RF AND MICROWAVE TRANSMITTER DESIGN

WILEY SERIES IN MICROWAVE AND OPTICAL ENGINEERING

KAI CHANG, Editor
Texas A&M University

A complete list of the titles in this series appears at the end of this volume.

RF AND MICROWAVE TRANSMITTER DESIGN

Andrei Grebennikov
Bell Labs, Alcatel-Lucent, Ireland

WILEY
A JOHN WILEY & SONS, INC., PUBLICATION

Published by John Wiley & Sons, Inc., Hoboken, New Jersey.
Published simultaneously in Canada.

Library of Congress Cataloging-in-Publication Data:

Grebennikov, Andrei
RF and microwave transmitter design/Andrei Grebennikov.
p. cm.
Includes bibliographical references and index.
ISBN 978-0-470-52099-4 (cloth)

Printed in Singapore

oBook ISBN: 978-0-470-92930-8
ePDF ISBN: 978-0-470-92929-1

10 9 8 7 6 5 4 3 2 1

CONTENTS

PREFACE

The main objective of this book is to present all relevant information required to design the transmitters in general and their main components in particular in different RF and microwave applications including well-known historical and recent novel architectures, theoretical approaches, circuit simulation results, and practical implementation techniques. This comprehensive book can be very useful for lecturing to promote the systematic way of thinking with analytical calculations and practical verification, thus making a bridge between theory and practice of RF and microwave engineering. As a result, this book is intended for and can be recommended to *university-level professors* as a comprehensive material to help in lecturing for graduate and postgraduate students, to *researchers and scientists* to combine the theoretical analysis with practical design and to provide a sufficient basis for innovative ideas and circuit design techniques, and to *practicing designers and engineers* as the book contains numerous well-known and novel practical circuits, architectures, and theoretical approaches with detailed description of their operational principles and applications.

Chapter 1 introduces the basic two-port networks describing the behavior of linear and nonlinear circuits. To characterize the nonlinear properties of the bipolar or field-effect transistors, their equivalent circuit elements are expressed through the impedance Z-parameters, admittance Y-parameters, or hybrid H-parameters. On the other hand, the transmission $ABCD$-parameters are very important for the design of the distributed circuits such as a transmission line or cascaded elements, whereas the scattering S-parameters are widely used to simplify the measurement procedure. The design formulas and curves are given for several types of transmission lines including stripline, microstrip line, slotline, and coplanar waveguide. Monolithic implementation of lumped inductors and capacitors is usually required at microwave frequencies and for portable devices. Knowledge of noise phenomena such as noise figure, additive white noise, low-frequency fluctuations, or flicker noise in active or passive elements is very important for the oscillator modeling in particular and entire transmitter design in general.

In Chapter 2, all necessary steps to provide an accurate device modeling procedure starting with the determination of the device small-signal equivalent circuit parameters are described and discussed. A variety of nonlinear models for MOSFET, MESFET, HEMT, and BJT devices including HBTs, which are very prospective for modern microwave monolithic integrated circuits, are given. In order to highlight the advantages or drawbacks of one over another nonlinear device model, a comparison of the measured and modeled volt–ampere and voltage–capacitance characteristics, as well as a frequency range of model application, are analyzed.

The main principles and impedance matching tools are described in Chapter 3. Generally, an optimum solution depends on the circuit requirements, such as the simplicity in practical realization, frequency bandwidth and minimum power ripple, design implementation and adjustability, stable operation conditions, and sufficient harmonic suppression. As a result, many types of the matching networks are available, including lumped elements and transmission lines. To simplify and visualize the matching design procedure, an analytical approach, which allows calculation of the parameters of the matching circuits using simple equations, and Smith chart traces are discussed. In addition, several examples of the narrowband and broadband power amplifiers using bipolar or MOSFET devices are given, including successive and detailed design considerations and explanations.

Chapter 4 describes the basic properties of the three-port and four-port networks, as well as a variety of different combiners, transformers, and directional couplers for RF and microwave power applications. For power combining in view of insufficient power performance of the active devices, it is best to use the coaxial-cable combiners with ferrite core to combine the output powers of RF power amplifiers intended for wideband applications. Since the device output impedance is usually too small for high power level, to match this impedance with a standard 50-Ω load, it is necessary to use the coaxial-line transformers with specified impedance transformation. For narrowband applications, the N-way Wilkinson combiners are widely used due to the simplicity of their practical realization. At the same time, the size of the combiners should be very small at microwave frequencies. Therefore, the commonly used hybrid microstrip combiners including different types of the microwave hybrids and directional couplers are described and analyzed.

Chapter 5 introduces the basic types of RF and microwave filters based on the low-pass or high-pass sections and bandpass or bandstop transformation. Classical filter design approaches using image parameter and insertion loss methods are given for low-pass and high-pass LC filter implementations. The quarterwave-line and coupled-line sections, which are the basic elements of microwave transmission-line filters, are described and analyzed. Different examples of coupled-line filters including interdigital, combline, and hairpin bandpass filters are given. Special attention is paid to microstrip filters with unequal phase velocities, which can provide unexpected properties because of different implementation technologies. Finally, the typical structures, implementation technology, operational principles, and band performance of the filters based on surface and bulk acoustic waves are presented.

Chapter 6 discusses the basic features of different types of analog modulation including amplitude, single-sideband, frequency, and phase modulation, and basic types of digital modulation such as amplitude shift keying, frequency shift keying, phase shift keying, or pulse code modulation and their variations. The principle of operations and various schematics of the modulators for different modulation schemes including Class S modulator for pulse-width modulation are described. Finally, the concept of time and frequency division multiplexing is introduced, as well as a brief description of different multiple access techniques.

A basic theory describing the operational principles of frequency conversion in receivers and transmitters is given in Chapter 7. The different types of mixers, from the simplest based on a single diode to a balanced and double-balanced type based on both diodes and transistors, are described and analyzed. The special case is a mixer based on a dual-gate transistor that provides better isolation between signal paths and simple implementation. The frequency multipliers that historically were a very important part of the vacuum-tube transmitters can extend the operating frequency range.

Chapter 8 presents the principles of oscillator design, including start-up and steady-state operation conditions, noise and stability of oscillations, basic oscillator configurations using lumped and transmission-line elements, and simplified equation-based oscillator analyses and optimum design techniques. An immittance design approach is introduced and applied to the series and parallel feedback oscillators, including circuit design and simulation aspects. Voltage-controlled oscillators and their varactor tuning range and linearity for different oscillator configurations are discussed. Finally, the basic circuits and operation principles of crystal and dielectric resonator oscillators are given.

Chapter 9 begins with description of the basic phase-locked loop concept. Then, the basic performance and structures of the analog, charge-pump, and digital phase-locked loops are analyzed. The basic loop components such as phase detector, loop filter, frequency divider, and voltage-controlled oscillator are discussed, as well as loop dynamic parameters. The possibility and particular realizations of the phase modulation using phase-locked loops are presented. Finally, general classes of frequency synthesizer techniques such as direct analog synthesis, indirect synthesis, and direct digital synthesis are discussed. The proper choice of the synthesizer type is based on the number of frequencies, frequency spacing, frequency switching time, noise, spurious level, particular technology, and cost.

Chapter 10 introduces the fundamentals of the power amplifier design, which is generally a complicated procedure when it is necessary to provide simultaneously accurate active device modeling, effective impedance matching depending on the technical requirements and operation conditions, stability in operation, and simplicity in practical implementation. The quality of the power amplifier design is evaluated by realized maximum power gain under stable operation condition with minimum amplifier stages, and the requirement of linearity or high efficiency can be considered where it is needed. For a stable operation, it is necessary to evaluate the operating frequency domains where the active device may be potentially unstable. To avoid the parasitic oscillations, the stabilization circuit technique for different frequency domains (from low frequencies up to high frequencies close to the device transition frequency) is discussed. The key parameter of the power amplifier is its linearity, which is very important for many wireless communication applications. The relationships between the output power, 1-dB gain compression point, third-order intercept point, and intermodulation distortions of the third and higher orders are given and illustrated for different active devices. The device bias conditions, which are generally different for linearity or efficiency improvement, depend on the power amplifier operation class and the type of the active device. The basic Classes A, AB, B, and C of the power amplifier operation are introduced, analyzed, and illustrated. The principles and design of the push–pull amplifiers using balanced transistors, as well as broadband and distributed power amplifiers, are discussed. Harmonic-control techniques for designing microwave power amplifiers are given with description of a systematic procedure of multiharmonic load–pull simulation using the harmonic balance method and active load–pull measurement system. Finally, the concept of thermal resistance is introduced and heatsink design issues are discussed.

Modern commercial and military communication systems require the high-efficiency long-term operating conditions. Chapter 11 describes in detail the possible circuit solutions to provide a high-efficiency power amplifier operation based on using Class D, Class F, Class E, or their newly developed subclasses depending on the technical requirements. In all cases, an efficiency improvement in practical implementation is achieved by providing the nonlinear operation conditions when an active device can simultaneously operate in pinch-off, active, and saturation regions, resulting in nonsinusoidal collector current and voltage waveforms, symmetrical for Class D and Class F and asymmetrical for Class E (DE, FE) operation modes. In Class F amplifiers analyzed in frequency domain, the fundamental-frequency and harmonic load impedances are optimized by short-circuit termination and open-circuit peaking to control the voltage and current waveforms at the device output to obtain maximum efficiency. In Class E amplifiers analyzed in time domain, an efficiency improvement is achieved by realizing the on/off active device switching operation (the pinch-off and saturation modes) with special current and voltage waveforms so that high voltage and high current do not concur at the same time.

In modern wireless communication systems, it is very important to realize both high-efficiency and linear operation of the power amplifiers. Chapter 12 describes a variety of techniques and approaches that can improve the power amplifier performance. To increase efficiency over power backoff range, the Doherty, outphasing, and envelope-tracking power amplifier architectures, as well as switched multipath power amplifier configurations, are discussed and analyzed. There are several linearization techniques that provide linearization of both entire transmitter system and individual power amplifier. Feedforward, cross cancellation, or reflect forward linearization techniques are available technologies for satellite and cellular base station applications achieving very high linearity levels. The practical realization of these techniques is quite complicated and very sensitive to both the feedback loop imbalance and the parameters of its individual components. Analog predistortion linearization technique is the simplest form of power amplifier linearization and can be used for handset application, although significant linearity improvement is difficult to realize. Different types of the feedback linearization approaches, together with digital linearization techniques, are very attractive to be used in handset or base station transmitters. Finally, the potential semidigital and digital amplification approaches are discussed with their architectural advantages and problems in practical implementation.

Chapter 13 discusses the circuit schematics and main properties of the semiconductor control circuits that are usually characterized by small size, low power consumption, high-speed performance, and operating life. Generally, they can be built based on the *p–i–n* diodes, silicon MOSFET, or GaAs MESFET transistors and can be divided into two basic parts: amplitude and phase control circuits. The control circuits are necessary to protect high power devices from excessive peak voltage or dc current conditions. They are also used as switching elements for directing signal between different transmitting paths, as variable gain amplifiers to stabilize transmitter output power, as attenuators and phase shifters to change the amplitude and phase of the transmitting signal paths in array systems, or as limiters to protect power-sensitive components.

Finally, Chapter 14 describes the different types of radio transmitter architectures, history of radio communication, conventional types of radio transmission, and modern communication systems. Amplitude-modulated transmitters representing the oldest technique for radio communication are based on high- or low-level modulation methods, with particular case of an amplitude keying. Single-sideband transmitters as the next-generation transmitters could provide higher efficiency due to the transmission of a single sideband only. Frequency-modulated transmitters then became a revolutionary step to improve the quality of a broadcast transmission. TV transmitters include different modulation techniques for transmitting audio and video information, both analog and digital. Wireless communication transmitters as a part of the cellular technologies provide a worldwide wireless radio access. Radar transmitters are required for many commercial and military applications such as phased-array radars, automotive radars, or electronic warfare systems. Satellite transmission systems contribute to worldwide transmission of any communication signals through satellite transponders and offer communication for areas with any population density and location. Ultra-wideband transmission is very attractive for their low-cost and low-power communication applications, occupying a very wide frequency range.

ACKNOWLEDGMENTS

To Drs. Frederick Raab and Lin Fujiang for useful comments and suggestions in book organization and content covering.

To Dr. Frank Mullany from Bell Labs, Ireland, for encouragement and support.

The author especially wishes to thank his wife, Galina Grebennikova, for performing computer-artwork design, as well as for her constant support, inspiration, and assistance.

Andrei Grebennikov

INTRODUCTION

A vacuum-tube or solid-state radio transmitter is essentially a source of a radio-frequency (RF) signal to be transmitted through antenna in different radio systems such as wireless communication, television (TV) and broadcasting, navigation, radar, or satellite, the information format and electrical performance of which should satisfy the corresponding standard requirements. The transmission of radio signals is produced by modulation of different types, with different output power and transmission mode, and in different frequency ranges, from high frequencies to millimeter waves. Transmitters in which the power output is generated directly by the modulated source are considered as possessing high-level modulation systems. In contrast, arrangements in which the modulation takes place at a power level less than the transmitter output are referred to as low-level modulation systems.

Figure I.1 shows the simplified block diagram of a conventional radio transmitter intended to operate at radio and microwave frequencies, which consists of the following: the source of the information signal, which is usually amplified, filtered, or transformed to the intermediate frequency; the local oscillator and frequency multiplier, which establish the stabilized carrier frequency or some multiple of it; the RF modulator or mixer, which combines the signal and carrier frequency components to produce one of the varieties of the RF modulated waves; the power amplifier to deliver the RF modulated signal of required power level to the antenna; the antenna duplexer to separate and isolate transmitting and receiving paths. The power amplifier usually consists of cascaded gain stages, and each stage should have adequate linearity in the case of transmitting signal with variable amplitude corresponding to amplitude-modulated or multicarrier signals. In practice, there are many variations in transmitter architectures depending on the particular frequency bandwidth, output power, or modulation scheme.

Dr. Lee De Forest was an inventor who changed the world with electronics by inventing the vacuum tube, which he called the *audion*. In January 1907, De Forest filed a patent for an oscillation detector based on a three-electrode device representing a vacuum tube [1]. His pioneering innovation was the insertion of a third electrode (grid) in between the cathode (filament) and the anode (plate) of the previously invented diode. However, it was not until 1911 that De Forest built the first vacuum-tube amplifier based on three audions as amplifiers [2]. The original audion was capable of slightly amplifying received signals, but at this stage could not be used for more advanced applications such as radio transmitters. The inefficient design of the original audion meant that it was initially of little value to radio, and due to its high cost and short life it was rarely used. Eventually, vacuum-tube design was improved enough to make vacuum tubes more than novelties. Beginning in 1912, various researchers discovered that properly constructed (i.e., according to scientific and engineering principles) vacuum tubes could be employed in electrical circuits that made radio receivers and amplifiers thousands of times more powerful, and could also be used to make compact and efficient radio transmitters, which for the first time made radio broadcasting practical.

In 1914, the first vacuum-tube radio transmitters began to appear—a key technical development that would lead to the introduction of widespread broadcasting. Both amateurs and commercial firms started to experiment with the new vacuum-tube transmitters, employing them for a variety of purposes. Six years after suspending his efforts to make audio transmissions, when he had unsuccessfully

RF and Microwave Transmitter Design, First Edition. Andrei Grebennikov.

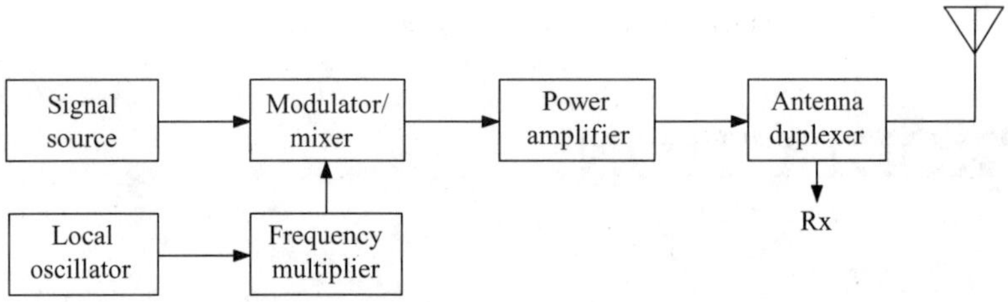

FIGURE I.1 Conventional transmitter architecture.

tried to use arc-transmitters, De Forest again took up developing radio to transmit sounds, including broadcasting news and entertainment, this time with much more success. He demonstrated that with this form of transmitter it was possible to telephone one to three miles, and by means of the small 3.5-V amplifier tube used with this apparatus, direct current could be transformed to alternating current at frequencies from 60 cycles per second to 1,000,000 cycles per second [3]. It is interesting that De Forest recognized the irony that he had overlooked the potential of developing his audion as a radio transmitter at the beginning. Reviewing his earlier arc-transmitter efforts, he wrote in his autobiography that he had been "totally unaware of the fact that in the little audion tube, which I was then using only as a radio detector, lay dormant the principle of oscillation which, had I but realized it, would have caused me to unceremoniously dump into the ash can all of the fine arc mechanisms which I had ever constructed, a procedure which a few years later actually took place all over the world" [4]. Meanwhile, in June 1915, the American Telephone & Telegraph Company installed a powerful experimental vacuum-tube transmitter in Arlington, Virginia, which quickly achieved remarkable distances for its audio transmissions [5]. The Marconi companies joined those experimenting with

PHOTO I.1 Lee De Forest.

the new vacuum-tube transmitters, achieving an oversea working range of 50 km between ship aerials [6]. With the capable assistance of engineers including H. Round and C. Franklin, Guglielmo Marconi began experimenting with shortwave vacuum-tube transmitters by about 1916 [7].

However, Alexander Meissner was the first to amplify high-frequency radio signals by using a regenerative (positive) feedback in a vacuum triode. This principle of reactive coupling, which was given in its general form by Meissner in March 1913, later became the basis of the radio transmitter development [8]. The first high-frequency vacuum tube transmitters of small power up to 15 W were built by the Telefunken Company at the beginning of 1915. To provide high operating efficiency of the transmitter, the plate current of the special wave form, or else an auxiliary voltage of triple frequency, was impressed on the grid; thus greatly reducing the losses because, at the time when the highest voltages are applied to the tube, the passage of current through it is prevented. At the very beginning of 1920s, 1-kW transmitters had already been in use for radio telephony and telegraphy in several of the larger German cities. Much higher transmitting power was achieved by connecting in parallel eight or more tubes, each delivering 1.5 kW. In Russia during 1920, the 5.5-kW radio transmitter, where the modulated oscillations were amplified by a tube and transferred to the grids of six tubes in parallel that fed the antenna of 120-meter height, covered long distances of more than 4500 km [9]. An attempt in the field of radio television had been tried out in 1920 in order to provide a radio transmission of photographs with two antennas: one of which sends the synchronizing signal while the other sends the actual picture.

Edwin H. Armstrong is widely regarded as one of the foremost contributors to the field of radio engineering, being responsible for the regenerative circuit (1912), the superheterodyne circuit (1918), and the complete frequency-modulation radio broadcasting system (1933). Armstrong studied the audion for several years, performed extensive measurements, and understood and explained its operation when he devised a circuit, in which part of the current at the plate was fed back to the grid to strengthen the incoming signal. This discovery led to the independent invention of regeneration (or feedback) principle and the vacuum-tube oscillator. He then disproved the currently accepted theory

PHOTO I.2 Alexander Meissner.

PHOTO I.3 Edwin Armstrong.

of the action of the triode (three-electrode vacuum tube), and published the correct explanation in 1914 [10].

At the same time, the theoretical development of quantum mechanics during the 1920s played an important role in driving solid-state electronics, with understanding of the differences between metals, insulators, and semiconductors [11]. These continuing theoretical efforts then quickly led to the discovery of new devices, when Julius Lilienfeld invented the concept of a field-effect transistor in 1926 [12]. He believed that applying a voltage to a poorly conducting material would change its conductivity and thereby achieve amplification, but no one was able to do anything practically with this device until much later time. In conjunction with this, it is worth mentioning that the oscillating crystal detector was described by W. Eccles in 1909 and then practically implemented as an oscillator and even a low-power transmitter (based on one-port negative resistance principle) by O. Losev in early 1920s [13,14]. All details needed to duplicate these circuits to make a tunnel-diode oscillator were reported in the September 1924 issue of *Radio News* and in the 1st and 8th October 1924 issues of *Wireless World*, with predictions that crystals would someday replace valves in electronics.

Shortly after the end of the war in 1945, Bell Labs formed a Solid State Physics Group led by William Shockley, with an assignment to seek a solid-state alternative to fragile glass vacuum-tube amplifiers. This group made a very important decision right at the beginning: that the simplest semiconductors were silicon and germanium and that their efforts would be directed at those two elements. The first attempts were based on Shockley's ideas about using an external electrical field on a semiconductor to affect its conductivity, but these experiments failed every time in all sorts of configurations and materials. The group was at a standstill until John Bardeen suggested a theory that invoked surface states that prevented the field from penetrating the semiconductor. In November 1947, John Bardeen and Walter Brattain, working without Shockley, succeeded in creating a point-contact transistor that achieved amplification when electrical field was applied to a crystal of germanium [15]. At the same time, Shockley secretly continued and successfully finished his own work to build a different sort of transistor based on *n–p* junctions that depended on the introduction of the minority carriers instead of point contacts, which he expected would be more

PHOTO I.4 William Shockley.

likely to be commercially viable [16]. In his seminal work *Electrons and Holes in Semiconductors with Applications to Transistor Electronics* (1955), Shockley worked out the critical ideas of drift and diffusion and the differential equations that govern the flow of electrons in solid-state crystals, where the Shockley ideal diode equation was also described. The term "transistor" was coined by John Pierce, who later recalled: "... at that time, it was supposed to be the dual of the vacuum tube. The vacuum tube had *transconductance*, so the transistor would have *transresistance*. And the name should fit in with the names of other devices, such as *varistor* and *thermistor*. And ... I suggested the name *transistor*."

The first to perceive the possibility of integrated circuits based upon semiconductor technology was Geoffrey Dummer, who said addressing the Electronic Components Conference in 1952: "With the advent of the transistor and the work in semiconductors generally, it seems now possible to envisage electronic equipment in a solid block with no connecting wires. The block may consist of layers of insulating, conducting, rectifying, and amplifying materials, the electrical functions being connected directly by cutting out areas of the various layers" [17]. In 1958, Jack Kilby of Texas Instruments developed the first integrated circuit consisting of a few mesa transistors, diffused capacitors, and bulk resistors on a piece of germanium using gold wires for interconnections [18]. The integrated circuit developed by Robert Noyce of Fairchild could resolve problems with wires by adding a final metal layer and then taking away some of it so that the wires required for the components to be connected were shaped so as to make the integrated circuit more suitable for mass production [19]. A year later, a planar process was developed by Jean Hoerni that utilized heat diffusion process, and oxide passivation of the surface protected the junctions and provided a reproducibility that assured more consistency than any previous manufacturing process. The first microwave gallium–arsenide (GaAs) Schottky-gate field-effect transistor (metal semiconductor field-effect transistor or MESFET), which had a maximum frequency $f_{\max} = 3$ GHz, was reported in 1967 [20]. But it was not until 1976 that the first fully monolithic single-stage GaAs MESFET X-band broadband amplifier was developed based on lumped matching elements [21]. Eight years later, on a GaAs chip of approximately the same area, an entire X-band transmit–receive (T/R) module was fabricated, consisting of two switches, a four-bit

TABLE I.1 IEEE Standard Letter Designations for Frequency Bands.

Band	Frequency	Wavelength
HF (high frequency)	3–30 MHz	100–10 m
VHF (very high frequency)	30–300 MHz	10–1.0 m
UHF (ultra high frequency)	300–1000 MHz	1.0 m to 30 cm
L	1–2 GHz	30–15 cm
S	2–4 GHz	15 cm to 7.5 cm
C	4–8 GHz	7.50–3.75 cm
X	8–12 GHz	3.75–2.50 cm
Ku	12–18 GHz	2.50–1.67 cm
K	18–27 GHz	1.67–1.11 cm
Ka	27–40 GHz	1.11 cm to 7.5 mm
V	40–75 GHz	7.5–4.0 mm
W	75–110 GHz	4.0–2.7 mm
Millimeter wave	110–300 GHz	2.7–1.0 mm

phase shifter, a three-stage low-noise amplifier, and a four-stage power amplifier [22]. Generally, a monolithic microwave integrated circuit (MMIC) was defined as an active or passive microwave circuit formed in situ on a semiconductor substrate by a combination of deposition techniques including diffusion, evaporation, epitaxy, and other means [23].

Table I.1 shows the IEEE standard letter designations for frequency bands [24]. In military radar band designations, millimeter-wave bandwidth occupies the frequency range from 40 to 300 GHz. The letter designations (L, S, C, X, Ku, K, Ka) were meant to be used for radar, but have become commonly used for other microwave frequency applications. The K-band is the middle band (18–27 GHz) that originated from the German word "Kurz", which means short, while Ku-band is lower in frequency (Kurz-under), and Ka-band is higher in frequency (Kurz-above). The 1984 revision defined the application of letters V and W to a portion of the millimeter-wave region each while retaining the previous letter designators for frequencies.

REFERENCES

1. L. De Forest, "Space Telegraphy," U.S. Patent 879,532, Feb. 1908 (filed Jan. 1907).
2. L. De Forest, "The Audion – Detector and Amplifier," *Proc. IRE*, vol. 2, pp. 15–29, Mar. 1914.
3. "High-Frequency Oscillating Transmitter for Wireless Telephony," *Electrical World*, p. 144, July 18, 1914.
4. T. H. White, United States Early Radio History, http://earlyradiohistory.us, sec. 11.
5. E. B. Craft and E. H. Colpitts, "Radio Telephony," *Trans. AIEE*, vol. 38, pp. 305–343, Feb. 1919.
6. "Wireless-Telephone Set," *Electrical World*, p. 487, Sept. 5, 1914.
7. J. E. Brittain, "Electrical Engineering Hall of Fame: Guglielmo Marconi," *Proc. IEEE*, vol. 92, pp. 1501–1504, Sept. 2004.
8. A. Meissner, "The Development of Tube Transmitters by the Telefunken Company," *Proc. IRE*, vol. 10, pp. 3–23, Jan. 1922.
9. V. Bashenoff, "Progress in Radio Engineering in Russia 1918–1922," *Proc. IRE*, vol. 11, pp. 257–270, Mar. 1923.
10. "Edwin Howard Armstrong," *Proc. IRE*, vol. 31, p. 315, July 1943.
11. W. F. Brinkman, D. E. Haggan, and W. W. Troutman, "A History of the Invention of the Transistor and Where It Will Lead Us," *IEEE J. Solid-State Circuits*, vol. SC-32, pp. 1858–1865, Dec. 1997.
12. J. E. Lilienfeld, "Method and Apparatus for Controlled Electric Currents," U.S. Patent 1745175, Jan. 1930 (filed Oct. 1926).

13. W. H. Eccles, "On an Oscillation Detector Actuated Solely by Resistance-Temperature Variations," *Proc. Phys. Society. London*, vol. 22, pp. 360–368, Jan. 1909.

14. M. A. Novikov, "Oleg Vladimirovich Losev: Pioneer of Semiconductor Electronics," *Physics of the Solid State*, vol. 46, pp. 1–4, Jan. 2004.

15. J. Bardin and W. H. Brattain, "Three-Electrode Circuit Element Utilizing Semiconductive Materials," U.S. Patent 2524035, Oct. 1950 (filed June 17, 1948).

16. W. Shockley, "Circuit Element Utilizing Semiconductive Material," U.S. Patent 2569347, Sept. 1951 (filed June 26, 1948).

17. J. S. Kilby, "Invention of the Integrated Circuit," *IEEE Trans. Electron Devices*, vol. ED-23, pp. 648–654, July 1976.

18. J. S. Kilby, "Miniaturized Electronic Circuits," U.S. Patent 3138743, June 1964 (filed Feb. 1959).

19. R. N. Noyce, "Semiconductor Device-and-Lead Structure," U.S. Patent 2981877, Apr. 1961 (filed July 1959).

20. D. N. McQuiddy, J. W. Wassel, J. Bradner LaGrange, and W. R. Wisseman, "Monolithic Microwave Integrated Circuits: An Historical Perspective," *IEEE Trans. Microwave Theory Tech.*, vol. MTT-32, pp. 997–1008, Sept. 1984.

21. R. S. Pengelly and J. A. Turner, "Monolithic Broadband GaAs F.E.T. Amplifiers," *Electronics Lett.*, vol. 12, pp. 251–252, May 1976.

22. E. C. Niehenke, R. A. Pucel, and I. J. Bahl, "Microwave and Millimeter-Wave Integrated Circuits," *IEEE Trans. Microwave Theory Tech.*, vol. MTT-50, pp. 846–857, Mar. 2002.

23. R. A. Pucel, "Design Considerations for Monolithic Microwave Circuits," *IEEE Trans. Microwave Theory Tech.*, vol. MTT-29, pp. 513–534, June 1981.

24. IEEE Standard 521-2002, *IEEE Standard Letter Designations for Radar-Frequency Bands*, 2003.

1 Passive Elements and Circuit Theory

The two-port equivalent circuits are widely used in radio frequency (RF) and microwave circuit design to describe the electrical behavior of both active devices and passive networks [1–4]. The two-port network impedance Z-parameters, admittance Y-parameters, or hybrid H-parameters are very important to characterize the nonlinear properties of the active devices, bipolar or field-effect transistors. The transmission $ABCD$-parameters of a two-port network are very convenient for designing the distributed circuits like transmission lines or cascaded elements. The scattering S-parameters are useful to characterize linear circuits, and are required to simplify the measurement procedure. Transmission lines are widely used in matching circuits in power amplifiers, in resonant circuits in the oscillators, filters, directional couplers, power combiners, and dividers. The design formulas and curves are presented for several types of transmission lines including stripline, microstrip line, slotline, and coplanar waveguide. Monolithic implementation of lumped inductors and capacitors is usually required at microwave frequencies and for portable devices. Knowledge of noise phenomena, such as the noise figure, additive white noise, low-frequency fluctuations, or flicker noise in active or passive elements, is very important for the oscillator modeling in particular and entire transmitter design in general.

1.1 IMMITTANCE TWO-PORT NETWORK PARAMETERS

The basic diagram of a two-port nonautonomous transmission system can be represented by the equivalent circuit shown in Figure 1.1, where V_S is the independent voltage source, Z_S is the source impedance, *LN* is the linear time-invariant two-port network without independent source, and Z_L is the load impedance. The two independent phasor currents I_1 and I_2 (flowing across input and output terminals) and phasor voltages V_1 and V_2 characterize such a two-port network. For autonomous oscillator systems, in order to provide an appropriate analysis in the frequency domain of the two-port network in the negative one-port representation, it is sufficient to set the source impedance to infinity. For a power amplifier or oscillator design, the elements of the matching or resonant circuits, which are assumed to be linear or appropriately linearized, can be found among the *LN*-network elements, or additional two-port linear networks can be used to describe their frequency domain behavior.

For a two-port network, the following equations can be considered to be imposed boundary conditions:

$$V_1 + Z_S I_1 = V_S \tag{1.1}$$

$$V_2 + Z_L I_2 = V_L. \tag{1.2}$$

Suppose that it is possible to obtain a unique solution for the linear time-invariant circuit shown in Figure 1.1. Then the two linearly independent equations, which describe the general two-port network

RF and Microwave Transmitter Design, First Edition. Andrei Grebennikov.

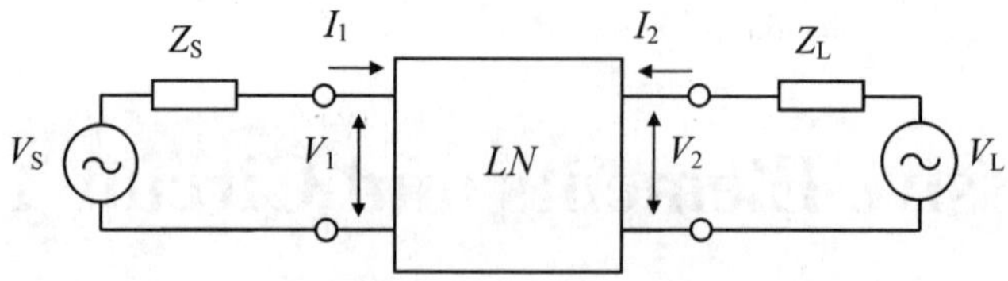

FIGURE 1.1 Basic diagram of two-port nonautonomous transmission system.

in terms of circuit variables V_1, V_2, I_1, and I_2, can be expressed in a matrix form as

$$[M][V]+[N][I]=0 \tag{1.3}$$

or

$$\left.\begin{aligned} m_{11}V_1+m_{12}V_2+n_{11}I_1+n_{12}I_2=0 \\ m_{21}V_1+m_{22}V_2+n_{21}I_1+n_{22}I_2=0 \end{aligned}\right\}. \tag{1.4}$$

The complex 2 × 2 matrices [*M*] and [*N*] in Eq. (1.3) are independent of the source and load impedances Z_S and Z_L and voltages V_S and V_L, respectively, and they depend only on the circuit elements inside the *LN* network.

If matrix [*M*] in Eq. (1.3) is nonsingular when $|M| \neq 0$, then this matrix equation can be rewritten in terms of [*I*] as

$$[V]=-[M]^{-1}[N][I]=[Z][I] \tag{1.5}$$

where [*Z*] is the open-circuit impedance two-port network matrix. In a scalar form, matrix Eq. (1.5) is given by

$$V_1 = Z_{11}I_1 + Z_{12}I_2 \tag{1.6}$$

$$V_2 = Z_{21}I_1 + Z_{22}I_2 \tag{1.7}$$

where Z_{11} and Z_{22} are the open-circuit driving-point impedances, and Z_{12} and Z_{21} are the open-circuit transfer impedances of the two-port network. The voltage components V_1 and V_2 due to the input current I_1 can be found by setting $I_2 = 0$ in Eqs. (1.6) and (1.7), resulting in an open-output terminal. Similarly, the same voltage components V_1 and V_2 are determined by setting $I_1 = 0$ when the input terminal becomes open-circuited. The resulting driving-point impedances can be written as

$$Z_{11} = \left.\frac{V_1}{I_1}\right|_{I_2=0} \qquad Z_{22} = \left.\frac{V_2}{I_2}\right|_{I_1=0} \tag{1.8}$$

whereas the two transfer impedances are

$$Z_{21} = \left.\frac{V_2}{I_1}\right|_{I_2=0} \qquad Z_{12} = \left.\frac{V_1}{I_2}\right|_{I_1=0}. \tag{1.9}$$

Dual analysis can be used to derive the short-circuit admittance matrix when the current components I_1 and I_2 are considered as outputs caused by V_1 and V_2. If matrix [*N*] in Eq. (1.3) is nonsingular when $|N| \neq 0$, this matrix equation can be rewritten in terms of $[V]$ as

$$[I] = -[N]^{-1}[M][V] = [Y][V] \tag{1.10}$$

where $[Y]$ is the short-circuit admittance two-port network matrix. In a scalar form, matrix Eq. (1.10) is written as

$$I_1 = Y_{11}V_1 + Y_{12}V_2 \tag{1.11}$$

$$I_2 = Y_{21}V_1 + Y_{22}V_2 \tag{1.12}$$

where Y_{11} and Y_{22} are the short-circuit driving-point admittances, and Y_{12} and Y_{21} are the short-circuit transfer admittances of the two-port network. In this case, the current components I_1 and I_2 due to the input voltage source V_1 are determined by setting $V_2 = 0$ in Eqs. (1.11) and (1.12), thus creating a short output terminal. Similarly, the same current components I_1 and I_2 are determined by setting $V_1 = 0$ when input terminal becomes short-circuited. As a result, the two driving-point admittances are

$$Y_{11} = \left.\frac{I_1}{V_1}\right|_{V_2=0} \qquad Y_{22} = \left.\frac{I_2}{V_2}\right|_{V_1=0} \tag{1.13}$$

whereas the two transfer admittances are

$$Y_{21} = \left.\frac{I_2}{V_1}\right|_{V_2=0} \qquad Y_{12} = \left.\frac{I_1}{V_2}\right|_{V_1=0}. \tag{1.14}$$

In some cases, an equivalent two-port network representation can be redefined in order to express the voltage source V_1 and output current I_2 in terms of the input current I_1 and output voltage V_2. If the submatrix

$$\begin{bmatrix} m_{11} & n_{12} \\ m_{21} & n_{22} \end{bmatrix}$$

given in Eq. (1.4) is nonsingular, then

$$\begin{bmatrix} V_1 \\ I_2 \end{bmatrix} = -\begin{bmatrix} m_{11} & n_{12} \\ m_{21} & n_{22} \end{bmatrix}^{-1} \begin{bmatrix} n_{11} & m_{12} \\ n_{21} & m_{22} \end{bmatrix} \begin{bmatrix} I_1 \\ V_2 \end{bmatrix} = [H] \begin{bmatrix} I_1 \\ V_2 \end{bmatrix} \tag{1.15}$$

where $[H]$ is the hybrid two-port network matrix. In a scalar form, it is best to represent matrix Eq. (1.15) as

$$V_1 = h_{11}I_1 + h_{12}V_2 \tag{1.16}$$

$$I_2 = h_{21}I_1 + h_{22}V_2 \tag{1.17}$$

where h_{11}, h_{12}, h_{21}, and h_{22} are the hybrid H-parameters. The voltage source V_1 and current component I_2 are determined by setting $V_2 = 0$ for the short output terminal in Eqs. (1.16) and (1.17) as

$$h_{11} = \left.\frac{V_1}{I_1}\right|_{V_2=0} \qquad h_{21} = \left.\frac{I_2}{I_1}\right|_{V_2=0} \tag{1.18}$$

where h_{11} is the driving-point input impedance and h_{21} is the forward current transfer function. Similarly, the input voltage source V_1 and output current I_2 are determined by setting $I_1 = 0$ when input terminal becomes open-circuited as

$$h_{12} = \left.\frac{V_1}{V_2}\right|_{I_1=0} \qquad h_{22} = \left.\frac{I_2}{V_2}\right|_{I_1=0} \tag{1.19}$$

where h_{12} is the reverse voltage transfer function and h_{22} is the driving-point output admittance.

The transmission parameters, often used for passive device analysis, are determined for the independent input voltage source V_1 and input current I_1 in terms of the output voltage V_2 and output current I_2. In this case, if the submatrix

$$\begin{bmatrix} m_{11} & n_{11} \\ m_{21} & n_{21} \end{bmatrix}$$

given in Eq. (1.4) is nonsingular, we obtain

$$\begin{bmatrix} V_1 \\ I_1 \end{bmatrix} = -\begin{bmatrix} m_{11} & n_{11} \\ m_{21} & n_{21} \end{bmatrix}^{-1} \begin{bmatrix} m_{12} & n_{12} \\ m_{22} & n_{22} \end{bmatrix} \begin{bmatrix} V_2 \\ -I_2 \end{bmatrix} = [ABCD] \begin{bmatrix} V_2 \\ -I_2 \end{bmatrix} \tag{1.20}$$

where $[ABCD]$ is the forward transmission two-port network matrix. In a scalar form, we can write

$$V_1 = AV_2 - BI_2 \tag{1.21}$$

$$I_1 = CV_2 - DI_2 \tag{1.22}$$

where A, B, C, and D are the transmission parameters. The voltage source V_1 and current component I_1 are determined by setting $I_2 = 0$ for the open output terminal in Eqs. (1.21) and (1.22) as

$$A = \left.\frac{V_1}{V_2}\right|_{I_2=0} \qquad C = \left.\frac{I_1}{V_2}\right|_{I_2=0} \tag{1.23}$$

where A is the reverse voltage transfer function and C is the reverse transfer admittance. Similarly, the input independent variables V_1 and I_1 are determined by setting $V_2 = 0$ when the output terminal is short-circuited as

$$B = \left.\frac{V_1}{I_2}\right|_{V_2=0} \qquad D = \left.\frac{I_1}{I_2}\right|_{V_2=0} \tag{1.24}$$

where B is the reverse transfer impedance and D is the reverse current transfer function. The reason a minus sign is associated with I_2 in Eqs. (1.20) to (1.22) is that historically, for transmission networks, the input signal is considered as flowing to the input port whereas the output current flowing to the load. The direction of the current $-I_2$ entering the load is shown in Figure 1.2.

The parameters describing the same two-port network through different two-port matrices (impedance, admittance, hybrid, or transmission) can be cross-converted, and the elements of each

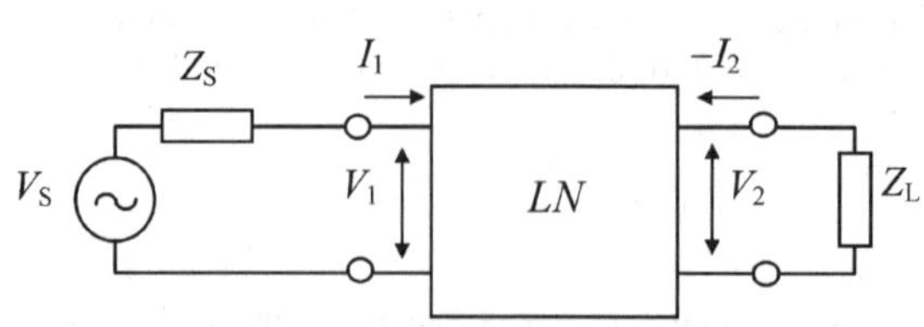

FIGURE 1.2 Basic diagram of loaded two-port transmission system.

TABLE 1.1 Relationships Between Z-, Y-, H- and $ABCD$-Parameters.

	[Z]	[Y]	[H]	[ABCD]
[Z]	$\begin{matrix} Z_{11} & Z_{12} \\ Z_{21} & Z_{22} \end{matrix}$	$\begin{matrix} \frac{Y_{22}}{\Delta Y} & -\frac{Y_{12}}{\Delta Y} \\ -\frac{Y_{21}}{\Delta Y} & \frac{Y_{11}}{\Delta Y} \end{matrix}$	$\begin{matrix} \frac{\Delta H}{h_{22}} & \frac{h_{12}}{h_{22}} \\ -\frac{h_{21}}{h_{22}} & \frac{1}{h_{22}} \end{matrix}$	$\begin{matrix} \frac{A}{C} & \frac{AD-BC}{C} \\ \frac{1}{C} & \frac{D}{C} \end{matrix}$
[Y]	$\begin{matrix} \frac{Z_{22}}{\Delta Z} & -\frac{Z_{12}}{\Delta Z} \\ -\frac{Z_{21}}{\Delta Z} & \frac{Z_{11}}{\Delta Z} \end{matrix}$	$\begin{matrix} Y_{11} & Y_{12} \\ Y_{21} & Y_{22} \end{matrix}$	$\begin{matrix} \frac{1}{h_{11}} & -\frac{h_{12}}{h_{11}} \\ \frac{h_{21}}{h_{11}} & \frac{\Delta H}{h_{11}} \end{matrix}$	$\begin{matrix} \frac{D}{B} & -\frac{AD-BC}{B} \\ -\frac{1}{B} & \frac{A}{B} \end{matrix}$
[H]	$\begin{matrix} \frac{\Delta Z}{Z_{22}} & \frac{Z_{12}}{Z_{22}} \\ -\frac{Z_{21}}{Z_{22}} & \frac{1}{Z_{22}} \end{matrix}$	$\begin{matrix} \frac{1}{Y_{11}} & -\frac{Y_{12}}{Y_{11}} \\ \frac{Y_{21}}{Y_{11}} & \frac{\Delta Y}{Y_{11}} \end{matrix}$	$\begin{matrix} h_{11} & h_{12} \\ h_{21} & h_{22} \end{matrix}$	$\begin{matrix} \frac{B}{D} & \frac{AD-BC}{D} \\ -\frac{1}{D} & \frac{C}{D} \end{matrix}$
[ABCD]	$\begin{matrix} \frac{Z_{11}}{Z_{21}} & \frac{\Delta Z}{Z_{21}} \\ \frac{1}{Z_{21}} & \frac{Z_{22}}{Z_{21}} \end{matrix}$	$\begin{matrix} -\frac{Y_{22}}{Y_{21}} & -\frac{1}{Y_{21}} \\ -\frac{\Delta Y}{Y_{21}} & -\frac{Y_{11}}{Y_{21}} \end{matrix}$	$\begin{matrix} -\frac{\Delta H}{h_{21}} & -\frac{h_{11}}{h_{21}} \\ -\frac{h_{22}}{h_{21}} & -\frac{1}{h_{21}} \end{matrix}$	$\begin{matrix} A & B \\ C & D \end{matrix}$

matrix can be expressed by the elements of other matrices. For example, Eqs. (1.11) and (1.12) for the Y-parameters can be easily solved for the independent input voltage source V_1 and input current I_1 as

$$V_1 = -\frac{Y_{22}}{Y_{21}} V_2 + \frac{1}{Y_{21}} I_2 \tag{1.25}$$

$$I_1 = -\frac{Y_{11}Y_{22} - Y_{12}Y_{21}}{Y_{21}} V_2 + \frac{Y_{11}}{Y_{21}} I_2. \tag{1.26}$$

By comparing the equivalent Eqs. (1.21) and (1.22) and Eqs. (1.25) and (1.26), the direct relationships between the elements of the transmission $ABCD$-matrix and admittance Y-matrix are written as

$$A = -\frac{Y_{22}}{Y_{21}} \quad B = -\frac{1}{Y_{21}} \tag{1.27}$$

$$C = -\frac{\Delta Y}{Y_{21}} \quad D = -\frac{Y_{11}}{Y_{21}} \tag{1.28}$$

where $\Delta Y = Y_{11}Y_{22} - Y_{12}Y_{21}$.

A summary of the relationships between the impedance Z-parameters, admittance Y-parameters, hybrid H-parameters, and transmission $ABCD$-parameters is shown in Table 1.1, where $\Delta Z = Z_{11}Z_{22} - Z_{12}Z_{21}$ and $\Delta H = h_{11}h_{22} - h_{12}h_{21}$.

1.2 SCATTERING PARAMETERS

The concept of incident and reflected voltage and current parameters can be illustrated by the one-port network shown in Figure 1.3, where the network impedance Z is connected to the signal source V_S

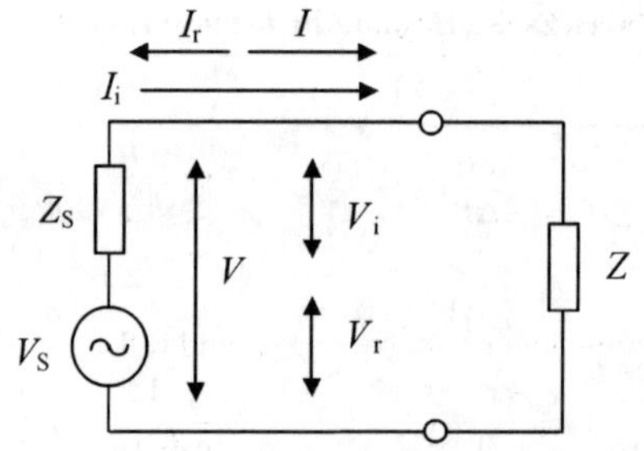

FIGURE 1.3 Incident and reflected voltages and currents.

with the internal impedance Z_S. In a common case, the terminal current I and voltage V consist of incident and reflected components (assume their root mean square [rms] values). When the load impedance Z is equal to the conjugate of source impedance expressed as $Z = Z_S^*$, the terminal current becomes the incident current. It is calculated from

$$I_i = \frac{V_S}{Z_S^* + Z_S} = \frac{V_S}{2\mathrm{Re}Z_S}. \tag{1.29}$$

The terminal voltage, defined as the incident voltage, can be determined from

$$V_i = \frac{Z_S^* V_S}{Z_S^* + Z_S} = \frac{Z_S^* V_S}{2\mathrm{Re}Z_S}. \tag{1.30}$$

Consequently, the incident power, which is equal to the maximum available power from the source, can be obtained by

$$P_i = \mathrm{Re}\left(V_i I_i^*\right) = \frac{|V_S|^2}{4\mathrm{Re}Z_S}. \tag{1.31}$$

The incident power can be presented in a normalized form using Eq. (1.30) as

$$P_i = \frac{|V_i|^2\, \mathrm{Re}Z_S}{\left|Z_S^*\right|^2}. \tag{1.32}$$

This allows the normalized incident voltage wave a to be defined as the square root of the incident power P_i by

$$a = \sqrt{P_i} = \frac{V_i \sqrt{\mathrm{Re}Z_S}}{Z_S^*}. \tag{1.33}$$

Similarly, the normalized reflected voltage wave b, defined as the square root of the reflected power P_r, can be written as

$$b = \sqrt{P_r} = \frac{V_r \sqrt{\mathrm{Re}Z_S}}{Z_S}. \tag{1.34}$$

The incident power can be expressed by the incident current I_i and the reflected power can be expressed by the reflected current I_r, respectively, as

$$P_i = |I_i|^2 \, \mathrm{Re}Z_S \tag{1.35}$$

$$P_r = |I_r|^2 \, \mathrm{Re}Z_S. \tag{1.36}$$

As a result, the normalized incident voltage wave a and reflected voltage wave b can be given by

$$a = \sqrt{P_i} = I_i\sqrt{\mathrm{Re}Z_S} \tag{1.37}$$

$$b = \sqrt{P_r} = I_r\sqrt{\mathrm{Re}Z_S}. \tag{1.38}$$

The parameters a and b can also be called the *normalized incident* and *reflected current waves*, or simply *normalized incident* and *reflected waves*, respectively, since the normalized current waves and the normalized voltage waves represent the same parameters.

The voltage V and current I, related to the normalized incident and reflected waves a and b, can be written as

$$V = V_i + V_r = \frac{Z_S^*}{\sqrt{\mathrm{Re}Z_S}}a + \frac{Z_S}{\sqrt{\mathrm{Re}Z_S}}b \tag{1.39}$$

$$I = I_i - I_r = \frac{1}{\sqrt{\mathrm{Re}Z_S}}a - \frac{1}{1\sqrt{\mathrm{Re}Z_S}}b \tag{1.40}$$

where

$$a = \frac{V + Z_S I}{2\sqrt{\mathrm{Re}Z_S}} \qquad b = \frac{V - Z_S^* I}{2\sqrt{\mathrm{Re}Z_S}}. \tag{1.41}$$

The source impedance Z_S is often purely real and, therefore, is used as the normalized impedance. In microwave design technique, the characteristic impedance of the passive two-port networks, including transmission lines and connectors, is considered as real and equal to 50 Ω. This is very important for measuring S-parameters when all transmission lines, source, and load should have the same real impedance. For $Z_S = Z_S^* = Z_0$, where Z_0 is the characteristic impedance, the ratio of the normalized reflected wave and the normalized incident wave for a one-port network is called the *reflection coefficient* Γ, defined as

$$\Gamma = \frac{b}{a} = \frac{V - Z_S^* I}{V + Z_S I} = \frac{V - Z_S I}{V + Z_S I} = \frac{Z - Z_S}{Z + Z_S} \tag{1.42}$$

where $Z = V/I$.

For a two-port network shown in Figure 1.4, the normalized reflected waves b_1 and b_2 can also be represented by the normalized incident waves a_1 and a_2, respectively, as

$$b_1 = S_{11}a_1 + S_{12}a_2 \tag{1.43}$$

$$b_2 = S_{21}a_1 + S_{22}a_2 \tag{1.44}$$

or, in a matrix form,

$$\begin{bmatrix} b_1 \\ b_2 \end{bmatrix} = \begin{bmatrix} S_{11} & S_{11} \\ S_{21} & S_{21} \end{bmatrix} \begin{bmatrix} a_1 \\ a_2 \end{bmatrix} \tag{1.45}$$

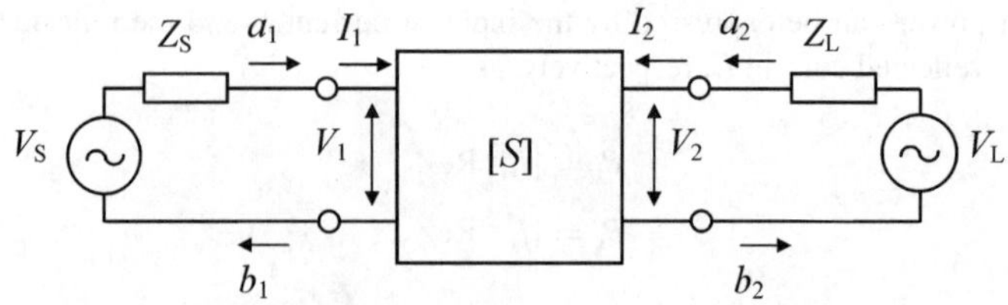

FIGURE 1.4 Basic diagram of S-parameter two-port network.

where the incident waves a_1 and a_2 and the reflected waves b_1 and b_2, for complex source and load impedances Z_S and Z_L, are given by

$$a_1 = \frac{V_1 + Z_S I_1}{2\sqrt{\text{Re}Z_S}} \qquad a_2 = \frac{V_2 + Z_L I_2}{2\sqrt{\text{Re}Z_L}} \tag{1.46}$$

$$b_1 = \frac{V_1 - Z_S^* I_1}{2\sqrt{\text{Re}Z_S}} \qquad b_2 = \frac{V_2 - Z_L^* I_2}{2\sqrt{\text{Re}Z_L}} \tag{1.47}$$

where S_{11}, S_{12}, S_{21}, and S_{22} are the S-parameters of the two-port network.

From Eq. (1.45) it follows that if $a_2 = 0$, then

$$S_{11} = \left.\frac{b_1}{a_1}\right|_{a_2=0} \qquad S_{21} = \left.\frac{b_2}{a_1}\right|_{a_2=0} \tag{1.48}$$

where S_{11} is the reflection coefficient and S_{21} is the transmission coefficient for ideal matching conditions at the output terminal when there is no incident power reflected from the load. Similarly,

$$S_{12} = \left.\frac{b_1}{a_2}\right|_{a_1=0} \qquad S_{22} = \left.\frac{b_2}{a_2}\right|_{a_1=0} \tag{1.49}$$

where S_{12} is the transmission coefficient and S_{22} is the reflection coefficient for ideal matching conditions at the input terminal.

To convert S-parameters to the admittance Y-parameters, it is convenient to represent Eqs. (1.46) and (1.47) as

$$I_1 = (a_1 - b_1)\frac{1}{\sqrt{Z_0}} \qquad I_2 = (a_2 - b_2)\frac{1}{\sqrt{Z_0}} \tag{1.50}$$

$$V_1 = (a_1 + b_1)\sqrt{Z_0} \quad V_1 = (a_2 + b_2)\sqrt{Z_0} \tag{1.51}$$

where it is assumed that the source and load impedances are real and equal to $Z_S = Z_L = Z_0$.

Substituting Eqs. (1.50) and (1.51) to Eqs. (1.11) and (1.12) results in

$$\frac{a_1 - b_1}{\sqrt{Z_0}} = Y_{11}(a_1 + b_1)\sqrt{Z_0} + Y_{12}(a_2 + b_2)\sqrt{Z_0} \tag{1.52}$$

$$\frac{a_2 - b_2}{\sqrt{Z_0}} = Y_{21}(a_1 + b_1)\sqrt{Z_0} + Y_{22}(a_2 + b_2)\sqrt{Z_0} \tag{1.53}$$

which can then be respectively converted to

$$-b_1(1+Y_{11}Z_0) - b_2Y_{12}Z_0 = -a_1(1-Y_{11}Z_0) + a_2Y_{12}Z_0 \tag{1.54}$$

$$-b_1Y_{21}Z_0 - b_2(1+Y_{22}Z_0) = a_1Y_{21}Z_0 - a_2(1-Y_{22}Z_0). \tag{1.55}$$

Eqs. (1.54) and (1.55) can be easily solved for the reflected waves b_1 and b_2 as

$$b_1\left[(1+Y_{11}Z_0)(1+Y_{22}Z_0) - Y_{12}Y_{21}Z_0^2\right] = a_1\left[(1-Y_{11}Z_0)(1+Y_{22}Z_0) + Y_{12}Y_{21}Z_0^2\right] - 2a_2Y_{12}Z_0 \tag{1.56}$$

$$b_2\left[(1+Y_{11}Z_0)(1+Y_{22}Z_0) - Y_{12}Y_{21}Z_0^2\right] = -2a_1Y_{21}Z_0 + a_2\left[(1+Y_{11}Z_0)(1-Y_{22}Z_0) + Y_{12}Y_{21}Z_0^2\right] \tag{1.57}$$

Comparing equivalent Eqs. (1.43) and (1.44) and Eqs. (1.56) and (1.57) gives the following relationships between the scattering S-parameters and admittance Y-parameters:

$$S_{11} = \frac{(1-Y_{11}Z_0)(1+Y_{22}Z_0) + Y_{12}Y_{21}Z_0^2}{(1+Y_{11}Z_0)(1+Y_{22}Z_0) - Y_{12}Y_{21}Z_0^2} \tag{1.58}$$

$$S_{12} = \frac{-2Y_{12}Z_0}{(1+Y_{11}Z_0)(1+Y_{22}Z_0) - Y_{12}Y_{21}Z_0^2} \tag{1.59}$$

$$S_{21} = \frac{-2Y_{21}Z_0}{(1+Y_{11}Z_0)(1+Y_{22}Z_0) - Y_{12}Y_{21}Z_0^2} \tag{1.60}$$

$$S_{22} = \frac{(1+Y_{11}Z_0)(1-Y_{22}Z_0) + Y_{12}Y_{21}Z_0^2}{(1+Y_{11}Z_0)(1+Y_{22}Z_0) - Y_{12}Y_{21}Z_0^2}. \tag{1.61}$$

Similarly, the relationships of S-parameters with Z-, H-, and $ABCD$-parameters can be obtained for the simplified case when the source impedance Z_S and the load impedance Z_L are equal to the characteristic impedance Z_0 [5].

1.3 INTERCONNECTIONS OF TWO-PORT NETWORKS

When analyzing the behavior of a particular electrical circuit in terms of the two-port network parameters, it is often necessary to define the parameters of a combination of the two or more internal two-port networks. For example, the general feedback amplifier circuit consists of an active two-port network representing the amplifier stage, which is connected in parallel with a passive feedback two-port network. In general, the two-port networks can be interconnected using parallel, series, series–parallel, or cascade connections.

To characterize the resulting two-port networks, it is necessary to take into account which currents and voltages are common for individual two-port networks. The most convenient set of parameters is one for which the common currents and voltages represent a simple linear combination of the independent variables. For the interconnection shown in Figure 1.5(a), the two-port networks Z_a and Z_b are connected in series for both the input and output terminals. Therefore, the currents flowing through these terminals are equal when

$$I_1 = I_{1a} = I_{1b} \quad I_2 = I_{2a} = I_{2b} \tag{1.62}$$

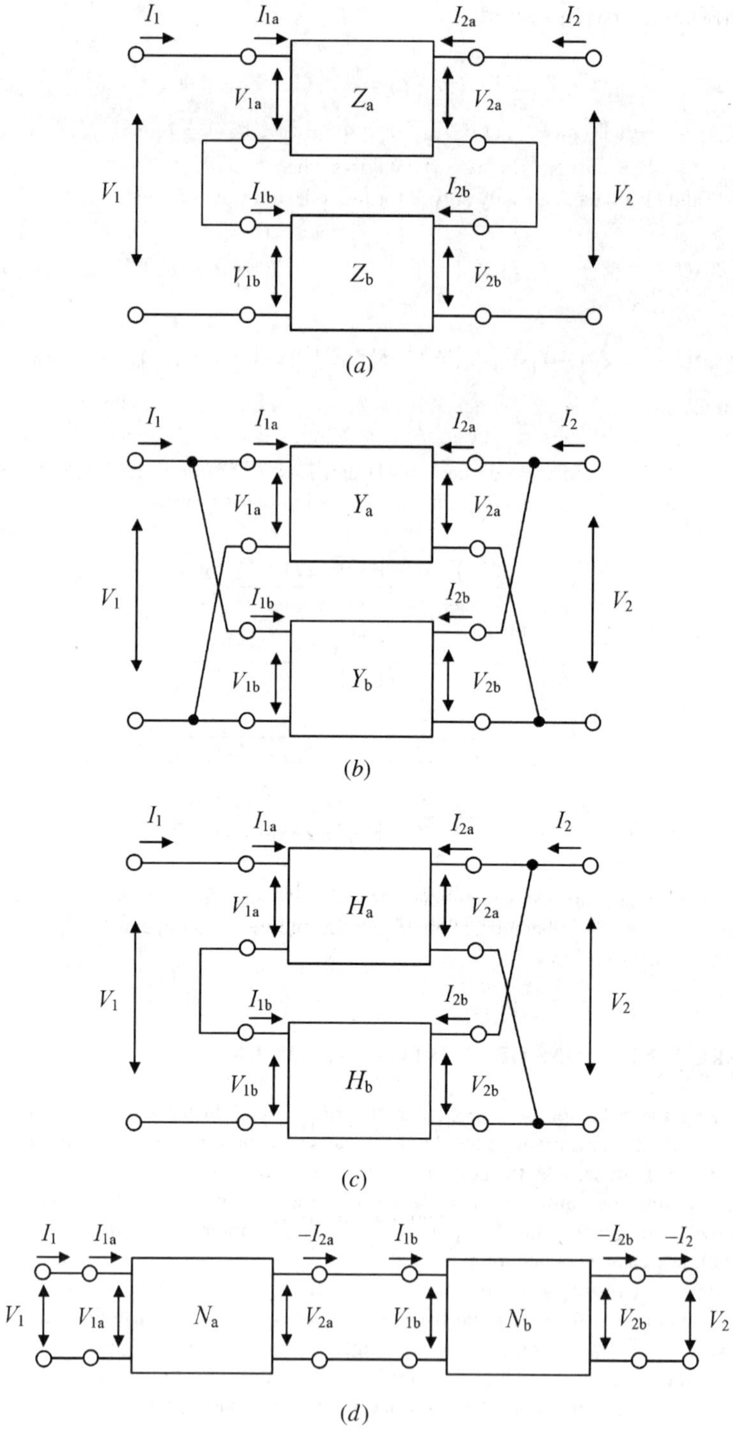

FIGURE 1.5 Different interconnections of two-port networks.

or, in a matrix form,

$$[I] = [I_a] = [I_b]\,. \tag{1.63}$$

The terminal voltages of the resulting two-port network represent the corresponding sums of the terminal voltages of the individual two-port networks when

$$V_1 = V_{1a} + V_{1b} \quad V_2 = V_{2a} + V_{2b} \tag{1.64}$$

or, in a matrix form,

$$[V] = [V_a] + [V_b]\,. \tag{1.65}$$

The currents are common components, both for the resulting and individual two-port networks. Consequently, to describe the properties of such a circuit, it is most convenient to use the impedance matrices. For each two-port network Z_a and Z_b, we can write using Eq. (1.62), respectively,

$$[V_a] = [Z_a]\,[I_a] = [Z_a]\,[I] \tag{1.66}$$

$$[V_b] = [Z_b]\,[I_b] = [Z_b]\,[I]\,. \tag{1.67}$$

Adding both sides of Eqs. (1.66) and (1.67) yields

$$[V] = [Z]\,[I] \tag{1.68}$$

where

$$[Z] = [Z_a] + [Z_b] = \begin{bmatrix} Z_{11a} + Z_{11b} & Z_{12a} + Z_{12b} \\ Z_{21a} + Z_{21b} & Z_{22a} + Z_{22b} \end{bmatrix}. \tag{1.69}$$

The circuit shown in Figure 1.5(*b*) is composed of the two-port networks Y_a and Y_b connected in parallel, where the common components for both resulting and individual two-port networks are input and output voltages, respectively,

$$V_1 = V_{1a} = V_{1b} \quad V_2 = V_{2a} = V_{2b} \tag{1.70}$$

or, in a matrix form,

$$[V] = [V_a] = [V_b]\,. \tag{1.71}$$

Consequently, to describe the circuit properties in this case, it is convenient to use the admittance matrices that give the resulting matrix equation in the form

$$[I] = [Y]\,[V] \tag{1.72}$$

where

$$[Y] = [Y_a] + [Y_b] = \begin{bmatrix} Y_{11a} + Y_{11b} & Y_{12a} + Y_{12b} \\ Y_{21a} + Y_{21b} & Y_{22a} + Y_{22b} \end{bmatrix}. \tag{1.73}$$

The series connection of the input terminals and parallel connection of the output terminals are characterized by the circuit in Figure 1.5(*c*), which shows a series–parallel connection of two-port networks. The common components for this circuit are the input currents and the output voltages. As a result, it is most convenient to analyze the circuit properties using hybrid matrices. The resulting two-port hybrid matrix is equal to the sum of the two individual hybrid matrices written as

$$[H] = [H_a] + [H_b] = \begin{bmatrix} h_{11a} + h_{11b} & h_{12a} + h_{12b} \\ h_{21a} + h_{21b} & h_{22a} + h_{22b} \end{bmatrix}. \tag{1.74}$$

Figure 1.5(*d*) shows the cascade connection of the two individual two-port networks. For such an approach using the one-by-one interconnection of the two-port networks, the output voltage and the output current of the first network is equal to the input voltage and the input current of the second one, respectively, when

$$V_1 = V_{1a} \quad I_1 = I_{1a} \tag{1.75}$$

$$V_{2a} = V_1 \quad -I_{2a} = I_{1b} \tag{1.76}$$

$$V_{2b} = V_2 \quad -I_{2b} = -I_2 \tag{1.77}$$

In this case, it is convenient to use a system of *ABCD*-parameters given by Eqs. (1.21) and (1.22). As a result, for the first individual two-port network shown in Figure 1.5(*d*),

$$\begin{bmatrix} V_{1a} \\ I_{1a} \end{bmatrix} = \begin{bmatrix} A_a & B_a \\ C_a & D_a \end{bmatrix} \begin{bmatrix} V_{2a} \\ -I_{2a} \end{bmatrix} \tag{1.78}$$

or, using Eqs. (1.75) and (1.76),

$$\begin{bmatrix} V_1 \\ I_1 \end{bmatrix} = \begin{bmatrix} A_a & B_a \\ C_a & D_a \end{bmatrix} \begin{bmatrix} V_{1b} \\ I_{1b} \end{bmatrix}. \tag{1.79}$$

Similarly, for the second individual two-port network,

$$\begin{bmatrix} V_{1b} \\ I_{1b} \end{bmatrix} = \begin{bmatrix} A_b & B_b \\ C_b & D_b \end{bmatrix} \begin{bmatrix} V_{2b} \\ -I_{2b} \end{bmatrix} = \begin{bmatrix} A_b & B_b \\ C_b & D_b \end{bmatrix} \begin{bmatrix} V_2 \\ -I_2 \end{bmatrix}. \tag{1.80}$$

Then, substituting matrix Eq. (1.80) to matrix Eq. (1.79) yields

$$\begin{bmatrix} V_1 \\ I_1 \end{bmatrix} = \begin{bmatrix} A_a & B_a \\ C_a & D_a \end{bmatrix} \begin{bmatrix} A_b & B_b \\ C_b & D_b \end{bmatrix} \begin{bmatrix} V_2 \\ -I_2 \end{bmatrix} = \begin{bmatrix} A & B \\ C & D \end{bmatrix} \begin{bmatrix} V_2 \\ -I_2 \end{bmatrix}. \tag{1.81}$$

Consequently, the transmission matrix of the resulting two-port network obtained by the cascade connection of the two or more individual two-port networks is determined by multiplying the transmission matrices of the individual networks. This important property is widely used in the analysis and design of transmission networks and systems.

1.4 PRACTICAL TWO-PORT NETWORKS

1.4.1 Single-Element Networks

The simplest networks, which include only one element, can be constructed by a series-connected admittance Y, as shown in Figure 1.6(*a*), or by a parallel-connected impedance Z, as shown in Figure 1.6(*b*).

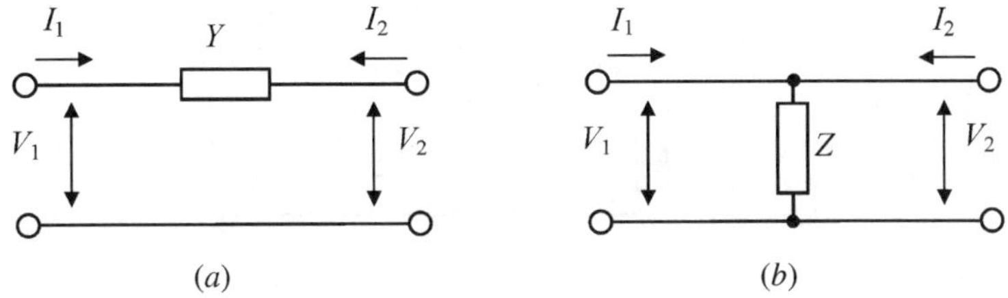

FIGURE 1.6 Single-element networks.

The two-port network consisting of the single series admittance Y can be described in a system of Y-parameters as

$$I_1 = YV_1 - YV_2 \tag{1.82}$$

$$I_2 = -YV_1 + YV_2 \tag{1.83}$$

or, in a matrix form,

$$[Y] = \begin{bmatrix} Y & -Y \\ -Y & Y \end{bmatrix} \tag{1.84}$$

which means that $Y_{11} = Y_{22} = Y$ and $Y_{12} = Y_{21} = -Y$. The resulting matrix is a singular matrix with $|Y| = 0$. Consequently, it is impossible to determine such a two-port network with the series admittance Y-parameters through a system of Z-parameters. However, by using H- and $ABCD$-parameters, it can be described, respectively, by

$$[H] = \begin{bmatrix} 1/Y & 1 \\ -1 & 0 \end{bmatrix} \quad [ABCD] = \begin{bmatrix} 1 & 1/Y \\ 0 & 1 \end{bmatrix}. \tag{1.85}$$

Similarly, for a two-port network with the single parallel impedance Z,

$$[Z] = \begin{bmatrix} Z & Z \\ Z & Z \end{bmatrix} \tag{1.86}$$

which means that $Z_{11} = Z_{12} = Z_{21} = Z_{22} = Z$. The resulting matrix is a singular matrix with $|Z| = 0$. In this case, it is impossible to determine such a two-port network with the parallel impedance Z-parameters through a system of Y-parameters. By using H- and $ABCD$-parameters, this two-port network can be described by

$$[H] = \begin{bmatrix} 0 & 1 \\ -1 & 1/Z \end{bmatrix} \quad [ABCD] = \begin{bmatrix} 1 & 0 \\ 1/Z & 1 \end{bmatrix}. \tag{1.87}$$

1.4.2 π- and T-Type Networks

The basic configurations of a two-port network that usually describe the electrical properties of the active devices can be represented in the form of a π-circuit shown in Figure 1.7(a) and in the form of a T-circuit shown in Figure 1.7(b). Here, the π-circuit includes the current source $g_m V_1$ and the T-circuit includes the voltage source $r_m I_1$.

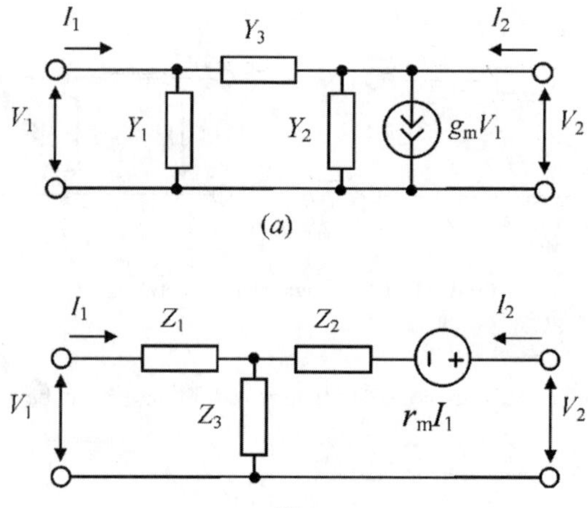

FIGURE 1.7 Basic diagrams of π- and T-networks.

By writing the two loop equations using Kirchhoff's current law or applying Eqs. (1.13) and (1.14) for the π-circuit, we obtain

$$I_1 - (Y_1 + Y_3)\,V_1 + Y_3 V_2 = 0 \tag{1.88}$$

$$I_2 + (g_m - Y_3)\,V_1 + (Y_2 + Y_3)\,V_2 = 0. \tag{1.89}$$

Eqs. (1.88) and (1.89) can be rewritten as matrix Eq. (1.3) with

$$[M] = \begin{bmatrix} 1 & 0 \\ 0 & 1 \end{bmatrix} \quad \text{and} \quad [N] = \begin{bmatrix} -(Y_1 + Y_3) & Y_3 \\ -g_m + Y_3 & -(Y_2 + Y_3) \end{bmatrix}.$$

Since matrix $[M]$ is nonsingular, such a two-port network can be described by a system of Y-parameters as

$$[Y] = -[M]^{-1}[N] = \begin{bmatrix} Y_1 + Y_3 & -Y_3 \\ g_m - Y_3 & Y_2 + Y_3 \end{bmatrix}. \tag{1.90}$$

Similarly, for a two-port network in the form of a T-circuit using Kirchhoff's voltage law or applying Eqs. (1.8) and (1.9), we obtain

$$[Z] = -[M]^{-1}[N] = \begin{bmatrix} Z_1 + Z_3 & Z_3 \\ r_m + Z_3 & Z_2 + Z_3 \end{bmatrix}. \tag{1.91}$$

If $g_m = 0$ for a π-circuit and $r_m = 0$ for a T-circuit, their corresponding matrices in a system of $ABCD$-parameters can be written as, for π-circuit,

$$[ABCD] = \begin{bmatrix} 1 + \dfrac{Y_2}{Y_3} & \dfrac{1}{Y_3} \\ Y_1 + Y_2 + \dfrac{Y_1 Y_2}{Y_3} & 1 + \dfrac{Y_1}{Y_3} \end{bmatrix} \tag{1.92}$$

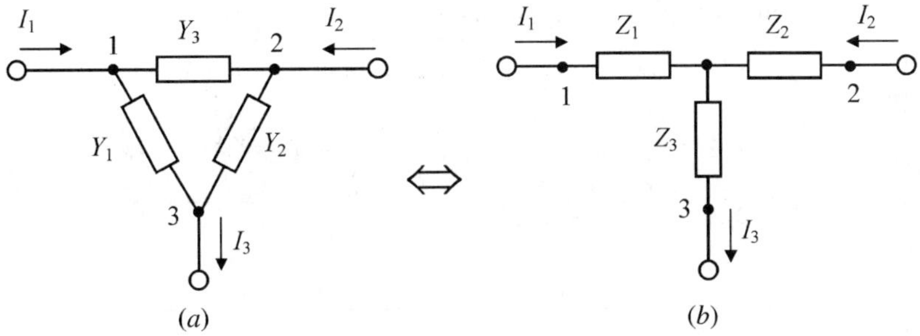

FIGURE 1.8 Equivalence of π- and T-circuits.

for T-circuit,

$$[ABCD] = \begin{bmatrix} 1 + \dfrac{Z_2}{Z_3} & Z_1 + Z_2 + \dfrac{Z_1 Z_2}{Z_3} \\ \dfrac{1}{Z_3} & 1 + \dfrac{Z_1}{Z_3} \end{bmatrix}. \quad (1.93)$$

For the appropriate relationships between impedances of a T-circuit and admittances of a π-circuit, these two circuits become equivalent with respect to the effect on any other two-port network. For a π-circuit shown in Figure 1.8(a), we can write

$$I_1 = Y_1 V_{13} + Y_3 V_{12} = Y_1 V_{13} + Y_3 (V_{13} - V_{23}) = (Y_1 + Y_3) V_{13} - Y_3 V_{23} \quad (1.94)$$

$$I_2 = Y_2 V_{23} - Y_3 V_{12} = Y_2 V_{23} - Y_3 (V_{13} - V_{23}) = -Y_3 V_{13} + (Y_2 + Y_3) V_{23}. \quad (1.95)$$

Solving Eqs. (1.94) and (1.95) for voltages V_{13} and V_{23} yields

$$V_{13} = \frac{Y_2 + Y_3}{Y_1 Y_2 + Y_1 Y_2 + Y_1 Y_2} I_1 + \frac{Y_3}{Y_1 Y_2 + Y_1 Y_2 + Y_1 Y_2} I_2 \quad (1.96)$$

$$V_{23} = \frac{Y_3}{Y_1 Y_2 + Y_1 Y_2 + Y_1 Y_2} I_1 + \frac{Y_1 + Y_3}{Y_1 Y_2 + Y_1 Y_2 + Y_1 Y_2} I_2. \quad (1.97)$$

Similarly, for a T-circuit shown in Figure 1.8(b),

$$V_{13} = Z_1 I_1 + Z_3 I_3 = Z_1 I_1 + Z_3 (I_1 + I_2) = (Z_1 + Z_3) I_1 + Z_3 I_2 \quad (1.98)$$

$$V_{13} = Z_1 I_1 + Z_3 I_3 = Z_1 I_1 + Z_3 (I_1 + I_2) = Z_3 I_1 + (Z_2 + Z_3) I_2 \quad (1.99)$$

and the equations for currents I_1 and I_2 can be obtained by

$$I_1 = \frac{Z_2 + Z_3}{Z_1 Z_2 + Z_1 Z_2 + Z_1 Z_2} V_{13} - \frac{Z_3}{Z_1 Z_2 + Z_1 Z_2 + Z_1 Z_2} V_{23} \quad (1.100)$$

$$I_2 = -\frac{Z_3}{Z_1 Z_2 + Z_1 Z_2 + Z_1 Z_2} V_{13} + \frac{Z_1 + Z_3}{Z_1 Z_2 + Z_1 Z_2 + Z_1 Z_2} V_{23}. \quad (1.101)$$

To establish a T- to π-transformation, it is necessary to equate the coefficients for V_{13} and V_{23} in Eqs. (1.100) and (1.101) to the corresponding coefficients in Eqs. (1.94) and (1.95). Similarly, to establish a π- to T-transformation, it is necessary to equate the coefficients for I_1 and I_2 in Eqs. (1.98)

TABLE 1.2 Relationships Between π- and T-Circuit Parameters.

T- to π-Transformation	π- to T-Transformation
$Y_1 = \dfrac{Z_2}{Z_1Z_2 + Z_2Z_3 + Z_1Z_3}$	$Z_1 = \dfrac{Y_2}{Y_1Y_2 + Y_2Y_3 + Y_1Y_3}$
$Y_2 = \dfrac{Z_1}{Z_1Z_2 + Z_2Z_3 + Z_1Z_3}$	$Z_2 = \dfrac{Y_1}{Y_1Y_2 + Y_2Y_3 + Y_1Y_3}$
$Y_3 = \dfrac{Z_3}{Z_1Z_2 + Z_2Z_3 + Z_1Z_3}$	$Z_3 = \dfrac{Y_3}{Y_1Y_2 + Y_2Y_3 + Y_1Y_3}$

and (1.99) to the corresponding coefficients in Eqs. (1.96) and (1.97). The resulting relationships between admittances for a π-circuit and impedances for a T-circuit are given in Table 1.2.

1.5 THREE-PORT NETWORK WITH COMMON TERMINAL

The concept of a two-port network with two independent sources can generally be extended to any multi-port networks. Figure 1.9 shows the three-port network where all three independent sources are connected to a common point. The three-port network matrix Eq. (1.3) can be described in a scalar form as

$$\left.\begin{aligned} m_{11}V_1 + m_{12}V_2 + m_{13}V_3 + n_{11}I_1 + n_{12}I_2 + n_{13}I_3 = 0 \\ m_{21}V_1 + m_{22}V_2 + m_{23}V_3 + n_{21}I_1 + n_{22}I_2 + n_{23}I_3 = 0 \\ m_{31}V_1 + m_{32}V_2 + m_{33}V_3 + n_{31}I_1 + n_{32}I_2 + n_{33}I_3 = 0 \end{aligned}\right\}. \tag{1.102}$$

If matrix $[N]$ in Eq. (1.102) is nonsingular when $|N| \neq 0$, this system of three equations can be rewritten in admittance matrix representation in terms of the voltage matrix $[V]$, similarly to a two-port network, by

$$\begin{bmatrix} I_1 \\ I_2 \\ I_3 \end{bmatrix} = \begin{bmatrix} Y_{11} & Y_{12} & Y_{13} \\ Y_{21} & Y_{22} & Y_{23} \\ Y_{31} & Y_{32} & Y_{33} \end{bmatrix} \begin{bmatrix} V_1 \\ V_2 \\ V_3 \end{bmatrix}. \tag{1.103}$$

The matrix $[Y]$ in Eq. (1.103) is the indefinite admittance matrix of the three-port network and represents a singular matrix with two important properties: the sum of all terminal currents entering the circuit is equal to zero, that is, $I_1 + I_2 + I_3 = 0$; and all terminal currents entering the circuit

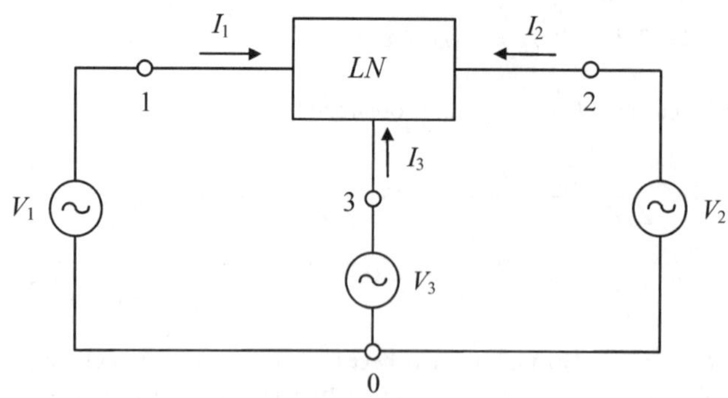

FIGURE 1.9 Basic diagram of three-port network with common terminal.

depend on the voltages between circuit terminals, which makes the sum of all terminal voltages equal to zero, that is, $V_{13} + V_{32} + V_{21} = 0$.

According to the first property, adding the left and right parts of matrix Eq. (1.103) results in

$$(Y_{11} + Y_{21} + Y_{31}) V_1 + (Y_{12} + Y_{22} + Y_{32}) V_2 + (Y_{13} + Y_{23} + Y_{33}) V_3 = 0. \tag{1.104}$$

Since all terminal voltages (V_1, V_2, and V_3) can be set independently from each other, Eq. (1.104) can be satisfied only if any column sum is identically zero,

$$\left.\begin{array}{l} Y_{11} + Y_{21} + Y_{31} = 0 \\ Y_{12} + Y_{22} + Y_{32} = 0 \\ Y_{13} + Y_{23} + Y_{33} = 0 \end{array}\right\}. \tag{1.105}$$

The neither terminal currents will neither decrease nor increase, with the simultaneous change of all terminal voltages, by the same magnitude. Consequently, if all terminal voltages are equal to a nonzero value when $V_1 = V_2 = V_3 = V_0$, a lack of the terminal currents occurs when $I_1 = I_2 = I_3 = 0$. For example, from the first row of the matrix Eq. (1.103) it follows that $I_1 = Y_{11}V_1 + Y_{12}V_2 + Y_{13}V_3$; then we can write

$$0 = (Y_{11} + Y_{12} + Y_{13}) V_0 \tag{1.106}$$

which results due to the nonzero value V_0 in

$$Y_{11} + Y_{12} + Y_{13} = 0. \tag{1.107}$$

Applying the same approach to other two rows results in

$$\left.\begin{array}{l} Y_{11} + Y_{12} + Y_{13} = 0 \\ Y_{21} + Y_{22} + Y_{23} = 0 \\ Y_{31} + Y_{32} + Y_{33} = 0 \end{array}\right\}. \tag{1.108}$$

Consequently, by using Eqs. (1.105) through (1.108), the indefinite admittance Y-matrix of three-port network can be rewritten by

$$[Y] = \begin{bmatrix} Y_{11} & Y_{12} & -(Y_{11} + Y_{12}) \\ Y_{21} & Y_{22} & -(Y_{21} + Y_{22}) \\ -(Y_{11} + Y_{21}) & -(Y_{12} + Y_{22}) & Y_{11} + Y_{12} + Y_{21} + Y_{22} \end{bmatrix}. \tag{1.109}$$

By selecting successively terminal 1, 2, and 3 as the datum terminal, the corresponding three two-port admittance matrices of the initial three-port network can be obtained. In this case, the admittance matrices will correspond to a common emitter configuration shown in Figure 1.10(a), a common base configuration shown in Figure 1.10(b), and a common collector configuration of the

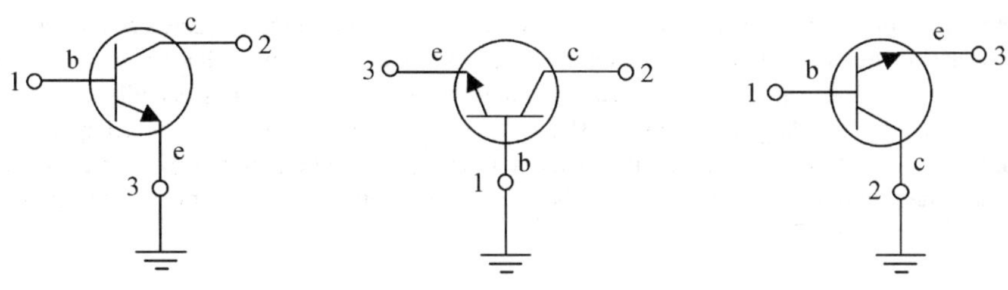

FIGURE 1.10 Bipolar transistors with different common terminals.

TABLE 1.3 *Y*- and *Z*-Parameters of Active Device with Different Common Terminal.

	Y-Parameters		*Z*-Parameters	
Common emitter (source)	Y_{11} Y_{21}	Y_{12} Y_{22}	Z_{11} Z_{21}	Z_{12} Z_{22}
Common base (gate)	$Y_{11}+Y_{12}+Y_{21}+Y_{22}$ $-(Y_{21}+Y_{22})$	$-(Y_{12}+Y_{22})$ Y_{22}	$Z_{11}+Z_{12}+Z_{21}+Z_{22}$ $-(Z_{21}+Z_{22})$	$-(Z_{12}+Z_{22})$ Z_{22}
Common collector (drain)	Y_{11} $-(Y_{11}+Y_{21})$	$-(Y_{11}+Y_{12})$ $Y_{11}+Y_{12}+Y_{21}+Y_{22}$	Z_{11} $-(Z_{11}+Z_{21})$	$-(Z_{11}+Z_{12})$ $Z_{11}+Z_{12}+Z_{21}+Z_{22}$

bipolar transistor shown in Figure 1.10(*c*), respectively. If the common emitter device is treated as a two-port network characterized by four *Y*-parameters (Y_{11}, Y_{12}, Y_{21}, and Y_{22}), the two-port matrix of the common collector configuration with grounded collector terminal is simply obtained by deleting the second row and the second column in matrix Eq. (1.109). For the common base configuration with grounded base terminal, the first row and the first column should be deleted because the emitter terminal is considered the input terminal.

A similar approach can be applied to the indefinite three-port impedance network. This allows the *Z*-parameters of the impedance matrices of the common base and the common collector configurations through known impedance *Z*-parameters of the common emitter configuration of the transistor to be determined. Parameters of the three-port network that can describe the electrical behavior of the three-port bipolar or field-effect transistor configured with different common terminals are given in Table 1.3.

1.6 LUMPED ELEMENTS

Generally, passive RF and microwave lumped or integrated circuits are designed based on the lumped elements, distributed elements, or combination of both types of elements. Distributed elements represent any sections of the transmission lines of different lengths, types, and characteristic impedances. The basic lumped elements are inductors and capacitors that are small in size in comparison with the transmission-line wavelength λ, and usually their linear dimensions are less than $\lambda/10$ or even $\lambda/16$. In applications where lumped elements are used, their basic advantages are small physical size and low production cost. However, their main drawbacks are lower quality factor and power-handling capability compared with distributed elements.

1.6.1 Inductors

Inductors are lumped elements that store energy in a magnetic field. Lumped inductors can be realized using several different configurations, such as a short-section of a strip conductor or wire, a single loop, or a spiral. The printed high-impedance microstrip-section inductor is usually used for low inductance values, typically less than 2 nH, and often meandered to reduce the component size. The printed microstrip single-loop inductors are not very popular due to their limited inductance per unit area. The approximate expression for the microstrip short-section inductance in free space is given by

$$L\,(\text{nH}) = 0.2 \times 10^{-3} l \left[\ln\left(\frac{l}{W+t}\right) + 1.193 + \frac{W+t}{3l} \right] K_{\text{g}} \tag{1.110}$$

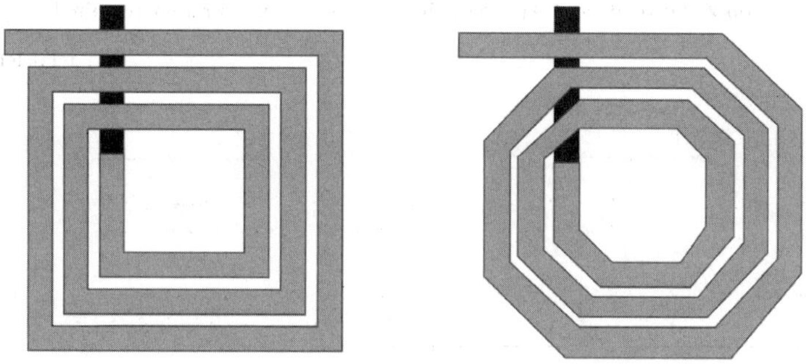

(a) (b)

FIGURE 1.11 Spiral inductor layouts.

where the conductor length l, conductor width W, and conductor thickness t are in microns, and the term K_g accounts for the presence of a ground plane, defined as

$$K_g = 0.57 - 0.145 \ln \frac{W}{h}, \quad \text{for } \frac{W}{h} > 0.05 \tag{1.111}$$

where h is the spacing from ground plane [6,7].

Spiral inductors can have a circular configuration, a rectangular (square) configuration shown in Figure 1.11(*a*), or an octagonal configuration shown in Figure 1.11(*b*), if the technology allows 45° routing. The circular geometry is superior in electrical performance, whereas the rectangular shapes are easy to lay out and fabricate. Printed inductors are based on using thin-film or thick-film Si or GaAs fabrication processes, and the inner conductor is pulled out to connect with other circuitry through a bondwire, an air bridge, or by using multilevel crossover metal. The general expression for spiral inductor, which is also valid for its planar integration within accuracy of around 3%, is based on Wheeler formula and can be obtained as

$$L\,(\text{nH}) = \frac{K_1 n^2 d_{avg}}{1 + K_2 \rho} \tag{1.112}$$

where n is the number of turns, $d_{avg} = (d_{out} + d_{in})/2$ is the average diameter, $\rho = (d_{out} + d_{in})/(d_{out} - d_{in})$ is the fill ratio, d_{out} is the outer diameter in μm, d_{in} is the inner diameter in μm, and the coefficients K_1 and K_2 are layout-dependent as follows: square—$K_1 = 2.34$, $K_2 = 2.75$; hexagonal—$K_1 = 2.33$, $K_2 = 3.82$; octagonal—$K_1 = 2.25$, $K_2 = 3.55$ [8,9].

In contrast to the capacitors, high-quality inductors cannot be readily available in a standard complementary metal-oxide-semiconductor (CMOS) technology. Therefore, it is necessary to use special techniques to improve the inductor electrical performance. By using a standard CMOS technology with only two metal layers and a heavily doped substrate, the spiral inductor will have a large series resistance, compared with three–four metal layer technologies, and the substrate losses become a very important factor due to a relatively low resistivity of silicon. A major source of substrate losses is the capacitive coupling when current is flowing not only through the metal strip, but also through the silicon substrate. Another important source of substrate losses is the inductive coupling when, due to the planar inductor structure, the magnetic field penetrates deeply into the silicon substrate, inducing current loops and related losses. However, the latter effects are particularly important for large-area inductors and can be overcome by using silicon micromachining techniques [10].

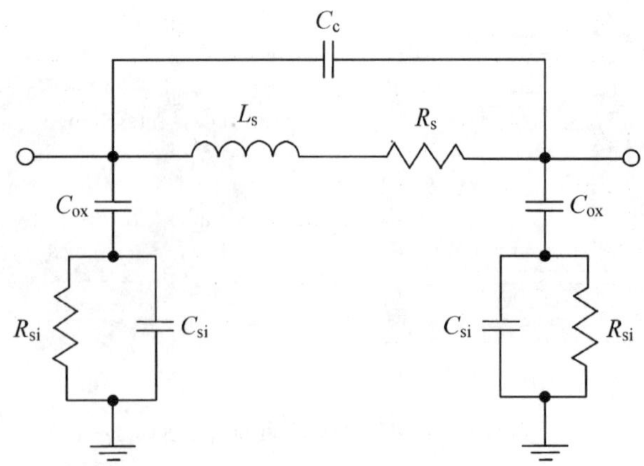

FIGURE 1.12 Equivalent circuit of a square spiral inductor.

The simplified equivalent circuit for the CMOS spiral microstrip inductor is shown in Figure 1.12, where L_s models the self and mutual inductances, R_s is the series coil resistance, C_{ox} is the parasitic oxide capacitance from the metal layer to the substrate, R_{si} is the resistance of the conductive silicon substrate, C_{si} is the silicon substrate parasitic capacitance, and C_c is the parasitic coupling capacitance [11]. The parasitic silicon substrate capacitance C_{si} is sufficiently small, and in most cases it can be neglected. Such a model shows an accurate agreement between simulated and measured data within 10% across a variety of inductor geometries and substrate dopings up to 20 GHz [12]. At frequencies well below the inductor self-resonant frequency ω_{SRF}, the coupling capacitance C_c between metal segments due to fringing fields in both the dielectric and air regions can also be neglected since the relative dielectric constant of the oxide is small enough [13]. In this case, if one side of the inductor is grounded, the self-resonant frequency of the spiral inductor can approximately be calculated from

$$\omega_{SRF} = \frac{1}{\sqrt{L_s C_{ox}}} \sqrt{\frac{L_s - R_s^2 C_{ox}}{L_s - R_{si}^2 C_{ox}}}. \tag{1.113}$$

At frequencies higher than self-resonant frequency ω_{SRF}, the inductor exhibits a capacitive behavior. The self-resonant frequency ω_{SRF} is limited mainly by the parasitic oxide capacitance C_{ox}, which is inversely proportional to the oxide thickness between the metal layer and substrate. The frequency at which the inductor quality factor Q is maximal can be obtained as

$$\omega_Q = \frac{1}{\sqrt{L_s C_{ox}}} \sqrt{\frac{R_s}{2R_{si}}} \left(\sqrt{1 + \frac{4R_{si}}{3R_s}} - 1 \right)^{0.5}. \tag{1.114}$$

The inductor metal conductor series resistance R_s can be easily calculated at low frequencies as the product of the sheet resistance and the number of squares of the metal trace. However, at high frequencies, the skin effect and other magnetic field effects will cause a nonuniform current distribution in the inductor profile. In this case, a simple increase in the diameter of the inductor metal turn does not necessarily reduce correspondingly the inductor series resistance. For example, for the same inductance value, the difference in resistance between the two inductors, when one of which has a two times wider metal strip, is only a factor of 1.35 [14]. Moreover, at very high frequencies, the largest contribution to the series resistance does not come from the longer outer turns, but from the inner turns. This phenomenon is a result of the generation of circular eddy currents in the inner conductors, whose direction is such that they oppose the original change in magnetic field. On the inner side of the inner turn, coil current and eddy current flow in the same direction, so the current density is larger than average. On the outer side, both currents cancel, and the current density is smaller than average. As a result, the current in the inner turn is pushed to the inside of the conductor.

In hybrid or monolithic applications, bondwires are used to interconnect different components such as lumped elements, planar transmission lines, solid-state devices, and integrated circuits. These bondwires, which are usually made of gold or aluminium, have 0.5- to 1.0-mil diameters, and their lengths are electrically shorter compared with the operating wavelength. To characterize the electrical behavior of the bondwires, simple formulas in terms of their inductances and series resistances can be used. As a first-order approximation, the parasitic capacitance associated with bondwires can be neglected. When $l \gg d$, where l is the bondwire length in μm and d is the bondwire diameter in μm,

$$L\,(\text{nH}) = 0.2 \times 10^{-3} l \left(\ln \frac{4l}{d} + 0.5 \frac{d}{l} - 1 + C \right) \tag{1.115}$$

where $C = 0.25 \tanh\,(4\delta/d)$ is the frequency-dependent correction factor, which is a function of bondwire diameter and its material's skin depth δ [8,15].

1.6.2 Capacitors

Capacitors are lumped elements that store energy due to an electric field between two electrodes (or plates) when a voltage is applied across them. In this case, charge of equal magnitude but opposite sign accumulates on the opposing capacitor plates. The capacitance depends on the area of the plates, separation, and dielectric material between them. The basic structure of a chip capacitor shown in Figure 1.13(a) consists of two parallel plates, each of area $A = W \times l$ and separated by a dielectric material of thickness d and permittivity $\varepsilon_0\varepsilon_r$, where ε_0 is the free-space permittivity (8.85×10^{-12} farads/m) and ε_r is the relative dielectric constant.

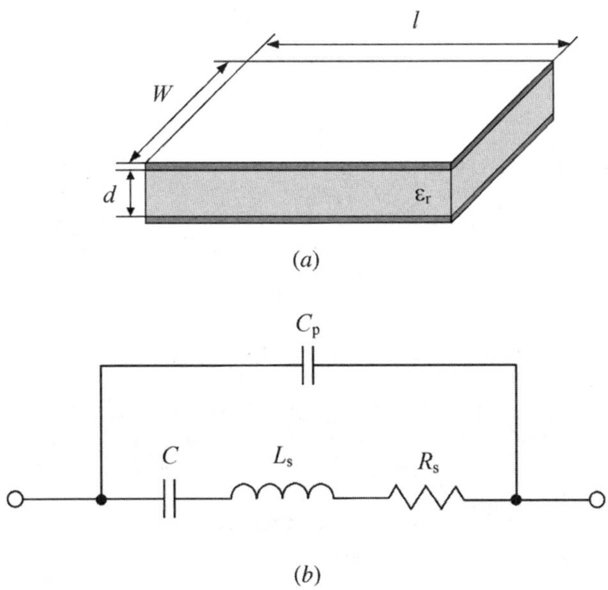

FIGURE 1.13 Parallel capacitor topology and its equivalent circuit.

Chip capacitors are usually used in hybrid integrated circuits when relatively high capacitance values are required. In the parallel-plate configuration, the capacitance is commonly expressed as

$$C\,(\text{pF}) = 8.85 \times 10^{-3}\varepsilon_r \frac{Wl}{d} \qquad (1.116)$$

where W, l, and d are dimensions in millimeters. Generally, the low-frequency bypass capacitor values are expressed in microfarads and nanofarads, high-frequency blocking and tuning capacitors are expressed in picofarads, and parasitic or fringing capacitances are written in femtofarads. This basic formula given by Eq. (1.116) can also be applied to capacitors based on a multilayer technique [7]. The lumped-element equivalent circuit of a capacitor is shown in Figure 1.13(*b*), where L_s is the series plate inductance, R_s is the series contact and plate resistance, and C_p is the parasitic parallel capacitance. When $C \gg C_p$, the frequency ω_{SRF}, at which the reactances of series elements C and L_s become equal, is called the *capacitor self-resonant frequency*, and the capacitor impedance is equal to the resistance R_s.

For monolithic applications, where relatively low capacitances (typically less than 0.5 pF) are required, planar series capacitances in the form of microstrip or interdigital configurations can be used. These capacitors are simply formed by gaps in the center conductor of the microstrip lines, and they do not require any dielectric films. The gap capacitor, shown in Figure 1.14(*a*), can be equivalently represented by a series coupling capacitance and two parallel fringing capacitances [16]. The interdigital capacitor is a multifinger periodic structure, as shown in Figure 1.14(*b*), where the capacitance occurs across a narrow gap between thin-film transmission-line conductors [17]. These gaps are essentially very long and folded to use a small amount of area. In this case, it is important to keep the size of the capacitor very small relative to the wavelength, so that it can be treated as a lumped element. A larger total width-to-length ratio results in the desired higher shunt capacitance and lower series inductance. An approximate expression for the total capacitance of interdigital structure, with $s = W$ and length l less than a quarter wavelength, can be given by

$$C\,(\text{pF}) = (\varepsilon_r + 1)\,l\,[(N-3)\,A_1 + A_2] \qquad (1.117)$$

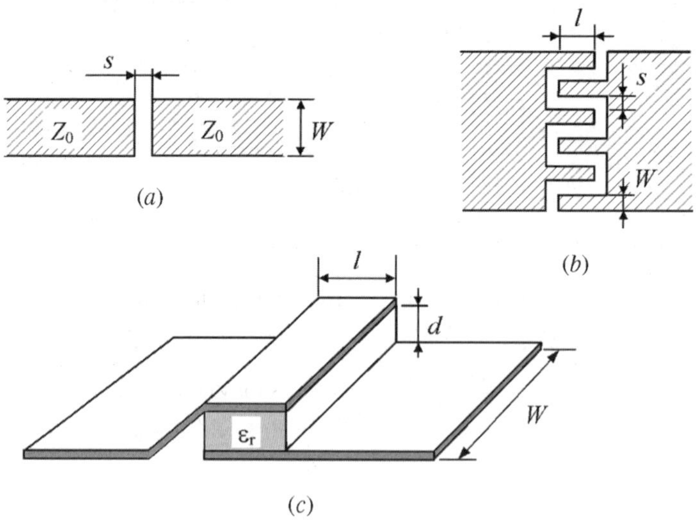

FIGURE 1.14 Different series capacitor topologies.

where N is the number of fingers and

$$A_1\ (\text{pF}/\mu\text{m}) = 4.409 \tanh\left[0.55\left(\frac{h}{W}\right)^{0.45}\right] \times 10^{-6} \tag{1.118}$$

$$A_2\ (\text{pF}/\mu\text{m}) = 9.92 \tanh\left[0.52\left(\frac{h}{W}\right)^{0.5}\right] \times 10^{-6} \tag{1.119}$$

where h is the spacing from the ground plane.

Series planar capacitors with larger values, which are called the metal-insulator-metal (MIM) capacitors, can be realized by using an additional thin dielectric layer (typically less than 0.5 μm) between two metal plates, as shown in Figure 1.14(c) [7]. The bottom plate of the capacitor uses a thin unplated metal, and typically the dielectric material is silicon nitride (Si_3N_4) for integrated circuits on GaAs and SiO_2 for integrated circuits on Si. The top plate uses a thick-plated conductor to reduce the loss in the capacitor. These capacitors are used to achieve higher capacitance values in small areas (10 pF and greater), with typical tolerances from 10% to 15%. The capacitance can be calculated according to Eq. (1.116).

1.7 TRANSMISSION LINE

Transmission lines are widely used in matching circuits in power amplifiers, in resonant and feedback circuits in the oscillators, filters, directional couplers, power combiners, and dividers. When the propagated signal wavelength is compared to its physical dimension, the transmission line can be considered as a two-port network with distributed parameters, where the voltages and currents vary in magnitude and phase over length.

Schematically, a transmission line is often represented as a two-wire line, as shown in Figure 1.15(a), where its electrical parameters are distributed along its length. The physical properties of a transmission line are determined by four basic parameters:

1. The series inductance L due to the self-inductive phenomena of two conductors.
2. The shunt capacitance C in view of the close proximity between two conductors.
3. The series resistance R due to the finite conductivity of the conductors.
4. The shunt conductance G that is related to the dielectric losses in the material.

As a result, a transmission line of length Δx represents a lumped-element circuit shown in Figure 1.15(b), where ΔL, ΔC, ΔR, and ΔG are the series inductance, the shunt capacitance, the series resistance, and the shunt conductance per unit length, respectively. If all these elements are distributed uniformly along the transmission line, and their values do not depend on the chosen position of Δx, this transmission line is called the *uniform transmission line*. Any finite length of the uniform transmission line can be viewed as a cascade of section length Δx.

To define the distribution of the voltages and currents along the uniform transmission line, it is necessary to write the differential equations using Kirchhoff's voltage law for instantaneous values of the voltages and currents in the line section of length Δx, distant x from its beginning. For the sinusoidal steady-state condition, the telegrapher equations for $V(x)$ and $I(x)$ are given by

$$\frac{d^2V(x)}{dx^2} - \gamma^2 V(x) = 0 \tag{1.120}$$

$$\frac{d^2I(x)}{dx^2} - \gamma^2 I(x) = 0 \tag{1.121}$$

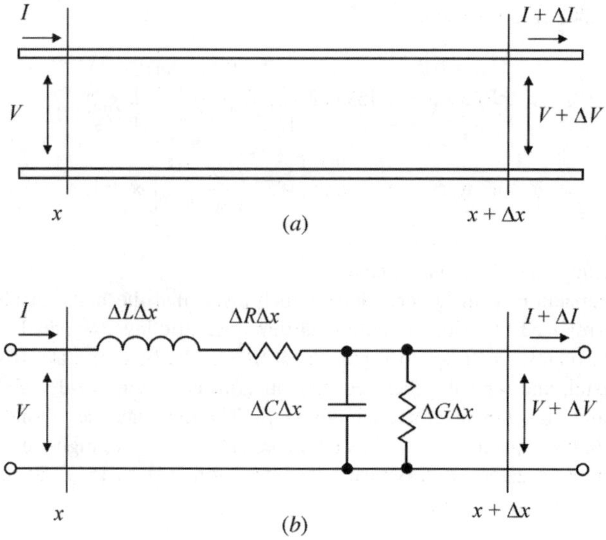

FIGURE 1.15 Transmission line schematics.

where $\gamma = \alpha + j\beta = \sqrt{(\Delta R + j\omega\Delta L)(\Delta G + j\omega\Delta C)}$ is the complex propagation constant (which is a function of frequency), α is the attenuation constant, and β is the phase constant. The general solutions of Eqs. (1.120) and (1.121) for voltage and current of the traveling wave in the transmission line can be written as

$$V(x) = A_1 \exp(-\gamma x) + A_2 \exp(\gamma x) \tag{1.122}$$

$$I(x) = \frac{A_1}{Z_0}\exp(-\gamma x) - \frac{A_2}{Z_0}\exp(\gamma x) \tag{1.123}$$

where $Z_0 = \sqrt{(\Delta R + j\omega\Delta L)/(\Delta G + j\omega\Delta C)}$ is the characteristic impedance of the transmission line, $V_i = A_1\exp(-\gamma x)$ and $V_r = A_2\exp(\gamma x)$ represent the incident voltage and the reflected voltage, respectively, and $I_i = A_1\exp(-\gamma x)/Z_0$ and $I_r = A_2\exp(\gamma x)/Z_0$ are the incident current and the reflected current, respectively. From Eqs. (1.122) and (1.123) it follows that the characteristic impedance of the transmission line Z_0 represents the ratio of the incident (reflected) voltage to the incident (reflected) current at any position on the line as

$$Z_0 = \frac{V_i(x)}{I_i(x)} = \frac{V_r(x)}{I_r(x)}. \tag{1.124}$$

For a lossless transmission line, when $R = G = 0$ and the voltage and current do not change with position, the attenuation constant $\alpha = 0$, the propagation constant $\gamma = j\beta = j\omega\sqrt{\Delta L \Delta C}$, and the phase constant $\beta = \omega\sqrt{\Delta L \Delta C}$. Consequently, the characteristic impedance is reduced to $Z_0 = \sqrt{L/C}$ and represents a real number. The wavelength is defined as $\lambda = 2\pi/\beta = 2\pi/\omega\sqrt{\Delta L \Delta C}$ and the phase velocity as $v_p = \omega/\beta = 1/\sqrt{\Delta L \Delta C}$.

Figure 1.16 represents a transmission line of characteristic impedance Z_0 terminated with a load Z_L. In this case, the constants A_1 and A_2 are determined at the position $x = l$ by

$$V(l) = A_1 \exp(-\gamma l) + A_2 \exp(\gamma l) \tag{1.125}$$

$$I(l) = \frac{A_1}{Z_0}\exp(-\gamma l) - \frac{A_2}{Z_0}\exp(\gamma l) \tag{1.126}$$

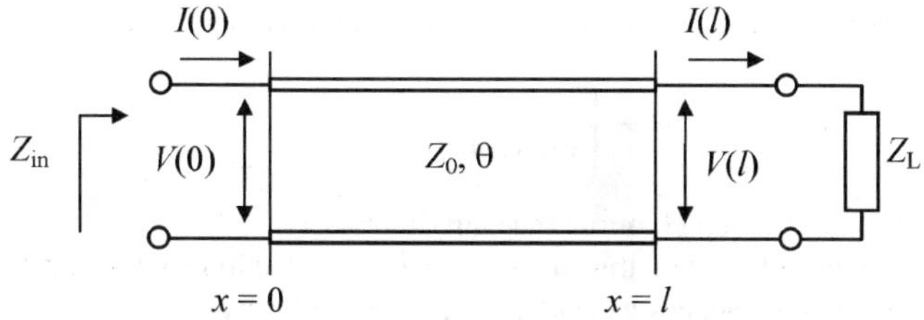

FIGURE 1.16 Loaded transmission line.

and equal to

$$A_1 = \frac{V(l) + Z_0 I(l)}{2} \exp(\gamma l) \tag{1.127}$$

$$A_2 = \frac{V(l) - Z_0 I(l)}{2} \exp(-\gamma l). \tag{1.128}$$

As a result, wave equations for voltage $V(x)$ and current $I(x)$ can be rewritten as

$$V(x) = \frac{V(l) + Z_0 I(l)}{2} \exp[\gamma(l - x)] + \frac{V(l) - Z_0 I(l)}{2} \exp[-\gamma(l - x)] \tag{1.129}$$

$$I(x) = \frac{V(l) + Z_0 I(l)}{2Z_0} \exp[\gamma(l - x)] - \frac{V(l) - Z_0 I(l)}{2Z_0} \exp[-\gamma(l - x)] \tag{1.130}$$

which allows their determination at any position on the transmission line.

The voltage and current amplitudes at $x = 0$ as functions of the voltage and current amplitudes at $x = l$ can be determined from Eqs. (1.129) and (1.130) as

$$V(0) = \frac{V(l) + Z_0 I(l)}{2} \exp(\gamma l) + \frac{V(l) - Z_0 I(l)}{2} \exp(-\gamma l) \tag{1.131}$$

$$I(0) = \frac{V(l) + Z_0 I(l)}{2Z_0} \exp(\gamma l) - \frac{V(l) - Z_0 I(l)}{2Z_0} \exp(-\gamma l). \tag{1.132}$$

By using the ratios $\cosh x = [\exp(x) + \exp(-x)]/2$ and $\sinh x = [\exp(x) - \exp(-x)]/2$, Eqs. (1.131) and (1.132) can be rewritten in the form

$$V(0) = V(l) \cosh(\gamma l) + Z_0 I(l) \sinh(\gamma l) \tag{1.133}$$

$$I(0) = \frac{V(l)}{Z_0} \sinh(\gamma l) + I(l) \cosh(\gamma l) \tag{1.134}$$

which represents the transmission equations of the symmetrical reciprocal two-port network expressed through the $ABCD$-parameters when $AD - BC = 1$ and $A = D$. Consequently, the transmission $ABCD$-matrix of the lossless transmission line with $\alpha = 0$ can be given by

$$[ABCD] = \begin{bmatrix} \cos\theta & jZ_0 \sin\theta \\ \dfrac{j\sin\theta}{Z_0} & \cos\theta \end{bmatrix}. \tag{1.135}$$

Using the formulas to transform *ABCD*-parameters into *S*-parameters yields

$$[S] = \begin{bmatrix} 0 & \exp(-j\theta) \\ \exp(-j\theta) & 0 \end{bmatrix} \tag{1.136}$$

where $\theta = \beta l$ is the electrical length of the transmission line.

In the case of the loaded lossless transmission line, the reflection coefficient Γ is defined as the ratio between the reflected voltage wave and the incident voltage wave, given at x as

$$\Gamma(x) = \frac{V_r}{V_i} = \frac{A_2}{A_1} \exp(2j\beta x). \tag{1.137}$$

By taking into account Eqs. (1.127) and (1.128), the reflection coefficient for $x = l$ can be defined as

$$\Gamma = \frac{Z - Z_0}{Z + Z_0} \tag{1.138}$$

where Γ represents the load reflection coefficient and $Z = Z_L = V(l)/I(l)$. If the load is mismatched, only part of the available power from the source is delivered to the load. This power loss is called the *return loss* (*RL*), and is calculated in decibels as

$$RL = -20\log_{10}|\Gamma|. \tag{1.139}$$

For a matched load when $\Gamma = 0$, a return loss is of ∞ dB. A total reflection with $\Gamma = 1$ means a return loss of 0 dB when all incident power is reflected.

According to the general solution for voltage at a position x in the transmission line,

$$V(x) = V_i(x) + V_r(x) = V_i[1 + \Gamma(x)]. \tag{1.140}$$

Hence, the maximum amplitude (when the incident and reflected waves are in phase) is

$$V_{max}(x) = |V_i|\,[1 + |\Gamma(x)|] \tag{1.141}$$

and the minimum amplitude (when these two waves are out of phase) is

$$V_{min}(x) = |V_i|\,[1 - |\Gamma(x)|]. \tag{1.142}$$

The ratio of V_{max} to V_{min}, which is a function of the reflection coefficient Γ, represents the *voltage standing wave ratio* (*VSWR*). The *VSWR* is a measure of mismatch and can be written as

$$VSWR = \frac{V_{max}}{V_{min}} = \frac{1 + |\Gamma|}{1 - |\Gamma|} \tag{1.143}$$

which can change from 1 to ∞ (where $VSWR = 1$ implies a matched load). For a load impedance with zero imaginary part when $Z_L = R_L$, the *VSWR* can be calculated using $VSWR = R_L/Z_0$ when $R_L \geq Z_0$ and $VSWR = Z_0/R_L$ when $Z_0 \geq R_L$.

From Eqs. (1.133) and (1.134) it follows that the input impedance of the loaded lossless transmission line can be obtained as

$$Z_{in} = \frac{V(0)}{I(0)} = Z_0 \frac{Z_L + jZ_0 \tan(\theta)}{Z_0 + jZ_L \tan(\theta)} \tag{1.144}$$

which gives an important dependence between the input impedance, the transmission-line parameters (electrical length and characteristic impedance), and the arbitrary load impedance.

1.8 TYPES OF TRANSMISSION LINES

Several types of transmission lines are available when designing RF and microwave active and passive circuits. Coaxial lines have very high bandwidth and high power-handling capability, and are widely used for impedance transformers and power combiners. Planar transmission lines as an evolution of the coaxial and parallel-wire lines are compact and readily adaptable to hybrid and monolithic integrated circuit fabrication technologies at RF and microwave frequencies [18]. If coaxial line is deformed in such a manner that both the center and outer conductors are square or rectangular in cross-section, and then if side walls of the rectangular coaxial system are extended to infinity, the resultant flat-strip transmission system would have a form factor that is adaptable to the printed-circuit technique. Similarly, if the parallel-wire line is replaced by its equivalent of a single wire and its image in a conducting ground plane, and if this single wire is, in turn, progressively distorted into a flat strip, the resulting transmission system is again a planar structure. There is an important aspect that differ flat-strip transmission lines from coaxial lines. In a coaxial line, an impedance discontinuity acts as a shunt capacitance, while a discontinuity in a flat strip has a series inductance in its equivalent circuit. Holes and gaps in center conductor strips also represent discontinuities that can be utilized in many applications to microwave circuitry.

1.8.1 Coaxial Line

A main type of wave propagated along a coaxial line shown in Figure 1.17 is the *transverse electromagnetic* (TEM) wave. When the transverse fields of a TEM wave are the same as the static fields that can exist between the conductors, the electromagnetic properties of a coaxial line can be characterized by the following parameters [19]: the shunt capacitance per unit length

$$C = 2\pi\varepsilon_0\varepsilon_r/\ln\left(\frac{b}{a}\right) \tag{1.145}$$

where a is the radius of inner conductor and b is the inner radius of outer conductor; the series inductance per unit length

$$L = \frac{\mu_0\mu_r}{2\pi}\ln\left(\frac{b}{a}\right) \tag{1.146}$$

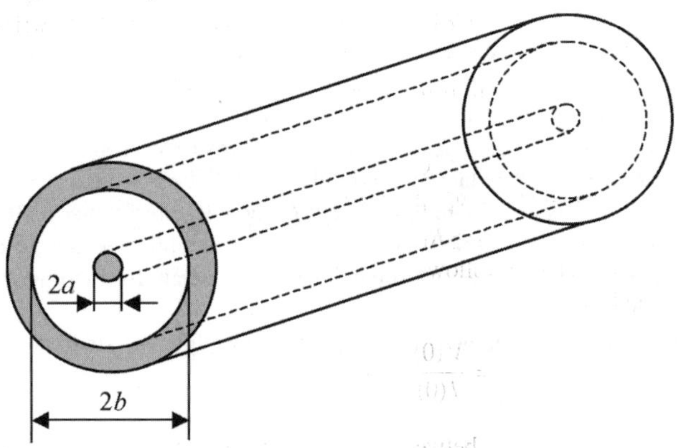

FIGURE 1.17 Coaxial line structure.

where $\mu_0 = 4\pi \times 10^{-7}$ H/m is the permeability of free space and μ_r is the relative magnetic constant or substrate permeability; the series resistance per unit length

$$R = \frac{R_s}{2\pi}\left(\frac{1}{b} + \frac{1}{a}\right) \tag{1.147}$$

where $R_s = \rho/\Delta(f) = \sqrt{\pi\mu_0\rho f}$ is the surface resistivity, ρ is the metallization electrical resistivity, $\Delta(f)$ is the penetration depth, and f is the frequency; the shunt conductance per unit length

$$G = 2\pi\sigma/\ln\left(\frac{b}{a}\right) = 2\pi\omega\varepsilon_0\varepsilon_r \tan\delta/\ln\left(\frac{b}{a}\right) \tag{1.148}$$

where σ is the dielectric conductivity and $\tan\delta$ is the dielectric loss tangent; the characteristic impedance

$$Z_0 = \frac{\eta}{2\pi}\ln\left(\frac{b}{a}\right) \tag{1.149}$$

where $\eta = \sqrt{\mu/\varepsilon}$ is the wave impedance of the lossless coaxial line identical to the intrinsic impedance of the medium.

The conductor loss factor (in Np/m) can be written as

$$\alpha_c = \frac{R}{2Z_0} \tag{1.150}$$

whereas the dielectric loss factor (in Np/m) can be written as

$$\alpha_d = \frac{GZ_0}{2} = \frac{\sigma\eta}{2} = \pi\sqrt{\varepsilon_r}\frac{\tan\delta}{\lambda} \tag{1.151}$$

where λ is the free-space wavelength.

1.8.2 Stripline

The geometry of a commonly used stripline is shown in Figure 1.18. The strip conductor of width W is placed between two flat dielectric substrates with the same dielectric constant. The outer surfaces of these substrates are metallized and serve as a ground conductor. In practice, the strip conductor is etched on one of the dielectric substrates by photolithography process. Since the stripline has two conductors and a homogeneous dielectric, it can support a pure TEM propagation mode, which is the usual mode of operation. The advantages of striplines are good electromagnetic shielding and low attenuation losses, which make them suitable for high-Q and low-interference applications. However, striplines require strong symmetry that makes their tuning complicated due to difficult access to center conductor. As a result, the stripline structure is not convenient for incorporating chip elements and associated bias circuitry.

The exact expression for the characteristic impedance of a lossless stripline of zero thickness is given by

$$Z_0 = \frac{30\pi}{\sqrt{\varepsilon_r}}\frac{K(k)}{K(k')} \tag{1.152}$$

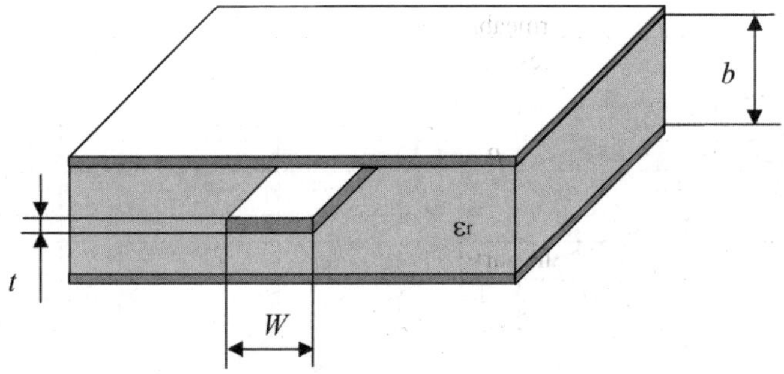

FIGURE 1.18 Stripline structure.

where

$$K(k) = \int_0^{\pi/2} \frac{d\varphi}{\sqrt{1 - k^2 \sin^2\varphi}} \tag{1.153}$$

is the complete elliptic integral of the first kind, $k = \text{sech}(\pi W/2b)$, and $k' = \sqrt{1 - k^2}$ [20,21].

An expression for the ratio $K(k)/K(k')$ can be simplified to

$$\frac{K(k)}{K(k')} = \begin{cases} \pi/\ln\left(2\dfrac{1+\sqrt{k'}}{1-\sqrt{k'}}\right) & \text{for } 0 \le k \le \dfrac{1}{\sqrt{2}} \\ \dfrac{1}{\pi}\ln\left(2\dfrac{1+\sqrt{k}}{1-\sqrt{k}}\right) & \text{for } \dfrac{1}{\sqrt{2}} \le k \le 1 \end{cases} \tag{1.154}$$

which provides the relative error lower than 3×10^{-6} [22].

In practice, it makes sense to use a sufficiently simple formula without complicated special functions [23]. In this case, the formula for Z_0 can be written within 1% of the exact results as

$$Z_0 = \frac{30\pi}{\sqrt{\varepsilon_r}} \frac{b}{W_e + 0.441b} \tag{1.155}$$

where W_e is the effective width of the center conductor defined by

$$\frac{W_e}{b} = \frac{W}{b} - \begin{cases} 0 & \text{for } \dfrac{W}{b} > 0.35b \\ \left(0.35 - \dfrac{W}{b}\right)^2 & \text{for } \dfrac{W}{b} < 0.35b \end{cases}. \tag{1.156}$$

For a stripline with a TEM propagation mode, the dielectric loss factor α_d is the same as for coaxial line, which is determined by Eq. (1.151). An approximation result for the conductor loss factor α_c (in Np/m) can be obtained by

$$\alpha_c = \begin{cases} A\dfrac{2.7 \times 10^{-3} R_s \varepsilon_r Z_0}{30\pi(b-t)} & \text{for } Z_0\sqrt{\varepsilon_r} < 120 \\ B\dfrac{0.16 R_s}{Z_0 b\pi} & \text{for } Z_0\sqrt{\varepsilon_r} > 120 \end{cases} \tag{1.157}$$

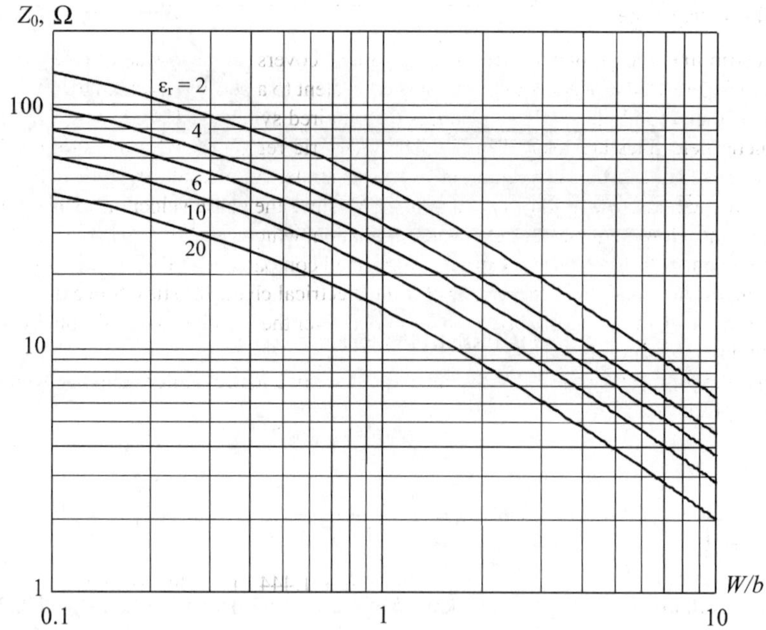

FIGURE 1.19 Stripline characteristic impedance versus W/b.

with

$$A = 1 + \frac{2W}{b-t} + \frac{1}{\pi}\frac{b+t}{b-t}\ln\frac{2b-t}{t} \tag{1.158}$$

$$B = 1 + \frac{b}{0.5W + 0.7t}\left(0.5 + \frac{0.414t}{W} + \frac{1}{2\pi}\ln\frac{4\pi W}{t}\right) \tag{1.159}$$

where t is the thickness of the strip [4].

Figure 1.19 shows the characteristic impedance Z_0 of a stripline as a function of the normalized strip width W/b for various ε_r according to Eqs. (1.155) and (1.156). Typical values of the main electrical and thermal properties of some substrate materials are listed in Table 1.4.

TABLE 1.4 Electrical and Thermal Properties of Substrate Materials.

Typical Substrate	Dielectric Constant, ε_r @ 10 GHz	Loss Tangent, $\tan\delta$ @ 10 GHz	Coefficient of Thermal Expansion (ppm/°C)
Alumina 99.5%	9.8	0.0003	6.7
Aluminum nitride	8.7	0.001	4.5
Barium tetratitanade	37	0.0002	8.3
Beryllia 99.5%	6.6	0.0003	7.5
Epoxy glass FR-4	4.7	0.01	3.0
Fused quartz	3.78	0.0001	0.5
Gallium arsenide	13.1	0.0006	6.5
Silicon	11.7	0.004	4.2
Teflon	2.5	0.0008	15

1.8.3 Microstrip Line

In a microstrip line, the grounded metallization surface covers only one side of dielectric substrate, as shown in Figure 1.20. Such a configuration is equivalent to a pair-wire system for the image of the conductor in the ground plane which produces the required symmetry [24]. In this case, the electric and magnetic field lines are located in both the dielectric region between the strip conductor and the ground plane and in the air region above the substrate. As a result, the electromagnetic wave propagated along a microstrip line is not a pure TEM, since the phase velocities in these two regions are not the same. However, in a quasistatic approximation, which gives sufficiently accurate results as long as the height of the dielectric substrate is very small compared with the wavelength, it is possible to obtain the explicit analytical expressions for the electrical characteristics. Since microstrip line is an open structure, it has a major fabrication advantage over the stripline due to simplicity of practical realization, interconnection, and adjustments.

The exact expression for the characteristic impedance of a lossless microstrip line with finite strip thickness is given by [25,26]

$$Z_0 = \begin{cases} \dfrac{60}{\sqrt{\varepsilon_{re}}} \ln\left(\dfrac{8h}{W_e} + \dfrac{W_e}{4h}\right) & \text{for } \dfrac{W}{h} \le 1 \\ \dfrac{120\pi}{\sqrt{\varepsilon_{re}}} \left[\dfrac{W_e}{h} + 1.393 + 0.667 \ln\left(\dfrac{W_e}{h} + 1.444\right)\right]^{-1} & \text{for } \dfrac{W}{h} \ge 1 \end{cases} \tag{1.160}$$

where

$$\frac{W_e}{h} = \frac{W}{h} + \frac{\Delta W}{h} \tag{1.161}$$

$$\frac{\Delta W}{h} = \begin{cases} \dfrac{1.25}{\pi}\dfrac{t}{h}\left(1 + \ln\dfrac{4\pi W}{t}\right) & \text{for } \dfrac{W}{h} \le 1/2\pi \\ \dfrac{1.25}{\pi}\dfrac{t}{h}\left(1 + \ln\dfrac{2h}{t}\right) & \text{for } \dfrac{W}{h} \ge 1/2\pi \end{cases} \tag{1.162}$$

$$\varepsilon_{re} = \frac{\varepsilon_r + 1}{2} + \frac{\varepsilon_r - 1}{2} \frac{1}{\sqrt{1 + 12h/W}} - \frac{\varepsilon_r - 1}{4.6} \frac{t}{h} \sqrt{\frac{h}{W}}. \tag{1.163}$$

Figure 1.21 shows the characteristic impedance Z_0 of a microstrip line with zero strip thickness as a function of the normalized strip width W/h for various ε_r according to Eqs. (1.160) to (1.163).

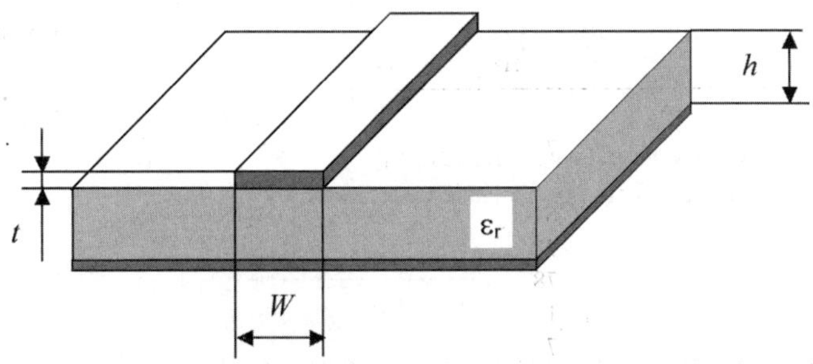

FIGURE 1.20 Microstrip line structure.

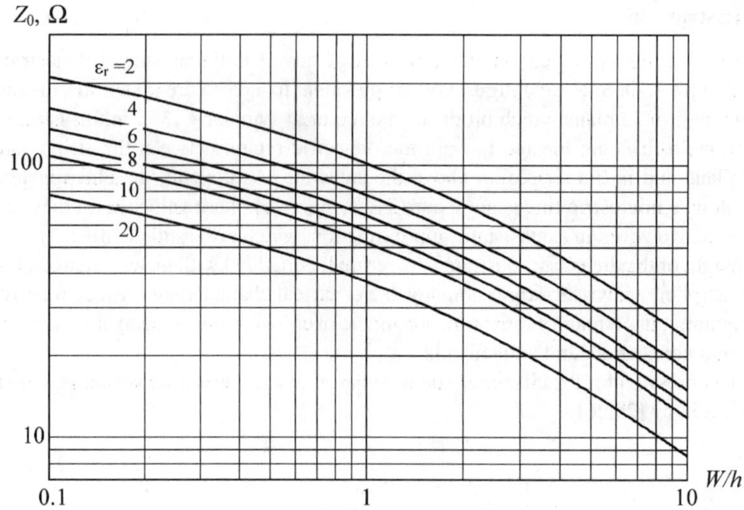

FIGURE 1.21 Microstrip characteristic impedance versus W/h.

In practice, it is possible to use a sufficiently simple formula to estimate the characteristic impedance Z_0 of a microstrip line with zero strip thickness written as [27]

$$Z_0 = \frac{120\pi}{\sqrt{\varepsilon_r}} \frac{h}{W} \frac{1}{1 + 1.735\varepsilon_r^{-0.0724} (W/h)^{-0.836}}. \tag{1.164}$$

For a microstrip line in a quasi-TEM approximation, the conductor loss factor α_c (in Np/m) as a function of the microstrip-line geometry can be obtained by

$$\alpha_c = \begin{cases} 1.38A \dfrac{R_s}{hZ_0} \dfrac{32 - (W_e/h)^2}{32 + (W_e/h)^2} & \text{for } \dfrac{W}{h} \le 1 \\ 6.1 \cdot 10^{-5} A \dfrac{R_s Z_0 \varepsilon_{re}}{h} \left(\dfrac{W_e}{h} + \dfrac{0.667\, W_e/h}{1.444 + W_e/h} \right) & \text{for } \dfrac{W}{h} \ge 1 \end{cases} \tag{1.165}$$

with

$$A = 1 + \frac{h}{W_e} \left(1 + \frac{1}{\pi} \ln \frac{2B}{t} \right) \tag{1.166}$$

$$B = \begin{cases} 2\pi W & \text{for } \dfrac{W}{h} \le 1/2\pi \\ h & \text{for } \dfrac{W}{h} \ge 1/2\pi \end{cases} \tag{1.167}$$

where W_e/h is given by Eqs. (1.161) and (1.162) [28].

The dielectric loss factor α_d (in Np/m) can be calculated by

$$\alpha_d = 27.3 \frac{\varepsilon_r}{\varepsilon_r - 1} \frac{\varepsilon_{re} - 1}{\sqrt{\varepsilon_{re}}} \frac{\tan\delta}{\lambda}. \tag{1.168}$$

Conductor loss is a result of several factors related to the metallic material composing the ground plane and walls, among which are conductivity, skin effect, and surface ruggedness. For most microstrip lines (except some kinds of semiconductor substrate such as silicon), the conductor loss is

TABLE 1.5 Electrical Resistivity of Conductor Materials.

Material	Symbol	Electrical Resistivity (μΩ cm)	Material	Symbol	Electrical Resistivity (μΩ cm)
Aluminum	Al	2.65	Palladium	Pd	10.69
Copper	Cu	1.67	Platinum	Pt	10.62
Gold	Au	2.44	Silver	Ag	1.59
Indium	In	15.52	Tantalum	Ta	15.52
Iron	Fe	9.66	Tin	Sn	11.55
Lead	Pb	21.0	Titanium	Ti	55.0
Molybdenum	Mo	5.69	Tungsten	W	5.6
Nickel	Ni	8.71	Zinc	Zn	5.68

much more significant than the dielectric loss. The conductor losses increase with increasing characteristic impedance due to greater resistance of narrow strips. The electrical resistivity of some conductor materials is given in Table 1.5.

1.8.4 Slotline

Slotlines are usually used when it is necessary to realize a high value of the characteristic impedance Z_0 [29,30]. A slotline is dual to a microstrip line and represents a narrow slot between two conductive surfaces, one of which is grounded. Changing the width of the slot can easily change the characteristic impedance of the slotline. The transverse electric H-mode wave propagates along the slotline. Three basic types of slotlines are unilateral, antipodal, and bilateral. The geometry of a unilateral slotline is shown in Figure 1.22, with a narrow gap in the conductive coating on one side of the dielectric substrate and being bare on the other side of substrate. Slotline can be used either alone or with microstrip line on the opposite side of substrate.

It is difficult to provide exact analytical expressions to calculate the slotline parameters. However, an equation for Z_0 can be obtained for a quasi-TEM approximation with zero conductor thickness and infinite width of the entire slotline system as

for $0.02 \leq W/h \leq 0.2$

$$Z_0 = 72.62 - 15.283\ln\varepsilon_r + 50\left(1 - 0.02\frac{h}{W}\right)\left(\frac{W}{h} - 0.1\right) + (19.23 - 3.693\ln\varepsilon_r)\ln\left(10^2\frac{W}{h}\right)$$
$$- \left(11.4 - 2.636\ln\varepsilon_r - 10^2\frac{h}{\lambda}\right)^2 \times \left[0.139\ln\varepsilon_r - 0.11 + \frac{W}{h}(0.465\ln\varepsilon_r + 1.44)\right] \quad (1.169)$$

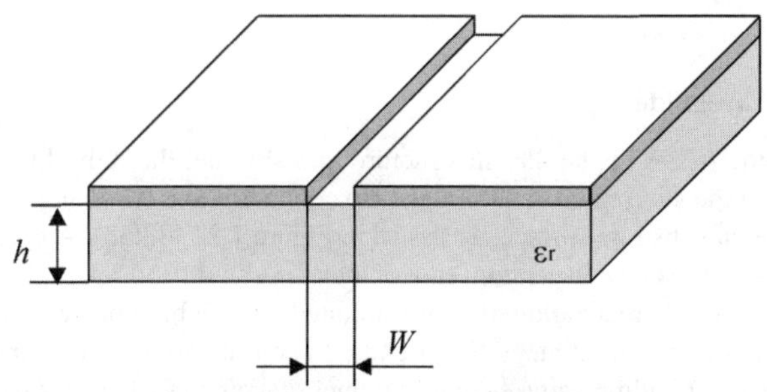

FIGURE 1.22 Slotline structure.

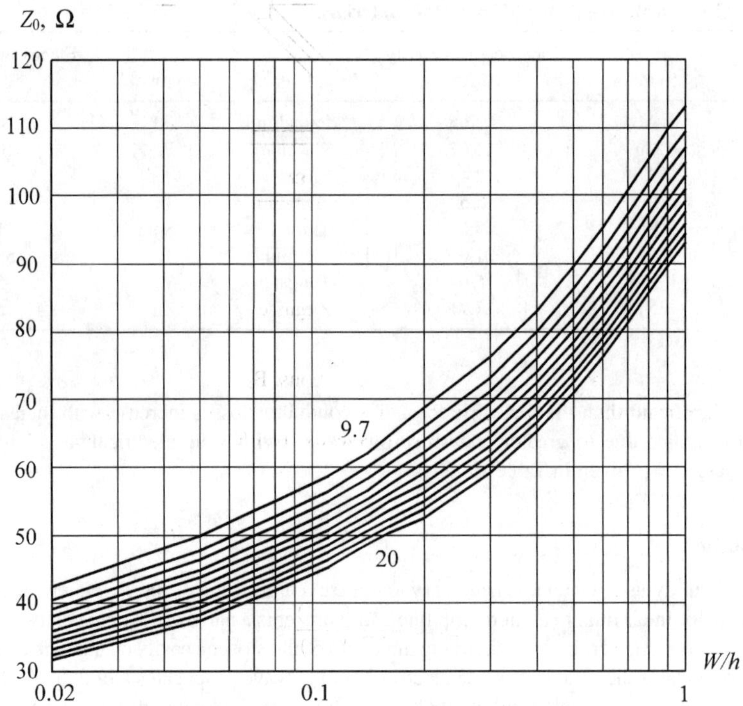

FIGURE 1.23 Slotline characteristic impedance versus W/h.

for $0.2 \leq W/h \leq 1.0$

$$Z_0 = 113.19 - 23.257 \ln\varepsilon_r + 1.25\frac{W}{h}(114.59 - 22.531 \ln\varepsilon_r) + 20\left(1 - \frac{W}{h}\right)\left(\frac{W}{h} - 0.2\right)$$

$$- \left[0.15 + 0.1 \ln\varepsilon_r + \frac{W}{h}(0.899 \ln\varepsilon_r - 0.79)\right]$$

$$\times \left[10.25 - 2.171 \ln\varepsilon_r + \frac{W}{h}(2.1 - 0.617 \ln\varepsilon_r) - 10^2\frac{h}{\lambda}\right]^2 \quad (1.170)$$

where $0.01 \leq h/\lambda \leq 0.25/\sqrt{\varepsilon_r - 1}$ and λ is the free-space wavelength [31].

Figure 1.23 shows the characteristic impedance Z_0 of a slotline within the error of 2% as a function of the normalized slot width W/h for $h/\lambda = 0.02$ and various $\varepsilon_r = 9.7, 11, 12, \ldots, 20$ calculated by Eqs. (1.169) and (1.170).

1.8.5 Coplanar Waveguide

A coplanar waveguide (CPW) is similar in structure to a slotline, the only difference being a third conductor centered in the slot region. The center strip conductor and two outer grounded conductors lie in the same plane on substrate surface, as shown in Figure 1.24 [32,33]. A coplanar configuration has some advantages such as low dispersion, ease of attaching shunt and series circuit components, no need for via holes, or simple realization of short-circuited ends, which makes a coplanar waveguide suitable for hybrid and monolithic integrated circuits. In contrast to the microstrip and stripline, the coplanar waveguide has shielding between adjacent lines that creates a better isolation between them. However, like microstrip and stripline, the coplanar waveguide can be also described by a quasi-TEM

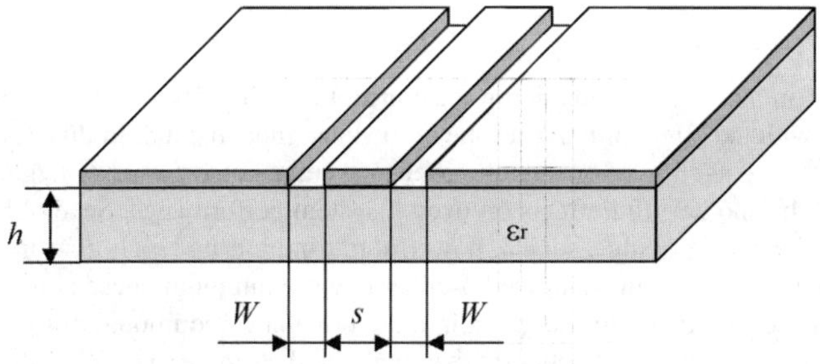

FIGURE 1.24 Coplanar waveguide structure.

approximation for both numerical and analytical calculations. Because of the high dielectric constant of the substrate, most of the RF energy is stored in the dielectric and the loading effect of the grounded cover is negligible if it is more than two slot widths away from the surface. Similarly, the thickness of the dielectric substrate with higher relative dielectric constants is not so critical, and practically it should be one or two times the width W of the slots.

The approximate expression of the characteristic impedance Z_0 for zero metal thickness which is satisfactory accurate in a wide range of substrate thicknesses can be written as

$$Z_0 = \frac{30\pi}{\sqrt{\varepsilon_{\rm re}}} \frac{K\,(k')}{K\,(k)} \tag{1.171}$$

where

$$\varepsilon_{\rm re} = 1 + \frac{\varepsilon_{\rm r} - 1}{2} \frac{K\,(k')}{K\,(k)} \frac{K\,(k_1)}{K\,(k_1')} \tag{1.172}$$

$k = s/(s + 2W)$, $k_1 = (\sinh \pi s/4h)/(\sinh \pi\,(s + 2W)/4h)$, $k' = \sqrt{1 - k^2}$, $k_1' = \sqrt{1 - k_1^2}$, and K is the complete elliptic integral of the first kind [34]. The values of ratios $K(k)/K(k')$ and $K(k_1)/K(k_1')$ can be defined from Eq. (1.154). Figure 1.25 shows the characteristic impedance Z_0 of a coplanar waveguide as a function of the parameter $s/(s + 2W)$ for various $\varepsilon_{\rm r}$ according to Eqs. (1.171) and (1.172).

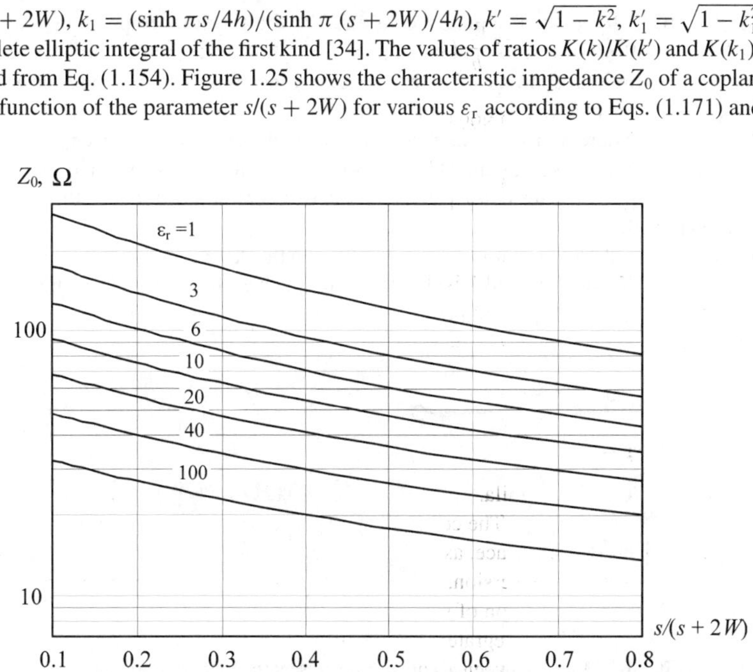

FIGURE 1.25 Coplanar waveguide characteristic impedance versus $s/(s + 2W)$.

1.9 NOISE

The electrical performance of RF and microwave transmitters of different applications can be affected by many factors, with the effect of noise as one of the most fundamental. In this case, it is necessary to keep the ratio of average (or peak) signal power to average noise power so large that the noise in a transmitter path has no harmful effects on overall system performance. Several basic approaches can provide this where it is possible, such as powerful transmitters and high-gain antennas to develop large signals at the receiver input, stabilized oscillators with minimum phase noise, power amplifier and mixer circuits so that they introduce a minimum amount of additional noise when processing signals, and modulation and coding schemes that facilitate the separation of signal and noise.

1.9.1 Noise Sources

There are several primary noise sources in the electrical circuit. *Thermal* or *white noise* is created by random motion of charge carriers due to thermal excitation, being always found in any conducting medium whose temperature is above absolute zero whatever the nature of the conduction process or the nature of the mobile charge carriers [35]. This random motion of carriers creates a fluctuating voltage on the terminals of each resistive element which increases with temperature. However, if the average value of such a voltage is zero, then the noise power on its terminal is not zero being proportional to the resistance of the conductor and to its absolute temperature. The resistor as a thermal noise source can be represented by either of the noise sources shown in Figure 1.26. The noise voltage source and noise current source can be respectively described by Nyquist equations through their mean-square noise voltage and noise current values as

$$\overline{e_n^2} = 4kTR\Delta f \tag{1.173}$$

$$\overline{i_n^2} = \frac{4kT\Delta f}{R} \tag{1.174}$$

where $k = 1.38 \times 10^{-23}$ J/K is the Boltzmann constant, T is the absolute temperature, and $kT = 4 \times 10^{-21}$ W/Hz $= -174$ dBm/Hz at ambient temperature $T = 290$ K. The thermal noise is proportional to the frequency bandwidth Δf, and it can be represented by the voltage source in series with resistor R, or by the current source in parallel to the resistor R. The maximum noise power can be delivered to the load when $R = R_L$, where R_L is the load resistance, being equal to $kT\Delta f$. Hence, the noise power density when the noise power is normalized by Δf is independent of frequency and is considered as white noise. The root-mean-square noise voltage and current are proportional to the square root of the frequency bandwidth Δf.

Shot noise is associated with the carrier injection through the device *p-n* junction, being generated by the movement of individual electrons within the current flow. In each forward biased junction,

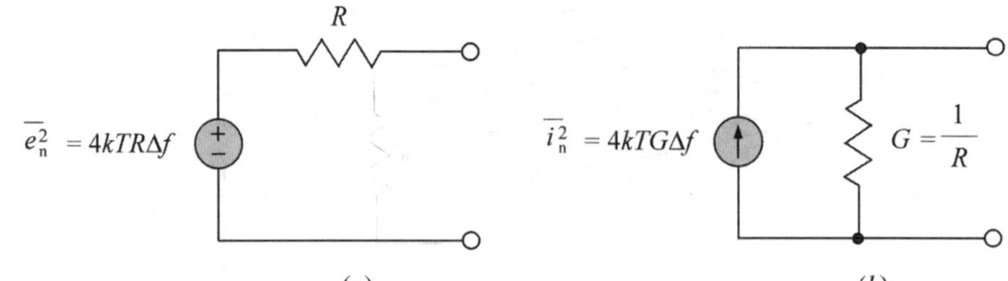

FIGURE 1.26 Equivalent circuits to represent thermal noise sources.

there is a potential barrier that can be overcome by the carriers with higher thermal energy. Such a process is random and mean-square noise current can be given by

$$\overline{i_n^2} = 2qI\Delta f \tag{1.175}$$

where q is the electron charge and I is the direct current flowing through the p-n junction. The shot noise depends on the thermal energy of the carriers near the potential barrier and its power density is independent of frequency. It has essentially a flat spectral distribution and can be treated as the thermal or white type of noise with current source $\overline{i_n^2}$ connected in parallel to the small-signal junction resistance. In a voltage noise representation, when the noise voltage source is connected in series with such a resistor, it can be written as

$$\overline{e_n^2} = 2kTr\Delta f \tag{1.176}$$

where $r = kT/qI$ is the junction resistance.

Circuits containing more than one resistor can be analyzed by reducing their number to the only one (Thevenin) equivalent resistance to obtain the mean-square noise voltage in the form of Eq. (1.173) [36]. As an example, the noise equivalent of a circuit shown in Figure 1.27(a), where a signal source V_S is driving a hypothetical noise-free load resistor R_{in} (which can be considered an input of the power amplifier) through three noise resistors R_1, R_2, and R_3, is a noise voltage source $\overline{e_n^2} = 2kTR_T\Delta f$ connected in series with an ideal (noise-free) resistor equal to the Thevenin resistance R_T, as shown in Figure 1.27(b).

Consider now a simple parallel RC circuit shown in Figure 1.28(a), where the thermal noise due to the parallel resistor is represented by a parallel noise current source i_n. Nyquist has determined the thermal noise output of a two-port network containing both resistive and reactive elements, as shown in Figure 1.28(b). In this case, the mean-square thermal noise voltage is given by

$$\overline{e_n^2} = 4kT\int_{\Delta f} R(f)\,df \tag{1.177}$$

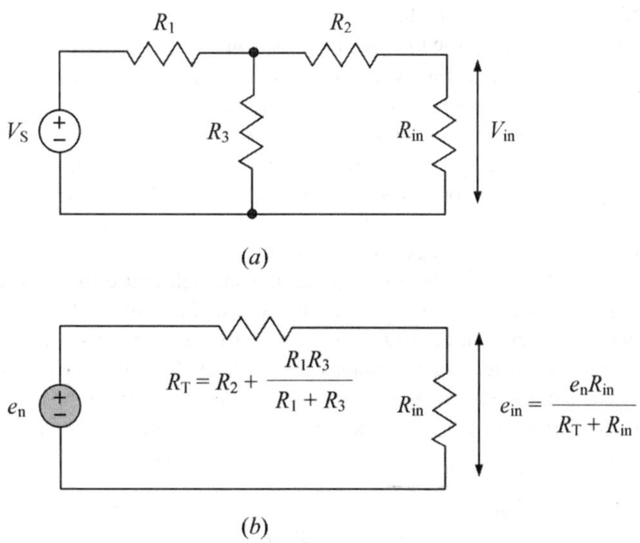

FIGURE 1.27 Circuit with three resistors and its equivalent with noise voltage source.

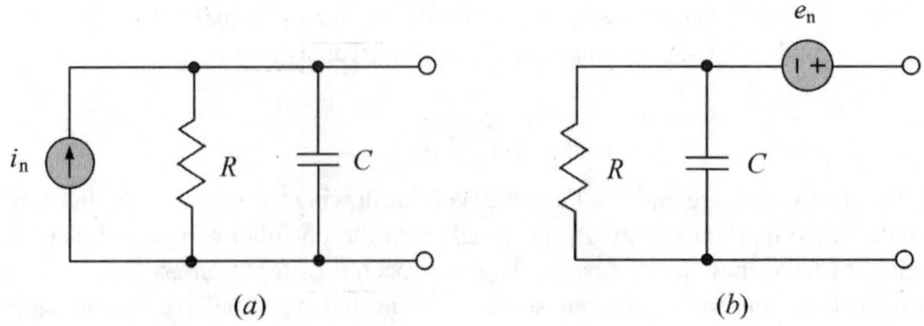

FIGURE 1.28 Noise characterization of two-port *RC* network.

where integration is performed over the frequency bandwidth of interest Δf and

$$R(f) = \frac{R}{1 + (2\pi f C R)^2} \tag{1.178}$$

is the real part of the output circuit impedance at frequency f.

Hence, the parallel current noise source can be equivalently transformed to the series noise voltage source by integration over infinite frequency bandwidth with the total mean-square noise voltage given by

$$\overline{e_n^2} = \frac{4kT}{2\pi} \int_0^\infty \frac{R d\omega}{1 + (\omega C R)^2} = \frac{2kTR}{\pi} \int_0^\infty \frac{d\omega}{1 + (\omega C R)^2} = \frac{kT}{C} \tag{1.179}$$

where the resistance R has no effect on the noise voltage which depends on the value of the capacitance C and temperature T only [36,37].

1.9.2 Noise Figure

It is well-known that any linear noisy two-port network can be represented as a noise-free two-port part with noise sources at the input and the output connected in different way [38,39]. For example, the noisy linear two-port network with internal noise sources shown in Figure 1.29(*a*) can be redrawn, either in the impedance form with external series voltage noise sources shown in Figure 1.29(*b*) or in the admittance form with external parallel current noise sources shown in Figure 1.29(*c*).

However, to fully describe the noise properties of the two-port network at fixed frequency, sometimes it is convenient to represent it through the noise-free two-port part and the noise sources equivalently located at the input. Such a circuit is equivalent to the configurations with noise sources located at the input and the output [40]. In this case, it is enough to use four parameters: the noise spectral densities of both noise sources and the real and imaginary parts of its correlation spectral density. These four parameters can be defined by measurements at the two-port network terminals. The two-port network current and voltage amplitudes are related to each other through a system of two linear algebraic equations. By taking into account the noise sources at the input and the output, these equations in the impedance and admittance forms can be respectively written as

$$V_1 = Z_{11} I_1 + Z_{12} I_2 - V_{n1} \tag{1.180}$$

$$V_2 = Z_{21} I_1 + Z_{22} I_2 - V_{n2} \tag{1.181}$$

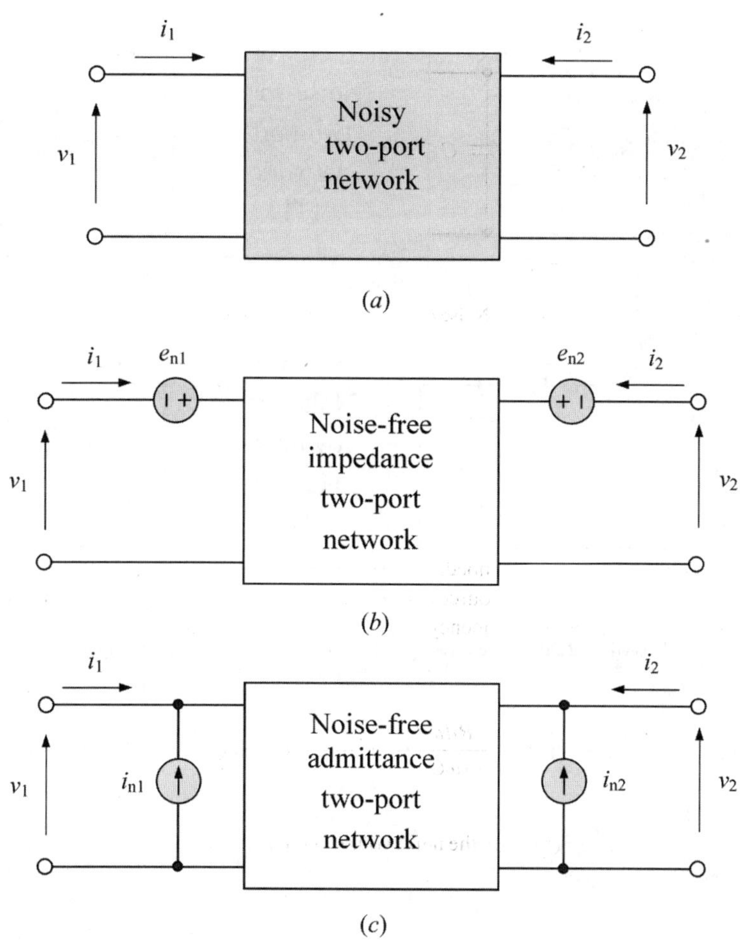

FIGURE 1.29 Linear two-port network with noise sources.

and

$$I_1 = Y_{11}V_1 + Y_{12}V_2 - I_{n1} \tag{1.182}$$

$$I_2 = Y_{21}V_1 + Y_{22}V_2 - I_{n2} \tag{1.183}$$

where the voltage and current noise amplitudes represent the Fourier transforms of noise fluctuations.

The equivalent two-port network with voltage and current noise sources located at its input is shown in Figure 1.30(*a*), where $[Y]$ is the two-port network admittance matrix and ratios between current and voltage amplitudes can be written as

$$I_1 = Y_{11}(V_1 + V_{ni}) + Y_{12}V_2 - I_{ni} \tag{1.184}$$

$$I_2 = Y_{21}(V_1 + V_{ni}) + Y_{22}V_2. \tag{1.185}$$

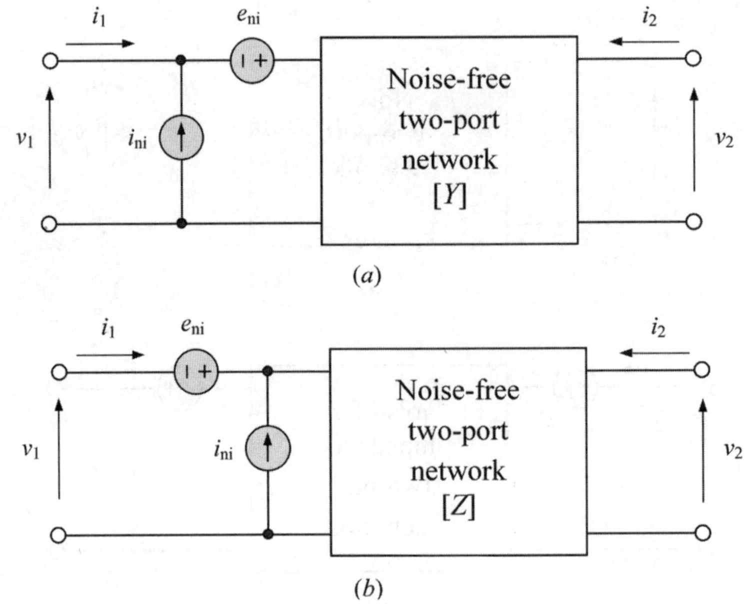

FIGURE 1.30 Linear two-port network with noise sources at input.

From comparison of Eqs. (1.182) and (1.183) with Eqs. (1.184) and (1.185) it follows that

$$V_{\text{ni}} = -\frac{I_{\text{n2}}}{Y_{21}} \tag{1.186}$$

$$I_{\text{ni}} = I_{\text{n1}} - \frac{Y_{11}}{Y_{21}} I_{\text{n2}} \tag{1.187}$$

representing the relationships between the current noise sources at the input and the output corresponding to the circuit shown in Figure 1.29(*c*) and the voltage and current noise sources at the input only corresponding to the circuit shown in Figure 1.30(*a*). In this case, Eqs. (1.186) and (1.187) are valid only if $Y_{21} \neq 0$ that always takes place in practice. Similar equations can be written for the circuit with the series noise voltage source followed by a parallel noise current source shown in Figure 1.30(*b*) in terms of impedance Z-parameters to represent the relationships between the voltage noise sources at the input and the output corresponding to the circuit shown in Figure 1.29(*b*). The use of voltage and current noise sources at the input enables the combination of all internal two-port network noise sources.

To evaluate the quality of a two-port network, it is important to know the amount of noise added to a signal passing through it. Usually, this can be done by introducing an important parameter such as a *noise figure* or *noise factor*. The noise figure of the two-port network is intended as an indication of its noisiness. The lower the noise figure, the less is the noise contributed by the two-port network. The noise figure is defined as

$$F = \frac{S_{\text{in}}/N_{\text{in}}}{S_{\text{out}}/N_{\text{out}}} \tag{1.188}$$

where $S_{\text{in}}/N_{\text{in}}$ is the signal-to-noise ratio available at the input and $S_{\text{out}}/N_{\text{out}}$ is the signal-to-noise ratio available at the output.

For a two-port network characterizing by the available power gain G_A, the noise figure can be rewritten as

$$F = \frac{S_{in}/N_{in}}{G_A S_{in}/G_A(N_{in} + N_{add})} = 1 + \frac{N_{add}}{N_{in}} \tag{1.189}$$

where N_{add} is the additional noise power added by the two-port network referred to the input. From Eq. (1.189) it follows that the noise figure depends on the source impedance Z_S shown in Figure 1.31(*a*), but not on the circuit connected to the output of the two-port network.

Hence, if the two-port network is driven from the source with impedance $Z_S = R_S + jX_S$, the noise figure F of this two-port network in terms of the model shown in Figure 1.31(*b*) with input voltage and current noise sources and noise-free two-port network can be obtained by

$$F = 1 + \frac{\overline{|e_n + Z_S i_n|^2}}{4kTR_S\Delta f} = 1 + \frac{R_n + |Z_S|^2 G_n + 2\sqrt{R_n G_n}\mathrm{Re}\,(CZ_S)}{R_S} \tag{1.190}$$

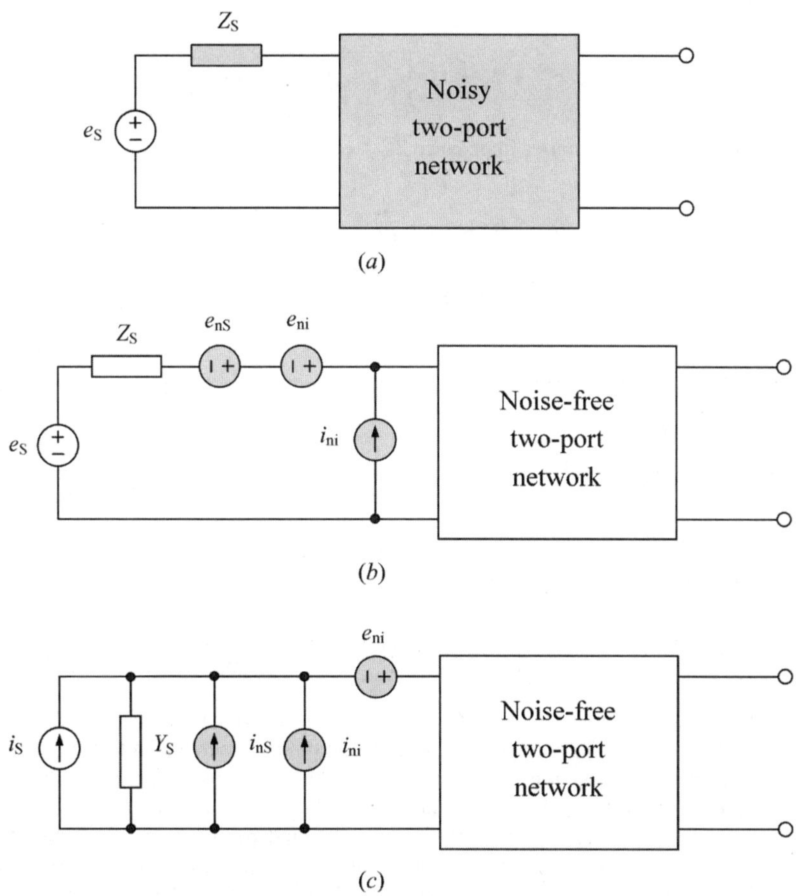

FIGURE 1.31 Linear two-port networks to calculate noise figure.

where

$$R_{\rm n} = \frac{\overline{e_{\rm ni}^2}}{4kT\Delta f} \tag{1.191}$$

is the equivalent input-referred noise resistance corresponding to the noise voltage source,

$$G_{\rm n} = \frac{\overline{i_{\rm ni}^2}}{4kT\Delta f} \tag{1.192}$$

is the equivalent input-referred noise conductance corresponding to the noise current source, and

$$C = \frac{\overline{i_{\rm ni}e_{\rm ni}^*}}{\sqrt{\overline{i_{\rm ni}^2}\,\overline{e_{\rm ni}^2}}} \tag{1.193}$$

is the correlation coefficient representing a complex number less than or equal to unity in magnitude [39]. Here, $G_{\rm n}$ and $R_{\rm n}$ generally do not represent the particular circuit immitances but depend on the bias level resulting in a dependence of the noise figure on the operating bias point of the active device.

As the source impedance $Z_{\rm S}$ is varied over all values with positive $R_{\rm S}$, the noise figure F has a minimum value of

$$F_{\rm min} = 1 + 2\sqrt{R_{\rm n}G_{\rm n}}\left[\sqrt{1-({\rm Im}C)^2} + {\rm Re}C\right] \tag{1.194}$$

which occurs for the optimum source impedance $Z_{\rm Sopt} = R_{\rm Sopt} + jX_{\rm Sopt}$ given by

$$\left|Z_{\rm Sopt}\right|^2 = \frac{R_{\rm n}}{G_{\rm n}} \tag{1.195}$$

$$X_{\rm Sopt} = \sqrt{\frac{R_{\rm n}}{G_{\rm n}}}{\rm Im}C. \tag{1.196}$$

As a result, the noise figure F for the input impedance $Z_{\rm S}$ which is not optimum can be expressed in terms of $F_{\rm min}$ as

$$F = F_{\rm min} + \left|Z_{\rm S} - Z_{\rm Sopt}\right|^2 \frac{G_{\rm n}}{R_{\rm S}} = F_{\rm min} + \left[\left(R_{\rm S} - R_{\rm Sopt}\right)^2 + \left(X_{\rm S} - X_{\rm Sopt}\right)^2\right]\frac{G_{\rm n}}{R_{\rm S}}. \tag{1.197}$$

Similarly, the noise figure F can be equivalently expressed using a model shown in Figure 1.31(*c*) with source admittance $Y_{\rm S} = G_{\rm S} + jB_{\rm S}$ as

$$F = F_{\rm min} + \left|Y_{\rm S} - Y_{\rm Sopt}\right|^2 \frac{R_{\rm n}}{G_{\rm S}} = F_{\rm min} + \left[\left(G_{S} - G_{\rm Sopt}\right)^2 + \left(B_{S} - B_{\rm Sopt}\right)^2\right]\frac{R_{\rm n}}{G_{\rm S}} \tag{1.198}$$

where $F_{\rm min}$ is the minimum noise figure of the two-port network which can be realized with respect to the source admittance $Y_{\rm S}$, $Y_{\rm Sopt} = G_{\rm Sopt} + jB_{\rm Sopt}$ is the optimal source admittance, and $R_{\rm n}$ is the equivalent noise resistance which measures how rapidly the noise figure degrades when the source admittance $Y_{\rm S}$ deviates from its optimum value $Y_{\rm Sopt}$ [41]. Since the admittance $Y_{\rm S}$ is generally complex, then its real and imaginary parts can be controlled independently. To obtain the minimum value of the noise figure, the two matching conditions of $G_{\rm S} = G_{\rm Sopt}$ and $B_{\rm S} = B_{\rm Sopt}$ must be satisfied.

The physical interpretation of the noise sources which are assumed to be stationary random processes is given by their self- and cross-power spectral densities which are defined as the Fourier transform of their auto- and cross-correlation function. These spectral densities in two-port matrix form leads to the so-called correlation matrices with their admittance, impedance, or chain representations [42]. The correlation matrix C belonging to the noise sources s_{n1} and s_{n2} can be written as

$$C = \frac{1}{\Delta f}\begin{bmatrix} \overline{s_{n1}s_{n1}^*} & \overline{s_{n1}s_{n2}^*} \\ \overline{s_{n2}s_{n1}^*} & \overline{s_{n2}s_{n2}^*} \end{bmatrix} \tag{1.199}$$

where the asterisk denotes the complex conjugate. For example, the admittance correlation matrix for the circuit shown in Figure 1.29 (*c*) with two parallel current noise sources is obtained as

$$C_Y = \frac{1}{\Delta f}\begin{bmatrix} \overline{i_{n1}i_{n1}^*} & \overline{i_{n1}i_{n2}^*} \\ \overline{i_{n2}i_{n1}^*} & \overline{i_{n2}i_{n2}^*} \end{bmatrix}. \tag{1.200}$$

Determination of the noise correlation matrix is based on the following procedure:

- Each element in the diagonal matrix is equal to the sum of the noise current of each element connected to the corresponding node: the first diagonal element is the sum of noise currents connected to the node 1, while the second diagonal element is the sum of noise currents connected to node 2.
- The off-diagonal elements are the negative noise current of the element connected to the pair of the corresponding node; therefore, a noise current source between nodes 1 and 2 goes into the matrix at locations (1, 2) and (2, 1).
- If a noise current source is grounded, it will only contribute to one entry in the noise correlation matrix—at the appropriate location on the diagonal; if it is not grounded, it will contribute to four entries in the matrix—two diagonal entries corresponding to the two nodes and two off-diagonal entries.

By applying these rules for the circuit with two current noise sources shown in Figure 1.32, the admittance noise correlation matrix C_Y can be defined as

$$C_Y = \frac{1}{\Delta f}\begin{bmatrix} \overline{i_{n1}^2} & -\overline{i_{n1}^2} \\ -\overline{i_{n1}^2} & \overline{i_{n1}^2} + \overline{i_{n2}^2} \end{bmatrix}. \tag{1.201}$$

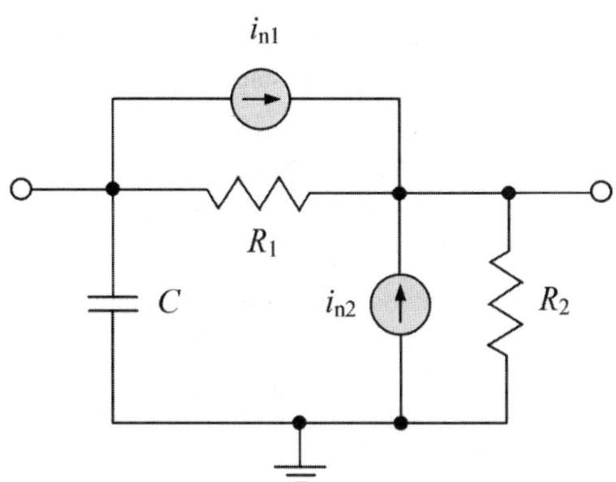

FIGURE 1.32 Circuit with two noise current sources.

To form the impedance noise correlation matrix with voltage noise sources, we can write

$$C_Z = \frac{1}{\Delta f}\begin{bmatrix} \overline{e_{n1}e_{n1}^*} & \overline{e_{n1}e_{n2}^*} \\ \overline{e_{n2}e_{n1}^*} & \overline{e_{n2}e_{n2}^*} \end{bmatrix} = [Z]\,[C_Y]\,[Z]^T \tag{1.202}$$

where $[Z]$ is the impedance Z-matrix of the two-port network and T denotes the Hermitian or transposed conjugation.

In the case where the correlation matrix cannot be theoretically derived, the measurements can be used for its determination. Such measurements are usually done by defining the equivalent noise resistance R_n, the optimal source admittance Y_{Sopt}, and the minimum noise figure F_{min}. As a result, the chain representation of the noise correlation matrix is obtained as

$$C_A = 4kT\begin{bmatrix} R_n & \dfrac{F_{min}-1}{2} - R_n Y_{Sopt} \\ \dfrac{F_{min}-1}{2} - R_n Y_{Sopt}^* & R_n\left|Y_{Sopt}\right|^2 \end{bmatrix} \tag{1.203}$$

where T is the absolute temperature [42]. If the correlation matrix has been determined, the noise parameters can be calculated analytically from

$$R_n = \frac{C_{11}^A}{4kT} \tag{1.204}$$

$$Y_{Sopt} = \sqrt{\frac{C_{22}^A}{C_{11}^A} - \mathrm{Im}^2\left(\frac{C_{12}^A}{C_{11}^A}\right)} - j\mathrm{Im}\left(\frac{C_{12}^A}{C_{11}^A}\right) \tag{1.205}$$

$$F_{min} = 1 + \frac{C_{12}^A + C_{11}^A Y_{Sopt}}{2kT}. \tag{1.206}$$

where C_{11}^A, C_{12}^A, C_{21}^A, and C_{22}^A are the elements of the chain correlation matrix C_A.

In a multistage transmitter system, the input signal travels through a cascade of many different components, each of which may degrade the signal-to-noise ratio to some degree. For a cascade of two stages having available gains G_{A1} and G_{A2} and noise figures F_1 and F_2, using Eq. (1.189) results in the output-to-input noise power ratio N_{out}/N_{in} written as

$$\frac{N_{out}}{N_{in}} = G_{A2}\left[G_{A1}\left(1 + \frac{N_{add1}}{N_{in}}\right) + \frac{N_{add2}}{N_{in}}\right] = G_{A1}G_{A2}\left(F_1 + \frac{F_2 - 1}{G_{A1}}\right) \tag{1.207}$$

where N_{add1} and N_{add2} are the additional noise powers added by the first and second stages, respectively. Consequently, an overall noise figure $F_{1,2}$ for a two-stage system based on Eq. (1.188) can be given by

$$F_{1,2} = F_1 + \frac{1}{G_{A1}}(F_2 - 1). \tag{1.208}$$

Eq. (1.208) can be generalized to a multistage transmitter system with n stages as

$$F_{1,n} = F_1 + \frac{F_2 - 1}{G_{A1}} + \cdots + \frac{F_n - 1}{G_{A1}G_{A2}\ldots G_{A(n-1)}} \tag{1.209}$$

which means that the noise figure of the first stage has the predominant effect on the overall noise figure, unless G_{A1} is small or F_2 is large [43].

1.9.3 Flicker Noise

The flicker or 1/*f* noise is a low-frequency noise associated with a fluctuation in the conductance with a power spectral density proportional to $f^{-\gamma}$, where $\gamma = 1.0 \pm 0.1$ in a wide frequency range, usually measured from 1 Hz to 10 kHz [44]. Its spectrum cannot be exactly f^{-1} at offset frequencies from $f = 0$ to $f \to \infty$, since neither the integral over the power density nor the Fourier transform would be able to have finite values. Unlike the thermal or shot noise sources, the origin of the 1/*f* noise is not exactly clear and open to debate despite its predictable behavior. Generally, it is a result of both surface and bulk effects in the semiconductor material and is not generated by the current. In series experiments it was shown that there is a type of 1/*f* noise that is a fluctuation in the carrier mobility due to lattice scattering.

Significant contribution to the low-frequency noise is made by the generation-recombination and burst noises [45]. The generation-recombination noise associated with the fluctuations in the number of the carriers rather than their mobility is due to trap centers within the bandgap of a semiconductor. It may have any frequency behavior between f^{0} and f^{-2}. If not masked by thermal noise, the low-frequency noise generated from these trap centers becomes f^{-2} at very high frequencies. However, if the lifetime of the carriers in the semiconductor is finite, the noise spectral density reaches a plateau at very low frequencies. Burst noise (random telegraph noise) is a special kind of generation-recombination noise due to a single trap in the active device region. It is often observed in submicron devices or in devices with very poor crystalline quality. In such devices, a trap level with certain energy and at a specific location in the active device region (a single localized trap) traps and detraps the carriers causing an on-off time-dependent signal similar to a telegraph signal [46].

The physical origin of a low-frequency 1/*f* noise for any type of the metal-oxide-semiconductor field-effect transistor (MOSFET) devices including CMOS transistors is based on two dominant processes: random fluctuation of the carriers in the channel due to fluctuations in the surface potential caused by trapping and releasing of the carriers by traps located near the Si–SiO_2 interface, and mobility fluctuations due to carrier interactions with lattice fluctuations [47]. However, for a CMOS transistor depending on its type, one effect can prevail over the other. For example, flicker noise in *n*-channel devices is mostly attributed to carrier number fluctuations, while flicker noise in *p*-channel devices is often attributed to mobility fluctuations. It was observed that pMOS transistors have significantly lower 1/*f* noise than nMOS transistors of the same size and fabricated with the same CMOS process (by one order of magnitude or more). This is because, when an n^+-polysilicon gate layer is used for both the nMOS and pMOS devices, nMOS transistors have a surface channel while pMOS transistors have a buried channel [48].

REFERENCES

1. L. O. Chua, C. A. Desoer, and E. S. Kuh, *Linear and Nonlinear Circuits*, New York: McGraw-Hill, 1987.
2. D. R. Cunningham and J. A. Stuller, *Circuit Analysis*, New York: John Wiley & Sons, 1995.
3. G. D. Vendelin, A. M. Pavio, and U. L. Rohde, *Microwave Circuit Design Using Linear and Nonlinear Techniques*, New York: John Wiley & Sons, 2005.
4. D. M. Pozar, *Microwave Engineering*, New York: John Wiley & Sons, 2004.
5. D. A. Frickey, "Conversions between S, Z, Y, h, ABCD, and T Parameters which are Valid for Complex Source and Load Impedances," *IEEE Trans. Microwave Theory Tech.*, vol. MTT-42, pp. 205–211, Feb. 1994.
6. E. F. Terman, *Radio Engineer's Handbook*, New York: McGraw-Hill, 1945.
7. I. J. Bahl, *Lumped Elements for RF and Microwave Circuits*, Boston: Artech House, 2003.
8. H. A. Wheeler, "Simple Inductance Formulas for Radio Coils," *Proc. IRE*, vol. 16, pp. 1398–1400, Oct. 1928.
9. S. S. Mohan, M. del Mar Hershenson, S. P. Boyd, and T. H. Lee, "Simple Accurate Expressions for Planar Spiral Inductances," *IEEE J. Solid-State Circuits*, vol. SC-34, pp. 1419–1424, Oct. 1999.

10. J. M. Lopez-Villegas, J. Samitier, C. Cane, P. Losantos, and J. Bausells, "Improvement of the Quality Factor of RF Integrated Inductors by Layout Optimization," *IEEE Trans. Microwave Theory Tech.*, vol. MTT-48, pp. 76–83, Jan. 2000.

11. J. R. Long and M. A. Copeland, "The Modeling, Characterization, and Design of Monolithic Inductors for Silicon RF IC's," *IEEE J. Solid-State Circuits*, vol. SC-32, pp. 357–369, Mar. 1997.

12. N. A. Talwalkar, C. P. Yue, and S. S. Wong, "Analysis and Synthesis of On-Chip Spiral Inductors," *IEEE Trans. Electron Devices*, vol. ED-52, pp. 176–182, Feb. 2005.

13. N. M. Nguyen and R. G. Meyer, "Si IC-Compatible Inductors and *LC Passive Filters*," *IEEE J. Solid-State Circuits*, vol. SC-25, pp. 1028–1031, Aug. 1990.

14. J. Craninckx and M. S. J. Steyaert, "A 1.8-GHz Low-Phase-Noise CMOS VCO Using Optimized Hollow Spiral Inductors," *IEEE J. Solid-State Circuits*, vol. SC-32, pp. 736–744, May 1997.

15. S. L. March, "Simple Equations Characterize Bond Wires," *Microwaves & RF*, vol. 30, pp. 105–110, Nov. 1991.

16. P. Benedek and P. Silvester, "Equivalent Capacitances of Microstrip Gaps and Steps," *IEEE Trans. Microwave Theory Tech.*, vol. MTT-20, pp. 729–733, Nov. 1972.

17. G. D. Alley, "Interdigital Capacitors and Their Applications in Lumped Element Microwave Integrated Circuits," *IEEE Trans. Microwave Theory Tech.*, vol. MTT-18, pp. 1028–1033, Dec. 1970.

18. R. M. Barrett, "Microwave Printed Circuits – The Early Years," *IEEE Trans. Microwave Theory Tech.*, vol. MTT-32, pp. 983–990, Sept. 1984.

19. S. Ramo, J. R. Whinnery, and T. Van Duzer, *Fields and Waves in Communication Electronics*, New York: John Wiley & Sons, 1993.

20. S. B. Kohn, "Characteristic Impedance of the Shielded-Strip Transmission Line," *IRE Trans. Microwave Theory Tech.*, vol. MTT-2, pp. 52–55, July 1954.

21. H. Howe, *Stripline Circuit Design*, Dedham: Artech House, 1974.

22. W. Hilberg, "From Approximations to Exact Relations for Characteristic Impedances," *IEEE Trans. Microwave Theory Tech.*, vol. MTT-17, pp. 259–265, May 1969.

23. I. J. Bahl and R. Garg, "A Designer's Guide to Stripline Circuits," *Microwaves*, vol. 17, pp. 90–96, Jan. 1978.

24. D. D. Grieg and H. F. Engelmann, "Microstrip – A New Transmission Technique for the Kilomegacycle Range," *Proc. IRE*, vol. 40, pp. 1644–1650, Dec. 1952.

25. E. O. Hammerstad, "Equations for Microstrip Circuit Design," *Proc. 5th Europ. Microwave Conf.*, pp. 268–272, Sept. 1975.

26. I. J. Bahl and R. Garg, "Simple and Accurate Formulas for Microstrip with Finite Strip Thickness," *Proc. IEEE*, vol. 65, pp. 1611–1612, Nov. 1977.

27. R. S. Carson, *High-Frequency Amplifiers*, New York: John Wiley & Sons, 1975.

28. K. C. Gupta, R. Garg, and R. Chadha, *Computer-Aided Design of Microwave Circuits*, Dedham: Artech House, 1981.

29. S. B. Cohn, "Slot Line on a Dielectric Substrate," *IEEE Trans. Microwave Theory Tech.*, vol. MTT-17, pp. 768–778, Oct. 1969.

30. E. A. Mariani, C. P. Heinzman, J. P. Agrios, and S. B. Cohn, "Slot Line Characteristics," *IEEE Trans. Microwave Theory Tech.*, vol. MTT-17, pp. 1091–1096, Dec. 1969.

31. R. Garg and K. C. Gupta, "Expression for Wavelength and Impedance of a Slotline," *IEEE Trans. Microwave Theory Tech.*, vol. MTT-24, p. 532, Aug. 1976.

32. C. P. Weng, "Coplanar Waveguide: A Surface Strip Transmission Line Suitable for Nonreciprocal Gyromagnetic Device Applications," *IEEE Trans. Microwave Theory Tech.*, vol. MTT-17, pp. 1087–1090, Dec. 1969.

33. R. N. Simons, *Coplanar Waveguide Circuits, Components, and Systems*, New York: John Wiley & Sons, 2001.

34. G. Ghione and C. Naldli, "Analytical Formulas for Coplanar Lines in Hybrid and Monolithic," *Electronics Lett.*, vol. 20, pp. 179–181, Feb. 1984.

35. A. van der Ziel, *Noise*, Englewood Cliffs: Prentice-Hall, 1954.

36. H. L. Krauss, C. W. Bostian, and F. H. Raab, *Solid State Radio Engineering*, New York: John Wiley & Sons, 1980.

37. B. L. Buckingham, *Noise in Electronic Devices and Systems*, New York: John Wiley & Sons, 1983.
38. H. C. Montgomery, "Transistor Noise in Circuit Applications," *Proc. IRE*, vol. 40, pp. 1461–1471, 1952.
39. H. Rothe and W. Dahlke, "Theory of Noise Fourpoles," *Proc. IRE*, vol. 44, pp. 811–818, June 1956.
40. A. G. T. Becking, H. Groendijk, and K. S. Knol, "The Noise Factor of Four-Terminal Networks," *Philips Res. Rep.*, vol. 10, pp. 349–357, 1955.
41. IRE Subcommittee 7.9 on Noise (H. A. Haus, Chairman), "Representation of Noise in Linear Twoports," *Proc. IRE*, vol. 48, pp. 69–74, Jan. 1960.
42. H. Hillbrand and P. H. Russer, "An Efficient Method for Computer Aided Noise Analysis of Linear Amplifier Networks," *IEEE Trans. Circuits and Systems*, vol. CAS-23, pp. 235–238, Apr. 1976.
43. H. T. Friis, "Noise Figures of Radio Receivers," *Proc. IRE*, vol. 32, pp. 419–422, July 1944.
44. F. N. Hooge, "1/*f* Noise Sources," *IEEE Trans. Electron Devices*, vol. ED-41, pp. 1926–1934, Nov. 1996.
45. S. Mohammadi and D. Pavlidis, "A Nonfundamental Theory of Low-Frequency Noise in Semiconductor Devices," *IEEE Trans. Electron Devices*, vol. ED-47, pp. 2009–2017, Nov. 2000.
46. X. N. Zhang, A. van der Ziel, K. H. Duh, and H. Morkoc, "Burst and Low-Frequency Generation-Recombination Noise in Double-Heterojunction Bipolar Transistors," *IEEE Trans. Electron Device Lett.*, vol. 5, pp. 277–279, July 1984.
47. Y. P. Tsividis, *Operation and Modeling of the MOS Transistor*, New York: McGraw-Hill, 1987.
48. K. K. O, N. Park, and D.-J. Yang, "1/*f* Noise of NMOS and PMOS Transistors and Their Implications to Design of Voltage Controlled Oscillators," *2002 IEEE RFIC Symp. Dig.*, pp. 59–62, June 2002.

2 Active Devices and Modeling

Accurate device modeling is an extremely important procedure in circuit design, especially in monolithic integrated circuits. Better approximations of the final design can only be achieved if the nonlinear device behavior is described accurately. Once a device model has been incorporated into a circuit simulator, it requires the parameters to specify the device characteristics according to the model equations. The next step is to provide a procedure for adequate extraction of S-parameter data. This poses the problem of how to extract the device parameters from the measurement results accurately, rapidly, and without unnecessary measurement complexity. The best way is to minimize the device bias points under S-parameter measurements and to combine the analytical approach with a final optimization procedure to provide the best fitting of the experimental curves and empirical equation-based model curves. Numerical optimization is often used to fit the model S-parameters to the measured parameters since the resulting device element values depend on the starting values and are not unique. The analytical approach incorporates a derivation of the basic intrinsic device parameters from its equivalent circuit based on S- to Y- or Z-parameter transformations using sufficiently simple equations. However, it is crucial to choose an appropriately large-signal model that is most suitable for a specific active device, accurately describes the nonlinear behavior of its equivalent circuit parameters, and contains a reasonable number of model parameters to be determined.

This chapter describes all necessary steps for an accurate device modeling procedure, beginning with determining the small-signal equivalent circuit parameters. A variety of nonlinear models, including noise models, for metal-oxide-semiconductor field-effect transistors (MOSFETs), metal-semiconductor field-effect transistors (MESFETs), high electron mobility transistors (HEMTs), bipolar devices including heterojunction bipolar transistors (HBTs), which are very prospective to design modern microwave hybrid and monolithic integrated circuits, are given and discussed.

2.1 DIODES

Most modern diodes are based on semiconductor p–n junctions. In a p–n diode, conventional current can flow from the p-type side (the anode) to the n-type side (the cathode), but cannot flow in the opposite direction. Another type of semiconductor diode, the Schottky diode, is formed from the contact between a metal and a semiconductor rather than by a p–n junction.

2.1.1 Operation Principle

A semiconductor diode current–voltage (I–V) characteristic curve is ascribed to the behavior of the so-called *depletion layer* or *depletion zone* that exists at the p–n junction between the differing semiconductors. When a p–n junction is first created, conduction band (mobile) electrons from the n-doped region diffuse into the p-doped region, where there is a large population of holes (places for electrons in which no electron is present) with which the electrons "recombine." When a mobile electron recombines with a hole, the hole vanishes and the electron is no longer mobile. Thus, two

RF and Microwave Transmitter Design, First Edition. Andrei Grebennikov.

charge carriers have vanished. The region around the p–n junction becomes depleted of charge carriers and thus behaves as an insulator.

However, the depletion width cannot grow without limit. For each electron–hole pair that recombines, a positively charged dopant ion is left behind in the n-doped region, and a negatively charged dopant ion is left behind in the p-doped region. As recombination proceeds and more ions are created, an increasing electric field develops through the depletion zone that acts to slow and then finally stops recombination. At this point, there is a built-in potential across the depletion zone. This built-in potential is positive because the n-side is at a higher potential than the p-side, which is proper to obtain a balance between drift and diffusion across the junction.

If an external voltage is placed across the diode with the same polarity as the built-in potential, the depletion zone continues to act as an insulator preventing a significant electric current. This is the *reverse bias* phenomenon. However, if the polarity of the external voltage opposes the built-in potential, recombination can once again proceed resulting in substantial electric current through the p–n junction. For silicon diodes, the built-in potential is approximately 0.6 V. Thus, if an external current is passed through the diode, about 0.6 V will be developed across the diode such that the p-doped region is positive with respect to the n-doped region and the diode is said to be "turned on" as it has a *forward bias*.

A diode I–V characteristic shown in Figure 2.1 can be approximated by two regions of operation. Below a certain difference in potential between the two leads, the depletion layer has significant width, and the diode can be thought of as an open (nonconductive) circuit. As the potential difference is increased, at some stage the diode will become conductive and allow charges to flow, at which point it can be thought of as a connection with zero (or at least very low) resistance. In a normal silicon diode at rated currents, the voltage drop across a conducting diode is approximately 0.6 to 0.7 V. In the reverse bias region for a normal p–n rectifier diode, the current through the device is

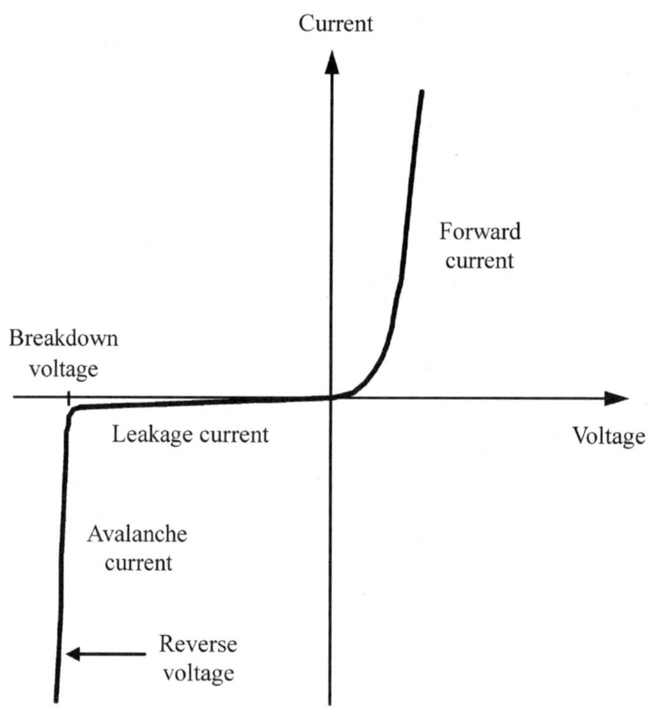

FIGURE 2.1 Diode current–voltage characteristic.

very low (in the μA range) for all reverse voltages up to a point called the *peak inverse voltage* (PIV). Beyond this point a process called *reverse breakdown* occurs that causes the device to be damaged along with a large increase in current. For special purpose diodes like the avalanche or Zener diodes, the concept of PIV is not applicable since they have a deliberate breakdown beyond a known reverse current such that the reverse voltage is "clamped" to a known value (called the *Zener voltage* or breakdown voltage). These devices however have a maximum limit to the current and power in the Zener or avalanche region.

The *Shockley ideal diode equation* or the *diode law* is the current–voltage (I–V) characteristic of an ideal diode in either forward or reverse bias (or no bias):

$$I = I_{\text{sat}} \left[\exp\left(\frac{V}{nV_{\text{T}}} \right) - 1 \right] \tag{2.1}$$

where I is the diode current, I_{sat} is a scale factor called the reverse saturation current, V is the voltage across the diode, V_{T} is the thermal voltage, and n is the emission coefficient, also known as the ideality factor. The term "saturation" indicates that the current I approaches an asymptote and becomes independent of the voltage V.

It is derived with the assumption that the only processes giving rise to current in the diode are drift (due to electrical field), diffusion, and thermal recombination–generation. It also assumes that the carrier recombination–generation current in the depletion region is insignificant. Additionally, it does not describe the "leveling off" of the I–V curve at high forward bias due to internal resistance, nor does it explain the practical deviation from the ideal at very low forward bias due to recombination–generation current in the depletion region.

The emission coefficient n varies from about 1 to 2 depending on the fabrication process and semiconductor material and in many cases is assumed to be approximately equal to 1 (thus omitted). The thermal voltage V_{T} is approximately 25.7 mV at room temperature 25°C or 298 K defined by

$$V_{\text{T}} = \frac{kT}{q} \tag{2.2}$$

where q is the electronic charge (or elementary charge), k is Boltzmann constant, and T is the absolute temperature of the p–n junction.

The diode small-signal electrical behavior can be described by the equivalent representation of the p–n junction with the differential resistance at the operating point

$$r = \left. \frac{\partial V}{\partial I} \right|_{\text{op}} = \left[\frac{I_{\text{sat}}}{nV_{\text{T}}} \exp\left(\frac{V}{nV_{\text{T}}} \right) \right]^{-1} \tag{2.3}$$

and by two forms of charge storage: charge storage in the depletion region due to the dopant concentrations and charge storage due to minority-carrier charges injected into the neutral p- and n-type regions resulting in the junction and diffusion capacitances, respectively [1]. There are several types of semiconductor junction diodes, which either emphasizes a different physical aspects of a diode often by geometric scaling, doping level, choosing the right electrodes, or just an application of a diode in a special circuit.

2.1.2 Schottky Diodes

Schottky diodes are constructed from a structure made by forming a contact between a metal and a semiconductor defined as a *metal–semiconductor junction*. Each metal–semiconductor junction is characterized by a potential barrier called the *Schottky barrier*, which depends only on the two materials and temperature, and it is not a function of the semiconductor doping. The Schottky barrier blocks the flow of electrons from the metal toward the semiconductor. Schottky diodes have a lower forward voltage drop than any *p–n* junction diode and its electrical behavior is described by similar Shockley diode current–voltage equation. Their forward voltage drop at forward currents of about 1 mA is in the range 0.15 to 0.45 V, which makes them useful in voltage clamping applications and prevention of transistor saturation.

Reverse-bias breakdown in Schottky diodes formed on lightly doped substrates occurs through avalanching in high-field regions, just as in a *p–n* junction diode. However, Schottky diodes tend to present lower breakdown voltage than *p–n* junction diodes. For Schottky diodes formed on more heavily doped substrates, breakdown may occur through tunneling, which is quite comparable to that of Zener breakdown in *p–n* junction diodes. Schottky diodes are usually referred to as majority-carrier devices and the forward-bias current in these devices is characterized by electron emission from *n*-type semiconductor into the metal, rather than by hole injection into the semiconductor. Therefore, they do not suffer from minority-carrier storage problems that slow down most normal diodes, thus resulting in a faster "reverse recovery" than any *p–n* junction diode. This contributes toward their high switching speed and their suitability in high speed circuits and RF devices such as a switched-mode power supply, mixers, and detectors.

Operating as a mixer diode, the Schottky diode is used as a variable-resistance diode with non-linear incremental small-signal conductance of the junction. The Mott diode is another variant of the conventional Schottky-diode structure which differs in its epitaxial layer. Being very thin and lightly doped, the epilayer is fully depleted and the depletion region extends into the buffer layer located between the epilayer and the substrate. As a result, the junction capacitance consists of two components that can be modeled by two capacitances connected in series: the capacitance of the depleted epilayer, which is very small and fixed, and the larger voltage-variable capacitance, arising from the part of the depletion region that extends into the buffer layer. But because the smaller fixed capacitance dominates, the Mott-diode capacitance, therefore, is only weakly dependent on voltage. The nonlinear Mott-diode model is shown in Figure 2.2, which includes the current source $I(V)$, two junction capacitances in series, C and C_j, and series parasitic resistance R_s [2]. Because of the Mott diode lower doping density, its reverse saturation current I_{sat} is lower than that of a conventional Schottky diode having the same diameter. The cutoff frequency f_c for both Mott and Schottky diodes is defined as

$$f_c = \frac{1}{2\pi R_s C_j(0)} \tag{2.4}$$

where the junction capacitance $C_j(0)$ defined at zero bias voltage is calculated at the operating bias point and the series resistance R_s includes skin effect.

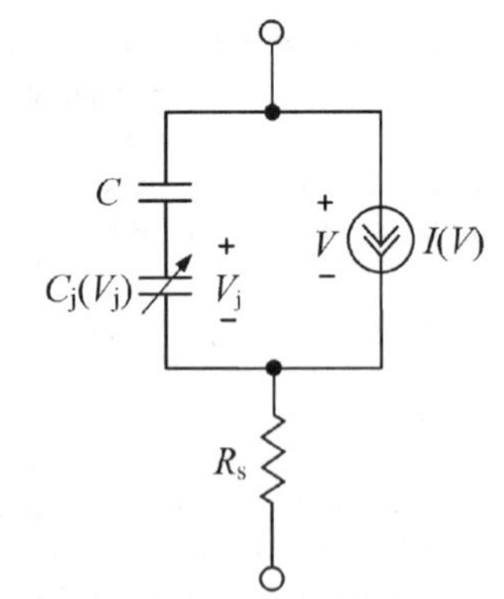

FIGURE 2.2 Mott diode equivalent circuit.

The most important difference between p–n junction diode and Schottky diode is reverse recovery time, when the diode switches from nonconducting to conducting state and vice versa. Where in p–n junction diode the reverse recovery time can be in the order of hundreds of nanoseconds, and less than 100 ns for fast diodes, Schottky diodes do not have a recovery time, because there is nothing to recover from. The switching time is of about 100 ps for the small signal diodes, up to tens of nanoseconds for special high-capacity power diodes. With p–n junction switching, there is a reverse recovery current associated with it, which among other things with high-power semiconductors brings increased electromagnetic interference (EMI) noise. With Schottky diodes switching instantly with only slight capacitive loading, this is of much lesser concern.

The most evident limitations of Schottky diodes are the relatively low reverse voltage rating for silicon–metal Schottky diodes, 50 V and below, and a relatively high reverse leakage current. The reverse leakage current, increasing with temperature, leads to a thermal instability issue. This often limits the useful reverse voltage to well below the actual rating. Luckily, times are changing and the diodes are becoming better and better and the voltage ratings are now up at 200 V. The silicon carbide Schottky diodes have about 40 times lower reverse leakage current compared to silicon counterparts and are made as 300-V, 600-V, and 1200-V variants. Silicon carbide has high thermal conductivity and temperature has little influence on switching and thermal characteristics, plus diodes have no thermal runaway. With special packing it is possible to have operating junction temperatures of over 500 K, which allows passive radiation cooling in aerospace applications.

2.1.3 *p–i–n* Diodes

A p–i–n diode is a device that operates as a variable resistor at RF and microwave frequencies. It is a silicon semiconductor diode in which a high-resistivity intrinsic i-layer is sandwiched between the p-type and n-type layers. When a p–i–n diode is forward biased, holes and electrons from p- and n-layers are injected into the i-layer, respectively. These charges do not recombine immediately; instead, a finite quantity of charge always remains stored for an average time, called the *carrier life time* τ, thus lowering the resistivity of the i-region. The quantity of stored charge Q is directly proportional to the carrier life time (recombination time) τ and the forward bias current I_f as $Q = I_f\tau$. The resistance R_s of the i-layer under forward bias condition is inversely proportional to Q and directly proportional to W^2, where W is the i-layer thickness. When the p–i–n diode is at zero or reverse bias, there is no stored charge in the i-layer and the diode appears as a junction capacitance C_j shunted by a parallel resistance R_p. The p–i–n diode equivalent circuits for (a) forward bias and (b) zero or reverse modes are shown in Figure 2.3, where L_s is the parasitic series inductance [3]. By varying

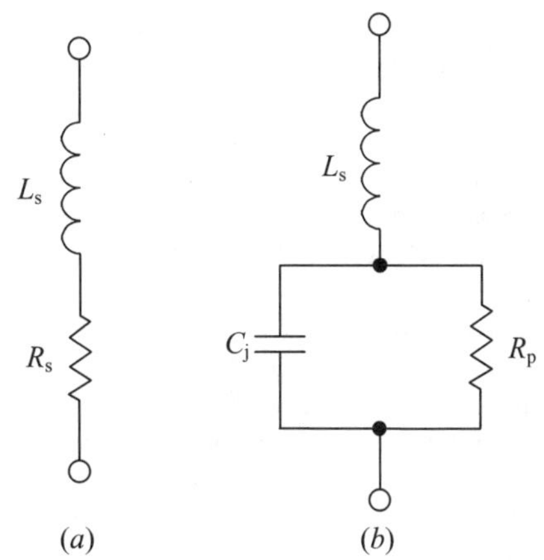

FIGURE 2.3 p–i–n diode equivalent circuits.

the *i*-layer thickness and diode area, it is possible to construct *p–i–n* diodes of different geometries to result in the same R_s and C_j values. These devices may have similar small-signal characteristics. However, a diode with thicker *i*-layer would have a higher breakdown voltage and better distortion properties. On the other hand, a diode with thinner *i*-layer would have faster switching speed.

At dc and very low frequencies being forward biased, the *p–i–n* diode behaves similar to a *p–n* junction diode and the diode resistance is described by the dynamic (or differential) resistance of the diode *I–V* characteristic at any forward bias point. The dc dynamic resistance point is not, however, valid for the *p–i–n* diodes at frequencies above which the period is shorter than the transit time of the *i*-region. The frequency at which this occurs is called the *transit time frequency* f_T, and this lowest frequency limit is primarily a function of the *i*-layer thickness W expressed by

$$f_T = \frac{1300}{W^2}\text{MHz}. \tag{2.5}$$

At high frequencies when a *p–i–n* diode is at zero and reverse bias, it appears as a parallel plate capacitor, essentially independent of reverse voltage being a function of the junction capacitance and thickness of *i*-layer. However, at frequencies much lower than f_T, the *p–i–n* diode acts as a varactor, and the junction capacitance C_j is determined by applying a sufficiently large reverse voltage, which fully depletes the *i*-region of carriers. At low reverse bias voltages, the finite resistivity of the *i*-region results in a lossy junction capacitance. As the reverse voltage is increased, carriers are depleted from the *i*-layer, resulting in an essentially lossless silicon capacitor. Under reverse bias, the *p–i–n* diode should not be biased beyond its dc rating voltage V_R. The diode avalanche or bulk breakdown voltage V_{BD} is proportional to the *i*-layer thickness W and is always higher than V_R. In a typical application, maximum negative voltage swing should never exceed V_{BD}. An instantaneous excursion of the RF signal into the positive bias direction generally does not cause the diode to go into conduction because of slow reverse-to-forward switching speed.

A step recovery diode (SRD) is a semiconductor junction diode having the ability to recover extremely quickly from a strong forward-conduction state to a cutoff state. It is also called the *snap-off diode* or *charge-storage diode*, and has a variety of uses in microwave electronic devices as pulse generator or parametric amplifier. The SRD must have high charge storage in the forward direction, low capacitance in the reverse direction, low series resistance, and high reverse breakdown voltage for power applications. Its switching time must be adequately short because switching speed establishes its high-frequency limit of operation. To meet these requirements, the SRD must have a relatively long charge-storage time, and the charge that is injected into the junction while it is forward-biased must not travel so far that it cannot be removed during the reverse-bias interval. Also, the depletion region must not be too wide, otherwise transit-time effects will reduce the efficiency at high frequency.

The term "step recovery" relates to the form of the reverse recovery characteristic of these devices. After a forward current has been passing in an SRD and the current is interrupted or reversed, the reverse conduction will cease very abruptly (as in a step waveform). The SRD has the *p–i–n* structure, in which the *i*-region is a layer of undoped or intrinsic semiconductor or is very lightly doped [2]. The *i*-region is formed by the overlap of the *p*- and *n*-layers, both of which have a steep doping profile. This profile gives a narrow depletion region in the *p*- and *n*-layers and a strong built-in electric field, which opposes the diffusion of charge into the unction. During forward conduction, holes and electrons are injected into the *i*-layer, where they recombine very slowly. The *i*-layer thus becomes a region in which charge is stored. When the SRD is reverse-biased, the *i*-layer is fully depleted, because of the wide depletion width, which includes the entire *i*-layer, resulting in a very low reverse capacitance. Under forward bias, the SRD can be modeled as a conventional *p–n* junction diode, however, when the SRD is reverse-biased, its capacitance is nearly constant and very low.

2.1.4 Zener Diodes

The effect when the *p–n* diodes can be made to conduct backward is called the *Zener breakdown*. It occurs at a precisely defined voltage, allowing the diode to be used as a precision voltage reference.

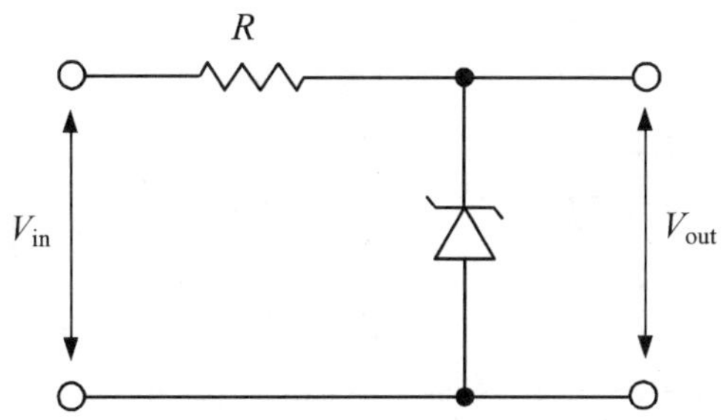

FIGURE 2.4 Circuit with Zener diode.

A conventional *p–n* diode will not let significant current flow if it is reverse-biased below its reverse breakdown voltage. By exceeding the reverse bias breakdown voltage, a conventional diode is subject to high current flow due to avalanche breakdown. Unless this current is limited by external circuitry, the diode will be permanently damaged. A Zener diode exhibits almost the same properties, except the device is specially designed so as to have a greatly reduced breakdown or Zener voltage. A Zener diode contains a heavily doped *p–n* junction allowing electrons to tunnel from the valence band of the *p*-type material to the conduction band of the *n*-type material. A reverse-biased Zener diode will exhibit a controlled breakdown and let the current flow to keep the voltage across the Zener diode at the Zener voltage. For example, a diode with a Zener breakdown voltage of 3.2 V will exhibit a voltage drop of 3.2 V if reverse bias voltage applied across it is more than its Zener voltage. However, the current is not unlimited, so the Zener diode is typically used to generate a reference voltage for a power amplifier stage, or as a voltage stabilizer for low-current applications. The breakdown voltage can be controlled quite accurately in the doping process. Tolerances to within 0.05% are available, though the most widely used tolerances are 5% and 10%.

Zener diodes are widely used to regulate the voltage across a circuit. When connected in parallel with a variable voltage source so that it is reverse biased, a Zener diode conducts when the voltage reaches the diode reverse breakdown voltage. From that point it keeps the voltage at that value. In the circuit shown in Figure 2.4, the bias resistor R provides the voltage drop between the input voltage V_{in} and the output voltage V_{out}. The value of R must satisfy two conditions: it must be small enough that the current through diode keeps diode in reverse breakdown and it must be large enough that the current through diode does not destroy the device. A Zener diode used in this way is known as a *shunt voltage regulator* (*shunt* means connected in parallel, and *voltage regulator* is a class of circuit that produces a stable voltage across any load).

2.2 VARACTORS

The term "varactor" comes from the phrase "variable reactor" and means a device whose reactance can be varied in a controlled manner, in this case, with dc bias voltage [4]. The discovery of the capacitance of a rectifying contact extends back to 1929 when the first comprehensive investigation of this phenomenon was made [5]. In 1949 Shockley published his classic paper on the theory of *p–n* junctions where not only the expressions for the current–voltage relationships in a diode are given, but the capacitance effect is considered as well [6]. The possibility of tuning diodes with exponents greater than one-half was firstly described in 1958, with the term "hyperabrupt" coined for such sensitive devices [7].

2.2.1 Varactor Modeling

A simplified varactor equivalent circuit is shown in Figure 2.5, where C_v is the variable depletion layer capacitance, C_p is the package capacitance, R_s is the series contact and bulk resistance, L_s is the

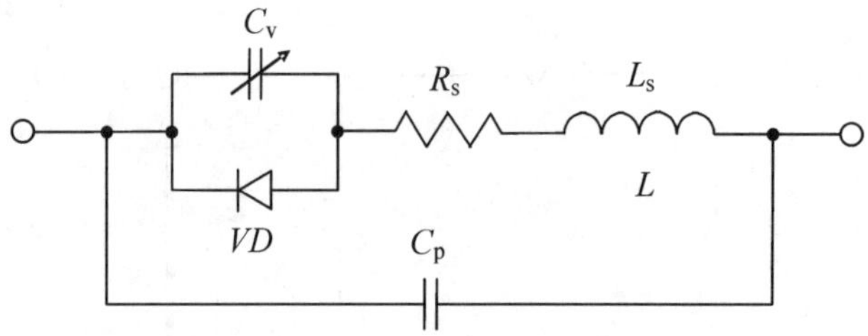

FIGURE 2.5 Simplified varactor equivalent circuit.

series inductance incorporating package inductance, and *VD* is the diode junction [8,9]. The diode is necessary to take into account because of the rectifying effect during a positive voltage swinging. The series resistance R_s is a function of applied voltage and operating frequency, although in most practical cases it can be considered constant. Such a model neglects some parasitic linear components, which should be taken into account for microwave applications including distributed-line package model and some capacitances due to the ground proximity. However, for most high frequency applications up to 2.5 GHz, these parasitics would not be significant unless higher order harmonics due to the varactor nonlinearity affect VCO performance.

The varactor junction capacitance C_v as a function of the reverse dc bias voltage V_v can be expressed by

$$C_v(V_v) = C_{v0}\left(1 + \frac{V_v}{\varphi}\right)^{-\gamma} \tag{2.6}$$

where

$$\gamma = -\frac{dC_v}{C_v} \bigg/ \frac{dV_v}{(V_v + \varphi)} \tag{2.7}$$

is the varactor junction sensitivity ($\gamma = 0.5$ for abrupt varactors, $1 \le \gamma \le 2$ for hyperabrupt varactors), φ is the contact potential which value depends on a doping profile of varactor; and $C_{v0} = C_v(0)$ is the varactor junction capacitance at $V_v = 0$. The voltage V_v is positive since it is assumed a reverse connection of the varactor.

When the junction is reverse-biased, a large electric field exists in the depletion region. As the bias voltage is increased, this field increases to a sufficiently high value where the thermally generated carriers traversing the depletion layer will generate additional hole–electron pairs by collision. These hole–electron pairs will in turn generate additional pairs, thus causing a multiplication effect or avalanche of carriers. The breakdown voltage at which avalanche occurs is the upper limit for the varactor voltage-control range and determines the minimum varactor junction capacitance $C_{v\min}$. The reverse breakdown voltage is relatively insensitive to temperature variations.

The doping concentration level in the varactor depletion region defines the difference between an abrupt junction varactor and a hyperabrupt junction one. For abrupt junction shown in Figure 2.6(*a*), the doping density is constant across the depletion region, whereas, for hyperabrupt junction shown in Figure 2.6(*b*), the doping density is a nonlinear function [10]. In the latter case, ion implantation or nonlinear epitaxial growing techniques can accomplish this. As a result, as the reverse voltage is increased, the higher doping density contributes to a greater capacitance change in the hyperabrupt varactor than in the abrupt varactor with a constant doping density. However, the averaged doping concentration of the undepleted epitaxial region for abrupt varactor is higher approximately by a factor of two, which gives the higher value of the series resistance R_s for hyperabrupt varactor.

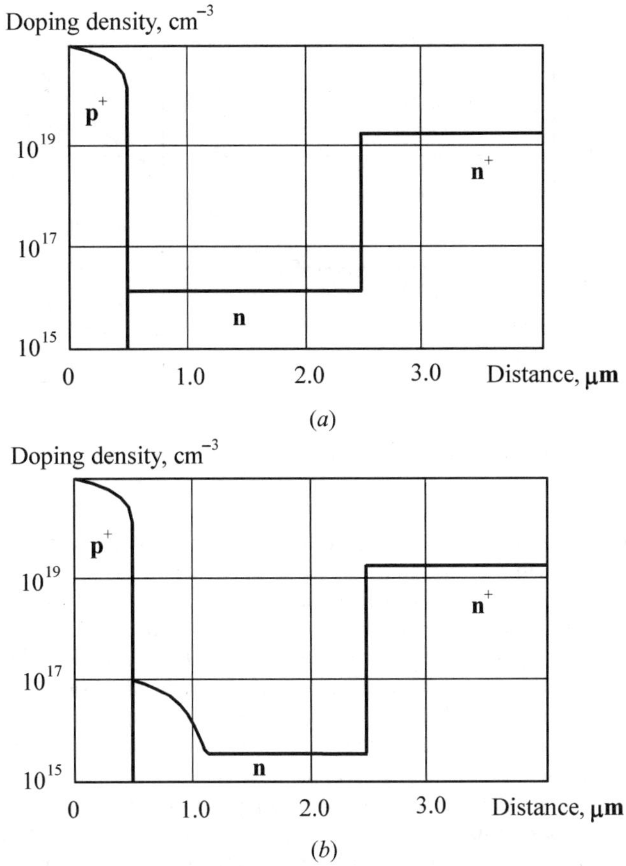

FIGURE 2.6 Doping density for abrupt and hyperabrupt varactors.

The quality factor of a varactor Q_v (taken into account that the varactor junction capacitance C_v is substantially higher than the package capacitance C_p) is defined as

$$Q_v(V_v, \omega) = \frac{1}{\omega R_s C_v(V_v)} \tag{2.8}$$

being a function of operating frequency and applied voltage. Since with the increase of R_s the varactor quality factor decreases, the Q_v of the abrupt varactor is higher than that of the hyperabrupt varactor at low reverse bias voltage. However, at higher reverse bias voltages, the quality factor of the hyperabrupt varactor becomes higher due to the more rapid decrease in the hyperabrupt varactor capacitance. As shown in Figure 2.7, usually over the linear tuning range for reverse bias voltage ranging from 1 to 10 V, the Q_v for hyperabrupt varactor is lower. As a result, the output power of such a voltage-controlled oscillator (VCO) with hyperabrupt varactor should be lower due to higher power losses in the varactor.

2.2.2 MOS Varactor

It is well known that an MOS transistor can be used as a variable capacitance due to the dependence of its total charge (representing the charge on the gate, effective interface charge, and charge in the

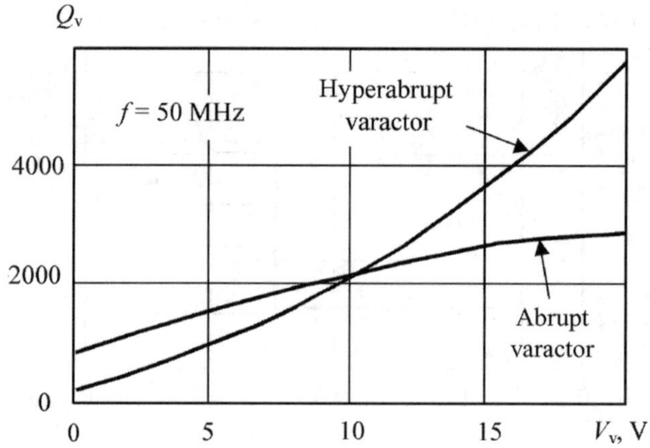

FIGURE 2.7 Varactor quality factor versus bias voltage.

semiconductor under the oxide) on the applied voltage between bulk and gate terminals [11]. In the case of *p*MOS varactor with the source, drain, and bulk terminals together connected, an inversion channel with mobile holes will be realized for gate–bulk voltages V_{gb} greater than the threshold voltage of the MOS transistor V_{th}. Further increase in V_{gb} results in a strong inversion of the MOS transistor operation. On the other hand, for voltages V_{gb} lower than V_{th}, an MOS transistor operates in accumulation region, where the voltage at the interface between gate oxide and semiconductor is positive and high enough to allow electrons to move freely.

Figure 2.8(*a*) shows the cross-section of the MOS transistor, where the movement behavior of the majority-charge carriers in the inversion, depletion, and accumulation regions is also shown [12]. A metal oxide structure is build on the top of a lightly doped *n*-well diffusion layer with the gate and the two n^+ contacts inside *n*-well. The device bias-dependent capacitance C can be modeled as oxide capacitance C_{ox} in series with the parallel connection of the capacitance C_b owing to the depletion region charge and C_i owing to the inversion layer charge at the gate–oxide interface. If C_b or C_i dominates, the MOS transistor is operated in a depletion or strong-inversion region, respectively. Otherwise, if neither capacitance dominates, the MOS transistor is operated in a weak-inversion region. The overall behavior of the MOS capacitance C versus gate–bulk voltage V_{gb} is qualitatively shown in Figure 2.8(*b*), where the charge carrier movement in strong- and moderate-inversion regions is indicated by solid lines shown in Figure 2.8(*a*).

When V_g approaches V_{th} in the moderate-inversion region, the concentration of holes at the oxide interface decreases steadily, but C_i continues to be much larger than C_b. However, when the MOS transistor enters the weak-inversion region, modulation of the depletion region becomes of the same importance as hole injection when $C_i \approx C_b$, being dominating effect in the depletion region when $C_i \ll C_b$. Thus, both in the strong-inversion and accumulation region, the overall device capacitance C approaches the oxide capacitance C_{ox}. Physically, an abundance of holes exists at high V_{gb} immediately below the oxide and provides bottom plate of the oxide capacitor, just as abundance of electrons provided that plate in the case of accumulation. In this case, the parasitic resistance is associated with the resistive losses of electrons moving from the bulk contact to the interface between the bulk and depletion regions, as shown in Figure 2.8(*a*) by dashed lines. Such a parasitic resistance can be reduced by using scaled technology with shorter device gate length L [13]. With the technology scaling, the oxide thickness is also reduced with correspondent increase in the oxide capacitance. Ideally, this should result not only in the better quality factor of a MOS varactor, but also in a wider tuning range since a minimum depletion capacitance increases at a lower rate.

To obtain monotonic dependence for C, it is necessary to provide the device operation without entering the accumulation region for a very wide range of gate voltage values. This can be

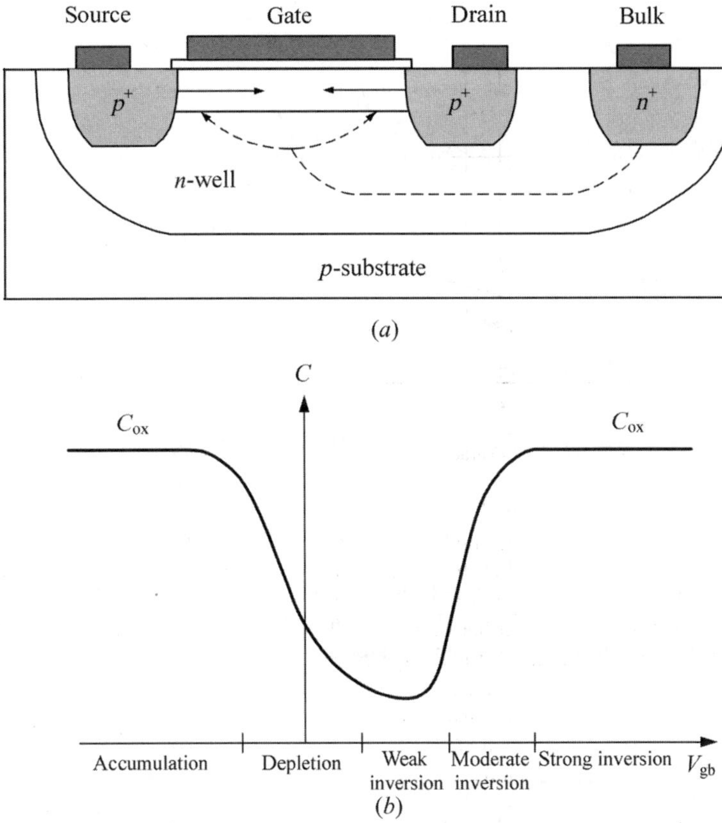

FIGURE 2.8 MOS varactor and its voltage–capacitance dependence.

accomplished by removing the connection between drain and source with bulk, by connecting the bulk terminal to the supply voltage as the highest dc voltage in the circuit. Figure 2.9 shows the voltage–capacitance dependence of such an inversion-mode MOS varactor operating in the strong-, moderate-, or weak-inversion regions only. However, a more attractive approach is to use the *p*MOS device in the depletion and accumulation regions only, resulting in a wider tuning range and better quality factor due to lower parasitic resistance since the electrons have mobility approximately three times higher than holes [14].

The voltage–capacitance dependence of such an accumulation-mode MOS varactor operating in the depletion and accumulation regions only is shown in Figure 2.10(*a*). To realize an accumulation-mode MOS varactor, the formation of the strong-, moderate-, and weak-inversion regions must be inhibited that requires the suppression of any injection of holes in the MOS channel. This can be accomplished by removing the source and drain diffusion p^+-doped layers and implementing the bulk n^+-doped contacts instead of them, thus minimizing the parasitic-well resistance, as shown in Figure 2.10(*b*).

Since physical models describing the behavior of a MOS device in the accumulation and depletion regions are different, it normally comprises separate models for these regions, the integration of which into a common circuit simulator such as SPICE is complicated. Figure 2.11(*a*) shows the cross-section of an accumulation-mode MOS varactor, where bulk represents shorted *n*-well contacts. Its single equivalent circuit is shown in Figure 2.11(*b*), where C_f represents the fringing capacitance mainly associated with sidewall of the gate; L_g and R_{poly} are the parasitic inductance and resistance of the

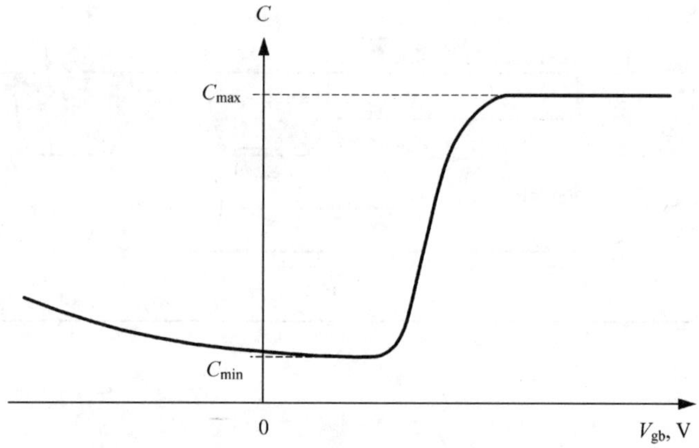

FIGURE 2.9 Inversion-mode MOS varactor tuning curve.

gate electrode, respectively [15]. The resistances R_{well}, R_{sub} and capacitances C_{sub1}, C_{sub2} are the substrate-related components. The resistance R_{sd} represents the source/drain regions. The channel resistance R_{ch} is the only bias-dependent resistance. It can be modeled as $R_{ch} = R_s + R_{acc}/R_p$, where R_s is the bias-independent n-well resistance between n^+ contact and accumulation or depletion regions underneath the gate, R_{acc} is the bias-dependent resistance of the accumulation layer, and R_p is the effective resistance along the edge of the depletion region.

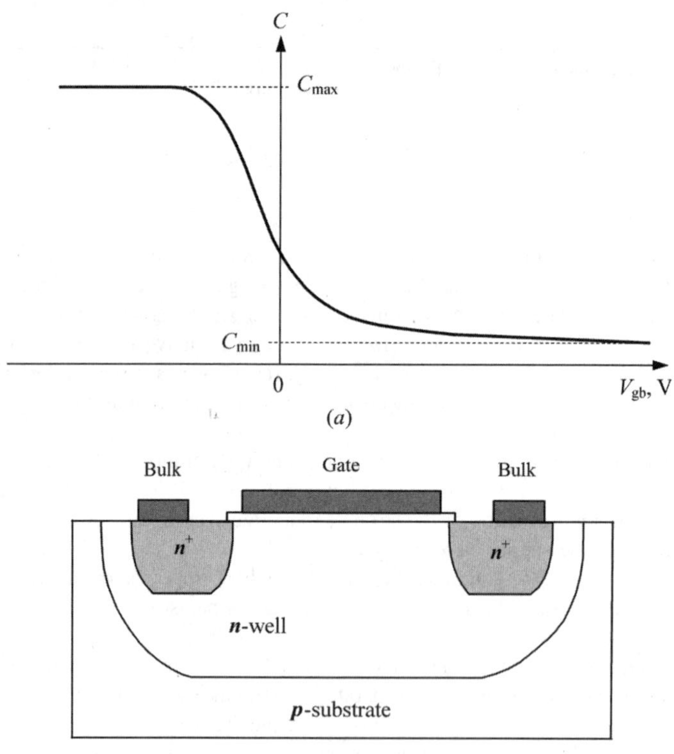

FIGURE 2.10 Accumulation-mode varactor and its tuning curve.

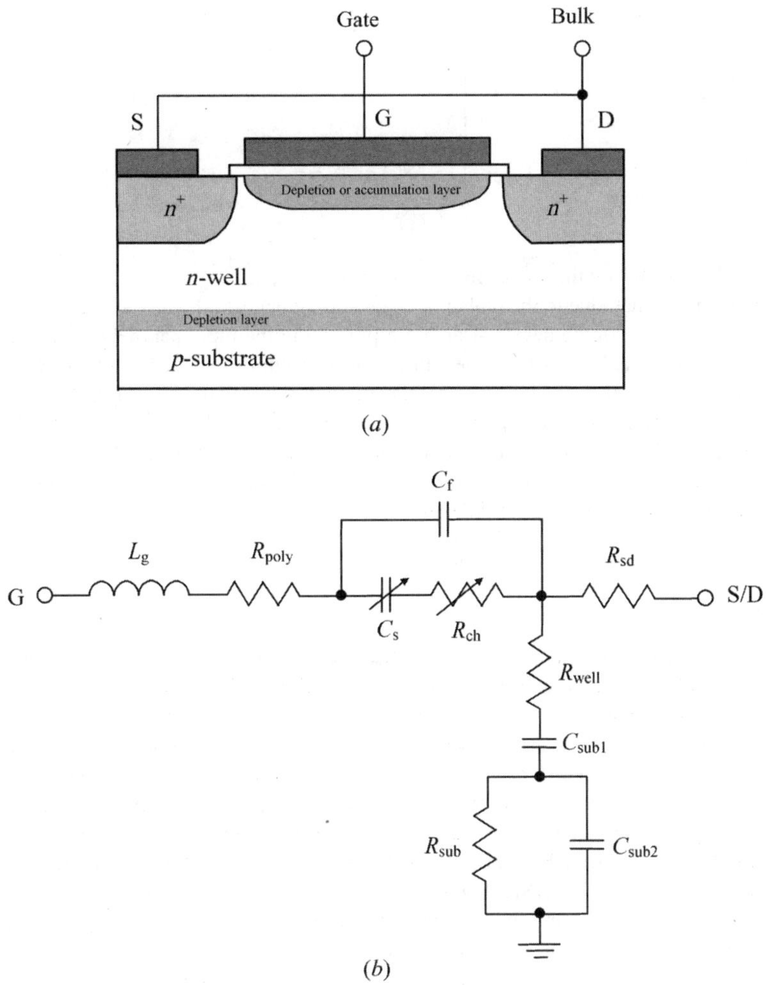

FIGURE 2.11 MOS varactor equivalent circuit.

In the accumulation region, R_{acc} becomes much smaller than R_p, and R_{ch} is approximately equal to $R_s + R_{acc}$. In the depletion region, R_{acc} can be considered infinite, and R_{ch} approaches a constant value of $R_s + R_p$. The gate bias-dependence model of R_{acc} based on the measurement results can be empirically given by

$$R_{acc}(V_{gb}) = \frac{1}{K_{acc}} \frac{1}{V_{gb} - V_{dep}} \quad \text{for } V_{gb} > V_{dep} \tag{2.9}$$

$$R_{acc}(V_{gb}) = \infty \quad \text{elsewhere} \tag{2.10}$$

where K_{acc} is the fitting parameter related to the device geometry and mobility of electrons in the accumulation region, and V_{dep} is the fitting parameter related to the flat-band voltage in the depletion region.

Accurate model of the MOS varactor capacitance C based on the description of its bias-dependent behavior in two regions separately, available in SPICE simulator and valid under different bias conditions for a frequency range up to 10 GHz, is given by

$$C(V_g) = C_j\left(1 - \frac{V_{gb}}{V_j}\right)^{M_j} \qquad \text{for } V_{gb} \leq F_c V_j \tag{2.11}$$

$$C(V_g) = C_j\frac{1 - F_c\left(1 + M_j\right) + M_j\dfrac{V_{gb}}{V_j}}{(1 - F_c)^{-1-M_j}} \qquad \text{for } V_{gb} > F_c V_j \tag{2.12}$$

where C_j, V_j, F_c, and M_j are the model-fitting parameters [16].

It should be noted that, due to the higher average doping beneath the gate and enhanced parasitic interconnect capacitance in the device structure representing the more parallel connected segments for the same gate area and smaller gate length L, the minimum capacitance will increase resulting in a lower capacitance ratio [17]. However, the minimum gate-length devices have the highest minimum quality factors since the polysilicon gate resistance at lower L dominates the overall parasitic resistance, which is several magnitudes smaller than the n-well resistance at large L. As a result, there is a tradeoff between the quality factor and capacitance tuning ratio that can be achieved, for example, for medium 0.65-μm gate-length MOS varactors.

2.3 MOSFETs

Personal wireless communication services have been driving the development of silicon MOSFET worldwide to provide reliable low-cost and high-performance technology. For example, the laterally diffused MOSFET (LDMOSFET) device structures have proven to be highly efficient, high gain, and linear for both high-power and low-voltage microwave and RF applications, including power amplifiers, low-noise amplifiers, mixers, and voltage-controlled oscillators. To develop low-cost silicon integrated circuits using CMOS technology for higher speed and higher frequency integrated circuits and subsystems within shorter design time, it is necessary to create accurate device models to allow efficient CAD simulation. Several well-known physically based MOSFET models can describe the device electrical behavior [1,18]. However, some of them such as Level 1, Level 2, or Level 3 large-signal models are very simple and cannot describe the current–voltage and voltage–capacitance characteristics with acceptable accuracy. Other popular models, as the BSIM3v3 or higher version models, are too much complicated and quite formal in general so that, for better learning of the device basic properties, it is useful to consider its intrinsic nonlinear core circuit only. Also, BSIM3v3 may not be as accurate for RFIC simulation due to their derivative discontinuity. Moreover, microwave parasitic effects in silicon MOSFET are not easy physically predictable. Table-based models, such as the HP Root model, are only accurate for the characterized structures and measurement conditions. An empirical analytical modeling approach is an explicit and valid compromise between physical models and data-based models.

2.3.1 Small-Signal Equivalent Circuit

To describe accurately the nonlinear properties of the large MOSFET devices, it is necessary to take into account the distributed nature of the gate capacitor, because the channel resistance is not equal to zero. In this case, the channel of such a device can be modeled as a bias-dependent RC distributed transmission line along the channel length, as shown in Figure 2.12. This one-dimensional approach assumes a gradual channel approximation when the quantity of charge in the channel is controlled completely by the gate electrode, only fields in the vertical dimension influence the depletion region, and channel current is provided entirely by drift with a constant mobility. Despite some drawbacks related to short-channel devices, this approach allows a compromise between accuracy and simplicity of a model derivation.

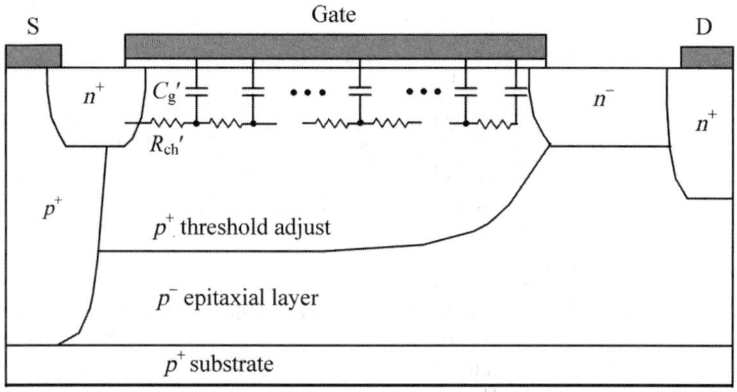

FIGURE 2.12 Schematic representation of MOSFET distributed channel structure.

The *ABCD*-matrix of the given *RC* transmission line can be written as

$$[ABCD] = \begin{bmatrix} \cosh\gamma L & Z_0 \sinh\gamma L \\ \dfrac{\sinh\gamma L}{Z_0} & \cosh\gamma L \end{bmatrix} \tag{2.13}$$

where $\gamma = \sqrt{j\omega\, R'_{ch} C'_g}$ is the propagation constant, $Z_0 = R'_{ch}/\gamma$ is the characteristic transmission line impedance, L is the channel length, $R'_{ch} = R_{ch}/L$, $C'_g = C_g/L$, R_{ch} is the channel charging resistance, which is a result of noninstantaneous respond to the changes of the gate–source voltage, and C_g is the total gate capacitance. In this case, the equivalent gate–source impedance Z_{gs} can be written using Eq. (2.13) as

$$Z_{gs} = \frac{A}{C} = R_{ch}\frac{\coth\gamma L}{\gamma L}. \tag{2.14}$$

The first-order approximation of Z_{gs} obtained from a power series expansion of Eq. (2.14) yields

$$Z_{gs} = R_{ch}\frac{\coth\gamma L}{\gamma L} \cong \frac{R_{ch}}{\gamma L}\left(\frac{1}{\gamma L} + \frac{\gamma L}{3}\right) = \frac{R_{ch}}{3} + \frac{1}{j\omega C_g}. \tag{2.15}$$

As we see from Eq. (2.15), the MOSFET intrinsic gate–source circuit can be modeled using a simple series circuit with the resistance $R_{gs} = R_{ch}/3$ and the capacitance $C_{gs} = C_g$. Figure 2.13 shows the intrinsic transistor equivalent circuit corresponding to the first-order channel approximation.

For a high-power MOSFET device whose channel width is significantly larger than its channel length, the distributed nature of the total gate resistance $R_{tot} = R_{sh}W/L$ across the width W (where R_{sh} is the sheet resistance of the gate material) has to be taken into consideration. In this case, silicon MOSFET can be decomposed into n devices, each with a width of W/n and a gate resistance of R_{tot}/n. For $n \to \infty$, it will be viewed as array of the small transistors distributed along the gate of the device. Commonly, it is necessary to consider a two-dimensional power MOSFET distributed model because it shows a distributed-gate nature along both the channel length and channel width. However, for a short-channel MOSFET, the distributed-gate effect along the channel length can be taken into account only in the frequency range close to the transition frequency f_T and higher. When $\omega R_{gs}C_{gs} \ll 1$, an analysis of the distributed-gate model along the channel width based on transmission

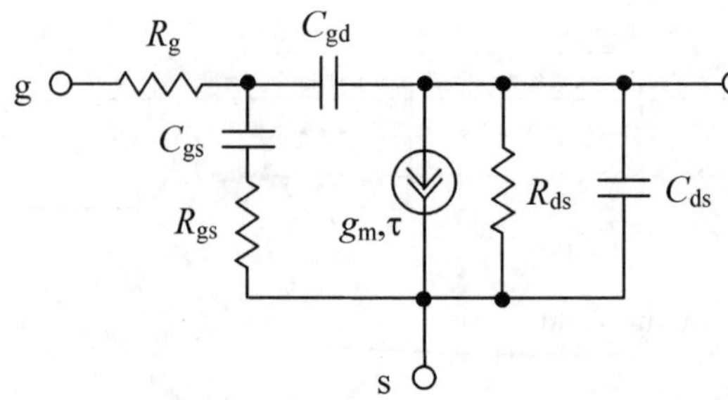

FIGURE 2.13 First-order approximation of intrinsic MOSFET equivalent circuit.

line theory shows that all transistor Y-parameters should be modified by the term tanh $(\gamma W)/(\gamma W)$ [19]. However, a linear power series expansion of this term

$$\frac{\tanh(\gamma W)}{\gamma W} = 1 - j\omega C_g \frac{R_{tot}}{3} \cong \frac{1}{1 + j\omega C_g \frac{R_{tot}}{3}} \tag{2.16}$$

leads to only the additional use of a series lumped gate resistance $R_{tot}/3$ that does not alter the structure of the transistor equivalent circuit. Consequently, the overall gate resistance R_g can generally be divided in two consecutive series resistances as $R_g = R_{ge} + R_{gi}$, where R_{ge} is the extrinsic contact and ohmic gate electrode resistance and $R_{gi} = R_{tot}/3$ is the intrinsic gate resistance due to the distributed-gate structure of the power MOSFET.

The complete small-signal MOSFET equivalent circuit with the extrinsic parasitic elements is shown in Figure 2.14. Here, R_{ds} is the differential channel resistance as a result of the channel length modulation by the drain voltage, C_{ds} is the drain–source capacitance, L_g is the gate lead inductance, R_s and L_s are the source bulk and ohmic resistance and lead inductance, R_d and L_d are the drain bulk and ohmic resistance and lead inductance, C_{gp} and C_{dp} are the gate and source pad capacitances, respectively.

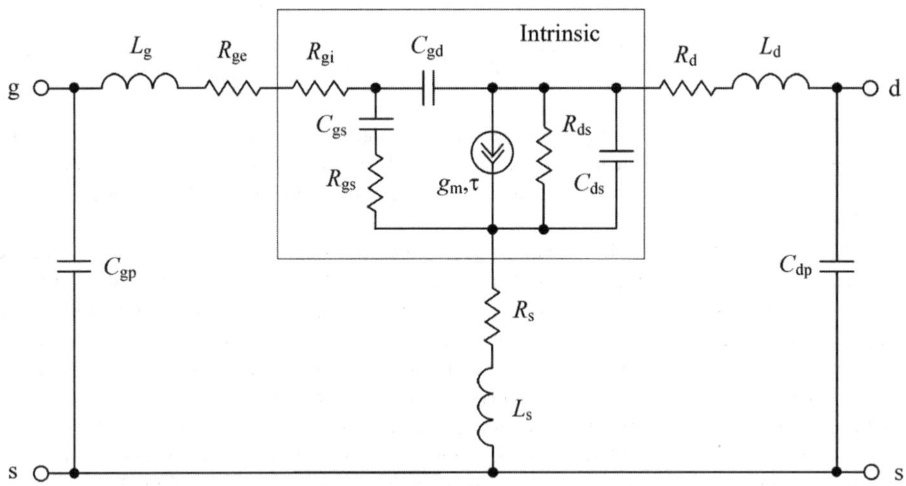

FIGURE 2.14 Nonlinear MOSFET equivalent circuit with extrinsic linear elements.

To characterize the transistor nonlinear electrical properties, it is sufficient to use the grounded-source intrinsic Y-parameters. Their two-port admittance matrix is

$$Y = \begin{bmatrix} \dfrac{j\omega C_{gs}}{1 + j\omega\tau_g} + j\omega C_{gd} & -j\omega C_{gd} \\ \dfrac{g_m \exp(-j\omega\tau)}{1 + j\omega\tau_g} - j\omega C_{gd} & \dfrac{1}{R_{ds}} + j\omega\left(C_{ds} + C_{gd}\right) \end{bmatrix} \quad (2.17)$$

where τ is the effective channel carrier transit time and $\tau_g = R_{gs}C_{gs}$. The intrinsic gate resistance R_{gi} can be considered an external gate element. In this case, the MOSFET intrinsic Y-matrix is the same as for the MESFET or HEMT devices. Consequently, to determine the elements of the intrinsic MOSFET small-signal equivalent circuit, it is possible to use similar analytical approach, which allows the determination of its elements through the real and imaginary parts of the device intrinsic admittance Y-parameters.

2.3.2 Nonlinear *I–V* Models

An empirical approach to approximate the nonlinear behavior of the drain current source I_{ds} (V_{gs}, V_{ds}) of a JFET device is described in [20]. Instead of using separate equations for the triode and pinch-off regions, irrespective of the device geometry and material parameters, a general expression based on hyperbolic functions was proposed:

$$I_{ds} = I_{dss}\left(1 - \frac{V_{gs}}{V_p}\right)^2 \tanh\alpha \left|\frac{V_{ds}}{V_p - V_{gs}}\right| \quad (2.18)$$

where I_{dss} is the saturation drain current for $V_{gs} = 0$, α is the saturation voltage parameter, and V_p is the pinch-off voltage. As a result, good agreement was obtained between the predicted and experimental results, which showed a promising prospect of such a simple empirical model.

A similar and sufficiently simple nonlinear model using a hyperbolic function and incorporating self-heating effect was later proposed to describe the *I–V* characteristics of a MOSFET device:

$$I_{ds} = \beta_{eff} V_{gst}^{\mathrm{VG\,exp}} (1 + \lambda V_{ds}) \tanh\left(\frac{\alpha V_{ds}}{V_{gst}^{\mathrm{SATexp}}}\right) \quad (2.19)$$

where

$$\beta_{eff} = \beta \Big/ \left(1 + \mu_{crit} V_{gst}^{\mathrm{GMexp}}\right)$$

$$V_{gst} = V_{st} \ln\left[1 + \exp\left(\frac{V_{gst1}}{V_{st}}\right)\right]$$

$$V_{gst1} = V_{gs} - V_{th0} - \gamma V_{ds}$$

where GMexp, SATexp, and μ_{crit} are the channel current parameters [21].

An empirical nonlinear model, which is single-piece and continuously differentiable, developed for silicon LDMOS transistors is given by

$$I_{ds} = \beta V_{gst}^{\mathrm{VG\,exp}} (1 + \lambda V_{ds}) \tanh\left(\frac{\alpha V_{ds}}{V_{gst}}\right) \left[1 + K_1 \exp(V_{BReff1})\right] + I_{ss} \exp\left(\frac{V_{ds} - V_{BR}}{V_T}\right) \quad (2.20)$$

where

$$V_{gst} = V_{st} \ln\left[1 + \exp\left(\frac{V_{gst2}}{V_{st}}\right)\right]$$

$$V_{gst2} = V_{gst1} - \frac{1}{2}\left(V_{gst1} + \sqrt{\left(V_{gst1} - V_K\right)^2 + \Delta^2} - \sqrt{V_K^2 + \Delta^2}\right)$$

$$V_{gst1} = V_{gs} - V_{th0} - \gamma V_{ds}$$

$$V_{BReff1} = \frac{V_{ds} - V_{BReff}}{K_2} + M_3 \frac{V_{ds}}{V_{BReff}}$$

$$V_{BReff} = \frac{V_{BR}}{2}\left[1 + \tanh\left(M_1 - V_{gst} M_2\right)\right]$$

where λ is the drain current slope parameter, β is the transconductance parameter, V_{th0} is the forward threshold voltage, V_{st} is the subthreshold slope coefficient, V_T is the temperature voltage, I_{ss} is the forward diode leakage current, V_{BR} is the breakdown voltage, K_1, K_2, M_1, M_2, and M_3 are the breakdown parameters, V_K, VGexp, Δ, and γ are the gate–source voltage parameters [22].

In many applications, it is necessary to take into consideration the MOSFET operation in the weak-inversion region when the gate–source voltage V_{gs} is smaller than the threshold voltage V_{th}. For example, to improve the conversion gain of a mixer or reduce the intermodulation distortion (IMD) of a class AB power amplifier when device is biased around the onset of the strong-inversion region from the weak-inversion region for low drain quiescent current. The drain current in the weak-inversion region is mainly dominated by the diffusion component that increases exponentially with the gate voltage [11]. On the other hand, in the strong-inversion saturation region when the gate–source voltage is greater than the threshold voltage, the drain current is proportional to the square of $(V_{gs} - V_{th})$.

To obtain continuous behavior from weak-inversion region to strong-inversion region for the drain current and a compromise between accurate device modeling and ease of circuit analysis, we can use

$$I_{ds}\left(V_{gs}\right) = A\left\{\ln\left[1 + \exp\left(B\left(V_{gs} - V_{th}\right)\right)\right]\right\}^2 \tag{2.21}$$

where A and B are the approximation parameters. In this case, the drain current is effectively proportional to the square of $(V_{gs} - V_{th})$ when V_{gs} is larger than V_{th} and exponentially decreases with the gate–source voltage when V_{gs} is smaller than V_{th}. The approximation parameters A and B are defined from the following conditions:

$$I_{ds}|_{V_{gs}=V_{th}} = I_{th} \qquad \left.\frac{\partial I_{ds}}{\partial V_{gs}}\right|_{V_{gs}=V_{th}} = S_{th} \tag{2.22}$$

where I_{th} is the threshold drain current, as shown in Figure 2.15, and S_{th} is a slope of the current–voltage transfer characteristic in the threshold point. Then,

$$A = \frac{I_{th}}{(\ln 2)^2} \qquad B = \frac{S_{th}}{I_{th}} \ln 2. \tag{2.23}$$

Consequently, the transfer characteristic can be defined in terms of only two physical parameters I_{th} and S_{th}, which are easily calculated from the device measurements. By using similar analytical approach, the EKV MOST model has been successfully applied to low-voltage and low-current analog circuit design and simulation, referring voltage V_g, V_d, and V_s to the device local substrate [23].

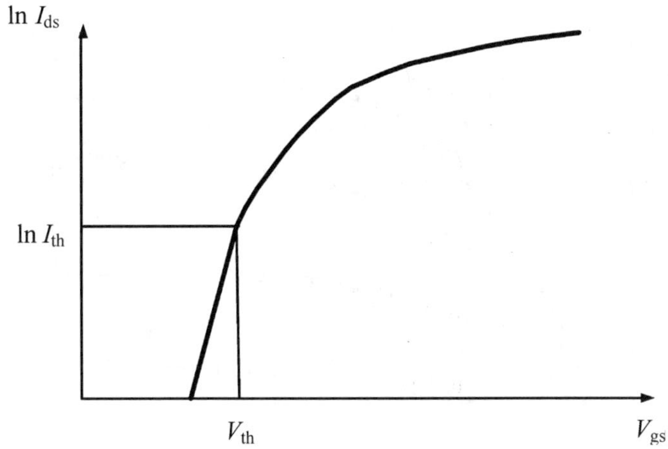

FIGURE 2.15 Drain current versus gate–source voltage.

In view of the monotonous behavior of $I_{ds}-V_{ds}$ curves, the entire drain current–voltage characteristics of a MOSFET device can be described as:

$$I_{ds}\left(V_{gs}, V_{ds}\right) = I_0 \Big/ \left[1 + \left(\frac{I_0}{I_{max}}\right)^{n}\right]^{\frac{1}{n}} \tag{2.24}$$

where

$$I_0 = \frac{I_{th}}{(\ln 2)^2\left(1 - \beta V_{gs}\right)}\left\{\ln\left[1 + \exp\left(\frac{S_{th}\ln 2}{I_{th}}\left(V_{gs} - V_{th}\right)\right)\right]\right\}^2$$

$$I_{max} = I_{sat}\left(1 + \lambda V_{ds}\right)\tanh\left(\alpha V_{ds}\right)$$

where I_{sat} is the saturated drain current, α is the saturation voltage parameter (which affects a slope of the $I_{ds}-V_{ds}$ characteristics in the linear region), λ is the parameter that determines a slope of the same drain characteristics in the saturation region, $V_{th} = V_{th0} - \sigma V_{ds}$, n and β are the fitting parameters that determine a slope of the transfer characteristics under large values of V_{gs}, σ is the parameter that expresses empirically the dependence of the threshold voltage on V_{ds} [24].

To verify that this model is applicable not only to high-power LDMOSFET devices but also to low-voltage MOSFETs, the appropriate low-voltage MOSFET power device with a gate width of $W =$ 2 mm was selected [25]. Figure 2.16(*a*) shows the transistor theoretical and experimental drain current curves $I_{ds}(V_{ds})$. The resulting current mean-square error of a family of the output current–voltage $I_{ds}-V_{ds}$ characteristics is 0.42%. The transfer $I_{ds}-V_{gs}$ characteristics were compared with the same characteristics calculated from the BSIM3v3 model, which was developed for modeling of deep submicrometer devices. The results shown in Figure 2.16(*b*) demonstrate a good agreement with the experimental curves and practically the same as in the case of the BSIM3v3 approximation.

2.3.3 Nonlinear *C–V* Models

The input capacitance C_{gs} normally influences the IMD level especially when the frequency increases in the microwave region [26]. Generally, the calculation of the gate–source capacitance C_{gs} or the gate–drain capacitance C_{gd} from the charges corresponding to the strong-inversion model only results in a mathematically complicated expression [11]. Therefore, in most cases when it is necessary to

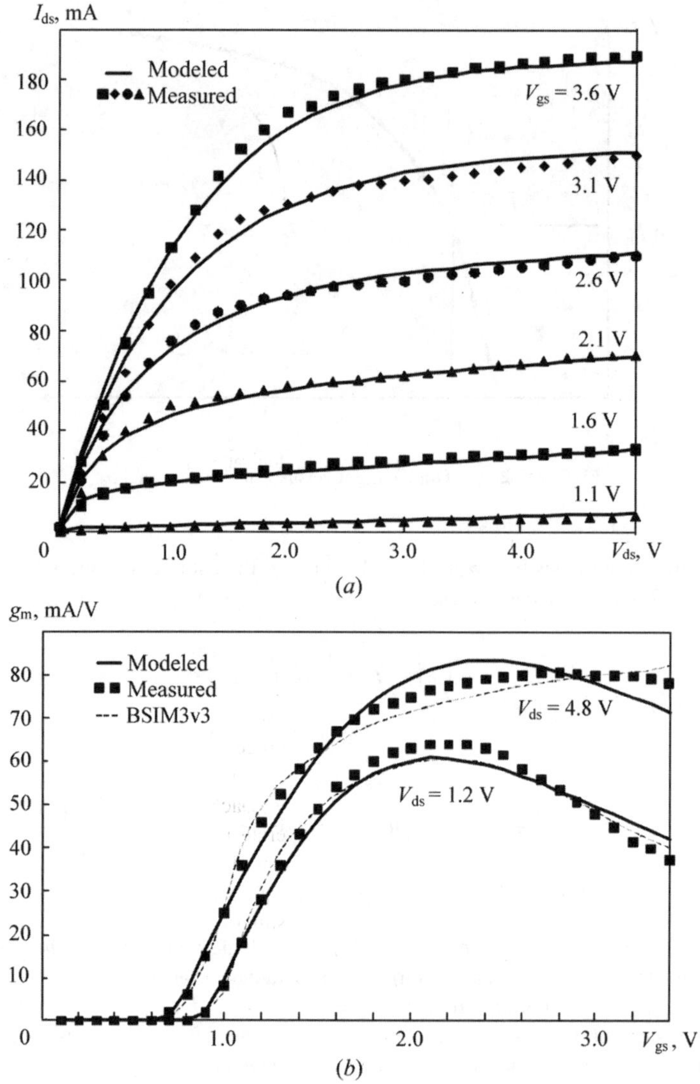

FIGURE 2.16 Measured and modeled current–voltage characteristics of low voltage MOSFET.

predict efficiency, gain, or output power of the power amplifier or oscillator, the capacitances C_{gs} and C_{gd} can be modeled as the fixed capacitances measured at the quiescent bias voltage, and the *p–n* junction diode capacitance model can be applied to the capacitance C_{ds} [27]. The gate–drain capacitance C_{gd} can also be considered as the bias-dependent junction capacitance [28]. With the increase in the drain bias voltage, a depletion region is formed under the oxide in the lightly doped drain region. Therefore, the capacitance C_{gd} can be considered as a junction capacitance, which strongly depends on the drain–source bias voltage V_{ds}. According to the accurate charge model calculations, the gate–drain capacitance C_{gd} has a strong dependence on V_{gs} only in the moderate-inversion region when $V_{gs} - V_{th} < 1$ V [11]. In this region, the behavior of C_{gd} is similar to C_{gs}, and can be evaluated using the same hyperbolic tangent functions. However, for high-voltage LDMOSFET devices, since the dependence of C_{gd} on V_{gs} is quite small, it seems sufficient to limit the dependence to V_{ds} only.

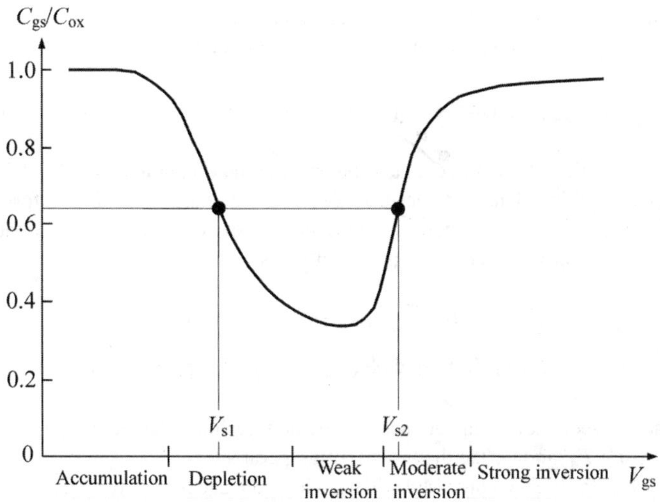

FIGURE 2.17 Gate–source capacitance versus gate–source voltage.

The drain–source capacitance C_{ds} varies due to the change in the depletion region, which is mainly determined by the value of V_{ds}.

The gate–source capacitance C_{gs} can be described as a function of the gate–source bias voltage. First, we consider an appropriate behavior of each of its main composite part: the intrinsic gate–source capacitance C_{gsi}, including both the gate–source and the source–substrate charge fluctuations, and the gate–substrate capacitance C_{gbi}. The gate–source voltage dependence of these components is substantially different [11]. The intrinsic gate–substrate capacitance C_{gbi} is constant in the accumulation region where it is equal to the total intrinsic oxide capacitance C_{ox}, slightly decreases in the weak-inversion region, significantly reduces in the moderate-inversion region, and becomes practically constant in the strong-inversion or saturation region. The intrinsic gate–source capacitance C_{gsi} grows rapidly in the moderate-inversion region and equals $2C_{ox}/3$ in the saturation region. The dependence of the total gate–source capacitance C_{gs} as a sum of its components C_{gsi} and C_{gbi} on V_{gs} is shown in Figure 2.17.

A hyperbolic tangent function can be used for each of two parts of the dependence $C_{gs}(V_{gs})$, where the gate–source capacitance can be approximated by

$$C_{gs} = C_{gsmin} + C_s \left\{ 1 + \tanh\left[\frac{S}{C_s} \left(V_{gs} - V_s \right) \right] \right\} \tag{2.25}$$

where $C_s = (C_{gsmax} - C_{gsmin})/2$, C_{gsmax} is the maximum gate–source capacitance, C_{gsmin} is the minimum gate–source capacitance, $S = (S_1, S_2)$ is the slope of $C_{gs}(V_{gs})$ at each bend point $V_{gs} = (V_{s1}, V_{s2})$, as shown in Figure 2.17,

$$S_1 = \frac{\partial C_{gs}}{\partial V_{gs}} \Big|_{V_{gs}=V_{s1}} \qquad S_2 = \frac{\partial C_{gs}}{\partial V_{gs}} \Big|_{V_{gs}=V_{s2}}. \tag{2.26}$$

The total gate–source capacitance C_{gs} as a function of V_{gs} can be described by

$$C_{gs} = C_{gsmax} - C_{gso} \left\{ 1 + \tanh\left[\frac{S_1}{C_s} \left(V_{gs} - V_{s1} \right) \right] \right\} \times \left\{ 1 + \tanh\left[\frac{S_2}{C_s} \left(V_{gs} - V_{s2} \right) \right] \right\} \tag{2.27}$$

where $C_{gsmax} = C_{ox}$ and C_{gso} is the model fitting parameter [24].

The approximation function for the gate–source capacitance C_{gs} as the dependence of V_{gs} can also be expressed by using the two components, both containing the hyperbolic functions:

$$C_{gs} = C_{gs1} + C_{gs2}\left\{1 + \tanh\left[C_{gs6}\left(V_{gs} + C_{gs3}\right)\right]\right\} + C_{gs4}\left[1 - \tanh\left(C_{gs5}V_{gs}\right)\right] \quad (2.28)$$

where C_{gs1}, C_{gs2}, C_{gs3}, C_{gs4}, C_{gs5}, and C_{gs6} are the approximation parameters [22].

If we consider the dependence of the gate–source capacitance C_{gs} on V_{ds} for submicrometer MOSFET devices when C_{gs} slightly increases with the increase of V_{ds}, the approximation expression for C_{gs} as a function of both V_{gs} and V_{ds} can be written as

$$\begin{aligned} C_{gs} &= C_{gs0} + C_{gs1}\left\{A_s + B_s\tanh\left[C_s\left(V_{gs} - V_{th}\right)\right]\right\} \\ &\times \left\{D_s + E_s\left[1 + \tanh\left(V_{gs} - V_{ds}\right)\right]\tanh\left[F_sV_{ds} - G_sV_{gs}\right]\right\} \end{aligned} \quad (2.29)$$

where C_{gs0} is the bias-dependent capacitance, V_{th} is the threshold voltage, C_{gs1} is the scaling factor, and A_s, B_s, C_s, D_s, E_s, F_s, and G_s are the model fitting parameters [29].

On the other hand, for high-voltage MOSFET devices, the dependencies of gate–drain capacitance C_{gd} and drain–source capacitance C_{ds} on V_{ds} can be accurately evaluated by the junction capacitance model as

$$C_{gd(ds)} = C_{gdo(dso)}\left(\frac{\varphi + V_{dso}}{\varphi + V_{ds}}\right)^{m} \quad (2.30)$$

where m (m_1 for C_{gd} or m_2 for C_{ds}) is the junction sensitivity depending on a doping profile ($m = 1/3$ for the linearly graded junction, $m = 1/2$ for the abrupt junction, and $m > 1/2$ for the hyperabrupt junction), φ is the contact potential, C_{gdo} and C_{dso} are the junction capacitances when $V_{ds} = V_{dso}$. For practical profiles of the junction, which are neither exactly abrupt nor exactly linearly graded, one often chooses the parameters m and φ to obtain the best matching between the theoretical model and the measurements.

For submicrometer MOSFET devices, to take into account the dependence of C_{gd} both on V_{gs} and V_{ds}, the approximation expression for the C_{gd} is written as

$$C_{gd} = C_{gd0} + C_{gd1}\left\{A_d + B_d\tanh\left[C_d\left(D_dV_{gs} - V_{ds}\right) - V_{th}\right]\right\} \quad (2.31)$$

where C_{gd0} is the bias-dependent capacitance, V_{th} is the threshold voltage, C_{gd1} is the scaling factor, while A_d, B_d, C_d, and D_d are the model fitting parameters [29].

2.3.4 Charge Conservation

To describe the small- and large-signal device models, it is necessary to satisfy the charge conservation condition. For a three-terminal MOSFET device, the matrix equation for the small-signal charging circuit in frequency domain is given by

$$\begin{bmatrix} I_g \\ I_d \\ I_s \end{bmatrix} = j\omega \begin{bmatrix} C_{gg} & -C_{gd} & -C_{gs} \\ -C_{dg} & C_{dd} & -C_{ds} \\ -C_{sg} & -C_{sd} & C_{ss} \end{bmatrix} \begin{bmatrix} V_g \\ V_d \\ V_s \end{bmatrix} \quad (2.32)$$

where I_g, I_d, and I_s are the terminal current amplitudes, V_g, V_d, and V_s are the terminal voltage amplitudes, and the capacitance between any two device terminals (k, l) is described as $C_{kl} = \partial Q_k/\partial V_l$ [11]. To transform a three-terminal device into a two-port network with a common source terminal, the current and voltage terminal conditions of $I_g = I_{gs}$, $I_d = I_{ds}$, $I_s = -(I_{gs} + I_{ds})$,

$V_g - V_s = V_{gs}$, and $V_d - V_s = V_{ds}$ should be taken into account. In addition, the following relationships between the terminal capacitances for three-terminal devices are valid:

$$\begin{aligned} C_{gg} &= C_{gd} + C_{gs} = C_{dg} + C_{sg} \\ C_{dd} &= C_{dg} + C_{ds} = C_{gd} + C_{sd} \\ C_{ss} &= C_{sg} + C_{sd} = C_{gs} + C_{ds}. \end{aligned} \tag{2.33}$$

The admittance Y_c-matrix for such a capacitive two-port network is

$$Y_c = \begin{bmatrix} j\omega\left(C_{gs} + C_{gd}\right) & -j\omega C_{gd} \\ -j\omega\left(C_{gd} + C_m\right) & j\omega\left(C_{ds} + C_m + C_{gd}\right) \end{bmatrix} \tag{2.34}$$

where $C_m = C_{dg} - C_{gd}$ is the transcapacitance, C_{gd} represents the effect of the drain on the gate, and C_{dg} represents the effect of the gate on the drain, and these effects are different [11]. Similarly to the *I–V* characteristics, there is no reason to expect that the effect of the drain voltage on the gate current, which is zero assuming no leakage current, is the same as the effect of the gate voltage on the drain current, which is significantly large.

Therefore, for power MOSFET devices, because the transcapacitance C_m is substantially less than C_{gs}, it can be translated to an additional delay time τ_c in a frequency range up to f_T by its combining with the transconductance g_m according to

$$g_m - j\omega C_m = g_m\sqrt{1 + \left(\frac{\omega}{\omega_T}\frac{C_m}{C_{gs}}\right)^2}\exp\left[-j\tan^{-1}\left(\frac{\omega}{\omega_T}\frac{C_m}{C_{gs}}\right)\right] \cong g_m\exp\left(-j\omega\tau_c\right) \tag{2.35}$$

where $\tau_c = C_m/(\omega_T C_{gs})$. To satisfy the charge conservation condition, the total delay time τ, shown in the MOSFET equivalent circuit in Figure 2.14, represents both the ideal transit time and delay time due to the transcapacitance. The transcapacitance C_m can be easily added to the drain–source capacitance C_{ds} under parameter extraction procedure.

2.3.5 Gate–Source Resistance

The gate–source resistance R_{gs} is determined by the effect of the channel inertia in responding to rapid changes of the time varying gate–source voltage, and varies in such a manner that the charging time $\tau_g = R_{gs}C_{gs}$ remains approximately constant. Thus, the increase of R_{gs} in the velocity saturation region (when the channel conductivity decreases) is partially compensated by the decrease of C_{gs} due to nonuniform channel charge distribution [30]. The effect of R_{gs} becomes significant at higher frequencies close to the transition frequency f_T of the MOSFET and may not be taken into account when designing RF circuits operating below 2 GHz, as used for commercial wireless applications [25,31]. For example, for the MOSFET with the depletion region doping concentration value $N_A = 1700\ \mu m^{-3}$, the phase of the small-signal transconductance g_m near f_T reaches the value only of $-15°$ [11].

2.3.6 Temperature Dependence

Silicon MOSFET devices are very sensitive to the operation temperature T and their characteristics are strongly temperature dependent [11]. The main parameters responsible for this are the effective carrier mobility μ and the threshold voltage V_{th}, resulting in the increase of the drain current through $V_{th}(T)$ and the decrease of the drain current through $\mu(T)$ with temperature. Increasing temperature decreases the slope of the $I_{ds}(V_{gs})$ curves. A certain value of V_{gs} can be found, at which the drain current becomes practically temperature independent over a large temperature range. The variation of V_{th} with temperature in a wide range from −50 to +200°C represents a nonlinear function, which is slowly decreased with temperature and can be approximated by

$$V_{th}(T) = V_{th}(T_{nom}) + V_{T1}\Delta T + V_{T2}\Delta T^2 \tag{2.36}$$

where $\Delta T = T - T_{nom}$, $T_{nom} = 300$ K (27°C), and V_{T1} and V_{T2} are the linear and quadratic temperature coefficients for threshold voltage [27].

The variation of μ with temperature can be taken into account by introducing the appropriate temperature variation of I_{sat} in Eq. (2.24). The temperature variation of I_{sat} represents an almost straight line, which decreases with temperature [27]. The temperature dependence $I_{sat}(T)$ can be approximated by the linear function as

$$I_{sat}(T) = I_{sat}(T_{nom}) + I_T\Delta T \tag{2.37}$$

where I_T is the linear temperature coefficient for the saturation current.

The temperature dependencies of the MOSFET capacitances and series resistances can be described by the following linear equations [1,32]:

$$C(T) = C(T_{nom}) + C_T\Delta T \tag{2.38}$$

$$R(T) = R(T_{nom}) + R_T\Delta T \tag{2.39}$$

where $C = (C_{gs}, C_{ds}, C_{gd})$, $R = (R_g, R_s, R_d)$, and R_T and C_T are the linear temperature coefficients for the capacitances and resistances, respectively.

At high value of V_{gs} and V_{ds} under dc measurement, the slope of $I_{ds} - V_{ds}$ curves can be negative that occurs due to the self-heating effect in a highly dissipated power region. For the drain current model given by Eq. (2.19), this effect can be taken into account by adding a linear component describing the temperature dependence as

$$\beta(T) = \beta(T_{nom}) + \beta_T\Delta T_j \tag{2.40}$$

$$\gamma(T) = \gamma(T_{nom}) + \gamma_T\Delta T_j \tag{2.41}$$

where $\Delta T_j = R_{th}P_{dis} + \Delta T$, R_{th}(°C/W) is the thermal resistance, P_{dis} is the dc power consumption in watts caused by dc biasing, and β_T and γ_T are the linear temperature coefficients with negative values, respectively [21].

Another way of taking into account the effect of the negative conductance at high biasing is to write the nonlinear $I_{ds} - V_{ds}$ model as follows:

$$I_{ds}(T, p_T) = \frac{I_{ds}(T)}{1 + p_T V_d I_{ds}(T)} \tag{2.42}$$

where the drain current source $I_{ds}(T)$ is given by Eq. (2.24), $V_d = V_{ds}/\sqrt{1 + (\omega\tau_{th})^2}$, p_T is the self-heating temperature coefficient, V_{ds} is the drain–source supply voltage, $\tau_{th} = R_{th}C_{th}$ is the thermal time constant, R_{th} is the thermal resistance, C_{th} is the thermal capacitance, and T_j is the function of ambient temperature T and p_T. A thermal equivalent circuit can be added to the large-signal MOSFET model as a parallel $R_{th}C_{th}$ circuit [22]. The thermal resistance R_{th} can be extracted from the temperature measurement of the dc characteristics. Since the slope of the dc measured $I_{ds}(V_{ds})$ curves changes its sign from positive to negative, the temperature coefficient p_T can be evaluated under the condition of

$$\frac{dI_{ds}(T, p_T)}{dV_{ds}} = 0 \tag{2.43}$$

As a result,

$$p_T = \frac{1}{I_{ds}^2(T)} \frac{dI_{ds}(T)}{dV_{ds}} \tag{2.44}$$

where $I_{ds}(V_{ds})$ curves are determined by measurement of the pulsed $I_{ds}(V_{ds})$ curves at ambient temperature T, and the value of $I_{ds}(T)$ is fixed the same as for zero slope of $I_{ds}(T, p_T)$.

The thermal time constant τ_{th} can be extracted by comparing pulsed $I_{ds}(V_{ds})$ curves calculated under different pulse widths and duty factors. A plot of I_{ds} as a function of pulse width under the fixed gate–source and drain–source bias voltages gives an appropriate value of τ_{th}.

2.3.7 Noise Model

The noise behavior of the MOSFET device can be described based on its equivalent circuit representation, which includes the main elements responsible for the device electrical behavior and noise sources. The noise generated by a circuit element can be modeled as a result of a small-signal electrical excitation. Each noise source is considered as statistically uncorrelated to the other noise sources in the circuit and the contribution of each noise source to the total noise is determined on the individual basis. The total noise represents the root-mean-square sum of these individual noise contributions. Since a device channel material is resistive, it exhibits thermal noise as a major source of noise, which can be represented by a noise current source $\overline{i_{nd}^2}$ connected between the drain and the source in the MOSFET small-signal equivalent circuit shown in Figure 2.18(a), where the flicker

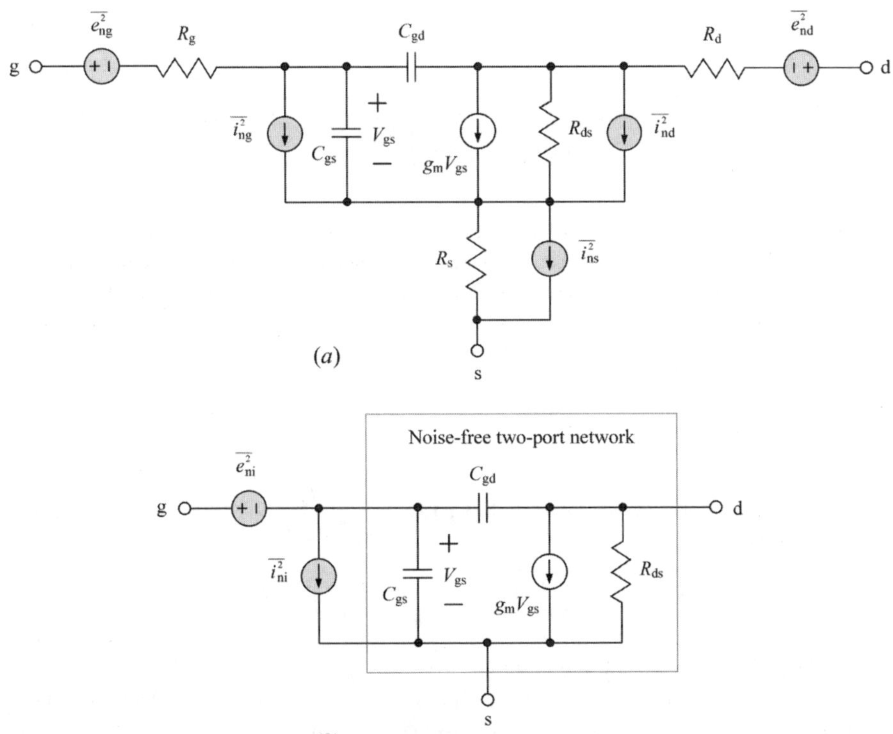

FIGURE 2.18 MOSFET equivalent circuits with noise sources.

noise is also included. The induced gate current noise is modeled by the gate noise current source $\overline{i_{\rm ng}^2}$ connected across the gate–source capacitance $C_{\rm gs}$.

The noise voltage and current sources can be given through their mean-square values as

$$\overline{e_{\rm ng}^2} = 4kTR_{\rm g}\Delta f \tag{2.45}$$

$$\overline{i_{\rm ng}^2} = \frac{4kT\eta\left(\omega C_{\rm gs}\right)^2\Delta f}{g_{\rm m}} \tag{2.46}$$

$$\overline{i_{\rm ns}^2} = \frac{4kT\Delta f}{R_{\rm s}} \tag{2.47}$$

$$\overline{e_{\rm nd}^2} = 4kTR_{\rm d}\Delta f \tag{2.48}$$

$$\overline{i_{\rm nd}^2} = \frac{4kT\gamma}{R_{\rm ds0}}\Delta f + K_{\rm F}\frac{I_{\rm d}^{A_{\rm F}}}{C_{\rm ox}L_{\rm eff}^2 f}\Delta f \tag{2.49}$$

where η is the induced gate noise coefficient, $g_{\rm m}$ is the device transconductance, $I_{\rm d}$ is the drain bias current, $A_{\rm F}$ is the flicker noise exponent, $K_{\rm F}$ is the flicker noise coefficient, $C_{\rm ox}$ is the oxide capacitance per unit area, $L_{\rm eff}$ is the effective channel length, $R_{\rm ds0}$ is the differential drain–source resistance at $V_{\rm ds} = 0$, γ is the channel noise coefficient [1,33]. In the long-channel devices, $R_{\rm ds0} = 1/g_{\rm m}$ and $\gamma = 2/3$, while both $R_{\rm ds0}$ and γ are complicated functions of the device parameters in the short-channel MOSFETs [34]. The equation for $\overline{i_{\rm nd}^2}$ (with excluded flicker noise) valid for both short-channel and long-channel devices expressed through the device parameters can be written in a simple form as

$$\overline{i_{\rm nd}^2} = 4kT\beta I_{\rm d}\Delta f \tag{2.50}$$

where

$$\beta = \frac{1}{V_{\rm dsat}} + \frac{\alpha^2 V_{\rm dsat}}{3\left(V_{\rm gs} - V_{\rm th}\right)^2}$$

$V_{\rm dsat}$ is the saturation drain–source voltage, and α is the bulk-charge effect coefficient [35].

The required minimum noise figure $F_{\rm min}$, noise resistance $R_{\rm n}$, and optimum source admittance $Y_{\rm Sopt}$ using the noise correlation $C_{\rm A}$-matrix parameters as functions of the input-referred noise voltage $\overline{v_{\rm ni}^2}$, noise current $\overline{i_{\rm ni}^2}$, and the parameters of the simplified noise-free two-port network shown in Figure 2.18(*b*) can be approximately estimated by

$$R_{\rm n} = R_{\rm g} + \frac{\beta I_{\rm d}}{g_{\rm m}^2} \tag{2.51}$$

$$Y_{\rm Sopt} = \frac{\sqrt{\beta I_{\rm d} R_{\rm g}}}{R_{\rm n}}\left(\frac{f}{f_{\rm T}}\right)\left(1 - j\frac{1}{g_{\rm m}}\sqrt{\frac{\beta I_{\rm d}}{R_{\rm g}}}\right) \tag{2.52}$$

$$F_{\rm min} = 1 + \left(\frac{f}{f_{\rm T}}\right)\sqrt{\beta I_{\rm d} R_{\rm g}}\left[1 + \left(\frac{f}{f_{\rm T}}\right)\sqrt{\beta I_{\rm d} R_{\rm g}}\right] \tag{2.53}$$

where f is the operation frequency and $f_{\rm T} = g_{\rm m}/2\pi(C_{\rm gs} + C_{\rm gd})$ [35].

There are two major theories to explain the physical origin of 1/*f* noise in MOSFET devices, one is based on the carrier number fluctuation theory when the flicker noise is attributed to the random trapping and detrapping processes of charges near the Si–SiO_2 interface, the other is based on mobility fluctuation theory considering the flicker noise as a result of the fluctuations in bulk mobility

[36–38]. Assuming that the channel can exchange charges with the oxide traps through tunneling, the charge fluctuation results in fluctuation of the surface potential, which in turn modulates the channel carrier density. At the same time, it is considered that the fluctuation of bulk mobility is induced by fluctuations in phonon population through phonon scattering. Generally, the measured noise power in MOSFET devices has a more complicated dependence on the gate bias and oxide thickness than due to the predictions based on the number or bulk mobility fluctuation theory only. Also, the surface mobility fluctuation mechanism should be taken into account attributed to the scattering effect of fluctuating oxide charge [39]. The implementation of oxide and interface trapping noise into a partial differential equation-based semiconductor device simulator shows the correct prediction of 1/*f* noise spectral densities for submicrometer MOSFET devices operating in subthreshold and strong inversion in saturation [40].

The dependence of flicker noise power on gate bias and oxide thickness for *n*-channel MOSFET in terms of the equivalent gate noise power $\overline{e_{ng}^2}$ can be modeled by the following empirical expression:

$$\overline{e_{ng}^2} = \frac{\Delta f}{LWf}\left[K_1\left(\frac{q}{C_{ox}}\right)^m (V_{gs} - V_{th}) + K_2\left(\frac{q}{C_{ox}}\right)^2\right] \tag{2.54}$$

where V_{gs} is the gate–source voltage, V_{th} is the threshold voltage, L is the channel length, W is the channel width, C_{ox} is the gate–oxide capacitance per unit area, q is the electron charge, K_1 and K_2 are empirical constants, and $m = 0.7$–1.2 [38]. In this case, it was proposed that the term with K_1 represents the contribution from the mobility fluctuation, whereas the term with K_2 represents the contribution from the number fluctuation.

Generally, the noise behavior of the *n*-channel and *p*-channel MOSFET devices is different, since the pMOS transistor being less noisy usually has a channel at a larger distance from the interface [41]. Unlike MOSFET device, the flicker noise of its JFET counterpart is negligible. At low temperatures the noise spectrum of JFET indicates the presence of the several types of the generation–recombination processes, but the 1/*f* noise component is absent [42]. However, the component related to 1/*f* noise can appear at high temperatures more than 200 K [43]. Such a situation can be explained by the fact that the noise behavior of a JFET device having the *p–n* junction cannot be characterized by the semiconductor–oxide surface effects since its channel is separated by the depletion region localized along the device channel.

2.4 MESFETs AND HEMTs

2.4.1 Small-Signal Equivalent Circuit

The small-signal equivalent circuit shown in Figure 2.19 proves to be an adequate representation for MESFETs and HEMTs in a frequency range up at least to 25 GHz. Here, the extrinsic elements R_g, L_g, R_d, L_d, R_s, and L_s are the bulk and ohmic resistances and lead inductances associated with the gate, drain, and source, while C_{gp} and C_{dp} are the gate and source pad capacitances, respectively. The capacitance C_{dsd} and resistance R_{dsd} model the dispersion of the MESFET or HEMT *I–V* characteristics due to the trapping effect in the device channel, which leads to discrepancies between dc measurement and *S*-parameter measurement at high frequencies [44–46]. The intrinsic model is described by the channel charging resistance R_{gs}, which represents the resistive path for charging of the gate–source capacitance C_{gs}, the feedback gate–drain capacitance C_{gd}, the output differential resistance R_{ds}, the drain–source capacitance C_{ds} and the transconductance g_m. The gate–source capacitance C_{gs} and gate–drain capacitance C_{gd} represent the charge depletion region and are nonlinear functions. The influence of the drain–source capacitance C_{ds} on the device behavior is insignificant and its value is practically bias independent. To model the transit time of electrons along the channel, the transconductance g_m usually includes the time constant τ.

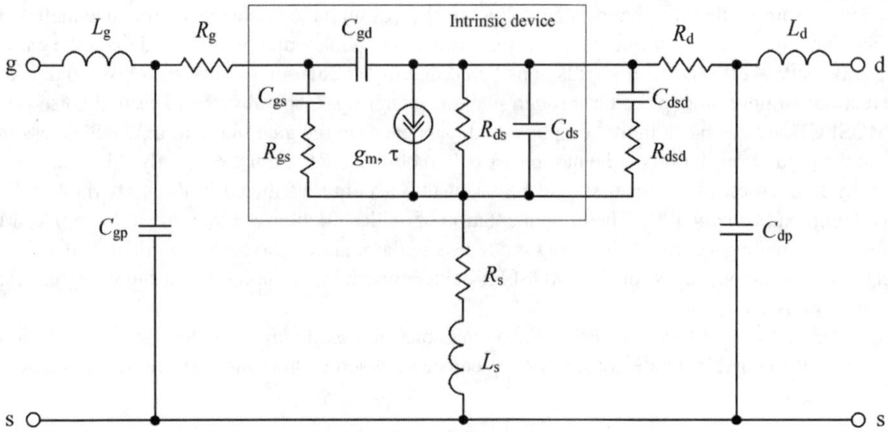

FIGURE 2.19 Small-signal equivalent circuit of field-effect transistor.

To describe accurately the small-signal and large-signal device models, it is necessary to satisfy charge conservation condition. The models for the device gate–source capacitance C_{gs} and gate–drain capacitance C_{gd} should be derived from the charge model. There are commonly four partial derivatives of the two device terminal charges, the gate charge Q_g and the drain charge Q_d with regard to V_{gs} and V_{ds}, which appropriately represent totally four capacitances, as shown Figure 2.20(*a*) [45]. However,

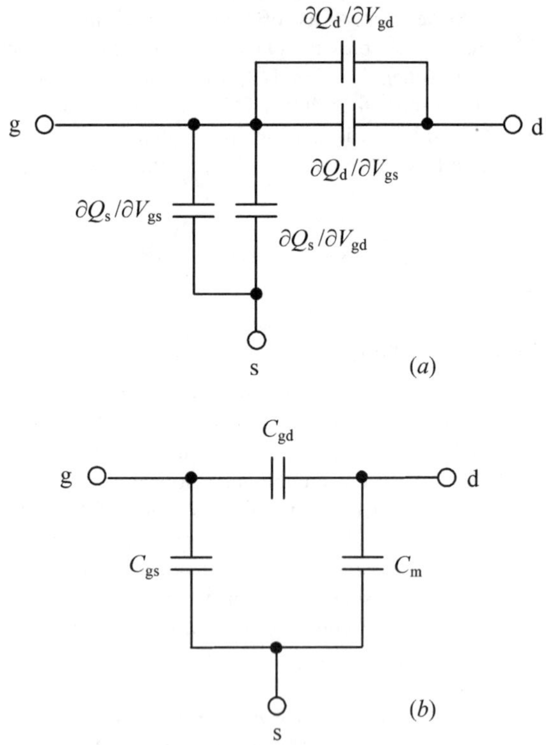

FIGURE 2.20 Capacitance equivalent circuits consistent with charge conservation.

the intrinsic small-signal equivalent circuit contains only two capacitances. Consequently, in this case, they can be defined as

$$C_{\rm gs} = \frac{\partial\left(Q_{\rm g}+Q_{\rm d}\right)}{\partial V_{\rm gs}} \qquad C_{\rm gd} = \frac{\partial\left(Q_{\rm g}+Q_{\rm d}\right)}{\partial V_{\rm gd}}. \tag{2.55}$$

The admittance $Y_{\rm c}$-matrix for such a capacitive two-port network is

$$Y_{\rm c} = \begin{bmatrix} j\omega\left(C_{\rm gs}+C_{\rm gd}\right) & -j\omega C_{\rm gd} \\ -j\omega\left(C_{\rm gs}-C_{\rm m}\right) & j\omega C_{\rm gd} \end{bmatrix} \tag{2.56}$$

where $C_{\rm m}$ is an additional transcapacitance, which is determined by

$$C_{\rm m} = \frac{\partial Q_{\rm g}}{\partial V_{\rm gd}} - \frac{\partial Q_{\rm d}}{\partial V_{\rm gs}}. \tag{2.57}$$

By adding the transcapacitance $C_{\rm m}$, the capacitance equivalent circuit becomes consistent with the charge conservation condition, as shown in Figure 2.20(*b*). Given that, for the MESFET devices, the transcapacitance $C_{\rm m}$ is substantially less than the gate–source capacitance $C_{\rm gs}$ in a frequency range where $\omega \le \omega_{\rm T}$, it can be translated to an additional delay time $\tau_{\rm c}$ by its combining with the small-signal transconductance $g_{\rm m}$ according to $g_{\rm m} - j\omega C_{\rm m} \approx g_{\rm m}\exp(-j\omega\tau_{\rm c})$, where $\tau_{\rm c} = C_{\rm m}/(\omega_{\rm T} C_{\rm gs})$.

2.4.2 Determination of Equivalent Circuit Elements

To characterize the nonlinear device electrical properties, first we consider the admittance Y-parameters derived from the intrinsic small-signal equivalent circuit as

$$Y_{11} = \frac{j\omega C_{\rm gs}}{1+j\omega C_{\rm gs}R_{\rm gs}} + j\omega C_{\rm gd} \tag{2.58}$$

$$Y_{12} = -j\omega C_{\rm gd} \tag{2.59}$$

$$Y_{21} = \frac{g_{\rm m}\exp\left(-j\omega\tau\right)}{1+j\omega C_{\rm gs}R_{\rm gs}} + j\omega C_{\rm gd} \tag{2.60}$$

$$Y_{22} = \frac{1}{R_{\rm ds}} + j\omega\left(C_{\rm ds}+C_{\rm gd}\right). \tag{2.61}$$

By dividing these equations into their real and imaginary parts, the parameters of the small-signal equivalent circuit can be determined as [47]

$$C_{\rm gd} = -\frac{{\rm Im}\,Y_{12}}{\omega} \tag{2.62}$$

$$C_{\rm gs} = -\frac{{\rm Im}\,Y_{11}-\omega C_{\rm gd}}{\omega}\left[1+\left(\frac{{\rm Re}Y_{11}}{{\rm Im}\,Y_{11}-\omega C_{\rm gd}}\right)^2\right] \tag{2.63}$$

$$R_{\rm gs} = \frac{{\rm Re}Y_{11}}{\left({\rm Im}\,Y_{11}-\omega C_{\rm gd}\right)^2+({\rm Re}Y_{11})^2} \tag{2.64}$$

$$g_{\rm m} = \sqrt{({\rm Re}Y_{21})^2+\left({\rm Im}\,Y_{21}+\omega C_{\rm gd}\right)^2}\sqrt{1+\left(\omega C_{\rm gs}R_{\rm gs}\right)^2} \tag{2.65}$$

$$\tau = \frac{1}{\omega}\sin^{-1}\left(\frac{-\omega C_{\rm gd}-{\rm Im}Y_{21}-\omega C_{\rm gs}R_{\rm gs}{\rm Re}Y_{21}}{g_{\rm m}}\right) \tag{2.66}$$

$$C_{ds} = \frac{\text{Im}\,Y_{22} - \omega C_{gd}}{\omega} \tag{2.67}$$

$$R_{ds} = \frac{1}{\text{Re}Y_{22}}. \tag{2.68}$$

Equations (2.62) to (2.68) are valid for the entire frequency range and for the drain voltages of $V_{ds} > 0$. If we assume that all extrinsic parasitic elements are already known, the only remaining problem is to determine the admittance Y-parameters of the intrinsic two-port network from experimental data. Consecutive stages shown in Figure 2.21 can represent such a determination procedure [48]:

- Measurement of the S-parameters of the extrinsic device.
- Transformation of the S-parameters to the impedance Z-parameters with subtraction of the series inductances L_g and L_d.
- Transformation of the impedance Z-parameters to the admittance Y-parameters with subtraction of the parallel capacitances C_{gp} and C_{dp}
- Transformation of the admittance Y-parameters to the impedance Z-parameters with subtraction of the series resistances R_g, R_s, R_d, and inductance L_s.
- Transformation of the impedance Z-parameters to the admittance Y-parameters of the intrinsic device two-port network.

The device extrinsic parasitic elements can be directly determined from measurements performed at $V_{ds} = 0$. Figure 2.22 shows the distributed RC channel network under the device gate for zero drain bias condition, where ΔC_g is the distributed gate capacitance, ΔR_{diode} is the distributed Schottky diode resistance, and ΔR_c is the distributed channel resistance. For any gate bias voltages, by taking into account the negligible influence of C_{gp} and C_{dp}, the extrinsic impedance Z-parameters are

$$Z_{11} = R_s + R_g + \frac{R_c}{3} + \frac{nkT}{qI_g} + j\omega\left(L_s + L_g\right) \tag{2.69}$$

$$Z_{12} = Z_{21} = R_s + \frac{R_c}{2} + j\omega L_s \tag{2.70}$$

$$Z_{22} = R_s + R_d + R_c + j\omega\left(L_s + L_d\right) \tag{2.71}$$

where R_c is the total channel resistance under the gate, nkT/qI_g is the differential resistance of the Schottky diode, n is the ideality factor, k is the Boltzmann constant, T is the Kelvin temperature, q is the electron charge, and I_g is the dc gate current. As a result, if the parasitic inductance L_s can be determined directly from measured ImZ_{12}, the parasitic inductances L_g and L_d are calculated from measured ImZ_{11} and ImZ_{22}, respectively. The resistance R_c is the channel technological parameter, which is usually known. The measured real parts of Z-parameters yield the values of R_s, R_g, and R_d.

At zero drain bias and for the gate voltages lower than the pinch-off voltage V_p, the small-signal equivalent circuit can be simplified to the one shown in Figure 2.23. Here, the capacitance C_b represents the fringing capacitance due to the depleted layer extension at each side of the gate. For low-frequency measurements usually up to a few gigahertz, when the extrinsic parasitic resistances and inductances have no influence on the device behavior, the imaginary parts of the Y-parameters can be written as

$$\text{Im}Y_{11} = j\omega\left(C_{gp} + 2C_b\right) \tag{2.72}$$

$$\text{Im}Y_{12} = \text{Im}Y_{21} = -j\omega C_b \tag{2.73}$$

$$\text{Im}Y_{22} = j\omega\left(C_b + C_{dp}\right). \tag{2.74}$$

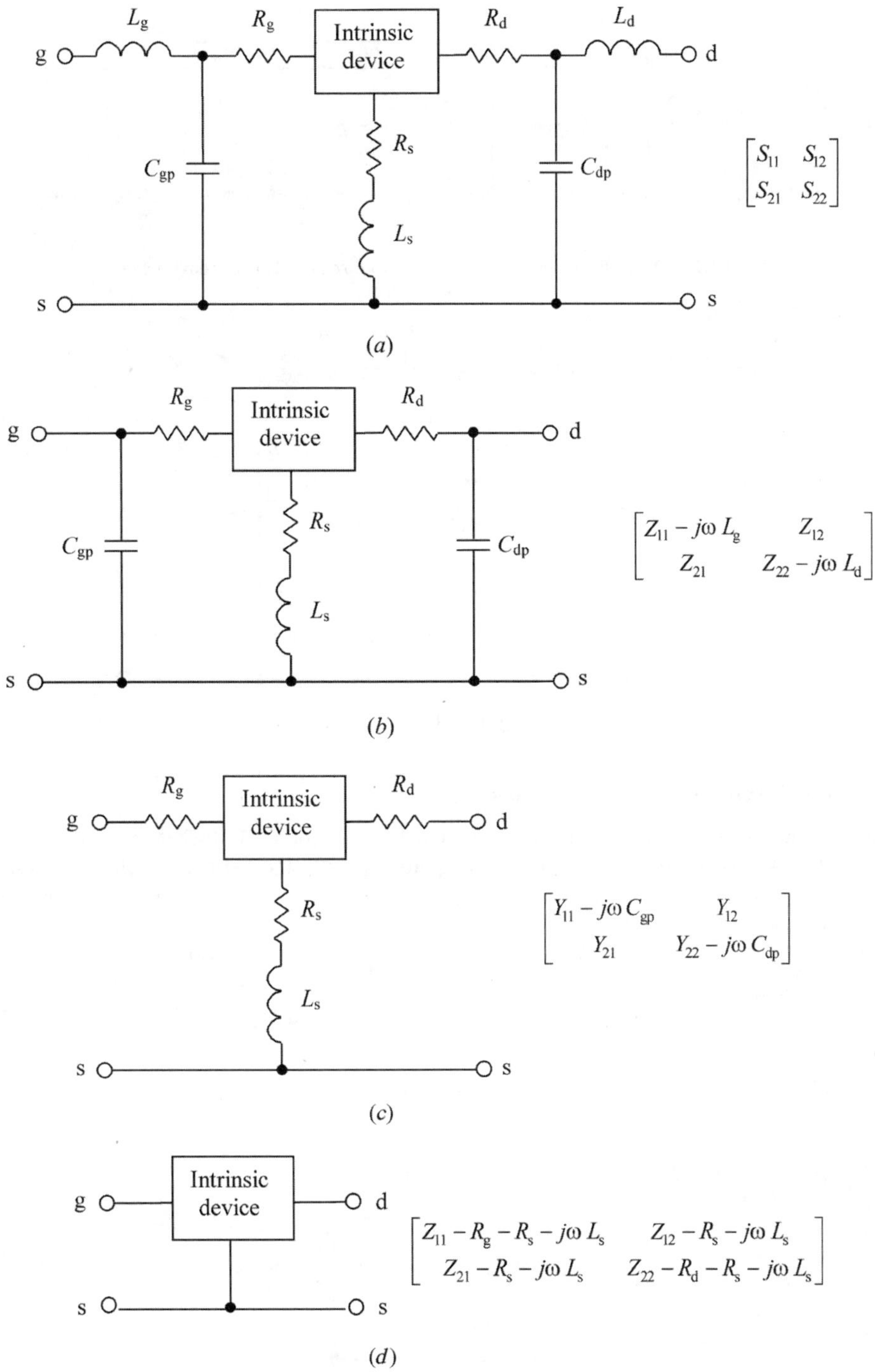

FIGURE 2.21 Method for extracting device intrinsic Z-parameters.

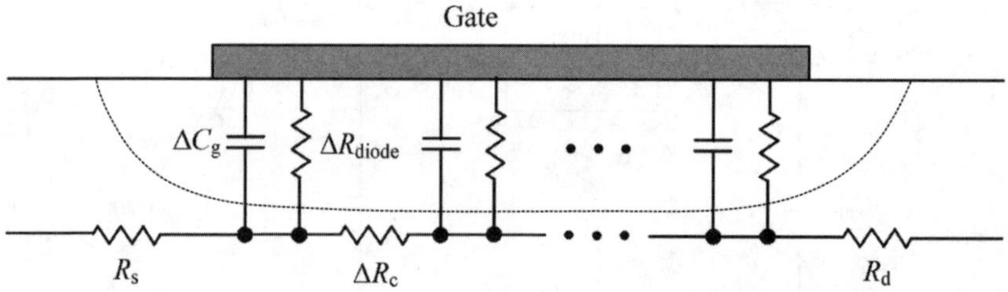

FIGURE 2.22 Distributed *RC* channel network schematic under device gate.

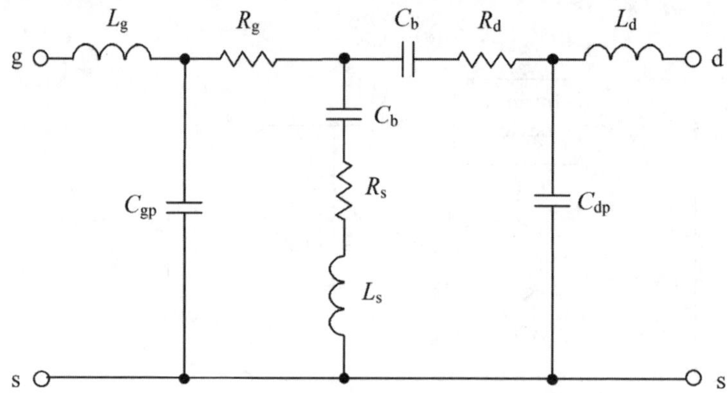

FIGURE 2.23 Small-signal FET circuit at zero drain bias voltage.

2.4.3 Curtice Quadratic Nonlinear Model

One of the first simple nonlinear intrinsic large-signal models for a MESFET device for use in the design of GaAs integrated circuits is shown in Figure 2.24 [49]. It consists of a voltage-controlled source $I_{ds}(V_{gs}, V_{ds})$, the gate–source capacitance $C_{gs}(V_{gs})$, and a clamping diode between gate and source. The gate–drain capacitance C_{gd} is assumed to be constant.

An analytical function proposed to describe the nonlinear current source behavior is

$$I_{ds} = \beta \left(V_{gs} - V_p\right)^2 (1 + \lambda V_{ds}) \tanh(\alpha V_{ds}) \tag{2.75}$$

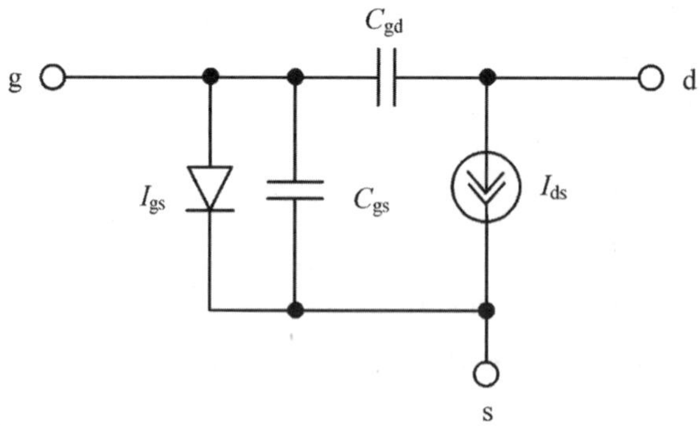

FIGURE 2.24 Curtice quadratic nonlinear intrinsic model.

where β is the transconductance parameter determined from experimental data, V_p is the pinch-off voltage, and λ is the slope of the drain characteristic in the saturated region. Due to the finite value of maximum charge velocity of about 10^7 cm/s during transient operation, change in the gate voltage does not cause an instantaneous change in the drain current. For example, it takes the order of 10 ps to change the drain current after the gate voltage is changed in a 1-μm gate-length MESFET. Consequently, the most important result of this effect is a time delay between gate–source voltage and drain current. Therefore, the current source given by Eq. (2.75) as $I_{ds}[V_{gs}(t), V_{ds}(t)]$ should be altered to be $I_{ds}[V_{gs}(t-\tau), V_{ds}(t)]$, where τ is equal to the transit time under the gate. To more accurate approximate the drain current behavior, a cubic nonlinear model can be used [50].

The time delay effect is not easily added to most circuit analysis program. Therefore, a simple and sufficiently accurate way to solve this problem is to assume the current source to be of the form

$$I_{ds}\left(V_{gs}\right) - \tau \frac{dI_{ds}\left(V_{gs}\right)}{dt} \tag{2.76}$$

where the second term is considered the first term of the Taylor series expansion of $I_{ds}(t-\tau)$ in time when for small τ the error is quite small,

$$\frac{dI_{ds}\left(V_{gs}\right)}{dt} = \left.\frac{dI_{ds}\left(V\right)}{dV_{gs}}\right|_{V_{gs}} \frac{dV_{gs}}{dt}.$$

The gate–source capacitance C_{gs} and gate–drain capacitance C_{gd} can be treated the voltage-dependent Schottky-barrier diode capacitances. For the negative gate–source and small drain–source voltages, these capacitances are practically equal. However, when the drain–source voltage is increased beyond the current saturation point, the gate–drain capacitance C_{gd} is much more heavily back-biased than the gate–source capacitance C_{gs}. Therefore, the gate–source capacitance C_{gs} is significantly more important and usually dominates the input impedance of the MESFET device. In this case, an analytical expression to approximate the gate–source capacitance C_{gs} is

$$C_{gs} = C_{gs0} / \sqrt{1 - \frac{V_{gs}}{V_{gsi}}} \tag{2.77}$$

where C_{gs0} is the gate–source capacitance for $V_{gs} = 0$ and V_{gsi} is the built-in gate voltage. When V_{gs} approaches V_{gsi}, the denominator in Eq. (2.77) must not be allowed to approach zero since C_{gs} will continue to increase as the depletion width reduces, so that a forward bias condition occurs, and the diffusion gate–source capacitance becomes of great importance. The built-in voltage V_{gsi} should be equal to the built-in voltage of the Schottky-barrier junction plus some part of the voltage drop along the channel under the gate.

2.4.4 Parker–Skellern Nonlinear Model

The comprehensive large-signal Parker–Skellern model, which intrinsic nonlinear part is shown in Figure 2.25, is a realistic description of measured characteristics of the MESFET device over all its operation regions [51]. The model maintains strict continuity in high-order derivatives, describes subthreshold and breakdown regions, self-heating effect, and frequency dependence of the output conductance and transconductance. The model parameters provide independent fitting to all operation regions. Both the gate–source capacitance C_{gs} and the gate–drain capacitance C_{gd} are treated as the standard voltage-dependent depletion capacitances with sensitivity $\gamma = 0.5$.

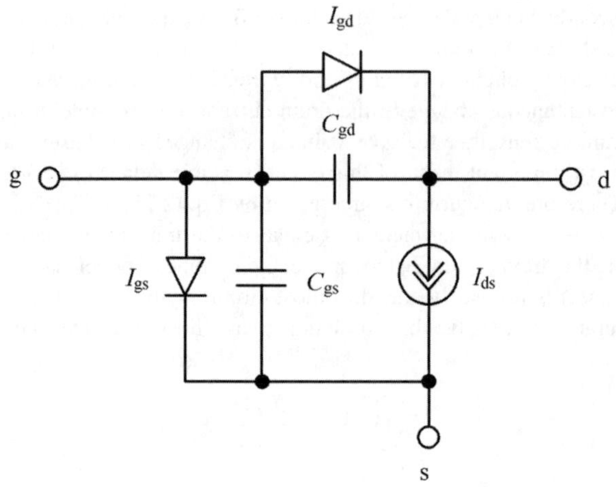

FIGURE 2.25 Parker–Skellern nonlinear intrinsic model.

An analytical function to describe the nonlinear current source behavior is proposed as

$$I_{ds} = \frac{\beta V_{gT}^{Q}}{1+\delta X}\left[1-\left(1-\frac{V_{dT}}{V_{gT}}\right)^{Q}\right] \tag{2.78}$$

where

$$X = \beta V_{gT}^{Q} V_{ds}\left[1-\left(1-\frac{V_{dT}}{V_{gT}}\right)^{Q}\right] - \tau_d \frac{dX}{dt} \tag{2.79}$$

with

$$V_{gT} = V_{sT}\left(1+MV_{ds}\right)\ln\left[1+\exp\left(\frac{V_{gsT}}{V_{sT}}\frac{1}{1+MV_{ds}}\right)\right]$$

$$V_{gsT} = V_{gs} - V_{th} - \gamma_1 \overline{V_{gd}} - \gamma_2\left(V_{gd}-\overline{V_{gd}}\right) - \gamma_3\left(V_{gs}-\overline{V_{gs}}\right)$$

$$\overline{V_{gs}} = V_{gs} - \tau_g \frac{d\overline{V_{gs}}}{dt} \qquad \overline{V_{gd}} = V_{gd} - \tau_g \frac{d\overline{V_{gd}}}{dt}$$

$$V_{dT} = \frac{1}{2}\sqrt{\left(V_{dP}\sqrt{1+Z}+V_{sat}\right)^2 + ZV_{sat}^2} - \frac{1}{2}\sqrt{\left(V_{dP}\sqrt{1+Z}-V_{sat}\right)^2 + ZV_{sat}^2}$$

$$V_{dP} = V_{ds}\frac{P}{Q}\left(\frac{V_{gT}}{\varphi - V_{th}}\right)^{P-Q}$$

$$V_{sat} = \frac{\xi\left(\varphi - V_{th}\right)V_{gT}}{\xi\left(\varphi - V_{th}\right)+V_{gT}}$$

where β is the linear-region transconductance scaling coefficient, δ is the thermal reduction coefficient, τ_d is the relaxation time for thermal reduction, τ_g is the relaxation time for gamma feedback, M is the subthreshold modulation, P is the linear-region power-law exponent, Q is the saturated-region power-law exponent, V_{th} is the threshold voltage, V_{sT} is the subthreshold potential, Z is the knee transition parameter, φ is the gate-junction potential, ξ is the saturation-knee potential factor, and γ_1, γ_2, and γ_3 are the fitting parameters.

The model must select V_{sat} to be less than V_{gT} and the velocity saturation potential that is set by the parameter ξ and the channel depletion potential $\varphi - V_{th}$. For a 1-μm GaAs MESFET, ξ is approximately 0.3. A single set of dual power-law model parameters P and Q can describe both the controlled-resistance and controlled-current regions. The drain conductance in a controlled-current region is fitted with parameter P, whereas the transconductance in a controlled-current region is fitted with parameter Q. To describe the frequency dispersion effects producing differences between the drain conductance at dc and at high frequencies, the model determines the bias condition by a continuous calculation of the average value of the thermal potentials $\overline{V_{gs}}$ and $\overline{V_{gd}}$, where the average is accumulated over a user-defined time constant τ_g which is normally between 0.1 to 1 ms. The average potential at time t is calculated by a weighted sum of the instantaneous potential at time t and previously calculated average at time $(t - \Delta t)$ as

$$\overline{V(t)} = V(t) + \left[\overline{V(t - \Delta t)} - V(t)\right] \exp\left(-\frac{\Delta t}{\tau}\right) \tag{2.80}$$

where $V(t)$ is replaced by $V_{gs}(t)$ or $V_{gd}(t)$ and τ is replaced by τ_g or τ_d. At very low frequencies, the instantaneous and average potentials are identical.

The gate junction current is implemented with identical gate–source and gate–drain diodes between the intrinsic nodes. The term δ is a model parameter related to the product of the temperature coefficient of current reduction and the channel resistance. The model calculates average dissipated power rather than using instantaneous power over a time constant τ_d using algorithm of Eq. (2.80).

2.4.5 Chalmers (Angelov) Nonlinear Model

A simple and accurate large-signal model for different submicrometer gate-length HEMT devices and commercially available MESFETs, which is capable of modeling the drain current–voltage characteristics and its derivatives, as well as the gate–source and gate–drain capacitances, is shown in Figure 2.26 [52,53]. This model can be used not only for large-signal analysis of power amplifier and oscillators, but also for predicting the performance of multipliers and mixers including intermodulation simulation.

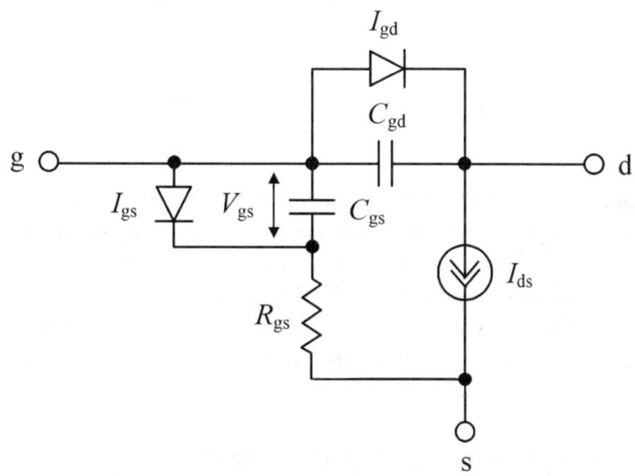

FIGURE 2.26 Angelov nonlinear intrinsic model.

The drain current source is described by using the hyperbolic functions as

$$I_{ds} = I_{pk}\,(1 + \tanh\psi)\,(1 + \lambda V_{ds})\tanh(\alpha V_{ds}) \tag{2.81}$$

where I_{pk} is the drain current at maximum transconductance with the contribution from the output conductance subtracted, λ is the channel length modulation parameter, and $\alpha = \alpha_0 + \alpha_1\tanh\psi$ is the saturation voltage parameter, where α_0 is the saturation voltage parameter at pinch-off and α_1 is the saturation voltage parameter at $V_{gs} > 0$.

The parameter ψ is a power series function centered at V_{pk} with the bias voltage V_{gs} as a variable,

$$\psi = P_1\left(V_{gs} - V_{pk}\right) + P_2\left(V_{gs} - V_{pk}\right)^2 + P_3\left(V_{gs} - V_{pk}\right)^3 + \cdots \tag{2.82}$$

where V_{pk} is the gate voltage for maximum transconductance g_{mpk}. The model parameters as a first approximation can be easily obtained from the experimental $I_{ds}(V_{gs}, V_{ds})$ dependencies at a saturated channel condition when all higher terms in ψ are assumed to be zero, and λ is the slope of the I_{ds}–V_{ds} curves.

The intrinsic maximum transconductance g_{mpk} is calculated from the measured maximum transconductance g_{mpkm}, by taking into account feedback effect due to the source resistance R_s, as

$$g_{mpk} = \frac{g_{mpkm}}{1 - g_{mpkm}R_s}. \tag{2.83}$$

To evaluate the gate voltage V_{pk} and parameter P_1, it is necessary to define the derivatives of the drain current. If higher order terms of ψ are neglected, the transconductance g_m becomes equal to

$$g_m = \frac{\partial I_{ds}}{\partial V_{gs}} = I_{pk}P_1\,\mathrm{sech}\left[P_1\left(V_{gs} - V_{pk}\right)\right]^2(1 + \lambda V_{ds})\tanh(\alpha V_{ds}). \tag{2.84}$$

The gate voltage V_{pk} that depends on the drain voltage can be extracted by finding the gate voltages for maximum transconductance, at which the second derivative of the drain current is equal to zero. In this case, it is advisable to use the simplified expression

$$V_{pk}\left(V_{ds}\right) = V_{pk0} + \left(V_{pks} - V_{pk0}\right)(1 + \lambda V_{ds})\tanh(\alpha V_{ds}) \tag{2.85}$$

where V_{pk0} is measured at V_{ds} closely to zero and V_{pks} is measured at V_{ds} in the saturation region.

A good fitting of P_1 and good results in harmonic balance simulations are obtained using

$$P_1 = P_{1sat}\left[1 + \left(\frac{P_{10}}{P_{1sat}} - 1\right)\frac{1}{\cosh^2(BV_{ds})}\right] \tag{2.86}$$

where $P_{10} = g_{m0}/I_{pk0}$ at V_{ds} close to zero and B is the fitting parameter ($B \approx 1.5\alpha$). The parameter P_2 makes the derivative of the drain current asymmetric, whereas the parameter P_3 changes the drain current values at voltages V_{gs} close to pinch-off voltage V_p. Three terms in Eq. (2.82) are usually enough to describe the behavior of the different MESFET or HEMT devices with acceptable accuracy.

The same hyperbolic functions can be used to model the intrinsic device capacitances. When 5–10% accuracy is sufficient, the gate–source capacitance C_{gs} and gate–drain capacitance C_{gd} can be described by

$$C_{gs} = C_{gs0}\left[1 + \tanh\left(P_{1gsg}V_{gs}\right)\right]\left[1 + \tanh\left(P_{1gsd}V_{ds}\right)\right] \tag{2.87}$$

$$C_{gd} = C_{gd0}\left[1 + \tanh\left(P_{1gdg}V_{gs}\right)\right]\left[1 - \tanh\left(P_{1gdd}V_{ds} + P_{1cc}V_{gs}V_{ds}\right)\right] \tag{2.88}$$

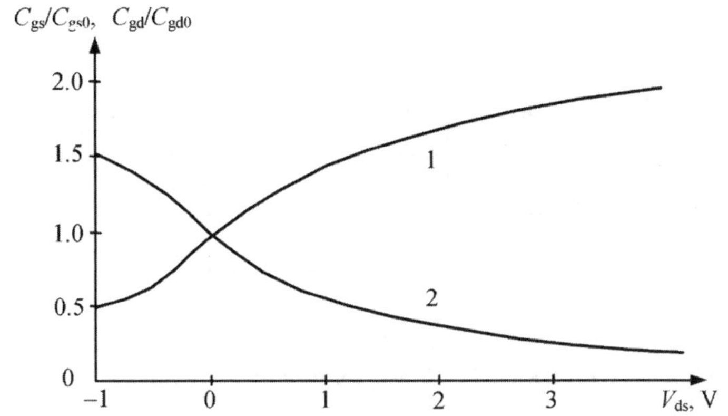

FIGURE 2.27 Gate–source and gate–drain capacitances versus drain–source voltage.

where the product $P_{1cc}V_{gs}V_{ds}$ reflects the cross-coupling of V_{gs} and V_{ds} on C_{gd} and the coefficients P_{1gsg}, P_{1gsd}, P_{1gdg}, and P_{1gdd} are the fitting parameters. These dependencies, unlike the commonly diode-like models, are suitable for HEMT devices with an undoped AlGaAs spacer-layer in view of the saturation effect of the gate–source capacitance C_{gs} for increasing V_{gs}. This is due to the absence of parasitic MESFET channel formation in the undoped AlGaAs layer, found in HEMTs with a doped AlGaAs layer. The approximate behavior of normalized gate–source capacitance C_{gs}/C_{gs0} (curve 1) and gate–drain capacitance C_{gd}/C_{gd0} (curve 2) as functions of V_{ds} for zero gate–source voltage is shown in Figure 2.27, where C_{gs0} and C_{gd0} are the gate–source and gate–drain capacitances at $V_{ds} = 0$, respectively. The character of the curves is the same for both positive and negative V_{gs}, except that the capacitance range decreases for the same range of V_{ds} with the decrease of V_{gs}.

The drain–source dispersive resistance R_{dsd} as a nonlinear function of the gate–source voltage V_{gs} can be defined by

$$R_{dsd} = R_{dsd0} + \frac{R_{dsdp}}{1 + \tanh\psi} \tag{2.89}$$

where R_{dsd0} is the minimum value of R_{dsd} and R_{dsdp} determines the value of R_{dsd} at the pinch-off [46].

2.4.6 IAF (Berroth) Nonlinear Model

An analytical charge conservative large signal model for HEMT devices, which is valid for a frequency range up to 60 GHz, is shown in Figure 2.28 [54]. In this model, the current sources I_{gs}, I_{gd}, and I_{ds} and capacitances C_{gs} and C_{gd} are considered as the nonlinear elements.

The drain current source is represented by the following nonlinear equation with 10 fitting parameters:

$$I_{ds} = f\left(V_{gs}\right)\left[1 + \frac{\lambda}{1 + \Delta_\lambda\left(V_{gs} - V_c + 2/\beta\right)}V_{ds}\right]\tanh\left(\alpha V_{ds}\right) \tag{2.90}$$

where

$$f\left(V_{gs}\right) = CD_{vc}\left\{1 + \tanh\left[\beta\left(V_{gs} - V_c\right) + \gamma\left(V_{gs} - V_c\right)3\right]\right\} + CD_{vsb}\left\{1 + \tanh\left[\delta\left(V_{gs} - V_{sb}\right)\right]\right\}$$

α is the slope of drain current in the pinch-off region, β is the slope parameter of drain current, γ is the slope parameter of drain current in the pinch-off region, λ is the slope of drain current in the

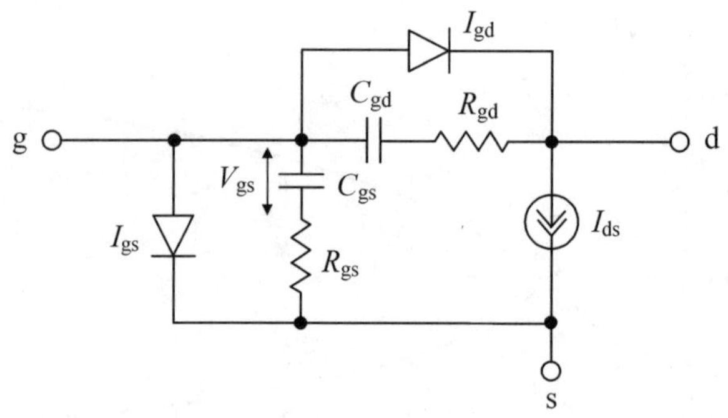

FIGURE 2.28 Berroth nonlinear intrinsic model.

saturation region, Δ_λ is the gate voltage parameter for slope of drain current, δ is the drain current slope parameter correction term, V_c is the gate voltage for maximum transconductance, V_{sb} is the gate voltage for maximum transconductance correction term, CD_{vc} is the drain current multiplication factor, and CD_{vsb} is the drain current multiplication factor correction term.

To describe the I_{gs} and I_{gd} current sources, the diode model for both forward and reverse bias operation modes was used in the form

$$I_{diode} = I_{dsat}\left[\exp\left(\frac{V_{diode}}{nV_T}\right) - 1\right] \tag{2.91}$$

where I_{dsat} is the forward bias fitting parameter, V_T is the temperature voltage, and n is the diode ideality factor.

The nonlinear capacitances C_{gs} and C_{gd} are calculated by differentiating the voltage-depending charge function $Q_g(V_{gs}, V_{ds})$ with respect to V_{gs} and V_{ds}, which leads to the input capacitance $C_{11} = C_{gs} + C_{gd}$ and transcapacitance $C_{12} = -C_{gd}$, respectively,

$$Q_g\left(V_{gs}, V_{ds}\right) = Af_1f_2 + E\left(V_{gs} - 0.5V_{ds}\right) \tag{2.92}$$

where

$$f_1 = \frac{1}{B}\text{lncosh}\left\{B\left[\left(V_{gs} - V_1\right) - 0.5\tanh\left(CV_{ds}\right)\right]\right\} + \left(V_{gs} - V_1\right) - 0.5\tanh\left(CV_{ds}\right)$$

$$f_2 = 1 + D\,\text{lncosh}\left(FV_{ds}\right)$$

V_1 is the transition voltage, and A, B, C, D, E, and F are the model-fitting parameters.

2.4.7 Noise Model

The noise properties of a MESFET device can be described based on both its physical and equivalent circuit models. The dominant intrinsic noise of a microwave GaAs MESFET device is the diffusion noise introduced by electrons experiencing velocity saturation. In a device two-zone model, a portion of the channel near the source end is assumed to be in the constant mobility operation mode (zone I), while the remaining portion near the drain end is postulated to be in velocity saturation (zone II). The position of the boundary between these zones is a strong function of the source–drain bias with weak dependence on the gate–source bias. It is assumed that the noise in zone I is thermal enhanced by hot electron effects [55,56]. However, zone II cannot be treated as an ohmic conductor. Its contribution must be represented as a high-field diffusion noise, being dominant in microwave devices [57]. This diffusion noise is proportional to the high-field diffusion coefficient and is linearly dependent on drain current. On the other hand, the thermal noise of zone I decreases with increasing drain current. As a result, a strong correlation exists between the drain noise and the induced gate noise which leads to a high degree of cancellation in the noise output of the GaAs MESFET [58].

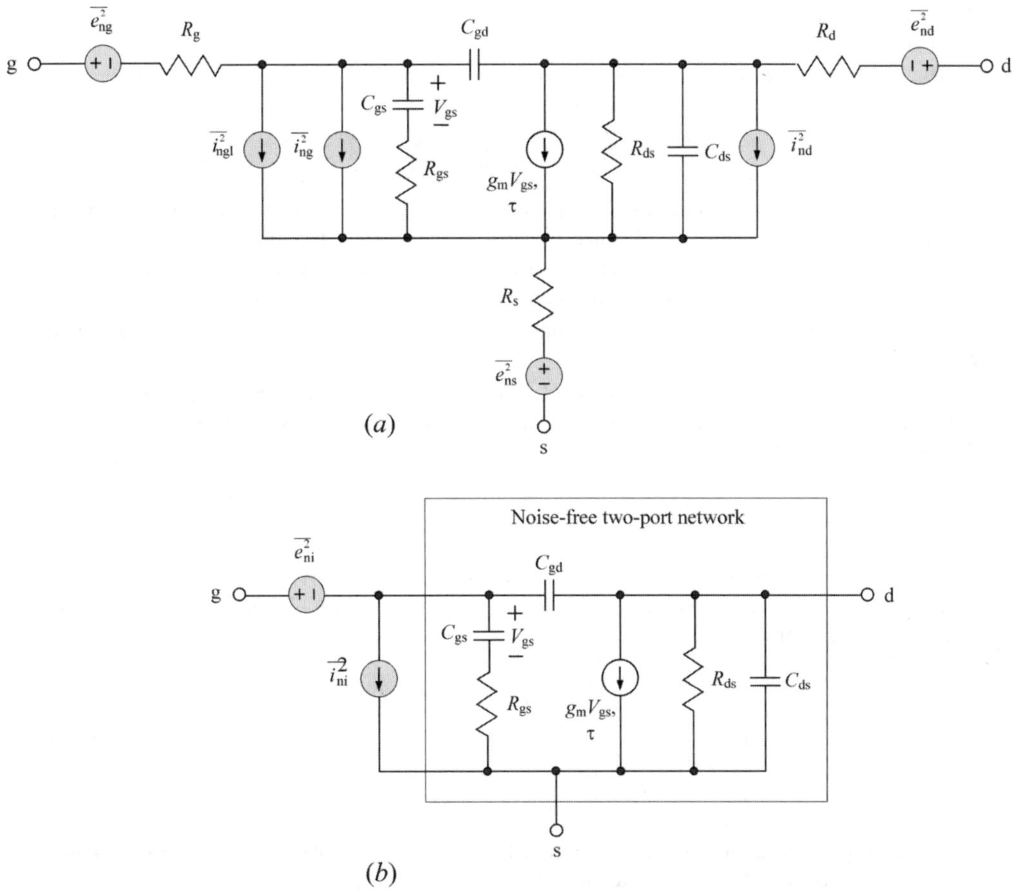

FIGURE 2.29 MESFET equivalent circuits with noise sources.

The noise equivalent circuit of the MESFET device with both intrinsic and extrinsic noise sources is shown in Figure 2.29(*a*) [55,58]. The noise source $\overline{i_{\rm ng}^2}$ represents the noise induced on the gate electrode by the passing thermal fluctuations in the drain current. The intrinsic drain noise source $\overline{i_{\rm nd}^2}$ has a flat spectrum. The resistance $R_{\rm gs}$ represents the resistive charging path for the gate–source capacitance $C_{\rm gs}$, and noise associated with this resistor is imbedded in the gate noise source. The series gate, source and drain resistances are represented by the voltage thermal noise sources $\overline{e_{\rm ng}^2}$, $\overline{e_{\rm ns}^2}$, and $\overline{e_{\rm nd}^2}$, respectively.

The noise voltage and current sources can be given through their mean-square values as

$$\overline{e_{\rm ng}^2} = 4kTR_{\rm g}\Delta f \tag{2.93}$$

$$\overline{i_{\rm ng}^2} = \frac{4kT\left(\omega C_{\rm gs}\right)^2 R\Delta f}{g_{\rm m}} \tag{2.94}$$

$$\overline{e_{\rm ns}^2} = 4kTR_{\rm s}\Delta f \tag{2.95}$$

$$\overline{e_{\rm nd}^2} = 4kTR_{\rm d}\Delta f \tag{2.96}$$

$$\overline{i_{\rm nd}^2} = 4kTg_{\rm m}P\Delta f \tag{2.97}$$

where R and P are the gate and drain noise model parameters depending upon the implementation technology and biasing conditions [59,60].

The cross-correlation between the gate–drain noise current sources $\overline{i_{ng}^2}$ and $\overline{i_{nd}^2}$ can be written as

$$\overline{i_{nd} i_{ng}} = 4kT\omega C_{gs} C\sqrt{PR}\Delta f \tag{2.98}$$

where C is the correlation coefficient. The quantities P, R, and C are the bias-dependent empirical correction factors, which may be obtained by noise de-embedding techniques [61]. Their typical values based on measurement and calculation of the noise figure for different devices can be chosen as $P = 1$, $R = 0.5$, and $C = 0.9$ [62]. It should be noted that $\overline{i_{nd}^2}$ and C increase in ohmic region and tend to saturate at high drain voltage, while $\overline{i_{ng}^2}$ increases with a near constant slope versus drain voltage.

The noise current source $\overline{i_{ngl}^2}$ is responsible for the effect of the gate leakage current, which should be taken into account when using a submicrometer gate-length HEMT device. It can be written as a shot noise source in the form

$$\overline{i_{ngl}^2} = 2\alpha q I_{gl}\Delta f \tag{2.99}$$

where I_{gl} is the gate leakage current and α is the fitting parameter [63,64].

The admittance Y-parameters of the MESFET intrinsic small-signal equivalent circuit can be written in matrix form

$$Y = \begin{bmatrix} \dfrac{j\omega C_{gs}}{1 + j\omega R_{gs} C_{gs}} + j\omega C_{gd} & -j\omega C_{gd} \\ \dfrac{g_m \exp(-j\omega\tau)}{1 + j\omega R_{gs} C_{gs}} - j\omega C_{gd} & \dfrac{1}{R_{ds}} + j\omega\left(C_{ds} + C_{gd}\right) \end{bmatrix}. \tag{2.100}$$

The corresponding admittance noise correlation matrix to calculate the equivalent noise resistance R_n, optimal source admittance Y_{Sopt}, and minimum noise figure F_{min} is given by

$$C_Y = 4kT \begin{bmatrix} \dfrac{\alpha I_{gl}}{2V_T} + \dfrac{\left(\omega C_{gs}\right)^2 R}{g_m} & j\omega C_{gs} C\sqrt{PR} \\ -j\omega C_{gs} C\sqrt{PR} & g_m P \end{bmatrix} \tag{2.101}$$

where $V_T = kT/q$ is the thermal voltage. However, if the correlation matrix has been determined, the noise parameters can be analytically calculated by using the elements of the chain correlation matrix [65]

$$C_{11}^A = \frac{C_{22}^Y}{|Y_{21}|^2} \tag{2.102}$$

$$C_{12}^A = \frac{Y_{11}^* C_{22}^Y}{|Y_{21}|^2} - \frac{C_{12}^Y}{Y_{12}} \tag{2.103}$$

$$C_{22}^A = C_{11}^Y + \frac{|Y_{11}|^2 C_{22}^Y}{|Y_{21}|^2} - \frac{Y_{11} C_{21}^Y}{Y_{21}} - \frac{Y_{11}^* C_{12}^Y}{Y_{21}^*}. \tag{2.104}$$

In a first approximation, the gate noise source $\overline{i_{ng}^2}$, feedback capacitance C_{gd}, and series drain resistance R_d can be neglected. As a result, the simple approximate expressions based on measurements can be obtained in terms of the device equivalent circuit elements as

$$F_{\min} = 1 + 0.016 f C_{gs} \sqrt{\frac{R_g + R_s}{g_m}} \tag{2.105}$$

$$R_n = \frac{0.8}{g_m} \tag{2.106}$$

$$R_{Sopt} = 2.2 \left(\frac{1}{4g_m} + R_g + R_s \right) \tag{2.107}$$

$$X_{Sopt} = \frac{160}{f C_{gs}} \tag{2.108}$$

provided that R_n, R_{Sopt}, X_{Sopt}, R_g, and R_s are in ohms, transconductance g_m in mhos, capacitance C_{gs} in picofarads, and operating frequency f is in gigahertz [66,67].

The minimum noise figure $F_{\min}$ given by Eq. (2.105) can also be expressed in terms of the device geometrical parameters as

$$F_{\min} = 1 + 0.27 L f \sqrt{g_m \left(R_g + R_s\right)} \tag{2.109}$$

where the effective gate length L is in micrometers [68].

A comparison of the noise performance of both HEMT and conventional MESFET devices demonstrates the HEMT superiority, mainly related to its higher transition frequency and correlation coefficient [62]. The transition frequency of HEMT device is greater for two main reasons: higher carrier mobility results in a higher average velocity and, therefore, a higher transconductance, whereas the small epilayer thickness yields higher transconductance and less effect of the parasitic capacitances. In addition, the correlation coefficient C is close to 0.7–0.8 for short-gate-length MESFETs, but becomes close to 0.8–0.95 for HEMTs.

The GaAs MESFET devices are characterized by a significant value of the flicker noise. This is a combined result of the gate leakage current, fluctuations of the Schottky barrier space charge region, and carrier number fluctuations in the channel and at the interface between the channel and substrate due to trapping phenomena [69]. The trapping mechanism is especially pronounced in GaAs material where the trapping centers can arise from a variety of causes such as trace impurities and crystal defects [70]. It is shown that the low-frequency noise in GaAs MESFET devices on InP substrate is directly related to the structural quality of GaAs active layers when increasing the buffer layer thickness for GaAs lattice mismatched on InP substrate improves the noise performance [71].

2.5 BJTs AND HBTs

2.5.1 Small-Signal Equivalent Circuit

The complete bipolar transistor small-signal equivalent circuit with extrinsic parasitic elements is shown in Figure 2.30. Based on this hybrid π-type representation, the electrical properties of the bipolar transistors, in particularly HBT devices, can be described with sufficient accuracy up to 30 GHz [72,73]. Here, the extrinsic elements R_b, L_b, R_c, L_c, R_e, and L_e are the series resistances and lead inductances associated with the base, collector, and emitter; and C_{pbe}, C_{pbc}, and C_{pce} are the parasitic capacitances associated with the contact pads, respectively. The lateral resistance with the base semiconductor resistance underneath the base contact and the base semiconductor resistance underneath the emitter are combined into a base-spreading resistance r_b. The intrinsic model is described by the dynamic diode resistance r_π, the total base–emitter junction capacitance and base charging capacitance C_π, the transconductance g_m, and the output Early resistance r_{ce} that model the

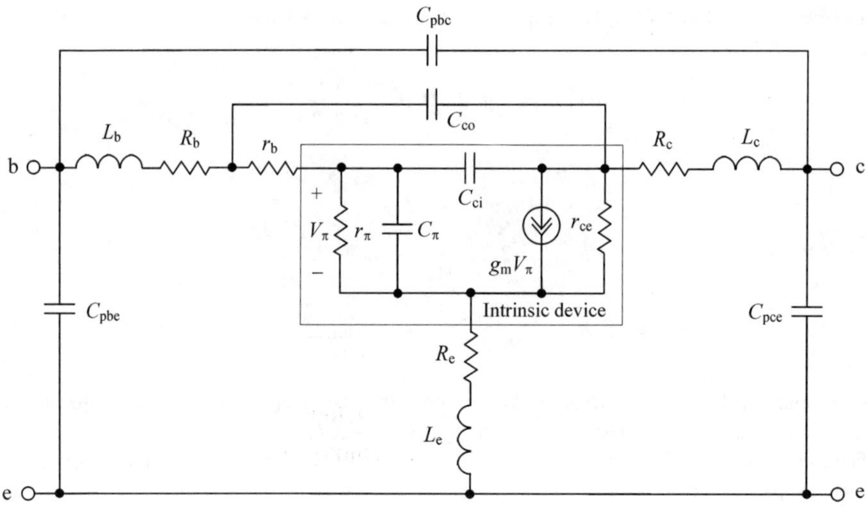

FIGURE 2.30 Small-signal equivalent circuit of bipolar device.

effect of base-width modulation on the transistor characteristics due to variations in the collector–base depletion region. To increase the usable operating frequency range of the device up to 50 GHz, it is necessary to properly include the collector-current delay time in the current source as $g_m\exp(-j\omega\tau_\pi)$, where τ_π is the transit time [74].

2.5.2 Determination of Equivalent Circuit Elements

If all extrinsic parasitic elements of the device equivalent circuit shown in Figure 2.30 are known, the intrinsic two-port network parameters can be embedded with the following determination procedure [72]:

- Measurement of the S-parameters of the transistor with extrinsic elements.
- Transformation of the S-parameters to the admittance Y-parameters with subtraction of the parasitic shunt capacitances C_{pbe}, C_{pbc}, and C_{pce}.
- Transformation of the new Y-parameters to the impedance Z-parameters with subtraction of the parasitic series elements L_b, R_b, L_e, R_e, L_c, and R_c.
- Transformation of the new Z-parameters to the Y-parameters with subtraction of the parasitic shunt capacitance C_{co}.
- Transformation of the new Y-parameters to the Z-parameters with subtraction of the parasitic series resistance r_b.
- Transformation of the new Z-parameters to the Y-parameters of the intrinsic device two-port network.

The bipolar transistor intrinsic Y-parameters can be written as

$$Y_{11} = \frac{1}{r_\pi} + j\omega(C_\pi + C_{ci}) \tag{2.110}$$

$$Y_{12} = -j\omega C_{ci} \tag{2.111}$$

$$Y_{21} = g_m\exp(-j\omega\tau_\pi) + j\omega C_{ci} \tag{2.112}$$

$$Y_{22} = \frac{1}{r_{ce}} + j\omega C_{ci}. \tag{2.113}$$

After separating Eqs. (2.110) to (2.113) into their real and imaginary parts, the elements of the intrinsic small-signal equivalent circuit can be determined analytically as

$$C_\pi = \frac{\mathrm{Im}(Y_{11} + Y_{12})}{\omega} \tag{2.114}$$

$$r_\pi = \frac{1}{\mathrm{Re}Y_{11}} \tag{2.115}$$

$$C_{\mathrm{ci}} = -\frac{\mathrm{Im}Y_{12}}{\omega} \tag{2.116}$$

$$g_{\mathrm{m}} = \sqrt{(\mathrm{Re}Y_{21})^2 + (\mathrm{Im}\,Y_{21} + \mathrm{Im}Y_{12})^2} \tag{2.117}$$

$$\tau_\pi = \frac{1}{\omega}\cos^{-1}\frac{\mathrm{Re}Y_{21} + \mathrm{Re}Y_{12}}{\sqrt{(\mathrm{Re}Y_{21})^2 + (\mathrm{Im}Y_{21} + \mathrm{Im}Y_{12})^2}} \tag{2.118}$$

$$r_{\mathrm{ce}} = \frac{1}{\mathrm{Re}Y_{22}}. \tag{2.119}$$

The parasitic capacitances associated with the pads can be determined by the measurement of the open test structure with the corresponding circuit model shown in Figure 2.31(*a*). When the values of these pad capacitances are known, it is easy to determine the values of the parasitic series inductances by measuring the shorted test structure with corresponding circuit model shown in Figure 2.31(*b*).

The values of series parasitic resistances can be calculated on the basis of physical parameters of the device or by adding them to the intrinsic device parameters (with the appropriate solution of a nonlinear system of eight equations with eight independent variables using iterative technique) [73]. In the latter case, it is assumed that influence of the transit time τ_π on the bipolar electrical properties in a frequency range up to 30 GHz is negligible.

The external parasitic parallel capacitance C_{co}, as well as the other device capacitances, can be estimated from the device behavior at low frequencies and cutoff operating conditions [75,76]. For

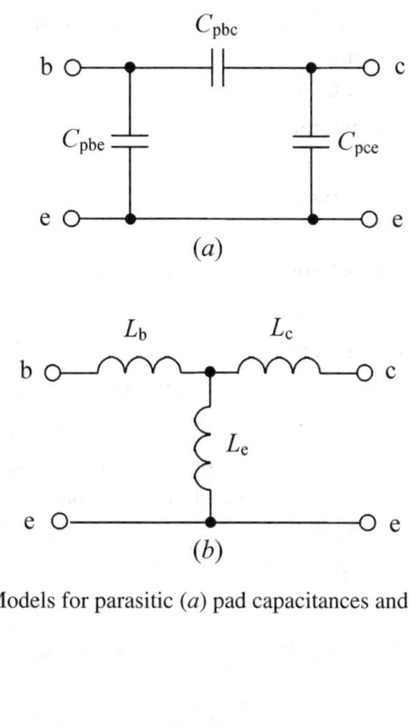

FIGURE 2.31 Models for parasitic (*a*) pad capacitances and (*b*) lead inductances.

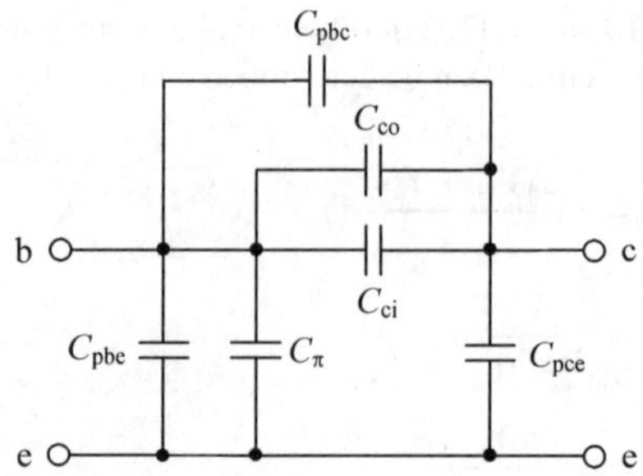

FIGURE 2.32 Small-signal model at low frequencies and cutoff operation mode.

such conditions, the device small-signal equivalent circuit is reduced to capacitive elements, as shown in Figure 2.32.

The device capacitances can be directly calculated from measured Y-parameters by

$$C_{pbe} + C_\pi = \frac{\mathrm{Im}\,(Y_{11} + Y_{12})}{\omega} \tag{2.120}$$

$$C_{pce} = \frac{\mathrm{Im}\,(Y_{11} + Y_{12})}{\omega} \tag{2.121}$$

$$C_{pbc} + C_{co} + C_{ci} = -\frac{\mathrm{Im}Y_{12}}{\omega}. \tag{2.122}$$

Since C_{pbc} and C_{co} are the bias-independent capacitances and C_{ci} is the base–collector junction capacitance, the extraction of $C_{pbc} + C_{co}$ can be carried out by fitting the sum $C_{pbc} + C_{co} + C_{ci}$ to the expression for junction capacitance at different base–collector voltages. If an approximation expression for C_{co} is given by

$$C_{co} = C_{jco}/\sqrt{1 + \frac{V_{bc}}{\varphi_c}} \tag{2.123}$$

then the extraction of the parameters C_{jco}, φ_c, and C_{ci} can be performed using the linear equation

$$\left(\frac{C_{jco}}{\frac{\mathrm{Im}Y_{12}}{\omega} + C_{pbc} + C_{ci}} \right)^2 = 1 + \frac{1}{\varphi_c} V_{bc}. \tag{2.124}$$

As a result, linearizing this equation by choosing a proper value for C_{ci} with known value of C_{pbc} gives the values for remaining two parameters C_{co} and φ_c from the slope and intercept point of the final linearized dependence.

2.5.3 Equivalence of Intrinsic π- and T-Type Topologies

The small-signal equivalent circuit of a bipolar transistor can be represented by both π-type and T-type topologies. The T-type equivalent circuit representation is appealing because all the model parameters can be directly tied to the physics of the device and an excellent fit between measured and simulated S-parameters in the frequency range up to 30–40 GHz [74–77]. The HBT small-signal equivalent circuit with T-type topology is shown in Figure 2.33.

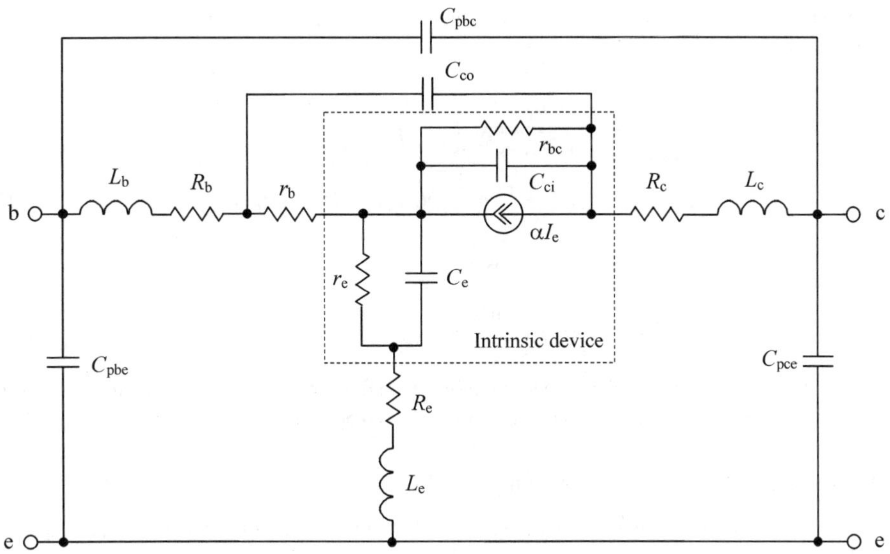

FIGURE 2.33 Small-signal T-model of bipolar device.

There is one-to-one correspondence between π-type and T-type device models. By comparing both small-signal equivalent circuits shown in Figures 2.30 and 2.33, the difference, which can be found in the representation of both intrinsic device models, are just enclosed in dashed boxes. From Figure 2.34 it follows that the admittances $Y_e = I_e/V_{be}$ for π- and T-type models are defined by

$$Y_e = \frac{1}{r_e} + j\omega C_e = \frac{1}{r_\pi} + j\omega C_\pi + g_m \exp(-j\omega\tau_\pi). \tag{2.125}$$

The collector source currents for both models are the same,

$$\alpha \exp(-j\omega\tau_{tee}) I_e = g_m \exp(-j\omega\tau_\pi) V_\pi \tag{2.126}$$

where τ_{tee} is the transit time for a T-type model, $\alpha = \alpha_0/(1 + j\omega\tau_\alpha)$, $\tau_\alpha = 1/(2\pi f_\alpha)$, f_α is the alpha cutoff frequency, and α_0 is the low frequency collector-to-emitter current gain.

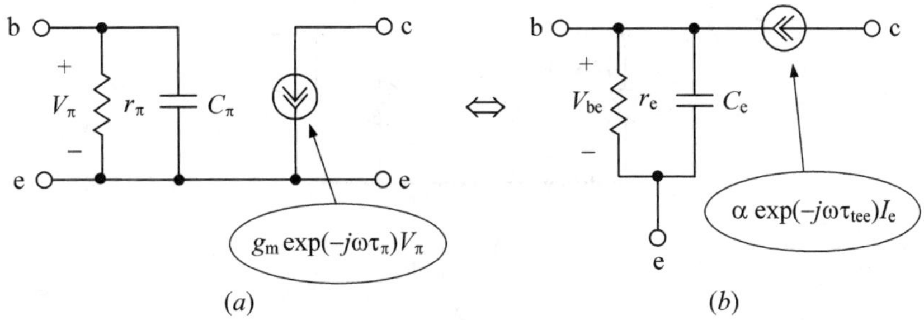

FIGURE 2.34 Intrinsic (a) π-type and (b) T-type device topologies.

The expressions for intrinsic π-model parameters can be derived through the intrinsic T-model parameters as [74]

$$g_{\rm m} = \frac{\alpha_0}{\sqrt{1+(\omega\tau_\alpha)^2}}\sqrt{\left(\frac{1}{r_{\rm e}}\right)^2+(\omega C_{\rm e})^2} \tag{2.127}$$

$$\tau_\pi = \tau_{\rm tee} - \frac{1}{\omega}\left[\tan^{-1}(\omega C_{\rm e} r_{\rm e}) + \tan^{-1}(\omega\tau_\alpha)\right] \tag{2.128}$$

$$\frac{1}{r_\pi} = \frac{1}{r_{\rm e}} - g_{\rm m}\cos(\omega\tau_\pi) \tag{2.129}$$

$$C_\pi = C_{\rm e} - g_{\rm m}\frac{\sin(\omega\tau_\pi)}{\omega}. \tag{2.130}$$

Both π-type and T-type bipolar device topologies can describe the transistor electrical properties in a very wide frequency range and, when optimized, up to 50 GHz.

2.5.4 Nonlinear Bipolar Device Modeling

Since the bipolar transistor can be considered to be an interacting pair of *p–n* junctions, the approach to model its nonlinear properties is the same as that used for the diode modeling. The simple large-signal Ebers–Moll model with a single current source between the collector and the emitter is shown in Figure 2.35 [1]. The collector–emitter source current $I_{\rm ce}$ is defined by

$$I_{\rm ce} = I_{\rm sat}\left[\exp\left(\frac{V_\pi}{V_{\rm T}}\right) - \exp\left(\frac{V_{\rm bc}}{V_{\rm T}}\right)\right] \tag{2.131}$$

where $I_{\rm sat}$ is the bipolar transistor saturation current and $V_{\rm T}$ is the thermal voltage calculated from Eq. (2.2).

The device terminal currents are defined as $I_{\rm c} = I_{\rm ce} - I_{\rm bc}$, $I_{\rm e} = -I_{\rm ce} - I_{\rm be}$, and $I_{\rm b} = I_{\rm be} + I_{\rm bc}$, where the diode currents are given by

$$I_{\rm be} = \frac{I_{\rm sat}}{\beta_{\rm F}}\left[\exp\left(\frac{V_\pi}{V_{\rm T}}\right) - 1\right] \tag{2.132}$$

$$I_{\rm bc} = \frac{I_{\rm sat}}{\beta_{\rm R}}\left[\exp\left(\frac{V_{\rm bc}}{V_{\rm T}}\right) - 1\right] \tag{2.133}$$

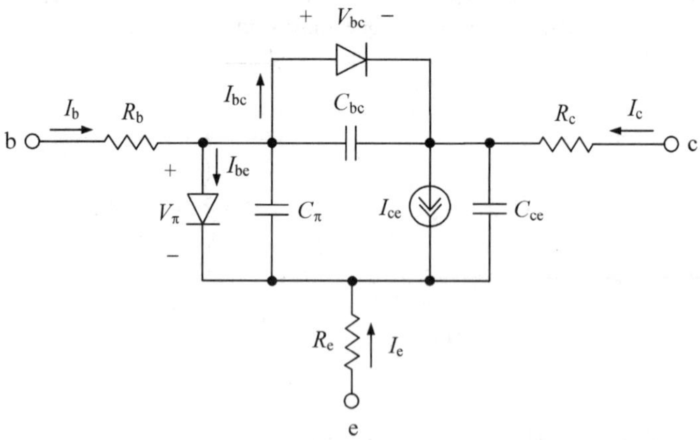

FIGURE 2.35 Large-signal Ebers–Moll model.

where β_F and β_R are the large-signal forward current gain and reverse current gain of a common emitter bipolar transistor, respectively.

The device capacitances C_π and C_{bc} each consist of two components and are modeled by the diffusion capacitance and junction capacitance, respectively, as

$$C_\pi = \tau_F \frac{dI_{be}}{dV_\pi} + C_{jeo}\left(1 - \frac{V_\pi}{\varphi_e}\right)^{-m_e} \tag{2.134}$$

$$C_{bc} = \tau_R \frac{dI_{bc}}{dV_{bc}} + C_{jco}\left(1 - \frac{V_{bc}}{\varphi_c}\right)^{-m_c} \tag{2.135}$$

where τ_F and τ_R are the ideal total forward time and reverse transit time, C_{jeo} and C_{jco} are the base–emitter and base–collector zero-bias junction capacitances, and m_e and m_c are the base–emitter and base–collector junction grading factors, respectively.

The collector–emitter substrate capacitance C_{ce} should be taken into account when designing integrated circuits. Its representation is adequate for many cases, since the epitaxial-layer–substrate junction is reversed-biased for isolation purposes, and usually it is modeled as a capacitance with constant value.

The Ebers–Moll model cannot describe the second-order effects due to low current and high-level injection, such as a base-width modulation (Early effect) and variation of the large-signal forward current gain β_F with collector current I_{ce}. In addition, to find a better approximation of the distributed structure of the base–collector junction at microwave frequencies, the junction capacitance C_{bc} should be divided into two separate capacitances: internal C_{ci} and external C_{co}. The lateral resistance and the base semiconductor resistance underneath the base contact and the base semiconductor resistance underneath the emitter are combined into a base-spreading resistance r_b. Figure 2.36 shows the modified π-type Gummel–Poon nonlinear model of the bipolar transistor that can describe the nonlinear electrical behavior of bipolar transistors, in particularly HBT devices, with sufficient accuracy up to 20 GHz [78,79].

For the Gummel–Poon large-signal model, the collector source current I_{ce} is determined by

$$I_{ce} = \frac{I_{ss}}{q_b}\left[\exp\left(\frac{V_\pi}{n_F V_T}\right) - \exp\left(\frac{V_{bc}}{n_R V_T}\right)\right] \tag{2.136}$$

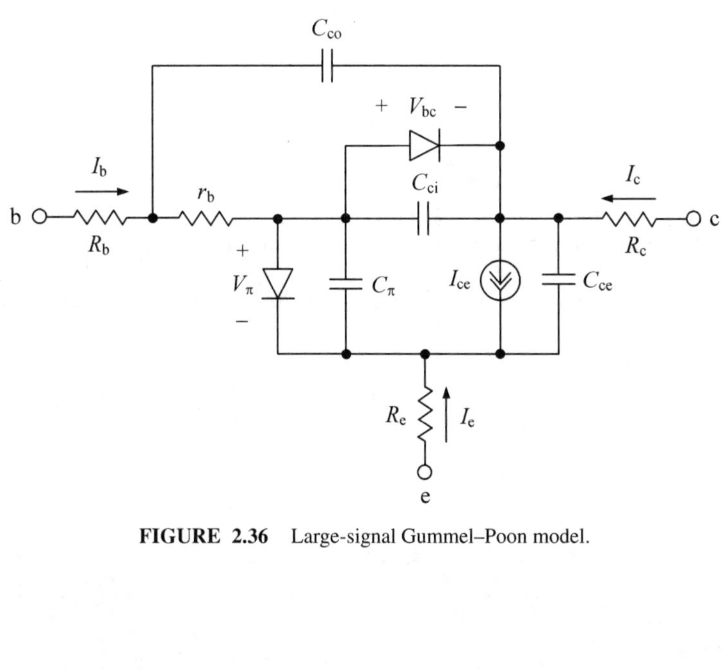

FIGURE 2.36 Large-signal Gummel–Poon model.

where I_{ss} is the BJT fundamental constant defined at zero-bias condition, n_F is the forward current emission coefficient, n_R is the reverse current emission coefficient, and q_b is the variable parameter defined by

$$q_b = \frac{q_1}{2} + \sqrt{\left(\frac{q_1}{2}\right)^2 + q_2} \tag{2.137}$$

where

$$q_1 = 1 + \frac{V_\pi}{V_B} + \frac{V_{bc}}{V_A}$$

$$q_2 = \frac{I_{ss}}{I_{KF}}\left[\exp\left(\frac{V_\pi}{n_F V_T}\right) - 1\right] + \frac{I_{ss}}{I_{KR}}\left[\exp\left(\frac{V_{bc}}{n_R V_T}\right) - 1\right]$$

V_A is the forward Early voltage, V_B is the reverse Early voltage, I_{KF} is the knee current for high-level injection in the normal active region, and I_{KR} is the knee current for low-level injection in inverse region [1].

The currents through model diodes are defined by

$$I_{be} = \frac{I_{sat}(0)}{\beta_{FM}(0)}\left[\exp\left(\frac{V_\pi}{n_F V_T}\right) - 1\right] + C_2 I_{sat}(0)\left[\exp\left(\frac{V_\pi}{n_{EL} V_T}\right) - 1\right] \tag{2.138}$$

$$I_{bc} = \frac{I_{sat}(0)}{\beta_{RM}(0)}\left[\exp\left(\frac{V_{bc}}{n_R V_T}\right) - 1\right] + C_4 I_{sat}(0)\left[\exp\left(\frac{V_{bc}}{n_{CL} V_T}\right) - 1\right] \tag{2.139}$$

where $I_{sat}(0)$ is the saturation current for $V_{bc} = 0$, $\beta_{FM}(0)$ and $\beta_{RM}(0)$ are the large-signal forward and reverse current gains of a common–emitter BJT in mid-current region for $V_{bc} = 0$, C_2 and C_4 are the forward and reverse low-current nonideal base current coefficients, respectively, n_{EL} is the nonideal low-current base–emitter emission coefficient, and n_{CL} is the nonideal low-current base–collector emission coefficient.

The nonlinear behavior of the intrinsic base resistance r_b can be described by

$$r_b = r_{bm} + 3(r_{b0} - r_{bm})\frac{\tan z - z}{z \tan^2 z} \tag{2.140}$$

where

$$z = \frac{\sqrt{1 + \dfrac{144}{\pi^2}\dfrac{I_b}{I_{rb}}} - 1}{\dfrac{24}{\pi^2}\sqrt{\dfrac{I_b}{I_{rb}}}}$$

r_{bm} is the minimum base resistance that occurs at high-current level, r_{b0} is the base resistance at zero bias with small base current level, and I_{rb} is the current where the base resistance falls halfway to its minimum value [1].

The intrinsic device capacitances C_π, C_{ci}, and C_{co} are modeled by the diffusion capacitance and junction capacitance, respectively, as

$$C_\pi = \frac{d}{dV_\pi}\left(\tau_{FF}\frac{I_{cc}}{q_b}\right) + C_{jeo}\left(1 - \frac{V_\pi}{\varphi_e}\right)^{-m_e} \tag{2.141}$$

$$C_{ci} = \tau_R \frac{dI_{bc}}{dV_{bc}} + k_c C_{jco}\left(1 - \frac{V_{bc}}{\varphi_c}\right)^{-m_c} \tag{2.142}$$

$$C_{co} = C_{jco}(1 - k_c)\left(1 - \frac{V_{bco}}{\varphi_c}\right)^{-m_c} \tag{2.143}$$

where k_c is the fraction of the base–collector junction capacitance connected to the base resistance r_b, V_{bco} is the voltage through the capacitance C_{co}, and τ_{FF} is the modulated transit time defined by

$$\tau_{FF} = \tau_F \left[1 + X_{\tau F} \left(\frac{I_{cc}}{I_{cc} + I_{\tau F}} \right)^2 \exp \left(\frac{V_{bc}}{1.44\, V_{\tau F}} \right) \right] \tag{2.144}$$

where $X_{\tau F}$ is the transit time bias dependence coefficient, $I_{\tau F}$ is the high-current parameter for effect on τ_F, $V_{\tau F}$ is the value of V_{bc} where the exponential equals to 0.5, and

$$I_{cc} = \frac{I_{ss}}{q_b} \left[\exp \left(\frac{V_\pi}{n_F V_T} \right) - 1 \right]. \tag{2.145}$$

As it follows from Eq. (2.141), the nonlinear behavior of capacitance C_π strongly depends on the effect of transit time modulation characterized by τ_{FF}. This transit charge variation results in significant changes of the transition frequency f_T at various operation conditions. For example, at medium currents, f_T reaches its peak value and is practically constant. Here, the ideal transit time is defined by $\tau_F = 1/(2\pi f_T)$ and the dominated base–emitter diffusion capacitance increases linearly with collector current. At low currents, f_T is dominated by the junction capacitance and increases with the increase in collector current. At high currents, the widening of the charge-neutral base region and pushing of the entire space-charge region toward the heavily doped collector region (Kirk effect) degrades the frequency response of the transistor by increasing the transit time and decreasing f_T. In this case, the transit time is modeled by τ_{FF}.

The more complicated models, such as VBIC, HICUM, or MEXTRAM, include the effects of self-heating of a bipolar transistor, take into account the parasitic *p–n–p* transistor formed by the base, collector, and substrate regions, provide an improved description of depletion capacitances at large-forward bias, take into account avalanche and tunneling currents, and other nonlinear effects corresponding to distributed high-frequency effects [80].

2.5.5 Noise Model

The noise in a bipolar transistor is assumed to arise from three basic sources: diffusion fluctuations, recombination fluctuations in the base region, and thermal noise in the base resistance [81]. The noise behavior of the bipolar transistor can be described based on its equivalent circuit representation shown in Figure 2.37(*a*), which includes the main elements responsible for the device electrical behavior and noise sources. Since the process of the carrier drifting into the collector–base depletion region is a random process, the collector current I_c demonstrates shot noise and is represented by a shot-noise collector current source $\overline{i_{nc}^2}$. The base current I_b is a result of the carrier injection from the base to the emitter and generation–recombination effect in the base and base–emitter depletion regions. Because all these components are independent, representing a random process, the base current also demonstrates a shot-noise behavior and is represented by a shot-noise base current source $\overline{i_{nb}^2}$. Flicker noise is represented by a current source across the internal base–emitter junction combined with the base current source $\overline{i_{nb}^2}$. The series base, emitter and collector resistances are represented by the voltage and current thermal noise sources $\overline{e_{nb}^2}$, $\overline{i_{ne}^2}$, and $\overline{e_{nc}^2}$, respectively.

The noise voltage and current sources can be given through their mean-square values as

$$\overline{e_{nb}^2} = 4kTr_b\Delta f \tag{2.146}$$

$$\overline{i_{ne}^2} = \frac{4kT\Delta f}{r_e} \tag{2.147}$$

$$\overline{i_{nc}^2} = 2qI_c\Delta f \tag{2.148}$$

$$\overline{e_{nc}^2} = 4kTr_c\Delta f \tag{2.149}$$

$$\overline{i_{nb}^2} = 2qI_b\Delta f + K_F\frac{I_b^{A_F}}{f}\Delta f \tag{2.150}$$

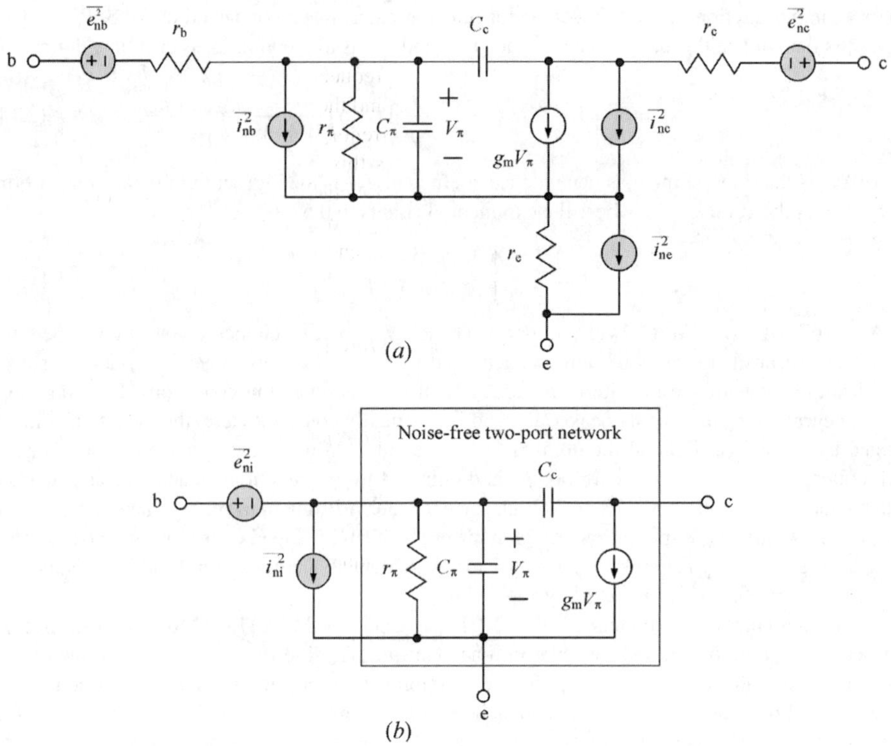

FIGURE 2.37 Bipolar equivalent circuits with noise sources.

where q is the electron charge, A_F is the flicker noise exponent, K_F is the flicker noise coefficient calculated as $2qf_c$, and f_c is the flicker noise corner frequency [1,82].

The required minimum noise figure F_{min}, noise resistance R_n, and optimum source admittance Y_{Sopt} using the noise correlation C_A-matrix parameters as functions of the input-referred noise voltage $\overline{v_{ni}^2}$ and noise current $\overline{i_{ni}^2}$, as well as Y-parameters of the simplified noise-free two-port network shown in Figure 2.37(b), can be calculated for a sufficiently high value of the low-frequency current gain $\beta = g_m r_\pi$ from

$$R_n = r_b \left(1 + \frac{r_b}{2r_\pi}\right) + \frac{r_\pi}{2\beta} + \frac{r_b^2}{2\beta r_\pi}\left(\frac{f}{f_T}\right)^2 \tag{2.151}$$

$$Y_{Sopt} = \sqrt{\frac{1}{2r_\pi R_n}\left[1 + \frac{1}{\beta}\left(\frac{f}{f_T}\right)^2\right] - \left(\frac{1}{2\beta R_n}\frac{f}{f_T}\right)^2} - j\frac{1}{2\beta R_n}\frac{f}{f_T} \tag{2.152}$$

$$F_{min} = 1 + \frac{r_b}{r_\pi}\left[1 + \frac{1}{\beta}\left(\frac{f}{f_T}\right)^2\right] + \sqrt{\frac{2r_b}{r_\pi}\left(1 + \frac{r_b}{2r_\pi}\right) + \frac{2r_b}{\beta r_\pi}\left(1 + \frac{r_b}{r_\pi}\right)\left(\frac{f}{f_T}\right)^2} \tag{2.153}$$

where f is the operation frequency and $f_T = g_m/(2\pi C_\pi)$ is the bipolar transition frequency (the effect of the feedback collector capacitance C_c is not taken into account) [83]. It should be noted that if the optimum noise conductance G_{Sopt} is insensitive to the collector capacitance C_c, then it can severely affect the other noise parameters at microwave frequencies, mostly due to the reduction of the device gain capability.

A noise model for HBT device operated at very high frequencies should include the contribution of both space-charge layers (at the emitter–base junction and the base–collector junction) to the shot noise. These two noise sources related to the collector current I_c are the result of the same electrons, which are injected from the emitter into the base, cross this layer, and then reach the collector. Therefore, their correlation can be given by a time delay function $\exp(-j\omega\tau)$, where τ is the transit time through the base and the collector–base junction which is τ_π for a π-type model [84–86]. Thus, to extend the HBT noise model valid up to its transition frequency, the base and collector noise current sources are rewritten as

$$\overline{i_{\mathrm{nb}}^2} = 2q\left(I_{\mathrm{b}} + |1 - \exp(-j\omega\tau)|^2 I_{\mathrm{c}}\right)\Delta f \tag{2.154}$$

$$\overline{i_{\mathrm{nc}}^2} = 2qI_{\mathrm{c}}\Delta f \tag{2.155}$$

$$\overline{i_{\mathrm{nb}}i_{\mathrm{nc}}^*} = 2q\left[\exp(-j\omega\tau) - 1\right]I_{\mathrm{c}}\Delta f. \tag{2.156}$$

Flicker noise in bipolar transistors is associated mainly with generation–recombination centers, which contribute to random trapping and release of free carriers [87,88]. This relatively slow process is always associated with flowing current by which mean-square value $\overline{i_{\mathrm{n}}^2}$ as a function of the offset frequency f can be approximated

$$\overline{i_{\mathrm{n}}^2} = K_{\mathrm{F}}I^{\mathrm{k}}\frac{\Delta f}{f} \tag{2.157}$$

where K_{F} is the flicker noise coefficient and k is the flicker noise exponent. Both these coefficients are device dependent. In conventional high-quality silicon bipolar devices, the low-frequency noise is determined by 1/f noise in the base current due to carriers injected from the base into the emitter, since the emitter series resistance can be neglected, the base series resistance is low and the collector current has an ideality factor of 1. In downscaled polysilicon bipolar transistors with lower emitter area, at low bias currents the 1/f noise in the base current is dominant, while at higher bias currents the influence of the series resistances on the noise becomes noticeable. In GaAs/AlGaAs HBT devices, the 1/f noise in the base current can be described by Eq. (2.157) with $K_{\mathrm{F}} \approx 10^{-10}$ $\mathrm{A}^{2-\mathrm{k}}$ and $k \approx 1.6$, whereas, for the 1/f noise in the collector current, $K_{\mathrm{F}} \approx 10^{-12}$ $\mathrm{A}^{2-\mathrm{k}}$ and $k \approx 1.3$ [89]. The contribution of the extrinsic base resistance noise becomes more important with scaling, especially for the device with very high transition frequency f_{T} [90].

REFERENCES

1. G. Massobrio and P. Antognetti, *Semiconductor Device Modeling with SPICE*, New York: McGraw-Hill, 1993.
2. S. A. Maas, *Nonlinear Microwave Circuits*, New York: IEEE Press, 1997.
3. Application Note 1002, "Design with PIN Diodes," Skyworks Solutions, 2005.
4. M. H. Norwood and E. Shatz, "Voltage Variable Capacitor Tuning: A Review," *Proc. IEEE*, vol. 56, pp. 788–798, May 1968.
5. W. Schottky and W. Deutschmann, "Zum Mechanismus der Richtwirkung in Kupferoxydulgleichrichten," *Phys. Z.*, vol. 30, pp. 839–846, Nov. 1929.
6. W. Shockley, "The Theory of *p–n* Junctions in Semiconductors," *Bell Syst. Tech. J.*, vol. 28, pp. 435–450, July 1949.
7. M. E. McMahon and G. F. Straube, "Voltage-Sensitive Semiconductor Capacitor," *IRE WESCON Conv. Rec.*, pt. 3, pp. 72–82, Aug. 1958.

8. P. Penfield and R. P. Rafuse, *Varactor Applications*, Cambridge: The M.I.T Press, 1962.
9. Application Note 1004, "Varactor SPICE Models for RF VCO Applications," Skyworks Solutions, 2005.
10. R. Beach, "High Speed Linear Oscillators," *Microwave J.*, vol. 21, pp. 59–65, Dec. 1978.
11. Y. P. Tsividis, *Operation and Modeling of the MOS Transistor*, New York: McGraw-Hill, 1987.
12. P. Andreani and S. Mattisson, "On the Use of MOS Varactors in RF VCO's," *IEEE J. Solid-State Circuits*, vol. SC-35, pp. 905–910, June 2000.
13. F. Svelto, P. Erratico, and R. Castello, "A Metal-Oxide-Semiconductor Varactor," *IEEE Electron Device Lett.*, vol. 20, pp. 164–166, Apr. 1999.
14. A.-S. Porret, T. Melly, C. C. Enz, and E. A. Vittoz, "Design of High-Q Varactors for Low-Power Wireless Applications Using a Standard CMOS Process," *IEEE J. Solid-State Circuits*, vol. SC-35, pp. 337–345, Mar. 2000.
15. S.-S. Song and H. Shin, "An RF Model of the Accumulation-Mode MOS Varactor Valid in Both Accumulation and Depletion Regions," *IEEE Trans. Electron Devices*, vol. ED-50, pp. 1997–1999, Sep. 2003.
16. B. Senapati, K. Ehwald, W. Winkler, and F. Furnhammer, "High Performance MOS Varactor SPICE Model," *Electronics Lett.*, vol. 38, pp. 1416–1417, Nov. 2002.
17. K. Molnar, G. Rappitsch, Z. Huszka, and E. Seebacher, "MOS Varactor Modeling With a Subcircuit Utilizing the BSIM3v3 Model," *IEEE Trans. Electron Devices*, vol. ED-49, pp. 1206–1211, July 2002.
18. Y. P. Tsividis and K. Suyama, "MOSFET Modeling for Analog Circuit CAD: Problems and Prospects," *IEEE J. Solid-State Circuits*, vol. SC-29, pp. 210–216, Mar. 1994.
19. E. Abou-Allam and T. Manku, "A Small-Signal MOSFET Model for Radio Frequency IC Applications," *IEEE Trans. Computer-Aided Design*, vol. 16, pp. 437–447, May 1997.
20. T. Taki, "Approximation of Junction Field-Effect Transistor Characteristics by a Hyperbolic Function," *IEEE J. Solid-State Circuits*, vol. SC-13, pp. 724–726, Oct. 1978.
21. D. Heo, E. Chen, E. Gebara, S. Yoo, J. Lascar, and T. Anderson, "Temperature Dependent MOSFET RF Large Signal Model Incorporating Self Heating Effects," *IEEE MTT-S Int. Microwave Symp. Dig.*, vol. 2, pp. 415–418, June 1999.
22. W. R. Curtice, J. A. Pla, D. Bridges, T. Liang, and E. E. Shumate, "A New Dynamic Electro-Thermal Nonlinear Model for Silicon RF LDMOS FETs," *IEEE MTT-S Int. Microwave Symp. Dig.*, vol. 2, pp. 419–422, 1999.
23. C. C. Enz and E. A. Vittoz, *Charge-Based MOS Transistor Modeling: The EKV Model for Low Power and RF IC Design*, Chichester: John Wiley & Sons, Ltd. 2006.
24. A. V. Grebennikov and F. Lin, "An Efficient CAD-Oriented Large-Signal MOSFET Model," *IEEE Trans. Microwave Theory Tech.*, vol. MTT-48, pp. 1732–1742, Oct. 2000.
25. M. C. Ho, K. Green, R. Culbertson, J. Y. Yang, D. Ladwig, and P. Ehnis, "A Physical Large Signal Si MOSFET Model for RF Circuit Design," *IEEE MTT-S Int. Microwave Symp. Dig.*, vol. 2, pp. 391–394, June 1997.
26. J. Angel Garcia, A. Mediavilla, J. Carlos Pedro, N. B. Carvalho, A. Tazon, and J. L. Garcia, "Characterizing the Gate to Source Nonlinear Capacitor Role on FET IMD Performance," *IEEE MTT-S Int. Microwave Symp. Dig.*, vol. 3, pp. 1635–1638, 1998.
27. P. Perugupalli, M. Trivedi, K. Shenai, and S. K. Leong, "Modeling and Characterization of an 80 V Silicon LDMOSFET for Emerging RFIC Applications," *IEEE Trans. Electron Devices*, vol. ED-45, pp. 1468–1478, July 1998.
28. D. Moncoqut, D. Farenc, P. Rossel, G. Charitat, H. Tranduc, J. Victory, and I. Pages, "LDMOS Transistor for Smart Power Circuits: Modeling and Design," *Proc. IEEE Bipolar/BiCMOS Circuits and Technology Meeting (BCTM)*, pp. 216–219, Sept./Oct. 1996.
29. C. E. Biber, M. L. Schmatz, T. Morf, U. Lott, and W. Bachtold, "A Nonlinear Microwave MOSFET Model for SPICE Simulators," *IEEE Trans. Microwave Theory Tech.*, vol. MTT-46, pp. 604–610, May 1998.
30. R. Sung, P. Bendix, and M. B. Das, "Extraction of High-Frequency Equivalent Circuit Parameters of Submicron Gate-Length MOSFET's," *IEEE Trans. Electron Devices*, vol. ED-45, pp. 1769–1775, Aug. 1998.
31. B. J. Cheu and P. K. Ko, "Measurement and Modeling of Short-Channel MOS Transistor Gate Capacitances," *IEEE J. Solid-State Circuits*, vol. SC-22, pp. 464–472, June 1987.
32. R. E. Anholt and S. E. Swirhun, "Experimental Investigations of the Temperature Dependence of GaAs FET Equivalent Circuits," *IEEE Trans. Electron Devices*, vol. ED-39, pp. 2029–2036, Sept. 1992.

33. C. Enz, "An MOS Transistor Model for RF IC Design Valid in All Regions of Operation," *IEEE Trans. Microwave Theory Tech.*, vol. MTT-50, pp. 342–359, Jan. 2002.

34. S. Tedja, J. van der Spiegel, and H. H. Williams, "Analytical and Experimental Studies of Thermal Noise in MOSFETs," *IEEE Trans. Electron Devices*, vol. ED-41, pp. 2069–2075, Nov. 1994.

35. S. Asgaran, M. J. Deen, and C.-H. Chen, "Analytical Modeling of MOSFET's Channel Noise and Noise Parameters," *IEEE Trans. Electron Devices*, vol. ED-51, pp. 2109–2114, Dec. 2004.

36. B. L. Buckingham, *Noise in Electronic Devices and Systems*, New York: John Wiley & Sons, 1983.

37. O. Jantsch, "Flicker (1/*f*) Noise Generated by a Random Walk of Electrons in Interfaces," *IEEE Trans. Electron Devices*, vol. ED-34, pp. 1100–1115, May 1987.

38. H. Mikoshiba, "1/*f* Noise in n-Channel Silicon-Gate MOS Transistors," *IEEE Trans. Electron Devices*, vol. ED-29, pp. 965–969, June 1982.

39. K. K. Hung, P. K. Ko, C. Hu, and Y. C. Cheng, "A Unified Model for the Flicker Noise in Metal-Oxide-Semiconductor Field-Effect Transistors," *IEEE Trans. Electron Devices*, vol. ED-37, pp. 654–664, Mar. 1990.

40. F.-C. Hou, G. Bosman, and M. E. Law, "Simulation of Oxide Trapping Noise in Submicron n-Channel MOSFETs," *IEEE Trans. Electron Devices*, vol. ED-50, pp. 846–852, Mar. 2003.

41. L. K. J. Vandamme, X. Li, and D. Rigaud, "1/*f* Noise in MOS Devices, Mobility or Number Fluctuations?" *IEEE Trans. Electron Devices*, vol. ED-41, pp. 1936–1944, Nov. 1994.

42. C. F. Hiatt, A. van der Ziel, and K. M. van Viet, "Generation-Recombination Noise Produced in the Channel of JFETs," *IEEE Trans. Electron Devices*, vol. ED-22, pp. 614–616, 1975.

43. F. M. Klaassen and J. R. Robinson, "Anomalous Noise Behaviour of the Junction Gate Field Effect Transistor at Low Temperatures," *IEEE Trans. Electron Devices*, vol. ED-17, pp. 852–857, 1970.

44. F. Ellinger, J. Kucera, and W. Baechtold, "Improvements on a GaAs MESFET Model for Nonlinear RF Simulations," *IEEE MTT-S Int. Microwave Symp. Dig.*, vol. 3, pp. 1623–1626, June 1998.

45. C.-J. Wei, Y. Tkachenko, and D. Bartle, "An Accurate Large-Signal Model of GaAs MESFET Which Accounts for Charge Conservation, Dispersion, and Self-Heating," *IEEE Trans. Microwave Theory Tech.*, vol. MTT-46, pp. 1638–1644, Nov. 1998.

46. I. Angelov, L. Bengtsson, and M. Garcia, "Extensions of the Chalmers Nonlinear HEMT and MESFET Model," *IEEE Trans. Microwave Theory Tech.*, vol. MTT-44, pp. 1664–1674, Oct. 1996.

47. M. Berroth and R. Bosch, "Broad-Band Determination of the FET Small-Signal Equivalent Circuit," *IEEE Trans. Microwave Theory Tech.*, vol. MTT-38, pp. 891–895, July 1990.

48. G. Dambrine, A. Cappy, F. Heliodore, and E. Playez, "A New Method for Determining the FET Small-Signal Equivalent Circuit," *IEEE Trans. Microwave Theory Tech.*, vol. MTT-36, pp. 1151–1159, July 1988.

49. W. R. Curtice, "A MESFET Model for Use in the Design of GaAs Integrated Circuits," *IEEE Trans. Microwave Theory Tech.*, vol. MTT-25, pp. 448–456, May 1980.

50. W. R. Curtice and M. Ettenberg, "A Nonlinear GaAs FET Model for Use in the Design of Output Circuits for Power Amplifiers," *IEEE Trans. Microwave Theory Tech.*, vol. MTT-33, pp. 1383–1394, Dec. 1985.

51. A. E. Parker and D. J. Skellern, "A Realistic Large-Signal MESFET Model for SPICE," *IEEE Trans. Microwave Theory Tech.*, vol. MTT-45, pp. 1563–1571, Sept. 1997.

52. I. Angelov, H. Zirath, and N. Rorsman, "A New Empirical Nonlinear Model for HEMT and MESFET Devices," *IEEE Trans. Microwave Theory Tech.*, vol. MTT-40, pp. 2258–2266, Dec. 1992.

53. I. Angelov, H. Zirath, and N. Rorsman, "Validation of a Nonlinear Transistor Model by Power Spectrum Characteristics of HEMT's and MESFET's," *IEEE Trans. Microwave Theory Tech.*, vol. MTT-43, pp. 1046–1051, May 1995.

54. R. Osorio, M. Berroth, W. Marsetz, L. Verweyen, M. Demmler, H. Massler, M. Neumann, and M. Schlechtweg, "Analytical Charge Conservative Large Signal Model for MODFETs Validated up to MM-Wave Range," *IEEE MTT-S Int. Microwave Symp. Dig.*, vol. 2, pp. 595–598, June 1998.

55. H. Statz, H. A. Haus, and R. A. Pucel, "Noise Characteristics of Gallium Arsenide Field-Effect Transistors," *IEEE Trans. Electron Devices*, vol. ED-21, pp. 549–562, Sept. 1974.

56. W. Baechtold, "Noise Behavior of GaAs Field-Effect Transistors with Short Gate Lengths," *IEEE Trans. Electron Devices*, vol. ED-19, pp. 674–680, May 1972.

57. A. van der Ziel, "Thermal Noise in the Hot Electron Regime in FET's," *IEEE Trans. Electron Devices*, vol. ED-18, p. 977, Oct. 1971.
58. R. A. Pucel, D. F. Masse, and C. F. Krumm, "Noise Performance of Gallium Arsenide Field-Effect Transistors," *IEEE Trans. Solid-State Circuits*, vol. SC-11, pp. 243–255, Apr. 1976.
59. A. van der Ziel, "Thermal Noise in Field Effect Transistor," *Proc. IRE*, vol. 50, pp. 1808–1812, Aug. 1962.
60. A. van der Ziel, "Gate Noise in Field Effect Transistors at Moderately High Frequencies," *Proc. IRE*, vol. 51, pp. 461–467, Mar. 1963.
61. V. Rizzoli, F. Mastri, and C. Cecchetti, "Computer-Aided Noise Analysis of MESFET and HEMT Mixers," *IEEE Trans. Microwave Theory Tech.*, vol. MTT-37, pp. 1401–1410, Sept. 1989.
62. A. Cappy, "Noise Modeling and Measurement Techniques," *IEEE Trans. Microwave Theory Tech.*, vol. MTT-36, pp. 1–9, Jan. 1988.
63. P. Heymann and H. Prinzler, "Improved Noise Model for MESFETs and HEMTs in Lower Gigahertz Frequency Range," *Electronics Lett.*, vol. 28, pp. 611–612, Mar. 1992.
64. D.-S. Shin, J. B. Lee, H. S. Min, J.-E. Oh, Y.-J. Park, W. Jung, and D. S. Ma, "Analytical Noise Model with the Influence of Shot Noise Induced by the Gate Leakage Current for Submicrometer Gate-Length High-Electron Mobility Transistors," *IEEE Trans. Electron Devices*, vol. ED-44, pp. 1883–1887, Nov. 1997.
65. J. Gao, C. L. Law, H. Wang, S. Aditya, and G. Boeck, "A New Method for pHEMT Noise-Parameter Determination Based on 50-Ω Noise Measurement System," *IEEE Trans. Microwave Theory Tech.*, vol. MTT-51, pp. 2079–2089, Oct. 2003.
66. H. Fukui, "Design of Microwave GaAs MESFET's for Broad-Band Low-Noise Amplifiers," *IEEE Trans. Microwave Theory Tech.*, vol. MTT-27, pp. 643–650, July 1979.
67. H. Fukui, "Addendum to Design of Microwave GaAs MESFET's for Broad-Band Low-Noise Amplifiers," *IEEE Trans. Microwave Theory Tech.*, vol. MTT-29, p. 1119, Oct. 1981.
68. H. Fukui, "Optimal Noise Figure of Microwave GaAs MESFET's," *IEEE Trans. Electron Devices*, vol. ED-26, pp. 1032–1037, July 1979.
69. A. van der Ziel, "Flicker Noise in Electronic Devices," *Advances in Electronics and Electron Physics*, vol. 49, pp. 225–297, 1979.
70. R. A. Pucel, "The GaAs FET Oscillator – Its Signal and Noise Performance," *Proc. 40th Annual Frequency Control Symp.*, pp. 385–391, 1986.
71. M. Chertouk and A. Chovet, "Origins and Characterization of Low-Frequency Noise in GaAs MESFET's Grown on InP Substrates," *IEEE Trans. Electron Devices*, vol. ED-43, pp. 123–129, 1996.
72. D. Costa, W. U. Liu, and J. S. Harris, "Direct Extraction of the AlGaAs/GaAs Heterojunction Bipolar Transistor Small-Signal Equivalent Circuit," *IEEE Trans. Electron Devices*, vol. ED-38, pp. 2018–2024, Sep. 1991.
73. H. Ghaddab, F. M. Ghannouchi, and F. Bouallegue, "Small-Signal Modeling of HBT's Using a Hybrid Optimization/Statistical Technique," *IEEE Trans. Microwave Theory Tech.*, vol. MTT-46, pp. 292–298, March 1998.
74. D. A. Teeter and W. R. Curtice, "Comparison of Hybrid Pi and Tee HBT Circuit Topologies and Their Relationship to Large Signal Modeling," *IEEE MTT-S Int. Microwave Symp. Dig.*, vol. 2, pp. 375–378, 1997.
75. A. Samelis and D. Pavlidis, "DC to High-Frequency HBT-Model Parameter Evaluation Using Impedance Block Conditioned Optimization," *IEEE Trans. Microwave Theory Tech.*, vol. MTT-45, pp. 886–897, June 1997.
76. B. Li, F. S. Prasad, L.-W. Yang, and S. C. Wang, "A Semianalytical Parameter-Extraction Procedure for HBT Equivalent Circuit," *IEEE Trans. Microwave Theory Tech.*, vol. MTT-46, pp. 1427–1435, Oct. 1998.
77. U. Schaper and B. Holzapfl, "Analytical Parameter Extraction of the HBT Equivalent Circuit with T-like Topology from Measured S-parameters," *IEEE Trans. Microwave Theory Tech.*, vol. MTT-43, pp. 493–498, March 1995.
78. N. M. Rohringer and P. Kreuzgruber, "Parameter Extraction for Large-Signal Modeling of Bipolar Junction Transistors," *Int. J. Microwave and Millimeter-Wave Computer-Aided Eng.*, vol. 5, pp. 161–272, Sept. 1995.
79. J. P. Fraysee, D. Floriot, P. Auxemery, M. Campovecchio, R. Quere, and J. Obregon, "A Non-Quasi-Static Model of GaInP/AlGaAs HBT for Power Applications," *IEEE MTT-S Int. Microwave Symp. Dig.*, vol. 2, pp. 377–382, June 1997.

80. M. Reisch, *High-Frequency Bipolar Transistors*, Berlin: Springer, 2003.
81. E. G. Nielsen, "Behavior of Noise Figure in Junction Transistors," *Proc. IRE*, vol. 45, pp. 957–963, July 1957.
82. H. Fukui, "The Noise Performance of Microwave Devices," *IEEE Trans. Electron Devices*, vol. ED-13, pp. 329–341, Mar. 1966.
83. L. Escotte, J.-P. Roux, R. Plana, J. Graffeuil, and A. Gruhle, "Noise Modeling of Microwave Heterojunction Bipolar Transistors," *IEEE Trans. Electron Devices*, vol. ED-42, pp. 883–889, May 1995.
84. M. Rudolph, R. Doerner, L. Klapproth, and P. Heymann, "An HBT Noise Model Valid Up to Transit Frequency," *IEEE Electron Device Lett.*, vol. 20, pp. 24–26, Jan. 1999.
85. G. Niu, J. D. Cressler, S. Zhang, W. E. Ansley, C. S. Webster, and D. L. Harame, "A Unified Approach to RF and Microwave Noise Parameter Modeling in Bipolar Transistors," *IEEE Trans. Electron Devices*, vol. ED-48, pp. 2568–2574, Nov. 2001.
86. J. Gao, X. Li, H. Wang, and G. Boeck, "Microwave Noise Modeling for InP-InGaAs HBTs," *IEEE Trans. Microwave Theory Tech.*, vol. MTT-52, pp. 1264–1272, Apr. 2004.
87. J. L. Plumb and E. R. Chenette, "Flicker Noise in Transistors," *IEEE Trans. Electron Devices*, vol. ED-10, pp. 304–308, Sep. 1963.
88. R. C. Jaeger and A. J. Broderson, "Low-Frequency Noise Sources in Bipolar Junction Transistors," *IEEE Trans. Electron Devices*, vol. ED-17, pp. 128–134, Feb. 1970.
89. T. G. M. Kleinpenning, "Low-Frequency Noise in Modern Bipolar Transistors: Impact of Intrinsic Transistor and Parasitic Series Resistances," *IEEE Trans. Electron Devices*, vol. ED-41, pp. 1981–1991, Nov. 1994.
90. G. Niu, J. Tang, Z. Feng, A. Joseph, and D. L. Harame, "SiGe HBT Scaling Implications on 1/*f* Noise and Oscillator Phase Noise," *IEEE RFIC Symp. Dig.*, pp. 299–302, June 2004.

3 IMPEDANCE MATCHING

This chapter begins by describing the main principles and impedance matching tools. Generally, an optimum solution depends on the circuit requirement, such as the simplicity in practical realization, the frequency bandwidth and minimum power ripple, design implementation and adjustability, stable operation conditions, and sufficient harmonic suppression. As a result, many types of the matching networks are available, including lumped elements and transmission lines. To simplify and visualize the matching design procedure, an analytical approach, which allows calculation of the parameters of the matching circuits using simple equations, and Smith chart traces are discussed. In addition, several examples of the narrowband and broadband power amplifiers using bipolar or MOSFET devices are given, including successive and detailed design considerations and explanations.

3.1 MAIN PRINCIPLES

Impedance matching is necessary to provide maximum delivery to the load of the RF power available from the source. This means that, when the electrical signal propagates in the circuit, a portion of this signal will be reflected at the interface between the sections with different impedances. Therefore, it is necessary to establish the conditions that allow to fully transmit the entire electrical signal without any reflection. To determine an optimum value of the load impedance Z_L, at which the power delivered to the load is maximal, consider the equivalent circuit shown in Figure 3.1(*a*).

The power delivered to the load can be defined as

$$P = \frac{1}{2}V_{in}^2 \text{Re}\left(\frac{1}{Z_L}\right) = \frac{1}{2}V_S^2 \left|\frac{Z_L}{Z_S + Z_L}\right|^2 \text{Re}\left(\frac{1}{Z_L}\right) \tag{3.1}$$

where $Z_S = R_S + jX_S$ is the source impedance, $Z_L = R_L + jX_L$ is the load impedance, V_S is the source voltage amplitude, and V_{in} is the load voltage amplitude. Substituting the real and imaginary parts of the source and load impedances Z_S and Z_L, into Eq. (3.1) yields

$$P = \frac{1}{2}V_S^2 \frac{R_L}{(R_S + R_L)^2 + (X_S + X_L)^2}. \tag{3.2}$$

Assume the source impedance Z_S is fixed and it is necessary to vary the real and imaginary parts of the load impedance Z_L until maximum power is delivered to the load. To maximize the output power, the following analytical conditions in the form of derivatives with respect to the output power are written:

$$\frac{\partial P}{\partial R_L} = 0 \quad \frac{\partial P}{\partial X_L} = 0. \tag{3.3}$$

RF and Microwave Transmitter Design, First Edition. Andrei Grebennikov.

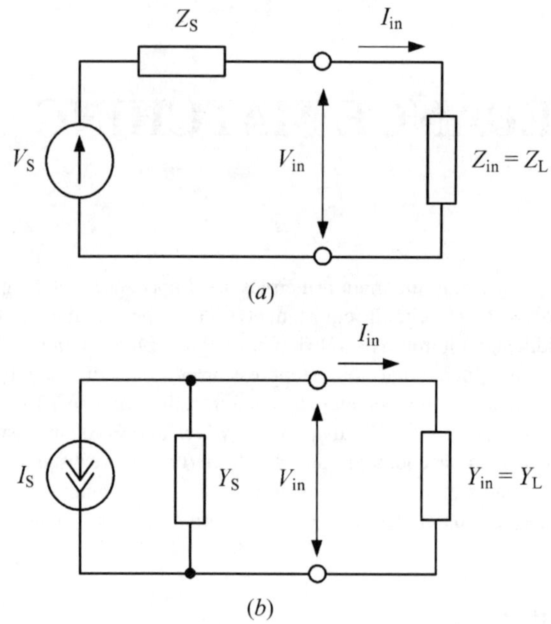

FIGURE 3.1 Equivalent circuits with (*a*) voltage and (*b*) current sources.

Applying these conditions and taking into consideration Eq. (3.2), the following system of two equations can be obtained:

$$\frac{1}{(R_L + R_S)^2 + (X_L + X_S)^2} - \frac{2R_L\,(R_L + R_S)}{\left[(R_L + R_S)^2 + (X_L + X_S)^2\right]^2} = 0 \tag{3.4}$$

$$\frac{2X_L\,(X_L + X_S)}{\left[(R_L + R_S)^2 + (X_L + X_S)^2\right]^2} = 0. \tag{3.5}$$

Simplifying Eqs. (3.4) and (3.5) results in

$$R_S^2 - R_L^2 + (X_L + X_S)^2 = 0 \tag{3.6}$$

$$X_L\,(X_L + X_S) = 0. \tag{3.7}$$

By solving Eqs. (3.6) and (3.7) simultaneously for R_S and X_S, one can obtain

$$R_S = R_L \tag{3.8}$$

$$X_L = -X_S \tag{3.9}$$

or in a common impedance case

$$Z_L = Z_S^* \tag{3.10}$$

where * denotes the complex-conjugate value [1].

Eq. (3.10) is called an *impedance conjugate matching condition*, and its fulfillment results in maximum power delivered to the load for fixed source impedance.

Maximum power delivered to the load must be equal to

$$P = \frac{V_S^2}{8R_S}. \tag{3.11}$$

The *admittance conjugate matching condition*, applied to the equivalent circuit presented in Figure 3.1(*b*),

$$Y_L = Y_S^* \tag{3.12}$$

can be readily obtained in the same way. Maximum power delivered to the load in this case can be written as

$$P = \frac{I_S^2}{8G_S} \tag{3.13}$$

where $G_S = \mathrm{Re}Y_S$ is the source conductance and I_S is the source current amplitude.

Thus, the conjugate matching conditions in a common case can be determined through the immittance parameters, that is any system of impedance Z-parameters or admittance Y-parameters, in the form of

$$W_L = W_S^*. \tag{3.14}$$

The matching circuit is connected between the source and the input of an active device shown in Figure 3.2(*a*) and between the output of an active device and the load shown in Figure 3.2(*b*).

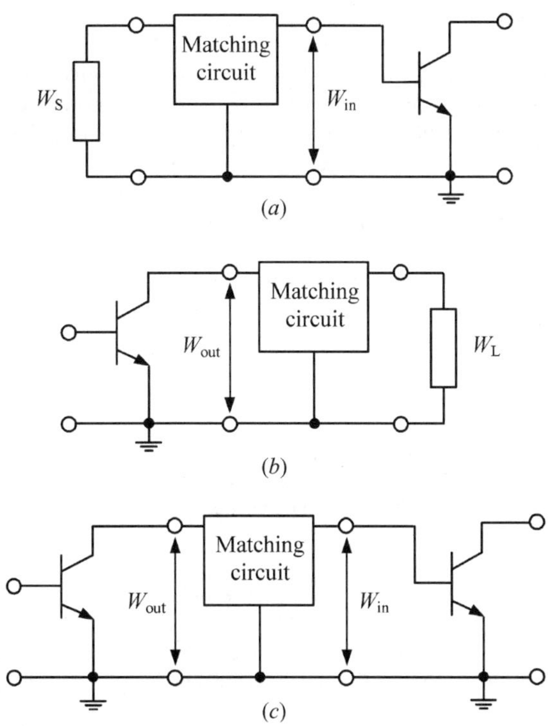

FIGURE 3.2 Matching circuit arrangements.

For a multistage power amplifier, the load represents an input circuit of the next stage. Therefore, the matching circuit is connected between the output of the active device of the previous amplifier stage and the input of the active device of the subsequent stage of the power amplifier shown in Figure 3.2(*c*). The main objective is to properly transform the load immittance W_L to the optimum device output immittance W_{out}, the value of which is properly determined by the supply voltage, the output power, the saturation voltage of the active device, and the selected class of the active device operation to maximize the operating efficiency and output power of the power amplifier. It should be noted that Eq. (3.14) is given in a general immittance form without indication of whether it is used in a small-signal or large-signal application. In the latter case, this only means that the device immittance W-parameters are fundamentally averaged over large signal swing across the device equivalent circuit parameters and that the conjugate-matching principle is valid in both the small-signal application and the large-signal application, where the optimum equivalent device output resistance (or conductance) is matched to the load resistance (or conductance) and the effect of the device output reactive elements is eliminated by the conjugate reactance of the load network. In addition, the matching circuits should be designed to provide the required voltage and current waveforms at the device output and the stability of operation conditions, and according to the requirements for the power amplifier amplitude and phase characteristics. The losses in the matching circuits must be as small as possible to deliver the output power to the load with maximum efficiency. Finally, it is desirable that the matching circuit be easy to tune.

3.2 SMITH CHART

The hemisphere Smith chart is one of the tools most widely used to match circuit designs because it gives a clear and simple graphical representation of the consecutive matching design procedure [2,3]. However, another hemisphere chart with a standing-wave indicator to measure the impedance ratio of a load on line where a grid of circular lines was marked with magnitude and angle of impedance, corresponding to latitude and longitude on a hemisphere, was simultaneously proposed by Carter [4,5]. The Smith chart can be applied for matching using both lumped elements and transmission lines. The Smith chart is particularly useful for matching circuit designs that use the transmission lines because analytical calculations in this case are very complicated. Also, when using the complete Smith chart, the circuit parameters such as voltage standing wave ratio (*VSWR*), reflection coefficient, return loss, or losses in the transmission line can be directly calculated. The Smith chart is a very important tool, being a part of modern computer-aided design software and test equipment and providing a useful visual way to understand the circuit behavior.

The Smith chart represents a relationship between the load impedance Z and the reflection coefficient Γ that can be written in the normalized form of

$$\frac{Z}{Z_0} = \frac{1+\Gamma}{1-\Gamma} \tag{3.15}$$

where the normalized impedance Z/Z_0 can also be written as

$$\frac{Z}{Z_0} = \frac{R}{Z_0} + j\frac{X}{Z_0} \tag{3.16}$$

and represents the reflection coefficient Γ through its real and imaginary parts as

$$\Gamma = \Gamma_r + j\Gamma_i. \tag{3.17}$$

Then, substituting Eqs. (3.16) and (3.17) into Eq. (3.15) results in

$$\frac{R}{Z_0} + j\frac{X}{Z_0} = \frac{1 + \Gamma_r + j\Gamma_i}{1 - \Gamma_r - j\Gamma_i}. \tag{3.18}$$

By equating the real and imaginary parts of Eq. (3.18), one can obtain

$$\left(\Gamma_r - \frac{R}{R + Z_0}\right)^2 + \Gamma_i^2 = \left(\frac{Z_0}{R + Z_0}\right)^2 \tag{3.19}$$

$$(\Gamma_r - 1)^2 + \left(\Gamma_i - \frac{Z_0}{X}\right)^2 = \left(\frac{Z_0}{X}\right)^2. \tag{3.20}$$

As a result, in the Γ_r–Γ_i coordinate plane, Eq. (3.19) represents a family of circles centered at points $\Gamma_r = R/(R + Z_0)$ and $\Gamma_i = 0$ with radii of $Z_0/(R + Z_0)$, which are called *constant-(R/Z_0) circles*. Eq. (3.20) represents a family of circles at points $\Gamma_r = 1$ and $\Gamma_i = Z_0/X$ with radii of Z_0/X, which are called *constant-(X/Z_0) circles*. These constant-(R/Z_0) and constant-(X/Z_0) circles with different normalized parameters are shown in Figure 3.3(*a*), where the points $\Gamma_r = -1$ and $\Gamma_r = 1$ are also indicated. The plot of such circles is called the *impedance Smith chart* or the *Z Smith chart*. The curve from the point *A* to the point *C* represents the impedance transformation from the pure resistance of 25 Ω to the inductive impedance of (25 + *j*25) Ω, which can be provided by using the inductance connected in series with the resistance.

Eqs. (3.19) and (3.20) can be rewritten easily in the admittance form with the real part *G* and imaginary part *B* when a relationship between the impedance *Y* and the reflection coefficient Γ can be written as

$$\frac{G}{Y_0} + j\frac{B}{Y_0} = \frac{1 - \Gamma_r - j\Gamma_i}{1 + \Gamma_r + j\Gamma_i} \tag{3.21}$$

where $Y_0 = 1/Z_0$. Then,

$$\left(\Gamma_r + \frac{G}{G + Y_0}\right)^2 + \Gamma_i^2 = \left(\frac{Y_0}{G + Y_0}\right)^2 \tag{3.22}$$

$$(\Gamma_r + 1)^2 + \left(\Gamma_i + \frac{Y_0}{B}\right)^2 = \left(\frac{Y_0}{B}\right)^2. \tag{3.23}$$

From Eqs. (3.22) and (3.23) it follows that the constant-(G/Y_0) circles are centered at $\Gamma_r = -G/(G + Y_0)$ and $\Gamma_i = 0$ with radii of $Y_0/(G + Y_0)$. The constant-(B/Y_0) circles are centered at points $\Gamma_r = -1$ and $\Gamma_i = -Y_0/B$ with radii of Y_0/B, which is shown in Figure 3.3(*b*). These circles are centered at antisymmetric points in contrast to the impedance Smith chart. The *admittance Smith chart* or the *Y Smith chart*, the admittances of which coincide with the appropriate impedances plotted at the *Z* Smith chart, is the mirror-reflected impedance Smith chart as a result of its rotation by 180°. The curve from the point *C* to the point *D* shows the admittance transformation from the inductive admittance of (20 − *j*20) mS to the pure conductance of 20 mS or resistance of 50 Ω, which can be provided by using the capacitance connected in parallel with the initial admittance. The impedances and admittances can be determined by any impedance or admittance Smith chart where the normalized parameters are indicated. As a general case, this diagram is called *the immittance Smith chart*. However, in this case, the impedance point and its corresponding admittance value are located one against another at the same distance from center (1, 0).

Therefore, sometimes it is advisable to use the combined impedance–admittance Smith chart shown in Figure 3.3(*c*). This is the case when, for any point, we can read the normalized impedance from the *Z* Smith chart and normalized admittance from the *Y* Smith chart. This *Z*–*Y* Smith chart

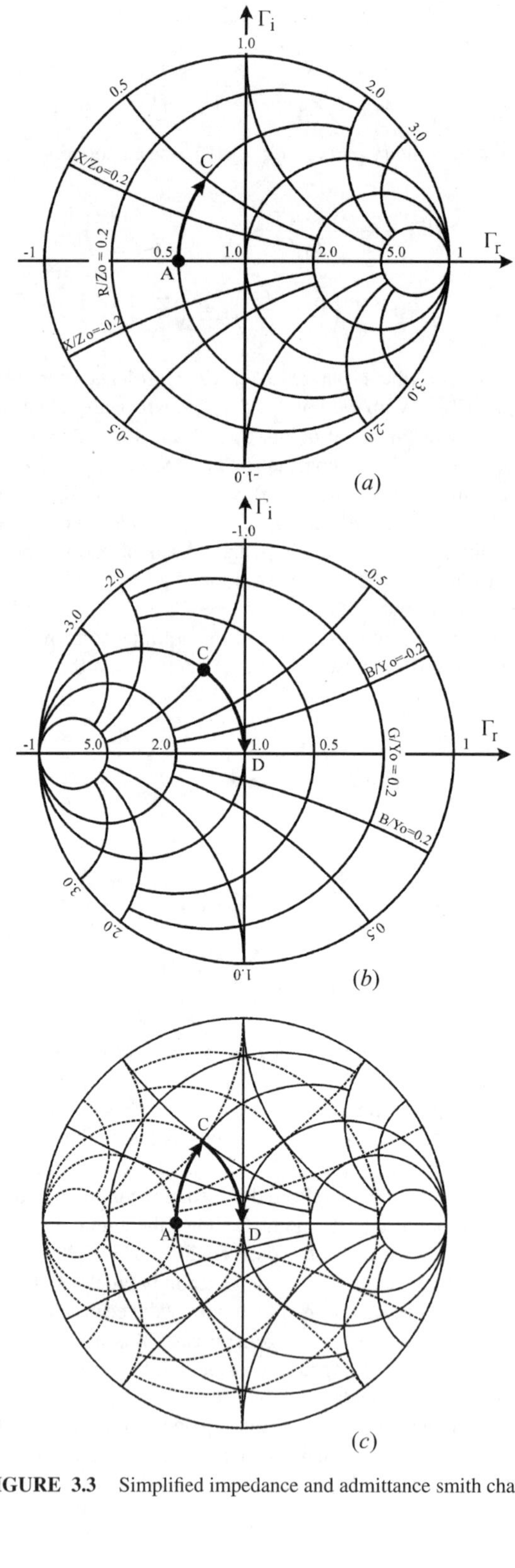

FIGURE 3.3 Simplified impedance and admittance smith charts.

avoids the necessity of impedance rotating by 180° to find the corresponding admittance. A combined impedance–admittance Smith chart is very convenient for matching using lumped elements. For example, it is necessary to convert the source active impedance of 25 Ω (point *A*) into the load resistance of 50 Ω (point *D*). First, the series inductance plotted at the *Z* Smith chart changes the source impedance by moving along constant-(X/Z_0) circle from point *A* as far as point *C*. Then, it is converted to the *Y* Smith chart. As a result, the parallel capacitance starting at this point changes the given admittance by moving along constant-(B/Y_0) circle from point *C* as far as point *D*. Such a transformation between these two resistances for normalizing impedance $Z_0 = 50$ Ω is shown in Figure 3.3(*c*).

When designing RF and microwave power amplifiers, it is very important to determine such parameters as *VSWR*, reflection coefficient, or losses in the matching circuits based on the lumped elements or using the transmission lines. The linear reference scales, indicating these additional characteristics, are added underneath the Smith chart, as shown in Figure 3.4. Scales around its

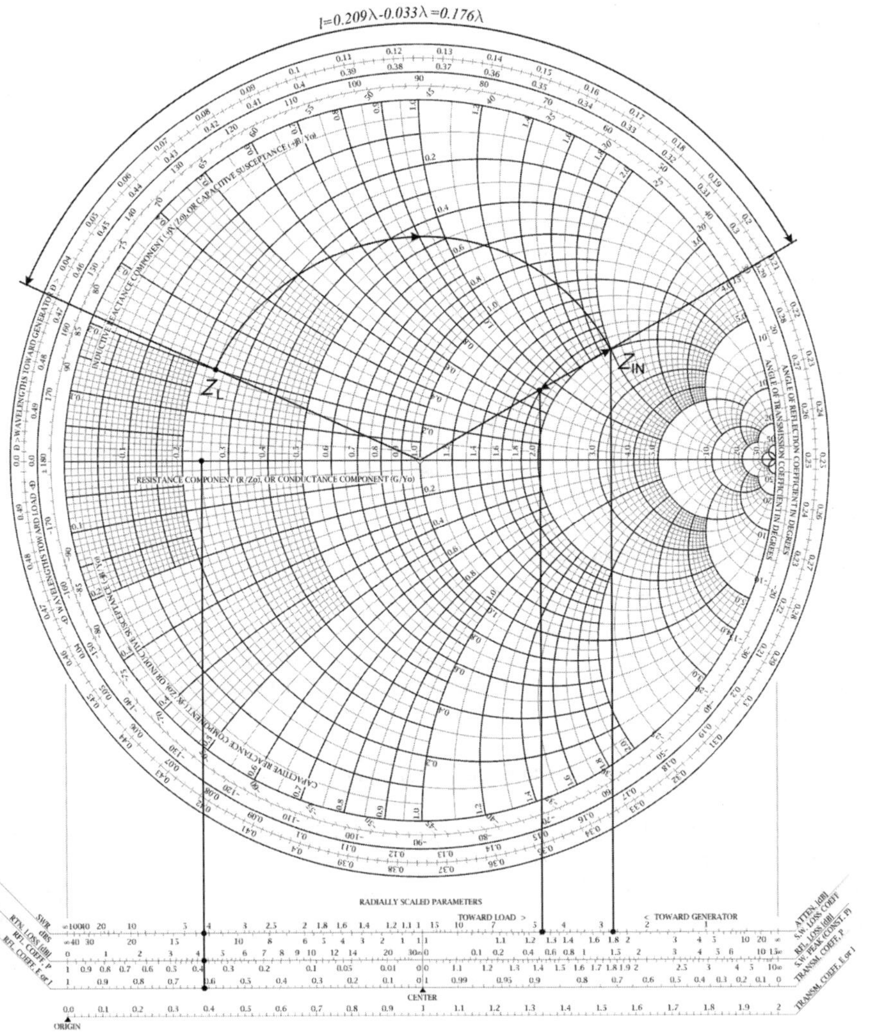

FIGURE 3.4 Smith chart.

periphery show the calibrated electrical wavelength and the angles of the reflection coefficients. The Smith chart can be easily implemented for the graphical design using the transmission lines. Expression for the input impedance of the lossless transmission line can be written in terms of the reflection coefficient as

$$\frac{Z_{in}}{Z_0} = \frac{1 + \Gamma \exp(-2j\theta)}{1 - \Gamma \exp(-2j\theta)}. \tag{3.24}$$

This equation differs from Eq. (3.15) by the added phase angle only. This means that the normalized input impedance, seen looking into the transmission line with electrical length θ, can be found by rotating this impedance point clockwise by 2θ about the center of the Smith chart with the same radius $|\Gamma|$. By subtracting 2θ from the phase angle of the reflection coefficient, its value decreases in the clockwise direction, according to the periphery scale. In this case, the half-wave transmission line provides a clockwise rotation of 2π or 360° about the center, returning the point to its original position.

For example, a typical 50 Ω transmission line provides a transformation of load impedance from $Z_L = (12 + j10)$ Ω to a new impedance of $Z_{in} = (100 + j100)$ Ω. The normalized load impedance is $Z_L/Z_0 = (0.24 + j0.2)$ Ω, which is plotted on the Smith chart, as shown in Figure 3.4. By using a compass to draw the circle from this point to the intersection point with a real axis, we obtain $|\Gamma| = 0.61$ at the reflection coefficient scale, $RL = 4.3$ dB at the return loss scale and $VSWR = 4.2$ at the standing wave ratio (*SWR*) scale. To determine the angle of the reflection coefficient, it needs to draw a radial line through the load impedance point to the intersection of the periphery circle (an angle of 83° can be read). If a radial line is drawn through the point of the input impedance at the outer wavelength scale, the difference between points of 0.209λ and 0.033λ gives the length of the transmission line as 0.176λ. However, in the case of the transmission line with losses of 2 dB, the point obtained should be moved to the point of $Z_{in} = (90 + j40)$ Ω, according to the attenuation scale.

3.3 MATCHING WITH LUMPED ELEMENTS

3.3.1 Analytic Design Technique

Generally, there is a variety of configurations for matching networks that can be used to connect a generating system efficiently to its useful load, but in order to obtain high transmission efficiency, any of these networks should be properly designed. The lumped matching circuits in the form of (*a*) *L*-transformer, (*b*) π-transformer, or (*c*) *T*-transformer shown in Figure 3.5 have proved for a long time to be effective for power amplifier design [6]. The simplest and most popular matching circuit is the matching circuit in the form of the *L*-transformer. The transforming properties of this matching circuit can be analyzed by using the equivalent transformation of a parallel into a series representation of *RX* circuit [7].

Consider the parallel *RX* circuit shown in Figure 3.6(*a*), where R_1 is the real (resistive) and X_1 is the imaginary (reactive) parts of the circuit impedance. $Z_1 = jX_1R_1/(R_1 + jX_1)$ and the series *RX* circuit shown in Figure 3.6(*b*), where R_2 is the resistive and X_2 is the reactive parts of the circuit impedance $Z_2 = R_2 + jX_2$. These two circuits, series and parallel, can be considered equivalent at some frequency if $Z_1 = Z_2$ resulting in

$$R_2 + jX_2 = \frac{R_1X_1^2}{R_1^2 + X_1^2} + j\frac{R_1^2X_1}{R_1^2 + X_1^2}. \tag{3.25}$$

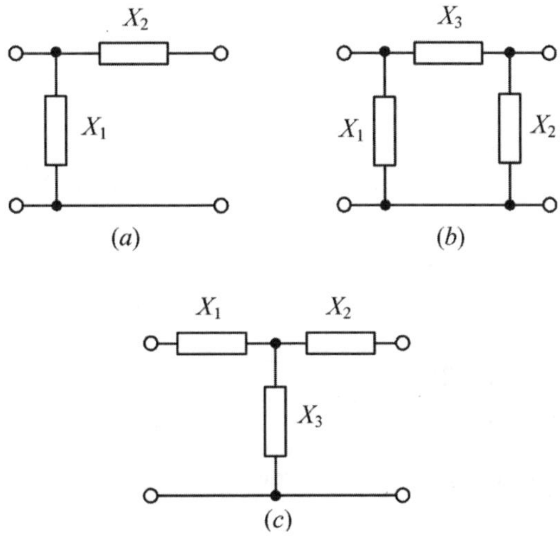

FIGURE 3.5 Matching circuits in the form of (*a*) *L*-, (*b*) π- and (*c*) *T*-transformers.

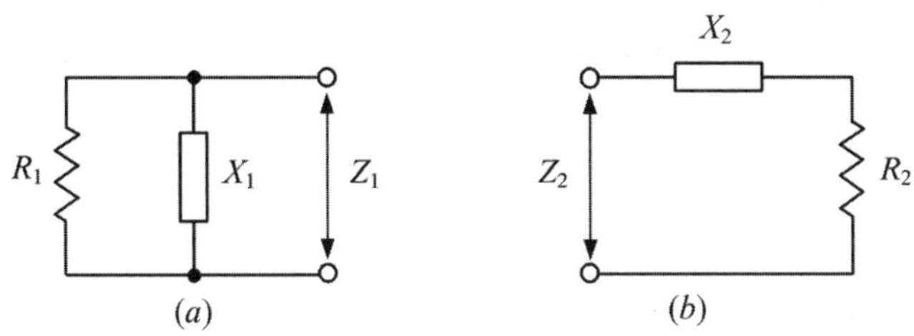

FIGURE 3.6 Impedance (*a*) parallel and (*b*) series equivalent circuits.

Eq. (3.25) can be rearranged by two separate equations for real and imaginary parts in the form of

$$R_1 = R_2\left(1 + Q^2\right) \tag{3.26}$$

$$X_1 = X_2\left(1 + Q^{-2}\right) \tag{3.27}$$

where $Q = R_1/|X_1| = |X_2|/R_2$ is the circuit quality factor, which is equal for both the series and parallel *RX* circuits.

Consequently, if the reactive impedance $X_1 = -X_2\left(1 + Q^{-2}\right)$ is connected in parallel to the series circuit R_2X_2, it allows the reactive impedance (or reactance) of the series circuit to be compensated. In this case, the input impedance of such a two-port network shown in Figure 3.7 will be resistive

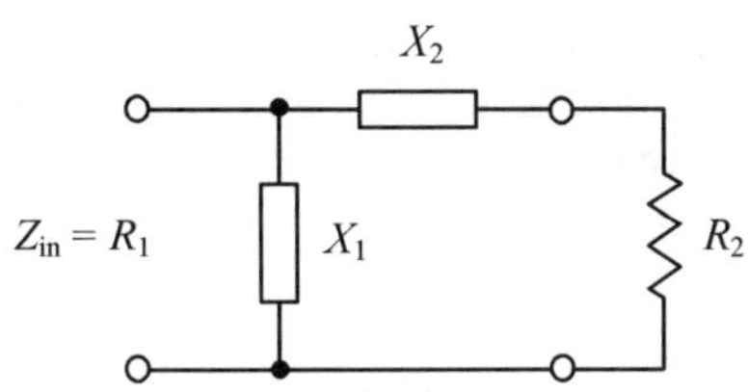

FIGURE 3.7 Input impedance of two-port network.

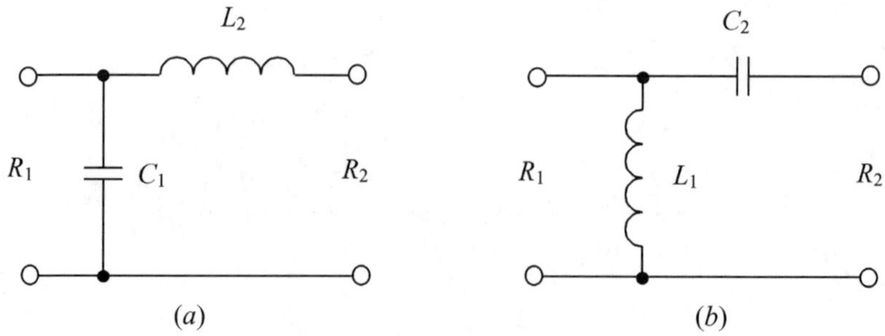

$$\left.\begin{array}{l}\omega C_1 = Q/R_1\\ \omega L_2 = Q R_2\end{array}\right\} \quad Q = \sqrt{\frac{R_1}{R_2} - 1} \quad \left\{\begin{array}{l}\omega L_1 = R_1/Q\\ \omega C_2 = 1/(Q R_2)\end{array}\right.$$

FIGURE 3.8 *L*-type matching circuits and relevant equations.

only and equal to R_1. Consequently, to transform the resistance R_1 into the other resistance R_2 at the given frequency, it is sufficient to connect between them a two-port *L*-transformer with the opposite signs of the reactances X_1 and X_2, the parameters of which can be easily calculated from the following simple equations:

$$|X_1| = \frac{R_1}{Q} \tag{3.28}$$

$$|X_2| = R_2 Q \tag{3.29}$$

where

$$Q = \sqrt{\frac{R_1}{R_2} - 1} \tag{3.30}$$

is the circuit quality factor expressed through the resistances to be matched. Thus, to design a matching circuit with fixed resistances to be matched, first we need to calculate the circuit quality factor Q according to Eq. (3.30) and then to define the reactive elements, according to Eqs. (3.28) and (3.29).

Due to the opposite signs of the reactances X_1 and X_2, the two possible circuit configurations (one in the form of a low-pass filter section and another in the form of a high-pass filter section) with the same transforming properties can be realized, as shown in Figure 3.8 together with the design equations.

The lumped matching circuits in the form of the *L*-transformer loaded on the resistance R_2 can also be considered as the parallel resonant circuit shown in Figure 3.9. In this case, the series inductance L_2 and series resistance R_2 transform to the parallel inductance L_2' and parallel resistance R_2', respectively, which become the frequency-dependent functions:

$$R_1 = R_2' = R_2\left(1 + Q^2\right) \tag{3.31}$$

$$L_2' = L_2\left(1 + Q^{-2}\right) \tag{3.32}$$

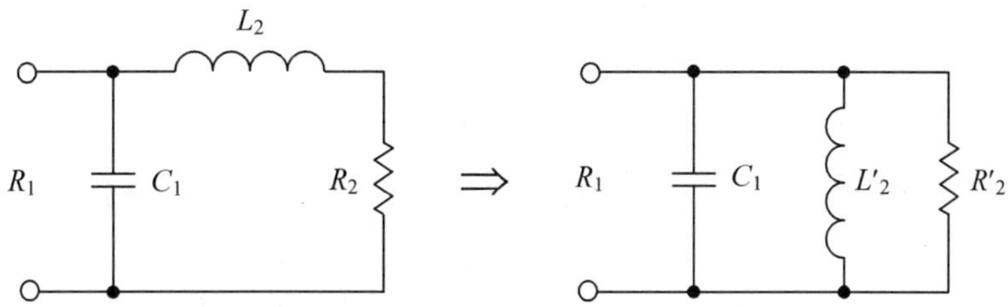

FIGURE 3.9 Parallel resonant circuit equivalent of loaded L-transformer.

where $Q = \omega L_2/R_2$. The resonant frequency f_0 of such an equivalent parallel resonant circuit is determined from

$$\omega_0 = 2\pi f_0 = \sqrt{\frac{1}{L_2 C_1} - \left(\frac{R_2}{L_2}\right)^2}. \tag{3.33}$$

If this matching circuit has small values of Q, wider frequency bandwidth but poor out-of-band suppression can be achieved. However, with large values of Q, the frequency bandwidth is substantially reduced. For the case of $R_1/R_2 \geq 10$, which corresponds to the condition of $Q \geq 3$, the frequency bandwidth Δf and out-of-band suppression factor F_n of such an L-transformer can be approximately evaluated by the same formulas as for a parallel resonant circuit:

$$\Delta f \cong \frac{f_0}{Q} \tag{3.34}$$

$$F_n \cong Q\left(n^2 - 1\right) \tag{3.35}$$

where f_0 is the operating frequency and n is the harmonic number [8]. The out-of-band suppression factor F_n of the matching circuit represents the ratio of the selected harmonic component in the input (collector) current to the same harmonic component in the output (load) current. Figure 3.10 shows the frequency behavior of the input impedance magnitude $|Z_{in}|$ of the parallel resonant circuit.

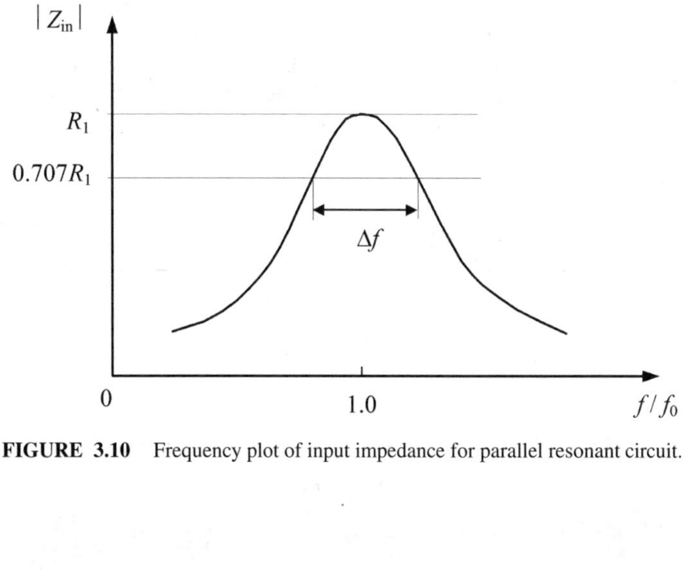

FIGURE 3.10 Frequency plot of input impedance for parallel resonant circuit.

The transformer efficiency η_T is determined by the ratio of P_L/P_{in}, where P_{in} is the power at the input of the transformer and P_L is the load transformer power. The efficiency for the L-transformer with negligible losses in the capacitor can be calculated from

$$\eta_T \cong 1 \bigg/ \left(1 + \frac{Q}{Q_{ind}}\right) \tag{3.36}$$

where Q_{ind} is the inductor quality factor. From Eq. (3.36) it follows that, with the increase of Q, the efficiency of the impedance transformer decreases. This means that, for the same R_1 and series parasitic resistance of the circuit inductance, the lower resistance R_2 provides the higher current flowing through this inductance, which leads to an additional power dissipation. An analysis of Eqs. (3.28) and (3.29) shows that, for the given resistances R_1 and R_2, each element of the L-transformer can have only one value for a fixed frequency. Consequently, it is difficult to satisfy simultaneously such contradictory requirements as efficiency, frequency bandwidth and out-of-band suppression.

To avoid the parasitic low-frequency oscillations and to increase the level of the harmonic suppression, it may be necessary to connect an additional L_0C_0 series circuit, which resonant frequency is equal to the operating frequency of the power amplifier, as shown in Figure 3.11. In this case, the out-of-band suppression factor F_n defined by Eq. (3.35) is written as

$$F_n \cong Q_\Sigma \left(n^2 - 1\right) \tag{3.37}$$

where $Q_\Sigma = Q + Q_0$ and $Q_0 = \omega L_0/R_2$. Better harmonic suppression is achieved at the expense of the frequency bandwidth narrowing.

In practice, it makes sense to use the single two-port L-transformers in power amplifiers as the interstage matching circuits, where the requirements for out-of-band suppression and efficiency are not as high as for the output matching circuits. In this case, the main advantage of such an L-transformer is in its simplicity when the only two reactive elements with fast tuning are needed. For larger values of $Q \geq 10$, it is possible to use a cascade connection of L-transformers, which allows wider frequency bandwidth and transformer efficiency to be realized.

The matching circuits in the form of (*a*) π-transformer and (*b*) T-transformer can be realized by appropriate connection of two L-transformers, as shown in Figure 3.12. For each L-transformer, the resistances R_1 and the resistance R_2 are transformed to some intermediate resistance R_0 with the value of $R_0 < (R_1, R_2)$ for a π-transformer and the value of $R_0 > (R_1, R_2)$ for a T-transformer. The value of R_0 is not fixed and can be chosen arbitrary depending on the frequency bandwidth. This means that, compared to the simple L-transformer with fixed parameters for the same ratio of R_2/R_1, the parameters of the π-transformer or T-transformer can be different. However, they provide narrower frequency bandwidths due to higher quality factors because the intermediate resistance R_0 is either greater or smaller than each of the resistances R_1 and R_2. By taking into account the two possible

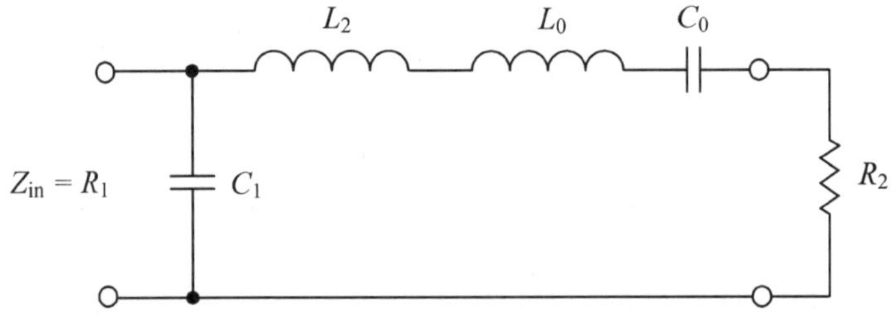

FIGURE 3.11 *L*-transformer with additional *LC*-filter.

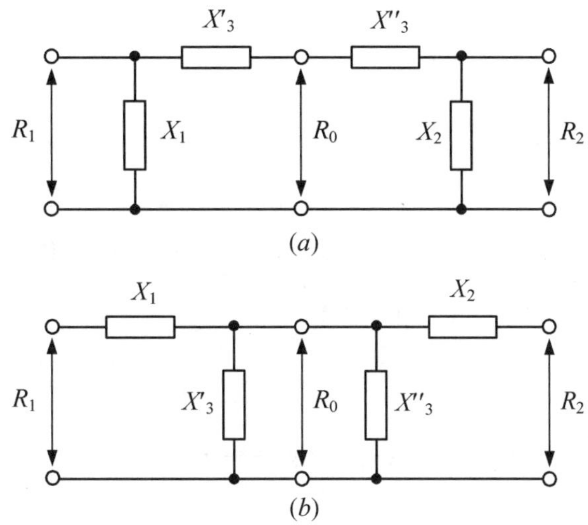

FIGURE 3.12 Matching circuits developed by connecting two L-transformers.

circuit configurations of the L-transformer shown in Figure 3.8, it is possible to develop the different circuit configurations of such two-port transformers shown in Figure 3.12(a), where $X_3 = X_3' + X_3''$, and in Figure 3.12(b), where $X_3 = X_3'X_3''/\left(X_3' + X_3''\right)$.

Several of the most widely used two-port π-transformers, together with the design formulas, are shown in Figure 3.13 [7,9]. The π-transformers are usually used as output matching circuits of high-power amplifiers in class B operation when it is necessary to achieve a sinusoidal drain (or collector) voltage waveform by appropriate harmonic suppression. In addition, it is convenient to use some of them as interstage matching circuits in low-power and medium-power amplifiers when it is necessary to provide sinusoidal voltage waveforms both at the drain (or collector) of the previous transistor and at the gate (or base) of the subsequent transistor. In this case, for the π-transformer with shunt capacitors, the input and output capacitances of these transistors can be easily included into the matching circuit elements C_1 and C_2, respectively. Finally, a π-transformer can be directly used as the load network for a high-efficiency class E mode with proper calculation of its design parameters.

Figure 3.14 shows the π-transformer with shunt capacitor and additional L_0C_0 series circuit in which elements are defined from

$$\omega L_0 = \frac{1}{\omega C_0} = \frac{R_1 Q_0}{1 + Q_1^2} \tag{3.38}$$

where Q_0 is the arbitrary value depending on the specification requirements to the harmonic suppression. The other elements are defined by the design equations given in Figure 3.13(a).

The π-transformer with the two shunt capacitors, shown in Figure 3.13(a), represents a face-to-face connection of two simple low-pass L-transformers. As a result, there is no special requirement for the resistances R_1 and R_2 in which ratio R_1/R_2 can be greater or smaller than unity. As an example, the design equations correspond to the case of $R_1/R_2 > 1$. However, as it will be further derived, the π-transformer with a series capacitor shown in Figure 3.13(b) can only be used for impedance matching when $R_1/R_2 > 1$. Such a π-transformer represents a face-to-face connection of the high-pass and low-pass L-transformers, as shown in Figure 3.15(a).

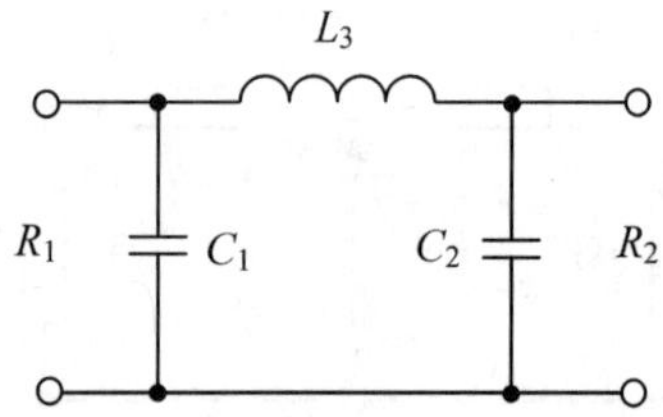

$$\omega C_1 = Q_1 / R_1 \qquad \omega C_2 = Q_2 / R_2 \qquad \omega L_3 = R_1(Q_1 + Q_2)/(1 + Q_1^2)$$

$$Q_2 = \sqrt{\frac{R_2}{R_1}\left(1 + Q_1^2\right) - 1} \qquad Q_1^2 > \frac{R_1}{R_2} - 1$$

(*a*)

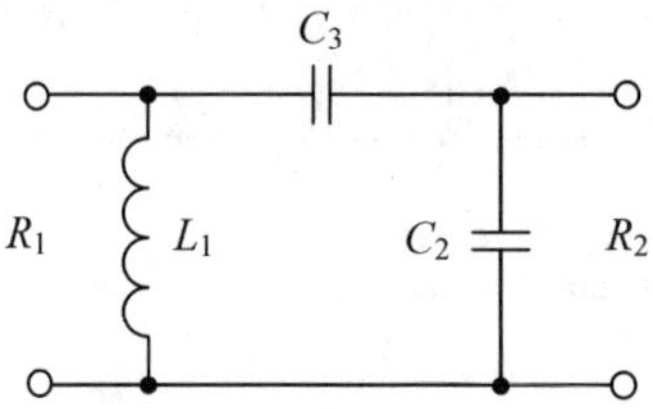

$$\omega L_1 = R_1 / Q_1 \qquad \omega C_2 = Q_2 / R_2 \qquad \omega C_3 = (1 + Q_2^2)/[R_2(Q_1 - Q_2)]$$

$$Q_2 = \sqrt{\frac{R_2}{R_1}\left(1 + Q_1^2\right) - 1} \qquad Q_1^2 > \frac{R_1}{R_2} - 1$$

(*b*)

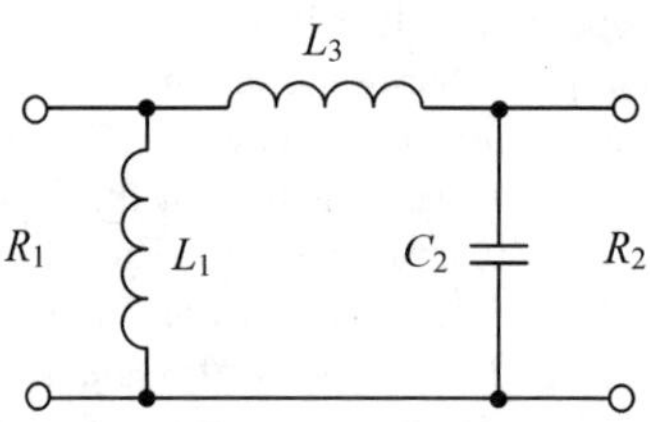

$$\omega L_1 = R_1 / Q_1 \qquad \omega C_2 = Q_2 / R_2 \qquad \omega L_3 = R_2(Q_2 - Q_1)/(1 + Q_2^2)$$

$$Q_1 = \sqrt{\frac{R_1}{R_2}\left(1 + Q_2^2\right) - 1} \qquad Q_2^2 > \frac{R_2}{R_1} - 1$$

(*c*)

FIGURE 3.13 π-transformers and relevant equations.

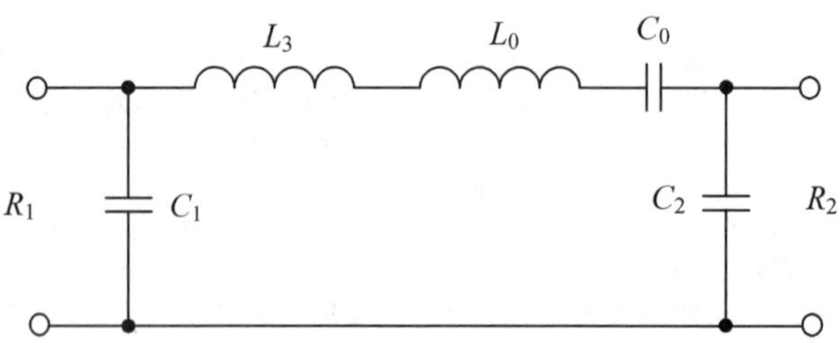

FIGURE 3.14 π-transformer with additional *LC*-filter.

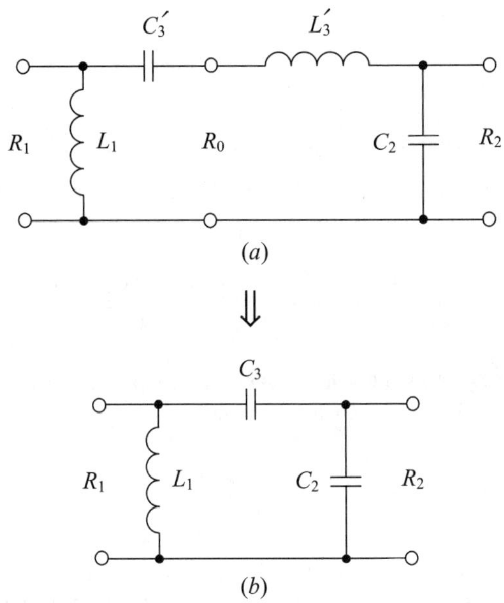

FIGURE 3.15 π-transformer with series capacitor.

The design equations for a high-pass section are written using Eqs. (3.28) to (3.30) as

$$\omega L_1 = \frac{R_1}{Q_1} \tag{3.39}$$

$$\omega C_3' = \frac{1}{Q_1 R_0} \tag{3.40}$$

$$Q_1^2 = \frac{R_1}{R_0} - 1 \tag{3.41}$$

where Q_1 is the circuit quality factor and R_0 is the intermediate resistance.

Similarly, for a low-pass section,

$$\omega C_2 = \frac{Q_2}{R_2} \tag{3.42}$$

$$\omega L_3' = Q_2 R_0 \tag{3.43}$$

$$Q_2^2 = \frac{R_2}{R_0} - 1. \tag{3.44}$$

Since it is assumed that $R_1 > R_2 > R_0$, from Eq. (3.41) it follows that the quality factor Q_1 of a high-pass L-transformer can be chosen from the condition of

$$Q_1^2 > \frac{R_1}{R_2} - 1. \tag{3.45}$$

Substituting Eq. (3.41) into Eq. (3.44) results in

$$Q_2 = \sqrt{\frac{R_2}{R_1}\left(1 + Q_1^2\right) - 1}. \tag{3.46}$$

Combining the reactances of two series elements (capacitor C_3' and inductor L_3') given by Eqs. (3.40) and (3.43) yields

$$\omega L_3' - \frac{1}{\omega C_3'} = R_0\left(Q_2 - Q_1\right) = \frac{R_2\left(Q_2 - Q_1\right)}{\left(1 + Q_2^2\right)}. \tag{3.47}$$

As a result, since $Q_1 > Q_2$, the total series reactance is negative, which can be provided by a series capacitance C_3, as shown in Eq. 3.15(b), with susceptance

$$\omega C_3 = \frac{1 + Q_2^2}{R_2\left(Q_1 - Q_2\right)}. \tag{3.48}$$

On the other hand, if $Q_2 > Q_1$ when $R_2 > R_1 > R_0$, then the total series reactance is positive, which can be provided by a series inductance L_3, and all matching circuit parameters can be calculated according to the design equations given in Figure 3.13(c). In this case, it needs first to choose the value of Q_2 for fixed resistances R_1 and R_2 to be matched, then to calculate the value of Q_1, and finally the values of the shunt inductance L_1, shunt capacitance C_2, and series inductance L_3.

Some of the matching circuit configurations of two-port T-transformers, together with the design formulas, are shown in Figure 3.16 [7,9]. The T-transformers are usually used in the high-power amplifiers as input, interstage, and output matching circuits, especially the matching circuit with shunt and series capacitors shown in Figure 3.16(b). In this case, if a high value of the inductance L_2 is chosen, then the current waveform at the input of the transistor with a small input resistive part will be close to sinusoidal. By using such a T-transformer for the output matching of a power amplifier, it is easy to realize a high-efficiency class F operation mode, because the series inductance connected to the drain (or collector) of the active device creates an open-circuit harmonic impedance conditions.

The T-transformer with the two-series inductors, shown in Figure 3.16(a), represents a back-to-back connection of two simple low-pass L-transformers. In this case, the resistance ratio R_1/R_2 can be greater or smaller than unity, similar to a π-transformer with the two shunt capacitors shown in Figure 3.13(a). As an example, the design equations correspond to the case of $R_1/R_2 > 1$. However, the T-transformer with series and shunt capacitors shown in Figure 3.16(b) can only be used for impedance matching when $R_1/R_2 > 1$. Such a T-transformer represents a back-to-back connection of the high-pass and low-pass L-transformers, as shown in Figure 3.17(a).

The design equations for a high-pass section of such a T-transformer are written using Eqs. (3.28) to (3.30) as

$$\omega C_1 = \frac{1}{R_1 Q_1} \tag{3.49}$$

$$\omega L_3' = \frac{R_0}{Q_1} \tag{3.50}$$

$$Q_1^2 = \frac{R_0}{R_1} - 1 \tag{3.51}$$

where Q_1 is the circuit quality factor and R_0 is the intermediate resistance.

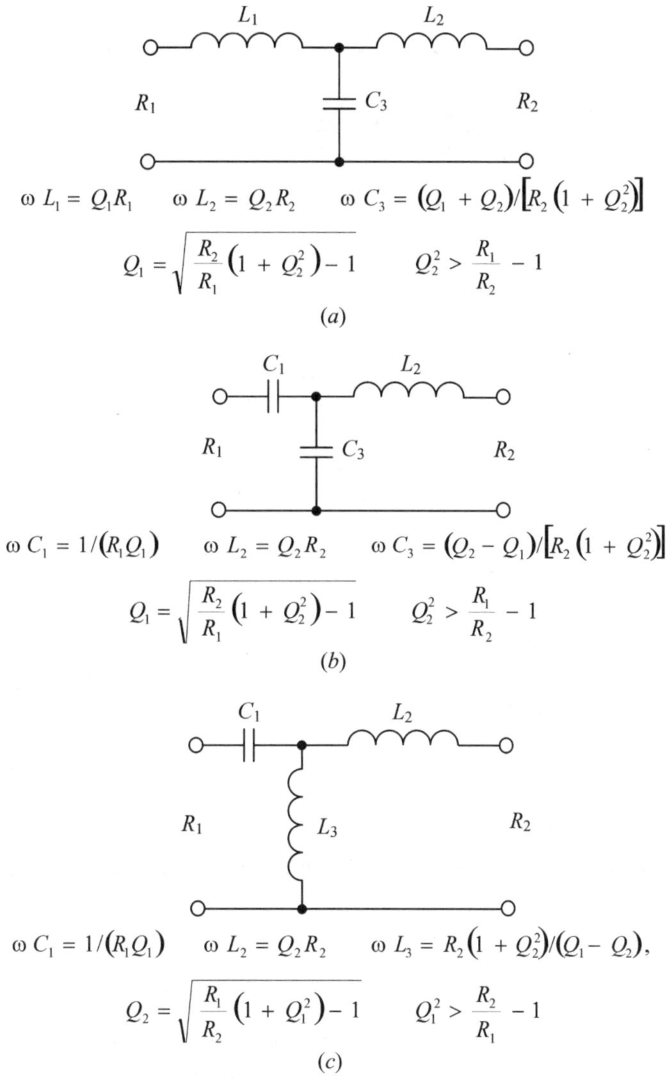

FIGURE 3.16 *T*-transformers and relevant equations.

Similarly, for a low-pass section,

$$\omega L_2 = Q_2 R_2 \tag{3.52}$$

$$\omega C_3' = Q_2 / R_0 \tag{3.53}$$

$$Q_2^2 = \frac{R_0}{R_2} - 1. \tag{3.54}$$

Since it is assumed that $R_0 > R_1 > R_2$, from Eq. (3.54) it follows that the quality factor Q_2 of a low-pass *L*-transformer can be chosen from the condition of

$$Q_2^2 > \frac{R_1}{R_2} - 1. \tag{3.55}$$

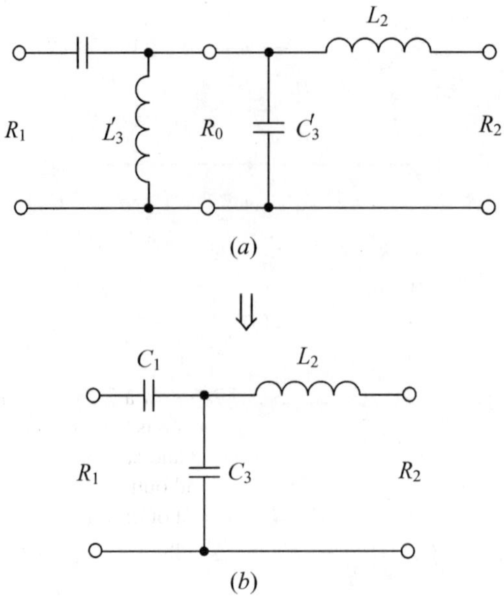

FIGURE 3.17 *T*-transformer with series and shunt capacitors.

Substituting Eq. (3.54) into Eq. (3.51) results in

$$Q_1 = \sqrt{\frac{R_2}{R_1}\left(1 + Q_2^2\right) - 1}. \tag{3.56}$$

Combining the susceptances of two shunt elements (inductor L_3' and capacitor C_3') given by Eqs. (3.50) and (3.53) yields

$$\omega C_3' - \frac{1}{\omega L_3'} = \frac{Q_2 - Q_1}{R_0} = \frac{Q_2 - Q_1}{R_2\left(1 + Q_2^2\right)}. \tag{3.57}$$

As a result, since $Q_2 > Q_1$, the total shunt susceptance is positive, which can be provided by a shunt capacitance C_3, as shown in Figure 3.17(*b*), with susceptance

$$\omega C_3 = \frac{Q_2 - Q_1}{R_2\left(1 + Q_2^2\right)}. \tag{3.58}$$

On the other hand, if $Q_1 > Q_2$ when $R_0 > R_2 > R_1$, then the total shunt susceptance is negative that can be provided by a shunt inductance L_3, and all matching circuit parameters can be calculated according to the design equations given in Figure 3.16(*c*). In this case, it needs first to choose the value of Q_1 for fixed resistances R_1 and R_2 to be matched, then to calculate the value of Q_2, and finally the values of the series capacitance C_1, series inductance L_2, and shunt inductance L_3.

If the elements of π- and T-transformers are chosen according to the condition $X_1 = X_2 = -X_3$, then the input impedance Z_{in} of the transformer loaded by the resistance R_L (from any side) is equal to

$$Z_{in} = R_{in} = \frac{X^2}{R_L} \tag{3.59}$$

where $X = |X_i|$, $i = 1, 2, 3$. As a result, the input impedance Z_{in} will be resistive, regardless of the value of the load resistance R_L. For example, setting $R_L = R_2$ for the transformers shown in Figure 3.13(*a*) and Figure 3.16(*a*) yields

$$R_1 = \frac{X^2}{R_2}. \tag{3.60}$$

When $X_1 \neq X_2 \neq -X_3$, the input impedances of the π- and T-transformers will be resistive for only one particular value of R_L.

3.3.2 Bipolar UHF Power Amplifier

Consider a design example of a 10 W 300 MHz bipolar power amplifier with supply voltage 12.5 V providing a power gain of at least 10 dB. The first design step is to select an appropriate active device that allows both simplifying the circuit design procedure and satisfying the specified requirements. Usually, the manufacturer states the values of the input and output impedances or admittances at the nominal operation point on the data sheet for the device. For example, the above requirements can be realized by an *n–p–n* silicon bipolar transistor operating at 300 MHz with the input impedance $Z_{in} = (1.3 + j0.9)\ \Omega$ and output admittance $Y_{out} = (150 - j70)$ mS, which is intended for power amplification in class AB with nominal supply voltages up to 13.5 V.

In this case, Z_{in} is expressed as a series combination of the transistor input resistance and inductive reactance, whereas Y_{out} is represented by a parallel combination of the transistor output resistance and inductive reactance. This means that, at required operating frequency, effect of the series parasitic collector lead inductance exceeds the effect of the shunt collector capacitance, which gives a net inductive reactance to the equivalent output circuit of an active device. To match the series input inductive impedance to the standard 50 Ω input source impedance, it is advisable to use a matching circuit in the form of a T-transformer shown in Figure 3.16(*b*), where the series capacitance C_1 can serve also as a blocking capacitor. Figure 3.18 shows the complete input network including input device elements and matching circuit. From the reactive part of the input impedance Z_{in} it follows that, at the operating frequency of 300 MHz, the input inductance will be equal approximately to 0.5 nH.

It is necessary first to calculate the quality factor Q_2, which is needed to determine the parameters of the matching circuit:

$$Q_2 > \sqrt{\frac{R_1}{R_{in}} - 1} = 6.1.$$

Consequently, the value of Q_2 must be larger than 6.1. For example, $Q_2 = 6.5$ provides a 3-dB bandwidth of 300 MHz/6.5 = 46 MHz. In this case, a value of Q_1 will be equal to 0.35. As a result,

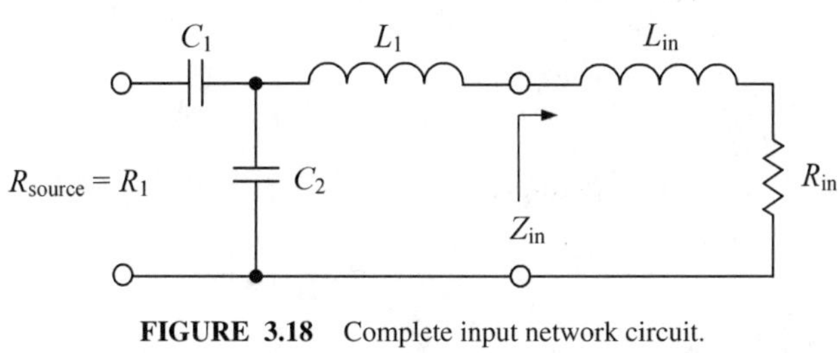

FIGURE 3.18 Complete input network circuit.

the values of the input matching circuit elements are as follows:

$$C_1 = \frac{1}{\omega Q_1 R_1} = 30\,\text{pF}$$

$$L_1 + L_{\text{in}} = \frac{Q_2 R_{\text{in}}}{\omega} = 4.5\,\text{nH} \Rightarrow L_1 = 4.0\,\text{nH}$$

$$C_2 = \frac{Q_2 - Q_1}{\omega R_{\text{in}}\left(1 + Q_2^2\right)} = 59\,\text{pF}.$$

This type of a T-transformer is widely used in practical matching circuit design because of its simplicity and convenience. In addition, the sufficiently small value of a series capacitance C_1 can contribute to the elimination of the low-frequency parasitic oscillations in the case of a multistage power amplifier. The function of each element can be graphically traced on the Smith chart as shown in Figure 3.19.

The easiest and most convenient way to plot the traces of the matching circuit elements is by plotting initially the traces of $Q_2 = 6.5$ and $Q_1 = 0.35$. The circle of equal Q is plotted, taking into account that, for each point located at this circle, a ratio of X/R or B/G must be the same. The trace of the series inductance L_1 must be plotted as far as the intersection point with Q_2-circle. This means

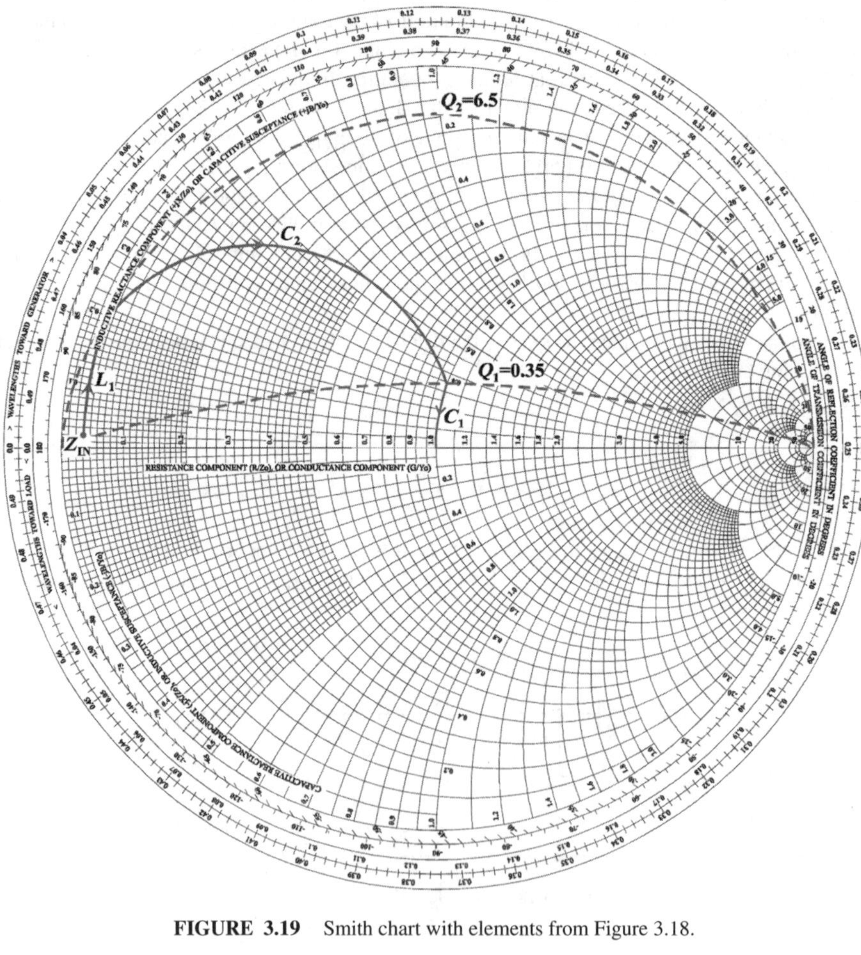

FIGURE 3.19 Smith chart with elements from Figure 3.18.

that, beginning at Z_{in}, a curve of increasing inductive reactance must be plotted up to Q_2-circle. The value of L_1 is determined from the normalized inductive impedance at this intersection point. Then, due to a normalization to 50 Ω, the chart value must be multiplied by factor 50.

The trace of the parallel capacitance C_2 must be plotted using admittance circles. The previous impedance point located at Q_2-circle is converted to its appropriate admittance one. This point is symmetrical to the impedance point regarding the center point and is located on a straight line from the intersection point drawn through the center of the Smith chart into its lower half at the same distance from the center point. A curve from this point with constant conductance and increasing capacitive susceptance is plotted. These points are transformed to appropriate impedances using a line through the center point extended at equal distance on the other side and stop when the transformed curve reaches the $Q_1 = 0.35$ circle. In other words, it is necessary to transform mentally or to use a transparent admittance Smith chart (impedance Smith chart rotated on 180°) to plot a curve for C_2 on the upper half of the impedance Smith chart. The difference between the susceptances at the beginning and the end of this curve determines the value of C_2. Then, a curve of reducing inductive reactance is plotted down to the center point to determine a value of the series capacitance C_1.

A similar design philosophy can be applied to the design of the output matching circuit shown in Figure 3.20. However, taking into account the presence of the output shunt inductance, it makes sense to use a matching circuit in the form of a π-transformer shown in Figure 3.13(*c*). The equivalent output resistance can be analytically evaluated by

$$R_{out} = \frac{\left[V_{cc} - V_{cc(sat)}\right]^2}{2P_{out}} \cong \frac{(0.9\,V_{cc})^2}{2P_{out}} \cong 6.3\ \Omega$$

where V_{cc} is the supply voltage, $V_{cc(sat)}$ is the saturation voltage, and P_{out} is the output power. Its value is very close to the measurement value given by the real part of the device output admittance calculated as $R_{out} = 1/0.15 = 6.7\ \Omega$. A value of L_{out} is approximately equal to 7.6 nH.

The quality factor Q_2 can be chosen as

$$Q_2 > \sqrt{\frac{R_2}{R_{out}} - 1} = 2.5.$$

The calculated quality factor Q_1 of the device output circuit is

$$Q_1 = \frac{R_{out}}{\omega L_{out}} = 0.47.$$

This value of L_{out} allows the output device admittance to be matched to 50 Ω load using such a π-transformer because

$$Q_2 = \sqrt{\frac{R_2}{R_{out}}\left(1 + Q_1^2\right) - 1} = 2.8 > 2.5.$$

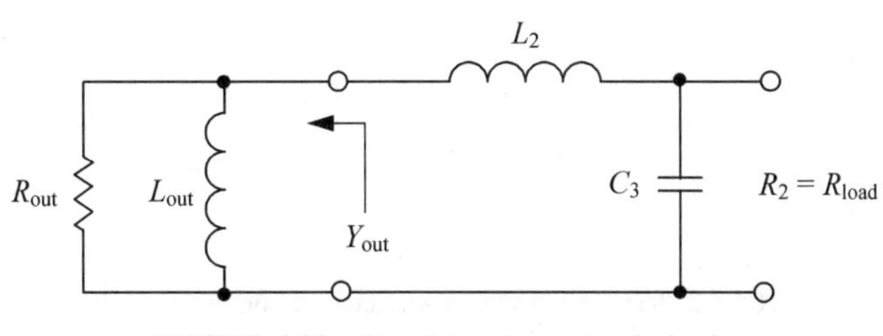

FIGURE 3.20 Complete output network circuit.

As a result, the values of the other two elements of the output matching circuit are

$$C_3 = \frac{Q_2}{\omega R_2} = 31\ \text{pF}$$

$$L_2 = \frac{R_2\,(Q_2 - Q_1)}{\omega\left(1 + Q_2^2\right)} = 6.8\ \text{nH}.$$

A blocking capacitor that performs dc supply decoupling can be connected after the π-transformer with sufficiently high value of its capacitance to avoid any negative influence on the matching circuit. Alternatively, it can be used in series with the inductor L_2, in order to form a series resonant circuit, as shown in Figure 3.11. The design of the output circuit using the Smith chart is given in Figure 3.21. Initially, it is necessary to transform the output admittance Y_{out} to the output impedance Z_{out} using the straight line of the Smith chart, putting the Z_{out} point at the same distance from the center point as for the Y_{out} point. Then, the effect of increasing series inductance L_2, by moving from the Z_{out} point along the curve of the constant R and increasing X until intersection with the $Q_2 = 2.8$ circle, is plotted. To determine the parallel capacitance C_3, it is necessary to transform this point to the corresponding

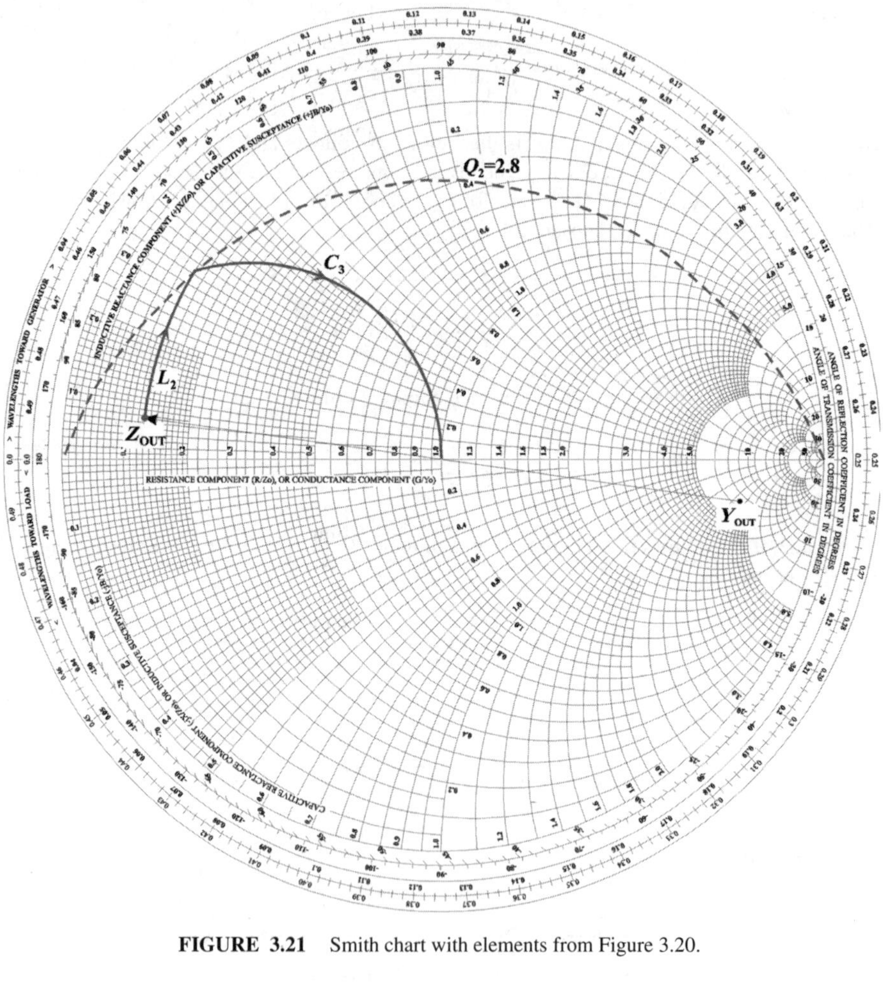

FIGURE 3.21 Smith chart with elements from Figure 3.20.

admittance one and plot the curve of the constant G and increasing B, which must intersect the center point of the Smith chart.

3.3.3 MOSFET VHF High-Power Amplifier

Now let us demonstrate lumped matching circuit technique to design a 150 W MOSFET power amplifier with the supply voltage 50 V operating in a frequency bandwidth of 132 to 174 MHz and providing a power gain greater than 10 dB. These requirements can be satisfied using a silicon n-channel enhancement mode VDMOS transistor designed for power amplification in the VHF frequency range. In this case, the center bandwidth frequency is equal to $f_c = \sqrt{132 \cdot 174} = 152$ MHz. For this frequency, the manufacturer states the following values of the input and output impedances: $Z_{in} = (0.9 - j1.2)\ \Omega$ and $Z_{out} = (1.8 + j2.1)\ \Omega$. Both Z_{in} and Z_{out} represent the series combination of an input or output resistance and a capacitive or inductive reactance, respectively. To cover the required frequency bandwidth, the low-Q matching circuits should be used that allows reduction of the in-band amplitude ripple and improvement of the input *VSWR*. The value of a quality factor for 3-dB bandwidth level must be less than $Q = 152/(174 - 132) = 3.6$. As a result, it is very convenient to design the input and output matching circuits using the simple L-transformers in the form of low-pass and high-pass filter sections with a constant value of Q [10].

To match input series capacitive impedance to the standard 50 Ω source impedance in a sufficiently wide frequency bandwidth, it is preferable to use three filter sections shown in Figure 3.22. From the negative reactive part of the input impedance Z_{in} it follows that the input capacitance at the operating frequency of 152 MHz is equal to approximately 873 pF. To compensate at the center bandwidth frequency for this capacitive reactance, it is sufficient to connect in series to it an inductance of 1.3 nH. Now, when the device input capacitive reactance is compensated, the design of the input matching circuit can proceed. To simplify the matching design procedure, it is best to cascade L-transformers with equal value of Q. Although equal Q values are not absolutely necessary, this provides a convenient guide for both analytical calculation of the matching circuit parameters and the Smith chart graphical design.

In this case, the following ratio can be written for the input matching circuit:

$$\frac{R_1}{R_2} = \frac{R_2}{R_3} = \frac{R_3}{R_{in}} \tag{3.61}$$

resulting in $R_2 = 13\ \Omega$ and $R_3 = 3.5\ \Omega$ for $R_{source} = R_1 = 50\ \Omega$ and $R_{in} = 0.9\ \Omega$. Consequently, a quality factor of each L-transformer is equal to $Q = 1.7$. The elements of the input matching circuit using the formulas given in Figure 3.8 can be calculated as $L_1 = 31$ nH, $C_1 = 47$ pF, $L_2 = 6.2$ nH, $C_2 = 137$ pF, $L_3 = 1.6$ nH, $C_3 = 509$ pF.

This equal-Q approach significantly simplifies the matching circuit design using the Smith chart. When calculating a value of Q, it is necessary to plot a circle of equal Q values on the Smith chart. Then, each element of the input matching circuit can be readily determined, as it is shown in Figure 3.23. Each trace for the series inductance must be plotted until the intersection point with

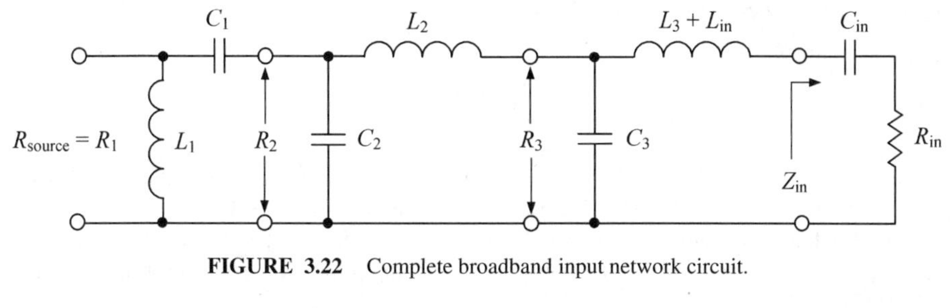

FIGURE 3.22 Complete broadband input network circuit.

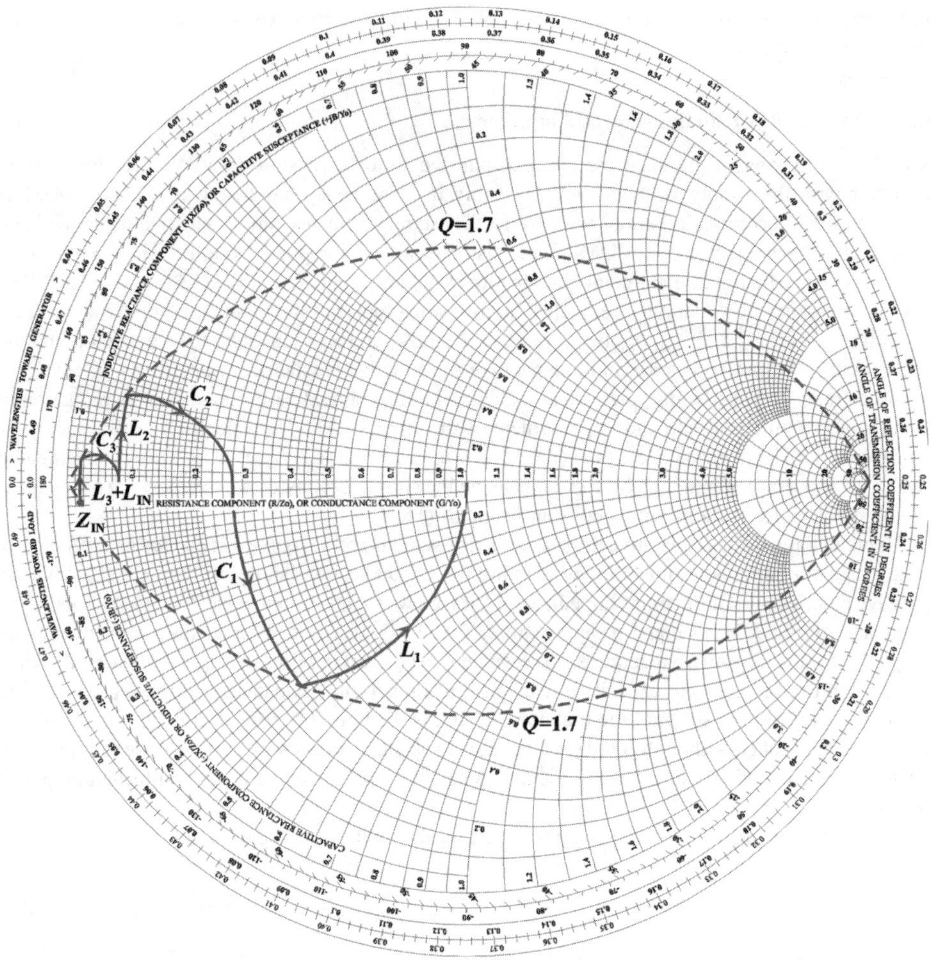

FIGURE 3.23 Smith chart with elements from Figure 3.22.

Q-circle, whereas each trace for the parallel capacitance should be plotted until intersection with horizontal real axis.

To match output series inductive impedance to the standard 50 Ω load impedance, it is sufficient to use only two filter sections, as shown in Figure 3.24. At the operating frequency of 152 MHz, the transistor series output inductance is equal to approximately 2.2 nH. This inductance can be used as a part of L-transformer in the form of a low-pass filter section. For an output matching circuit, the condition of equal-Q values gives the following ratio:

$$\frac{R_2}{R_1} = \frac{R_1}{R_{out}} \tag{3.62}$$

with the value of $R_1 = 9.5\ \Omega$ for $R_{load} = R_2 = 50\ \Omega$ and $R_{out} = 1.8\ \Omega$. Consequently, a quality factor of each L-transformer is equal to $Q = 2.1$, which is substantially smaller than a value of Q for 3-dB bandwidth level. Now it is necessary to check a value of a series inductance of the low-pass section, which must exceed the value of 2.2 nH for correct matching procedure. The appropriate calculation

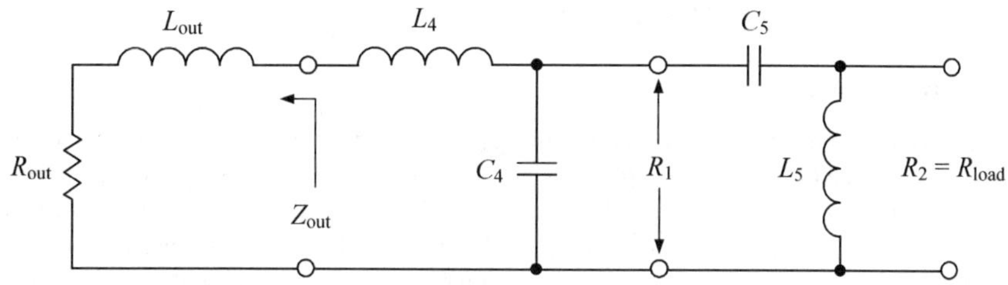

FIGURE 3.24 Complete broadband output network circuit.

gives a value of total series inductance $L_4 + L_{out}$ of approximately 4 nH. As a result, the values of the output matching circuit elements are $L_4 = 1.8$ nH, $C_4 = 231$ pF, $C_5 = 52$ pF, $L_5 = 25$ nH.

The output matching circuit design using the Smith chart with constant Q-circle is shown in Figure 3.25. For the final high-pass section, a trace for the series capacitance C_5 must be plotted until the intersection with $Q = 2.1$ circle, whereas a trace for the shunt inductance L_5 should be plotted until the intersection with the center point of the Smith chart.

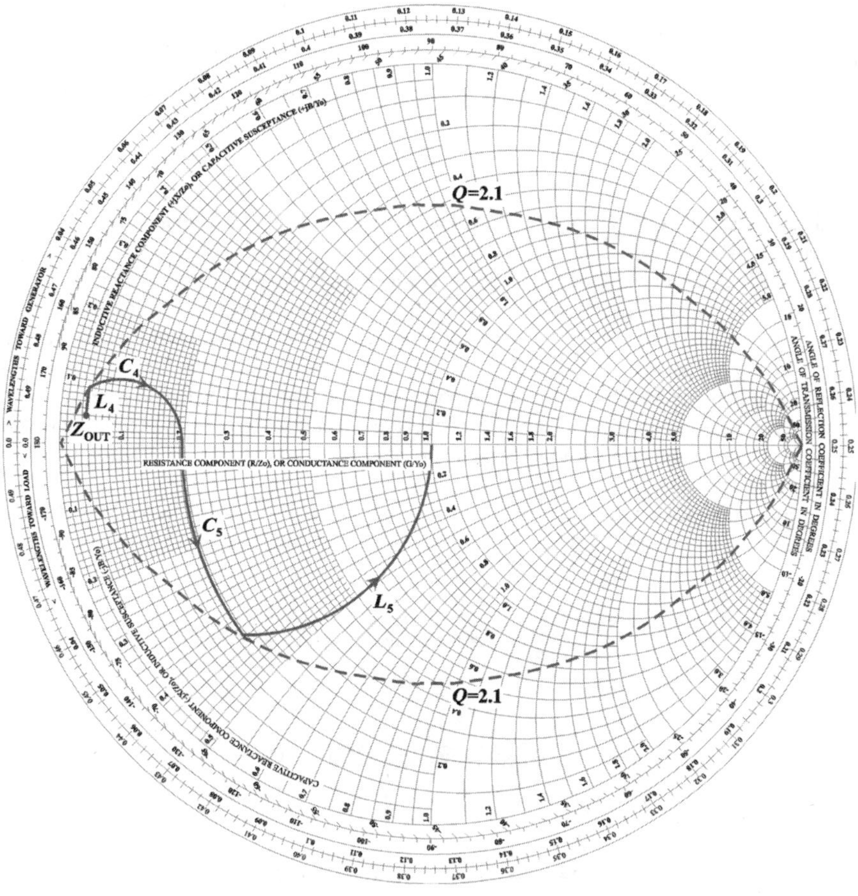

FIGURE 3.25 Smith chart with elements from Figure 3.24.

3.4 MATCHING WITH TRANSMISSION LINES

3.4.1 Analytic Design Technique

At very high frequencies, it is very difficult to implement lumped elements with predefined accuracy in view of a significant effect of their parasitic parameters, for example, the parasitic interturn and direct-to-ground capacitances for lumped inductors and the stray inductance for lumped capacitors. However, these parasitic parameters become a part of a distributed *LC* structure such as a transmission line. In this case, for a microstrip line, the series inductance is associated with the flow of current in the conductor and the shunt capacitance is associated with the strip separated from the ground by the dielectric substrate. If the line is wide, the inductance is reduced but the capacitance is large. However, for a narrow line, the inductance is increased but the capacitance is small.

Figure 3.26 shows an impedance matching circuit in the form of a transmission-line transformer connected between the source impedance Z_S and load impedance Z_L. The input impedance as a function of the length of the transmission line with arbitrary load impedance is

$$Z_{in} = Z_0 \frac{Z_L + jZ_0 \tan\theta}{Z_0 + jZ_L \tan\theta} \tag{3.63}$$

where Z_0 is the characteristic impedance, $\theta = \beta l$ is the electrical length of the transmission line, $\beta = \frac{\omega}{c}\sqrt{\mu_r \varepsilon_r}$ is the phase constant, c is the speed of light in free-space, μ_r is the substrate permeability, ε_r is the substrate permittivity, ω is the radial frequency, l is the geometrical length of the transmission line [3,11].

For a quarter-wavelength transmission line when $\theta = \pi/2$, the expression for Z_{in} simplified to

$$Z_{in} = Z_0^2/Z_L \tag{3.64}$$

from which it follows that, for example, a 50 Ω load is matched to a 12.5 Ω source with characteristic impedance of 25 Ω.

Usually, such a quarter-wavelength impedance transformer is used for impedance matching in a narrow-frequency bandwidth of 10 to 20%, and its length is chosen at the bandwidth center frequency. However, using a multisection quarterwave transformer widens the bandwidth and expands the choice of the substrate to include materials with high dielectric permittivity, which reduces the transformer's size. For example, by using a transformer composed of seven quarter-wavelength transmission lines of different characteristic impedances, whose lengths are selected at the highest bandwidth frequency, the power gain flatness of ±1 dB was achieved over frequency range of 5 to 10 GHz for a 15 W GaAs MESFET power amplifier [12].

To provide a complex-conjugate matching of the input transmission line impedance Z_{in} with the source impedance $Z_S = R_S + jX_S$ when $R_S = \mathrm{Re}Z_{in}$ and $X_S = -\mathrm{Im}Z_{in}$, Eq. (3.63) can be rewritten as

$$R_S - jX_S = Z_0 \frac{R_L + j(X_L + Z_0 \tan\theta)}{Z_0 - X_L \tan\theta + jR_L \tan\theta}. \tag{3.65}$$

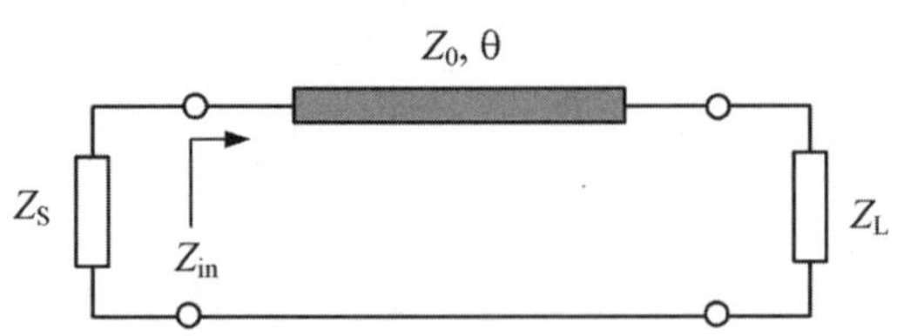

FIGURE 3.26 Transmission-line impedance transformer.

For a quarter-wavelength transformer, Eq. (3.65) can be divided into two separate equations representing the real and imaginary parts of the source impedance Z_S as

$$R_S = Z_0^2 \frac{R_L}{R_L^2 + X_L^2} \tag{3.66}$$

$$X_S = -Z_0^2 \frac{X_L}{R_L^2 + X_L^2}. \tag{3.67}$$

For a pure real load impedance with $X_L = 0$, a quarter-wavelength transmission line with the characteristic impedance Z_0 can provide impedance matching for a purely active source and load only when

$$Z_0 = \sqrt{R_S R_L}. \tag{3.68}$$

Generally, Eq. (3.65) can be divided into two equations representing the real and imaginary parts,

$$R_S (Z_0 - X_L \tan\theta) - R_L (Z_0 - X_S \tan\theta) = 0 \tag{3.69}$$

$$X_S (X_L \tan\theta - Z_0) - Z_0 (X_L + Z_0 \tan\theta) + R_S R_L \tan\theta = 0. \tag{3.70}$$

Solving Eqs. (3.69) and (3.70) for the two independent variables Z_0 and θ yields

$$Z_0 = \sqrt{\frac{R_S \left(R_L^2 + X_L^2\right) - R_L \left(R_S^2 + X_S^2\right)}{R_L - R_S}} \tag{3.71}$$

$$\theta = \tan^{-1} \left(Z_0 \frac{R_S - R_L}{R_S X_L - X_S R_L} \right). \tag{3.72}$$

As a result, the transmission line with the characteristic impedance Z_0 and electrical length θ, determined by Eqs. (3.71) and (3.72), can match any source and load impedances when the impedance ratio gives a positive value inside the square root in Eq. (3.71).

For a particular case of a purely active source when $Z_S = R_S$, the ratio between the load and transmission-line parameters can be expressed by

$$X_L Z_0 \left(1 - \tan^2\theta\right) + \left(Z_0^2 - X_L^2 - R_L^2\right) \tan\theta = 0. \tag{3.73}$$

Then, for the electrical length of the transmission line having $\theta = \pi/4$, the expression for the characteristic impedance Z_0 given by Eq. (3.71) can be simplified to

$$Z_0 = |Z_L| = \sqrt{R_L^2 + X_L^2} \tag{3.74}$$

whereas the required real source impedance R_S should be equal to

$$R_S = R_L \frac{Z_0}{Z_0 - X_L}. \tag{3.75}$$

Consequently, any load impedance can be transformed to a real source impedance defined by Eq. (3.75) using a $\lambda/8$ transformer, the characteristic impedance of which is equal to the magnitude of the load impedance [13].

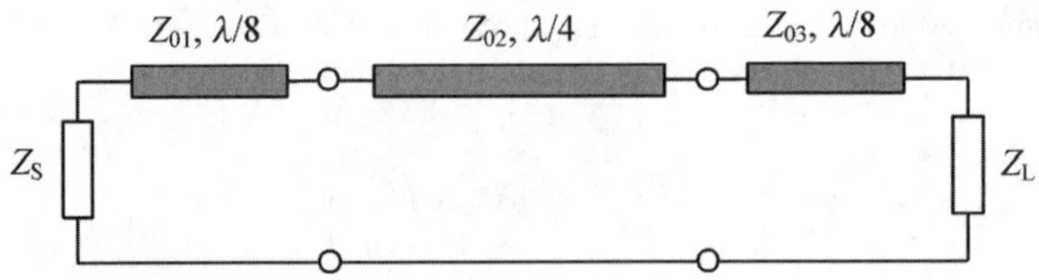

FIGURE 3.27 Transmission-line transformer for any source and load impedances.

Applying the same approach to match purely resistive load with the source impedance, the total matching circuit that includes two $\lambda/8$ transformers and a $\lambda/4$ transformer can provide impedance matching between any complex source impedance Z_S and load impedance Z_L, as shown in Figure 3.27.

In practice, to simplify the power amplifier design at microwaves, the simple matching circuits are very often used, including an *L*-transformer with a series transmission line as the basic matching section. It is convenient to analyze the transforming properties of this matching circuit by substituting the equivalent transformation of the parallel *RX* circuit to the series one. For example, R_1 is the resistance and $X_1 = -1/\omega C$ is the reactance of the impedance $Z_1 = jR_1X_1/(R_1 + jX_1)$ for a parallel *RC*-circuit, and $R_{in} = \text{Re}Z_{in}$ is the resistance and $X_{in} = \text{Im}Z_{in}$ is the reactance of the impedance $Z_{in} = R_{in} + jX_{in}$ for the series transmission-line circuit shown in Figure 3.28. For a conjugate matching when $Z_1 = Z_{in}^*$, we obtain

$$\frac{R_1X_1^2}{R_1^2 + X_1^2} + j\frac{R_1^2X_1}{R_1^2 + X_1^2} = R_{in} - jX_{in}. \tag{3.76}$$

The solution of Eq. (3.76) can be written in the form of two expressions for real and imaginary impedance parts by

$$R_1 = R_{in}\left(1 + Q^2\right) \tag{3.77}$$

$$X_1 = -X_{in}\left(1 + Q^{-2}\right) \tag{3.78}$$

where $Q = R_1/|X_1| = X_{in}/R_{in}$ is a quality factor equal for both parallel capacitive and series transmission-line circuits. Using Eq. (3.63), the real and imaginary parts of the input impedance Z_{in} can be written as

$$R_{in} = Z_0^2R_2\frac{1 + \tan^2\theta}{Z_0^2 + (R_2\tan\theta)^2} \tag{3.79}$$

$$X_{in} = Z_0\tan\theta\frac{Z_0^2 - R_2^2}{Z_0^2 + (R_2\tan\theta)^2}. \tag{3.80}$$

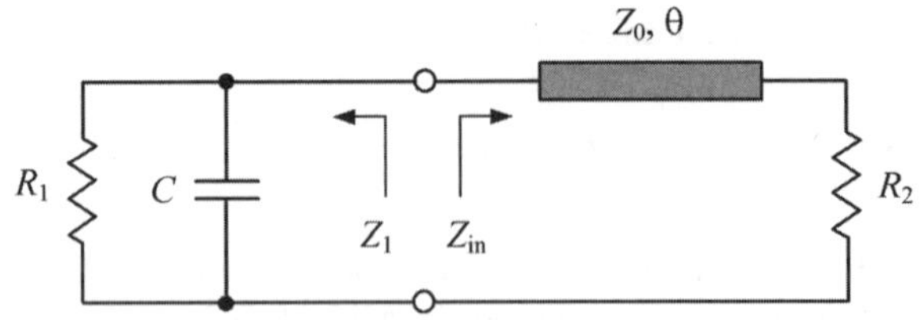

FIGURE 3.28 *L*-transformer with series transmission line.

From Eq. (3.80) it follows that an inductive input impedance (necessary to compensate for the capacitive parallel component) is provided when $Z_0 > R_2$ for $\theta < \pi/2$ and $Z_0 < R_2$ for $\pi/2 < \theta < \pi$. As a result, to transform the resistance R_1 into the other resistance R_2 at the given frequency, it is necessary to connect a two-port *L*-transformer (including a parallel capacitance and a series transmission line) between them. When one parameter (usually the characteristic impedance Z_0) is known, the matching circuit parameters can be calculated from the following two equations:

$$C = \frac{Q}{\omega R_1} \tag{3.81}$$

$$\sin 2\theta = \frac{2Q}{\dfrac{Z_0}{R_2} - \dfrac{R_2}{Z_0}} \tag{3.82}$$

where

$$Q = \sqrt{\frac{R_1}{R_2}\left[\cos^2\theta + \left(\frac{R_2}{Z_0}\right)^2 \sin^2\theta\right] - 1} \tag{3.83}$$

is the circuit quality factor defined as a function of the resistances R_1 and R_2 and the parameters of the transmission line, the characteristic impedance Z_0 and electrical length θ.

It follows from Eqs. (3.82) and (3.83) that the electrical length θ can be calculated as a result of the numerical solution of a transcendental equation with one unknown parameter. However, it is more convenient to combine these two equations and to rewrite them in the implicit form of

$$\frac{R_1}{R_2} = \frac{1 + \left(\dfrac{Z_0}{R_2} - \dfrac{R_2}{Z_0}\right)^2 \sin^2\theta\cos^2\theta}{\cos^2\theta + \left(\dfrac{R_2}{Z_0}\right)^2 \sin^2\theta}. \tag{3.84}$$

Figure 3.29 shows the resistance ratio of R_1/R_2 as a function of the normalized parameter Z_0/R_2 and electrical length θ in the form of two nomographs: for the case of $Z_0/R_2 > 0$ shown in Figure 3.29(*a*) and for the case of $Z_0/R_2 < 0$ shown in Figure 3.29(*b*). When the input resistance R_1 and output resistance R_2 are known in advance, it is easy to evaluate the required value of θ for a fixed transmission-line characteristic impedance Z_0 using these nomographs. The graphical results show that, in contrast to a lumped *L*-transformer, a transmission-line *L*-transformer can match purely resistive source and load impedances with any ratio of R_1/R_2.

A π-transformer can be realized by back-to-back connection of the two *L*-transformers, as shown in Figure 3.30(*a*), where resistances R_1 and R_2 are transformed to some intermediate resistance R_0. In this case, to minimize the length of a transmission line, the value of R_0 should be smaller than that of both R_1 and R_2, that is, $R_0 < (R_1, R_2)$. The same procedure for a *T*-transformer shown in Figure 3.30(*b*) gives a value of R_0 that is larger than that of both R_1 and R_2, that is $R_0 > (R_1, R_2)$. Then, for a *T*-transformer, two shunt adjacent capacitances are combined. For a π-transformer, two adjacent series transmission lines are combined into a single transmission line with total electrical length.

For a π-transformer, the lengths of each part of the combined transmission line can be calculated by equating the imaginary parts of the impedances from both sides at the reference plane A–A' to zero, which means that the intermediate impedance R_0 is real. This leads to two quadratic equations

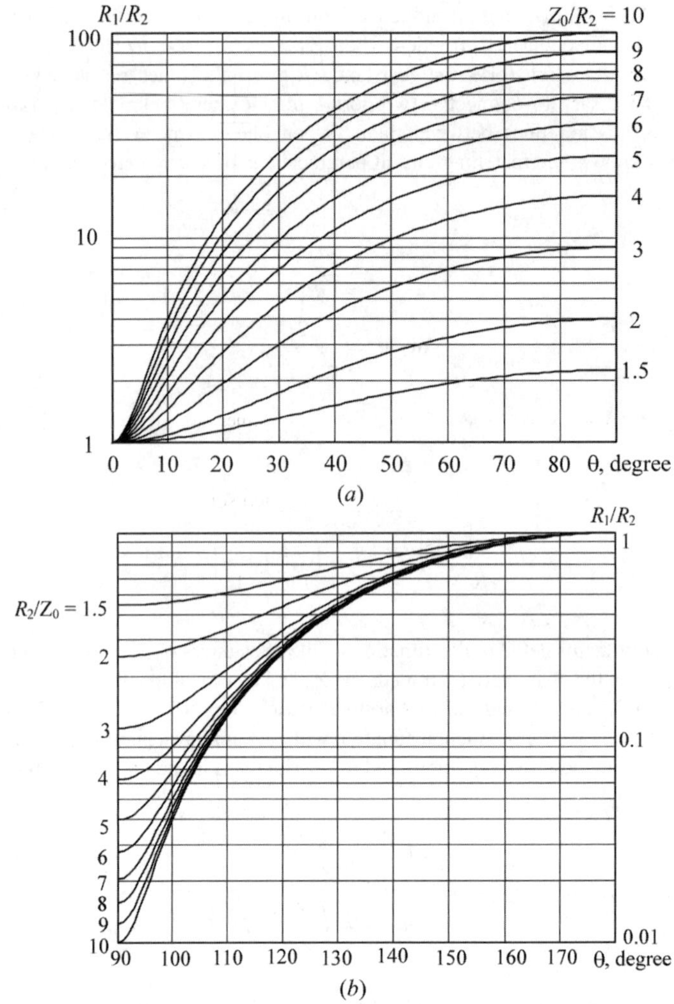

FIGURE 3.29 Nomographs for calculating transmission-line L-transformer.

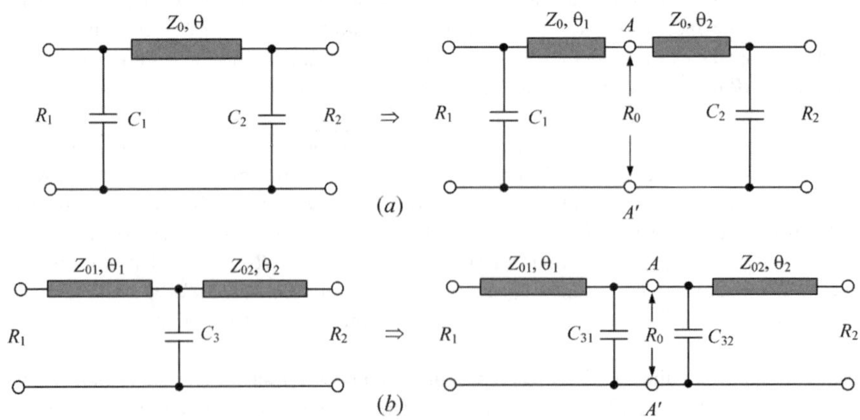

FIGURE 3.30 π- and T-transformers with transmission lines.

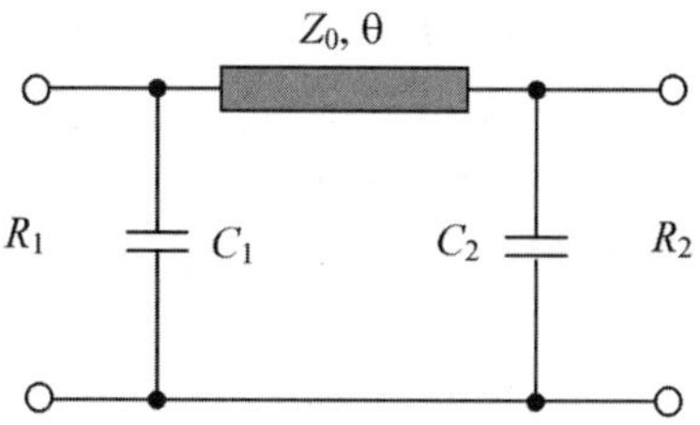

$$\omega C_1 = \frac{Q_1}{R_1} \qquad \omega C_2 = \frac{Q_2}{R_2} \qquad \theta = \theta_1 + \theta_2$$

$$Q_2 = \sqrt{\frac{R_2}{R_1}\left(1 + Q_1^2\right) - 1} \qquad Q_1 > \sqrt{\frac{R_1}{R_2} - 1}$$

FIGURE 3.31 Transmission-line π-transformer and relevant equations.

to calculate the electrical lengths θ_1 and θ_2 of the combined series transmission line:

$$\tan^2\theta_1 - \frac{R_1}{Z_0 Q_1}\left[1 - \left(1 + Q_1^2\right)\left(\frac{Z_0}{R_1}\right)^2\right]\tan\theta_1 - 1 = 0 \tag{3.85}$$

$$\tan^2\theta_2 - \frac{R_2}{Z_0 Q_2}\left[1 - \left(1 + Q_2^2\right)\left(\frac{Z_0}{R_2}\right)^2\right]\tan\theta_2 - 1 = 0 \tag{3.86}$$

where $Q_1 = \omega C_1 R_1$ and $Q_2 = \omega C_2 R_2$.

However, to simplify the matching circuit design procedure, it is very helpful to use the nomographs shown in Figure 3.29. If the values of R_1 and R_2 are selected in advance to set the intermediate resistance R_0, and the characteristic impedance of the transmission line Z_0 is known, the values of θ_1 and θ_2 can be easily determined from one of these nomographs.

A widely used two-port π-transformer with two shunt capacitors along with its design formulas is presented in Figure 3.31 [7,14]. Such a transformer can be conveniently used as an output matching circuit when the device collector or drain–source capacitance can be considered as a first shunt capacitance and parasitic series lead inductance can easily be added to a series transmission line. Also, it is convenient to use this transformer as the matching circuits in balanced power amplifiers, where the shunt capacitors can be connected between series transmission lines due to effect of virtual grounding. The schematic of a transmission-line two-port T-transformer with the series and shunt capacitors along with the design formulas is given in Figure 3.32.

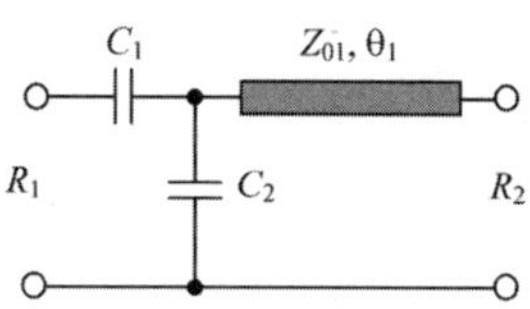

$$\omega C_1 = \frac{1}{R_1 Q_1} \qquad \sin 2\theta_1 = 2Q_2 \Big/ \left(\frac{Z_{01}}{R_2} - \frac{R_2}{Z_{01}}\right) \qquad \omega C_2 = \frac{Q_2 - Q_1}{R_1\left(1 + Q_1^2\right)}$$

$$Q_1 = \sqrt{\frac{R_2}{R_1}\,\frac{1 + Q_2^2}{\cos^2\theta_1 + \left(R_2/Z_{01}\right)\sin^2\theta_1} - 1} \qquad Q_2 > \sqrt{\frac{R_1}{R_2} - 1}$$

FIGURE 3.32 Transmission-line T-transformer and relevant equations.

3.4.2 Equivalence Between Circuits with Lumped and Distributed Parameters

Generally, the input impedance of the transmission line at a particular frequency can be expressed as that of a lumped element, the equivalence of which for a shunt inductor L is shown in Figure 3.33(a) and for a shunt capacitor C is shown in Figure 3.33(b). If load represents a short when $Z_L = 0$, from Eq. (3.63) it follows that

$$Z_{in} = jZ_0 \tan\theta \tag{3.87}$$

which corresponds to the inductive input impedance for $\theta < \pi/2$. The equivalent inductance at the design frequency ω is calculated from

$$L = \frac{X_{in}}{\omega} = \frac{Z_0 \tan\theta}{\omega} \tag{3.88}$$

where $X_{in} = \mathrm{Im}Z_{in}$ is the input reactance. This means that the network shunt inductor can equivalently be replaced with an open-circuited transmission line of characteristic impedance Z_0 and electrical length θ.

Similarly, when $Z_L = \infty$,

$$Z_{in} = -jZ_0 \cot\theta \tag{3.89}$$

which corresponds to the capacitive input impedance for $\theta < \pi/2$. The equivalent capacitance at the design frequency ω is determined from

$$C = -\frac{1}{\omega X_{in}} = \frac{\tan\theta}{\omega Z_0}. \tag{3.90}$$

These equivalences between lumped elements and transmission lines are exact only at the design frequency. The reactance of the inductor increases linearly with increasing frequency, while the reactance of a shorted line increases as tanθ. For a short transmission line when $\theta \ll 90^\circ$, the input impedance increases linearly with frequency since tan$\theta \approx \theta$. Therefore, the short-circuit line behaves like an inductor and open-circuit line behaves like a capacitor over a range of frequencies where the electrical lengths of these transmission lines are much less than 90°.

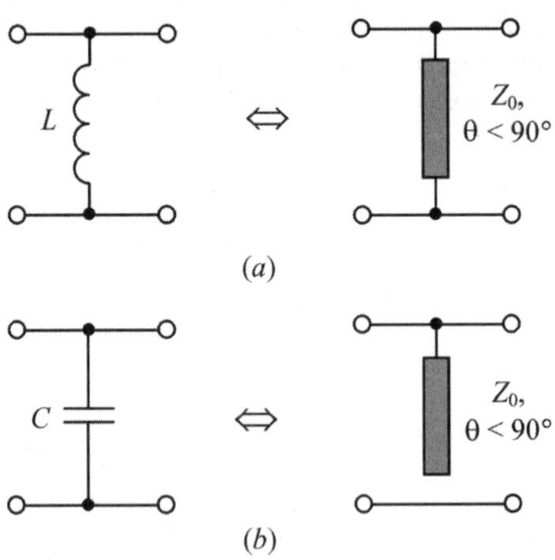

FIGURE 3.33 Equivalence between lumped element and transmission line.

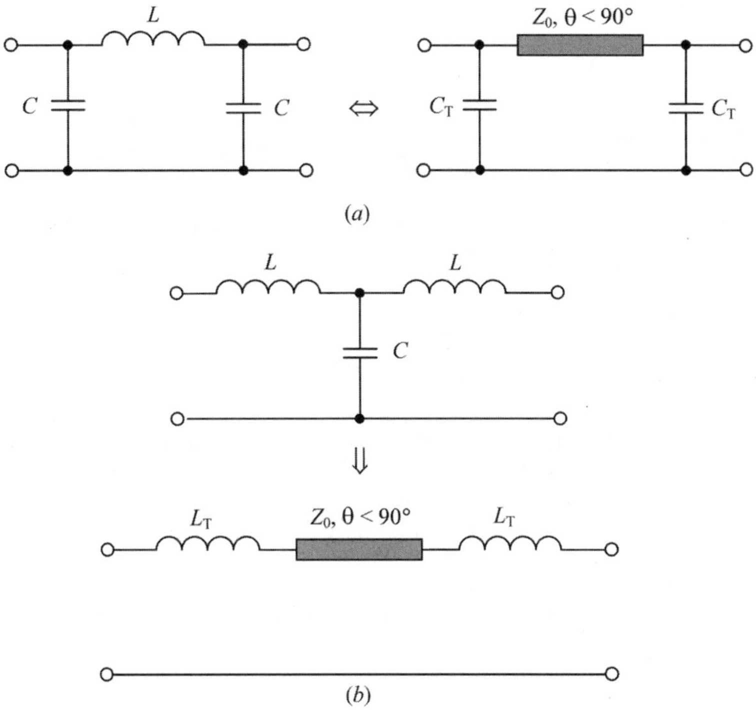

FIGURE 3.34 Lumped and transmission line matching-circuit equivalence.

To define the single-frequency equivalence between π-transformers using lumped and distributed elements, it is convenient to use two-port transmission *ABCD*-parameters. The transmission *ABCD*-matrices of these π-transformers shown in Figure 3.34(*a*) can respectively be given by

$$[ABCD]_L = \begin{bmatrix} 1 & 0 \\ j\omega C & 1 \end{bmatrix} \begin{bmatrix} 1 & j\omega L \\ 0 & 1 \end{bmatrix} \begin{bmatrix} 1 & 0 \\ j\omega C & 1 \end{bmatrix} = \begin{bmatrix} 1-\omega^2 LC & j\omega L \\ j\omega C\left(2-\omega^2 LC\right) & 1-\omega^2 LC \end{bmatrix} \tag{3.91}$$

$$[ABCD]_T = \begin{bmatrix} 1 & 0 \\ j\omega C & 1 \end{bmatrix} \begin{bmatrix} \cos\theta & jZ_0\sin\theta \\ j\dfrac{\sin\theta}{Z_0} & \cos\theta \end{bmatrix} \begin{bmatrix} 1 & 0 \\ j\omega C & 1 \end{bmatrix}$$

$$= \begin{bmatrix} \cos\theta - \omega C_T Z_0 \sin\theta & jZ_0\sin\theta \\ \dfrac{j}{Z_0}\left(2\omega C_T Z_0\cos\theta + \sin\theta - \omega^2 C_T^2 Z_0^2 \sin\theta\right) & \cos\theta - \omega C_T Z_0\sin\theta \end{bmatrix} \tag{3.92}$$

where θ is the electrical length of a transmission line at the design frequency ω.

Since equivalent circuits must have equal matrix elements, then for $A_L = A_T$ and $B_L = B_T$ we can write

$$1-\omega^2 LC = \cos\theta - \omega C_T Z_0 \sin\theta \tag{3.93}$$

$$j\omega L = jZ_0\sin\theta. \tag{3.94}$$

As a result, the equivalent series inductance can be expressed through the parameters of the transmission line as

$$L = \frac{Z_0 \sin\theta}{\omega} \tag{3.95}$$

whereas the relationship between the shunt capacitances C from a lumped π-transformer and the shunt capacitance C_T from a transmission-line π-transformer can be obtained by

$$C_T = \frac{\cos\theta + \omega C Z_0 \sin\theta - 1}{\omega Z_0 \sin\theta}. \tag{3.96}$$

From Eq. (3.96) it follows that $C_T \approx C$ for high values of the characteristic impedance Z_0 when $\omega C Z_0 \gg 1$ or small values of the electrical length θ when $\cos\theta \approx 1$. Consequently, the series lumped inductor L can be replaced by a short transmission line, the parameters of which can be calculated according to Eq. (3.95). The characteristic impedance of the series transmission line is chosen to be sufficiently high to provide better accuracy.

Similarly, to define single-frequency equivalence between a lumped T-transformer and a transmission-line transformer shown in Figure 3.34(b), their transmission $ABCD$-matrices can respectively be given by

$$[ABCD]_L = \begin{bmatrix} 1 & j\omega L \\ 0 & 1 \end{bmatrix} \begin{bmatrix} 1 & 0 \\ j\omega C & 1 \end{bmatrix} \begin{bmatrix} 1 & j\omega L \\ 0 & 1 \end{bmatrix} = \begin{bmatrix} 1-\omega^2 LC & j\omega L\left(2-\omega^2 LC\right) \\ j\omega C & 1-\omega^2 LC \end{bmatrix} \tag{3.97}$$

$$[ABCD]_T = \begin{bmatrix} 1 & j\omega L_T \\ 0 & 1 \end{bmatrix} \begin{bmatrix} \cos\theta & jZ_0\sin\theta \\ j\dfrac{\sin\theta}{Z_0} & \cos\theta \end{bmatrix} \begin{bmatrix} 1 & j\omega L_T \\ 0 & 1 \end{bmatrix}$$

$$= \begin{bmatrix} \cos\theta - \dfrac{\omega L_T \sin\theta}{Z_0} & jZ_0\left(2\dfrac{\omega L_T}{Z_0}\cos\theta + \sin\theta - \dfrac{\omega^2 L_T^2}{Z_0^2}\sin\theta\right) \\ j\dfrac{\sin\theta}{Z_0} & \cos\theta - \dfrac{\omega L_T \sin\theta}{Z_0} \end{bmatrix} \tag{3.98}$$

where θ is the electrical length of a transmission line at the design frequency ω.

For equivalent matrix elements $A_L = A_T$ and $C_L = C_T$,

$$1 - \omega^2 LC = \cos\theta - \frac{\omega L_T}{Z_0}\sin\theta \tag{3.99}$$

$$j\omega C = j\frac{\sin\theta}{Z_0}. \tag{3.100}$$

As a result, the equivalent shunt capacitance can be expressed through the parameters of the transmission line as

$$C = \frac{\sin\theta}{\omega Z_0} \tag{3.101}$$

whereas the relationship between the series inductance L from a lumped T-transformer and the series inductance L_T from a transmission-line transformer can be obtained by

$$L_T = \frac{\cos\theta + \dfrac{\omega L}{Z_0}\sin\theta - 1}{\dfrac{\omega}{Z_0}\sin\theta}. \tag{3.102}$$

From Eq. (3.102) it follows that $L_T \approx L$ for small values of the characteristic impedance Z_0 when $\omega L/Z_0 \gg 1$ or small values of the electrical length θ when $\cos\theta \approx 1$. Consequently, the shunt lumped capacitor C can be replaced by a short transmission line, the parameters of which can be calculated according to Eq. (3.101). In this case, the characteristic impedance of the series transmission line is chosen to be sufficiently low to provide better accuracy.

3.4.3 Narrowband Microwave Power Amplifier

As an example, consider the design of a transmission-line output matching circuit for a 5 W 1.6 GHz bipolar power amplifier that operates at a supply voltage of 24 V and provides a power gain of about 10 dB. These requirements can be satisfied by using an *n–p–n* silicon microwave transistor intended to operate in class AB power amplifiers for a frequency range of 1.5 to 1.7 GHz. At the operating frequency of 1.6 GHz, let the output device impedance is $Z_{out} = (5.5 - j6.5)\ \Omega$, which corresponds to a series combination of the transistor output resistance and capacitance. To match this capacitive impedance to the standard 50 Ω load, it is best to use a matching circuit in the form of a T-transformer shown in Figure 3.32. Figure 3.35 shows the complete two-port network including the output device impedance and matching circuit.

The circuit should compensate for the series capacitance inherent in the output impedance. For the small electrical length when $\tan\theta_{in} \approx \theta_{in}$ and characteristic impedance $Z_0 \gg R_{out}$, Eqs. (3.79) and (3.80) in terms of the output resistance R_{out} and reactance X_{out} can respectively be simplified to

$$R_{out} \cong R_2 \tag{3.103}$$

$$\theta_2 \cong -\frac{X_{out}}{Z_0} = \frac{1}{\omega C_{out} Z_0} \tag{3.104}$$

where θ_2 is a part of the total transmission line that is required to compensate for the output capacitive reactance. If Z_0 is chosen to be 50 Ω, then $\theta_2 = 6.5/50 = 0.13$ radians, which is equal to the electrical length of approximately 7.5°. Then, the value of quality factor Q_2 is defined by

$$Q_2 > \sqrt{\frac{R_1}{R_2} - 1} = 2.84.$$

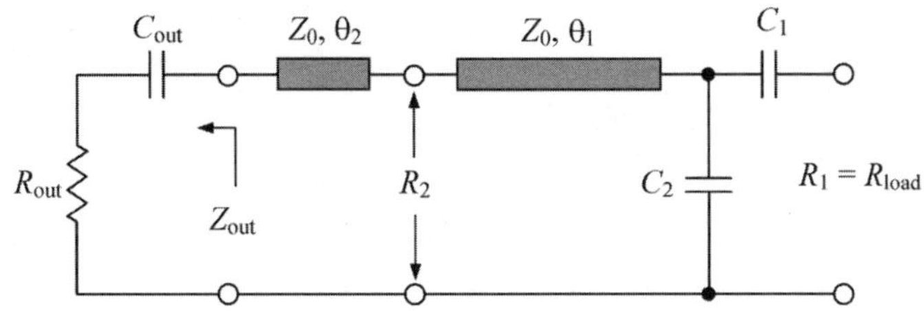

FIGURE 3.35 Complete output two-port network circuit.

Consequently, Q_2 must be larger than 2.84. For example, a value of $Q_2 = 3$ can be chosen to yield a 3-dB frequency bandwidth of 1.6 GHz/3 = 533 MHz. The values of the output matching circuit parameters are

$$\theta_1 = \frac{1}{2}\sin^{-1}\left[2Q_2 \Big/ \left(\frac{Z_0}{R_2} - \frac{R_2}{Z_0}\right)\right] = 21^\circ$$

$$Q_1 = \sqrt{\frac{R_2}{R_1}\frac{1+Q_2^2}{\cos^2\theta_1 + (R_2/Z_0)^2\sin^2\theta_1} - 1} = 0.5$$

$$C_2 = \frac{Q_2 - Q_1}{\omega R_1\left(1+Q_1^2\right)} = 4\,\text{pF}$$

$$C_1 = \frac{1}{\omega Q_1 R_1} = 4\,\text{pF}.$$

The function of each element for visual effect can be traced on the Smith chart, as shown in Figure 3.36. The easiest and most convenient way to plot the traces of the matching circuit elements is to first plot the traces of Q_1 and Q_2, then plot the trace of the series transmission line as far as the intersection point with Q_2-circle, followed by the plot the trace of the shunt capacitance C_2 as far as the intersection point with Q_1-circle, and finally plot the trace of the series capacitance C_1 to the

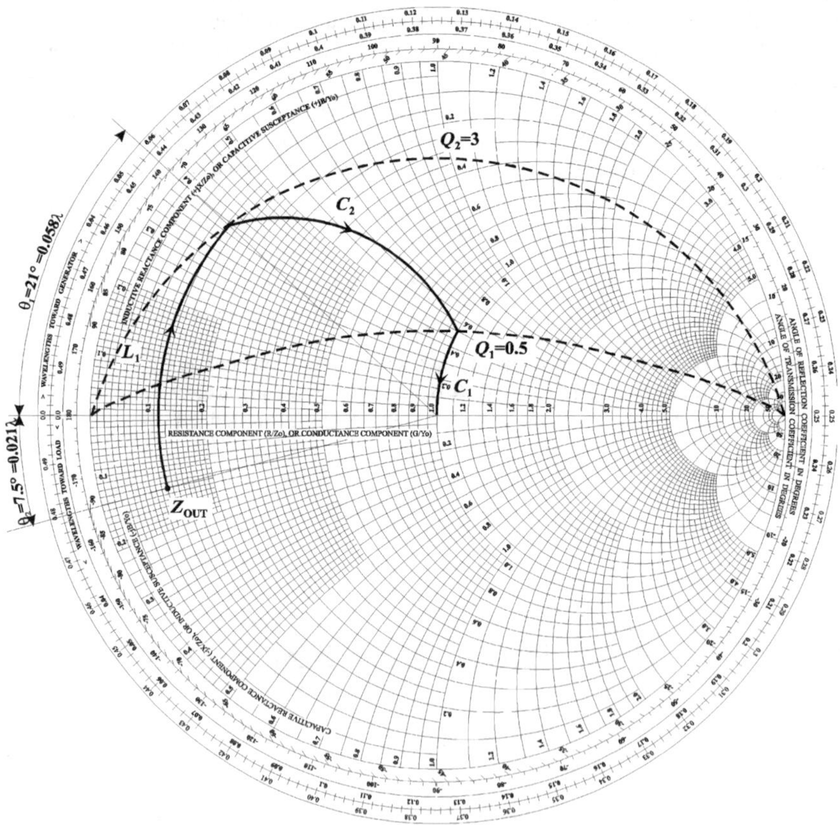

FIGURE 3.36 Smith chart with elements from Figure 3.35.

center 50-Ω point. The plot of the series transmission line represents an arc of the circle with the center point at the center of the Smith chart.

3.4.4 Broadband UHF High-Power Amplifier

Now consider the design example of a broadband 150 W power amplifier that operates over a frequency bandwidth of 470 to 860 MHz, uses a 28 V supply voltage, and provides a power gain of more than 10 dB. In this case, it is convenient to use a high power balanced LDMOS transistor specially designed for UHF TV transmitters as the active device. Let us assume that the manufacturer states the value of input impedance for each transistor-balanced part of $Z_{in} = (1.7 + j1.3)\ \Omega$ at the center bandwidth frequency $f_c = \sqrt{470 \cdot 860} = 635$ MHz. The input impedance Z_{in} represents a series combination of the input resistance and inductive reactance. To cover the required frequency bandwidth, low-Q matching circuits should be used to reduce in-band amplitude ripple and improve input *VSWR*.

To achieve a 3-dB frequency bandwidth, the value of a quality factor must be less than $Q = 635/(860 - 470) = 1.63$. Based on this value of Q, the next step is to define a number of matching sections. For example, for a single-stage input lumped matching circuit, the value of quality factor Q is

$$Q > \sqrt{\frac{50}{1.7} - 1} = 5.33$$

which means that the entire frequency range can be appropriately cover using a multistage matching circuit only.

In this case, it is important that the device input quality factor is smaller than 1.63, being equal to $Q_{in} = 1.3/1.7 = 0.76$. It is very convenient to design the input matching circuit (as well as the output matching circuit) by using simple low-pass L-transformers composed of the series transmission line and shunt capacitance each, with a constant value of Q, for each balanced part of the active device. Then, these two input matching circuits are combined by inserting capacitances, the values of which are reduced twice, between the two series transmission lines.

To match the series input inductive impedance to the standard 50-Ω source, we use three low-pass L-transformers, as shown in Figure 3.37. In this case, the input resistance R_{in} can be assumed to be constant over entire frequency range. At the center bandwidth frequency of 635 MHz, the input inductance is equal approximately to 0.3 nH. Taking this inductance into account, it is necessary to subtract the appropriate value of electrical length θ_{in} from the total electrical length θ_3. Due to the short size of this transmission line when $\tan\theta_{in} \approx \theta_{in}$, a value of θ_{in} can be easily calculated in accordance with

$$\theta_{in} \cong \frac{X_{in}}{Z_0} = \frac{\omega L_{in}}{Z_0}. \tag{3.105}$$

According to Eq. (3.83), there are two simple possibilities to provide input matching using a technique with equal quality factors of L-transformers. One option is to use the same values of characteristic impedance for all transmission lines; the other is to use the same electrical lengths for

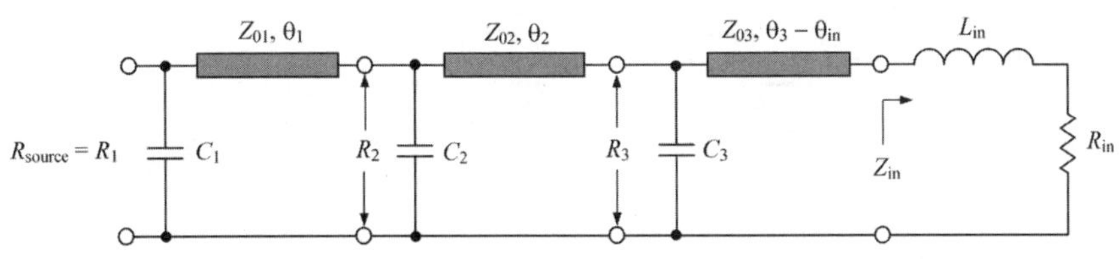

FIGURE 3.37 Complete broadband input two-port network circuit.

all transmission lines. Consider the first approach, which also allows direct use of Smith chart, and choose the value of the characteristic impedance $Z_0 = Z_{01} = Z_{02} = Z_{03} = 50\ \Omega$. The ratio of input and output resistances can be written as

$$\frac{R_1}{R_2} = \frac{R_2}{R_3} = \frac{R_3}{R_{\rm in}} \tag{3.106}$$

which gives the values of $R_2 = 16.2\ \Omega$ and $R_3 = 5.25\ \Omega$ for $R_{\rm source} = R_1 = 50\ \Omega$ and $R_{\rm in} = 1.7\ \Omega$. The values of electrical lengths are determined from the nomograph shown in Figure 3.29(a) as $\theta_1 = 30°, \theta_2 = 7.5°, \theta_3 = 2.4°$.

To calculate the quality factor Q, equal for each L-transformer, from Eq. (3.83), it is enough to know the electrical length θ_1 of the first L-transformer. The remaining two electrical lengths can be directly obtained from Eq. (3.82). As a result, the quality factor of each L-transformer is equal to a value of $Q = 1.2$. The values of the shunt capacitances using Eq. (3.81) are $C_1 = 6$ pF, $C_2 = 19$ pF, $C_3 = 57$ pF.

For a constant Q, we can simplify significantly the design of the matching circuit by using the Smith chart. After calculating the value of Q, it is necessary to plot a constant Q-circle on the Smith chart. Figure 3.38 shows the input matching circuit design using the Smith chart with a constant Q-circle, where the curves for the series transmission lines represent the arcs of the circles with center

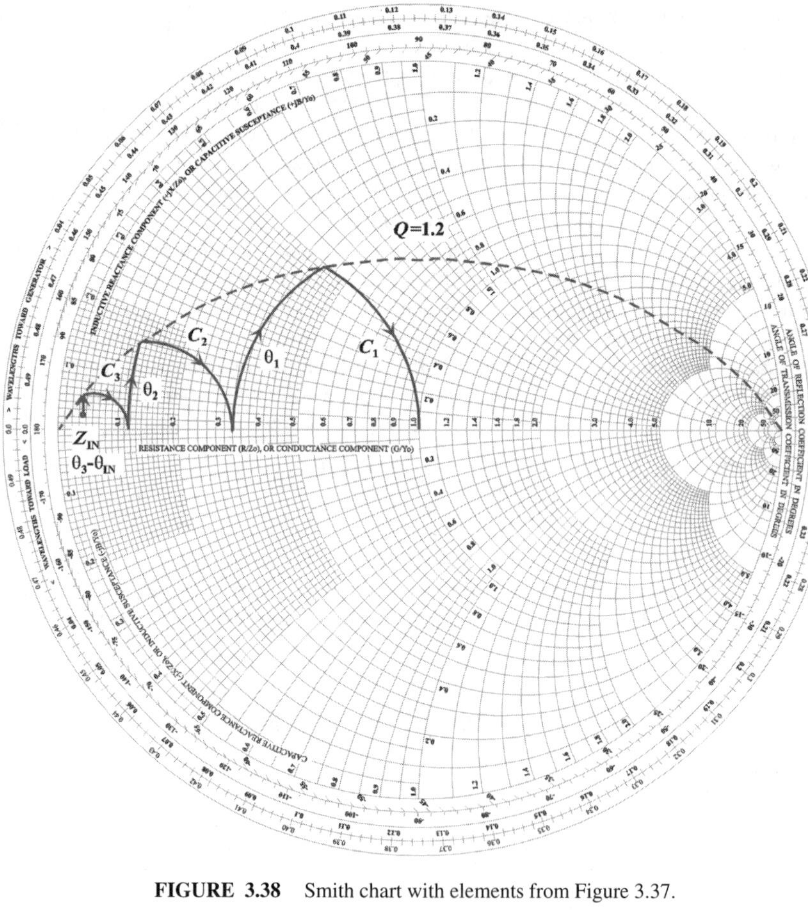

FIGURE 3.38 Smith chart with elements from Figure 3.37.

point at the center of the Smith chart. The capacitive traces are moved along the circles with the increasing susceptances and constant conductances.

Another approach assumes the same values of electrical lengths $\theta = \theta_1 = \theta_2 = \theta_3$ and calculates the characteristic impedances of series transmission lines from Eq. (3.83) at equal ratios of the input and output resistances according to Eq. (3.106). Such an approach is more convenient in practical design, because, when using the transmission lines with standard characteristic impedance $Z_0 = 50\ \Omega$, the electrical length of the transmission line adjacent to the active device input terminal is usually too short. In this case, it makes sense to set the characteristic impedance of the first transmission line to $Z_{01} = 50\ \Omega$. Then, the value of $\theta = 30^\circ$ is determined directly from the nomograph shown in Figure 3.29(*a*). Further calculation of Q from Eq. (3.81) for fixed θ and Z_0/R_2 yields $Q = 1.2$. The characteristic impedances of the remaining two transmission lines are then calculated easily from Eq. (3.81) or Eq. (3.83). Their values are $Z_{02} = 15.7\ \Omega$ and $Z_{03} = 5.1\ \Omega$ with the same values of the shunt capacitances.

3.5 MATCHING NETWORKS WITH MIXED LUMPED AND DISTRIBUTED ELEMENTS

The matching circuits, which incorporate mixed lumped and transmission line elements, are widely used both in hybrid and monolithic design technique. Such matching circuits are especially very convenient when designing the push–pull power amplifiers where the shunt capacitances are simply connected between two series microstrip lines. In this case, the lumped matching circuits can be easily transformed to the transmission-line matching circuits based on the single-frequency equivalence between lumped and distributed elements. First, consider a periodic *LC*-structure in the form of the low-pass ladder π-network which is used as a basis for the lumped matching prototype. Then, the lumped prototype should be split up into individual π-type sections with equal capacitances by consecutive step-by-step process and replaced by their equivalent distributed network counterparts. Finally, the complete mixed matching structure is optimized to improve the overall performance by employing standard nonlinear optimization routine on the element values. For the single frequency equivalence between lumped and distributed elements, the low-pass lumped π-type ladder section with equal shunt capacitances can be made equivalent to a symmetrically loaded transmission line at the single frequency, as shown in Figure 3.39(*a*), where relationships between the circuit parameters are given by Eqs. (3.95) and (3.96).

To provide the design method using a single frequency equivalent technique, the following consecutive design steps can be performed [15]:

- Designate the lumped π-type C_1–L_1–C_2 section to be replaced.
- From the chosen π-type C_1–L_1–C_2 section, form the symmetrical with equal capacitances C shown in Figure 3.39(*b*), the choice of which is arbitrary, however the values cannot exceed the minimum of (C_1, C_2).
- Calculate the parameters of the equivalent symmetrical C_T–TL–C_T section using the parameters of the lumped equivalent π-section by setting the transmission-line electrical length θ in accordance with Eqs. (3.95) and (3.96), assuming that the minimum of the capacitances C_1 and C_2 should be greater or equal than C_T, so that C_T can be readily embedded in the new C_T–TL–C_T section.
- Finally, replace the π-type C_1–L_1–C_2 ladder section by the equivalent symmetrical C_T–TL–C_T section and combine adjacent shunt capacitances, as shown in Figure 3.39(*b*), where the loaded parallel capacitances C_A and C_B are given as $C_A = C_1' + C_T$ and $C_B = C_2' + C_T$.

Figure 3.40(*a*) shows the electrical schematic of a broadband VHF high-power LDMOSFET amplifier. To provide an output power of about 15 W with a power gain of more than 10 dB in a

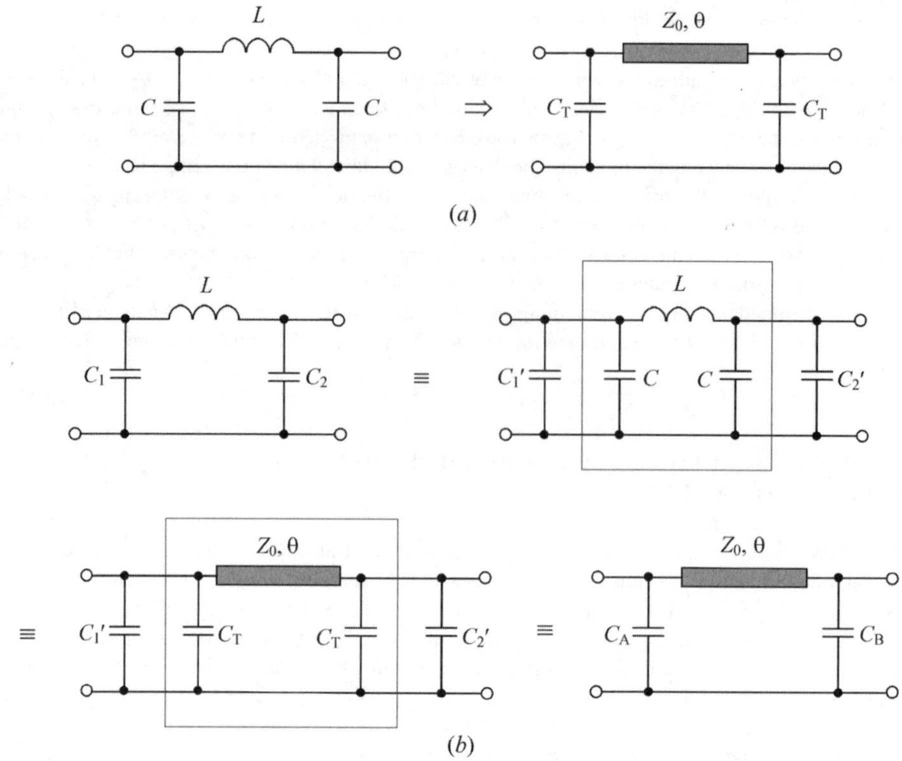

FIGURE 3.39 Transforming design procedure for lumped and distributed matching circuits.

frequency range of 225 to 400 MHz, an LDMOSFET device with the gate geometry of 1.25 μm × 40 mm and a supply voltage of 28 V was chosen. In this case, the matching design technique is based on the multisection low-pass networks with the series transmission line and shunt capacitances, two π-type sections for input matching circuit, and one π-type section for output matching circuit. The sections adjacent to the device input and output terminals incorporate the corresponding internal input gate–source and output drain–source device capacitances. Since a difference between the device equivalent output resistance at the fundamental for several tens of watts of output power and the load resistance of 50 Ω is not significant in this case, it is sufficient to be confined to only one section for output matching network.

Once a matching network structure is chosen, based on the requirements to the electrical performance and frequency bandwidth, the simplest and fastest way is to apply an optimization procedure using CAD simulators to satisfy certain criteria over a wide frequency range. For such a broadband power amplifier, these criteria can be, for example, the minimum output power variation and input return loss with the maximum power gain and efficiency. To minimize the overall dimensions of the power amplifier board, the parallel microstrip line in the drain circuit can be treated as an element of the output matching circuit and its electrical length can be considered as a variable to be optimized as well. Applying a nonlinear broadband CAD optimization technique implemented in any high-level circuit simulator and setting the ranges of electrical length of the transmission lines between 0 and 90° and parallel capacitances from 0 to 100 pF, we obtain the parameters of the input and output matching circuits. The characteristic impedances of all transmission lines can be set to 50 Ω for simplicity and convenience of the circuit implementation. However, to speed up this procedure, it is best to optimize circuit parameters separately for input and output matching circuits with the device equivalent input and output impedances: a series RC-circuit for the device input and a parallel

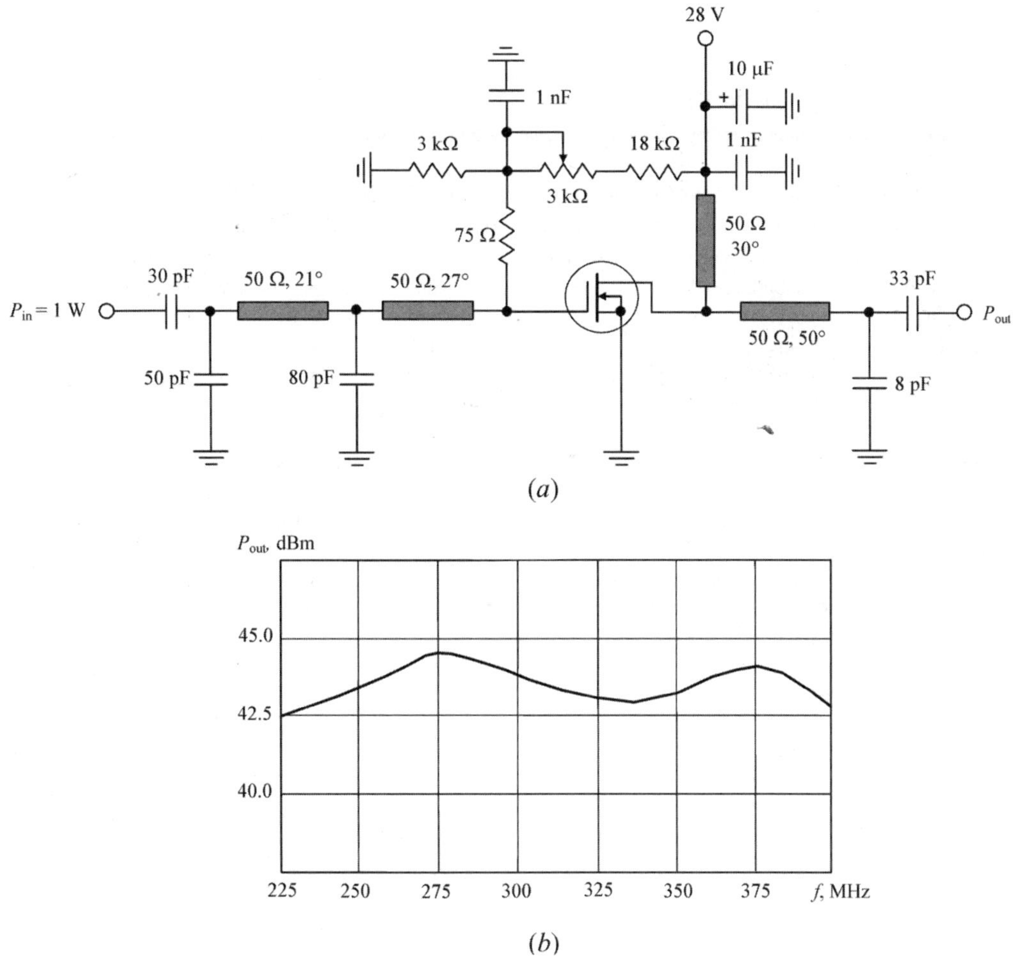

FIGURE 3.40 Circuit schematic and performance of broadband LDMOSFET power amplifier.

RC-circuit for the device output. It is sufficient to use a fast linear optimization process, which will take only a few minutes to complete the matching circuit design. Then, the resulting optimized values are incorporated into the overall power amplifier circuit for each element and final optimization is performed using a large-signal active device model. In this case, the optimization process is finalized by choosing the nominal level of input power with optimizing elements in much narrower ranges of their values of about 10–20% from nominal for most critical elements. For practical convenience all transmission lines might have the characteristic impedances of 50 Ω. Figure 3.40(*b*) illustrates the simulated broadband power amplifier performance, where the output power is within the range of 42.5 to 44.5 dBm with a power gain of 13.5 ± 1 dB in a frequency bandwidth of 225 to 400 MHz [7].

REFERENCES

1. S. Roberts, "Conjugate-Image Impedances," *Proc. IRE*, vol. 34, pp. 198–204, Apr. 1946.
2. P. H. Smith, "Transmission Line Calculator," *Electronics*, vol. 12, pp. 29–31, Jan. 1939.
3. P. H. Smith, *Electronic Applications of the Smith Chart*, New York: Noble Publishing, 2000.

4. P. S. Carter, "Charts for Transmission-Line Measurements and Computations," *RCA Rev*, vol. 3, pp. 355–368, Jan. 1939.
5. H. A. Wheeler, "Reflection Charts Relating to Impedance Matching," *IEEE Trans. Microwave Theory Tech*, vol. MTT-32, pp. 1008–1021, Sept. 1984.
6. W. L. Everitt, "Output Networks for Radio-Frequency Power Amplifiers," *Proc. IRE*, vol. 19, pp. 725–737, May. 1931.
7. A. Grebennikov, *RF and Microwave Power Amplifier Design*, New York: McGraw-Hill, 2004.
8. L. F. Gray and R. Graham, *Radio Transmitters*, New York: McGraw-Hill, 1961.
9. V. M. Bogachev and V. V. Nikiforov, *Transistor Power Amplifiers* (in Russian), Moskva: Energiya, 1978.
10. A. Tam, "Network Building Blocks Balance Power Amp Parameters," *Microwaves & RF*, vol. 23, pp. 81–87, July. 1984.
11. D. M. Pozar, *Microwave Engineering*, New York: John Wiley & Sons, 2004.
12. Y. Ito, M. Mochizuki, M. Kohno, H. Masuno, T. Takagi, and Y. Mitsui, "A 5–10 GHz 15-W GaAs MESFET Amplifier with Flat Gain and Power Responses," *IEEE Microwave and Guided Wave Lett*, vol. 5, pp. 454–456, Dec. 1995.
13. D. H. Steinbrecher, "An Interesting Impedance Matching Network," *IEEE Trans. Microwave Theory Tech*, vol. MTT-15, p. 382, June. 1967.
14. A. V. Grebennikov, "Create Transmission-Line Matching Circuits for Power Amplifiers," *Microwaves & RF*, vol. 39, pp. 113–122, Oct. 2000.
15. B. S. Yarman and A. Aksen, "An Integrated Design Tool to Construct Lossless Matching Networks with Mixed Lumped and Distributed Elements," *IEEE Trans. Circuits and Systems – I: Fundamental Theory Appl*, vol. CAS-I-39, pp. 713–723, Sep. 1992.

4 Power Transformers, Combiners, and Couplers

It is critical, particularly at higher frequencies, that the special types of combiners and dividers are used to avoid insufficient power performance of the individual active devices. The methods of configuration of the combiners or dividers differ depending on the operating frequency, frequency bandwidth, output power, and size requirements. Coaxial cable combiners with ferrite cores are used to combine the output powers of power amplifiers intended for wideband applications. The device output impedance is usually sufficiently small for high power levels; so, to match this impedance with a standard 50-Ω load, coaxial-line transformers with specified impedance transformation are used. For narrow-band applications, the N-way Wilkinson combiners are widely used due to their simple practical realization. For microwaves, the size of combiners should be very small and, therefore, the hybrid microstrip combiners (including different types of the microwaves hybrids and directional couplers) are commonly used to combine output powers of power amplifiers or oscillators. In this chapter, the basic properties of three-port and four-port networks are presented, as well as a variety of different combiners, transformers, and directional couplers for application in RF and microwave transmitters.

4.1 BASIC PROPERTIES

Basic three-port or four-port networks can be used to divide the output power of a single power source or combine the output powers of two or more power amplifiers or oscillators. Generally, the multiport network required to combine the output powers of N identical power sources is based on these basic networks. In this case, it is very important to match the output impedances of all power amplifiers or oscillators with the load to provide the overall output power in N times larger than the output power of a single power amplifier or oscillator. Changes in the operation condition of one power amplifier or oscillator should not affect the operation conditions of the remaining power amplifiers or oscillators. To satisfy this requirement, all input ports of the combiner should be decoupled or mutually independent. When one of the power sources is eliminated, the total output power must decrease by as small as possible, being within the limits of a maximum permissible level. In addition, the combiners can effectively be used for both narrowband and broadband transmitters.

4.1.1 Three-Port Networks

The simplest devices used for power division and combining are the three-port networks with one input and two outputs in the power divider, as shown in Figure 4.1(a), and two inputs and one output in the power combiner, as shown in Figure. 4.1(b). The scattering S-matrix of an arbitrary three-port

RF and Microwave Transmitter Design, First Edition. Andrei Grebennikov.

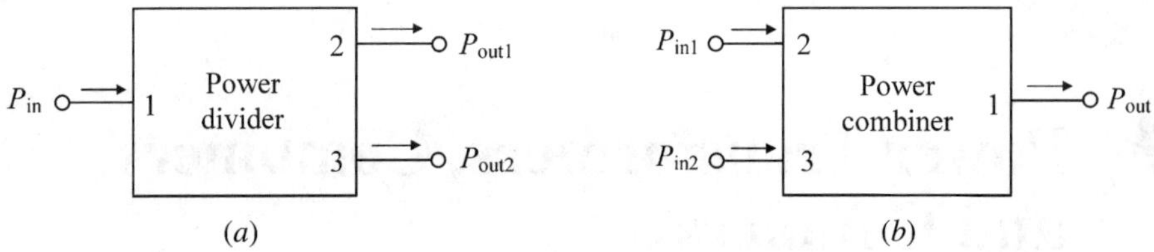

FIGURE 4.1 Schematic diagrams of (*a*) power divider and (*b*) power combiner.

network can be written by

$$[S] = \begin{bmatrix} S_{11} & S_{12} & S_{13} \\ S_{21} & S_{22} & S_{23} \\ S_{31} & S_{32} & S_{33} \end{bmatrix} \tag{4.1}$$

where $S_{ij} = S_{ji}$ for the symmetric scattering *S*-matrix when all components are passive and reciprocal. In this case, if all ports are appropriately matched (when $S_{ii} = 0$), the scattering *S*-matrix is reduced to

$$[S] = \begin{bmatrix} 0 & S_{12} & S_{13} \\ S_{12} & 0 & S_{23} \\ S_{13} & S_{23} & 0 \end{bmatrix}. \tag{4.2}$$

A lossless condition applied to a fully matched *S*-matrix given by Eq. (4.2) requires it to be a unitary matrix when

$$[S]^* \, [S] = 1 \tag{4.3}$$

where $[S]^*$ is the matrix, complex-conjugated to the original *S*-matrix [1,2].

As a result of multiplying two matrices,

$$|S_{12}|^2 + |S_{13}|^2 = |S_{12}|^2 + |S_{23}|^2 = |S_{13}|^2 + |S_{23}|^2 = 1 \tag{4.4}$$

$$S_{13}^* S_{23} = S_{23}^* S_{12} = S_{12}^* S_{13} = 0. \tag{4.5}$$

From Eq. (4.5) it follows that at least two of three available parameters S_{12}, S_{13}, and S_{23} should be zero, which is inconsistent with at least one condition given by Eq. (4.4). This means that a three-port network cannot be lossless, reciprocal, and matched at all ports. However, if any one of these three conditions is not fulfilled, a physically realizable device is possible for practical implementation. A lossless and reciprocal three-port network can be realized if only two of its ports are matched, or it is lossy being reciprocal and fully matched at all three ports, as in the case of the resistive divider.

If a reciprocal three-port network represents the 3-dB power divider when, for a given input power at the port 1, the output powers at the ports 2 and 3 are equal, then, according to Eq. (4.4),

$$|S_{12}| = |S_{13}| = \frac{1}{\sqrt{2}}. \tag{4.6}$$

4.1.2 Four-Port Networks

The four-port networks are used for directional power coupling when, for a given input signal at the port 1, the output signals are delivered to the ports 2 and 3, and no power is delivered to the port 4 (ideal case), as shown in Figure 4.2. The scattering *S*-matrix of a reciprocal four-port network

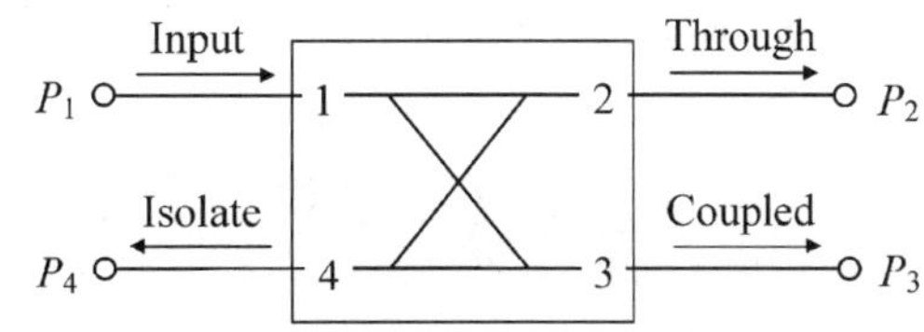

FIGURE 4.2 Schematic diagram of directional coupler.

matched at all its ports is given by

$$[S] = \begin{bmatrix} 0 & S_{12} & S_{13} & S_{14} \\ S_{12} & 0 & S_{23} & S_{24} \\ S_{13} & S_{23} & 0 & S_{34} \\ S_{14} & S_{24} & S_{34} & 0 \end{bmatrix} \tag{4.7}$$

where $S_{ij} = S_{ji}$ for the symmetric scattering S-matrix when all components are passive and reciprocal. In this case, the power supplied to the input port 1 is coupled to the coupled port 3 with coupling factor $|S_{13}|^2$, whereas the remainder of the input power is delivered to the through port 2 with coupling factor $|S_{12}|^2$.

For a lossless four-port network, the unitary condition of the fully matched S-matrix given by Eq. (4.7) results in

$$|S_{12}|^2 + |S_{13}|^2 = |S_{12}|^2 + |S_{24}|^2 = |S_{13}|^2 + |S_{34}|^2 = |S_{24}|^2 + |S_{34}|^2 = 1 \tag{4.8}$$

which implies a full isolation between ports 2 and 3 and ports 1 and 4, respectively, when

$$S_{14} = S_{41} = S_{23} = S_{32} = 0 \tag{4.9}$$

and that

$$|S_{13}| = |S_{24}| \quad |S_{12}| = |S_{34}|. \tag{4.10}$$

The scattering S-matrix of such a directional coupler, matched at all its ports with two decoupled two-port networks, reduces to

$$[S] = \begin{bmatrix} 0 & S_{12} & S_{13} & 0 \\ S_{12} & 0 & 0 & S_{24} \\ S_{13} & 0 & 0 & S_{34} \\ 0 & S_{24} & S_{34} & 0 \end{bmatrix}. \tag{4.11}$$

The directional coupler can be classified according to the phase shift ϕ between two output ports 2 and 3 as the in-phase coupler with $\phi = 0$, quadrature coupler with $\phi = 90^\circ$ or $\pi/2$, and out-of-phase coupler with $\phi = 180^\circ$ or π. The following important quantities are used to characterize the directional coupler:

- The power-split ratio or power division ratio K^2, which is calculated as the ratio of powers at the output ports when all ports are nominally (reflectionless) terminated,

$$K^2 = \frac{P_2}{P_3}$$

- The insertion loss C_{12}, which is calculated as the ratio of powers at the input port 1 relative the output port 2,

$$C_{12} = 10\log_{10}\frac{P_1}{P_2} = -20\log_{10}|S_{12}|$$

- The coupling C_{13}, which is calculated as the ratio of powers at the input port 1 relative to the output port 3,

$$C_{13} = 10\log_{10}\frac{P_1}{P_3} = -20\log_{10}|S_{13}|$$

- The directivity C_{34}, which is calculated as the ratio of powers at the output port 3 relative to the isolated port 4,

$$C_{34} = 10\log_{10}\frac{P_3}{P_4} = 20\log_{10}\frac{|S_{13}|}{|S_{14}|}$$

- The isolation C_{14} and C_{23}, which are calculated as the ratios of powers at the input port 1 relative to the isolated port 4 and between the two output ports (output port 2 is considered an input port), respectively,

$$C_{14} = 10\log_{10}\frac{P_1}{P_4} = -20\log_{10}|S_{14}|$$

$$C_{23} = 10\log_{10}\frac{P_2}{P_3} = -20\log_{10}|S_{23}|$$

- The voltage standing wave ratio at each port or $VSWR_i$, where $i = 1, 2, 3, 4$, calculated as

$$VSWR_i = \frac{1 + |S_{ii}|}{1 - |S_{ii}|}.$$

In an ideal case, the directional coupler would have $VSWR_i = 1$ at each ports, insertion loss $C_{12} = 3$ dB, coupling $C_{13} = 3$ dB, infinite isolation and directivity $C_{14} = C_{23} = C_{34} = \infty$.

4.2 TRANSMISSION-LINE TRANSFORMERS AND COMBINERS

The transmission-line transformers and combiners can provide very wide operating bandwidths and operate up to frequencies of 3 GHz and higher [3,4]. They are widely used in matching networks for antennas and power amplifiers in the HF and VHF bands, in mixer circuits, and their low losses make them especially useful in high-power circuits [5,6]. Typical structures for transmission-line transformers consist of parallel wires, coaxial cables, or bifilar twisted wire pairs. In the latter case, the characteristic impedance can easily determined by the wire diameter, the insulation thickness, and, to some extent, the twisting pitch [7,8]. For coaxial cable transformers with correctly chosen characteristic impedance, the theoretical high-frequency bandwidth limit is reached when the cable length comes in order of a half wavelength, with the overall achievable bandwidth being about a decade. By introducing the low-loss high permeability ferrites alongside a good quality semirigid coaxial or symmetrical strip cable, the low frequency limit can be significantly improved providing bandwidths of several or more decades.

The concept of a broadband impedance transformer consisted of a pair of interconnected transmission lines was first disclosed and described by Guanella [9,10]. Figure 4.3(a) shows a Guanella

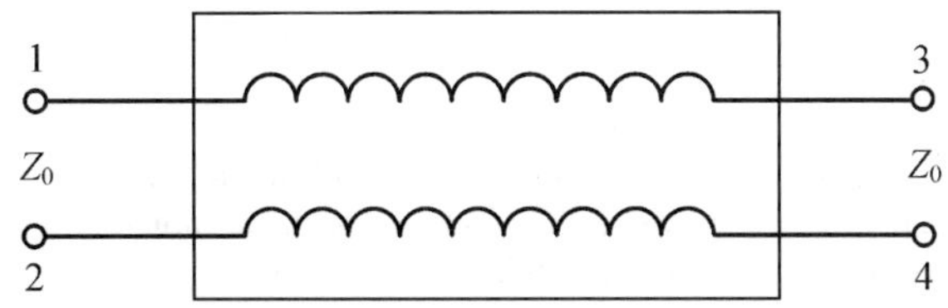

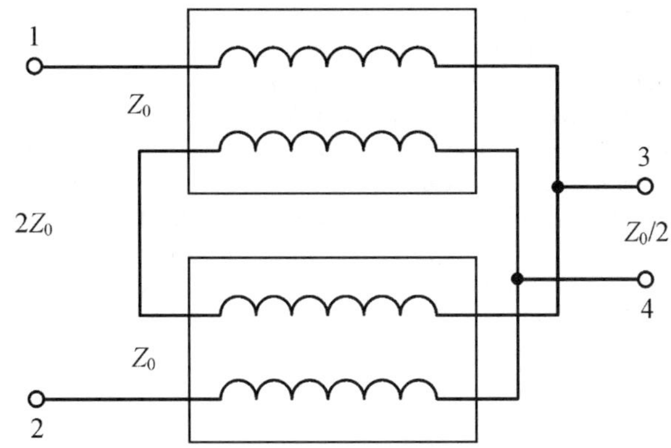

FIGURE 4.3 Schematic configurations of Guanella 1:1 and 4:1 transformers.

transformer system with transmission-line character achieved by an arrangement comprising one pair of cylindrical coils that are wound in the same sense and are spaced a certain distance apart by an intervening dielectric. In this case, one cylindrical coil is located inside the insulating cylinder and the other coil is located on the outside of this cylinder. For the currents flowing through both windings in opposite directions, the corresponding flux in the coil axis is negligibly small. However, for the currents flowing in the same direction through both coils, the latter may be assumed to be connected in parallel, and a coil pair represents a considerable inductance for such currents and acts like a choke coil. With terminal 4 being grounded, such a 1:1 transformer provides matching of the balanced source to unbalanced load and is called the *balun* (*bal*anced-to-*un*balanced transformers). In this case, if terminal 2 is grounded, it represents simply a delay line. In a particular case, when terminals 2 and 3 are grounded, the transformer performs as a phase inverter.

A series-parallel connection of a plurality of coil pairs can produce a match between unequal source and load resistances. Figure 4.3(*b*) shows a 4:1 impedance (2:1 voltage) transmission-line transformer where the two pairs of cylindrical transmission line coils are connected in series at the input and in parallel at the output. For the characteristic impedance Z_0 of each transmission line, this results in the two times higher impedance $2Z_0$ at the input and two times lower impedance $Z_0/2$ at the output. By grounding terminal 4, such a 4:1 impedance transformer provides impedance matching of the balanced source to the unbalanced load. In this case, when terminal 2 is grounded, it performs as a 4:1 *unun* (*un*balanced-to-*un*balanced transformer). With a series-parallel connection of n coil pairs with the characteristic impedance Z_0 each, the input impedance is equal to nZ_0 and the output impedance is equal to Z_0/n. Since Guanella adds voltages that have equal delays through the transmission lines, such a technique results in the so called *equal-delay* transmission-line transformers.

The simplest transmission-line is a quarterwave transmission line whose characteristic impedance is chosen to give the correct impedance transformation. However, this transformer provides a narrowband performance valid only around frequencies for which the transmission line is odd multiples of a quarter wavelength. If a ferrite sleeve is added to the transmission line, common-mode currents flowing in both transmission line inner and outer conductors in phase and in the same direction are suppressed and the load may be balanced and floating above ground [11,12]. If the characteristic impedance of the transmission line is equal to the terminating impedances, the transmission is inherently broadband. If not, there will be a dip in the response at the frequency at which the transmission line is a quarter-wavelength long.

A coaxial cable transformer, the physical configuration, and equivalent circuit representation of which are shown in Figures 4.4(*a*) and 4.4(*b*), respectively, consists of the coaxial line arranged inside the ferrite core or wound around the ferrite core. Due to its practical configuration, the coaxial cable transformer takes a position between the lumped and distributed systems. Therefore, at lower frequencies its equivalent circuit represents a conventional low-frequency transformer shown in Figure 4.4(*c*), while at higher frequency it is a transmission line with the characteristic impedance Z_0 shown in Figure 4.4(*d*). The advantage of such a transformer is that the parasitic interturn capacitance determines its characteristic impedance, whereas in the conventional wire-wound transformer with discrete windings this parasitic capacitance negatively contributes to the transformer frequency performance.

When $R_S = R_L = Z_0$, the transmission line can be considered a transformer with a 1:1 impedance transformation. To avoid any resonant phenomena, especially for complex loads, which can contribute to the significant output power variations, as, a general rule, the length l of the transmission line is kept to no more than an eight of a wavelength λ_{min} at the highest operating frequency,

$$l \leq \frac{\lambda_{min}}{8} \tag{4.12}$$

where λ_{min} is the minimum wavelength in the transmission line corresponding to the high operating frequency f_{max}.

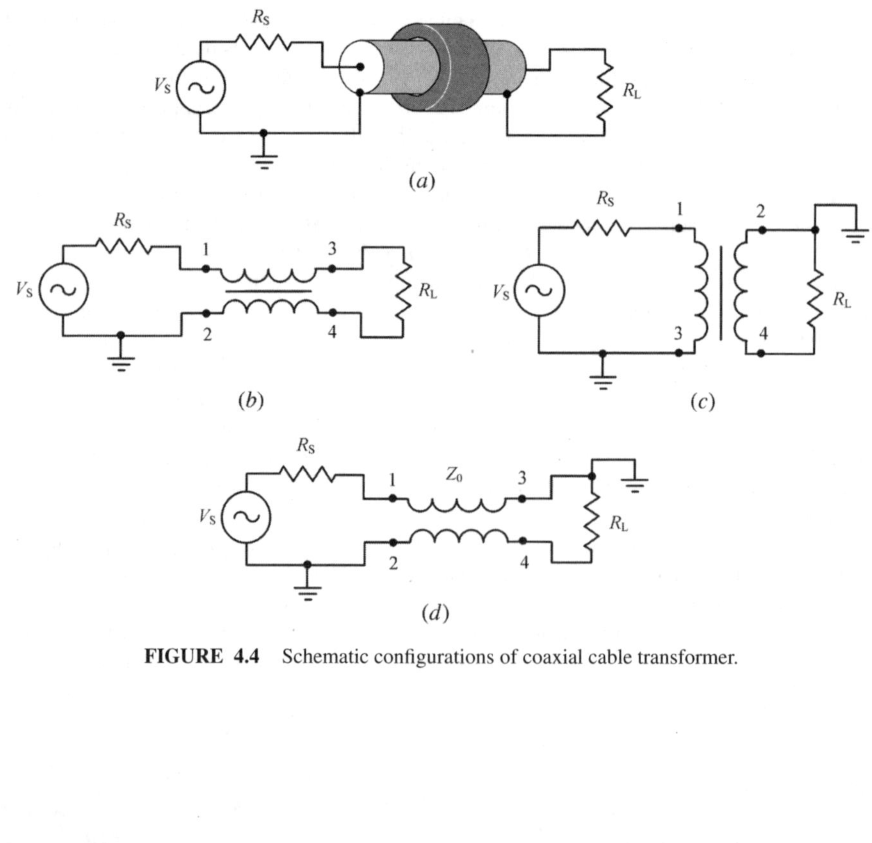

FIGURE 4.4 Schematic configurations of coaxial cable transformer.

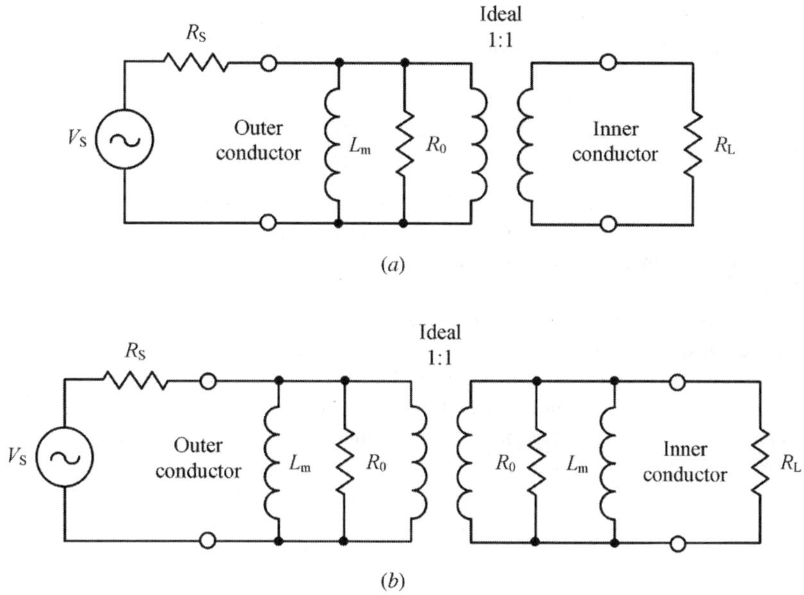

FIGURE 4.5 Low-frequency models of 1:1 coaxial cable transformer.

The low-frequency bandwidth limit of a coaxial cable transformer is determined by the effect of the magnetizing inductance L_m of the outer surface of the outer conductor according to the equivalent low-frequency transformer model shown in Figure 4.5(*a*), where the transmission line is represented by the ideal 1:1 transformer [6]. The resistance R_0 represents the losses of the transmission line. An approximation to the magnetizing inductance can be made by considering the outer surface of the coaxial cable to be the same as that of a straight wire (or linear conductor), which at higher frequencies where the skin effect causes the current to be concentrated on the outer surface, would have the self-inductance of

$$L_m = 2l\left[\ln\left(\frac{2l}{r}\right) - 1\right]\text{nH} \tag{4.13}$$

where l is the length of the coaxial cable in cm and r is the radius of the outer surface of the outer conductor in cm [6].

High permeability of core materials results in shorter transmission lines. If a toroid is used for the core, the magnetizing inductance L_m is obtained by

$$L_m = 4\pi n^2 \mu \frac{A_e}{L_e}\ \text{nH} \tag{4.14}$$

where n is the number of turns, μ is the core permeability, A_e is the effective cross-sectional area of the core in cm^2, and L_e is the average magnetic path length in cm [13].

Considering the transformer equivalent circuit shown in Figure 4.4(*a*), the ratio between the power delivered to the load P_L and power available at the source $P_S = V_S^2/8R_S$ when $R_S = R_L$ can be obtained from

$$\frac{P_L}{P_S} = \frac{(2\omega L_m)^2}{R_S^2 + (2\omega L_m)^2} \tag{4.15}$$

which gives the minimum operating frequency $f_{\min}$ for a given magnetizing inductance L_m, taking into account the maximum decrease of the output power by 3 dB, as

$$f_{\min} \geq \frac{R_S}{4\pi L_m}. \tag{4.16}$$

A similar low-frequency model for a coaxial cable transformer using twisted or parallel wires is shown in Figure 4.5(*b*) [6]. Here, the model is symmetrical as both conductors are exposed to any magnetic material and therefore contribute identically to the losses and low-frequency performance of the transformer.

An approach using a transmission line based on a single bifilar wound coil to realize a broadband 1:4 impedance transformation was introduced by Ruthroff [14,15]. In this case, by using a core material of sufficiently high permeability, the number of turns can be significantly reduced. Figure 4.6(*a*) shows the circuit schematic of an unbalanced-to-unbalanced 1:4 transmission-line transformer where terminal 4 is connected to the input terminal 1. As a result, for $V = V_1 = V_2$, the output voltage is twice the input voltage, and the transformer has a 1:2 voltage step-up ratio. As the ratio of input voltage to input current is one-fourth the load voltage to load current, the transformer is fully matched for maximum power transfer when $R_L = 4R_S$, and the characteristic impedance of the transmission-line Z_0 is equal to the geometric mean of the source and load impedances:

$$Z_0 = \sqrt{R_S R_L} \tag{4.17}$$

where R_S is the source resistance and R_L is the load resistance. Figure 4.6(*b*) shows an impedance transformer acting as a phase inverter, where the load resistance is included between terminals 1 and 4 to become a 1:4 balun. This technique is called the *bootstrap effect*, which does not have the same high-frequency response as Guanella equal-delay approach because it adds a delayed voltage to a

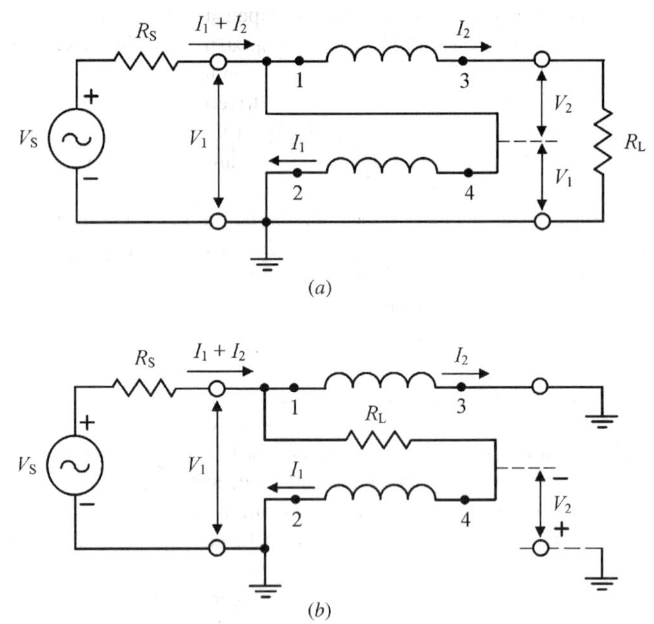

FIGURE 4.6 Schematic configurations of Ruthroff 1:4 impedance transformer.

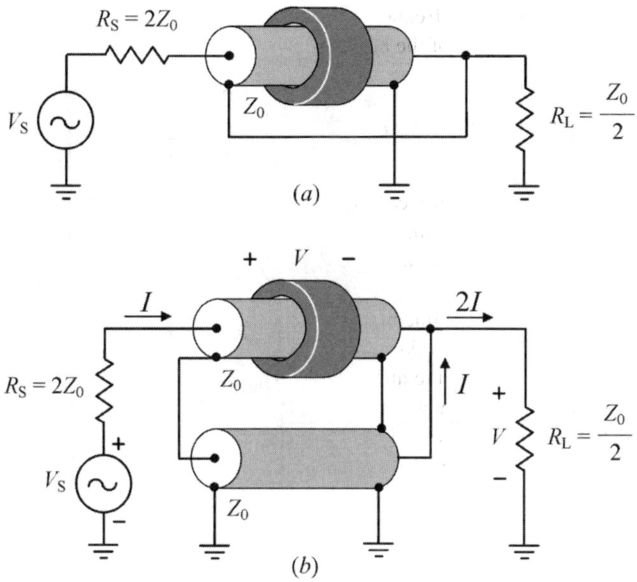

FIGURE 4.7 Schematic configurations of 4:1 coaxial cable transformer.

direct one [16]. The delay becomes excessive when the transmission line reach a significant fraction of a wavelength.

Figure 4.7(*a*) shows the physical implementation of the 4:1 impedance Ruthroff transformer using a coaxial cable arranged inside the ferrite core. At lower frequencies, such a transformer can be considered an ordinary 2:1 voltage autotransformer. To improve the performance at higher frequencies, it is necessary to add an additional phase-compensating line of the same length shown in Figure 4.7(*b*), resulting in a Guanella ferrite-based 4:1 impedance transformer. In this case, a ferrite core is necessary only for the upper line because the outer conductor of the lower line is grounded at both ends, and no current is flowing through it. A current I driven into the inner conductor of the upper line produces a current I that flows in the outer conductor of the upper line, resulting in a current $2I$ flowing into the load R_L. Because the voltage $2V$ from the transformer input is divided in two equal parts between the coaxial line and the load, such a transformer provides impedance transformation from $R_S = 2Z_0$ into $R_L = Z_0/2$, where Z_0 is the characteristic impedance of each coaxial line. The bandwidth extension for the Ruthroff transformers can also be achieved by using transmission lines with step-function and exponential changes in their characteristic impedances [17,18]. To adopt this transmission-line transformer for microwave planar applications, the coaxial line can be replaced by a pair of stacked strip conductors or coupled microstrip lines [19,20].

Figure 4.8 shows similar arrangements for the 3:1 voltage coaxial cable transformers, which produce 9:1 impedance transformation. A current I driven into the inner conductor of the upper line in Figure 4.8(*a*) will cause a current I to flow in the outer conductor of the upper line. This current then produces a current I in the outer conductor of the lower line, resulting in a current $3I$ flowing into the load R_L. The lowest coaxial line can be removed, resulting in a 9:1 impedance coaxial cable transformer shown in Figure 4.8(*b*). The characteristic impedance of each transmission line is specified by the voltage applied to the end of the line and the current flowing through the line and is equal to Z_0.

By using the transmission-line baluns with different integer-transformation ratios in certain connection, it is possible to obtain the fractional-ratio baluns and ununs [4,21,22]. Figure 4.9 shows a transformer configuration for obtaining an impedance ratio of 2.25:1, which consists of a 1:1 Guanella

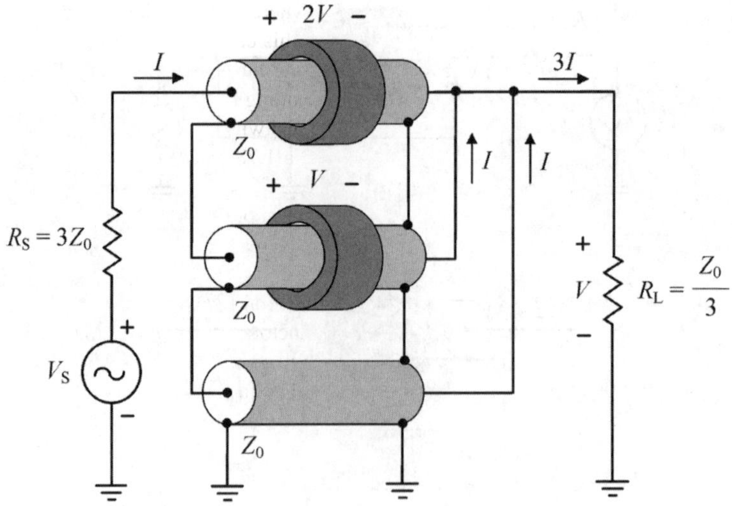

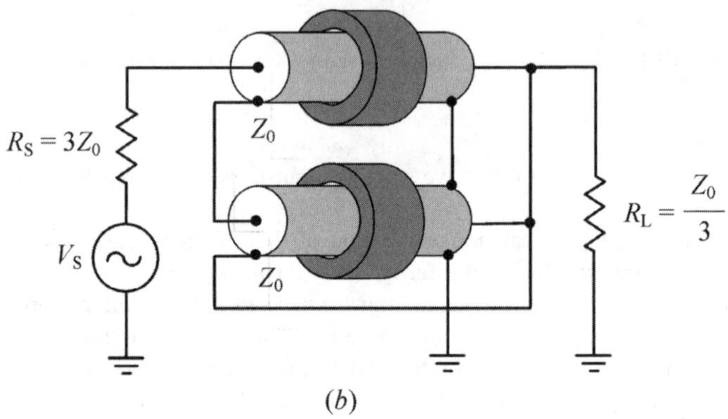

FIGURE 4.8 Schematic configurations of 9:1 coaxial cable transformer.

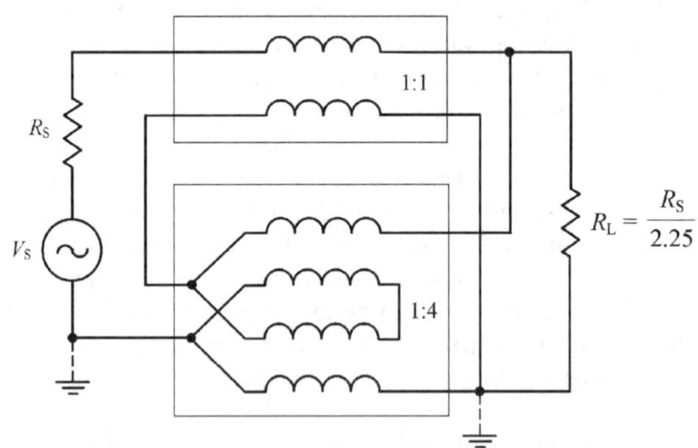

FIGURE 4.9 Schematic configuration of equal-delay 2.25:1 unun.

balun on the top combined with a 1:4 Guanella balun where voltages on the left-hand side are in series and on the right-hand side are in parallel [21]. In this case, the left-hand side has the higher impedance. In a matched condition, this transformer should have a high frequency response similar to a single transmission line. By grounding the corresponding terminals (shown by dashed line), it becomes a broadband unun. Different ratios can be obtained with other configurations. For example, using a 1:9 Guanella balun below the 1:1 unit results in a 1.78:1 impedance ratio, whereas, with a 1:16 balun, the impedance ratio becomes 1.56:1.

On the other hand, the overall 1:1.5 voltage transformer configuration can be achieved by using the cascade connection of a 1:3 voltage transformer to increase the impedance by 9 times, and a 2:1 voltage transformer to decrease the impedance by 4 times, which block schematic is shown in Figure 4.10(*a*) [22]. The practical configuration using coaxial cables and ferrite cores is shown in Figure 4.10(*b*). Here, the currents $I/3$ in the inner conductors of two lower lines cause an overall current $2I/3$ in the outer conductor of the upper line, resulting in a current $2I/3$ flowing into the load R_L. A load voltage $3V/2$ is out of phase with a longitudinal voltage $V/2$ along the upper line, resulting in a voltage V at the transformer input. The lowest line also can be eliminated with direct connection of the points at both ends of its inner conductor, as in the case of the 2:1 and 3:1 Ruthroff voltage

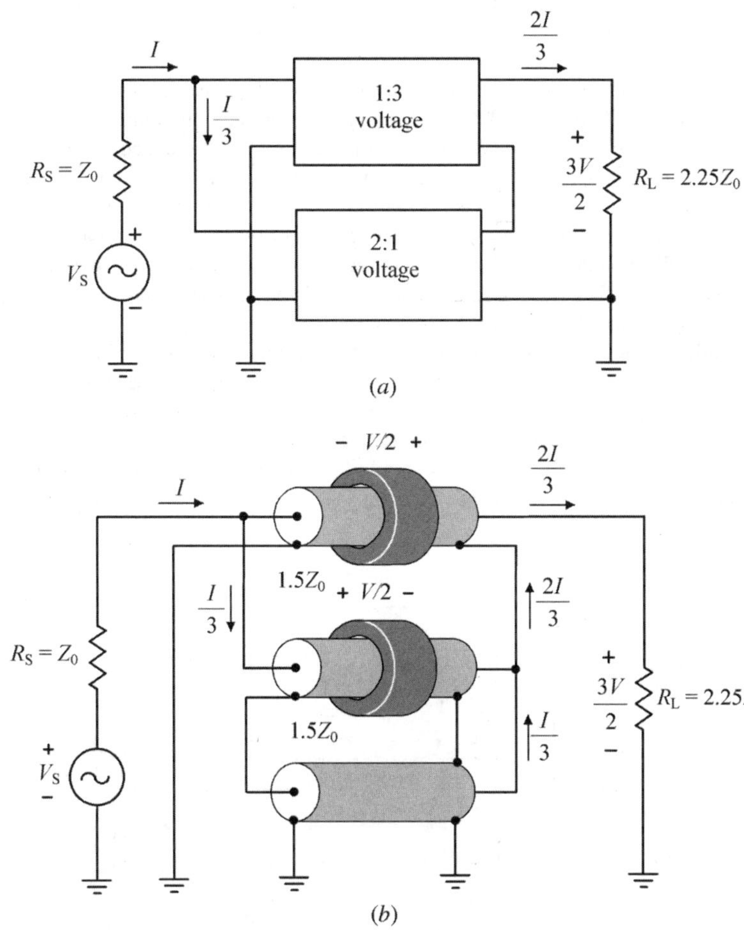

FIGURE 4.10 Schematic configurations of fractional 1:2.25 impedance transformer.

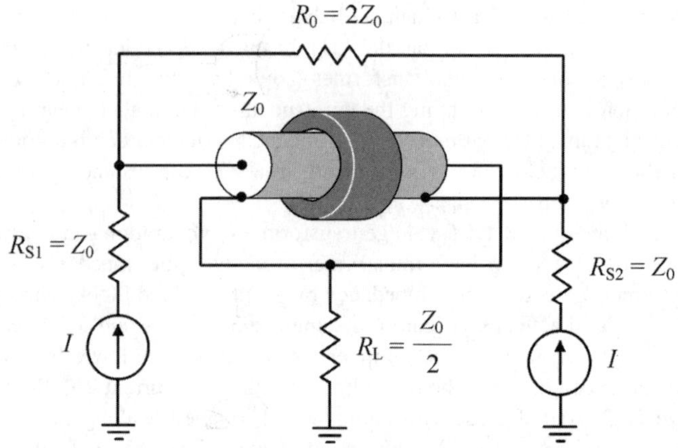

FIGURE 4.11 Coaxial cable combiner.

transformers shown in Figures 4.7(*a*) and 4.8(*b*), respectively. If the source impedance is 50 Ω, then the characteristic impedance of all three transmission lines should be 75 Ω. In this case, the matched condition corresponds to a load impedance of 112.5 Ω.

By using the coaxial cable transformers, the output powers from two or more power sources can be combined. Figure 4.11 shows an example of such a transformer, combining two in-phase signals when both signal are delivered to the load R_L and no signal will be dissipated in the ballast resistor R_0 if their amplitudes are equal [14]. The main advantage of this transformer is the zero longitudinal voltage along the line for equal input powers; as a result, no losses occur in the ferrite core. When one input signal source (for example, power amplifier) defaults or disconnects, the longitudinal voltage becomes equal to half a voltage of another input source. For this transformer, it is possible to combine two out-of-phase signals when the ballast resistor is considered the load, and the load resistor in turn is considered the ballast resistor.

The schematic of another hybrid coaxial cable transformer using as a combiner is shown in Figure 4.12. The advantage of this combiner is that both the load R_L and the ballast resistor R_0 are grounded. These transformer-based combiners can also be used for the power division when the output power from single source is divided and delivered into two independent loads. In this case, the original load and the two signal sources should be switched. These hybrid transformer-based combiners can also be used for the power division when the output power from single source is

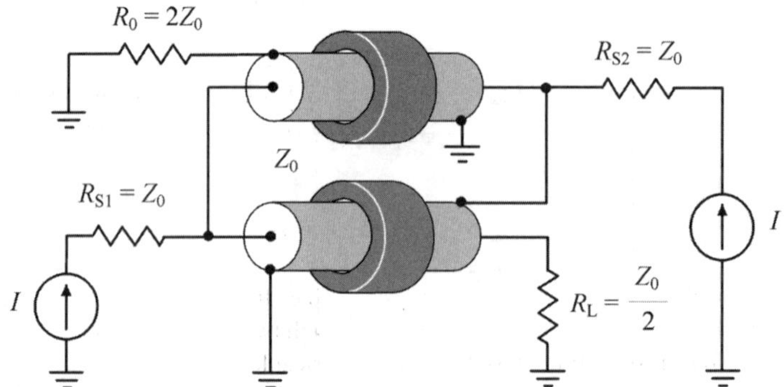

FIGURE 4.12 Two-cable hybrid combiner.

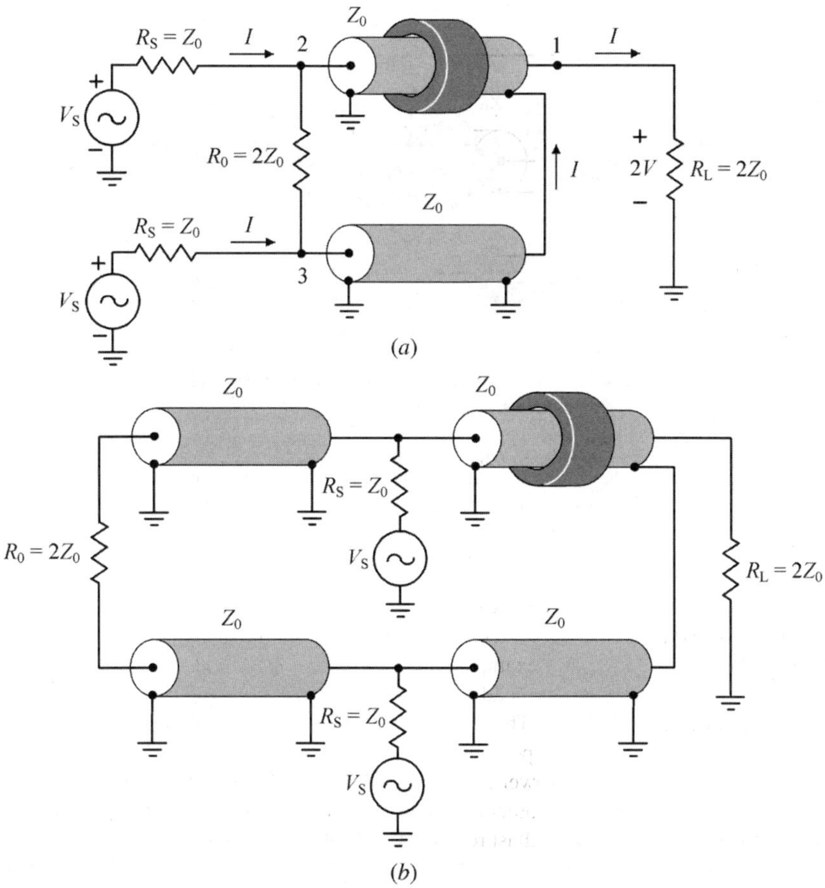

FIGURE 4.13 Coaxial cable combiners with increased isolation.

divided and delivered into two independent loads. In this case, the original load and the two signal sources should be switched. As it turns out, the term "hybrid" comes not from the fact that the transformer might be constructed of two different entities (for example, cable and resistor), but just because it is being driven by two signals as opposed to only one. Consequently, the hybrid transformer represents a four-port device having two input ports, one sum port and one difference port. The unique characteristic of the hybrid transformer is ability to isolate the two input signal sources.

Figure 4.13(*a*) shows a coaxial cable two-way combiner, where the input signals having the same amplitudes and phases at ports 2 and 3 are matched at higher frequencies when all lines are of the same lengths and $R_S = Z_0 = R_L/2 = R_0/2$ [4]. In this case, the isolation between these input ports can be calculated by

$$C_{23}(\text{dB}) = 10\log_{10}\left[4\left(1 + 4\cot^2\theta\right)\right] \tag{4.18}$$

where θ is the electrical length of each transmission line. In order to improve the isolation, the symmetrical ballast resistor R_0 should be connected through two additional lines, as shown in Figure 4.13(*b*), where all transmission lines have the same electrical lengths.

Figure 4.14 shows a coaxial cable two-way combiner, which is fully matched and isolated in pairs [4]. Such combiners can be effectively used in high-power broadcasting VHF FM and VHF–UHF

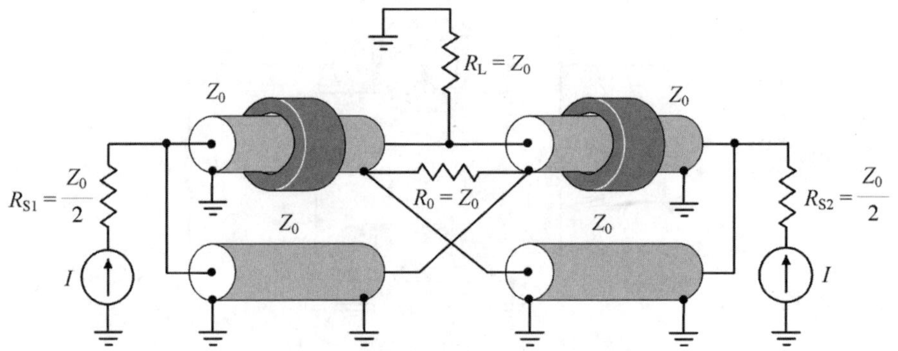

FIGURE 4.14 Fully matched and isolated coaxial cable combiner.

TV transmitters. In this case, for power amplifiers with the identical output impedances R_{S1} and R_{S2} when $R_{S1} = R_{S2} = Z_0/2$, it is necessary to choose the values of the ballast resistor R_0 and the load R_L of $R_0 = R_L = Z_0$, where Z_0 is the characteristic impedance of the each transmission line of the same length.

4.3 BALUNS

Baluns are very important elements in the design of mixers, push–pull amplifiers, or oscillators to link a symmetrical (balanced) circuit to asymmetrical (unbalanced) circuit. Therefore, it makes sense to discuss their circuit configurations and performance in details separately. The main requirements to baluns are to provide an accurate 180-degree phase shift over required frequency bandwidth, with minimum loss and equal balanced impedances. In power amplifiers and oscillators, lack of symmetry will degrade output power and efficiency. Besides, the symmetrical port must be well isolated from ground to minimize an unwanted effect of parasitic capacitances.

A wire-wound transformer whose simplified equivalent schematic is shown in Figure 4.15(*a*) provides an excellent broadband balun covering in commercial applications frequencies from low kHz to beyond 2GHz. They are usually realized with a center-tapped winding that provides a short circuit to even-mode (common-mode) signals while having no effect on the differential (odd-mode) signal. Wire-wound transformers are more expensive than the printed or lumped *LC* baluns, which are more suitable in practical mixer designs. However, unlike wire-wound transformers, the lumped *LC* baluns are narrowband as containing the resonant elements.

Figure 4.15(*b*) shows the circuit schematic of a lattice-type *LC* balun that was proposed a long ago for combining powers in push–pull amplifier and their delivery to antenna [23]. It consists of two capacitors and two inductors, which produce the ±90-degree phase shifts at the output ports. The values of identical inductances *L* and capacitances *C* can be obtained by

$$L = \frac{\sqrt{R_{out} R_L}}{\omega_0} \tag{4.19}$$

$$C = \frac{1}{\omega_0 \sqrt{R_{out} R_L}} \tag{4.20}$$

where ω_0 is the center bandwidth frequency, R_{out} is the balanced output resistance, and R_L is the unbalanced load resistance. When designing this circuit, need to be confident that the operating frequency is well below the self-resonant frequencies of their components.

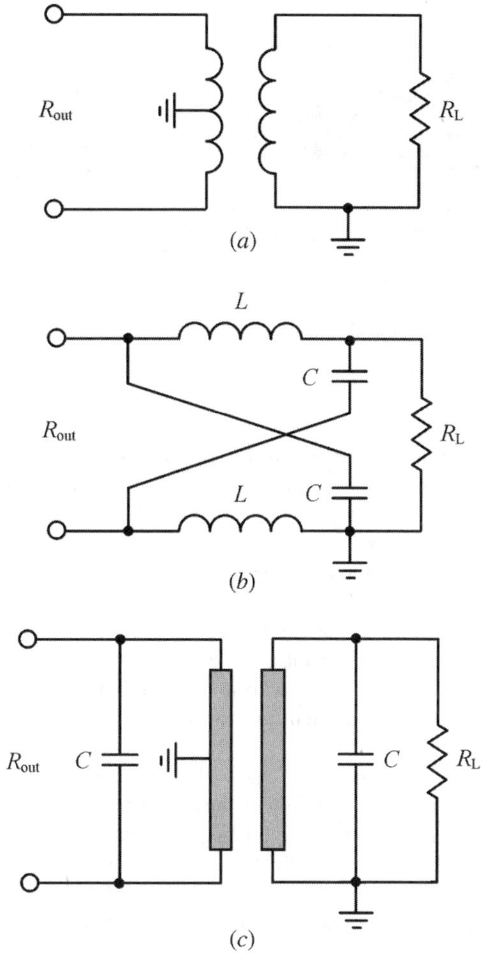

FIGURE 4.15 Different circuit configurations of 1:1 balun.

In monolithic microwave applications where the lumped inductances are usually replaced by transmission lines, the designs with microstrip coupled lines, Lange couplers, or multilayer coupled structures are very popular. However, the electrical length of the transmission lines at center bandwidth frequency is normally set to a quarter-wavelength, which is too large for applications in wireless communication systems. Therefore, it is very attractive to use the lumped-distributed balun structures, which can significantly reduce the balun size and, at the same time, can satisfy the required electrical characteristics. Figure 4.15(*c*) shows such a compact balun with lumped-distributed structure consisting of the two coupled planar microstrip lines and two parallel capacitors, where the input transmission line is grounded at midpoint and the output transmission line is grounded at its one port [24]. Without these capacitors, it is necessary a very small spacing between quarterwave microstrip lines to achieve a 3-dB coupling between them. However, by optimizing the balun elements around the center bandwidth of 900 MHz, the planar structure of approximately one-sixteenth the size of the conventional quarter-wavelength structure was realized, with spacing $S = 8$ mils (note that 1 mil $=$ 0.0254 mm) using an FR4 board with substrate thickness of 300 mils.

Figure 4.16 shows the circuit arrangement with two coaxial-line transformers combined to provide a push–pull operation of the power amplifier by creating a balanced-to-unbalanced impedance

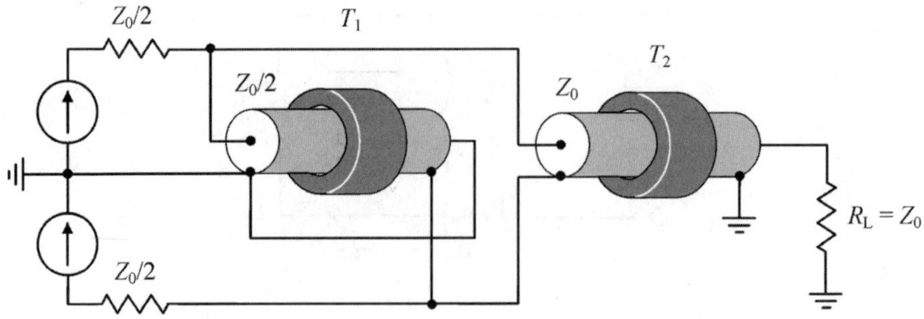

FIGURE 4.16 Circuit arrangement with two cable transformers for push–pull operation.

transformation with higher spectral purity. Ideally, the out-of-phase RF signals from both active devices will have pure half-sinusoidal waveforms, which contain (according to the Fourier series expansion) only fundamental and even harmonic components. This implies a 180-degree shift between fundamental components from both active devices and in-phase condition for remaining even harmonic components. In this case, the transformer T_1 representing a phase inverter is operated as a filter for even harmonics because currents flow through its inner and outer conductors in opposite directions. For each fundamental flowing through its inner and outer conductors in the same directions, it works as an RF choke, the impedance of which depends on the core permeability. Consequently, since the transformer T_2 represents a 1:1 balun, in order to provide maximum power delivery to the load R_L, the output equivalent resistance of each active device should be two times smaller.

For a simple 1:1 transmission-line balun realized with a twisted wire pair or coaxial cable, the balanced end is isolated from ground only at the center bandwidth frequency. To compensate for the short-circuited line reactance over certain frequency bandwidth around center frequency, a series open-circuited transmission line was introduced by Marchand, resulting in a compensated balun, the simplified schematic of which is shown in Figure 4.17(*a*) [25]. In this case, at center bandwidth frequency when the electrical length of the compensated line is a quarter-wavelength, the load resistance R_L is seen unchanged. When this structure is realized with coaxial cables, to eliminate unwanted current existing in the outer conductor and corresponding radiation, it is necessary to additionally provide the certain coupling between the coaxial cables forming a transmission line with two outer conductors, as shown in Figure 4.17(*b*) [26]. Generally, the shunting reactance of this compensating line can reduce the overall balun reactance about center frequency or reverse it sign depending on the balanced load resistance, characteristic impedance of the compensating line, and coupling (characteristic impedance) between the outer conductors of two lines. Hence, a compensating line can create a complementary reactance to a balanced load and provide an improved match over broader frequency range. At microwaves, wire-wound transformers are usually replaced by a pair of the coupled transmission lines shown in Figure 4.17(*c*), thus resulting in a compact planar structure. It should be noted that generally the characteristic impedances of the coaxial or coupled transmission lines can be different to optimize the frequency–bandwidth response.

Multilayer configurations make Marchand balun even more compact and can provide wide bandwidths due to the tight coupling between coupled-line sections. Modeling and synthesis result of a two-layer monolithic Marchand balun configuration with two-coupled lines, the basic structure of which is shown in Figure 4.18(*a*), is discussed in [27]. In this configuration, the unbalanced terminal is connected to the microstrip line located at the upper metallization level, whereas the balanced load is connected to the microstrip lines located at the lower metallization level. The transmission-line sections in different layers are not isolated from each other. It should be noted that, for a given set

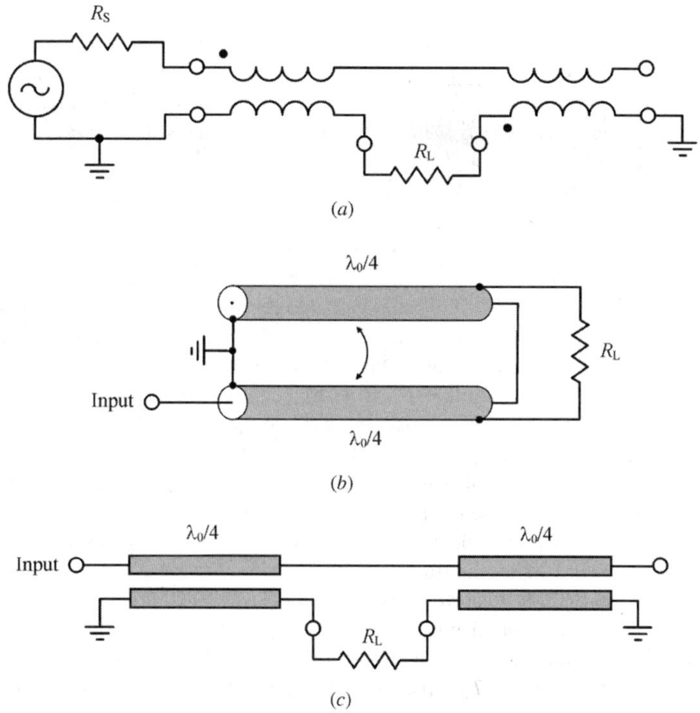

FIGURE 4.17 Schematic configurations of Marchand balun.

of the output balanced and load unbalanced resistances R_{out} and R_L, the characteristic impedances of the outer and inner microstrip lines Z_{01} and Z_{02} are not unique, and they can be calculated from

$$C = \frac{1}{2}\sqrt{\frac{Z_{02}}{Z_{01}}} \tag{4.21}$$

$$Z_{02} = 4Z_{01} - \frac{R_{out}R_L}{Z_{01}} \tag{4.22}$$

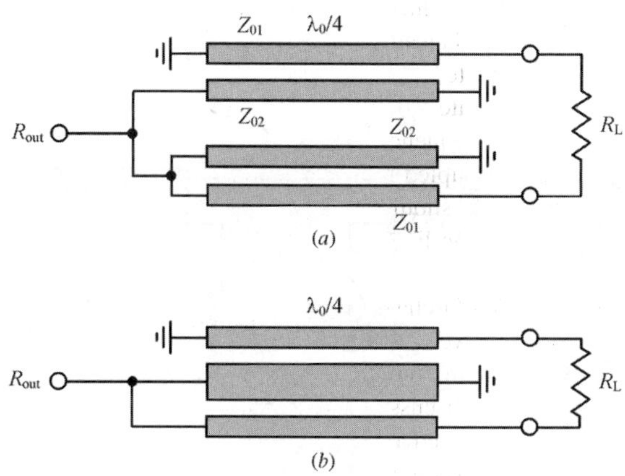

FIGURE 4.18 Schematic configurations of coupled-line Marchand balun.

where C is the coupling factor [28]. However, a different choice of Z_{01} and Z_{02} leads to a different frequency bandwidth. For example, for $R_{out} = 50\ \Omega$ and $R_L = 100\ \Omega$, it was found that using the symmetrical directional coupler with $Z_{01} = Z_{02} = 40.825\ \Omega$ results in a frequency bandwidth of 48.4% with $|S_{11}| < -10$ dB and amplitude imbalance within 0.91 dB, whereas the frequency bandwidth of 20.9% with amplitude imbalance of less than 1.68 dB will be realized for the nonsymmetrical case when $Z_{01} = 38\ \Omega$ and $Z_{02} = 20.42\ \Omega$.

The design of a three-line microstrip balun, the basic schematic of which is shown in Figure 4.18(*b*), is based on the equivalence between a six-port section of three coupled lines and a six-port combination of two couplers [28]. The results of circuit analysis and optimization show that the spacings between adjacent microstrip lines are so narrow that it is difficult to fabricate a single-layer three-line balun. For a two-layer three-line balun with two coupled outer lines on the top metallization level, the spacing between these lines is significantly wider than in a single-layer case. However, wider frequency range can be achieved using a two-layer three-line balun with two coupled outer lines at the lower metallization level. For example, the measurements results for this balun show that, being fabricated on the Duroid RT5880 substrate, it can provide frequency range of 2.13 to 3.78 GHz with amplitude imbalance within 2.12 dB and phase error of less than 4.51°.

To improve the performance of multilayer Marchand balun based on microstrip-line technology over frequency range, a short transmission line can be included connecting the two couplers, as shown in Figure 4.19(*a*) [29]. This additional short microstrip line effectively compensates for the amplitude and phase imbalance caused by the difference in even- and odd-mode phase velocities. Besides, to minimize the balun size, the transmission lines of the coupler can be implemented in meander form that can give up to 90% reduction in size. As a result, the phase and amplitude differences of the compensated balun were within 180 ± 10° and 0 ± 1 dB over the frequency range of 5 to 30 GHz. The compensation can also be implemented by employing capacitors at each end of the coupled lines, as

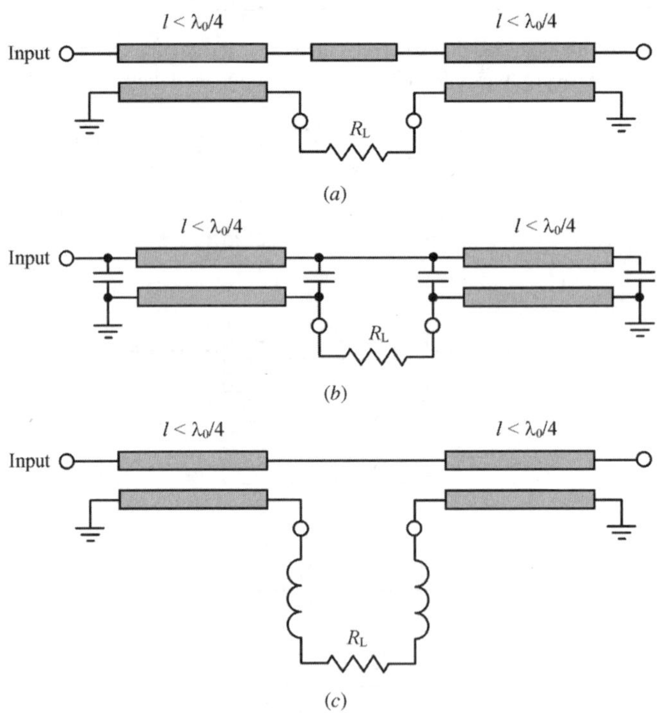

FIGURE 4.19 Schematic configurations of planar Marchand balun.

shown in Figure 4.19(*b*) [30]. In this case, the capacitor will not affect the even-mode but effectively increases odd-mode phase length, thus resulting in a minimum amplitude and phase imbalance over certain frequency bandwidth. An exact synthesis technique that is widely used in filter design can be applied to develop and analyze new classes of miniaturized mixed lumped-distributed planar Marchand baluns using microstrip lines and lumped capacitors [31]. As an alternative, by employing two additional inductors at each balanced output and optimum coupling between the grounded strips shown in Figure 4.19(*c*), a frequency bandwidth of 53% centered around 6.2 GHz with size reduction of 64% over a conventional coupled-line Marchand balun is achieved [32]. A combined compensation technique uses a series capacitor at the unbalanced input port to improve the matching bandwidth and inductors at the ground connections to minimize amplitude and phase imbalance [33].

Figure 4.20(*a*) shows the broadband parallel-connected coaxial cable balun as an alternative to a series-connected Marchand balun [34]. It consists of an unbalanced input coaxial cable connected to a dummy cable that maintains symmetry. On the opposite side of the balun, the output inner and outer conductors are connected in parallel to each other, while the input inner and outer conductors of coaxial cables are cross-connected. The right-hand portion of the balun forms a high impedance balanced load. By means of the cross connection, the high impedance is reduced to a low impedance showing a 4:1 impedance transformation ratio, for example, from a balanced load of 200 Ω to a single-ended 50 Ω. The frequency bandwidth of the balun is limited by the shunting effect at lower frequencies and near half-wave resonance. These parallel-connected baluns can provide approximately four times the operating frequency bandwidth of their series-connected counterparts as covering in the experiment the frequency range from 160 to 4000 MHz, a 25:1 bandwidth [35]. The parallel-connected balun may be realized in a variety of configurations, some of which are shown in Figures 4.20(*b*) and Figure 4.20(*c*) [34,35].

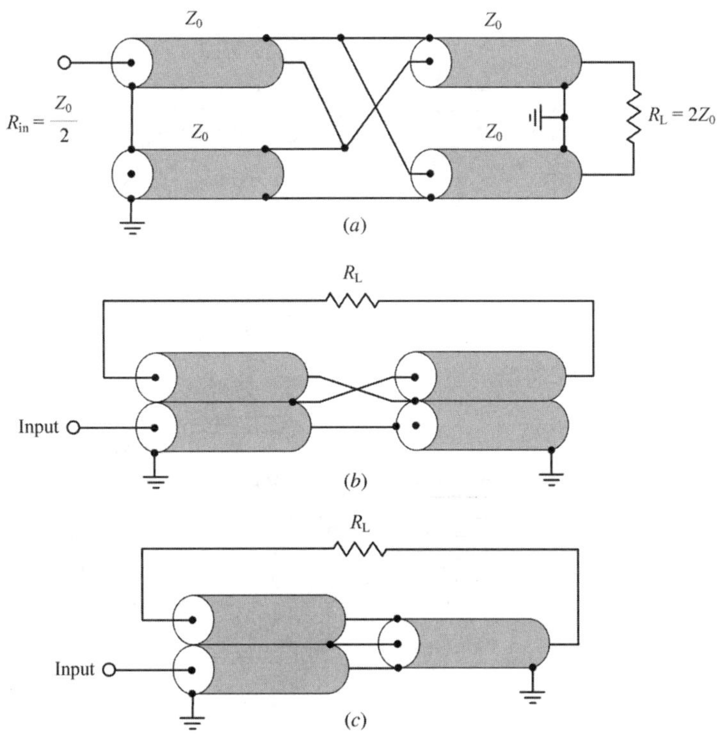

FIGURE 4.20 Broadband parallel-connected cable 1:4 baluns.

4.4 WILKINSON POWER DIVIDERS/COMBINERS

The in-phase power combiners and dividers are the important components of the RF and microwave transmitters when it is necessary to deliver a high level of the output power to antenna, especially in phased-array systems. In this case, it is required to provide a high degree of isolation between output ports over some frequency range for identical in-phase signals with equal amplitudes. Figure 4.21(*a*) shows a planar structure of the basic parallel beam N-way divider/combiner, which provides a combination of powers from the N signal sources. Here, the input impedance of the N transmission lines (connected in parallel) with the characteristic impedance of Z_0 each is equal to Z_0/N. Consequently, an additional quarterwave transmission line with the characteristic impedance $Z_0/\sqrt{N}$ is required to convert the input impedance Z_0/N to standard impedance Z_0. However, this N-way combiner cannot provide sufficient isolation between input ports. The impedances are matched only when all input signals have the same amplitudes and phases at any combiner input. The effect of any input on the remaining ones becomes smaller for combiners with greater number of inputs. For example, if the input signal is delivered into the input port 2 and all other $(N - 1)$ input ports and output port 1 are matched, then the power dissipated at any load connected to the matched input ports will be decreased by $\left(1 - 1/N^2\right) / (2N - 1)$ times and isolation between any two input ports expressed through S-parameters is obtained by

$$S_{ij} = -10\log_{10}\left(\frac{1}{N^2}\frac{N^2-1}{2N-1}\right) \tag{4.23}$$

where N is a number of the input ports and $i, j = 2, \ldots, N + 1$.

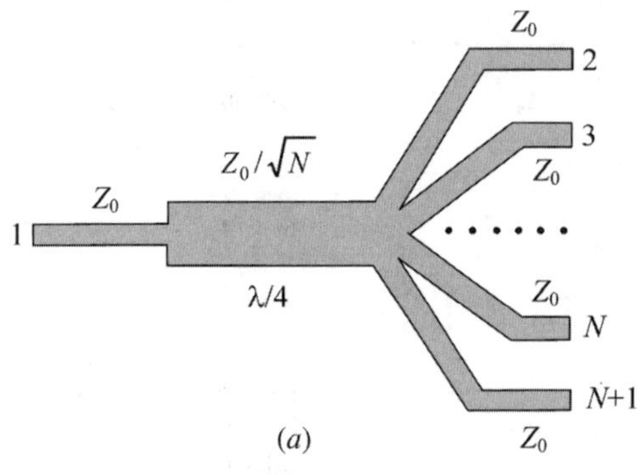

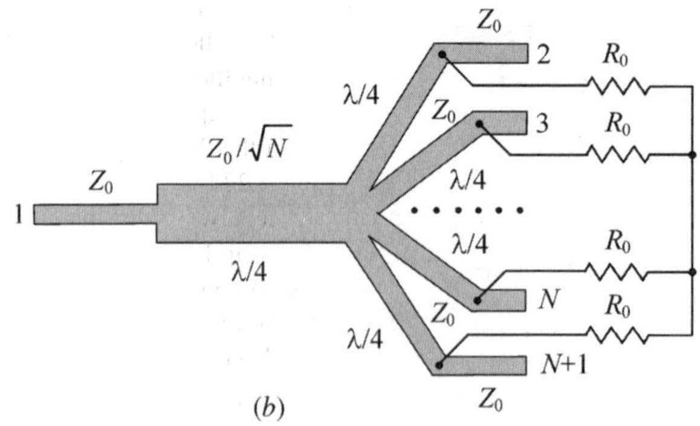

FIGURE 4.21 Circuit topologies of N-way in-phase combiners/dividers.

In most cases, better isolation is required than obtained by Eq. (4.23). The simplest way to provide full isolation between the input and output ports of the combiner is to connect the ferrite isolators (circulators) at the input ports $2, \ldots, N+1$. In this case, the lengths of the transmission lines connected between each ferrite isolator and a quarterwave transmission line should be equal. Although the ferrite isolators increase the overall weight and dimensions of the combiner and contribute to additional insertion losses, nevertheless they provide a very simple combiner realization and protect the connected power amplifiers from the load variations. By using such a 12-way parallel beam combiner, the continuous output power of 1 kW for the L-band transmitter was obtained at the operating frequency of 1.25 GHz [36].

When one or more power amplifiers are destroyed for some reasons, the overall output power P_{out} and efficiency η_c of the combiner can be calculated, respectively, by

$$P_{out} = \frac{(N-M)^2}{N} P_1 \tag{4.24}$$

$$\eta_c = \frac{P_{out}}{P_{in}} = 1 - \frac{M}{N} \tag{4.25}$$

where $P_{in} = (N-M)P_1$, P_1 is the output power from a single power amplifier, N is the number of the input ports, and M is the number of the destroyed power amplifiers. Part of the output power of the remaining power amplifiers will be dissipated within the ferrite isolators (in ballast resistors of circulators). For each ferrite isolator connected to the operating power amplifier, the dissipated power P_{do} can be defined as

$$P_{do} = \left(\frac{M}{N}\right)^2 P_1 \tag{4.26}$$

whereas for each isolator connected to the destroyed power amplifier, the dissipated power P_{dd} can be calculated from

$$P_{dd} = \left(\frac{N-M}{N}\right)^2 P_1. \tag{4.27}$$

In this case, by adding the ballast resistors $R_0 = Z_0$, the right-hand side terminals of which are combined together in a common junction as shown in Figure 4.21(b), matching of all ports, low loss and high isolation between input and output ports can be provided. Such kind of a simple N-way hybrid power divider is known as a Wilkinson power divider [37]. However, it should be mentioned that, historically, this divider/combiner was reported a little bit earlier [38–40]. Originally, a Wilkinson power divider was composed of a coaxial line in which the hollow inner conductor has been split into N splines of length $\lambda/4$, with shorting plate connecting the splines at the input end and resistors connected in a radial manner between each spline at the output end and a common junction. The frequency response of the voltage standing wave ratio at the divider input port, $VSWR_{in}$, depending on the number of the output ports N, is shown in Figure 4.22 [41].

The hybrid planar microstrip realization of the simplest two-way Wilkinson divider is shown in Figure 4.23(a). It consists of the two quarterwave microstrip lines connected in parallel at the input end and the planar ballast resistor connected between the output ports of the microstrip lines. Despite its small dimensions and simple construction, such a divider provides a sufficient isolation between output ports over sufficiently wide frequency bandwidth when equal power division is provided due to a symmetrical configuration with $R_0 = 2Z_0$ and $Z_1 = Z_0\sqrt{2}$. However, in practice, it is necessary to take into account the distributed RC structure of the ballast resistor when its size is sufficiently large, as well as manufacturing tolerances and discontinuities. As a result, in a frequency bandwidth

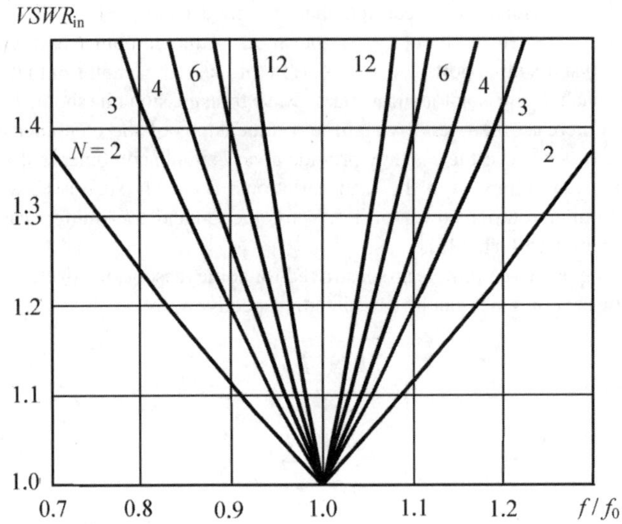

FIGURE 4.22 Frequency performance of N-way Wilkinson divider.

of 30% with $VSWR_{in} \leq 1.2$ at input port 1 and $VSWR_{out} \leq 1.03$ at output ports 2 and 3, the isolation between the divider outputs can be better than 20 dB [42].

In millimeter-wave integrated circuits, in order to increase a self-resonant frequency of the ballast chip resistor, the overall MMIC dimensions must be very small. This means that the two branches of the power divider are very close to each other, which leads to strong mutual coupling between the output microstrip lines and, as a result, upsets the desired power-split ratio. A possible solution is to use the branches with the electrical lengths of $3\lambda/4$ instead of $\lambda/4$ and to include the two additional branches into a semicircle, as shown in Figure 4.23(b) [43]. These additional branches should be of the half-wave electrical lengths with the characteristic impedances equal to Z_0. In this case, isolation can be better than 17 dB between all ports with the insertion loss of about 1.3 dB at the operating frequency of 30.4 GHz.

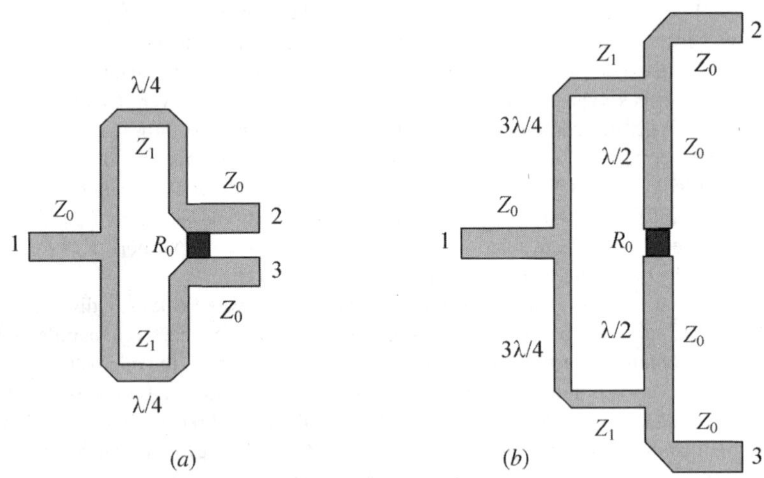

FIGURE 4.23 Microstrip realization of two-way Wilkinson dividers.

However, the ballast resistors of the conventional N-way Wilkinson combiners/dividers cannot be designed to be a planar structure when their physical lengths and connecting wires are minimal to provide sufficient isolation among output ports over the required frequency range. For example, the radial and fork N-way hybrids have reasonably wide frequency bandwidth, of about 20% and higher, but their match and isolation are not perfect even at the center bandwidth frequency [44]. Besides, due to small size of the ballast resistor compared to the wavelength and its balanced structure, it is difficult to heat-sink it in the case of high power combining. In order to provide higher output power capability, it is possible to modify the N-way Wilkinson combiner/divider by replacing the ballast resistive star with a combination of quarterwave transmission lines and shunt-connected resistors [45]. In this case, each ballast resistor is connected to corresponding output port through a transmission line. At the same time, all ballast resistors are connected to a common floating starpoint by the transmission lines. Such a modification has an advantage of external isolation loads (high-power ballast resistors) and easy monitoring capability for imbalances at the output ports. For a two-way planar power combiner/divider, the circuit topology of which is shown in Figure 4.24, the balanced 100-Ω ballast resistor is replaced by a transmission-line network and two 50-Ω resistors are connected to ground acting as the out-of-phase load, where $Z_1 = Z_0\sqrt{2}$, $Z_2 = Z_0/\sqrt{2}$, and $Z_0 = R_0 = 50\ \Omega$ [46].

The cascade connection of two-way Wilkinson power combiners/dividers can provide a multiway power division or power combining. The simplest practical realization is the binary power divider/combiner, composed of the n stages when each consecutive stage of which contains an increasing by 2^N number of two-way dividers/combiners. For a single destroyed power amplifier, the power dissipated in the ballast resistors is equal to

$$P_{\rm db} = \left(1 - \frac{1}{N}\right)^2 P_1. \tag{4.28}$$

The output power of $P_1/2$ is dissipated in the ballast resistor adjacent to the destroyed power amplifier; the output power of $P_1/4$ is dissipated in the ballast resistor of the next stage, and so on. It should be mentioned that the power divider with a number of outputs multiple to 4^N represents the convenient case when the characteristic impedance of the transmission line are of the same impedance. Figure 4.25 shows the four-way microstrip Wilkinson divider/combiner fabricated on alumina substrate with six 50-Ω quarterwave microstrip lines and two 100 Ω and one 50-Ω thin-film resistors. This microstrip Wilkinson power divider/combiner can provide the insertion loss of less

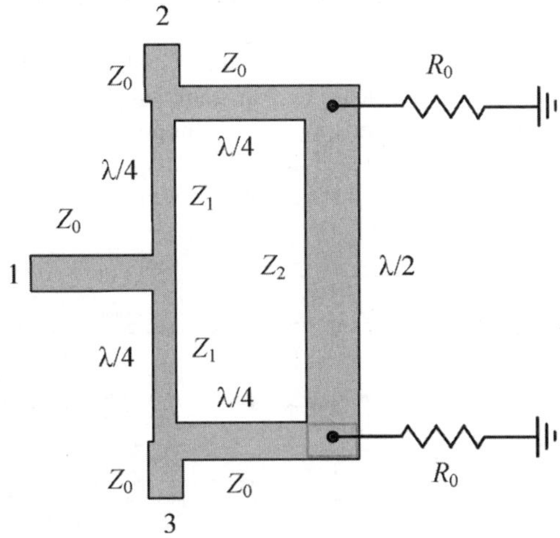

FIGURE 4.24 Gysel high-power in-phase planar combiner/divider.

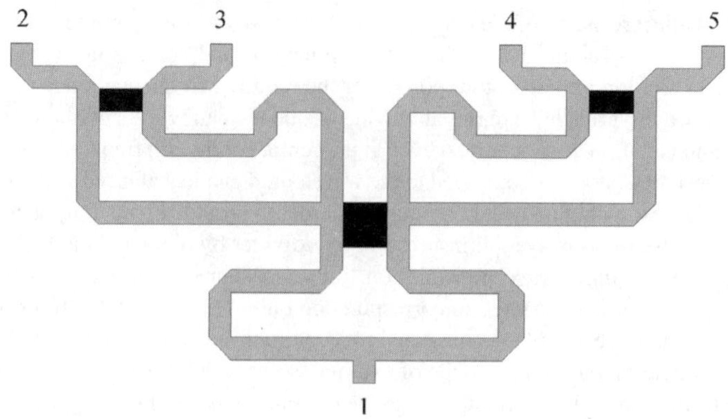

FIGURE 4.25 Practical four-way microstrip wilkinson power combiner/divider.

than 0.3 dB and isolation between any outputs of about 20 dB in a frequency bandwidth of ±10% in decimeter frequency band.

The frequency bandwidth property of a Wilkinson power divider/combiner can be improved with an increasing number of its sections [42]. Generally, a broadband two-way Wilkinson power divider can contain N pairs of equal-length transmission lines and N bridging resistors distributed from input port 1 to output ports 2 and 3. For example, for $N = 2$, the theoretical minimum isolation in an octave band between ports 2 and 3 can achieve 27.3 dB with *VSWR* at each port better than 1.1. In monolithic microwave integrated circuits (MMICs), by using a two-metal layer GaAs HBT process when the bottom metal layer can realize a coplanar waveguide (CPW) transmission line and the top metal layer can realize a microstrip transmission line, the size of a two-section two-way power divider/combiner can be reduced. In this case, an isolation of 15 dB and a return loss of 15 dB can be achieved in a frequency bandwidth from 15 to 45 GHz [47].

Figure 4.26 shows the equivalent circuit representation of a three-way modified Wilkinson power divider/combiner [48]. Assuming all the impedances of the input and three output ports be 50 Ω, the characteristic impedances of the quarterwave transmission lines are selected for a maximally flat performance as $Z_1 = 114\ \Omega$ and $Z_2 = 65.8\ \Omega$. To match circuit at the center frequency, the values of the ballast planar resistors should be chosen as $R_1 = 64.95\ \Omega$ and $R_2 = 200\ \Omega$. In this case, the isolation between output ports of such a three-way divider demonstrates more than 20 dB in an octave frequency bandwidth.

Generally, high characteristic impedance values (usually higher than 100 Ω) for the transmission lines can create a problem in their practical microstrip implementation, since their narrow widths increase the insertion loss. In this case, using a recombinant power divider, the topology of which is shown in Figure 4.27, provides an isolation of 20 dB in a frequency range of 72% for a maximum line impedance of 80 Ω and requires only three isolation resistors [49]. This three-way recombinant

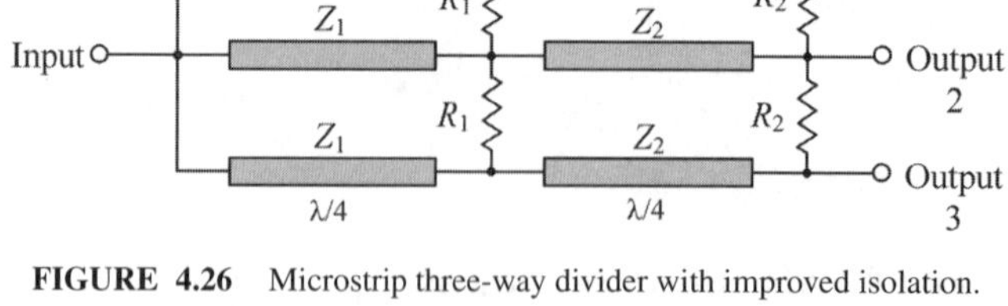

FIGURE 4.26 Microstrip three-way divider with improved isolation.

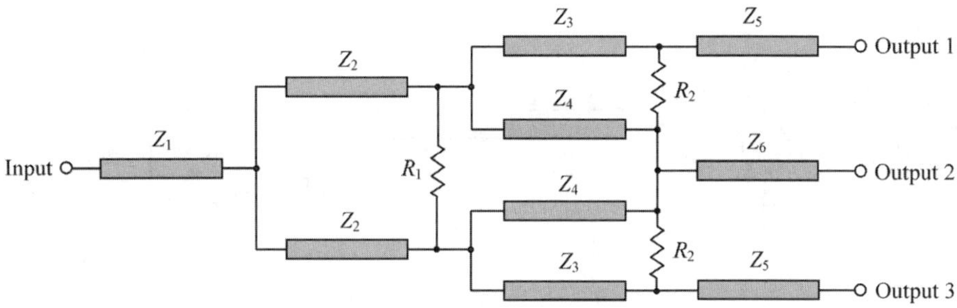

FIGURE 4.27 Microstrip three-way recombinant divider with improved isolation.

divider is characterized by the insertion loss of about 1 dB and return loss of more than 12 dB in a frequency range of 6 to 14 GHz, fabricated in 25-mil thick 99.6% alumina substrate. The design values for the quarterwave transmission lines were $Z_1 = 36\ \Omega$, $Z_2 = Z_3 = 40\ \Omega$, $Z_4 = 80\ \Omega$, and $Z_5 = Z_6 = 40\ \Omega$ with the ballast resistors $R_1 = 50\ \Omega$ and $R_2 = 100\ \Omega$, respectively. Over a 2:1 bandwidth, the center-to-side and side-to-side isolations exceed 20 dB.

The divider broadband properties can also be improved by using the more complicated phase-shifting circuit instead of a simple microstrip line. The phase shift between two output ports 2 and 3 will be close to 90° in an octave frequency range if a Schiffman element based on the coupled microstrip lines is connected to one output port [50]. At the same time, an additional microstrip line with the electrical length of 270° at the center bandwidth frequency should be connected to the second output port.

In the design of a microwave distributed network, a power divider providing two equal-phase outputs with unequal power division is often required. The split-tee power divider is a simple compact and broadband device. It provides two isolated equal-phase unequal-amplitude outputs with a good match at each port. Since a split-tee power divider is similar to the N-way equiphase equiamplitude power divider, it can be developed from this N-way divider as follows: connect M of the output ports together to form one port and the remaining $(N - M)$ output ports together to form the other port, connect quarterwave transformers to the resulting output ports to adjust their impedance level, and a power divider with two equiphase outputs and power ratio of $N/(N - M)$ is derived.

The basic schematic of a power divider with unequal output load impedances is shown in Figure 4.28(a) [51]. This power divider is designed so that, when fed from input port 1, the perfect match will be achieved at the center bandwidth frequency when the output power at port 3 is K^2 times the output power at port 2, and the voltage between port 2 and ground is equal to the voltage between port 3 and ground when measured at equal distances from port 1. To satisfy these conditions, the characteristic impedances Z_1 and Z_2 for unequal loads $R_2 = KZ_0$ and $R_3 = Z_0/K$ are calculated from

$$Z_1 = KZ_0\sqrt{K + \frac{1}{K}} \tag{4.29}$$

$$Z_2 = \frac{Z_0}{K}\sqrt{K + \frac{1}{K}} \tag{4.30}$$

where both transmission lines are of a quarter wavelength at the center bandwidth frequency.

Since the voltages at port 2 and port 3 are equal with this design, a resistor may be placed between these two ports without causing any power dissipation. However, isolation between output ports and a good match seen looking in at any ports is obtainable because of this resistor. Finally, to transform the two unequal output impedances to output impedance Z_0 equal for each output port, the characteristic

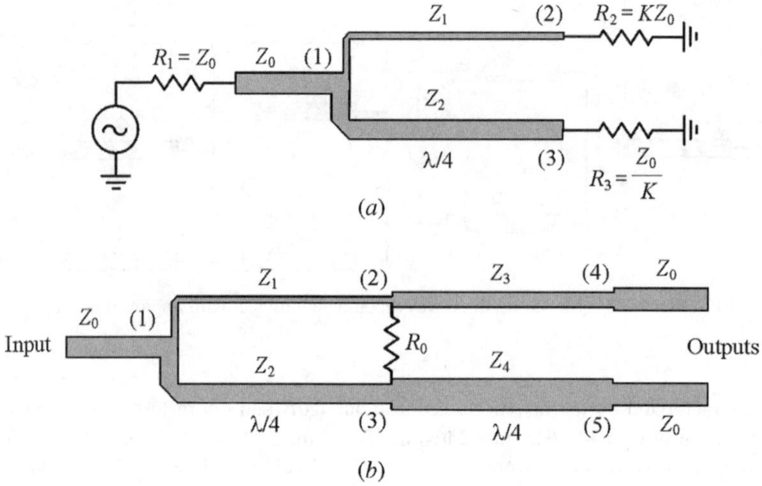

FIGURE 4.28 Split-tee power divider.

impedances of additional quarterwave transformers Z_3 and Z_4 and ballast resistor R_0 shown in Figure 4.28(*b*) are determined from

$$Z_3 = Z_0\sqrt{K} \tag{4.31}$$

$$Z_4 = \frac{Z_0}{\sqrt{K}} \tag{4.32}$$

$$R_0 = Z_0\left(K + \frac{1}{K}\right). \tag{4.33}$$

The three-way power divider with various output power ratios, which represents a planar structure and can be easily realized using microstrip lines with reasonable characteristic impedances, is shown in Figure 4.29 [52]. When port 1 is an input port, the input power is divided by a ratio of M:N:K at corresponding output ports 2, 4, 6 with isolated ports 3 and 5. The electrical lengths of the transmission lines must be 90° except for the half-wave middle horizontal line. The characteristic impedances of the transmission lines can be calculated from

$$Z_1 = Z_0\sqrt{\frac{\Delta_1}{\Delta_2}} \tag{4.34}$$

$$Z_2 = Z_0\sqrt{\frac{\Delta_1}{M}} \tag{4.35}$$

$$Z_3 = Z_0 \tag{4.36}$$

$$Z_4 = Z_0\sqrt{\frac{\Delta_2}{N}} \tag{4.37}$$

$$Z_5 = Z_0\sqrt{\frac{\Delta_2}{K}} \tag{4.38}$$

where $\Delta_1 = M + N + K$ and $\Delta_2 = N + K$. For example, for a three-way divider with $M = 3$, $N = 2$ and $K = 1$, it follows that $Z_1 = Z_2 = 1.41Z_0$, $Z_4 = 1.22Z_0$ and $Z_5 = 1.73Z_0$. The same

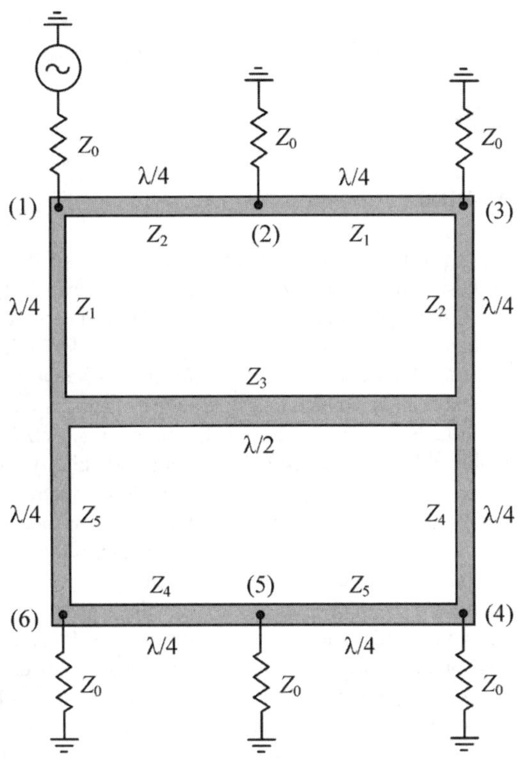

FIGURE 4.29 New type of three-way power divider.

characteristic impedances are required for a 1:1:1 equal-power three-way divider, only the input port must be changed to port 4 in this case.

Figure 4.30 shows the compact microstrip three-way Wilkinson power divider designed to operate over a frequency range of 1.7 to 2.1 GHz, with minimum combining efficiency of 93.8%, maximum amplitude imbalance of 0.35 dB, and isolation better than 15 dB [53]. To avoid any amplitude and phase imbalances between the divider 50 Ω output ports, the ballast resistor connected to its middle branch should be split into two equal parallel resistors. To obtain an ideal floating node, these two

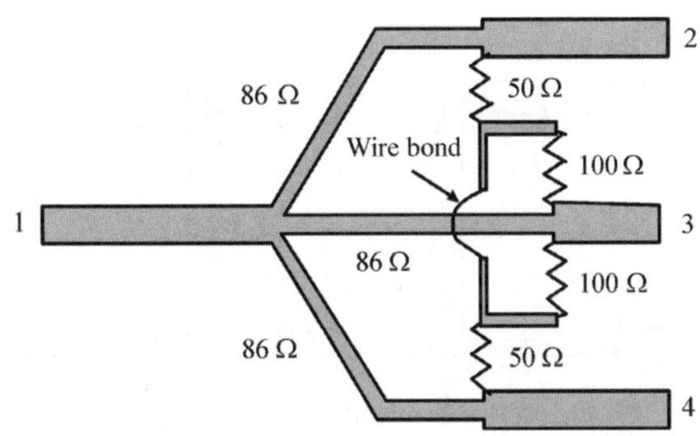

FIGURE 4.30 Compact microstrip three-way Wilkinson power divider.

resistors are connected together with narrow microstrip lines that are as short as possible. Finally, to connect the resistors from both sides of the middle branch, a copper wire of 7-mil diameter is used. The most critical parameter is the isolation between port 2 and port 4, which can be improved by shortening the bondwire length.

4.5 MICROWAVE HYBRIDS

The branch-line couplers or hybrids were firstly described more than six decades ago; however, the problem of their exact synthesis remained a puzzle for a number of years [54]. Initially, the branch-line hybrid was analyzed as a four-arm symmetrical network based on a superposition of the results obtained in the even and odd modes [55]. By writing the even- and odd-mode matrices together, the characteristic impedances of the branch lines and coupling into different ports can be obtained. A general synthesis procedure that can be applied to any structure of a multibranch hybrid, based on an invariance of the Richard's variable $S = j\tan\theta$ to the transformation of $S \rightarrow 1/S$ apart from a 180° phase change, had become available a decade later [56]. As a result, with highly precise computer-design techniques available for branch-line hybrids, it became possible to generate any coupling value in the useful 0–15-dB coupling range. Their waveguide designs that have been used in large complex feeds for phase-array radars are compact, highly predictable in amplitude and phase characteristics, and handle very high power. Coaxial, microstrip, or stripline implementations of branch-line hybrids provide simple planar structures of moderate bandwidth capability, up to about 2/3 of an octave.

For a fully matched case with standard 50-Ω source and load impedances, when the characteristic impedances of its transverse branches are 50 Ω and the characteristic impedances of its longitudinal main lines are $50/\sqrt{2} = 35.4$ Ω, the microstrip branch-line hybrid shown in Figure 4.31 represents a 3-dB directional coupler, for which power in arm 1 divides evenly between arms 2 and 3 with the phase shift of 90°. No power is delivered to arm 4, because the signal flowing through different paths (lengths of λ/4 and 3λ/4) have the same amplitude and opposite phases at this port. The branch-line hybrid does not depend on the load mismatch level for equal reflected coefficients from the outputs when all reflected power is dissipated in the 50-Ω ballast resistor. However, in practice, due to the quarter-wavelength transmission-line requirement, the bandwidth of such a single-stage quadrature branch-line hybrid is limited to 10–20%.

Figure 4.32 shows the calculated frequency bandwidth characteristics of a single-section branch-line hybrid matched at the center bandwidth frequency with the load impedance $Z_L = Z_0 = 50\,\Omega$ [57]. At millimeter-wave frequencies, the lengths of the microstrip lines can actually get shorter than the widths and the mutual coupling between the input lines and discontinuities at the input increases significantly. This has a direct effect on the input/output match, frequency bandwidth, and isolation. To minimize the effect of these problems, the branch-line hybrid can be designed as a two-section

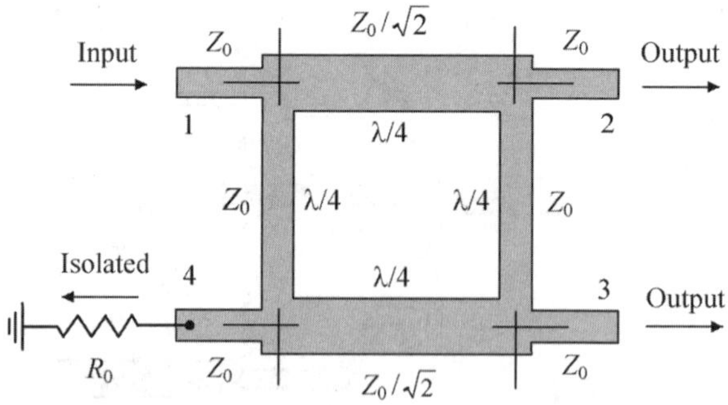

FIGURE 4.31 Microstrip branch-line quadrature hybrid.

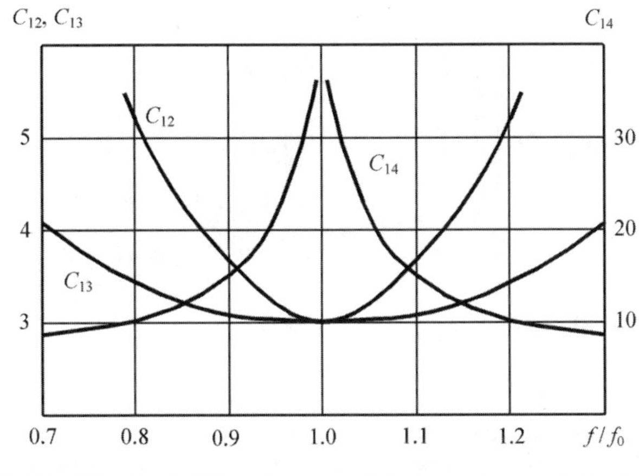

FIGURE 4.32 Bandwidth performance of single-section branch-line hybrid.

hybrid using three-quarterwave lines for the series main lines and quarterwave lines for the shunt branch lines, with all inputs/outputs orthogonal to each other [58]. As a result, the return loss can achieve 10 dB or better over 90% of the band, the isolation was 10 dB or better over the whole band, and the difference in the coupling can be equal or less than 1 dB over about 75% of the frequency band from 26 to 40 GHz.

If one pair of terminating resistors has different values compared to the other pair, the resulting branch-line hybrid can operate as a directional coupler and an impedance transformer simultaneously [59]. Design values of the branch- and main-line characteristic impedances for a single-section branch-line hybrid shown in Figure 4.33, related to the input source impedance Z_{0S} and output load impedance Z_{0L}, can be calculated by

$$Z_1 = \frac{Z_{0S}}{K} \tag{4.39}$$

$$Z_2 = \sqrt{\frac{Z_{0S} Z_{0L}}{1 + K^2}} \tag{4.40}$$

$$Z_3 = \frac{Z_1 Z_{0L}}{Z_{0S}} \tag{4.41}$$

where K is the voltage-split ratio between output ports 2 and 3, and $R_0 = Z_{0S}$ [60]. Such a hybrid with a 2-to-1 (50- to 25-Ω) impedance transformation ratio can provide approximately 20-percent

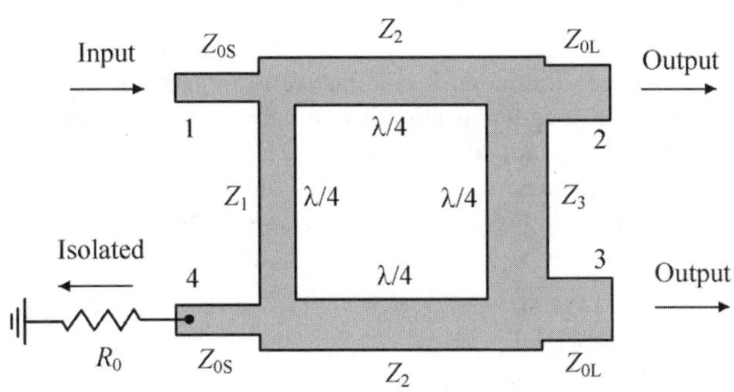

FIGURE 4.33 Microstrip branch-line quadrature impedance-transforming hybrid.

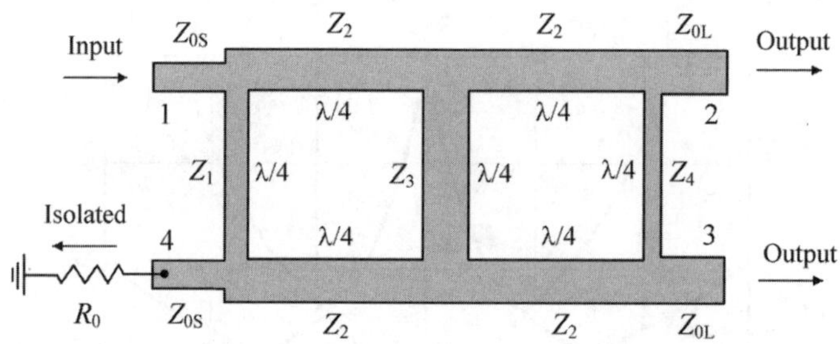

FIGURE 4.34 Broadband microwave branch-line quadrature hybrid.

frequency bandwidth with ±0.25-dB amplitude imbalance. However, for a fixed directivity, the frequency bandwidth of branch-line impedance-transforming hybrid increases as the output-to-input impedance ratio is reduced [59].

The operating bandwidth can be significantly increased using multistage impedance-transforming hybrids. A two-section branch-line impedance-transforming quadrature hybrid is shown in Figure 4.34. To design this hybrid with the given impedance transformation ratio r and power-split ratio K^2, the branch- and main-line characteristic impedances should be chosen according to

$$Z_1 = Z_{0S}\sqrt{r\frac{t^2 - r}{t - r}} \tag{4.42}$$

$$\frac{Z_2^2}{Z_3} = Z_{0S}\sqrt{r - \left(\frac{r}{t}\right)^2} \tag{4.43}$$

$$Z_4 = Z_{0S}\frac{\sqrt{r\left(t^2 - r\right)}}{t - 1} \tag{4.44}$$

where $t = r\sqrt{1 + K^2}$ [61]. The condition of $Z_2 = Z_3$ gives maximum bandwidth when the best performance at the center bandwidth frequency is specified.

For an equal power division when $K = 1$, the condition of $t = r\sqrt{2}$ specifies a minimum value of r, which is equal to 0.5. However, in practice, it is better to choose r in the range of 0.7 to 1.3, in order to provide the physically realizable branch-line characteristic impedances for 50-Ω input impedance. For example, for the 50- to 35-Ω impedance transformation using a two-stage hybrid, the impedances are as follows: $Z_1 = 72.5\ \Omega$, $Z_2 = Z_3 = 29.6\ \Omega$, and $Z_4 = 191.25\ \Omega$. This gives the power balance between the output ports better than 0.5 dB with the return loss and isolation better than 20 dB over a frequency bandwidth of 25% for a 2-GHz hybrid.

For MMIC applications, the overall size of the quadrature branch-line hybrids with quarterwave transmission lines is too large. Therefore, it is attractive to replace each quarterwave branch line with the combination of a short-length transmission line and two shunt capacitors providing the same bandwidth properties. Consider the admittance matrix $[Y_a]$ for a quarterwave transmission line shown in Figure 4.35(a) and the admittance matrix $[Y_b]$ for a circuit consisting of a short transmission line with two shunt capacitors shown in Figure 4.35(b), which are given by

$$[Y_a] = \frac{1}{jZ_0}\begin{bmatrix} 0 & -1 \\ -1 & 0 \end{bmatrix} \tag{4.45}$$

$$[Y_b] = \frac{1}{jZ\sin\theta}\begin{bmatrix} \cos\theta - \omega CZ\sin\theta & -1 \\ -1 & \cos\theta - \omega CZ\sin\theta \end{bmatrix} \tag{4.46}$$

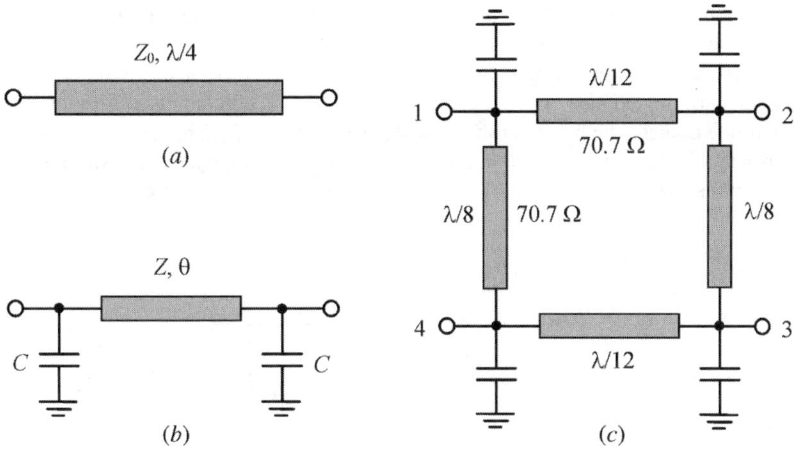

FIGURE 4.35 Reduced-size branch-line quadrature hybrid.

where Z_0 is the characteristic impedance of a quarterwave line, Z and θ are the characteristic impedance and electrical length of a shortened line, and C is the shunt capacitance. By equating the corresponding Y-parameters in Eqs. (4.45) and (4.46), the simple ratios between the circuit parameters can be obtained in the form of

$$Z = \frac{Z_0}{\sin\theta} \tag{4.47}$$

$$C = \frac{1}{\omega Z_0 \cos\theta} \tag{4.48}$$

from which it follows that the lengths of the hybrid transmission lines can be made much shorter by increasing their characteristic impedance Z. For example, when choosing the electrical length of $\theta = 45^\circ$, the characteristic impedance of the transmission line increases by a factor of $\sqrt{2}$.

A circuit schematic of the reduced-size branch-line quadrature hybrid is shown in Figure 4.35(*c*) [62]. Compared to the conventional branch-line hybrid with the characteristic impedances of its branch- and main-lines of Z_0 and $Z_0/\sqrt{2}$, respectively, the circuit parameters of the reduced-size branch-line hybrid are obtained from

$$\theta_1 = \sin^{-1}\left(\frac{Z_0}{Z}\right) \tag{4.49}$$

$$\theta_2 = \sin^{-1}\left(\frac{Z_0}{Z\sqrt{2}}\right) \tag{4.50}$$

$$\omega C Z_0 = \sqrt{1-\left(\frac{Z_0}{Z}\right)^2} + \sqrt{2-\left(\frac{Z_0}{Z}\right)^2} \tag{4.51}$$

where θ_1 and θ_2 are the electrical lengths of the shunt branch-line and series main-line, respectively. For a particular case of the standard characteristic impedance $Z_0 = 50\ \Omega$, the characteristic impedance and electrical lengths of the transmission lines are defined as $Z = Z_0\sqrt{2}, \theta_1 = 45^\circ$, and $\theta_2 = 30^\circ$, as shown in Figure 4.35(*c*). Experimental results for a 25-GHz reduced-size branch-line quadrature hybrid show that its bandwidth performance is slightly narrower than that of the conventional quarterwave hybrid, but its overall size is more than 80% smaller.

To further reduce the MMIC size, the transmission lines can be fully replaced by the lumped planar inductors. Such an approach becomes possible because the symmetrical lumped LC-type π- or T-section is equivalent at a single frequency to the transmission-line section with the appropriate characteristic impedance and electrical length. The lumped-element equivalent circuit of a transmission-line branch-line hybrid is shown in Figure 4.36(a) [63]. This circuit has also some additional advantages when each its section can work as a separate impedance transformer, a low-pass filter, and a phase shifter. The circuit can be diced into four separate sections and cascaded for the

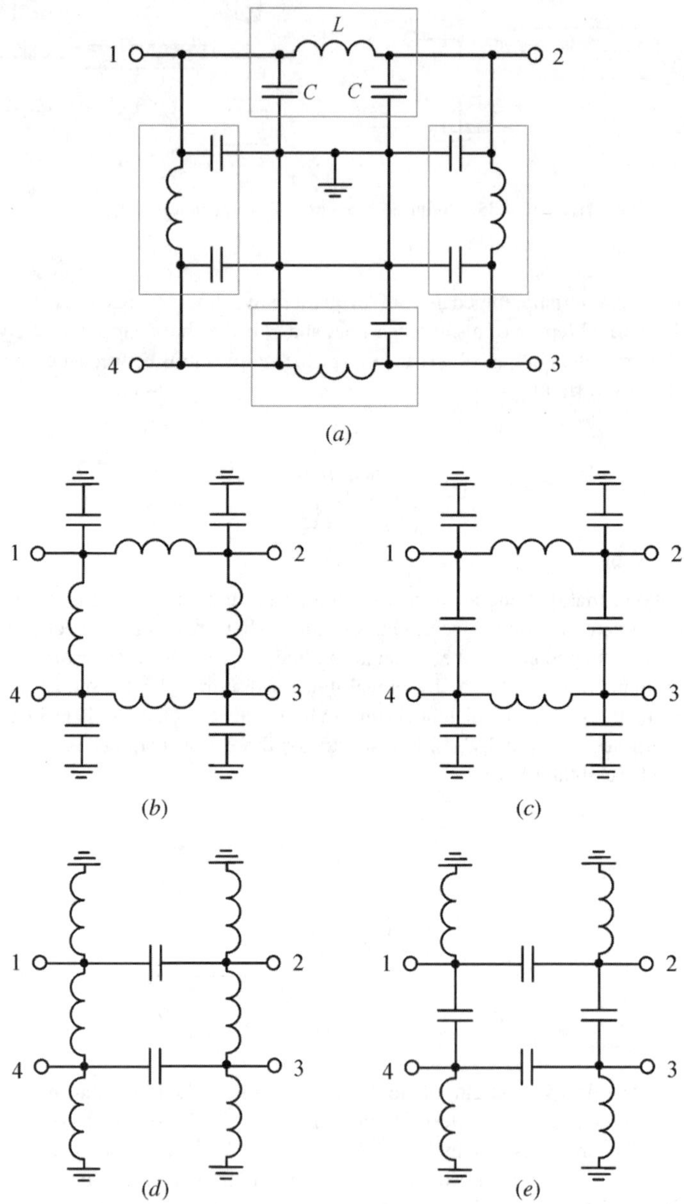

FIGURE 4.36 Equivalent circuits of lumped LC-type hybrid.

desired transmission characteristics. The circuit analysis indicates that various types of networks fulfill the conditions required for an ideal hybrid. Therefore, greater design flexibility in the choice of the hybrid structure and performance is possible.

Several possible single-section two-branch hybrid options are shown in Figure 4.36 [64]. In this case, it should be noted that not only a low-pass section but also a high-pass section can be respectively used. In the latter case, the high-pass *LC* section is considered an equivalent single-frequency replacement for a 270-degree transmission line [65]. Figures 4.36(*c*) and 4.36(*d*) illustrate the use of both low-pass and high-pass sections simultaneously, while only high-pass sections compose the hybrid shown in Figure 4.36(*e*). The performances of the hybrids shown in Figures 4.36(*b*) and 4.36(*e*) are very similar to that of the classical single-section branch-line hybrid. The bandwidth performances of the hybrids shown in Figures 4.36(*c*) and 4.36(*d*) are narrower because the power balance between their output ports is much narrower. Broader bandwidth and lower output impedances can be provided with a two-section three-branch lumped-element hybrid.

Figure 4.37(*a*) shows the equivalent circuit of a capacitively coupled lumped-element hybrid, which is used for monolithic design of variable phase shifters [66]. However, the power and phase balance bandwidth at the output ports of this hybrid is very narrow, in the limits of a few percent. An alternative design of an inductively coupled lumped-element hybrid is shown in Figure 4.37(*b*) [67]. As a basic element, it includes lumped multiturn mutually coupled spiral inductors with the coupling coefficient, which can be realized using a bifilar (a sandwich of two multiturn spiral inductors with inner and outer windings) spiral transformer to achieve a coupling coefficient $k = 0.707$. In this case, this basic lumped-element configuration is completely equivalent to a transmission line in the vicinity of the center bandwidth frequency. The inductively coupled hybrid can provide a power balance within 0.2 dB and phase balance within 1° in a frequency bandwidth of ±10% in 2-GHz wireless applications. However, during the design procedure, some parasitic effects should be taken

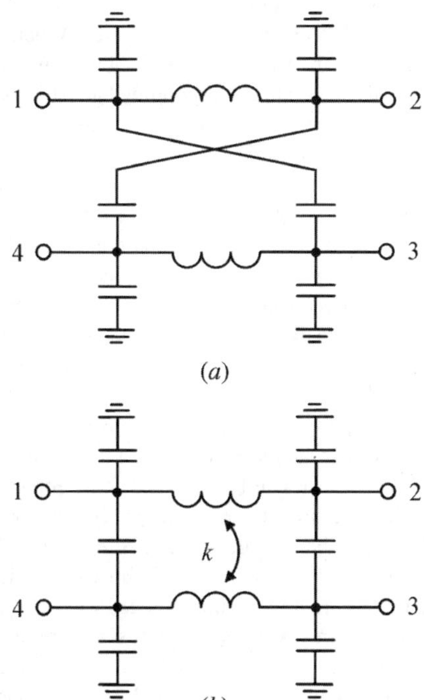

FIGURE 4.37 Equivalent circuits of lumped hybrid with capacitive and inductive coupling.

into account. For example, the coupled inductor itself has a sufficient value of internal capacitance. Also, the finite value of the inductor quality factor results in a modest amplitude imbalance, but it leads to a significant phase deviation from ideal quadrature 90-degree difference. In this case, to compensate for the resulting performance degradation, the electromagnetic simulation of the structure and optimization of the values of the added shunt capacitors on both sides of the circuit are required. Two of these inductively coupled hybrids can be cascaded in tandem to significantly extend the frequency bandwidth. As a result, the phase shift of $93 \pm 6^\circ$, the insertion loss between 1 and 1.5 dB, the return loss better than 16 dB, and the isolation between the output ports better than 18 dB were measured over the frequency range from 2 to 6 GHz [67].

A hybrid-ring directional coupler or rat-race, which is one of the fundamental components used in microwave circuits was described and analyzed more than six decades ago [68]. Its operation principle is based on an assumption that the voltage at any point along the transmission line is a superposition of the forward and backward propagating waves. Signal from the excitation source spreads out in the driving point and propagates along the line. As the forward wave reaches the far-end termination, it reflects, propagates backward, reflects from the near-end termination, propagates forward again, and continues in a loop. According to this wave behavior, a pure standing wave is set up within the ring when there is no mechanism for dissipation other than the minor ohmic losses associated with wave transmission. The point of voltage minimum (zero) corresponds to the case when the waves have phase shift of 180° with respect to each other. As a result, the standing wave within the ring can be mapped by marking off alternate voltage maxima and minima at quarterwave intervals. Consequently, if standing waves are set up in the main arm, the side arm receives maximum power when a voltage minimum of the standing-wave pattern coincides with the center of this arm and minimum power when a voltage maximum located in this point.

The $3\lambda/2$ ring hybrid, whose microstrip topology is shown in Figure 4.38(*a*), can be used to divide the driving power or to combine the powers from two sources. In the case of power division, a signal applied to the port 1 will be evenly split into two in-phase components at ports 2 and 3, and port 4 will be isolated. If the input signal is applied at port 4, it will be equally split into two components with 180-degree difference at ports 2 and 3, and port 1 will be isolated. When operated as a combiner, with input signals applied at ports 2 and 3, the sum of the input signals will be delivered to port 1. Their difference will flow to port 4 and dissipated in corresponding ballast resistor R_0. For equal signals and matched ring hybrid with $Z_0 = R_0$, there is zero-power dissipation at port 4 [55]. The ideal rat-race or ring hybrid has the bandwidth of approximately 27.6% at the tolerance limits of the deviation of 0.43 dB for split and of 20 dB for the maximum return loss and isolation.

Generally, a hybrid-ring directional coupler can provide arbitrary power divisions when the power-split ratio is adjusted by varying the impedances between the arms [69]. Figure 4.38(*b*) shows the planar microstrip topology of a ring hybrid where the characteristic impedances of four arms are equal to the standard characteristic impedance Z_0. The variable parameters are the two characteristic impedances Z_1 and Z_2, which determine the degree of coupling of the output arms and the impedance matching condition for the input arm. The analysis of this ring hybrid consists of the usual procedure of reducing the four-terminal network to a two-terminal network by taking advantage of the symmetry about the plane A–A'. For example, when two in-phase waves of equal amplitude are applied to terminals 1 and 3, the current is zero at the plane A–A'. As a result, the ring can be open-circuited at this plane and only one-half of the circuit can be analyzed. This condition is called the even mode. On the other hand, the odd-mode condition is a result of applying the opposite-phase waves of equal amplitude to terminals 1 and 3 when the voltage at plane A–A' is zero. In this case, the ring can be short-circuited at this plane and only one half of the circuit can be analyzed. Once the scattering matrices for the even and odd modes are known, the reflected waves in each arm can be determined. Then, by superimposing the waves of the two modes, the resultant reflected waves in each arm and a single incident wave in one arm is obtained. Thus, if arm 1 is input, the output voltage ratio between arm 2 and 3 is equal to Z_1/Z_2, and no power is delivered to the isolated port 4 where the ballast resistor $R_0 = Z_0$ is connected.

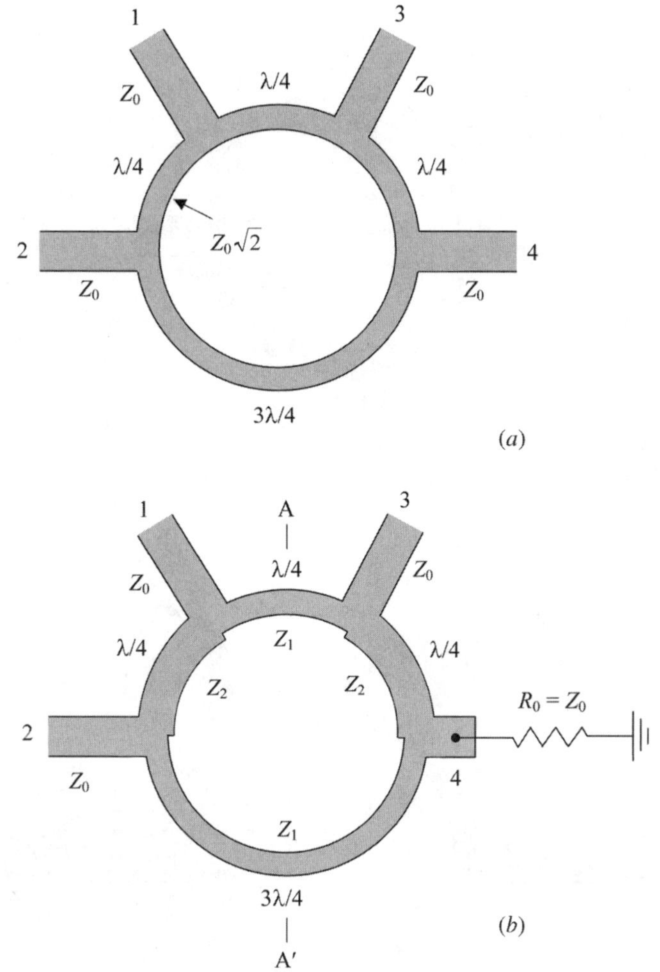

FIGURE 4.38 Microstrip rate-race ring hybrids.

Although usually considered a narrowband device, the rat-race hybrid can provide much broader performance if its three-quarter wavelength section (a main limiting factor in a conventional configuration) is replaced by one having the same characteristic impedance, but whose electrical length is realized by a quarter wavelength of line exhibiting the characteristics of an ideal phase-reversing network [70]. It may be a coaxial cable with a crossover of its inner and outer conductors at one end, thus resulting in an isolation of more than 20 dB in a frequency bandwidth of 30% [57]. As a compact planar alternative, a pair of equilateral broadside-coupled segments of strip transmission lines having diametrically opposing ends short-circuited, as shown in Figure 4.39(*a*), approximates a phase-reversing network over a wide frequency range.

The characteristic impedance of the coupled section Z_c is given by

$$Z_c = \frac{2Z_{0e}Z_{0o}\sin\theta}{\sqrt{(Z_{0e} - Z_{0o})^2 - (Z_{0e} + Z_{0o})^2\cos^2\theta}} \tag{4.52}$$

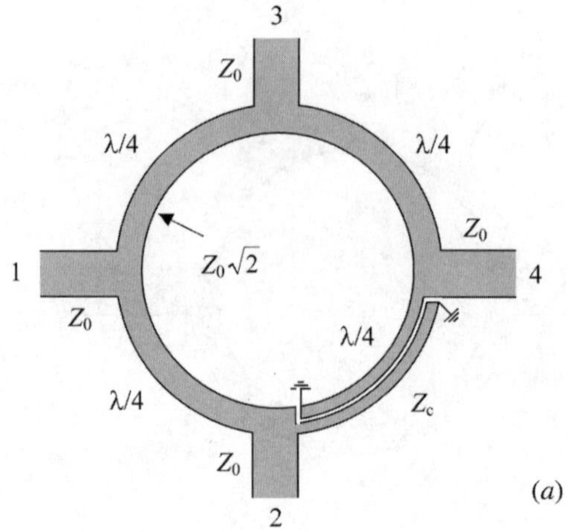

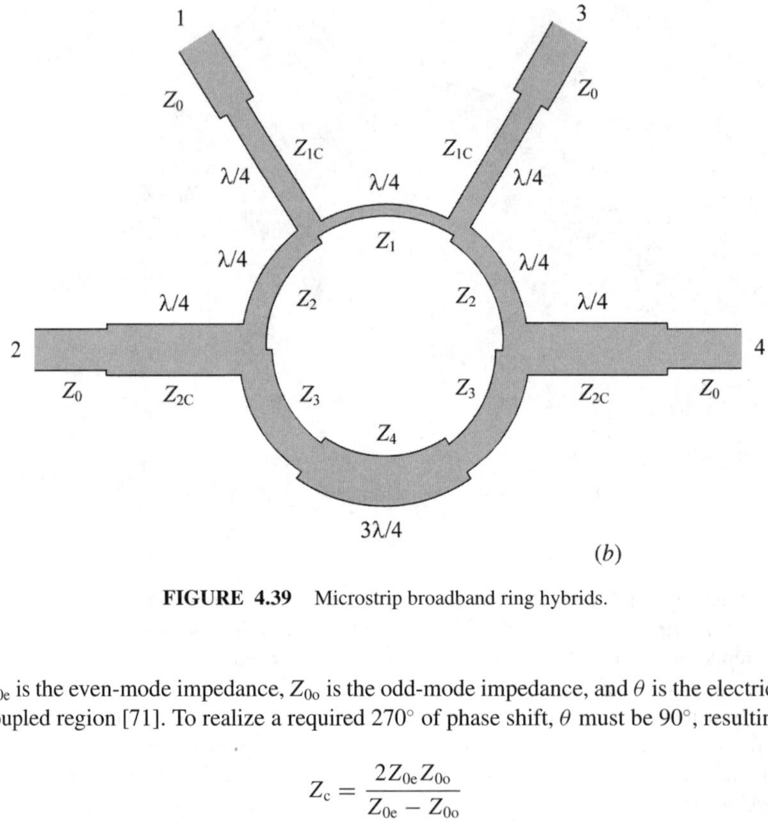

FIGURE 4.39 Microstrip broadband ring hybrids.

where Z_{0e} is the even-mode impedance, Z_{0o} is the odd-mode impedance, and θ is the electrical length of the coupled region [71]. To realize a required 270° of phase shift, θ must be 90°, resulting in

$$Z_c = \frac{2Z_{0e}Z_{0o}}{Z_{0e} - Z_{0o}} \tag{4.53}$$

which for proper operation should be equal at midband to $Z_0\sqrt{2}$. Since $Z_c = \sqrt{Z_{0e}Z_{0o}}$,

$$Z_{0e} = \left(2 + \sqrt{2}\right) Z_0 \tag{4.54}$$

$$Z_{0o} = \left(2 - \sqrt{2}\right) Z_0 \tag{4.55}$$

which means that, for a 50-Ω transmission system and a median 3-dB coupling, Z_{0e} must be 170.7 Ω and Z_{0o} must be 29.3 Ω.

To extend the frequency bandwidth and simplify fabrication process with a uniplanar structure, the ring hybrid can represent a slotline ring with one slotline and three CPW arms [72]. The design technique substitutes one reverse-phase slotline T-junction for the conventional rat-race phase delay section. Since the phase reverse of the slotline T-junction is frequency independent, the resulting slotline ring hybrid provides a broad bandwidth. Experimental results show that this uniplanar crossover ring-hybrid coupler has a bandwidth from 2 to 4 GHz with ±0.4 dB power dividing balance and ±1° phase balance. By using a microcoplanar stripline ring with a broadband coplanar stripline phase inverter and four CPW arms, the coupling between the output ports is within 3.5 ± 0.5 dB over the frequency range from 1.43 to 2.95 GHz [73].

However, it is possible to increase the operating bandwidth of the conventional rat-race without a quarterwave phase-reversing section by applying the concept of hypothetical ports and adding the proper quarterwave transformer to each port [74]. The bandwidth will be significantly broadened by replacing a three-quarterwave equal-impedance section with symmetrical transmission-line structure, the middle quarterwave section of which has different characteristic impedance. To optimally realize this, the numerical optimization procedure to minimize an objective evaluation function is required. In this case, to improve coupling, matching, and isolation, it is necessary to connect either the open-circuited compensating stubs at the hypothetical ports (connecting points of the quarterwave lines with different characteristic impedances), or to add the quarterwave transformers (transmission lines with different characteristic impedances) to the hybrid ports, as shown in Figure 4.39(b). As a result, the frequency bandwidth of the improved ring hybrid becomes 1.84 times as wide as the conventional rat-race.

To significantly reduce the overall size of the conventional ring hybrid with the 90-degree and 270-degree transmission line sections, which is crucial for MMIC design, the principle of the single-frequency equivalence between the circuits with the lumped and distributed parameters can be applied [65]. Let us compare the elements of the transmission matrix $[ABCD_a]$ for a 270-degree transmission line and the transmission matrix $[ABCD_b]$ for a high-pass π-type lumped circuit, consisting of the series capacitor and two shunt inductors, given by

$$[ABCD_a] = \begin{bmatrix} \cos\theta & jZ_0\sin\theta \\ j\dfrac{\sin\theta}{Z_0} & \cos\theta \end{bmatrix}_{\theta=3\pi/2} = \begin{bmatrix} 0 & -jZ_0 \\ \dfrac{-j}{Z_0} & 0 \end{bmatrix} \tag{4.56}$$

$$[ABCD_b] = \begin{bmatrix} 1 & 0 \\ -j/\omega L & 1 \end{bmatrix}\begin{bmatrix} 1 & -j/\omega C \\ 0 & 1 \end{bmatrix}\begin{bmatrix} 1 & 0 \\ -j/\omega L & 1 \end{bmatrix} = \begin{bmatrix} 1-\dfrac{1}{\omega^2 LC} & -j\dfrac{1}{\omega C} \\ -j\dfrac{1}{\omega L}\left(2-\dfrac{1}{\omega^2 LC}\right) & 1-\dfrac{1}{\omega^2 LC} \end{bmatrix}. \tag{4.57}$$

Equating the corresponding elements of both matrices yields the simple equations to determine the circuit elements in the form of

$$Z_0\omega C = \frac{Z_0}{\omega L} = 1. \tag{4.58}$$

Hence, to reduce the size of a ring hybrid, a 270-degree transmission line shown in Figure 4.40(a) is replaced by a high-pass lumped section. Then, the three quarterwave-line sections are replaced by shortened ones with two shunt capacitors, as shown in Figure 4.35(b) for a branch-line coupler. Finally, the characteristic impedance of the transmission line sections is chosen so that the shunt inductors of the high-pass section shown in Figure 4.40(b) could resonant with the shunt capacitors of the

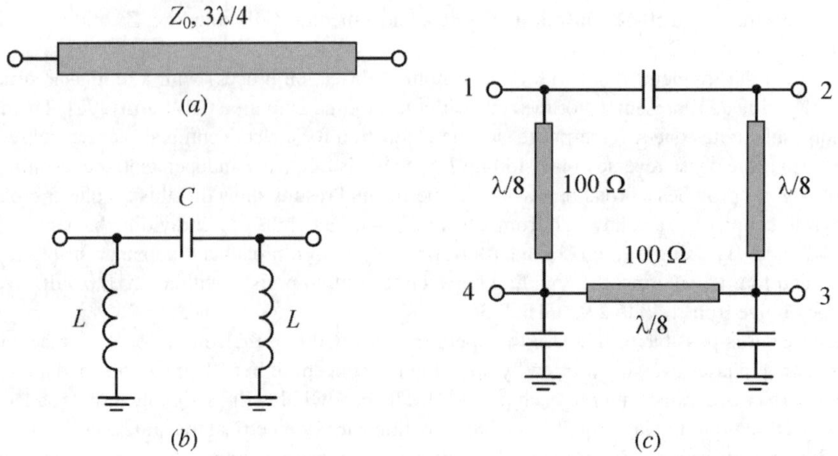

FIGURE 4.40 Reduced-size ring hybrid.

low-pass section at the center bandwidth frequency, in order to completely remove these components. Figure 4.40(*c*) shows the circuit diagram of the reduced-size ring hybrid with the characteristic impedances of the transmission-line sections of 100 Ω and their electrical lengths of 45° [62]. As a result, the overall reduced hybrid size is more than 80% smaller than that of the conventional hybrid.

4.6 COUPLED-LINE DIRECTIONAL COUPLERS

The first directional couplers consisted of either a two-wire balanced line coupled to a second balanced line along a distance of quarter wavelength, or a pair of rods a quarter wavelength long between ground planes [54]. Although the propagation of waves on systems of parallel conductors was investigated many decades ago in connection with the problem of crosstalk between open wire lines or cable pairs in order to eliminate the natural coupling rather than use it, the first exact design theory for transverse electromagnetic (TEM) transmission-line couplers was introduced by Oliver [75]. In terms of the even and odd electric-field modes describing a system of the coupled conductors, it can be stated that the coupling is backward with coupled wave on the secondary line propagating in the direction opposite to the direction of the wave on the primary line, the directivity will be perfect with *VSWR* equal to unity if $Z_0^2 = Z_{0e} Z_{0o}$ at all cross sections along the directional coupler, and the midband voltage coupling coefficient C of the directional coupler is defined as

$$C = \frac{Z_{0e} - Z_{0o}}{Z_{0e} + Z_{0o}} \tag{4.59}$$

where $C = 0$ for zero coupling and $C = 1$ for completely superposed transmission lines.

A coupled-line directional coupler, the stripline single-section topology of which is shown in Figure 4.41(*a*), can be used for broadband power division or combining. Its electrical properties are described using a concept of two types of excitations for the coupled lines in TEM approximation. In this case, for the even mode, the currents flowing in the strip conductors are equal in amplitude and flow in the same direction. The electric field has even symmetry about the center line, and no current flows between the two strip conductors. For the odd mode, the currents flowing in the strip conductors are equal in amplitude, but flow in opposite directions. The electric field lines have an odd symmetry about the center line, and a voltage null exists between these two strip conductors.

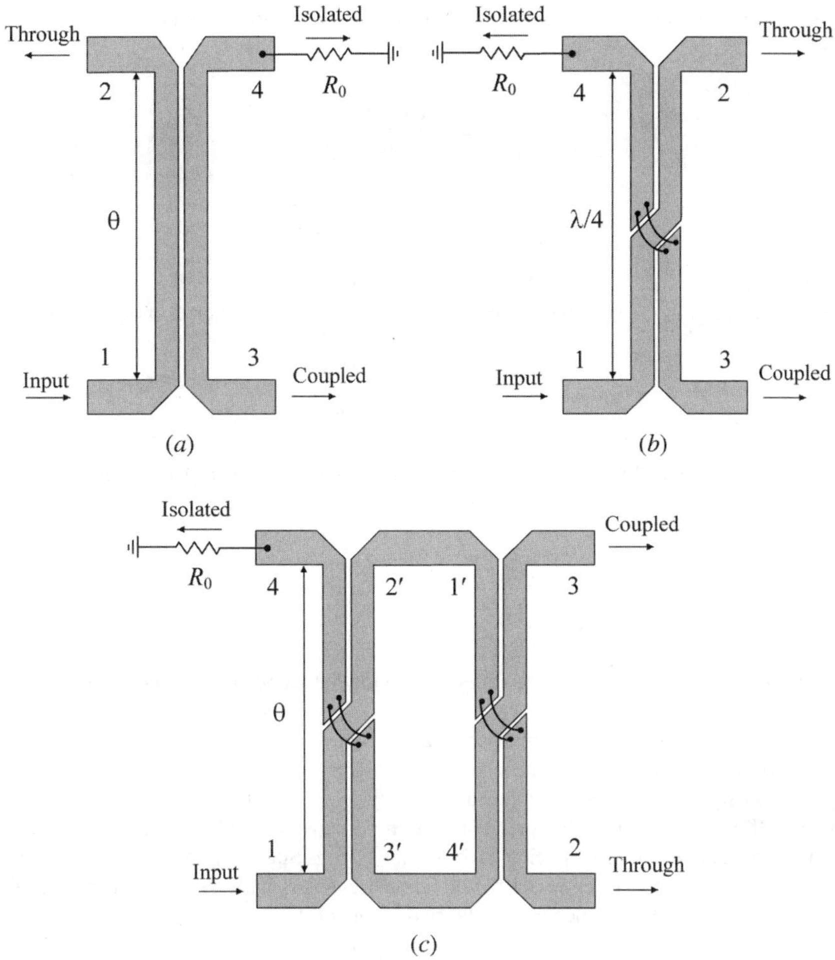

FIGURE 4.41 Coupled-line directional couplers.

An arbitrary excitation of the coupled lines can always be treated as a superposition of appropriate amplitudes of even and odd modes. Therefore, the characteristic impedance for even excitation mode Z_{0e} and the characteristic impedance for the odd excitation mode Z_{0o} characterize the coupled lines. When the two coupled equal-strip lines are used in a standard system with characteristic impedance of Z_0, $Z_0^2 = Z_{0e}Z_{0o}$ and

$$Z_{0e} = Z_0\sqrt{\frac{1+C}{1-C}} \tag{4.60}$$

$$Z_{0o} = Z_0\sqrt{\frac{1-C}{1+C}}. \tag{4.61}$$

An analysis in terms of scattering S-parameters gives $S_{11} = S_{14} = 0$ for any electrical lengths of the coupled lines and the output port 4 is isolated from the matched input port 1. Changing the coupling between the lines and their widths can change the characteristic impedances Z_{0e} and Z_{0o}. In this case,

$$S_{12} = \frac{\sqrt{1-C^2}}{\sqrt{1-C^2}\cos\theta + j\sin\theta} \tag{4.62}$$

$$S_{13} = \frac{jC\sin\theta}{\sqrt{1-C^2}\cos\theta + j\sin\theta} \tag{4.63}$$

where θ is the electrical length of the coupled-line section.

The voltage-split ratio K is defined as the ratio between voltages at port 2 and port 3 by

$$K = \left|\frac{S_{12}}{S_{13}}\right| = \frac{\sqrt{1-C^2}}{C\sin\theta} \tag{4.64}$$

where K can be controlled by changing the coupling coefficient C and electrical length θ.

For a quarter-wavelength-long coupler when $\theta = 90°$, Eqs. (4.62) and (4.63) reduce to

$$S_{12} = -j\sqrt{1-C^2} \tag{4.65}$$

$$S_{13} = C \tag{4.66}$$

from which it follows that equal voltage split between the output ports 2 and 3 can be provided with $C = 1/\sqrt{2}$.

If it is necessary to provide the output ports 2 and 3 at one side, it is best to use a construction of a microstrip directional coupler with crossed bondwires, as shown in Figure 4.41(*b*). The strip crossover for a stripline directional coupler can be easily achieved with the three-layer sandwich. The microstrip 3-dB directional coupler fabricated on alumina substrate for idealized zero strip thickness should have the calculated strip spacing of less than 10 μm. Such a narrow value easily explains the great interest to the constructions of the directional couplers with larger spacing. The effective solution is to use a tandem connection of the two identical directional couplers, which alleviates the physical problem of tight coupling, since two individual couplers need only 8.34-dB coupling to achieve a 3-dB coupler [76,77]. The tandem coupler shown in Figure 4.41(*c*) has the electrical properties of the individual coupler when the output ports 1, 4 and 2, 3 are isolated in pairs, and the phase difference between the output ports 2 and 3 is of 90°.

From an analysis of the signal propagation from input port 1 to output ports 2 and 3 of the tandem coupler, when the signal from the input port 1 propagates to the output port 2 through the traces 1-2′-1′-2 and 1-3′-4′-2 while the signal flowing through the traces 1-2′-1′-3 and 1-3′-4′-3 is delivered to the output port 3, the ratio of the scattering parameters S_{12}^{T} and S_{13}^{T} of a tandem coupler can be expressed through the corresponding scattering parameters S_{12} and S_{13} of the individual coupler as

$$\frac{S_{12}^{T}}{S_{13}^{T}} = \frac{S_{12}^2 + S_{13}^2}{2S_{12}S_{13}} = -j\frac{1-C^2\left(1+\sin^2\theta\right)}{2C\sqrt{1-C^2}\sin\theta}. \tag{4.67}$$

As a result, the signal at the port 2 overtakes the signal at the port 3 by 90°. In this case, for a 3-dB tandem coupler with $\theta = 90°$, the magnitude of Eq. (4.67) must be equal to unity. Consequently, the required voltage coupling coefficient is calculated as

$$C = 0.5\sqrt{2-\sqrt{2}} = 0.3827$$

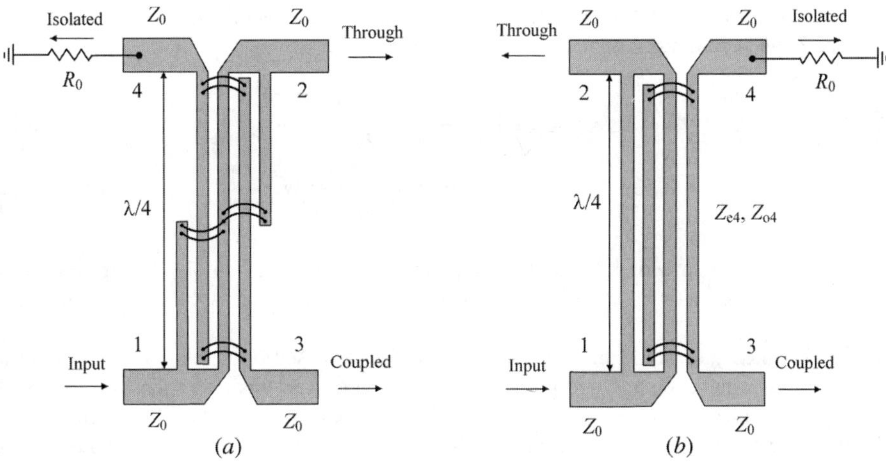

FIGURE 4.42 Lange directional couplers.

or

$$C_{12} = C_{13} = 8.34 \text{ dB}.$$

As an example, a tandem 8.34-dB directional coupler has the dimensions of $W/h = 0.77$ and $S/h = 0.18$ for alumina substrate with $\varepsilon_r = 9.6$, where W is the strip width, S is the strip spacing, and h is the substrate thickness [57].

Another way to increase the coupling between the two edge-coupled microstrip lines is to use several parallel narrow microstrip lines interconnected with each other by the bondwires, as shown in Figure 4.42. For a Lange coupler shown in Figure 4.42(a), four coupled microstrip lines are used, achieving a 3-dB coupling over an octave or more bandwidth [78]. In this case, the signal flowing to the input port 1 is distributed between the output ports 2 and 3 with the phase difference of 90°. However, this structure is quite complicated for practical implementation when, for alumina substrate with $\varepsilon_r = 9.6$, the dimensions of a 3-dB Lange coupler are $W/h = 0.107$ and $S/h = 0.071$, where W is the width of each strip and S is the spacing between adjacent strips.

Figure 4.42(b) shows the unfolded Lange coupler with four strips of equal length; it offers the same electrical performance but is easier for circuit modeling [79]. The even-mode characteristic impedance Z_{e4} and odd-mode characteristic impedance Z_{o4} of the Lange coupler with $Z_0^2 = Z_{e4}Z_{o4}$ in terms of the characteristic impedances of a two-conductor line (which is identical to any pair of adjacent lines in the coupler) can be obtained by

$$Z_{e4} = \frac{Z_{0o} + Z_{0e}}{3Z_{0o} + Z_{0e}} Z_{0e} \tag{4.68}$$

$$Z_{o4} = \frac{Z_{0e} + Z_{0o}}{3Z_{0e} + Z_{0o}} Z_{0o} \tag{4.69}$$

where Z_{0e} and Z_{0o} are the even- and odd-mode characteristic impedances of the two-conductor pair [80].

The midband voltage coupling coefficient C is given by

$$C = \frac{Z_{e4} - Z_{o4}}{Z_{e4} + Z_{o4}} = \frac{3\left(Z_{0e}^2 - Z_{0o}^2\right)}{3\left(Z_{0e}^2 + Z_{0o}^2\right) + 2Z_{0e}Z_{0o}}. \tag{4.70}$$

The even- and odd-mode characteristic impedances Z_{0e} and Z_{0o}, as functions of the characteristic impedance Z_0 and coupling coefficient C, are determined by

$$Z_{0e} = Z_0\sqrt{\frac{1+C}{1-C}\frac{4C-3+\sqrt{9-8C^2}}{2C}} \tag{4.71}$$

$$Z_{0o} = Z_0\sqrt{\frac{1-C}{1+C}\frac{4C+3-\sqrt{9-8C^2}}{2C}}. \tag{4.72}$$

For alumina substrate with $\varepsilon_r = 9.6$, the dimensions of such a 3-dB unfolded Lange coupler are $W/h = 0.112$ and $S/h = 0.08$, where W is the width of each strip and S is the spacing between the strips.

The design theory for TEM transmission-line couplers is based on an assumption of the same phase velocities of the even and odd propagation mode. However, this is not the case for coupled microstrip lines, since they have unequal even- and odd-mode phase velocities. In this case, the odd mode has more fringing electric field in the air region rather than the even mode with electrical field concentrating mostly in the substrate underneath the microstrip lines. As a result, the effective dielectric permittivity in the latter case is higher, thus indicating a smaller phase velocity for the even mode. Consequently, it is required to apply the phase velocity compensation techniques to improve the coupler directivity that decreases with increasing frequency. Figure 4.43(*a*) shows a typical wiggly-line coupler (with sawtooth shape of coupled lines), where wiggling the adjacent edges of the microstrip lines, which makes their physical lengths longer, slows the odd-mode wave without much affecting the even-mode wave [81]. High directivity can also be achieved by using capacitive compensation. Figure 4.43(*b*) shows the capacitively compensated microstrip directional coupler where the two identical lumped capacitors are connected between coupled lines at their edges. Physically, these edge capacitors affect the odd mode by equivalent extension of the transmission-line electrical lengths, with almost no effect for even mode. For an ideal lossless operation condition at 12 GHz using standard alumina substrate, the compensated coupled-line microstrip directional coupler can improve directivity from 13.25 dB to infinity [82]. Capacitive compensation can be performed by a gap coupling of the open-circuit stub formed in a subcoupled line [83]. In this case, the coupler directivity can be improved by 23 dB in a frequency range from 1 to 2.5 GHz compared to the directivity of the conventional uncompensated microstrip coupler.

At radio frequencies and low microwaves, the conventional quarter-wavelength directional coupler has very large dimensions that limit their practical application especially in monolithic circuits. Figure 4.44 shows a reduced-size directional coupler consisting of the two coupled microstrip lines,

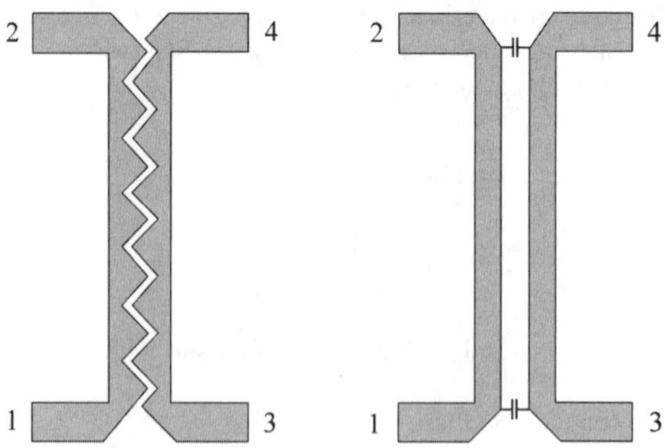

(*a*) (*b*)

FIGURE 4.43 Coupled-line directional couplers.

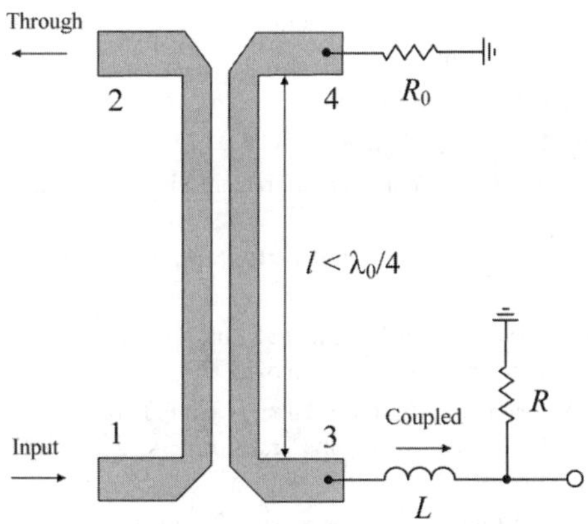

FIGURE 4.44 Coupled-line reduced-size directional coupler.

the electrical lengths of which are much smaller than quarter wavelength. The main problem of the coupler at frequencies, where the electrical length of its coupled lines is smaller than quarter wavelength, is that the degree of coupling linearly varies with frequency. To compensate this frequency behavior, the output port 3 can be connected to a series inductor L followed by a shunt resistor R [38,84]. The inductance value depends on the coupling value and flatness, and midband frequency, while the resistance value depends on the impedance of the secondary line and inductance value. Such a microstrip reduced-size directional coupler with $L = 180$ nH and $R = 62\ \Omega$ can provide the coupling of about 30 dB with flatness of ± 0.1 dB, directivity greater than 20 dB, insertion lossless than 0.25 dB, and *VSWR* less than 1.15 in a frequency bandwidth of 60% around 200 MHz. Tuning of the center bandwidth frequency and coupling can be simply realized by varying the inductance value.

REFERENCES

1. D. M. Pozar, *Microwave Engineering*, New York: John Wiley & Sons, Ltd. 2004.
2. R. E. Collin, *Foundation for Microwave Engineering*, New York: McGraw-Hill, 1992.
3. H. L. Krauss, C. W. Bostian, and F. H. Raab, *Solid State Radio Engineering*, New York: John Wiley & Sons, 1980.
4. Z. I. Model, *Networks for Combining and Distribution of High Frequency Power Sources* (in Russian), Moskva: Sov. Radio, 1980.
5. J. Sevick, *Transmission Line Transformers*, Norcross: Noble Publishing, 2001.
6. C. Trask, "Transmission Line Transformers: Theory, Design and Applications," *High Frequency Electronics*, vol. 4, pp. 46–53, Dec. 2005, vol. 5, pp. 26–33, Jan. 2006.
7. E. Rotholz, "Transmission-Line Transformers," *IEEE Trans. Microwave Theory Tech.*, vol. MTT-29, pp. 148–154, Apr. 1981.
8. J. Horn and G. Boeck, "Design and Modeling of Transmission Line Transformers," *Proc. IEEE SBMO/MTT-S Int. Microwave and Optoelectronics Conf.*, vol. 1, pp. 421–424, Sept. 2003.
9. G. Guanella, "New Method of Impedance Matching in Radio-Frequency Circuits," *The Brown Boveri Rev.*, vol. 31, pp. 327–329, Sept. 1944.
10. G. Guanella, "High-Frequency Matching Transformer," U.S. Patent 2,470,307, May 1949.
11. R. K. Blocksome, "Practical Wideband RF Power Transformers, Combiners, and Splitters," *Proc. RF Technology Expo*, vol. 86, pp. 207–227, 1986.
12. J. L. B. Walker, D. P. Myer, F. H. Raab, and C. Trask, *Classic Works in RF Engineering: Combiners, Couplers, Transformers, and Magnetic Materials*, Norwood: Artech House, 2005.

13. J. Sevick, "Magnetic Materials for Broadband Transmission Line Transformers," *High Frequency Electronics*, vol. 4, pp. 46–52, Jan. 2005.
14. C. L. Ruthroff, "Some Broad-Band Transformers," *Proc. IRE*, vol. 47, pp. 1337–1342, Aug. 1959.
15. C. L. Ruthroff, "Broadband Transformers," U.S. Patent 3,037,175, May 1962.
16. J. Sevick, "A Simplified Analysis of the Broadband Transmission Line Transformer," *High Frequency Electronics*, vol. 3, pp. 48–53, Feb. 2004.
17. R. T. Irish, "Method of Bandwidth Extension for the Ruthroff Transformer," *Electronics Lett.*, vol. 15, pp. 790–791, Nov. 1979.
18. S. C. Dutta Roy, "Optimum Design of an Exponential Line Transformer for Wide-Band Matching at Low Frequencies," *Proc. IEEE*, vol. 67, pp. 1563–1564, Nov. 1979.
19. M. Engels, M. R. Jansen, W. Daumann, R. M. Bertenburg, and F. J. Tegude, "Design Methodology, Measurement and Application of MMIC Transmission Line Transformers," *IEEE MTT-S Int. Microwave Symp. Dig.*, vol. 3, pp. 1635–1638, May 1995.
20. S. P. Liu, "Planar Transmission Line Transformer Using Coupled Microstrip Lines," *IEEE MTT-S Int. Microwave Symp. Dig.*, vol. 2, pp. 789–792, June 1998.
21. J. Sevick, "Design of Broadband Ununs with Impedance Ratios Less than 1:4," *High Frequency Electronics*, vol. 3, pp. 44–50, Nov. 2004.
22. S. E. London and S. V. Tomashevich, "Line Transformers with Fractional Transformation Factor," *Telecommunications and Radio Engineering*, vol. 28/29, pp. 129–130, Apr. 1974.
23. C. Lorenz, "Schaltungsanordnung zum Ubergang von Einer symmetrischen, Elektrischen Anordnung zu Einer Unsymmetrischen, Insbesondere bei Hochfrequenzanordnungen," German Patent DE603816, Sept. 1934.
24. B. P. Kumar and G. R. Branner, "Optimized Design of Unique Miniaturized Planar Baluns for Wireless Applications," *IEEE Microwave and Wireless Comp. Lett.*, vol. 13, pp. 134–136, Feb. 2003.
25. N. Marchand, "Transmission-Line Conversion Transformers," *Electronics*, vol. 17, pp. 142–145, Dec. 1944.
26. G. Oltman, "The Compensated Balun," *IEEE Trans. Microwave Theory Tech.*, vol. MTT-14, pp. 112–119, March. 1966.
27. C.-M. Tsai and K. C. Gupta, "A Generalized Model for Coupled Lines and its Applications to Two-Layer Planar Circuits," *IEEE Trans. Microwave Theory Tech.*, vol. MTT-40, pp. 2190–2199, Dec. 1992.
28. C. Cho and K. C. Gupta, "A New Design Procedure for Single-Layer and Two-Layer Three-Line Baluns," *IEEE Trans. Microwave Theory Tech.*, vol. MTT-46, pp. 2514–2119, Dec. 1998.
29. K. Nishikawa, I. Toyoda, and T. Tokumitsu, "Compact and Broad-Band Three-Dimensional MMIC Balun," *IEEE Trans. Microwave Theory Tech.*, vol. MTT-47, pp. 96–98, Jan. 1999.
30. C. Y. Ng, M. Chongcheawchamnan, and I. D. Robertson, "Analysis and Design of a High-Performance Planar Marchand Balun," *IEEE MTT-S Int. Microwave Symp. Dig.*, vol. 1, pp. 113–116, June 2002.
31. W. M. Fathelbab and M. B. Steer, "New Classes of Miniaturized Planar Marchand Baluns," *IEEE Trans. Microwave Theory Tech.*, vol. MTT-53, pp. 1211–1220, Apr. 2005.
32. V. Govind, W. Yun, S. Dalmia, V. Sundaram, G. White, and M. Swaminathan, "Analysis and Design of Compact Wideband Baluns on Multilayer Liquid Crystalline Polymer (MLP) Based Structures," *IEEE MTT-S Int. Microwave Symp. Dig.*, pp. 1–4, June 2005.
33. R. Phromloungsri, S. Srisathit, M. Chongcheawchamnan, and I. D. Robertson, "Novel Technique for Performance Improvement in Impedance-Transforming Planar Marchand Baluns," *Proc. 35th Europ. Microwave Conf.*, pp. 533–536, 2005.
34. H. Jasik, "Broad Band Balun Transformer," U.S. Patent 2925566, Feb. 1960.
35. H. R. Phelan, "A Wide-Band Parallel-Connected Balun," *IEEE Trans. Microwave Theory Tech.*, vol. MTT-18, pp. 259–263, May 1970.
36. S. V. Bearse, "New Combining and Cooling Techniques Developed for 1 kW L-Band Transmitter," *Microwaves*, vol. 16, pp. 9–10, Sept. 1977.
37. E. J. Wilkinson, "An *N-Way Hybrid Power Divider*," *IRE Trans. Microwave Theory Tech.*, vol. MTT-8, pp. 116–118, Jan. 1960.
38. L. G. Maloratsky, *Passive RF & Microwave Integrated Circuits*, New York: Elsevier, 2004.
39. V. M. Katushkina and Z. I. Model, "*N*-Way Power Combiner of VHF-UHF Transmitters (in Russian)," *Elektrosvyaz*, vol. 7, pp. 17–25, July 1959.

40. S. Y. London, "Independent Operation of High Power VHF Amplifiers on Common Load (in Russian)," *Voprosy Radioelektroniki, Ser. X, Tekhnika Svyazi*, vol. 6, pp. 87–96, June 1959.
41. J. J. Taub and G. P. Kurpis, "A More General *N*-Way Hybrid Power Divider," *IEEE Trans. Microwave Theory Tech.*, vol. MTT-17, pp. 406–408, July 1969.
42. S. B. Cohn, "A Class of Broad-Band Three-Port TEM-Mode Hybrids," *IEEE Trans. Microwave Theory Tech.*, vol. MTT-16, pp. 110–116, Feb. 1968.
43. D. Antsos, R. Crist, and L. Sukamto, "A Novel Wilkinson Power Divider with Predictable Performance at K and Ka-Band," *IEEE MTT-S Int. Microwave Symp. Dig.*, vol. 2, pp. 907–910, May 1994.
44. A. A. M. Saleh, "Planar Electrically Symmetric *n*-Way Hybrid Power Dividers/Combiners," *IEEE Trans. Microwave Theory Tech.*, vol. MTT-28, pp. 555–563, June 1980.
45. U. H. Gysel, "A New N-Way Power Divider/Combiner Suitable for High-Power Applications," *IEEE MTT-S Int. Microwave Symp. Dig.*, pp. 116–118, May 1975.
46. R. Knochel and B. Mayer, "Broadband Printed Circuit 0°/180° Couplers and High Power Inphase Power Dividers," *IEEE MTT-S Int. Microwave Symp. Dig.*, vol. 1, pp. 471–474, May 1990.
47. Y. Sun and A. P. Freundorfer, "Broadband Folded Wilkinson Power Combiner/Splitter," *IEEE Microwave and Wireless Components Lett.*, vol. 14, pp. 295–297, June 2004.
48. N. Nagai, E. Maekawa, and K. Ono, "New *N*-way Hybrid Power Dividers," *IEEE Trans. Microwave Theory Tech.*, vol. MTT-25, pp. 1008–1012, Dec. 1977.
49. M. E. Goldfarb, "A Recombinant, In-Phase Power Divider," *IEEE Trans. Microwave Theory Tech.*, vol. MTT-39, pp. 1438–1440, Aug. 1991.
50. B. M. Schiffman, "A New Class of Broad-Band Microwave 90-Degree Phase Shifter," *IRE Trans. Microwave Theory Tech.*, vol. MTT-6, pp. 232–237, Apr. 1958.
51. L. I. Parad and R. L. Moynihan, "Split-Tee Power Divider," *IEEE Trans. Microwave Theory Tech.*, vol. MTT-13, pp. 91–95, Jan. 1965.
52. J.-S. Lim and S.-Y. Eom, "A New 3-Way Power Divider with Various Output Power Ratios," *IEEE MTT-S Int. Microwave Symp. Dig.*, vol. 2 pp. 785–788, June 1996.
53. D. Maurin and K. Wu, "A Compact 1.7-2.1 GHz Three-Way Power Combiner Using Microstrip Technology with Better Than 93.8% Combining Efficiency," *IEEE Microwave and Guided Wave Lett.*, vol. 6, pp. 106–108, Feb. 1996.
54. S. B. Cohn and R. Levy, "History of Microwave Passive Components with Particular Attention to Directional Couplers," *IEEE Trans. Microwave Theory Tech.*, vol. MTT-32, pp. 1046–1054, Sept. 1984.
55. J. Reed and G. J. Wheeler, "A Method of Analysis of Symmetrical Four-Port Networks," *IRE Trans. Microwave Theory Tech.*, vol. MTT-4, pp. 246–252, Oct. 1956.
56. R. Levy and L. Lind, "Synthesis of Symmetrical Branch-Guide Directional Couplers," *IEEE Trans. Microwave Theory Tech.*, vol. MTT-16, pp. 80–89, Feb. 1968.
57. A. Grebennikov, *RF and Microwave Power Amplifier Design*, New York: McGraw-Hill, 2004.
58. P. Meaney, "A Novel Branch-Line Coupler Design for Millimeter-Wave Applications," *IEEE MTT-S Int. Microwave Symp. Dig.*, vol. 1, pp. 585–588, May 1990.
59. L. F. Lind, "Synthesis of Asymmetrical Branch-Guide Directional Coupler-Impedance Transformers," *IEEE Trans. Microwave Theory Tech.*, vol. MTT-17, pp. 45–48, Jan. 1969.
60. R. K. Gupta, S. E. Anderson, and W. J. Getsinger, "Impedance-Transforming 3-dB 90° Hybrids," *IEEE Trans. Microwave Theory Tech.*, vol. MTT-35, pp. 1303–1307, Dec. 1987.
61. S. Kumar, C. Tannous, and T. Danshin, "A Multisection Broadband Impedance Transforming Branch-Line Hybrid," *IEEE Trans. Microwave Theory Tech.*, vol. MTT-43, pp. 2517–2523, Nov. 1995.
62. T. Hirota, A. Minakawa, and M. Muraguchi, "Reduced-Size Branch-Line and Rat-Race Hybrid for Uniplanar MMIC's," *IEEE Trans. Microwave Theory Tech.*, vol. MTT-38, pp. 270–275, March 1990.
63. M. Caulton, B. Hershenov, S.P. Knight, and R. E. DeBrecht, "Status of Lumped Elements in Microwave Integrated Circuits – Present and Future," *IEEE Trans. Microwave Theory Tech.*, vol. MTT-19, pp. 588–599, July 1971.
64. R. W. Vogel, "Analysis and Design of Lumped- and Lumped-Distributed-Element Directional Couplers for MIC and MMIC Applications," *IEEE Trans. Microwave Theory Tech.*, vol. MTT-40, pp. 253–262, Feb. 1992.
65. S. J. Parisi, "180° Lumped Element Hybrid," *IEEE MTT-S Int. Microwave Symp. Dig.*, vol. 3, pp. 1243–1246, June 1989.

66. R. C. Frye, S. Kapur and R. C. Melville, "A 2-GHz Quadrature Hybrid Implemented in CMOS Technology," *IEEE J. Solid-State Circuits*, vol. SC-38, pp. 550–555, March 2003.
67. F. Ali and A. Podell, "A Wide-Band GaAs Monolithic Spiral Quadrature Hybrid and its Circuit Applications," *IEEE J. Solid-State Circuits*, vol. SC-26, pp. 1394–1398, Oct. 1991.
68. W. A. Tyrrell, "Hybrid Circuits for Microwaves," *Proc. IRE*, vol. 35, pp. 1294–1306, Nov. 1947.
69. C. Y. Pon, "Hybrid-Ring Directional Coupler for Arbitrary Power Divisions," *IRE Trans. Microwave Theory Tech.*, vol. MTT-9, pp. 529–535, Nov. 1961.
70. S. March, "A Wideband Stripline Hybrid Ring," *IEEE Trans. Microwave Theory Tech.*, vol. MTT-16, pp. 361, June 1968.
71. G. L. Matthaei, E. M. T. Jones, and L. Young, *Microwave Filters, Impedance-Matching Networks and Coupling Structures*, New York: Artech House, 1980.
72. C.-H. Ho, L. Fan, and K. Chang, "Broad-Band Uniplanar Hybrid-Ring and Branch-Line Couplers," *IEEE Trans. Microwave Theory Tech.*, vol. MTT-41, pp. 2116–2124, Dec. 1993.
73. T. Wang and K. Wu, "Size-Reduction and band-Broadening Design Technique of Uniplanar Hybrid Ring Coupler Using Phase Inverter for M(H)MIC's," *IEEE Trans. Microwave Theory Tech.*, vol. MTT-42, pp. 198–206, Feb. 1999.
74. D. I. Kim and Y. Naito, "Broad-Band Design of Improved Hybrid-Ring 3-dB Directional Couplers," *IEEE Trans. Microwave Theory Tech.*, vol. MTT-30, pp. 2040–2046, Nov. 1982.
75. B. M. Oliver, "Directional Electromagnetic Couplers," *Proc. IRE*, vol. 42, pp. 1686–1692, Nov. 1954.
76. G. D. Monteath, "Coupled Transmission Lines as Symmetrical Directional Couplers," *IEE Proc.*, vol. 102, part B, pp. 383–392, May 1955.
77. T. P. Shelton and J. A. Mosko, "Synthesis and Design of Wide-Band Equal-Ripple TEM Directional Couplers and Fixed Phase Shifters," *IEEE Trans. Microwave Theory Tech.*, vol. MTT-14, pp. 462–473, Oct. 1966.
78. J. Lange, "Interdigitated Stripline Quadrature Hybrid," *IEEE Trans. Microwave Theory Tech.*, vol. MTT-17, pp. 1150–1151, Dec. 1969.
79. R. Waugh and D. LaCombe, "Unfolding the Lange Coupler," *IEEE Trans. Microwave Theory Tech.*, Vol. MTT-20, pp. 777–779, Nov. 1972.
80. W. P. Ou, "Design Equations for an Interdigitated Directional Coupler," *IEEE Trans. Microwave Theory Tech.*, vol. MTT-23, pp. 253–255, Feb. 1975.
81. A. Podell, "A High Directivity Microstrip Coupler Technique," *G-MTT Int. Microwave Symp. Dig.*, pp. 33–36, May 1970.
82. M. Dydyk, "Accurate Design of Microstrip Directional Couplers with Capacitive Compensation," *IEEE MTT-S Int. Microwave Symp. Dig.*, vol. 1, pp. 581–584, May 1990.
83. C.-S. Kim, J.-S. Lim, D.-J. Kim, and D. Ahn, "A Design of Single and Multi-Section Microstrip Directional Coupler with the High Directivity," *IEEE MTT-S Int. Microwave Symp. Dig.*, vol. 3, pp. 1895–1898, June 2004.
84. L. Maloratsky, "Miniature Directional Coupler," U.S. Patent 5424694, June 1995.

5 Filters

Historically, an introduction of bandpass electrical filters invented independently by Karl W. Wagner and George A. Campbell played a vital role in their successful development in high-frequency wireless telegraphy, wireless telephony, or multiplex wire telephony transmitting systems when a maximum number of channels could simultaneously operate on a given circuit with a minimum frequency interval between adjacent channels [1,2]. In addition to separating the various channels from each other in multiplex line, it was found convenient from an operating standpoint to separate, within the toll offices, the carrier frequencies as a group (using a high-pass filter) from the frequencies intended for ordinary telephony and telegraphy (using a low-pass filter), and transmit them by separate lines [3].

This chapter discusses the basic types of radio frequency (RF) and microwave filters based on the low-pass or high-pass sections, and bandpass or bandstop transformation. Classical filter design approaches, using image parameter and insertion loss methods, are given for low-pass and high-pass *LC* filter implementations. The quarterwave-line and coupled-line section, which are the basic elements of microwave transmission-line filters, are described and analyzed. Different examples of coupled-line filters including interdigital, combline, and hairpin bandpass filters are given. Special attention is paid to microstrip filters with unequal phase velocities that can provide unexpected properties due to different implementation technologies. Finally, the typical structures, implementation technology, operational principles, and band performance of the filters, based on surface and bulk acoustic waves, are presented.

5.1 TYPES OF FILTERS

The most general types of electrical filters in terms of the circuit configuration of the basic elements can be represented by a ladder structure in the form of a T-network shown in Figure 5.1(a) and a π-network shown in Figure 5.1(b), or a lattice structure shown in Figure 5.1(c), where the corresponding reactive elements are represented schematically by conventional blocks [4]. A lattice network can be realized with any desired amplitude characteristic, and its characteristic impedance is independent of its transmission properties. According to the classification in terms of the character of their elements, they can be considered as the *LC* filters, transmission-line filters, or resonator-type filters.

The low-pass filter (LPF) with the frequency response shown in Figure 5.2(a) passes the wave energy in a passband region from zero frequency up to a determined cutoff frequency ideally without attenuation and rejects all energy beyond that limit in a stopband region, with a finite transition between passband and stopband regions. The high-pass filter (HPF) with the frequency response shown in Figure 5.2(b) prevents the transmission of frequencies in a stopband region below a determined point (cutoff frequency) and appears to be electrically transparent to frequencies beyond this point ideally without attenuation, with a finite transition between stopband and passband regions.

The bandpass filter (BPF) with the frequency response shown in Figure 5.2(c) passes the wave energy in a passband region from certain lower to upper frequency limits ideally without attenuation and stops all energy outside these two limits in the stopband regions. The band-reject filter (BRF)

RF and Microwave Transmitter Design, First Edition. Andrei Grebennikov.

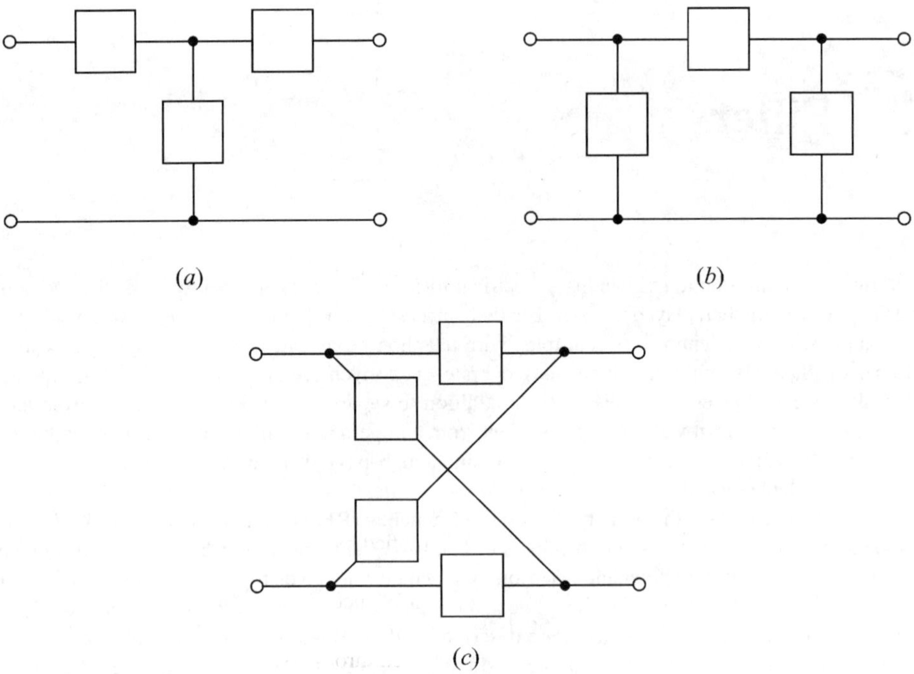

FIGURE 5.1 Filter configurations.

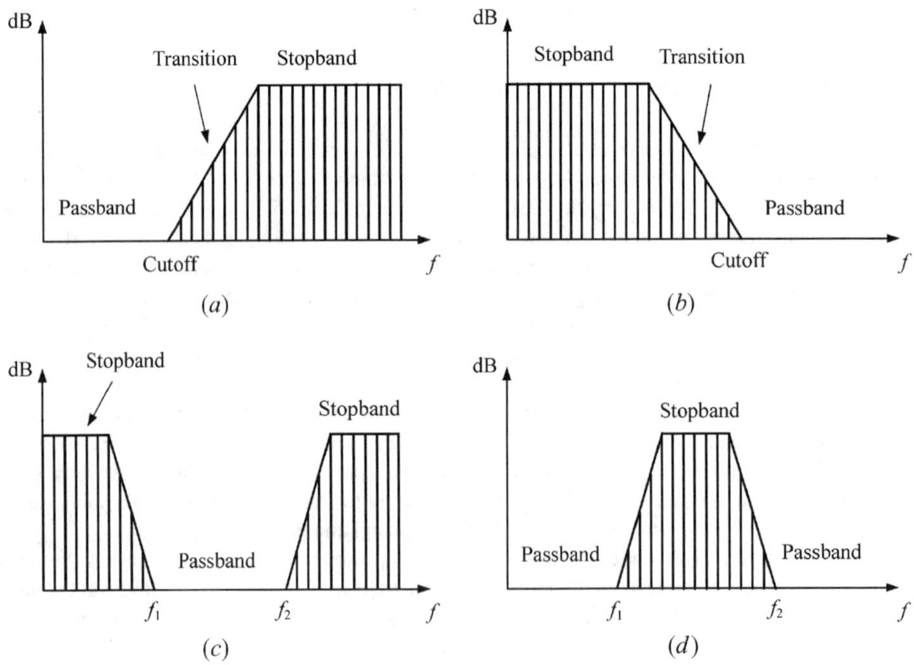

FIGURE 5.2 Filter frequency responses.

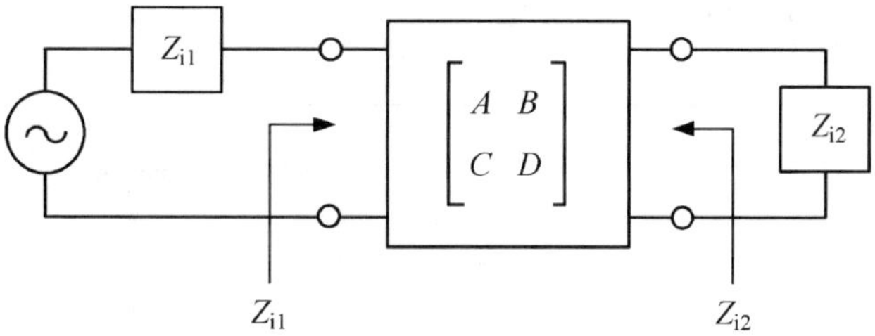

FIGURE 5.3 Voltage-driven two-port network terminated in its image impedance.

with the frequency response shown in Figure 5.2(*d*) is used when a certain unwanted frequency or band of frequencies has to be rejected. Outside of the rejection band or stopband, all frequencies will pass ideally without attenuation. All-pass filters pass all frequency components of the input signal, but introduce a predictable phase shift for its different components. A short impulse on the input side of such a filter is modified into a longer frequency-modulated signal at the output. As an example, an all-pass filter can be represented by a matched transmission line or *LC* lattice structure of a different order.

For an arbitrary voltage-driven two-port network specified by its transmission *ABCD*-parameters, as shown in Figure 5.3, the image impedance Z_{i1} is defined as the input impedance at input port 1 when output port 2 is terminated with the image impedance Z_{i2}, which is defined as the input impedance at output port 2 when port 1 is terminated with Z_{i1}. Thus, both ports are matched when terminated in their image impedances that can be expressed through the *ABCD*-parameters as

$$Z_{i1} = \sqrt{\frac{AB}{CD}} \tag{5.1}$$

$$Z_{i2} = \sqrt{\frac{BD}{AC}} \tag{5.2}$$

or $Z_{i2} = DZ_{i1}/A$, where the factor $\sqrt{D/A}$ can be interpreted as a transformer turns ratio [5]. If the network is symmetric, then $A = D$ and $Z_{i2} = Z_{i1}$.

The propagation factor for the two-port network can be defined as

$$e^{\gamma} = \sqrt{AD} - \sqrt{BC} \tag{5.3}$$

where $\gamma = \alpha + j\beta$ is the propagation constant, α is the attenuation constant, and β is the phase constant. Eq. (5.3) can be rewritten as

$$\cosh\gamma = \sqrt{AD}. \tag{5.4}$$

Table 5.1 lists the image impedances and propagation factors (along with other important parameters) for the two-port T- and π-networks.

The basic network unit for realizing all-pass prototype filters is a lattice structure, as shown in Figure 5.4, where there is a conventional abbreviated representation on the right-hand side. This lattice structure is not only symmetric with respect to the two ports, but also balanced with respect to ground. If the impedance of the series arm of the lattice is equal to Z_1 and the impedance of the parallel arm is equal to Z_2, the image impedance of the lattice structure is defined as

$$Z_i = \sqrt{Z_1 Z_2} \tag{5.5}$$

TABLE 5.1 Image Parameters for T- and π-Networks.

T-Network	π-Network
ABCD-parameters	*ABCD*-parameters
$A = 1 + \dfrac{Z_1}{2Z_2}$	$A = 1 + \dfrac{Z_1}{2Z_2}$
$B = Z_1 + \dfrac{Z_1^2}{4Z_2}$	$B = Z_1$
$C = \dfrac{1}{Z_2}$	$C = \dfrac{1}{Z_2} + \dfrac{Z_1}{4Z_2^2}$
$D = 1 + \dfrac{Z_1}{2Z_2}$	$D = 1 + \dfrac{Z_1}{2Z_2}$
Z-parameters	Y-parameters
$Z_{11} = Z_{22} = Z_2 + \dfrac{Z_1}{2}$	$Y_{11} = Y_{22} = \dfrac{1}{Z_1} + \dfrac{1}{2Z_2}$
$Z_{12} = Z_{21} = Z_2$	$Y_{12} = Y_{21} = \dfrac{1}{Z_1}$
Image impedance	Image impedance
$Z_{iT} = \sqrt{Z_1 Z_2}\sqrt{1 + \dfrac{Z_1}{4Z_2}}$	$Z_{i\pi} = \dfrac{\sqrt{Z_1 Z_2}}{\sqrt{1 + \dfrac{Z_1}{4Z_2}}} = \dfrac{Z_1 Z_2}{Z_{iT}}$
Propagation factor	Propagation factor
$e^{\gamma} = 1 + \dfrac{Z_1}{2Z_2} + \sqrt{\dfrac{Z_1}{Z_2} + \left(\dfrac{Z_1}{2Z_2}\right)^2}$	$e^{\gamma} = 1 + \dfrac{Z_1}{2Z_2} + \sqrt{\dfrac{Z_1}{Z_2} + \left(\dfrac{Z_1}{2Z_2}\right)^2}$

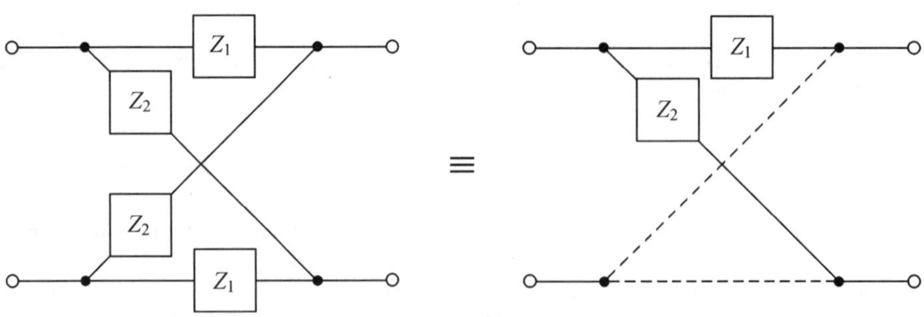

FIGURE 5.4 Lattice filter configuration.

and the propagation constant γ is derived from

$$\tanh\frac{\gamma}{2} = \sqrt{\frac{Z_1}{Z_2}}. \tag{5.6}$$

The amplitude response of the lattice filter depends on only the ratio of the branch reactances and is independent of the input and output impedances. When $Z_1 = Z_2$, then tanh $(\gamma/2) = \pm 1$, and a peak of attenuation occurs. The position of this peak can be modified without impedance-change by multiplying Z_1 by a real positive factor and dividing Z_2 by the same factor. Lattice networks must be designed and build with great care in order to balance their impedances. Once reactance values are chosen to resonate at strategic positions in the passband (such as the cutoffs and the center of passband), the level of the reactances between these strategic positions must be maintained in both arms in an exactly prescribed fashion. Moreover, the peak of attenuation outside the passband is controlled by the level of the impedances of both arms when they are of the same nature (capacitive or inductive), whereas the position of the attenuation peak depends on the sharpness of the reactance curve at cutoff, which is effectively within the passband.

5.2 FILTER DESIGN USING IMAGE PARAMETER METHOD

5.2.1 Constant-*k* Filter Sections

The image attenuation of the filter is completely defined by its poles with their specified orders. If the image impedance is of the first order, called the *constant-k*, the filter design leads to the structure where the filter consists of as many sections as there are poles, and each section is designed separately. First consider the low-pass T-network shown in Figure 5.5(a) where the series inductors and shunt capacitors tend to block high-frequency signals while passing low-frequency signals. For $Z_1 = j\omega L$ and $Z_2 = 1/(j\omega C)$, from Table 5.1 it follows that the image impedance Z_{iT} can be written as

$$Z_{iT} = \sqrt{\frac{L}{C}}\sqrt{1 - \frac{\omega^2 LC}{4}}. \tag{5.7}$$

If we introduce and define a cutoff frequency ω_c as

$$\omega_c = \frac{2}{\sqrt{LC}} \tag{5.8}$$

and a characteristic impedance Z_0 as

$$Z_0 = \sqrt{\frac{L}{C}} = k \tag{5.9}$$

where k is a constant, then Eq. (5.7) can be rewritten as

$$Z_{iT} = Z_0\sqrt{1 - \left(\frac{\omega}{\omega_c}\right)^2} \tag{5.10}$$

from which it follows that $Z_{iT} = Z_0$ for $\omega = 0$.

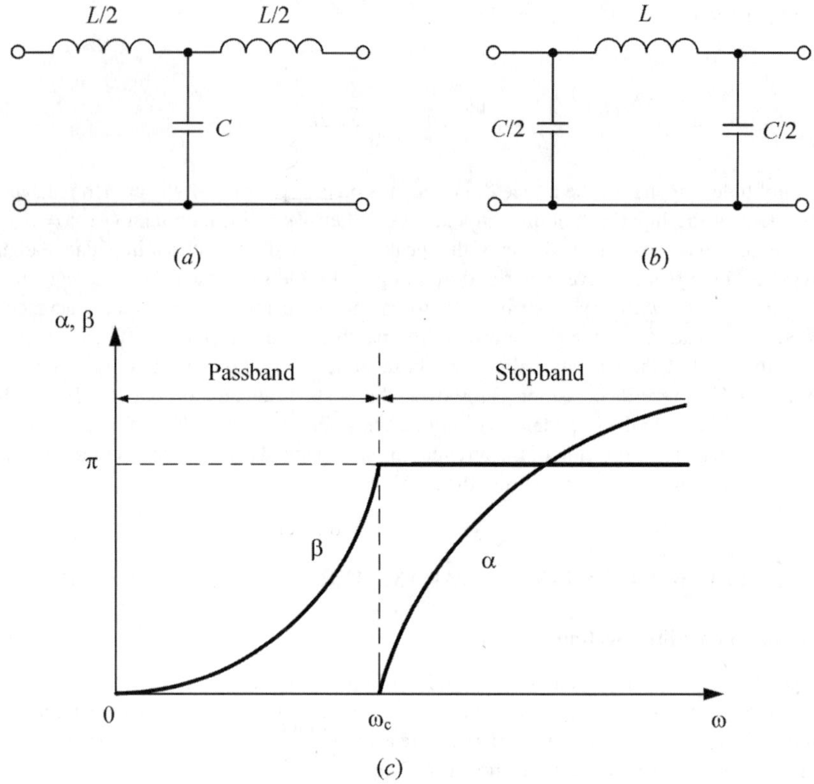

FIGURE 5.5 Low-pass constant-k filter sections and their typical frequency response.

In this case, the propagation factor given in Table 5.1 can be written as

$$e^{\gamma} = 1 - 2\left(\frac{\omega}{\omega_c}\right)^2 + 2\frac{\omega}{\omega_c}\sqrt{\frac{L}{C}}\sqrt{\left(\frac{\omega}{\omega_c}\right)^2 - 1}. \tag{5.11}$$

Hence, each filter section is characterized by its Zobel parameter m related to the attenuation pole $\Omega = \omega/\omega_c$, defined as

$$m = \sqrt{1 - \left(\frac{\omega}{\omega_c}\right)^2}. \tag{5.12}$$

For the low-pass π-network shown in Figure 5.5(b) when $Z_1 = j\omega L$ and $Z_2 = 1/(j\omega C)$, the cutoff frequency ω_c, characteristic impedance Z_0, and propagation factor exp (γ) are the same as for a low-pass T-network given by Eqs. (5.8), (5.9), and (5.11), respectively. At $\omega = 0$, we have that $Z_{iT} = Z_{i\pi} = Z_0$, where $Z_{i\pi}$ is the image impedance of the low-pass π-network, but Z_{iT} and $Z_{i\pi}$ are generally not equal at other frequencies.

Typical phase and attenuation constants for a low-pass constant-k filter section are shown in Figure 5.5(c). In this case, the attenuation constant α is zero in the passband when $\omega < \omega_c$, and tends to infinity in the stopband as $\omega \to \infty$. The attenuation rate for $\omega \gg \omega_c$ is 40 dB/decade. According to Eq. (5.11), the phase constant β increases from 0 to π in the passband where the image impedance is

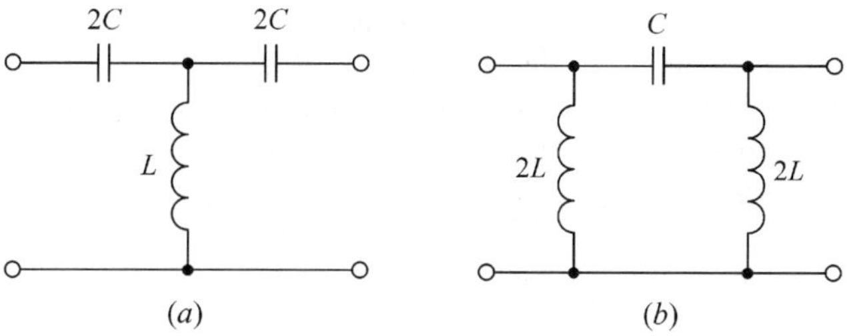

FIGURE 5.6 High-pass constant-k filter sections.

real, and becomes constant and equal to π in the stopband where the image impedance is imaginary. This type of filter is known as a constant-k low-pass prototype, with only two parameters to choose (L and C), which are determined by the cutoff frequency ω_c and characteristic impedance Z_0.

For high-pass constant-k sections, the positions of the inductors and capacitors are reversed from those in the low-pass prototype. As a result, the characteristic impedance Z_0 of the high-pass constant-k filter sections in the form of T- and π-sections, shown in Figure 5.6, is defined similarly to the low-pass constant-k filter sections given by Eq. (5.9), and its cutoff frequency ω_c is calculated from

$$\omega_c = \frac{1}{2\sqrt{LC}}. \tag{5.13}$$

5.2.2 *m*-Derived Filter Sections

The main disadvantage of the constant-k filter is that the image impedance, which should terminate the filter section at both ports, is a function of frequency that is not likely to match a given source or load impedance. In addition, its attenuation is sufficiently small near cutoff frequency and slowly increases. The problems can be overcome with the modified m-derived filter sections.

In this case, as shown in Figure 5.7(a), the impedances Z_1 and Z_2 in a constant-k T-section are replaced with the modified impedances Z_{1m} and Z_{2m} as

$$Z_{1m} = mZ_1 \tag{5.14}$$

$$Z_{2m} = \frac{1-m^2}{4m}Z_1 + \frac{Z_2}{m} \tag{5.15}$$

the values of which are chosen to obtain the same value of the image impedance Z_{iT} as for the constant-k section given in Table 5.1 [6]. Because the impedances Z_1 and Z_2 represent reactive elements, the

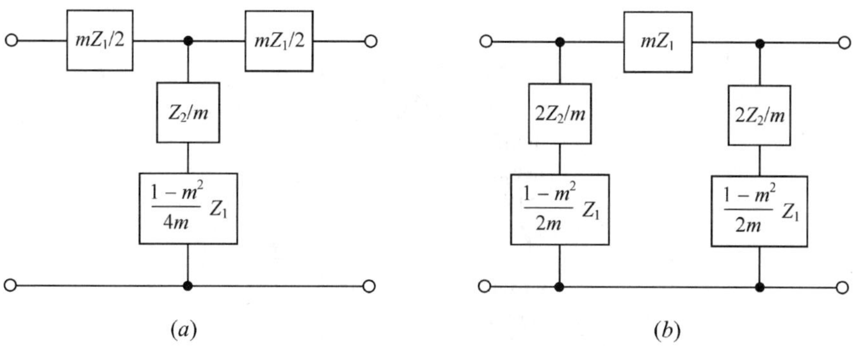

FIGURE 5.7 m-derived filter sections.

modified impedance Z_{2m} represents two elements in series. Note that $m = 1$ reduces the m-derived filter section to the original constant-k- filter section.

For a low-pass T-section when $Z_1 = j\omega L$ and $Z_2 = 1/(j\omega C)$, the modified impedances Z_{1m} and Z_{2m}, given by Eqs. (5.14) and (5.15), can be rewritten as

$$Z_{1m} = jm\omega L \tag{5.16}$$

$$Z_{2m} = j\frac{1-m^2}{4m}\omega L + \frac{1}{jm\omega C} \tag{5.17}$$

which results in the filter circuit of Figure 5.8(a). In this case, the propagation factor for the m-derived section can be written using Table 5.1 as

$$e^{\gamma} = 1 + \frac{Z_{1m}}{2Z_{2m}} + \sqrt{\frac{Z_{1m}}{Z_{2m}}\left(1 + \frac{Z_{1m}}{4Z_{2m}}\right)}. \tag{5.18}$$

For the low-pass m-derived T-section,

$$\frac{Z_{1m}}{Z_{2m}} = -\frac{\left(2m\dfrac{\omega}{\omega_c}\right)^2}{1-(1-m^2)\left(\dfrac{\omega}{\omega_c}\right)} \tag{5.19}$$

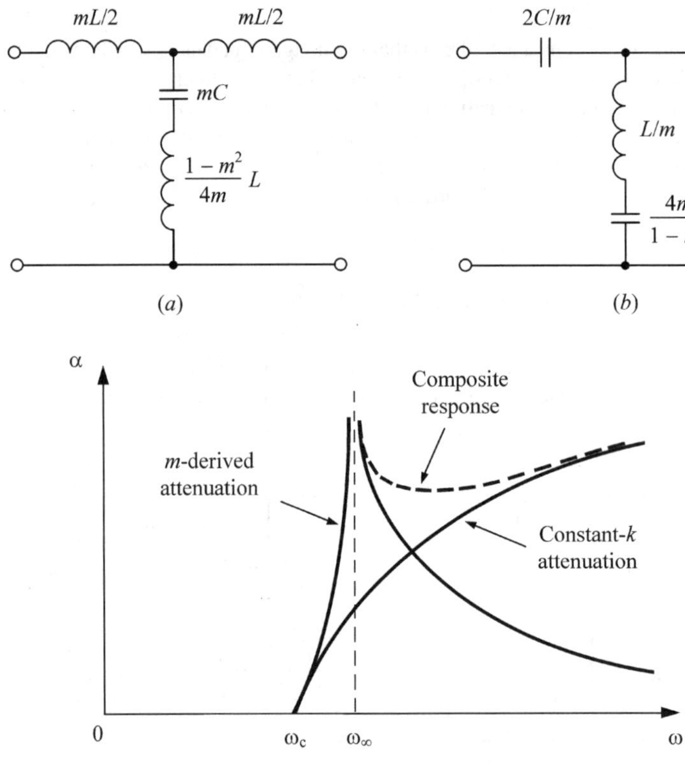

FIGURE 5.8 Low-pass and high-pass m-derived T-sections and frequency response.

where $\omega_c = 2/\sqrt{LC}$ as for the low-pass constant-k filter section, and

$$1 + \frac{Z_{1m}}{4Z_{2m}} = \frac{1 - \left(\dfrac{\omega}{\omega_c}\right)^2}{1 - \left(1 - m^2\right)\left(\dfrac{\omega}{\omega_c}\right)}. \tag{5.20}$$

Substituting Eqs. (5.19) and (5.20) into Eq. (5.18) shows that when $0 < m < 1$, then the propagation factor e^{γ} is real and $|e^{\gamma}| > 1$ for $\omega > \omega_c$. Hence, the filter stopband begins at $\omega = \omega_c$, similarly to the constant-k section. However, when $\omega = \omega_\infty$ and the filter Zobel parameter m is defined as

$$m = \sqrt{1 - \left(\frac{\omega_c}{\omega_\infty}\right)^2} \tag{5.21}$$

the denominators in both Eqs. (5.19) and (5.20) vanish and e^{γ} becomes infinite, thus implying infinite attenuation. Physically, this pole in the attenuation characteristic is caused by the resonance of the series LC resonator at ω_∞ in the shunt arm of the T-section. The infinite attenuation occurs after the cutoff frequency ω_c, and the position of pole can be controlled with the value of m. The best result with minimum variation of the image impedance Z_{iT} over the filter passband is achieved when $m =$ 0.6. The high-pass m-derived T-section is shown in Figure 5.8(b), for which $\omega_c = 1/(2\sqrt{LC})$ as for the high-pass constant-k filter section and

$$m = \sqrt{1 - \left(\frac{\omega_\infty}{\omega_c}\right)^2}. \tag{5.22}$$

The low-pass m-derived section shown in Figure 5.7(b) can be obtained from the corresponding two m-derived T-sections connected in cascade. In this case, the impedances Z_{1m} and Z_{2m} in the series and shunt arms must be multiplied by a factor of 2. Figure 5.9 shows the (a) low-pass and (b) high-pass m-derived π-sections, where m for sharp cutoff is defined by Eqs. (5.21) and (5.22), respectively, and equals to 0.6 for optimum matching. The inductor L and capacitor C are the same as for the corresponding low-pass and high-pass constant-k π-sections.

From Figure 5.8(c) it follows that the filter attenuation of the m-derived section decreases for $\omega > \omega_\infty$. Since it is often desirable to have infinite attenuation as $\omega \to \infty$, the m-derived section can be cascaded with a constant-k section to give the composite attenuation response. By combining in cascade the constant-k T-section, the m-derived sharp-cutoff T-section, and the m-derived bisected-π matching sections at the filter input and output, the four-stage composite filter with desired

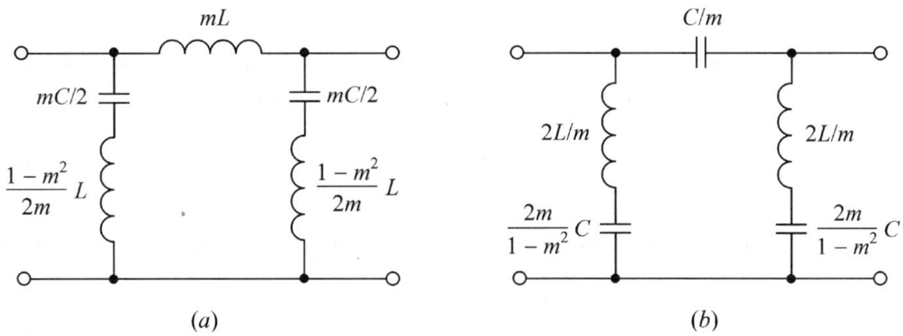

FIGURE 5.9 Low-pass and high-pass m-derived π-sections.

attenuation and matching properties can be realized [5]. The sharp-cutoff section with $m < 0.6$ places an attenuation pole near the cutoff frequency to provide a sharp attenuation response, whereas the constant-k section provides high attenuation further into the stopband. The bisected-π sections at the ends of the filter match the nominal source and load impedance Z_0 to the internal image impedances Z_{iT} of the constant-k and m-derived sections.

5.3 FILTER DESIGN USING INSERTION LOSS METHOD

Unlike the image parameter method, the insertion loss method allows a high degree of the filter performance control over the passband and stopband when the desired amplitude and phase characteristic can be synthesized with a systematic way. For example, if a minimum insertion loss is most important, a binomial filter response could be used. However, a requirement for the sharpest cutoff can be satisfied with a Chebyshev filter response, or better phase response can be obtained by using a linear phase filter design. In all cases, the insertion loss method allows filter performance to be improved in a straightforward manner, at the expense of a higher order filter.

In the insertion loss method, a filter response is defined by its attenuation A (or insertion loss in decibels equal to $10\log_{10}A$), which is defined as the ratio of power available from the source P_S to power delivered to the load P_L,

$$A = \frac{P_S}{P_L} = \frac{1}{1 - |\Gamma(\omega)|^2} \tag{5.23}$$

where $\Gamma(\omega)$ is the reflection coefficient. For the input impedance $Z_{in}(\omega) = R_{in}(\omega) + jX_{in}(\omega)$, the reflection coefficient $\Gamma(\omega)$ can be written as

$$\Gamma(\omega) = \frac{Z_{in}(\omega) - Z_0}{Z_{in}(\omega) + Z_0} = \frac{R_{in}(\omega) - Z_0 + jX_{in}(\omega)}{R_{in}(\omega) + Z_0 + jX_{in}(\omega)} \tag{5.24}$$

where Z_0 is the characteristic impedance. From Eq. (5.24) it follows that the magnitude of the reflection coefficient is even function of ω because

$$|\Gamma(\omega)|^2 = |\Gamma(-\omega)|^2 \tag{5.25}$$

and it can be expressed as a polynomial of ω^2. As a result, Eq. (5.23) can be rewritten as

$$A = 1 + \frac{M(\omega^2)}{N(\omega^2)}. \tag{5.26}$$

5.3.1 Maximally Flat Low-Pass Filter

The low-pass binomial or Butterworth response of the filter (named in honor of S. Butterworth who described this response in 1930), which is completely defined by its poles, is specified as

$$A = 1 + k^2\left(\frac{\omega}{\omega_c}\right)^{2n} \tag{5.27}$$

where n is the order of the filter, which corresponds to the number of required reactive elements, and ω_c is the cutoff frequency. The insertion-loss function of this filter in the passband has the flattest possible shape in the passband (maximally flat), having the maximum number of $(2n - 1)$ zero derivatives at $\omega = 0$, and is a monotonically increasing function. The passband extends from $\omega = 0$

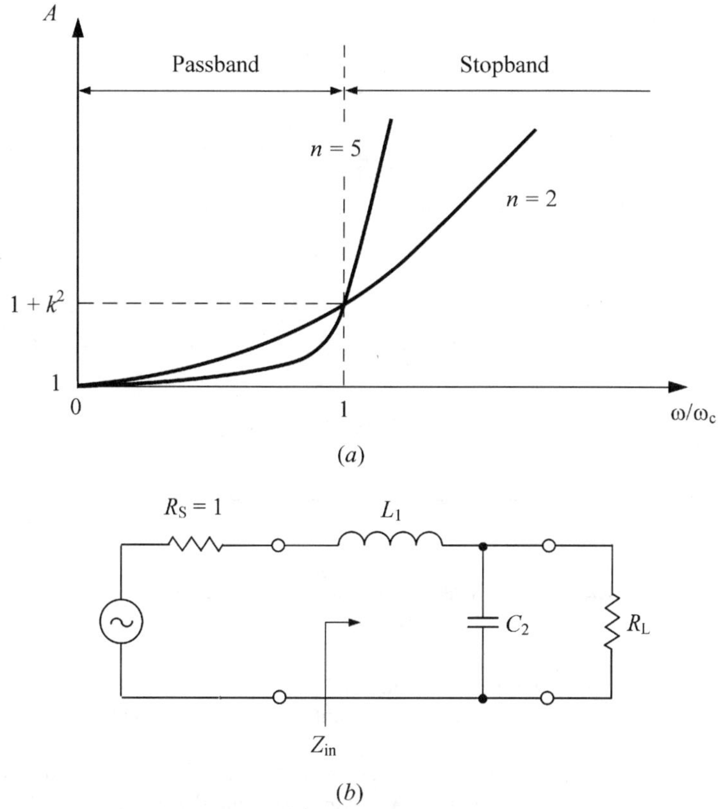

FIGURE 5.10 Maximally flat responses and low-pass filter prototype.

to $\omega = \omega_c$, with the power loss ratio of $(1 + k^2)$ at the band edge and $k = 1$ for a commonly used maximum passband attenuation of 3 dB. Above the cutoff frequency, the attenuation increases almost linearly that gives a rate of $6n$ dB/octave on a logarithmic frequency scale. The attenuation functions of the low-pass Butterworth filter are sketched in Figure 5.10(a) for small ($n = 2$) and large ($n = 5$) filter orders. It is clearly seen that the larger order provides a better approximation to the ideal low-pass response.

As a design example, consider the two-element low-pass filter prototype shown in Figure 5.10(b) and derive the normalized element values of an inductor L_1 and a capacitor C_2 for a maximally flat response, assuming a source impedance $R_S = Z_0 = 1\ \Omega$ and a cutoff frequency $\omega_c = 1$ radian/s. For $n = 2$, the input impedance of this filter can be written as

$$Z_{in} = j\omega L_1 + \frac{1 - j\omega C_2 R_L}{1 + (\omega C_2 R_L)^2} R_L. \tag{5.28}$$

Then, substituting Eq. (5.28) to Eq. (5.23), and using Eq. (5.24), results in

$$A = 1 + \frac{(1 - R_L)^2 + \left(R_L^2 C_2^2 + L_1^2 - 2L_1 C_2 R_L^2\right)\omega^2 + (L_1 C_2 R_L)^2\,\omega^4}{4R_L} \tag{5.29}$$

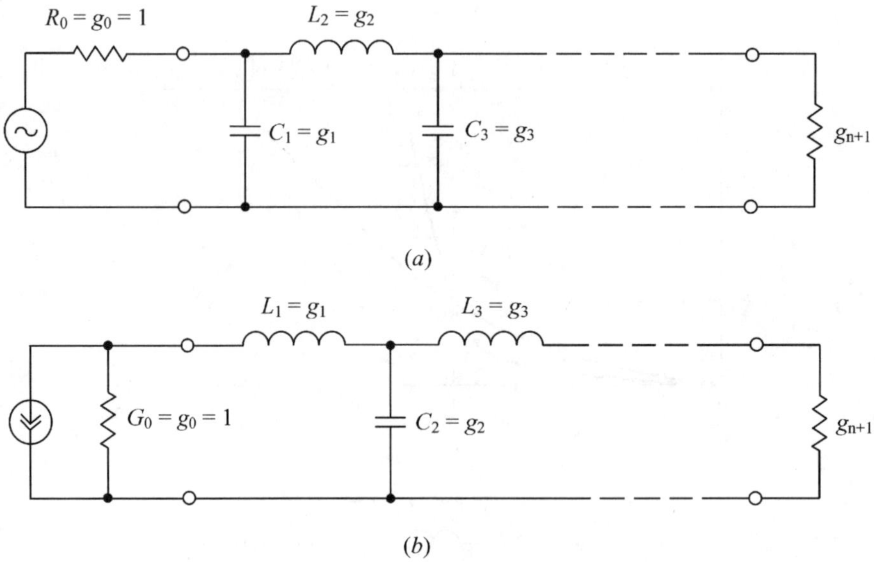

FIGURE 5.11 Ladder circuits for low-pass filter prototypes.

which represents a polynomial of ω^2. Comparing to Eq. (5.27) for $k = 1$ and $n = 2$ shows that $R_L = 1\ \Omega$, since $A = 1$ for $\omega = 0$. In this case, the coefficient of ω^2 must be equal to zero, which results in $L_1 = C_2$. Then, for the coefficient of ω^4 to be unity, $L_1 = C_2 = \sqrt{2}$.

In principle, this design procedure can be extended to find the element values for filters with an arbitrary numbers of elements, but analytically it becomes too complicated. Generally, by constructing the rational transfer function from Eq. (5.27), the values of elements (normalized capacitors and inductors) shown in Figure 5.11 are calculated from

$$\begin{aligned} g_0 &= 1.0 \\ g_i &= 2\sin\left(\frac{2i-1}{2n}\pi\right) \qquad \text{for } i = 1, \ldots, n \\ g_{n+1} &= 1.0 \end{aligned} \tag{5.30}$$

which are given in numerical form in Table 5.2 [7,8]. For convenience, the element values for maximally flat low-pass filter prototypes are given for $n = 1$ to 8, and these data can be used with

TABLE 5.2 Element Values for Maximally Flat Low-Pass Filter Prototypes.

n	g_1	g_2	g_3	g_4	g_5	g_5	g_7	g_8	g_9
1	2.0000	1.0000							
2	1.4142	1.4142	1.0000						
3	1.0000	2.0000	1.0000	1.0000					
4	0.7654	1.8478	1.8478	0.7654	1.0000				
5	0.6180	0.6180	2.0000	0.6180	0.6180	1.0000			
6	0.5176	1.4142	1.9318	1.9318	1.4142	0.5176	1.0000		
7	0.4450	1.2470	1.8019	2.0000	1.8019	1.2470	0.4450	1.0000	
8	0.3902	1.1111	1.6629	1.9615	1.9615	1.6629	1.1111	0.3902	1.0000

either of the ladder circuits of Figure 5.11, where $R_S = R_0$ and $G_S = G_0$. The two-port Butterworth filters considered here represent the symmetrical network structure when $g_0 = g_{n+1}$, $g_1 = g_n$, and so on. If a termination resistance higher than 1 Ω is desired, the reactances are multiplied accordingly. The reactances are then scaled so that they have the same value at the new desired cutoff frequency ω_c.

The elements alternate between series and shunt connections, and g_0 and g_{n+1} represent the source and load resistances for a network in Figure 5.11(*a*) and the source and load conductances for a network in Figure 5.11(*b*); the reactive components are calculated from

$$C_i = \frac{g_i}{\omega_c R_0} \qquad L_i = \frac{g_i R_0}{\omega_c} \tag{5.31}$$

and both low-pass ladder circuits of Figure 5.11 give the same frequency response.

The degree of a Butterworth low-pass filter prototype defined by a minimum stopband attenuation A_{min} at $\omega = \omega_s$ and a maximum passband attenuation A_{max} at $\omega = \omega_c$ can be derived from Eq. (5.27) in the form

$$n \geq \frac{\log_{10} \dfrac{A_{min} - 1}{A_{max} - 1}}{2 \log_{10} \dfrac{\omega_s}{\omega_c}}. \tag{5.32}$$

For example, if a minimum attenuation in the stopband $A_{min} = 10^4$ (or 40 dB) at $\omega_s/\omega_c = 2$ for $A_{max} = 2$ (or 3 dB) is required, then $n \geq 6.644$, and a 7-pole ($n = 7$) Butterworth prototype should be chosen.

In the high-pass ladder filters, which are dual of the low-pass ladder filters, the series inductors are replaced with series capacitors, while shunt capacitors are replaced by shunt inductors. The reactances are the same at the normalized cutoff frequency $\omega_c = 1$ radian/s and $R_0 = 1$ Ω. The Butterworth high-pass filter prototype is described by the following denormalization of the normalized low-pass data from Table 5.2:

$$C_i = \frac{1}{g_i \omega_c R_0} \qquad L_i = \frac{R_0}{g_i \omega_c}. \tag{5.33}$$

The Butterworth approximation is useful for many applications; however, its main advantage is its mathematical simplicity. The low-pass Butterworth response was derived on the assumption that behavior at zero frequency was far more important than behavior at any other frequency. This leads to a class of filters with good phase response and tolerably good amplitude response but with very poor characteristics around the cutoff frequency. The Butterworth function is unsuitable for applications that require uniform transmission of frequencies in the passband and sharp rise at cutoff.

5.3.2 Equal-Ripple Low-Pass Filter

By changing the approximation conditions, it is possible to obtain much better characteristics near cutoff frequency. Besides, it may be required that all frequencies in the passband are equally important, and that it is desirable to minimize the maximum deviation from the ideal response. The Chebyshev approximation can exhibit the equal-ripple passband and maximally flat stopband when the attenuation of the low-pass filter is written as

$$A = 1 + k^2 T_n^2 \left(\frac{\omega}{\omega_c} \right) \tag{5.34}$$

where k is the ripple constant and $T_n(\omega/\omega_c)$ is the a Chebyshev function of the first kind of order n, which is defined as

$$T_n\left(\frac{\omega}{\omega_c}\right) = \begin{cases} \cos\left(n\cos^{-1}\dfrac{\omega}{\omega_c}\right) & \text{for } \dfrac{\omega}{\omega_c} \leq 1 \\ \cosh\left(n\cosh^{-1}\dfrac{\omega}{\omega_c}\right) & \text{for } \dfrac{\omega}{\omega_c} \geq 1 \end{cases}. \tag{5.35}$$

From Eq. (5.35) is not clear that $T_n(\omega/\omega_c)$ is actually the nth-order polynomial. However, by applying simple trigonometric identities, the first four Chebyshev polynomials are written as

$$T_1(x) = x \tag{5.36}$$

$$T_2(x) = 2x^2 - 1 \tag{5.37}$$

$$T_3(x) = 4x^3 - 3x \tag{5.38}$$

$$T_4(x) = 8x^4 - 8x^2 + 1 \tag{5.39}$$

with higher order polynomials found using general recurrence relation,

$$T_n(x) = 2xT_{n-1}(x) - T_{n-2}(x) \tag{5.40}$$

where $x = \omega/\omega_c$.

Since the Chebyshev polynomials have the property that $T_n(0) = 0$ for odd n and $T_n(0) = 1$ for even n, from Eq. (5.34) it follows that the equally-ripple low-pass filter have a unity attenuation at $\omega = 0$ for odd n, but an attenuation of $(1 + k^2)$ at $\omega = 0$ for even n. In the passband, the attenuation response varies between the values of zero and $A_{max} = 1 + k^2$, as seen from Figure 5.12(a) for $n = 2$, 3. Above the cutoff frequency ω_c, the attenuation curve rises monotonically. The rate of increase depends not only upon the number of poles or resonators but also upon such a design parameter as the height of the ripples: the attenuation rate is higher for larger passband ripples.

For the two-element low-pass filter prototype, shown in Figure 5.12(b) with a source impedance $R_S = R_0 = 1\ \Omega$ and a cutoff frequency $\omega_c = 1$ radian/s, equating Eq. (5.34) to Eq. (5.29) and using Eq. (5.37) result in

$$1 + k^2\left(4\omega^4 - 4\omega^2 + 1\right) = 1 + \frac{(1 - R_L)^2 + \left(R_L^2C_2^2 + L_1^2 - 2L_1C_2R_L^2\right)\omega^2 + (L_1C_2R_L)^2\omega^4}{4R_L} \tag{5.41}$$

which can be solved to define R_L, L_1, and C_2 if the ripple determined by k^2 is known. Since $k^2 = (1 - R_L)^2/(4R_L)$ at $\omega = 0$, then equating coefficients of ω^2 and ω^4 yields the additional equations,

$$4k^2 = \frac{(L_1C_2R_L)^2}{4R_L} \tag{5.42}$$

$$-4k^2 = \frac{(C_2R_L)^2 + L_1^2 - 2L_1C_2R_L^2}{4R_L} \tag{5.43}$$

which are necessary to find L_1 and C_2. Note that a value for R_L is not unity for even n, thus resulting in an impedance mismatch if the load represents a unity normalized impedance.

Because it is too complicated to find analytically element values for filters with arbitrary numbers of elements, by constructing the rational transfer function from Eq. (5.34), the values of elements

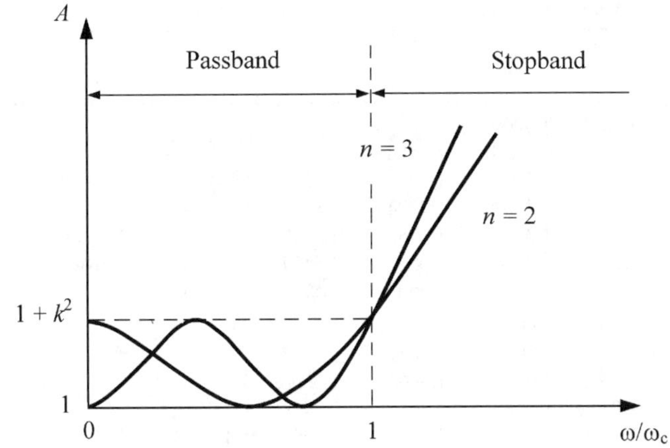

(a)

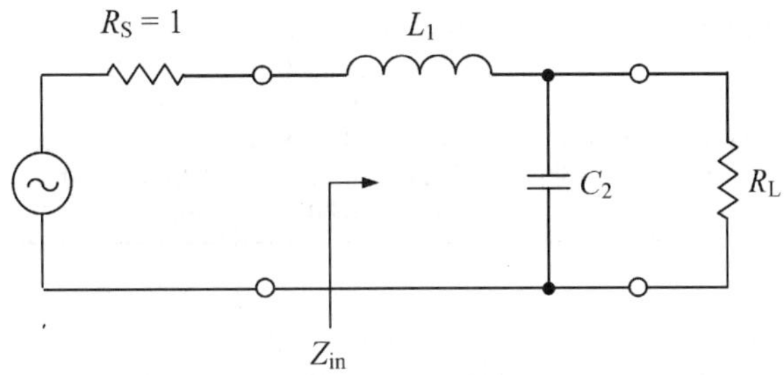

(b)

FIGURE 5.12 Equal-ripple responses and low-pass filter prototype.

(normalized capacitors and inductors) shown in Figure 5.11 can be calculated from

$$
\begin{aligned}
g_0 &= 1.0 \\
g_1 &= \frac{2}{\gamma}\sin\left(\frac{\pi}{2n}\right) \\
g_i &= \frac{1}{g_{i-1}}\,\frac{4\sin\left(\dfrac{2i-1}{2n}\pi\right)\sin\left(\dfrac{2i-3}{2n}\pi\right)}{\gamma^2} \qquad \text{for } i = 2, 3, \ldots, n \\
g_{n+1} &= 1.0 \text{ for odd } n \text{ and } g_{n+1} = \coth^2\frac{\beta}{4} \text{ for even } n
\end{aligned}
\tag{5.44}
$$

with

$$\gamma = \sinh\frac{\beta}{2n} \qquad \beta = \ln\left[\coth\frac{A\,(\text{dB})}{17.3718}\right]$$

where A(dB) is the passband ripple in decibels [7]. Some typical element values for such filters are given in numerical form in Table 5.3 for various passband ripples A(dB) and for the filter degree of $n = 1$ to 7.

TABLE 5.3 Element Values for Equal-Ripple Low-Pass Filter Prototypes.

Ripple (dB)	n	g_1	g_2	g_3	g_4	g_5	g_6	g_7	g_8
0.1	1	0.3052	1.0000						
	2	0.8431	0.6220	1.3554					
	3	1.0316	1.1474	1.0316	1.0000				
	4	1.1088	1.3062	1.7704	0.8181	1.3554			
	5	1.1468	1.3712	1.9750	1.3712	1.1468	1.0000		
	6	1,1681	1.4040	2.0562	1.5171	1.9029	0.8618	1.3554	
	7	1.1812	1.4228	2.0967	1.5734	2.0967	1.4228	1.1812	1.0000
0.5	1	0.6986	1.0000						
	2	1.4029	0.7071	1.9841					
	3	1.5963	1.0967	1.5963	1.0000				
	4	1.6703	1.1926	2.3661	0.8419	1.9841			
	5	1.7058	1.2296	2.5408	1.2296	1.7058	1.0000		
	6	1.7254	1.2479	2.6064	1.3137	2.4758	0.8696	1.9841	
	7	1.7372	1.2583	2.6381	1.3444	2.6381	1.2583	1.7372	1.0000
3.0	1	1.9953	1.0000						
	2	3.1013	0.5339	5.8095					
	3	3.3487	0.7117	3.3487	1.0000				
	4	3.4389	0.7483	4.3471	0.5920	5.8095			
	5	3.4817	0.7618	4.5381	0.7618	3.4817	1.0000		
	6	3.5045	0.7685	4.6061	0.7929	4.4641	0.6033	5.8095	
	7	3.5182	0.7723	4.6386	0.8039	4.6386	0.7723	3.5182	1.0000

The degree of a Chebyshev low-pass filter prototype, defined by a minimum stopband attenuation $A_{\min}$ at $\omega = \omega_s$ and a maximum passband ripple $A_{\max}$ at $\omega = \omega_c$, can be derived from Eqs. (5.34) and (5.35) in the form

$$n \geq \frac{\cosh^{-1}\sqrt{\dfrac{A_{\min}-1}{A_{\max}-1}}}{\cosh^{-1}\dfrac{\omega_s}{\omega_c}}. \tag{5.45}$$

For instance, using the same example as given for the Butterworth low-pass filter prototype with a minimum attenuation in the stopband $A_{\min} = 10^4$ (or 40 dB) at $\omega_s/\omega_c = 2$, but for a passband ripple $A_{\max} = 1.0233$ (or 0.1 dB) for the Chebyshev response, results in $n \geq 5.45$. Hence, a 6-pole ($n = 6$) Chebyshev prototype should be chosen to meet this specification that demonstrates the superiority of the Chebyshev design over the Butterworth design.

The Chebyshev function is extremely useful in applications where the magnitude of the transfer function is of primary concern. This approximation gives more constant magnitude response throughout the passband and faster rate of cutoff outside the passband. As a consequence, the transition range for reaching a prescribed $A_{\min}$ is a minimum, and the attenuation in the stopband is never less than that prescribed attenuation. However, the phase response of the Chebyshev filter tends to be poor, with a rapid increase in the group-delay variations at the band edges.

5.3.3 Elliptic Function Low-Pass Filter

The maximally flat and equal-ripple responses both have monotonically increasing attenuation in stopband. However, in many applications it is required to specify a minimum stopband attenuation, in which case a better cutoff rate can be obtained. By changing the approximation conditions, it is possible to obtain much better characteristics near cutoff frequency. This can be achieved by using elliptic function filters having equal-ripple responses both in the passband and the stopband, as shown

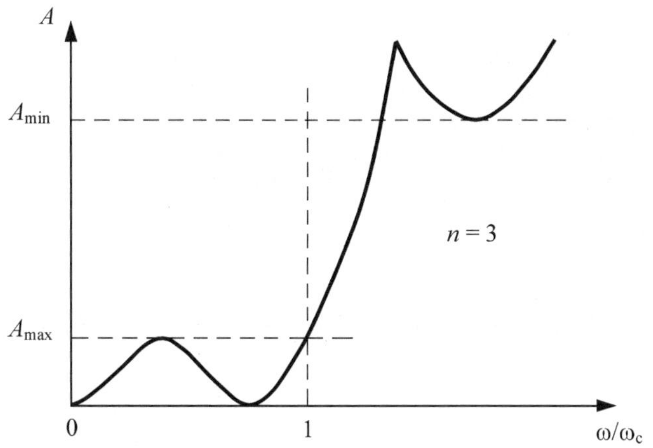

FIGURE 5.13 Elliptic function low-pass filter response.

in Figure 5.13, where A_{max} is the maximum attenuation in the passband and A_{min} is the minimum attenuation in the stopband. Elliptic function filters are also known as Cauer or Zolotarev filters, which achieve the smallest filter order for the same specifications, or the narrowest transition width for the same filter order, as compared to other filter types [4,9].

The attenuation of the elliptic function low-pass filter is written as

$$A = 1 + k^2 T_n^2 \left(\frac{\omega}{\omega_c} \right) \tag{5.46}$$

where k is the ripple constant and $T_n(\omega/\omega_c)$ is chosen so that it has an equal-ripple attenuation in the passband and the stopband. Depending on whether it is even or odd, the polynomial function $T_n(\omega/\omega_c)$ represents one of two following forms:

$$T_n\left(\frac{\omega}{\omega_c}\right) = \begin{cases} M \dfrac{\prod_{i=1}^{n/2} \left[\left(\dfrac{\omega_i}{\omega_c}\right)^2 - \left(\dfrac{\omega}{\omega_c}\right)^2 \right]}{\prod_{i=1}^{n/2} \left[\left(\dfrac{\omega_s}{\omega_i}\right)^2 - \left(\dfrac{\omega}{\omega_c}\right)^2 \right]} & \text{for even } n \\[2ex] N \dfrac{\dfrac{\omega}{\omega_c} \prod_{i=1}^{(n-1)/2} \left[\left(\dfrac{\omega_i}{\omega_c}\right)^2 - \left(\dfrac{\omega}{\omega_c}\right)^2 \right]}{\prod_{i=1}^{(n-1)/2} \left[\left(\dfrac{\omega_s}{\omega_i}\right)^2 - \left(\dfrac{\omega}{\omega_c}\right)^2 \right]} & \text{for odd } n \geq 3 \end{cases} \tag{5.47}$$

where $0 < \omega_i/\omega_c < 1$ and $\omega_s/\omega_c > 1$ represent some critical frequencies (stopband begins with frequency ω_s), M and N are constants to be defined, and $T_n(\omega/\omega_c)$ must lie between the limits -1 and $+1$ in the passband and should take the maximum possible absolute values for the given degree of n in the stopband [10,11].

Figure 5.14 shows two commonly used network structures for elliptic function low-pass filter prototypes. The series–parallel resonant circuits in Figure 5.14(a) are introduced for realizing the finite-frequency zeroes to block the signal transmission at these frequencies by means of infinite open-circuit conditions. In this case, g_i represents the normalized capacitance of a shunt capacitor for odd i, while g_i is the normalized inductance of an inductor and the primed g_i' is the normalized capacitance of a capacitor in a parallel resonant circuit for even n. For the dual network structure shown in Figure 5.14(b), the shunt branches of series resonant circuits are used for implementing

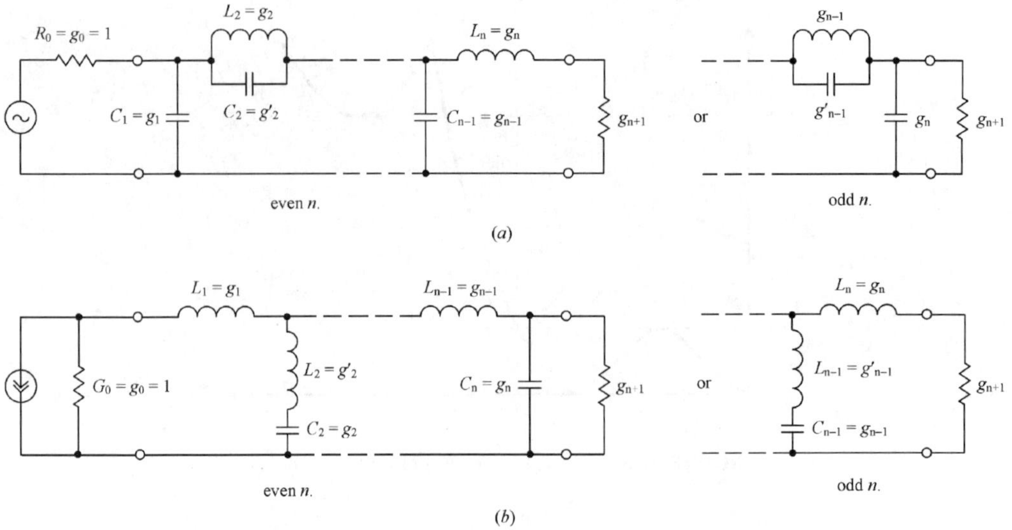

FIGURE 5.14 Elliptic function low-pass filter prototypes.

the finite-frequency zeroes to short out the signal transmission at these frequencies by means of the short-circuit conditions. In this case, g_i represents the normalized inductance of a series inductor for odd i, while g_i is the normalized capacitance of a capacitor and the primed g'_i is the normalized inductance of an inductor in a series resonant circuit for even n. Note that either low-pass filter structure form can be used because both give the same response.

Unlike Butterworth and Chebyshev low-pass filter prototypes, there is no simple formula available for determining element values of the elliptic function low-pass filter prototypes. Table 5.4 tabulates

TABLE 5.4 Element Values for Elliptic Function Low-Pass Filter Prototypes.

n	ω_s/ω_c	A_{min}, dB	g_1	g_2	g'_2	g_3	g_4	g'_4	g_5
3	1.4493	13.5698	0.7427	0.7096	0.5412	0.7427			
	1.6949	18.8571	0.8333	0.8439	0.3252	0.8333			
	2.0000	24.0012	0.8949	0.9375	0.2070	0.8949			
	2.5000	30.5161	0.9471	1.0173	0.1205	0.9471			
4	1.2000	12.0856	0.3714	0.5664	1.0929	1.1194	0.9244		
	1.2425	14.1259	0.4282	0.6437	0.8902	1.1445	0.9289		
	1.2977	16.5343	0.4877	0.7284	0.7155	1.1728	0.9322		
	1.3962	20.3012	0.5675	0.8467	0.5261	1.2138	0.9345		
	1.5000	23.7378	0.6282	0.9401	0.4073	1.2471	0.9352		
	1.7090	29.5343	0.7094	1.0688	0.2730	1.2943	0.9348		
	2.0000	36.0438	0.7755	1.1765	0.1796	1.3347	0.9352		
5	1.0500	13.8785	0.7081	0.7663	0.7357	1.1276	0.2014	4.3812	0.0499
	1.1000	20.0291	0.8130	0.9242	0.4934	1.2245	0.3719	2.1350	0.2913
	1.1494	24.5451	0.8726	1.0083	0.3845	1.3097	0.4991	1.4450	0.4302
	1.2000	28.3031	0.9144	1.0652	0.3163	1.3820	0.6013	1.0933	0.5297
	1.2500	31.4911	0.9448	1.1060	0.2694	1.4415	0.6829	0.8827	0.6040
	1.2987	34.2484	0.9681	1.1366	0.2352	1.4904	0.7489	0.7425	0.6613
	1.4085	39.5947	1.0058	1.1862	0.1815	1.5771	0.8638	0.5436	0.7578
	1.6129	47.5698	1.0481	1.2416	0.1244	1.6843	1.0031	0.3540	0.8692
	1.8182	54.0215	1.0730	1.2741	0.0919	1.7522	1.0903	0.2550	0.9367
	2.000	58.9117	1.0876	1.2932	0.0732	1.7939	1.1433	0.2004	0.9772

some useful design data for equally terminated ($g_0 = g_{n+1} = 1$) elliptic function low-pass filter prototypes representing two equivalent structures shown in Figure 5.14, where normalized element values are given for various ω_s/ω_c and minimum stopband attenuation A_{min} for a maximum passband ripple $A_{max} = 0.1$ dB [12]. In this case, a smaller ω_s/ω_c means a higher selectivity of the filter at the expense of reducing stopband rejection.

Considering the same example as used for the Butterworth and Chebyshev low-pass filter prototypes with a minimum attenuation in the stopband $A_{min} = 40$ dB at $\omega_s/\omega_c = 2$ and a maximum passband ripple $A_{max} = 0.1$ dB, from Table 5.4 it follows that a 5-pole ($n = 5$) elliptic function filter prototype should be chosen that demonstrates its superiority over both the Butterworth and Chebyshev designs for this type of specification. Note that the even-degree Zolotarev low-pass filter has a mismatch at zero frequency and is realizable as an impedance-transforming *LC* ladder network with unequal terminations [13].

5.3.4 Maximally Flat Group-Delay Low-Pass Filter

A maximally flat group-delay low-pass filter prototype is characterized by its transfer (or gain) function $G(j\omega/\omega_c)$ that is written in the form

$$G\left(j\frac{\omega}{\omega_c}\right) = \frac{a_0}{\sum_{k=0}^{n} a_k \left(j\frac{\omega}{\omega_c}\right)^k} \tag{5.48}$$

where

$$a_k = \frac{(2n-k)!}{2^{n-k} k!\,(n-k)!} \tag{5.49}$$

represents the coefficients of the reverse Bessel polynomials and ω_c is a frequency chosen to give the desired cutoff frequency. This transfer function results in better responses in the time domain and approaches the ideal Gaussian curve as the degree of approximation is increased ($n \to \infty$). It also contributes to a group delay that has maximum possible number of zero derivatives with respect to ω at $\omega = 0$, resulting in a maximally flat group delay around $\omega = 0$, and is in a sense complementary to the Butterworth response, which has a maximally flat amplitude. This maximally flat group-delay approximation was originally derived by W. E. Thompson, and for these reasons the filter prototypes of this type are also called Bessel–Thompson filters [14]. The first four Bessel polynomials with respect to a loss function $H(j\omega/\omega_c) = 1/G(j\omega/\omega_c)$ are written as

$$H(s) = \begin{cases} s+1 & \text{for } n = 1 \\ s^2 + 3s + 1 & \text{for } n = 2 \\ s^3 + 6s^2 + 15s + 15 & \text{for } n = 3 \\ s^4 + 10s^3 + 45s^2 + 105s + 105 & \text{for } n = 4 \end{cases} \tag{5.50}$$

where $s = j\omega/\omega_c$ is the normalized complex frequency variable.

Figure 5.15 shows the typical maximally flat group-delay responses in terms of attenuation and group delay for $n = 3$ and $n = 5$ that are properly obtained from Eq. (5.48). In general, the Bessel–Thompson filters have a poor selectivity, as can be seen from the attenuation curves in Figure 5.15(*a*). However, its selectivity improves according to

$$A(\text{dB}) = 10\log_{10}\exp\left[\frac{1}{2n-1}\left(\frac{\omega}{\omega_c}\right)^2\right] \tag{5.51}$$

with increasing order n of the filter [15]. From Eq. (5.51), the 3-dB frequency bandwidth can be determined as

$$\frac{\omega_{3dB}}{\omega_c} = \sqrt{(2n-1)\ln 2} \tag{5.52}$$

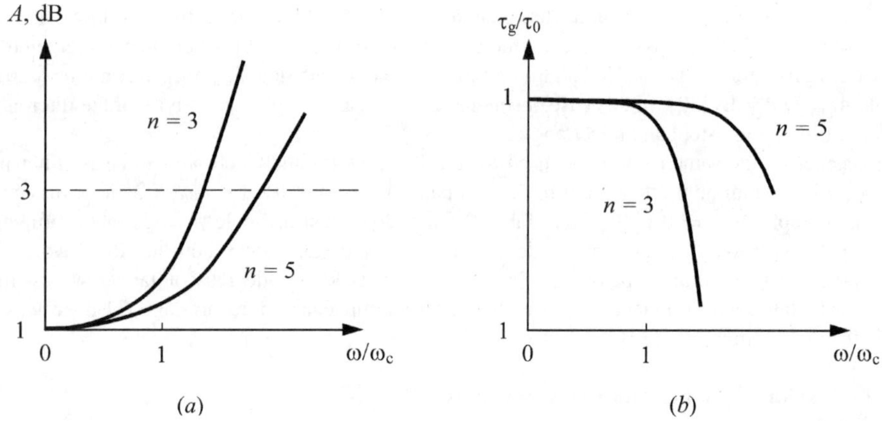

FIGURE 5.15 Maximally flat group-delay response.

approximation of which is sufficiently accurate for $n \geq 3$. Hence, unlike the Butterworth response, the 3-dB bandwidth of a Bessel–Thompson filter is a function of the filter order when the higher the filter order, the wider the 3-dB bandwidth.

However, the Bessel–Thompson filters have a quite flat group delay in the passband, as shown in Figure 5.15(*b*) for a normalized group delay τ_g/τ_0, where τ_0 is the delay at zero frequency, and this normalized group delay is inversely proportional to the bandwidth of the passband. With increasing the filter order *n*, the group delay is flat over a wider frequency range. Therefore, a high-order Bessel–Thompson filter is usually used for achieving a flat group delay over a large passband.

As an example, consider the gain function $G(s)$ for a third-order Bessel–Thompson low-pass filter prototype derived from Eq. (5.48) as

$$G(s) = \frac{15}{s^3 + 6s^2 + 15s + 15}. \tag{5.53}$$

In this case, the attenuation $A = |H(s)|^2 = 1/|G(s)|^2$ can be written by assuming $\omega_c = 1$ radian/s as

$$A = 1 + \frac{\omega^2}{5} + \frac{\omega^4}{75} + \frac{\omega^6}{225} \tag{5.54}$$

and the phase $\phi = \arg[G(s)]$ as

$$\phi(\omega) = -\tan^{-1}\left(\frac{15\omega - \omega^3}{15 - 6\omega^2}\right). \tag{5.55}$$

The group delay, which is a measure of the transit time of a signal through a particular transmitting structure versus frequency, is then

$$\tau_g = -\frac{\phi(\omega)}{\omega} = \frac{6\omega^4 + 45\omega^2 + 225}{\omega^6 + 6\omega^4 + 45\omega^2 + 225}. \tag{5.56}$$

Finally, the Taylor series expansion of the group delay is

$$\tau_g = 1 - \frac{\omega^6}{225} + \frac{\omega^8}{1125} + \cdots \tag{5.57}$$

TABLE 5.5 Element Values for Maximally Flat Group-Delay Low-Pass Filter Prototypes.

n	g_1	g_2	g_3	g_4	g_5	g_5	g_7	g_8	g_9
1	2.0000	1.0000							
2	1.5774	0.4226	1.0000						
3	1.2550	0.5528	0.1922	1.0000					
4	1.0598	0.5116	0.3181	0.1104	1.0000				
5	0.9303	0.4577	0.3312	0.2090	0.0718	1.0000			
6	0.8377	0.4116	0.3158	0.2364	0.1480	0.0505	1.0000		
7	0.7677	0.3744	0.2944	0.2378	0.1778	0.1104	0.0375	1.0000	
8	0.7125	0.3446	0.2735	0.2297	0.1867	0.1387	0.0855	0.0289	1.0000

where the two terms in ω^2 and ω^4 are zero, resulting in a very flat group delay at $\omega = 0$. This is the greatest number of terms that can be set to zero, since there are a total of four coefficients in the third-order Bessel polynomial, requiring four equations in order to be defined. One equation specifies that the gain be unity at $\omega = 0$ and a second equation specifies that the gain be zero at $\omega = \infty$, leaving two equations to specify two terms in the series expansion to be zero. This is a general property of the group delay for a Bessel–Thompson filter of order n: the first $(n - 1)$ terms in the series expansion of the group delay will be zero, thus maximizing the flatness of the group delay at $\omega = 0$.

Design element values for maximally flat group-delay low-pass filter prototypes with $g_0 = 1$ derived for the ladder circuits of Figure 5.11 for different values of $n = 1$ to 8 are given in Table 5.5.

Figure 5.16 shows the frequency transfer characteristics of the Butterworth, Chebyshev, and Bessel–Thompson low-pass filter prototypes based on the ladder circuit shown in Figure 5.11(a) and the elliptic low-pass filter prototype based on the circuit structure of Figure 5.14(a). Each filter configuration was chosen to have an order of $n = 5$, with a passband ripple of 0.5 dB for the Chebyshev

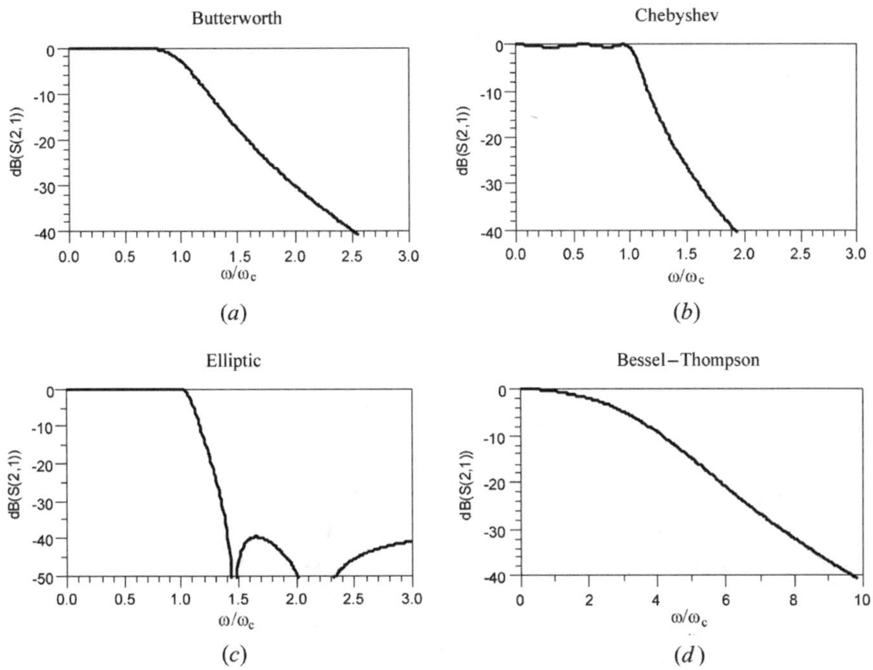

FIGURE 5.16 Filter transfer frequency responses for $n = 5$.

TABLE 5.6 Filter Ranking.

Filter Type	Gain Roll-Off	Linear Phase
Bessel–Thompson	Worst	Best
Butterworth	Poor	Better
Chebyshev	Better	Poor
Elliptic	Best	Worst

low-pass filter and passband ripple of 0.1 dB for the elliptic low-pass filter. Note that attenuation of the elliptic filter can be increased to more than 50 dB starting from $\omega_s/\omega_c = 1.7$, as it follows from Table 5.4.

Table 5.6 shows a rank ordering of the filters in terms of both gain and phase performance. In this case, the elliptic filter offers the very best standard approximation to the ideal low-pass filter amplitude behavior, but its group delay deviates considerably from a constant. On the other end, the Bessel–Thompson filter provides excellent phase performance, but a quite high order is required to achieve a reasonable gain characteristic. If both excellent gain and phase characteristics are absolutely necessary, two approaches are possible. One either uses computer optimization techniques to simultaneously approximate gain and phase, or one uses elliptic filter with excellent gain performance followed by the phase compensation circuit with an inverse phase characteristic.

5.4 BANDPASS AND BANDSTOP TRANSFORMATION

Low-pass filter prototypes can be transformed to have the bandpass or bandstop responses. This transformation into a passband response can be obtained using the frequency substitution in the form

$$\omega \to \frac{\omega_0}{\Delta\omega}\left(\frac{\omega}{\omega_0} - \frac{\omega_0}{\omega}\right) \tag{5.58}$$

where ω_0 is the center bandwidth frequency, $\Delta\omega = \omega_{c+} - \omega_{c-}$ is the passband, and ω_{c-} and ω_{c+} are the low and high cutoff edges of the passband. In this case, the passband Chebyshev filter is realized with the same equal-ripple level, as shown in Figure 5.17.

As a result, a series inductor L_i is transformed into a series resonant circuit with inductor L_i' and capacitor C_i' according to

$$\omega L_i = \frac{\omega_0}{\Delta\omega}\left(\frac{\omega}{\omega_0} - \frac{\omega_0}{\omega}\right) L_i = \omega L_i' - \frac{1}{\omega C_i'} \tag{5.59}$$

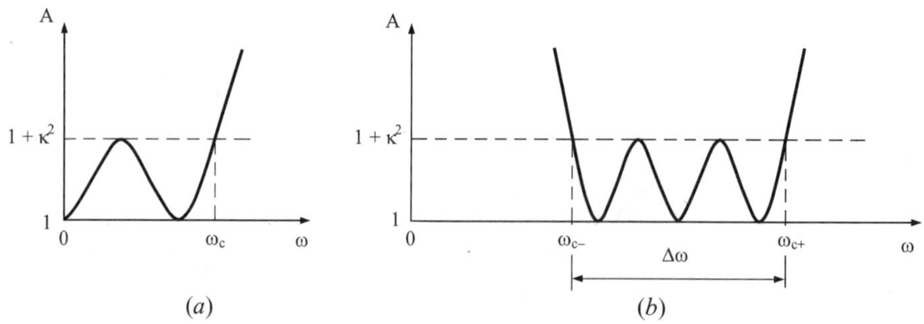

FIGURE 5.17 Bandpass frequency transformation for equal-ripple low-pass filter prototype.

where

$$L'_i = \frac{L_i}{\Delta\omega} \quad C'_i = \frac{\Delta\omega}{\omega_0^2 L_i}. \tag{5.60}$$

Similarly, a shunt capacitor C_i is transformed into a shunt resonant circuit with inductor L'_i and capacitor C'_i as

$$\omega C_i = \frac{\omega_0}{\Delta\omega}\left(\frac{\omega}{\omega_0} - \frac{\omega_0}{\omega}\right) C_i = \omega C'_i - \frac{1}{\omega L'_i} \tag{5.61}$$

where

$$C'_i = \frac{C_i}{\Delta\omega} \quad L'_i = \frac{\Delta\omega}{\omega_0^2 C_i}. \tag{5.62}$$

The low-pass filter prototype will be transformed to the bandpass filter when all its series elements are replaced by the series resonant circuits and all its parallel elements are replaced by the parallel resonant circuits, where each of them are tuned to the center bandwidth frequency ω_0. In this case, the elements in the series resonant circuit of a bandpass filter can be calculated from Eq. (5.60) with impedance scaling as

$$L'_i = \frac{g_i R_0}{\Delta\omega} \quad C'_i = \frac{\Delta\omega}{\omega_0^2 g_i R_0} \tag{5.63}$$

whereas the elements in the shunt resonant circuit of a bandpass filter can be calculated from Eq. (5.62) with impedance scaling as

$$C'_i = \frac{g_i}{\Delta\omega R_0} \quad L'_i = \frac{\Delta\omega R_0}{\omega_0^2 g_i} \tag{5.64}$$

where i is an element serial number for low-pass prototype filter and g_i is the appropriate coefficient given by Table 5.2 for maximally flat low-pass filter prototype or Table 5.3 for equal-ripple low-pass filter prototype.

As an example, let us design a third-order ($n = 3$) bandpass filter with the center frequency of 2 GHz and bandwidth of 20%, having a 0.5-dB equal-ripple response and an impedance $R_0 = 50\ \Omega$. From Table 5.3, the element values for the low-pass prototype circuit shown in Figure 5.18(a) are given by

$$g_1 = 1.5963 \quad g_2 = 1.0967$$
$$g_3 = 1.5963 \quad g_4 = 1.0000.$$

Then, the impedance-scaled and frequency-transformed element values for the passband filter circuit shown in Figure 5.18(b) using Eqs. (5.63) and (5.64) are

$$L'_1 = L'_3 = 3.76\,\text{nH} \quad C'_1 = C'_3 = 0.1994\,\text{pF}$$
$$L'_2 = 0.7256\,\text{nH} \quad C'_2 = 8.727\,\text{pF}.$$

Finally, the resulting amplitude response of the Chebyshev passband filter in a frequency domain is shown in Figure 5.18(c) where the out-of-band suppression of 40 dB is achieved below 1.3 GHz and above 3.0 GHz.

The inverse transformation that can be applied to obtain a bandstop response is written as

$$\omega \to \frac{\Delta\omega}{\omega_0}\left(\frac{\omega}{\omega_0} - \frac{\omega_0}{\omega}\right)^{-1} \tag{5.65}$$

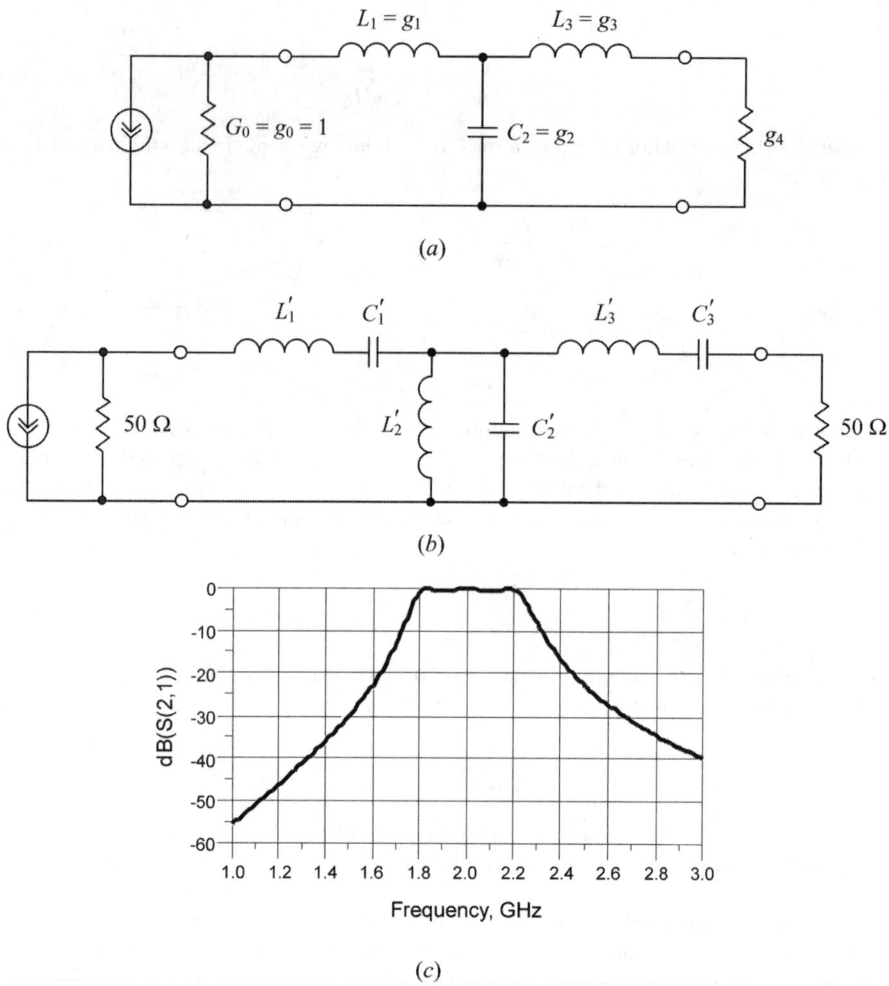

FIGURE 5.18 Bandpass filter circuit and its amplitude response.

where ω_0 and $\Delta\omega$ have the same definition as in Eq. (5.58). In this case, a series inductor L_i of the low-pass filter prototype is transformed into a parallel resonant circuit with inductor L_i' and capacitor C_i' according to

$$L_i' = \frac{\Delta\omega L_i}{\omega_0^2} \quad C_i' = \frac{1}{\Delta\omega L_i} \tag{5.66}$$

and a shunt capacitor C_i of the low-pass filter prototype is transformed into a series resonant circuit with inductor L_i' and capacitor C_i' as

$$C_i' = \frac{\Delta\omega C_i}{\omega_0^2} \quad L_i' = \frac{1}{\Delta\omega C_i} \tag{5.67}$$

with further denormalization similar to that given by Eqs. (5.63) and (5.64) for a passband filter.

5.5 TRANSMISSION-LINE LOW-PASS FILTER IMPLEMENTATION

The design approaches for the filters with lumped elements generally work well at sufficiently low frequencies or in small-size monolithic integrated circuits. However, the lumped elements such as inductors and capacitors are difficult to implement at microwave frequencies where they can be treated as distributed elements, and distances between the filter components are not negligible. The other problem is that the quality factors for inductors are sufficiently small, thus contributing to additional losses.

5.5.1 Richards's Transformation

Generally, the design of a practical distributed filter is based on some approximate equivalence between lumped and distributed elements, which can be established by applying a Richards's transformation [16]. This implies that the distributed circuits composed of equal-length open- and short-circuited transmission lines can be treated as lumped elements under the transformation

$$s = j\tan\frac{\pi\omega}{2\omega_0} \tag{5.68}$$

where $s = j\omega/\omega_c$ is the conventional normalized complex frequency variable and ω_0 is the radian frequency for which the transmission lines are a quarter wavelength [12].

As a result, the one-port impedance of a short-circuited transmission line corresponds to the reactive impedance of a lumped inductor Z_L as

$$Z_L = sL = j\omega L = jL\tan\frac{\pi\omega}{2\omega_0}. \tag{5.69}$$

Similarly, the one-port admittance of an open-circuited transmission line corresponds to the reactive admittance of a lumped capacitor Y_C as

$$Y_C = sC = j\omega C = jC\tan\frac{\pi\omega}{2\omega_0}. \tag{5.70}$$

The results given in Eqs. (5.69) and (5.70) show that an inductor can be replaced with a short-circuited stub of electrical length $\theta = \pi\omega/(2\omega_0)$ and characteristic impedance $Z_0 = L$, while a capacitor can be replaced with an open-circuited stub of electrical length $\theta = \pi\omega/(2\omega_0)$ and characteristic impedance $Z_0 = 1/C$ when a unity filter characteristic impedance is assumed.

From Eq. (5.68) it follows that, for a low-pass filter prototype, the cutoff occurs when $\omega = \omega_c$, resulting in

$$\tan\frac{\pi\omega_c}{2\omega_0} = 1 \tag{5.71}$$

that gives a stub length $\theta = 45^\circ$ (or $\pi/4$) with $\omega_c = \omega_0/2$. Hence, the inductors and capacitors of a lumped-element filter can be replaced with short-circuited and open-circuited stubs, as shown in Figure 5.19. Since the lengths of all stubs are the same and equal to $\lambda/8$ at the cutoff frequency ω_c, these lines are called the *commensurate lines*. At the frequency $\omega = \omega_0$, the transmission lines will be a quarter-wavelength long, resulting in an attenuation pole. However, at any frequency away from ω_c, the impedance of each stub will no longer match the original lumped-element impedances, and the filter response will differ from the desired filter prototype response. Note that the response will be periodic in frequency, repeating every $4\omega_c$.

Since the transmission line generally represents a four-port network, it is very convenient to use a matrix technique for a filter design. In the case of cascade of several networks, the rule is that the overall matrix of the new network is simply the matrix product of the matrices for the individual

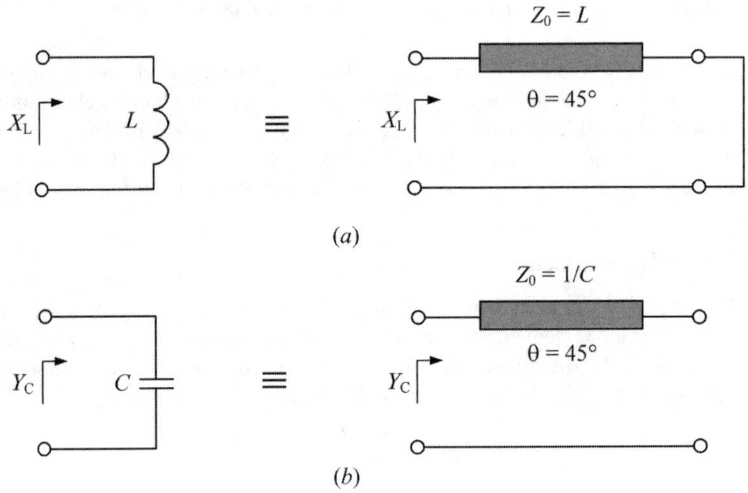

FIGURE 5.19 Equivalence between lumped elements and transmission lines.

networks taken in the order of connection [17]. In terms of a Richards's variable, an *ABCD*-matrix for a transmission line with the characteristic impedance Z_0 can be written as

$$\begin{bmatrix} A & B \\ C & D \end{bmatrix} = \frac{1}{\sqrt{1-s^2}} \begin{bmatrix} 1 & sZ_0 \\ \dfrac{s}{Z_0} & 1 \end{bmatrix} \tag{5.72}$$

representing a unit element which has a half-order transmission zero at $s = \pm 1$. The matrix of the unit element is the same as that of a transmission line of electrical length θ and characteristic impedance Z_0. Unit elements are usually introduced to separate the circuit elements in transmission-line filters, which are otherwise located at the same physical point.

5.5.2 Kuroda Identities

The application of a Richards's transformation provides a sequence of short-circuited and open-circuited stubs, which are then converted to a more practical circuit implementation. This can be done based on a series of equivalent circuits known as Kuroda identities, which allows these stubs to be physically separated, transforming the series stub into the shunt and changing impractical characteristic impedances into more realizable impedances [18]. The Kuroda identities use the unit elements, and these unit elements are thus commensurate with the stubs used to implement inductors and the capacitors of the prototype design. Connecting the unit element with characteristic impedance Z_0 to the same load impedance Z_0 does not change the input impedance. The four Kuroda identities are illustrated in Figure 5.20, where the combinations of unit elements with the characteristic impedance Z_0 and electrical length $\theta = 45°$, the reactive elements, and the relationships between them are given.

To prove the equivalence, consider two circuits of identity at the first row in Figure 5.20 when *ABCD*-matrix for the entire left-hand circuit can be written as

$$\begin{bmatrix} A & B \\ C & D \end{bmatrix}_L = \frac{1}{\sqrt{1-s^2}} \begin{bmatrix} 1 & 0 \\ sC & 1 \end{bmatrix} \begin{bmatrix} 1 & sZ_1 \\ \dfrac{s}{Z_1} & 1 \end{bmatrix} = \frac{1}{\sqrt{1-s^2}} \begin{bmatrix} 1 & sZ_1 \\ s\left(C + \dfrac{1}{Z_1}\right) & 1 + s^2 Z_1 C \end{bmatrix} \tag{5.73}$$

where Z_1 is the characteristic impedance of the left-hand unit element.

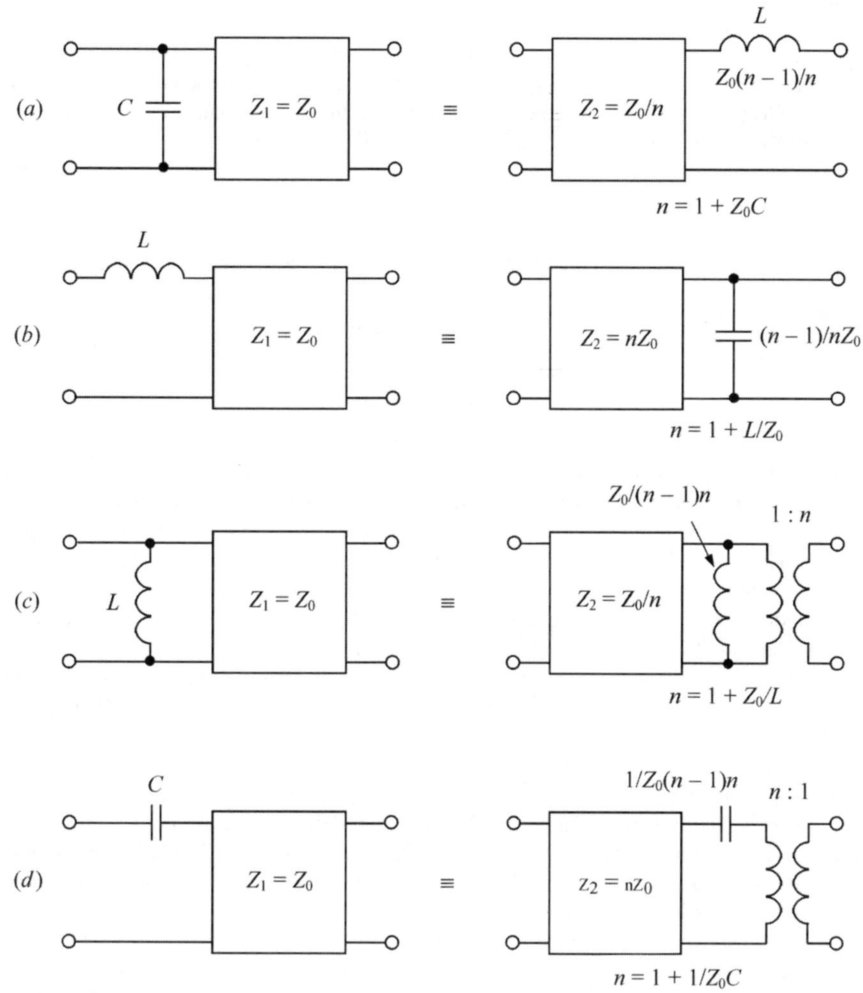

FIGURE 5.20 Four Kuroda identities.

Similarly, for the right-hand circuit,

$$\begin{bmatrix} A & B \\ C & D \end{bmatrix}_R = \frac{1}{\sqrt{1-s^2}} \begin{bmatrix} 1 & sZ_2 \\ \dfrac{s}{Z_2} & 1 \end{bmatrix} \begin{bmatrix} 1 & sL \\ 0 & 1 \end{bmatrix} = \frac{1}{\sqrt{1-s^2}} \begin{bmatrix} 1 & s(Z_2+L) \\ \dfrac{s}{Z_2} & 1+\dfrac{s^2L}{Z_2} \end{bmatrix} \tag{5.74}$$

where Z_2 is the characteristic impedance of the right-hand unit element.

The results in Eqs. (5.73) and (5.74) are identical if

$$Z_1 = Z_2 + L \quad \frac{1}{Z_1} + C = \frac{1}{Z_2} \quad \frac{L}{Z_2} = Z_1 C$$

or

$$Z_2 = \frac{Z_1}{n} \quad L = \frac{n-1}{n} Z_1 \tag{5.75}$$

where $n = 1 + Z_1 C$.

5.5.3 Design Example

To design a low-pass microwave filter based on the transmission lines, let us choose a lumped third-order low-pass Chebyshev filter prototype shown in Figure 5.18(*a*) with the cutoff frequency $f_c = \omega_c/(2\pi) = 2$ GHz, having a 0.5-dB equal-ripple response and an impedance $R_0 = 50\ \Omega$. For unit element with $R_0 = Z_0 = 1$ and normalized inductance $L_3 = g_3 = 1.5963$,

$$n = 1 + L/Z_0 = 2.5963.$$

Using Table 5.3 and Eq. (5.31) results in the denormalized circuit parameters given by

$$L_1 = \frac{g_1 R_0}{\omega_c} = L_3 = 6.375\,\text{nH}$$

$$C_2 = \frac{g_2}{\omega_c R_0} = 1.745\,\text{pF}.$$

Figure 5.21 shows the design transformation of a lumped low-pass filter prototype to a transmission-line filter using the Kuroda identities. The first step shown in Figure 5.21(*a*) is to add a unit element at the right end of the circuit and convert a series inductor into a shunt capacitor using the second Kuroda identity, as shown in Figure 5.21(*b*), resulting in the shunt capacitor having a normalized value of $(n - 1)/n = 0.6148$. Then, adding another unit element at the left end of the circuit and applying the first Kuroda identity lead to the same result for the symmetrical filter structure, as shown in Figure 5.21(*c*) with resulting two unit elements and three shunt capacitors. To keep the same physical dimensions during the calculation of the circuit parameters, the inductance should be taken in nanohenri. The capacitance is measured in nanofarad if the cutoff frequency is measured in gigahertz.

Finally, a Richards's transformation is used to convert the shunt capacitors to open-circuited stubs. In this case, the normalized characteristic impedance of an open-circuited stub equal to $1/C_i$ is necessary to be multiplied by $Z_0 = 50\ \Omega$. Thus, the series transmission-line stubs have the characteristic impedance of $2.5963 \times 50 = 129.3\ \Omega$, the left- and right-hand side open-circuited stubs have the characteristic impedance of $50/0.6148 = 81.3\ \Omega$ each, and the centre open-circuited stub has the characteristic impedance of $50/1.0967 = 45.6\ \Omega$.

Figure 5.21(*d*) shows the transmission-line structure of the final low-pass Chebyshev filter. The lengths of the series and shunt open-circuited stubs are $\lambda/8$ at 2 GHz. The simulated amplitude response of the designed transmission-line low-pass filter is shown in Figure 5.21(*e*), along with the response of its lumped-circuit prototype. It is clearly seen that the passband characteristics are very similar; however, the transmission-line filter has a sharper cutoff transition and its response repeats every 8 GHz as a result of the periodic nature of a Richards's transformation.

5.6 COUPLED-LINE FILTERS

The filter design can also be based on the parallel coupled transmission lines that can be easily implemented in microstrip or stripline form. First, let us consider the basic properties of a quarterwave line and a coupled-line section and then describe the most popular designs of the passband and stopband filter configurations.

5.6.1 Impedance and Admittance Inverters

The impedance and admittance inverters are useful to transform series-connected elements to shunt-connected elements, or vice versa. An ideal impedance inverter represents a two-port network shown in Figure 5.22(*a*) having a unique property at all frequencies which means that, if it is terminated in

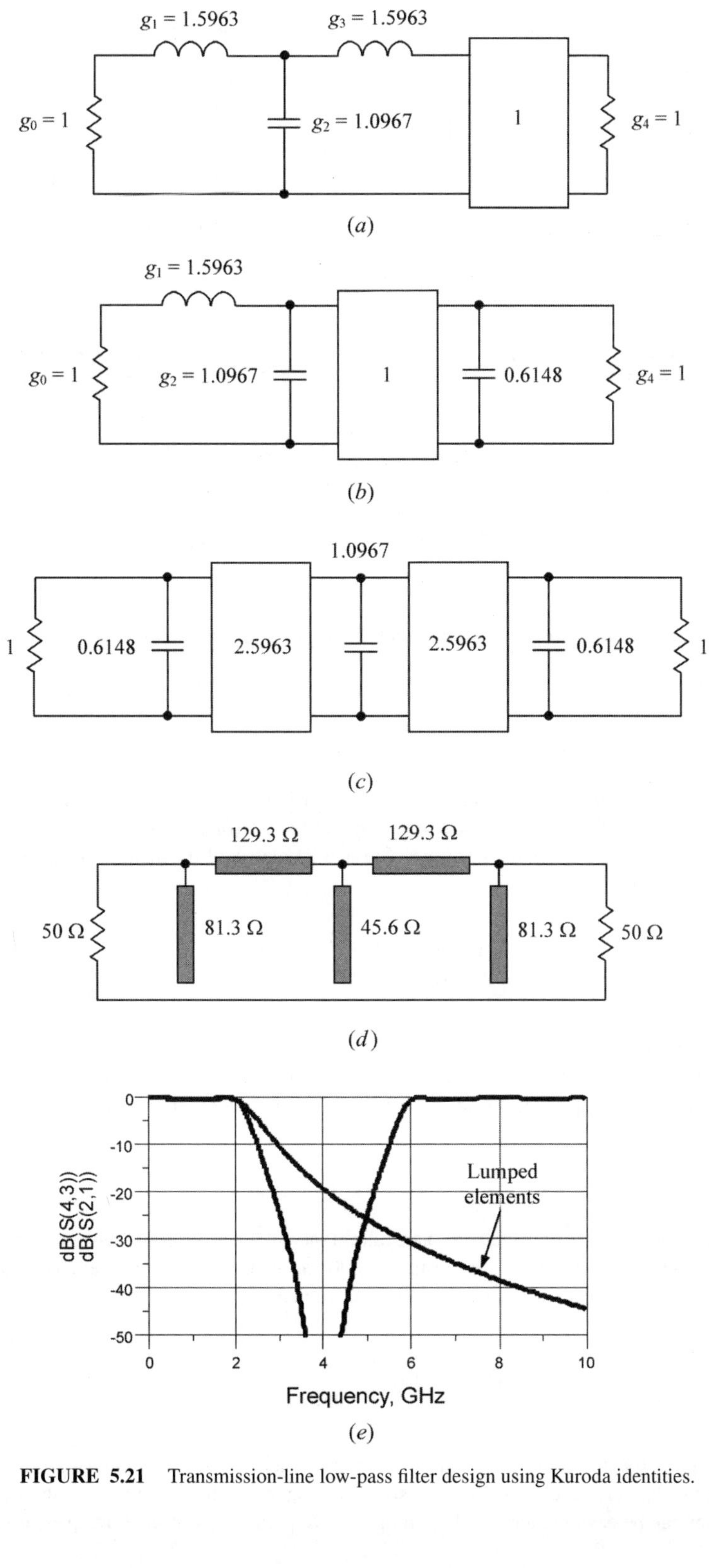

FIGURE 5.21 Transmission-line low-pass filter design using Kuroda identities.

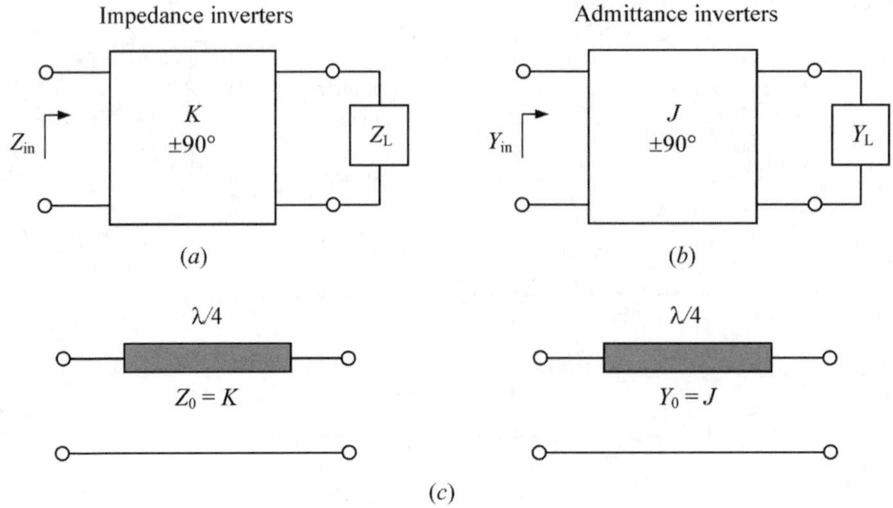

FIGURE 5.22 Impedance and admittance inverters.

an impedance Z_L on one port considered a load, the input impedance Z_{in} seen looking into the other port is

$$Z_{in} = \frac{K^2}{Z_L} \tag{5.76}$$

where K is a constant image impedance of the inverter known as a K-inverter [19]. In this case, if Z_L is capacitive (or inductive), then Z_{in} will become inductive (or capacitive), and hence the inverter has a phase shift of $\pm 90^\circ$ or its odd multiple. The $ABCD$-matrix of the ideal impedance inverters can generally be written as

$$\begin{bmatrix} A & B \\ C & D \end{bmatrix} = \begin{bmatrix} 0 & \mp jK \\ \pm \frac{1}{jK} & 0 \end{bmatrix}. \tag{5.77}$$

Similarly, an idealized admittance inverter is a two-port network shown in Figure 5.22(*b*) that exhibits such a unique property at all frequencies that, if an impedance Y_L is connected at on one port considered a load, the input admittance Y_{in} seen looking into the other port is

$$Y_{in} = \frac{J^2}{Y_L} \tag{5.78}$$

where J is real and defined as the characteristic admittance of the inverter known as a J-inverter. In this case, the admittance inverter has a phase shift of $\pm 90^\circ$ or its odd multiple as well. The $ABCD$-matrix of the ideal admittance inverters can generally be written as

$$\begin{bmatrix} A & B \\ C & D \end{bmatrix} = \begin{bmatrix} 0 & \pm \frac{1}{jJ} \\ \mp jJ & 0 \end{bmatrix}. \tag{5.79}$$

In its simplest form, a K- or J-inverter can be implemented using a quarterwave transformer of the appropriate characteristic impedance, as shown in Figure 5.22(*c*). In this case, the $ABCD$-matrix of the inverter can be easily determined from the $ABCD$-parameters of the transmission line.

5.6.2 Coupled-Line Section

A parallel-coupled stripline section representing a four-port network with port voltages and currents is shown in Figure 5.23(*a*). It is well known that the even (symmetric) and odd (antisymmetric) excitation modes exist in a system of the two-coupled transmission lines. The transverse electromagnetic (TEM) even- and odd-mode electric field distributions in a homogeneous dielectric medium are shown in Figures 5.23(*b*) and 5.23(*c*), respectively. In this case, stronger coupling between the transmission lines takes place at an odd excitation mode, and $Z_{0e} > Z_0 > Z_{0o}$, where the characteristic impedance Z_{0e} corresponds to even mode, the characteristic impedance Z_{0o} corresponds to odd mode, and $Z_0 = \sqrt{Z_{0e} Z_{0o}}$ is the characteristic impedance of a single transmission line.

The four-port impedance *Z*-matrix for a lossless parallel coupled-line section is derived based on a superposition of the results at both types of excitation in a homogeneous medium, which can be written as

$$\begin{bmatrix} V_1 \\ V_2 \\ V_3 \\ V_4 \end{bmatrix} = \begin{bmatrix} Z_{11} & Z_{12} & Z_{13} & Z_{14} \\ Z_{21} & Z_{22} & Z_{23} & Z_{24} \\ Z_{31} & Z_{32} & Z_{33} & Z_{34} \\ Z_{41} & Z_{42} & Z_{43} & Z_{44} \end{bmatrix} \begin{bmatrix} I_1 \\ I_2 \\ I_3 \\ I_4 \end{bmatrix} \tag{5.80}$$

where

$$Z_{11} = Z_{22} = Z_{33} = Z_{44} = -j\left(Z_{0e} + Z_{0o}\right)\frac{\cot\theta}{2}$$

$$Z_{12} = Z_{21} = Z_{34} = Z_{43} = -j\left(Z_{0e} - Z_{0o}\right)\frac{\cot\theta}{2}$$

$$Z_{13} = Z_{31} = Z_{24} = Z_{42} = -j\left(Z_{0e} - Z_{0o}\right)\frac{\csc\theta}{2}$$

$$Z_{14} = Z_{41} = Z_{23} = Z_{32} = -j\left(Z_{0e} + Z_{0o}\right)\frac{\csc\theta}{2}$$

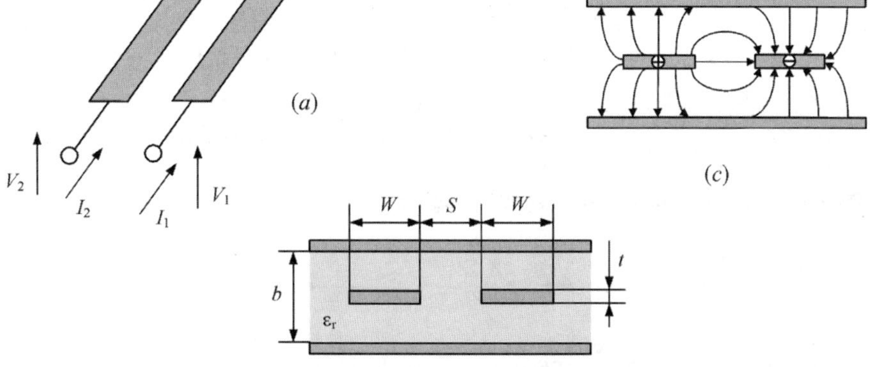

FIGURE 5.23 Parallel-coupled stripline four-port network, even- and odd-mode electric fields.

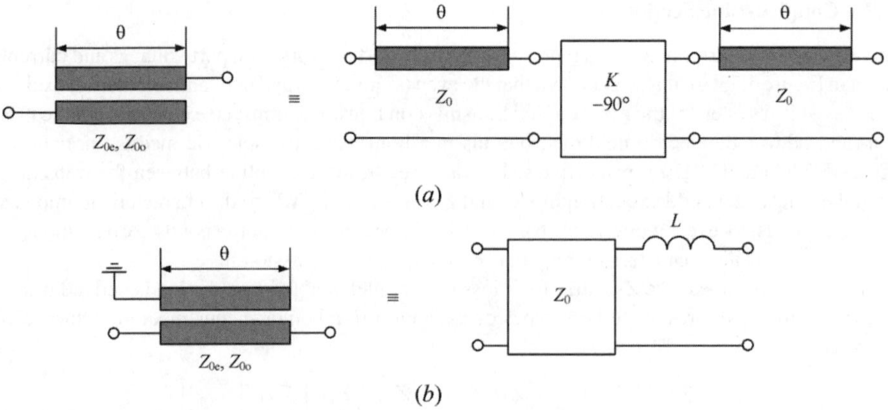

FIGURE 5.24 Equivalent circuits of coupled-line sections.

where θ is the electrical length of the transmission lines [20]. Equation (5.80) holds for a simplest case when both striplines have equal widths. However, by using unequal strip widths, an additional degree of freedom can be obtained as required in many practical cases [21].

Any two-port network with different boundary conditions can be built from the coupled-line section by terminating two of the four ports in either open or short circuits. In this case, the various circuits have different frequency responses, including low-pass, bandpass, all-pass, and all-stop. The most popular configuration of a bandpass filter can be realized with open-circuited conditions when $I_2 = I_4 = 0$, as open circuits are easier to fabricate than short circuits. From Table 5.1, the image impedance for the coupled-line section in terms of Z-parameters shown in the left-hand side of Figure 5.24(a) is

$$Z_i = \sqrt{Z_{11}^2 - \frac{Z_{11}Z_{13}^2}{Z_{33}}} = \frac{\sqrt{(Z_{0e} - Z_{0o})^2 - (Z_{0e} + Z_{0o})^2 \cos^2\theta}}{2\sin\theta}. \tag{5.81}$$

When an electrical length of the coupled-line section is $\theta = \pi/2$ (or it is $\lambda/4$ long), the image impedance reduces to

$$Z_i = \frac{Z_{0e} - Z_{0o}}{2} \tag{5.82}$$

which is real and positive since $Z_{0e} > Z_{0o}$. However, when $\theta \to 0$ or π, then $Z_i \to \infty$, thus indicating a stopband, and the passband cutoff frequencies can be obtained from Eq. (5.81) when $Z_i = 0$ as

$$\cos\theta_c = \pm\frac{Z_{0e} - Z_{0o}}{Z_{0e} + Z_{0o}}. \tag{5.83}$$

The propagation constant can be defined using the results of Table 5.1 as

$$\cos\beta = \sqrt{\frac{Z_{11}Z_{33}}{Z_{13}^2}} = \frac{Z_{0e} + Z_{0o}}{Z_{0e} - Z_{0o}}\cos\theta \tag{5.84}$$

which shows that β is real for $\theta_{c-} < \theta < \theta_{c+} = \pi - \theta_{c-}$, where θ_{c-} and θ_{c+} correspond to the low and high cutoff edges of the passband, respectively, and $\cos\theta_{c-} = (Z_{0e} - Z_{0o})/(Z_{0e} + Z_{0o})$ [20].

The design procedure for a transmission-line passband filter is based on an equivalence of a single coupled-line section and an equivalent circuit that includes an ideal impedance inverter with a phase shift of -90°, as shown in Figure 5.24(a). The $ABCD$-matrix of the entire equivalent circuit can be obtained from the multiplication of the $ABCD$-matrices of the ideal transmission lines and the impedance inverter as

$$\begin{bmatrix} A & B \\ C & D \end{bmatrix} = \begin{bmatrix} \cos\theta & jZ_0\sin\theta \\ j\dfrac{\sin\theta}{Z_0} & \cos\theta \end{bmatrix} \begin{bmatrix} 0 & -jK \\ -j\dfrac{1}{K} & 0 \end{bmatrix} \begin{bmatrix} \cos\theta & jZ_0\sin\theta \\ j\dfrac{\sin\theta}{Z_0} & \cos\theta \end{bmatrix}$$
$$= \begin{bmatrix} \left(\dfrac{K}{Z_0} - \dfrac{Z_0}{K}\right)\sin\theta\cos\theta & j\left(\dfrac{Z_0^2}{K}\sin^2\theta - K\cos^2\theta\right) \\ j\left(\dfrac{K}{Z_0^2}\sin^2\theta - \dfrac{\cos^2\theta}{K}\right) & \left(\dfrac{K}{Z_0} - \dfrac{Z_0}{K}\right)\sin\theta\cos\theta \end{bmatrix}. \tag{5.85}$$

Then, the image impedance of the equivalent circuit can be found using Eq. (5.1) as

$$Z_i = \sqrt{\frac{B}{C}} = Z_0\sqrt{\frac{\dfrac{Z_0}{K}\sin^2\theta - \dfrac{K}{Z_0}\cos^2\theta}{\dfrac{K}{Z_0}\sin^2\theta - \dfrac{Z_0}{K}\cos^2\theta}} \tag{5.86}$$

while the propagation constant can be calculated from Eq. (5.4) as

$$\cos\beta = A = \left(\frac{K}{Z_0} - \frac{Z_0}{K}\right)\sin\theta\cos\theta. \tag{5.87}$$

By equating Eqs. (5.81) and (5.86) for the image impedances, Eqs. (5.84) and (5.87) for the propagation constants, and setting the electrical length of the coupled-line section to $\theta = 90^\circ$, which corresponds to the center frequency of the bandpass response and for which a single-frequency equivalence is fully established, the separate equations for even- and odd-mode characteristic impedances can be derived as

$$\frac{Z_{0e}}{Z_0} = 1 + \frac{Z_0}{K} + \left(\frac{Z_0}{K}\right)^2 \tag{5.88}$$

$$\frac{Z_{0o}}{Z_0} = 1 - \frac{Z_0}{K} + \left(\frac{Z_0}{K}\right)^2. \tag{5.89}$$

Similarly, the equivalence between the coupled-line section with open-circuited and short-circuited conditions for a coupled line and the cascade of a unit element and a series inductor shown in Figure 5.24(b) can be established [22]. In this case, the transmission-line low-pass filter can be fabricated using coupled-line sections with such a configuration where even- and odd-mode characteristic impedances can be derived as

$$Z_{0e} = Z_0 + L + \sqrt{L(Z_0 + L)} \tag{5.90}$$

$$Z_{0o} = \frac{Z_{0e}}{1 + \dfrac{2L}{Z_0} + \dfrac{2}{Z_0}\sqrt{L(Z_0 + L)}} \tag{5.91}$$

where L is the series inductance of the equivalent circuit after application of Kuroda identities to the low-pass filter-prototype. Generally, there are ten possible combinations of boundary conditions for a coupled-line section [22].

In a practical implementation of the symmetric coupled stripline structure shown in Figure 5.23(d), the required strip widths and spacing for a given set of the characteristic impedances Z_{0e} and Z_{0o} and the dielectric constant ε_r for an idealized case of $t = 0$ can be calculated from

$$Z_{0e} = \frac{94.15}{\sqrt{\varepsilon_r}\left[\frac{\ln 2}{\pi} + \frac{W}{b} + \frac{1}{\pi}\ln\left(1 + \tanh\frac{\pi S}{2b}\right)\right]} \tag{5.92}$$

$$Z_{0o} = \frac{94.15}{\sqrt{\varepsilon_r}\left[\frac{\ln 2}{\pi} + \frac{W}{b} + \frac{1}{\pi}\ln\left(1 + \coth\frac{\pi S}{2b}\right)\right]} \tag{5.93}$$

where W is the strip width, S is the strip spacing, and b is the distance between ground conductors [23]. Sufficiently accurate closed-form expressions for the effective dielectric constants and the characteristic impedances of coupled microstrip lines can be found in [24].

5.6.3 Parallel-Coupled Bandpass Filters Using Half-Wavelength Resonators

Narrowband bandpass filters can be designed using the cascaded coupled-line sections with open ends, as shown in Figure 5.25(a). In this case, the sections are of equal length (one-quarter wavelength at the center frequency), and their electrical design is completely specified by the even- and odd-mode characteristic impedances Z_{0e} and Z_{0o}, respectively. The total filter structure always will be symmetrical for maximally flat or equal-ripple response. In the design of a parallel-coupled with half-wavelength resonators, it is necessary first to select the type of response function and the number of resonators that will yield the desired transfer or attenuation function in the pass and stop bands. Then, the even- and odd-mode characteristic impedances for each coupled-line section are calculated based on Eqs. (5.88) and (5.89), respectively, from

$$Z_{0e_i} = Z_0\left[1 + \frac{Z_0}{K_i} + \left(\frac{Z_0}{K_i}\right)^2\right] \tag{5.94}$$

$$Z_{0o_i} = Z_0\left[1 - \frac{Z_0}{K_i} + \left(\frac{Z_0}{K_i}\right)^2\right] \tag{5.95}$$

where

$$\begin{aligned} \frac{Z_0}{K_1} &= \sqrt{\frac{\Delta\omega}{2\omega_0}\frac{\pi}{g_1}} \\ \frac{Z_0}{K_i} &= \frac{\Delta\omega}{2\omega_0}\frac{\pi}{\sqrt{g_{i-1}g_i}} \qquad \text{for } i = 2, 3, \ldots, n \\ \frac{Z_0}{K_{n+1}} &= \sqrt{\frac{\Delta\omega}{2\omega_0}\frac{\pi}{g_n g_{n+1}}} \end{aligned} \tag{5.96}$$

$g_1, g_2, \ldots, g_{n+1}$ are the normalized values for the ladder-type low-pass prototype elements from Tables 5.2 and 5.3, and $\omega_0 = (\omega_{c+} + \omega_{c-})/2$ is the center bandwidth frequency [8,19]. Finally, the transmission-line physical dimensions in each section should be designed to yield these characteristic impedances. In general, the strip widths and spacing will differ from section to section, and hence the width of the resonators will not be constant. The design equations have excellent accuracy for

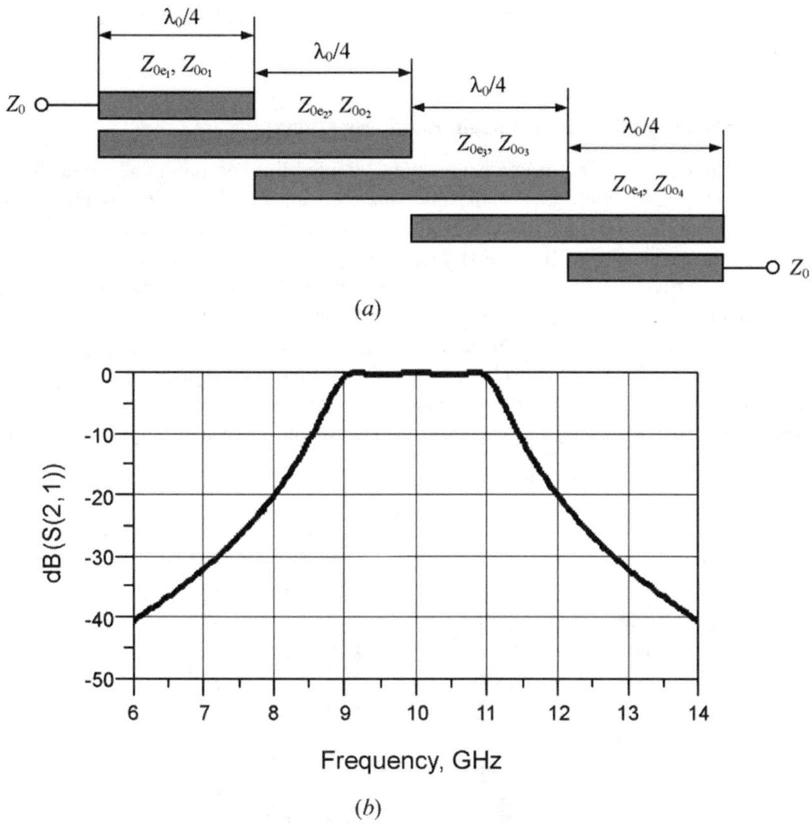

FIGURE 5.25 Coupled-line bandpass filter and its response.

bandwidths up to 20% in the case of maximally flat response and 30% in the case of equal-ripple response.

As an example, let us design a third-order ($n = 3$) coupled-line bandpass filter with the center frequency of 10 GHz and bandwidth of 20%, having a 0.5-dB equal-ripple response and a characteristic impedance $Z_0 = 50\ \Omega$. From Table 5.3, the element values for the low-pass prototype circuit are given by

$$g_1 = 1.5963 \quad g_2 = 1.0967$$

$$g_3 = 1.5963 \quad g_4 = 1.0000.$$

Then, the expressions from Eq. (5.96) are used to determine the normalized impedance inverter constants in the form of Z_0/K_i and finally the even- and odd-mode characteristic impedances Z_{0e} and Z_{0o} can be calculated from Eq. (5.94) and (5.95), respectively, as

$$Z_{oe} = 82.02\ \Omega \quad Z_{0o} = 37.66\ \Omega \quad \text{for } i = 1, 4$$

$$Z_{oe} = 64.69\ \Omega \quad Z_{0o} = 40.95\ \Omega \quad \text{for } i = 2, 3.$$

The resulting amplitude response of the designed microwave third-order coupled-line passband filter in a frequency domain is shown in Figure 5.25(*b*), where the out-of-band suppression of 40 dB is achieved below 6 GHz and above 14 GHz with a minimum suppression at 20 GHz. Note that passbands also occur at higher frequencies, having the center bandwidth frequencies of 30 GHz, 50 GHz, and so on.

5.6.4 Interdigital, Combline, and Hairpin Bandpass Filters

An interdigital bandpass filter represents an array of coupled-line resonators where each resonator element is a quarter-wavelength long at the midband frequency and is short-circuited at one end and open-circuited at the other end. A typical interdigital filter construction is realized by stripline suspending resonators in an air-filled metal case, and generally the physical dimensions of these resonators can be different. Coupling is achieved by way of the fields fringing between adjacent resonator elements. Coupled-stripline self- and mutual capacitances provide the starting point for the determination of the resonator width and spacing [25,26]. The mutual coupling between the resonators causes the resonator width to be less than the width of uncoupled lines.

Figure 5.26(a) shows the interdigital bandpass filter structure with short-circuited lines at the ends where each element serves as a resonator, except for the input and output lines which have

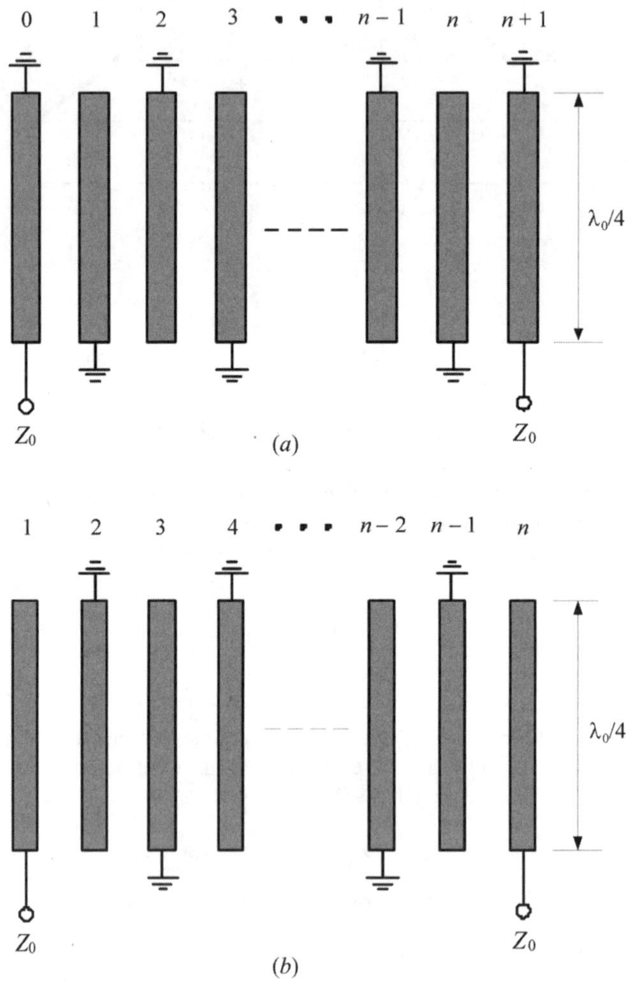

FIGURE 5.26 Interdigital passband filter structures.

an impedance-matching function. This type of design is most practical for filters having narrow or moderate bandwidths. The main drawback in applying the design procedure to filters of wider bandwidths is that the gaps between lines 0 and 1 and between lines n and $n + 1$ tend to become inconveniently small when the bandwidth is large, and the thickness of lines 1 and n tend to become very small. The other interdigital filter structure is shown in Figure 5.26(b) where the terminating lines are open-circuited that gives filter structural dimensions most suitable for moderate and wide bandwidths. In this case, all of the line elements serve as resonators. The main drawback of this filter structure for a narrowband design is that lines 1 and n will attain extremely high impedance.

Interdigital passband filters have a number of attractive features: they are compact; the manufacturing tolerances are relatively relaxed due to the relatively large spacing; the second passband is centered at three times the center frequency of the first passband, and there is no possibility of spurious responses in between; the rates of cutoff and the strength of the stopbands are enhanced by multiple-order poles of attenuation at dc and even multiples of the center frequency of the first passband [25]. In microstrip implementation, the interdigital filter is superior to the parallel coupled bandpass filter based on half-wavelength resonators because the microstrip interdigital filter occupies less space at the expense of shorts through the substrate [27]. However, its microstrip implementation suffers from severe asymmetry of the filter response due to effect of coupling between nonadjacent resonators.

Figure 5.27 shows the tapped combline bandpass filter which consists of an array of coupled resonators that are short-circuited at one end and loaded by a grounded lumped capacitance C_i at the other end. In this case, the resonator lines will be less than a quarter-wavelength long at resonance each, and the coupling between resonators is predominantly magnetic [28]. However, if the capacitors C_i were not present, the resonator lines would be a full $\lambda_0/4$ long at resonance, and the structure would have no passband. Without some kind of reactive loading at the ends of the resonator lines, the magnetic and electric coupling effects would cancel each other, and the combline structure would become an all-stop structure. The larger the loading capacitances, the shorter the resonator lines, which result in a more compact filter structure with a wider stopband between the first desired passband and the second unwanted passband. For example, when resonator length is equal to $l = \lambda_0/8$, then the second passband will appear at over four times the midband frequency, and when $l = \lambda_0/16$, the second passband will be located at over eight times the midband frequency [29].

In practice, the minimum resonator line length is limited by the decrease of unloaded quality factor of resonator and a requirement of heavy capacitive loading. The bandwidth of combline filters is a function of the ground-plane spacing b and spacing S between resonators when the greater bandwidth is achieved for greater b and S. The spacing b determines the resonator impedance and length, as well as the maximum power rating and quality factor. The loading capacitance for each resonator can be

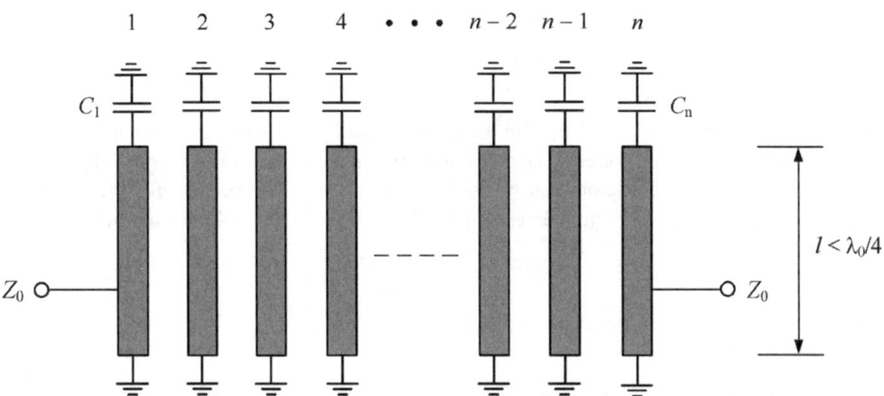

FIGURE 5.27 Structure of tapped combline passband filter.

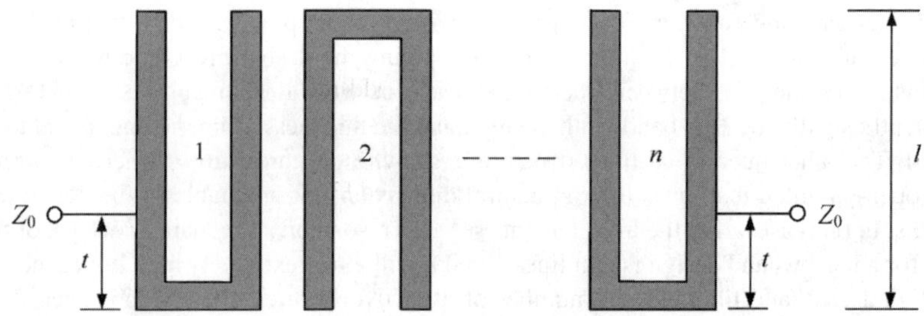

FIGURE 5.28 Structure of tapped hairpin-line passband filter.

calculated as $C_i = \cot\theta_0/(\omega_0 Z_{0i})$ where $\theta_0 = 2\pi l/\lambda_0$ is the electrical length at the midband frequency and Z_{0i} is the characteristic impedance of the ith resonator in view of the adjacent grounded lines $i-1$ and $i+1$. In the physical realization of narrow- and moderate-bandwidth interdigital and combline filters, the conventional input and output transformer couplings shown in Figure 5.26 are sometimes replaced with direct tapped connections shown in Figure 5.27 [30,31].

Hairpin-line passband filters, the basic tapped structure of which is shown in Figure 5.28, are preferred structure for microstrip realization because they offer small size and need no ground connection for resonators. This type of filters conceptually is obtained by folding the resonators of a parallel-coupled half-wavelength resonator filter into a U-shape. As a result, the same design equations for the parallel-coupled passband filter with half-wavelength resonators can be used [32]. However, to fold the resonators, it is necessary to take into account the reduction of the coupled-line lengths, which reduces the coupling between resonators. The line between two bends tends to shorten the physical length of the coupling sections, and the coupled section is slightly less than a quarter-wavelength [33]. Open-circuit resonators reduce free-space radiation due to phase cancellation of fields at the ends. Also, the radiation decreases with decreasing space between the folded lines of the hairpin. However, when this space is small, self-resonator coupling causes a decrease in filter bandwidth and an increase in losses. A reasonable spacing is two to three times the inter-resonator spacing, or five times the substrate thickness [34]. Microstrip hairpin filters require a sufficiently large spacing between resonators to achieve the desired narrow passband.

The hairpin-line bandpass filter design parameters can be calculated from the low-pass filter prototype parameters g_i by

$$\begin{aligned} Q_{e1} &= \frac{g_0 g_1}{\Delta\omega} \\ C_{i,i+1} &= \frac{\pi\,\Delta\omega}{\sqrt{g_i g_{i+1}}} \qquad \text{for } i = 1, 2, \ldots, n-1 \\ Q_{en} &= \frac{g_n g_{n+1}}{\Delta\omega} \end{aligned} \tag{5.97}$$

where $\Delta\omega$ is the passband, Q_{e1} and Q_{en} are the external quality factors of the resonators at the input and output, and $C_{i,\,i+1}$ are the coupling coefficients between the adjacent resonators [11].

The input and output resonators can be slightly shortened to compensate for the effect of the tapping line and the adjacent coupled resonator. The design equation for estimating the tapping length t can be written as

$$t = \frac{2l}{\pi}\sin^{-1}\sqrt{\frac{\pi}{2Q_e}\frac{Z_0}{Z_{0i}}} \tag{5.98}$$

where Z_{0i} is the characteristic impedance of the hairpin line, Z_0 is the terminating impedance, and l is the length of the coupled section [35]. Although the design equation ignores the effects of discontinuity at the tapped point and coupling between two folded arms, such estimation is sufficiently accurate. In the microstrip hairpin-line bandpass filter, a spurious mode occurs at approximately twice the passband frequency due to the different even- and odd-mode propagation velocities of coupled resonators. To resolve this problem, the tapped length t can be chosen close to $\lambda_0/8$. In addition, to minimize the overall size and make microstrip bandpass hairpin-line filters more compact, their structures in the cross-coupled or zigzag form can be used [36,37].

5.6.5 Microstrip Filters with Unequal Phase Velocities

The conventional approach to design parallel-coupled microstrip passband filters results in the spurious response at twice the passband frequency, which causes passband response to be asymmetric and reduces the width of the upper stopband [38]. This is a result of the inequality of the even- and odd-mode phase constants of the coupled line for each stage, which becomes more severe with high relative permittivity of the dielectric substrate. In this case, to partially solve this problem, the different lengths for the even- and odd-modes or overcoupling of the end stages with high image impedance of the filter can be used [39,40]. The lumped capacitors can be connected to the resonators to extend the traveling path of the odd mode [41,42]. Besides, the optimization of the resonator length and characteristic impedances can improve the return loss in the passband and out-of-band rejection [43,44].

However, the effect of unequal even and odd modes with high velocity ratios can alternatively be used to introduce some novel properties of the filters based on an inhomogeneous coupled-line structure. For example, a single microwave type-C filter section based on inhomogeneous broadside-coupled strips with phase velocity ratio of 2, which is an all-pass circuit in the homogeneous case, achieves an equal-ripple 3-peak stopband and passband response [45,46].

The TEM even- and odd-mode electric field distributions in an inhomogeneous dielectric medium with relative permittivity ε_r are shown in Figure 5.29(a), where the stronger coupling between the transmission lines takes place at an odd excitation mode. For each excitation mode, the characteristic impedance Z_{0e} corresponds to even mode, while the characteristic impedance Z_{0o} corresponds to odd mode. In this case,

$$Z_{0e} > Z_0 > Z_{0o} \tag{5.99}$$

where $Z_0 = \sqrt{Z_{0e}Z_{0o}}$ is the characteristic impedance of a single transmission line. The equivalent representation of a single open-circuit comb-line section can be given by a lattice filter section with semilattice impedances Z_a and Z_b, as shown in Figure 5.29(b), where Z_a is the impedance of the equivalent two-port network defined as the input impedance of the original two-port network under its odd-mode excitation, while Z_b is the corresponding impedance under even-mode excitation. In this case, an independent control of the odd mode only, resulting in high phase velocity ratio, can be provided by adjusting the position of the short circuit between the conductors along the coupled-line section, as shown in Figure 5.29(c), with no odd-mode field existing between the short circuit and the open ends. The overall result is defined by a superposition of the results at both types of excitation.

The four-port impedance Z-matrix for a single comb-line section shown in Figure 5.29(b) in an inhomogeneous dielectric medium can be written as

$$[Z] = \frac{1}{2}\begin{bmatrix} Z_b + Z_a & Z_b - Z_a \\ Z_b - Z_a & Z_b + Z_a \end{bmatrix} \tag{5.100}$$

where

$$Z_a = jX_a = -jZ_{0o}\cot\left(\frac{\pi}{2}\frac{f}{f_{0o}}\right) \tag{5.101}$$

$$Z_b = jX_b = -jZ_{0e}\cot\left(\frac{\pi}{2}\frac{f}{f_{0e}}\right) \tag{5.102}$$

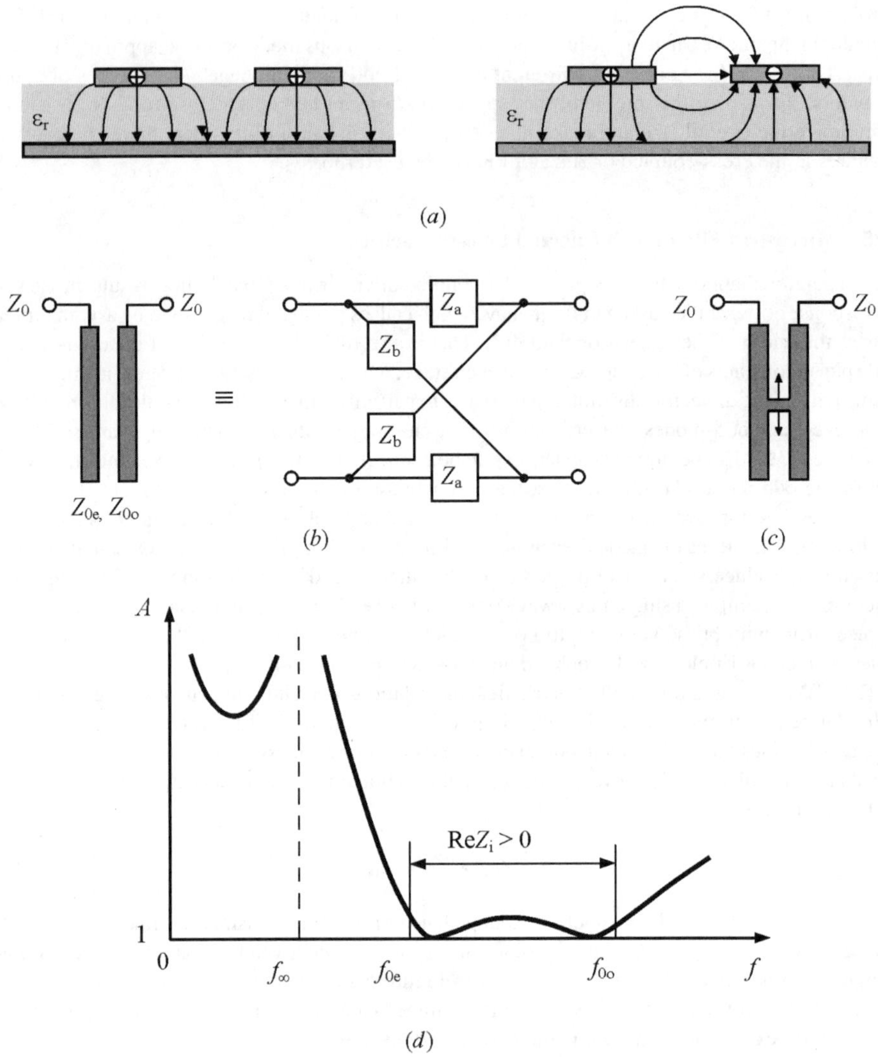

FIGURE 5.29 Parallel-coupled microstrip structure with even and odd electric field modes, its equivalent representation, and frequency response.

where f_{0e} and f_{0o} are the frequencies at which a comb-line section are a quarter-wavelength long electrically when excited in the even- and odd mode, respectively [47,48].

The effective attenuation of a comb-line section in terms of the equivalent lattice circuit parameters can be written as

$$A = 1 + \left(\frac{1 + X'_a X'_b}{X'_a - X'_b} \right)^2 \tag{5.103}$$

where X'_a and X'_b represent the corresponding reactances X_a and X_b normalized to Z_0 [4,7].

The image impedance can be obtained from Eq. (5.5) as

$$Z_i = \sqrt{Z_a Z_b} = j\sqrt{X_a X_b} \tag{5.104}$$

From Eqs. (5.103) and (5.104) it follows that, in a inhomogeneous dielectric medium with unequal phase velocities when $f_{0e} \neq f_{0o}$, the image impedance Z_i of a comb-line section becomes real in a frequency region with boundary frequencies f_{0e} and f_{0o}, as shown in Figure 5.29(*d*). This is because the reactances X_a and X_b have different signs in this region, negative for X_a and positive for X_b, with zeros of the attenuation function A defined by $Z_i(f_0) = Z_0$ [49,50]. In a homogeneous dielectric medium, the electrical lengths for even and odd modes are equal, and a comb-line section represents an all-stop circuit when the frequency region of real image impedances degenerates to $f_{0e} = f_{0o} = f_\infty$, with an attenuation pole at this frequency.

5.6.6 Bandpass and Bandstop Filters Using Quarter-Wavelength Resonators

The bandpass or bandstop response can be achieved by using quarter-wavelength open- or short-circuited transmission-line stubs which can be placed in shunt along a transmission line, as shown in Figure 5.30 for $\theta = \pi/2$. In this case, each open-circuit stub acts as a series resonant circuit, whereas each short-circuit stub operates as a parallel resonant circuit. Quarter-wavelength sections of the transmission line between stubs can be considered as impedance inverters to convert alternate shunt resonators to series resonators. For narrow bandwidths, the characteristic impedance of the series quarter-wavelength transmission lines is equal to Z_0. Bandstop filters can be used to eliminate a band of frequencies, or to separate such a band from other frequencies.

The design equations for the required stub characteristic impedances Z_{0n} is derived based on the element values of a low-pass filter prototype. Generally, the input impedance of an open-circuit stub with the characteristic impedance Z_0 and electrical length θ, which can be approximated as a series *LC* resonator shown in Figure 5.31(*a*), can be written as $Z_{in} = -jZ_0 \cot\theta$, whereas the input impedance of a short-circuit stub with the characteristic impedance Z_0 and electrical length θ, which can be approximated as a parallel *LC* resonator shown in Figure 5.31(*b*), can be written as

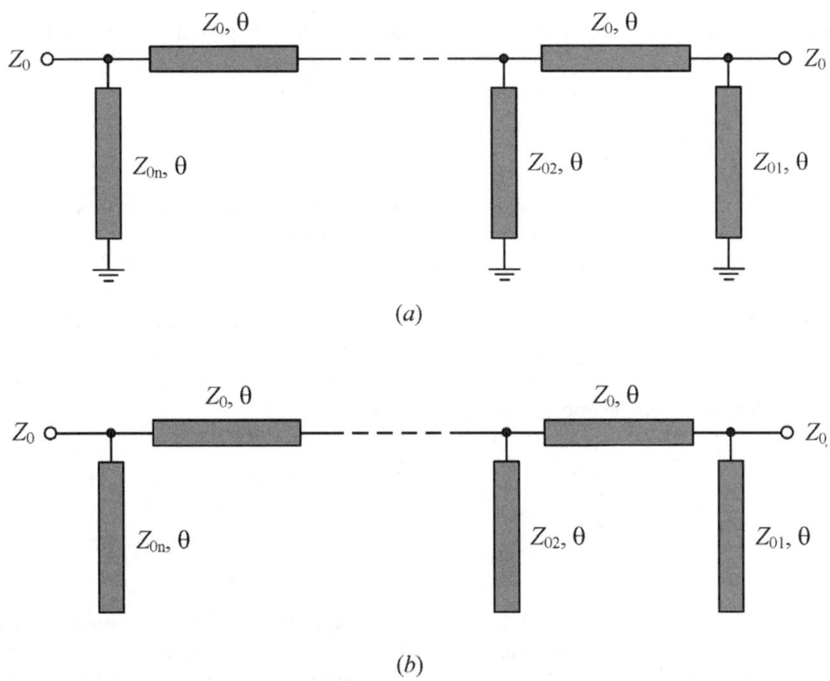

FIGURE 5.30 Bandpass and bandstop filters with shunt quarterwave resonators.

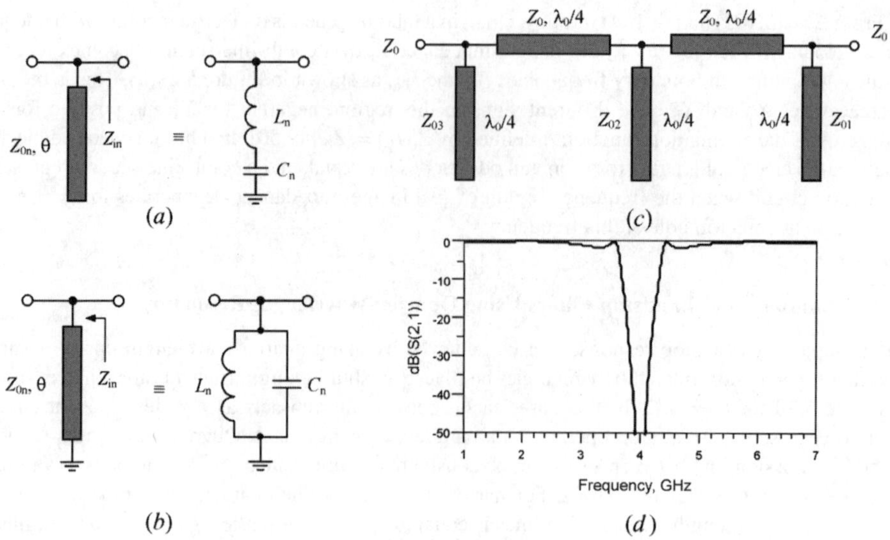

FIGURE 5.31 Third-order bandstop filter with quarterwave resonators and its response.

$Z_{in} = jZ_0 \tan\theta$. As a result, the characteristic impedance Z_{0i} of each open-circuit transmission-line stub used in the n-order bandstop filters is calculated from

$$Z_{0i} = \frac{4Z_0\omega_0}{\pi g_i \Delta\omega} \tag{5.105}$$

whereas the characteristic impedance Z_{0i} of each short-circuit transmission-line stub is determined by

$$Z_{0i} = \frac{\pi Z_0 \Delta\omega}{4g_i\omega_0} \tag{5.106}$$

where g_i are the element values for maximally flat or equal-ripple low-pass filters prototypes, $\Delta\omega$ is the frequency bandwidth, and ω_0 is the center bandwidth frequency, $i = 1, 2, \ldots, n$ [5,51]. Note that these results can only be applied to filters having input and output impedances of Z_0, and cannot be used for equal-ripple designs with even n.

As an example, let us design a third-order ($n = 3$) coupled-line bandstop filter with the center frequency of 4 GHz and bandwidth of 25%, having a 0.5 dB equal-ripple response and a characteristic impedance $Z_0 = 50\ \Omega$. For the corresponding element value of the low-pass prototype circuit tabulated in Table 5.3, the characteristic impedance Z_{0i} is calculated as

$$g_1 = 1.5963 \quad Z_{01} = 159.5\ \Omega \quad \text{for}\ i = 1$$

$$g_2 = 1.0967 \quad Z_{02} = 232.2\ \Omega \quad \text{for}\ i = 2$$

$$g_3 = 1.5963 \quad Z_{03} = 159.5\ \Omega \quad \text{for}\ i = 3$$

resulting in the transmission-line filter topology shown in Figure 5.31(c) and amplitude response shown in Figure 5.31(d). In this case, the ripple in the passband is somewhat greater than expected, which is the result of a narrowband assumption corresponding to the stop-band bandwidths up to a few percent.

For better accuracy over wider bandwidths, the characteristic impedances of the interconnecting quarter-wavelength lines must be variable calculated according to the corresponding design equations [8,52]. These distributed microwave bandstop filters are characterized by their second harmonic response centered at no more than three times the fundamental bandstop center frequency. However, such a limitation can be removed by the use of compound stub resonators, each composed of a pair of transmission lines with different characteristic impedances with each line being of the commensurate length when the first upper stopband center frequency becomes as high as six times the fundamental bandstop center frequency [53]. By providing a parallel association of two different bandstop structures, the narrow bandpass filters with short- and open-circuit stubs of different lengths and characteristic impedances can be realized which allow an independent control of each attenuated band on either side of one bandpass [54].

5.7 SAW AND BAW FILTERS

Serious interest in surface acoustic wave (SAW) technology commenced in 1967 with the development of the interdigital electrode transducer (IDT), which permitted efficient transduction between electromagnetic acoustic energy [55]. The IDT consists of a set of interleaved metal electrodes and is fabricated by photoetching a deposited metal film. In the simplest form, the width and spacing of electrodes is equal and uniform throughout the pattern. In this case, electrical excitation of the transducer with a sinusoidal voltage produces a periodic electrical field which penetrates into the piezoelectric substrate. In this case, suitable choice of substrate orientation produces two surface acoustic wave beams propagating normal to the IDT electrodes. As a result, peak output occurs at the synchronous frequency $f_0 = v/l$, where v is the SAW velocity and l is the transducer periodic length. Under this condition of acoustic synchronism, the stress contributions of all electrodes add in phase.

The lower SAW filter frequency limit is set by practical size limits on piezoelectric substrates and the range of velocities of available surface wave materials. At the high-frequency end, the limit is set by the resolution capabilities of state-of-the-art patterning techniques, substrate velocities, and frequency-dependent loss mechanisms. The essential feature in changing the operating frequency of surface-wave filters is simply a matter of changing the interdigital electrode spacing. In this case, the width of acoustic beam is set by electrode stripe length w which can be adjusted to obtain a convenient impedance level. With increasing frequency, however, electrode thickness must decrease to avoid larger distortion from reflections due to increased parasitic resistive loss. Several filter applications were found at both ends of available surface-wave frequency range, from the frequency modulated (FM) radio filters at 10.7 MHz with a 200-kHz signal bandwidth to the 800-MHz filters for mobile telephone transceivers and L-band radar and satellite transponder filters operating at 1.5 GHz and higher [56–58]. To achieve higher SAW operating frequency, it is necessary either to use substrate with higher velocities than can be achieved in quartz and lithium niobate, or to develop higher resolution fabrication. An important relationship in SAW filter design is that between fractional bandwidth ($\Delta f/f_0$) and insertion loss, with benefit of lithium niobate substrate over quartz substrate due to stronger piezoelectric coupling despite its moderate temperature sensitivity. The out-of-band rejection better than 60 dB over a wide range of bandwidths is available with weighting techniques such as finger overlap, finger withdrawal, phase weighting, and multistrip couplers. Principle obstacles to achieve or improve such a performance include diffraction and spurious responses.

Figure 5.32 shows the equivalent electric circuit for IDT, where G represents the radiation conductance, C_T represents the capacitance between transducer fingers, and B represents the acoustic susceptance due to vibration [59]. This circuit corresponds to a shunt representation for transducer input immittance in a "crossed-field" model configuration when the applied electric field is normal to the acoustic propagation vector. However, it can equivalently be illustrated by a series representation for transducer input immittance in an "in-line" model configuration characterized by parallel electric field and propagation vectors. For a given piezoelectric with given cut and orientation, the choice

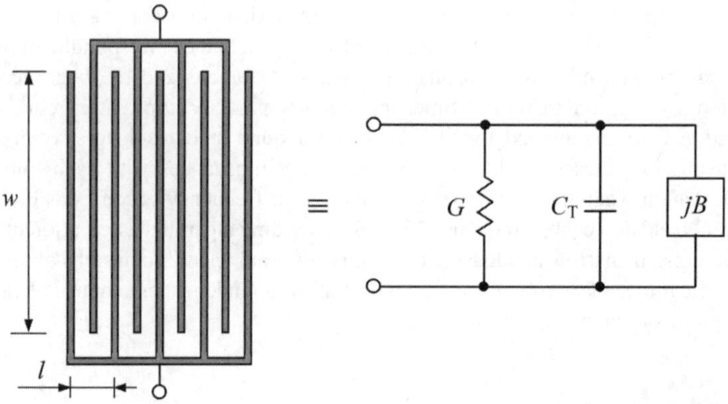

FIGURE 5.32 Equivalent electric circuit for IDT.

between these models is made by evaluating the contribution to Rayleigh wave (RW) excitation of the perpendicular and parallel to the surface electric field components.

From computer simulation and practical experiments, it was found that the input impedance (or admittance) of an IDT with quarter-wavelength-solid fingers can be classified into three types, depending on the number of finger pairs N and electromechanical coupling coefficient defined as $k^2 = 2\Delta v/v$, where Δv is the perturbation of phase velocity due to nonidentical field distributions for interdigital and metal-coated configurations [60]. In this case, for an optimized $N = 1.5/k^2$, a wide frequency range can be achieved. The total susceptance ($\omega C_T + B$) becomes very small over this range due to the cancellation of electric and acoustic susceptance components, being capacitive for smaller N and inductive for larger N.

Figure 5.33 shows the configuration with sharp cutoff frequency response representing an image-impedance connection where one pair of electrically connected interdigital transducers with an optimum number of finger pairs is introduced. Input and output transducers with a broader frequency response are arranged on both sides where SAW reflectors are also placed [61]. With this nonweighting configuration, a sharp cutoff response is achieved because, in the passband where the susceptance is cancelled, the current from an upper image-impedance IDT flows into a lower one without reflection. If the frequency is outside passband, almost all current is reflected at the image-impedance connected point because the susceptance component is larger than the conductance component. In this case, an insertion loss of as low as 3.5–4.0 dB is achieved for the ultra high frequency (UHF) filter, and 1.7–2.0 dB for the very high frequency (VHF) filter.

The loss mechanism in a UHF SAW filter is basically defined by the leakage (bidirectionality), weighting, and thin Al electrode conductivity losses. Since bidirectionality loss is the largest in conventional transversal SAW filters, an introduction of the resonant structure with lateral repetitions in addition to reflectors can minimize this loss. To reduce weighting loss, a new phase weighting, which excites SAW structure with uniform wavefront distribution in the passband, can be included. Figure 5.34 shows the new repetition structure for interdigital transducers with a tapering number of finger pairs where one pair of input and output transducers is laterally repeated [61]. Since there is a larger number of finger pairs in the center of the filter and a smaller number at both sides, thus the energy of acoustic vibrations is confined within the filter. As a result, six repetitions are enough to achieve 0.7 dB of insertion loss using a simple taper in the number of finger pairs (half the number of finger pairs at both sides). To reduce the conductivity loss, the filter aperture should be small and the impedance of each repeated IDT must be increased.

The device will usually be hermetically packaged to protect the sensitive surface from contamination. Often, one or two reactive components must be added at each end, outside the package. An inductor may be needed because the interdigital transducers have capacitive reactance which

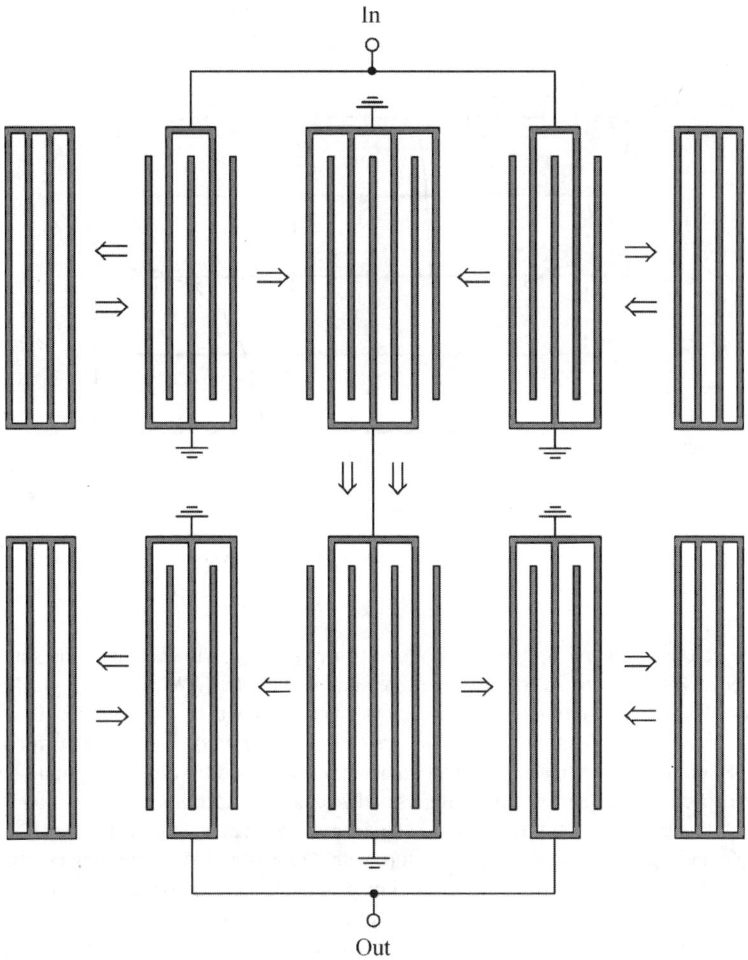

FIGURE 5.33 SAW filter with sharp cutoff frequency response.

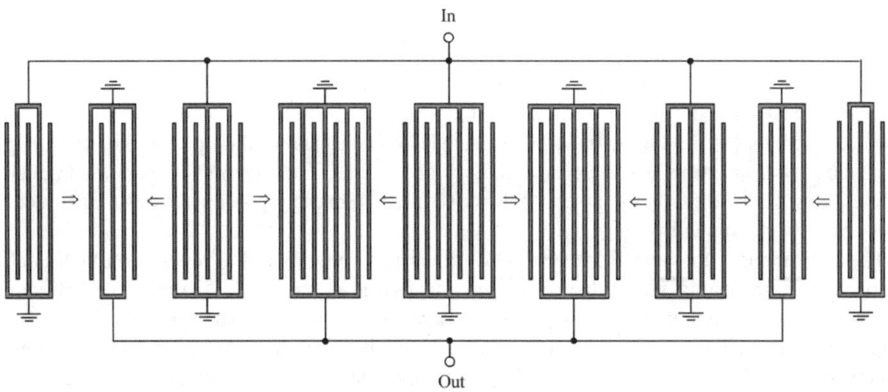

FIGURE 5.34 SAW filter with tapered number of finger pairs.

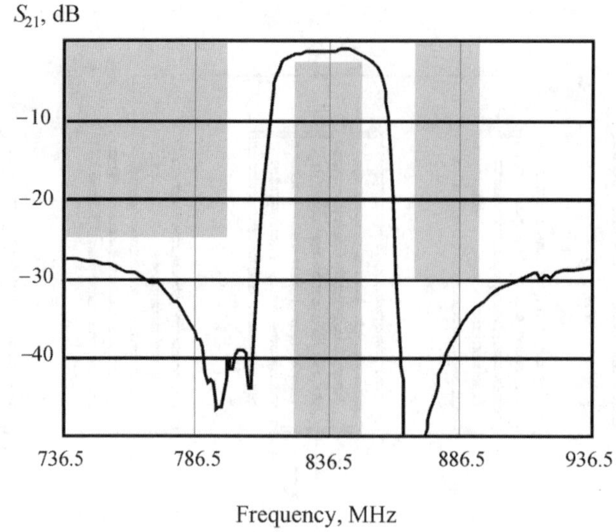

FIGURE 5.35 Typical SAW filter response.

may need to be tuned out. Also, lumped *LC* circuits are often used to transform the source or load impedance (usually 50 Ω) to an impedance more suitable for the SAW device. Typical SAW filter response for an 800-MHz mobile telephone transceiver is shown in Figure 5.35. To improve its power capability, insertion loss, and out-of-band rejection satisfying the requirements for a miniature antenna duplexer, the transmitter SAW filter can be realized in a ladder-type configuration with additional shunt capacitors between electrode fingers and ground [62]. Extremely low-loss and ultra-steep cutoff SAW filters used in a 1.9-GHz antenna duplexer can be achieved with IDT metallization ratio less than 50% and optimized ladder-type structure with a reduction in the coupling coefficient k^2 for only selected series resonators [63,64]. To minimize SAW filter size for multimode mobile phone applications, a chip-sized SAW package, which is flip-chip mounted onto a chip carrier serving as bottom of the package with electrical connections to the chip provided by solder bumps, represents a key packaging technology [65].

The traditional SAW filter technology demonstrates good selectivity and very small size, but operation frequencies of SAW filters is limited below 3 GHz due to their high insertion losses at high input power levels of above 1 W. On the other hand, despite the recent advances in ceramic material technology which have made dielectric filters very largely employed as duplexers since they can handle high power levels and demonstrate high selectivity, their large dimensions limit their application in mobile RF front-end modules. At the same time, the radio-frequency bulk acoustic wave (BAW) filters and duplexers based on film bulk acoustic resonator (FBAR) filters or solidly mounted resonator (SMR) technologies are expected to replace these traditional RF filter technologies due to their lower insertion loss, better selectivity, higher power handling, higher operation frequency. Besides, they can be directly integrated with active circuits, making possible of fully-integrated RF front-ends at very competitive costs.

The core element for both BAW and SAW filters technologies is the resonator which is characterized by high quality factor due to low propagation loss and small size due to much shorter acoustic wavelength than that of the electromagnetic wave. Besides, acoustic waves can be easily and efficiently excited and converted back to electrical signal by using transducers and the piezoelectric property of the SAW and BAW materials. Though the impedance responses of these filters are similar, the resonances are realized quite differently in SAW and BAW filter technologies. In the former case, a surface acoustic wave is an elastic wave that travels along the surface of a crystal substrate, however

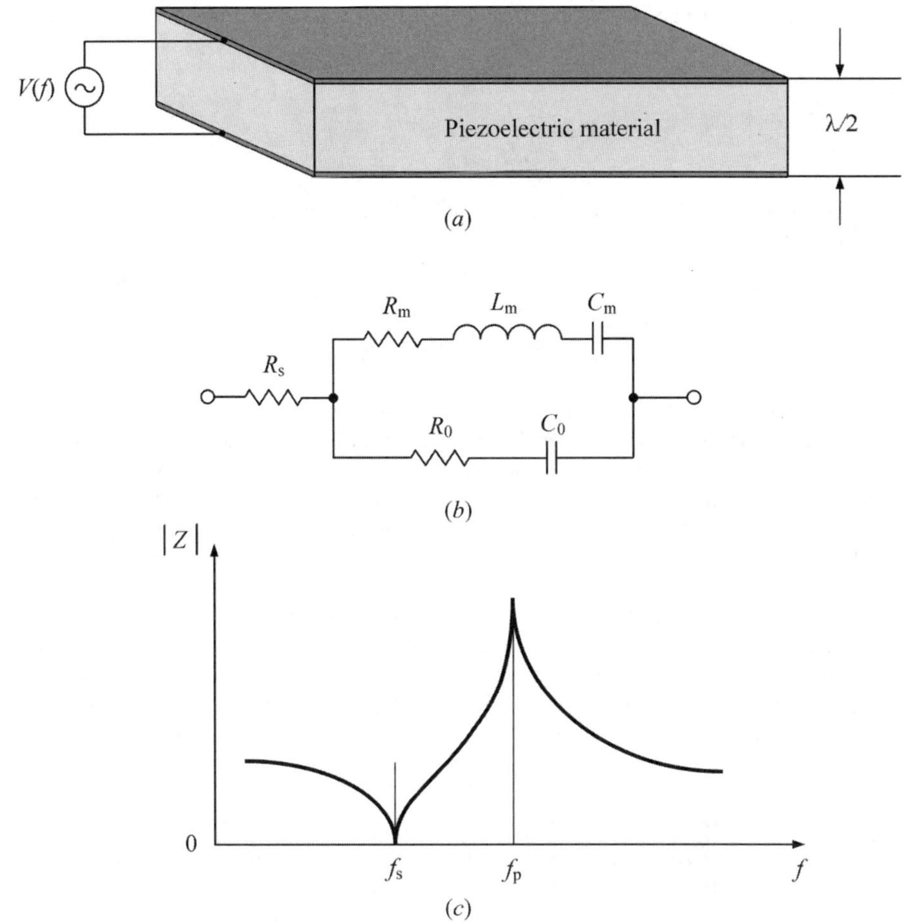

FIGURE 5.36 BAW resonator structure, its equivalent circuit, and frequency response.

in the latter case, a bulk acoustic wave represents an elastic wave that travels inside solid material. In a SAW resonator, the wave amplitude decays rapidly with depth, with 90% or more energy of acoustic energy being within one wavelength of the surface material. In a BAW resonator whose structure is shown in Figure 5.36(*a*) together with a signal source, the acoustic wave which is called the *longitudinal wave* propagates in a vertical direction between top and bottom metal electrodes [66,67].

Generally, the equivalent circuits of BAW and SAW resonators valid near the fundamental resonance are the same and are shown in Figure 5.36(*b*), where C_0 is the static (or plate) capacitance, C_m is the motional (or acoustic) capacitance which is much smaller than C_0, L_m is the motional inductance, R_m is the motional resistance, and the resistance R_s and R_0 characterize the electrode ohmic loss and undesired acoustic wave loss, respectively [68,69]. Being a second-order circuit, it has both a zero which defines the series resonant frequency f_s and pole which defines the parallel resonant frequency f_p, as shown in Figure 5.36(*c*). In this case, the motional inductance L_m forms a series resonance with C_m and, at a slightly higher frequency, a parallel resonance with C_0, with the inductive impedance in between f_s and f_p.

The most important characteristic of a BAW resonator is the strength of the piezoelectric coupling which depends on the piezoelectric and electrode materials and the resonator structure. With a single piezoelectric plate without any layers, the coupling coefficient can be determined from the series and parallel resonance frequencies as

$$k^2 = \frac{\pi}{2}\frac{f_s}{f_p}\tan\left(\frac{\pi}{2}\frac{f_p - f_s}{f_p}\right) \tag{5.107}$$

However, in a real FBAR resonator, there are other layers and, then, the effective piezoelectric coupling coefficient is defined by

$$k_{\mathrm{eff}}^2 = 1 - \left(\frac{f_s}{f_p}\right)^2 = \frac{C_m}{C_m + C_0} \approx \frac{8}{\pi^2}k^2 \tag{5.108}$$

which means that one must bear in mind that $k = 1.1k_{\mathrm{eff}}$ when comparing these coupling coefficients [70]. For a BAW resonator, the only piezoelectric thin-film material that has been established as the best balance of performance and manufacturability is aluminum nitride (AlN). With the best layer stack design and with optimized AlN thin-film deposition system, the maximum achievable resonator relative bandwidth $BW = (f_p - f_s)/f_p$ and k_{eff}^2 are 2.8% and 6.9%, respectively [66].

The BAW (or SAW) resonator filters can be generally divided in two basic topologies: ladder shown in Figure 5.37(*a*) and lattice shown in Figure 5.37(*b*), or the combination of them [71,72]. Ladder-type BAW filters have a very steep rejection response at passband ends followed by a lower rejection at further frequencies. On the other hand, lattice-type BAW filters demonstrate slower roll-off coefficient and higher rejection at both stopbands which are more effective to for rejecting undesired frequency bands. As a result, a combined ladder-lattice BAW filters can provide high in-band selectivity and improved out-of-band isolation simultaneously. Using FBAR technology with acoustic loading layer,

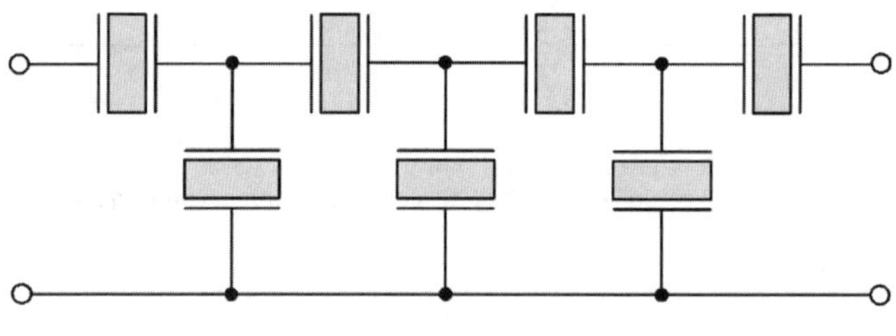

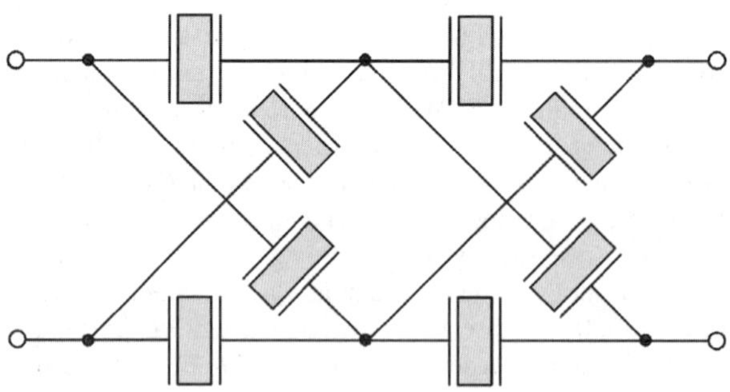

FIGURE 5.37 Ladder and lattice BAW filter configurations.

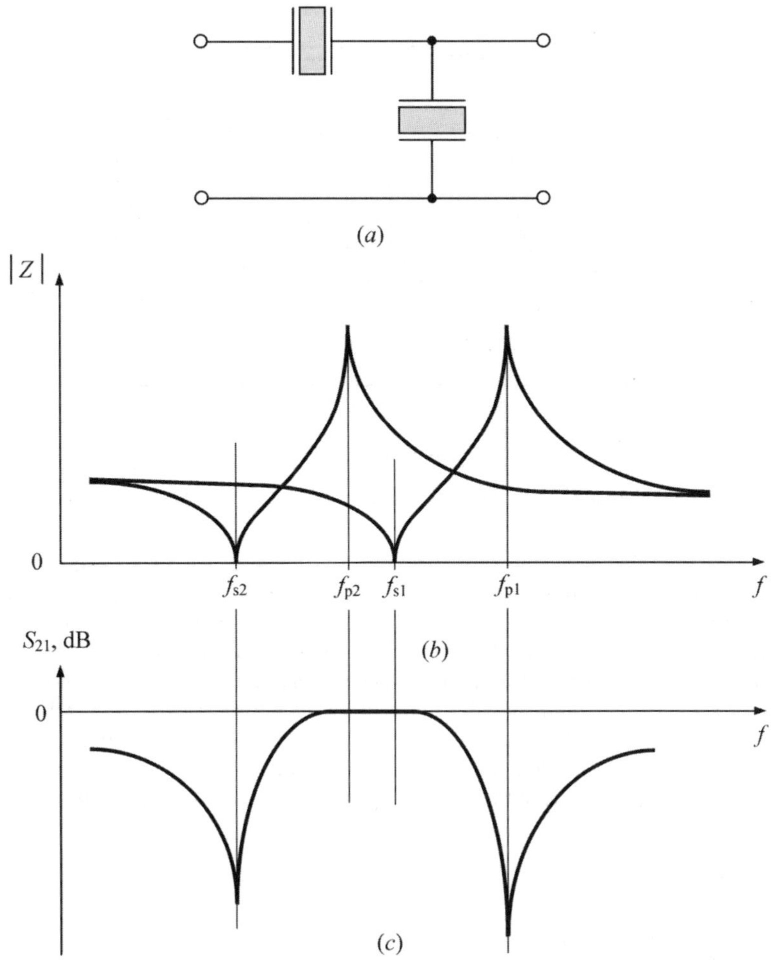

FIGURE 5.38 Operation principle of filter *L*-section.

the double lattice BAW filter can achieve an insertion loss of 2.5 dB, a bandwidth of 250 MHz, and selectivity around 56 at *X*-band (11.325 GHz) [73].

The operation principle of a ladder-type filter can be explained based on its basic *L*-section which is composed of the consecutive series and shunt resonators, as shown in Figure 5.38(*a*). The series resonator is characterized by the series resonant frequency f_{s1} and parallel resonant frequency f_{p1}. At the same time, the shunt resonator is tuned to operate at a slightly lower frequency by increasing the period of the interdigital transducer for SAW resonator or by mass loading the electrode for BAW resonator. When its parallel resonant frequency f_{p2} is chosen to be equal to or slightly lower than its series resonant frequency f_{s1}, as shown in Figure 5.38(*b*), a passband is formed at frequencies around f_{p2} and f_{s1}, as shown in Figure 5.38(*c*), where there is a low impedance between the input and output terminals of the *L*-section and a high impedance between the output terminal and ground. At frequency f_{s2}, any current flowing into the *L*-section is shorted to ground by the low impedance of the shunt resonator, and a zero is seen in the transition response below the passband. At frequency f_{p1}, the impedance between the input and output terminals of the *L*-section is high and another zero occurs in the transmission response above the passband. Far below f_{s2} and above f_{p1}, the *L*-section

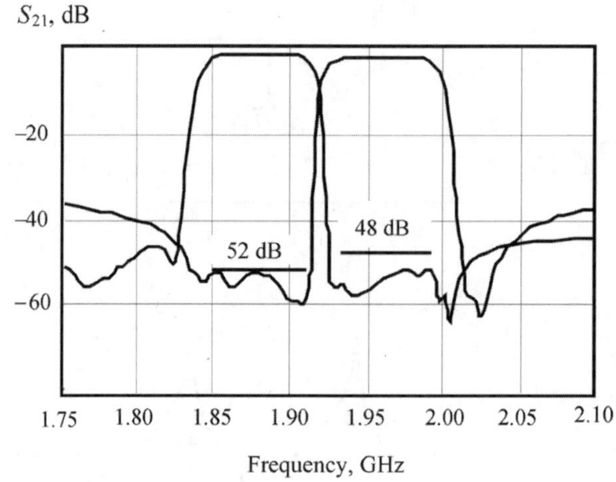

FIGURE 5.39 Frequency response of antenna BAW duplexer.

behaves as a two-capacitor network characterizing by insufficient out-of-band rejection. However, better rejection can be achieved by cascading more L-sections and by trading off with insertion loss.

The typical frequency response measured in the antenna BAW duplexer is shown in Figure 5.39, with an insertion loss in the passbands of 2.5 dB for transmitter path to the antenna and 3.0 dB for receiver path. In the path from transmitter to antenna, the insertion loss within the passband is definitely not the limiting factor because that role is played by the excessive roll-off at the filter edges. The total attenuation level in the signal path from antenna to receiver is greater due to the additional half-stage in the ladder-type topology of the receive filter. Typical suppression of the receive signal in the transmit band is 52 dB, that of the transmit signal in the receive band is 48 dB.

REFERENCES

1. G. A. Campbell, "Electrical Receiving, Translating, or Repeating Circuit," U.S. Patent 1227114, May 1917 (filed June 1916).
2. V. Belevitch, "Summary of the History of Circuit Theory," *Proc. IRE*, vol. 50, pp. 848–852, May 1962.
3. E. H. Colpitts and O. B. Blackwell, "Carrier Current Telephony and Telegraphy," *AIEE Trans.*, vol. 40, pp. 205–300, 1921; reprinted in *Proc. IEEE*, vol. 52, pp. 340–359, Apr. 1964.
4. A. I. Zverev, *Handbook of Filter Synthesis*, New York: John Wiley & Sons, 2005.
5. D. M. Pozar, *Microwave Engineering*, New York: John Wiley & Sons. 2004.
6. O. J. Zobel, "Wave Filter," U.S. Patent 1538964, May 1925 (filed Jan. 1921).
7. J. D. Rhodes, *Theory of Electrical Filters*, New York: John Wiley & Sons, 1976.
8. G. L. Matthaei, L. Young, and E. M. T. Jones, *Microwave Filters, Impedance-Matching Networks, and Coupling Structures*, Dedham: Artech House, 1980.
9. W. Cauer, *Synthesis of Linear Communication Circuits*, New York: McGraw-Hill, 1958.
10. L. Weinberg, *Network Analysis and Synthesis*, New York: McGraw-Hill, 1962.
11. J.-S. Hong and M. J. Lancaster, *Microstrip Filters for RF/Microwave Applications*, New York: John Wiley & Sons, 2001.
12. R. Saal and E. Ulbrich, "On the Design of Filters by Synthesis," *IRE Trans. Circuit Theory*, vol. CT-5, pp. 284–327, Dec. 1958.
13. H. C. Clark, "Zolotarev Bandpass Filters," *IEEE Trans. Microwave Theory Tech.*, vol. MTT-49, pp. 2357–2362, Dec. 2001.

14. W. E. Thomson, "Delay Network Having Maximally Flat Frequency Characteristics," *Proc. IEE*, pt. 3, vol. 96, pp. 487–490, Nov. 1949.

15. G. C. Temes and S. K. Mitra, *Modern Filter Theory and Design*, New York: John Wiley & Sons, Ltd. 1973.

16. P. I. Richards, "Resistor-Transmission-Line Circuits," *Proc. IRE*, vol. 36, pp. 217–220, Feb. 1948.

17. P. I. Richards, "Applications of Matrix Algebra to Filter Theory," *Proc. IRE*, vol. 34, pp. 145–150, Mar. 1946.

18. H. Ozaki and J. Ishii, "Synthesis of Transmission-Line Networks and the Design of UHF Filters," *IRE Trans. Circuit Theory*, vol. CT-2, pp. 325–336, Dec. 1955.

19. S. B. Cohn, "Parallel-Coupled Transmission-Line-Resonator Filters," *IRE Trans. Microwave Theory Tech.*, vol. MTT-6, pp. 145–153, Apr. 1958.

20. E. M. T. Jones and J. T. Bolljahn, "Coupled-Strip-Transmission-Line Filters and Directional Couplers," *IRE Trans. Microwave Theory Tech.*, vol. MTT-4, pp. 75–81, Apr. 1956.

21. H. Ozaki and J. Ishii, "Synthesis of a Class of Strip-Line Filters," *IRE Trans. Circuit Theory*, vol. CT-5, pp. 104–109, June 1958.

22. R. J. Wenzel, "Exact Design of TEM Microwave Networks Using Quarter-Wave Lines," *IEEE Trans. Microwave Theory Tech.*, vol. MTT-12, pp. 94–111, Jan. 1964.

23. S. B. Cohn, "Shielded Coupled-Strip Transmission Line," *IRE Trans. Microwave Theory Tech.*, vol. MTT-3, pp. 29–38, Oct. 1955.

24. M. Kirschning and R. H. Jansen, "Accurate Wide-Range Design Equations for Parallel Coupled Microstrip Lines," *IEEE Trans. Microwave Theory Tech.*, vol. MTT-32, pp. 83–90, Jan. 1984; vol. MTT-33, p. 288, Mar. 1985.

25. G. L. Matthaei, "Interdigital Band-Pass Filters," *IRE Trans. Microwave Theory Tech.*, vol. MTT-10, pp. 479–491, Nov. 1962.

26. R. J. Wenzel, "Exact Theory of Interdigital Bandpass Filters and Related Structures," *IEEE Trans. Microwave Theory Tech.*, vol. MTT-13, pp. 559–575, Sept. 1965.

27. T. A. Milligan, "Dimensions of Microstrip Coupled Lines and Interdigital Structures," *IEEE Trans. Microwave Theory Tech.*, vol. MTT-25, pp. 405–410, May 1977.

28. G. L. Matthaei, "Combline Band-Pass Filters of Narrow or Moderate Bandwidth," *Microwave J.*, vol. 6, pp. 82–91, Aug. 1963.

29. R. J. Wenzel, "Synthesis of Combline and Capacitively Loaded Interdigital Filters of Arbitrary Bandwidth," *IEEE Trans. Microwave Theory Tech.*, vol. MTT-19, pp. 678–686, Aug. 1971.

30. E. G. Cristal, "Tapped-Line Coupled Transmission Lines with Applications to Interdigital and Combline Filters," *IEEE Trans. Microwave Theory Tech.*, vol. MTT-23, pp. 1007–1113, Dec. 1975.

31. S. Caspi and J. Adelman, "Design of Combline and Interdigital Filters with Tapped-Line Input," *IEEE Trans. Microwave Theory Tech.*, vol. MTT-36, pp. 759–763, Apr. 1988.

32. E. G. Cristal and S. Frankel, "Design of Hairpin-Line and Hybrid Hairpin-Parallel-Coupled-Line Filters," *IEEE Trans. Microwave Theory Tech.*, vol. MTT-20, pp. 719–728, Nov. 1972.

33. U. H. Gysel, "New Theory and Design for Hairpin-Line Filters," *IEEE Trans. Microwave Theory Tech.*, vol. MTT-22, pp. 523–531, May 1974.

34. L. G. Maloratsky, *Passive RF & Microwave Integrated Circuits*, New York: Newnes, 2004.

35. J. S. Wong, "Microstrip Tapped-Line Filter Design," *IEEE Trans. Microwave Theory Tech.*, vol. MTT-27, pp. 44–50, Jan. 1979.

36. J.-S. Hong and M. J. Lancaster, "Cross-Coupled Microstrip Hairpin-Resonator Filters," *IEEE Trans. Microwave Theory Tech.*, vol. MTT-46, pp. 118–122, Jan. 1998.

37. G. L. Matthaei, "Narrow-Band, Fixed-Tuned, and Tunable Bandpass Filters with Zig-Zag Hairpin-Comb Resonators," *IEEE Trans. Microwave Theory Tech.*, vol. MTT-51, pp. 1214–1219, Apr. 2003.

38. C.-Y. Chang and T. Itoh, "A Modified Parallel-Coupled Filter Structure That Improves the Upper Stopband and Response Symmetry," *IEEE Trans. Microwave Theory Tech.*, vol. MTT-39, pp. 310–314, Feb. 1991.

39. A. Riddle, "High Performance Parallel Coupled Microstrip Filters," *IEEE MTT-S Int. Microwave Symp. Dig.*, vol. 1, pp. 427–430, May 1988.

40. J.-T. Kuo, S.-P. Chen, and M. Jiang, "Parallel-Coupled Microstrip Filters with Over-Coupled End Stages for Suppression of Spurious Responses," *IEEE Microwave and Wireless Components Lett.*, vol. 13, pp. 440–442, Oct. 2003.

41. S. L. March, "Phase Velocity Compensation in Parallel-Coupled Microstrip," *IEEE MTT-S Int. Microwave Symp. Dig.*, pp. 410–412, June 1982.

42. I. J. Bahl, "Capacitively Compensated High Performance Parallel Coupled Microstrip Filters," *IEEE MTT-S Int. Microwave Symp. Dig.*, vol. 2, pp. 679–682, June 1989.

43. D. Kajfez and S. Govind, "Effect of Difference in Odd- and Even-Mode Wavelengths on a Parallel-Coupled Bandpass Filter," *Electronics Lett.*, vol. 11, pp. 117–118, Mar. 1975.

44. H.-M. Lee and C.-M. Tsai, "Improved Coupled-Microstrip Filter Design Using Effective Even-Mode and Odd-Mode Characteristic Impedances," *IEEE Trans. Microwave Theory Tech.*, vol. MTT-53, pp. 2812–2818, Sept. 2005.

45. G. I. Zysman and A. Matsumoto, "Properties of Microwave *C*-Sections," *IRE Trans. Circuit Theory*, vol. CT-12, pp. 74–82, Mar. 1965.

46. J. L. Allen, "Inhomogeneous Coupled-Line Filters with Large Mode-Velocity Ratios," *IEEE Trans. Microwave Theory Tech.*, vol. MTT-22, pp. 1182–1186, Dec. 1974.

47. G. Zysman and A. Johnson, "Coupled Transmission Line Networks in an Inhomogeneous Dielectric Medium," *IEEE Trans. Microwave Theory Tech.*, vol. MTT-17, pp. 753–759, Oct. 1969.

48. G. M. Aristarkhov and Y. P. Vershinin, "Analysis of Coupled-Line Filters with Unequal Phase Velocities (in Russian)," *Radiotekhnika i Elektronika*, vol. 28, pp. 1714–1724, Sept. 1983.

49. G. M. Aristarkhov and V. P. Chernyshev, "Equivalent Mode Representation of Microstrip Filters Based on Multiconductor Transmission Lines with Unequal Phase Velocities (in Russian)," *Radiotekhnika i Elektronika*, vol. 30, pp. 2289–2297, Dec. 1985.

50. G. M. Aristarkhov, A. N. Alekseev, P. S. Mikhnevich, and V. P. Chernyshev, "Functional Possibilities of Structures Based on Slotlines and Coupled Microstrip Lines (in Russian)," *Radiotekhnika i Elektronika*, vol. 33, pp. 7–13, Jan. 1989.

51. L. Young, G. L. Matthaei, and E. M. T. Jones, "Microwave Band-Stop Filters with Narrow Stop Bands," *IRE Trans. Microwave Theory Tech.*, vol. MTT-10, pp. 416–427, Nov. 1962.

52. B. M. Schiffman and G. L. Matthaei, "Exact Design of Band-Stop Microwave Filters," *IEEE Trans. Microwave Theory Tech.*, vol. MTT-12, pp. 6–15, Jan. 1964.

53. R. Levy, R. V. Snyder, and S. Shin, "Bandstop Filters with Extended Upper Passbands," *IEEE Trans. Microwave Theory Tech.*, vol. MTT-54, pp. 2503–2515, June 2006.

54. C. Quendo, E. Rius, and C. Person, "Narrow Bandpass Filters Using Dual-Behavior Resonators Based on Stepped-Impedance Stubs and Different-Length Stubs," *IEEE Trans. Microwave Theory Tech.*, vol. MTT-52, pp. 1034–1044, Mar. 2004.

55. P. M. Grant, J. D. Adam, and J. H. Collins, "Surface Wave Device Applications in Microwave Communication Systems," *IEEE Trans. Commun.*, vol. COM-22, pp. 1410–1419, Sept. 1974.

56. R. D. Hays and C. S. Hartmann, "Surface-Acoustic-Wave Devices for Communications," *Proc. IEEE*, vol. 64, pp. 652–671, May 1976.

57. Y. Kinoshita, M. Hikita, M. Toya, T. Toyama, and Y. Fujiwara, "Low Loss and High Performance SAW Filters for 800 MHz Mobile Telephone Transceiver," *IEEE Trans. Vehicular Technol.*, vol. VT-32, pp. 225–229, Aug. 1983.

58. H. Shinonaga and Y. Ito, "Microwave SAW Bandpass Filters for Spacecraft Applications," *IEEE Trans. Microwave Theory Tech.*, vol. MTT-40, pp. 1110–1116, June 1992.

59. W. R. Smith, H. M. Gerard, J. H. Collins, T. M. Reeder, and H. J. Shaw, "Analysis of Interdigital Surface Wave Transducers by Use of an Equivalent Circuit Model," *IEEE Trans. Microwave Theory Tech.*, vol. MTT-17, pp. 856–864, Nov. 1969.

60. M. Hikita, H. Kojima, T. Tabuchi, and Y. Kinoshita, "800-MHz High-Performance SAW Filter Using New Resonant Configuration," *IEEE Trans. Microwave Theory Tech.*, vol. MTT-33, pp. 510–518, June 1985.

61. M. Hikita, T. Tabuchi, and A. Sumioka, "Miniaturized SAW Devices for Radio Communication Transceivers," *IEEE Trans. Vehicular Technol.*, vol. VT-38, pp. 2–8, Feb. 1989.

62. M. Hikita, Y. Ishida, T. Tabuchi, and K. Kurosawa, "Miniature SAW Antenna Duplexer for 800-MHz Portable Telephone Used in Cellular Radio Systems," *IEEE Trans. Microwave Theory Tech.*, vol. MTT-36, pp. 1047–1056, June 1988.

63. J. Tsutsumi, S. Inoue, Y. Iwamoto, T. Matsuda, M. Miura, Y. Satoh, M. Ueda, and O. Ikata, "Extremely Low-Loss SAW Filter and Its Application to Antenna Duplexer for the 1.9 GHz PCS Full-Band," *Proc. IEEE Int. Frequency Control Symp.*, pp. 861–867, May 2003.

64. S. Inoue, J. Tsutsumi, T. Matsuda, M. Ueda, O. Ikata, and Y. Satoh, "Ultra-Steep Cut-Off Double Mode SAW Filter and Its Application to a PCS Duplexer," *IEEE Trans. Ultrasonic Ferroelectric Frequency Control*, vol. 54, pp. 1882–1887, Sept. 2007.

65. H. Meier, T. Baier, and G. Riha, "Miniaturization and Advanced Functionalities of SAW Devices," *IEEE Trans. Microwave Theory Tech.*, vol. MTT-49, pp. 743–748, Apr. 2001.

66. F. Z. Bi and B. P. Barber, "Bulk Acoustic Wave RF Technology," *IEEE Microwave Mag.*, vol. 9, pp. 65–80, Oct. 2008.

67. M. Ueda, M. Iwaki, T. Nishihara, Y. Satoh, and K. Hashimoto, "A Circuit Model for Nonlinear Simulation of Radio-Frequency Filters Using Bulk Acoustic Wave Resonators," *IEEE Trans. Ultrason. Ferroelectric Frequency Control*, vol. 55, pp. 849–856, Apr. 2008.

68. J. F. Rosenbaum, *Bulk Acoustic Wave Theory and Devices*, London: Artech House, 1988.

69. M. Ylilammi, J. Ella, M. Partanen, and J. Kaitila, "Thin Film Bulk Acoustic Wave Filter," *IEEE Trans. Ultrasonic Ferroelectric Frequency Control*, vol. 49, pp. 535–539, Apr. 2002.

70. H. Jaffe and D. A. Berlincourt, "Piezoelectric transducer materials," *Proc. IEEE*, vol. 53, pp. 1372–1386, Oct. 1965.

71. G. G. Fattinger, J. Kaitila, R. Aigner, and W. Nessler, "Thin Film Bulk Acoustic Wave Devices for Applications at 5.2 GHz," *Proc. IEEE Ultrasonic Symp.*, vol. 1, pp. 174–177, Oct. 2003.

72. A. A. Shirakawa, J.-M. Pham, P. Jarry, E. Kerherve, F. Dumont, J.-B. David, and A. Cathelin, "A High Isolation and High Selectivity Ladder-Lattice BAW-SMR Filter," *Proc. 36th Europ. Microwave Conf.*, pp. 905–908, Sept. 2006.

73. A. A. Shirakawa, J.-M. Pham, P. Jarry, and E. Kerherve, "Design of FBAR Filters at High Frequency Bands," *Int. J. RF and Microwave Computer-Aided Eng.*, vol. 17, pp. 115–122, Jan. 2007.

6 Modulation and Modulators

Modulation is the process that allows the information content of an audio, video, or data signal to be transferred to a region of higher frequencies where the transmission can be effective. In this case, the spectrum bandwidth of the modulating signal must be much smaller compared to the carrier frequency being modulated. During the modulation process, one or several parameters of the carrier signal such as amplitude, frequency, or phase vary according to the modulating signal when the modulator changes the signal into a form suitable for transmission over the proper radio channel. More complex modulation process includes digitization and encoding of the modulating signals. Depending on the communication systems with corresponding requirements on transmitting power, signal quality, frequency bandwidth, power consumption, system complexity, or cost, different types of the modulation scheme can be used based on analog and digital modulation techniques.

This chapter discusses the basic features of different types of analog modulation including amplitude, single-sideband, frequency, and phase modulation and basic types of digital modulation such as amplitude shift keying, frequency shift keying, phase shift keying, or pulse code modulation and their variations. The principle of operations and various schematics of the modulators for different modulation schemes including Class-S modulator for pulse-width modulation are described. Finally, the concept of time and frequency division multiplexing is introduced, as well as brief description of different multiple access techniques is given.

6.1 AMPLITUDE MODULATION

The amplitude modulation (AM) concept invented in 1902 by Reginald Fessenden and followed by his audio radio broadcast demonstrations in 1906 by means of the alternator–transmitter became the original method used for audio radio transmissions [1]. The basic design idea was to produce a steady radio signal when connected to an aerial by simply placing a carbon microphone in the transmission line, and the strength of the radio signal could be varied in order to add sounds to the transmission. This means that AM would be used to impress audio on the radio frequency carrier wave. However, it would take many years of expensive development before even a prototype alternator–transmitter would be ready, and a few more years beyond that for high-power versions to become available. Later in 1920s, audio radio broadcasting became widespread by using the vacuum-tube transmitters rather than the alternator.

6.1.1 Basic Principle

The basic single-frequency carrier signal $v = V\cos(\omega_0 t + \phi_0)$ is characterized by the constant amplitude V, carrier frequency ω_0, and initial phase ϕ_0 that can be set to zero. Such a signal does not contain any information since it repeats over and over again with the same parameters. However, when this signal is modulated, either in amplitude or frequency, it is no longer a simple sine wave, but is instead a mixture of several waves of slightly different frequencies, superimposed upon each other.

RF and Microwave Transmitter Design, First Edition. Andrei Grebennikov.

Thus, in the common case of modulating signal v_m, the AM signal with time-varying amplitude $V = V_0 + v_m(t)$ can be written as

$$v(t) = [V_0 + v_m(t)] \cos \omega_0 t \tag{6.1}$$

where V_0 represents the carrier amplitude without modulation. Generally, such a transmitting signal may represent any periodical and non-periodical processes, the spectral components of which occupy very wide frequency range. However, the main signal energy is concentrated in a sufficiently narrow frequency bandwidth that gives an opportunity to make the communication channel bandwidth much narrower to transmit signal with required quality characterized by acceptable distortion level.

In a simple case of a sinusoidal modulating signal $v_m(t) = V_m \cos\Omega t$, where V_m is the modulating amplitude and Ω is the modulating frequency, Eq. (6.1) can be rewritten as

$$v(t) = V_0 [1 + m \cos \Omega t] \cos \omega_0 t \tag{6.2}$$

where m is the modulation index or modulation factor defined as

$$m = \frac{V_m}{V_0} \tag{6.3}$$

In this case, the maximum and minimum values of the AM signal amplitude is obtained from Eq. (6.2) as $V_{max} = V_0(1 + m)$ and $V_{min} = V_0(1 - m)$, respectively, and modulation factor can be characterized through the relative amplitude variation as

$$m = \frac{V_{max} - V_{min}}{V_{max} + V_{min}} \tag{6.4}$$

as shown in Figure 6.1(*a*).

Equation (6.2) can be rearranged by employing a trigonometric identity into the form

$$v(t) = V_0 \cos\omega_0 t + \frac{m V_0}{2} \cos(\omega_0 + \Omega) t + \frac{m V_0}{2} \cos(\omega_0 - \Omega) t \tag{6.5}$$

where the first term is called the *carrier*, which is unaffected by modulation process and the amplitude of which is equal to V_0. The other terms at frequencies $\omega_0 + \Omega$ and $\omega_0 - \Omega$ are called the *upper* and *lower sideband frequencies*, respectively. Their amplitudes are proportional to the modulation factor and their frequencies differ from the carrier frequency by an amount equal to the modulating frequency. Figure 6.1(*b*) shows the frequency spectrum of such a single-frequency AM signal with a bandwidth of 2Ω. The relative phase relationships of these three components are illustrated in Figure 6.1(*c*), with the vector of the carrier frequency rotating counterclockwise in a circle with angular velocity ω_0 and the side-frequency vectors rotating with angular velocities $\omega_0 + \Omega$ and $\omega_0 - \Omega$, respectively. Thus, the side-frequency vectors are rotating in opposite directions relatively to the carrier vector with the angular velocity Ω, and their vector sum will always lie along the carrier vector, indicating amplitude variation but no frequency deviation. This means that, in the complex plane, both the carrier and the sum of the side-frequency phasors are always located on the real horizontal axis. Hence, in AM, carrier and modulation vectors are in phase [2]. Consequently, the resulting AM-signal vector is rotating with the constant angular velocity ω_0, and its length varies periodically with the angular velocity Ω.

Figure 6.2 shows the time-domain behavior of the AM signal for fixed values of carrier and modulating frequencies and different modulation factors: (*a*) $m = 50\%$, (*b*) $m = 100\%$, and (*c*) $m = 150\%$. In this case, for $m < 1$, the carrier time-varying amplitude or *envelope* of the modulated wave varies in proportion to the time-varying modulating signal $v_m(t) = V_m \cos\Omega t$. However, for $m > 1$, the modulation symmetry is violated since the negative peaks of v_m when $V_m > V_0$ reduce the total

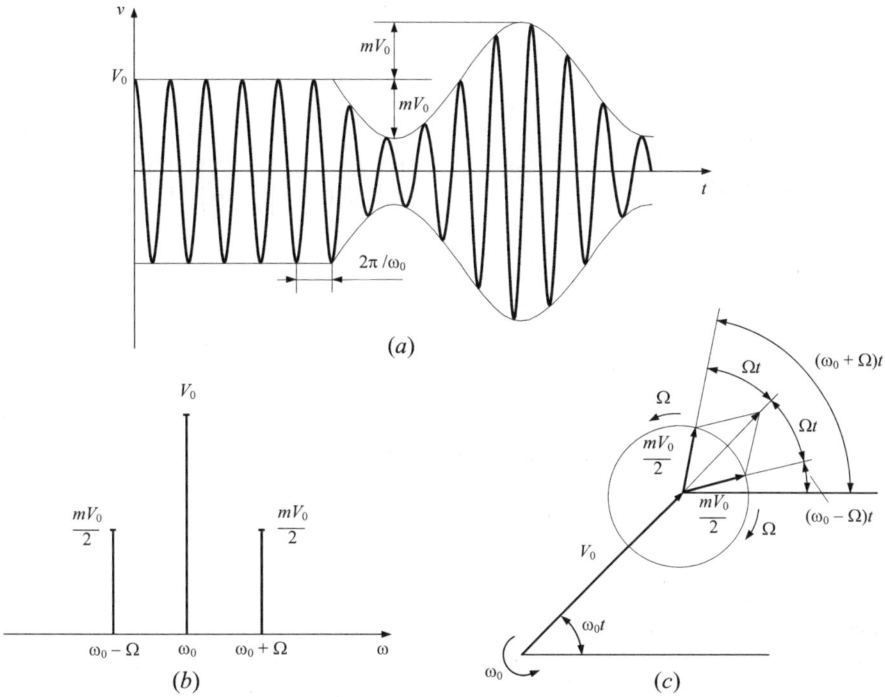

FIGURE 6.1 Time-domain, frequency-domain, and vector representations of AM wave.

voltage to the points less than zero. In this case, the resulting AM signal envelope do not corresponds anymore to the modulating signal waveform. This condition is called the *overmodulation*, resulting in a significant nonlinear distortion for the transmitting signal.

In a common case of the complicated modulating signal shown in Figure 6.3, the upper and lower frequency sidebands appear, each of which corresponds to the modulating signal spectrum. If Ω_{max} is the highest modulating angular frequency present and ω_0 is the carrier angular frequency, the complete spectrum of the amplitude-modulated waveform extends from the bottom of the lower sideband at $\omega_0 - \Omega_{max}$ to the top of the upper sideband $\omega_0 + \Omega_{max}$, with a total bandwidth of $(\omega_0 + \Omega_{max}) - (\omega_0 - \Omega_{max}) = 2\Omega_{max}$.

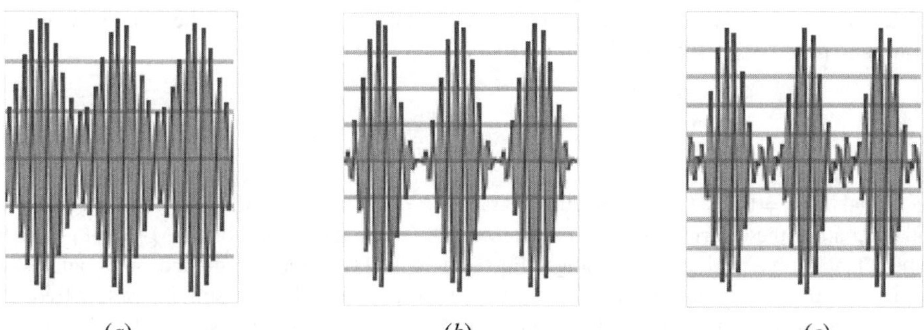

FIGURE 6.2 AM waveforms with different modulation factors.

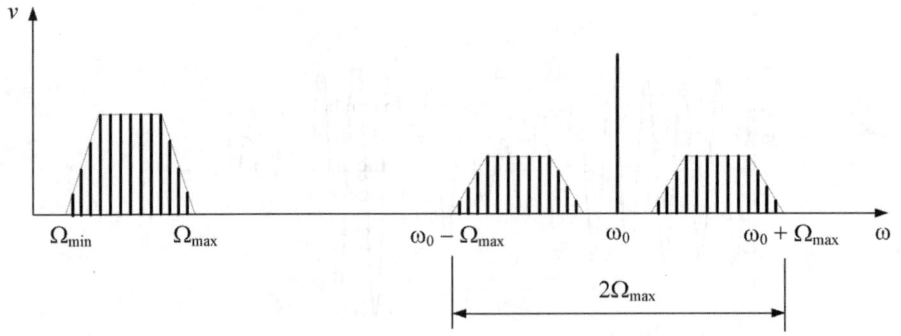

FIGURE 6.3 Amplitude modulation spectrum representation.

The carrier power without modulation can be written as

$$P_0 = \frac{V_0^2}{2R_L} \tag{6.6}$$

where R_L is the load resistance. At the same time, the maximum power P_{max} corresponding to the maximum AM signal amplitude V_{max} is defined as

$$P_{max} = \frac{V_{max}^2}{2R_L} = \frac{V_0^2 (1+m)^2}{2R_L} = P_0 (1+m)^2 \tag{6.7}$$

which is also called the *peak envelope power* (PEP).

From Eq. (6.7) it follows that, for $m = 1$ when maximum possible linear transmission mode without overmodulation can be realized, the maximum power P_{max} (or peak envelope power P_{PEP}) is four times greater than the carrier power P_0, that is, $P_{max} = 4P_0$. The average power P_{avr} over complete cycle of the modulating signal can be found as

$$P_{avr} = \frac{1}{2\pi} \int_0^{2\pi} P_0 (1 + m \cos \Omega t)^2 \, d(\Omega t) = P_0 \left(1 + \frac{m^2}{2}\right) \tag{6.8}$$

where the total sideband power is equal to $P_{sb} = (m^2/2)P_0$. In this case, the bandwidth of the output resonant circuit tuned to the carrier frequency must be much wider than the modulating frequency bandwidth to present purely R_L for complete AM signal. Thus, the useful information within the complete AM signal, which is represented by P_{sb}, defined as a ratio to the maximum possible transmitting power P_{max}, is a function of the modulation factor m according to

$$\frac{P_{sb}}{P_{max}} = \frac{m^2}{2(1+m)^2} \tag{6.9}$$

being equal to $P_{sb} = 0.125P_{max}$ for modulation factor $m = 1$. This means that the total sideband power containing all of the information does not exceed 12.5% in a maximum total AM-signal power that can be potentially transmitted. According to Eq. (6.8), each of two sideband components contains one-fourth as much power as does the carrier, so that the power required to transmit the carrier-frequency signal is at least twice greater than the useful power corresponding to the sidebands. Moreover, for complicated AM signals, the long-time average modulating factor is usually significantly smaller than its maximum short-time value, thus resulting in a significant reduction of the sideband power in an overall transmitting AM-signal power. Therefore, despite the simplicity of the AM approach, the overall AM transmitter efficiency is sufficiently small, especially when the linear signal transmission with negligible nonlinear distortion is required.

An amplitude-modulated wave possesses distortion to the extent that the modulation envelope fails to reproduce exactly the modulating signal. This distortion can be classified as amplitude (nonlinear), frequency, or phase (time-delay) distortion [3]. Amplitude distortion exists when the modulation envelope contains frequency components not present in the modulating signal due to the inherent nonlinearity of the active device being modulated. This results in harmonics of the modulating signal, which in turn denotes the presence of higher order sideband components in the output spectrum. Frequency distortion arises when the degree of modulation produced by a modulating signal of given amplitude varies with the frequency of the modulating signal, when the sideband components of different frequencies do not have the correct relative amplitudes. Time-delay distortion occurs when the phase relationships between different frequency components of the modulation envelope differ from the phase relationships that exist in the modulating signal.

Distortion in an AM signal can arise either from imperfections in the modulation system that produce this signal, or from the action of the active and passive circuits that transmit AM signal. Thus, when an AM signal is applied to a tuned circuit resonant at the carrier frequency, the upper and lower sideband frequencies can be reduced in amplitude symmetrically by an amount that increases the higher the modulating frequency. At the same time, the sideband frequencies undergo symmetrical phase shifts that introduce a time delay. If the carrier frequency does not coincide with resonant frequency of the tuned circuit, then the upper and lower sidebands are characterized by unequal transmission because of the asymmetrical phase shifts with respect to the carrier. This introduces quite severe amplitude distortion of the modulation envelope.

6.1.2 Amplitude Modulators

Generally, an AM can be generated by multiplication of the carrier and modulating signals in a nonlinear circuit. For example, if the transfer volt–ampere characteristic of a nonlinear device (diode or transistor) is represented by a second order polynomial

$$i(v) = a_0 + a_1 v + a_2 v^2 \tag{6.10}$$

then, by neglecting the effect of the output voltage on the output current and substituting the sum of the carrier and modulating signals

$$v = v_0 + v_m = V_0 \cos\omega_0 t + V_m \cos \Omega t \tag{6.11}$$

into Eq. (6.10), the output current $i(v)$ can be written as

$$i(v) = a_0 + a_1 \left(V_0 \cos \omega_0 t + V_m \cos \Omega t\right) + a_2 \left(V_0^2 \cos^2 \omega_0 t + 2 V_m V_0 \cos \Omega t \cos \omega_0 t + V_m^2 \cos^2 \Omega t\right). \tag{6.12}$$

The output current given by Eq. (6.12) can be rewritten as a sum of the harmonic components with different frequencies using trigonometric identities by

$$\begin{aligned} i(v) = a_0 &+ \frac{a_2}{2}\left(V_0^2 + V_m^2\right) + a_1 \left(V_0 \cos \omega_0 t + V_m \cos \Omega t\right) + a_2 V_m V_0 \left[\cos\left(\omega_0 + \Omega\right) + \cos\left(\omega_0 - \Omega\right)\right] \\ &+ \frac{a_2}{2}\left(V_0^2 \cos 2\omega_0 t + V_m^2 \cos 2\Omega t\right) \end{aligned} \tag{6.13}$$

whose spectrum, containing the carrier frequency ω_0 and its second harmonic component $2\omega_0$, the modulating frequency Ω and its second harmonic component 2Ω, and the sum and difference intermodulation products $\omega_0 \pm \Omega$, is shown in Figure 6.4(*a*) [4].

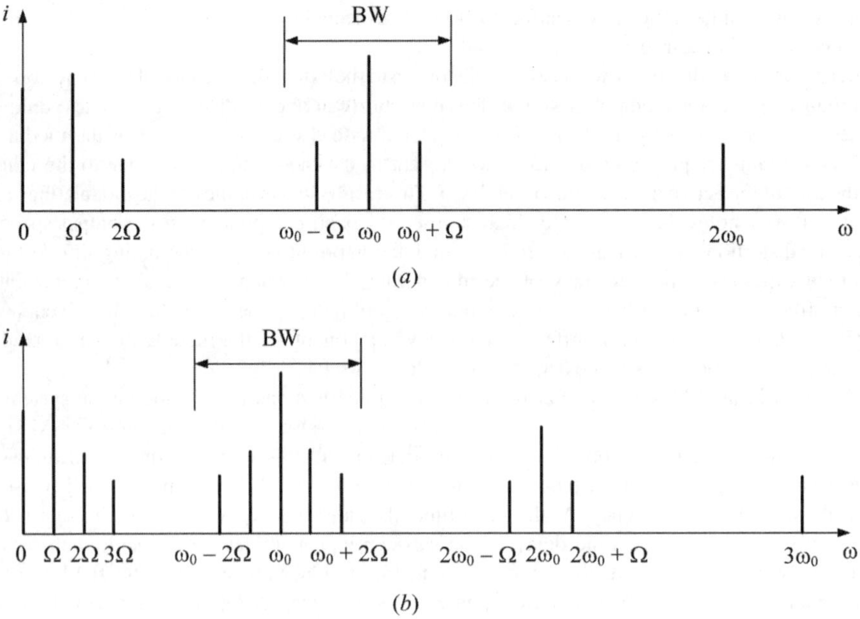

FIGURE 6.4 Output AM current spectra.

In order to obtain the required AM signal, it is necessary to represent only the frequency components ω_0, $\omega_0 + \Omega$, and $\omega_0 - \Omega$ from the whole spectrum. This can be achieved by using a resonant circuit tuned to the carrier frequency ω_0 with a certain frequency bandwidth (BW) when the portion of the output current flowing to the load can be obtained from Eq. (6.12) as

$$i_{\mathrm{L}} = a_1 V_0 \cos \omega_0 t + 2a_2 V_{\mathrm{m}} V_0 \cos \Omega t \cos \omega_0 t. \tag{6.14}$$

As a result, if the bandwidth of the resonant circuit is such that the load is seen as pure resistance R_{L} for frequencies ω_0, $\omega_0 + \Omega$, and $\omega_0 - \Omega$ and is zero for the rest frequency components, then the AM voltage across the resonant circuit is defined by

$$v_{\mathrm{L}}^{\mathrm{AM}} = a_1 R_{\mathrm{L}} V_0 \left(1 + \frac{2a_2 V_{\mathrm{m}}}{a_1} \cos \Omega t\right) \cos \omega_0 t \tag{6.15}$$

which can be simply rewritten as

$$v_{\mathrm{L}}^{\mathrm{AM}} = V_{\mathrm{L}} \left(1 + m \cos \Omega t\right) \cos \omega_0 t \tag{6.16}$$

where $V_{\mathrm{L}} = a_1 R_{\mathrm{L}} V_0$ is the carrier amplitude and $m = 2(a_2/a_1)V_{\mathrm{m}}$ is the modulation factor. In this case, larger values of the modulation factor m correspond to stronger device nonlinearity defined by a_2 and greater modulating amplitude V_{m}. Since the modulation factor m is directly proportional to the modulating-signal amplitude V_{m}, then ideally the modulation process is distortion free.

However, if the transfer volt–ampere characteristic of a nonlinear device is represented by a third-order polynomial $i(v) = a_0 + a_1 v + a_2 v^2 + a_3 v^3$, then the output current spectrum will contain the three harmonic components of the carrier and modulating frequency, ω_0 and Ω, and the intermodulation components of the second and third orders, $\omega_0 \pm \Omega$, $\omega_0 \pm 2\Omega$, and $2\omega_0 \pm \Omega$, as shown in Figure 6.4(*b*). To achieve a distortion-free AM signal, it is necessary to be confined only to the frequency components ω_0 and $\omega_0 \pm \Omega$. However, in this case the bandwidth of the resonant circuit is not narrow enough to suppress the second order intermodulation products $\omega_0 \pm 2\Omega$, resulting in a parasitic AM of the carrier frequency by the second harmonic of the modulating frequency 2Ω.

An AM can be easily realized by variation of the base- or gate-bias voltage of the transistor, which produces a corresponding variation in the amplitude of the transistor output voltage in accordance with the modulating voltage. However, despite the simplicity of implementation, some distortion (5% to 10%) of the envelope is inherent with this method because of the nonlinearities of the device input capacitance and transconductance as nonlinear functions of a bias voltage [5]. In addition, it is difficult to provide a sufficient isolation between the carrier and modulating signal paths, especially if the spectrum of the modulating frequencies is wide enough, which is the case for modern telecommunication modulation formats. To improve the linearity performance of the AM transmitting signal without additional linearization envelope feedback loops, it is best to apply an AM at low power levels using either cascode or differential-pair transistor stage, followed then by a linear power AM amplifier.

Figure 6.5(*a*) shows the circuit schematic of an amplitude modulator based on a cascode where the bottom transistor in a common emitter configuration is used for carrier signal and the top transistor in a

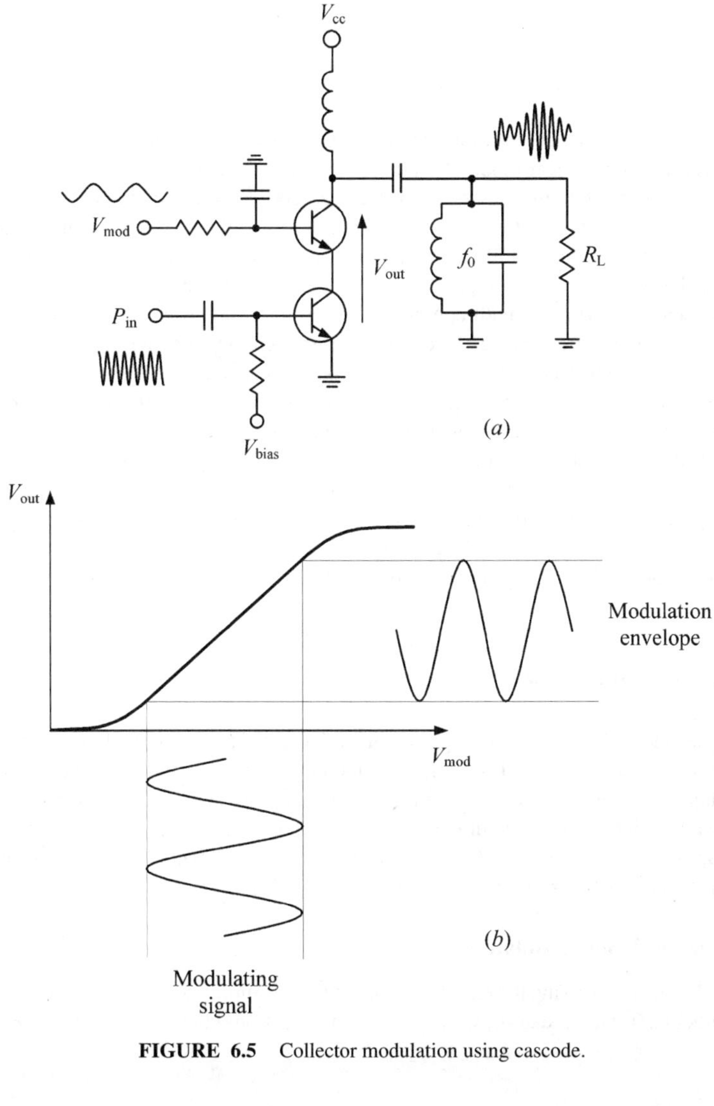

FIGURE 6.5 Collector modulation using cascode.

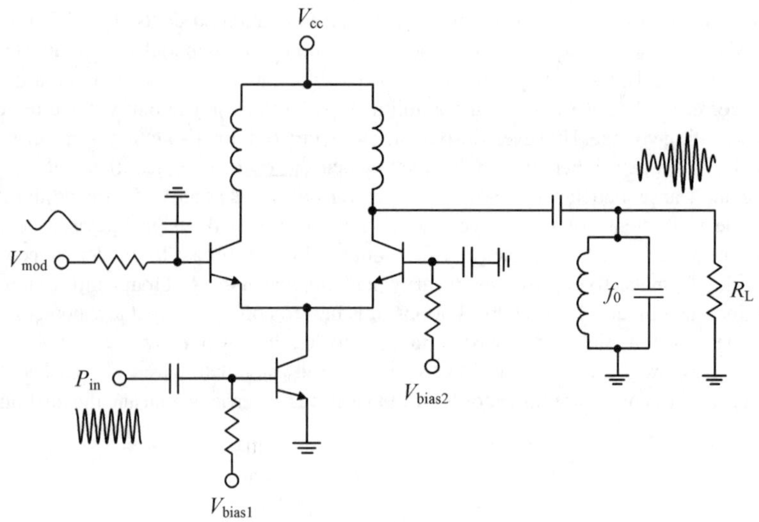

FIGURE 6.6 Collector modulation using differential pair.

common base configuration is used for modulating signal. Such a cascode amplifier can operate as an amplitude modulator with linear modulation range of about 15–20 dB, because of the collector–bias voltage variation of the bottom transistor when the cascode amplitude characteristic $V_{out}(V_{mod})$ is linear over a significant range of the modulating amplitudes applied to the base of the top transistor, as shown in Figure 6.5(*b*). The significant isolation of the modulating and carrier paths can be easily achieved by using a proper *RC* low-pass filter connected to the base of the top transistor. As an alternative, the carrier frequency path can be connected to the top transistor, while the modulating signal is applied to the bottom device, providing a linear emitter modulation of the carrier signal. However, due to a sufficiently high collector–base device capacitance, the significant feedthrough carrier power flows to the load in this case. Similar or even better linearity of the output AM signal can be achieved using a differential-pair transistor stage where the carrier signal is applied to the current–source device, as shown in Figure 6.6. In this case, the modulating signal can be applied to one transistor of the differential pair, while the load for the resulting AM signal can be connected to the collector of the other differential-pair transistor. The modulating signal can also be applied to both differential-pair devices to obtain the differential output AM signals at the outputs of the transistors relatively to each other. The voltage range of the modulating amplitudes is set by a series resistor in the modulating signal path.

6.2 SINGLE-SIDEBAND MODULATION

The single-sideband (SSB) method of signal transmission was proposed by John Carson in 1915 as a result of pure mathematical studies related to modulation of a continuous-wave carrier by means of vacuum tubes. He was the first to recognize that the suppression of the carrier either without or with suppression of one sideband would uniquely define the transmitting message that eventually led to the development of the *double-sideband amplitude modulation* (DSB-AM) and *single-sideband amplitude modulation* (SSB-AM), respectively [6,7].

6.2.1 Double-Sideband Modulation

The usual method of suppressing the carrier component of an amplitude-modulated signal is to employ a balanced modulator, which can provide carrier compensation when combining two AM waves in

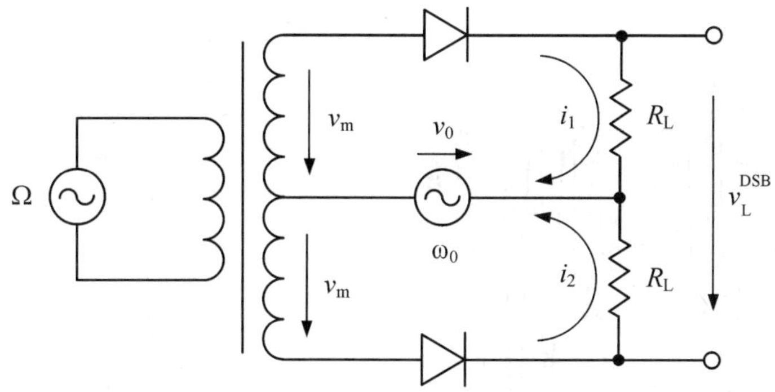

FIGURE 6.7 Schematic of diode balanced modulator.

a common load. In this case, the carrier voltage is applied with the same phase to the inputs of two single amplitude modulators, while the modulating signal is applied in opposite phases, usually by means of the center-tapped transformer. If the transfer volt–ampere characteristic of the nonlinear device is defined by Eq. (6.10), then the amplitude-modulated voltage in the load $v_{\mathrm{L}}^{\mathrm{AM}}$ is defined by Eq. (6.16) for $v = v_0 + v_{\mathrm{m}}$. However, for $v = v_0 - v_{\mathrm{m}}$, it can be rewritten as

$$v_{\mathrm{L}}^{\mathrm{AM}} = V_{\mathrm{L}} (1 - m \cos \Omega t) \cos \omega_0 t. \tag{6.17}$$

Thus, subtracting the load voltage given in Eqs. (6.16) and (6.17) due to the back-to-back connected nonlinear devices, as shown in Figure 6.7 for a diode balanced modulator, results in a double-sideband output signal in the form

$$\begin{aligned} v_{\mathrm{L}}^{\mathrm{DSB}} &= V_{\mathrm{L}} (1 + m \cos \Omega t) \cos \omega_0 t - V_{\mathrm{L}} (1 - m \cos \Omega t) \cos \omega_0 t = 2m V_{\mathrm{L}} \cos \Omega t \cos \omega_0 t \\ &= m V_{\mathrm{L}} [\cos (\omega_0 + \Omega) t + \cos (\omega_0 - \Omega) t] \end{aligned} \tag{6.18}$$

whose resulting envelope varying at twice the modulation frequency with amplitude $|2mV_{\mathrm{L}}\cos\Omega t|$ is shown in Figure 6.8(*a*). In this case, the modulated waveform consists of two side-frequency components with no carrier. With complicated modulating signal, there will be upper and lower sidebands. Therefore, it is called the *double-sideband* or *suppressed-carrier wave*.

The vector representation of this wave is shown in Figure 6.8(*b*), with zero carrier vector and the side-frequency vectors rotating with angular velocities $\omega_0 + \Omega$ and $\omega_0 - \Omega$, respectively. Here, similarly to AM signal, the side-frequency vectors are rotating in opposite directions with the angular velocity Ω, and, in the complex plane, their phasor sum will always lie along the real horizontal axis, indicating amplitude variation but no frequency deviation. Since there is no carrier, all the transmitted power fully corresponds to the information transmission, whereas in ordinary AM less than one-third of the total power carries the modulating information. Both systems require a bandwidth equal to twice the highest modulation frequency $\Omega_{\max}$. However, because the envelope of the DSB-AM wave is a full-wave-rectified replica of the modulating signal, the simple envelope detector cannot be used without first reintroducing the carrier.

For a balanced modulator shown in Figure 6.7 with identical diode volt–ampere characteristics approximated by the third-order polynomial where the voltages across the top and bottom diodes are $v_0 + v_{\mathrm{m}}$ and $v_0 - v_{\mathrm{m}}$, respectively (for simplicity, the voltage drops across the resistors R_{L} are neglected), the corresponding diode currents i_1 and i_2 can be written as

$$i_1 (v_0 + v_{\mathrm{m}}) = a_0 + a_1 (v_0 + v_{\mathrm{m}}) + a_2 (v_0 + v_{\mathrm{m}})^2 + a_3 (v_0 + v_{\mathrm{m}})^3 \tag{6.19}$$

$$i_2 (v_0 - v_{\mathrm{m}}) = a_0 + a_1 (v_0 - v_{\mathrm{m}}) + a_2 (v_0 - v_{\mathrm{m}})^2 + a_3 (v_0 - v_{\mathrm{m}})^3 \tag{6.20}$$

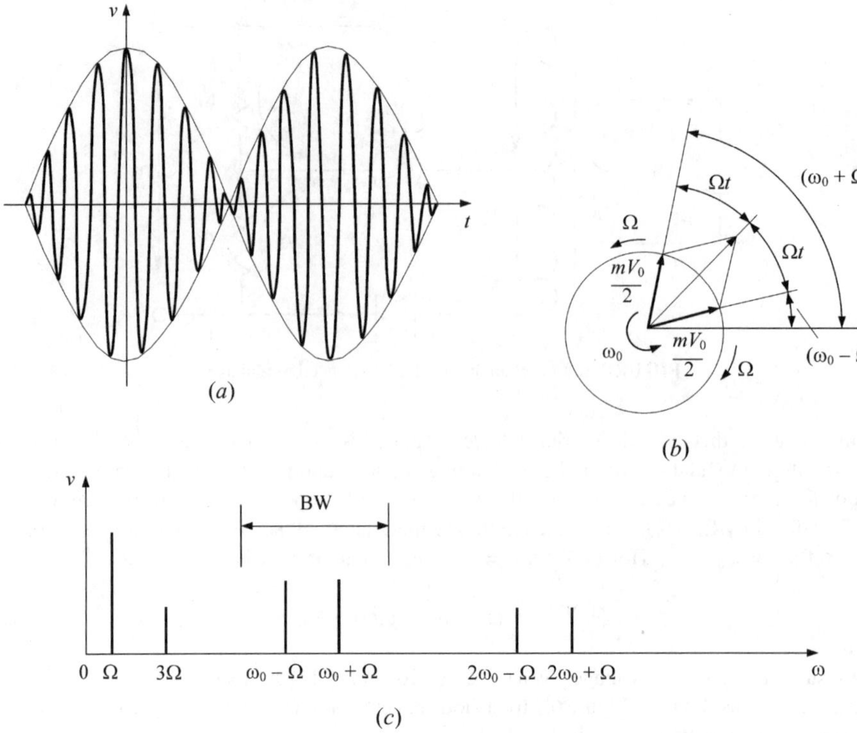

FIGURE 6.8 Time-domain, frequency-domain, and vector representation of DSB-AM wave.

resulting in an output voltage

$$v_{\mathrm{L}}^{\mathrm{DSB}} = R_{\mathrm{L}}\left[i_1\left(v_0 + v_{\mathrm{m}}\right) - i_2\left(v_0 - v_{\mathrm{m}}\right)\right] = 2R_{\mathrm{L}}\left(a_1 v_{\mathrm{m}} + 2a_2 v_0 v_{\mathrm{m}} + 3a_3 v_0^2 v_{\mathrm{m}} + a_3 v_{\mathrm{m}}^3\right). \tag{6.21}$$

From Eq. (6.21) it follows that the output voltage spectrum obtained by using trigonometric identities will contain much less frequency components than that of a single-diode modulator. For example, the dc, second modulating (2Ω) and third carrier ($3\omega_0$) harmonic components will be suppressed, as well as the intermodulation components of the second order $\omega_0 \pm 2\Omega$, as shown in Figure 6.8(*c*), compared to the output AM spectrum shown in Figure 6.4(*b*). In this case, the double-sideband signal can be easily filtered by the output resonant circuits connected in parallel to each load resistor R_{L}.

Moreover, the further and significantly better harmonic suppression can be achieved by adding two additional diodes, thus resulting in a double-balanced diode modulator shown in Figure 6.9. In this case, the currents $i_1(v_0 + v_{\mathrm{m}})$ and $i_2(v_0 - v_{\mathrm{m}})$ are defined by Eqs. (6.19) and (6.20), while the currents $i_3(-v_0 - v_{\mathrm{m}})$ and $i_4(-v_0 + v_{\mathrm{m}})$ differ from the currents i_1 and i_2 by opposite signs of the voltages v_0 and v_{m}, respectively. Hence, the resulting output voltage at the output of a double-balanced modulator by adding the two output voltages, one caused by current i_1 and i_2 defined by Eq. (6.21) and another caused by currents i_3 and i_4 defined by Eq. (6.21) with opposite signs for voltages v_0 and v_{m}, can be written as

$$\begin{aligned} v_{\mathrm{L}}^{\mathrm{DSB}} &= R_{\mathrm{L}}\left(i_1 - i_2\right) + R_{\mathrm{L}}\left(i_3 - i_4\right) = 2R_{\mathrm{L}}\left(a_1 v_{\mathrm{m}} + 2a_2 v_0 v_{\mathrm{m}} + 3a_3 v_0^2 v_{\mathrm{m}} + a_3 v_{\mathrm{m}}^3\right) \\ &\quad + 2R_{\mathrm{L}}\left(-a_1 v_{\mathrm{m}} + 2a_2 v_0 v_{\mathrm{m}} - 3a_3 v_0^2 v_{\mathrm{m}} - a_3 v_{\mathrm{m}}^3\right) = 8R_{\mathrm{L}} a_2 v_0 v_{\mathrm{m}} \end{aligned} \tag{6.22}$$

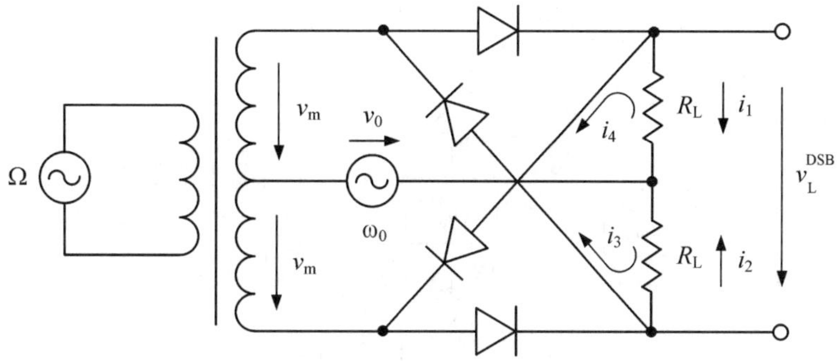

FIGURE 6.9 Schematic of diode double-balanced modulator.

which demonstrates the behavior of a double-balanced modulator as an ideal mixer-multiplier of two input signals with output spectrum containing the side-frequency components $\omega_0 + \Omega$ and $\omega_0 - \Omega$ only.

6.2.2 Single-Sideband Generation

An SSB signal can be obtained by passing the output of a carrier–suppression system through the sideband filter that is sufficiently selective to transmit one sideband while significantly suppressing the other. For example, if the lowest modulating frequency for speech transmission is typically 300 Hz, the filter characteristic must change from full transmission in its passband to very effective rejection in an interval of 2×300 Hz $= 600$ Hz. Even with well-defined filters such sharpness of the filter cutoff characteristic is possible only if the carrier frequency is low enough compared to the modulating frequency. Moreover, as the sharpness of cutoff is increased, the filter will introduce more phase and amplitude distortion in the portion of the transmitted signal near the cutoff region. Therefore, multiple-pole crystal or ceramic filters with adequately sharp cutoff characteristic can be used and frequency upconversion action can be applied if necessary for an SSB transmission at high radio frequencies [5].

The need for sharp cutoff filters can be avoided using a second method generally called the *phasing method*, which provides elimination of the unwanted sideband by phase cancelation. Its block diagram consists of the two balanced modulators and 90° phase shifters, which are necessary to deliver the input carrier and modulating signals for lower modulator with a phase difference of 90° each compared to the upper modulator, as shown in Figure 6.10(*a*) [8,9]. As a result, if the phase requirements are satisfied exactly and the balanced modulators are identical, the combined outputs cancel for one sideband and add for the other. As a practical matter, it is quite easy to realize 20-dB suppression, reasonable to expect 30 dB, and quite difficult to go beyond 40 dB. The main problem of this system is to provide an exact phase shift of 90° over entire bandwidth of modulating frequencies. To approximate the desired result physically, the modulating signal is passed through two networks having phase shifts that differ by 90° over the frequency range of interest while the attenuation difference is substantially constant [10].

The third method does not need either sharp cutoff filters or wideband 90° phase shifters. Figure 6.10(*b*) shows the circuit diagram of such an SSB system with the corresponding basic frequency relationships [11]. In addition, imperfections in the phasing or balancing do not result in the presence of the unwanted sideband in its usual location. Instead, the unwanted sideband occupies the same frequency band as the desired sideband, except that it is inverted. As it follows from Figure 6.10(*b*), the modulating signal flows to balanced modulators along with the corresponding quadrature components of a carrier frequency ω_0 followed by the low-pass filters that remove the upper sidebands, respectively. The filter output components are combined in the other pair of balanced modulators with corresponding quadrature signals from a source of frequency ω_c, the band center of the final

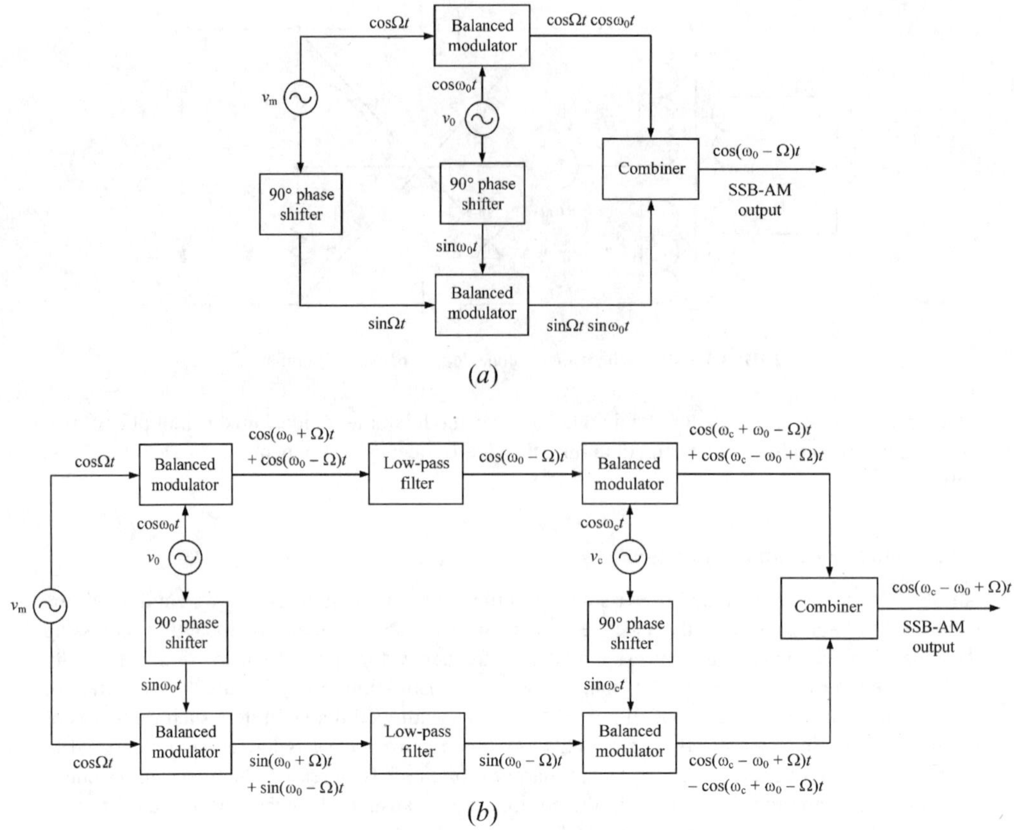

FIGURE 6.10 Methods of single-sideband modulation.

SSB frequency. As a result, the combined outputs from these balanced modulators cancel one pair of sideband components and add the other pair of sideband components to produce the desired SSB signal. As the entire circuit is bilateral, it can be used in demodulation as well as in generation of the SSB signals.

6.2.3 Single-Sideband Modulator

SSB modulators have numerous applications in a variety of existing and new communication systems including optical and millimeter waves. Basically, a practical SSB modulator represents an upconverter that generates an SSB suppressed-carrier (SSB/SC) output signal without the use of filters. Figure 6.11(*a*) shows the block diagram of a typical high-frequency SSB modulator where either the lower sideband or the upper sideband can be selected by exchanging the in-phase (I) and quadrature (Q) modulating inputs [12]. Generally, the system should include two balanced modulators, an I/Q generator to generate output quadrature modulating signals, a 90° branch-line hybrid combiner to input the carrier signals, and an in-phase Wilkinson combiner for summing the two output modulator signals. As an alternative, the upper sideband can be achieved by using an out-of-phase rat-race 180° subtracter at the output instead of a Wilkinson combiner. In microwave monolithic implementation, both the transmission-line branch-line coupler and Wilkinson combiner can be replaced by their lumped low-pass and high-pass filter equivalents based on monolithic spiral inductors and MIM capacitors [13]. To further minimize size and phase imbalance between modulator paths, the GaAs MESFETs or HEMTs as nonlinear elements rather than diodes can be used.

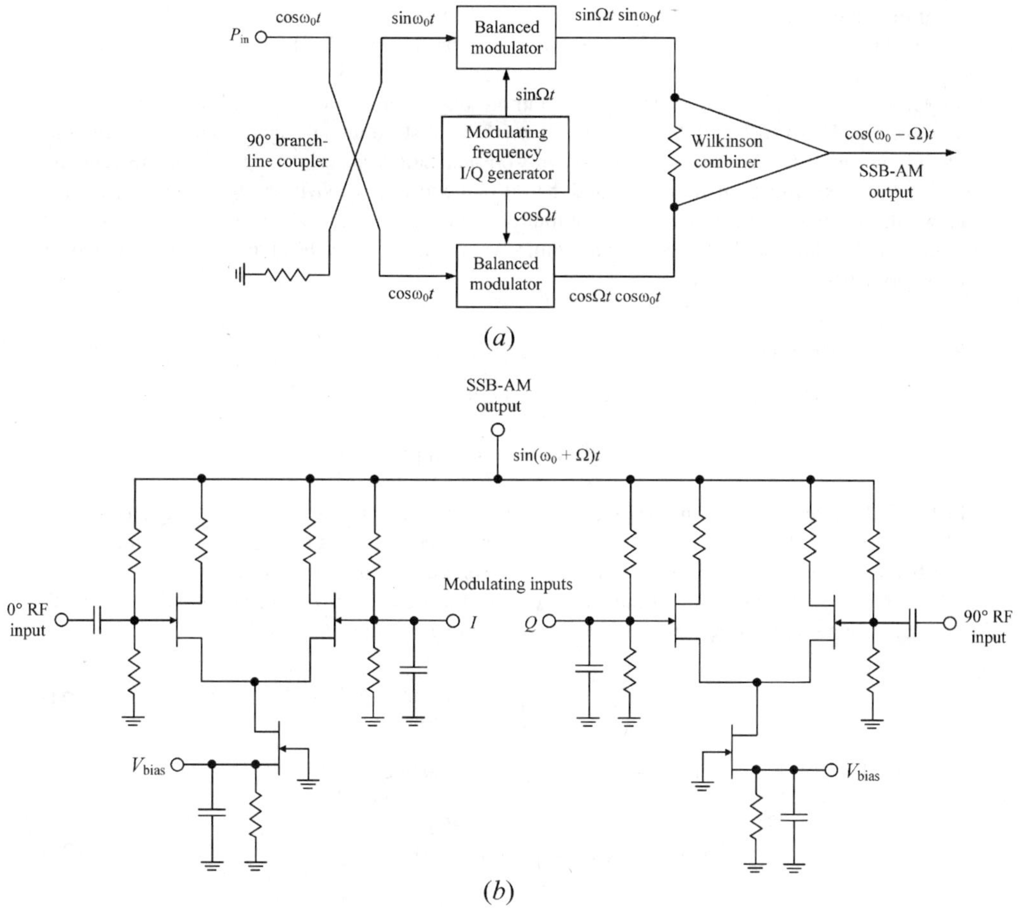

FIGURE 6.11 Microwave double-balanced single-sideband modulators.

Figure 6.11(*b*) shows the microwave monolithic GaAs MESFET SSB modulator circuit, operating with a 7-GHz carrier and providing better than a 17-dB suppression of any spectral products including carrier, undesired sideband, and third harmonic [12]. The circuit consists of two differential MESFET pairs, with the common source node of each pair biased through additional MESFET that acts as a constant current source. The drain node of each pair is connected together to add output signals from both differential pairs in phase. The quadrature carrier signals applied to the gate of one MESFET in each pair can be provided by a Lange coupler. The modulating audio signal is split into the *I* and *Q* components and applied to the gate of the second MESFET in each pair. The constant current source devices are necessary to balance the drain currents in both differential pairs, thus creating a double-sideband spectrum at the MESFET drains of each pair, connected together in parallel. Then by connecting in parallel both differential pairs, the SSB spectrum with upper sideband is realized. If the *I* and *Q* inputs are reversed in phase, the lower sidebands combine and the upper sidebands cancel.

6.3 FREQUENCY MODULATION

The subject of frequency modulation (FM) is well known a long time ago, though the early experiments in radio transmission appeared interested only in AM. In 1902, Fessenden proposed to use a condenser type of microphone in circuits that appear to produce FM, but in his publication remarks about tests he seemed interested in using the frequency variation to produce amplitude variation by throwing the frequency in and out of resonance with the antenna or other tuned circuit to modulate the amplitude [14]. Moreover, in 1922, Carson published his famous paper pointing out that FM would not provide a narrower band than AM, and that it inherently produces distortion in the signal, and hence, the FM system is inferior to the AM system [15]. And only one and half decades later, Armstrong theoretically explained and experimentally demonstrated that, if one widened the swing in FM much beyond the bandwidth used in AM and used an effective limiter in the receiver, noise lower than the carrier could be substantially eliminated, thus proving a significant superiority of the FM broadcasting system over its AM counterpart [16].

6.3.1 Basic Principle

Generally, the modulated signal can be written in the form

$$v(t) = V(t)\cos\phi(t) \tag{6.23}$$

which is fully described by a time-varying amplitude $V(t)$ and a time-varying phase angle $\phi(t)$. If in AM the carrier envelope $V(t)$ is varied while $\phi(t)$ remains constant, then in angle modulation $V(t) = V_0$ is fixed and the modulating signal varies $\phi(t)$. Angular modulation can be either FM or phase modulation (PM), depending upon the exact relationship between $\phi(t)$ and the modulating signal. The instantaneous angular velocity $\omega(t)$ can be written in a common case as

$$\omega(t) = 2\pi f(t) = \frac{d\phi(t)}{dt} \tag{6.24}$$

from which the instantaneous phase can be found by integrating according to

$$\phi(t) = \int_0^t \omega(t)dt + \phi_0 \tag{6.25}$$

where ϕ_0 is the initial phase at $t = 0$.

A frequency-modulated signal with cosine modulation when the instantaneous modulating signal varies in accordance with $v_{\rm m}(t) = V_{\rm m}\cos\Omega t$ represents by definition a signal in which the instantaneous angular velocity is varied according to

$$\omega(t) = \omega_0 + \frac{v_{\rm m}(t)}{V_{\rm m}}\Delta\omega = \omega_0 + 2\pi\,\Delta f\,\cos\Omega t \tag{6.26}$$

where ω_0 is the carrier or average radian frequency, $\Omega = 2\pi f_{\rm m}$ is the modulating radian frequency, and Δf is the maximum deviation of instantaneous frequency from average obtained at time moments when $v_{\rm m} = V_{\rm m}$. A fundamental characteristic of a frequency-modulated signal is that the maximum frequency deviation Δf is proportional to the peak amplitude of the modulating signal $V_{\rm m}$ and is independent of the modulating frequency. Integrating Eq. (6.26) gives

$$\phi(t) = \omega_0 t + \frac{\Delta f}{f_{\rm m}}\sin\Omega t + \phi_0 \tag{6.27}$$

where for simplicity it can be assumed that $\phi_0 = 0$.

Consequently, substituting Eq. (6.27) in Eq. (6.23) results in

$$v(t) = V_0\cos\left(\omega_0 t + \frac{\Delta f}{f_{\rm m}}\sin\Omega t\right) \tag{6.28}$$

which is commonly written for FM as

$$v(t) = V_0 \cos(\omega_0 t + m \sin \Omega t) \tag{6.29}$$

where

$$m = k\frac{V_m}{f_m} = \frac{\Delta f}{f_m} \tag{6.30}$$

is called the *modulation index* for frequency modulation, where k is the factor of proportionality [4]. Note that, for a given frequency deviation Δf, the modulation index m varies inversely as the modulating frequency f_m. Time-domain representation of the FM process with a sinusoidal modulating wave is shown in Figure 6.12. As the modulating signal swings positive, the carrier frequency is increased, reaching its highest frequency at the positive peak of the modulating signal. When the signal swings in the negative direction, the carrier frequency is lowered, reaching a minimum when the signal passes through its peak negative value.

Equation (6.29) can be rewritten through trigonometric expansion as

$$v(t) = V_0 \cos(m \sin \Omega t) \cos \omega_0 t - V_0 \sin(m \sin \Omega t) \sin \omega_0 t \tag{6.31}$$

which represents a sum of two quadrature signals with a carrier frequency ω_0, each of which is amplitude-modulated with a modulating frequency Ω. Generally, two different classes of FM signals can be distinguished, depending on the values of the modulation index m [17]. For $m \leq 0.5$, it is

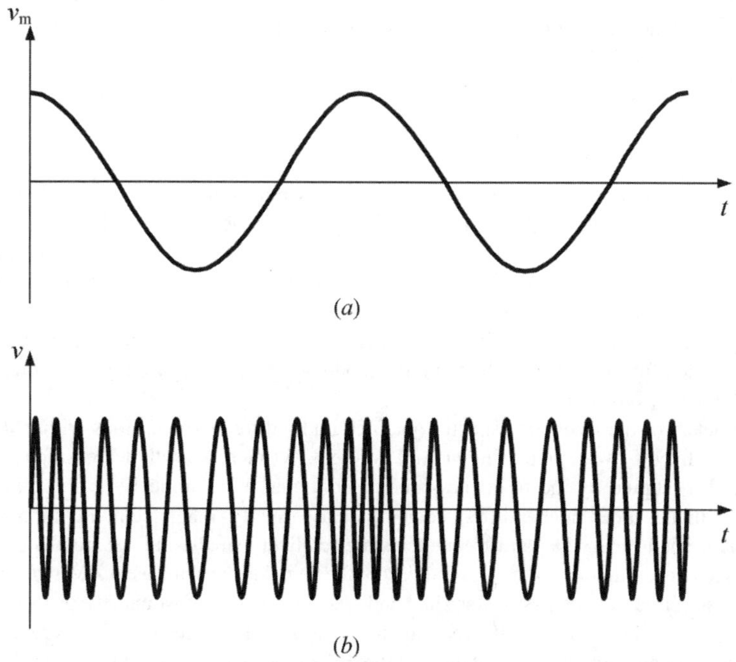

FIGURE 6.12 Frequency modulation process in time domain.

called the *narrowband frequency modulation*, and Eq. (6.31) can be approximated as

$$v(t) \cong V_0 \cos \omega_0 t - m V_0 \sin\Omega t \sin \omega_0 t = V_0 \cos \omega_0 t + \frac{m V_0}{2} \cos(\omega_0 + \Omega) t - \frac{m V_0}{2} \cos(\omega_0 - \Omega) t \tag{6.32}$$

since cos(m sinΩt) $\cong$ 1 and sin(m sinΩt) $\cong$ m sinΩt. Hence, the bandwidth of a narrowband FM is approximately 2Ω, which is the same as for AM signal containing the carrier frequency ω_0 and two sideband frequencies $\omega_0 + \Omega$ and $\omega_0 - \Omega$. In this case, the amplitudes of the sideband frequencies are defined by a modulation index m. However, the narrowband FM spectrum differs from the AM spectrum due to opposite sign in the second term, resulting in its phase shift by 180°.

For $m > 0.5$, the frequency-modulated process is called the *wideband frequency modulation*, which is the type of FM most often used in analog communication systems. To determine its bandwidth requirements, it is necessary to expand the terms cos(m sinΩt) and sin(m sinΩt) in Eq. (6.31) as the even and odd periodic functions in a Fourier series according to

$$\cos(m\sin\Omega t) = J_0(m) + 2J_2(m)\cos(2\Omega t) + 2J_4(m)\cos(4\Omega t) + \cdots \tag{6.33}$$

$$\sin(m\sin\Omega t) = 2J_1(m)\sin(\Omega t) + 2J_3(m)\sin(3\Omega t) + \cdots \tag{6.34}$$

where $J_n(m)$ are the Bessel functions of the first kind.

Substituting Eqs. (6.33) and (6.34) into Eq. (6.31) and using trigonometric identities result in

$$\begin{aligned} v(t) &= V_0 \{J_0(m)\cos\omega_0 t + J_1(m)[\cos(\omega_0 + \Omega)t - \cos(\omega_0 - \Omega)t] \\ &\quad + J_2(m)[\cos(\omega_0 + 2\Omega)t + \cos(\omega_0 - 2\Omega)t] \\ &\quad + J_3(m)[\cos(\omega_0 + 3\Omega)t - \cos(\omega_0 - 3\Omega)t] + \cdots\} \\ &= V_0 \left\{ J_0(m)\cos\omega_0 t + \sum_{n=1}^{\infty} J_n(m) J_1(m) \left[\cos(\omega_0 + n\Omega)t + (-1)^n \cos(\omega_0 - n\Omega)t\right] \right\} \end{aligned} \tag{6.35}$$

from which it follows that the FM signal spectrum is discrete, symmetrical relatively to the carrier frequency ω_0, and contains an infinite number of sideband frequencies spaced at radian frequencies $\pm\,\Omega$, $\pm\,2\Omega$, ..., about the carrier with amplitudes $V_n = V_0 J_n(m)$. The plots and numbers of the relative carrier and sideband-frequency amplitudes for different modulation indexes m are shown in Figure 6.13. It is clearly seen that, as the modulation index varies, a carrier frequency or a sideband-frequency pair may vanish entirely. This phenomenon can be used to set the frequency deviation of an FM transmitter, for example, by choosing the modulating frequency f_m for a specified frequency deviation Δf so that $J_0(m) = 0$.

The characters of the frequency spectra obtained under different conditions with FM are shown in Figure 6.14 (fixed f_m with increasing m) and Figure 6.15 (fixed Δf with decreasing f_m). Although the bandwidth occupied by the frequency-modulated signal is theoretically infinite, in reality the amplitudes of higher order sideband frequencies decrease rapidly. When the modulation index is less than 0.5, the second-order sideband frequencies are relatively small, and the frequency bandwidth (BW) required to accommodate the essential part of the signal is the same as in AM when BW $= 2f_m$. Besides, the amplitude of the first-order sideband frequencies is almost exactly proportional to the modulation index. On the other hand, when the modulation index exceeds unity, there are important higher order sideband components contained in the FM signal, and, for very large m, the approximate

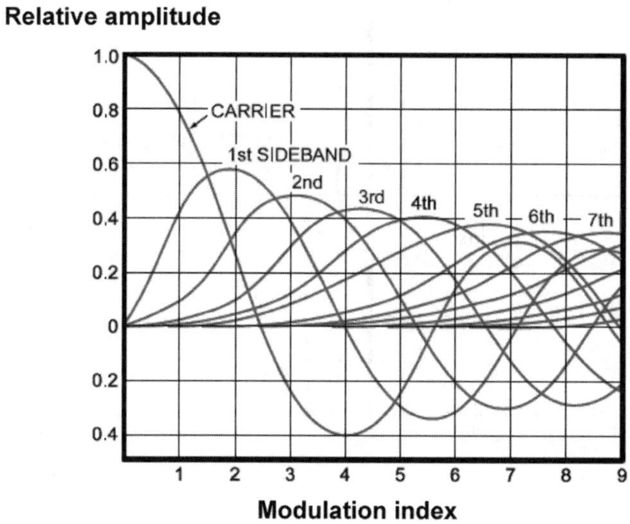

Modulation index	Carrier	1	2	3	4	5	6	7	8	9	10	11	12	13	14
0	1.00														
0.25	0.98	0.12													
0.5	0.94	0.24	0.03												
1.0	0.77	0.44	0.11	0.02											
1.5	0.51	0.56	0.23	0.06	0.01										
2.0	0.22	0.58	0.35	0.13	0.03										
2.41	0	0.52	0.43	0.20	0.06	0.02									
2.5	-0.05	0.50	0.45	0.22	0.07	0.02	0.01								
3.0	-0.26	0.34	0.49	0.31	0.13	0.04	0.01								
4.0	-0.40	-0.07	0.36	0.43	0.28	0.13	0.05	0.02							
5.0	-0.18	-0.33	0.05	0.36	0.39	0.26	0.13	0.05	0.02						
5.53	0	-0.34	-0.13	0.25	0.40	0.32	0.19	0.09	0.03	0.01					
6.0	0.15	-0.28	-0.24	0.11	0.36	0.36	0.25	0.13	0.06	0.02					
7.0	0.30	0	-0.3	-0.17	0.16	0.35	0.34	0.23	0.13	0.06	0.02				
8.0	0.17	0.23	-0.11	-0.29	-0.10	0.19	0.34	0.32	0.22	0.13	0.06	0.03			
8.65	0	0.27	0.06	-0.24	-0.23	0.03	0.26	0.34	0.28	0.18	0.10	0.05	0.02		
9.0	-0.09	0.25	0.14	-0.18	-0.27	-0.06	0.20	0.33	0.31	0.21	0.12	0.06	0.03	0.01	
10.0	-0.25	0.04	0.25	0.06	-0.22	-0.23	-0.01	0.22	0.32	0.29	0.21	0.12	0.06	0.03	0.01

FIGURE 6.13 Carrier and sideband amplitudes for FM signals.

bandwidth requirements can be defined as $\mathrm{BW} = 2mf_{\mathrm{m}} = 2\Delta f$. Figure 6.16 illustrates the analyzer display of an FM signal with $m = 95$ when it is required to use the entire spectrum width.

Thus, a general rule of thumb for the approximate bandwidth of a frequency-modulated signal for any m is

$$\mathrm{BW} = 2(\Delta f + f_{\mathrm{m}}) = 2f_{\mathrm{m}}(m+1) \tag{6.36}$$

which is called the *Carson's bandwidth rule*. In other words, any frequency-modulated signal will have an infinite number of sidebands but in practice all significant sideband power (98% or more) is concentrated within the bandwidth defined by Eq. (6.36). However, setting the arbitrary definition of occupied bandwidth at 98% of the total power still means that the power outside the band is only by $10 \times \log_{10}(0.02/0.98) = 16.9$ dB less than the in-band power. For voice communications a higher

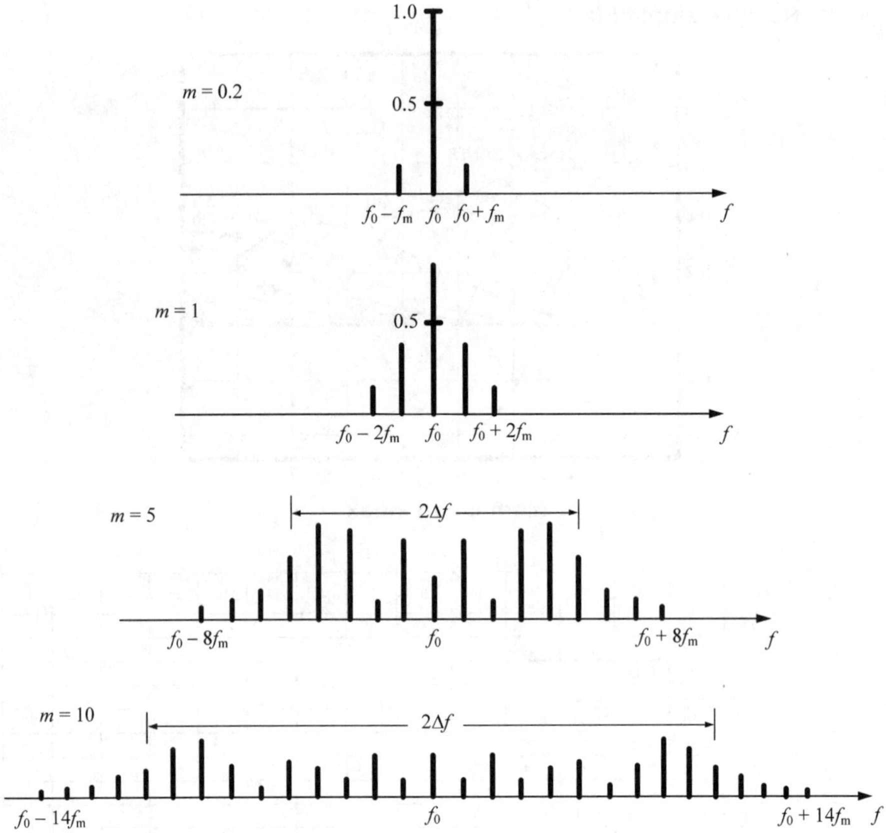

FIGURE 6.14 Frequency spectra of FM signal with fixed modulating frequency.

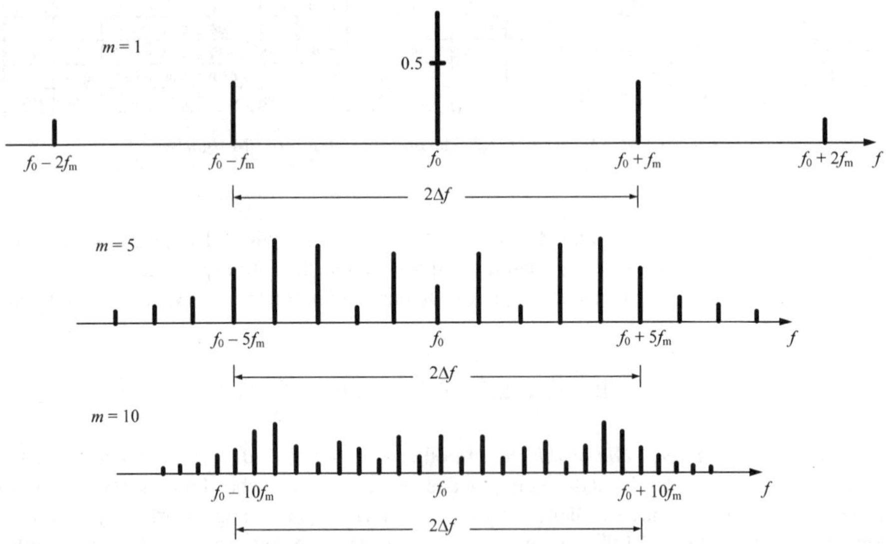

FIGURE 6.15 Frequency spectra of FM signal with fixed frequency deviation.

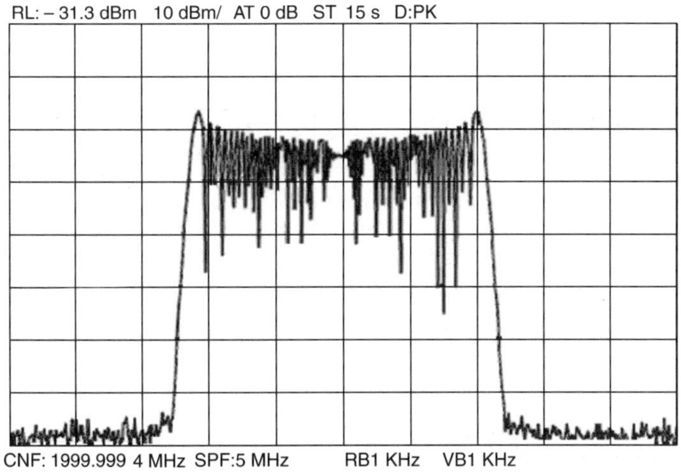

FIGURE 6.16 Analyzer display of FM signal with $m = 95$.

degree of distortion can be tolerated, and it is possible to ignore all sidebands with less than 10% of the carrier voltage. In broadcast FM, where Δf is restricted to be 75 kHz and maximum f_m should be less than or equal to 15 kHz, the smallest modulation index is $m = 75$ kHz/15 kHz $= 5$. As a result, from Eq. (6.36) it follows that BW = 180 kHz, which is safely within the total bandwidth of 200 kHz allocated to each FM station.

More accurate approximation of the frequency-modulated signal bandwidth for $m = 0$ to 24 can be calculated for sinusoidal modulation as

$$\mathrm{BW} = 2f_{\mathrm{m}}\left(m + \sqrt{m} + 1\right) \tag{6.37}$$

and for square-wave modulation as

$$\mathrm{BW} = 2f_{\mathrm{m}}\sqrt{\frac{200}{\pi}m + m^2} \tag{6.38}$$

with amplitudes less than 1% of the non-modulated carrier amplitude [18].

Generally, the relative carrier and sideband frequency amplitudes in a frequency-modulated signal will vary with amplitude and frequency of the modulating signal, but the total power contained in the modulated waveform remains constant. This is in contrast to AM, where the sideband amplitudes and the total power are controlled by the modulation, but the carrier amplitude is not. Besides, contrary to the situation in AM when each frequency component in the modulating signal produces a single pair of sideband frequencies, the superposition principle does not hold in a frequency-modulated signal for complex modulating signal, resulting in asymmetrical arrays of sideband frequencies, even with harmonically related modulating signals [4].

6.3.2 Frequency Modulators

Generally, there are two basic methods to provide an FM process: direct method by varying the reactive element of the frequency oscillator and indirect method by proper phase varying in the phase modulator according to Eq. (6.27), which follows the fixed-frequency oscillator. Usually, varactor whose junction capacitance varies in accordance with applied reverse bias voltage is used as a frequency-tuning element in the oscillators. By providing a proper reverse bias voltage and low-frequency modulating

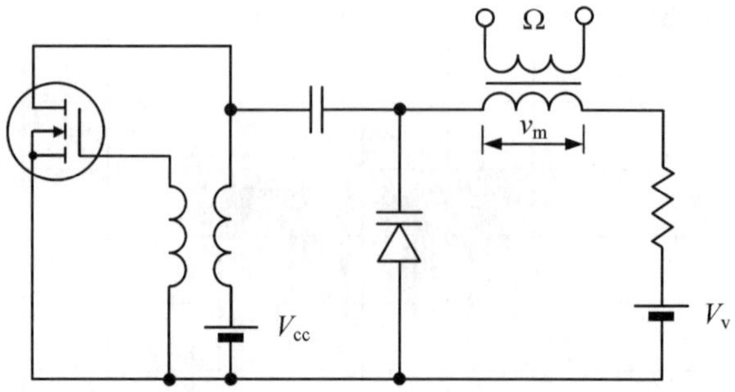

FIGURE 6.17 Schematic of varactor-tuned transformer-coupled oscillator.

signal to the varactor that is a part of entire oscillator resonant circuit, the frequency-modulated signal at the output port of the oscillator is achieved. To evaluate the quality of such a frequency-modulated signal, it is very important to know the behavior of the varactor voltage–capacitance characteristic and its influence on the oscillator modulation characteristic, representing a dependence of the oscillation frequency on a slowly varying varactor bias voltage. Figure 6.17 shows the simplified transformer-coupled MOSFET oscillator circuit where the modulating signal $v_m = V_m \cos\Omega t$ is applied to the varactor included into the oscillator resonant circuit.

The voltage–capacitance characteristic of the abrupt-junction varactor can be written as

$$C_v(V_v) = \frac{C_{v0}}{\sqrt{1 + \frac{V_v}{\varphi}}} \tag{6.39}$$

where V_v is the fixed bias voltage, C_{v0} is the varactor capacitance when $V_v = 0$, and φ is the contact potential. Substituting the voltage increment $v = V_v + \Delta v$ instead of V_v into Eq. (6.39) gives

$$\frac{\Delta C_v}{C_v} = \frac{1}{\sqrt{1+\xi}} - 1 \tag{6.40}$$

where $\xi = \Delta v/(\varphi + V_v)$. Equation (6.40) can be expanded into a Taylor series as

$$x = -\frac{\xi}{2} + \frac{3\xi^2}{8} - \frac{5\xi^3}{16} + \cdots \tag{6.41}$$

where $x = \Delta C_v/C_v$. By assuming that the normalized voltage increment ξ represents the value of the modulating voltage as $\xi = \Delta v/(\varphi + V_v) = V_m \cos\Omega t$, then Eq. (6.41) can be rewritten as

$$\frac{\Delta C_v}{C_v} = -\frac{3V_m^2}{16} - \frac{V_m}{2}\cos\Omega t - \frac{3V_m^2}{16}\cos 2\Omega t - \cdots. \tag{6.42}$$

For a single resonant circuit oscillator with the radian frequency deviation $\Delta\omega$ around the oscillation frequency ω_0 when varactor represents the only circuit capacitance,

$$\Delta\omega = \frac{1}{\sqrt{L(C_v + \Delta C_v)}} - \frac{1}{\sqrt{LC_v}} = \omega_0 \left[\left(1 + \frac{\Delta C_v}{C_v}\right)^{-1/2} - 1 \right] \tag{6.43}$$

where $\omega_0 = 1/\sqrt{LC_v}$. As a result, from Eq. (6.43) it follows that

$$\frac{\Delta f}{f_0} = \frac{1}{\sqrt{1+x}} - 1 = -\frac{x}{2} + \frac{3x^2}{8} - \cdots. \tag{6.44}$$

Substituting Eq. (6.41) with two first factors only into Eq. (6.44) results in

$$\frac{\Delta f}{f_0} = \frac{\xi}{4} - \frac{3\xi^2}{32} + \cdots \tag{6.45}$$

which represents a nonlinear dependence of the frequency deviation Δf versus voltage variation Δv on the varactor. Then, substituting $\xi = V_m \cos\Omega t$ into Eq. (6.45) gives

$$\frac{\Delta f}{f_0} = -\frac{3V_m^2}{64} + \frac{V_m}{4}\cos\Omega t - \frac{3V_m^2}{64}\cos 2\Omega t + \cdots \tag{6.46}$$

with the frequency deviation

$$\Delta f = \frac{V_m}{4} f_0 \tag{6.47}$$

the second-harmonic nonlinear coefficient

$$k_2 = \frac{3V_m}{16} = \frac{3}{4}\frac{\Delta f}{f_0} \tag{6.48}$$

and the relative shift of the average frequency

$$\frac{\Delta f_{avr}}{f_0} = \frac{3V_m^2}{64} = \frac{3}{4}\left(\frac{\Delta f}{f_0}\right)^2. \tag{6.49}$$

Generally, the function $x(\xi)$ in Eq. (6.41) depends on the voltage–capacitance characteristic of the particular varactor, whereas the function $\Delta f(x)$ is characterized by the particular type of the varactor connection to the oscillator resonant circuit. Figure 6.18(*a*) shows the general case of the partial varactor connection to the oscillator resonant circuit. When $C_1 \gg C_2 + C_v$, it can represent a parallel resonant circuit with variable capacitance C_v, as shown in Figure 6.18(*b*). When $C_2 \ll C_v$, the oscillator can be represented by a series resonant circuit with variable capacitance C_v, as shown in Figure 6.18(*c*).

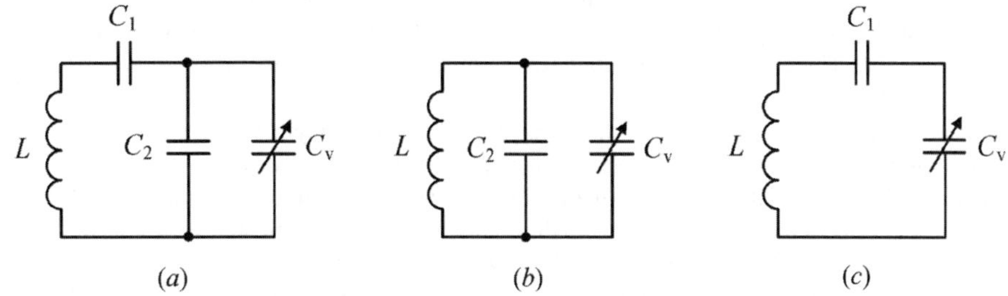

FIGURE 6.18 Varactor connection to resonant circuit.

For a parallel resonant circuit with $C_0 = C_2 + C_v$ and varactor capacitance variation ΔC_v due to Δv,

$$y = \frac{\Delta C}{C_0} = \frac{C_2 + C_v + \Delta C_v}{C_2 + C_v} - 1 = \frac{C_v}{C_2 + C_v}\frac{\Delta C_v}{C_v} = px \tag{6.50}$$

where the total capacitance variation ΔC corresponds to the varactor voltage variation Δv and $p = C_v/(C_2 + C_v)$ is the coefficient of partial varactor connection into the parallel oscillator resonant circuit.

For a series resonant circuit with $C_0 = C_v C_1/(C_v + C_1)$,

$$y = \frac{\Delta C}{C_0} = \frac{(1+x)(C_1 + C_v)}{C_1 + C_v(1+x)} - 1 = \frac{px}{1 + (1-p)x} = px - p(1-p)x^2 + \cdots \tag{6.51}$$

where $p = C_1/(C_1 + C_v)$ is the coefficient of partial varactor connection into the series oscillator resonant circuit.

The general-type resonant circuit shown in Figure 6.18(a) represents a combination of the series and parallel resonant circuits. The relative deviation y_1 of the capacitance $C'_v = C_2 + C_v$ can be defined from Eq. (6.50) as

$$x_1 = \frac{\Delta C'_v}{C'_v} = p_1 x \tag{6.52}$$

where $p_1 = C_v/(C_2 + C_v)$. In this case, the capacitance C'_v is the variable capacitance for the circuit including the capacitance C_1 and inductance L. Therefore, using Eq. (6.51),

$$y(x) = p_2 x_1 - p_2(1 - p_2)x_1^2 + \cdots \tag{6.53}$$

where $p_2 = C_1/(C_1 + C'_v) = C_1/(C_1 + C_2 + C_v)$. Substituting Eq. (6.52) into Eq. (6.53) and representing Eq. (6.53) as $y(x) = a_1 x + a_2 x^2 + \cdots$ allow the first two coefficients to be determined as $a_1 = p_1 p_2$ and $a_2 = -p_1^2 p_2(1 - p_2)$.

As a result, by choosing the coefficients p_1 and p_2 with proper values of C_1, C_2, and, C_v, the behavior of the oscillator modulation characteristic can be optimized for linear frequency tuning for the particular varactor voltage–capacitance characteristic [19].

Generally, the voltage v_v across the varactor shown in Figure 6.19 consists of the dc bias voltage V_v, low-frequency modulating component v_m with amplitude V_m, and high-frequency carrier component v_0 with amplitude V_0, which can be written as

$$v_v = V_v + V_m \cos \Omega t + V_0 \cos \omega_0 t \tag{6.54}$$

where Ω is the modulating frequency and ω_0 is the carrier frequency. The bias voltage V_v is usually chosen at the center point of the varactor voltage–capacitance characteristic to maximize the frequency tuning bandwidth without performance degradation due to the forward conduction of the active device or its breakdown. Under these assumptions, the borders of voltage variations across the varactor junction should satisfy the condition $V_m + V_0 = 0.5V_{vmax}$.

Figure 6.20 shows the schematic of the linearized direct frequency modulator based on a crystal transistor oscillator with audio-frequency input port, which is intended to use in the land mobile radio of low VHF band [20]. Assuming that the crystal fundamental frequency may be selected on a cost–size basis only, a multiplier ratio of 3 will require a crystal fundamental frequency of 10 to 17 MHz to cover the 30 to 50 MHz VHF band. To provide a 5 kHz deviation with a multiplication ratio of 3, the varactor-tuned crystal oscillator must be capable of a ±1.7 kHz deviation at the fundamental. In this case, the bias voltage must be large enough to insure the varactor diode remains reverse biased.

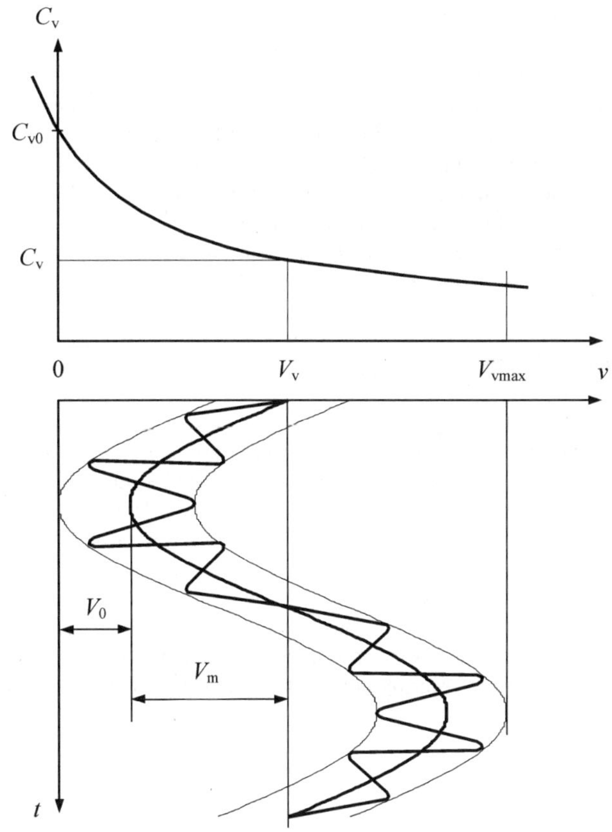

FIGURE 6.19 Varactor capacitance and frequency modulation.

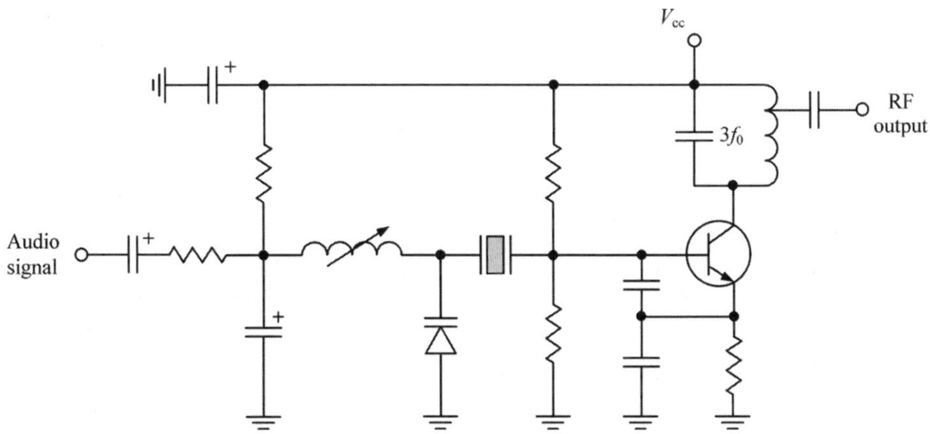

FIGURE 6.20 Schematic of linearized direct frequency modulator.

The tank circuit in the collector is tuned to the third harmonic of the fundamental crystal frequency, which is then amplified without further multiplication straight through antenna. The linearization of the modulation characteristic can be achieved by optimizing the value of the variable inductor connected in parallel to the varactor diode. The value of the shunt capacitor at the audio input is chosen to look like an RF short to the inductor but still have relatively high reactance over the audio band up to 3000 Hz. By proper choice of both the inductance and the varactor capacitance, the undesirable modulation nonlinearity can be considerably reduced. The linearizing network for a microwave frequency modulator based on the varactor-tuned cavity or dielectric resonator transistor oscillator can include the optimized combination of a shunt variable capacitor and the high-impedance and low-impedance transmission lines [21].

6.4 PHASE MODULATION

In order to introduce the PM principle, first consider the carrier cycle as the projection of a point rotating counterclockwise in a circle where the phase at any given point is the angle between the start point and the point on the waveform, as shown in Figure 6.21. This is called the *vector representation*, which can also be plotted as the amplitude against the number of degrees of the point rotation. For each cycle of the carrier, the point rotates in one complete circle, or period of 360°: starting from 0°, rotating through 180°, and coming back to 0°, where the next cycle begins. The position of the point at any instant can be indicated by a vector drawn from the center of the circle, showing the particular case of 45° in Figure 6.21. Note that this vector is rotating at the carrier frequency ω_0.

A phase-modulated signal with cosine modulation (instantaneous modulating signal varies in accordance with $\cos\Omega t$) will represent by definition a signal in which the instantaneous phase is varied according to

$$\phi(t) = \omega_0 t + \frac{v_m(t)}{V_m}\Delta\phi + \phi_0 = \omega_0 t + \Delta\phi \cos\Omega t + \phi_0 \tag{6.55}$$

where ω_0 is the carrier or average radian frequency, $\Omega = 2\pi f_m$ is the modulating radian frequency, ϕ_0 is the initial phase, and $\Delta\phi$ is the maximum deviation of instantaneous phase from its average value. Similarly to a frequency-modulated signal represented by Eq. (6.27), Eq. (6.55) for a phase-modulated signal can be rewritten as

$$\phi(t) = \omega_0 t + m \cos\Omega t + \phi_0 \tag{6.56}$$

where $m = kV_m = \Delta\phi$ is called the *modulation index* for phase modulation [4].

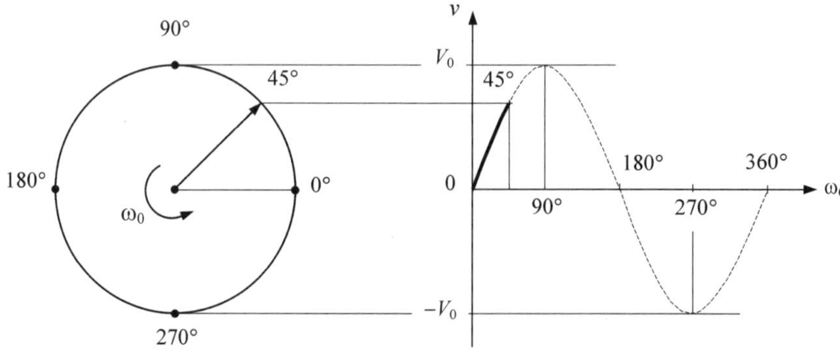

FIGURE 6.21 Projection of circulating vector point to form sine-wave cycle.

Substituting Eq. (6.56) into Eq. (6.23) with constant carrier amplitude V_0 results in a phase-modulated signal with cosine modulation written as

$$v(t) = V_0 \cos(\omega_0 t + m\cos\Omega t) \tag{6.57}$$

where it is assumed that $\phi_0 = 0$ for the sake of simplicity. A comparison of Eqs. (6.29) and (6.57) shows that the phase-modulated signal contains similar sideband components as does the frequency-modulated signal, and if the modulation indexes in these two cases are the same, the relative amplitudes of their corresponding components will also be the same. The frequency of the phase-modulated signal obtained from Eq. (6.56) varies according to

$$\omega = \frac{d\phi(t)}{dt} = \omega_0 - m\Omega\sin\Omega t. \tag{6.58}$$

As a result, from a comparison of Eqs. (6.56) and (6.58) it follows that the cosine varying of the signal phase causes the sinusoidal varying of the signal frequency. In this case, the phase modulation with maximum phase deviation $\Delta\phi$ is accompanied by the frequency modulation with maximum frequency deviation $\Delta f = f_m \Delta\phi$. At the same time, as it follows from a comparison of Eqs. (6.26) and (6.27), the FM with maximum deviation Δf is equivalent to the PM with maximum phase deviation $\Delta\phi$ as

$$\Delta\phi = \frac{\Delta f}{f_m}. \tag{6.59}$$

Consequently, varying the modulating frequency f_m affects differently the frequency spectra of FM and PM signals. For example, the modulation index of FM signal increases by reducing the modulating frequency, as shown in Figure 6.15; however, the spectrum bandwidth remains almost the same. On the contrary, the spectrum bandwidth of FM signal reduces with increasing modulating frequency, as shown in Figure 6.22 where the modulating frequency f_m becomes three times smaller,

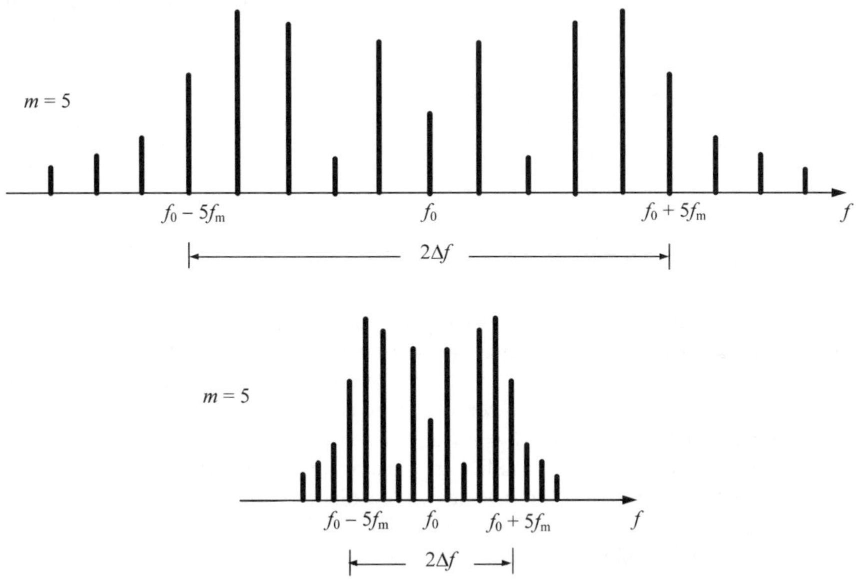

FIGURE 6.22 Frequency spectra of PM signal with fixed modulation index.

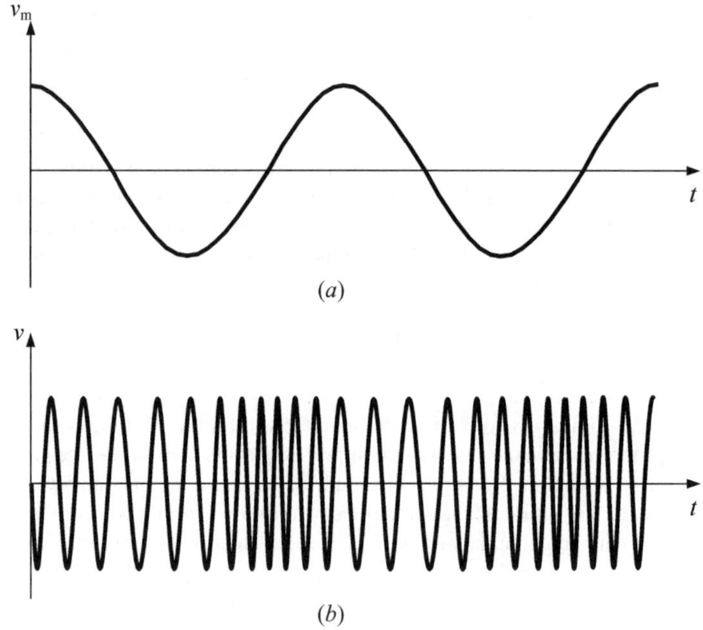

FIGURE 6.23 Phase modulation process in time domain.

but has no effect on the modulation index of PM signal. Time-domain representation of the PM process with a cosine modulating wave is shown in Figure 6.23. If the phase-modulated wave is shifted by 90°, it becomes looking alike the frequency-modulated wave shown in Figure 6.12.

Thus, the common feature of both FM and PM is that the frequency varying inevitably results in a corresponding phase varying and vice versa, and, to produce a phase-modulated signal that is identical in phase with a frequency-modulated signal, the modulating voltages must differ in phase by 90°. However, the difference between FM and PM is that the maximum phase deviation for a harmonic FM is reversely proportional to frequency of the modulating signal according to Eq. (6.57), whereas the maximum frequency deviation for harmonic PM is directly proportional to frequency of the modulating signal. For constant amplitude of the modulating frequency, the maximum frequency deviation for a harmonic FM is constant, whereas it is linearly proportional to the modulating frequency for a harmonic PM, increasing with velocity of 6 dB per octave.

The PM process can therefore be used to generate a true frequency-modulated signal by arranging so that the amplitude of the modulating voltage actually used to produce the phase variation is the signal modified by passing through the integrating resistance-capacitance network, in which the transmission is inversely proportional to frequency, instead of being the actual modulating signal. If the voltage appearing at the output of such a network is then used to control the instantaneous phase, the result is exactly the same as if the original signal were employed to control the instantaneous frequency [3]. Figure 6.24(*a*) shows the block schematic of an FM modulator where the frequency-modulated output signal is achieved by applying the modulating signal to the phase modulator input through the integrating network to reduce modulating amplitude versus increasing modulating frequency. Similarly, Figure 6.24(*b*) shows the block schematic of a PM modulator where the phase-modulated output signal is a result of applying the modulating signal through the differential network to increase the modulating amplitude versus increasing modulating frequency.

An important practical consideration is the fact that the average power in an angle-modulated signal is constant. The average power in a band-limited signal is found by adding the power contributions

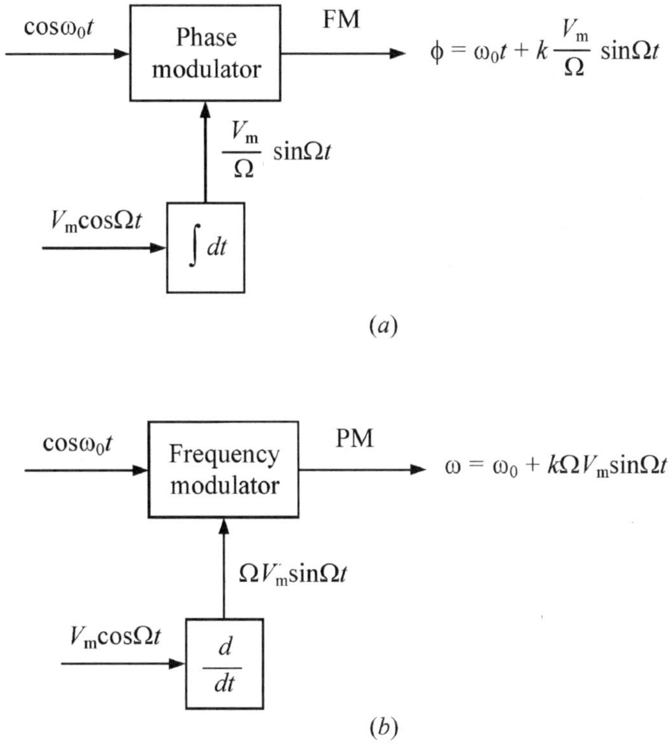

FIGURE 6.24 Indirect methods of frequency and phase modulations.

of the various frequency components delivered to the load resistance R_L as

$$P = \frac{V_0^2}{2R_L} \sum_{n=0}^{\infty} J_n^2(m) \tag{6.60}$$

by using Parseval's theorem according to which the average power is proportional to the sum of the squares of the individual Fourier components of the modulated wave.

Figure 6.25(*a*) shows the circuit schematic of the phase modulator based on a three-section low-pass filter having identical resonant circuits, each tuned by the varactor. For a proper tuning, such a phase modulator can provide the phase deviation up to 30° for each section when the parasitic AM is negligible and nonlinear distortions are sufficiently small [22]. The phase modulator, whose basic reactive elements compose a second-order phase-shifting four-pole network, is shown in Figure 6.25(*b*). Here, the low-frequency modulating voltage applied directly to the varactors changes their capacitances so that, for a proper dc biasing and sufficiently small capacitance variation, the amplitude variations are negligible with the phase deviation up to 70°.

Figure 6.26 shows the simplified schematics of the reflection-type varactor-tuned phase modulator using either (*a*) quadrature branch-line hybrid or (*b*) 3-dB coupled-line directional coupler. In both cases, the input carrier signal divides equally between the two opposite ports of the hybrid or coupler, each of which is connected to the varactor. The varactor diodes are both biased in the same forward or reverse biased state, so that the resulting modulated waves reflected from these two reactive terminations will add in phase at the output port, which originally represents the isolated port. To isolate the high-frequency carrier and low-frequency modulating paths, the quarterwave transmission

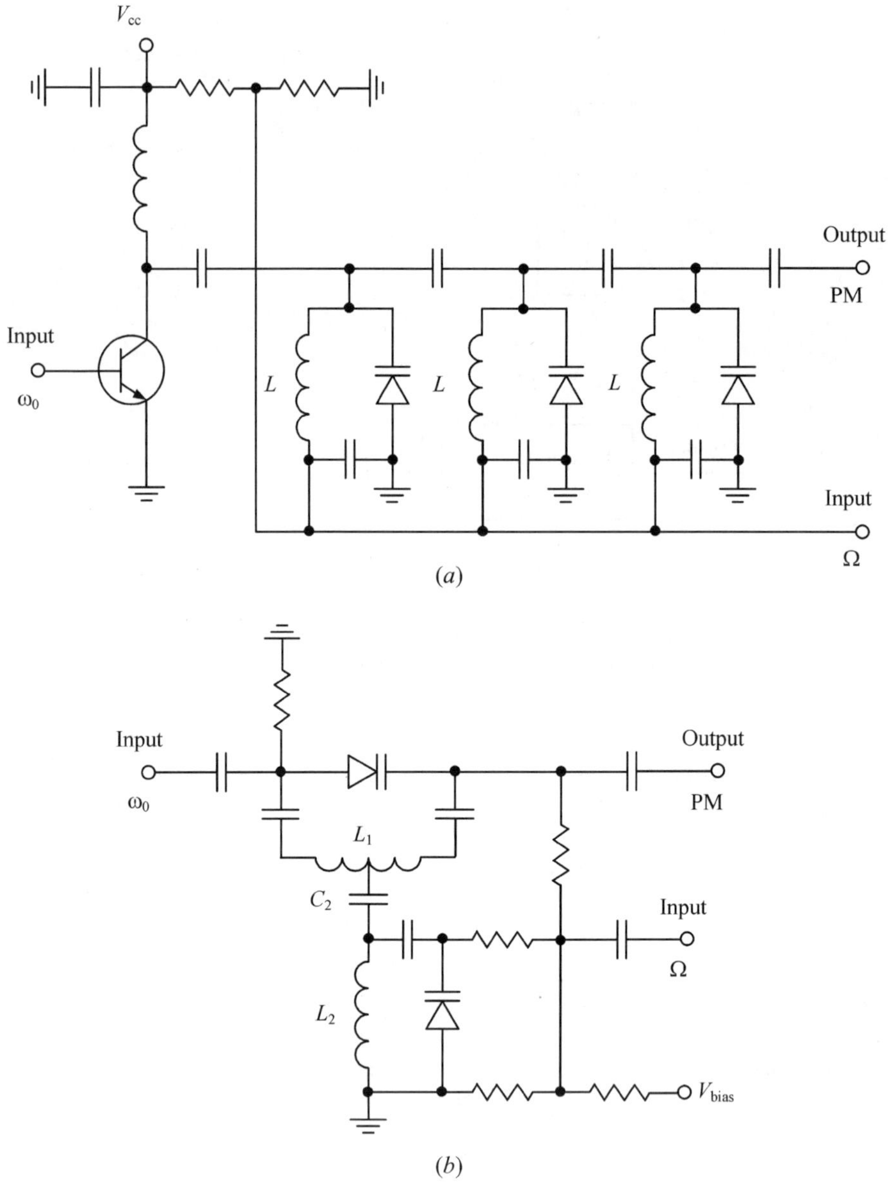

FIGURE 6.25 Schematics of varactor-tuned phase modulators.

lines can be used instead of lumped *LC* circuits. To improve the modulation linearity, a pair of varactors that are reversely biased with different voltages and separated by microstrip lines with optimized lengths to maximize reflection can be configured instead of a single varactor [23]. In this case, the phase deviation of 90° with linearity of 5% had been achieved within the frequency bandwidth of 0.8 to 1.2 GHz [24]. Much broader bandwidth can be achieved by using a 3-dB tandem or multisection Lange-type coupler. Three cascaded sections (with tapered varactor size), each of which is based on the lumped reflection-type quadrature hybrid and two varactors, could provide an *X*-band phase modulation with peak phase deviation of 300° and linearity of ±2° [25].

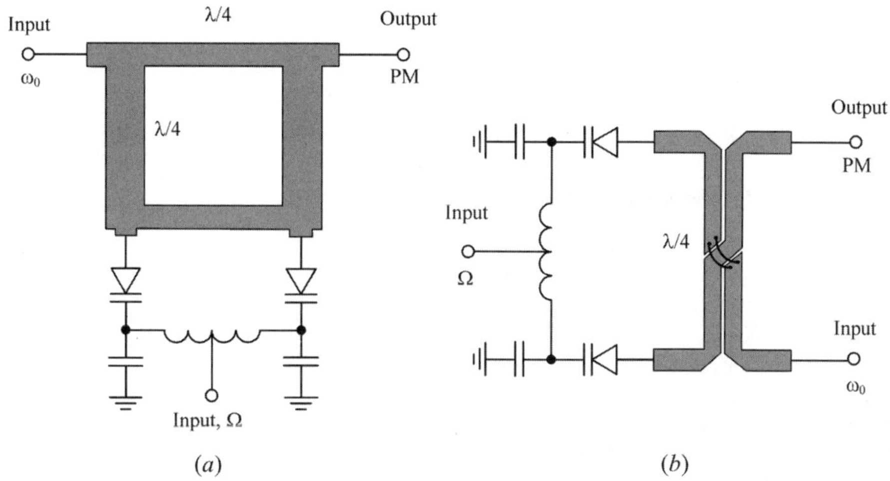

FIGURE 6.26 Microstrip reflection-type phase modulators.

6.5 DIGITAL MODULATION

Any modulated signal described by Eq. (6.23) can be represented in polar form by

$$v(t) = V(t)\cos\left[\omega_0 t + \varphi(t)\right] \tag{6.61}$$

where $V(t)$ is the time-varying amplitude (or envelope), ω_0 is the carrier radian frequency, and $\varphi(t)$ is the time-varying phase. By using the trigonometric identities, Eq. (6.61) can be rewritten in rectangular form by

$$v(t) = I(t)\cos\omega_0 t - Q(t)\sin\omega_0 t \tag{6.62}$$

where $I(t) = V(t)\cos\varphi(t)$ is the in-phase time-varying signal component and $Q(t) = V(t)\sin\varphi(t)$ is the quadrature time-varying signal component, as shown in Figure 6.27(a). These components are orthogonal and do not interfere with each other. On a polar diagram, the I-axis lies on the zero degree

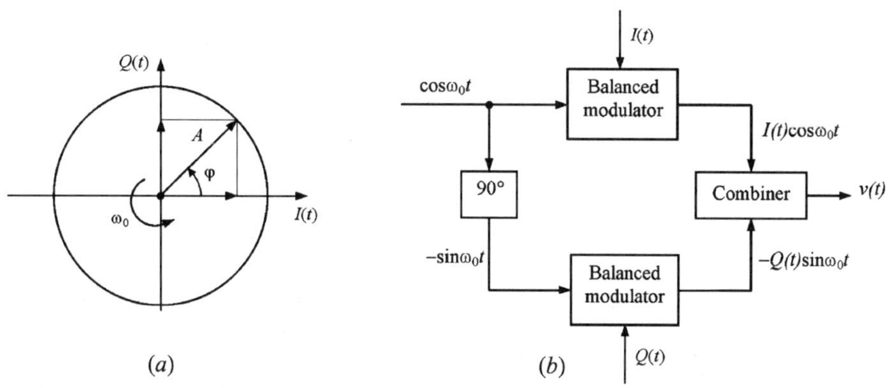

FIGURE 6.27 Vector representation and block implementation of I/Q modulator.

phase reference, and the Q-axis is rotated by 90 degrees. The signal vector projection onto the I-axis is its in-phase (I) component, while the projection onto the Q-axis is its quadrature (Q) component. In this case, the corresponding signal magnitude is defined as

$$A = \sqrt{I^2 + Q^2} \tag{6.63}$$

whereas the signal phase is obtained by

$$\varphi = \tan^{-1}\frac{Q}{I}. \tag{6.64}$$

Since two baseband signals $I(t)$ and $Q(t)$ modulate two exactly 90° out-of-phase carriers $\cos\omega_0 t$ and $-\sin\omega_0 t$, respectively, then the system operating according to Eq. (6.62) is called the *I/Q modulator*. The operating principle of an I/Q modulator, the block schematic of which is shown in Figure 6.27(*b*), is based on the splitting of the carrier signal into two equal signals when one flows directly to the balanced modulator to form the I-channel and the other flows into the other balanced modulator via 90° phase shifter to form the Q-channel. The baseband $I(t)$ and $Q(t)$ signals (either analog or digital) modulate appropriately the carrier to produce the I (in-phase) and Q (quadrature) frequency components, which are finally combined to produce the desired radio transmitting signal. Since any RF signal can be represented in the I/Q form, any modulation scheme can be implemented by an I/Q modulator. Digital modulation is easy to accomplish with the I/Q modulators because most digital modulation schemes map the data to a number of discrete points on the I/Q plane, resulting in a simultaneous amplitude and phase modulation.

To compare different modulation format efficiencies, it is necessary to first introduce the terms "bit" and "symbol," and then to understand the difference between bit rate and symbol rate. A *bit* is a binary digit as a basic unit of information capacity, taking a value of either 0 or 1. A *symbol* is a state or significant condition of the communication channel that persists for a fixed period of time. A sending device places symbols on the channel at a fixed and known *symbol rate*, which is measured in baud (Bd) or symbols per second. The baud unit is named in honor of Emile Baudot, the inventor of the Baudot code for telegraphy. If the transmitted pulses take on only two possible levels, each pulse represents 1 bit, and the bit rate = symbol rate = $1/\tau$ bits/s, where τ is the width of rectangular pulse. However, when raised cosine pulses are used for data transmission, then the pulse rate or symbol rate = $1/\tau$ symbols/s. Generally, the basic difference is that the signal bandwidth needed for the communication channel depends on the symbol rate, and not on the bit rate. The bit rate is the frequency of a system bit stream. The symbol rate is the bit rate divided by the number of bits transmitted with each symbol. If one bit is transmitted per symbol, then the symbol rate would be the same as the bit rate. However, if two bits are transmitted per symbol, then the symbol rate would be half of the bit rate. If more bits can be sent with each symbol, then the same amount of data can be sent in a narrower spectral bandwidth. This is why modulation formats that are more complex and use higher number of states can send the same information over a narrower frequency bandwidth. For example, for 8-state phase-shift keying modulation, there are eight possible states that the signal can transmit at any time. Since $8 = 2^3$, there are three bits per symbol, which means that the symbol rate is one-third of the bit rate.

6.5.1 Amplitude Shift Keying

Amplitude shift keying (ASK) is a form of digital modulation that represents digital data as pulsed variations in the amplitude of a carrier wave. The amplitude of an analog sinusoidal carrier signal varies in accordance with the bit stream (modulating signal), keeping frequency and phase constant. In the modulated signal, logic 0 is represented by the absence of a carrier, while logic 1 is represented by some carrier amplitude, thus giving on-off keying operation. The ASK technique is also commonly used

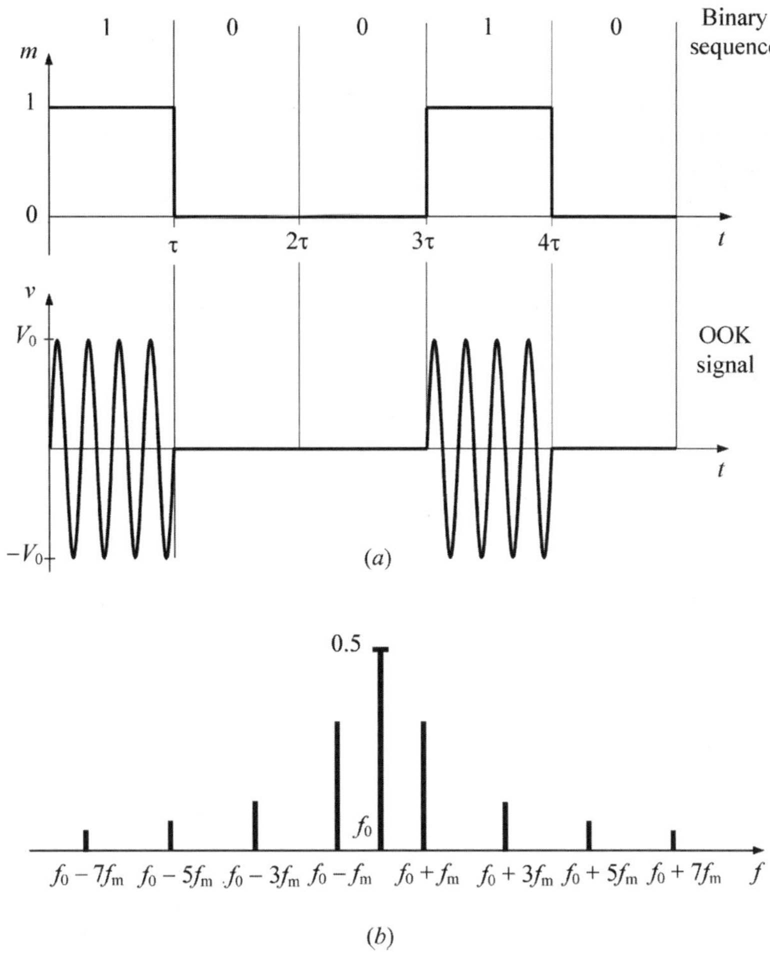

FIGURE 6.28 Binary modulating signal, OOK signal and spectrum.

to transmit digital data over optical fiber. For the fiber optic LED (light-emitting diode) transmitters, binary 1 is represented by a short pulse of light and binary 0 by the absence of light. *On-off keying* (OOK) is the simplest type of ASK modulation representing digital data as the presence (peak value) or absence (zero value) of a carrier wave. OOK is most commonly used to transmit Morse code over radio frequencies; the term "keying" is a historical remnant of telegraph transmission days. When more than two-level pulses are used, called the *m-ary transmission* for *m*-level signals, the modulation process is usually called the *m-ary amplitude shift keying* (MASK).

Figure 6.28(*a*) shows the binary sequence 10010 that generates the OOK signal in the form of an RF pulse train. The OOK modulation can be considered as a modulation of the carrier by the periodic rectangular pulses with the symbol rate $f_m = 1/(2\tau) = 0.5$ Bd, where τ is the symbol (bit) duration time. Figure 6.28(*b*) shows the OOK frequency spectrum where its spectral components are defined using a Fourier transform of the OOK signal with unit carrier amplitude as

$$a_0 = 0.5,\ a_n = \frac{1}{n\pi}\sin\frac{n\pi}{2}, \qquad n = 1,\ 3,\ 5, \ldots .$$

The OOK modulators can be used in modern digital communication systems where it is necessary to transmit an RF pulse-width train instead of the signal with non-constant envelope for highly efficient operation, and they play an especially important role in subscriber radio systems employing *time-division multiple access* (TDMA) technology. The simplest implementation of the OOK modulators at microwave and millimeter-wave frequencies is based on using the different types of quarter-wavelength transmission lines and *p–i–n* diodes [26]. Figure 6.29(*a*) shows the typical circuit schematic of the OOK modulator based on a single-ended amplifying transistor stage that can be used at radio frequencies. Here, both the carrier signal and binary pulse sequence are delivered to the common transistor base, being isolated from each other by a low-pass second-order *RLC* circuit.

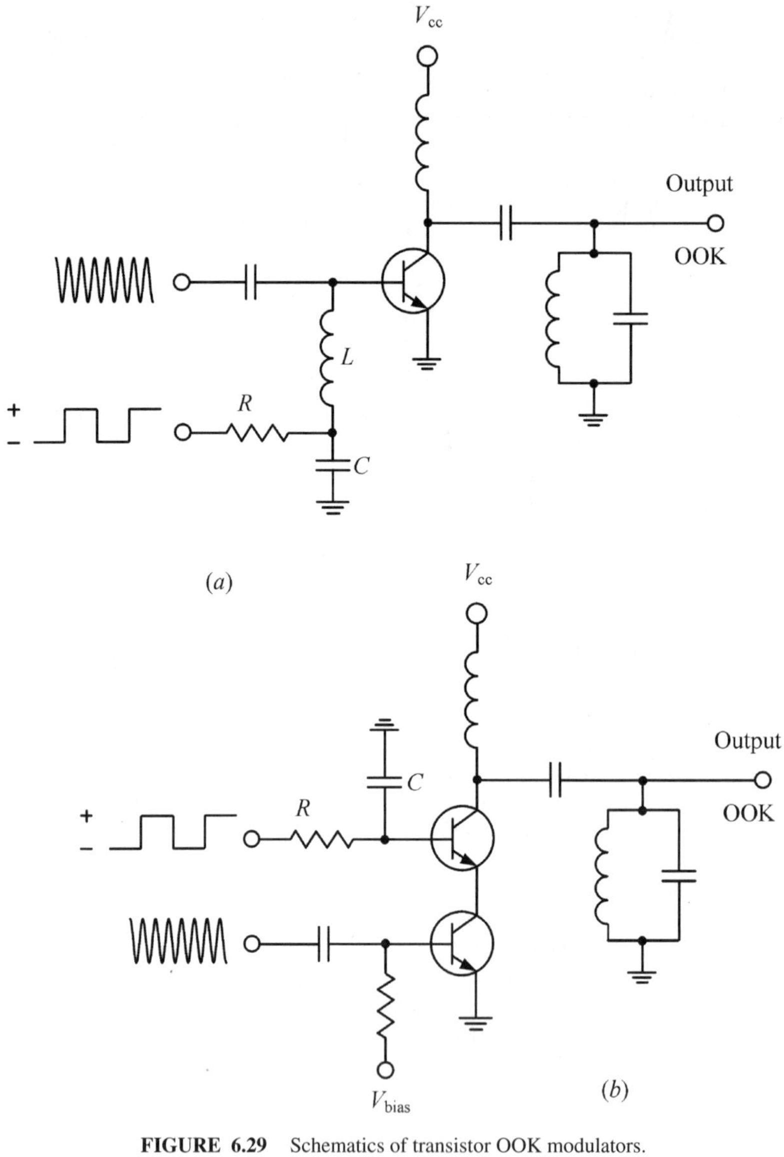

FIGURE 6.29 Schematics of transistor OOK modulators.

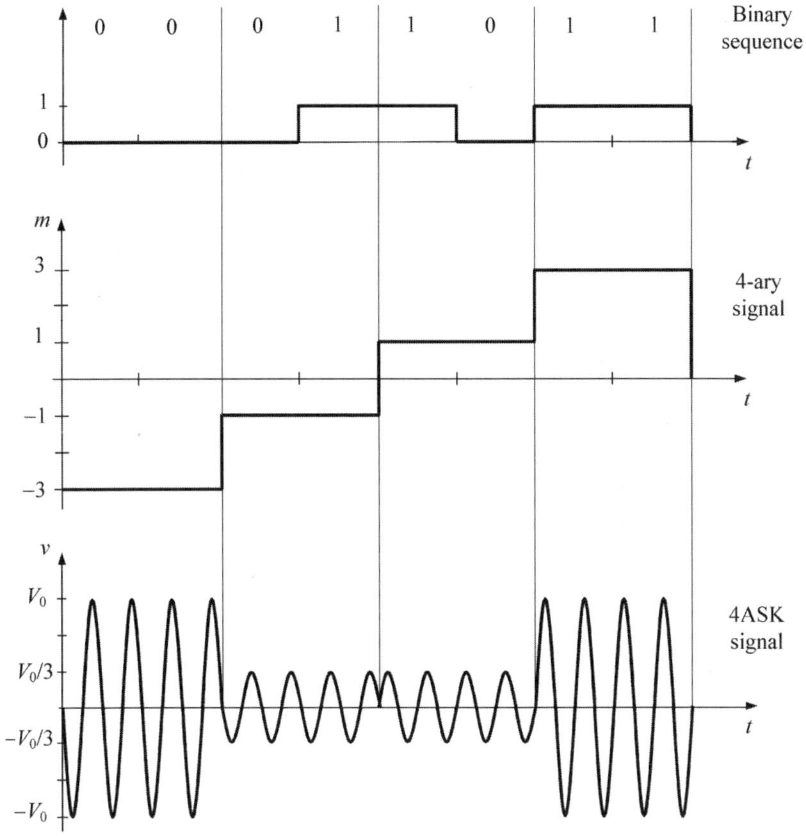

FIGURE 6.30 Binary sequence and 4ASK modulation.

In this case, it is necessary to optimally choose its elements to minimize the overall circuit transient response caused by the finite pulse fall and rise times and periodic damped oscillation process. A very convenient practical implementation, especially for monolithic circuits, can be based on a cascode circuit shown in Figure 6.29(*b*) where the bottom transistor in a common emitter configuration is used for carrier signal and the top transistor in a common base configuration is used for binary pulse sequence. The significant isolation of the pulse-modulating and carrier paths can be easily achieved by using an optimized low-pass *RC* filter connected to the base of the top transistor.

Figure 6.30 shows the binary sequence, 4-ary digital signal, and 4ASK signal with the amplitudes that are defined as $V_i = V_0[2i - (m - 1)]$ for $i = 0, 1, 2, \ldots, m$ and $m \geq 4$. It is interesting to note that simply by observing the resulting 4ASK signal alone, it is difficult to directly distinguish this signal from the conventional binary ASK (BASK) signal with a modulation index $m = 0.5$. However, each RF pulse of the 4ASK signal, corresponding to a particular binary code, is characterized by different initial phase compared to the RF pulse with the same amplitude in a conventional binary case.

6.5.2 Frequency Shift Keying

Frequency shift keying (FSK) is an FM scheme in which digital information is transmitted through discrete frequency changes of a carrier wave. The simplest FSK is a *binary frequency shift keying*

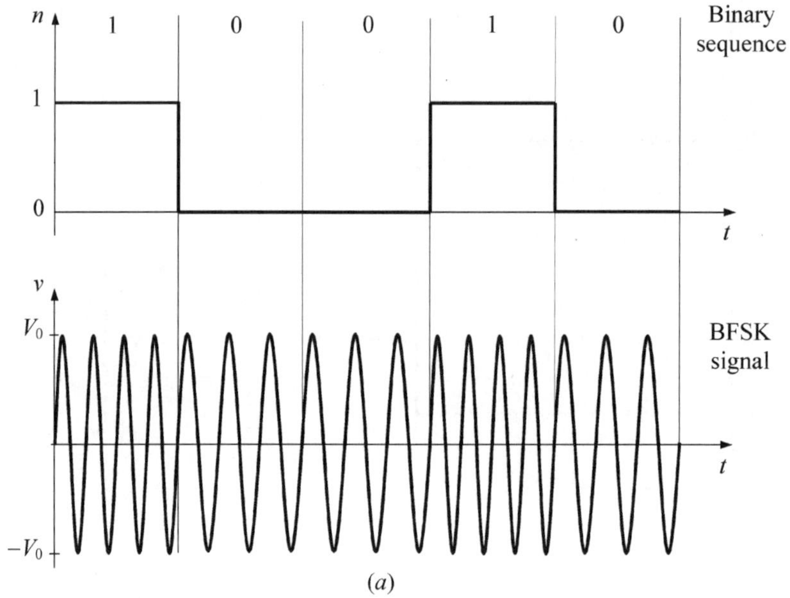

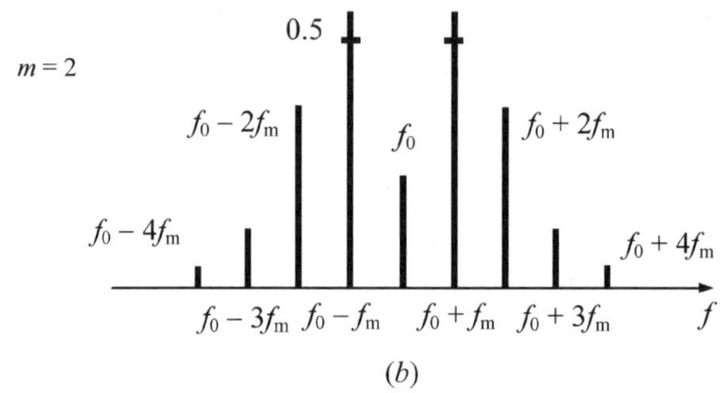

FIGURE 6.31 Binary modulating signal, BFSK signal, and spectrum.

(BFSK). For a binary sequence, FSK simply consists of a single-frequency sinusoidal pulse for logic 1 and a different frequency sinusoidal pulse for logic 0. The FSK is used in many applications including paging and cordless systems. Figure 6.31(a) shows the binary sequence 10010 that generates a BFSK waveform. The FSK spectrum depends both on the symbol rate f_m and maximum frequency deviation Δf. For a sufficiently small frequency deviation when $\Delta f \leq 3.4 f_m$, the FSK spectrum shown in Figure 6.31(b) occupies less frequency bandwidth required for the receiver according to Eq. (6.36) than the ASK spectrum shown in Figure 6.28(b). This is because higher FSK frequency components decrease faster than the corresponding higher ASK frequency components.

The simple way to implement BFSK is to use a crystal oscillator with a switching capacitance. Figure 6.32 show the circuit schematic of a crystal transistor-controlled FSK modulator where the capacitance C_2 is turned on or off by the transistor in accordance with the binary sequence delivered to its input to transmit different frequencies. The frequency shift corresponding to the required BFSK can be controlled by proper selection of the capacitances C_1 and C_2. Typically, the frequency deviations of about few megahertz are possible for such an FSK modulator, because larger deviation reduces the resonator quality factor and thus increases the phase noise.

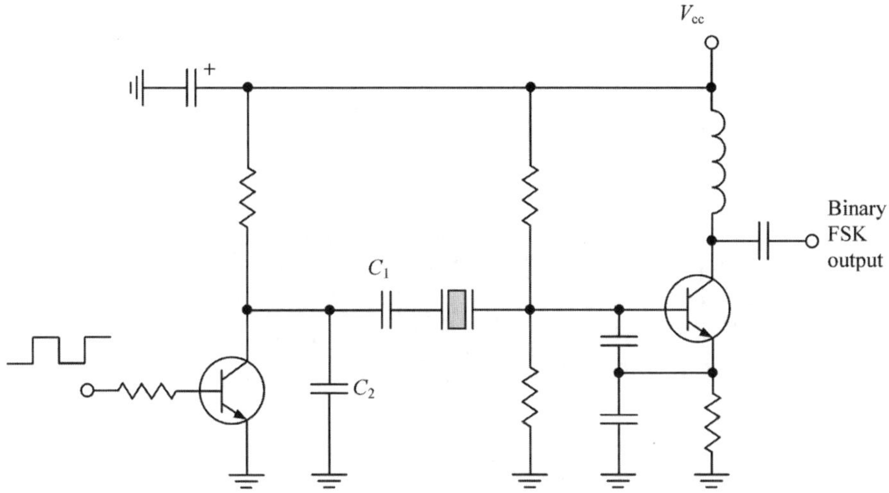

FIGURE 6.32 Schematic of binary FSK modulator.

The BFSK signal can be represented as a sum of two OOK signals generated by two binary sequences n_1 and n_2, as shown in Figure 6.33. In this case, logics 1 and 0 for binary sequences should alternate with each other when logic 1 for binary sequence n_1 corresponds to logic 0 for binary sequence n_2 and vice versa. The BFSK can be provided using an I/Q modulator with quadrature components of the carrier frequency f_0 when the I and Q inputs are driven by a low-frequency oscillator, set at the frequency offset f_m. The output of the oscillator is split into quadrature components and applied to the modulator, thus generating an SSB output at the frequency $f_0 - f_m$ (or $f_0 + f_m$, depending on the phase relationships of I and Q inputs). The low-frequency oscillator can be on-to-off keyed by a binary sequence to toggle between the fixed frequency f_0 representing a carrier feedthrough mode and the modulated frequency $f_0 - f_m$ representing an SSB mode.

In *m-ary frequency shift keying* (MFSK), *n* bits are combined into a symbol before transmission, and each symbol is represented by a separate frequency of the signal. There can be a total of *m* different frequencies, with each change in frequency occurring every symbol period equal to *n*-bit periods. For example, in quaternary or 4FSK, the data is transmitted in symbols of two bits, as shown in Figure 6.34, where each symbol is represented by a different frequency of the transmitted signal. The signal can have any of four frequencies depending on the type of symbol (00, 01, 10, or 11). The frequencies are spaced equidistant from each other and the spacing generally is even multiples of the symbol frequency f_s. The MFSK signal requires less bandwidth compared to BFSK; however, more resolution is required in the receiver.

6.5.3 Phase Shift Keying

Phase shift keying (PSK) modulation scheme is used to transmit digital information by shifting the phase of a carrier among several discrete values. When a binary sequence is to be transmitted, the phase is usually switched between 0° and 180°, and the *binary phase shift keying* (BPSK) signal can be defined as

$$v(t) = \begin{cases} V_0 \cos(\omega_0 t + \phi_0) & \text{for logic 1} \\ -V_0 \cos(\omega_0 t + \phi_0) & \text{for logic 0} \end{cases} \tag{6.65}$$

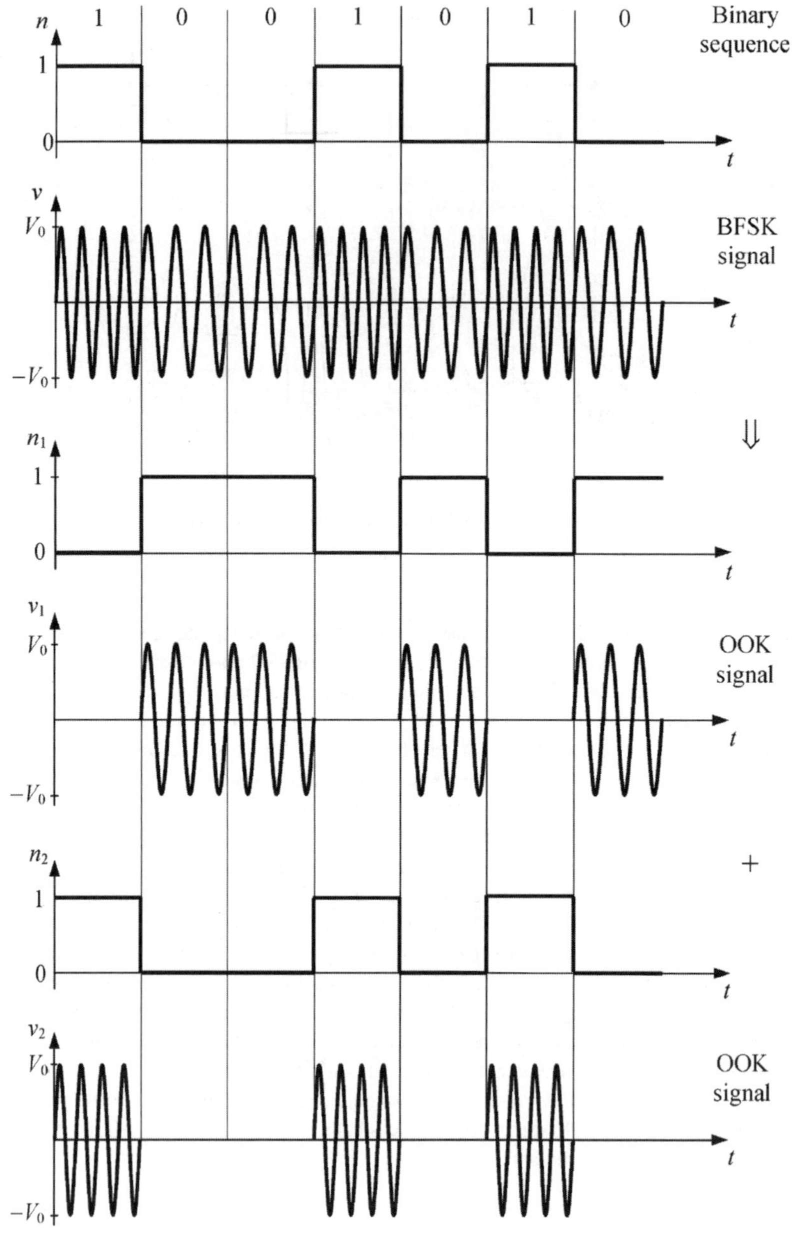

FIGURE 6.33 BFSK signal as sum of two OOK signals.

where ϕ_0 is the initial phase. Hence, there are two possible locations on an I-axis of the I/Q diagram, so a binary one or zero can be sent, and the symbol rate is one bit per symbol. Figure 6.35(a) shows the binary sequence 10010 that generates a BPSK waveform. The PSK spectrum generally contains the carrier and sideband components located symmetrically at offset frequencies multiple to a bit rate f_m. However, for a phase keying of $\pm$ 180°, the carrier component is absent, and BPSK spectrum shown in Figure 6.35(b) becomes similar to an ASK spectrum with suppressed carrier, with zero

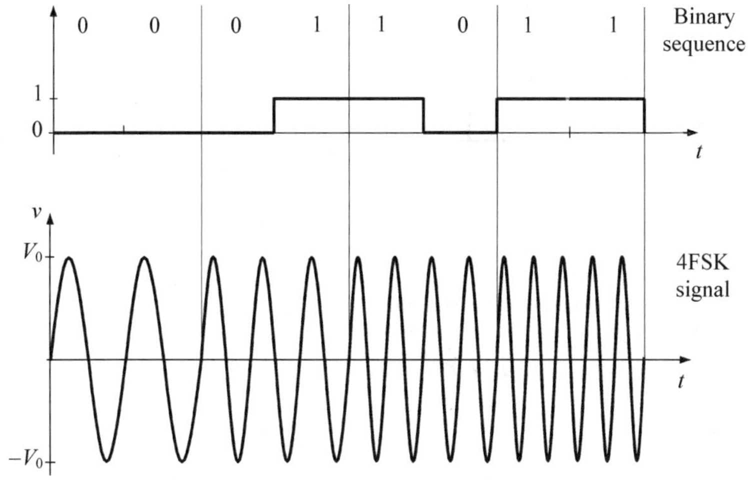

FIGURE 6.34 Binary sequence and 4FSK modulation.

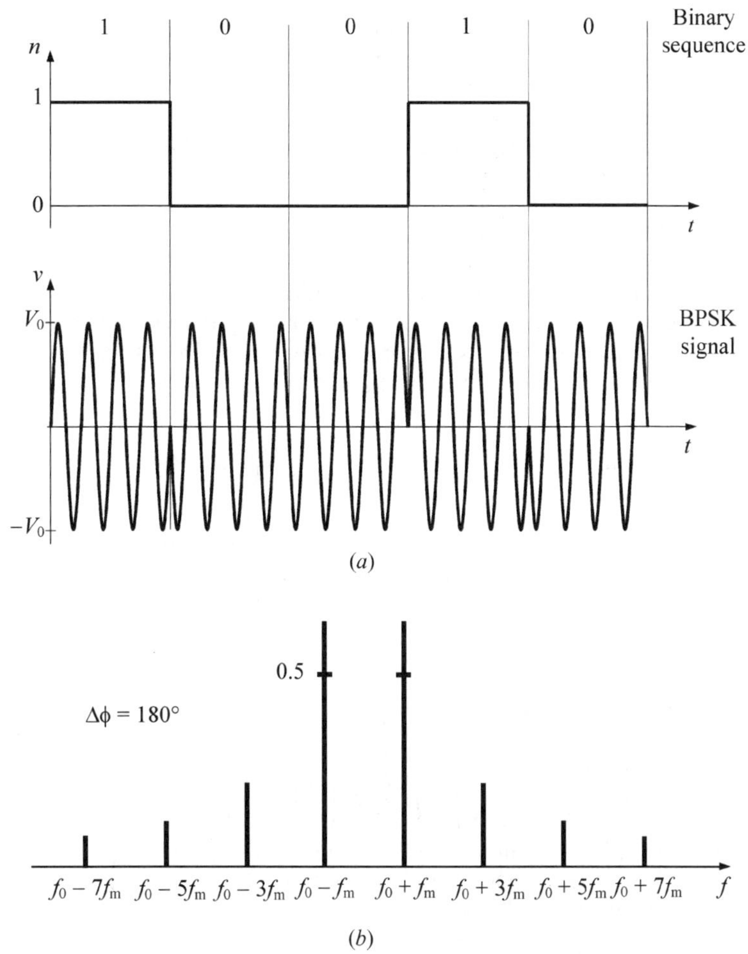

FIGURE 6.35 Binary modulating signal, BPSK signal, and spectrum.

crossing corresponding to the even-order sidebands and increased level of the odd-order sideband components, both having approximately the same bandwidths.

An *m-ary phase shift keying* (MPSK) signal can be defined as

$$v(t) = V_0 \cos\left(\omega_0 t + 2\pi \frac{i}{m} + \phi_0\right) \tag{6.66}$$

where $i = 1, 2, \ldots, m$. In MPSK, n bits are combined and transmitted as a symbol, with a total of $m = 2^n$ different symbols, and each symbol is a different combination of n bits. Each transmitted symbol is represented by a different phase of the signal. Therefore, the signal can have up to m phase changes occurring every symbol period. The transmitted signal in Eq. (6.66) can be rewritten using a trigonometric identity as

$$v(t) = V_0 \cos\left(2\pi \frac{i}{m}\right) \cos\left(\omega_0 t + \phi_0\right) - V_0 \sin\left(2\pi \frac{i}{m}\right) \sin\left(\omega_0 t + \phi_0\right). \tag{6.67}$$

Since PSK requires coherent demodulation with phase reference, *differential phase shift keying* (DPSK) avoids the requirement for an accurate local oscillator phase by using the phase during the immediately preceding symbol interval as the phase reference [17]. As long as the preceding phase is received correctly, the phase reference is accurate. There are a variety of encoding techniques for implanting a DPSK system. One technique, which can be used for one-bit-at-a-time transmission, is to obtain a differential binary sequence from the input binary sequence and then assign phases to the bits in the differential sequence. The differential binary sequence is generated by repeating the preceding bit in the differential sequence if the message bit corresponds to logic 1 or by changing to the opposite bit if the message bit corresponds to logic 0. Phases are then assigned to the differential binary sequence by transmitting 0° for logic 1 and 180° for logic 0. Another common DPSK encoding scheme is to group bits into singles, pairs (dibits), or triples (tribits), and then associate a particular phase change with each group.

In *quadrature phase shift keying* (QPSK) or 4PSK with $n = 2$ and $m = 4$, two bits are combined to form a symbol before transmission when each symbol is represented by a different phase of the transmitted signal and phase change occurs every symbol period. The phases are usually spaced equally so that the transmitted phases for 4PSK signal are 90° apart. Figure 6.36 shows the binary

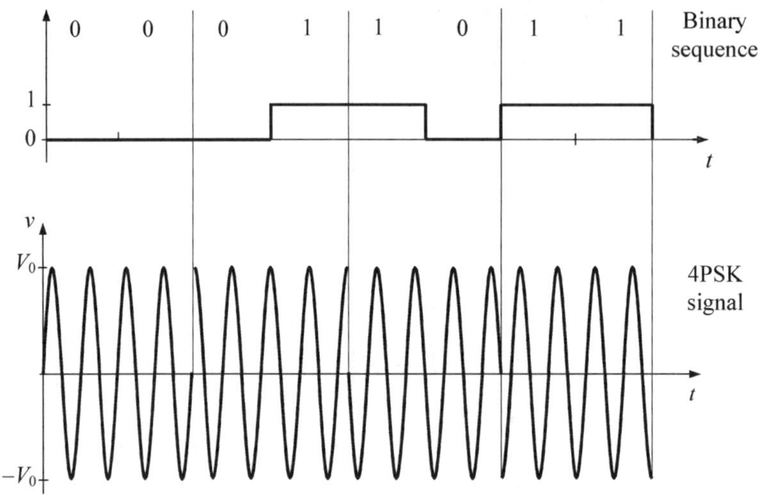

FIGURE 6.36 Binary sequence and 4PSK modulation.

sequence and 4PSK modulation according to Eq. (6.67) with $\phi_0 = -90^\circ$. Any sudden phase change results in non-constant envelope. The more abrupt the phase change, the more envelope changes and the wider the spectrum is, with the worst case of a 180° phase change. To minimize the power in spectral adjacent channels, pulse shaping is usually applied. In 8PSK, there are eight states of points spaced equally with 45°, 16PSK corresponds to the signal with equal phase spacing of 22.5°, and so forth.

It is convenient to represent a PSK signal, as well as some other types of the modulated signal, in the form of a phasor or constellation diagram. Using this scheme, the phase of the signal is represented by the angle around the circle, and the amplitude by the distance from the origin or center of the circle. In this way, the signal can be resolved into quadrature components representing the sine for in-phase (I) component and the cosine for the quadrature (Q) component. Most PSK systems use constant amplitude, and therefore, points appear on one circle with constant amplitude and the changes in state being represented by movement around the circle. For a BPSK using phase reversals, the constellation diagram represents two points appearing at opposite points on the I axis, as shown in Figure 6.37(a). Other forms of PSK may use different points on the circle, and there can be more points on the circle. The constellation diagram for a QPSK signal with initial phase started at 45° is shown in Figure 6.37(b). Since a rotation of this constellation diagram has no effect on modulation performance, it can also start at zero phase, as shown in Figure 6.37(c). Figure 6.37(d) shows the trajectories for an 8PSK signal, where the bits are mapped to possible phases according to the *Gray coding*. In this case, the bits are numbered such that each adjacent phase means just bit difference so that when a phase mistake is made and the most likely one is the nearest phase, then only one bit is decoded incorrectly. Using a constellation view of the signal enables quick fault-finding in a system.

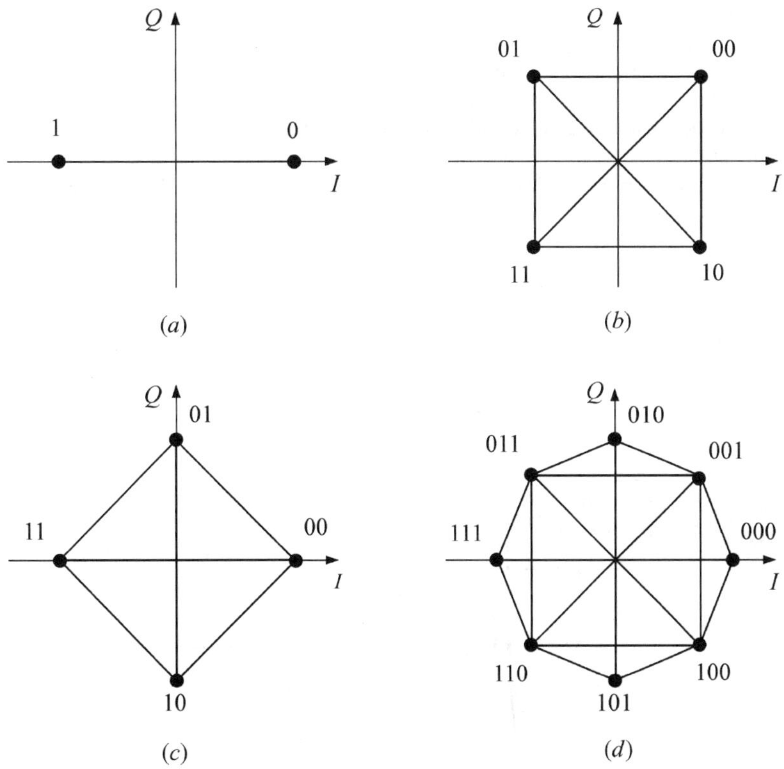

FIGURE 6.37 Trajectories for (a) BPSK, (b) QPSK, (c) shifted QPSK, (d) 8PSK.

If the problem is related to phase, the constellation will spread around the circle. If the problem is related to magnitude, the constellation will spread off the circle, either towards or away from the origin. These graphical techniques assist in isolating problems much faster than when using other methods.

A form of QPSK is used in the Digital Advanced Mobile Phone System (DAMPS) cellular architecture, as well as for the forward link from the base station to the handset in the IS-95 Code Division Multiple Access (CDMA) cellular system. On the reverse link from the handset to the base station, *offset-quadrature phase shift keying* (OQPSK) is used to prevent transitions through the origin. Figure 6.38 shows the block diagram of an OQPSK modulator where a serial bit stream is transformed by a serial-to-parallel (S/P) converter to separate I and Q bit streams. In OQPSK, one vector component is delayed by one bit period (one half of a symbol period), so the vector will move down first and then over, thus avoiding moving through the origin. In this case, the range of phase transitions is 0° and 90° when the possibility of a phase of 180° is eliminated and occurs twice as often, but with half intensity of the QPSK. If a QPSK modulated signal undergoes filtering to reduce the spectral side lobes, the resulting waveform will no longer have a constant envelope, and the occasional 180° shifts in phase will cause the envelope to go to zero momentarily. Moreover, a high nonlinearity of the active device transfer characteristic in this operation region close to pinch-off tends to introduce a significant nonlinear distortion in the signal transmission through the power amplifier. However, when an OQPSK signal undergoes bandlimiting, the resulting intersymbol interference causes the phase transition of ±90°, and the envelope will never go to zero.

Another method to avoid the transitions through the origin is to use the *differential phase shift keying* (DPSK) system when the information is not carried by the absolute value, but is carried by the transitions between states. A DPSK transmission system can have transitions from any symbol position to any other symbol position. The $\pi/4$-DPSK modulation format where the carrier trajectory does not go through the origin is widely used in many wireless applications such as IS-54 North American Digital Cellular (NADC) standard, cordless Personal Handyphone System (PHS), or trunked radio system TETRA (Trans European Trunked Radio). It is based on two QPSK constellations, being offset between each other by 45°, where transitions must occur from one constellation to the other. The data are encoded in the magnitude and direction of the phase shift, but not in the absolute position on the constellation.

This is achieved by using the two QPSK modulators running in parallel and using one symbol from each one each time (2τ), as shown in Figure 6.39(a). From the signal constellation diagram

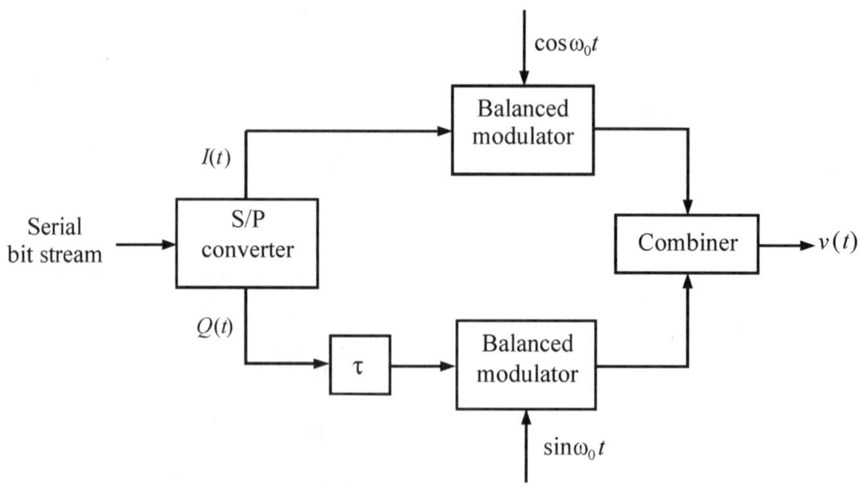

FIGURE 6.38 Block diagram of OQPSK modulator.

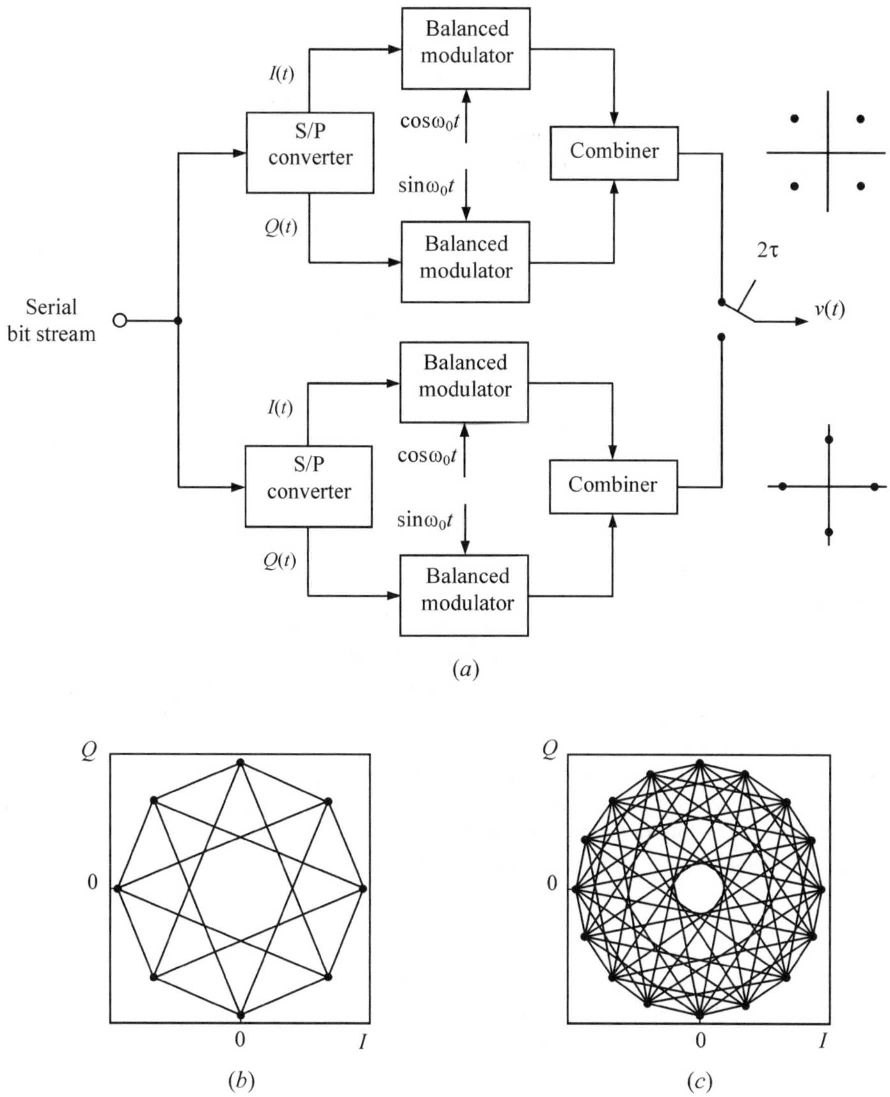

FIGURE 6.39 Block and constellation diagrams for $\pi/4$-QPSK and $3\pi/8$-8PSK.

shown in Figure 6.39(*b*), it is possible to obtain that the maximum phase change is 135°. The $3\pi/8$-8PSK modulation scheme is similar to $\pi/4$-DPSK in the sense that rotation of the constellation occurs from one time interval to the next. This time, however, the rotation of the constellation from one symbol to the next is $3\pi/8$. This modulation scheme is used in the EDGE (Enhanced Data Rates for GSM Evolution) system and provides 3 bits per symbol. Its constellation diagram where the continuous rotation of the symbols by a $3\pi/8$ offset prevents the signal from crossing through the origin is shown in Figure 6.39(*c*). This condition can be viewed as having two 8PSK constellation planes offset by $3\pi/8$ and swapping from one plane to the other at every consecutive symbol time.

Generally, filtering allows the transmitted bandwidth to be significantly reduced without losing the content of the digital data. There are many different varieties of filtering techniques, among which

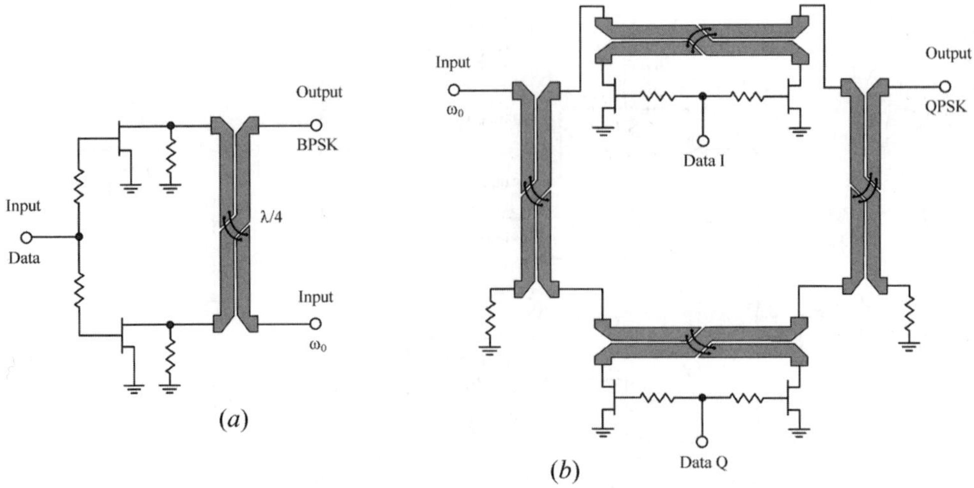

FIGURE 6.40 Circuit topologies of single-stage and balanced BPSK modulators.

the most common are the raised-cosine, the square-root raised-cosine and the Gaussian filters [27]. The raised-cosine filters, being a class of Nyquist filters, have the property that their impulse response rings at the symbol rate. This means that time response of the filter goes through zero with a period that exactly corresponds to the symbol spacing. Nyquist filters heavily filter the signal without blurring the symbols together at the symbol times. This is important for transmitting information without errors caused by inter-symbol interference. Usually, the filter is split, with one half in the transmit path and the other half in the receiver path, being commonly called the square-root raised-cosine filters, so that their combined response is that of a Nyquist filter. The Gaussian filters used in GSM (Global System for Mobile communications or Global Speciale Mobile) represents a Gaussian shape in both the time and frequency domains, without ringing inherent in the raised-cosine filter. As a result, its effects in the time domain are relatively short and each symbol interacts significantly with only preceding and succeeding symbols.

Accurate bi-phase modulation with equal amplitudes and 180° phase shift can be obtained in a reflection-type modulator if the switched device (*p–i–n* diode, MESFET, or HEMT) provides zero impedance when it is turned on and infinite impedance when it is turned off. Figure 6.40(*a*) shows the simplified schematic of a bi-phase BPSK modulator based on a hybrid quadrature Lange directional coupler and microwave MESFET devices in a switching-mode operation [28]. To make equal reflections from MESFETs in two states by minimizing the device parasitic capacitive and inductive elements, a thin-film resistor can be placed between drain and source terminals of each MESFET. To achieve a broadband modulator performance with a reflected power balance of ±1 dB and a bi-phase difference of 180° ± 10° over the full octave range from 4 to 8 GHz, a tandem directional coupler can be used. A quadri-phase QPSK modulator can be designed using two bi-phase BPSK modulators, a 90° hybrid input splitter, and a 90° output hybrid combiner, as shown in Figure 6.40(*b*). Near perfect amplitude and phase performance in millimeter-wave applications can be achieved by using pHEMT devices acting as switches. In this case, a multifunctional *I*/*Q* vector modulator for multilevel modulation schemes can be built using two quadri-phase QPSK modulators with a 90° hybrid input splitter and an output in-phase Wilkinson combiner [29].

6.5.4 Minimum Shift Keying

The power spectral density of a PSK signal represents a $\text{sinc}^2(x)$ spectrum with the main lobe centered at the carrier frequency and side lobes of decreasing amplitudes extending on the neighboring channels, the most important of which are adjacent and alternate, creating interference with these channels, which is called the *spectral regrowth*. Spectral regrowth can be reduced by using a modulation format called the *minimum shift keying* (MSK), which occurs when the phase changes of the transmitted signal are smooth as opposed to instantaneous changes in QPSK. Smooth phase changes in MSK are provided by filtering the digital data before modulation, thus resulting in the side lobes much lower in amplitude compared to BPSK or QPSK. However, the addition of a filter in the data path creates symbol distortion, and therefore, the choice of filtering must take into consideration the tradeoff between symbol distortion and side lobe suppression [30].

MSK can be mathematically understood as an OQPSK with a special pulse shaping in such a way that the transitions between symbols are smoothed and there are no abrupt changes in phase or frequency. If sinusoidal pulses are employed instead of rectangular shapes, the modified signal can be defined as MSK and equals to

$$v(t) = a_I(t)V_0 \cos\left(\frac{\pi}{2\tau}t\right) \cos \omega_0 t + a_Q(t)V_0 \sin\left(\frac{\pi}{2\tau}t\right) \sin \omega_0 t \tag{6.68}$$

where a_I and a_Q are two data streams consisting of even and odd bits, being separated from the binary input bit stream arriving at a rate of $1/\tau$ baud. Figure 6.41(*a*) shows the modified in-phase bit stream waveform with the corresponding amplitudes $\pm$ 1 shown inside waveforms. The in-phase carrier, representing by the first term in Eq. (6.68), is obtained by multiplying the waveform in Figure 6.41(*a*) by $\cos\omega_0 t$, as shown in Figure 6.41(*b*). Similarly, the sinusoidally shaped odd-bit stream and the quadrature carrier, multiplied by $\sin\omega_0 t$, are shown in Figures 6.41(*c*) and 6.41(*d*), respectively. The composite MSK signal $v(t)$, the addition of Figures 6.41(*b*) and 6.41(*d*), is shown in Figure 6.41(*e*).

To better understand the waveform shown in Figure 6.41(*e*), Eq. 6.68 can be rewritten using a well-known trigonometric identity as

$$v(t) = V_0 \cos\left[\omega_0 t + b_k(t)\frac{\pi}{2\tau}t + \phi_k\right] \tag{6.69}$$

where b_k is equal to $+1$ when a_I and a_Q have opposite signs and b_k is equal to -1 when a_I and a_Q have the same sign, and ϕ_k is equal to 0 or π corresponding to $a_I = +1$ or -1, respectively. Note that b_k(t) can also be written as $-a_I(t)a_Q(t)$.

From Eq. (6.69) and Figure 6.41(e), the following basic properties of MSK can be derived:

1. It has constant envelope.
2. There is phase continuity in the RF carrier at the bit transition instants.
3. The signal represents an FSK signal with signaling frequencies $f_+ = f_0 + 1/(4\tau)$ and $f_- = f_0 - 1/(4\tau)$, where $f_0 = \omega_0/(2\pi)$, with the frequency deviation equals to half the bit rate, that is, $\Delta f = f_+ - f_- = 1/(2\tau)$. This is the minimum frequency spacing that allows the two FSK signals to be coherently orthogonal. Since the frequency spacing is only half as much as the conventional $1/\tau$ spacing used in noncoherent detection of FSK signals, MSK is also referred to as the *fast frequency shift keying* (FFSK).
4. The excess phase of the MSK signal, referenced to the carrier phase, is given by the term

$$\varphi(t) = b_k(t)\frac{\pi}{2\tau}t = \pm\frac{\pi}{2\tau}t$$

in Eq. (6.69), which increases or decreases linearly during each bit period of τ seconds. A bit b_k of $+1$ corresponds to an increase of the carrier phase by 90° and corresponds to an FSK signal at the higher frequency f_+. Similarly, $b_k = -1$ implies a linear decrease of phase by 90° over τ seconds, corresponding to the lower frequency f_-. In order to make the phase continuous at bit transitions, the carrier frequency f_0 should be chosen such that f_0 is an integral multiple of $1/(4\tau)$, one-fourth the bit rate.

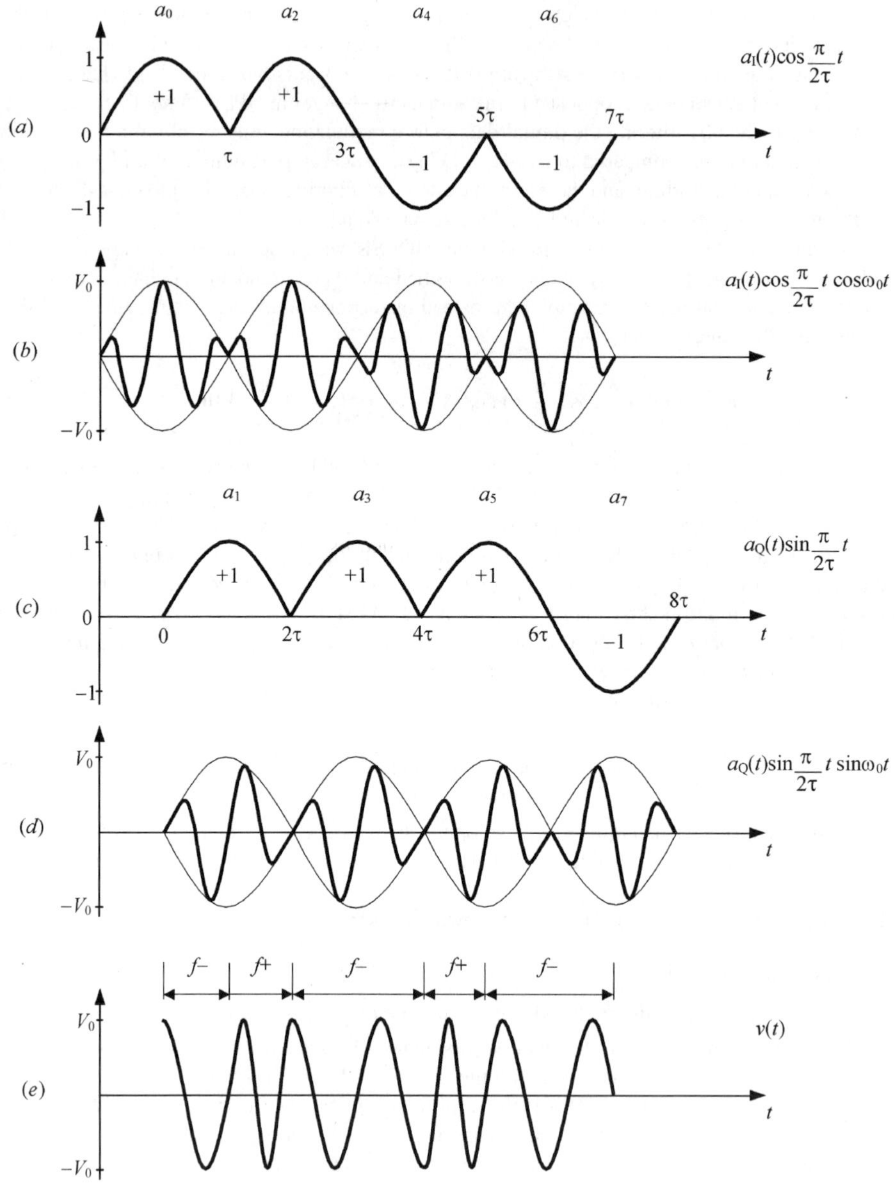

FIGURE 6.41 Binary modulating signal and MSK signal.

Thus, MSK can be viewed either as an OQPSK signal with cosine pulse weighting or as a *continuous phase frequency shift keying* (CPFSK) signal with a frequency separation equal to one-half the bit rate. Figure 6.42 shows the block diagram of an MSK modulator corresponding to Eq. (6.68).

By using a Gaussian pulse shaping, the MSK spectral performance can be improved even further in terms of lower side lobes as a result of smoother transitions between symbols with optimum filter roll-off factor to compromise between a bit rate and out-of-band interference. The modulation obtained this way is called the *Gaussian minimum shift keying* (GMSK). In GMSK modulation, a data stream is passed through a Gaussian filter and the filtered response drives an I/Q modulator,

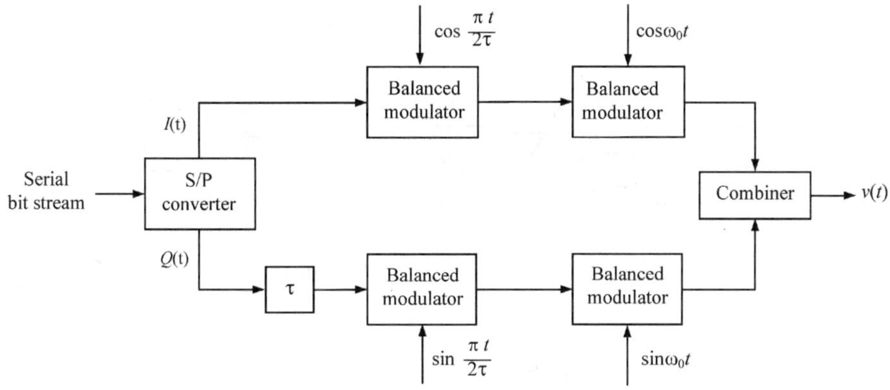

FIGURE 6.42 Block diagram of MSK modulator.

with similar architecture as shown in Figure 6.42. GMSK can also be achieved by using a $\pi/2$-shift BPSK or CPFSK modulator driving the voltage-controlled oscillator (VCO) included in a phase-locked loop together with Gaussian filter [31,32]. GSMK is used in such communication systems as GSM850/900, DECT (Digital European Cordless Telephone), CDPD (Cellular Digital Packet Data), DCS1800 (Digital Communication System in the 1800 MHz band), or PCS1900 (Personal Communications Services in the 1900 MHz band).

6.5.5 Quadrature Amplitude Modulation

The above described digital modulation schemes use either amplitude or phase carrier changes to transmit binary data stream, with the constellation points lying on a circle of constant amplitude. However, much more information can be transmitted if both amplitude and phase are varied, as it can be done using *quadrature amplitude modulation* (QAM), based on a proper digital processing. An *m-ary quadrature amplitude modulation* (MQAM) signal can be defined as

$$v(t) = V_i \cos(\omega_0 t + \phi_i + \phi_0) = V_i \cos(\phi_i)\cos(\omega_0 t + \phi_0) - V_i \sin(\phi_i)\sin(\omega_0 t + \phi_0) \tag{6.70}$$

where V_i are the amplitudes, ϕ_i are the phases, and ϕ_0 is the initial phase, $i = 1, 2, \ldots, m$ [33,34]. Figure 6.43(a) shows the block diagram of the MQAM modulator transmitting a bit stream with n bits and m symbols. The QAM constellation is produced by separately amplitude modulating the I and Q carriers, which have the same frequency but are 90° out of phase. The most common form of QAM is a square QAM, or a rectangular QAM with equal numbers of I and Q states, where the constellation points are usually arranged in a square grid with equal vertical and horizontal spacing.

In 16-state QAM (16QAM), there are four I values and four Q values resulting in a total of 16 possible states for the signal. It can transit from any state to any other state at every symbol time. Since $16 = 2^4$, four bits per symbol can be sent, with symbol rate equal to one-fourth of the bit rate. Note that 2QAM is in fact BPSK, as well as 4QAM is the same as 4PSK or QPSK, which follows from the constellation diagrams shown in Figures 6.37(b) and 6.43(b). Also, the error-rate performance of 8QAM is close to that of 16QAM, but its data rate is only three-quarters that of 16QAM. Therefore, the most common forms are 16QAM, 64QAM, 128QAM, and 256QAM. By moving to a higher order constellation, it is possible to transmit even more bits per symbol. However, the symbols become very close to each other, and are thus more subject to errors due to noise and distortion. Figure 6.43(c) shows the constellation diagram for 16QAM with a Gray coded bit assignment. 64-QAM and 256-QAM are often used in digital cable television and cable modem applications.

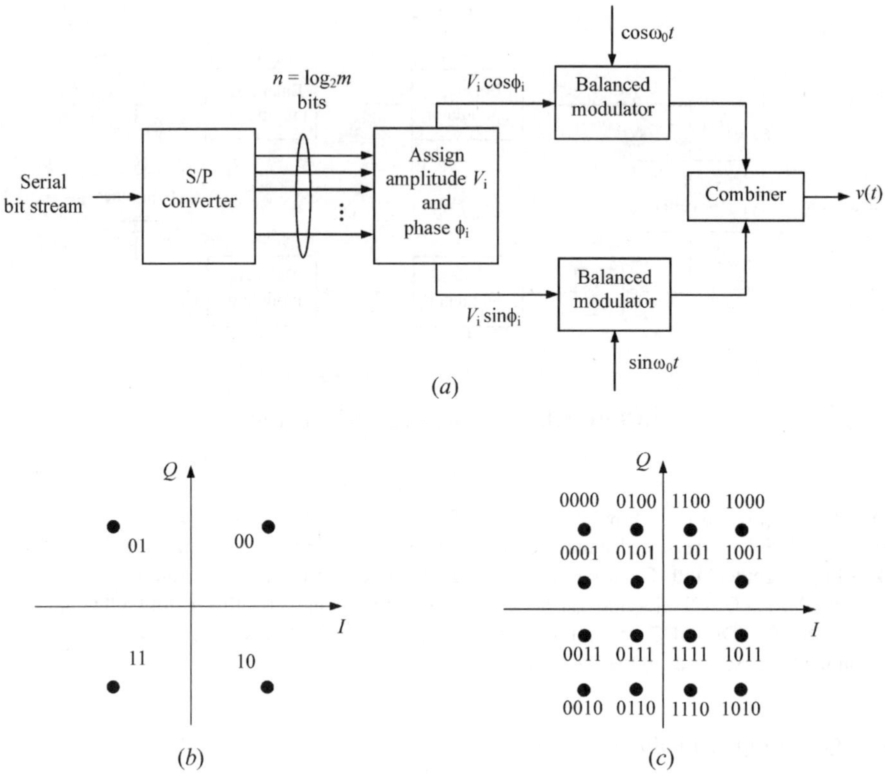

FIGURE 6.43 Block diagram of QAM modulator and constellation diagrams.

6.5.6 Pulse Code Modulation

Pulse code modulation (PCM) was originally developed in 1939 by Alec Reeves for voice communication as a method for transmitting digital signals over analog communications channels [35]. This is generally a method of transmitting continuous signals in which the amplitude of the signal is sampled regularly at uniform intervals and then quantized to a series of symbols in a numeric (usually binary) code. Pulse-code modulation is very popular because of many advantages it offers such as inexpensive digital circuitry required in the system, all-digital transmission, possible encryption with further digital processing, or error minimization by appropriate coding of the signals. PCM variants are based on different mathematical techniques for quantization, including linear, logarithmic, and adaptive. Quantization is the process of converting a continuous amplitude into one of a finite number of discrete amplitudes.

Figure 6.44 shows how an analog signal in the form of a sine wave is converted (sampled) into 16 amplitude levels with equal spacing by a 16-level quantizer. For each sample, one of the available values shown on the *Y*-axis is chosen by some algorithm (in this case, the floor function is used when floor(x) is the largest integer not greater than x). This produces a fully discrete representation of the input signal (shaded area) that can be easily encoded as digital data for storage or manipulation. For this sine wave signal, we can verify that the quantized values at the sampling moments are 7, 9, 11, 12, 13, 14, 14, 15, 15, 15, 14, and so on. Encoding these values as binary numbers would result in the following set of nibbles (or half-octets): 0111, 1001, 1011, 1100, 1101, 1110, 1110, 1111, 1111, 1111, 1110, and so on. These digital values could then be further processed or analyzed by a purpose-specific digital signal processor or general purpose central processing unit (CPU). Several PCM data streams

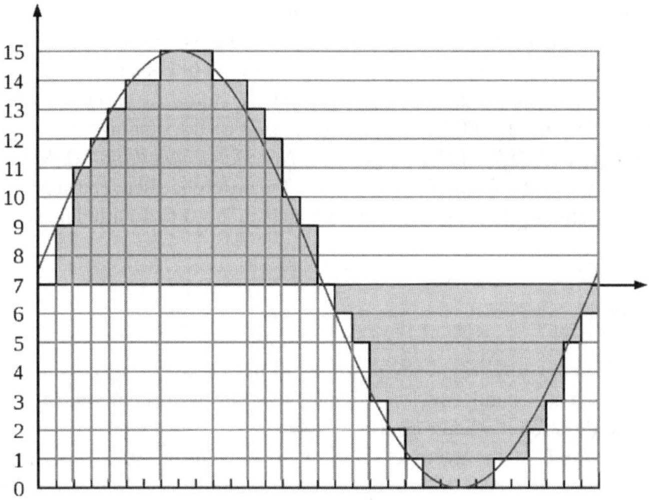

Digit	Binary code	Gray code
0	0000	0000
1	0001	0001
2	0010	0011
3	0011	0010
4	0100	0110
5	0101	0111
6	0110	0101
7	0111	0100
8	1000	1100
9	1001	1101
10	1010	1111
11	1011	1110
12	1100	1010
13	1101	1011
14	1110	1001
15	1111	1000

FIGURE 6.44 Signal sampling and quantization for 4-bit PCM.

could also be multiplexed into a larger aggregate data stream, generally for transmission of multiple streams over a single physical link. There are many ways to implement a real device that performs this task. In real systems, such a device is commonly implemented on a single integrated circuit that lacks only the clock necessary for sampling, and is generally referred to as an ADC (analog-to-digital converter). These devices will produce on their output a binary representation of the input whenever they are triggered by a clock signal, which would then be read by a processor of some sort.

Generally, quantization reduces the degree of accuracy of representation of the sampled signal, which is characterized by a quantization error or noise. To evaluate the quantizer performance, one commonly used measure is the quantizer signal-to-noise ratio (*SQNR*). Assuming a uniform distribution of input signal values, the quantization noise is a uniformly-distributed random signal with peak-to-peak amplitude of one quantization level, making the amplitude ratio $2^n/1$. Then, if the input to the quantizer is a full-scale sine wave signal (that is, the quantizer is designed such that it has the same minimum and maximum values as the input signal), the quantization noise approximates a sawtooth wave with peak-to-peak amplitude of one quantization level and uniform distribution [17]. As a result, for a n-bit quantizer,

$$SQNR\,(\text{dB}) = 20n \log_{10} 2 + 1.76 = 6.02n + 1.76 \tag{6.71}$$

Thus, for an 8-bit quantizer, the *SQNR* for a full-scaled sine wave is about 50 dB, and each extra quantization bit increases the *SQNR* by roughly 6 dB. In practical digital telephone systems, $256 = 2^8$ levels are used to keep quantization error to an acceptable level, whereas $65{,}536 = 2^{16}$ levels are used for the CD digital system.

If the quantized samples are transmitted directly over a channel, such a process is simply called the *pulse amplitude modulation* (PAM) process in the transmission system. In PCM system, the quantized signal is coded into a block of digits for transmission. The decimal-to-binary conversion can be done in various ways, for example, using two simple coding rules based on the natural binary and Gray codes, as shown in Figure 6.44 for a 16-level sample into 4 binary digits. In the Gray coding, adjacent levels differ by only a single bit, and hence, for equally likely bit errors in each position, a single channel bit error is more likely to produce an adjacent output level than in other codes.

Some forms of PCM combine signal processing with coding. *Differential pulse code modulation* (DPCM) encodes the PCM values as differences between the current value and the predicted value.

An algorithm predicts the next sample based on the previous samples, and the encoder stores only the difference between this prediction and the actual value. If the prediction is reasonable, fewer bits can be used to represent the same information. For audio signals, this type of encoding reduces the number of bits required per sample by about 25% compared to PCM. *Adaptive differential pulse code modulation* (ADPCM) is a variant of DPCM that varies the size of the quantization step, to allow further reduction of the required bandwidth for a given *SQNR*. *Delta modulation* (DM or Δ-modulation), another variant, is the simplest form of DPCM where the difference between successive samples is encoded into n-bit data streams, and the transmitted data is reduced to a 1-bit data stream.

6.6 CLASS-S MODULATOR

A class-S modulator is a high-efficiency low-frequency power amplifier based on *pulse-width modulation* (PWM), and its output is a baseband envelope signal including a dc component [5,36]. Pulse-width modulation (or pulse-duration modulation) uses a square-wave signal whose pulse width is modulated resulting in the variation of the average value of the waveform. Such a modulator produces only positive output current and, therefore, requires only a single transistor and a single diode that is required to provide a suitable reverse-direction path, as shown in Figure 6.45. The output voltage can have any value between nearly zero voltage and the supply voltage V_{cc}. The low-pass filter (LPF) should have high impedance to the switching (sampling, or clock) frequency and its harmonics to provide a high level of spectral purity of the output RF signal. Usually, it is preferable to use a first-order LPF, since further increasing the number of its elements improves the modulator performance insignificantly, but significantly increases size and complexity of a Class-S modulator. When the transistor is turned on, the supply voltage V_{cc} is connected to the input of the LPF, and the load current flows through the device. However, when transistor is turned off, the load current flows through the diode, and the voltage at the LPF input is equal to the voltage across the diode.

In an ideal operation mode, the active device never experiences simultaneous nonzero voltage and nonzero current resulting in 100% efficiency. However, in practical implementation, the losses in the diode and transistor degrade efficiency due to the finite values of their on-resistances and finite rise and fall times, especially at higher operation frequencies. To minimize the intermodulation distortion of the envelope signal, it is necessary to choose correctly the LPF parameters for a maximally flat frequency response, which can be calculated from

$$L = 0.7\frac{R_L}{\pi f_c} \qquad C = \frac{1}{2.8\pi f_c R_L} \tag{6.72}$$

where f_c is the cutoff frequency of the low-pass filter.

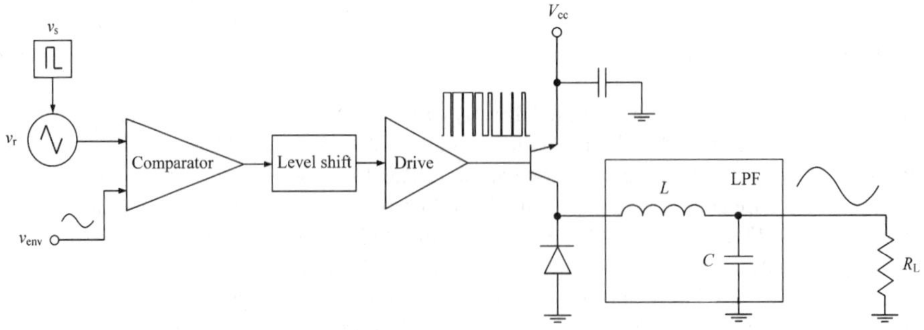

FIGURE 6.45 Schematic of Class-S modulator.

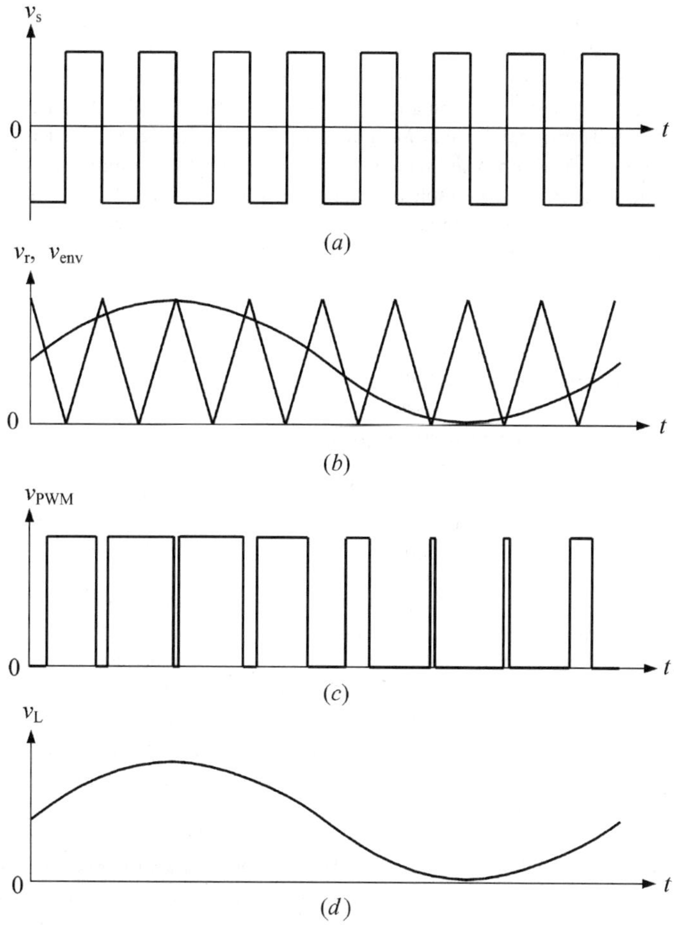

FIGURE 6.46 Waveforms of pulse-width modulation of envelope signal.

Generally, the PWM signal can be accomplished by several techniques, the most popular of which is a comparator method shown in Figure 6.45. In this case, a comparison of the envelope input to a triangular reference wave shown in Figure 6.46(*b*) by a comparator produces a PWM switching signal, the width of which varies with envelope amplitude, as shown in Figure 6.46(*c*). The comparator produces maximum output when the input signal is larger than the triangular wave and zero output when the input signal is smaller than the triangular wave. The triangular wave can be supplied directly from a function-generator circuit or obtained by integration of the output of a switching generator with rectangular-wave signal, which is shown in Figure 6.46(*a*). The maximum modulation frequency depends strongly on the switching frequency required for a PWM process. To keep the spurious products in the output spectrum at least 30–40 dB below the carrier, the switching frequency must four or five times exceed the highest modulation frequency. In practice, to restore the input envelope with minimum level of intermodulation components using an LPF, as shown in Figure 6.46(*d*), it is better to use a factor of 10. An efficiency of about 90% can be achieved for a Class-S modulator with an envelope bandwidth up to 150 kHz [36,37]. However, a PWM is an inherently nonlinear process that generates intermodulation components with as higher level as wider RF signal bandwidth [38].

Therefore, for the envelope-varying systems with wider bandwidths and stronger requirements for the intermodulation distortion levels, it is preferable to use a Class-S modulator based on a low-pass delta–sigma modulation ($\Delta\Sigma$-modulation or DSM) scheme. For the same switching frequency (or oversampling ratio), a system based on DSM can provide wider bandwidth and lower distortion than a PWM system. The signal in a delta–sigma modulator is digitized by a quantizer (single-bit comparator), the output of which is subtracted from the input signal through a digital feedback loop acting as a discrete-time sharp low-pass filter and quantization noise is forced outside of the band of interest [39,40]. The degree of suppression of the quantization noise depends on an oversampling ratio, which is the ratio of a sampling frequency to twice the bandwidth or highest frequency of the signal being sampled.

6.7 MULTIPLE ACCESS TECHNIQUES

6.7.1 Time and Frequency Division Multiplexing

In the history of electrical communications, the earliest reason for sampling a signal was to interlace samples from different telegraphy sources, and transmit them over a single telegraph cable. The electrical engineer W. Miner used an electromechanical commutator for time-division multiplex of multiple telegraph signals in 1903, and also applied this technology to telephony where he obtained intelligible speech from channels sampled at a rate above 3500–4300 Hz by means of a pulse-amplitude modulation rather than PCM. Thus, *time division multiplexing* (TDM) is alternate process in which a number of signals or bit streams are transmitted simultaneously in one communication channel by letting the signals occupy the same frequency band on a time-sharing basis. In such systems the transmitter is switched or commutated to each signal channel sequentially. In this case, the time domain is divided into several recurrent timeslots of fixed length, one for each subchannel, when a sample byte or data block of subchannel 1 is transmitted during timeslot 1, subchannel 2 during timeslot 2, and so on. The receiving system must then be switched in synchronism with the transmitter to separate the various signals prior to final demodulation. In circuit switched networks such as the Public Switched Telephone Network (PSTN) there exists the need to transmit multiple subscribers' calls along the same transmission medium, and to accomplish this, network designers make use of TDM that allows switches to create channels within a transmission stream.

Frequency division multiplexing (FDM) is a form of signal multiplexing that involves assigning non-overlapping frequency ranges to different signals. Since the portion of the spectrum to be utilized is determined by the carrier frequency, different signals can modulate carriers of different frequencies and all of them can be transmitted simultaneously. The receiver can choose the desired signal band by means of selective filters. FDM can be also used to combine multiple signals before final modulation onto a carrier wave. In this case, the carrier signals are referred to as subcarriers. For example, in stereo FM transmission, a 38-kHz subcarrier is used to separate the left–right difference signal from the central left–right sum channel, prior to the frequency modulation of the composite signal. A television channel is divided into subcarrier frequencies for video, color, and audio. Digital Subscribe Line (DSL) uses different frequencies for voice and for upstream and downstream data transmission on the same conductors.

Orthogonal frequency division multiplexing (OFDM) represents an FDM scheme utilized as a digital multicarrier modulation method where a large number of closely-spaced orthogonal subcarriers are used for data transmission [41]. In this case, the data is divided into several parallel data streams or channels, one for each subcarrier, and the subcarrier is modulated with a conventional modulation scheme such as QAM or PSK at a low symbol rate, maintaining total data rates similar to conventional single-carrier modulation schemes in the same bandwidth. OFDM has developed into a popular scheme for wideband digital communication used in applications such as digital television and audio broadcasting, wireless networking and broadband internet access. In OFDM, the subcarriers are chosen so that they are orthogonal to each other, meaning that cross-talk between the subchannels is eliminated and intercarrier guard bands are not required. This greatly simplifies the design of both the transmitter and the receiver when a separate filter for each subchannel is not required. As a result, an OFDM carrier signal is the sum of a number of orthogonal subcarriers, with baseband data on each subcarrier being independently modulated commonly using some type of QAM or PSK.

6.7.2 Frequency Division Multiple Access

When FDM is used as to allow multiple users to share a physical communication channel, it is called the *frequency division multiple access* (FDMA), which is the traditional way of separating radio signals from different transmitters. FDMA gives users an individual allocation of one or several frequency bands, allowing them to utilize the allocated radio spectrum without interfering with each other and coordinate access between multiple users. In FDMA, a predetermined frequency band is available for the entire period of communication, with stream data that may not be packetized. In the commercial radio broadcast bands, 535–1705 kHz for AM and 88–108 MHz for FM, each local broadcast station (user) is assigned to a specific slice of spectrum within the frequency band allocated for that purpose. As long as the station broadcasts, no other radio station in the same area can use that radio frequency bandwidth to send a signal. Another broadcast station can use that same bandwidth only when the distance between the stations is sufficient to reduce the risk of interference. Therefore, FDMA requires high-performing filters in the radio systems. In addition, because adjacent channel interference is an important factor in channel quality, frequency planning is a key consideration when selecting fixed or base station locations.

It is important to distinguish between FDMA and *frequency division duplexing* (FDD). While FDMA allows multiple users simultaneous access to a certain system, FDD refers to how the radio channel is shared between the uplink and downlink (for instance, the traffic going back and forth between a mobile phone and a base station). Similarly, FDM should not be confused with FDMA. The former is a physical layer technique that combines and transmits low-bandwidth channels through a high-bandwidth channel. FDMA, on the other hand, is an access method in the data link layer. FDMA also supports demand assignment in addition to fixed assignment. *Demand assignment* allows all users apparently continuous access of the radio spectrum by assigning carrier frequencies on a temporary basis using a statistical assignment process.

Orthogonal frequency division multiple access (OFDMA) is a multi-user version of the OFDM digital modulation scheme. Multiple access is achieved in OFDMA by assigning subsets of subcarriers to individual users. This allows simultaneous low data rate transmission from several users.

6.7.3 Time Division Multiple Access

Time division multiple access (TDMA) represents a channel access method for shared radio networks. TDMA systems can improve spectrum efficiency compared to FDMA because they allow not only assigning users into an available pair of channels, but also allowing several users to share the same frequency channel by dividing into different timeslots. The users transmit in rapid succession, one after the other, each using own timeslot. This allows multiple stations to share the same radio channel, while using only a part of its channel capacity. However, the user can only send or receive information at that time, regardless of the availability of other timeslots. Information flow is not continuous for any user, but rather is sent and received in bursts that are then reassembled at the receiving end and appear to provide continuous signal due to fast process. TDMA is used in the digital cellular systems such as GSM or DECT (Digital Enhanced Cordless Telecommunications) standard for portable phones, as well as in the satellite systems. Figure 6.47 shows the TDMA frame structure based on a data stream divided into frames with those frames divided into timeslots.

TDMA is a type of TDM when multiple transmitters are used instead of a single transmitter connected to a single receiver. Because TDMA systems also split an allotted portion of the frequency spectrum into smaller slots (channels), they require the same level of frequency planning as FDMA systems. Generally, FDMA can provide extended battery life and talk time, more efficient use by

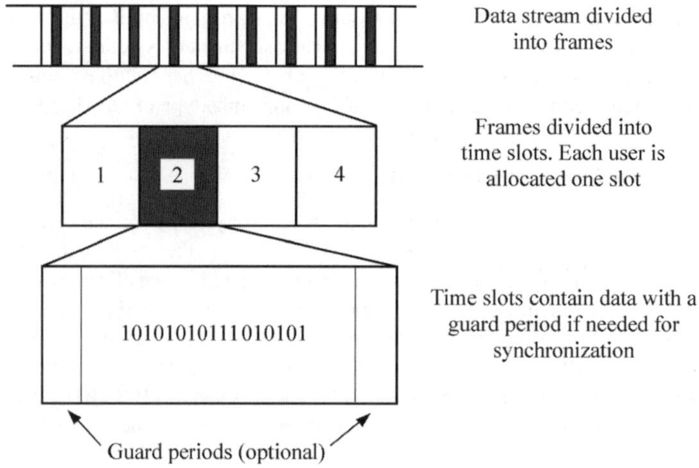

FIGURE 6.47 TDMA frame structure.

accommodating more users in the same spectrum space, efficient utilization of hierarchical cell structures from macrocells to picocells and can efficiently handle video and audio data. However, in TDMA systems, the network and spectrum planning are intensive; they create interference at a frequency that is directly connected to the timeslot length; the frequency guard bands limit the potential bandwidth of a TDMA channel; and dropped calls are possible when users switch in and out of different cells. Handsets that are moving will need to constantly adjust their timings to ensure their transmission is received at precisely the right time, because as they move further from the base station, their signal will take longer to arrive, which also means that the major TDMA systems have hard limits on cell sizes in terms of range.

In the GSM system, the synchronization of the mobile phones is achieved by sending timing advance commands from the base station that instructs the mobile phone to transmit earlier and by how much. The mobile phone is not allowed to transmit for its entire timeslot, but there is a guard interval at the end of each timeslot. As the transmission moves into the guard period, the mobile network adjusts the timing advance to synchronize the transmission.

6.7.4 Code Division Multiple Access

Code division multiple access (CDMA) is a class of modulation that uses specialized codes to provide multiple communication channels in a designated frequency range [42]. It is based on a spread spectrum technique used to increase spectrum efficiency over current FDMA and TDMA systems. In this case, CDMA spreads the information contained in a signal over the entire available bandwidth and not simply through one frequency. Due to the wide bandwidth of a spread-spectrum signal, it is very difficult to cause jamming, difficult to interfere with, and difficult to identify. CDMA systems are widely used in military and cellular applications.

The use of CDMA for terrestrial cellular communications was first conceived in 1985 and was then standardized as IS-95 to become the first commercialized CDMA system. In North American cellular system, CDMA starts with a basic rate of 9600 b/s, and then it spreads to a transmitted bit rate, or chip rate, of 1.2288 MHz [43]. Spreading consists of applying digital codes to the data bits that increase the data rate while adding redundancy to the system. The chips (transmitted bits) are transmitted using a form of a QPSK modulation that is filtered to limit the bandwidth of the signal. When the signal is received, the coding is removed from the desired signal, returning it to a rate of 9600 b/s. The ratio of transmitted bits or chips to data bits is the coding gain, which is equal to

128 or 21 dB and because of which an interference of up to 18 dB above the signal level (3 dB below the signal strength after coding gain) can be tolerated. The CDMA codes are designed to have very low cross-correlation.

The basic differences of CDMA communication systems from the others are:

1. Multiple users share one frequency. In a fully loaded CDMA system, there are about 35 users on each carrier frequency. (Note that there are actually two carrier frequencies per channel, 45 MHz away from each other. One is for the base-mobile link, when it is called the *forward direction*, while the other is for the mobile-base link, which is called the *reverse direction*).
2. The channel is defined by a code. There is a carrier frequency assignment, with a frequency bandwidth of 1.25 MHz.
3. The capacity limit is soft. Additional users add more interference to the system, which can cause a higher data error rate for all users, but this limit is not set by the number of physical channels.

CDMA makes use of multiple forms of diversity such as spatial, frequency, and time diversities. The traditional form of spatial diversity, using multiple antennas, is used for the cell site receiver. Another form of spatial diversity called the *soft handoff* is used during the process of handling off a call from one cell to the next. Soft handoffs allow the mobile telephone to communicate simultaneously with two or more cells, and the best signal quality is selected until the handoff is complete. Frequency diversity is provided in the bandwidth of the transmitted signal. With time diversity for multipath signals, the multiple correlative receiver elements can be assigned to different time-delayed copies of the same signal when the time-diverse signals are combined in optimal manner. One important aspect of CDMA is the use of Walsh codes based on the Walsh matrix, which is a square matrix with binary elements that always has a dimension that is power of two.

Each user in a CDMA system uses a different code to modulate their signal. Choosing the codes used to modulate the signal is very important in the performance of CDMA systems. The best performance will occur when there is good separation between the signal of a desired user and the signals of other users. The separation of the signals is made by correlating the received signal with the locally generated code of the desired user. If the signal matches the desired user's code, then the correlation function will be high, and the system can extract that signal. If the desired user's code has nothing in common with the signal, the correlation should be as close to zero as possible (thus eliminating the signal), which is referred to as *cross-correlation*. If the code is correlated with the signal at any time offset other than zero, the correlation should be as close to zero as possible. This is referred to as *auto-correlation* and is used to reject multipath interference.

In general, CDMA belongs to two basic categories: synchronous (orthogonal codes) and asynchronous (pseudorandom codes). Most modulation schemes try to minimize the bandwidth of the transmitted signal since bandwidth is a limited resource. However, spread-spectrum techniques use a transmission bandwidth that is several orders of magnitude greater than the minimum required signal bandwidth. One of the initial reasons for doing this was military applications, including guidance and communication systems. These systems were designed using spread spectrum because of its security and resistance to jamming. Asynchronous CDMA has some level of privacy built in because the signal is spread using a pseudorandom code; this code makes the spread-spectrum signals appear random or have noise-like properties. CDMA can also effectively reject narrowband interference. Since narrowband interference affects only a small portion of the spread-spectrum signal, it can easily be removed through notch filtering without much loss of information. Convolution encoding and interleaving can be used to assist in recovering this lost data. CDMA signals are also resistant to multipath fading. Since the spread-spectrum signal occupies a large bandwidth, only a small portion of this will undergo fading due to multipath at any given time. Like the narrowband interference, this will result in only a small loss of data and can be overcome.

Frequency reuse is the ability to reuse the same radio channel frequency at other cell sites within a cellular system. In the FDMA and TDMA systems, frequency planning is an important consideration. The frequencies used in different cells need to be planned carefully in order to ensure that the signals from different cells do not interfere with each other. In a CDMA system, the same frequency can be used in every cell because channelization is done using the pseudorandom codes. Reusing the same frequency in every cell eliminates the need for frequency planning in a CDMA system; however, planning of the different pseudorandom sequences must be done to ensure that the received signal from one cell does not correlate with the signal from a nearby cell.

WCDMA (Wideband Code Division Multiple Access) is a wideband spread-spectrum channel access method providing higher speeds and support more users compared to CDMA and TDMA schemes and represents an air interface implemented in 3G (Third Generation) mobile telecommunications networks [44]. The structures of the WCDMA physical channels, as well as the modulation and spreading techniques, differs in some respects from that of IS-95 CDMA. In WCDMA systems, the QPSK nodulation is employed, with baseband filters representing the root raised-cosine filters having a roll-off factor of 0.22. WCDMA standards define a wide variety of data and chip rates. In this case, data rates for the traffic channels are between 32 and 1024 kSymbols/s. Chip rate of 3.84 MChips/s is chosen for a channel bandwidth of 5 MHz, with frame length of 10 ms and each frame is divided into 15 slots. A strong difference in the WCDMA structure with respect to IS-95 is the use of a pilot signal. This is a special data pattern that is identical for all code channels and is transmitted every 0.625 ms. WCDMA traffic channel structure is based on a single code transmission for small data rates and multicode for higher data rates. WCDMA supports FDD and TDD modes, using one carrier frequency for uplink and one for downlink and fifteen slots per radio frame in UMTS (Universal Mobile Telecommunications System) for time division. Multiuser detection and smart antennas can be used to increase capacity and coverage. Modulation schemes such as OQPSK or GMSK do not prevent zero-crossing for multiple channels at different amplitude levels with complex scrambling in 3G spread-spectrum systems. In this case, HPSK (Hybrid Phased Shift Keying), representing a complex spreading modulation technique that is based on Walsh coding and specific orthogonal spreading functions, can minimize zero-crossing and 0° phase shift transitions, thus improving the peak-to-average power ratio of the signal by approximately 1.0 to 1.5 dB [45].

REFERENCES

1. R. A. Fessenden, "Apparatus for Signaling by Electromagnetic Waves," U.S. Patent 706,747, Aug. 1902 (filed Sept. 1901).
2. H. Roder, "Amplitude, Phase, and Frequency Modulation," *Proc. IRE*, vol. 19, pp. 2145–2176, Dec. 1931.
3. F. E. Terman, *Electronic and Radio Engineering*, New York: McGraw-Hill, 1955.
4. H. S. Black, *Modulation Theory*, New York: Van Nostrand, 1953.
5. H. L. Krauss, C. W. Bostian, and F. H. Raab, *Solid State Radio Engineering*, New York: John Wiley & Sons, 1980.
6. J. R. Carson, "Method and Means for Signaling with High Frequency Waves," U.S. Patent 1,449,382, Mar. 1923 (filed Dec. 1915).
7. A. A. Oswald, "Early History of Single-Sideband Transmission," *Proc. IRE*, vol. 44, pp. 1676–1679, Dec. 1956.
8. R. V. L. Hartley, "Modulation System," U.S. Patent 1,666,206, Apr. 1928 (filed Jan. 1925).
9. O. G. Villard, "A High-Level Single-Sideband Transmitter," *Proc. IRE*, vol. 36, pp. 1419–1425, Nov. 1948.
10. S. D. Bedrosian, "Normalized Design of 90 Degree Phase Difference Networks," *IEEE Trans. Circuit Theory*, vol. CT-7, pp. 128–136, June 1960.
11. D. K. Weaver, "A Third Method of Generation and Detection of Single-Sideband Signals," *Proc. IRE*, vol. 44, pp. 1703–1705, Dec. 1956.

12. S. D. Thompson and A. M. Pavio, "A Monolithic Double Balanced Single Sideband Modulator," *1987 IEEE MTT-S Int. Microwave Symp. Dig.*, vol. 2, pp. 899–902.

13. S. J. Parisi, "Monolithic, Lumped Element, Single Sideband Modulator," *1992 IEEE MTT-S Int. Microwave Symp. Dig.*, vol. 2, pp. 1047–1050.

14. R. A. Heising, "Modulation Methods," *Proc. IRE*, vol. 50, pp. 896–901, May 1962.

15. J. R. Carson, "Notes on the Theory of Modulation," *Proc. IRE*, vol. 10, pp. 57–67, Feb. 1922; reprinted in *Proc. IEEE*, vol. 51, pp. 893–896, June 1963.

16. E. H. Armstrong, "A Method of Reducing Disturbances in Radio Signaling by a System of Frequency Modulation," *Proc. IRE*, vol. 24, pp. 689–740, May 1936; reprinted in *Proc. IEEE*, vol. 52, pp. 1042–1062, Aug. 1964.

17. J. D. Gibson, *Principles of Digital and Analog Communications*, New York: Prentice Hall, 1993.

18. E. I. Manaev, "About Bandwidth to Receive Frequency-Modulated Signals without Nonlinear Distortions (in Russian)," *Radiotekhnika*, vol. 3, pp. 54–61, May 1948.

19. A. Grebennikov, *RF and Microwave Transistor Oscillator Design*, London: John Wiley & Sons, 2007.

20. S. J. Lipoff, "Linearization of Direct FM Frequency Modulators," *IEEE Trans. Vehicular Technol.*, vol. VT-27, pp. 7–17, Feb. 1978.

21. E. Marazzi and V. Rizzoli, "The Design of Linearizing Networks for High-Power Varactor-Tuned Frequency Modulators," *IEEE Trans. Microwave Theory Tech.*, vol. MTT-28, pp. 767–773, July 1980.

22. K. I. Kukk and V. G. Sokolinsky, *Transmitter Devices of Multichannel Radio Relay Systems* (in Russian), Moskva: Svyaz, 1968.

23. R. V. Garver, "360° Varactor Linear Phase Modulator," *IEEE Trans. Microwave Theory Tech.*, vol. MTT-17, pp. 137–147, Mar. 1969.

24. T. Morawaki, J. Zborowska, and P. Miazga, "Multi-Octave Phase Modulators," *1984 IEEE MTT-S Int. Microwave Symp. Dig.*, pp. 378–380.

25. F. Ali and N. Mysoor, "An Analog MMIC Phase Modulator for X-Band Satellite Transponder Applications," *IEEE Microwave and Guided Wave Lett.*, vol. 2, pp. 445–446, Nov. 1992.

26. Y. Tarusawa, H. Ogawa, and T. Hirota, "A New Constant-Resistance ASK Modulator Using Double-Sided MIC," *IEEE Trans. Microwave Theory Tech.*, vol. MTT-35, pp. 819–822, Sept. 1987.

27. Application Note 1298, *Digital Modulation in Communication Systems – An Introduction*, Agilent Technologies, 2001.

28. B. E. Dobratz, N. J. Ho, G. E. Lee, and H. T. Suyematsu, "Microwave Analog and Digital Processors," *1979 IEEE Int. Solid-State Conf. Dig.*, pp. 76–77.

29. A. E. Ashtiani, S.-I. Nam, A. d'Espona, S. Lucyszyn, and I. D. Robertson, "Direct Multilevel Carrier Modulation Using Millimeter-Wave Balanced Vector Modulators," *IEEE Trans. Microwave Theory Tech.*, vol. MTT-46, pp. 2611–2619, Dec. 1998.

30. S. Pasupathy, "Minimum Shift Keying: A Spectrally Efficient Modulation," *IEEE Commun. Mag.*, vol. 17, pp. 14–22, July 1979.

31. K. Murota and K. Hirade, "GMSK Modulation for Digital Radio Telephony," *IEEE Trans. Commun.*, vol. COM-29, pp. 1044–1050, July 1981.

32. D. M. Klymyshyn, S. Kumar, and A. Mohammadi, "Direct GMSK Modulation with a Phased-Locked Power Oscillator," *IEEE Trans. Vehicular Technol.*, vol. VT-48, pp. 1616–1625, Sept. 1999.

33. T. Noguchi, Y. Daido, and J. A. Nossek, "Modulation Techniques for Microwave Digital Radio," *IEEE Commun. Mag.*, vol. 24, pp. 21–30, Oct. 1986.

34. F. Xiong, *Digital Modulation Techniques*, Norwood: Artech House, 2006.

35. W. M. Waggener, *Pulse Code Modulation Techniques*, Cambridge: Springer, 2007.

36. F. H. Raab and D. J. Rupp, "Class-S High-Efficiency Amplitude Modulator," *RF Design*, vol. 17, pp. 3–11, May 1994.

37. F. H. Raab, B. E. Sigmon, R. G. Myers, and R. M. Jackson, "*L*-Band Transmitter Using Kahn EER Technique," *IEEE Trans. Microwave Theory Tech.*, vol. MTT-46, pp. 2220–2225, Dec. 1996.

38. F. H. Raab, "Intermodulation Distortion in Kahn-Technique Transmitters," *IEEE Trans. Microwave Theory Tech.*, vol. MTT-44, pp. 2273–2278, Dec. 1996.

39. I. Galton, "Delta-Sigma Data Conversion in Wireless Transceivers," *IEEE Trans. Microwave Theory Tech.*, vol. MTT-50, pp. 302–315, Jan. 2002.

40. J. Choi, J. Yim, J. Yang, J. Kim, J. Cha, D. Kang, D. Kim, and B. Kim, "A $\Delta\Sigma$-Digitized Polar RF Transmitter," *IEEE Trans. Microwave Theory Tech.*, vol. MTT-55, pp. 2679–2690, Dec. 2007.

41. L. Litwin and M. Pugel, "The Principles of OFDM," *RF Design*, vol. 24, pp. 30–48, Jan. 2001.

42. A. J. Viterbi, *CDMA: Principles of Spread Spectrum Communication*, Boston: Addison-Wesley, 1995.

43. D. P. Whipple, "The CDMA Standard," *Applied Microw. & Wireless*, vol. 6, pp. 24–37, Winter/Spring 1994.

44. K.-D. Tiepermann, "Understand the Basics of WCDMA Signal Generation," *Microwaves & RF*, vol. 37, pp. 83–86, Dec. 1998.

45. Application Note 1335, *HPSK Spreading for 3G*, Agilent Technologies, 2000.

7 Mixers and Multipliers

This chapter begins with a basic theory describing the operational principles of frequency conversion in receivers and transmitters. The different types of mixers, from the simplest based on a single diode to a balanced and double-balanced type based on both diodes and transistors, are described and analyzed. The special case is a mixer based on a dual-gate transistor that provides better isolation between signal paths and simple implementation. The frequency multipliers that historically were a very important part of the vacuum-tube transmitters can extend the operating frequency range.

7.1 BASIC THEORY

In wireless transmitters, it is generally required to convert a low-frequency baseband information signal to high frequency or frequencies at which the resulting signal can be effectively transmitted via antenna by electromagnetic propagation to the desired destination. This can be done either by a direct modulation of the high-frequency signal source or using a frequency converter traditionally called the *mixer*. Generally, any device (diode, vacuum tube, or transistor) that exhibits amplitude-nonlinear behavior can serve as a mixer, as nonlinear distortion results in the production of frequencies not present in the input. However, mixer as itself is fundamentally a linear device, which is shifting a signal from one frequency to another, keeping the properties of the initial signal (amplitude and phase), thus generally providing a linear operation. At the same time, mixers have relatively high levels of intermodulation distortion, spurious responses, and other undesirable nonlinear phenomena. Although mixers are equally important in wireless transmission and reception, traditional mixer terminology favors the receiving case because mixing was first applied as such in receiving applications [1,2].

The basic characteristic of a frequency-converter stage is its conversion transconductance, which is determined by the device transconductance of the input-electrode voltage to the output-electrode current. The general analysis of a frequency converter or mixer is applicable to all types of mixers no matter how or to what electrodes the oscillator and signal voltages are applied. Under assumption that the signal voltage is very small and the local oscillator (LO) voltage is large, the transconductance g_m of the three-element device (or conductance g of the two-element device) may be considered as a periodically varying function of the oscillator sinusoidal voltage only, which can be written as a Fourier series

$$g_m = a_0 + \sum_{n=1}^{\infty} a_n \cos n\omega_0 t \tag{7.1}$$

where ω_0 is the angular frequency of the LO, n is the harmonic number,

$$a_0 = \frac{1}{2\pi} \int_0^{2\pi} g_m d\left(\omega_0 t\right) \tag{7.2}$$

RF and Microwave Transmitter Design, First Edition. Andrei Grebennikov.

$$a_n = \frac{1}{\pi} \int_0^{2\pi} g_m \cos n\omega_0 t \, d(\omega_0 t). \tag{7.3}$$

Use of the cosine series implies that the device transconductance is a single-valued function of the oscillator voltage that varies as $\cos\omega_0 t$. When a small signal $v_s = V_s \sin\omega_s t$ is applied to the device input, the resulting output current i in a first-order approximation with the oscillator-frequency terms only may be written as

$$i = g_m V_s \sin\omega_0 t = a_0 V_s \sin\omega_s t + V_s \sum_{n=1}^{\infty} a_n \sin\omega_s t \cos n\omega_0 t = a_0 V_s \sin\omega_s t$$

$$+ \frac{V_s}{2} \sum_{n=1}^{\infty} a_n \sin(\omega_s + n\omega_0) t + \frac{V_s}{2} \sum_{n=1}^{\infty} a_n \sin(\omega_s - n\omega_0) t. \tag{7.4}$$

Examination of Eq. (7.4) shows that the output current contains a component having a sum frequency $(\omega_s + \omega_0)$ and a magnitude $a_1 V_s/2$. The term $a_1/2$ is called the *conversion transconductance* g_c. It is analogous to the device transconductance g_m representing the factor that, when multiplied by the amplitude of the applied signal, will give the amplitude of the sum-frequency component of the output current. The conversion transconductance g_c for the sum frequency is seen from Eq. (7.3) to have the value

$$g_c = \frac{1}{2\pi} \int_0^{2\pi} g_m \cos\omega_0 t \, d(\omega_0 t). \tag{7.5}$$

In a common case, if a circuit tuned to the intermediate frequency $(\omega_s - n\omega_0)$ is inserted in the output port, then the mixer operates as a *downconverter* converting the incoming signal frequency to a lower intermediate frequency. However, if a circuit inserted in the output port is tuned to the intermediate frequency $(\omega_s + n\omega_0)$, then the mixer operates as an *upconverter* converting the incoming signal frequency to a higher intermediate frequency. For a given small-signal frequency ω_s, an ideal mixer with a perfect LO with no harmonics and no noise sidebands would produce only two intermediate-frequency outputs: one at the sum frequency $(\omega_s + \omega_0)$ and another at the frequency difference $(\omega_s - \omega_0)$. Filtering can be used to select the desired intermediate-frequency product and reject the unwanted one, which is sometimes referred to as the *intermediate-frequency* (IF) *image*. Since n is an integer, it is evident that the intermediate frequency, in general, may be chosen to be the difference between the signal frequency and any integral multiple of the LO frequency. The conversion transconductance at the nth harmonic of the LO is given by

$$g_{cn} = \frac{I_{\omega_s \pm n\omega_0}}{V_s} = \frac{a_n}{2} \tag{7.6}$$

where $I_{\omega_s \pm n\omega_0}$ is the amplitude of the intermediate-frequency output current component $(\omega_s \pm n\omega_0)$.

All mixers are *multipliers* in the sense that the various new output components they produce can be described mathematically as the multiplicative products of their input frequencies. Mixing is achieved by the application of two signals to a nonlinear device whose transfer characteristic may differ depending upon the particular device. An active device with square-law transfer characteristic is ideal for mixer performance, since the least number of undesired frequencies is produced. In this case, if a device has the transfer characteristic

$$i(t) = a_0 v(t) + a_1 v^2(t) \tag{7.7}$$

then, with a two-tone input signal

$$v(t) = V_1 \cos \omega_1 t + V_2 \cos \omega_2 t \tag{7.8}$$

the output current becomes

$$i(t) = a_0 V_1 \cos \omega_1 t + a_0 V_2 \cos \omega_2 t + a_1 V_1^2 \cos^2 \omega_1 t + a_1 V_2^2 \cos^2 \omega_2 t + 2a_1 V_1 V_2 \cos\omega_1 t \cos\omega_2 t \tag{7.9}$$

where the first two terms are not necessary for mixer performance and must be filtered out in practical circuit. By the use of the trigonometric identity when

$$a_1 V_1^2 \cos^2 \omega_1 t = \frac{a_1}{2} a_1 V_1^2 (1 + \cos 2\omega_1 t) \tag{7.10}$$

the third and fourth terms are seen to represent a dc component and second harmonics of the input frequencies. Consequently, the final term in Eq. (7.9), which is called the *product term*, represents the desired output spectral components derived as

$$2a_1 V_1 V_2 \cos \omega_1 t \ \cos \omega_2 t = a_1 V_1 V_2 [\cos (\omega_1 - \omega_2) + \cos (\omega_1 + \omega_2)] \tag{7.11}$$

where the amplitudes of the sum- and difference-frequency components are proportional to the product $V_1 V_2$ of the input-signal amplitudes. The sum and difference frequencies generated by the squared term in Eq. (7.7) are called the *second-order intermodulation products*, whereas the spectral frequency components generated by the cubed term such as $2\omega_1 - \omega_2$ and $2\omega_2 - \omega_1$ are called the *third-order intermodulation products*.

In a heterodyne receiver application, a low-level RF signal and an RF LO signal are mixed together to produce an intermediate difference frequency $f_{IF} = f_{RF} - f_{LO}$ (low-side conversion) or $f_{IF} = f_{LO} - f_{RF}$ (high-side conversion) and a much higher sum frequency $f_{RF} + f_{LO}$ that is filtered out, as shown in Figure 7.1(*a*). In a particular transmitter operation, an RF LO signal and IF signal can be mixed together to produce an output RF signal $f_{RF} = f_{LO} + f_{IF}$ (sum mixer) or $f_{RF} = f_{LO} - f_{IF}$ (difference mixer) with corresponding filtering, as shown in Figure 7.1(*b*). A single LO can provide the mixing function in the transmitter (upconversion) and receiver (downconversion), when both are used in radar or transceiver system.

7.2 SINGLE-DIODE MIXERS

The simplest mixer type that was for many years a key element in receiving systems was a diode mixer. Although other semiconductor junctions, such as the *p–n* junction, also exhibit nonlinear behavior, the metal–semiconductor Schottky-barrier diodes are widely used in modern mixers because of inherently low junction capacitance and high switching speed. As a result, Schottky diodes operate well into the millimeter-wave frequency with cutoff frequencies exceeding 200 GHz due to their much higher carrier mobility. Despite the fact that GaAs diodes are more expensive than those in silicon, they can provide better conversion loss and noise performance, especially at high frequencies. Besides, GaAs devices generally have higher breakdown voltages than silicon ones and better resistance to ionizing radiation. However, their advantage at lower frequencies is minimal and their cost is much greater.

At low frequencies, the relationship between the current i and voltage v across the mixer diode can generally be written as

$$i = I_{sat} \left[\exp \left(\frac{v}{nV_T} \right) - 1 \right] \tag{7.12}$$

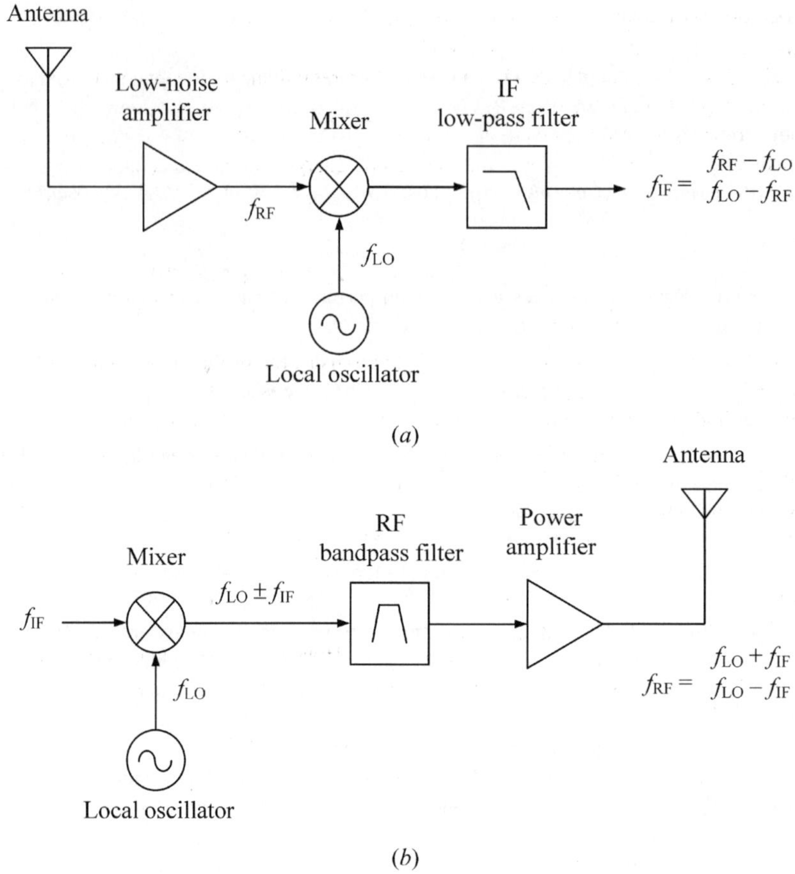

FIGURE 7.1 Frequency conversion in receiver and transmitter.

where I_{sat} is the reverse saturation current, V_T is the thermal voltage, and n is the ideality factor. The diode differential conductance at the operating point is therefore given by

$$g = \frac{di}{dv} = \frac{I_{sat}}{nV_T} \exp\left(\frac{v}{nV_T}\right) = \frac{i + I_{sat}}{nV_T} \cong \frac{i}{nV_T} \quad i \gg I_{sat} \tag{7.13}$$

If a cosine oscillator voltage is assumed to apply to the mixer diode,

$$v = V_0 + V \cos \omega_0 t \tag{7.14}$$

where V_0 is the dc bias voltage, then its substitution into Eq. (7.12) allows us to rewrite the diode conductance g as a time-varying function by

$$g(t) = \frac{I_{sat}}{nV_T} \exp\left(\frac{V_0}{nV_T}\right) \exp\left(\frac{V \cos \omega_0 t}{nV_T}\right) = \frac{I_{sat}}{nV_T} \exp\left(\frac{V_0}{nV_T}\right) \sum_{k=0}^{\infty} I_k \left(\frac{V}{nV_T}\right) \cos k\omega_0 t \tag{7.15}$$

where $I_k[V/(nV_T)]$ are the modified Bessel functions of the first kind of order k with argument $V/(nV_T)$ [3].

If there were no charge storage in the diode, it would be infinitely fast because of absence any charge inertia, and the diode current could be changed in zero time instantly. However, there are two forms of charge storage: charge storage in the depletion region due to dopant concentration resulting in a junction capacitance C_j and charge storage due to minority-carrier charges injected into the neutral regions (holes stored in the *n*–type region and electrons stored in the *p*–type region) resulting in a diffusion capacitance C_{diff}. Thus, the total diode charge-storage capacitance for forward-bias voltages $V_0 \leq 0.5\varphi$ is defined as

$$C(v) = C_{diff}(v) + C_j(v) = \tau \frac{di}{dv} + C_j(V_0)\left(\frac{\varphi - V_0}{\varphi - v}\right)^{\gamma} \tag{7.16}$$

where γ is the junction grading coefficient, τ is the transit time of the diode, and φ is the junction potential [4].

When the diode has finite series resistance r_s, then both the forward and the reverse impedance at microwave frequencies are affected. In the forward direction, increasing the pumped voltage amplitude V across the diode junction causes the diode conductance to rise until it finally reaches maximum conductance $g_{max} = 1/r_s$, as shown in Figure 7.2. In this case, the diode capacitance is represented by the diffusion capacitance C_{diff} as a linear function of current i and its value depends on the diode transit time, according to Eqs. (7.13) and (7.16). If the forward excursion is arbitrarily restricted, for instance, to the point where the diode conductance rises to $g_{max}/5$, the only way to reduce the conversion loss and noise figure with increased sinusoidal pumping is to use the reverse bias, resulting in a smaller pulsed current duty cycle. However, if the voltage swing is too far in the reverse direction, the loss will also increase rapidly due to the series resistance r_s and junction capacitance C_j that lead to a residual diode conductance $g_{min} = r_s(\omega C_j)^2$, as shown in Figure 7.2. In this region, the diode is operated as a varactor with a quality factor $Q_v = 1/(\omega\, r_s C_j)$.

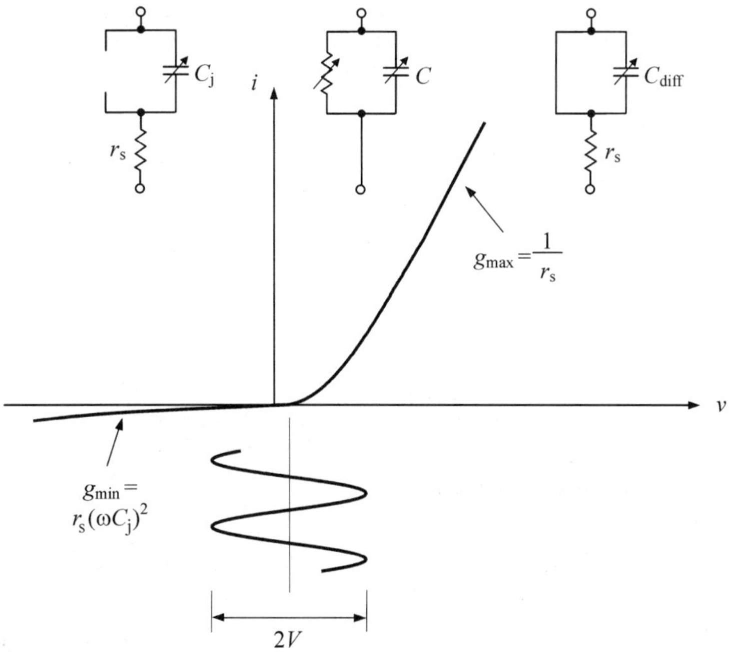

FIGURE 7.2 Effect of diode parasitics on diode conductance.

In the past, it was common to use varactor frequency converters to generate moderate to high levels of RF power. Their major advantage as the reactive devices over other types of the frequency converters was a little noise. The general relations between the real powers at all mixing frequencies, which were derived by Manley and Rowe, apply to any nonlinear reactance, including varactor as a nonlinear capacitor, driven by one or two incommensurable frequencies. Generally, if a nonlinear capacitor is excited by two frequencies f_1 and f_2, the sideband frequencies of the form $mf_1 + nf_2$ are generated because of the varactor nonlinearity, resulting in the following two equations:

$$\sum_{m=0}^{\infty} \sum_{n=-\infty}^{\infty} \frac{m P_{\rm mn}}{mf_1 + nf_2} = 0 \tag{7.17}$$

$$\sum_{n=0}^{\infty} \sum_{m=-\infty}^{\infty} \frac{n P_{\rm mn}}{mf_1 + nf_2} = 0 \tag{7.18}$$

where $P_{\rm mn}$ is the real power produced in the circuit at the frequency $mf_1 + nf_2$ [5,6]. These equations imply that the nonlinear capacitor is lossless and can be applied to parametric amplifiers, frequency upconverters and multipliers.

In a frequency multiplier, there is only a single excitation signal with frequency f_1 when $f_2 = 0$, and n can be set to zero. Consequently, the summation over n in Eq. (7.17) can be omitted, all terms in Eq. (7.18) become zero, and Eq. (7.17) is simplified to

$$\sum_{m=0}^{\infty} P_{\rm m} = 0 \tag{7.19}$$

where $P_{\rm m}$ represents the power at the frequency mf_1. Equation (7.19) states that applying the input signal with power P_1 and frequency f_1 to a lossless nonlinear capacitor must be converted to the output powers at the harmonics of f_1. In particular, for mth-harmonic multiplier, the highest possible value of $P_{\rm m}$ occurs when only P_1 and $P_{\rm m}$ are not zero, and the input power flowing to the capacitor is equal to the output power flowing from the capacitor, resulting in a multiplier efficiency of 100%. To achieve such an idealized condition, it is necessary to provide the varactor termination by pure reactances at all other harmonics.

However, varactor mixers and multipliers are narrowband and very sensitive to slight mistuning that makes difficult their performance optimization. In addition, they can generate parasitic parametric oscillations due to capacitor nonlinearity depending on how strong varactor is pumped. Also, the small variation of pumping level results in a significant variation of the desired output power because the average varactor capacitance strongly depends on the voltage amplitude across the varactor junction. Thermal noise of the varactor series capacitance provides a fundamental limit on the noise performance obtainable from the frequency converters. The upper-sideband (USB) upconverter is generally stable, but its gain is limited, and for high-frequency signals no gain is possible. At the same time, the lower-sideband (LSB) upconverter is potentially unstable, and can have negative input and output resistances [6]. Therefore, the resistive diode mixers based on Schottky-barrier diodes are widely used in modern designs due to the circuit simplicity, performance stability and predictability, broadband design capability, and applicability up to millimeter-wave frequencies [7].

Figure 7.3(*a*) shows the microwave upconversion single-diode mixer (SDM) schematic where the signal from the LO flows through the directional coupler to the mixing diode connected in parallel, and the low-pass and bandpass filters are included into the IF and RF signal paths, respectively. In this case, *conversion gain* (*loss*) is the ratio of the output (RF) signal power to the input (IF) signal power. *Isolation* represents the amount of "leakage" or "feedthrough" between the mixing ports. *Dynamic range* is the amplitude range over which the mixer can operate without performance degradation. *Harmonic intermodulation distortion* results from the mixing of mixer-generated harmonics of the input signals having frequencies $mf_{\rm LO} \pm nf_{\rm IF}$, where m and n represent the harmonic order.

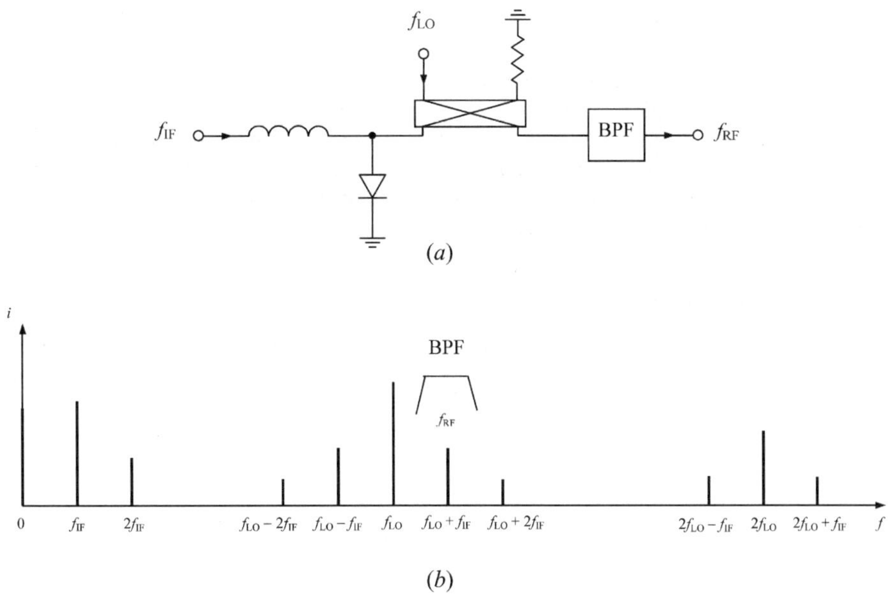

FIGURE 7.3 Single-diode upconversion mixer and its spectrum.

Cross-modulation distortion is the amount of modulation transferred from a modulated input signal to an unmodulated input signal when both signals are applied to the IF port.

In upconversion mixers, the signal that is necessary to transfer to higher frequency band is applied to the IF port, the power level of the LO signal is increased, and then the high-side RF signal is filtered out, as shown in Figure 7.3(*b*). In this case, when both the LO and IF signals may have comparably high power levels, the spectrum of the output RF signal is rich with intermodulation products. For instance, if voltage applied to the mixing diode with exponential characteristic given by Eq. (7.12) represents the sum of two voltages,

$$v = V_0 + V_1 \cos \omega_1 t + V_2 \cos \omega_2 t \tag{7.20}$$

where V_0 is the dc bias voltage, then the currents corresponding to any intermodulation product in a diode mixer can be written as

$$i_{\text{kl}} = I_{\text{sat}} \exp\left(\frac{V_0}{nV_{\text{T}}}\right) I_{\text{k}}\left(\frac{V_1}{nV_{\text{T}}}\right) I_{\text{l}}\left(\frac{V_2}{nV_{\text{T}}}\right) \cos\left(k\omega_1 t \pm l\omega_2 t\right) \tag{7.21}$$

where $I_{\text{k}}[V_1/(nV_{\text{T}})]$ and $I_{\text{l}}[V_2/(nV_{\text{T}})]$ are the modified Bessel functions of the first kind of orders k and l with argument $V_1/(nV_{\text{T}})$ and $V_2/(nV_{\text{T}})$, respectively [8]. Here, the quantities k and l designate the current harmonic components produced in the diode mixer. As a result, the conversion loss of the SDM is sufficiently high, of about 10 dB. However, some improvement of 1–2 dB in the RF output power level can be achieved if to provide an optimum impedance for the second LO harmonic ($2f_{\text{LO}}$) at the RF output when its combining with a low-side RF frequency creates an intermodulation product of the secondary conversion $f_{\text{RF}} = 2f_{\text{LO}} - (f_{\text{LO}} - f_{\text{IF}})$ which is in phase with the correspondent product of the primary upconversion $f_{\text{RF}} = f_{\text{LO}} + f_{\text{IF}}$, similarly to that kind of process in a downconverted image-rejection diode mixer having optimum image-frequency impedance [9].

7.3 BALANCED DIODE MIXERS

Balanced mixers are generally divided into two classes, called the *single-balanced mixers* and the *double-balanced mixers*, which combine two or more identical single-diode mixers. The advantage of balanced upconversion mixers over single-diode mixers includes both the rejection of spurious responses and intermodulation products and significantly better LO-to-RF, RF-to-IF, and LO-to-IF isolation. The level of rejection is dependent on the amplitude and phase balance of the baluns, providing the balanced drive and the matching between the diodes.

7.3.1 Single-Balanced Mixers

In a single-balanced mixer (SBM) circuit shown in Figure 7.4 where the two diodes as nonlinear elements are excited in phase by one signal and 180° out of phase by the other signal, no intermodulation outputs that include even-order harmonics of the out-of-phase signal exist [10]. In a general case when the transfer characteristic of the nonlinear element can be described as a polynomial

$$i = \sum_{n=0}^{\infty} A_{\mathrm{n}} v^{\mathrm{n}} \tag{7.22}$$

the current in the upper nonlinear element can be calculated as

$$\begin{aligned} i_1 &= \sum_{n=0}^{\infty} A_{\mathrm{n}} \left(V_2 \cos\omega_2 t + V_1 \cos\omega_1 t\right)^{\mathrm{n}} = \sum_{n=0}^{\infty} B_{\mathrm{n}} \cos^{\mathrm{n}} \omega_2 t + \sum_{n=0}^{\infty} C_{\mathrm{m}} \cos^{\mathrm{m}} \omega_1 t \\ &+ \left(\sum_{p=1}^{\infty} D_{\mathrm{p}} \cos^{\mathrm{p}} \omega_2 t\right)\left(\sum_{q=1}^{\infty} E_{\mathrm{q}} \cos^{\mathrm{q}} \omega_1 t\right) = \left(\sum_{r=0}^{\infty} F_{\mathrm{r}} \cos^{\mathrm{r}} \omega_2 t\right)\left(\sum_{s=0}^{\infty} G_{\mathrm{s}} \cos^{\mathrm{s}} \omega_1 t\right). \end{aligned} \tag{7.23}$$

By expanding each term into the Fourier series, Eq. (7.23) can be equivalently rewritten in frequency domain with harmonic representation as

$$i_1 = \left(\sum_{r=0}^{\infty} H_{\mathrm{r}} \cos r\omega_2 t\right)\left(\sum_{s=0}^{\infty} I_{\mathrm{s}} \cos s\omega_1 t\right). \tag{7.24}$$

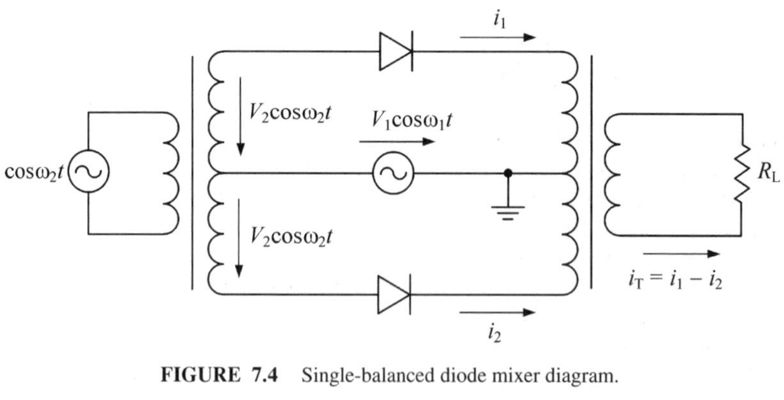

FIGURE 7.4 Single-balanced diode mixer diagram.

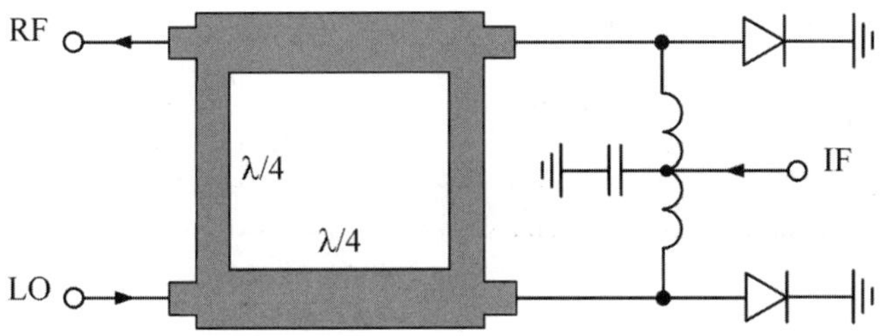

FIGURE 7.5 Microstrip reflection-type branch-line diode mixer.

Since both nonlinear elements are assumed to be fully identical, the current in the lower nonlinear element excited by the same input signal but with a 180° out-of-phase shift can be written as

$$i_2 = \left[\sum_{r=0}^{\infty} H_r \cos r\,(\omega_2 t + 180^\circ) \right] \left(\sum_{s=0}^{\infty} I_s \cos s\omega_1 t \right). \tag{7.25}$$

As a result, the total current i_T that flows through the load is equal to

$$i_T = i_1 - i_2 = \left(\sum_{s=0}^{\infty} I_s \cos s\omega_1 t \right) \times \left\{ \left(\sum_{r=0}^{\infty} H_r \cos r\omega_2 t \right) - \left[\sum_{r=0}^{\infty} H_r \cos r\,(\omega_2 t + 180^\circ) \right] \right\}. \tag{7.26}$$

From Eq. (7.26) it follows that, for all r where r is an even integer, the current i_T becomes equal to zero. Therefore, all intermodulation products defined by frequencies of the form

$$\omega_{IM} = |\pm s\omega_1 \pm r\omega_2| \tag{7.27}$$

exist for all s but only for odd r.

The circuit for a microwave balanced upconversion mixer based on a branch-line hybrid is shown in Figure 7.5. Although not shown, generally any single-diode mixer (SDM) requires matching and bias circuits. Hybrid microwave combiners become practical for upconversion mixers when IF frequency is much smaller than both LO and RF frequencies, and both LO and RF frequencies differ insignificantly from each other to fit the hybrid operating bandwidth. For a fully symmetrical SBM circuit, the resulting current i_{RF} corresponding to the USB RF frequency $f_{RF} = f_{LO} + f_{IF}$ can be written as

$$\begin{aligned} i_{RF} = -i_{RF1} - i_{RF2} = -I_{RF} \left\{ \cos \left[\left(\omega_{LO} t - \frac{\pi}{2} \right) + \omega_{IF} t + \pi \right] \right. \\ \left. + \cos \left(\omega_{LO} t + \omega_{IF} t + \frac{\pi}{2} \right) \right\} = 2 I_{RF} \sin\,(\omega_{LO} + \omega_{IF})\, t. \end{aligned} \tag{7.28}$$

where $\omega_{IF} = 2\pi f_{IF}$ and $\omega_{LO} = 2\pi f_{LO}$.

Instead of a branch-line hybrid combiner, the rate-race ring hybrid combiner or coupled-line 3-dB coupler can be used [11,12]. The hybrid ring is more easily fabricated and controlled, but it is larger than the hybrid coupler and its operating frequency bandwidth is nowhere as good. Due to suppression of the even LO harmonics, the SBM provides less level of the intermodulation components in the transmitted RF signal spectrum than the SDM. When one mixing diode of an SBM fails, the circuit continues to operate, although the RF output power level drops by about 3 dB.

To increase the operating frequency, a subharmonic mixer can be used that provides an ability to operate with halved LO frequency, thus avoiding signal leakage problems inherent with higher frequency LO sources. Figure 7.6(a) shows the simplified schematic of a subharmonic mixer with an

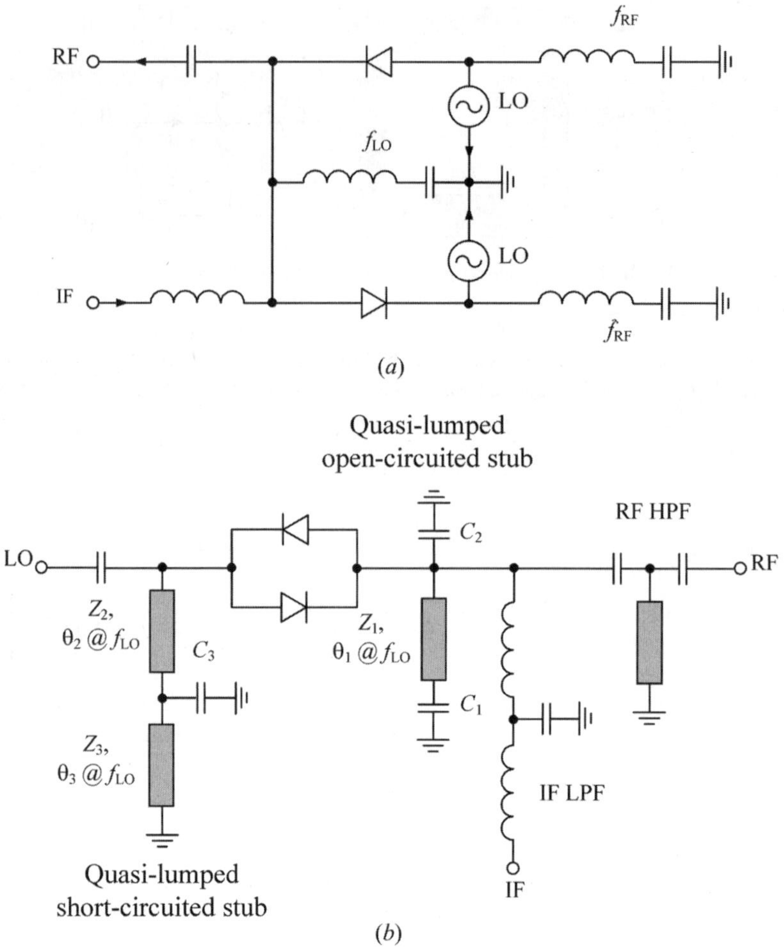

FIGURE 7.6 Schematics of subharmonically pumped mixers.

antiparallel mixing diode pair that presents an antisymmetric current–voltage characteristic [13]. In this case, the subharmonic mixer is equivalent to two SDMs in parallel, and each mixer is excited by a carrier LO signal at $f_{RF}/2$. For in-phase LO signals with reversed polarity of two parallel-connected diodes, as shown in Figure 7.6(*a*), the total current i_T that flows through the RF load is equal to $i_T = i_1 + i_2$ and, as it follows from Eq. (7.26) where the negative sign must be replaced by the positive one, this current becomes equal to zero for all r where r is an odd integer, including the fundamental component. Similar result at an RF load can also be achieved for 180° out-of-phase LO signals with the same polarity of two parallel-connected diodes. At microwave frequencies, the series resonant circuits tuned to f_{RF} and f_{LO} to isolate the LO and RF signal paths from each other can be replaced by the short-circuited and open-circuited microstrip stubs. In this case, the short-circuited stub at the LO port is quarter wavelength long at f_{LO}, and becomes a half wavelength long at f_{RF}, thus providing a short circuit to the RF signal. On the other hand, the open-circuit stub at the RF output presents an open circuit to the RF but is a quarter wavelength long at f_{LO}, presenting a short circuit for LO signal. The frequency of the IF signal to be upconverted is normally far enough from f_{RF} to allow easy realization of an IF filter presenting an open-circuit output to the RF port.

Figure 7.6(*b*) shows the circuit schematic of the subharmonically pumped (SHP) mixer MMIC [14]. The mixer includes an IF low-pass T-type lumped filter and RF high-pass T-type quasi-lumped filter to isolate RF and IF ports. To isolate RF and LO ports, the quasi-lumped open-circuited and short-circuited stubs are used. The former stub provides a short circuit at f_{LO} with an open circuit at f_{RF}, whereas the latter stub introduces a short circuit at f_{RF}, providing an open circuit at f_{LO}. The capacitances C_1 and C_2 of the quasi-lumped open-circuited stub are defined as

$$C_1 = \frac{1}{\omega_{LO} Z_1 \tan\theta_1}. \tag{7.29}$$

$$C_2 = \frac{C_1}{3 + \tan^2\theta_1}. \tag{7.30}$$

where Z_1 is the characteristic impedance and θ_1 ($0 < \theta_1 < 90^\circ$) is the electrical length of the stub at f_{LO}. To effectively convert the IF signal to the RF, the quasi-lumped short-circuited stub should be grounded at IF frequency. In this case, the shunt capacitance C_3 and electrical length θ_3 can be calculated from

$$\tan\theta_3 = \frac{Z_3}{Z_2}\frac{1 + 3\tan^2\theta_2}{2\tan\theta_2} - \sqrt{\left(\frac{Z_3}{Z_2}\right)^2\left(\frac{1 + 3\tan^2\theta_2}{2\tan\theta_2}\right)^2 - 3}. \tag{7.31}$$

$$C_3 = \frac{1}{\omega_{LO} Z_3}\left(\frac{1}{\tan\theta_3} - \frac{Z_3}{Z_2}\tan\theta_2\right). \tag{7.32}$$

where the electrical lengths θ_2 and θ_3, which are set at f_{LO}, must vary within the range of $0 < (\theta_2, \theta_3) < 45^\circ$ to short RF signal at the input of the short-circuited stub.

7.3.2 Double-Balanced Mixers

A double-balanced mixer (DBM) normally makes use of four diodes, which can be connected in different configurations, with both LO and RF signal paths being balanced. In this case, all ports of the mixer are inherently isolated from each other, and isolation is achieved by means of center-tapped transformers. The advantages of a double-balanced design over a single-balanced scheme are the increased linearity, improved suppression of spurious products, and inherent isolation between all ports. However, they require a higher level LO drive to pump the diodes and two additional baluns. The DBMs have higher conversion loss and lower limit in maximum frequency, but they provide broader bandwidth.

Figure 7.7(*a*) shows the bridge double-balanced diode upconversion mixer schematic with two center-tapped transformers to provide isolation between three ports: LO, IF, and RF [15]. The LO voltage is assumed to be large enough to properly control on/off cycle of the diodes. The RF currents i_1 and i_3, which are shown in Figure 7.7(*b*), flow through the diodes D_1 and D_3, respectively, during the time when voltage v_{LO} makes point *a* positive with respect to point *b* and voltage v_{IF} makes point *c* positive with respect to point *d*. Diodes D_1 and D_3 are turned on by v_{LO}, and current i_{LO} flows around the loop *a*-*c*-*b*-*a*. In this case, diodes D_2 and D_4 are turned off because they are reverse-biased. The RF currents i_1 and i_3 add in the load resistance R_L to produce the RF voltage with indicated polarity. Note that i_{LO} does not flow through the IF path between the center points of the transformers because these points have the same potential at f_{LO} if the diodes and the LO transformer are perfectly balanced. The polarity of v_{IF} in Figure 7.7(*c*) is the same as in Figure 7.7(*b*), but the polarity of v_{LO} is reversed, thus making point *b* positive with respect to point *a*. In this case, diodes D_2 and D_4 are turned on, while diodes D_1 and D_3 are turned off. As a result, the RF currents i_2 and i_4 flows into the RF load, producing the RF voltage of opposite polarity to that in Figure 7.7(*b*). The LO current flows through the loop involving diodes D_2 and D_4, but not in the RF transformer. This type of mixer can operate over a wide frequency range determined primarily by the design of transformers. If toroidal-cored transmission-line transformers are used, bandwidths of 1000:1 can be achieved. The balance of the

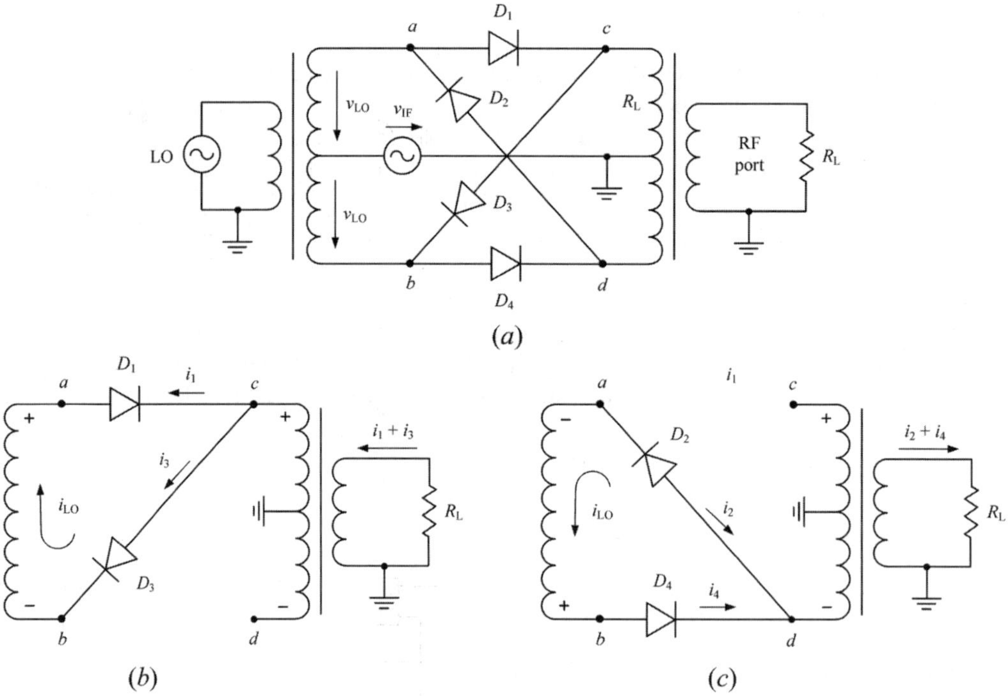

FIGURE 7.7 Equivalent schematics of diode double-balanced mixer.

mixer and isolation between ports is determined by the accuracy of the transformer windings and careful matching of the diode characteristics.

Generally, the two most common types of DBMs are the ring mixer (RM) and the star mixer [7]. The RM is more relevant to low-frequency applications, in which wire-wound or transmission-line ferrite transformers can be used. The star mixer is used primarily in microwave applications, as it is more amenable to operate with microwave baluns. However, there is no significant difference in their performances. Figure 7.8(*a*) shows the classical diode RM circuit where the secondary windings of the transformers are connected to the nodes of the diode ring composed by four diodes. The LO signal alternately turns the top diode pair D_1 and D_2 and bottom diode pair D_3 and D_4 on and off in anti-phase, thus providing the corresponding closed loops for IF currents through the diodes and the grounded center tap of the LO transformer in opposite directions. In this case, if the diodes are identical, the points *b* and *d* represent virtual grounding to the RF signal, hence no RF voltage appears across the secondary winding of the LO transformer. Similarly, the points *a* and *c* are virtual ground to the LO signal, thus no LO voltage appears across the secondary winding of the RF transformer. If diode ring is fabricated on a single chip, the isolation between LO and RF ports in microwave mixers can be achieved from 30 to 40 dB. The IF signal can also be delivered through the center tap of the LO transformer rather than through the center tap of the RF transformer. The RM concept can be extended to include double-ring mixer topologies that have the added advantage of allowing the IF frequency response to overlap the RF and LO frequency bands, thus making easier the process of frequency selection [16]. In this case, to reduce the conversion loss and to extend the frequency upconversion to higher operating frequencies, the broadband active center-tapped balun can be implemented in microwave monolithic design.

The upper-frequency limit of the DBMs using tapered baluns and low parasitic diode packages can also be extended if a balun structure is formed by two coupled transmission-line pairs, each having a quarter wavelength at the center bandwidth frequency, as shown in Figure 7.8(*b*) [17]. Without

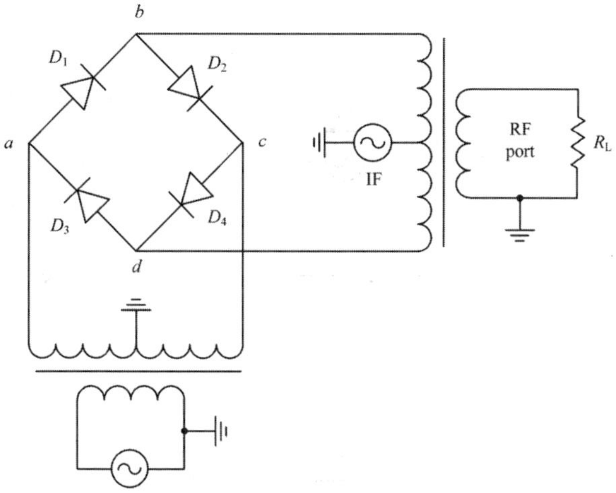

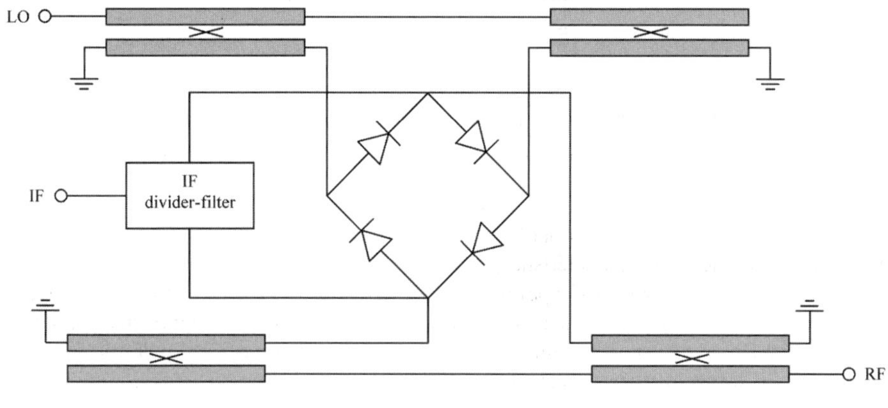

FIGURE 7.8 Schematics of diode ring double-balanced mixers.

any frequency compensation, this structure can exhibit greater than octave of bandwidth and can be realized in a variety of media, such as coaxial cable, microstrip, or coplanar waveguide. As a result, the upper-frequency limit of such a DBM occurs when the total length of the balun becomes 360°. To minimize the overall mixer size for monolithic microwave applications above 10 GHz, the mixer architecture for frequency upconversion can include a miniature spiral balun for IF path and a 180° hybrid formed with an interdigitated Lange microstrip coupler with two (+45° and −45°) phase shifters. This allows us to split an LO incoming signal into two paths with equal amplitudes and 180° relative phase difference and to provide an output port for the RF extraction of upconverter application [18].

In a star DBM architecture shown in Figure 7.9, one terminal of each four diodes is connected to a common node used as the IF terminal [7,19]. Similarly to the RM, this mixer operates as a polarity-reversing switch. For the LO transformer T_2, when the dotted sides of its secondary windings T_{2a} and T_{2b} are positive, diodes D_1 and D_2 are turned on, diodes D_3 and D_4 are turned off, and the dotted

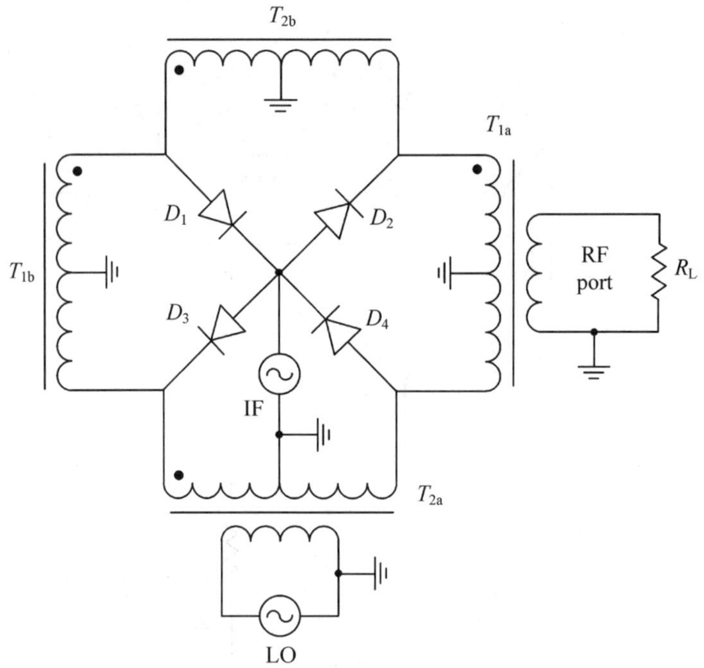

FIGURE 7.9 Schematic of diode double-balanced star mixer.

sides of the secondary windings T_{1a} and T_{1b} of the RF transformer T_1 are connected to the IF port. The RF port is thus connected to the IF port through the RF transformer T_1 and the diodes. However, when the LO polarity reverses, diodes D_1 and D_2 are turned off, diodes D_3 and D_4 are turned on, and then the undotted sides of both secondary windings T_{1a} and T_{1b} are connected to the IF port. In this case, the RF port is again connected to the IF port, but with reversed polarity. As a result, the RF polarity is therefore always applied to the IF port, but its polarity is reversed periodically at the LO frequency. Because generally the operational principles of the star and RMs are the same, it should expect similar spurious-response properties, and the only difference may be defined by the convenience and complexity of the implementation in a particular design and technology.

Single-sideband (SSB) mixer architectures are useful in discriminating and removing the LSB and USB generated during frequency upconversion, especially when sidebands are very close in frequency and attenuation of one of the sidebands cannot be achieved with filtering. Conventional SSB mixers generally include a Wilkinson divider to provide two LO signals, well balanced both in amplitude and phase, a 90° phase shifter to provide quadrature IF signals, and a quadrature hybrid coupler to combine the extracted RF signal at USB, as shown in Figure 7.10(*a*). Unlike the low-noise receiver downconverter, the transmitter upconverter operates with sufficiently high power levels (in limits of 5 to 10 dBm) at both IF and LO inputs, and therefore, the sidebands are rich with IF harmonics. However, since the even IF harmonics and their intermodulation components are well suppressed in each SBM, only the sidebands with the odd-order components such as $f_{LO} \pm (f_{IF}, 3f_{IF}, 5f_{IF}, \ldots)$ are delivered to the output quadrature coupler of the DBM. The carrier LO frequency is also suppressed in the SBMs, typically by about 20 dB, due to the corresponding isolation of their output hybrids.

Consider the propagation paths of the main signal components $\omega_{LO} \pm \omega_{IF}$ from both SSB mixers through the output quadrature coupler–combiner in Figure 7.10(*a*), by assuming that the phase delay in directions 1–3 and 2–4 is equal to 90° and there is no phase delay in diagonal directions. As a

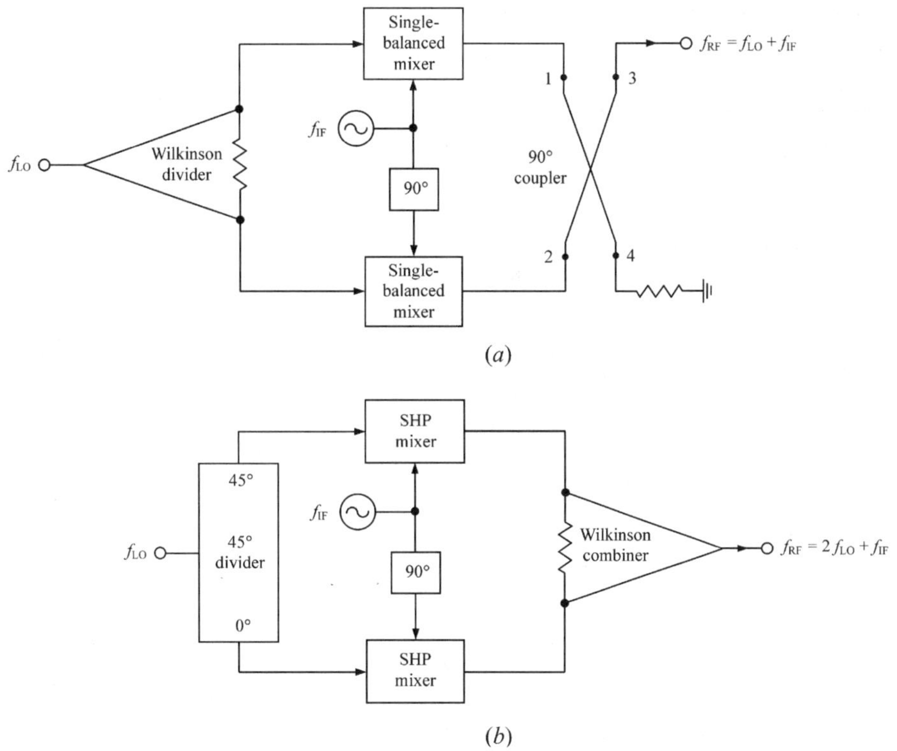

FIGURE 7.10 Block diagrams of single-sideband mixer-upconverters.

result, for an USB RF signal at port 3, the phase delay from the top mixer is equal to $\phi_{1,3} = \phi_{LO} + \phi_{IF} + 90^\circ$, whereas the phase delay from the bottom mixer is equal $\phi_{2,3} = \phi_{LO} + (\phi_{IF} + 90^\circ) + 0^\circ$, which are in phase, and the USB RF signal $f_{LO} + f_{IF}$ flows to the RF load. At the same time, for a LSB RF signal, the in-phase condition is achieved at port 4, when the phase delay from the top mixer is equal to $\phi_{1,4} = \phi_{LO} - \phi_{IF} + 0^\circ$, whereas the phase delay from the bottom mixer is equal $\phi_{2,4} = \phi_{LO} - (\phi_{IF} + 90^\circ) + 90^\circ$. As a result, the LSB RF signal $f_{LO} - f_{IF}$ is fully dissipated in the ballast resistor and does not appear in the RF load.

The suppression level of an unwanted sideband strongly depends on a circuit symmetry level (equality of pumped LO signals, identity of voltage–ampere diode characteristics, equality of phase shifts to nominal values, etc.). For instance, if two RF signals having equal amplitudes and being orthogonal in phase flow from both top and bottom SSB mixer outputs, then their suppression level at the output of the non-ideal quadrature coupler–combiner can be evaluated according to

$$L = \frac{1 + 2\Delta A \cos \Delta\phi + \Delta A^2}{1 - 2\Delta A \cos \Delta\phi + \Delta A^2} \tag{7.33}$$

where ΔA is the ratio of amplitudes and $\Delta\phi$ is the phase difference produced by the coupler. Consequently, in order to achieve the suppression level of an unwanted sideband of more than 20 dB, it is necessary to provide ΔA and $\Delta\phi$ to be better than ± 0.2 dB and 10°, respectively. As an example, in a microwave SSB diode mixer operated as a frequency upconverter with IF power $P_{IF} =$ 5 dBm and LO power $P_{LO} = 10$ dBm, an upconversion of the modulated IF to RF can occur with the conversion loss of 6 dB and LO suppression of 9 dB, relatively to the USB $f_{RF} = f_{LO} + f_{IF}$ [20].

TABLE 7.1 Intermodulation Components in RF Spectrum for Different Mixer Types.

	Harmonics LO						
Harmonics IF	1	2	3	4	5	6	Mixer Type
1	+	+	+	+	+	+	SDM
	+	−	+	−	+	−	SBM
	+	−	+	−	+	−	DBM, RM
2	+	+	+	+	+	+	SDM
	+	−	+	−	+	−	SBM
	−	−	−	−	−	−	DBM, RM
3	+	+	+	+	+	+	SDM
	+	−	+	−	+	−	SBM
	+	−	+	−	+	−	DBM, RM
4	+	+	+	+	+	+	SDM
	+	−	+	−	+	−	SBM
	−	−	−	−	−	−	DBM, RM
5	+	+	+	+	+	+	SDM
	+	−	+	−	+	−	SBM
	+	−	+	−	+	−	DBM, RM
6	+	+	+	+	+	+	SDM
	+	−	+	−	+	−	SBM
	−	−	−	−	−	−	DBM, RM

To simplify filtering problem by mixing with a harmonic of a lower frequency LO signal, which is very important when IF frequency is not so high, and to increase an operating frequency bandwidth for RF signal towards millimeter waves, an SSB upconverter based on subharmonically pumped mixers with antiparallel pair of diodes can be used [14,21]. Figure 7.10(*b*) shows the block schematic of the wideband SSB SHP upconversion mixer consisting of two SHP mixers, an LO power divider with a phase shift of 45°, and an RF in-phase power combiner. In this case, the IF signals are fed into each SHP mixers having the phase difference of 90°, the LO signal is divided into two signals with the phase difference of 45°, and then the SHP mixers generate the USBs and LSBs of the upconverted signal $2f_{\rm LO} \pm f_{\rm IF}$. Since upconverted USB signals $2f_{\rm LO} + f_{\rm IF}$ are in phase with each other at the input of the Wilkinson combiner, hence the desired upconverted USB resulting signal appears at the RF output, while out-of-phase LSB resulting signal is canceled.

To compare different diode mixer architectures, the presence or absence of various intermodulation component in an RF spectrum due to IF and LO harmonic components (from fundamental to sixth) is shown in Table 7.1, where the harmonic availability is marked by the "+" symbol. Thus, the spectrum of an SDM contains all possible harmonic components, unlike an SBM, where the even-order LO harmonics are suppressed. However, in a DBM or RM, the number of intermodulation components reduces even more, in total by four times, due to the corresponding suppression of both IF and LO even harmonics.

7.4 TRANSISTOR MIXERS

Transistor mixers can be built based on both bipolar transistor and FET device; however, GaAs FETs are commonly used at microwave frequencies due to their superior noise and gain performance. For a bipolar mixer, it was found that the frequency conversion, which is strongly dependent on the forward collector–emitter current gain of the device in a common base configuration in the low-frequency frequency range, primarily depends on the base resistance in the medium-frequency range, and is mainly dependent on the base–emitter capacitance [22]. Due to its highly nonlinear characteristic, the FET device turns out to be a very efficient mixer, and its conversion transconductance exhibits a flat maximum as a function of the gate bias voltage. In this case, an optimum voltage range can be obtained where the ratio between mixing current and distortion product currents is most favorable. Generally, FET mixers exhibit smaller transconductance than mixer stages with bipolar transistors [23]. However, the FET input impedance is higher and thus does allow the use of better and more selective input resonance circuits. The more effective mixer performance can be achieved with lower gate capacitance and bulk resistances at higher frequencies where FET devices are preferred over bipolar transistors as having less intermodulation and cross-modulation distortion and their lower feedback capacitance provides better circuit stability.

In a general case, if a device has the transfer characteristic

$$i(t) = a_0 v(t) + a_1 v^2(t) + a_2 v^3(t) \tag{7.34}$$

then applying a two-tone input signal

$$v(t) = V_1 \cos \omega_1 t + V_2 \cos \omega_2 t \tag{7.35}$$

results in the following components of the output current provided correspondingly by first-order term

$$a_0 V_1 \cos \omega_1 t + a_0 V_2 \cos \omega_2 t \tag{7.36}$$

second-order term

$$\frac{a_1}{2}\left(V_1^2 + V_2^2\right) + \frac{a_1}{2} V_1^2 \cos 2\omega_1 t + \frac{a_1}{2} V_2^2 \cos 2\omega_2 t + a_1 V_1 V_2 \left[\cos(\omega_1 + \omega_2) t + \cos(\omega_1 - \omega_2) t\right] \tag{7.37}$$

and third-order term

$$\begin{aligned}
&\left(\frac{3a_2}{4} V_1^3 + \frac{3a_2}{2} V_1 V_2^2\right) \cos \omega_1 t + \left(\frac{3a_2}{4} V_2^3 + \frac{3a_2}{2} V_1^2 V_2\right) \cos \omega_2 t + \frac{a_2}{4} V_1^3 \cos 3\omega_1 t \\
&+ \frac{a_2}{4} V_2^3 \cos 3\omega_2 t + \frac{3a_2}{4} V_1^2 V_2 \left[\cos(2\omega_1 + \omega_2) t + \cos(2\omega_1 - \omega_2) t\right] \\
&+ \frac{3a_2}{4} V_1 V_2^2 \left[\cos(2\omega_2 + \omega_1) t + \cos(2\omega_2 - \omega_1) t\right]
\end{aligned} \tag{7.38}$$

where the term $a_0 v$ represents linear mixer action, reproducing the input signals at the output, the term $a_1 v^2$ gives rise to a dc component and second harmonics of the input signals as well as the product terms at frequencies $f_1 \pm f_2$, and the term $a_2 v^3$ produces components at frequencies f_1, f_2, $3f_1$, $3f_2$, $2f_1 \pm f_2$, and $2f_2 \pm f_1$ [15].

The basic block schematic of a single FET mixer, where an active device is treated as a three-port network, is shown in Figure 7.11(*a*), with each port generally intended to be connected to LO, IF, or RF signal path, respectively. However, conventional FET mixers usually employ a gate mixing to produce a desired frequency component when LO signal shares the same gate terminal with IF modulated signal. In this case, independent matching for each signal is impossible, and, to achieve a sufficiently high isolation between both LO and IF signals, it is necessary to use hybrid circuits such as Lange couplers, power dividers, or baluns that increase the mixer circuit complexity.

Figure 7.11(*b*) shows the circuit schematic of a monolithic upconversion heterojunction FET mixer MMIC based on coplanar waveguides (CPW) as the transmission lines that can be used in microwave

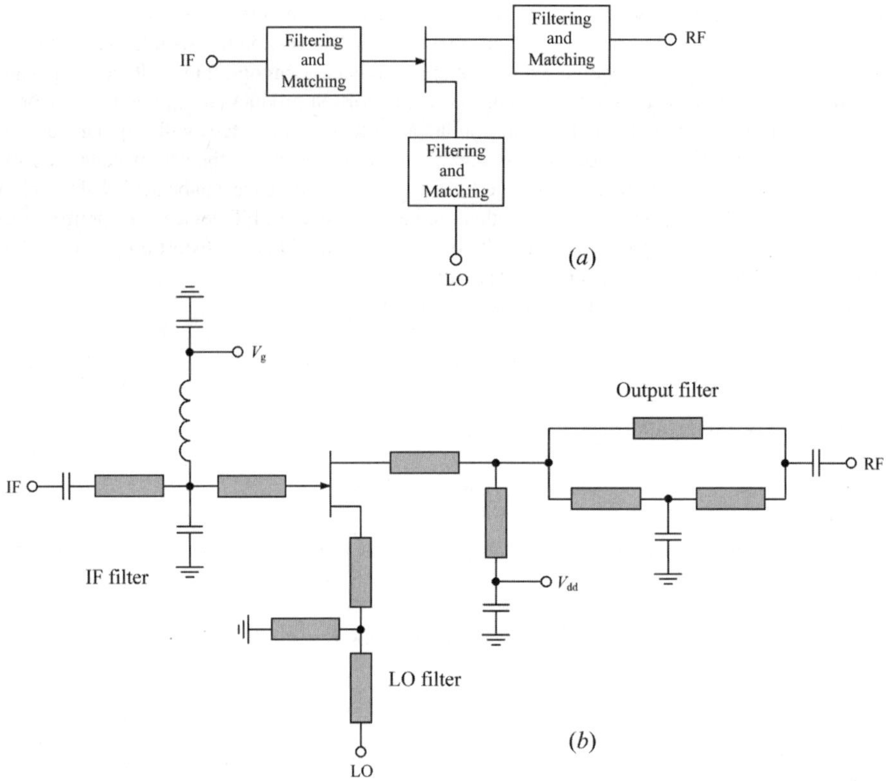

FIGURE 7.11 Block diagram and circuit schematic of source-injection FET mixer.

and millimeter-wave applications [24,25]. In this case, a source injection concept was introduced in which the IF signal with a frequency f_{IF} is applied to the gate terminal through the IF filtering and matching network, the LO signal with a frequency f_{LO} is applied to the source terminal through the LO filtering and matching network, and the resultant USB RF signal having a frequency $f_{RF} = f_{IF} + f_{LO}$ is extracted from the drain terminal, after passing through the RF matching circuit and output filter. The RF output filter consists of two T-type networks connected in parallel for suppressing the LSB frequency $f_{IF} - f_{LO}$ and LO frequency f_{LO}, but passing the USB frequency $f_{IF} + f_{LO}$. The LO matching network is essentially responsible to provide an unconditionally stable operation for the upconverter, both in presence and absence of an LO signal. In this structure, since the LO signal is directly applied to the source terminal, it has the largest effect on the device transconductance modulation and, consequently, on nonlinear mixing enhancement, which facilitates a low LO power operation. As a result, the upconverter can operate with an LO power level as low as −16 dBm for IF signals in 1.5–2.5 GHz band, and LO signals in 20–23 GHz band. In this case, LO suppression at IF and RF ports is better than 20 dB, whereas IF suppression at RF port is better than 35 dB.

Figure 7.12 shows the circuit schematic of a monolithic gate-injection HEMT mixer that shifts a signal in the 16 GHz band up to the V-band using a 48 GHz LO signal [26]. Such a mixer topology was chosen because it is simple in implementation and can achieve high conversion gain for low LO power levels. The network at the gate side of the mixer was designed to provide 50-Ω matching at both IF (16 GHz) and LO (48 GHz) frequencies. However, since two signals are applied to the same port, the external diplexer is required. The gate bias network also includes a small (5 Ω) resistor, which is needed to ensure unconditional stability in the circuit. The network at the drain side of

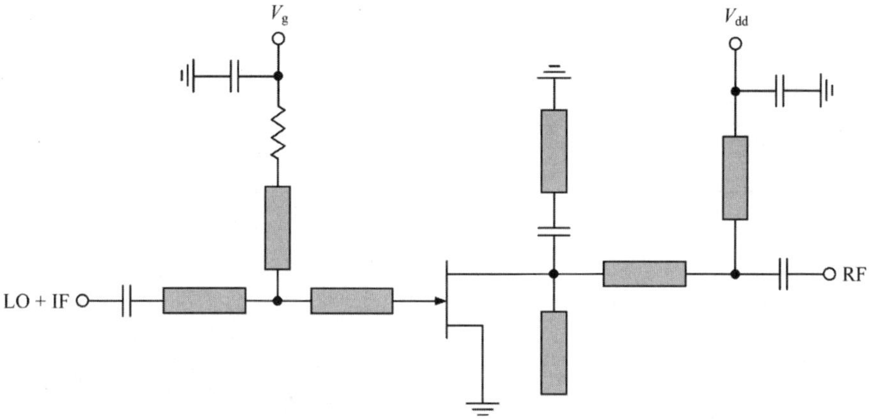

FIGURE 7.12 Circuit schematic of upconverting HEMT mixer.

the device provides both the suppression of LO and IF signals and 50-Ω matching at RF (64 GHz) frequency. Obtaining a good suppression is important because it allows the HEMT device to remain in the saturation region over the entire LO cycle. This achieved by an open-circuit stub with an electrical length of 90° at 48 GHz. Measurements of the fabricated circuit demonstrated a peak conversion gain of 1 dB at 64.5 GHz for −1.7 dBm LO power, an LO suppression better than 30 dB, and an input third-order intercept point of −1.6 dBm.

7.5 DUAL-GATE FET MIXER

Dual-gate FET devices have a major advantage over their single-gate counterparts so that the LO and IF (or RF) signal paths can be connected to separate gates, thus providing a good LO–IF (or LO–RF) isolation because the capacitance between the gates is very low. Therefore, single-gate FET devices are mostly used in realizing balanced mixers since a balanced dual-gate FET mixer requires separate hybrids or baluns for splitting input signals that makes the mixer design too complicated.

The dual-gate MESFET device and its decomposition into an equivalent cascode connection of two single-gate MESFET parts with RF grounded second gate is shown Figure 7.13(*a*) and Figure 7.13(*b*), respectively [27,28]. In this case, the first device is operated in a common source configuration, whereas the second device is operated in a common gate configuration, and the drain current I_{d1} of the first device represents the input current of the second device. This means that the first transistor can operate as a current source for the second transistor, and the drain–source voltage V_{ds1} across the first transistor can be controlled by the gate bias voltage V_{gs2} of the second transistor. The other advantage of the dual-gate FET device is that it can provide operation stability over wider frequency range due to much lower feedback capacitance between the second-device drain and the first-device gate.

The main nonlinear operation regions of a dual-gate MESFET can be identified using its bidimensional transfer and output voltage–ampere transfer characteristics shown in Figure 7.13(*c*) [29]. Here, the vertical traces correspond to the different gate voltages V_{gs2} of the FET2 for constant $V_{ds} = 5$ V and must be divided between the channels of the two single-gate FETs because $V_{ds} = V_{ds1} + V_{ds2}$. In this case, there are mainly three nonlinear regions of the dual-gate MESFET to operate as a mixer:

- Region I where FET1 operates in a linear region of its output voltage–ampere characteristics and FET2 operates in a current saturation region with low V_{gs1} and V_{gs2} close to pinch-off voltages.

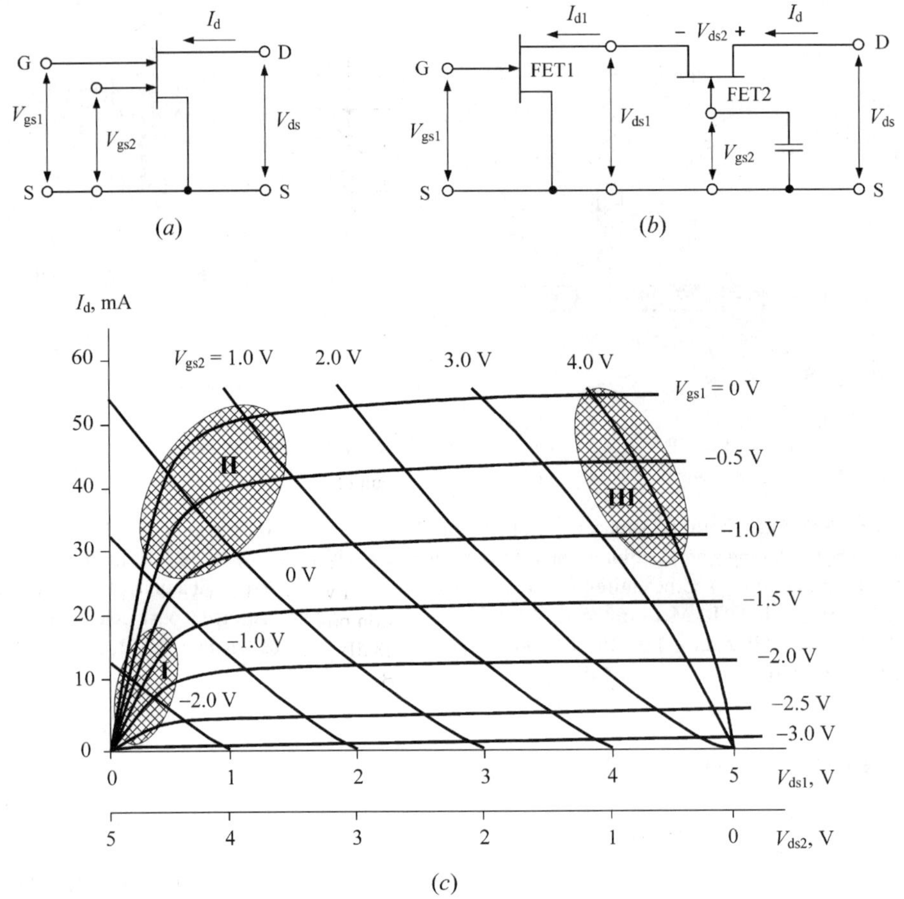

FIGURE 7.13 Dual-gate MESFET, its equivalent cascode representation, and voltage-ampere characteristics.

- Region II where the FET1 operates in a linear region close to saturation and the FET2 operates in a current saturation for high V_{gs1} and V_{gs2} far from pinch-off voltages.
- Region III where the FET1 operates in a saturation region and the FET2 operates in a linear region with high V_{gs1} and V_{gs2} far from pinch-off voltages.

Figure 7.14 shows the equivalent circuit of a dual-gate MESFET showing the elements responsible for nonlinear operation in different bias regions. It is obvious that in bias regions I and II, mixing takes place inside of FET1, while FET2 acts as a linear amplifier. In this case, if the mixing in region I is mainly provided at low drain current by the nonlinearity of the device transconductance g_{m1} and output differential resistance R_{ds1} as the functions of V_{gs1}, then, in region II, the nonlinear effect of the gate–source capacitance C_{gs1} at high drain current becomes significant as well. Due to the low drain current, the mixer operating in region I requires smaller LO power and provides low-noise performance. However, if the dual-gate MESFET operates as a mixer in region III, the nonlinearities are located in FET2, and FET1 acts now as a linear amplifier.

Figure 7.15 shows the block schematic of a microwave upconversion mixer based on a dual-gate GaAs MESFET operating at X-band with an IF input signal varying over 700 ± 250 MHz applied to the second gate [30]. Such an upconverter pumping by the LO signal at 7.4 GHz can offer the

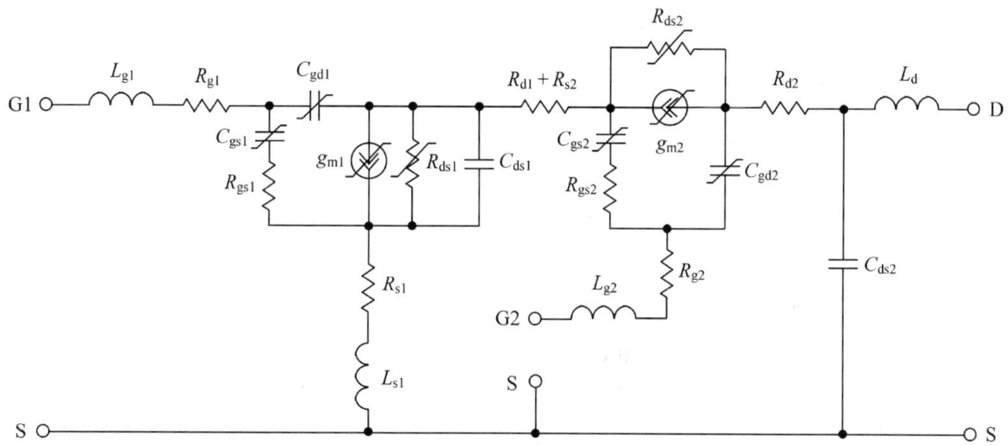

FIGURE 7.14 Nonlinear equivalent circuit of dual-gate MESFET.

conversion gains up to 15 dB, noise figure of 3.2 dB and saturated RF output power up to 12 dBm. For a particular case of the saturated output power of 9 dBm with the LO level set at 12 dBm, the third-order intermodulation product at the 1-dB compression point was −22 dBc. The isolation from the input LO port to the input IF port was measured to be 28 dB, whereas the isolation from the output RF port to either of the input ports was greater than 20 dB.

7.6 BALANCED TRANSISTOR MIXERS

7.6.1 Single-Balanced Mixers

A pair of single-gate MESFET mixers can be combined into a single-balanced MESFET mixer using different hybrid combiners or simply parallel connection at the output. In this case, the properties of balanced MESFET mixers with regard to intermodulation products and isolation of input and output ports are essentially the same as in balanced diode mixers. Figure 7.16 shows the typical block schematic of a balanced MESFET mixer where the LO signals being 180° out of phase are mixed with IF signals fed from a 180° hybrid [31]. As a result, the output upconverted signals are combined at the RF port as being in phase, while the LO signals are canceled as being out of phase. In the practical layout of such a 20-GHz balanced upconversion mixer, a 180° phase difference between the LO signals can be obtained through the slotline Y-junction formed on the other side of substrate. The

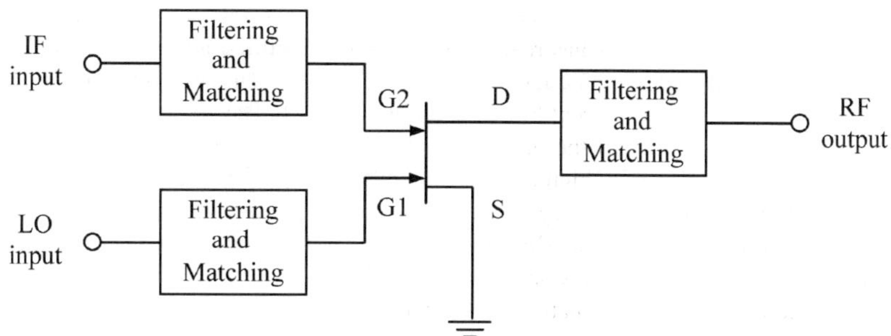

FIGURE 7.15 Dual-gate MESFET upconverter configuration.

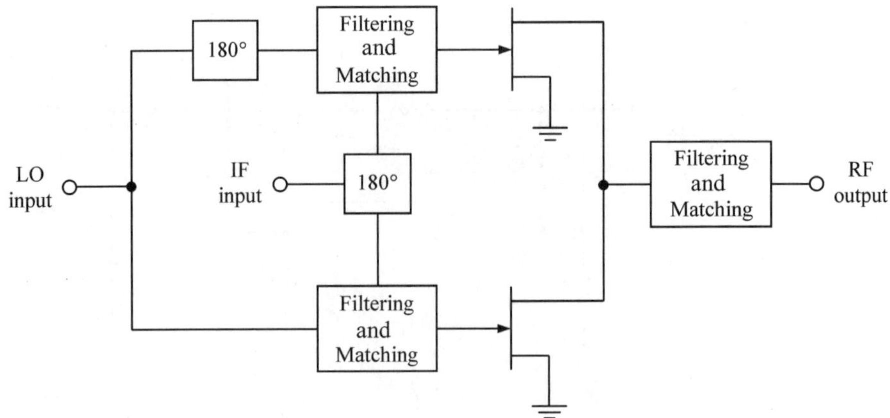

FIGURE 7.16 Balanced MESFET upconverter configuration.

combiner can consist of a coupled microstrip lines with a connecting gold ribbon whose position is properly adjusted to optimize the LO reflection phase. As a result, the LO level measured at the output RF port was lower by 12 dB than the upconverted output power of 15.9 dBm without external filter.

Since the input MESFET impedance for IF frequency is high, the gate voltage swing of the IF signal is considerably large. Consequently, a simplified analysis shows that the MESFET transconductance $g_m(t)$ in time domain can be approximated by its saturation value g_{m0} at positive half cycles of the IF signals and zero at negative half cycles. A Fourier series expansion of $g_m(t)$ results in

$$g_m(t) = g_{m0}\left(\frac{1}{2} + \frac{2}{\pi}\sin\omega_{IF}t + \cdots\right) \tag{7.39}$$

where $\omega_{IF} = 2\pi f_{IF}, f_{IF}$ is the IF frequency. When the LO voltage of a radian frequency $\omega_{LO} = 2\pi f_{LO}$ and an amplitude V_{LO} is applied, the drain current $i_d(t)$ can be expressed as

$$\begin{aligned} i_d(t) &= g_m(t)V_{LO}\sin\omega_{LO}t \\ &= g_{m0}V_{LO}\left\{\frac{1}{2}\sin\omega_{LO}t - \frac{1}{\pi}\left[\cos\left(\omega_{LO}+\omega_{IF}\right)t - \cos\left(\omega_{LO}-\omega_{IF}\right)t\right] + \cdots\right\} \end{aligned} \tag{7.40}$$

which means that the RF output signal is lower by the factor of $1/\pi$ (or -9.9 dB) corresponding to both low-side upconversion at $\omega_{LO} - \omega_{IF}$ and high-side upconversion at $\omega_{LO} + \omega_{IF}$ than the maximum RF output signal when the MESFET is used in a power amplifier application.

For millimeter-wave applications, a concept of subharmonic balanced mixer with the LO frequency about half the RF output frequency can be used. Figure 7.17 shows the simplified circuit schematic of a 60-GHz balanced CMOS upconversion mixer where a miniature transformer-coupler is introduced to provide two LO input signals with 90° phase difference [32]. The IF signal feeds through the low-pass filter and injects to transistor M_1. By using the property of a common source transistor, the IF signals at the drain and gate of M_2 have 180° phase difference at low frequency. Therefore, the IF signals flowing to the mixing transistors M_3 and M_4 are out of phase. The drains of M_3 and M_4 are connected together to cancel the fundamental IF and second harmonic LO. As a result, the measured LO–RF and 2LO–RF isolations for RF frequency from 58 to 66 GHz were better than 40 dB.

Figure 7.18 shows the circuit schematic of a balanced *Ku*-band upconverter using HEMT technology for low-noise monolithic application [33]. The mixer consists of an HEMT pair with the common source and common gate transistor configurations. For a common gate HEMT, the LO

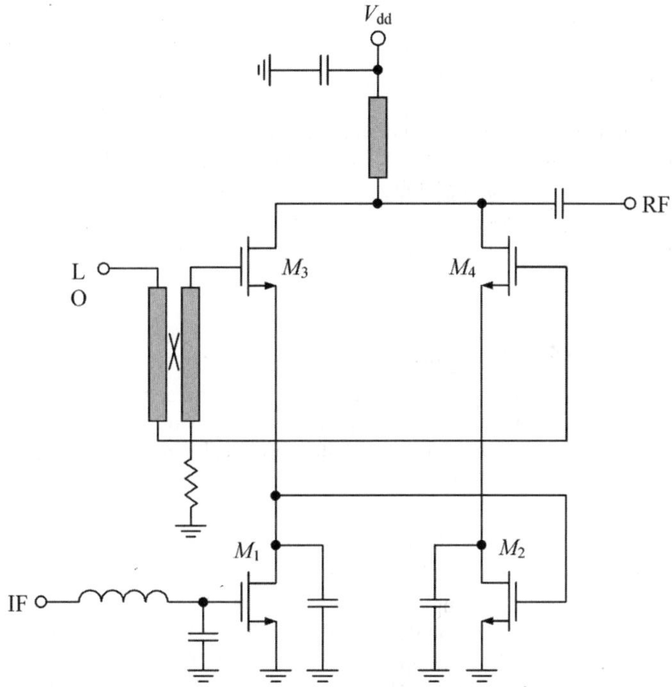

FIGURE 7.17 Schematic of balanced CMOS upconversion mixer MMIC.

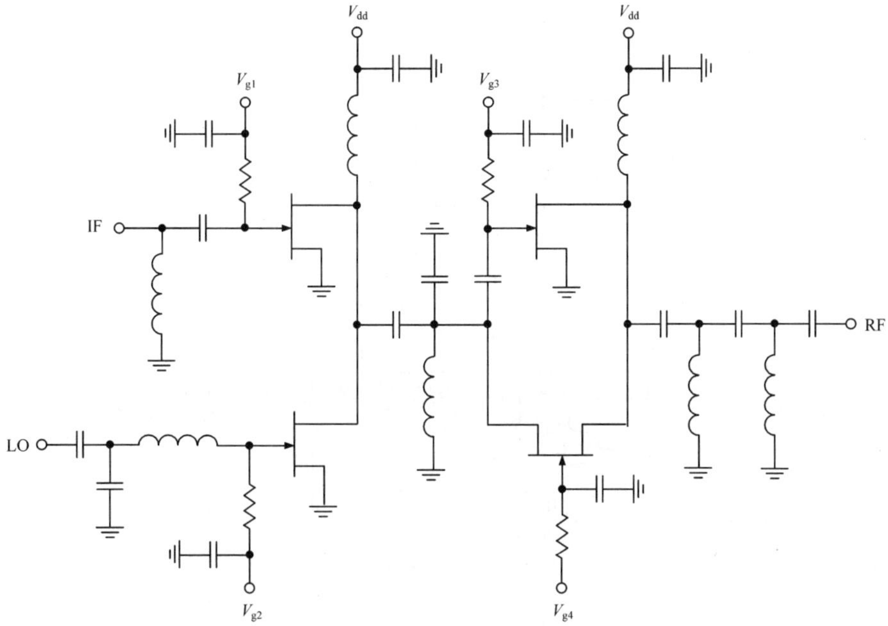

FIGURE 7.18 Circuit schematic of balanced HEMT upconversion mixer MMIC.

signal is injected between the gate and source ports. In both configurations, the time-varying device transconductance $g_m(t)$ is the dominant contributor to frequency conversion, and the effect of other nonlinearities is minimal. By properly selecting the gate width of the common source transistor, it is possible to have the same gain in both devices. As a result, the IF and LO signals will be canceled at the output of the parallel-connected transistors due to the phase shift (ideally 180°) in the common source transistor. To combine the LO and IF signals, two active circuits based on the single-stage amplifying stages were used, as well as a high-pass filter at the RF output was included to improve isolation. As a result, a conversion gain over 4.2 dB at RF frequency of 14 GHz (IF = 1.885 GHz) was achieved with 3 dBm of LO power and isolation between any ports over 27 dB.

7.6.2 Double-Balanced Mixers

The double-balanced transistor mixers have essentially the same characteristics as the double-balanced diode mixers in terms of isolation, spurious responses, or bandwidth but substantially higher conversion gain and they do not require hybrids or baluns that make them very attractive for compact monolithic implementation. The DBM in the form of a pair of the transistorized differential amplifiers was originally invented by Howard Jones to use as a dual output synchronous detector [34]. However, Alberto Bilotti was the first who described and analyzed the fully balanced arrangement of the three differential pairs shown in Figure 7.19 with regard to its different monolithic applications such as balanced product mixer (downconverter or upconverter) and suppressed carrier modulator [35]. In this case, a DBM requires low-level operation at both inputs, whereas suppressed carrier modulation is obtained by low-level modulating signal $v_1(t)$ and high-level carrier signal $v_2(t)$. In the latter case, carrier high-level operation provides a modulated output independent from the carrier level and should be preferred unless the harmonic spectral response is acceptable. Note that the amount of carrier suppression is a function of the matching accuracy of the differential pairs.

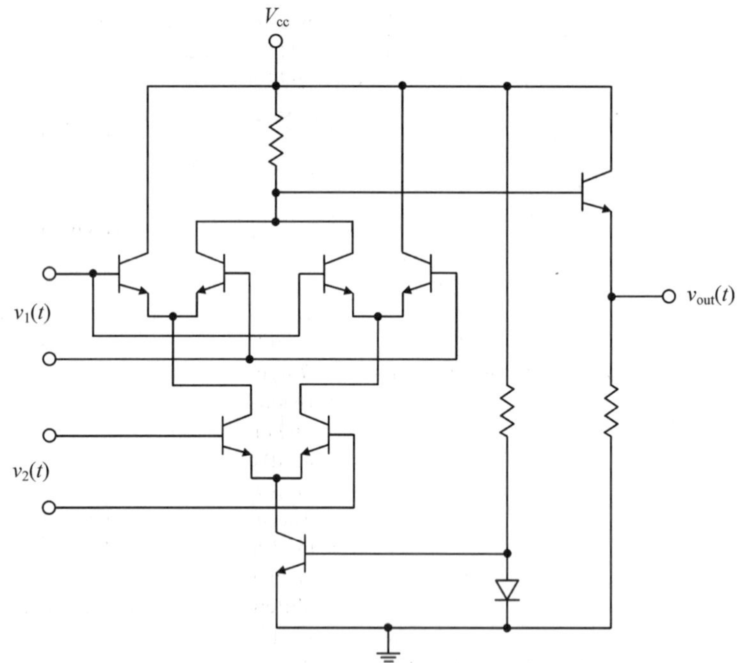

FIGURE 7.19 Monolithic double-balanced mixer configuration.

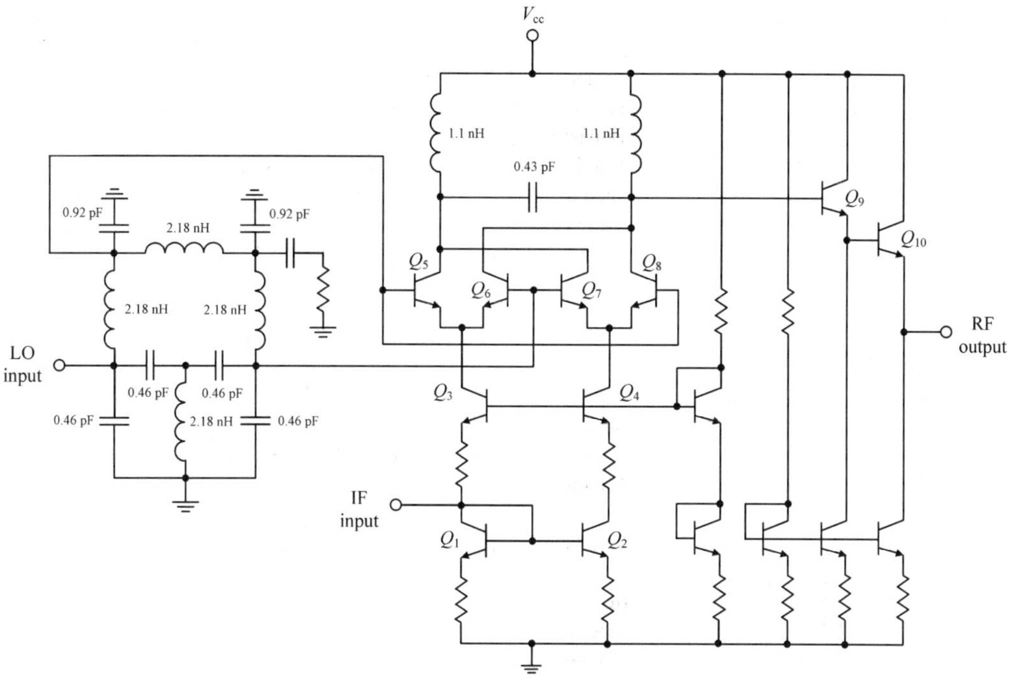

FIGURE 7.20 Schematic of monolithic double-balanced mixer with lumped hybrid.

For an upconversion balanced mixer, both the input signal $v_1(t)$ and the input signal $v_2(t)$ can respectively be considered either an LO signal or an IF signal to be converted into the output RF signal $v_{out}(t)$. Being implemented with a GaAs HBT technology, such a double balanced upconverter when $v_1(t) = v_{IF}(t)$ and $v_2(t) = v_{LO}(t)$ can provide the conversion gain greater than 20 dB (with buffer amplifiers) up to an RF output frequency of 5.5 GHz [36]. At the same time, a 20-GHz AlGaAs/GaAs HBT frequency upconverter can exhibit the conversion gain of 5 dB with the RF–LO isolation of 23 dB for RF output up to 8.5 GHz when $v_1(t) = v_{LO}(t)$ and $v_2(t) = v_{IF}(t)$ [37].

Since generally the differential-pair DBM requires external baluns for the LO and IF input ports resulting in its bulky and expensive implementation, one way to solve this problem is to use a lumped equivalence of the hybrid baluns. GaAs material has the semi-insulating substrate with low conductivity, and therefore, a sufficiently high quality factor of the lumped-element inductors can be achieved. Figure 7.20 shows the circuit schematic of a fully integrated double-balanced GaInP/GaAs HBT upconversion mixer, consisting of a mixer core based on two differential pairs (Q_5 to Q_8) with a 180° lumped rat-race hybrid for the balanced LO signal, an active IF balun (Q_1 to Q_4), and an RF output *LC* current combiner with an emitter–follower output buffer (Q_9 and Q_{10}) [38]. The rat-race hybrid employs the π-type *LC* low-pass network and the *T*-type *LC* high-pass network to replace the quarter-wavelength and three-quarter-wavelength transmission-line sections, respectively. The output *LC* current combiner formed by two inductors of 1.1 nH and one capacitor of 0.43 pF is necessary to perform the differential-to-single (or balanced-to-unbalanced) conversion at the RF output port. As a result, the conversion gain of 1 dB when $f_{IF} = 300$ MHz, $f_{LO} = 4.9$ GHz, and $f_{RF} = 5.2$ GHz can be achieved, with the measured LO–RF isolation of 38 dB. The single-to-differential conversion for the LO signal can be provided by an active balun in the form of the differential HBT pair reducing the overall die size but consuming extra power with limited dynamic range and conversion gain. In this case, in order to increase the conversion gain, the output differential amplifier is used providing also the differential-to-single conversion of the output RF signal [39].

Figure 7.21 shows the simplified schematic of a monolithic double-balanced MESFET mixer that upconverts an input 100 to 500 MHz IF signals to an output 0.6 to 1.75 GHz RF signals with a

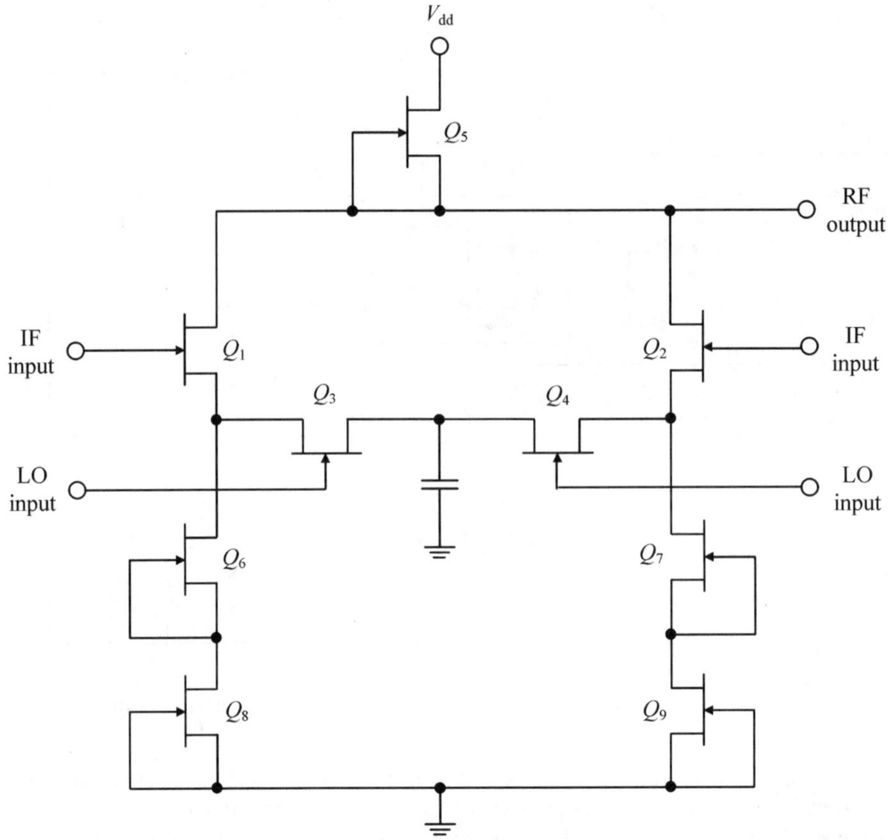

FIGURE 7.21 Simplified schematic of double-balanced MESFET mixer.

conversion gain of 8 dB [40]. The circuit consists of two mixers stages that are excited by IF and LO signals that are 0° and 180° out of phase with respect to one another. The outputs of the two mixers are then combined and amplified to produce an RF output, with the conversion gain depending on the LO level. Mixing action is accomplished by alternating amplification of the two 180° out-of-phase IF signals at the LO frequency. When the LO transistor is turned on, ideally an ac ground is created across it, which makes an effective common-source amplifier for the IF signal. Conversely, when the LO transistor is turned off, no amplification of the IF signal occurs at the output. Therefore, the LO signal alternately turns each side of the mixer into an IF amplifier, thus creating an output RF signal that is a combination of the two signals. To achieve the proper mixing action, the dc components of the LO and IF transistors must be identical to ensure in the quiescent state that the drain-to-source voltages of the devices Q_3 and Q_4 are zero, thereby preventing any LO feedthrough in the absence of an IF signal.

For low supply voltage and low power consumption requirements, RF CMOS technology can be used in the high microwave range in phased-array radars, wireless local networks, local multipoint distribution services (LMDS), and other industrial-scientific-medical (ISM) band applications. Although passive mixers have the inherent advantage of better frequency and linearity response, they have conversion loss and also require a high LO drive level. A typical circuit schematic of the double-balanced CMOS mixer based on a pair of differential amplifiers is shown in Figure 7.22(*a*) where the gate and source degeneration inductors are used to match LO and IF inputs and RF output to standard 50 Ω [41]. Having source degeneration inductors also helps to improve the linearity at some

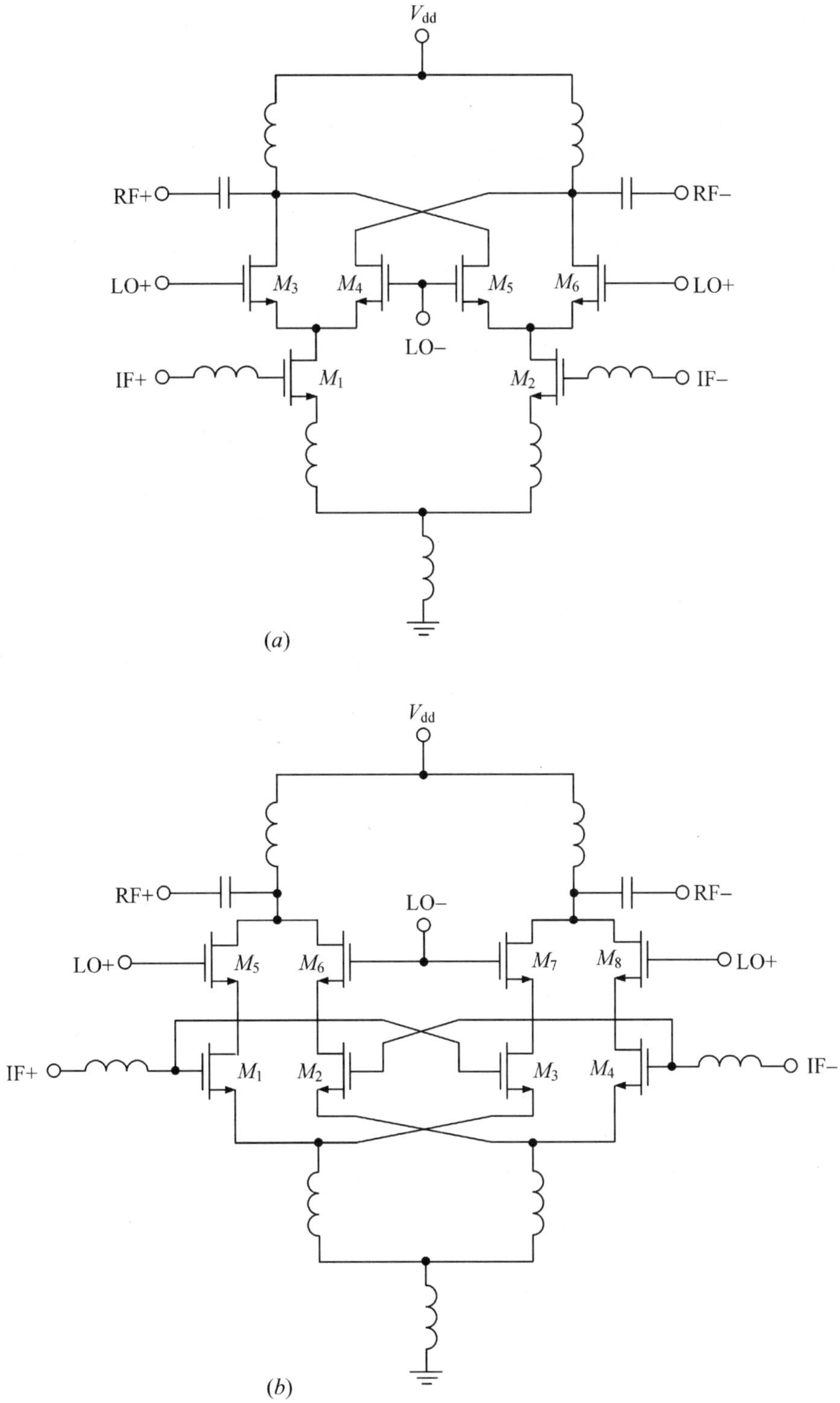

FIGURE 7.22 Simplified schematics of double-balanced CMOS mixer.

expense of conversion gain. To improve conversion gain of the upconversion mixer that is intended to operate in a superheterodyne 60-GHz transmitter, the transmission-line stubs in the output circuit and between each differential pair and current-source transistors can be used [42].

However, even lower available LO drive can be achieved in a dual-gate DBM, the simplified circuit schematic of which is shown in Figure 7.22(*b*) [41]. In a dual-gate mixer, a finite LO signal must be presented at the drains of lower devices (M_1 to M_4), which causes the devices to operate between linear and saturation regions, thus resulting in the modulated transconductance of these devices. This is because the lower devices are operated in linear region during most part of the LO cycle, unlike a conventional double-differential mixer where switching action of the LO transistor pair occurs between the cutoff and saturation regions. As a result, the conversion gain of a dual-gate mixer is slightly lower, but its linearity performance is moderately better. The mixer can achieve a conversion gain of 1.7 dB and an output 1-dB compression point of −3.4 dBm at a supply voltage of 1.2 V with a current consumption of 9 mA when $f_{IF} = 2{,}3$ GHz and $f_{RF} = 23$ GHz.

7.7 FREQUENCY MULTIPLIERS

The frequency multipliers historically were the very important part of the vacuum-tube transmitters to achieve a high-frequency transmission of a stable modulated signal. In this case, a high-frequency transmitter chain was comprised of a crystal-controlled vacuum-tube circuit, oscillating on the crystal frequency, and a non-oscillating power amplifier or several power amplifiers, the output resonant circuit of which is tuned to a small integral multiple of the crystal frequency [43,44]. It was found that, for any given excitation voltage, there is a fairly critical adjustment of negative grid voltage corresponding to Class C mode for maximum efficiency and another critical adjustment for maximum output harmonic power. As the order of multiplication is increased, these adjustments become more critical and the points of the maximum harmonic efficiency and power more nearly coincide. However, both higher efficiency and greater output power is obtained for higher values of excitation voltage.

For a piecewise-linear approximation of the device transfer characteristic shown in Figure 7.23 with cosine driving voltage, an effect of the duty cycle (or conduction angle) of the pulsed output current on the operation mode of a frequency multiplier can be readily understood based on the ratio of the *n*th-harmonic current amplitude I_n and maximum amplitude I_{max} of the output current, which can written as

$$\alpha_n = \frac{I_n}{I_{max}} \tag{7.41}$$

where maximum current amplitude I_{max} as a function of a half-conduction angle θ is equal to

$$I_{max} = I\,(1 - \cos\theta) \tag{7.42}$$

where I is the pulsed current amplitude, and $I = I_{max}$ when $\theta = 90^\circ$, which corresponds to 50% duty cycle. The amplitude I_n of *n*th current harmonic is defined as $I_n = \gamma_n(\theta)I$, where $\gamma_n(\theta)$ is the *n*-harmonic current coefficient defined as

$$\gamma_n(\theta) = \frac{1}{\pi}\left[\frac{\sin(n-1)\theta}{n(n-1)} - \frac{\sin(n+1)\theta}{n(n+1)}\right]. \tag{7.43}$$

As a result,

$$\alpha_n = \frac{\gamma_n(\theta)}{1 - \cos\theta} \tag{7.44}$$

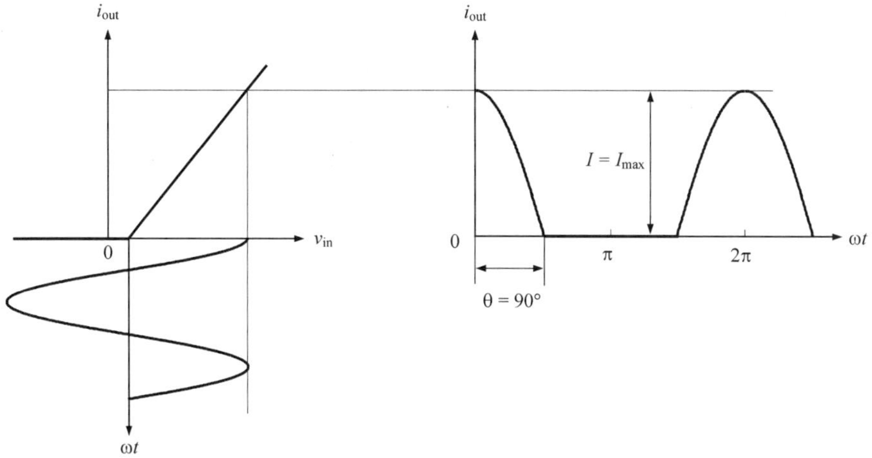

FIGURE 7.23 Piecewise-linear approximation and pulsed output current waveform.

which shows that, for each harmonic component, there is a particular angle of current flow, for which α_n is a maximum, and hence where the maximum harmonic output is obtainable. In this case, the maximum value of $\alpha_n(\theta)$ is achieved when $\theta = 120°/n$, where n is the harmonic order. The best angle of current flow is almost exactly inversely proportional to the order of the harmonic and differs slightly for the piecewise-linear and square-law cases [45].

The possibility of a frequency multiplication in the transistor amplifiers can be explained on the basis of two physical principles. The first principle is based on the transistor operation with a certain conduction angle that results in the harmonic generation at the device output. In this case, it is enough to employ idler circuit at the collector of the bipolar transistor tuned to the selected harmonic component, as shown in Figure 7.24(*a*), where the tank idler circuit tuned to the *n*th harmonic is used to provide low impedance conditions at the collector for the rest harmonic components including fundamental. Such a circuit configuration of the transistor frequency multiplier is usually used for low orders of frequency multiplication and high-power levels [46]. The second multiplication principle is based on the nonlinearity of the device collector-junction capacitance that is used in the varactor parametric multipliers [6]. In this case, due to the load-network configuration with series idler harmonically-tuned circuits with finite loaded quality factors, as shown in Figure 7.24(*b*), the collector current also flows through the feedback collector abrupt-junction capacitance, thus resulting in a frequency multiplication due to parametric effect. The parametric frequency multiplication is a very effective approach when it is necessary to achieve the output frequency exceeding the transistor transition frequency f_T by 2–3 times. In addition, for lower power transistors and higher order multiplication, the idler circuit can be placed at the base of the transistor that exhibits the abrupt change of base-spreading resistance as the input voltage increases that may result in an additional improvement of the multiplier conversion gain [47].

With introduction of GaAs MESFET devices operated as frequency multipliers, the high conversion gain and good isolation between the input and output ports due to significantly lower feedback gate-drain capacitance can be achieved. Besides, the MESFET frequency multiplier provides high conversion efficiency and circuit stability, requires low driving power, and the lack of an idler circuit at the input makes it reasonable broadband [48]. Generally, the major nonlinearities in the MESFET causing harmonic generation are the nonlinear gate-source and gate-drain capacitances, the drain current clipping due to the gate-source Schottky diode, and the nonlinear transconductance and output differential resistance due to the drain current source operation in the pinch-off and saturation regions. However, the largest contributor to the MESFET frequency doubler is the drain-current clipping effect when the device is biased near close to 0 V, just below the forward conduction point

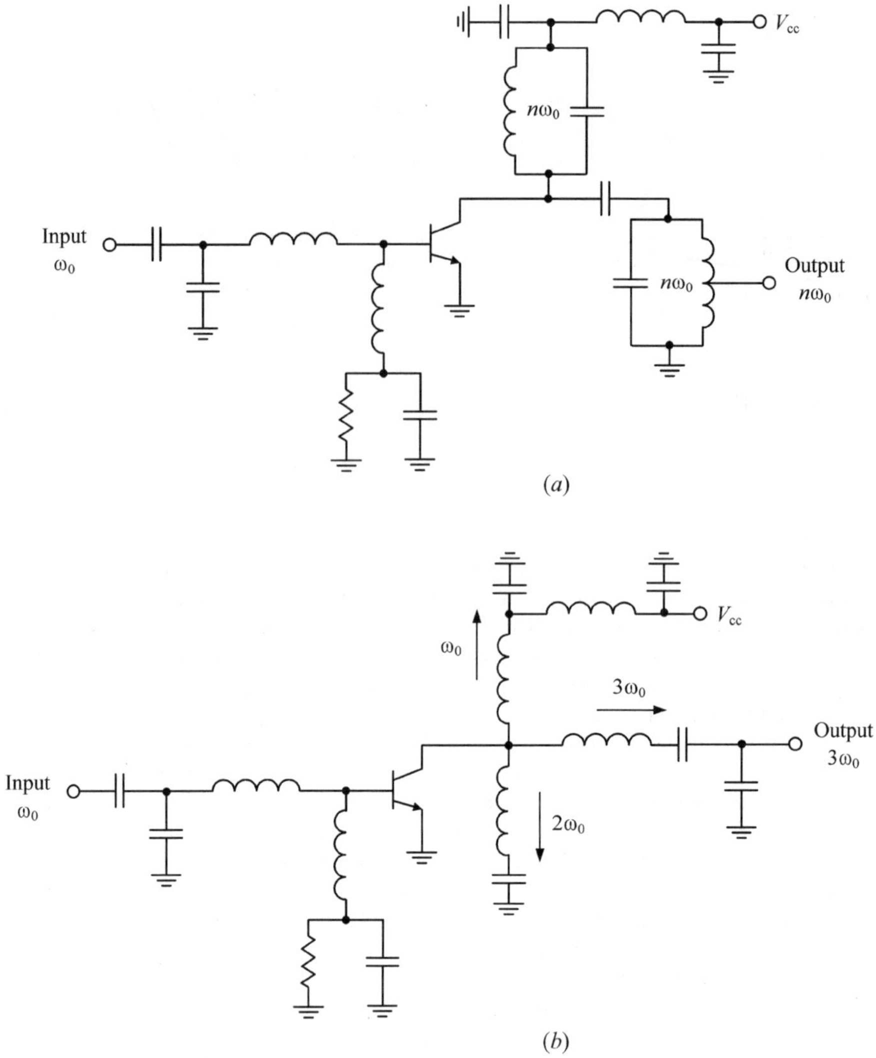

FIGURE 7.24 Circuit schematics of bipolar frequency multipliers.

of the gate-source junction, resulting in a half-rectified waveform and a maximum conversion gain for the MESFET frequency doubler [49]. This suggests that the device is principally a resistive doubler and therefore harmonic conversion gain is expected to fall as $(1/n)^2$. In a dual-gate MESFET, the frequency multiplication is provided primarily by the combining effect of the input diode and drain current source nonlinearities [50]. In this case, the first gate transconductance is periodically modulated by an input signal and the amplified signal swing at the second gate is modulated in turn by its nonlinearity, thereby generating harmonics. Then, the resultant harmonics are further amplified and extracted from the drain. As a result, the dual-gate MESFET multiplier offers the advantages of better efficiency and easier gain control over its single-gate counterpart.

Figure 7.25 shows the simplified schematic of a reflector-type MESFET frequency doubler where the fundamental-frequency matching circuit and second-harmonic reflector are placed at the input, and the second-harmonic matching circuit and fundamental-frequency reflector are placed at the

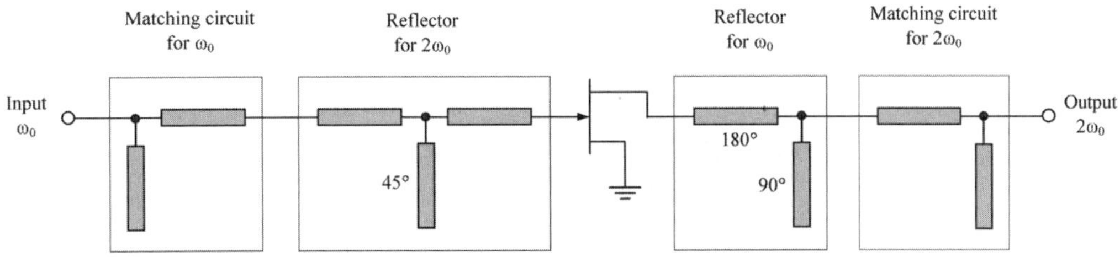

FIGURE 7.25 Schematic of reflector-type MESFET frequency doubler.

output [51,52]. The open-circuit stub with an electrical length of 45° is located inside the input reflector to provide the short-circuit condition for the second harmonic, while the open-circuit stub with an electrical length of 90° and series stub with an electrical length of 180° represent the output reflector to provide the short-circuit condition for the fundamental at the device drain. As a result, a conversion gain of 6 dB for the MESFET frequency doubler was achieved at an output frequency of 24 GHz. Using the separate fundamental-frequency rejection network at the device drain when it can be adjusted independently from the frequency-selective load network can improve significantly the conversion gain for the 11.5 to 34.5 GHz HEMT frequency tripler [53]. The fundamental-frequency termination at the device output is especially useful for a millimeter-wave HEMT multiplier when the frequency ranges from 180 to 220 GHz for a doubler and from 130 to 150 GHz for a tripler with acceptable conversion gain can be achieved [54].

To provide high spectral purity and broadband operation, the balanced frequency-doubler configurations are preferred because they provide efficient rejection of the fundamental and odd-harmonic components. In this case, the bandwidth of the doubler is mainly limited by the phase and amplitude imbalance of the hybrid that increases with frequency. To overcome this limitation, it is necessary to apply broadband MMIC approaches for combining scheme or hybrid implementation. Figure 7.26(*a*) shows the miniaturized broadband balanced frequency doubler MMIC that consists of a common-gate and common-source MESFETs connected in parallel electrically and in series by electrode configuration such as source-gate-drain-gate-source [55]. Here, the gate of a common-gate device is grounded through a capacitor, and the source of a common-gate device and the gate of a common-source device are connected in phase at the input port through the phase shifter and dc-blocking capacitor, respectively. The phase shifter compensates for a phase error between the device outputs, and the output matching circuit is required for load matching at the second harmonic. The doubler operates near the gate-source pinch-off voltage, and the amplitudes of the output signals from both MESFETs are made equal by adjusting each gate bias. Due to different device configurations, their fundamental components are 180° out of phase, whereas the second-harmonic components are in phase, thus resulting in a cancelation of the fundamental frequency signals and an in-phase combining of the second-harmonic signals. The conversion gain from −8 to −10 dB was achieved with fundamental-signal suppression better than 17 dB from 6 to 16 GHz.

Figure 7.26(*b*) shows the basic configuration of a balanced frequency doubler consisting of the two identical MESFETs and a 180° hybrid divider [56]. Since a practical rat-race hybrid was characterized by amplitude imbalance of about 1.1 dB and phase imbalance of about 2° from 28.8 to 39.4 GHz, the bias compensation approach when each transistor is biased differently to compensate for the hybrid asymmetry was applied. As a result, the fundamental-frequency rejection achieved was better than 35 dB over second-harmonic output frequency bandwidth of 31.5 to 37.5 GHz. A planar Marchand balun is very effective to divide the incoming signal into two equal parts with 180° phase difference in millimeter-wave applications to provide a conversion gain of −10 dB for a balanced HEMT doubler covering the frequency range from 150 to 220 GHz [57]. To achieve frequency doubling over an octave bandwidth, good conversion gain and fundamental-frequency rejection, the microwave MESFET frequency doubler can be built based on a balanced structure with Lange couplers followed

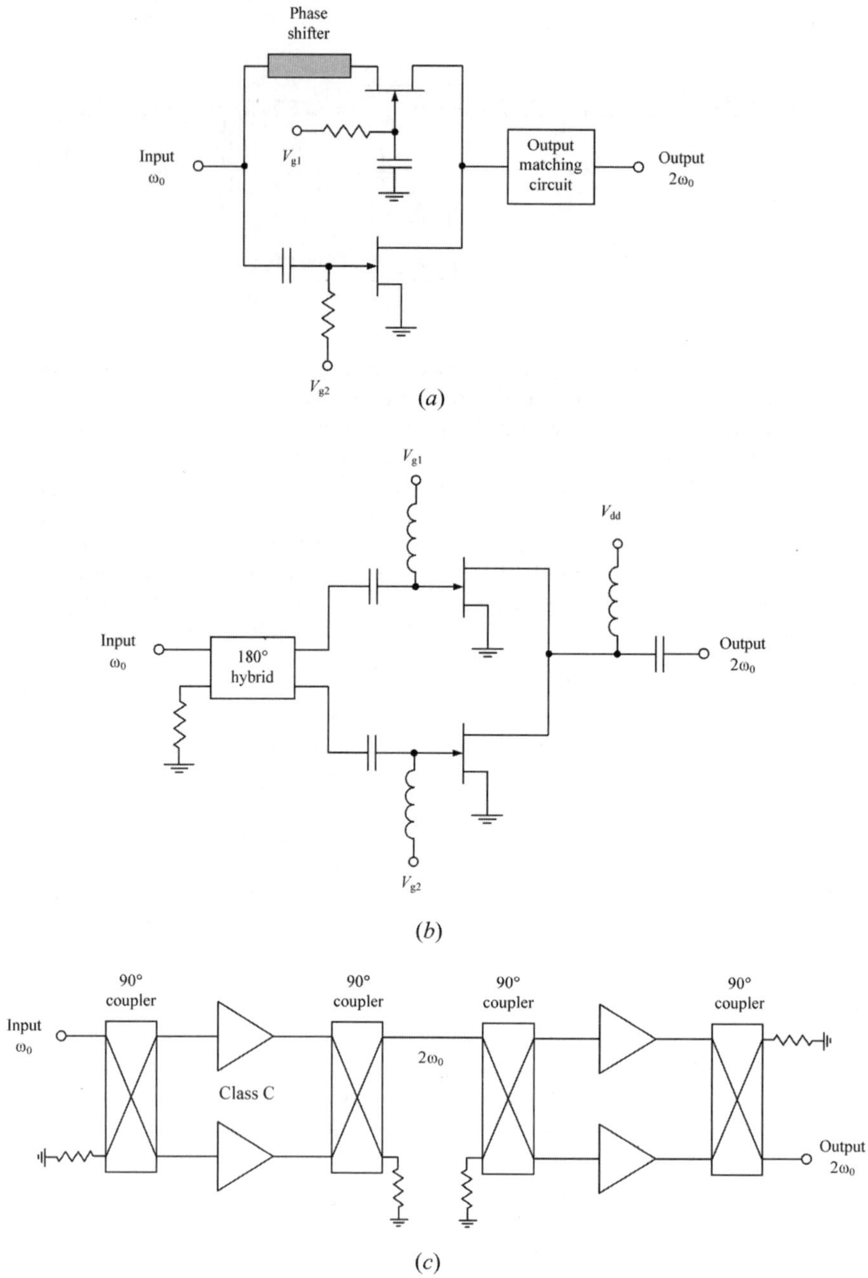

FIGURE 7.26 Schematics of balanced MESFET frequency doublers.

by a broadband balanced amplifier, as shown in Figure 7.26(*c*) [58]. In this case, the input and output Lange couplers in a frequency doubler are connected symmetrically, rather than in complementary symmetry, as for a conventional amplifier, thereby creating a frequency doubled waveform with rejected fundamental signal to be then amplified. The bandwidth of the doubler is controlled by the bandwidth over which the Lange couplers can maintain a 180° phase-difference path. An average

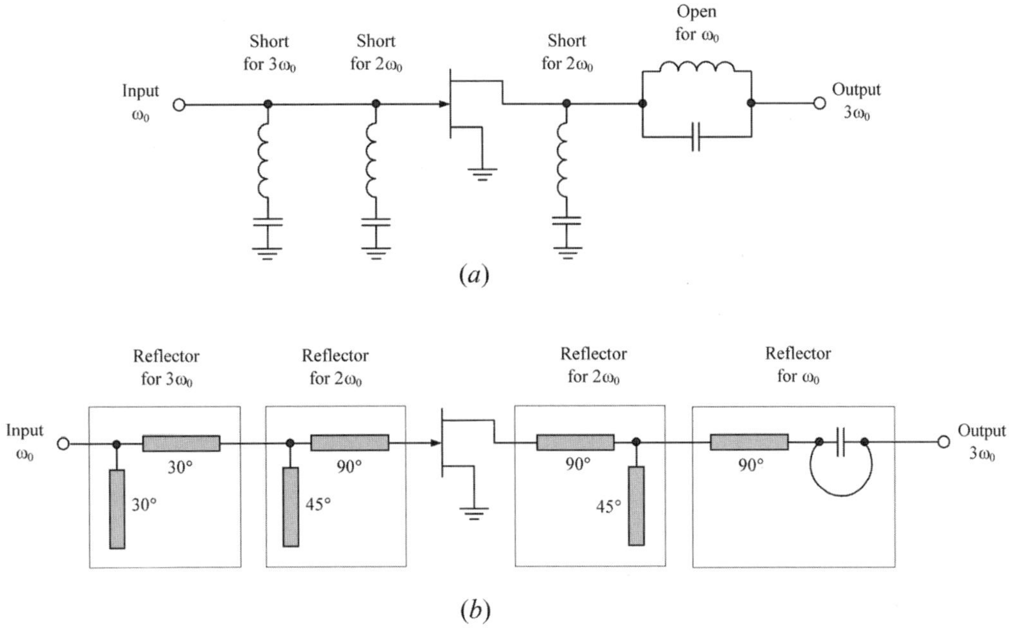

FIGURE 7.27 Circuits schematics of reflector-type HEMT frequency tripler.

conversion gain of −3 dB with fundamental-frequency rejection greater than 12 dB was achieved in the 8–16 GHz output band.

Figure 7.27(a) shows the simplified circuit schematic of a lumped HEMT tripler where the architecture of the input and output circuits is chosen to exploit the fact that large amounts of unwanted harmonic signals exist at both the transistor input and output [59]. For a tripler, the schematic includes resonant circuits for $2\omega_0$ and $3\omega_0$ at the gate and for ω_0 and $2\omega_0$ at the drain to reflect these unwanted harmonics back into the transistor gate and drain, causing nonlinear interactions that can be utilized to maximize the resultant $3\omega_0$ output signal. Optimization of the idealized input and output circuits shows that maximum conversion gain is provided when both input resonant circuits represent a short circuit for $2\omega_0$ and $3\omega_0$, whereas the output resonant circuits should represent an open circuit for ω_0 and a short circuit for $2\omega_0$, respectively. Besides, inserting the offset transmission lines between the harmonic loads and the transistor from both sides can further improve the tripler performance by varying the phase of the harmonic signals. In practical microwave implementation, the short circuit elements (input $2\omega_0$ and $3\omega_0$ and output $2\omega_0$) are realized using corresponding quarter-wavelength open-circuit stubs, as shown in Figure 7.27(b). The open circuit for ω_0 was implemented with a parallel combination of chip capacitor and wire-loop inductor, allowing reflection of the fundamental-frequency signal without affecting $3\omega_0$ output. As a result, for such a reflector-type HEMT tripler, the measured conversion gain greater than 3.5 dB for the input fundamental-frequency signal at 2.94 GHz was achieved.

Figure 7.28 shows the simplified circuit schematic of a balanced MESFET tripler with quadrature input and output couplers [60]. In this case, the input Ku-band signal is amplified in the first stage and feeds the tripler MESFETs with 90° out of phase. At the output ports of the second stage, the fundamental-frequency signal remains with 90° out of phase, whereas the second and third harmonic signals are 180° and 270° out of phase, respectively. When combined in the output quadrature coupler, only third-harmonic Q-band signals are in phase. The fundamental-frequency components are 180° out of phase and cancel each other at the output. The fundamental suppression of 40 dB over frequency range from 42 to 51 GHz was achieved using Lange couplers.

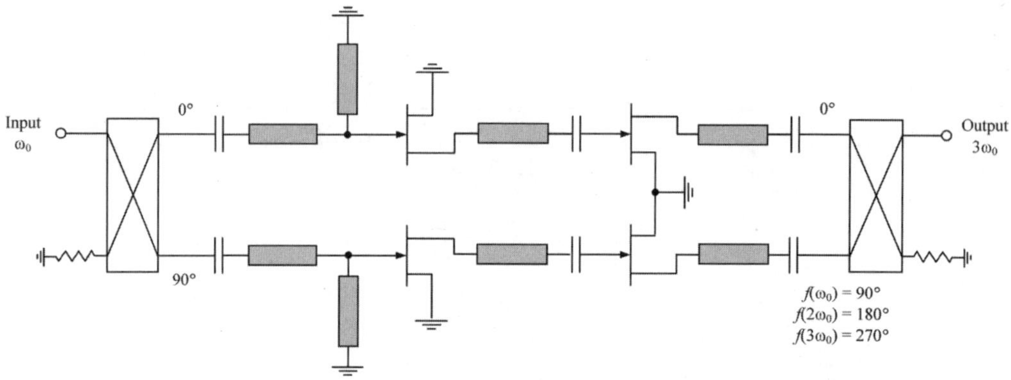

FIGURE 7.28 Schematic of balanced MESFET frequency tripler.

REFERENCES

1. E. W. Herold, "The Operation of Frequency Converters and Mixers for Superheterodyne Reception," *Proc. IRE.*, vol. 30, pp. 84–103, Feb. 1942.
2. E. W. Herold and L. Matter, "Some Aspects of Radio Reception at Ultra-High Frequencies," *Proc. IRE.*, vol. 31, pp. 575–582, Oct. 1943.
3. M. R. Barber, "Noise Figure and Conversion Loss of the Schottky Barrier Mixer Diode," *IEEE Trans. Microwave Theory Tech.*, vol. MTT-15, pp. 629–635, Nov. 1967.
4. P. R. Gray and R. G. Meyer, *Analysis and Design of Analog Integrated Circuits*, New York: John Wiley & Sons, 1977.
5. J. M. Manley and H. E. Rowe, "Some General Properties of Nonlinear Elements," *Proc. IRE.*, vol. 44, pp. 904–913, July 1956.
6. P. Penfield and R. P. Rafuse, *Varactor Applications*, Cambridge: The M.I. T Press, 1962.
7. S. A. Maas, *Nonlinear Microwave Circuits*, New York: IEEE Press, 1997.
8. L. M. Orloff, "Intermodulation Analysis of Crystal Mixer," *Proc. IEEE.*, vol. 52, pp. 173–179, Feb. 1964.
9. E. W. Herold, R. R. Bush, and W. R. Ferris, "Conversion Loss of Diode Mixers Having Image-Frequency Impedance," *Proc. IRE.*, vol. 33, pp. 603–609, Sept. 1945.
10. D. Kaye, "Derivation of Intermodulation Outputs of a Pair of General Nonlinear Elements in a Balanced Mixer Array," *Proc. IEEE.*, vol. 53, pp. 1783–1784, Nov. 1965.
11. C. M. Howell, "Integrated Balanced Mixers for S- and X-Band," *1966 IEEE G-MTT Int. Microwave Symp. Dig.*, pp. 151–155.
12. D. M. Pozar, *Microwave Engineering*, New York: John Wiley & Sons, 2004.
13. F. L. Di Alessio, A. D'Orazio, and R. Lavolpe, "Even-Harmonic *C*-Band Modulator," *IEEE Trans. Microwave Theory Tech.*, vol. MTT-53, pp. 1203–1210, Apr. 2005.
14. H. Okazaki and Y. Yamaguchi, "Wide-Band SSB Subharmonically Pumped Mixer MMIC," *IEEE Trans. Microwave Theory Tech.*, vol. MTT-45, pp. 2375–2379, Dec. 1997.
15. H. L. Krauss, C. W. Bostian, and F. H. Raab, *Solid State Radio Engineering*, New York: John Wiley & Sons, 1980.
16. A. M. Pavio, R. H. Halladay, S. D. Bingham, and C. Sapashe, "Double Balanced Mixers Using Active and Passive Techniques," *IEEE Trans. Microwave Theory Tech.*, vol. MTT-36, pp. 1948–1957, Dec. 1988.
17. G. D. Vendelin, A. M. Pavio, and U. L. Rohde, *Microwave Circuit Design Using Linear and Nonlinear Techniques*, New York: John Wiley & Sons, 1990.
18. C.-M. Lin, H.-K. Lin, C.-F. Lin, Y.-A. Lai, C.-H. Lin, and Y.-H. Wang, "A 16–44 GHz Compact Doubly Balanced Monolithic Ring Mixer," *IEEE Microwave and Wireless Components Lett.*, vol. 18, pp. 620–622, Sept. 2008.

19. R. B. Mouw, "A Broad-Band Hybrid Junction and Application to the Star Modulator," *IEEE Trans. Microwave Theory Tech.*, vol. MTT-16, pp. 911–918, Nov. 1968.

20. D. Neuf and S. Spohrer, "Conventional and New Applications for the Quadrature IF Microwave Mixer," *Microwave J.*, vol. 26, pp. 99–109, Jan. 1983.

21. K. Hettak, G. A. Morin, and M. G. Stubbs, "Size Reduction of a MMIC Direct Up-Converter at 44 GHz in Multilayer CPW Technology Using Thin-Film Microstrip Stubs Loading," *IEEE Trans. Microwave Theory Tech.*, vol. MTT-54, pp. 3453–3461, Sept. 2006.

22. J. Zawels, "The Transistor as a Mixer," *Proc. IRE*, vol. 42, pp. 542–548, Mar. 1954.

23. J. S. Vogel, "Nonlinear Distortion and Mixing Processes in Field-Effect Transistors," *Proc. IEEE.*, vol. 55, pp. 2109–2116, Dec. 1967.

24. M. Madihian, L. Desclos, K. Maruhashi, K. Onda, and M. Kuzuhara, "A Monolithic AlGaAs/InGaAs Upconverter IC for K-Band Wireless Networks," *IEEE Trans. Microwave Theory Tech.*, vol. MTT-43, pp. 2773–2777, Dec. 1995.

25. M. Madihian, L. Desclos, K. Maruhashi, K. Onda, and M. Kuzuhara, "60-GHz Monolithic Down- and Up-Converters Utilizing a Source-Injection Concept," *IEEE Trans. Microwave Theory Tech.*, vol. MTT-46, pp. 1003–1006, July 1998.

26. A. Orzati, F. Robin, H. Meier, H. Benedikter, and W. Bachtold, "A V-Band Up-Converting InP HEMT Active Mixer with Low LO-Power Requirements," *IEEE Microwave and Wireless Components Lett.*, vol. 13, pp. 202–204, June 2003.

27. S. Asai, F. Murai, and H. Kodera, "GaAs Dual-Gate Schottky-Barrier FET's for Microwave Frequencies," *IEEE Trans. Electron Devices*, vol. ED-22, pp. 897–904, Oct. 1975.

28. C. A. Liechti, "Performance of Dual-Gate GaAs MESFET's as Gain-Controlled Low-Noise Amplifiers and High-Speed Modulators," *IEEE Trans. Microwave Theory Tech.*, vol. MTT-23, pp. 461–469, June 1975.

29. C. Tsironis, R. Meierer, and R. Stahlmann, "Dual-Gate MESFET Mixers," *IEEE Trans. Microwave Theory Tech.*, vol. MTT-32, pp. 248–255, Mar. 1984.

30. W. S. Tsai, S. F. Paik, and B. S. Hewitt, "An X-Band Dual-Gate FET Up-Converter," *1979 IEEE MTT-S Int. Microwave Symp. Dig.*, pp. 495–497.

31. T. Hirota and H. Ogawa, "A Novel K-Band Balanced FET Up-Converter," *IEEE Trans. Microwave Theory Tech.*, vol. MTT-32, pp. 679–683, July 1984.

32. P.-S. Wu, C.-H. Wang, C.-S. Lin, K.-Y. Lin, and H. Wang, "A Compact 60 GHz Integrated Up-Converter Using Miniature Transformer Couplers with 5 dB Conversion Gain," *IEEE Microwave and Wireless Components Lett.*, vol. 18, pp. 641–643, Sept. 2008.

33. M. L. de la Fuente, J. Portilla, J. P. Pascual, and E. Artal, "Low-Noise Ku-Band MMIC Balanced P-HEMT Upconverter," *IEEE J. Solid-State Circuits*, vol. SC-34, pp. 259–263, Feb. 1999.

34. H. E. Jones, "Dual Output Synchronous Detector Utilizing Transistorized Differential Amplifiers," U.S. Patent 3,241,078, Mar. 1966 (filed June 1963).

35. A. Bilotti, "Applications of a Monolithic Analog Multiplier," *IEEE J. Solid-State Circuits*, vol. SC-3, pp. 373–380, Dec. 1968.

36. A. Y. Umeda, C. T. Matsuno, A. K. Oki, G. S. Dow, K. W. Kobayashi, D. K. Umemoto, and M. E. Kim, "A Monolithic GAAs HBT Upconverter," *1990 IEEE Microwave and Millimeter-Wave Monolithic Circuits Symp. Dig.*, pp. 77–80.

37. K. Osafune and Y. Yamauchi, "20-GHz 5-dB-Gain Analog Multipliers with AlGaAs/GaAs HBT's," *IEEE Trans. Microwave Theory Tech.*, vol. MTT-42, pp. 518–520, Mar. 1994.

38. C. Meng, T.-H. Wu, and M. Lin, "Compact 5.2-GHz GaInP/GaAs HBT Gilbert Upconverter Using Lumped Rat-Race Hybrid and Current Combiner," *IEEE Microwave and Wireless Components Lett.*, vol. 15, pp. 688–690, Oct. 2005.

39. K. W. Kobayashi, L. T. Tran, S. Bui, A. K. Oki, D. C. Streit, and M. Rosen, "InAlAs/InGaAs HBT *X*-Band Double-Balanced Upconverter," *IEEE J. Solid-State Circuits*, vol. SC-29, pp. 1238–1243, Oct. 1994.

40. M. West and M. Kumar, "Monolithic Integrated Blanking Up-Converter," *IEEE Trans. Microwave Theory Tech.*, vol. MTT-38, pp. 1227–1231, Sept. 1990.

41. A. Verma, K. K. O, and J. Lin, "A Low-Power Up-Conversion CMOS Mixer for 22–29-GHz Ultra-Wideband Applications," *IEEE Trans. Microwave Theory Tech.*, vol. MTT-54, pp. 3295–3300, Aug. 2006.

42. A. Valdes-Garcia, S. Reynolds, and J.-O. Plouchart, "60 GHz Transmitter Circuits in 65 nm CMOS," *Proc. 2008 IEEE RFIC Symp.*, pp. 641–644.

43. R. S. Ohl, "High Frequency Oscillation Generator," U.S. Patent 1,756,131, Apr. 1930 (filed Aug. 1925).

44. R. M. Page, "An Investigation of the Phenomena of Frequency Multiplication as Used in Tube Transmitters," *Proc. IRE*, vol. 17, pp. 1649–1655, Sept. 1929.

45. F. E. Terman and J. H. Ferns, "The Calculation of Class C Amplifier and Harmonic Generator Performance of Screen-Grid and Similar Tubes," *Proc. IRE*, vol. 22, pp. 359–373, Mar. 1934.

46. H. C. Lee and R. Minton, "Designing Transistor Multipliers," *Microwaves*, vol. 4, pp. 18–28, Nov. 1965.

47. G. D. O'Clock and R. J. Dauphinee, "High-Gain Transistor Frequency Multiplier," *Proc. IEEE*, vol. 58, pp. 1363–1365, Sept. 1970.

48. J. J. Pan, "Wideband MESFET Microwave Frequency Multiplier," *1978 IEEE MTT-S Int. Microwave Symp. Dig.*, pp. 306–308.

49. A. Gopinath and J. B. Rankin, "Single-Gate MESFET Frequency Doublers," *IEEE Trans. Microwave Theory Tech.*, vol. MTT-30, pp. 869–875, June 1982.

50. P. T. Chen, C.-T. Li, and P. H. Wang, "Dual-Gate GaAs FET as a Frequency Multiplier at Ku-Band," *1978 IEEE MTT-S Int. Microwave Symp. Dig.*, pp. 309–311.

51. C. Rauscher, "High-Frequency Doubler Operation of GaAs Field-Effect Transistors," *IEEE Trans. Microwave Theory Tech.*, vol. MTT-31, pp. 462–473, June 1983.

52. Y. Iyama, A. Iida, T. Takagi, and S. Urasaki, "Second-Harmonic Reflector Type High-Gain FET Frequency Doubler Operating in K-Band," *1989 IEEE MTT-S Int. Microwave Symp. Dig.*, pp. 1291–1294.

53. G. Zhang, R. D. Pollard, and C. M. Snowden, "A Novel Technique for HEMT Tripler Design," *1996 IEEE MTT-S Int. Microwave Symp. Dig.*, pp. 663–666.

54. Y. Campos-Roca, C. Schworer, A. Leuther, and M. Seelmann-Eggebert, "*G*-Band Metamorphic HEMT-Based Frequency Multipliers," *IEEE Trans. Microwave Theory Tech.*, vol. MTT-54, pp. 2983–2992, July 2006.

55. T. Hiraoka, T. Tokumitsu, and M. Akaike, "A Miniaturized Broad-Band MMIC Frequency Doubler," *IEEE Trans. Microwave Theory Tech.*, vol. MTT-38, pp. 1932–1937, Dec. 1990.

56. B. Piernas, K. Nishikawa, T. Nakagawa, H. Hayashi, and K. Araki, "Analysis of Balanced Active Doubler for Broad-Band Operation – The Frequency-Tuning Concept," *IEEE Trans. Microwave Theory Tech.*, vol. MTT-50, pp. 1120–1126, Apr. 2002.

57. C. Schworer, Y. Campos-Roca, A. Leuther, A. Tessmann, M. Seelmann-Eggebert, H. Massler, M. Schlechtweg, and G. Weimann, "A 150 to 220 GHz Balanced Doubler MMIC Using a 50 nm Metamorphic HEMT Technology," *Proc. 35th Europ. Microwave Conf.*, vol. 3, pp. 1–4, 2005.

58. R. Gilmore, "Octave-Bandwidth Microwave FET Doubler," *Electronics Lett.*, vol. 21, pp. 532–533, June 1985.

59. J. E. Johnson, G. R. Branner, and J.-P. Mima, "Design and Optimization of Large Conversion Gain Active Microwave Frequency Triplers," *IEEE Microwave and Wireless Components Lett.*, vol. 15, pp. 457–459, July 2005.

60. H. Fudem and E. C. Niehenke, "Novel Millimeter Wave Active MMIC Triplers," *1998 IEEE MTT-S Int. Microwave Symp. Dig.*, pp. 387–390.

8 Oscillators

This chapter presents the principles of oscillator design, including start-up and steady-state operation conditions, noise and stability of oscillations, basic oscillator configurations using lumped and transmission-line elements, and simplified equation-based oscillator analyses and optimum design techniques. An immittance design approach is introduced and applied to the series and parallel feedback oscillators, including circuit design and simulation aspects. Voltage-controlled oscillators (VCOs) and their varactor tuning range and linearity for different oscillator configurations are discussed. Finally, the basic circuits and operation principles of crystal and dielectric resonator oscillators are given.

8.1 OSCILLATOR OPERATION PRINCIPLES

8.1.1 Steady-State Operation Mode

A simple feedback oscillator model is shown in Figure 8.1(*a*) where an oscillator circuit is decomposed into a forward nonlinear network and a feedback linear network, both of which are two-port networks. Figure 8.1(*b*) shows an example of a transformer-coupled metal-oxide-semiconductor field-effect transistor (MOSFET) oscillator without bias circuit to illustrate common features of the feedback oscillator. Because an oscillator is an autonomous circuit, electronic noise in the active device or power supply turn-on transient leads to the self-excitation of the oscillations. This provides the initial oscillation build-up. As the oscillation amplitude grows, the active device displays larger nonlinearity and then limits the amplitude increase.

In a steady-state operation mode, the following complex equation also known as the *Barkhausen criterion* can be written for the parallel feedback oscillator:

$$\mathbf{T}(V_{\text{in}}, \omega) = \mathbf{A}(V_{\text{in}}, j\omega)\,\beta(j\omega) = 1 \tag{8.1}$$

where $\boldsymbol{A} = \boldsymbol{I}_{\text{out}}/\boldsymbol{V}_{\text{in}}$ is the forward transfer function and $\beta = \boldsymbol{V}_{\text{in}}/\boldsymbol{I}_{\text{out}}$ is the feedback transfer function. This equation means that the oscillator complex loop gain is equal to unity [1,2]. The feedback transfer function can be represented as

$$\beta(j\omega) = K(j\omega)\,Z(j\omega) \tag{8.2}$$

where $\boldsymbol{K} = \boldsymbol{V}_{\text{in}}/\boldsymbol{V}_{\text{out}}$ is the voltage feedback coefficient and $\boldsymbol{Z} = \boldsymbol{V}_{\text{out}}/\boldsymbol{I}_{\text{out}}$ is the oscillator resonant circuit impedance. Presenting each of these complex quantities in the form of $\boldsymbol{A} = A\exp(j\phi_{\text{A}})$, $\boldsymbol{K} = K\exp(j\phi_{\text{K}})$, and $\boldsymbol{Z} = Z\exp(j\phi_{\text{Z}})$, the following equations for magnitudes and phases directly result from Eq. (8.1):

$$A(V_{\text{in}}, \omega)\,K(\omega)\,Z(\omega) = 1 \tag{8.3}$$

$$\phi_{\text{A}} + \phi_{\text{K}} + \phi_{\text{Z}} = 0, 2\pi, \ldots. \tag{8.4}$$

RF and Microwave Transmitter Design, First Edition. Andrei Grebennikov.

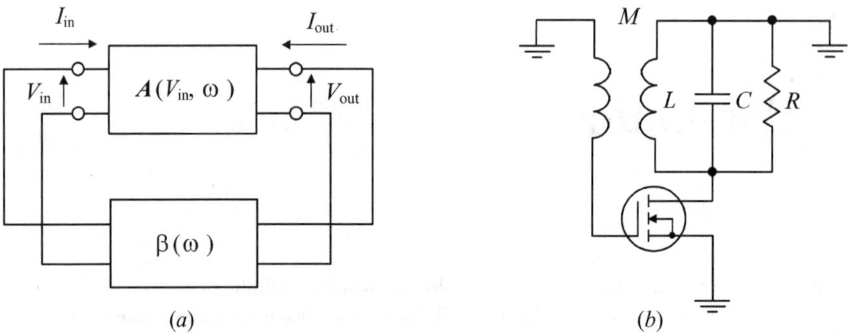

FIGURE 8.1 Schematics of (*a*) parallel feedback oscillator and (*b*) transformer-coupled MOSFET oscillator.

Equation (8.3), which is called the *amplitude balance condition*, means that the oscillator loop gain is equal to unity in the steady-state stationary operation mode. In this equation, it is assumed that two quantities, K and Z, depend on frequency only. Consequently, the amplitude balance condition is satisfied only under the appropriate value of input voltage amplitude V_{in}. To define the value of this amplitude, let us rewrite Eq. (8.3) in the form

$$A\left(V_{in}, \omega\right) = \frac{1}{K\left(\omega\right) Z\left(\omega\right)}. \tag{8.5}$$

In Figure 8.2, the amplitude dependence $A(V_{in})$ and feedback straight line $1/KZ$ are plotted. The intersection point of these dependencies determines the steady-state oscillation amplitude V_{in}^{o}.

Equation (8.4), which is called the *phase balance condition*, means that the sum of all oscillator loop phase shifts must be equal to zero or $2\pi n$, where $n = 1, 2, \ldots$. This equation defines the value of the oscillation frequency f_{osc}. In the simple case when $\phi_K = 0$ (transistor input admittance is equal to zero and feedback magnitude K depends only on a mutual-coupling coefficient M between primary and secondary inductances), and the active device does not produce the phase shift, that is $\phi_A = 0$, then $\phi_Z = 0$, and the oscillation frequency f_{osc} is equal to parallel resonant circuit frequency $f_0 = \omega_0/(2\pi)$, where $\omega_0 = 1/\sqrt{LC}$. If $\phi_A \neq 0$, the oscillation frequency f_{osc} will differ from f_0 in order to fully compensate for the available phase shift ϕ_Z according to

$$\phi_Z = -\tan^{-1}\left(2Q\frac{\Delta\omega}{\omega_0}\right) = -\phi_A \tag{8.6}$$

where $Q = 1/(\omega CR)$ is a quality factor of the oscillator resonant circuit, $\Delta\omega = 2\pi(f_{osc} - f_0)$ [2].

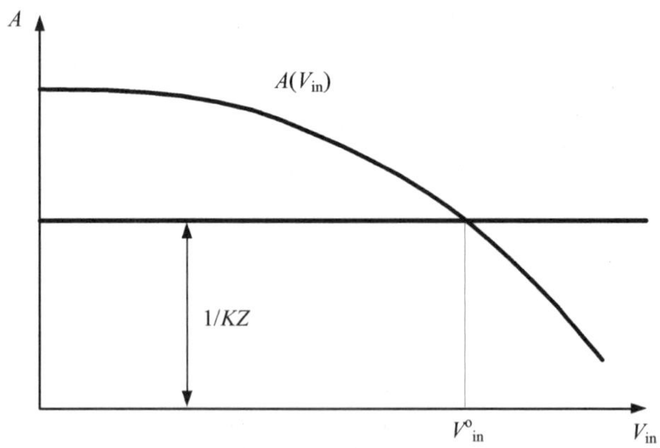

FIGURE 8.2 Graphic balance amplitude condition.

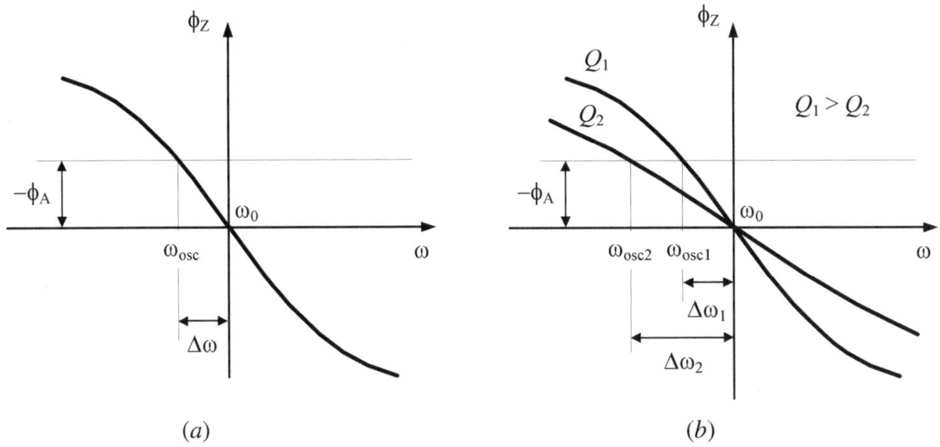

FIGURE 8.3 Deviation of oscillation frequency f_{osc} from resonant frequency f_0.

Equation (8.6) can be rewritten in the form

$$\frac{\Delta\omega}{\omega_0} = \frac{\tan\phi_{\mathrm{A}}}{2Q} \tag{8.7}$$

which determines the deviation of the oscillation frequency f_{osc} from resonant frequency f_0 as a function of the phase of the forward transfer function of the active device and oscillator quality factor.

From graphical representation of Eq. (8.6) shown in Figure 8.3(*a*), it follows that the oscillation frequency f_{osc} is smaller than the resonant frequency f_0. Furthermore, the lower the quality factor Q the smaller oscillation frequency f_{osc}, as shown in Figure 8.3(*b*), where the deviation $\Delta\omega_2$ from radian resonant frequency ω_0 becomes greater for the case of $Q_2 < Q_1$.

8.1.2 Start-Up Conditions

The start-up conditions can be illustrated on the plane of the transfer function $i_{\mathrm{out}} = f(v_{\mathrm{in}})$, as shown in Figure 8.4. Let us choose the bias voltage V_{g}' corresponding to the maximum small-signal transfer function when $\partial i_{\mathrm{out}}/\partial v_{\mathrm{in}} = 0$. Such a condition is adequate to the maximum forward transfer function $A(V_{\mathrm{in}} = 0)$ shown in Figure 8.2. According to Figure 8.2, the amplitude V_{in} of the oscillations grows monotonically, whereas the transfer function A decreases. The amplitude of the output current i_{out} is trying to reach its maximum value corresponding to the stable steady-state operation mode when $V_{\mathrm{in}} = V_{\mathrm{in}}^{\mathrm{o}}$. Such a behavior of the transfer function A means that the start-up oscillation conditions can be defined as

$$A(\omega)\, K(\omega)\, Z(\omega) > 1 \tag{8.8}$$

$$\phi_{\mathrm{A}} + \phi_{\mathrm{K}} + \phi_{\mathrm{Z}} = 0, 2\pi, \ldots. \tag{8.9}$$

Let the bias voltage V_{g}'' be chosen close to the device threshold voltage, as shown in Figure 7.4. Such a bias condition corresponds to the small initial value of transfer function $A(V_{\mathrm{in}} = 0)$, which is not enough to establish the stable steady-state stationary oscillations because $A < 1/KZ$. This is demonstrated by curve *II* in Figure 8.5(*a*). Therefore, the oscillation system with such a bias condition requires an impulse to initiate self-oscillations, as it cannot oscillate by itself. The result of this impulse should be output current amplitude resulting in a condition of $A > 1/KZ$. Hence, the system has hard operating conditions and the process of the oscillation establishment is called the

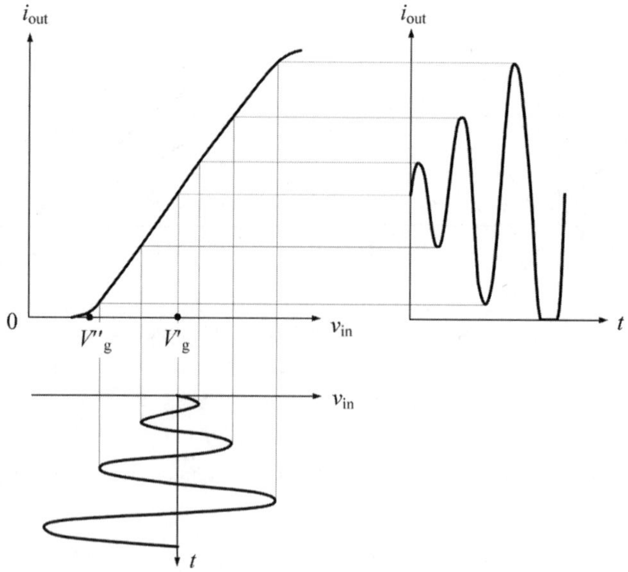

FIGURE 8.4 Start-up of oscillations.

hard build-up of self-oscillations. This means that oscillations of finite amplitude are established suddenly under some external influence.

Figure 8.4 shows the process of establishment of the stable self-oscillations as a result of the nonlinearity of the transfer function of the active device operating in pinch-off and active regions. This means that the start-up oscillation conditions correspond to a Class A operation mode of the active device having maximum or close to maximum value of the small-signal transconductance. The steady-state oscillation conditions are established when the active device operates in Class AB mode characterized by certain conduction angle whose particular value depends on the initial bias conditions.

Let us assume a high value of the resonant circuit quality factor when the input cosinusoidal voltage

$$v_{in} = V_{in} \cos \omega t \tag{8.10}$$

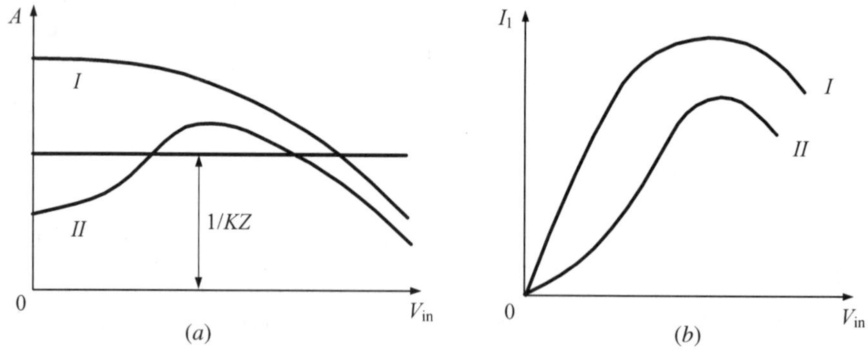

FIGURE 8.5 Balanced amplitude conditions for different biasing points.

is applied to the active device, representing an idealized nonlinear voltage-controlled current source. Then, the output current i_{out} contains the harmonic components and, being an even function, can be written as

$$i_{out} = I_0 + I_1 \cos \omega t + I_2 \cos 2\omega t + I_3 \cos 3\omega t + \cdots. \tag{8.11}$$

For the transfer function $i_{out} = f(v_{in})$, the fundamental component I_1 can be obtained from Eqs. (8.10) and (8.11) using a Fourier formula according to

$$I_1 = \frac{1}{\pi} \int_{-\pi}^{\pi} f\left(V_{in} \cos \omega t\right) \cos \omega t \, d\left(\omega t\right). \tag{8.12}$$

Consequently, dividing the output fundamental amplitude I_1 by the input voltage amplitude V_{in} provides an analytical expression to calculate the fundamentally averaged (average) value of the forward transfer function $A(V_{in})$ in a quasilinear approximation. The dependence of the output fundamental current amplitude I_1 on the input voltage amplitude V_{in}, expressed generally as

$$I_1 = f_1\left(V_{in}\right) \tag{8.13}$$

can be called the *amplitude characteristic* of the oscillator. Figure 8.5 shows the dependencies of (*a*) the amplitude characteristic and (*b*) average transfer function as functions of the input voltage amplitude V_{in} for different operation modes where curves *I* and curves *II* correspond to the *soft* and *hard* start-up conditions for self-oscillation establishment, respectively.

Consider the influence of the harmonic components on the process of the establishment of the self-oscillations for finite value of the quality factor Q. Then, the input voltage according to Figure 8.1(*b*) can be written as

$$v_{in} = V_{in1} \cos \omega t + V_{in2} \cos \left(2\omega t - \frac{\pi}{2}\right) + V_{in3} \cos \left(3\omega t - \frac{\pi}{2}\right) + \cdots \tag{8.14}$$

where the resonant circuit is tuned to the fundamental and has the capacitive reactance for the second-order and higher order harmonics. If we provide a second-order polynomial approximation of the nonlinear transfer function of the active device in the form

$$i_{out} = a_0 + a_1 v_{in} + a_2 v_{in}^2 \tag{8.15}$$

and to confine our attention to the first two components in Eq. (8.14), the fundamental component of the output current can be obtained by

$$i_1' = I_1 \cos \omega t + a_2 V_{in1} V_{in2} \cos \left(\omega t - \frac{\pi}{2}\right) \tag{8.16}$$

where $I_1 = a_1 V_{in1}$.

Equation (8.16) can be rewritten in the form

$$i_1' = I_1' \cos\left(\omega t + \phi_A\right) \tag{8.17}$$

where

$$I_1' = I_1 \sqrt{1 + \left(\frac{a_2}{a_1}\right)^2 V_{in2}^2} \tag{8.18}$$

and

$$\tan\phi_A = -\frac{a_2 V_{in2}}{a_1}. \tag{8.19}$$

Thus, the second component in Eq. (8.16) results in the variation of the output current fundamental amplitude when $I_1' \neq I_1$ and the appearance of a negative value of the phase ϕ_A of the device transfer function that means the phase shift between the fundamental voltage and fundamental current. In this case, an increase in the harmonic content of the output voltage spectrum (smaller Q-factor) causes a decrease in the frequency of self-oscillations, as follows from Figure 8.3(*b*).

Physically, the influence of the harmonic content on the frequency variation can be explained as follows: if the oscillations are purely sinusoidal, the energy distribution in both arms of the resonant LC-circuit is equal; when the harmonics appear, the currents corresponding to them flow mainly through the capacitive arm, and therefore they increase the electrostatic energy of this arm in comparison with the inductive arm; and in order to keep the energy equal in both arms, the fundamental frequency must slightly diminish itself in respect to the frequency given by the tank circuit only [3].

In Figure 8.6(*a*), for the example of the transformer-coupled MOSFET oscillator shown in Figure 8.1(*b*), the dependencies $I_1(V_{in})$ for a soft start-up condition with different values of mutual coupling coefficient M ($M_1 < M_2 < M_3 < M_4$) are shown [1,2]. As M increases from zero to M_2, the only stable equilibrium corresponds to the static operation mode at the point $V_{in} = 0$. When $M > M_2$, from two potentially existing equilibrium conditions the dynamic conditions at the points A_3 and A_4 are stable. To check the stable equilibrium at the point A_3 with amplitude $V_{in} = V_{in}^o$ and unstable at the $V_{in} = 0$, assume an oscillation rise of the small input amplitude V_{in}' due to some internal or external effect. This would cause an appearance of the output current with the amplitude I_1', which is determined according to the oscillator amplitude characteristic $I_1(V_{in})$. At that time this current makes conditional the input voltage with amplitude V_{in}''. As a result, the oscillation amplitude that has arisen accidentally increases up to the equilibrium value of V_{in}^o at the point A_3. With the growth of M, the output current amplitude I_1 changes smoothly, as shown in Figure 8.6(*b*). When M decreases, the amplitude I_1 changes in accordance with $I_1(M)$ curve and under $M = M_2$ the oscillations disappear.

In Figure 8.7(*a*), the dependencies $I_1(V_{in})$ for hard start-up condition with different values of mutual coupling coefficient M ($M_1 < M_2 < M_3 < M_4$) are shown [1,2]. At the point $M = M_3$

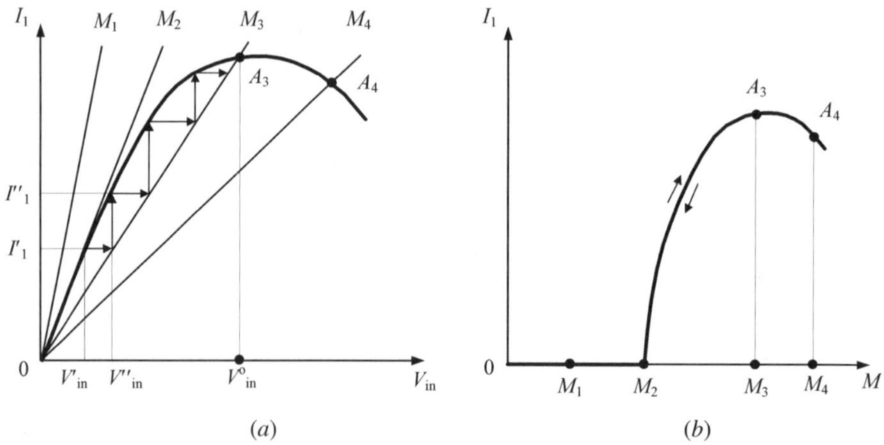

FIGURE 8.6 Start-up conditions with different values of mutual-coupling coefficient M.

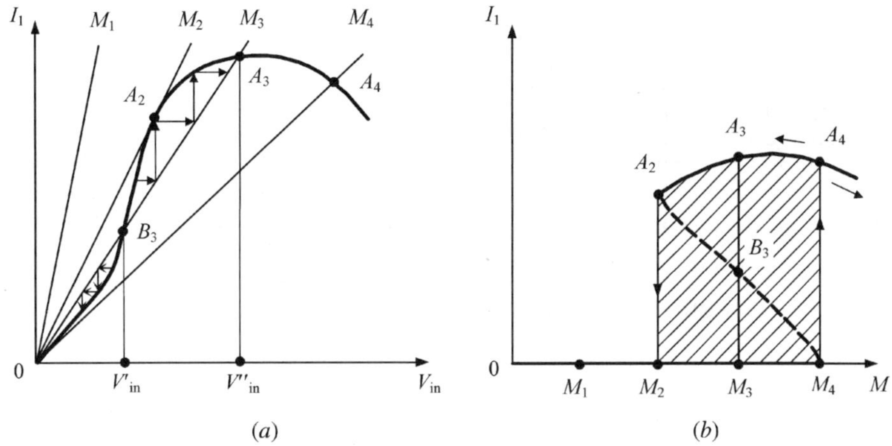

FIGURE 8.7 Hard start-up condition versus mutual-coupling coefficient M.

both curves intersect at three points, corresponding to three stationary modes: 0 is an equilibrium condition, B_3 and A_3 are the dynamic modes with amplitudes V'_{in} and V''_{in}, respectively. From the process of changing of the output fundamental current amplitude I_1 with initial voltage deviations from V'_{in} and V''_{in}, it follows that the stable conditions correspond to the points 0 and A_3, whereas the condition in the point B_3 is unstable.

Let us define the dependence of output fundamental current amplitude I_1 versus feedback coefficient M. As M increases from zero up to M_4 when the straight line corresponding to M_4 and the curve $I_1(V_{in})$ are tangents in the origin, the only stable equilibrium corresponds to the static operation mode at the point $V_{in} = 0$ when the small fluctuations can not produce an oscillation arise. At the point $M = M_4$, the above-mentioned operation mode becomes unstable and small oscillations grow up to large amplitude value corresponding to the point A_4. The subsequent increase of M leads to the amplitude change along the curve $I_1(M)$, as shown in Figure 8.7(b). If, then, to decrease M down to M_2, the collapse of the oscillation can occur only at the point A_2 because the dynamic operations modes corresponding to the points A_3 and A_4 are stable. This fact results from the qualitative process examination of Figure 8.7(a), where the points A_2 and B_3 are unstable at the same time. Hence, the *hysteresis region* takes place between the points A_2 and A_4 in limits of $M_2 < M < M_4$. Thus, the hard build-up of the self-oscillation is characterized by spasmodic development of the oscillation with large amplitude under smooth increase of the feedback coefficient M and spasmodic collapse of the oscillation under smooth decrease of M.

8.2 OSCILLATOR CONFIGURATIONS AND HISTORICAL ASPECT

The first oscillator configurations using vacuum tubes were based on the electromagnetic coupling between the output and input circuits. By providing a close enough coupling with the output, sufficient energy is supplied to the input circuit to keep the continuous self-sustaining oscillations. In this case, the oscillation frequency is approximately equal to the resonant frequency of the parallel LC-circuit if the coupling and active device parasitics are small enough.

There are two basic configurations of a transformer-coupled oscillator. The oscillator configuration with a parallel resonant circuit at the input electromagnetically coupled with the output, as shown in Figure 8.8(a), is called the *Armstrong oscillator* in honor of Edwin H. Armstrong who first developed and described the condition of obtaining the self-sustaining oscillations using such a configuration [4]. Approximately at the same time, Alexander Meissner in Germany described the transformer-coupled

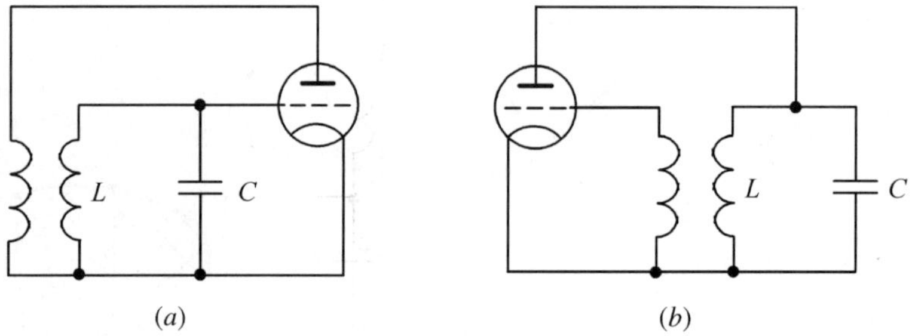

FIGURE 8.8 Schematics of (*a*) Armstrong and (*b*) Meissner oscillators.

oscillator with a parallel resonant circuit at the output electromagnetically coupled with the input, as shown in Figure 8.8(*b*), which is called the *Meissner oscillator* [5]. High-purity stable oscillations can be obtained by a fully balanced version of Meissner oscillator with a three-turn link providing an excellent isolation of the resonant circuit when very little energy is taken from the resonator to provide the voltages to drive the input and sustain the oscillations [6].

To satisfy a phase balance condition in the transformer-coupled oscillator, the transformer should provide a phase shift of 180°. If the primary and secondary windings of the transformer have the same direction of wind, it is necessary to connect the secondary winding in the opposite direction relatively the primary one, that is, to connect the end of the secondary winding where the voltage is in-phase with the anode (drain or collector) voltage to the ground. The coupling coefficient of the transformer is chosen to provide a soft start-up oscillation mode.

The schematic of the *Hartley* oscillator is very close to the schematic of the Meissner oscillator [7]. The difference is that the tank inductor L having an additional output replaces the transformer, as shown in Figure 8.9(*a*). The inductance ratio determines the feedback coefficient. The Hartley oscillator can be represented by a well-known inductive three-point configuration shown in Figure 8.9(*b*), where the feedback element from the output to the input is the capacitor C. The inductors L_1 and L_2 represent the output and input circuits, respectively.

Unlike the Hartley oscillator with electromagnetic coupling, the schematic of a *Colpitts* oscillator is based on electrostatic coupling using the capacitive divider, the capacitance ratio of which determines the feedback coefficient, as shown in Figure 8.10(*a*) [8]. The Colpitts oscillator can be represented by a well-known capacitive three-point configuration shown in Figure 8.10(*b*), where the feedback element from the output to the input is the inductor L [9]. The capacitors C_1 and C_2 represent the output and input circuits, respectively.

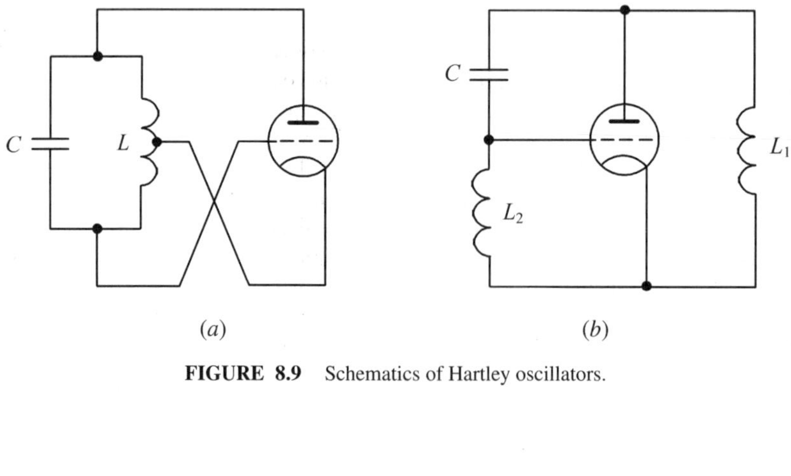

FIGURE 8.9 Schematics of Hartley oscillators.

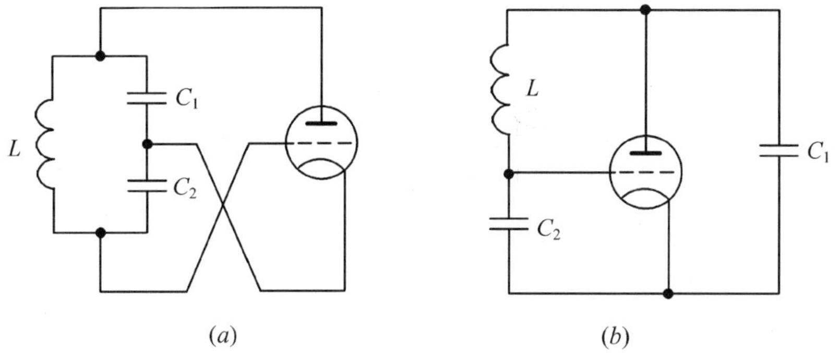

FIGURE 8.10 Schematics of Colpitts oscillators.

Depending on which electrode of the active device is grounded, we can distinguish three basic configurations representing a Colpitts family of oscillators based on the MOSFET devices, as shown in Figure 8.11. The Colpitts oscillator with grounded (common) source is shown in Figure 8.11(*a*), the Colpitts oscillator with grounded (common) drain is shown in Figure 8.11(*b*), and the Colpitts oscillators with grounded (common) gate is shown in Figure 8.11(*c*, *d*). The connection of the appropriate electrode of the tube (or transistor) to the ground have important effects upon the following:

- The manner in which the dc power is fed to the oscillator and the corresponding loading effects on the resonant circuit.
- The manner in which the output power is fed to the external load.
- The distribution of the stray elements to the ground plane, which has important effects, particularly at microwaves.

Calculating the output impedance Z_{out} for a common source oscillator shown in Figure 8.11(*a*) and a common drain oscillator shown in Figure 8.11(*b*), which are identical, shows that grounding of any terminal of the oscillator circuit does not change its electrical performance, provided there are no changes in the connection of the feedback elements and load to the active device. Therefore, the schematic with a common gate shown in Figure 8.11(*c*) is characterized by the same electrical behavior as the oscillator circuits with common source and common drain. However, in such a configuration there is no connection of any load terminal to the ground that makes its practical implementation more complicated. Figure 8.11(*d*) shows a common gate oscillator schematic with the load connected in parallel to the inductance *L*, that is, between the drain and gate terminals. However, the other load connection, such as this with one grounded port, results in different electrical properties of the oscillator compared with common source or common drain configuration.

The Gouriet–Clapp oscillator is a variation of the Colpitts oscillator with a tank inductor replaced by the series combination of an inductor *L* and a variable capacitor C_v, as shown in Figure 8.12(*a*). In this case, frequency stability is improved because the reactance of such a series circuit varies more rapidly with frequency than that of a single inductor. However, the possibility to improve the oscillator stability by connecting a capacitor in series to one or each circuit inductor was first found based on a Hartley type of the oscillator [10]. In modern practical oscillator design, the Gouriet–Clapp oscillator configuration is called the *Clapp oscillator*, although G. Gouriet and J. Clapp had developed it independently [11]. A parallel counterpart of the Clapp oscillator shown in Figure 8.12(*b*) was described by E. Seiler and, therefore, is called the *Seiler oscillator* [11]. Such an oscillator configuration is useful for wideband frequency tuning when the capacitance C_v in the parallel circuit is variable.

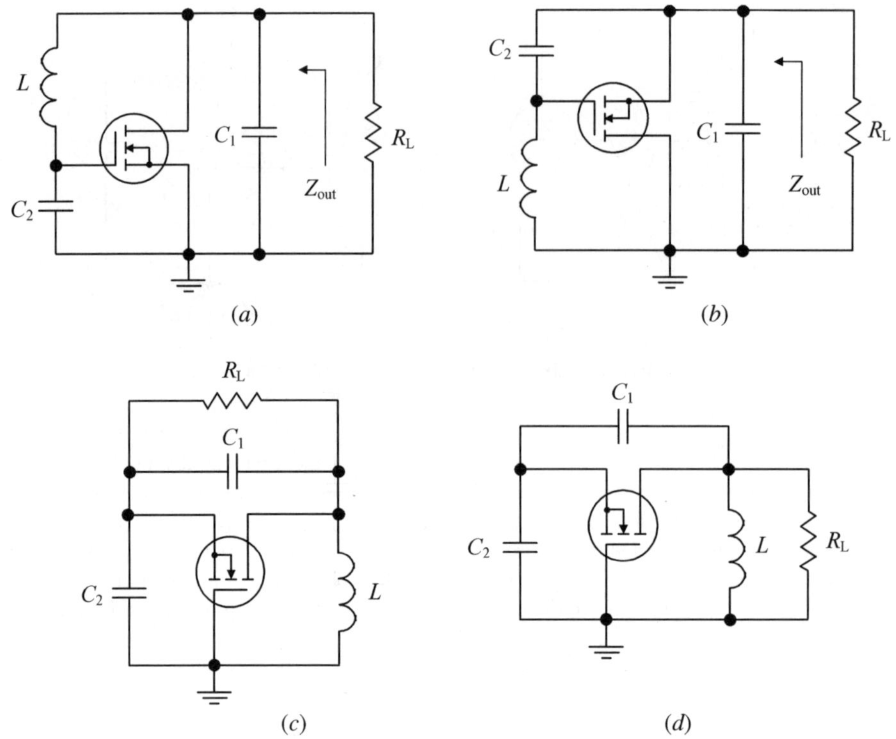

FIGURE 8.11 Colpitts family of oscillators.

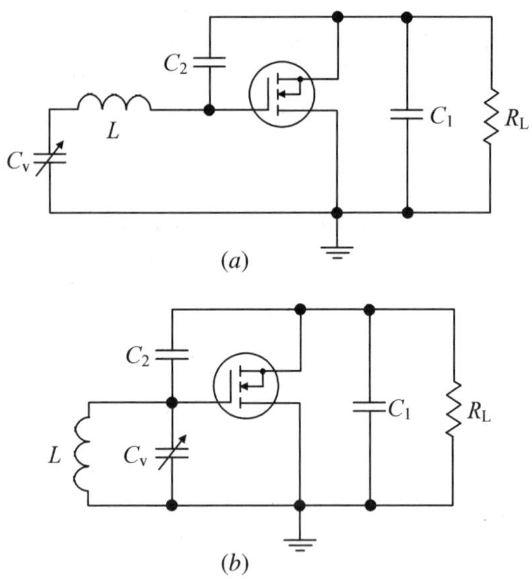

FIGURE 8.12 Schematics of (*a*) Gouriet–Clapp and (*b*) Seiler oscillators.

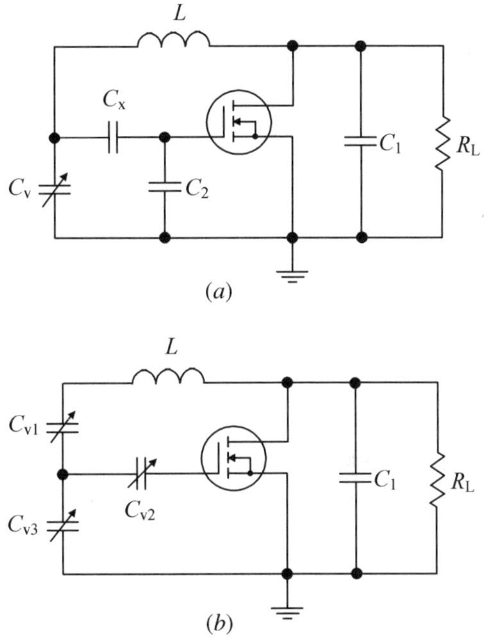

FIGURE 8.13 Schematics of Vackar oscillator.

The *Vackar oscillator* is the modified Clapp oscillator with additional variable capacitor C_v, whose simplified equivalent circuit representation is shown in Figure 8.13(*a*) [11,12]. It combines the features of the circuits with the series and parallel arrangements and is useful for very wideband frequency tuning. When $C_v = 0$, the schematic of Vackar oscillator becomes similar to the schematic of the Clapp oscillator shown in Figure 8.12(*a*). The configuration of the Vackar oscillator shown in Figure 8.13(*b*) demonstrates the main difference compared with the Clapp schematic. Here, the capacitance C_{v2} represents the phase-varying capacitance providing an additional phase shift. With this circuit, it is possible to utilize the maximum value of the oscillator quality factor over the complete tuning range, and the circuit has substantially constant output amplitude. Both oscillator schematics are equivalent, since the π-representation of the capacitances C_v, C_x, and C_2 are replaced by the equivalent T-representation of the capacitances C_{v1}, C_{v2}, and C_{v3}.

The transformer-coupled oscillator based on a differential amplifier is shown in Figure 8.14(*a*), where the coupling coefficient is chosen to provide a stable soft start-up condition. The simple oscillator configuration using a differential transistor pair is shown in Figure 8.14(*b*). Since the voltage at the gate of the transistor connected to the parallel resonant circuit is in-phase to the voltage at the drain of the transistor with grounded gate, the feedback in such a differential-pair oscillator is positive.

The push–pull connection of the transistors in power amplifiers is usually used to increase the resulting output power, simplify the output matching with load, and improve spectral performance by suppressing even-order harmonics. The same concept can also be applied to the oscillator design. The push–pull oscillator circuit shown in Figure 8.15(*a*) is based on the two single-ended Meissner oscillators, where the transistors are turned on and turned off alternately. Since the voltage at the gate of the one transistor is in-phase to the voltage at the drain of the other transistor, there is no need to invert phase using the secondary winding. Another push–pull oscillator configuration based on two single-ended Seiler oscillators is shown in Figure 8.15(*b*), where the positive feedback is formed by the capacitive divider based on the capacitors C_1 and C_2.

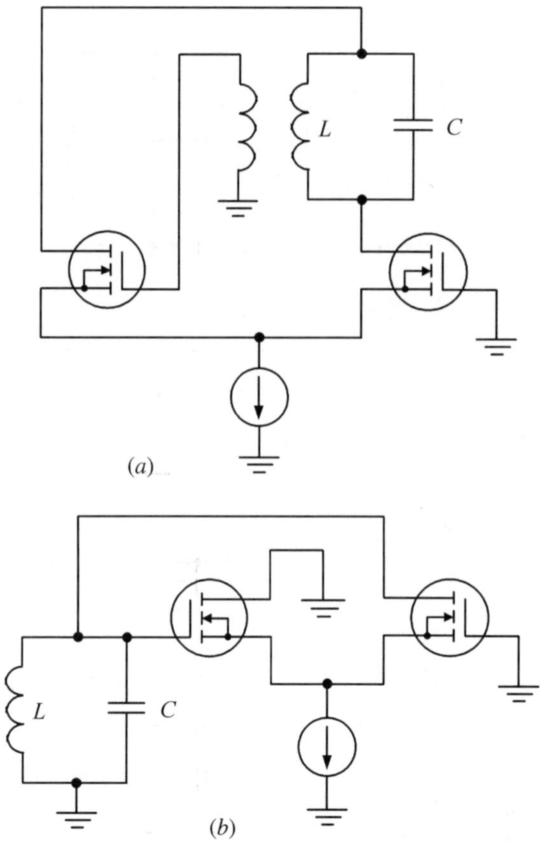

FIGURE 8.14 Schematics of differential-pair oscillators.

8.3 SELF-BIAS CONDITION

To provide more effective operation mode of the oscillator, a self-bias resistor is usually included in the oscillator circuit. Figure 8.16 illustrates the circuit principle of self-biasing by two examples. In a common source transformer-coupled oscillator shown in Figure 8.16(*a*), the self-bias resistor R_s is shunted by the capacitor C_s to minimize the RF signal losses. In a common gate oscillator whose schematic is shown in Figure 8.16(*b*), the RF choke is connected in series with the self-bias resistor R_s for the same purpose.

Let us consider the principle of the self-bias operation. The start-up conditions are satisfied under the initial gate–source bias corresponding to the large value of the small-signal transconductance when $V_s = 0$. As the oscillation amplitude grows, the dc bias gate–source voltage $V_{gs} = V_g - V_s$ decreases due to dc voltage drop across the self-bias resistor with the dc collector current increase ($V_s > 0$), as shown in Figure 8.17(*a*). The decrease of the gate–source bias voltage leads to an appropriate decrease of the large-signal transconductance and to the high-efficiency steady-state operation mode with the gate–source fundamental voltage amplitude V_{in}. As a result, the self-bias condition combines the soft self-excitation of the oscillation with high efficiency of the Class AB, Class B, or Class C operation mode under hard self-excitation of the oscillation with a certain value of the conduction angle.

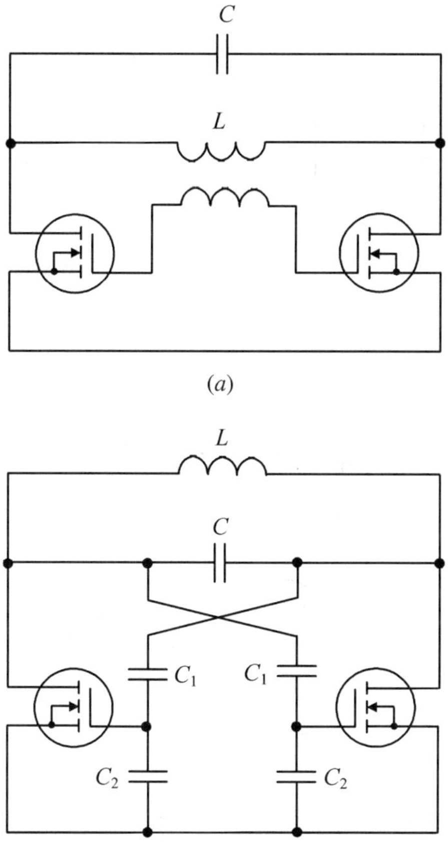

FIGURE 8.15 Schematics of push–pull oscillators.

Using a piecewise-linear approximation of the device transfer characteristic, the dc drain current I_0 as a function of the input gate voltage amplitude V_{in} can be determined by means of the conduction angle as

$$I_0 = g_m V_{in} \gamma_0(\theta) \tag{8.20}$$

where g_m is the device small-signal transconductance, $\gamma_0(\theta)$ is the dc current coefficient, and

$$\cos\theta = -\frac{V_{gs} - V_p}{V_{in}} \tag{8.21}$$

where 2θ is the conduction angle and V_p is the device pinch-off voltage.

Substituting Eq. (8.21) into Eq. (8.20) allows us to obtain the relationship between gate–source bias voltage V_{gs} and dc drain current I_0 in the form

$$V_{gs} = V_p - \frac{I_0}{g_m} \frac{\cos\theta}{\gamma_0(\theta)}. \tag{8.22}$$

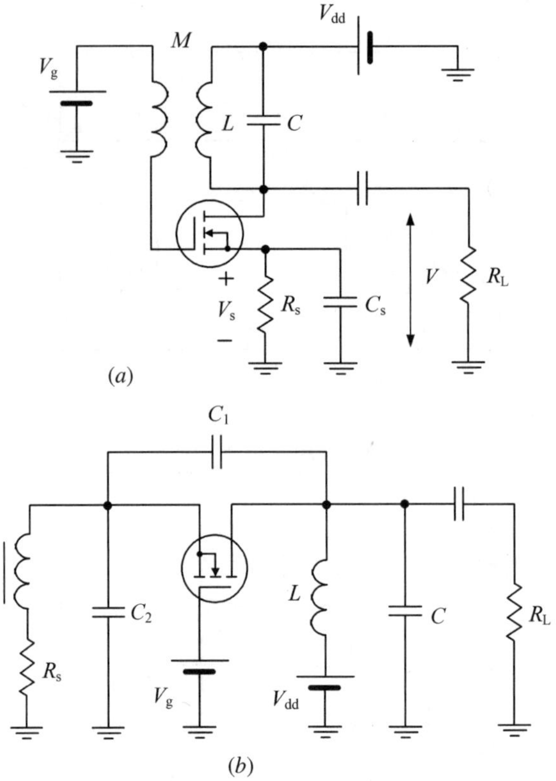

FIGURE 8.16 MOSFET oscillators with self-bias resistors.

On the other hand, a similar relationship can be written by

$$V_{gs} = V_g - I_0 R_s. \tag{8.23}$$

Figure 8.17(*b*) shows the graphical solution of a system of these two equations, where Eq. (8.22) with the conduction angle of $2\theta = 360^\circ$ is plotted by curve 1, Eq. (8.22) with the conduction angle of $2\theta < 180^\circ$ is plotted by curve 2, and Eq. (8.23) is plotted by curve 3. The intersection of curve 3 with curve 1 at the point *A* corresponds to the start-up oscillation mode, whereas the point *B* corresponds to the steady-state oscillation mode with constant amplitude. During the oscillation build-up, the dc drain current increases by the value of ΔI_0.

Consequently, during the process of self-oscillations build-up, the two separate processes occur simultaneously: the increase of the oscillation amplitude across the resonant circuit and the decrease of the bias gate–source voltage due to the presence of the self-bias resistor R_s. Generally, both these processes are interdependent. The velocity of the first process is defined by the time constant of the resonant circuit given by

$$T = \frac{Q}{\omega_0} \tag{8.24}$$

where $Q = \omega_0 C R_L$ is the oscillator quality factor at the resonant frequency ω_0. The velocity of the second process is defined by the time constant of the self-bias circuit as

$$T_s = R_s C_s. \tag{8.25}$$

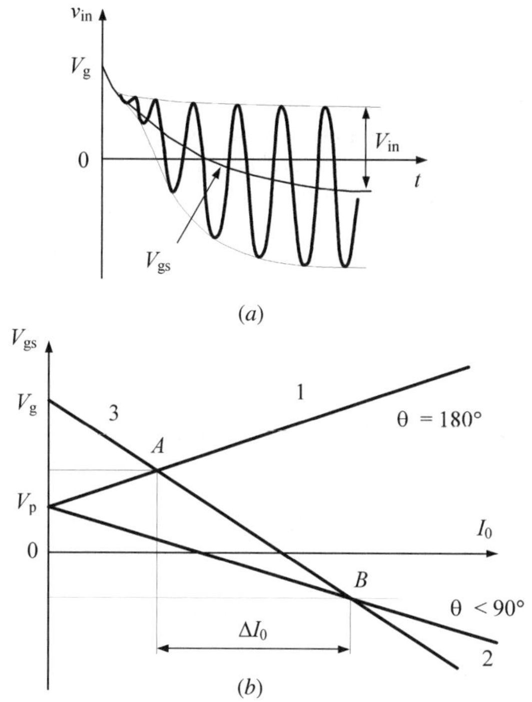

FIGURE 8.17 Self-bias operation.

If $T_s \gg T$, the self-oscillations grow very rapidly, corresponding to the soft start-up condition for the bias voltage of $V_g - V_s = V_g$ ($V_s = 0$). The intersection of the oscillator amplitude characteristic $I_1(V_{in})$ with $V_s = 0$ and feedback line provides the voltage amplitude across the resonant circuit corresponding to point 1 shown in Figure 8.18(*a*). The dc drain current grows with the same velocity, resulting in the slow growth of the self-bias voltage V_s due to the slow transient response of the self-bias circuit. This process contributes to the gradual transition to the points 2 and then 3, characterized

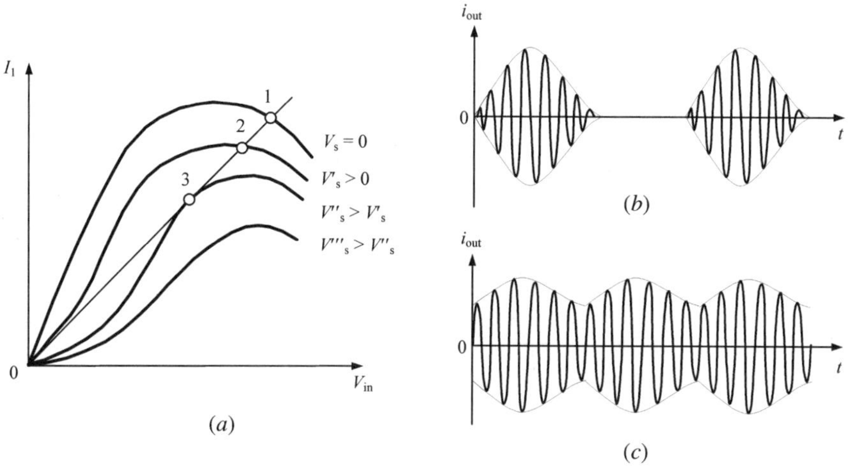

FIGURE 8.18 Self-pulse modulation and self-modulation phenomena.

by different types of the oscillator amplitude characteristic and smaller voltage amplitudes, with further collapse of the self-oscillations. The dc drain current becomes zero, resulting in the onset of the discharging process of the self-bias circuit. When the discharging process is finished, the process of the self-oscillations build-up will start once again. Thus, such a self-oscillation process is accompanied by the *self-pulse modulation*, as shown in Figure 8.18(*b*). For a smaller difference between T_s and T, the process of the oscillation amplitude decrease may not reach the collapse point when the oscillation amplitude starts to grow up. As a result, the self-oscillations have an amplitude modulation in the steady-state operation mode. This process is called the *self-modulation*. To eliminate the self-modulation effect, it is necessary to choose the condition when $T_s < T$.

8.4 PARALLEL FEEDBACK OSCILLATOR

The determination of the start-up and steady-state oscillation conditions is very often based upon a loop or nodal analysis of the circuit. However, the oscillator analysis using matrix techniques brings out the similarities between several types of the oscillators and results in one group of equations, which can be used to analyze different oscillator configurations [13]. In this case, a two-port network can represent both the active device and feedback element. Depending on the oscillator configuration with parallel or series feedback, using of admittance Y- or impedance Z-parameters can be respectively applied.

A generalized equivalent circuit of the parallel feedback oscillator is shown in Figure 8.19(*a*), where Y_1, Y_2, and Y_3 are the feedback elements. To calculate the steady-state stationary operation mode, it is convenient to add the matrix of the parallel feedback element Y_2 and the matrix of the active device Y-parameters according to

$$[Y] + [Y_2] = \begin{bmatrix} Y_{11} + Y_2 & Y_{12} - Y_2 \\ Y_{21} - Y_2 & Y_{22} + Y_2 \end{bmatrix}. \tag{8.26}$$

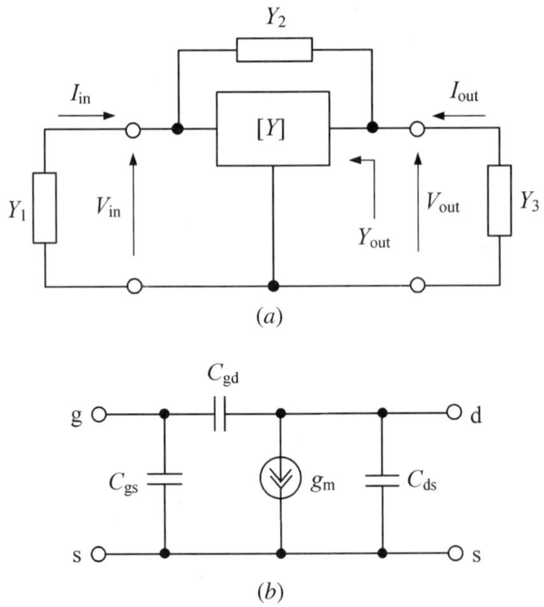

FIGURE 8.19 Equivalent circuit of parallel feedback oscillator.

In this case, a system of two equations for the two-port network input current I_{in}, output current I_{out}, input voltage V_{in}, and output voltage V_{out} can be written as

$$I_{\text{in}} = (Y_{12} + Y_2)\, V_{\text{in}} + (Y_{12} - Y_2)\, V_{\text{out}} \tag{8.27}$$

$$I_{\text{out}} = (Y_{21} - Y_2)\, V_{\text{in}} + (Y_{22} + Y_2)\, V_{\text{out}}. \tag{8.28}$$

Since, for the oscillator shown in Figure 8.19(a), the boundary conditions are obtained in the form

$$I_{\text{in}} = -Y_1 V_{\text{in}} \tag{8.29}$$

$$I_{\text{out}} = -Y_3 V_{\text{out}} \tag{8.30}$$

the following matrix equation can be written as

$$\begin{bmatrix} Y_{11} + Y_1 + Y_2 & Y_{12} - Y_2 \\ Y_{21} - Y_2 & Y_{22} + Y_2 + Y_3 \end{bmatrix} \begin{bmatrix} V_{\text{in}} \\ V_{\text{out}} \end{bmatrix} = 0. \tag{8.31}$$

Thus, the steady-state stationary condition can be derived by equating determinant to zero,

$$\begin{vmatrix} Y_{11} + Y_1 + Y_2 & Y_{12} - Y_2 \\ Y_{21} - Y_2 & Y_{22} + Y_2 + Y_3 \end{vmatrix} = 0. \tag{8.32}$$

After some simplification, Eq. (8.32) can be rewritten by

$$Y_{22} + Y_2 + Y_3 - \frac{(Y_{12} - Y_2)(Y_{21} - Y_2)}{Y_{11} + Y_1 + Y_2} = 0 \tag{8.33}$$

As a result, the steady-state oscillation condition for the parallel feedback oscillator represented as a one-port negative conductance oscillator can be written as

$$Y_{\text{out}} + Y_{\text{L}} = 0 \tag{8.34}$$

where $Y_3 = Y_{\text{L}}$, and

$$Y_{\text{out}} = Y_{22} + Y_2 - \frac{(Y_{12} - Y_2)(Y_{21} - Y_2)}{Y_{11} + Y_1 + Y_2}. \tag{8.35}$$

Consequently, the separate equations for real and imaginary parts of the admittances for a negative conductance oscillator, which are similar to Eqs. (8.3) and (8.4) for magnitude and phase of the loop gain of a parallel feedback oscillator corresponding to the steady-state oscillation process, can be obtained as

$$\text{Re}Y_{\text{out}} + \text{Re}Y_{\text{L}} = 0 \tag{8.36}$$

$$\text{Im}Y_{\text{out}} + \text{Im}Y_{\text{L}} = 0. \tag{8.37}$$

Similarly, the start-up conditions for the parallel feedback oscillator given by Eqs. (8.8) and (8.9) can be rewritten as

$$\text{Re}Y_{\text{out}} + \text{Re}Y_{\text{L}} < 0 \tag{8.38}$$

$$\text{Im}Y_{\text{out}} + \text{Im}Y_{\text{L}} = 0. \tag{8.39}$$

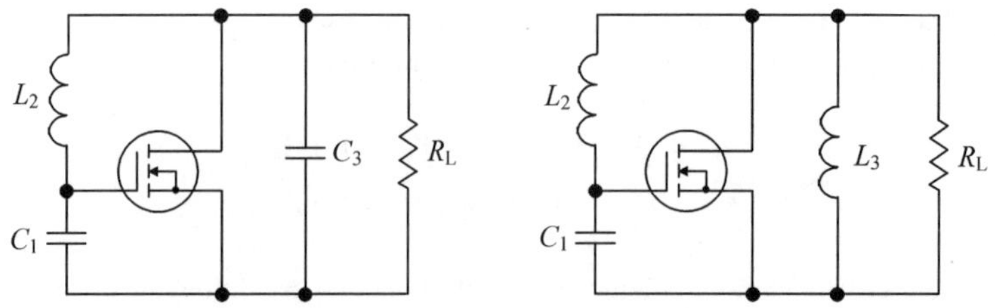

FIGURE 8.20 Electrical circuits of parallel feedback oscillators.

To obtain the analytical relationships between the active device and resonant circuit parameters, let us consider the simplified intrinsic MOSFET high-frequency small-signal equivalent circuit shown in Figure 8.19(*b*). The Y-parameters of the equivalent circuit are

$$\begin{aligned} Y_{11} &= j\omega\left(C_{gs} + C_{gd}\right) \quad & Y_{12} &= -j\omega C_{gd} \\ Y_{21} &= g_m - j\omega C_{gd} \quad & Y_{22} &= j\omega\left(C_{ds} + C_{gd}\right). \end{aligned} \tag{8.40}$$

An assumption of the lossless feedback elements allows us to provide a simple qualitative evaluation of the oscillator start-up conditions. For a Colpitts oscillator, whose basic circuit schematic is shown in Figure 8.20(*a*), the feedback and load admittances are $Y_1 = j\omega C_1$, $Y_2 = 1/(j\omega L_2)$, and $Y_L = 1/R_L + j\omega C_3$.

Then, by substituting all admittances in Eq. (8.35), the small-signal device transconductance g_m required to excite the self-oscillations will be determined by

$$g_m > \frac{1}{R_L}\frac{C_{gs} + C_1}{C_{ds} + C_3}. \tag{8.41}$$

The self-oscillation frequency that depends on both transistor equivalent circuit parameters and feedback elements can be defined from

$$\omega = \sqrt{\frac{1}{L_2}\frac{C_{gs} + C_{ds} + C_1 + C_3}{\left(C_{gs} + C_{gd} + C_1\right)\left(C_{ds} + C_3\right) + \left(C_{gs} + C_1\right)C_{gd}}}. \tag{8.42}$$

If the value of the MOSFET intrinsic feedback capacitance C_{gd} is negligible, the expression for the oscillation is simplified to

$$\omega = \sqrt{\frac{1}{L_2}\left(\frac{1}{C_{gs} + C_1} + \frac{1}{C_{ds} + C_3}\right)}. \tag{8.43}$$

From Eq. (8.41) it follows that, at lower frequencies compared with the device transition frequency f_T when the effect of the elements of the device equivalent circuit is not significant, in order to provide the soft build-up of the oscillation, it is necessary to choose the feedback elements, load resistance, and dc bias point to provide

$$g_m > \frac{1}{R_L}\frac{C_1}{C_3}. \tag{8.44}$$

Because it was assumed that the value of the feedback susceptance $B_3 = \mathrm{Im}Y_3$ is positive, consequently, to satisfy the start-up and steady-state oscillation conditions, it is necessary to use the capacitance C_3. By rewriting Eq. (8.43) as

$$B_3 = \omega \left(\frac{C_{gs} + C_1}{\omega^2 L_2 \left(C_{gs} + C_1 \right) - 1} - C_{ds} \right) \tag{8.45}$$

it is easy to show that when

$$L_2 < \frac{1}{\omega^2 \left(C_{gs} + C_1 \right)} \tag{8.46}$$

then the value of B_3 becomes negative. This means that such an oscillator feedback element must be inductive with the value of $L_3 = -1/(\omega B_3)$. The circuit configuration for this oscillator is shown in Figure 8.20(*b*).

8.5 SERIES FEEDBACK OSCILLATOR

A generalized equivalent circuit of the series feedback oscillator is shown in Figure 8.21(*a*), where Z_1, Z_2, and Z_3 are the feedback elements. To calculate the steady-state operation mode, it is convenient to add the matrix of the series feedback element Z_2 and the matrix of the active device Z-parameters according to

$$[Z] + [Z_2] = \begin{bmatrix} Z_{11} + Z_2 & Z_{12} + Z_2 \\ Z_{21} + Z_2 & Z_{22} + Z_2 \end{bmatrix}. \tag{8.47}$$

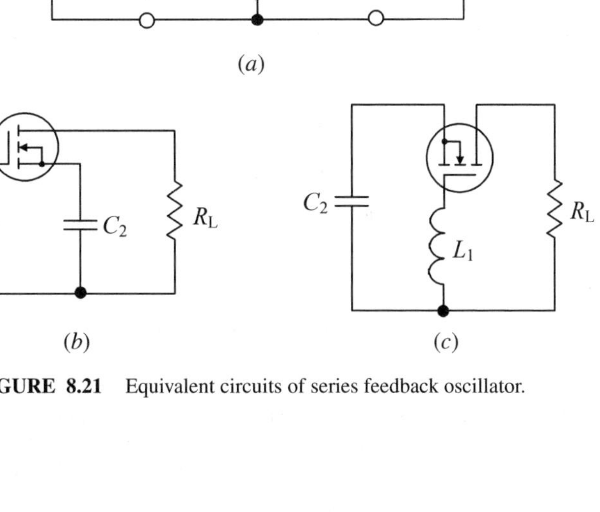

FIGURE 8.21 Equivalent circuits of series feedback oscillator.

In this case, a system of two equations for the two-port network input current I_{in}, output current I_{out}, input voltage V_{in}, and output voltage V_{out} can be written as

$$V_{\mathrm{in}} = (Z_{12} + Z_2)\, I_{\mathrm{in}} + (Z_{12} + Z_2)\, I_{\mathrm{out}} \tag{8.48}$$

$$V_{\mathrm{out}} = (Z_{21} + Z_2)\, I_{\mathrm{in}} + (Z_{22} + Z_2)\, I_{\mathrm{out}}. \tag{8.49}$$

Since, for the oscillator shown in Figure 8.21(a), the boundary conditions are obtained in the form

$$V_{\mathrm{in}} = -Z_1 I_{\mathrm{in}} \tag{8.50}$$

$$V_{\mathrm{out}} = -Z_3 I_{\mathrm{out}} \tag{8.51}$$

the following matrix equation can be written as

$$\begin{bmatrix} Z_{11} + Z_1 + Z_2 & Z_{12} + Z_2 \\ Z_{21} + Z_2 & Z_{22} + Z_2 + Z_3 \end{bmatrix} \begin{bmatrix} I_{\mathrm{in}} \\ I_{\mathrm{out}} \end{bmatrix} = 0. \tag{8.52}$$

Thus, the steady-state oscillation condition can be expressed by

$$\begin{vmatrix} Z_{11} + Z_1 + Z_2 & Z_{12} + Z_2 \\ Z_{21} + Z_2 & Z_{22} + Z_2 + Z_3 \end{vmatrix} = 0. \tag{8.53}$$

After some simplifications, Eq. (8.53) can be rewritten as

$$Z_{22} + Z_2 + Z_3 - \frac{(Z_{12} + Z_2)(Z_{21} + Z_2)}{Z_{11} + Z_1 + Z_2} = 0. \tag{8.54}$$

As a result, the steady-state oscillation condition for the series feedback oscillator represented as a one-port negative resistance oscillator can be written as

$$Z_{\mathrm{out}} + Z_{\mathrm{L}} = 0 \tag{8.55}$$

where $Z_3 = Z_{\mathrm{L}}$, and

$$Z_{\mathrm{out}} = Z_{22} + Z_2 - \frac{(Z_{12} + Z_2)(Z_{21} + Z_2)}{Z_{11} + Z_1 + Z_2}. \tag{8.56}$$

Consequently, the separate equations for real and imaginary parts of the negative resistance oscillator, which are similar to Eqs. (8.36) and (8.37) for the parallel feedback oscillator corresponding to the steady-state oscillation process, can be obtained by

$$\mathrm{Re}Z_{\mathrm{out}} + \mathrm{Re}Z_{\mathrm{L}} = 0 \tag{8.57}$$

$$\mathrm{Im}Z_{\mathrm{out}} + \mathrm{Im}Z_{\mathrm{L}} = 0. \tag{8.58}$$

Similarly, the start-up conditions for the series feedback oscillator can be rewritten as

$$\mathrm{Re}Z_{\mathrm{out}} + \mathrm{Re}Z_{\mathrm{L}} < 0 \tag{8.59}$$

$$\mathrm{Im}Z_{\mathrm{out}} + \mathrm{Im}Z_{\mathrm{L}} = 0. \tag{8.60}$$

Let us obtain the analytical relationships between the active device and resonant circuit parameters of the oscillator based on the MOSFET device, the admittance Y-parameters of the equivalent circuit of which are given by Eq. (8.40). The ratios between admittance Y-parameters and impedance Z-parameters can be expressed in the form

$$Z_{11} = \frac{Y_{22}}{\Delta Y} \quad Z_{12} = -\frac{Y_{12}}{\Delta Y} \quad Z_{21} = -\frac{Y_{21}}{\Delta Y} \quad Z_{22} = \frac{Y_{11}}{\Delta Y}. \tag{8.61}$$

Then, the steady-state condition for the series feedback oscillator can be written as

$$\frac{1 + Z_2 (Y_{11} + Y_{12} + Y_{21} + Y_{22}) + Z_1 (Y_{11} + Z_2 \Delta Y)}{Y_{22} + (Z_1 + Z_2) \Delta Y} + Z_3 = 0 \tag{8.62}$$

where $\Delta Y = 1/\Delta Z$ and $\Delta Y = Y_{11}Y_{22} - Y_{12}Y_{21}$. For a simple oscillator circuit configuration with external gate inductance shown in Figure 8.21(*b*), the feedback elements including load are defined as $Z_1 = j\omega L_1$, $Z_2 = 1/j\omega C_2$, and $Z_3 = R_L$.

Usually, in a wide frequency range up to $0.3f_T$, it is possible to neglect the intrinsic gate–drain capacitance C_{gd} without the substantial loss of accuracy. Then, the small-signal transconductance g_m corresponding to the soft start-up condition and the frequency of the self-oscillations can be evaluated, respectively, by

$$g_m > C_{gs} R_L \frac{(\omega C_{ds})^2}{C_{ds} + C_2} \tag{8.63}$$

$$\omega = \sqrt{\frac{1}{L_1}\left(\frac{1}{C_{gs}} + \frac{1}{C_{ds} + C_2}\right)}. \tag{8.64}$$

From Eq. (8.63) it follows that the oscillation build-up can be easily provided by choosing the MOSFET device with the smaller drain–source capacitance C_{ds} and using a sufficiently large value of the feedback capacitance C_2.

In a common case, different sets of two-port network parameters can be used for oscillator design including impedance Z-parameters, admittance Y-parameters, or scattering S-parameters. The choice of any type of these parameters depends on the operating frequency, required design accuracy and implementation technique, availability of the small- or large-signal active device parameters, or the possibility of using proper measurement and simulation tools. For example, in terms of one-port negative resistance approach, the conditions for self-oscillations expressed through the small-signal S-parameters can be written as

$$K = \frac{1 + |\Delta|^2 - |S_{11}|^2 - |S_{22}|^2}{2\,|S_{12}S_{21}|} < 1 \tag{8.65}$$

$$\Gamma_{in}\Gamma_S = 1 \tag{8.66}$$

$$\Gamma_{out}\Gamma_L = 1 \tag{8.67}$$

where K is the stability factor, $\Delta = S_{11}S_{22} - S_{12}S_{21}$, Γ_{in} and Γ_S are the input and source reflection coefficients, Γ_{out} and Γ_L are the output and load reflection coefficients, respectively [14]. The stability factor should be less than unity for any possibility of self-oscillations. The passive terminations Γ_S and Γ_L must be added to resonate input and output ports at the oscillation frequency. It should be noted that the conditions described by Eqs. (8.66) and (8.67) are inter-related, and if one condition is satisfied, then the other condition should be satisfied as well. The large-signal S-parameters of the transistor optimized for maximum output power can be measured by varying the input drive level and load impedance, or injecting signal into the output port [15]. By using the measured S-parameters, the required ratios of terminal voltages and currents can be calculated.

By converting S-parameters to Z-parameters, the active device impedance parameters can be obtained as

$$\begin{bmatrix} Z_{11} & Z_{12} \\ Z_{21} & Z_{22} \end{bmatrix} = \frac{Z_0}{(1-S_{11})(1-S_{22})-S_{12}S_{21}} \times \begin{bmatrix} (1+S_{11})(1-S_{22})+S_{12}S_{21} & 2S_{12} \\ 2S_{21} & (1-S_{11})(1+S_{22})+S_{12}S_{21} \end{bmatrix} \quad (8.68)$$

that allows us to calculate the input or output impedance of the loaded active device in the parallel feedback two-port network, or negative resistance series feedback one-port network oscillator configurations.

8.6 PUSH–PUSH OSCILLATORS

Using a common collector configuration of the transistors and a series resonant circuit located between the devices in the push–pull oscillators simplifies the load connection and allows the operation condition at twice the operating frequency. Figure 8.22(a) shows the general simplified equivalent circuit of the balanced oscillator having two load resistors, one connected in parallel to the series resonant circuit inductor and the other connected into the inductor center point. The series inductor located between the two active devices is common for both oscillators.

For a simplified circuit analysis, let us represent the oscillator schematic in the form of a general negative conductance oscillator with two active devices connected to a common two-port network, as shown in Figure 8.22(b). The two-port network is characterized by the admittance Y-parameters and terminal voltages represented by voltage phasors given as

$$\mathbf{V}_1 = V_1 \exp(j\phi_1) \quad \mathbf{V}_2 = V_2 \exp(j\phi_2) \quad (8.69)$$

where V_1, V_2, ϕ_1, and ϕ_2 are the magnitudes and phases of two voltage phasors, respectively.

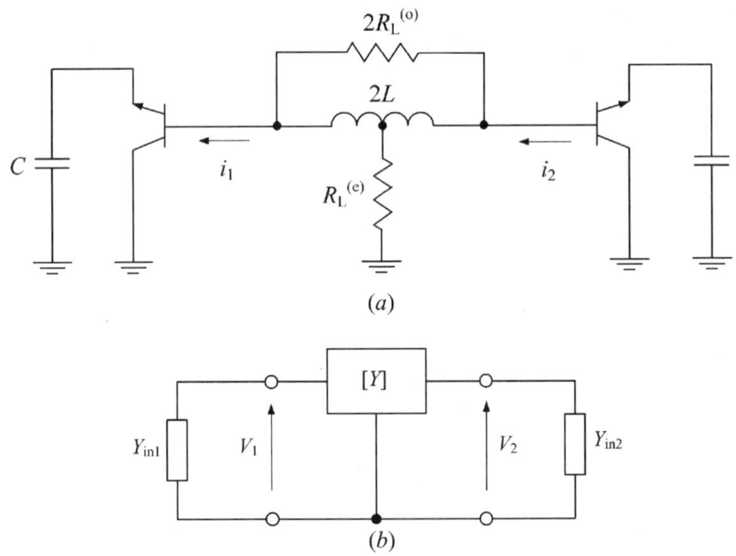

FIGURE 8.22 Simplified circuit schematics of bipolar push–pull oscillator.

The relationships between the circuit currents and voltages in a steady-state operation mode through the admittance parameters in a matrix form can be written by

$$\begin{bmatrix} -Y_{\text{in1}} & \mathbf{V}_1 \\ -Y_{\text{in2}} & \mathbf{V}_2 \end{bmatrix} = \begin{bmatrix} Y_{11} & Y_{12} \\ Y_{21} & Y_{22} \end{bmatrix} \begin{bmatrix} \mathbf{V}_1 \\ \mathbf{V}_2 \end{bmatrix} \tag{8.70}$$

where Y_{in1} and Y_{in2} represent the input admittances of the negative conductance devices.

Since the oscillator consists of two identical nonlinear active halves when $V_1 = V_2 = V$ and $V_{\text{in1}} = V_{\text{in2}} = V_{\text{in}}$ and symmetrical passive linear two-port network when $Y_{11} = Y_{22}$ and $Y_{12} = Y_{21}$, matrix Eq. (8.70) can be rewritten in the form of two equations as

$$-Y_{\text{in}} = Y_{11} + Y_{12} \exp(j\phi) \tag{8.71}$$

$$-Y_{\text{in}} = Y_{12} \exp(-j\phi) + Y_{11} \tag{8.72}$$

where $\phi = \phi_2 - \phi_1$ is the phase difference between voltage phasors.

Simplified analysis of the steady-state operation modes shows that, in such a balanced oscillator, there may exist two modes with equal amplitudes [2,16]:

- odd mode with

$$\phi = (2k + 1)\pi \quad \text{where } k = 0,\ 1,\ 2,\ \ldots$$

- even mode with

$$\phi = 2k\pi \quad \text{where } k = 0,\ 1,\ 2,\ \ldots.$$

The frequency and amplitude of each of the two oscillation modes can be determined by solving the following equation:

$$Y_{\text{in}}(V,\omega) + Y_{11}(\omega) \pm Y_{12}(\omega) = 0. \tag{8.73}$$

In a time domain, assuming symmetrical current waveforms flowing into the base of both transistors, their Fourier series expansions in a common case can be written by

$$i_1 = I_1^{(1)} \cos\omega t + I_2^{(1)} \cos 2\omega t + I_3^{(1)} \cos 3\omega t + \cdots + I_{\text{n}}^{(1)} \cos n\omega t \tag{8.74}$$

$$i_2 = I_1^{2)} \cos(\omega t + \phi) + I_2^{(2)} \cos 2(\omega t + \phi) \quad + I_3^{(2)} \cos 3(\omega t + \phi) + \cdots + I_{\text{n}}^{(2)} \cos n(\omega t + \phi) \tag{8.75}$$

where n is the harmonic number.

Consequently, in the odd mode, the currents are flowing in opposite directions providing a push–pull operation of both transistors having 180° out-of-phase base–emitter voltages. In this case, the circuit becomes antimetric and, at its center point, a virtual ground will be formed at the fundamental frequency f_0 with zero fundamental voltage at this point. The output current flowing into the load $2R_{\text{L}}^{(\text{o})}$ is a result of the out-of-phase summation of currents given by Eqs. (8.74) and (8.75) as

$$i^{(\text{o})} = 2I_1 \cos\omega t + 2I_3 \cos 3\omega t + \cdots + 2I_{2\text{k}+1} \cos(2k + 1)\omega t \tag{8.76}$$

which contains only odd current components, and there are no odd components flowing into the load $R_{\text{L}}{}^{(\text{e})}$. At the same time, the output current flowing into the load $R_{\text{L}}{}^{(\text{e})}$ is a result of the in-phase summation of currents as

$$i^{(\text{e})} = 2I_2 \cos 2\omega t + 2I_4 \cos 4\omega t + \cdots + 2I_{2\text{k}} \cos 2k\omega t \tag{8.77}$$

which contains only even current components.

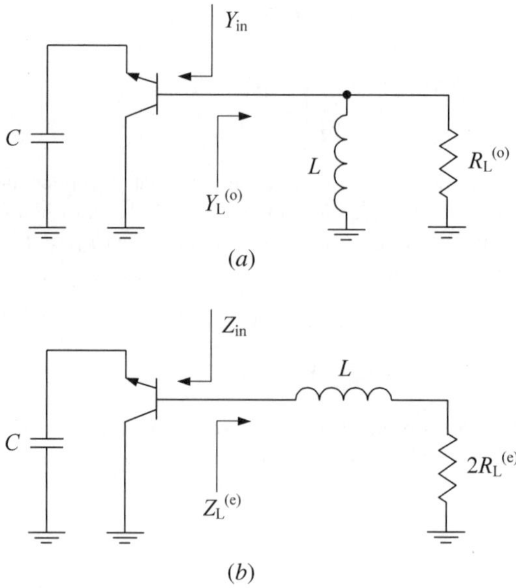

FIGURE 8.23 Equivalent oscillator circuits with (*a*) odd and (*b*) even modes.

In the even mode, the currents are flowing in the same direction providing a push–push operation of both transistors having in-phase base–emitter voltages. In this case, the circuit becomes symmetric with a load resistor $R_L^{(e)}$ at the center of symmetry. For such an operation, the odd current components obtained by Eq. (8.76) will flow to this load, whereas the even components obtained by Eq. (8.77) will be dissipated at the load $2R_L^{(o)}$.

The equivalent circuits of the oscillator under odd and even operation modes are shown in Figure 8.23. For oscillations to occur in the odd mode, the negative conductance generated by the oscillator, whose schematic is shown in Figure 8.23(*a*), should exceed the resistive losses in the resonant circuit, that is

$$\mathrm{Re}Y_{in} + \mathrm{Re}Y_L^{(o)} < 0. \tag{8.78}$$

For oscillations to occur in the even mode, the negative resistance generated by the oscillator, whose schematic is shown in Figure 8.23(*b*), should exceed the resistive losses in such a resonant circuit according to

$$\mathrm{Re}Z_{in} + \mathrm{Re}Z_L^{(e)} < 0. \tag{8.79}$$

Generally, both odd and even operation modes can exist in the oscillator simultaneously depending on the values of the load resistances $R_L^{(o)}$ and $R_L^{(e)}$. Under a start-up condition given by Eq. (8.78), the even mode will be stable for the case of $R_L^{(o)} = \infty$ corresponding to the situation of the summation of the output powers of individual oscillators with resistive coupling [17]. A simplified criterion of the odd mode stable operation for the oscillator circuit shown in Figure 8.22(*a*) can be obtained as

$$R_L^{(o)} < \frac{2R_L^{(e)}}{1 + \left(\dfrac{\omega L}{2R_L^{(e)}}\right)^2} \tag{8.80}$$

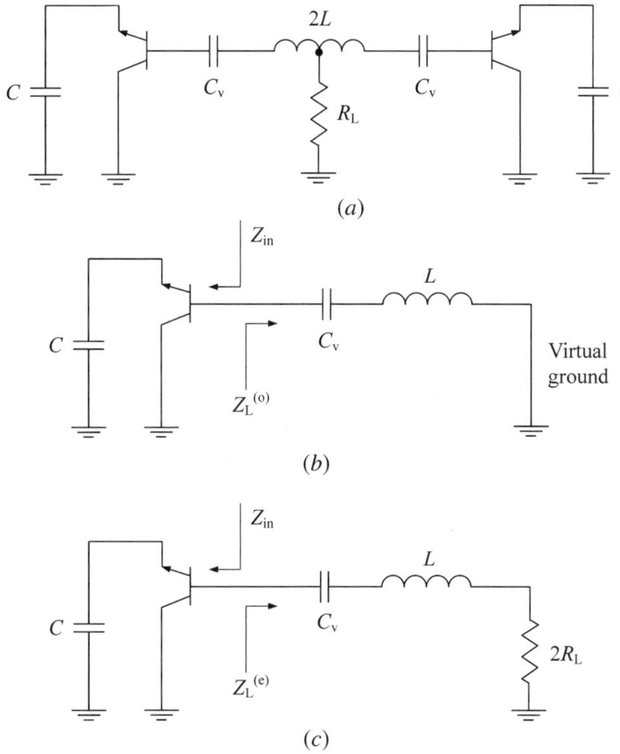

FIGURE 8.24 Simplified circuit schematics of bipolar push–push oscillator.

when the odd components will flow into the load $2R_L^{(o)}$, while the even components will dissipate at the load $R_L^{(e)}$. To improve stability of the odd operation mode, it is necessary to inhibit the oscillation condition given by Eq. (8.79).

Such a balanced oscillator configuration with a series resonant circuit creates a simple opportunity to double the oscillation frequency when each half-circuit oscillates at the resonant frequency f_0, while the output signal at the load oscillates at the frequency $2f_0$. In this case, it is necessary to provide stable operation in the odd mode and to inhibit the oscillations in the even mode by removing the load resistor $R_L^{(o)}$. This results in the oscillator circuit configuration shown in Figure 8.24, where the variable capacitors C_v can be used for frequency tuning. For a lossless tank inductor L, the start-up amplitude oscillation conditions can be rewritten as

$$\text{Re}Z_{\text{in}} < 0 \tag{8.81}$$

$$\text{Re}Z_{\text{in}} + 2R_L > 0 \tag{8.82}$$

when the even harmonic components dissipate at the load R_L. A significant margin in negative resistance can be achieved compared with an equivalent single-ended oscillator, even at very high frequencies [18]. The outputs for the out-of-phase odd components can be connected to the device emitters. Because, for the even output current components, the transistors are operated in phase, it is called the *push–push operation mode*. A push–push operation is the inverse to a push–pull operation, in that the load is either conductively or capacitively coupled to the center point of inductor or transmission line located between the bases (gates) of the active devices [19,20]. In this case, the currents flow into the load in the same direction during both half periods or 180° phases of the active

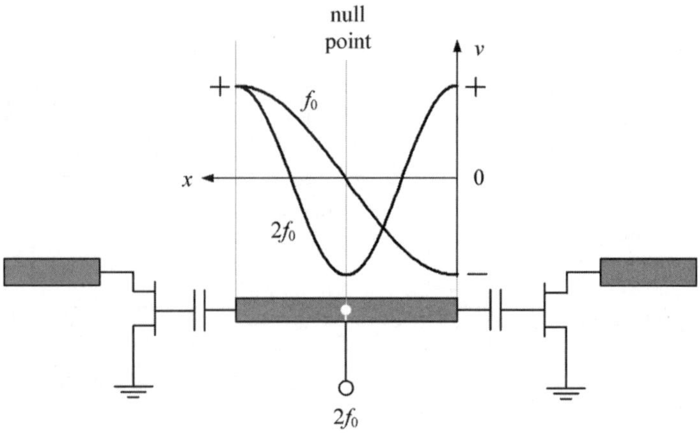

FIGURE 8.25 Push–push and balanced microstrip oscillator configurations.

device operation. Since there are two 180° phases during each cycle, the circuit acts as a frequency doubler. Such a push–push oscillator is a result of its odd operation mode with virtual ground; therefore, ideally this design is independent of the output load pulling compared to a fundamental oscillator approach.

At microwaves, the resonant circuits usually incorporate the transmission lines as inductive elements or resonators. Figure 8.25 shows the push–push oscillator using half wavelength or $\lambda/2$ microstrip resonator [21]. For the fundamental frequency f_0, this resonator has a null point at the center of microstrip line, being a point of the oscillator symmetry, which is considered as a virtual ground or short-circuited point. In this case, the resonance voltage has maximum values at both ends of the resonator with a phase difference of 180°, and the resonance voltage is zero at the center of the resonator. For the desired second-harmonic frequency $2f_0$, such a point could be considered as an open-circuited point. The null point is an ideal output port to combine the second-harmonic signals from both active devices effectively. To compromise the maximum second-harmonic output signal and significant harmonic suppression, a sufficiently small series capacitor is usually used. The operating bias points with zero gate voltages can provide nonlinear operation of the used high electron mobility transistor (HEMT) devices generating a high-power second-harmonic signal. The impedance of half-wavelength microstrip open-circuited resonator is optimized to make the impedance of the output port close to 50 Ω. For such a microwave push–push oscillator using two Fujitsu FHX35LG devices, a maximum power of 8.4 dBm at the second-harmonic frequency of 21.68 GHz with the fundamental frequency suppression of 26 dB was achieved.

8.7 STABILITY OF SELF-OSCILLATIONS

Applying dc bias to the active device does not generally result in the negative resistance or proper feedback condition. This condition has to be induced in these devices, and it is determined by the physical mechanism in the device and chosen circuit topology. The transistor in the oscillator circuits is mostly represented as the active two-port network, whose operation principle is reflected through its equivalent circuit. The influence of the circuit and transistor parameters can result in the hysteresis effect or oscillation instability in practical design. In high-frequency practical implementation, the presence of the parasitic device and circuit elements can contribute to the multiresonant circuits.

A steady-state free-running oscillation build-up is provided with the velocity due to dissipation factor $\delta = [R_{\text{out}}(I) + R_{\text{L}}]/L$ for a single series resonant circuit with an inductance L and $\delta = [G_{\text{out}}(V) +$

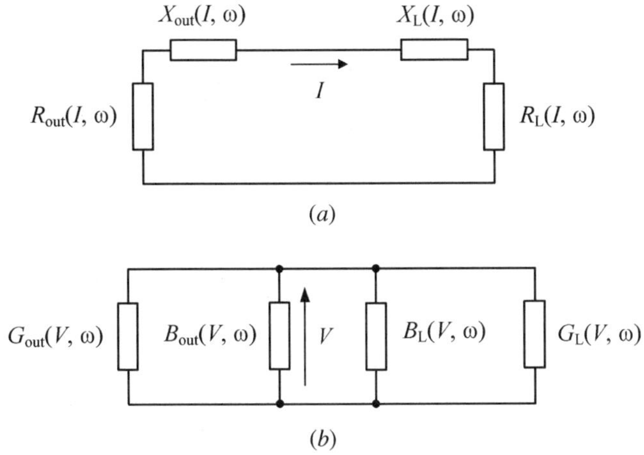

FIGURE 8.26 One-port (*a*) negative resistance and (*b*) negative conductance oscillator circuits.

$G_L]/C$ for a single parallel resonant circuit with a shunt capacitance C shown in Figures 8.26(*a*) and 8.26(*b*), respectively, the value of which is reduced with the decrease of negative output resistance $R_{out}(I)$ or conductance $G_{out}(V)$. It becomes zero in a steady-state oscillation mode when $R_{out}(I) + R_L = 0$ or $G_{out}(V) + G_L = 0$, which is written in a general immittance form as

$$\mathrm{Re}W_{out}(A_0) + \mathrm{Re}W_L = 0 \tag{8.83}$$

where A_0 represents a steady-state amplitude of the self-oscillation, voltage or current.

A stability criterion for such simple oscillator circuits, with only an amplitude dependence of output resistance or conductance, can be easily calculated through the perturbation method. Let us consider small perturbation $\Delta A > 0$ in the operating point $\mathrm{Re}W(A_0)$ when $A = A_0 + \Delta A$. Using the linear Taylor series expansion results in

$$\mathrm{Re}W(A_0 + \Delta A) \cong \mathrm{Re}W(A_0) + \frac{\partial \mathrm{Re}W(A_0)}{\partial A}\Delta A. \tag{8.84}$$

A steady-state oscillation mode will be stable if an amplitude of the oscillation dissipates according to $\exp(-\delta t)$ that is obtained from the solution of a second-order differential equation, describing the electrical behavior of each single resonant circuit shown in Figure 8.26 [2]. This takes place when $\mathrm{Re}W(A_0 + \Delta A) > \mathrm{Re}W(A_0)$. Therefore, a stability criterion is

$$\frac{\partial \mathrm{Re}W(A_0)}{\partial A} > 0 \tag{8.85}$$

which means that, as the active device negative resistance or conductance reduces with increase of the oscillation amplitude, stable oscillations are established in the oscillator.

In a common case of (*a*) negative resistance or (*b*) negative conductance one-port transistor oscillator circuit model shown in Figure 8.26, a complex equation consisting of the nonlinear immittances in the steady-state stationary operation mode can be expressed as

$$W(A_0, \omega_0) = \mathrm{Re}W\,(A_0, \omega_0) + j\mathrm{Im}W\,(A_0, \omega_0) = 0 \tag{8.86}$$

where

$$\mathrm{Re}W(A_0, \omega_0) = \mathrm{Re}W_{\mathrm{out}}(A_0, \omega_0) + \mathrm{Re}W_{\mathrm{L}}(A_0, \omega_0) \tag{8.87}$$

$$\mathrm{Im}W(A_0, \omega_0) = \mathrm{Im}W_{\mathrm{out}}(A_0, \omega_0) + \mathrm{Im}W_{\mathrm{L}}(A_0, \omega_0). \tag{8.88}$$

Assuming a solution of Eq. (8.86) as $a(t) = A_0 \cos\omega_0 t$ or $a(t) = A_0\mathrm{Re}[\exp(j\omega_0 t)]$, one can find the amplitude A_0 and frequency ω_0 of free-running oscillations. Then, consider the small perturbation from the steady-state stationary mode when the amplitude and frequency of the oscillation are turned out to be $A = A_0 + \Delta A$ and $\omega = \omega_0 + \Delta\omega$, respectively, where $\Delta A \ll A_0$ and $\Delta\omega \ll \omega_0$. Under conditions of small perturbations from the steady-state mode, a nonlinear system behaves as linear, and amplitude of the oscillation dissipates exponentially. Therefore, the expected solution can be written as

$$a(t) = (A_0 + \Delta A)\exp(-\Delta\delta t)\cos(\omega_0 + \Delta\omega)t \tag{8.89}$$

or

$$a(t) = (A_0 + \Delta A)\,\mathrm{Re}\left[\exp j(\omega_0 + \Delta\omega + j\Delta\delta)t\right]. \tag{8.90}$$

According to Eq. (8.90), the steady-state oscillation condition in the form of Eq. (8.86) can be rewritten as

$$\mathrm{Re}W(A_0 + \Delta A, \omega_0 + \Delta\omega + j\Delta\delta) + j\mathrm{Im}W(A_0 + \Delta A, \omega_0 + \Delta\omega + j\Delta\delta) = 0. \tag{8.91}$$

By applying the linear Taylor series expansion of Eq. (8.91) by degrees of the small parameters ΔA, $\Delta\omega$, and $\Delta\delta$ for each components and replacing one complex equation with two equations for real and imaginary parts, the following system of two equations in consideration of Eq. (8.86) can be written:

$$\frac{\partial \mathrm{Re}W}{\partial A}\Delta A + \frac{\partial \mathrm{Re}W}{\partial \omega}\Delta\omega + \frac{\partial \mathrm{Re}W}{\partial \delta}\Delta\delta = 0 \tag{8.92}$$

$$\frac{\partial \mathrm{Im}W}{\partial A}\Delta A + \frac{\partial \mathrm{Im}W}{\partial \omega}\Delta\omega + \frac{\partial \mathrm{Im}W}{\partial \delta}\Delta\delta = 0. \tag{8.93}$$

It is known that for analytic function $W(\omega + j\Delta\delta) = \mathrm{Re}W(\omega, \Delta\delta) + j\mathrm{Im}W(\omega, \Delta\delta)$, the Cauchy–Riemann equations can be written as

$$\frac{\partial \mathrm{Re}W}{\partial \omega} = \frac{\partial \mathrm{Im}W}{\partial \delta} \tag{8.94}$$

$$\frac{\partial \mathrm{Re}W}{\partial \delta} = -\frac{\partial \mathrm{Im}W}{\partial \omega}. \tag{8.95}$$

Then, excluding $\Delta\omega$ from Eqs. (8.92) and (8.93) and using Eqs. (8.94) and (8.95) result in

$$\frac{\Delta\delta}{\Delta A} = \frac{\dfrac{\partial \mathrm{Re}W}{\partial A}\dfrac{\partial \mathrm{Im}W}{\partial \omega} - \dfrac{\partial \mathrm{Im}W}{\partial A}\dfrac{\partial \mathrm{Re}W}{\partial \omega}}{\left(\dfrac{\partial \mathrm{Re}W}{\partial \omega}\right)^2 + \left(\dfrac{\partial \mathrm{Im}W}{\partial \omega}\right)^2}. \tag{8.96}$$

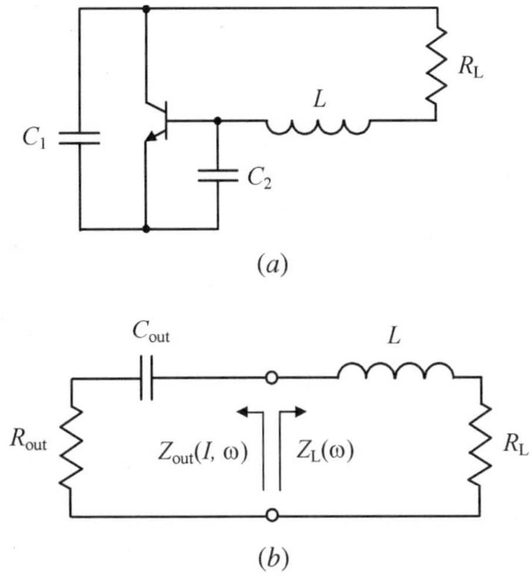

FIGURE 8.27 Parallel feedback (*a*) electrical and (*b*) equivalent oscillator circuits.

The steady-state stationary mode will be stable if a small perturbation of the oscillation amplitude ΔA dissipates in time and the values ΔA and $\Delta\delta$ have the same signs. Thus, the stable oscillations are established in the oscillator if

$$\frac{\partial \mathrm{Re}W}{\partial A}\frac{\partial \mathrm{Im}W}{\partial \omega} - \frac{\partial \mathrm{Im}W}{\partial A}\frac{\partial \mathrm{Re}W}{\partial \omega} > 0 \tag{8.97}$$

which corresponds in terms of immittance parameters to a general form of a stability condition obtained by Kurokawa [22].

Figure 8.27 shows (*a*) the electrical and (*b*) the equivalent circuits of the parallel feedback negative resistance oscillator with a common collector. In this case, a transistor has been configured and biased so that the output resistance is negative. The capacitors C_1 and C_2 are used to increase a regeneration factor and to provide steady-state oscillation conditions of the amplitude and phase.

In this series circuit configuration with the negative imaginary part of output transistor impedance, the load is constant and the negative output resistance $R_{out} = \mathrm{Re}Z_{out}(I, \omega)$ decreases with increase of the oscillation frequency. As a result,

$$\frac{\partial X}{\partial \omega} = \frac{\partial}{\partial \omega}\left(\omega L - \frac{1}{\omega C_{out}}\right) = L + \frac{1}{\omega^2 C_{out}} > 0 \tag{8.98}$$

and

$$\frac{\partial R}{\partial \omega} = \frac{\partial}{\partial \omega}(R_{out} + R_L) = \frac{\partial R_{out}}{\partial \omega} > 0. \tag{8.99}$$

Consequently, by taking into account Eq. (8.97), to obtain stable oscillations in a steady-state operation mode, it is required to satisfy the following sufficient conditions for the series resonant circuit negative resistance oscillator with constant load:

$$\frac{\partial R_{out}}{\partial I} > 0 \frac{\partial X_{out}}{\partial I} < 0. \tag{8.100}$$

8.8 OPTIMUM DESIGN TECHNIQUES

8.8.1 Empirical Approach

For optimum oscillator design with maximum output power, generally it is required extensive small- and large-signal measurements. The small-signal S-parameter measurements can be made at several frequencies, along with estimated device equivalent circuit parameters, including package parasitics. Then, a computer optimization program is used to match the measured S-parameters to the S-parameters computed for the active device from its equivalent circuit. The next step is to vary those elements of the equivalent circuit, which can vary under large signals. By varying the active device nonlinear equivalent circuit parameters, the set of large-signal S-parameters can be obtained corresponding to the saturation condition where maximum oscillator output power can be achieved.

The simplified design approach assumes that all of the S-parameters except the magnitude of S_{21} are constant under large signals. For example, for a metal-semiconductor field-effect transistor (MESFET) device up to saturation with small limitation in accuracy, it is possible to derive the large-signal behavior of the main device nonlinear elements as functions of the device transconductance g_m [23]. Consequently, the S-parameters become functions of g_m only and, at each incremental reduction of g_m, are recomputed and optimized along with power gain. If one were interested in the 1-dB compression point, the S-parameters used would be those at that point.

For a common source (emitter) power oscillator, it is helpful to characterize the maximum power through the saturated output power P_{sat} and small-signal transducer power gain G_T of the corresponding power amplifier. An empirical expression for output power of a common source MESFET power amplifier can be written by

$$P_{out} = P_{sat}\left[1 - \exp\left(-\frac{G_T P_{in}}{P_{sat}}\right)\right] \tag{8.101}$$

where P_{in} is the input power [23].

The objective is to maximize the oscillator output power $P_{osc} = P_{out} - P_{in}$ by providing a condition

$$\frac{dP_{osc}}{dP_{in}} = \frac{d(P_{out} - P_{in})}{dP_{in}} = 0. \tag{8.102}$$

Substituting Eq. (8.101) into Eq. (8.102) yields

$$\frac{dP_{out}}{dP_{in}} = G_T \exp\left(-\frac{G_T P_{in}}{P_{sat}}\right) = 1. \tag{8.103}$$

From Eq. (8.103) it follows that

$$\exp\left(\frac{G_T P_{in}}{P_{sat}}\right) = G_T \tag{8.104}$$

or

$$\frac{P_{in}}{P_{sat}} = \frac{\ln G_T}{G_T}. \tag{8.105}$$

As a result,

$$P_{out} = P_{sat}\left(1 - \frac{1}{G_T}\right) \tag{8.106}$$

and the maximum output power of the oscillator can be approximated by

$$P_{\text{osc}} = P_{\text{sat}}\left(1 - \frac{1}{G_{\text{T}}} - \frac{\ln G_{\text{T}}}{G_{\text{T}}}\right). \tag{8.107}$$

Consequently, the oscillator maximum power can be predicted from the saturated power and small-signal transducer power gain of a common source power amplifier. From Eq. (8.107) it follows that the oscillator output power P_{osc} will approach P_{sat} at low frequencies where the small-signal transducer power gain G_{T} is large. As the transducer power gain approaches unity, the oscillator output power approaches zero.

In this case, the maximum efficiency power gain G_{ME} can be calculated from

$$G_{\text{ME}} = \frac{G_{\text{T}} - 1}{\ln G_{\text{T}}}. \tag{8.108}$$

For example, a MESFET device, which is characterized by the small-signal transducer power gain $G_{\text{T}} = 7.5$ dB and saturated amplifier output power of 1 W, would be capable of a maximum oscillator power of 515 mW. The maximum efficiency power gain at this point calculated from Eq. (8.108) is equal to $G_{\text{ME}} = 4.3$ dB. The expression for maximum efficiency power gain G_{ME} given by Eq. (8.108) can be used to determine at what gain level the large-signal S-parameters were to be used for optimum oscillator design.

A quasilinear design approach based on the measured small-signal S-parameters and static output voltage–ampere device characteristic can be used to predict the oscillator output power [24]. However, this method assumes that the significant nonlinear effects in GaAs MESFET represent the rapid increase in the gate–source conductance due to the forward-biasing effect of the gate–source Schottky diode and output drain–source conductance [25]. To maximize the output power of a series feedback MESFET oscillator, an analytic procedure was derived using the input and output fundamental voltages as independent variables. The systematic approach to oscillator design using large-signal S-parameters, which may be measured under high drive conditions or obtained through the use of an active device nonlinear model, represents an alternative method [26]. The device can be modeled in a quasilinear approximation by its large-signal S-parameters, each assumed to be a function of a single variable when S_{11} and S_{21} are the functions of the input fundamental voltage amplitude, whereas S_{12} and S_{22} are the functions of the output fundamental voltage amplitude. As a result, the set of four equations describes the oscillation condition that requires standard computer-based nonlinear root-finding methods to determine the optimum feedback parameters providing a delivery of maximum output power into the load.

A technique for the design of microwave transistor oscillator, in which measurements made on an experimentally optimized power amplifier, has been presented in [15]. Generally, the transistor power amplifier is easily analyzed and optimized than the corresponding oscillator because, in quasilinear approximation at a given frequency and bias point, only two parameters need to be varied: the input RF drive level and the output load impedance. Once these parameters are experimentally optimized, measuring the input impedance and constructing an input matching circuit complete the design. In the case of the transistor oscillator, however, there are a multitude of possible oscillator configurations and a large number of interacting circuit elements, which must be varied to optimize an oscillator for maximum power. But since we know that a transistor operates under the same set of RF voltages and currents when delivering its maximum output power, it is possible to take information obtained from an easily optimized power amplifier and use this information to calculate optimum oscillator configurations.

The first step in this procedure is to experimentally optimize the large-signal behavior of the transistor by varying the load impedance and drive level for the maximum output power delivered into the load. The incident and reflected waves a_1, b_1, a_2, and b_2 shown in Figure 8.28 can be measured with calculation of the output power delivered into the load according to

$$P_L = |b_2|^2 \left(1 - \frac{1}{|S'_{22}|^2} - \frac{1 - |S'_{11}|^2}{|S'_{21}|^2}\right) \tag{8.109}$$

where

$$S'_{11} = \frac{b_1}{a_1} \quad S'_{22} = \frac{b_2}{a_2} \quad S'_{21} = \frac{b_2}{a_1}$$

S'_{11} is the input reflection coefficient Γ_1 and S'_{22} is the output reflection coefficient Γ_2 [14].

Using these three measured parameters S'_{11},S'_{21}, and S'_{22}, it is then possible to calculate the required ratios of the transistor terminal voltages and currents

$$\frac{V_2}{V_1} = \frac{S'_{21}\left(\frac{1}{S'_{22}} + 1\right)}{1 + S'_{11}} \tag{8.110}$$

$$\frac{I_2}{I_1} = \frac{S'_{21}\left(\frac{1}{S'_{22}} - 1\right)}{1 - S'_{11}} \tag{8.111}$$

$$\frac{V_1}{I_1} = R_S \frac{1 + S'_{11}}{1 - S'_{11}} \tag{8.112}$$

$$\frac{V_2}{I_1} = R_S \frac{S'_{21}\left(\frac{1}{S'_{22}} + 1\right)}{1 - S'_{11}} \tag{8.113}$$

$$\frac{I_2}{V_1} = \frac{1}{R_S} \frac{S'_{21}\left(\frac{1}{S'_{22}} - 1\right)}{1 + S'_{11}} \tag{8.114}$$

where V_1 and I_1 are the transistor input port voltage and current, V_2 and I_2 are the transistor output port voltage and current, and R_S is the source resistance considered as the characteristic impedance of the measuring system, as shown in Figure 8.28.

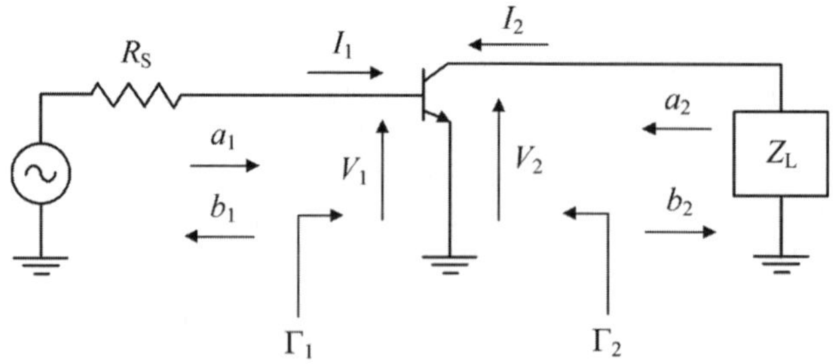

FIGURE 8.28 Power amplifier schematic to be experimentally optimized.

8.8.2 Analytic Approach

Depending on a type of the transistor used, operating frequencies, required output power and spectral characteristics, most of the oscillator schematics can be reduced to the two basic arrangements with external positive parallel or series feedback shown in Figures 8.29(*a*) and 8.29(*b*), respectively. One model may be preferred over another, depending on the oscillator configuration and characteristics. The two-port network representing an active device can be generally characterized by a system of immittance W-parameters. This describes parameter systems of a two-port network, such as a system of admittance Y-parameters for parallel feedback or a system of impedance Z-parameters for series feedback oscillator circuits, respectively. The oscillator circuit and the output load are characterized by immittances W_1, W_2, and $W_3 = W_L$, respectively.

Then, the steady-state oscillation condition for a single frequency of oscillation can be expressed as

$$W_{out} + W_L = 0 \tag{8.115}$$

where the active two-port network, together with the feedback elements W_1 and W_2, will be considered as a one-port negative resistance oscillator circuit.

To optimize the oscillator circuit in terms of the maximum value of the negative real part of the equivalent one-port network output immittance, the expression for output immittance W_{out} should be written as

$$W_{out} = W_{22} + W_2 - \frac{(W_{12} \mp W_2)(W_{21} \mp W_2)}{W_{11} + W_1 + W_2} \tag{8.116}$$

where the minus signs in the factors correspond to output admittance of the circuit shown in Figure 8.29(*a*) and the plus signs correspond to the output impedance of the circuit shown in Figure 8.29(*b*). Such an optimization approach was first applied to a bipolar transformer-coupled oscillator to define maximum output power [27]. Later it was used to determine optimum feedback elements with reference to a series feedback MESFET oscillator using a simplified device equivalent circuit without the intrinsic feedback gate–drain capacitance [28].

It is convenient to represent this optimum approach in a general form regardless of the type of the active device, for commonly used parallel and series feedback oscillators [2,29]. The first step in the design is to determine the optimum combination of the values of the feedback reactive elements $\text{Im}W_1$ and $\text{Im}W_2$, which maximize the absolute value of the real part of the output immittance $\text{Re}W_{out}$ at the desired frequency of oscillation. Such a condition will permit to obtain self-sustained oscillations with the largest amplitude that implies maximum output power delivered to the load [30]. Analyzing

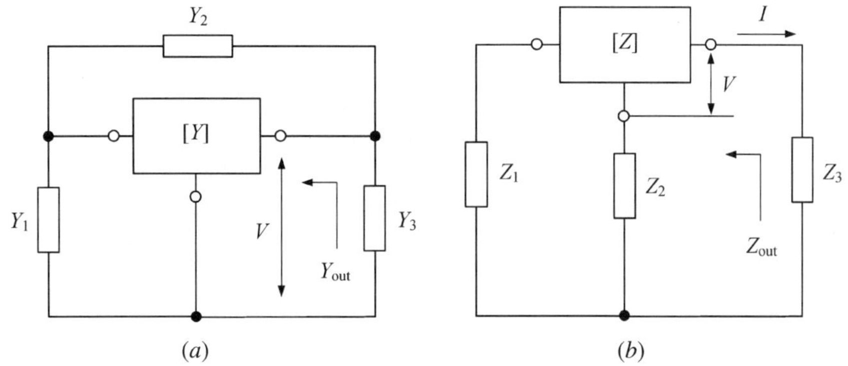

FIGURE 8.29 Two-port oscillator circuits with (*a*) parallel and (*b*) series feedback.

Eq. (8.116) in extremum, we can find the optimum values $\mathrm{Im}W_1^{\mathrm{o}}$ and $\mathrm{Im}W_2^{\mathrm{o}}$, at which the negative value $\mathrm{Re}W_{\mathrm{out}}$ is maximal, by solving

$$\frac{\partial\,\mathrm{Re}W_{\mathrm{out}}}{\partial\,\mathrm{Im}W_1}=0\qquad\frac{\partial\,\mathrm{Re}W_{\mathrm{out}}}{\partial\,\mathrm{Im}W_2}=0. \tag{8.117}$$

As a result, the optimum values $\mathrm{Im}W_1^{\mathrm{o}}$ and $\mathrm{Im}W_2^{\mathrm{o}}$ depend on the immittance parameters of the active device two-port network according to

$$\mathrm{Im}W_1^{\mathrm{o}}=\mp\frac{\mathrm{Re}\,(W_{21}-W_{12})}{\mathrm{Im}\,(W_{21}-W_{12})}\left[\frac{\mathrm{Re}\,(W_{12}+W_{21})}{2}\pm\mathrm{Re}\,(W_{11}+W_1)\right]-\mathrm{Im}W_{11}\mp\frac{\mathrm{Im}\,(W_{12}+W_{21})}{2} \tag{8.118}$$

$$\mathrm{Im}W_2^{\mathrm{o}}=\pm\frac{\mathrm{Re}\,(W_{21}+W_{12}\mp 2W_2)\,\mathrm{Re}\,(W_{21}-W_{12})}{2\mathrm{Im}\,(W_{21}-W_{12})}\pm\frac{\mathrm{Im}\,(W_{12}+W_{21})}{2}. \tag{8.119}$$

The next step is to determine the optimum load immittance W_{L} in a steady-state operation mode defined by Eq. (8.115). By substituting expressions for $\mathrm{Im}W_1^{\mathrm{o}}$ and $\mathrm{Im}W_2^{\mathrm{o}}$ into Eq. (8.116), the optimum real and imaginary parts of the output immittance $W_{\mathrm{out}}^{\mathrm{o}}$ will be respectively defined by

$$\mathrm{Re}W_{\mathrm{out}}^{\mathrm{o}}=\mathrm{Re}W_{22}+\mathrm{Re}W_2-\frac{\mathrm{Re}^2\,(W_{12}+W_{21}\mp 2W_1)+\mathrm{Im}^2\,(W_{21}-W_{12})}{4\mathrm{Re}\,(W_{11}+W_1+W_2)} \tag{8.120}$$

$$\mathrm{Im}W_{\mathrm{out}}^{\mathrm{o}}=\mathrm{Im}W_{22}+\mathrm{Im}W_2^{\mathrm{o}}-\frac{\mathrm{Re}\,(W_{21}-W_{12})}{\mathrm{Im}\,(W_{21}-W_{12})}\mathrm{Re}\left(W_{\mathrm{out}}^{\mathrm{o}}-W_{22}-W_2\right). \tag{8.121}$$

In a large-signal operation mode, all immittance parameters generally become the functions of the voltage amplitudes across the elements of the device equivalent circuit. For the negative resistance (conductance) one-port oscillators shown in Figure 8.29, the output power can be written as $P_{\mathrm{out}}=V^2G_{\mathrm{out}}(V)/2$ in terms of Y-parameters or $P_{\mathrm{out}}=I^2R_{\mathrm{out}}(I)/2$ in terms of Z-parameters, where V is voltage amplitude across the load resistance and I is the amplitude of the output current flowing into the load. Since, for the same output voltage or current amplitude, the combination of the optimum feedback parameters obtained by Eqs. (8.118) and (8.119) provides a maximum negative real part of the output immittance according Eq. (8.120), then the maximum power will be delivered to the load. Thus, such an analytic approach presumes that, once the nonlinear model of the active device is developed, the elements of the optimum feedback parameters and the load can be easily and explicitly calculated under both the small-signal and large-signal conditions corresponding to the start-up and steady-state operation modes, respectively.

An analytical evaluation of the start-up and steady-state oscillation conditions for a bipolar oscillator can be done using a simplified transistor equivalent circuit shown in Figure 8.30 [2]. In this case, the parasitic lead inductances and resistances can be considered amongst the external feedback circuit. The typical condition $r_\pi \gg 1/(\omega C_\pi)$ simplifies the analytical calculations substantially without a significant decrease in accuracy at high frequencies. Let us consider the transistor parasitic lead inductances L_{b}, L_{e}, and L_{c} among the external feedback elements Z_1, Z_2, and Z_{L}, respectively. In addition, loss in the feedback elements should be taken into account in elements r_{b} and r_{e}.

As a result, the optimum values of imaginary parts of the feedback elements X_1^{o} and X_2^{o}, expressed through the parameters of the bipolar transistor equivalent circuit, can be calculated from

$$X_1^{\mathrm{o}}=\frac{1}{2\omega C_{\mathrm{c}}}-r_{\mathrm{b}}\frac{\omega}{\omega_{\mathrm{T}}} \tag{8.122}$$

$$X_2^{\mathrm{o}}=-\frac{1}{2\omega C_{\mathrm{c}}}-r_{\mathrm{e}}\frac{\omega}{\omega_{\mathrm{T}}}. \tag{8.123}$$

where $\omega_{\mathrm{T}}=g_{\mathrm{m}}/C_\pi$ is the radian transition frequency.

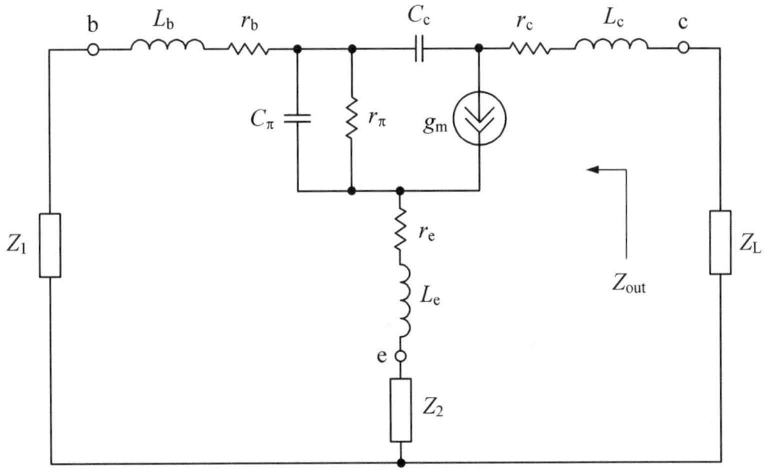

FIGURE 8.30 Simplified series feedback bipolar oscillator circuit.

The real and imaginary parts of the optimum output impedance $Z^o_{out} = R^o_{out} + jX^o_{out}$, including the collector series resistor r_c and lead inductance L_c, are given by

$$R^o_{out} = r_c + \frac{1}{r_b + r_e + R_{11}} \left[r_b \left(r_e + R_{11} + \frac{1}{a\omega_T C_c} \right) - \frac{1}{a} \left(\frac{1}{2\omega C_c} \right)^2 \right] \tag{8.124}$$

$$X^o_{out} = \omega L_c - \frac{1}{2\omega C_c} + \left(R^o_{out} - r_c \right) \frac{\omega}{\omega_T} \tag{8.125}$$

where $R_{11} = a/g_m$ and

$$a = \frac{1}{1 + \left(\dfrac{\omega}{\omega_T} \right)^2}.$$

From Eq. (8.124) it follows that, as frequency increases, the absolute value of the negative resistance R^o_{out} reduces and becomes zero at maximum oscillation frequency f_{max}. Without accounting for the parasitic parameters of the bipolar transistor equivalent circuit, the expression for f_{max} is

$$f_{max} = \sqrt{\frac{f_T}{8\pi r_b C_c}}. \tag{8.126}$$

This coincides with the well-known expression for a maximum oscillation frequency f_{max} of the bipolar transistor, on which a maximum available power gain becomes equal to unity and a steady-state oscillation condition can be established only for lossless feedback elements. Such a condition corresponds to the three-terminal mode of the transistor operation. However, the circuit may be capable of oscillating at frequencies greater than the f_{max} obtained by Eq. (8.126), when it is operated in a two-terminal mode, by making Z_1 an open circuit. In this configuration, no RF current is flowing into the base terminal to produce negative resistance (though the base terminal can still be used for biasing purposes). The transistor is now behaving as a two-terminal negative resistance, and the base resistance does not directly affect the maximum oscillation frequency, which will be determined by the parasitic resistances in the emitter and collector circuits [31].

TABLE 8.1 Microwave Bipolar Transistor Equivalent Circuit Parameters.

C_c, pF	r_π, Ω	C_π, pF	L_e, nH	L_b, nH	L_c, nH	r_e, Ω	r_b, Ω	r_c, Ω	f_T, GHz
0.5	30	40	0.3	0.3	0.5	0.3	4.0	1.75	6.0

The oscillator simulations were performed using a harmonic balance technique implemented in a Microwave Harmonica that is a part of the computer circuit simulator Ansoft Software, considering the oscillation frequency as an additional optimization variable [32]. The transistor equivalent circuit parameters are listed in Table 8.1. Accuracy of an analytic approach was verified based on a series feedback microwave bipolar oscillator shown in Figure 8.31. For a preliminary chosen oscillation frequency $f = 4$ GHz, the optimum oscillator feedback parameters according to the theoretical predictions given by Eqs. (8.122) and (8.123) must be equal to $L = 1.2$ nH and $C = 0.8$ pF.

Figure 8.32 shows the oscillation frequency and output power in a steady-state mode for various values of the emitter bias resistor R_e. In spite of the preliminary simplification, the simulation results obtained indicate minimum discrepancy between the chosen oscillation frequency and simulated oscillation frequency using an analytic calculation of the oscillator feedback parameters (taking

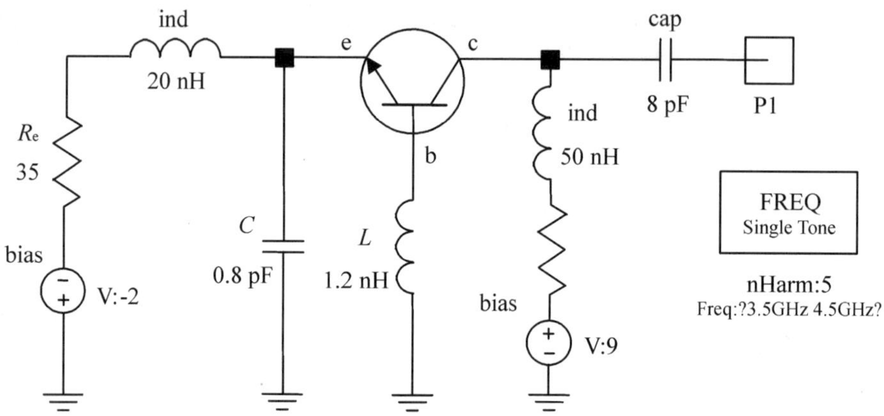

FIGURE 8.31 Simulated series feedback bipolar oscillator equivalent circuit.

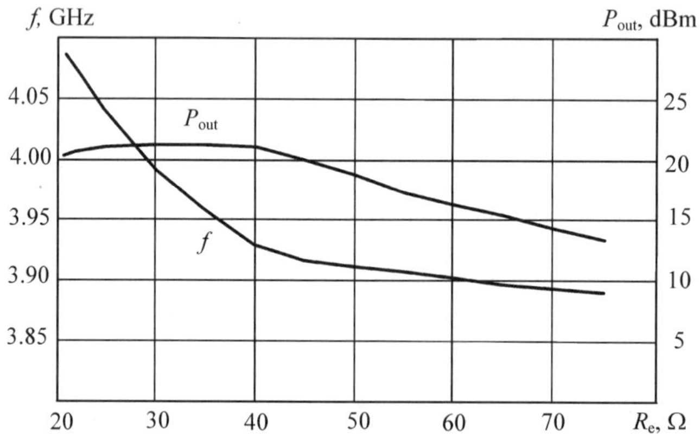

FIGURE 8.32 Output power and oscillation frequency versus emitter bias resistance.

into account the standard load resistance $R_L = 50\ \Omega$). An exact value of the oscillation frequency $f = 4$ GHz in a steady-state mode is realized for $R_e = 28\ \Omega$, when the output power P_{out} is close to the maximum value. Moreover, the output power $P_{out} = 21.5$ dBm is very close to a maximum value of 21.9 dBm for optimum load $R_L^o = 45\ \Omega$.

A common gate MESFET oscillator circuit with a series feedback between the gate and the ground is shown in Figure 8.33. Such a circuit configuration was selected because of its inherent broadband negative resistance. If the correct feedback reactance is added, oscillations can occur from very low frequencies to approximately maximum oscillation frequency f_{max}. Figure 8.33 also shows the small-signal MESFET equivalent circuit, which characterizes with good accuracy the device performance up to 50 GHz [33].

The results of microwave bipolar oscillator analytic design based on a quasilinear approach show the attractiveness and high effectiveness of preliminary analytical calculations of the oscillator feedback parameters according to the simple optimum analytical expressions. Therefore, one can assume that, to speedup the design process for microwave MESFET oscillators, a simplification of the analytical expressions for its feedback elements is possible [34]. For example, influence of the gate–drain capacitance C_{gd} and transit time τ can be ignored, and the transistor parasitic lead inductances L_g, L_s, and L_d can be considered among the external feedback elements Z_1, Z_2, and Z_L, respectively. Then, the optimum values of imaginary parts of the lossless feedback elements X_1^o, X_2^o, and output reactance X_{out}^o can be given by

$$X_1^0 = \frac{1}{\omega C_{gs}} + R_{ds}\left[-\omega C_{ds}\left(R_{gs} + R_g\right) + \frac{g_m}{2\omega C_{gs}}\right] \tag{8.127}$$

$$X_2^0 = -R_{ds}\left(\omega C_{ds} R_s + \frac{g_m}{2\omega C_{gs}}\right) \tag{8.128}$$

$$X_{out}^0 = -R_{ds}\left[\omega C_{ds} R_{out}^o + \frac{g_m}{2\omega C_{gs}}\right]. \tag{8.129}$$

As a result, for an optimum series feedback MESFET oscillator, according to Eqs. (8.127) and (8.128), the optimum values of the reactances X_1^o and X_2^o should be inductive and capacitive,

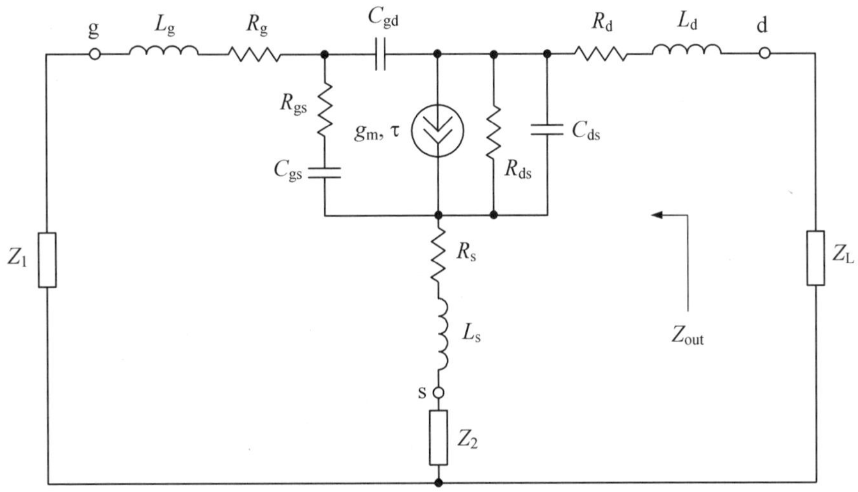

FIGURE 8.33 Series feedback microwave MESFET oscillator circuit.

TABLE 8.2 Small-Signal Parameters of GaAs MESFET Equivalent Circuit.

L_g, pH	L_s, pH	L_d, pH	R_g, Ω	R_{gs}, Ω	R_s, Ω	R_d, Ω	C_{gs}, pF	C_{gd}, pF	C_{ds}, pF	g_m, mS
50.4	0.1	60.1	2.0	2.0	0.93	1.1	1.2	0.087	0.199	97.4

respectively. An analytical equation to calculate the optimum output resistance $R^{\text{o}}_{\text{out}}$ in a small-signal operation mode can be written by

$$R^0_{\text{out}} \cong R_s + \frac{R_{\text{ds}}}{1 + (\omega C_{\text{ds}} R_{\text{ds}})^2}\left[1 - \frac{R_{\text{ds}}}{R_g + R_s + R_{gs}}\left(\frac{g_m}{2\omega C_{gs}}\right)^2\right]. \tag{8.130}$$

To determine the differential drain–source resistance R_{ds} as a function of the optimum output resistance $R^{\text{o}}_{\text{out}}$, it is enough to solve the quadratic equation for R_{ds} obtained from Eq. (8.130). As a result,

$$R_{\text{ds}} = \frac{1 + \sqrt{1 - 4\left(R^{\text{o}}_{\text{out}} - R_s\right) G_{\text{dso}}}}{2G_{\text{dso}}} \tag{8.131}$$

where

$$G_{\text{dso}} = \frac{1}{R_g + R_s + R_{gs}}\left(\frac{g_m}{2\omega C_{gs}}\right)^2 + \left(R^{\text{o}}_{\text{out}} - R_s\right)(\omega C_{\text{ds}})^2 .$$

Let us verify the accuracy of the analytic approach using the power microwave MESFET device with gate length $l = 1$ µm and gate width $w = 4 \times 200$ µm. To determine a large-signal value of the output resistance $R^{\text{o}}_{\text{out}}$ for a certain value of the load resistance R_L, it is necessary to use the amplitude balance equation $R^{\text{o}}_{\text{out}} + R_d + R_L = 0$. In such a situation, R_{ds} is considered as a fundamentally averaged drain–source resistance R_{ds1} under large-signal operation. The nonlinear circuit simulation was performed for a 12 GHz microwave series feedback MESFET oscillator, with the small-signal parameters of its equivalent circuit listed in Table 8.2 [33].

According to the theoretical predictions, the optimum oscillator feedback parameters must be equal to $L = 0.35$ nH and $C = 0.5$ pF (for load resistance $R_L = 50\ \Omega$), as shown in Figure 8.34. In this

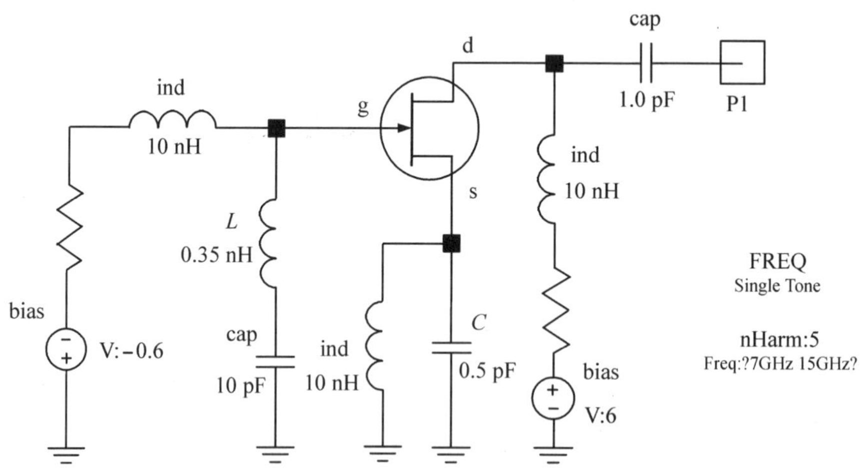

FIGURE 8.34 Simulated series feedback 12 GHz MESFET oscillator.

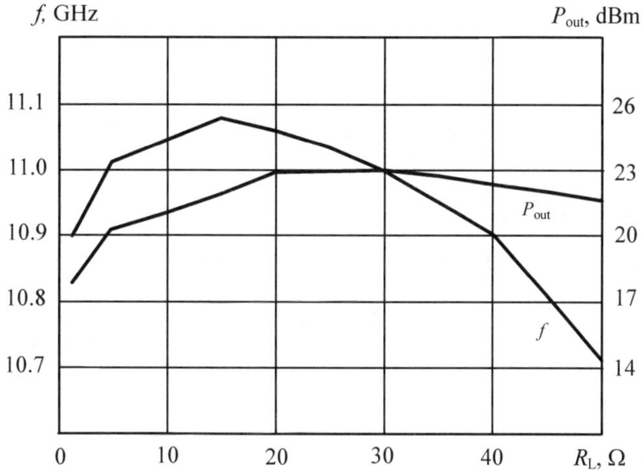

FIGURE 8.35 Output power and oscillation frequency versus load resistance.

case, to satisfy the phase balance condition $X^{o}_{out} + \omega L_d + X_L = 0$, the value of the load reactance X_L should be capacitive of $C_L = 1$ pF. A simulated value of the oscillation frequency is 10.72 GHz, which differs from the selected value by only 11%. Figure 8.35 shows the dependencies of the oscillation frequency f and output power P_{out} on the load resistance R_L. From this it follows that maximum output power $P_{out} = 22.9$ dBm can be obtained for load values in limits of 20–30 Ω.

8.9 NOISE IN OSCILLATORS

The instantaneous output signal of an oscillator can be represented by

$$v(t) = A_0 \left[1 + \frac{\Delta A(t)}{A_0} \right] \cos \left[2\pi f_0 t + \phi_0 + \Delta\phi(t) \right] \tag{8.132}$$

where A_0 is the voltage amplitude of the steady-state oscillations, f_0 is the oscillation frequency, ϕ_0 is the initial phase of the oscillations at $t = 0$, $\Delta A(t)$ and $\Delta\phi(t)$ are the amplitude and phase deviations of the corresponding amplitude and phase fluctuations, respectively. Generally, the nature of the fluctuations may be discrete or random where the discrete signals are called the *spurious*, appearing as distinct spectral components, while the random fluctuations are called the *phase noise*.

The instantaneous frequency as a function of time can be written as

$$f(t) = \frac{1}{2\pi} \frac{d}{dt} \left[2\pi f_0 t + \phi_0 + \Delta\phi(t) \right] = f_0 + \Delta f(t) \tag{8.133}$$

where

$$\Delta f(t) = \frac{1}{2\pi} \frac{d\Delta\phi(t)}{dt}.$$

Since the process associated with frequency fluctuations $\Delta f(t)$ is stationary, then

$$\Delta\phi(t) = 2\pi \int_0^t \Delta f(\tau) d\tau \tag{8.134}$$

which is in general a nonstationary process. However, as the phase fluctuation process is sufficiently slow during the natural period of the oscillations, the stationary model to describe the phase noise performance of free-running oscillators can be used. It should be noted that an autonomous free-running oscillator does not have a reference plane and the initial phase ϕ_0 in an autonomous system can be chosen arbitrary, for example, of zero value.

8.9.1 Parallel Feedback Oscillator

It is very important for the oscillator noise model to express a clear relationship between the oscillator spectral noise power density, resonant circuit, and active device parameters. The simple Leeson linear model for a feedback oscillator, which was derived empirically, is based on the expectations that the real oscillator has two basic components [35]. The first component is caused by the phase fluctuations due to the additive white noise at frequency offsets close to the carrier, as well as due to the noise having a mixing nature resulting from the circuit nonlinearities. The second component is a result of the low-frequency fluctuations or flicker noise upconverted to the carrier region because of the active device nonlinear effects.

The phase noise at the input of the power amplifier is added to a signal as the sum of every bandwidth $\Delta f = 1$ Hz, each producing an available noise power at the input of the noise-free amplifier, as shown in Figure 8.36(*a*). Maximum power delivery can be achieved when the source internal impedance is conjugate-matched to the input impedance of the amplifier. As a result, only one-half of the root-mean-square noise voltage e_n appears across the amplifier input and is equal to

$$e_{in} = \frac{e_n}{2} = \frac{\sqrt{4FkTR}}{2} = \sqrt{FkTR} \tag{8.135}$$

where F is the noise figure, k is the Boltzmann constant, T is the absolute temperature, and R is the equivalent resistance, which can be represented as the input resistance for the input root-mean-square noise voltage [36]. The input phase noise produces a root-mean-square phase deviation $\Delta\phi_{rms} = \Delta\phi/\sqrt{2}$ at each offset frequency $\pm f_m$ from the carrier f_0, as shown in Figure 8.36(*b*), for which a total power-wise sum can be written for a small phase perturbation as

$$\Delta\phi = \Delta\phi_{rms}\sqrt{2} = \frac{e_{in}\sqrt{2}}{V_{in}} = \sqrt{\frac{FkT}{P_{in}}} \tag{8.136}$$

where $V_{in} = \sqrt{2P_{in}R}$ is the signal voltage amplitude at the power amplifier input.

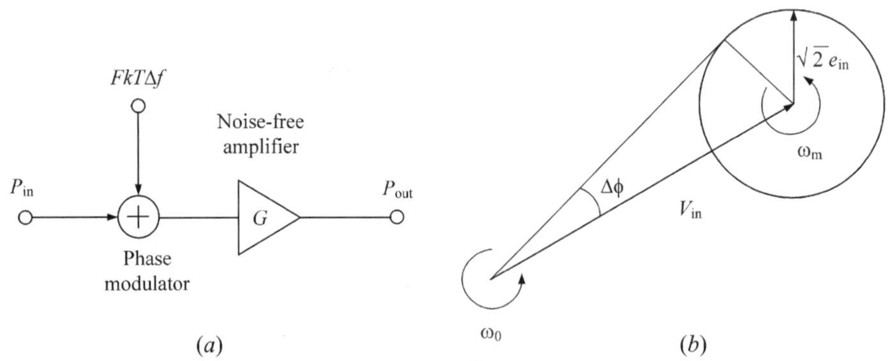

FIGURE 8.36 Simplified feedback oscillator noise model.

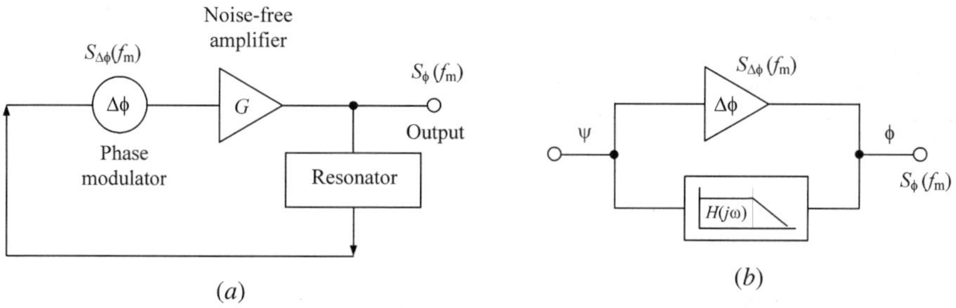

FIGURE 8.37 Equivalent model of parallel feedback oscillator.

As a result, the double-sideband spectral power density of the thermal phase noise in a frequency bandwidth $\Delta f = 1$ Hz can be written as

$$S_{\Delta\phi} = \Delta\phi^2 = \frac{FkT}{P_{in}}. \tag{8.137}$$

The Leeson model consists of an amplifier with a noise figure F and a resonator (or filter) in the feedback loop, as shown in Figure 8.37(*a*) [36]. The oscillator phase noise is modeled by assuming a noise-free power amplifier and adding a phase modulator to its input. Based on empirical predictions, the phase noise level of the oscillator at an offset frequency f_m from the carrier f_0 can be described by

$$S_\phi(f_m) = S_{\Delta\phi}(f_m)\left(\frac{f_0}{2Q_L f_m}\right)^2 \quad \text{for } f_m < \frac{f_0}{2Q_L} \tag{8.138}$$

$$S_\phi(f_m) = S_{\Delta\phi}(f_m) \quad \text{for } f_m > \frac{f_0}{2Q_L} \tag{8.139}$$

where $S_{\Delta\phi}(f_m)$ is determined using Eq. (8.137) as

$$S_{\Delta\phi}(f_m) = \frac{FkT}{P_{in}}\left(1 + \frac{f_c}{f_m}\right) \tag{8.140}$$

taking into the effect of the signal purity degradation due to the low-frequency flicker noise effect close to the carrier, described empirically by the corner frequency f_c. It should be noted that the empirical Leeson equation for $S_{\Delta\phi}(f_m)$ contains a multiplication factor of two in the numerator. Moreover, accurate agreement was achieved between the model and experiment results when the power density $S_{\Delta\phi}(f_m)$ was expressed in terms of the compressed (or large-signal) power gain G and output power P_{out} as $S_{\Delta\phi}(f_m) = 2GFkT/P_{out}$ [37].

The parameter F in Eq. (8.140) is associated with the active device noise figure and can be called an effective noise factor because, generally, it should represent the effects of the active device noise sources and the cyclostationary noise resulting from periodically varying processes in practical oscillators. Due to the inherent nonlinear nature of the active device, the effects of intermodulation between the wideband white noise and various harmonics of fundamental frequency (for example, nonlinear transformation of the noise near the third harmonic downconverted to the near carrier region due to mixing effect with second harmonic) must be included [35]. Also, the effect of low-frequency noise modulation of the current, resulting in the reactance modulation of the input impedance of the circuit (for example, variation of the phase angle of the device forward transfer function versus emitter current), cannot be neglected [38]. Hence, it is impossible to calculate F accurately without taking into account the effect of the oscillator resonant circuit. Therefore, for such a linear model, the effective noise factor F as well as the corner frequency f_c can be considered more like fitting parameters, based on measured data.

The corresponding combined expression to calculate the normalized double-sideband phase noise power spectral density or the double-sideband noise-to-carrier ratio at the input of the feedback oscillator can be obtained from

$$S_{\phi}(f_m) = \frac{FkT}{P_{in}}\left(1 + \frac{f_c}{f_m}\right)\left[1 + \left(\frac{f_0}{2Q_L f_m}\right)^2\right] \tag{8.141}$$

which gives an asymptotic model showing generally the noise reduction of 9 dB/octave in the offset region with predominant low-frequency 1/*f* noise, 6 dB/octave in the offset region due to feedback loop and 0 dB/octave representing the thermal or white noise spectrum.

The single-sideband noise-to-carrier ratio at the input of the feedback oscillator can be described by

$$L(f_m) = \frac{1}{2}\frac{FkT}{P_{in}}\left(1 + \frac{f_c}{f_m}\right)\left[1 + \left(\frac{f_0}{2Q_L f_m}\right)^2\right] \tag{8.142}$$

whose idealized sideband spectral behavior for different values of the loaded quality factors is illustrated in Figure 8.38. For the low-Q_L case, there are regions with $1/f_m^3$ and $1/f_m^2$ dependencies for spectral power density close to carrier, as shown in Figure 8.38(*a*). For the moderate-Q_L case, Figure 8.38(*b*) demonstrates only $1/f_m^3$ dependence as far as intersection with thermal noise floor. For the high-Q_L case, the regions with $1/f_m^3$ and $1/f_m$ dependencies are observed near the carrier, as shown in Figure 8.38(*c*). Closest to the carrier, $1/f_m^3$ phase noise behavior is a result of random frequency modulation of the oscillator by low frequency 1/*f* noise. In the region of $1/f_m^2$ phase noise behavior, the white noise causes random frequency modulation. The $1/f_m$ dependence is due to the mixing up of the 1/*f* noise with the oscillation frequency. Finally, the phase noise becomes constant, which is a result of the mixing up of white noise around the oscillation frequency.

To calculate the same phase noise power spectral density at the oscillator output, it is necessary to replace the input power P_{in} by the power available at the output P_{out} and to multiply the numerator of Eq. (8.142) by the power gain G. As a result, neglecting the effect of flicker noise and considering the case of $f_m \ll f_0$, one can obtain

$$L(f_m) = \frac{GFkT}{8Q_L^2 P_{out}}\left(\frac{f_0}{f_m}\right)^2 \tag{8.143}$$

where

$$G = \frac{1}{\left(1 - \frac{Q_L}{Q_0}\right)^2} \tag{8.144}$$

is considered the transducer power gain and Q_0 is the unloaded quality factor [39]. From Eqs. (8.142) and (8.143) it follows that, to minimize the oscillator phase noise, it is necessary to reduce the noise figure F and to increase the input power P_{in} (or the output power P_{out} for a fixed power gain G of the amplifier) as much as possible. In addition, for frequency offsets inside the resonator bandwidth, it is desirable to maximize the oscillator loaded quality factor Q_L. However, the resonator insertion loss and loaded Q_L are inter-related, and one cannot arbitrarily increase Q_L without increasing the insertion loss, otherwise a larger power gain G is needed. The two competing effects result in an optimum loaded Q_L of approximately one-half the unloaded Q_0 and insertion loss of about 6 dB [40,41]. Thus, the minimum noise occurs when $Q_L/Q_0 = 0.5$ resulting in

$$L(f_m) = \frac{2FkT}{P_{out}}\left(\frac{f_0}{2Q_L f_m}\right)^2. \tag{8.145}$$

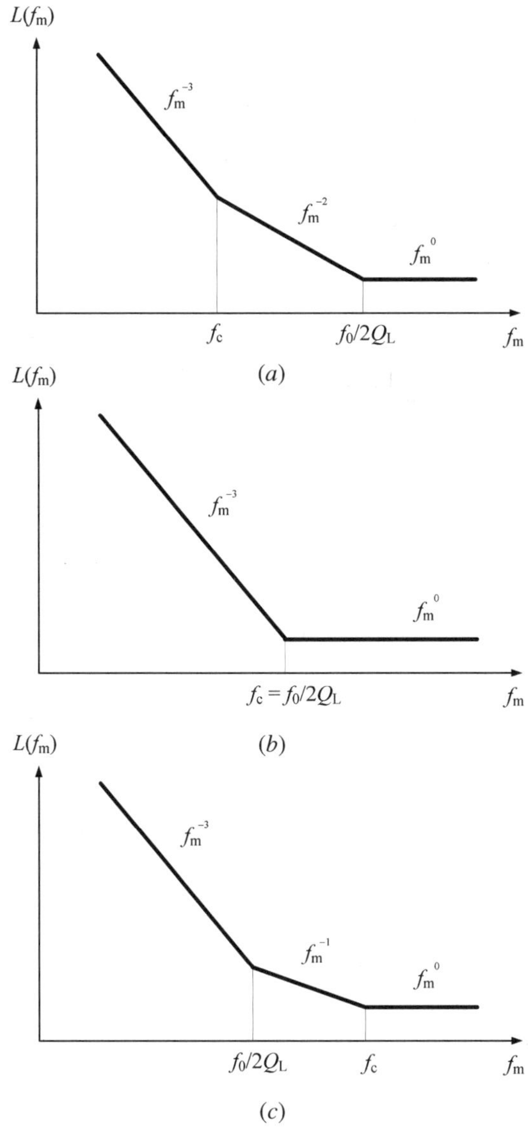

FIGURE 8.38 Single sideband oscillator phase noise behavior.

Note that the difference in the optimum noise performance predicted by different definitions of the output power (power dissipated in the resonant circuit or power available at the amplifier output) is small [39].

Figure 8.37(*b*) shows another representation of the Leeson model with a phase feedback loop. Suppose that the phase-noise modulation occurs in the oscillator active element as $\Delta\phi(t)$, and it is necessary to define how the oscillator reacts to this internal noise. As it is known from transmission theory, the transfer function of a modulated high-frequency signal, passing through a bandpass filter, equals to the transfer function of the modulating signal passing through an equivalent low-pass filter prototype. Thus, the phase relationship resulting from the null-phase condition on the loop and from the filtering of $\phi(t)$ by the selective filter can be written as

$$\psi(t) + \Delta\phi(t) = \phi(t) \tag{8.146}$$

where

$$\psi(t) = \int_{-\infty}^{+\infty} \phi(t)\, h(\tau - t)\, d\tau = \phi(t) \times h(t) \tag{8.147}$$

$h(t)$ is the impulse response of the equivalent low-pass filter, and the asterisk denotes a convolution product. The integral in Eq. (8.147) converges for nearly all samples of $\phi(t)$ provided that the filter is linear and time invariant, and that the stationary random process $\phi(t)$ possesses a finite second-order moment.

In this case, the Leeson formula for the double-sideband phase noise power spectral density of the feedback oscillator given by Eq. (8.141) can be expressed in a more general form

$$S_\phi(\omega_m) = \frac{S_{\Delta\phi}(\omega_m)}{[H(j\omega_m) - 1][H^*(j\omega_m) - 1]} \tag{8.148}$$

where $H(j\omega_m)$ is the equivalent low-pass transfer function and asterisk denotes the complex-conjugate value [42].

Thus, by representing the transfer function of the first order low-pass filter as

$$H(j\omega_m) = \frac{\alpha}{\alpha + j\omega_m} \tag{8.149}$$

where $\alpha = \omega_0/2Q_L$ is the half-bandwidth of the resonator and $\omega_m = 2\pi f_m$, Eq. (8.148) can be rewritten as

$$S_\phi(\omega_m) = S_{\Delta\phi}(\omega_m)\left[1 + \left(\frac{\alpha}{\omega_m}\right)^2\right] \tag{8.150}$$

which is similar to Eq. (8.141).

Now consider a case of the second-order low-pass filter based on the two coupled resonators having the transfer function

$$H(j\omega_m) = \frac{\alpha_1}{\alpha_1 + j\omega_m}\,\frac{\alpha_2}{\alpha_2 + j\omega_m} = \frac{1}{1 - \left(\frac{\omega_m}{\alpha}\right)^2 + j2\omega_m\delta} \tag{8.151}$$

where $\alpha = \sqrt{\alpha_1\alpha_2}$ and $\delta = (\alpha_1 + \alpha_2)/2\sqrt{\alpha_1\alpha_2}$ [43]. In this case, Eq. (8.148) can be rewritten as

$$S_\phi(\omega_m) = S_{\Delta\phi}(\omega_m)\left[1 + \frac{1 - 2\left(\frac{\omega_m}{\alpha}\right)^2}{\left(\frac{\omega_m}{\alpha}\right)^2 + 4\delta^2}\left(\frac{\alpha}{\omega_m}\right)^2\right] \tag{8.152}$$

which shows the substantially better phase noise performance in the near vicinity of the carrier frequency for the case of loosely coupled resonators when $\delta > 1$. However, since the available output power becomes low, it is necessary to provide post low-noise output signal amplification.

Let us quantitatively compare both cases of the Leeson phase noise models using the first- and second-order low-pass filters in the oscillator feedback loop for the same arbitrary chosen technical data:

- Oscillation (carrier) frequency $f_0 = 2$ GHz.
- Offset frequency $f_m = 10$ kHz.
- Oscillator resonant circuit loaded quality factor $Q_L = 10$.
- Noise figure of the active device $F = 6$ dB.
- Input power delivered to the device $P_{in} = 10$ mW.
- Corner frequency for flicker noise $f_c = 3$ kHz.

Substituting these parameters into Eq. (8.142) for a single-sideband phase noise spectral density of the oscillator with the first-order low-pass filter results in

$$L(f_m) = 10\log(1.05128 \times 10^{-10}) = -99.78\,\text{dBc/Hz}.$$

For the case of the oscillator with two coupled resonators, let us assume that the loaded Q_L of the second resonator is five times as much as the first one, that is $\alpha_1 = 6.283 \times 10^8$ and $\alpha_2 = 1.2566 \times 10^8$. Then,

$$L(f_m) = 10\log(2.918 \times 10^{-12}) = -115.35\,\text{dBc/Hz}$$

which clearly shows the significant phase noise improvement compared to the oscillator having the first-order low-pass filter in a feedback loop.

Figure 8.39 shows the typical oscillator output power spectrum. The noise distribution on each side of the oscillator signal is subdivided into a large number of strips of width Δf located at the distance f_m away from signal. It should be noted that, generally, the spectrum of the output signal consists of the amplitude and phase noise components. Hence, to measure the phase noise close to the carrier frequency, one needs to make sure that any contributions of parasitic amplitude modulation to the oscillator output noise spectrum are negligible compared with those from frequency modulation. The single-sideband phase noise $L(f_m)$ usually given logarithmically is defined as the ratio of a signal

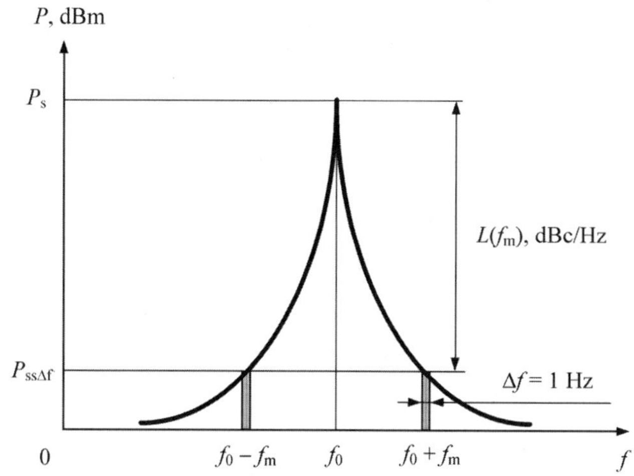

FIGURE 8.39 Oscillator output power spectrum.

power $P_{ss\Delta f}$ in one-phase modulation sideband per bandwidth $\Delta f = 1$ Hz, at an offset f_m away from the carrier, to the total signal power P_s.

Despite some limitations of the linear Leeson model when device is operated in a conduction angle large-signal mode and output signal is not purely sinusoidal, which have an effect on the active device noise factor and low-frequency flicker noise upconversion, such an approach gives a sense of the phase noise performance for oscillators with different resonant circuits. This applies especially, if the theoretical results can be supported by sufficiently accurate measurements of a loaded quality factor of the oscillator resonant circuit and simulations of the effective noise figure based on the modeled active device parameters and operation conditions.

In addition, such a simple model indicates the basic factors and provides the design rules that are necessary to follow in order to minimize the oscillator phase noise:

- Choose the resonator with maximum unloaded quality factor Q_0 and optimize the loaded quality factor Q_L of the oscillator resonant circuit by proper load coupling.
- Maximize the output power P_L delivered to the load by maximizing the RF voltage amplitude across the resonant circuit with limitations due to active device breakdown voltage and operation in the saturation mode.
- Choose a device with the lowest noise figure F and corner frequency f_c for low-frequency flicker noise.

8.9.2 Negative Resistance Oscillator

Now consider the equivalent circuit of a simple single-resonant negative resistance oscillator shown in Figure 8.40, where the available noise power is assumed to be totally from the active device. Here, R_n is the equivalent noise resistance associated with active device noise sources, the negative resistance $R_{out,}$ and the equivalent output capacitance C_{out} represent the device negative output impedance, L is the tank inductance, and R_L is the load resistance. The derivation of the power spectral density will be based on the fact that the available noise power in the active device is amplified in a frequency selective way, resulting at resonance in the output power P_L being dissipated in the load resistance R_L [44]. This will happen in a steady-state condition when the values of the negative resistance and the load resistance are close to each other.

Then, assuming that $R_L + \Delta R = -R_{out}$ and defining the magnitude of the mean-square noise current flowing into load from the mean-square noise voltage source, we can write

$$\overline{i_n^2} = \frac{\overline{e_n^2}}{(\Delta R)^2 + \left(\omega L - \dfrac{1}{\omega C_{out}}\right)^2}. \tag{8.153}$$

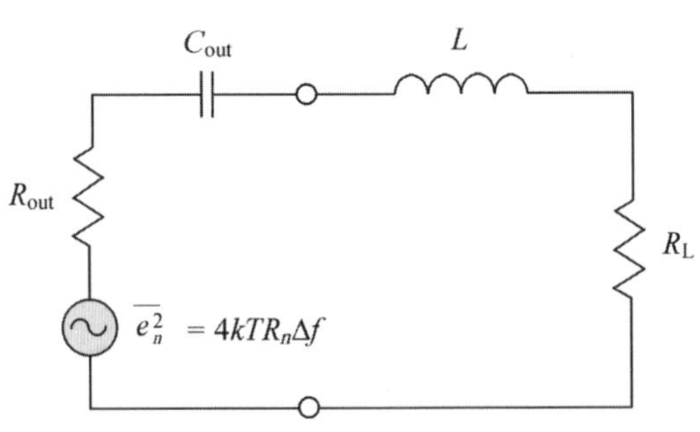

FIGURE 8.40 Simplified negative resistance oscillator noise model.

Equation (8.153) can be rewritten in the normalized general form as

$$\overline{i_n^2} = \frac{1}{\left(\dfrac{\Delta R}{R_\mathrm{L}}\right)^2 + Q_\mathrm{L}^2\left(\dfrac{\omega}{\omega_0} - \dfrac{\omega_0}{\omega}\right)^2} \frac{\overline{e_n^2}}{R_\mathrm{L}^2} \tag{8.154}$$

where ω_0 is the radian resonant frequency and Q_L is the oscillator loaded quality factor at the resonant frequency.

By normalizing to the power P_L dissipated in the load resistor R_L, Eq. (8.154) can be rewritten through the spectral power densities by

$$S_\phi = \frac{S_{\Delta\phi}}{\left(\dfrac{\Delta R}{R_\mathrm{L}}\right)^2 + Q_\mathrm{L}^2\left(\dfrac{\omega}{\omega_0} - \dfrac{\omega_0}{\omega}\right)^2} \tag{8.155}$$

where $S_{\Delta\phi} = 4kTR_n/(R_\mathrm{L}P_\mathrm{L})$, $S_\phi = \overline{i_n^2}R_\mathrm{L}/P_\mathrm{L}$ is the power spectral density of the noise current across the load resistor R_L and $\Delta f = 1$ Hz.

Since at small offset frequencies $\omega_\mathrm{m} = \omega - \omega_0$ close to the resonant frequency,

$$\frac{\omega}{\omega_0} - \frac{\omega_0}{\omega} \cong \frac{2\omega_\mathrm{m}}{\omega_0}$$

then Eq. (8.155) can be rewritten as

$$S_\phi = \frac{S_{\Delta\phi}}{\left(\dfrac{\Delta R}{R_\mathrm{L}}\right)^2 + \left(\dfrac{2Q_\mathrm{L}\omega_\mathrm{m}}{\omega_0}\right)^2} \tag{8.156}$$

which is similar to the power spectral density at frequency offsets close to the resonant frequency for the parallel feedback oscillator. Equation (8.156) represents a Lorentz function corresponding to an exponential decay of the autocorrelation function in the time domain [45].

Since the total output power delivered to the load is equal to P_L,

$$\frac{1}{2\pi}\int_0^\infty S_\phi(\omega)\mathrm{d}\omega = \left(\frac{\omega_0}{2Q_\mathrm{L}}\right)^2 \frac{S_{\Delta\phi}}{2\Delta\omega_n} = 1 \tag{8.157}$$

where

$$\Delta\omega_n = \frac{\Delta R}{R_\mathrm{L}} \frac{\omega_0}{2Q_\mathrm{L}}$$

is the Lorentzian linewidth (half-width at half-maximum), which is an oscillator spectrum linewidth characterized by the natural phase fluctuations due to the thermal and shot noises of the oscillator. However, in a common case, due to the variation of the oscillator resonant circuit parameters, flicker noise, pushing or pulling effects, the effective spectrum linewidth widens, especially close to the resonant frequency.

By using a widely used definition of the loaded quality factor of the passive resonator in the form

$$Q_\mathrm{L} = \frac{\omega_0}{\Delta\omega_{3\mathrm{dB}}} \tag{8.158}$$

where $\Delta\omega_{3\text{dB}}$ is the full linewidth at half-maximum level, one can write

$$2\Delta\omega_n = \frac{S_{\Delta\phi}}{4}\Delta\omega_{3\text{dB}}^2 = kT\frac{R_n}{R_\text{L}}\frac{\Delta\omega_{3\text{dB}}^2}{P_\text{L}}. \tag{8.159}$$

As a result, there is a complete analogy between Lorentzian linewidth defined by Eq. (8.159) and the expression for semiconductor laser homogeneous linewidth [46]. In this case, the characteristic energy kT corresponds to the photon energy, the oscillator output power P_L corresponds to the laser output power, and the oscillator noise-gain ratio R_n/R_L corresponds to the inversion factor representing the ratio between spontaneous emission rate and the optical gain rate. The widening of the oscillator spectral line due to the low-frequency fluctuations is similar to an inhomogeneous laser line broadening due to the Doppler effect.

The normalized power spectral density can be expressed through the Lorentzian linewidth as

$$S_\phi(\omega_\text{m}) = \frac{2\Delta\omega_n}{\Delta\omega_n^2 + \omega_\text{m}^2} \cong \frac{2\Delta\omega_n}{\omega_\text{m}^2} \tag{8.160}$$

showing a simple linear relationship between Lorentzian linewidth and oscillator phase noise spectrum at offset frequencies $\omega_\text{m} \gg \Delta\omega_n$. Substituting Eq. (8.159) into Eq. (8.160) and taking into account that $F = R_\text{n}/R_\text{L}$ result in the single-sideband noise-to-carrier ratio

$$L(f_\text{m}) = \frac{kTF}{2P_\text{L}}\left(\frac{f_0}{Q_\text{L} f_\text{m}}\right)^2 \tag{8.161}$$

which is similar to the Edson noise formula [44,47].

8.9.3 Colpitts Oscillator

As an example, let us define the linear phase noise model for a popular Colpitts oscillator, the simplified circuit schematic of which is shown in Figure 8.41(*a*). The power loss in the tank inductor L is included in the load resistance R_L. The transistor equivalent circuit with voltage and current noise sources in shown in Figure 8.41(*b*), where the resonator noise is modeled by a noise current i_nR. The noise voltage and current sources can be given through their mean-square values as

$$\overline{i_\text{nR}^2} = \frac{4kT\Delta f}{R_\text{L}} \tag{8.162}$$

$$\overline{v_\text{nb}^2} = 4kTr_\text{b}\Delta f \tag{8.163}$$

$$\overline{i_\text{nb}^2} = \frac{2qI_\text{c}\Delta f}{\beta} \tag{8.164}$$

$$\overline{i_\text{nc}^2} = 2qI_\text{c}\Delta f. \tag{8.165}$$

where the input noise current source $\overline{i_\text{nb}^2}$ is related to shot noise at the emitter–base junction due to electrons that recombine with holes inside the neutral base and $\overline{i_\text{nc}^2}$ represents the shot noise generated at the collector–base junction due to collector electrons. The mean-square values of these noise sources in a bandwidth Δf are given by

$$\overline{i_\text{nb}^2} = 2qI_\text{b}\Delta f = 2kTn_\text{b}g_\pi\Delta f \tag{8.166}$$

$$\overline{i_\text{nc}^2} = 2qI_\text{c}\Delta f = 2kTn_\text{c}g_\text{m}\Delta f. \tag{8.167}$$

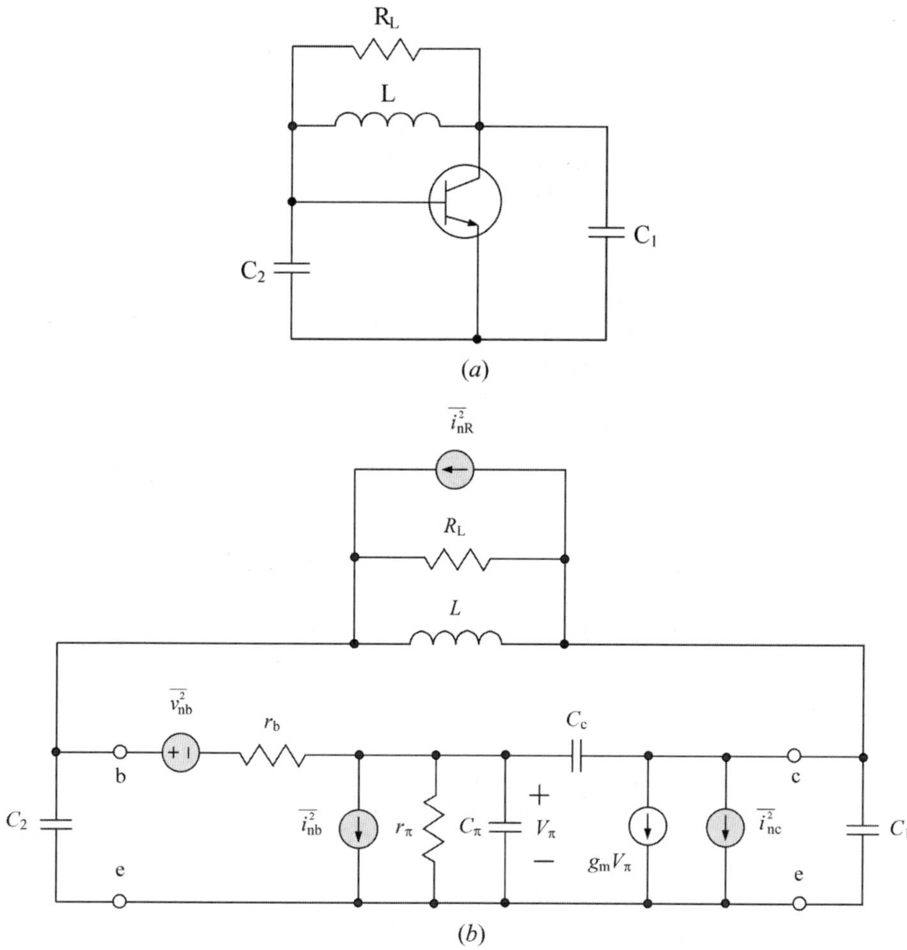

FIGURE 8.41 Equivalent circuits of Colpitts oscillator.

where $g_\pi = 1/r_\pi$, I_b and I_c are the dc base and collector currents, n_b and n_c are the ideality factors of the emitter–base and collector–base junctions, respectively [48].

The admittance Y-parameters of the internal transistor excluding the base resistance r_b can be obtained as

$$[Y] = \begin{bmatrix} \dfrac{1}{r_\pi} + j\omega(C_\pi + C_c) & -j\omega C_c \\ g_m - j\omega C_c & j\omega C_c \end{bmatrix} \tag{8.168}$$

where $g_m = \beta/r_\pi$ is the device transconductance.

For a Colpitts oscillator with parallel feedback capacitors C_1 and C_2, it is convenient to represent all noise sources in a parallel configuration with input and output noise current sources. Figure 8.42 shows the transformation of the device noise model with the series thermal voltage noise source due to the base resistance into the equivalent device noise model with two parallel current noise sources at the input and output only, using the transmission $ABCD$-parameters. By using formulas

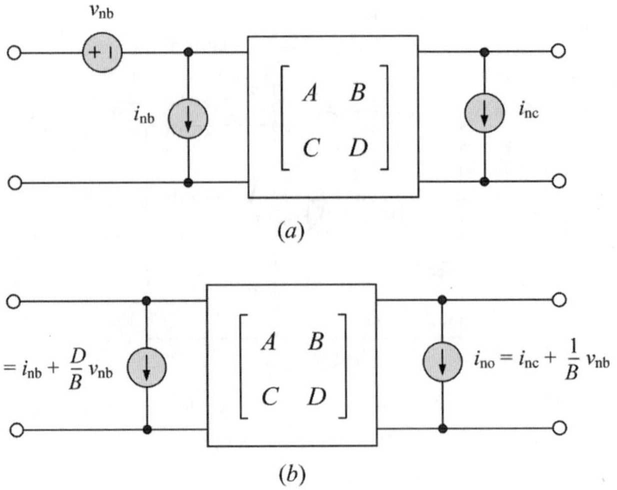

FIGURE 8.42 Transformation of base resistance thermal voltage noise.

of the transformation from the admittance Y-parameters to the transmission $ABCD$-parameters, one can write

$$\frac{1}{B} = -Y_{21} = -(g_m - j\omega C_c). \tag{8.169}$$

$$\frac{D}{B} = Y_{11} = \frac{1}{r_\pi} + j\omega(C_\pi + C_c). \tag{8.170}$$

Let us simplify the analytical calculations by taking into account that, as it follows from Eqs. (8.169) and (8.170), the contribution of the collector capacitance C_c is not significant. As a result, the mean-square input current source $\overline{i_{ni}^2}$ and output current source $\overline{i_{no}^2}$ can be written as

$$\overline{i_{ni}^2} = \frac{2qI_c\Delta f}{\beta} + \frac{4kT\Delta f}{r_b}\left(\frac{r_b}{r_\pi}\right)^2\left(1 + \beta^2\frac{f^2}{f_T^2}\right) \tag{8.171}$$

$$\overline{i_{n0}^2} = 2qI_c\Delta f + \frac{4kT\Delta f}{r_b}\left(\frac{\beta r_b}{r_\pi}\right)^2 \tag{8.172}$$

where f is the operating frequency, $f_T = g_m/(2\pi C_\pi)$ is the transition frequency, and it is assumed that the dc and small-signal values of the current gain β are equal.

To calculate the noise figure of the oscillator, the noise current sources $\overline{i_{ni}^2}$ and $\overline{i_{no}^2}$ should be transformed in parallel with $\overline{i_{nR}^2}$. By using Eqs. (8.166) and (8.167) for the case of ideal junctions when $n_b = n_c = 1$, the total equivalent noise source $\overline{i_{n\Sigma}^2}$ connected in parallel to the resonator and load can now be obtained as

$$\overline{i_{n\Sigma}^2} = \frac{4kT\Delta f}{R_L} + \frac{4kT\Delta f}{2r_\pi}\left[1 + \frac{2r_b}{r_\pi}\left(1 + \beta^2\frac{f^2}{f_T^2}\right)\right]\left(\frac{C_1}{C_1 + C_2}\right)^2 + \frac{4kT\beta\Delta f}{2r_\pi}\left(1 + \frac{2\beta r_b}{r_\pi}\right)\left(\frac{C_2}{C_1 + C_2}\right)^2. \tag{8.173}$$

The noise figure F of the oscillator is defined as the ratio of the total noise power due to all noise current sources and the noise power from the loaded resonator due to the noise source $\overline{i_{nR}^2}$. As a result,

$$F = 1 + \frac{R_L}{2r_\pi}\left[1 + \frac{2r_b}{r_\pi}\left(1 + \beta^2\frac{f^2}{f_T^2}\right)\right]\left(\frac{C_1}{C_1 + C_2}\right)^2 + \frac{R_L}{2r_\pi}\left(1 + \frac{2\beta r_b}{r_\pi}\right)\left(\frac{C_2}{C_1 + C_2}\right)^2. \quad (8.174)$$

In most cases, the contribution of the collector shot noise dominates over the contribution of the shot noise caused by the base current. Consequently, for $r_b \ll r_\pi$ and $f \ll f_T$, the expression for the oscillator noise figure is simplified to

$$F = 1 + \frac{R_L}{2r_\pi}\left(1 + \frac{2\beta r_b}{r_\pi}\right)\left(\frac{C_2}{C_1 + C_2}\right)^2. \quad (8.175)$$

Finally, the single-sideband noise-to-carrier ratio at the output of the Colpitts oscillator in a linear consideration, using Eq. (8.161), can be defined by

$$L(f_m) = \frac{kT}{2P_L}\left(\frac{f_0}{Q_L f_m}\right)^2\left[1 + \frac{R_L}{2r_\pi}\left(1 + \frac{2\beta r_b}{r_\pi}\right)\left(\frac{C_2}{C_1 + C_2}\right)^2\right]. \quad (8.176)$$

8.9.4 Impulse Response Model

The behavior of autonomous second-order weakly nonlinear oscillation systems with low damping factor close to linear conservative systems and small time-varying external force $f(t)$ can be described by

$$\frac{d^2x}{d\tau^2} + x = n(t) \quad (8.177)$$

where x is the time-dependent variable, voltage or current, and $\tau = \omega_0 t$ is the time normalized by the angular resonant frequency ω_0.

The phase plane method is one of the theoretical approaches that allows one to analyze qualitatively and quantitatively the dynamics of the oscillation systems described by the second-order differential equations such as Eq. (8.177) [1]. By setting the small external force equal to zero, the solution of the linear second-order differential equation takes the form

$$x = A\cos(\tau + \phi) = A\cos\psi \quad (8.178)$$

$$\frac{dx}{d\tau} = -A\sin(\tau + \phi) = -A\sin\psi \quad (8.179)$$

where A is the amplitude of the oscillations and ϕ is the phase of the oscillation. The phase portrait shown in Figure 8.43(a) represents the family of circular trajectories enclosing each other with radii $r = A$ (limit cycles) depending on energy stored in the system.

Let us define the variations of the amplitude $A(t)$ and phase $\phi(t)$ under the effect of the external force applied to the oscillation system [2,49]. Assuming that the effect of the external force is small and these variations are slow, the amplitude and phase can be considered constant during a natural period of the oscillation. Then, Eq. (8.177) can be rewritten in the form of two first-order equations by

$$\frac{dx}{d\tau} = y \quad (8.180)$$

$$\frac{dy}{d\tau} = -x + n(t). \quad (8.181)$$

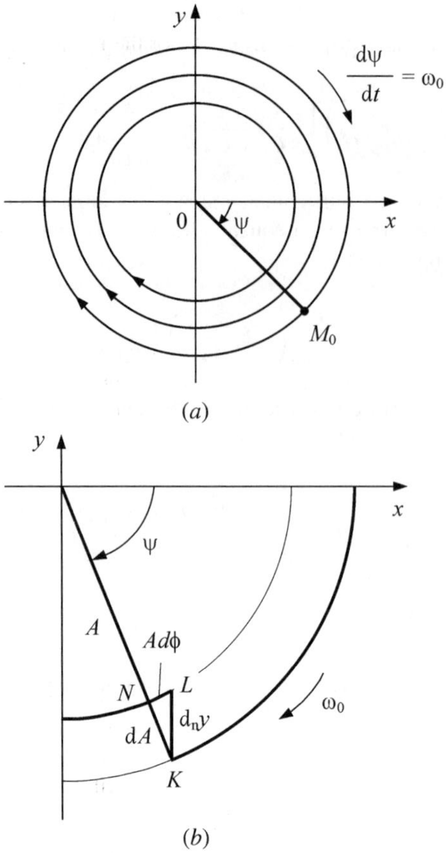

FIGURE 8.43 Phase portrait of second-order oscillation system and effect of injected impulse.

From Eqs. (8.180) and (8.181) it follows that the instantaneous change of the ordinate y by a value of $d_n y = n d\tau$ will occur, due to the small external force injected to the oscillation system. This corresponds to a step change of the representative point M_0 from the position K to the position L, thus resulting in the amplitude and phase changes shown in Figure 8.43(*b*). These changes can be determined from the consideration of a triangle *KLN* as

$$dA = -d_n y \sin\psi = -n \sin\psi\, d\tau \tag{8.182}$$

$$d\phi = -\frac{d_n y}{A}\cos\psi = -\frac{n}{A}\cos\psi\, d\tau. \tag{8.183}$$

Thus, separate first-order differential equations for the time-varying amplitude and phase can be obtained from Eqs. (8.182) and (8.183) as

$$\frac{dA}{d\tau} = -n \sin\psi \tag{8.184}$$

$$\frac{d\phi}{d\tau} = -\frac{n}{A}\cos\psi. \tag{8.185}$$

Since the right-hand sides of Eqs. (8.184) and (8.185) are small, time-averaged differential equations can be used instead of the differential equations for the instantaneous values of the amplitude and phase. Hence, the changes of the amplitude and phase for a time period $t \leq T$ are defined as

$$\Delta A(t) = -\int_0^{\omega_0 t} n(\tau)\sin\tau d\tau \tag{8.186}$$

$$\Delta\phi(t) = -\int_0^{\omega_0 t} \frac{n(\tau)}{A}\cos\tau d\tau. \tag{8.187}$$

Figure 8.44(a) shows the equivalent circuit of a negative resistance oscillator with injected small perturbation current $i(t)$. In a steady-state oscillation mode, when the losses in the resonant circuit are compensated by the energy inserted into the circuit by the active device, the second-order differential equation of the oscillator is written as

$$\frac{d^2v}{dt^2} + \omega_0^2 v = \frac{1}{C}\frac{di}{dt} \tag{8.188}$$

where $\omega_0 = 1/\sqrt{LC}$ is the resonant frequency and $v(t)$ is the voltage across the resonant circuit. If a current impulse $i(t)$ is injected, the amplitude and phase of the oscillator will have time-dependent responses. According to Eqs. (8.186) and (8.187), the resultant amplitude and phase changes have quadrature dependence with respect to each other. When an impulse is applied at the peak of the voltage across the capacitor, there will be a maximum amplitude deviation with no phase shift, as shown in Figure 8.44(b). On the other hand, if the current impulse is applied at zero crossing, it will result in a maximum phase deviation with no amplitude response, as shown in Figure 8.44(c).

Suppose that a perturbation current $i(t)$, injected into the oscillation circuit, is a periodical function that can generally be expanded into a Fourier series

$$i(t) = I_0\cos\Delta\omega t + \sum_{k=1}^{\infty}\{I_{kc}\cos[(k\omega_0 + \Delta\omega)t] + I_{ks}\sin[(k\omega_0 + \Delta\omega)t]\} \tag{8.189}$$

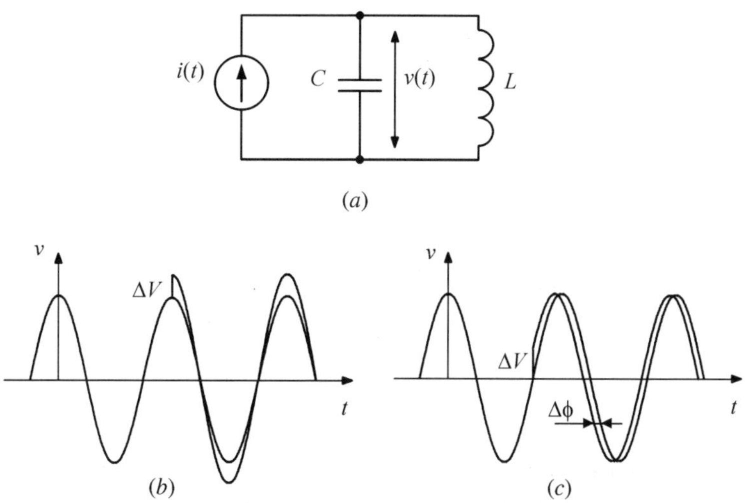

FIGURE 8.44 Second-order LC oscillator and effect of injected impulse.

where I_0 is the dc current, I_{kc} is the kth cosine current harmonic amplitude, I_{ks} is the kth sinusoidal current harmonic amplitude, and $\Delta\omega \ll \omega_0$. In this case, the small external force can be redefined as

$$n = \frac{1}{\omega_0 C}\frac{di}{d\tau} = L\frac{di}{dt}. \tag{8.190}$$

Consequently, substituting Eq. (8.189) into Eqs. (8.186) and (8.187) and taking into account that $\omega_0 + \Delta\omega \cong \omega_0$ result in

$$\Delta V(t) = -\frac{I_{1s}}{2C}\frac{\cos(\Delta\omega t) - 1}{\Delta\omega} \tag{8.191}$$

$$\Delta\phi(t) = -\frac{I_{1c}}{2CV}\frac{\sin(\Delta\omega t)}{\Delta\omega} \tag{8.192}$$

where V is the voltage amplitude across the capacitor C. In this case, only the fundamental components in the injected current $i(t)$ can contribute to the amplitude (sine amplitude) and phase (cosine amplitude) fluctuations given by Eqs. (8.191) and (8.192), because for the dc and kth-order current components, the arguments for all their integrals in Eqs. (8.186) and (8.187) are significantly attenuated by the averaging over integration period.

The output voltage of an ideal cosine oscillator with constant amplitude V and phase fluctuations $\Delta\phi$ can be written as

$$v(t) = V\cos[\omega_0 t + \Delta\phi(t)] = V\cos[\Delta\phi(t)]\cos\omega_0 t - V\sin[\Delta\phi(t)]\sin\omega_0 t \tag{8.193}$$

resulting in an output spectrum of the oscillator with sidebands close to the oscillation frequency ω_0. Since, for a narrowband phase modulation with small phase fluctuations, $\sin[\Delta\phi(t)] \cong \Delta\phi(t)$ and $\cos[\Delta\phi(t)] \cong 1$, the phase modulation spectrum given by Eq. (8.193) can be rewritten using Eq. (8.192) as

$$v(t) = V\cos\omega_0 t + \frac{I_{1c}}{4C\Delta\omega}\cos[(\omega_0 t - \Delta\omega)t] - \frac{I_{1c}}{4C\Delta\omega}\cos[(\omega_0 t + \Delta\omega)t] \tag{8.194}$$

which is similar to the single-tone amplitude modulation spectrum containing the spectral components corresponding to the carrier frequency ω_0 and two close sideband frequencies $\omega_0 - \Delta\omega$ and $\omega_0 + \Delta\omega$.

The injection of the current

$$i(t) = I_0\cos\Delta\omega_0 t + \sum_{k=1}^{\infty}\{I_{kc}\cos[(k\omega_0 - \Delta\omega)t] + I_{ks}\sin[(k\omega_0 - \Delta\omega)t]\} \tag{8.195}$$

has similar effect, resulting in twice the noise power at the sidebands. Therefore, an injected total current $i(t)$ results in a pair of equal sidebands at $\omega_0 \pm \Delta\omega$ with a sideband power P_{sb} relative to the carrier power P_c given by

$$\frac{P_{sb}(\omega_0 \pm \Delta\omega)}{P_c(\omega_0)} = 2\left(\frac{I_{1c}}{4CV\Delta\omega}\right)^2. \tag{8.196}$$

Now let us assume that a stationary thermal noise current with a white power spectral density $\overline{i_n^2}$ is injected into the oscillator circuit close to carrier. Then, by making a replacement between the amplitude and root-mean-square current values when $I_{1c}^2/2 = \overline{i_n^2}$, the single sideband power spectral density for the phase fluctuations at $\Delta\omega$ offset from the carrier ω_0 in $1/f^2$ region can be written using Eq. (8.196) as

$$L(f_m) = \frac{\overline{i_n^2}}{4C^2V^2\Delta\omega^2}. \tag{8.197}$$

Finally, taking into account that $\overline{i_n^2} = 4FkT/R_L$ for $\Delta f = 1$ Hz, $Q_L = \omega_0 CR_L$, and $P_L = V^2/2R_L$, where R_L is the tank parallel or load resistance, Eq. (8.197) can be rewritten as

$$L(f_m) = \frac{2FkT}{P_L}\left(\frac{\omega_0}{2Q_L\Delta\omega}\right)^2. \tag{8.198}$$

which is similar to Eq. (8.161) for the negative resistance oscillator.

Consideration of both cosine current and cosine voltage across the device output terminals implies the device operation in a linear active region only when the establishment of the oscillations can be achieved by using a separate diode as a nonlinear element. Therefore, the mixing effect from the nonlinear behavior of the active device is not taken into account. For a general case of the active device described by a two-port network equivalent circuit, in order to evaluate the output port noise voltage generator, it is necessary to provide a transformation of the noise source from the input port to the output port of the device.

As an example, consider an oscillator with a nonlinear output resistance dependent on the applied dc bias voltage and on the amplitude of the self-sustained oscillations. The basic oscillator circuit with the nonlinear negative output resistance R_{out}, capacitance C, inductance L, load resistance R_L, and noise current $i_n(t)$ is shown in Figure 8.45(*a*). The electrical behavior of such an oscillator, in terms of voltage $v(t)$ across the capacitance, can be represented by a second-order nonlinear differential equation

$$LC\frac{d^2v}{dt^2} + \frac{L}{R_L}\frac{dv}{dt} + v + L\frac{di}{dt} = e_n(t) \tag{8.199}$$

where

$$e_n(t) = L\frac{di_n(t)}{dt} \tag{8.200}$$

is the equivalent noise voltage and

$$i = I_0 + \frac{v}{R_{out}} + \sum_{k=2}^{\infty} G_k v^k \tag{8.201}$$

represents a power series expansion where I_0 is the dc current and G_k are the small coefficients.

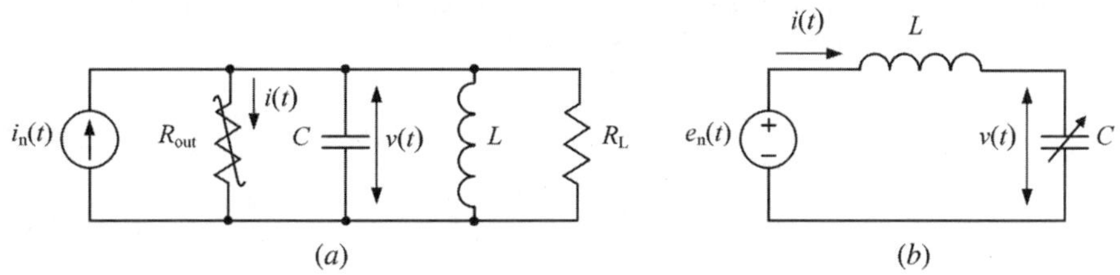

FIGURE 8.45 Second-order nonlinear oscillation systems.

In a steady-state operation mode, when the active device compensates for the losses in the load resistance according to $R_{out} + R_L = 0$, Eq. (8.199) can be rewritten as

$$LC\frac{d^2v}{dt^2} + v + L\frac{d}{dt}\left(\sum_{k=2}^{\infty} G_k v^k\right) = e_n(t) \tag{8.202}$$

Seeking the general solution of the inhomogeneous differential equation as the superposition of a general solution of the homogeneous (noise-free) and specific solutions of Eq. (8.202) as

$$v(t) = V(t)\cos[\omega_0 t + \phi(t)] + e_n(t) \tag{8.203}$$

and applying a van der Pol approach for the slowly time-varying amplitude $V(t)$ and phase $\phi(t)$ allow us to rewrite Eq. (8.202) in the form

$$2\omega_0\frac{dV}{dt}\sin(\omega_0 t + \phi) + 2\omega_0 V\frac{d\phi}{dt}\cos(\omega_0 t + \phi) = \frac{1}{C}\frac{d}{dt}\left(\sum_{k=2}^{\infty} G_k v^k\right) \tag{8.204}$$

where $\omega_0 = 1/\sqrt{LC}$. It is assumed that $e_n(t)$ is a small slowly time-varying low-frequency noise voltage, for which

$$LC\frac{d^2e_n(t)}{dt^2} \ll e_n(t).$$

As a result, substituting Eq. (8.203) into the right-hand side of Eq. (8.204) and using trigonometric identities yield

$$\frac{d\phi}{dt} = 0 \tag{8.205}$$

which means that the nonlinear output resistance has no impact on the phase fluctuations. However, the amplitude fluctuations are not equal to zero because all factors on the right-hand side of Eq. (8.204) have the first-order sine components. Thus, the resistive type of nonlinearities alone would cause amplitude noise only, since the reactive elements determining the oscillation frequency remain constant. However, if the high-frequency noise current is injected close to the carrier frequency ω_0 (for example, at small offset $\Delta\omega$), it will cause the phase fluctuations according to Eqs. (8.192) and (8.204).

Now consider a varactor-controlled oscillator with the varactor as a nonlinear element, whose capacitance depends not only on the applied dc bias voltage, but also on the amplitude of the self-sustained oscillations. The basic VCO circuit consists of the varactor with a nonlinear capacitance C, an inductance L, and a noise voltage $e_n(t)$, as shown in Figure 8.45(*b*) [50]. The voltage $e_n(t)$ can represent all the noise coming from both inside and outside the circuit, including any thermal noise from the resistors, flicker noise from the active device, and noise from the power supply. The electrical behavior of the oscillator can be described by

$$v + L\frac{di}{dt} = e_n(t) \tag{8.206}$$

$$i = \left(C + v\frac{dC}{dv}\right)\frac{dv}{dt} \tag{8.207}$$

where the nonlinear term vdC/dv is included in Eq. (8.207).

By expanding a nonlinear capacitance C into a power series

$$C = C_0 + \sum_{k=1}^{\infty} C_k v^k \tag{8.208}$$

with the small coefficients C_k, substituting Eq. (8.207) into Eq. (8.206) and applying an asymptotic perturbation procedure with decomposition of the perturbed and unperturbed equations, the first-order differential equation for phase fluctuations with the slowly time-varying noise voltage e_n can be derived as

$$\frac{d\phi}{dt} = -\frac{\omega_0}{C_0}\left[C_1 e_n + C_2\left(\frac{3}{4}V^2 + 3e_n^2\right) + C_3 e_n\left(3V^2 + 4e_n^2\right) + \cdots\right] \tag{8.209}$$

where V is the voltage amplitude across the varactor [51]. Note that a nonlinear capacitance has no impact on the amplitude noise of the oscillator.

From Eq. (8.209) it follows that

- the first-order capacitance nonlinearity described by the coefficient C_1 contributes to the up-conversion of the low-frequency noise $e_n(t)$ to the sideband noise near carrier ω_0;
- the second-order nonlinearity described by the coefficient C_2 generates a phase noise, due to both amplitude-to-phase conversion and low-frequency noise upconversion;
- the higher order nonlinearities described by the coefficients C_k, $k = 3, 4, 5, \ldots$, cause a more complicated noise behavior of the oscillator based on hybrid upconversion and amplitude-to-phase conversion due to the cross-terms of V and e_n.

In the case of a single-frequency LC oscillator, the main contributor to the phase noise is the nonlinear collector capacitance of the bipolar device or the gate–source capacitance of the field-effect transistor (FET) device. However, in a general case, the equivalent circuit of the active device is very complicated, including both nonlinear intrinsic and linear parasitic external elements. This means that it is difficult to evaluate analytically the impact of each nonlinear element on the upconversion mechanism. Moreover, the joint effect of different nonlinear circuit elements will result in both amplitude and phase fluctuations. For example, the phase noise can be significantly reduced by linearizing both the transconductance g_m and the gate–source capacitance C_{gs}, since both nonlinearities are important contributors to the phase noise [51]. The amplitude noise also depends on the capacitance and transconductance nonlinearities. However, the capacitance nonlinearity will not affect the output current if the series gate resistance R_g is set to zero. The nonlinearities of the gate–drain capacitance C_{gd} and drain–source resistance R_{ds} have negligible effect on the amplitude and phase noise.

Generally, the transition from soft start-up oscillation conditions to steady-state self-sustained oscillations is provided as a result of the degradation of the device transconductance in a large-signal mode, when the active device is operated in both pinch-off and active regions. As a result, for the cosine voltage across the resonant circuit, the output collector (or drain) current $i(t)$ represents a Fourier series expansion

$$i(t) = I_0 + \sum_{n=1}^{\infty} I_n \cos(n\omega_0 t) \tag{8.210}$$

where I_0 is the dc current and I_n is the amplitude of the nth harmonic component.

If the oscillation frequency is equal to the resonant circuit frequency, which means that the active device has no effect on the oscillation frequency, then the fundamental component of the collector voltage will be in phase with the fundamental component of the collector current. However, for all higher order voltage harmonics, the impedance of the resonant circuit will be capacitive since the

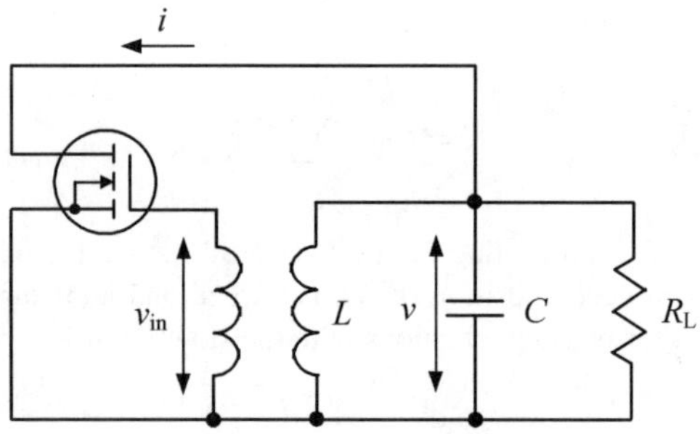

FIGURE 8.46 Schematic of parallel feedback oscillator.

collector current harmonics are mostly flowing through the shunt capacitance. Therefore, for the Meissner oscillator circuit shown in Figure 8.46, the voltage at the input of the active device can be written as

$$v_{in}(t) = V_{in1}\cos\omega_0 t + \sum_{k=2}^{\infty} V_{ink}\cos\left(k\omega_0 t - \frac{\pi}{2}\right) \tag{8.211}$$

where $V_{ink} \ll V_{in1}$ for a high value of the oscillator loaded quality factor.

As a result, when the device transfer characteristic is approximated by a polynomial, the presence of higher order voltage harmonic contributes, first, to the changes in the fundamental amplitude and, secondly, to the appearance of the phase shift between the fundamental voltage and the fundamental current. In a general form, the frequency deviation $\Delta\omega$ caused by the presence of the second-order and higher order harmonic components of the voltage v on the resonant circuit can be obtained from

$$\frac{\Delta\omega}{\omega_0} = -\frac{1}{2}\sum_{k=2}^{\infty}\left(k^2 - 1\right) m_k^2 \tag{8.212}$$

where k is the order of the harmonic component and $m_k = V_k/V_1$ is the ratio of the harmonic voltage component to the fundamental voltage amplitude [3].

In view of the multiharmonic representation of the oscillator output spectrum due to the device and resonant circuit nonlinearities, the total phase fluctuations can be represented by a superposition integral as a result of each harmonic contribution. This is similar to the Fourier harmonic expansion in a frequency domain of the voltage waveform, when the phase trajectory on the phase plane is a result of the phase trajectories with different radii and velocities corresponding to the dc shift and harmonic amplitudes. In this case, Eq. (8.187) can take a general form

$$\Delta\phi(t) = -\frac{1}{A}\int_0^{\omega_0 t} n(\tau)\Gamma(\tau)\,d\tau \tag{8.213}$$

where

$$\Gamma(t) = \frac{c_0}{2} + \sum_{n=1}^{\infty} c_n \cos\left(n\omega_0 t + \phi_n\right) \tag{8.214}$$

is a dimensionless periodic function characterizing the shape of the limit cycle or phase trajectory corresponding to the oscillation waveform and depending on the oscillator topology. It is called the *impulse sensitivity function* (ISF) for an approximate model for the oscillator phase behavior [52] and serves a similar role as the *perturbation projection vector* (PPV) for the exact model [53]. The initial phase ϕ_n in Eq. (8.214) is not important for random noise sources and can be neglected. For an ideal case of a purely sinusoidal oscillator, $c_1 = 1$ and $\Gamma(t) = \cos\omega_0 t$.

Now if any stationary noise current with a white power spectral density $\overline{i_n^2}/\Delta f$ is injected into the oscillator circuit close to any harmonic $n\omega_0 + \Delta\omega$ or $n\omega_0 - \Delta\omega$, it will result in a pair of equal sidebands at $\omega_0 \pm \Delta\omega$. Then, the total single-sideband power spectral density for the phase fluctuations in a bandwidth $\Delta f = 1$ Hz can be written, based on Eq. (8.197), as

$$L(f_m) = \frac{\overline{i_n^2} \sum_{n=0}^{\infty} c_n^2}{4C^2V^2\Delta\omega^2} = \frac{\overline{i_n^2}\Gamma_{\text{rms}}^2}{2C^2V^2\Delta\omega^2} \tag{8.215}$$

where Γ_{rms} is the root-mean-square value of $\Gamma(t)$ [54]. Thus, the total noise power near the carrier frequency of the oscillator is a result of the upconverted $1/f$ noise near dc, weighted by coefficient c_0, the noise near the carrier weighted by coefficient c_1, and the downconverted white noise near the second-order and higher order harmonics weighted by coefficients c_n, $n = 2, 3, \ldots$. The converted phase noise due to the conversion from one sideband to another can be of the order of 6 dB higher than the additive noise in the oscillator [55].

From Eq. (8.215) it follows that the effect of the converted phase noise can be reduced by minimizing the dc coefficient c_0 and the higher order harmonic coefficients c_n of the $\Gamma(t)$ approximating the cosine waveform of the injected node voltage. For a symmetrical flattened drain voltage waveform corresponding to a Class F operation mode, the ratio between the voltage fundamental and third harmonic should be equal to $V_1/V_3 = 9$ with the significantly suppressed even harmonic components. On the other hand, for a half-cosine drain voltage waveform corresponding to an inverse Class F operation mode, the ratio between the voltage fundamental and second harmonic should be equal to $V_1/V_2 = 4$ with the significantly suppressed odd harmonic components. Note that, in Class E operation with nonsymmetrical voltage waveform, the effect of the second-order and higher order harmonics is significant resulting in a high value of the voltage peak factor. The importance of the symmetry is necessary also to minimize the coefficient c_0 responsible for the low noise upconversion and amplitude-to-phase conversion [52]. As it is seen from Eq. (8.209), the phase noise improvement can be achieved by reducing the effect of the device and circuit nonlinear capacitances. Due to the amplitude-to-phase conversion, the phase for each higher order harmonic component changes with amplitude resulting in a generally asymmetric voltage waveform.

Equation (8.213) describes an approximate phase noise behavior, compared with the accurate equation where the phase $\phi(t)$ also appears in its right-hand side [53]. Such a simplified phase noise model is valid for the case of stationary noise sources such as white noise. However, when the noise sources are no longer stationary, it can be accurate only in the limits of an assumption of the small phase shifts for which $\cos\Delta\phi$ is close to unity. This implies that the approximate model is not accurate enough to analyze neither the injection-locking phenomenon nor the related issues such as behavior of phase differences of coupled oscillators [56].

The noise sources in an oscillator generally cannot be only modeled as purely stationary since the statistical properties of some of them may change with time in a periodic manner. Such types of noise sources are referred to as cyclostationary. If the thermal noise of the resistor has a stationary nature, then the collector shot noise of the transistor is an example of cyclostationary noise due to the time-varying nature of the collector current. The most important issue is that the collector shot noise is dominant, compared to the noise from base resistance or tank losses, and can achieve about 70% of total phase noise of the oscillator [57].

Figure 8.47(*a*) shows a simplified single-ended common gate complementary metal-oxide semiconductor (CMOS) Colpitts oscillator configuration where the required regeneration factor for the start-up oscillation conditions is chosen using a proper ratio of the feedback capacitances C_1 and C_2.

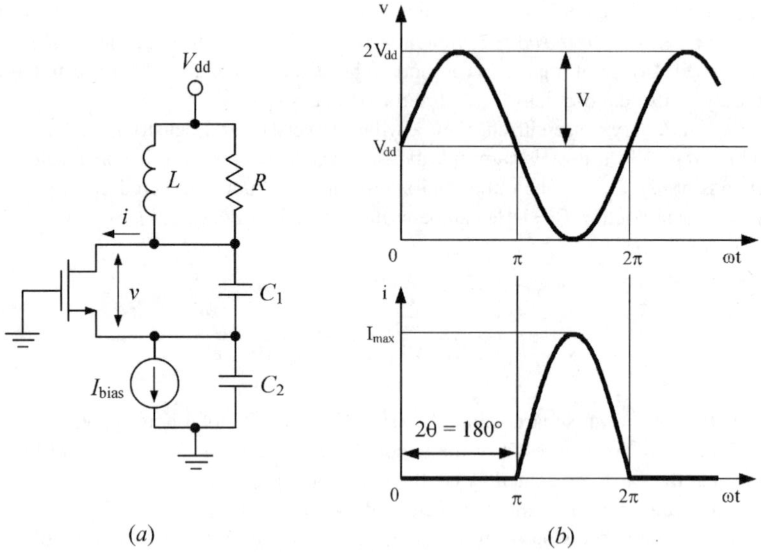

FIGURE 8.47 Simplified Colpitts oscillator and drain voltage and current waveforms.

The idealized voltage and current waveforms corresponding to the large-signal operation in idealized Class B with zero saturation voltage are shown in Figure 8.47(*b*).

Equation (8.210) for the drain time-varying current can be rewritten in the form

$$i\,(t) = I_{\max}\left[\alpha_0 + \sum_{n=1}^{\infty}\alpha_{\rm n}\cos\,(n\omega_0 t)\right] \tag{8.216}$$

where $\alpha_{\rm n}$ is the ratio of the nth current harmonic amplitude to the peak output current $I_{\max}$, expressed through a half-conduction angle θ as

$$\alpha_{\rm n} = \frac{I_{\rm n}}{I_{\max}} = \frac{\gamma_{\rm n}\,(\theta)}{1 - \cos\theta}. \tag{8.217}$$

where $\gamma_{\rm n}(\theta)$ are the current coefficients.

To account for the cyclostationary drain noise source as a result of total noise sources injected at frequencies $n\omega_0 \pm \Delta\omega$, Eq. (8.215) can be rewritten in a general form as

$$L\,(f_{\rm m}) = \frac{\overline{i_{\rm nd}^2}\sum_{n=0}^{\infty}(\alpha_{\rm n}c_{\rm n})^2}{4C^2V^2\Delta\omega^2} \tag{8.218}$$

where $\overline{i_{\rm nd}^2} = 2qI_{\max}$ is the drain current noise power density in a frequency bandwidth $\Delta f = 1$ Hz. The Fourier components for the current waveform close to a half-cosine show that the drain shot noise is mixed mostly with the fundamental and second harmonics to contribute to the total phase noise of the oscillator.

To minimize the oscillator phase noise, it is very important to choose the optimum value of a capacitance feedback ratio $k = C_2/C_1$ for the same total capacitance $C = C_1C_2/(C_1 + C_2)$. This is because different values of the conduction angle correspond to different harmonic contribution to the output spectrum. In the case of a Class B with $\theta = 90^\circ$, the third-order, fifth-order, and higher order harmonics can be eliminated since their current coefficients γ_n (for $n = 3, 5, \ldots$) become equal to zero. As a rule-of-thumb, the optimum capacitance feedback ratio for a Colpitts oscillator can be chosen to be approximately $k = 3.5$–4 [52,57].

8.10 VOLTAGE-CONTROLLED OSCILLATORS

The VCOs are key components in many applications, especially in wireless communication systems, measurement equipment, or military applications. A growing market of wireless applications requires highly integrated circuit solutions, where both high-performance transistors and passive elements with high-quality factors can be used. To analyze the tuning linearity of the VCO circuit, a general approach describing the oscillation circuit in terms of natural frequencies of a lossless two-port network when one of its ports is short circuited can be used [58]. Figure 8.48 shows the block diagram of the general VCO equivalent circuit, where a linear lossless network can incorporate one or several resonant circuits including active device reactive elements, the baseband modulation signal is brought to the varactor using port 3–3′, and the load is connected to the port 4–4′. Such a block representation enables the description of the VCO modulated curve in terms of poles and zeros, irrespective of any particular circuit diagram.

The oscillation frequency can be found form the phase balance condition of

$$X_{in}(\omega_0) = -X_v(\omega_0) \tag{8.219}$$

where $X_{in}(\omega_0)$ is the reactance seen by the varactor, $X_v(\omega_0)$ is the reactance of the varactor, and $\omega_0 = 2\pi f_0$ is the radian oscillation frequency. The reactance seen by the active device is equal to zero at the oscillation frequency.

Generally, the reactance X_{in} can be expressed in terms of poles and zeros. For instance, for the case shown in Figure 8.49 where the poles occur at the origin and ω_2 while the zeros occur at ω_1 and ω_3, the reactance X_{in} is written as

$$X_{in}(\omega) = K\frac{\left(\omega^2 - \omega_1^2\right)\left(\omega^2 - \omega_3^2\right)}{\omega\left(\omega^2 - \omega_2^2\right)}. \tag{8.220}$$

The reactance due to the varactor junction capacitance can be written using Eq. (2.6) in Chapter 2 in the form

$$X_v = -\frac{(v_v + \varphi)^{\gamma}}{\omega C_{vo}\varphi^{\gamma}} \tag{8.221}$$

where the voltage v_v on the varactor consists of the dc bias voltage V_v and of the modulation voltage v_m.

The modulation voltage can be represented in the normalized form

$$v = \frac{v_m}{V_v + \varphi}. \tag{8.222}$$

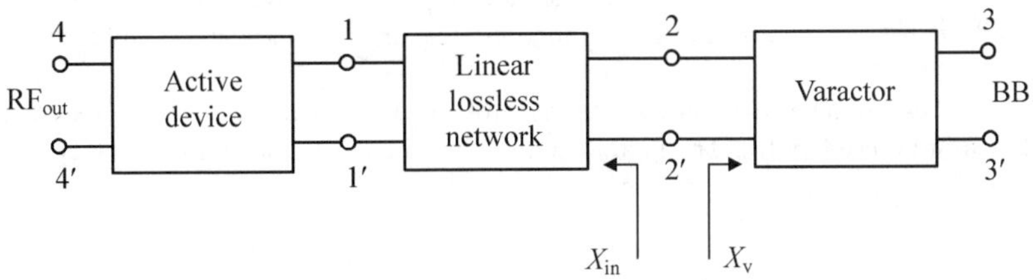

FIGURE 8.48 Block diagram of VCO with linear circuit.

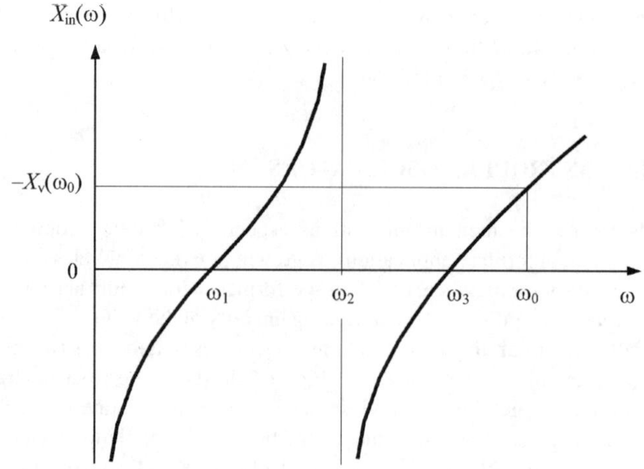

FIGURE 8.49 Reactance of lossless network.

For the initial conditions of $\omega = \omega_0$, v is equal to zero. This leads to

$$1 + v = \left[\frac{\left(\omega^2 - \omega_1^2\right)\left(\omega_0^2 - \omega_2^2\right)\left(\omega^2 - \omega_3^2\right)}{\left(\omega_0^2 - \omega_1^2\right)\left(\omega^2 - \omega_2^2\right)\left(\omega_0^2 - \omega_3^2\right)} \right]^{1/\gamma}. \tag{8.223}$$

Introducing the frequency normalization in the form of

$$\frac{\omega_i}{\omega_0} = \Omega_i \quad \text{for } i = 1, 2, 3 \tag{8.224}$$

and taking into account that, in the vicinity of the operating point

$$\frac{\omega}{\omega_0} = 1 + \eta \tag{8.225}$$

where $\eta = (\omega - \omega_0)/\omega_0$ is the normalized frequency change representing a very small number, the modulation curve can be rewritten by

$$1 + v = \left[\frac{(1+\eta)^2 - \Omega_1^2}{1 - \Omega_1^2} \frac{1 - \Omega_2^2}{(1+\eta)^2 - \Omega_2^2} \frac{(1+\eta)^2 - \Omega_3^2}{1 - \Omega_3^2} \right]^{1/\gamma}. \tag{8.226}$$

When the linear lossless circuit between the varactor and the active device has more than three finite natural frequencies, Eq. (8.226) can be appropriately expanded to include any additional zeros and poles.

The simplest two-port network between the varactor and the active device is a single series resonant circuit shown in Figure 8.50(a). The modulation curve for such a resonant circuit can be obtained from Eq. (8.226) as

$$1 + v = \left[1 + A_1\eta + \frac{A_1}{2}\eta^2 \right]^{1/\gamma} \tag{8.227}$$

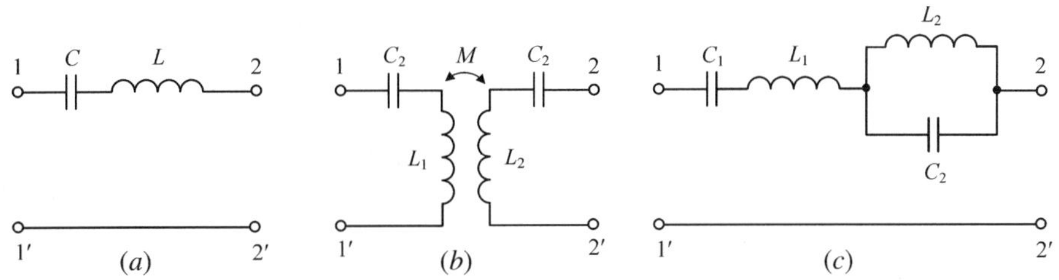

FIGURE 8.50 Schematics of two-port networks.

where

$$A_1 = \frac{2}{1-\Omega_1^2}.$$

Since η is a small number, it is convenient to use the binomial expansion to obtain first two terms of the Taylor series for voltage v by

$$v = C_1\eta + C_2\eta^2 \tag{8.228}$$

where

$$C_1 = \frac{A_1}{\gamma}$$

$$C_2 = \frac{A_1}{2\gamma}\left[1 + \left(\frac{1}{\gamma} - 1\right)A_1\right].$$

The slope coefficient C_1 is the inverse of the normalized circuit sensitivity S_n equal to

$$S_n = \left.\frac{d\eta}{dv}\right|_{\omega_0} = S\frac{V_v + \varphi}{f_0} \tag{8.229}$$

where S is the slope of the actual circuit sensitivity (or modulation curve) in hertz per volt. As a result, for the series resonant circuit shown in Figure 8.50(a), the normalized sensitivity must satisfy the condition of

$$S_n = \frac{\gamma}{2}\left(1 - \Omega_1^2\right). \tag{8.230}$$

The modulation curve is linearized by requesting $C_2 = 0$ resulting in

$$\gamma = \frac{2}{1+\Omega_1^2}. \tag{8.231}$$

From Eq. (8.231) it follows that the varactor junction sensitivity γ must be larger than unity because $\Omega_1 < 1$. Consequently, the only possibility to linearize the modulation curve for a series resonant circuit is to use a hyperabrupt varactor. However, such a resonant circuit for a fixed operating point does not provide enough flexibility to accommodate small variations in γ.

The inductively coupled pair of lumped resonant circuits shown in Figure 8.50(b) offers more degree of freedom to linearize the modulation curve, having the position of natural frequencies ω_1, ω_2, and ω_3 (when port 2–2′ is short-circuited), as shown in Figure 8.49. To achieve resonance with the varactor capacitance, reactance X_{in} must be positive with oscillation frequency ω_0 located above the zero ω_3.

The linearity of the modulation curve of the resonant circuit with two coupled resonators can be defined by expanding Eq. (8.226) in a three-term Taylor series

$$v = C_1\eta + C_2\eta^2 + C_3\eta^3 \tag{8.232}$$

where

$$C_1 = \frac{1}{\gamma}(A_1 - A_2 + A_3)$$

$$C_2 = \frac{1}{2}\left[C_1 + C_1^2 - \frac{1}{\gamma}(A_1^2 - A_2^2 + A_3^2)\right]$$

$$C_3 = -\frac{C_1^3}{3} - \frac{C_1^2}{2} - \frac{C_1}{2} + C_2\left(\frac{1}{2} + C_1\right) + \frac{1}{3\gamma}(A_1^3 - A_2^3 + A_3^3)$$

where

$$A_i = \frac{2}{1 - \Omega_i^2} \quad \text{for } i = 1, 2, 3.$$

Since the normalized sensitivity S_{n0} at the center frequency ω_0 can be written as $S_{n0} = 1/C_1$, and the linearity of the modulation curve requires $C_2 = C_3 = 0$, then the three nonlinear equations can be obtained as

$$A_1 - A_2 + A_3 = \frac{\gamma}{S_{n0}} \tag{8.233}$$

$$A_1^2 - A_2^2 + A_3^2 = \frac{\gamma}{S_{n0}}\left(1 + \frac{\gamma}{S_{n0}}\right) \tag{8.234}$$

$$A_1^3 - A_2^3 + A_3^3 = \frac{\gamma}{S_{n0}}\left[\frac{3}{2} + \frac{1}{S_{n0}}\left(\frac{3}{2} + \frac{1}{S_{n0}}\right)\right] \tag{8.235}$$

which should be solved numerically to find the optimum position of natural frequencies forcing elimination of both the second and third coefficients in Eq. (8.232).

The calculated values Ω_1, Ω_2, and Ω_3 are substituted in Eq. (8.226), and the relative frequency η is gradually varied so that normalized modulation curve $v(\eta)$ is evaluated [58]. The higher the chosen value of S_{n0}, the wider bandwidth within which the deviation from linearity stays within prescribed limits. For example, by using an abrupt varactor with $\gamma = 0.5$, for 1% deviation from linearity, the sensitivity $S_{n0} = 0.02$ results in the relative bandwidth of 0.0214, while the relative bandwidth is 0.0319 for $S_{n0} = 0.03$. To provide oscillation stability, the following approximate condition should be satisfied:

$$\Omega_2^2 \geq \Omega_r^2\left(1 + \frac{kQ_v\Omega_r}{2\Omega_v^2}\right) \tag{8.236}$$

where $\Omega_r = \omega_r/\omega_0$ is the normalized resonant frequency associated with the varactor side,

$$\omega_r = \sqrt{\frac{C_2 + C_v}{L_2 C_2 C_v}}$$

$\Omega_v = \omega_v/\omega_0$ is the normalized resonant frequency associated with the varactor only,

$$\omega_v = \frac{1}{\sqrt{L_2 C_v}}$$

$k^2 = M^2/L_1 L_2$ is the inductive coupling coefficient, and Q_v is the varactor quality factor.

To increase the frequency tuning bandwidth of a negative resistance oscillator, a reactance compensation technique based on the tandem connection of a series resonant circuit and parallel resonant circuit both tuned to the fundamental frequency, as shown Figure 8.50(*c*), can be used [2,59]. Such a reactance compensation technique can also provide a linearization of the frequency tuning characteristic for the certain ratios between the oscillator circuit parameters [60]. For example, a linear tuning bandwidth of 1.7 GHz with a minimum slop ratio of 1.08 can be achieved for an *X*-band negative resistance oscillator.

Depending on the operating frequency and application requirements, there is a variety of VCO implementation techniques based on using different types of the active devices, circuit schematic approaches, and hybrid or monolithic integrated circuit technologies. For example, their low cost implementation and low-phase noise performance are required in wireless communication systems. At microwaves, most of these VCOs use MESFET or heterojunction bipolar transistor (HBT) devices because of producing lower phase noise than VCOs based on high electron mobility transistor (HEMT) devices. However, high performance HEMT oscillators are essential if they can be integrated together with amplifiers and mixers for single-chip receivers or transmitters using the same technology.

The common gate VCO configuration is usually used to generate strong negative resistance over a wide frequency range. However, due to the series resistance of the varactor, generally the phase noise of such a VCO is relatively higher than that of a single-frequency oscillator. Nevertheless, the phase noise can be reduced by placing the varactor into the source rather than into the gate circuit. Figure 8.51 shows the circuit schematic of a common gate MESFET VCO MMIC where an output L_2C_2 matching network is incorporated in the form of a high-pass section to eliminate low-frequency parasitic oscillations [61]. Based on an analysis of the trajectories of the reflection coefficient lines of the resonator Γ_{in} and the device Γ_{out} as a function of the drain supply voltage, the oscillator loaded quality factor Q_L was maximized. As the drain voltage increases, the angle between these trajectories approaches 90° where Q_L becomes maximal [62]. Increasing the drain voltage also reduces the phase noise by extending the depletion region in the device channel to the drain side, thus reducing the sensitivity of the oscillator to the gate–source voltage. Being implemented in a commercial 0.6-μm GaAs MESFET process, such a VCO demonstrates the phase noise of −91 dBc/Hz at 100 kHz offset with an output power of 11.5 dBm and frequency tuning of 500 MHz around the center frequency of 11.5 GHz.

Wideband VCOs are used in a variety of RF and microwave systems, including broadband measurement equipment, wireless and TV applications and military electronic countermeasures

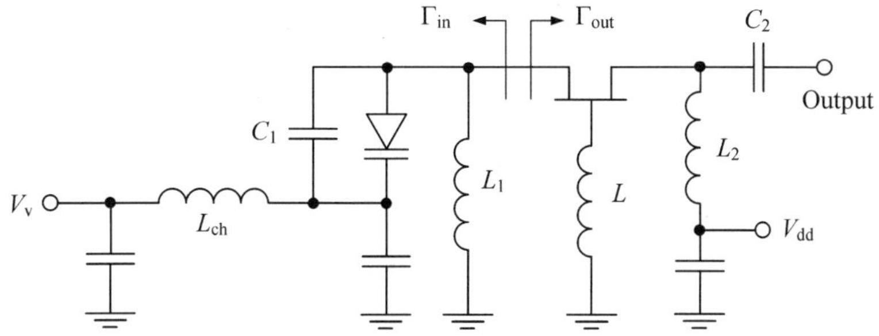

FIGURE 8.51 Circuit schematic of common gate MESFET VCO.

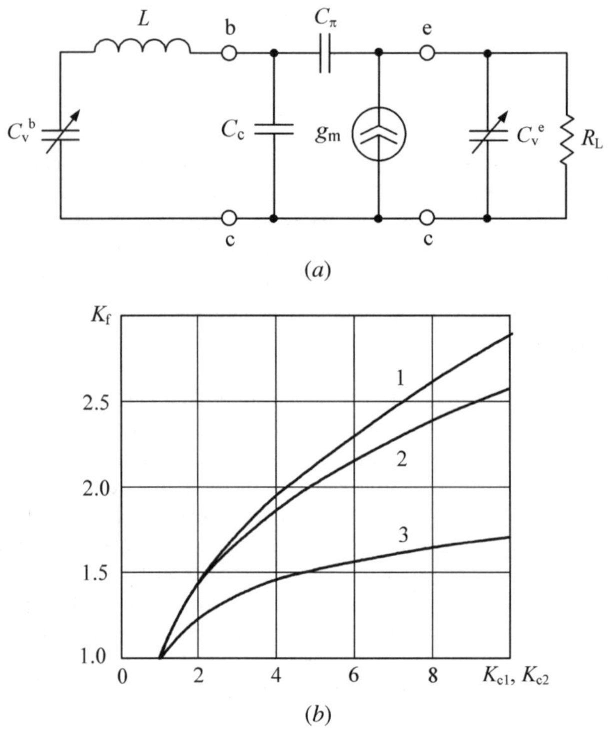

FIGURE 8.52 Equivalent circuit of common collector bipolar VCO.

(ECM) systems. In modern ECM systems, they serve as the frequency-agile local oscillators in receiver subsystems and fast-modulation noise sources in active jamming subsystems. Among wideband tunable signal sources such as YIG-tuned (yttrium iron garnet) oscillators, wideband VCOs are preferable because of their small size, low weight, high settling time speed and capability of fully monolithic integration. Therefore, modern radar and communication applications demand VCOs that are capable of being swept across a wide range of potential threat frequencies with a speed and settling time far beyond of the YIG-tuned oscillators.

The tuning possibility of a bipolar VCO resonant circuit can be evaluated by using the simple device equivalent circuit shown in Figure 8.52(*a*) [63]. Here, the collector terminal is common and usually RF grounded in the practical realization with a bypass capacitor. The simplified equivalent circuit includes the collector capacitance C_c, the base–emitter capacitance C_π (including diffusion and junction capacitances), and a current source described by the small-signal transconductance g_m. To provide wideband frequency tuning, the two varactors C_v^b and C_v^e are included into the base and emitter circuits, respectively.

For such a common collector bipolar VCO, the equation for resonant frequencies in a steady-state operation mode can be given by

$$\omega^2 L C_v^b \left(C_c + \frac{C_\pi C_v^e}{C_\pi + C_v^e} \right) = C_v^b + C_c + \frac{C_\pi C_v^e}{C_\pi + C_v^e}. \tag{8.237}$$

To characterize the VCO band properties, it is convenient to use the generalized dependencies $K_f(K_{c1}, K_{c2})$, where $K_{c1} = C_{vmax}^b / C_{vmin}^b$ and $K_{c2} = C_{vmax}^e / C_{vmin}^e$, and the normalized parameters to obtain the results regardless of the particular values of the circuit parameters. By using the normalized parameters $m_0 = \omega_T C_c/g_m$, $q_1 = C_c/C^b_{vmin}$, and $q_2 = C_c/C^e_{vmin}$, Eq. (8.237) can be rewritten in a general form

$$K_f = \sqrt{K_{c1}\frac{(1+q_1)(m_0+q_2)+q_1}{(q_1+K_{c1})(m_0K_{c2}+q_2)+q_1K_{c2}}\frac{K_{c2}(1+m_0)+q_2}{1+m_0+q_2}}. \tag{8.238}$$

Figure 8.52(*b*) shows the different dependencies $K_f(K_{c1}, K_{c2})$ for various values of q_1 and q_2 and fixed value $m_0 = 0.012$. Here, curve 1 is plotted for $q_1 = 1$ and $q_2 = 0.5$ with varactor tuning in the base and emitter circuits simultaneously. Curve 2 is characterized by $q_1 = 1$ and $q_2 = 0.05$ with varactor tuning only in the base circuit when $K_{c2} = 1$. Curve 3 is calculated for $q_1 = 0.1$ and $q_2 = 0.5$ with only varactor tuning in the emitter circuit when $K_{c1} = 1$. A comparison of the curves shows that, for varactor tuning in the base and emitter circuits simultaneously, maximum tuning bandwidth is achieved (curve 1). Using varactors only in the base circuit (curve 2) gives larger tuning bandwidth than in the case of the only varactor tuning in the emitter circuit (curve 3). In this case, decreasing q_2 and increasing q_1 can increase the tuning bandwidth. To increase the tuning bandwidth only by emitter varactor tuning, it is necessary to reduce the parameter q_1 significantly, provided $q_2 = 1$.

Figure 8.53 shows a typical common collector lumped VCO circuit where the two back-to-back varactors provide a wideband tuning [19]. In this circuit, the load is conductively connected to the resonant circuit inductor. The choke inductors L_{ch} and bypass capacitors C_b form the low-pass filters having high impedance at the fundamental frequency to isolate voltage supplies and low impedance for the modulation frequencies. By using abrupt varactors with $K_c = 3$ in a bias voltage range from 0 to 5 V and minimum capacitance $C_{vmin} = 0.6$ pF, it is possible to provide a wideband tuning in a frequency range with $K_f = 1.6$. Similar tuning bandwidth is predicted by curve 2 shown in Figure 8.52. Taking into account that the equivalent device input capacitance is equal to 1.5 pF, the tuning bandwidth from 5.5 to 8.0 GHz was achieved with a tank inductor of $L = 1.9$ nH.

However, for a common collector VCO, the conductive load connection is not the only way to obtain a maximum level of output power. In addition, it is very important to provide its minimum flatness over the entire tuning bandwidth. The load can also be connected to the emitter terminal, thus decreasing the influence of the load impedance on the resonant circuit that enables one to provide its higher quality factor. Figure 8.54 shows (*a*) the simplified common collector VCO schematic and (*b*) two possible combinations of the admittances in the base and emitter circuits. In the first case, the load is connected to the resonant circuit conductively or inductively, provided the impedance in the

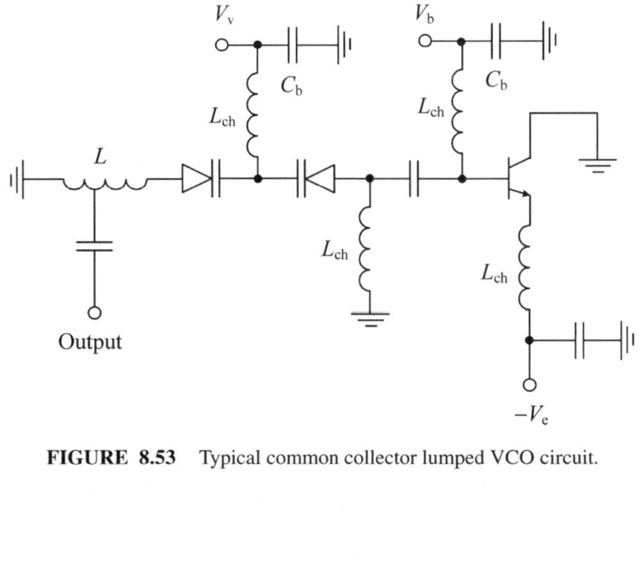

FIGURE 8.53 Typical common collector lumped VCO circuit.

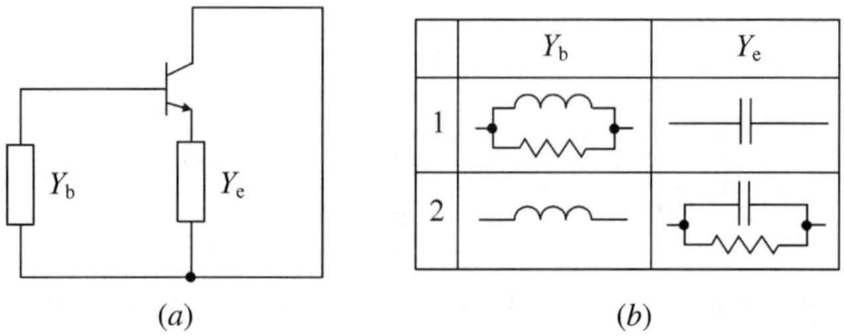

FIGURE 8.54 Simplified VCO schematic with two combinations of base and emitter circuits.

emitter circuit is capacitive. The second combination requires inductive impedance in the base circuit when the load is connected in parallel to the device emitter and collector terminals.

Figure 8.55 shows the calculated dependencies of the normalized output power versus normalized frequency for both above-mentioned cases [64]. For case 1, the load is connected to the base circuit and the output power has maximum when $\omega = \omega_\alpha$, where $\omega_\alpha = 2\pi f_\alpha, f_\alpha$ is the alpha cutoff frequency. For case 2, the output power comes from the device emitter and its level changes negligibly up to $\omega = 0.5\ \omega_\alpha$. As a result, for the latter case, VCO can be tuned easily in a very wide frequency range by a simple tuning of the value of the series inductance in the base circuit using a varactor diode in reverse bias operation. The series or parallel *RC*-circuit with constant capacitance and load resistance can provide the capacitive impedance in the emitter circuit.

Figure 8.56 shows the electrical circuit of a monolithic common drain MESFET VCO designed for a phase-locked loop application in the telecommunication system operating at 14 GHz [65]. A planar Schottky-barrier diode with $0.5 \times 280\ \mu m^2$ stripe suitable for monolithic integration with the MESFET device of the same gate geometry is used as a varactor. The varactor junction capacitance is tuned from 1.0 to 0.5 pF by applying a reverse bias voltage from 0 to 5 V. To provide inductive reactance in the gate circuit, the microstrip line is used as a quarterwave transformer between the varactor diode and the MESFET device. A 50-Ω resistor was inserted into the source circuit for self-biasing resulting

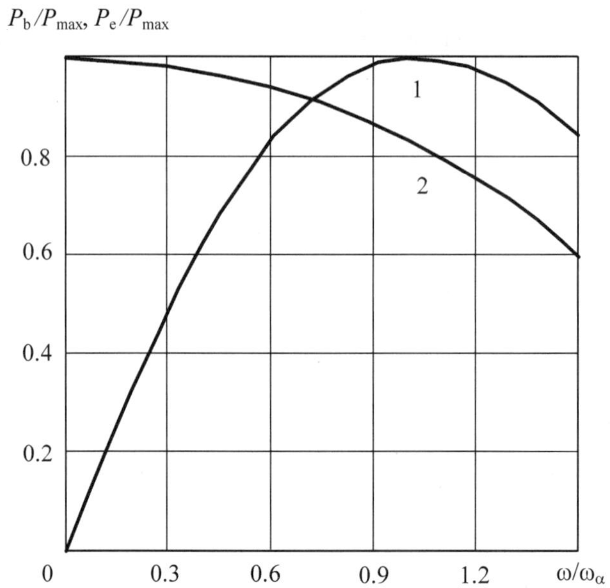

FIGURE 8.55 Output power versus oscillation frequency.

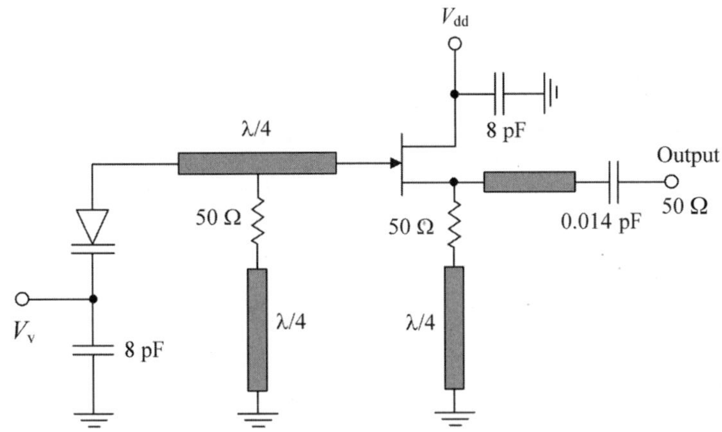

FIGURE 8.56 Schematics of microwave MESFET VCO.

in a drain current stabilization. Another 50-Ω resistor was connected to the microstrip impedance transformer to provide a gate zero bias point. These resistors are necessary to prevent low-frequency oscillations by contributing to the significant decrease of the loaded quality factor at frequencies much lower or higher than resonant frequency. An output series capacitance is small enough to reduce the pulling effect, compromising the output power. The frequency tuning bandwidth from 11.3 to 14.3 GHz was achieved by varactor tuning in a bias range from 0 to 7 V with an output power of −4 dBm.

Figure 8.57 shows the wideband VCO circuit that was used to design and fabricate four GaAs monolithic VCO chips that cover 2–4 GHz, 4–7 GHz, 7–12 GHz, and 12–18 GHz frequency ranges, respectively [66]. Each monolithic chip includes the MESFET device, two varactors, gate inductor L_g, source inductor L_s, and two bypass capacitors C_b. The varactor diode represents a single implanted structure into semi-insulating material and is formed with the same active region as for the MESFET device. Its capacitance ratio is typically of 8:1 or greater. The common gate VCO configuration using the gate inductor as a regenerative feedback element exhibits negative impedance at the source terminal across the entire tuning bandwidth for the equivalent load resistance connected to the drain

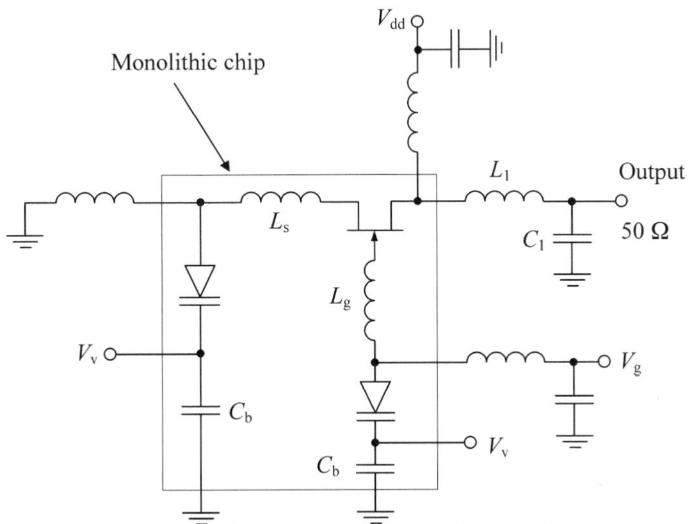

FIGURE 8.57 Microwave monolithic wideband VCO circuit schematic.

terminal of approximately 15 Ω. To provide the output matching with standard load of 50 Ω, the L-type matching circuit with a series inductor L_1 and a shunt capacitor C_1 is used. Due to the required additional area, the output matching circuit was not included on the chip.

The VCO performance in terms of phase noise and tuning range determines the basic characteristics of a whole transceiver. However, a limited frequency-tuning range is usually a serious problem for VCO fully based on CMOS technology. There are four basic candidate differential topologies shown in Figure 8.58 [67]. Figure 8.58(a) shows an nMOS differential VCO topology with tail current

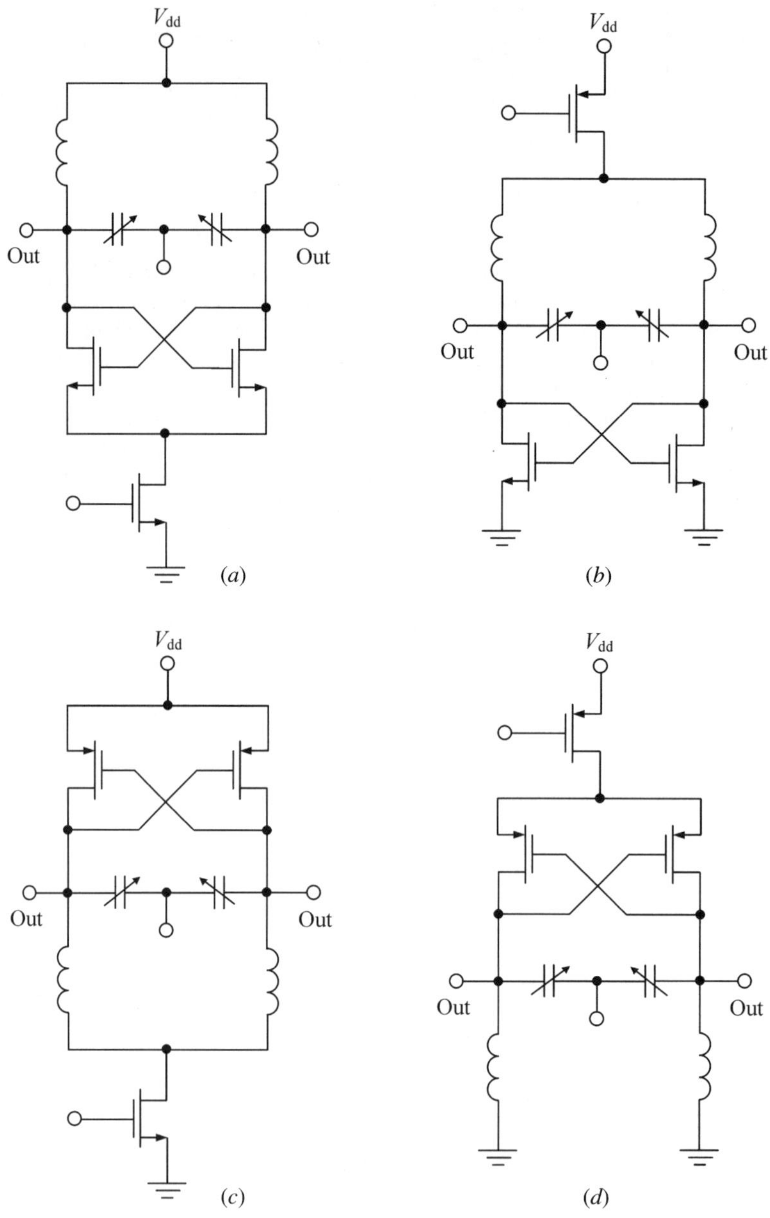

FIGURE 8.58 Single differential topologies of CMOS VCO.

source, while Figure 8.58(*b*) shows an nMOS differential topology with top current source. The pMOS differential topologies with tail and top current sources are shown Figures 8.58(*c*) and 8.58(*d*), respectively. As regards tuning range capability, topologies shown in Figures 8.58(*b*) and 8.58(*c*) are preferable. This is because the anode varactor bias voltage is fixed to V_{dd} for an nMOS oscillator with tail current source, and the anode varactor bias voltage is fixed to zero for a pMOS oscillator with top current source.

Since using pMOS devices results in a lower phase noise due to inherently smaller low-frequency $1/f$ noise, the pMOS differential topology with tail current source shown in Figure 8.58(*c*) represents the best choice for a low-noise wideband tuning. Being implemented in a 0.25-μm standard CMOS process, such a differential VCO with a Q-factor of the tank inductor of 7.5 provides the phase noise of −109 and −123 dBc/Hz at 100 and 500 kHz offset from carrier frequency of 1.3 GHz, respectively. The tuning range was 13.3% for $V_{\mathrm{dd}} = 1.4$ V and 20.1% for $V_{\mathrm{dd}} = 2.0$ V.

Using a proper configuration of the accumulation MOS varactors can significantly improve the tuning range. Such a varactor with high quality factor fabricated using the 0.13-μm CMOS silicon-on-insulator (SOI) technology demonstrates the capacitance ratio of 5 with ± 1 V voltage tuning that provides over 50% frequency-tuning range [68]. However, a high capacitance ratio implies a high varactor sensitivity or K_{VCO}, which makes the oscillator phase noise worse. A band switching solution can reduce the varactor sensitivity, but requires extra control circuitry. A simple and effective solution can be provided by differential varactor tuning to avoid the effect of high K_{VCO}.

8.11 CRYSTAL OSCILLATORS

The traditional and most common type of a piezoelectric resonator used in electronics is the quartz crystal. Its operation is based on a piezoelectric effect that converts the electrical signal applying to the opposite sides of the crystal to mechanical motion and reconverts the vibratory motion of the crystal back into an electrical signal at the resonator terminals. The amount of motion varies over wide extremes depending on how closely the applied signal frequency approaches a natural mechanical resonance of the crystal. In a properly designed resonator these regions of high-amplitude mechanical vibration are very narrow in frequency and are ideally suited for oscillator stabilization. Therefore, the oscillator circuits using a quartz crystal are called the *crystal oscillators*. The first crystal-controlled oscillator, using a crystal of Rochelle salt, was first built by Alexander M. Nicolson and the results were filed in the form of a patent application in 1918 [69]. In 1919, Walter G. Cady used the quartz to control the frequency of an oscillator and described the use of quartz bars and plates as frequency standards and wave filters [70]. Since then, it is generally accepted that Cady was the first to use the quartz to control the frequency of an oscillator circuit. However, subsequent litigation resulted in a legal decision in favor of Nicolson who is therefore considered to be the inventor of the piezoelectric oscillator.

The quartz crystal represents one of forms of silicone dioxide (SiO_2) that is found in nature and is ideal for use as a frequency-determining device because of its predictable thermal, mechanical, and electrical characteristics with a high quality factor. The quartz properties are primarily determined by the orientation angle at which the quartz wafers are cut, being dependent on the reference directions within the crystal that are referred to as *axes*. There are three axes in quartz, forwarded along the X, Y, and Z directions. Ideal quartz would consist of a hexagonal prism with six facets at each end, with a cross section shown in Figure 8.59 [71]. The Z-axis is known as the *optical axis*, repeating its physical properties every 120° as the crystal is rotated about the Z-axis. This axis is not anisotropic to light; therefore light passes readily. The X-axis called the *electrical axis* is parallel to a line bisecting the angles between adjacent prism faces. Electrical polarization occurs in this direction when mechanical pressure is applied. The Y-axis, which is called the *mechanical axis*, runs at right angles through the face of the prism, and at right angles to the X-axis.

The AT cut is the most popular of the Y-axis group because of its excellent temperature characteristic, and it is produced by cutting the quartz bar at an angle of approximately 35°15′ from Z-axis. The

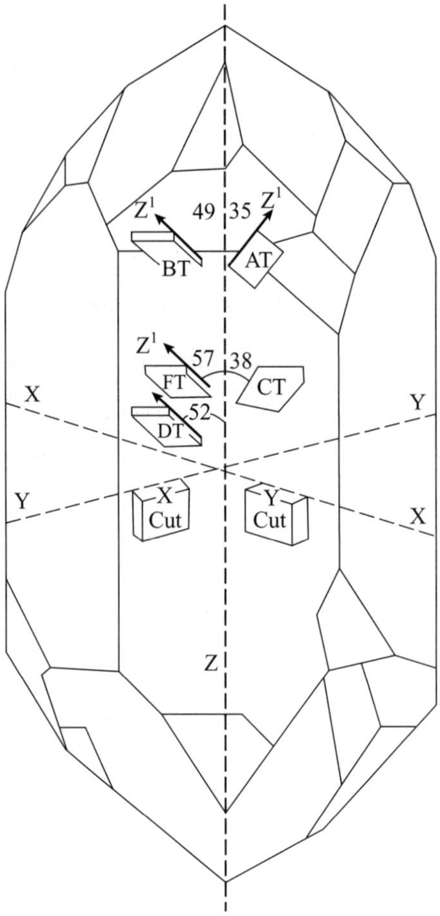

FIGURE 8.59 Typical crystal cuts from a doubly terminated quartz crystal.

crystal resonator is usually represents a round disc and its fundamental mode measured in megahertz is defined as

$$f = \frac{N}{d} \tag{8.239}$$

where d is the disc thickness in millimeters and N is the frequency constant for particular cut, for example, equal to 1.661 MHz/mm for AT cut, 1.797 MHz/mm for SC cut, or 2.536 MHz/mm for BT cut [72]. Because the frequency of the crystal is related to its thickness, there is a limitation in the manufacturing of high-frequency fundamental crystals. For example, a 100 MHz AT-cut crystal resonator has a thickness of only 16.61 microns. In this case, the higher the frequency, the thinner should be the crystal plate. Such a thin wafer is not only somewhat fragile but also impractical to fabricate by conventional means, which are normally limited to thicknesses of 30–35 microns (approximately 50 MHz for AT-cut resonators). However, by introducing chemical polishing methods using fluorides and composite double inverted mesa blank structures with air-gapped electrode have resulted in AT-cut crystal resonators with fundamental frequencies up to 1.6 GHz, corresponding to a resonator plate thickness of just less than 1 micron [73,74]. The measured quality factors of these units ranged from 73,000 at 100 MHz and 32,000 at 250 MHz to 5,000 at 950 MHz.

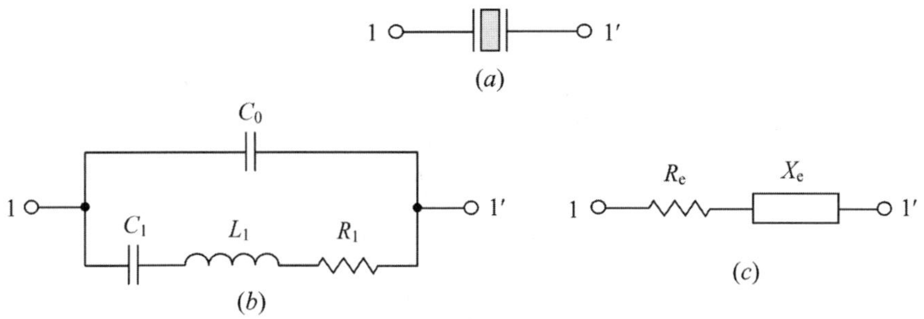

FIGURE 8.60 Simplified equivalent circuit of crystal resonator.

Figure 8.60(a) shows the symbol for a crystal resonator that should generally be treated as a two-port device together with its metallic enclosure, leads and supports. However, in some practical cases, especially at lower frequencies, the complete equivalent circuit of a crystal resonator can be simplified, as shown in Figure 8.60(b), where the crystal vibration portion is represented in the vicinity of a series resonance by dynamic parameters such as an inductance L_1, a capacitance C_1 and a resistance R_1, whereas C_0 is the static capacitance associated with the crystal and its adherent electrodes plus the stray capacitances internal to the crystal enclosure [74]. The impedance of a crystal resonator (between point $1-1'$) can be written as

$$Z(j\omega) = \frac{1}{j\omega C_0} \frac{\omega_s^2 + j\omega\dfrac{R_1}{L_1} - \omega^2}{\omega_p^2 + j\omega\dfrac{R_1}{L_1} - \omega^2}. \tag{8.240}$$

where

$$\omega_s = \frac{1}{\sqrt{L_1 C_1}} \tag{8.241}$$

is the series resonant frequency in radians per second and, for $C_0 \gg C_1$,

$$\omega_p = \sqrt{\frac{C_1 + C_0}{L_1 C_1 C_0}} = \omega_s \sqrt{1 + \frac{C_1}{C_0}} \cong \omega_s \left(1 + \frac{C_1}{2C_0}\right) \tag{8.242}$$

is the parallel resonant frequency in radians per second.

The simplified model of a crystal resonator shown in Figure 8.60(b) can be represented by an equivalent general network, as shown in Figure 8.60(c), with the impedance expressed as

$$Z(j\omega) = R_e(\omega) + jX_e(\omega) \tag{8.243}$$

where

$$R_e(\omega) = \frac{R_1}{(1 - ab)^2 + b^2} \tag{8.244}$$

$$X_e(\omega) = R_1 \frac{a(1 - ab) - b}{(1 - ab)^2 + b^2} \tag{8.245}$$

where $a = (\omega_s L_1/R_1)(\omega/\omega_s - \omega_s/\omega)$ and $b = \omega C_0 R_1$.

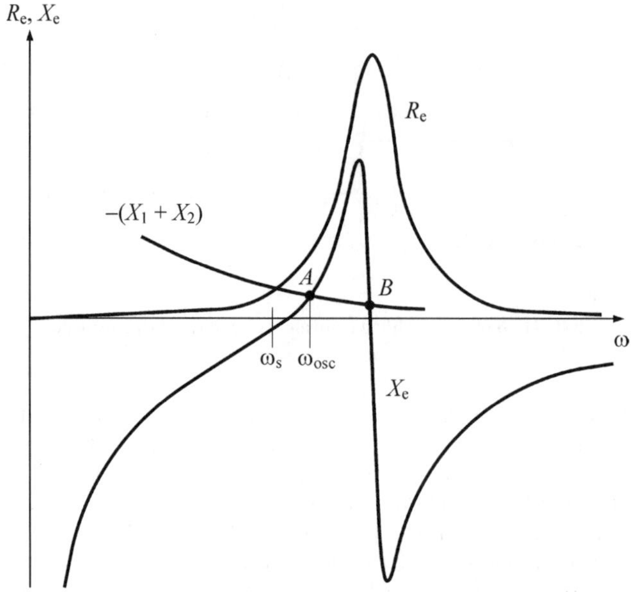

FIGURE 8.61 Resistive and reactive parts of crystal resonator network.

The equivalent-circuit resistance R_e and reactance X_e given by Eqs. (8.244) and (8.245), respectively, are plotted in Figure 8.61 as functions of frequency over its wide range. It is seen that reactance X_e is zero at two values of frequency where the lower frequency is just slightly above the series resonant frequency ω_s and upper frequency corresponds to a nearly maximum resistance R_e. Equation (8.245) can be simplified to

$$X_e(\omega) \approx \frac{2}{\omega C_1} \frac{\Delta f}{f} \tag{8.246}$$

which is used as a measure of the slope of the curve at the operating frequency f. The slope of X_e versus frequency is a measure of crystal resonator *stiffness* [72]. The resonator quality factor Q_x is defined as

$$Q_x = \frac{1}{\omega C_1 R_1} \tag{8.247}$$

the inverse value of which is a measure of loss in the resonator. Therefore, the resonator (or unloaded) quality factor is always higher than the oscillator (or loaded) quality factor.

Generally, the crystal oscillators can be divided into two basic groups. In the first group, the crystal resonator represents an inductive reactance, whereas, in the second group, it is used as a series resonant circuit included into the feedback loop. The basic and most widely used crystal oscillator circuit corresponding to the first group is shown in Figure 8.62(a), where the crystal resonator is included between collector and base terminals. It is called the *Pierce oscillator* in the honor of George W. Pierce, whose modified and simplified single-stage crystal oscillator circuit became the touchstone of the radio communication art. It is also possible to include crystal oscillator between the base and emitter terminals, as shown in Figure 8.62(b), and between the collector and emitter terminals, as shown in Figure 8.62(c). In both latter cases, the crystal oscillator represents an inductive three-point circuit, which is a crystal analog of a Hartley LC oscillator, unlike of a Colpitts LC analog

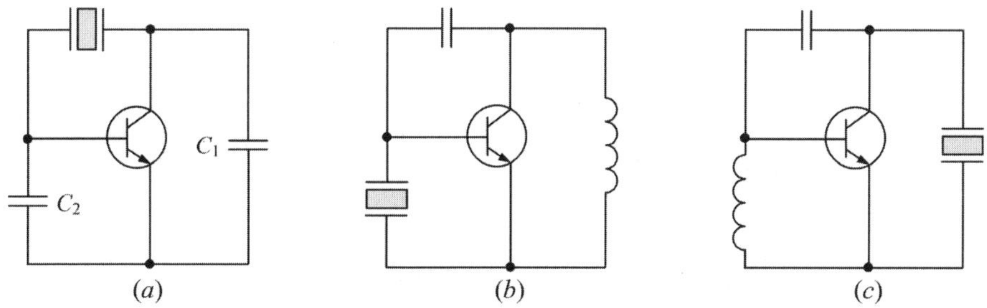

FIGURE 8.62 Simplified oscillator circuits with crystal resonator.

of a crystal Pierce oscillator. Adding additional capacitance across the crystal will cause the parallel resonance to shift downward. This can be used to adjust the frequency at which a crystal oscillator oscillates.

For an ideal Pierce oscillator shown in Figure 8.62(*a*) when the active device is regarded as an ideal and loss in crystal is neglected, the phase balance condition can be written as

$$X_1(\omega_{\rm osc}) + X_2(\omega_{\rm osc}) + X_{\rm e}(\omega_{\rm osc}) = 0 \tag{8.248}$$

where $\omega_{\rm osc}$ is the oscillation frequency, $X_1 = 1/(\omega_{\rm osc}C_1)$, $X_2 = 1/(\omega_{\rm osc}C_2)$, and $X_{\rm e}(\omega_{\rm osc})$ is defined by Eq. (8.245). The graphical solution of Eq. (8.248) is shown in Figure 8.61, where there are two points, in which a curve $-(X_1 + X_2)$ intersects a curve $X_{\rm e}$. In this case, point A corresponds to the oscillation frequency $\omega_{\rm osc}$ with a low resistance $R_{\rm e}$, whereas point B corresponds to the frequency with a high resistance $R_{\rm e}$ where the amplitude balance condition is not satisfied and build-up of self-oscillations can never happen. An analytical solution of Eq. (8.248) determines the frequency of self-oscillation equal to

$$\omega_{\rm osc} = \omega_{\rm s}\left[1 + \frac{1}{2}\frac{C_1}{C_0 + C_2C_3/(C_2 + C_3)}\right]. \tag{8.249}$$

Most types of the quartz vibration may be used on different overtones or harmonics. Among them, the most popular by far are the thickness vibrations of plates such as AT- and BT-cut of quartz, where the thickness determines the fundamental frequency according to Eq. (8.239). Odd overtones of the fundamental frequency may be excited in resonator, and Figure 8.63(*a*) represents the equivalent circuit of a family of overtones associated with the fundamental, where $C_{\rm N} = C_1/N^2$, $L_{\rm N} = L_1$, and $R_{\rm N} = R_1N^2$, $N = 1, 3, 5, \ldots$ is the overtone number [72]. It should be noted that overtones are used because the smaller $C_{\rm N}$ values lead to greater frequency stability due to increasing reactance slope with increasing N. In order to operate at one of the crystal overtones, the oscillator schematic should be slightly changed. In this case, an additional inductor L can be connected in parallel to a capacitor C_1, as shown in Figure 8.63(*b*). The resonant frequency of LC_1 circuit is chosen to be lower than the required operating frequency but higher than the nearest overtone. As a result, this circuit represents a capacitive reactance at the operating frequency, providing a conventional capacitive three-point oscillator configuration. However, at lower frequencies, the LC_1 circuit represents an inductive reactance when the phase balance condition is not satisfied. When the crystal is connected as the transistor emitter load impedance, such a harmonic emitter coupled crystal oscillator can provide very good short-term frequency stability [75,76].

To design crystal oscillator at higher frequencies, it is necessary to maximize its regeneration factor, which can be increased by minimizing the effect of the internal transistor feedback collector-base capacitance, thus resulting in a higher device maximum oscillation frequency. This can be done by using a cascode crystal oscillator configuration, simplified schematic of which is shown in

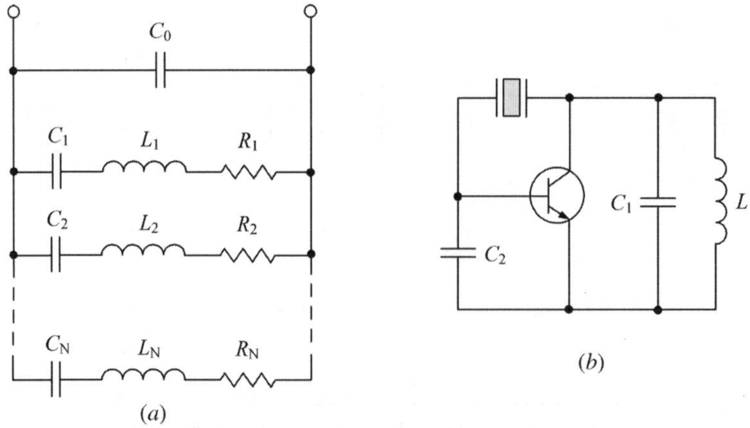

FIGURE 8.63 Crystal oscillator overtone operation.

Figure 8.64(*a*) [77]. In this case, the small-value collector–emitter capacitance of a common base transistor is connected in series with the high-value collector–base capacitance of a common emitter transistor. As a result, the operating frequencies of a crystal oscillator using the fundamental quartz or SAW resonator can be extended up to 1 GHz by using the same transistors.

In addition to the oscillator configuration where the crystal resonator operates as an inductance, there is a popular member of a family of the crystal oscillators where the crystal resonator is used as a series resonant circuit. Such a crystal oscillator, the circuit configuration of which is shown in Figure 8.64(*b*), is called the *Butler or Bridged-T oscillator* [72]. An operation principle of this oscillator is based on the fact that the real part of the crystal impedance at its series resonant frequency ω_s is minimal and drastically increases at the frequencies far from ω_s. Therefore, if a crystal resonator is included into the feedback loop, for example, between the external *LC* resonant circuit and the transistor emitter, as shown in Figure 8.64(*b*), then, at frequencies close to the series resonant frequency ω_s, the feedback loop becomes a closed loop and soft build-up of the self-oscillations occurs. At the same time, at frequencies different from ω_s, the real part of the series crystal impedance is high, resulting in a small regeneration factor, and the amplitude oscillation condition is not satisfied. Such an operating mode has the advantage that its amplitude is optimized since the regeneration factor is maximized. However, the frequency stability is not at maximum level in this case, and to maximize frequency stability, it is necessary to include an additional reactive element (capacitance or inductance) in series to the crystal resonator, depending on the character of

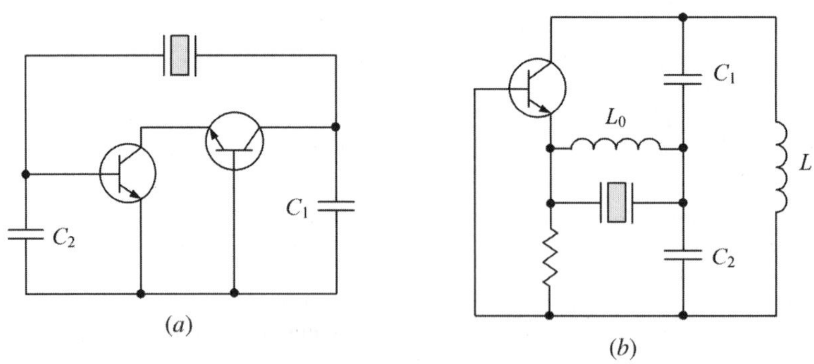

FIGURE 8.64 Schematic of cascode and Butler crystal oscillators.

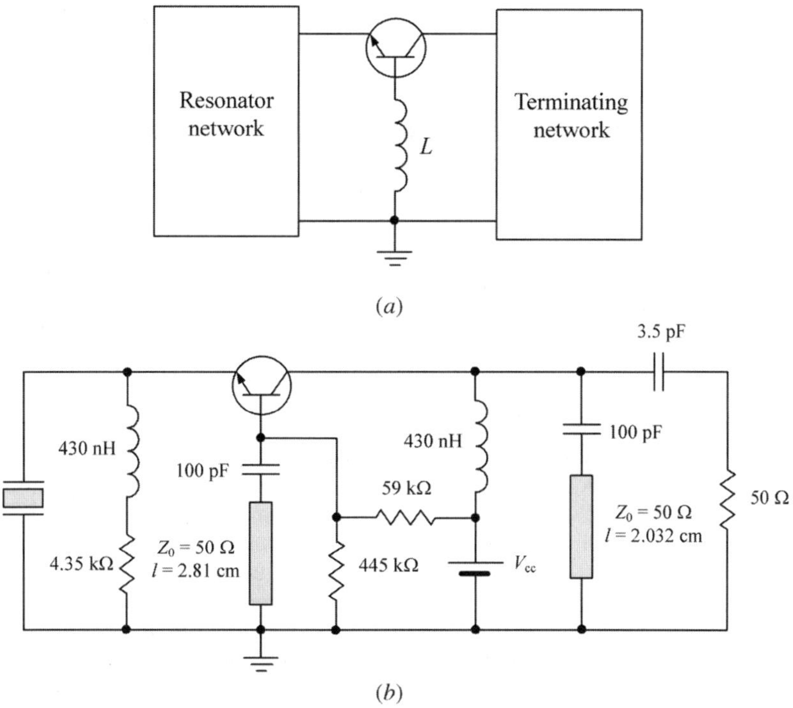

FIGURE 8.65 Schematics of negative resistance crystal oscillator.

the loading reactance of the external resonant circuit. It should be noted that the parasitic oscillations can be arisen at frequencies significantly higher than ω_s due to the crystal static capacitance C_0, whose reactance is small at these high frequencies and can result in a significant feedback with potential establishing of self-oscillations. To prevent this parasitic phenomenon, it is necessary to include an external neutralizing inductance L_0 in parallel to C_0, in order to form the parallel resonant circuit at very high frequencies.

The negative resistance approach to design a crystal oscillator can also be attractive, based on a common base transistor configuration with a series feedback inductor at the base, as shown in Figure 8.65(*a*) [78]. In this case, the parasitic capacitances of the transistor, together with the series inductive feedback obtained by a short-circuited stub, and a properly designed terminating network provides the negative resistance required at the input port, connected to the resonant network, to satisfy the oscillation conditions in a wide frequency range. Figure 8.65(*b*) shows the complete schematic of a microwave crystal oscillator operating at the fundamental resonance of the crystal resonator of 842.911 MHz. The oscillator was built using a substrate with relative dielectric constant of 4.8 and thickness of 1.5 mm. The experimental results showed that the negative resistance approach to design the crystal oscillator is well suited for its potential low-noise microwave application.

8.12 DIELECTRIC RESONATOR OSCILLATORS

It is known yet from 1930s that the dielectric object with free-space boundaries can resonate in various modes [79]. If the relative dielectric constant is high, the electric and magnetic fields of a given resonant mode will be confined in and near the resonator and will attenuate to negligible values at a distance small compared to free-space wavelength. Therefore, radiation loss is very small, and the unloaded Q of the resonance is limited mainly by losses inside the dielectric body. Since the magnetic permeability of the dielectric material used is unity, then magnetic losses are zero. Electrical losses occur as a result of the finite loss tangent ($\tan\delta$) of the dielectric material. Typical $\tan\delta$ values for dielectric materials used as resonators are about 0.0001–0.0002, thus resulting in the values of unloaded Q of about 5000–10,000 and even higher for high-purity TiO_2 disks [80]. For the fundamental-mode resonance, the dimensions of a dielectric resonator are on the order of one wavelength in the dielectric material. As a result, for the material with high relative dielectric constant, the size of the dielectric resonator can be very small. For example, for a polycrystalline TiO_2 ceramic with a relative dielectric constant ε_r of about 100, the effective wavelength in the dielectric will be one-tenth of the wavelength in air only. However, the disadvantage of dielectric resonators is significant variation of dielectric constant with temperature. For TiO2 material, the relative change of dielectric constant ε_r is about 1000 ppm/°C. Higher dielectric-constant materials such as strontium titanate ($\varepsilon_r > 250$) have much greater temperature sensitivities. Therefore, materials with dielectric constants $10 \leq \varepsilon_r \leq 100$ are generally used, such as barium tetratitanate and titanium dioxide.

The most practical shape of a dielectric resonator is a cylindrical disk whose length (thickness) L is less than its diameter D. With this shape, the lowest frequency resonant mode has a circular electric field distribution, whereas the magnetic field is strongest on the axis of the disk and at a sufficient distance outside the disk the field resembles that of an axial magnetic dipole, as shown in Figure 8.66 [81]. The axial-dipole mode offers the best separation from other resonances when the dielectric optimum ratio of L/D is about 0.4 [80]. The resonant frequency of the dielectric resonator depends on its shape, being a function of the relative dielectric constant ε_r, diameter D, and thickness L for a cylindrical resonator configuration [82].

In practical microwave resonant circuits, the dielectric resonator is placed on the alumina substrate and is confined within a metal case. In this case, the resonant circuit has a tripled-layered structure of alumina substrate, cylindrical dielectric resonator, and air gap. As a result, the resonant frequencies of such tripled-layered structures are higher than the values predicted theoretically based on the simplified assumptions. For example, the resonant frequency decreases with increasing resonator diameter and alumina substrate thickness for a fixed optimum ratio $L/D = 0.4$ [83]. Fine adjustment of resonant frequency is possible by changing air gap thickness between the resonator and the metal disk.

When a dielectric resonator is placed in the vicinity of a microstrip line on the alumina substrate, as shown in Figure 8.67(a), magnetic coupling between the resonator and microstrip line is caused, increasing if the distance l between the resonator edge and microstrip-line edge decreases [84]. In this case, the dielectric resonator operates like a reaction cavity that reflects the radiofrequency energy at

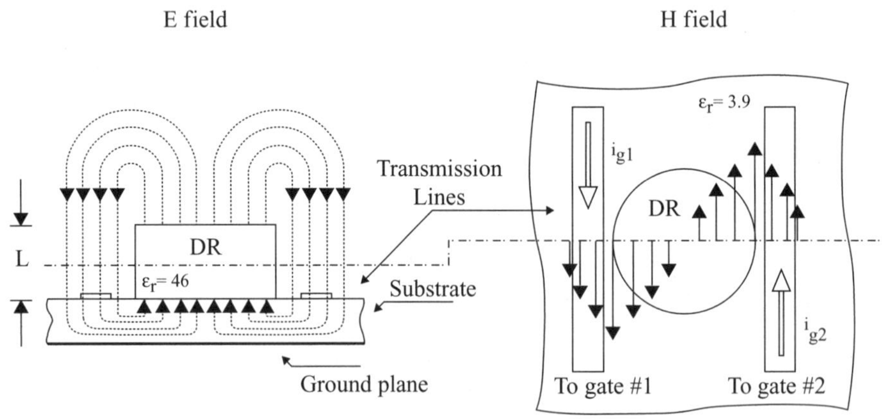

FIGURE 8.66 DR-microstrip coupling and electromagnetic fields.

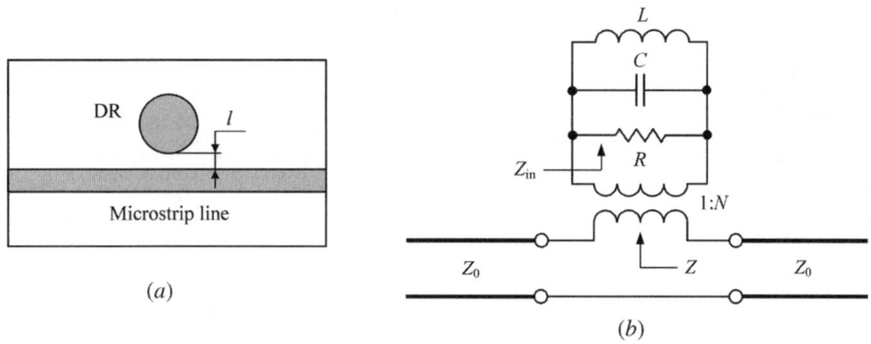

FIGURE 8.67 DR-microstrip coupling and equivalent circuit.

the resonant frequency, which is similar to an open circuit with a voltage maximum at the reference plane at the resonant frequency. Because coupling is via the magnetic filed, the dielectric resonator appears as a series load on the microstrip line, as shown in the equivalent circuit of Figure 8.67(*b*). The resonator is modeled as a parallel *RLC* circuit, and the coupling to the line is represented by the turns ratio of the equivalent transformer. Since the input impedance Z_{in} of a parallel *RLC* circuit can be written as

$$Z_{in} = \frac{R}{1 + 2jQ\dfrac{\Delta\omega}{\omega_0}} \tag{8.250}$$

where $Q = R/(\omega_0 L) = \omega_0 CR$ is the unloaded quality factor of the resonator, $\omega_0 = 1/\sqrt{LC}$ is the resonant frequency, and $\Delta\omega = \omega - \omega_0$, then the equivalent series impedance Z seen by the microstrip line is expressed as

$$Z = \frac{N^2 R}{1 + 2jQ\dfrac{\Delta\omega}{\omega_0}} \tag{8.251}$$

where N is the turns ratio of the equivalent transformer [85].

The coupling factor g between the resonator and the feedline represents the ratio of the unloaded Q to external Q_e, and can be written as

$$g = \frac{Q}{Q_e} = \frac{R}{\omega_0 L} / \frac{R_L}{N^2 \omega_0 L} = \frac{N^2 R}{2Z_0} \tag{8.252}$$

where $R_L = 2Z_0$ is the load resistance for a feedline with source and termination resistances Z_0 [86,87]. In some cases, the feedline is terminated with an open-circuit quarterwave line from the resonator to maximize the magnetic field at that point, thus resulting in $R_L = Z_0$, and the coupling factor is twice the value given in Eq. (8.252).

The reflection coefficient seen by the source on the terminated microstrip line toward the resonator can be written as

$$\Gamma = \frac{(Z_0 + N^2 R) - Z_0}{(Z_0 + N^2 R) + Z_0} = \frac{N^2 R}{2Z_0 + N^2 R} = \frac{g}{1+g} \tag{8.253}$$

which can be used to determine the coupling coefficient from the direct measurements as

$$g = \frac{1-\Gamma}{\Gamma}. \tag{8.254}$$

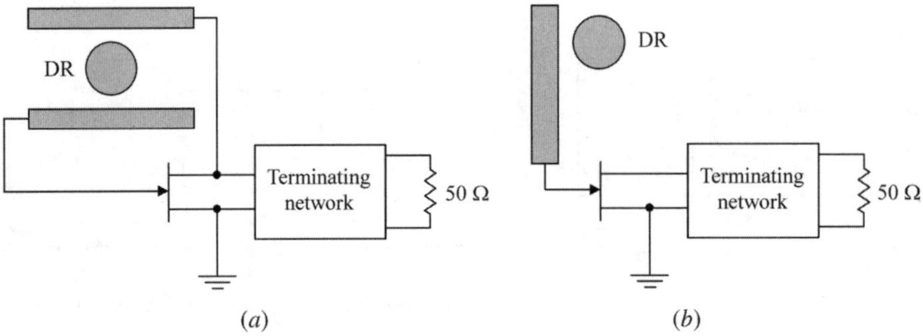

FIGURE 8.68 Schematics of dielectric resonator oscillator.

which is a strong function of the distance between the resonator and the microstrip line under fixed shielding conditions for given substrate thickness and dielectric constant [87,88]. For 60-GHz cylindrical dielectric resonators coupled to a microstrip line on a GaAs substrate, it was shown by the numerical calculation that the maximum coupling coefficient is provided when the distance between the resonator center and the microstrip line is approximately 3/5 of the resonator radius and this distance is almost independent of structural parameters [89].

A dielectric resonator can be incorporated into the oscillator circuit to improve its frequency stability using generally either the parallel feedback configuration shown in Figure 8.68(*a*) or the series feedback arrangement shown in Figure 8.68(*b*). In a parallel feedback configuration, the dielectric resonator is coupled to two microstrip lines, operating as a high-Q bandpass filter when a portion of the transistor output signal is coupled to its input. The amplitude level of coupling is controlled by the spacing between the dielectric resonator and the microstrip lines, while the phase is determined by the lengths of both microstrip lines. The series feedback configuration with a single microstrip line can be easily implemented, but typically does not have a tuning range as wide as that obtained with parallel feedback. However, placing the dielectric resonator on the gate port that is isolated from the output through the very small drain-gate capacitance minimizes interaction between the oscillator output and input circuits, thus resulting in very high loaded quality factors in a series feedback configuration. As a result, a series feedback dielectric resonator oscillator (DRO) demonstrates greater capability to achieve the lower phase noise level compared to its parallel feedback counterpart [90].

Figure 8.69(*a*) shows the series feedback bipolar oscillator circuit with a dielectric resonator ($f_0Q = 10^{14}$ Hz) coupled to the microstrip line, which is included into the base circuit [91]. To provide high-power and low-phase noise oscillator performance, a four finger InGaP/GaAs HBT device with a total emitter area of 240 μm^2 is used, resulting in a -124 dBc/Hz at 10 kHz offset from a carrier frequency of 6.7 GHz. This HBT DRO together with a 6-dB buffer amplifier was integrated on a 0.6-mm alumina substrate. At higher frequencies, the GaAs MMIC technology can be used for basic active circuit with external dielectric resonator. Figure 8.69(*b*) shows such a series feedback 10.7-GHz DRO with an oscillator chip size of 1.5×1.5 mm^2, where the dielectric resonator ($Q = 7400$ and $\varepsilon_r = 36.3$) is mounted on alumina substrate and coupled to a microstrip line terminated by a 50-Ω load [92]. The MESFET device with a gate length of 1 μm and a width of 300 μm was used to obtain an oscillator output power of 10 dBm. The value of a feedback capacitance connected to the source terminal is optimized for maximum reflection coefficient at the drain terminal. The resistor and quarterwave shunt microstrip line in the source circuit were used for a single power supply operation with chosen drain current. By using a 0.15-μm AlGaAs/InGaAs heterojunction field-effect transistor (HJFET) technology, the DRO oscillation frequency can be extended to a V band. In this case, the oscillation frequency of 59.6 GHz with an output power of 19 dBm, a phase noise of -90 dBc/Hz at 100-kHz offset and a temperature stability of 1.6 ppm/°C was achieved using a dielectric resonator with an unloaded Q of 5000 and a relative dielectric constant $\varepsilon_r = 23.8$ [93].

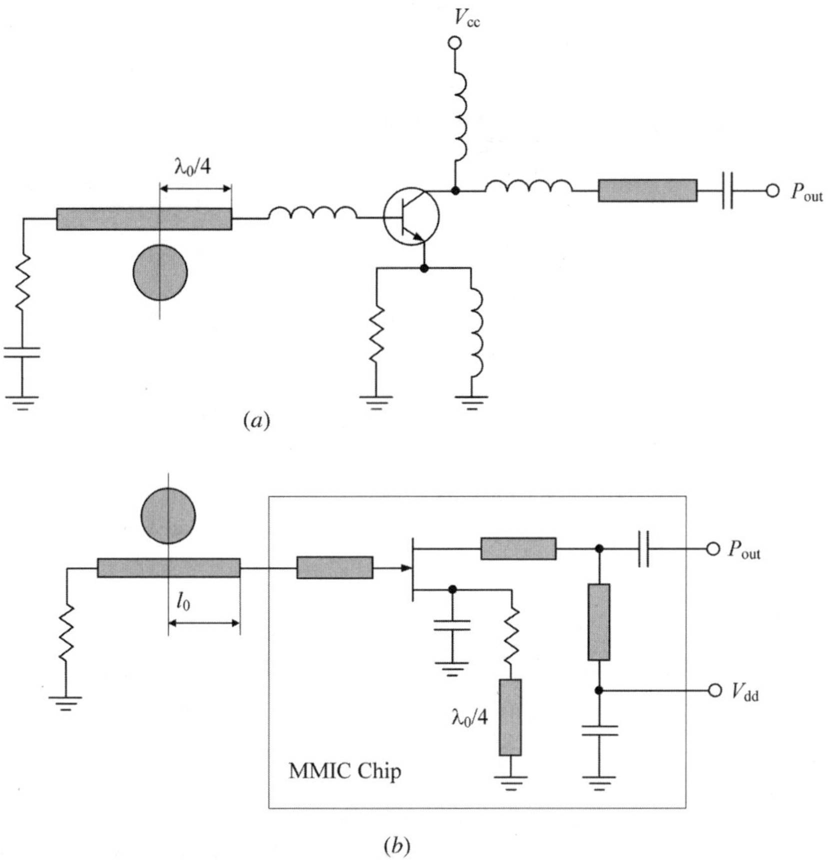

FIGURE 8.69 Schematics of series feedback dielectric resonator oscillator.

However, at high microwave frequencies, it is very difficult to effectively manufacture accurate high-Q dielectric resonators, as well as their physical handling, since a 2-GHz resonator is in the order of 1 mm in diameter. Also, at *Ka*-band frequencies and higher, the gain of the transistors for the same output power becomes marginal, and resonator coupling would need to be excessive, thus destroying the inherent Q and spectral purity of the oscillator. These problems can be partially eliminated by employing a push-push oscillator design approach. Figure 8.70 shows the push–push DRO configuration, where outputs from both transistors are combined in parallel using ordinary microstrip technique. In this case, each microstrip line has a characteristic impedance of 100 Ω and is of a half-wavelength at fundamental frequency f_0, in order to provide a parallel connection of the transistors with the output impedances of 100 Ω [94]. Since the gate circuits of each MESFET are on opposite sides of the resonator, the currents coupled at each gate will be exactly antiphased, as shown in Figure 8.66. Under these conditions, each MESFET is phase locked to the other, with their second-harmonic power combined in phase at the output of the oscillator and delivered to the 50-Ω load. A positive feedback by using a capacitive open-circuit stub is added to each device source. The level of the fundamental power at the load is usually of −25 dBc and less, relatively to the second-harmonic power, which can be maximized when MESFET is biased near pinch-off or forward conduction. The dielectric resonator must be located at the distance of $\theta = 180^\circ$ from the gates, however input device reactances should be taken into account, since they provides the effective transmission-line electrical length extension. The 50-Ω resistors at the ends of the gate transmission lines are necessary

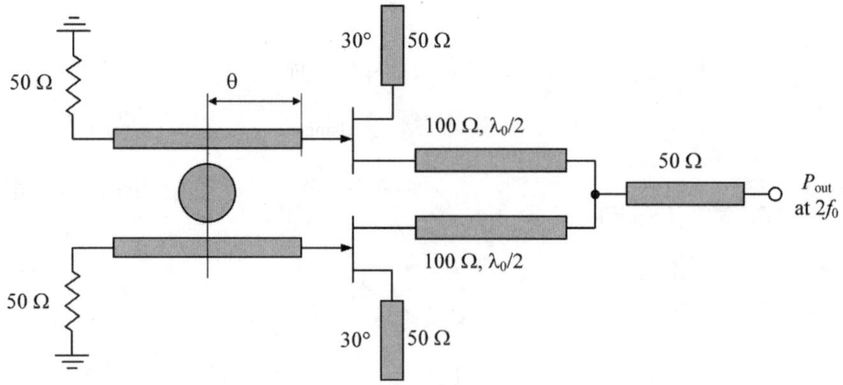

FIGURE 8.70 Schematic of push–push dielectric resonator oscillator.

to minimize unwanted parasitic oscillations. The single-sideband phase noise of such a push–push DRO at 12 GHz can reach the level of −110 dBc/Hz at 100-kHz offset from the carrier [81].

REFERENCES

1. A. A. Andronov, A. A. Vitt, and S. E. Khaikin, *Theory of Oscillations*, New York: Dover Publications, 1987.
2. A. Grebennikov, *RF and Microwave Transistor Oscillator Design*, London: John Wiley & Sons, Ltd. 2007.
3. J. Groszkowski, "The Interdependence of Frequency Variation and Harmonic Content, and the Problem of Constant-Frequency Oscillators," *Proc. IRE*, vol. 21, pp. 958–9813, July 1933.
4. E. H. Armstrong, "Some Recent Developments in the Audion Receiver," *Proc. IRE*, vol. 3, pp. 215–247, Sep. 1915; reprinted in *Proc. IEEE*, vol. 85, pp. 685–697, Apr. 1997.
5. A. Meissner, "Production of Waves by Cathode Ray Tubes," U. S. Patent 1,924,796, Aug. 1933 (filed Mar. 1914).
6. N. Demma, "Balanced Meissner Oscillator Circuits," *RF Des.*, pp. 72–74, Dec. 1993.
7. R. V. L. Hartley, "Oscillation Generator," U. S. Patent 1,356,763, Oct. 1920 (filed June 1915).
8. E. H. Colpitts, "Oscillation Generator," U. S. Patent 1,624,537, Apr. 1927 (filed Feb. 1918).
9. L. A. Hazeltine, "Oscillating Audion Circuits," *Proc. IRE*, vol. 6, pp. 63–97, Feb. 1918.
10. F. B. Llewellyn, "Constant Frequency Oscillator," U. S. Patent 1,896,781, Feb. 1933 (filed Mar. 1931).
11. J. K. Clapp, "Frequency Stable *LC* Oscillators," *Proc. IRE*, vol. 42, pp. 1295–1300, Aug. 1954.
12. J. Vackar, "Stabilization of Resonant Circuits," U. S. Patent 2,706,249, Apr. 1955 (filed Feb. 1950).
13. A. J. Cote, "Matrix Analysis of Oscillators and Transistor Applications," *IRE Trans. Circuit Theory*, vol. CT-5, pp. 181–188, Sept. 1958.
14. G. D. Vendelin, A. M. Pavio, and U. L. Rohde, *Microwave Circuit Design Using Linear and Nonlinear Techniques*, New York: John Wiley & Sons, 2005.
15. K. L. Kotzebue, "A Technique for the Design of Microwave Transistor Oscillators," *IEEE Trans. Microwave Theory Tech.*, vol. MTT-32, pp. 719–721, July 1984.
16. Z. Ding and K. Chang, "Modes and Their Stability of a Symmetric Two-Element Coupled Negative Conductance Oscillator Driven Spatial Power Combining Array," *IEEE Trans. Microwave Theory Tech.*, vol. MTT-44, pp. 1628–1636, Oct. 1996.
17. K. D. Stephan, "Inter-Injection-Locked Oscillators for Power Combining and Phased Arrays," *IEEE Trans. Microwave Theory Tech.*, vol. MTT-34, pp. 1017–1025, Oct. 1986.
18. K. W. Kobayashi, A. K. Oki, L. T. Tran, J. C. Cowles, A. Gutierrez-Aitken, F. Yamada, T. R. Block, and D. C. Streit, "A 108-GHz InP-HBT Monolithic Push-Push VCO with Low Phase Noise and Wide Tuning Bandwidth," *IEEE J. Solid-State Circuits*, vol. SC-34, pp. 1225–1231, Sept. 1999.

19. A. Shipow, "Linearity in Solid State Microwave Voltage Tuned Oscillators," *Microwave J.*, vol. 26, pp. 130–137, Feb. 1983.

20. J. R. Bender and C. Wong, "Push-Push Design Extends Bipolar Frequency Range," *Microwave & RF*, vol. 22, pp. 91–98, Oct. 1983.

21. H. Xiao, T. Tanaka, and M. Aikawa, "Push-Push Oscillator Using $\lambda_g/2$ Microstrip Resonator," *IEICE Trans. Electron.*, vol. E86-C, pp. 1438–1443, Aug. 2003.

22. K. Kurokawa, "Some Basic Characteristics of Broadband Negative Resistance Oscillator Circuits," *Bell Syst. Tech. J.*, vol. 48, pp. 1937–1955, July 1969.

23. K. M. Johnson, "Large Signal GaAs MESFET Oscillator Design," *IEEE Trans. Microwave Theory Tech.*, vol. MTT-27, pp. 217–227, Mar. 1979.

24. H. Abe, "A GaAs MESFET Oscillator Using Quasi-Linear Design Method," *IEEE Trans. Microwave Theory Tech.*, vol. MTT-34, pp. 19–25, Jan. 1986.

25. Y. Tajima, B. Wrona, and K. Mishima, "GaAs FET Large-Signal Model and Its Application to Circuit Design," *IEEE Trans. Electron Devices*, vol. ED-28, pp. 171–175, Feb. 1981.

26. R. J. Gilmore and F. J. Rosenbaum, "An Analytic Approach to Optimum Oscillator Design Using *S*-Parameters," *IEEE Trans. Microwave Theory Tech.*, vol. MTT-31, pp. 633–639, Aug. 1983.

27. D. F. Page, "A Design Basis for Junction Transistor Oscillator Circuits," *Proc. IRE*, vol. 46, pp. 1271–1280, Nov. 1958.

28. M. Madihian and T. Noguchi, "An Analytical Approach to Microwave GaAs FET Oscillator Design," *NEC Res. Dev.*, no. 82, pp. 49–56, July 1986.

29. A. V. Grebennikov and V. V. Nikiforov, "An Analytic Method of Microwave Transistor Oscillator Design," *Int. J. Electron.*, vol. 83, pp. 849–858, Dec. 1997.

30. M. Maeda, K. Kimura, and H. Kodera, "Design and Performance of *X*-Band Oscillators with GaAs Schottky-Gate Field-Effect Transistors," *IEEE Trans. Microwave Theory Tech.*, vol. MTT-23, pp. 661–667, Aug. 1975.

31. J. F. Gibbons, "An Analysis of the Modes of Operation of a Simple Transistor Oscillator," *Proc. IRE*, vol. 49, pp. 1383–1390, Sep. 1961.

32. A. V. Grebennikov, "Microwave Transistor Oscillators: Analytic Approach to Simplify Computer-Aided Design," *Microwave J.*, vol. 42, pp. 292–300, May 1999.

33. J.-M. Dortu, J.-E. Muller, M. Pirola, and G. Ghione, "Accurate Large-Signal GaAs MESFET and HEMT Modeling for Power MMIC Amplifier Design," *Int. J. Microw. Millimet. Wave Computer-Aided Eng.*, vol. 5, pp. 195–209, May 1995.

34. A. V. Grebennikov, "Microwave FET Oscillators: Analytic Approach to Simplify Computer-Aided Design," *Microwave J.*, vol. 43, pp. 100–110, Jan. 2000.

35. D. B. Leeson, "A Simple Model of Feedback Oscillator Noise Spectrum," *Proc. IEEE*, vol. 54, pp. 329–330, Feb. 1966.

36. D. Scherer, "Today's Lesson – Learn about Low-Noise Design," *Microwaves*, vol. 18, pp. 116–122, Apr. 1979.

37. T. E. Parker, "Characteristics and Sources of Phase Noise in Stable Oscillators," *Proc. 41st Annu. Freq. Control Symp.*, pp. 99–110, 1987.

38. D. J. Healy III, "Flicker of Frequency and Phase and White Frequency and Phase Fluctuations in Frequency Sources," *Proc. 26th Annu. Freq. Control Symp.*, pp. 43–49, 1972.

39. J. K. A. Everard, "A Review of Low Noise Oscillator: Theory and Design," *Proc. IEEE Int. Frequency Control Symp.*, pp. 909–918, 1997.

40. T. E. Parker, "Current Developments in SAW Oscillator Stability," *Proc. 31st Annu. Freq. Control Symp.*, pp. 359–364, 1977.

41. H. J. Siweris and B. Schiek, "Analysis of Noise Upconversion in Microwave FET Oscillators," *IEEE Trans. Microwave Theory Tech.*, vol. MTT-33, pp. 223–242, Mar. 1985.

42. G. Sauvage, "Phase Noise in Oscillators: A Mathematical Analysis of Leeson's Model," *IEEE Trans. Instrum. Meas.*, vol. IM-26, pp. 408–410, Dec. 1977.

43. T. Hajder, "Higher Order Loops Improve Phase Noise of Feedback Oscillators," *Appl. Microwave Wireless*, vol. 14, pp. 24–31, Oct. 2002.

44. S. Hamilton, "FM and AM Noise in Microwave Oscillators," *Microwave J.*, vol. 21, pp. 105–109, June 1978.

45. F. Herzel, "An Analytical Model for the Power Spectral Density of a Voltage-Controlled Oscillator and Its Analogy to the Laser Linewidth Theory," *IEEE Trans. Circuits Syst.—I: Fund. Theor. Appl.*, vol. CAS-I-45, pp. 904–908, Sept. 1998.
46. F. Herzel, K. Hennenberger, and W. Vogel, "The Semiconductor Laser Linewidth: A Green's Function Approach," *IEEE J. Quantum Electron.*, vol. QE-29, pp. 2891–2897, Dec. 1993.
47. W. A. Edson, "Noise in Oscillators," *Proc. IRE*, vol. 48, pp. 1454–1466, Aug. 1960.
48. A. van der Ziel, "Noise in Solid-State Devices and Lasers," *Proc. IEEE*, vol. 58, pp. 1178–1206, Aug. 1970.
49. K. A. Samoylo, *Method of Analysis of Second-Order Oscillation Systems* (in Russian), Moskva: Sov. Radio, 1976.
50. T. Ohira, "Higher-Order Analysis on Phase Noise Generation in Varactor-Tuned Oscillators – Baseband Noise Upconversion in GaAs MESFET Oscillators," *IEICE Trans. Electron.*, vol. E76-C, pp. 1851–1854, Dec. 1993.
51. S. Lee and K. J. Webb, "The Influence of Transistor Nonlinearities on Noise Properties," *IEEE Trans. Microwave Theory Tech.*, vol. MTT-53, pp. 1314–1321, Apr. 2005.
52. A. Hajimiri and T. H. Lee, "A General Theory of Phase Noise in Electrical Oscillators," *IEEE J. Solid-State Circuits*, vol. SC-33, pp. 179–194, Feb. 1998.
53. A. Demir, A. Mehrotra, and J. Roychowdhury, "Phase Noise in Oscillators: A Unifying Theory and Numerical Methods for Characterization," *IEEE Trans. Circuits Syst.—I: Fund. Theory Appl.*, vol. CAS-I-47, pp. 655–673, Apr. 2000.
54. T. H. Lee and A. Hajimiri, "Oscillator Phase Noise: A Tutorial," *IEEE J. Solid-State Circuits*, vol. SC-35, pp. 326–336, Mar. 2000.
55. J.-C. Nallatamby, M. Prigent, and J. Obregon, "On the Role of the Additive and Converted Noise in the Generation of Phase Noise in Nonlinear Oscillators," *IEEE Trans. Microwave Theory Tech.*, vol. MTT-53, pp. 901–906, Mar. 2005.
56. P. Vanassche, G. Gielen, and W. Sansen, "On the Difference between Two Widely Publicized Methods for Analyzing Oscillator Phase Behavior," *Proc. IEEE/ACM Int. Conf. Comput. Aided Des.*, pp. 229–233, 2002.
57. M. A. Margarit, J. L. Tham, R. G. Meyer, and M. J. Deen, "A Low-Noise, Low-Power VCO with Automatic Amplitude Control for Wireless Applications," *IEEE J. Solid-State Circuits*, vol. SC-34, pp. 761–771, June 1999.
58. D. Kajfez and E. J. Hwan, "Linearity Limits of the Varactor-Controlled Osc-Mod Circuits," *IEEE Trans. Microwave Theory Tech.*, vol. MTT-33, pp. 620–625, July 1985.
59. C. S. Aitchison and R. V. Gelsthorpe, "A Circuit Technique for Broadbanding the Electronic Tuning Range of Gunn Oscillators," *IEEE J. Solid-State Circuits*, vol. SC-12, pp. 21–28, Feb. 1977.
60. W. H. Lockyear, "Linearize VCOs by Reactance Compensation," *Microwave J.*, vol. 23, pp. 60–68, Feb. 1980.
61. C.-H. Lee, S. Han, B. Matinpour, and J. Laskar, "A Low Phase Noise X-band MMIC GaAs MESFET VCO," *IEEE Microwave Guided Wave Lett.*, vol. 10, pp. 325–327, Aug. 2000.
62. M. J. Howes and D. V. Morgan, *Microwave Devices*, New York: John Wiley & Sons, 1976.
63. A. V. Grebennikov, "Wideband VCO Designs Are Independent of Circuit Parameters," *Microwaves & RF*, vol. 40, pp. 147–155, Aug. 2001.
64. H.-N. Toussaint and P. Olfs, "Transistor Power Oscillator, Electronically Tunable from 250 to 500 MHz," *Proc. IEEE*, vol. 56, pp. 226–227, Feb. 1968.
65. T. Ohira, T. Hiraoka, and H. Kato, "MMIC 14-GHz VCO and Miller Frequency Divider for Low-Noise Local Oscillators," *IEEE Trans. Microwave Theory Tech.*, vol. MTT-35, pp. 657–661, July 1987.
66. B. N. Scott, M. Wurtele, and B. B. Creiger, "A Family of Four Monolithic VCO MIC's Covering 2–18 GHz," *1984 IEEE Microwave and Millimeter-Wave Monolithic Circuits Symp. Dig.*, pp. 58–61.
67. N. Itoh, "Low Voltage Low Phase Noise CMOS VCO and Its Flicker Noise Influence," *IEICE Trans. Electron.*, vol. E86-C, pp. 1062–1068, June 2003.
68. N. H. W. Fong, J.-O. Plouchart, N. Zamdmer, D. Liu, L. F. Wagner, C. Plett, and G. Tarr, "A Low-Voltage Multi-GHz VCO with 58% Tuning Range in SOI CMOS," *Proc. 2002 IEEE Custom Integrated Circuits Conf.*, pp. 423–426.
69. A. M. Nicolson, "Piezophony," U. S. Patent 1,495,429, May 1924 (filed 1918).

70. W. G. Cady, "Piezo Electric Resonator," U. S. Patent 1,450,246, Apr. 1923 (filed 1920).

71. Application Note 200–2, *Fundamentals of Quartz Oscillators*, Hewlett-Packard, 1997.

72. B. Parzen, *Design of Crystal and Other Harmonic Oscillators*, New York: John Wiley & Sons, 1983.

73. J. R. Hunt and R. C. Smythe, "Chemically Milled VHF and UHF AT-Cut Resonators," *Proc. 39th Annu. Freq. Control Symp.*, pp. 292–300, 1985.

74. M. Umeki, T. Sato, H. Uehara, and M. Okazaki, "622 MHz High Frequency Fundamental Composite Crystal Resonator with an Air-Gapped Electrode," *Proc. IEEE Int. Frequency Control Symp.*, pp. 365–368, 2004.

75. M. M. Driscoll, "Two-Stage Self-Limiting Series Mode Quartz-Crystal Oscillator Exhibiting Improved Short-Term Frequency Stability," *IEEE Trans. Instrum. Meas.*, vol. IM-22, pp. 130–138, June 1973.

76. R. J. Matthys, *Crystal Oscillator Circuits*, Malabar, FL: Krieger Publishing, 1992.

77. N. Nomura, Y. Aoyagi, C.-K. Chang, K. Asano, and Y. Sekine, "A Colpitts-Type Crystal Oscillator for Gigahertz Frequency," *Proc. IEEE Int. Freq. Control Symp.*, pp. 233–236, 2006.

78. G. Gonzalez and B. Avanic, "Design of Crystal Oscillators with Fundamental Frequencies in the UHF and Microwave Regions," *Electron. Lett.*, vol. 23, pp. 1379–1381, Dec. 1987.

79. R. D. Richtmeyer, "Dielectric Resonators," *J. Appl. Phys.*, vol. 10, pp. 391–398, June 1939.

80. S. B. Cohn, "Microwave Bandpass Filters Containing High-Q Dielectric Resonators," *IEEE Trans. Microwave Theory Tech.*, vol. MTT-16, pp. 218–227, Apr. 1968.

81. J.-F. Gravel and J. S. Wight, "On the Conception and Analysis of a 12-GHz Push-Push Phase-Locked DRO," *IEEE Trans. Microwave Theory Tech.*, vol. MTT-54, pp. 153–159, Jan. 2006.

82. J. C. Sethares and S. J. Naumann, "Design of Microwave Dielectric Resonators," *IEEE Trans. Microwave Theory Tech.*, vol. MTT-14, pp. 2–7, Jan. 1966.

83. H. Abe, Y. Takayama, A. Higashisaka, and H. Takamizawa, "A Highly Stabilized Low-Noise GaAs FET Integrated Oscillator with a Dielectric Resonators in the *C* Band," *IEEE Trans. Microwave Theory Tech.*, vol. MTT-26, pp. 156–162, Mar. 1978.

84. J. K. Plourde and C. L. Ren, "Applications of Dielectric Resonators in Microwave Components," *IEEE Trans. Microwave Theory Tech.*, vol. MTT-29, pp. 754–770, Aug. 1981.

85. Y. Komatsu and Y. Murakami, "Coupling Coefficient between Microstrip Line and Dielectric Resonator," *IEEE Trans. Microwave Theory Tech.*, vol. MTT-31, pp. 34–40, Jan. 1983.

86. D. Kajfez and P. Guillon, *Dielectric Resonators*, New York: Artech House, 1986.

87. D. M. Pozar, *Microwave Engineering*, New York: John Wiley & Sons, 2004.

88. A. Khanna and Y. Garault, "Determination of Loaded, Unloaded, and External Quality Factors of a Dielectric Resonator Coupled to a Microstrip Line," *IEEE Trans. Microwave Theory Tech.*, vol. MTT-31, pp. 261–264, Mar. 1983.

89. K. Hosoya, T. Inoue, M. Funabashi, and K. Ohata, "Systematic Evaluation and Analysis for 60-GHz Dielectric Resonators Coupled to a Microstrip Line on a GaAs Substrate," *IEEE Trans. Microwave Theory Tech.*, vol. MTT-46, pp. 352–358, Apr. 1998.

90. M. Regis, O. Llopis, and J. Graffeuil, "Nonlinear Modeling and Design of Bipolar Transistors Ultra-Low Phase-Noise Dielectric-Resonator Oscillators," *IEEE Trans. Microwave Theory Tech.*, vol. MTT-46, pp. 1589–1593, Oct. 1998.

91. S. Perez, D. Floriot, P. Maurin, P. M. Gutierrez, J. Obregon, and S. L. Delage, "Extremely Low Noise InGaP/GaAs HBT Oscillator at C-band," *Electron. Lett.*, vol. 34, pp. 813–814, Apr. 1998.

92. S. Hori, K. Kamei, K. Shibata, M. Tatematsu, K. Mishima, and S. Okano, "GaAs Monolithic MIC's for Direct Broadcast Satellite Receivers," *IEEE Trans. Microwave Theory Tech.*, vol. MTT-31, pp. 1089–1096, Dec. 1983.

93. K. Hosoya, K. Ohata, M. Funabashi, T. Inoue, and M. Kuzuhara, "V-Band HJFET MMIC DROs with Low Phase Noise, High Power, and Excellent Temperature Stability," *IEEE Trans. Microwave Theory Tech.*, vol. MTT-51, pp. 2250–2258, Nov. 2003.

94. A. M. Pavio and M. A. Smith, "A 20–40-GHz Push–Push Dielectric Resonator Oscillator," *IEEE Trans. Microwave Theory Tech.*, vol. MTT-33, pp. 1346–1349, Dec. 1985.

9 Phase-Locked Loops

This chapter begins with introduction of the basic phase-locked loop concept. Then, the basic performance and structures of the analog, charge-pump, and digital phase-locked loops are analyzed. The basic loop components such as phase detector, loop filter, frequency divider, and voltage-controlled oscillator are discussed, as well as loop dynamic parameters. The possibility and particular realizations of the phase modulation using phase-locked loops are presented. Finally, general classes of frequency synthesizer techniques such as direct analog synthesis, indirect synthesis, and direct digital synthesis are discussed. The proper choice of the synthesizer type is based on the number of frequencies, frequency spacing, frequency switching time, noise, spurious level, particular technology, and cost.

9.1 BASIC LOOP STRUCTURE

The basic phase-locked loop (PLL) concept has been well known and widely used since it was proposed and described by Appleton and later analyzed by Bellescize [1,2]. Starting with the rapid development of integrated circuits in the 1970s, the PLL had become an important part of modern communication systems, improving their performance and reliability. A PLL circuit responds to both the frequency and the phase of the input signal, automatically raising or lowering the frequency of a voltage-controlled oscillator until it is synchronized with the reference in both frequency and phase. In other words, the PLL controls the phase of its RF output signal in such a way that the phase error between output phase and reference phase reduces to a minimum. The basic structure of a PLL is shown in Figure 9.1, which consists of a phase detector (PD), a loop filter (LF), and a voltage-controlled oscillator (VCO). The PD compares the phase of the input reference signal with that of the VCO, and its output voltage is filtered and applied to the VCO whose output frequency moves in the direction so as to reduce the phase difference of both input and output signals. As a result, when the frequency is locked, the frequency of the VCO is exactly equal to the frequency of the reference oscillator.

To describe the general electrical behavior of the PLL, consider the input reference signal v_{in} and output signal v_{out} expressed as

$$v_{in}(t) = V_{in}\cos(\omega_{in}t + \theta_{in}) \tag{9.1}$$

$$v_{out}(t) = V_{out}\cos(\omega_{out}t + \phi_{out}) \tag{9.2}$$

where V_{in} and V_{out} are the amplitudes, ω_{in} and ω_{out} are the angular frequencies, and θ_{in} and ϕ_{out} are the phase constants of the input reference and output VCO signals, respectively [3,4]. If the loop is initially unlocked and the PD operating as a signal multiplier has a sinusoidal transfer characteristic, the error signal v_e at its output is written as

$$v_e(t) = K_d\left\{\cos\left[(\omega_{in} - \omega_{out})t + (\theta_{in} - \phi_{out})\right] + \cos\left[(\omega_{in} + \omega_{out})t + (\theta_{in} + \phi_{out})\right]\right\} \tag{9.3}$$

RF and Microwave Transmitter Design, First Edition. Andrei Grebennikov.

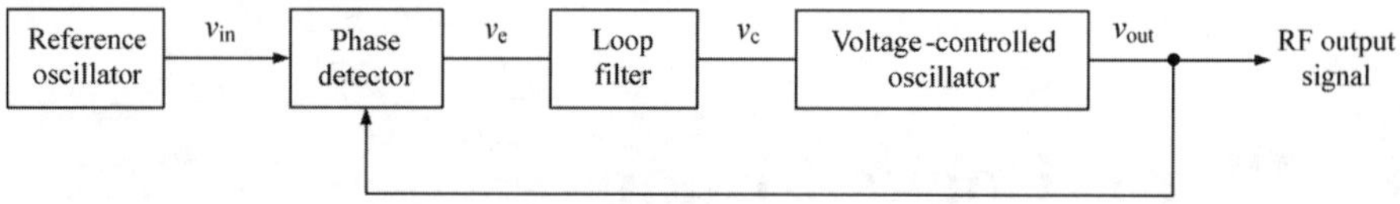

FIGURE 9.1 Basic structure of phase-locked loop.

where $K_d = V_{in}V_{out}/2$ is the gain factor of the PD. Since the higher frequency component $\omega_{in} + \omega_{out}$ is eliminated by the ideal loop low-pass filter (LPF) used, the signal v_c at the LPF output can be given by

$$v_c(t) = K_d \cos\left[(\omega_{in} - \omega_{out})\,t + (\theta_{in} - \phi_{out})\right]. \tag{9.4}$$

When the loop is locked after a period of time sufficiently long for transient, the VCO signal v_{out} becomes synchronous with the input reference signal v_{in}, and it can be written as

$$v_{out}(t) = V_{out} \cos(\omega_{in}t + \varphi_{out}) \tag{9.5}$$

where φ_{out} is the initial phase. Comparing Eqs. (9.2) and (9.5) shows that the phase ϕ_{out} in Eq. (9.2) is a linear function of time expressed by

$$\phi_{out} = (\omega_{in} - \omega_{out})\,t + \varphi_{out} \tag{9.6}$$

and the LPF output signal v_c becomes a dc signal given by

$$v_c(t) = K_d \cos(\theta_{in} - \varphi_{out}). \tag{9.7}$$

The instantaneous angular frequency ω_{osc} of the VCO operating as a frequency-modulated oscillator represents a linear function of the control signal v_c around the center angular frequency ω_{out} according to

$$\omega_{osc} = \frac{d}{dt}(\omega_{out}t + \phi_{out}) = \omega_{out} + K_{VCO}v_c(t) \tag{9.8}$$

and

$$\frac{d\phi_{out}}{dt} = K_{VCO}v_c(t) \tag{9.9}$$

where K_{VCO} is the VCO sensitivity measured in radian per second per volt and ω_{out} is the frequency that occurs when the PD is in the center of its output transfer characteristic or the frequency midway between the frequencies that occur at the edges of the PD range.

From Eqs. (9.6), (9.7), and (9.9) it follows that

$$\omega_{in} - \omega_{out} = K_d K_{VCO} \cos(\theta_{in} - \varphi_{out}) \tag{9.10}$$

resulting in

$$\varphi_{out} = \theta_{in} - \cos^{-1}\frac{\omega_{in} - \omega_{out}}{K_d K_{VCO}}. \tag{9.11}$$

Substituting Eq. (9.11) into Eq. (9.7) yields

$$v_c = \frac{\omega_{in} - \omega_{out}}{K_{VCO}} \tag{9.12}$$

which clearly shows that it is the dc signal v_c that changes the VCO frequency from its central value ω_{out} to the input signal angular frequency ω_{in} according to

$$\omega_{osc} = \omega_{out} + K_{VCO} v_c = \omega_{in}. \tag{9.13}$$

If the angular frequency difference $\omega_{in} - \omega_{out}$ is much lower than the loop gain $K = K_d K_{VCO}$, Eq. (9.11) is simplified to $\theta_{in} - \varphi_{out} \approx \pi/2$, thus indicating that the VCO output signal is almost in phase quadrature with the input reference signal when the loop is locked, where the phase quadrature condition corresponds to $\omega_{in} = \omega_{out}$. In this case, by letting $\theta_{out} = \varphi_{out} + \pi/2$, Eq. (9.7) can be rewritten as

$$v_c(t) = K_d \sin(\theta_{in} - \theta_{out}) \tag{9.14}$$

where $\theta_e = \theta_{in} - \theta_{out}$ is the phase error between the two signals, which becomes zero when the initial frequency offset equals to zero. It should be noted that Eq. (9.14) takes into account the phase shift of 180° around the loop. For a sufficiently small phase difference $\theta_e = \theta_{in} - \theta_{out}$, the following approximation can be used

$$v_c(t) \approx K_d(\theta_{in} - \theta_{out}) \tag{9.15}$$

with the PD gain factor K_d measured in volt per radian. When the difference $|\omega_{in} - \omega_{out}|$ exceeds the loop gain K in a PD with sinusoidal transfer characteristic, synchronization can no longer maintain according to Eq. (9.11), and the loop falls out of lock.

As one of the first practical implementations, the PLL structure was used to stabilize the operating frequency of a power ultra-high-frequency (UHF) VCO developed as part of a solid-state microwave radio relay system [5]. In this case, in order to increase the reference frequency provided by a crystal oscillator, the frequency doubler and tripler with buffer amplifiers were used. Integrated PLLs are now well developed and widely used in a variety of applications such as a reference source, signal modulator, frequency-selective demodulator, or frequency synthesizer.

9.2 ANALOG PHASE-LOCKED LOOPS

The basic analog PLL structure also includes a frequency divider (DIV) between the oscillator and the PD to match the frequency of the reference signal, as shown in Figure 9.2(*a*), since it creates enough flexibility in designing of both reference oscillator and VCO with improved performance. When the system is phased-lock, the VCO is N times that of the reference frequency, where N is the frequency division ratio. Changing the value of the divider will cause the PD to sense a frequency error. In this case, the feedback will respond with a correcting voltage. The operating range is set by the maximum frequency of the divider, the division ratio of the divider, and the VCO tuning range. The PD, LF, and VCO compose the feedforward path with the feedback path containing the DIV. Removal of the DIV produces unity gain in the feedback path for $N = 1$, and the VCO output frequency is equal to the PD input frequency.

An analog PLL is generally a nonlinear system since its PD is described by a nonlinear transfer characteristic in a common case. However, when its loop is locked, a PLL can be accurately approximated by a linear model when the phase error θ_e is small and analyzed using the concepts and terminology of feedback control systems based on Laplace transform. In this case, each component

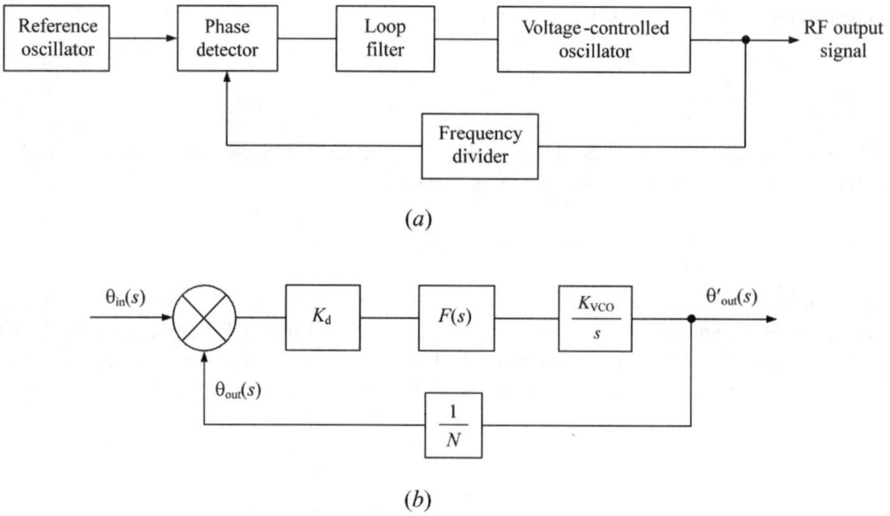

FIGURE 9.2 Typical structure of analog phase-locked loop.

of the PLL can be described by a linear transfer function. When the loop is closed, the response of any component and entire system can be expressed as the ratio of phases at corresponding output and input points. Figure 9.2(*b*) shows the typical structure of a linearized analog PLL where the PD is assumed to be sinusoidal, operating linearly over a phase range of ± 1 rad and represented by the frequency multiplier with separate gain block, $F(s)$ is the filter transfer function, and $s = j\omega$ is the complex frequency (Laplace) variable. The VCO is usually modeled by an integrator, since phase is the integral of instantaneous frequency with $K_{\rm VCO}$.

According to feedback control system analogy, the system forward gain is written as

$$G(s) = \frac{K_{\rm d} K_{\rm VCO} F(s)}{s} \tag{9.16}$$

with the open-loop transfer function or open-loop gain $GH(s) = G(s)/N$, whereas the closed-loop transfer function $H(s)$, which is a measure of the loop response to changes in the input phase or frequency, is obtained by

$$H(s) = \frac{\theta'_{\rm out}(s)}{\theta_{\rm in}(s)} = \frac{G(s)}{1 + G(s)/N} = \frac{K_{\rm d} K_{\rm VCO} F(s)}{s + K_{\rm d}(K_{\rm VCO}/N) F(s)}. \tag{9.17}$$

where the roots of $1 + G(s)/N$ are the poles of the system function, determining the transient behavior of the loop. This affects the ability of the loop to follow rapid changes in input frequency and phase.

By considering the change in output frequency produced by introducing a test frequency at various point in the PLL, all transfer functions from these points to output can be derived, which are shown in Table 9.1 [6,7]. Note that the transfer function of the LF is a major factor in loop performance. As the filter bandwidth is reduced, its response time is increased. This helps to keep the loop in lock through momentary losses of input signal, and minimizes the noise transmitted through the loop at the expense of a reduction in the capture range. In this case, the phase error function is written as

$$\theta_{\rm e}(s) = \theta_{\rm in}(s) - \theta_{\rm out}(s) = \frac{s\theta_{\rm in}(s)}{s + K_{\rm d} K_{\rm VCO} F(s)/N}. \tag{9.18}$$

TABLE 9.1 Transfer Functions for Various Analog PLL Ports.

Source to Output	Transfer Function
Reference oscillator	$\frac{G(s)}{1+G(s)/N}$
N divider	$\frac{G(s)}{1+G(s)/N}$
Phase detector	$\frac{1}{K_d}\frac{G(s)}{1+G(s)/N}$
Loop filter	$\frac{K_{VCO}}{s}\frac{1}{1+G(s)/N}$
VCO	$\frac{1}{1+G(s)/N}$

The order of the loop equals the number of poles in the open-loop transfer function $GH(s)$ defined by the highest power of s in its denominator [8]. Therefore, a PLL without LF when $F(s) = 1$ is called the *first-order PLL*, because the highest power of s in the denominator of $G(s)$ is 1. In this case, the open-loop gain of a first-order PLL is defined by the forward gain and equals to $K_d(K_{VCO}/N)/s$, where $K = K_d K_{VCO}/N$ represents a dc loop gain referred to the PD. This is a type 1 system when the phase error function is simplified to

$$\theta_e(s) = \frac{s\theta_{in}(s)}{s + K_d K_{VCO}/N} \tag{9.19}$$

where the dc loop gain is defined according to Figure 9.1 by taking into account Eqs. (9.12) and (9.15) as

$$K = \frac{v_e}{\theta_e}\frac{\Delta\omega}{v_e} = \frac{\Delta\omega}{\theta_e}. \tag{9.20}$$

Equation (9.17), which can simply be rewritten as $H(s) = 1/(1 + s/K)$ for $N = 1$, represents an LPF characteristic with a cutoff (−3 dB) frequency of $\omega = K$. Equation (9.18), rewritten for $\theta_e(s)/\theta_{in}(s)$ as $s/(s + K)$ for $N = 1$, shows that the relative error phase is characterized by a high-pass filter characteristic.

The steady-state phase error resulting from a step change of input phase of magnitude $\Delta\theta_{in}$ when $\theta_{in}(s) = \Delta\theta_{in}/s$ is derived in a stable feedback control system according to

$$\Delta\theta_e(s) = \lim_{s\to 0} s\theta_e(s) = \lim_{s\to 0}\frac{s\Delta\theta_{in}}{s + K_d K_{VCO}/N} = 0 \tag{9.21}$$

and the steady-state error resulting from a ramp input phase, or from a step change in reference frequency of magnitude $\Delta\omega$ when $\theta_{in}(s) = \Delta\omega/s^2$ is equal to

$$\Delta\theta_e(s) = \lim_{s\to 0} s\theta_e(s) = \lim_{s\to 0}\frac{\Delta\omega}{s + K_d K_{VCO}/N} = \frac{\Delta\omega N}{K_d K_{VCO}} \tag{9.22}$$

which is a constant [6]. In addition, it is important to know the first-order PLL response to a ramp change in frequency with time at a rate of $d\Delta\omega/dt$ when $\theta_{in}(s) = (1/s^3)\, d\Delta\omega/dt$. As a result,

$$\Delta\theta_e(s) = \lim_{s\to 0}\frac{d\Delta\omega/dt}{s^2 + sK_d K_{VCO}/N} = \infty \tag{9.23}$$

which means that above some critical value of change rate of reference frequency the loop will no longer stay locked or, conversely, if VCO frequency is linearly swept at a rate above a critical value to achieve locking, the latter will not occur.

These results indicate that a first-order PLL will eventually track out any step change in input phase that is within the system hold-in range and will follow a step frequency change with a phase error that is proportional to the magnitude of the frequency step and inversely proportional to the dc loop gain. The loop will behave in the same manner with respect to a phase or frequency change of the VCO signal. The time response of a first-order PLL to a step change in reference phase or frequency does not have overshoot. For example, it is associated with the exponential rise of the output phase due to a step change in the input phase with time constant equal to K. A high dc loop gain results in a fast loop response. Besides, the first-order PLL behaves as an LPF with respect to the phase noise of the reference signal and as a high-pass filter with respect to VCO noise. It should be noted that the basic first-order loop is unconditionally stable for all values of dc loop gain [3]. However, this simplified form of PLL is rarely used in frequency synthesis as having limited lock range. For narrow bandwidth, a second-order loop is most common, since a third-order loop is complicated and can lead to instability if not properly controlled [9].

To overcome some problems inherent with the second-order PLLs such as increasing the hold-in range without changing acquisition dynamics, aided acquisition loops can be used as an alternative, as shown in Figure 9.3(a) [10]. The basic principle here is to combine two feedback loops in such a way that only the frequency detector (FD) loop provides the error signal when input and output frequencies are different. If these two frequencies are sufficiently close, the PD loop begins to work, providing the required dc input to the VCO so that $\theta_{in} = \theta_{out.}$ Then, the FD loop gradually completes the acquisition and becomes inactive as the PD becomes activated. The exact dynamics of this

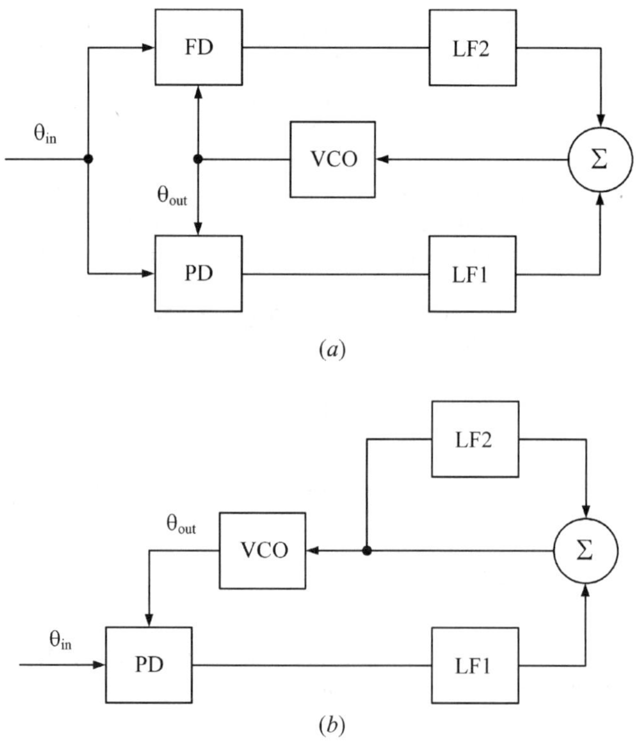

FIGURE 9.3 Phase-locked loops with aided acquisition loops.

process will be dependent on the actual implementation of detectors and LFs. However, since a PD implicitly contains the frequency information, a possible FD can be considered as a PD followed by a differentiator. Then, integration by the VCO and the differentiation operation cancel out, comprising the scaling factor that results in a PLL with an aided acquisition loop shown in Figure 9.3(*b*), where the LF1 and LF2 are LPFs, and only a conventional PD is used [11]. In this case, a second-order PLL with an aided acquisition loop can be seen as the equivalent to a third-order PLL.

9.3 CHARGE-PUMP PHASE-LOCKED LOOPS

Charge-pump PLLs based on sequential-logic phase frequency detectors are very popular due to their extended tracking range, frequency-aided acquisition, and low cost. The purpose of using charge pump (CP), which follows the phase and frequency detector (PFD), is to convert the logic states of the PFD into analog signals suitable for controlling the VCO. A CP simply represents a three-position electronic switch that is controlled by the three states of the PFD—it is either open or delivers positive and negative pump voltages or currents [12,13]. The current CP followed by a passive filter represents a convenient practical configuration that is easier to analyze. In this case, the low power performance and high integration level at microwave and millimeter-wave frequencies make complementary metal-oxide-semiconductor (CMOS) technology attractive in ultra-fast PLL designs. For example, a fully integrated PLL circuit in 90-nm CMOS technology can be designed for the 77-GHz automotive radars using a multistage DIV, a cross-coupled VCO, and a PFD based on the single-sideband mixer to suppress the reference feedthrough [14].

Figure 9.4(*a*) shows a typical block diagram of the charge-pump PLL with DIV in the feedback path. Here, the PFD compares the phase difference between the reference signal with phase θ_{in} and the VCO output signal (divided by DIV division ratio N) with phase θ_{out}, as shown in Figure 9.4(*b*). The PFD output signal represents a series of pulses whose duty cycle is proportional to this phase difference $\theta_{in} - \theta_{out}$, and the CP converts the voltage pulses into current pulses with a predetermined amplitude I_p. The LF converts in turn the current pulses into a low-pass filtered voltage signal that controls the frequency of the VCO. In this case, if the feedback is negative, the phase difference $\theta_{in} - \theta_{out}$ gradually becomes smaller and smaller approaching zero value until $\theta_{in} = \theta_{out}$, when the loop is referred to be locked. Once the loop is locked, the frequency of the VCO output signal is equal to the frequency of the reference signal multiplied by the feedback division factor N.

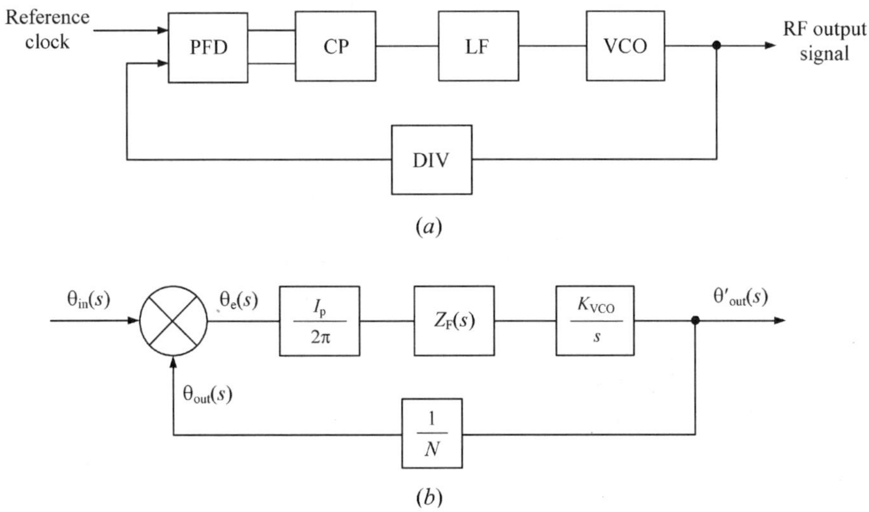

FIGURE 9.4 Charge-pump PLL block diagrams.

Since the operation of the PFD and CP is performed in the discrete-time domain, the complete transfer function of the loop becomes complicated due to the z-transform representation. A more intuitive equation can be obtained by assuming that the phase error is small when the PFD and CP can be modeled together by a simple combined gain block $I_p/2\pi$, as shown in Figure 9.4(b). As a result, the forward gain of a charge-pump PLL can be written in a linear approximation as

$$G(s) = \frac{I_p Z_F(s)}{2\pi} \frac{K_{VCO}}{s} \tag{9.24}$$

where $Z_F(s)$ is the filter transimpedance transfer function defined as a ratio of the filter output voltage to its input current, while the closed-loop transfer function H(s) is obtained by

$$H(s) = \frac{\theta'_{out}(s)}{\theta_{in}(s)} = \frac{G(s)}{1 + G(s)/N} = \frac{I_p Z_F(s) K_{VCO}}{2\pi s + I_p Z_F(s)(K_{VCO}/N)} \tag{9.25}$$

and the roots of $1 + G(s)/N$ are the poles of the system function, determining the transient behavior of the loop.

Considering the change in output frequency produced by introducing a test frequency at various points in the PLL results in the transfer function from these points to output, which are shown in Table 9.2. In this case, the phase error function is written as

$$\theta_e(s) = \theta_{in}(s) - \theta_{out}(s) = \frac{s\theta_{in}(s)}{s + \left(I_p Z_F/2\pi\right)(K_{VCO}/N)} \tag{9.26}$$

so that, by taking into account the analog PLL analogy, the steady-state phase error (or loop stress) for the charge-pump PLL, resulting from a step change of input phase of magnitude $\Delta\theta_{in}$, is defined by

$$\Delta\theta_e(s) = \lim_{s\to 0} \frac{s\Delta\theta_{in}}{s + \dfrac{I_p Z_F(s) K_{VCO}}{2\pi N}} = 0 \tag{9.27}$$

TABLE 9.2 Transfer Functions for Various Charge-Pump PLL Ports.

Source to Output	Transfer Function
Reference oscillator	$\dfrac{G(s)}{1+G(s)/N}$
N divider	$\dfrac{G(s)}{1+G(s)/N}$
Charge pump	$\dfrac{2\pi}{I_p}\dfrac{G(s)}{1+G(s)/N}$
Loop filter	$\dfrac{K_{VCO}}{s}\dfrac{1}{1+G(s)/N}$
VCO	$\dfrac{1}{1+G(s)/N}$

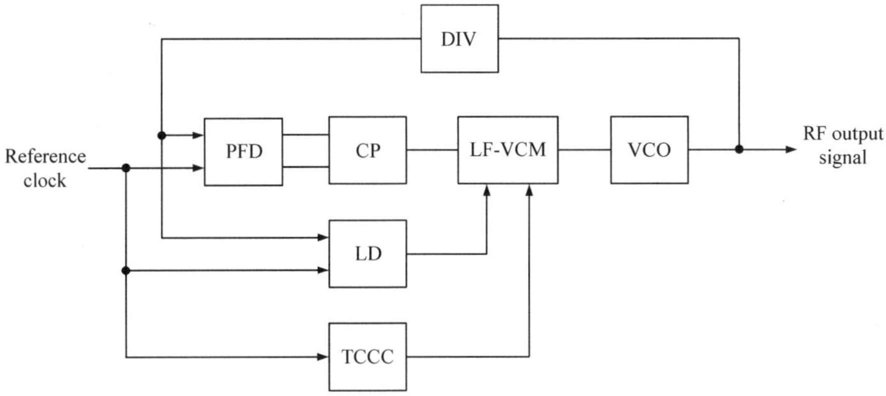

FIGURE 9.5 Charge-pump PLL with calibration circuit.

and the steady-state phase error resulting from a step change in reference frequency of magnitude $\Delta\omega$ when $\theta_{in}(s) = \Delta\omega/s^2$ is equal to

$$\Delta\theta_e(s) = \lim_{s\to 0} \frac{\Delta\omega}{s + \dfrac{I_d Z_F(s) K_{VCO}}{2\pi N}} = \frac{2\pi\,\Delta\omega N}{I_p Z_F(0)\,K_{VCO}} \tag{9.28}$$

having the same form as obtained for a conventional analog PLL [3,12]. The step response of the closed-loop transfer function shows the locking transient, with settling time performance determined from the transient waveform.

For a typical charge-pump PLL with high integration level, the process and temperature variations can alter its dynamic parameters away from the desired values. In this case, an analog time-constant calibration circuit can be used to precisely control the time constant of the LF, whereas the variable capacitance multiplier is introduced to adjust the equivalent capacitance in the LF. Figure 9.5 shows the charge-pump PLL block schematic that includes a time-constant calibration circuit (TCCC), LF with variable capacitance multiplier (LF-VCM), and a lock detector (LD) [15]. When the lock detector is turned off, the capacitance in the LF is not multiplied that equivalently enlarges the PLL loop bandwidth. However, when the lock detector is turned on, the LF capacitance is multiplied, thus resulting in a small loop bandwidth with low jitter performance. In this case, the lock time for phase acquisition can be reduced by about four times. As a result, the time constant of the LF and the dynamic parameters of the charge-pump PLL can effectively be tracked with the period of the reference clock, and a fast locking acquisition for a PLL is achieved.

9.4 DIGITAL PHASE-LOCKED LOOPS

In the early development of the PLL, the work was primarily on analog PLLs. However, with increasing emphasis on digital circuitry because of decreasing cost, increased reliability, and smaller size, significant efforts have been done to develop analog–digital (hybrid) PLL, discrete PLL, and finally digital PLL [9]. The hybrid PLL is a variation of analog PLL where one or more, but not all elements in the loop, are digital. A digital PLL is realized from analog PLL where all the analog components are replaced by digital devices. The first all-digital PLL with a second-order LF, in which all components were clocked synchronously by a common 1.097-MHz reference, was built and tested in 1967 [16]. The basic block diagram for a digital PLL is shown in Figure 9.6(*a*). Here, the input signal is sampled and compared with a reconstructed reference to produce digitized error samples that

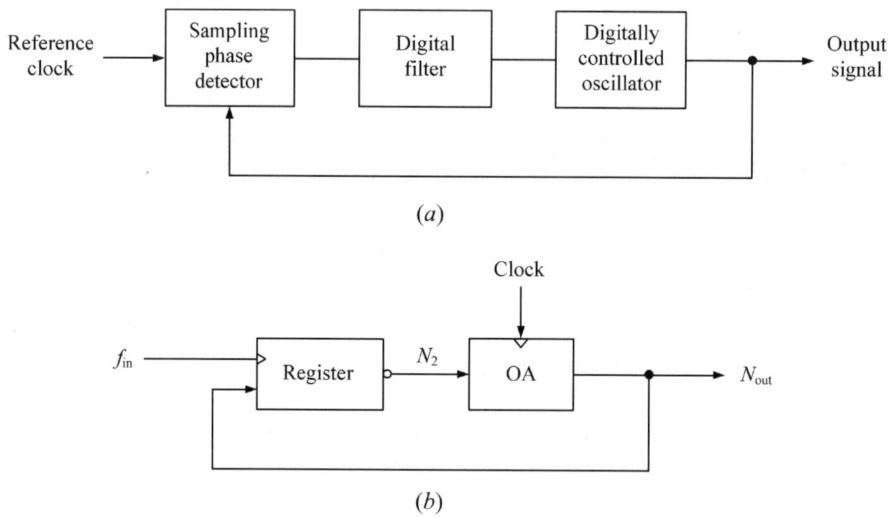

FIGURE 9.6 Simplified digital PLL block diagrams.

are proportional to their phase differences. These samples are then filtered by a numerical algorithm provided by a digital filter, and the corresponding output samples are used to control the period of the digitally controlled oscillator (DCO). If the loop is designed properly, the reconstructed reference will be forced to resemble the input signal.

Figure 9.6(*b*) shows the simplified block diagram of a digital PLL to derive the equivalence between digital and analog PLL blocks [8]. In this case, the input is a binary clock of frequency f_{in}, and the output represents a number that has an average repetition rate of f_{in} but follows the input with an accuracy provided by the loop parameters. The VCO is replaced by an output accumulator (OA) whose output represents a number that advances each clock cycle by an amount equal to its input N_2. Each time the OA reaches its capacity N_v, it recycles to 0, and $N_v - 1$ is the highest number that can be attained. Hence, one output cycle is represented by N_v and the output phase of the OA is

$$\theta_{out} = \frac{N_{out}}{N_v} \text{ cycles.} \tag{9.29}$$

The output frequency is written as

$$f_{out} = \frac{\Delta\theta_{out}}{\Delta t} = \frac{N_2}{N_v} f_{clock} \tag{9.30}$$

since the output is incremented by N_2 at each cycle of the OA clock with $T = 1/f_{clock}$. If N_2 is considered to be the tuning signal, the OA gain constant can be obtained from Eq. (9.30) as

$$K_v = \frac{df_{out}}{dN_2} = \frac{f_{clock}}{N_v}. \tag{9.31}$$

The register stores the value of N_{out} at each cycle where N_{out} represents the loop output phase. As a result, the stored number is the output phase at the last transition of f_{in}, and this can be considered the phase difference between the output and input signals. Consequently, the register operates as a PD and zero-order hold. However, since a phase inversion is required, one simple way to do this is to make N_2 the complement of the sampled value of N_{out}, which is represented by the small circle at the output of the register. In this case, during the nth input cycle (following the nth input transition), $N_2 = -N_{out}$ with $\theta_{in} = 2\pi n$, then the PD gain constant is defined as

$$K_d = -\frac{dN_2}{d\theta_{out}} = -\frac{dN_2}{dN_{out}}\frac{dN_{out}}{d\theta_{out}} = N_v \text{ cycle}^{-1} \tag{9.32}$$

and the loop gain constant is written as

$$K = K_d K_v = f_{clock} \text{ cycle}^{-1}. \tag{9.33}$$

The OA is the essential part of a digital PLL representing a combination of memory and adder, whose output phase advances at a rate proportional to the control signal that is the output from a PD (register). This simplified first-order digital PLL with the OA output N_{out} fed back to the PD performs similarly to an analog loop with a sawtooth PD, the mathematical block schematic of which is given by Figure 9.2(*b*) with $F(s) = N = 1$.

Generally, the PLL architecture for RF wireless transmitter applications includes an analog VCO circuit and therefore cannot be considered a fully digital PLL where all building blocks are defined as digital. However, an analog VCO can be implemented as a DCO that deliberately avoids any analog tuning voltage controls [17,18]. This allows for its loop control circuitry to be implemented in a fully digital manner. In this case, the DCO is analogous to a flip-flop whose internals are analog, but the analog nature does not propagate beyond its boundaries. Figure 9.7 shows the linear s-domain model for an all-digital PLL, where f_R is the reference frequency, N is frequency command word (FCW), and $\hat{K}_{DCO}$ is the DCO gain estimate. Here, the digital phase error is conditioned by a simple digital LF with loop gain parameters α and ρ, and then normalized by the DCO gain estimate $\hat{K}_{DCO}$. The DCO gain normalization with $f_R/\hat{K}_{DCO}$ multiplier is needed to precisely establish the loop bandwidth throughout the system (to be independent from the process, supply voltage, and temperature variations) and to perform a direct transmit frequency modulation. In a steady-state mode, such a PLL achieves a zero or constant averaged phase difference between the output variable θ_{out} and the reference input θ_{in} normalized timestamps. In this case, the clock quantization error must be added to the conventional definition of the phase error as a difference between θ_{in} and θ_{out}. Unlike conventional analog PLL architecture with frequency division in the feedback path, here the frequency multiplication by $N \equiv$ FCW is not a part of the open-loop transfer function and, hence, does not affect the loop bandwidth. On the other hand, phase deviation of the frequency reference needs to be multiplied by N, since it is measured by the same phase detection mechanism normalized to the DCO clock cycle.

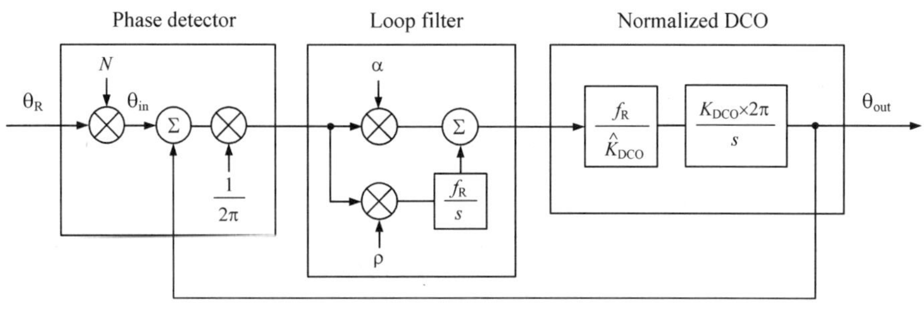

FIGURE 9.7 All-digital PLL block schematic.

9.5 LOOP COMPONENTS

9.5.1 Phase Detector

In the commonly used PDs, its output voltage v_e can be a sinusoidal, triangular, or sawtoothed function of the phase angle θ_e, where θ_e is shifted by $\pi/2$ from the VCO output phase according to Eq. (9.14) for the sinusoidal and triangular types, and by π for the sawtoothed type. As a result, it is convenient to provide a direct comparison between these three types of the PD transfer characteristics when the output voltage v_e is equal to zero for $\theta_e = 0$ in all cases, as shown in Figure 9.8. With the loop in lock, the phase angle θ_e stays within the limits $\pm\pi/2$ for sinusoidal and triangular types, and $\pm\pi$ for sawtoothed type. If the angular excursion is greater than this, the loop goes out of lock by skipping the cycles.

In a useful region where the phase excursions are sufficiently small compared with limiting values, the corresponding characteristic for each type of the PD can be written as

$$\text{sinusoidal:} \quad v_e = V_e \sin\theta_e$$

$$\text{triangular:} \quad v_e = \frac{2}{\pi} V_e \theta_e$$

$$\text{sawtooth:} \quad v_e = \frac{1}{\pi} V_e \theta_e$$

where V_e is the maximum output voltage. For the sinusoidal PD, a reasonable linear approximation of its characteristic for loop performance calculations with the loop in lock will be for $\theta_e \leq 0.2$ rad when $\sin\theta_e \approx \theta_e$.

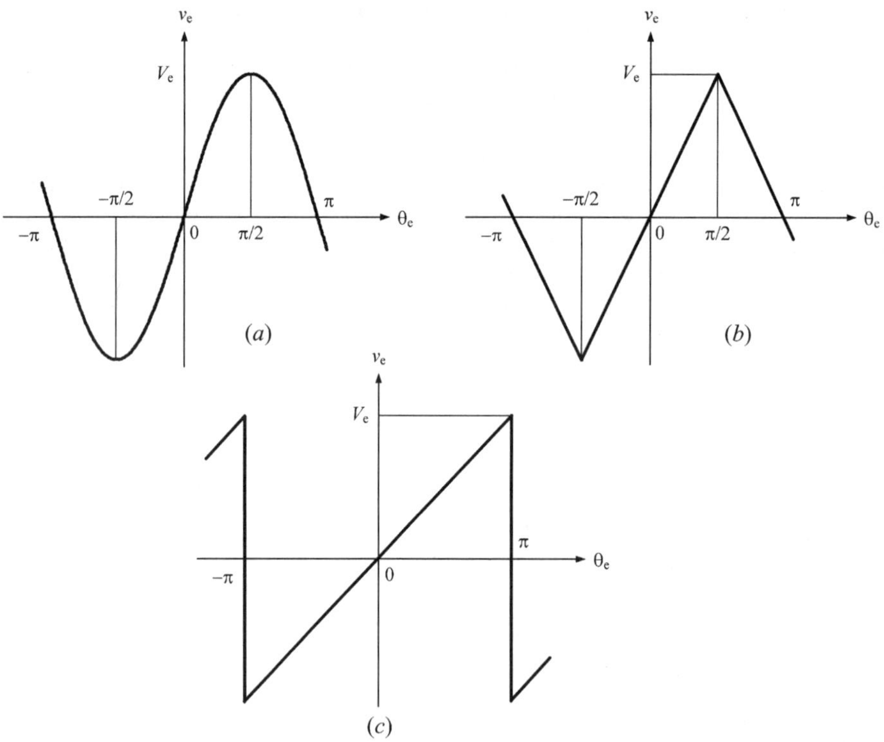

FIGURE 9.8 Three types of phase detector transfer characteristics.

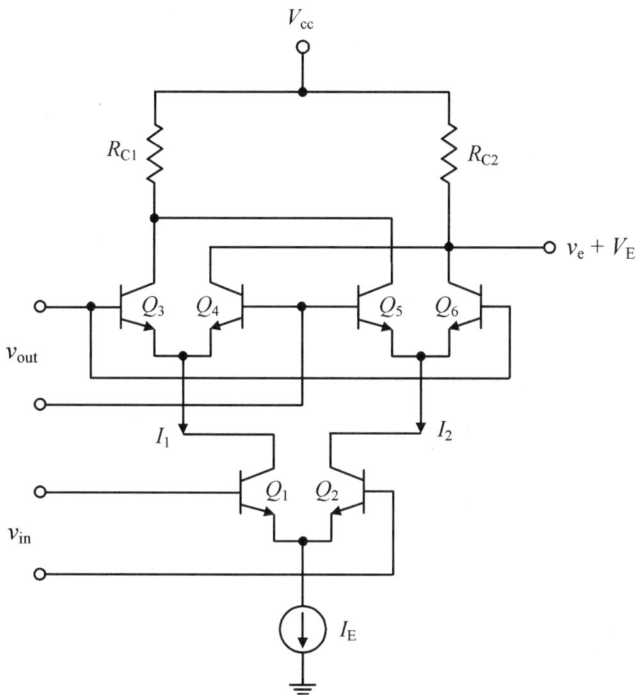

FIGURE 9.9 Double-balanced phase detector schematic.

The double-balanced PD circuit schematic based on bipolar transistors, which is widely used in a variety of analog PLLs, is shown in Figure 9.9, where I_E is a constant-current source for the differential amplifier on transistors Q_1 and Q_2 [19]. In this case, a change ΔI_1 in the collector current of Q_1 is accompanied by a corresponding change $\Delta I_2 = -\Delta I_1$ in the collector current of Q_2, since the total current $I_1 + I_2$ is equal to the constant current I_E. Hence, for the positive increment Δv_{in} of the input reference voltage v_{in}, the current in Q_1 increases while the current in Q_2 decreases. Then, in the case of the positive increment Δv_{out} of the output VCO voltage v_{out} when this signal is assumed to be a square wave of sufficient amplitude, the transistors Q_3 and Q_6 are turned completely on (saturation mode), and the collector current I_2 flows through transistor Q_6 and resistor R_{C2}. However, for the negative increment Δv_{out}, the transistors Q_4 and Q_5 are turned on, and the collector current I_1 flows through transistor Q_4 and resistor R_{C2}. Since the incremental components of the collector currents I_1 and I_2 due to Δv_{in} are of opposite phase and VCO output acts as a reversing switch on the amplified version of v_{in} that appears across R_{C2}, the output voltage v_e becomes the product of v_{in} times a square-wave switching function produced by v_{out}.

The time-domain voltage diagrams in Figure 9.10 illustrate the PD behavior for several combinations of v_{in} and v_{out}, both of which are assumed to be square waves. In Figure 9.10(*a*), the voltages v_{in} and v_{out} are 90° out of phase between each other, and the resulting v_e represents equal positive and negative areas around V_E, resulting in its average zero value. In Figure 9.10(*b*), the phase of v_{out} is shifted relative to that in Figure 9.10(*a*), resulting in a positive average value (dc component) due to longer positive areas. In this case, the polarity of v_e indicates the direction of phase shift, and its amplitude is proportional to the phase displacement, thus yielding the triangular PD function of Figure 9.8(*b*).

Such a double-balanced multiplier circuit will also operate as a PD even though the two input signals are not square waves, as long as one or both of them are large enough to cause the transistor

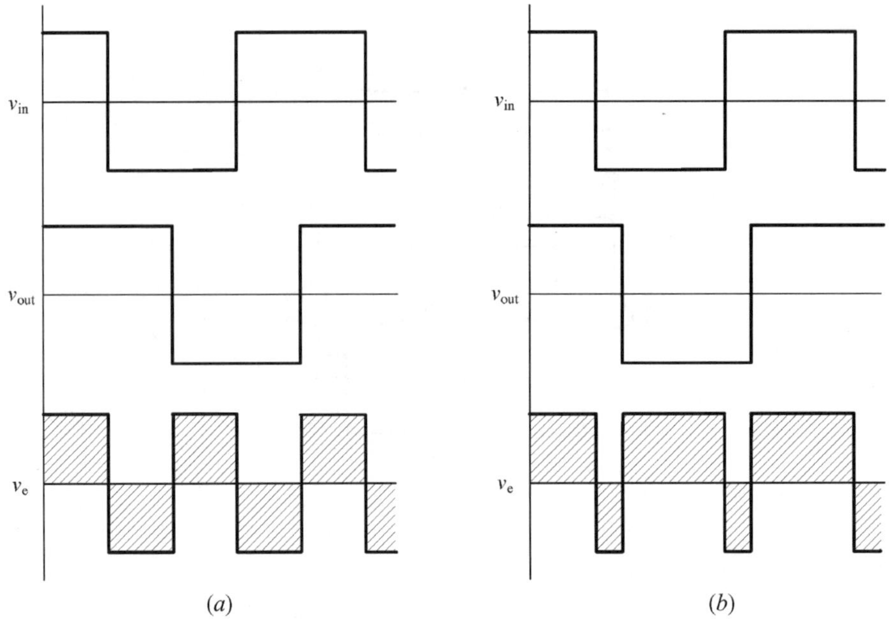

FIGURE 9.10 Phase detector voltage waveforms.

transfer function to be nonlinear. In this case, the resulting PD transfer function may be sinusoidal rather than triangular, especially for sine-wave inputs. The phase error signal amplitude V_e at the PD output versus its phase θ_e can be written as

$$V_e = \frac{1}{2\pi} \int_{\theta_e}^{\theta_e+\pi} \sin\theta d\theta = \frac{\cos\theta_e - \cos(\theta_e + \pi)}{2\pi} = \frac{1}{\pi}\cos\theta_e \tag{9.34}$$

showing that, for $\theta_e = 90°$, the amplitude V_e of the resulting signal is equal to zero.

Figure 9.11(*a*) shows the operation of a flip-flop PD and symbol for a set-reset (SR) flip-flop. It should be mentioned that the term "flip-flop" was derived long ago from the sound produced on a speaker connected to one of the back-coupled amplifiers' outputs during the trigger process within the circuit. The output state Q changes in response to state change at either S or R input, resulting in $Q = 1$ state when S is pulsed high, and then stays high even after S returns low. Similarly, if R is pulsed high while S is held low, then $Q = 0$, and stays low even after R returns low. When both inputs are high ($S = R = 1$), the output state is undefined, so the input waveforms are shown as very narrow pulses to avoid overlap of high states at the two inputs. The duration of the $Q = 1$ state, and thus the average output voltage, is proportional to the time difference between inputs, and hence to their phase difference, the average value of which represents the useful PD output. The average voltage has a sawtooth-shaped characteristic versus phase, as shown in Figure 9.8(*c*), where the phase varies linearly within the range of 2π radians.

The operation of the exclusive-OR (XOR) gate PD and its symbol are shown in Figure 9.11(*b*). The XOR gate is a digital logic gate that implements exclusive disjunction and behaves according to the truth table where a high output (1) results if one, and only one, of the inputs to the gate is high (1). If both inputs are low (0) or both are high (1), a low output (0) results. Since the part of the time when both inputs are the same (or different) indicates their relative phase, the XOR can be

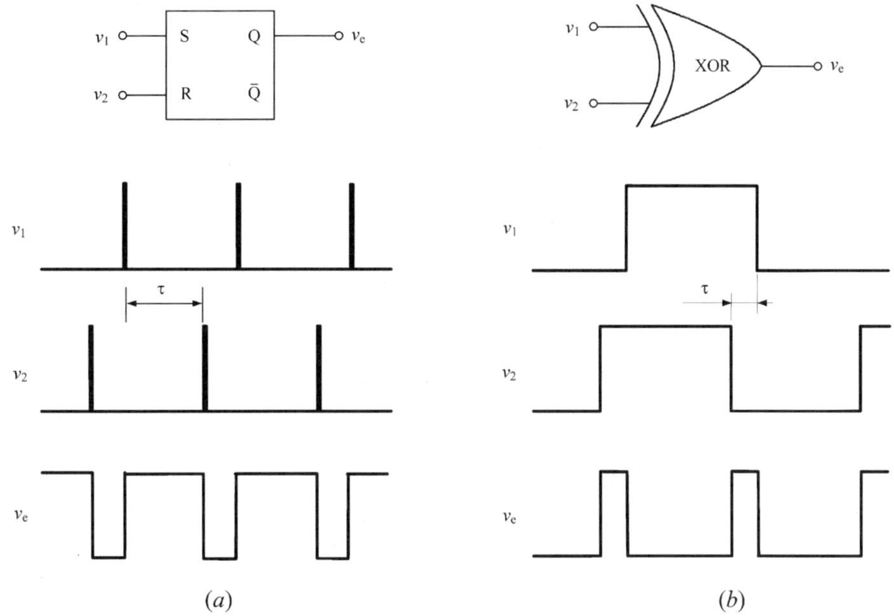

FIGURE 9.11 Flip-flop and exclusive-OR phase detectors.

used as a PD. However, deviation from a perfect in-phase (or out-of-phase) condition in either an increasing or a decreasing direction of phase change produces the same change in duty factor, and hence in average voltage. Therefore, the PD characteristic is triangular, as shown in Figure 9.8(*b*), with each linear slope extending for only half a cycle when the phase varies linearly within the range of π radians.

A PFD is an asynchronous sequential logic circuit based on two SR flip-flops, which produces an output even when the two input signals being compared differ not only in phase but also in frequency. The PFD prevents a false-lock condition, in which the VCO of the PLL synchronizes with the wrong phase of the input signal or with the wrong frequency. Figure 9.12 shows the basic circuit and operation waveforms of this PFD where the output v_{up} goes to 1 as before when triggered by the waveform of v_1 [8]. The difference is that the waveform of v_2 resets the upper flip-flop only indirectly, although with the same result. The voltage v_2 sets the lower flip-flop, which in turn provides a second 1 input to the AND gate. This causes a 1 at the AND output, which then resets both flip-flops. For positive $\tau > 0$, the results are the same, as for the PD in Figure 9.11(*a*). However, when τ goes to zero and then negative, with v_2 pulse preceding v_1 pulse, the results are considerably different. Now the output pulse appears for v_{down} being proportional to $-\tau$. The waveform of v_{down} will eventually be inverted relative to the waveform of v_{up}, so the pulse width of v_{down} represents negative phase relative to that of v_{up}. Such a PFD has a linearly changing output (average pulse width) from τ equal to -1 reference period to $+1$ period with a phase range of $\pm 360^\circ$. Its basic drawback is a dead zone in the phase characteristic at the equilibrium point. The dead zone generates phase jitter since the control system does not change the control voltage when the phase error is within dead zone. In this case, either variable inverter in the reset path or two inverters at both inputs for delays should be inserted to remove the dead zone [20,21].

The function of the CP is to alleviate any loading of the PFD in driving the rest of the circuit by converting the voltage pulses into current pulses with predetermined amplitude and delivering this current to a LPF, thus adjusting the control voltage of the VCO. Therefore, any noise generated by the

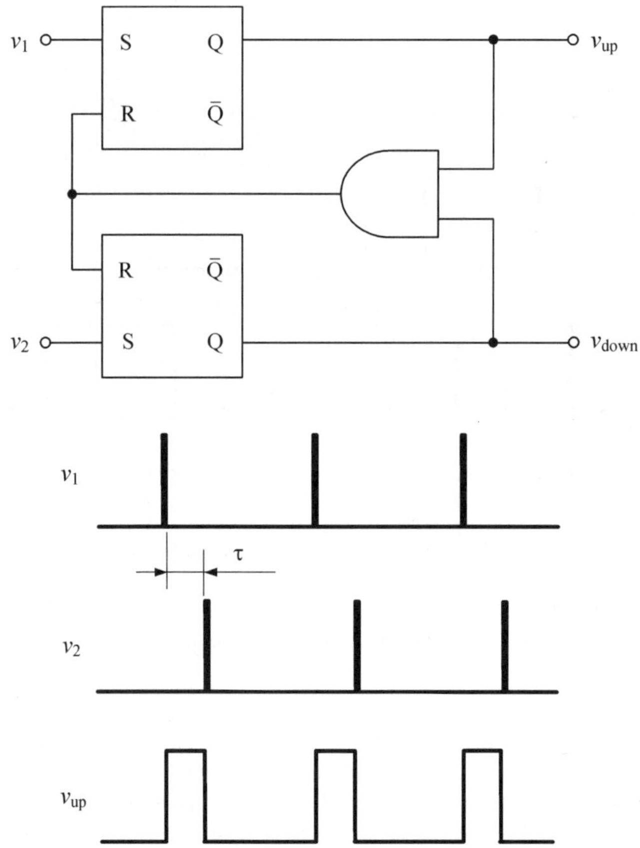

FIGURE 9.12 Phase and frequency detector and its waveforms.

CP will directly contribute to the VCO phase noise. The CP has a differential input and a single-ended output where the output PFD signals v_{up} and v_{down} are considered the CP inputs. Generally, the PFD and CP can represent the single-ended and differential configurations with different numbers of current sources [22,23]. Adding two complimentary current sources into a differential CP configuration provides an improved current matching, which results in a better performance with respect to spikes on the voltage controlled signal, less leakage current, and higher voltage swing [24].

Figure 9.13 shows a basic single-ended CP circuit implemented in CMOS technology where the output signal v_{up} is fed to a *p*-type CMOS transistor, whereas the output signal v_{down} is fed to an *n*-type CMOS transistor. This CP circuit is characterized by some nonideal effects that impair its performance. For example, the currents i_{up} and i_{down} should ideally be exactly equal. However, due to the finite output impedances, they are only equal for certain output voltages, resulting in some current mismatch over the other output voltages. The capacitors at the nodes *A* and *B* consist of the parasitic source–drain capacitances of the MOS devices. When the switching pMOS transistor M_1 and nMOS transistor M_2 are close, these parasitic capacitors store some charges and then transfer them to the LF, causing the voltage spikes on the VCO tuning curve. Similar effect occurs due to the parasitic gate–drain capacitances when voltages v_{up} and v_{down} change their logic levels (clock feedthrough). The common problem in submicron CMOS technology is a leakage current, resulting in the output leakage current in the CP that can also cause the VCO spikes. When PLL is in a locking state, all these nonideal effects contribute to the VCO phase noise.

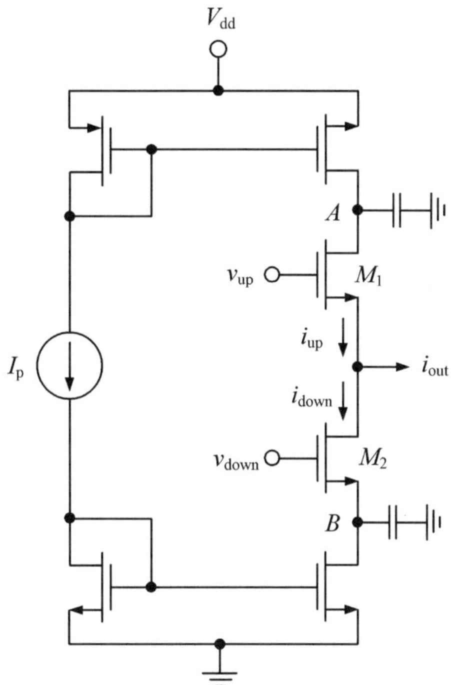

FIGURE 9.13 Schematic of basic single-ended CP CMOS circuit.

9.5.2 Loop Filter

The design of a PLL loop filter is a big challenge because it determines lock time, noise, stability, or reference spurious products, and is a biggest contributor to the chip area in monolithic integrated circuits. For example, fast lock times are possible with wide LF bandwidth, although this will lead to higher reference spurious products. A narrow LF bandwidth can reduce reference spurs, but leads to increased lock time and decreased stability. Therefore, the desired loop characteristics can be controlled by inserting the corresponding types of LFs. The simplest configuration is a passive one-pole low-pass *RC* circuit shown in Figure 9.14(*a*). The filter first-order transfer function that is produced by a series resistor *R* and a shunt capacitor *C* is derived as

$$F(s) = \frac{1}{1 + sRC} \tag{9.35}$$

where $s = j\omega$. The filter amplitude and phase response is shown in Figure 9.14(*b*), where $\omega_c = 1/RC$ is the cutoff frequency corresponding to the amplitude of $1/\sqrt{2}$ and phase shift of -45°.

From Eq. (9.17) for $N = 1$, the loop response becomes

$$\frac{\theta_{out}}{\theta_{in}}(s) = \frac{\dfrac{K_d K_{VCO}}{RC}}{s^2 + \dfrac{s}{RC} + \dfrac{K_d K_{VCO}}{RC}} \tag{9.36}$$

with the second-order denominator that can be written similarly to the second-order oscillator circuit as

$$s^2 + 2s\varsigma\omega_n + \omega_n^2$$

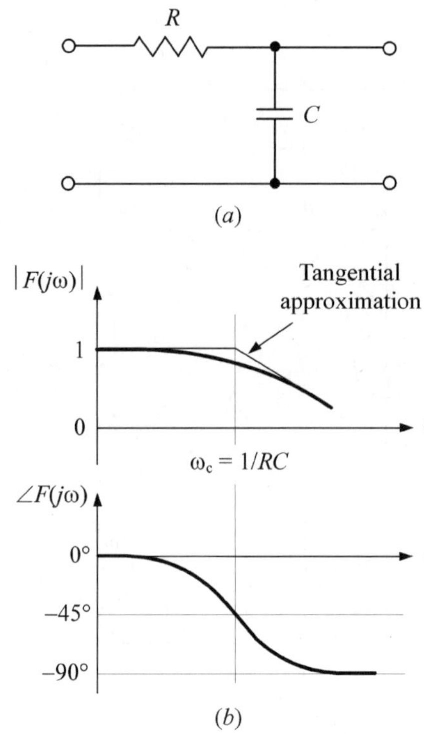

FIGURE 9.14 Passive first-order *RC* filter and its frequency response.

where

$$\varsigma = \frac{1}{2\sqrt{K_d K_{VCO} RC}} \tag{9.37}$$

is the damping ratio (or factor) and

$$\omega_n = \sqrt{\frac{K_d K_{VCO}}{RC}} \tag{9.38}$$

is the undamped natural frequency.

The loop undamped natural frequency is a measure of the response time of the loop, and the damping ratio is a measure of the overshoot and ringing. Ideally, the natural frequency should be high and the damping ratio should be near 0.707 (critical damping). However, with a single-pole filter, it is not possible to control the loop frequency and damping ratio independently.

The slightly more effective first-order *RC* filters in the form of proportional plus integral control networks with two resistors and one capacitor, the transfer functions of which include one pole and one zero, are shown in Figure 9.15 [25]. The loop frequency response for these types of filters is

$$\frac{\theta_{out}}{\theta_{in}}(j\omega) = \frac{1 + j2\varsigma\dfrac{\omega}{\omega_n}\left(1 - \dfrac{\omega_n}{2\varsigma K_d K_{VCO}}\right)}{1 + j2\varsigma\dfrac{\omega}{\omega_n} - \left(\dfrac{\omega}{\omega_n}\right)^2} \tag{9.39}$$

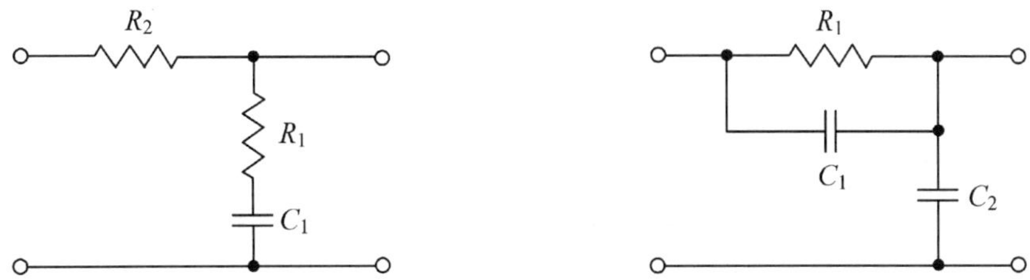

$\tau_1 = R_1C_1$

$\tau_2 = (R_1 + R_2)C_1$

$$F(s) = \frac{1 + s\tau_1}{1 + s\tau_2}$$

$\tau_1 = R_1C_1$

$\tau_2 = R_1(C_1 + C_2)$

(*a*) (*b*)

FIGURE 9.15 Passive low-pass filters for second-order loop.

showing that the cutoff frequency of the system for $\zeta = 0.5$ is approximately defined as $\omega_c \cong \omega_n$. The bandwidth and the gain constant of the system can be adjusted independently if a double time-constant control network is employed.

In the charge-pump PLL, the resistor in a single-pole *RC* LF introduces a stabilizing zero to improve the loop transient response. However, this resistor causes a ripple due to charge-pump output current I_p, which modulates the VCO frequency and introduces excessive jitter in the output. In order to suppress the ripple-induced jitter, a small shunt capacitor is added, resulting in a typical second-order filter circuit shown in Figure 9.16(*a*). However, this shunt capacitor introduces an additional pole, thus increasing the order of the system to three. Therefore, the phase degradation due to this pole has to be accounted for by a proper choice of the other loop parameters.

The transimpedance function of the second-order LF based on linear analysis can be written as

$$Z_F(s) = \frac{1 + s\tau_1}{s(C_1 + C_2)(1 + s\tau_2)} \tag{9.40}$$

where the time constants $\tau_1 = R_1C_1$ and

$$\tau_2 = R_1\left(\frac{1}{C_1} + \frac{1}{C_2}\right)^{-1}$$

determine the pole and zero frequencies of the filter transfer function and depend on the specified loop bandwidth and phase margin.

In this case, the third-order charge-pump PLL open-loop gain, using Eqs. (9.24) and (9.40), is obtained as

$$GH(s) = \frac{I_p K_{VCO}}{2\pi N} \frac{1 + s\tau_1}{s^2(C_1 + C_2)(1 + s\tau_2)} \tag{9.41}$$

where K_{VCO} is the VCO sensitivity, I_p is the charge-pump current, and N is the frequency division ratio.

This third-order loop with Bode plot shown in Figure 9.16 is described by a zero and three poles. The phase margin degradation due to the third pole can be expressed by

$$\phi_m(\omega_c) = \tan^{-1}(\omega_c\tau_1) + \tan^{-1}(\omega_c\tau_2) \tag{9.42}$$

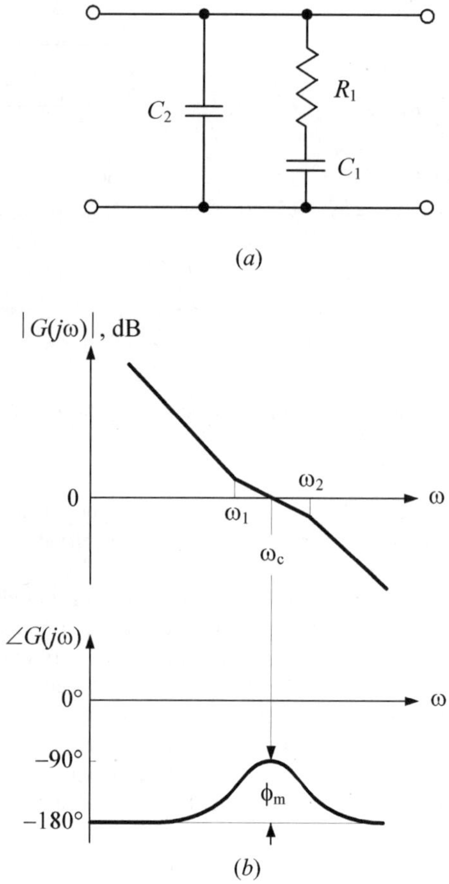

FIGURE 9.16 Passive filter for third-order loop and Bode plot.

where $\omega_1 = 1/\tau_1$ and $\omega_2 = 1/\tau_2$ [26,27]. The maximum phase margin ϕ_m can be analytically derived by setting the phase of the open-loop gain as derivative with respect to frequency at the cross-over frequency ω_c to zero when

$$\frac{1}{1+(\omega_c\tau_1)^2} + \frac{1}{1+(\omega_c\tau_2)^2} = 0 \tag{9.43}$$

which results in the values of the time constants τ_1 and τ_2 given by

$$\tau_1 = \frac{1}{\omega_c^2\tau_2} \qquad \tau_2 = \frac{1-\sin\phi_m}{\omega_c\cos\phi_m}. \tag{9.44}$$

As a result, the maximum phase margin occurs when

$$\omega_c = \frac{1}{\tau_1}\sqrt{\frac{C_1}{C_2}+1} \tag{9.45}$$

and the relationship between the two filter capacitances and maximum phase margin, by substituting Eq. (9.45) into Eq. (9.43), is described by

$$C_1 = 2C_2 \tan\phi_m \left(\tan\phi_m + \sqrt{\tan^2\phi_m + 1}\right) \tag{9.46}$$

where the choice of the basic loop parameters, such as loop bandwidth and phase margin, depends on a particular application. For example, in the case of clock generators with a poor phase noise VCO and a pure low-frequency reference, a relatively high bandwidth is required. However, it is necessary to choose a low loop bandwidth and a large phase margin if it is required to avoid any jitter peaking.

The cross-over frequency ω_c can also be approximated as a function of the loop parameters as

$$\omega_c \approx \frac{I_p K_{VCO} R_1}{2\pi N} \frac{C_1}{C_1 + C_2} \approx \frac{I_p K_{VCO} R_1}{2\pi N} \tag{9.47}$$

resulting in the simple design equations for the filter parameters

$$R_1 = \frac{2\pi N}{I_p K_{VCO}} \omega_c \tag{9.48}$$

$$C_1 = \frac{\alpha}{R_1 \omega_c} = \frac{I_p K_{VCO}}{2\pi N} \frac{\alpha}{\omega_c^2} \tag{9.49}$$

$$C_2 = \frac{1}{\beta R_1 \omega_c} = \frac{I_p K_{VCO}}{2\pi N} \frac{1}{\beta \omega_c^2} \tag{9.50}$$

where $\alpha = \omega_c/\omega_1$ and $\beta = \omega_2/\omega_c$ [28]. The factors α and β are typically equal to 4 each, which gives a capacitance ratio $C_1/C_2 = 16$, according to Eqs. (9.49) and (9.50), and results to a phase margin of approximately 60°, as it follows from Eq. (9.46).

An active LF employs an operational amplifier in its configuration, as shown in Figure 9.17(*a*) for a typical filter circuit. The transimpedance function of such a second-order active LF driven by a voltage, which is related to the passive lag–lead LF shown in Figure 9.15(*a*), is written as

$$Z_F(s) = \frac{1 + s\tau_1}{s\tau_2} \tag{9.51}$$

with the time constants $\tau_1 = R_1C_1$ and $\tau_2 = R_2C_1$, assuming an ideal operational amplifier with infinite gain. The response has a pole at the origin and a zero in the left half plane. The zero is vital in preserving system stability and in controlling the step response. It should be noted that the integrator, which is realized when $R_1 = 0$ for an active LF of Figure 9.17(*a*), is not a practical LF because there is no frequency at which it does not have a transfer of −90° that would inevitably put the loop on the verge of instability [8]. Also, the series capacitance at the filter input should be avoided.

If the active LF is driven by current, as it happens in a charge-pump PLL, there is no need for the series resistor R_2. The second-order passive LF shown in Figure 9.16(*a*) has some limitations; for example, one of them due to significant contribution of the filter resistance to the system output phase noise. In this case, a very low value of R_1 is needed, but it leads to extremely large capacitance values that are not feasible for integration. In addition, a small offset at the input of the PFD causing spurs in the output spectrum can occur due to the finite output impedance of the charge-pump current sources when the VCO input control voltage must vary over a wide range to use the full tuning range of the VCO. As a result, an extra pole must be inserted in the frequency response of the LF shown in Figure 9.17(*b*), which is placed on top of ω_2 by making $\tau_2 = R_3C_3$. To keep the phase margin high enough, the factor β must be increased from 4 to 6. Since the loop transfer function now falls off at a 60 dB/decade for frequencies beyond ω_2, the noise of the CP and the LF resistor R_1 will have a dependency on the offset frequency of ω_m^{-6}. However, this solution leads to a total capacitance that

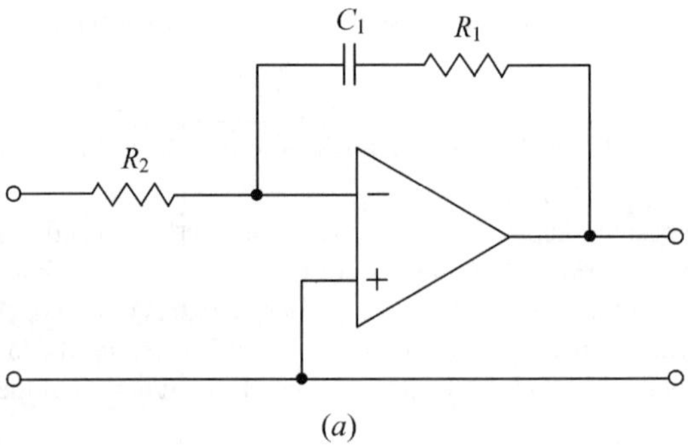

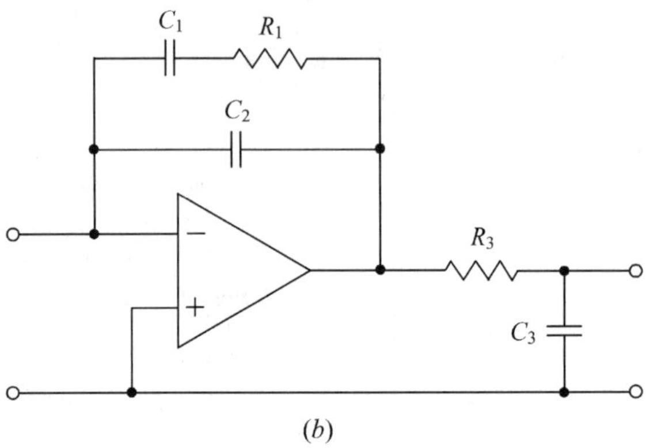

FIGURE 9.17 Active low-pass loop filters.

may be too large to be integrated. In this case, a special filter topology that splits up the LF in two paths, passive and active, can be employed [28].

9.5.3 Frequency Divider

There are different types of circuit that perform frequency division, with one of the simplest based on the oscillator injection-locking principle [29,30]. By injecting into an oscillator an RF signal whose frequency ω_{inj} is approximately a nth harmonic of the free-running oscillator frequency ω_0, one can force the oscillator to lock to subharmonic of the injected signal with $\omega_{inj} = N(\omega_0 + \Delta\omega)$, where $\Delta\omega \ll \omega_0$. In practice, the external signal is normally injected into the oscillator input. For a n-order polynomial transfer characteristic of the active nonlinear device expressed by $i_{out}(v) = a_0 + a_1 v + a_2 v^2 + \cdots + a_n v^n$, where $n = 1, 2, \ldots, N$, the basic contribution to the frequency division by N times is provided by the series component $a_N v^N = a_N \left(v_0 + v_{inj}\right)^N$. It contains the constituent element $N a_N v_0^{N-1} v_{inj}$, one of the components of which is directly proportional to $N a_N V_0^{N-1} V_{inj} \cos[(N-1)\omega_0 t] \cos \omega_{inj} t$. In this case, if its amplitude is sufficiently high, the signal component with difference frequency $\omega_{diff} = \omega_{inj} - (N-1)\omega_0 = \omega_0 + N\Delta\omega$, which is closed to ω_0, provides an injection locking. The initial oscillator frequency $\omega_{osc} = \omega_0$ starts approaching the difference frequency, which also changes in turn, and the transient process is completed by establishment of the steady-state mode, in which both frequencies become equal ($\omega_{diff} = \omega_{osc}$), which happens when $\omega_{osc} = \omega_{inj}/N$. Usually such an approach is used for frequency division with $N = 2$ or 3.

Figure 9.18 shows a typical oscillator circuit configured in a grounded-base Colpitts-type oscillator, where C_b is the bypass capacitor [6]. Here, if the reactance of the capacitor C_3 is selected so that

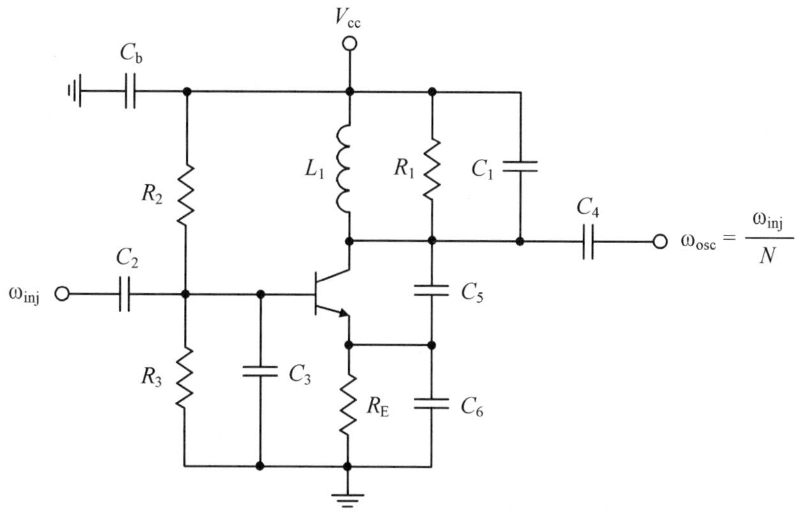

FIGURE 9.18 Locked-oscillator frequency divider circuit.

the capacitor no longer provides a good RF short from the base of the transistor to ground, and if an external signal whose frequency is harmonically related to the oscillator frequency is injected into the transistor base, the oscillator will lock in frequency to the injected signal. To increase the locking range, a load resistor R_1 is inserted across the oscillator-tuned L_1C_1 circuit, which effectively increases the bandwidth of the tuned circuit. In this case, the required frequency shift can be written as

$$\Delta\omega = \frac{\omega_0}{2Q_L}\frac{I_{inj}}{I_{osc}} \tag{9.52}$$

where Q_L is the oscillator quality factor, I_{inj} is the injected current amplitude, and I_{osc} is the oscillator fundamental current amplitude [30,31].

The injection-locking approach for frequency division is characterized by relatively low cost and has no lower or upper limit because, theoretically, an oscillator operating at any frequency can be locked to a signal source by injection locking. The circuit operates over a narrow frequency band (12–15%), though electronic tuning can be easily implemented. The major problem of the DIV based on the injection-locking principle is the presence of an output RF signal even if driving circuitry fails to generate an injection signal.

In the locked-oscillator dividers, the frequency division occurs due to availability of the difference frequency component $\omega_{diff} = \omega_{inj} - (N-1)\omega_{osc}$, the appearance of which can be considered as a result of two consecutive operations: multiplication of ω_{osc} by $(N-1)$ times resulting in a frequency $\omega' = (N-1)\omega_{osc}$, and then the frequency transformation by separating the component $\omega_{diff} = \omega_{inj} - \omega'$. Besides, signal amplification occurs due to the active device. As a result, the regenerative DIVs can be considered as modified locked-oscillator DIVs in which any from the above-mentioned operations (or some of them) is fulfilled by separate stages that allows the required characteristics of each stage to be independently controlled. As a result, the overall system performance can be improved. A regenerative DIV, also known as a Miller DIV, simplified block schematic of which is shown in Figure 9.19, provides mixing of the input injected signal with the feedback signal delivered from the output through the multiplier and bandpass filter [6,32]. For the oscillations to be sustained, the loop gain should be greater than unity, and the total phase shift must also be an integer multiple of 2π. To operate properly with correct frequency division, the multiplier has to be hard-driven and minimum required input power should be provided [33]. The advantage of using

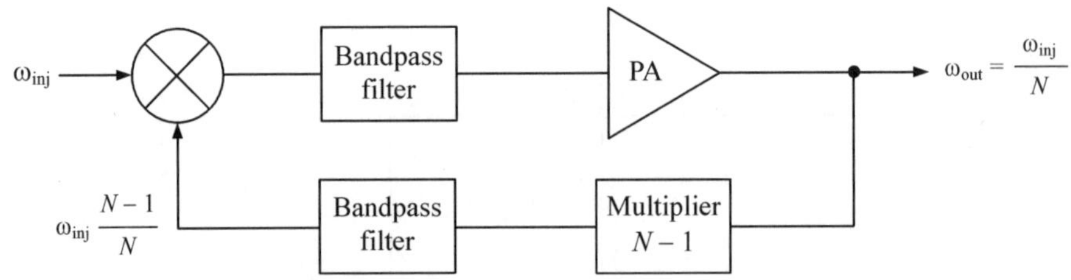

FIGURE 9.19 Regenerative frequency divider block schematic.

this circuit is that theoretically it can be made to divide at a very high frequency because both mixer and frequency multiplier can operate at very high frequencies depending on the implementation technology. Another advantage is that, in the absence of an input signal, no RF output is generated by the divider. However, the high cost, narrow bandwidth of operation, low division ratio, and difficulty of changing the division ratio make the regenerative divider inferior by far to other types of frequency division at sufficiently low operating frequencies. As a result, to accommodate the severe tradeoffs between the input frequency and operation range, a good strategy is to place the injection-locked and regenerative DIVs in descending order of frequency, and then they can be followed by the static dividers [34].

The typical digital DIV is based on a D-type flip-flop, as shown in Figure 9.20(a) [35]. The D-type flip-flop represents an edge-triggered device, which means that the change of state occurs on a clock transition (in this case the rising clock pulse as it goes from 0 to 1). The circuit operates in a simple way. The incoming pulse train acts as a clock for the device, and the data that is on the D input is then clocked through to the output Q. The Q output always takes on the state of the D input at the moment of a rising clock edge (or falling edge if the clock input is active low). It is called the D-type flip-flop for this reason, since the output takes the value of the D or *Data* input and delays it by one clock count. The $\bar{Q}$ output is opposite to that of Q and achieves the opposite state. In this case, when the logic voltage at the output Q is equal to level 1, this means that the output $\bar{Q}$ will be at level 0. Each positive edge occurs once every cycle of the input clock, but as the output of the D-type flip-flop requires two changes to complete a cycle, it means that the output from the D-type circuit changes

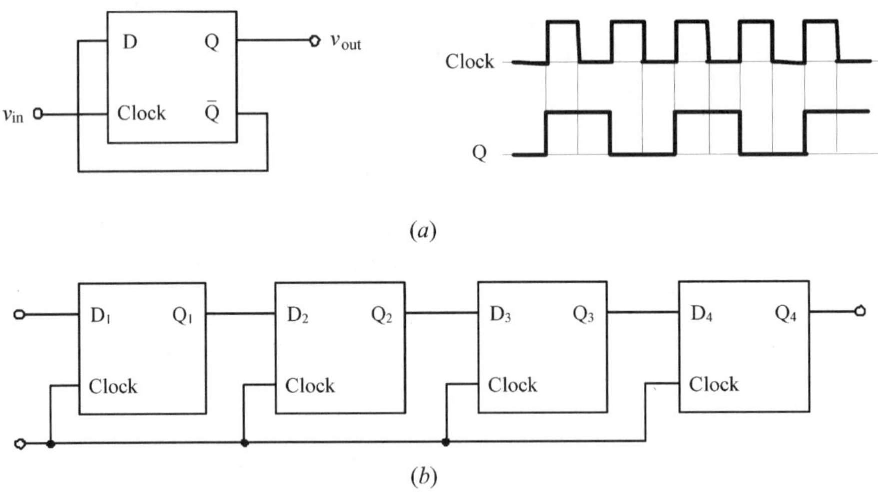

FIGURE 9.20 Frequency dividers with D-type flip-flops.

at half the rate of the incoming pulse train, thus representing a divide-by-2 circuit according to the timing diagram. It should be noted that the pulse train should have sharp edges. If the rising edges are insufficiently sharp, then there may be problems with the proper circuit operation. However, placing an inverter before the clock input has the effect of sharpening the edges on the incoming signal. The key building block in such a digital DIV is the bistable circuit based on a differential amplifier and a cross-coupled oscillator, although the other circuit configurations are also possible [36].

An arbitrary number of DIVs may be cascaded to produce very low frequencies. In this case, the input clock signal must be of the proper voltage level and must usually have a reasonably fast transition to ensure that the flip-flop is properly triggered. In cascade, a later flip-flop may be clocked by either the Q or $\bar{Q}$ output of the preceding stage because it does not matter which output is used if it is necessary to provide the frequency division by 2^n. Figure 9.20(*b*) shows the DIV based on a shift register, where a four-bit shift register is built from four cascaded D-type flip-flops [35]. With each clock pulse, the information at the input D of each flip-flop is transferred to the corresponding output Q. Assume that a logical 1 is applied to the input, and that initially all Q values are zero. When the first clock pulse arrives, Q_1 will be 1 while the rest remain zero. After the second clock pulse, Q_2 is 1 as well. Finally, after the fourth clock pulse, a logical 1 appears at the output. This may be used to trigger a one-shot multivibrator to reset all of the flip-flops. The result is a divide-by-4 circuit, and the division number corresponds to the number of stages in the register. Due to the propagation delay of several stages, uncertainty or jitter in the delay will lead to unwanted phase modulation of the output signal. This can be eliminated through the application of a D-type flip-flop at the output of a large divider chain.

In a differential CMOS implementation, a DIV is typically represented as an asynchronous cascade of divide-by-2 circuits, where each stage is clocked by the previous one. Figure 9.21(*a*) shows a common topology for a digital DIV with $N = 2$, where two D-latches are connected in master/slave configuration and the output $\bar{Q}$ of the second latch is fed back to the input D of the first latch [37]. The same clock is used to drive both D-latches with opposite logic, and either master or slave D-latch is activated in each clock, but not both at the same time. On the next cycle, inverted output is fed back to the input again, which causes the output to toggle. The same event repeats for every two input clocks. The jitter at the output of this DIV is not affected by the noise of the first latch, since the latter has no control on the output switching. Therefore, the noise sources of the second latch only need to be taken into account in the evaluation of the output jitter. The differential topology of the latch stage is shown in Figure 9.21(*b*), where the pMOS transistors biased in the triode region act as load resistors. The pairs of nMOS transistors are alternatively switched on and off.

9.5.4 Voltage-Controlled Oscillator

The VCO is considered as the heart of a phase-locked system generally determining its performance quality. For example, for a low phase noise VCO, it may be controlled with a narrow bandwidth loop with no compromise in overall noise performance. At low frequencies when it is possible, instead of the VCO in a PLL system some designs are used a current-controlled oscillator, representing a multivibrator with a square-wave output because it is desirable for proper operation of the PD and is compatible with logic circuitry. In this case, linear variation of frequency with control voltage (or current) is easier to obtain with a multivibrator than with a sine-wave LC oscillator. For high-frequency operation with better stability and spectral purity, or if sinusoidal waveform is required, a varactor–tuned LC oscillator is preferable. The VCO will convert the applied tuning voltage to an output frequency, and its sensitivity can vary drastically over its full frequency range. In general, the lower the tuning sensitivity of the VCO, the better the VCO phase noise will be.

Generally, different circuit topologies based on a Colpitts core circuit and including one or several varactors as the frequency-tuning elements can be used. The varactor diode is operated with a reverse bias on the junction, and its capacitance is a function of the reverse bias voltage. Figure 9.22(*a*) shows

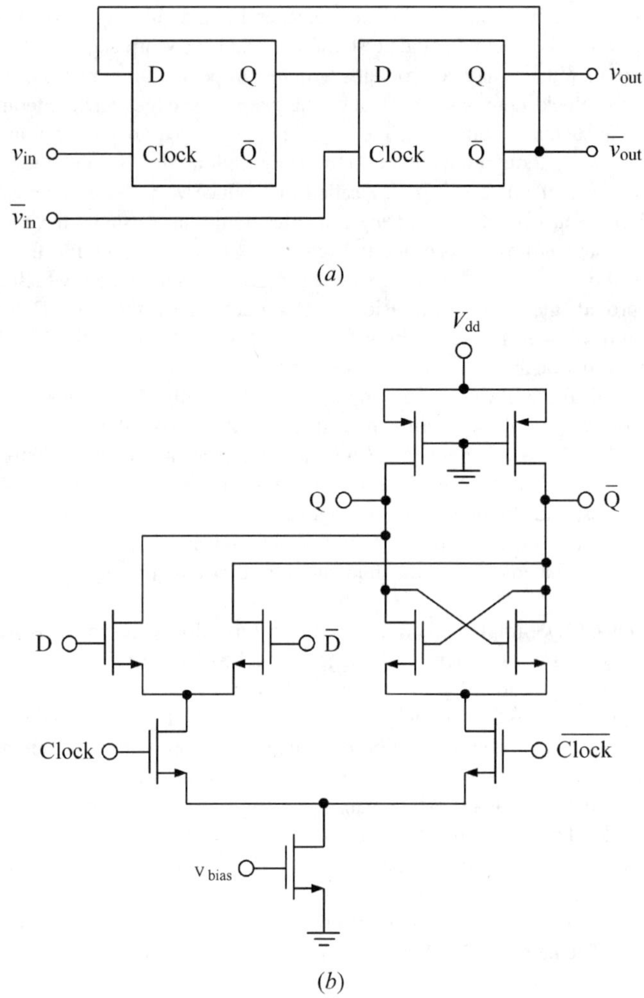

FIGURE 9.21 Differential digital frequency divider schematics.

a typical schematic diagram for a Seiler oscillator, where C_2 and C_3 are the feedback capacitors, and the varactor is connected in parallel to the resonant circuit inductor L_1 [6,38]. Grounded-base (or grounded-gate) oscillator configuration is often chosen due to its good temperature stability and ease to provide the self-oscillation conditions. An additional shunt capacitor C_1 is necessary to properly choose the inductance value that is evaluated at the upper edge of the frequency band to be tuned using the minimum varactor capacitance. Besides, both the shunt capacitance C_1 and series capacitance C_4 define the required tuning frequency range of the VCO. In this case, the smaller the tuning range of a VCO, the better the possible noise performance may be, and the varactor diode loss is of less significance in lowering resonator quality factor with a restricted tuning range. The linear tuning characteristic is very important because if the tuning is not linear with varactor bias voltage, the loop gain will vary that generally degrades the loop stability. One control circuit is employed when a VCO is tuned over a narrow frequency band. However, two or more separate controls are used to prevent a significant change in K_{VCO}, as the VCO is coarse-tuned, from affecting the PLL gain. Figure 9.22(*b*) shows a transmission-line VCO with both the control voltage and tuning voltage networks separated

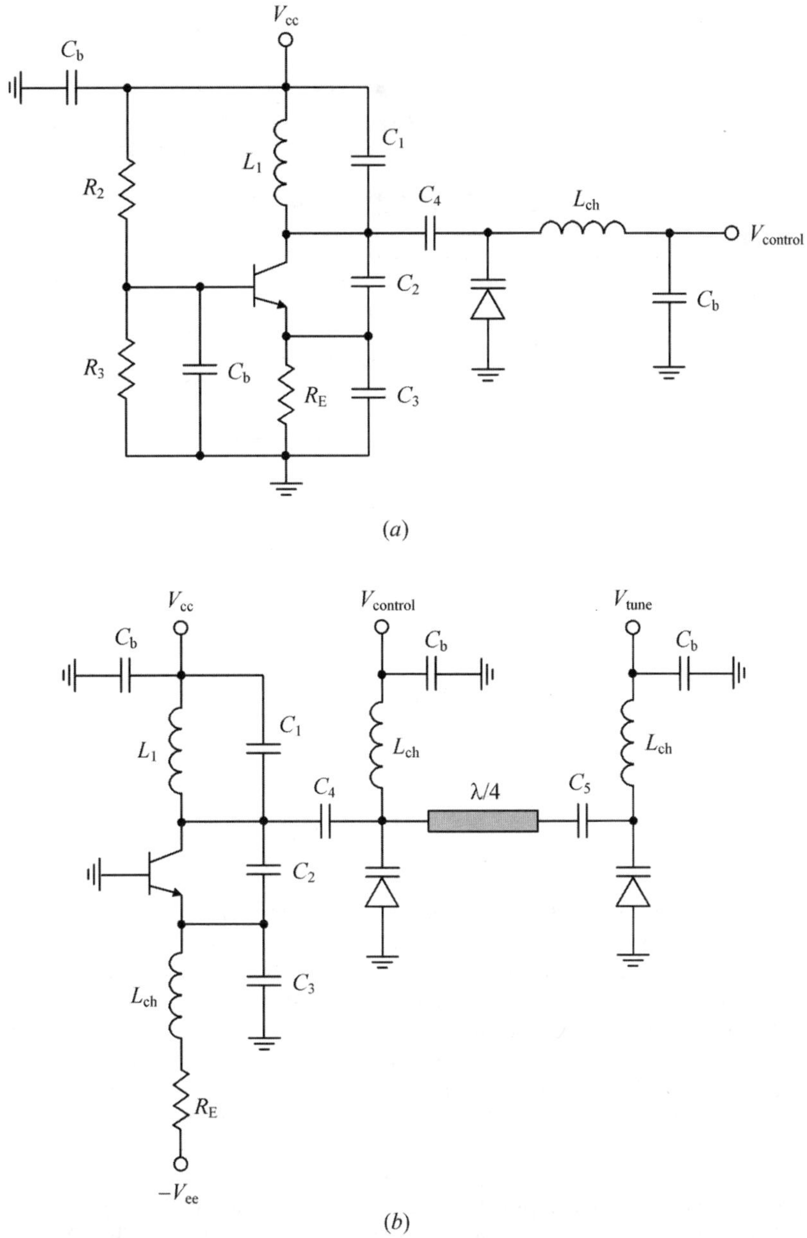

FIGURE 9.22 Schematics of bipolar *LC* voltage-controlled oscillators.

by a quarter-wavelength transmission line. The control voltage varactor is reversed-biased at a dc voltage that is generated by the phase detector (or phase comparator) of the PLL. It provides the fine-tuning control necessary to keep the VCO frequency-locked to a reference source, whereas the tuning voltage varactor coarse-tunes the VCO to within the PLL capture range. The overall linear tuning range can be expanded by using switching reactances into the circuit where capacitors can be switched with *p–i–n* diodes [6,35].

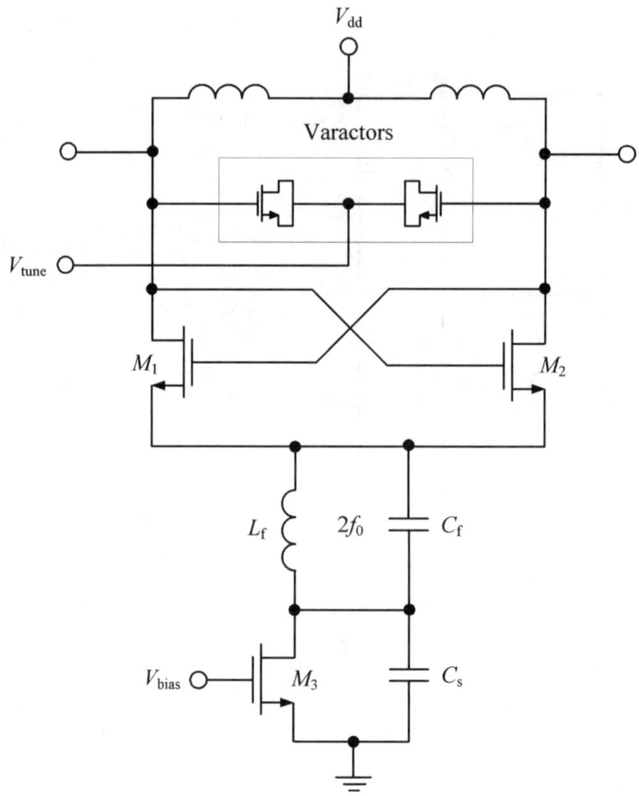

FIGURE 9.23 Differential tail-biased VCO with complete noise filter.

The popular cross-coupled VCO implemented in CMOS technology serves as the most suitable candidate for a very high-speed operation. Figure 9.23 shows the circuit schematic of such a VCO where the nMOS differential pair M_1 and M_2 provides a sufficient negative resistance, and the shunt capacitor C_s is necessary to reduce the upconversion of high-frequency noise from the source bias circuit on M_3. The parallel *LC* filter based on the series inductance L_f connected in parallel to the capacitance C_f resonates on the second harmonic. Its impedance is limited only to the quality factor of the inductor. As a result, the inserted inductor and two capacitors comprise a complete noise filter for a tail-biased differential *LC* oscillator [39]. The general guidelines for designing high-quality spiral inductors in silicon substrate include the requirements to limit the width of the metal conductors because of the skin effect, to use minimum spacing in between the conductors to maximize the inductance value, not to fill the inductor up to center because of eddy currents, and to limit the area occupied by the coil to minimize the losses due to conductive substrate [28]. In a DCO, which is used to perform the digital-to-frequency conversion and is based on an analog core circuit, varactors are implemented using an *n*-poly/*n*-well CMOS capacitor structures. In this case, gate terminals of the varactors are connected to oscillation nodes, and the digital inputs control the *n*-well nodes to eliminate the effects of parasitic capacitance from digital control bits [40]. All varactors are tuned in the same way between accumulation and depletion regions. The four DCO input digital-logic buses comprise the oscillator tuning word that is analogous to the tuning range of a VCO. Their bits individually control capacitive states of the *LC*-tank varactors, thus establishing the *LC*-tank resonant frequency.

9.6 LOOP PARAMETERS

9.6.1 Lock Range

The steady-state phase error resulting from a step change in reference frequency in a first-order PLL with sinusoidal PD, whose transfer characteristic is shown Figure 9.8(a), is defined by Eqs. (9.12) and (9.14), where $\Delta\omega$ is the difference between the free-running and the locked VCO frequencies. As a result, the maximum VCO frequency error approaching $\pm\pi/2$ rad that can be compensated for by a first-order PLL, which is called the *hold-in, lock, or synchronization range*, is equal to the dc loop gain, or

$$\Delta\omega_{\text{hold-in}} = K \tag{9.53}$$

where $K = K_{\text{d}}F(0)K_{\text{VCO}}$ for an analog PLL and $K = (I_{\text{p}}/2\pi)Z_{\text{F}}(0)K_{\text{VCO}}$ for a charge-pump PLL. However, the hold-in range $\Delta\omega_{\text{hold-in}}$, which is referred to the PD, should be divided by N. For the triangular and sawtooth PDs whose linear transfer characteristics are shown in Figures 9.8(b) and 9.8(c), respectively, the hold-in range is obtained by taking into account Eq. (9.20) as

$$\Delta\omega_{\text{hold-in}} = K\theta_{\text{e,max}} \tag{9.54}$$

where $\theta_{\text{e, max}}$ is equal to π for the triangular and 2π for the sawtooth detector characteristic.

If the input reference frequency is sufficiently close to the VCO free-running frequency, a PLL locks up with just a phase transient, without slipping cycle prior to lock. The frequency range over which the loop acquires phase to lock without slips is called the *seize or lock-in range* of the PLL. In a first-order loop, the lock-in range is equal to the hold-in range, however for the second and higher order loops the lock-in range is always less than the hold-in-range, which is defined by Eqs. (9.53) or (9.54). As an example for the second-order loop with a lag–lead filter shown in Figure 9.15(a), if the difference between the free-running and the locked VCO frequencies is less than 3-dB bandwidth of the closed-loop transfer function $H(s)$, the loop will lock up without slipping cycles and the maximum lock-in range can be written as

$$\Delta\omega_{\text{lock-in}} = \frac{K\tau_1}{\tau_1 + \tau_2} \tag{9.55}$$

or, in terms of the LF damping factor ς and natural frequency ω_{n},

$$\Delta\omega_{\text{lock-in}} \approx 2\varsigma\omega_{\text{n}} \tag{9.56}$$

where

$$\varsigma = \frac{\omega_{\text{n}}}{2}\left(\tau_1 + \frac{1}{K}\right) \tag{9.57}$$

$$\omega_{\text{n}} = \sqrt{\frac{K}{\tau_1 + \tau_2}}. \tag{9.58}$$

However, there is a frequency interval, smaller than the hold-in range and larger than the lock-in range, which is called the *pull-in or acquisition range*, over which the loop will acquire lock after slipping cycles for a while. The relationship between the hold-in range, lock-in range, and pull-in range is shown in Figure 9.24. The equalities in this apply to the first-order loops and highly damped second-order loops, which act like first-order loops. For the second-order PLL with a lag–lead filter,

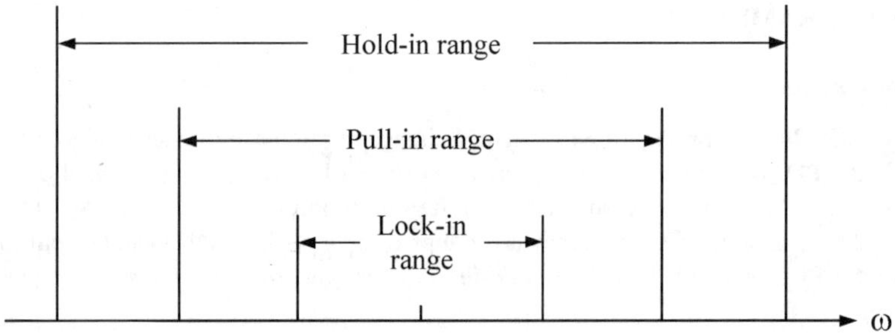

FIGURE 9.24 Ranges of dynamic limits of phase-locked loop.

locking is still possible with some cycles of VCO frequency skipped as long as $\Delta\omega < \Delta\omega_{\text{pull-in}}$, where the pull-in range is estimated as

$$\Delta\omega_{\text{pull-in}} \approx \sqrt{2}\sqrt{2\varsigma\omega_n K - \omega_n^2} \tag{9.59}$$

which is valid for $K/\omega_n > 2.5$ [6]. Once the loop is in lock, small loop bandwidth is desirable to minimize noise transmission. On the other hand, large bandwidth is desirable to aid initial capture. These incompatible requirements are sometimes met by reducing the bandwidth after lock is acquired or using small bandwidth but sweeping the VCO frequency until lock is acquired [3].

9.6.2 Stability

As in any feedback system, stability is one of the most important aspects of the PLL design since the condition for phase locking does not assure loop stability. A PLL can oscillate if at some frequency its open-loop gain is greater than unity and phase shift exceeds 180°. All PLLs have −90° phase shift due to the $1/s$ term, so excess phase is 90° even before any effect from the LF is considered. Therefore, the basic first-order PLL is stable unless the filtering provided in the forward and feedback path introduces a large amount of phase shift. Whether a loop will oscillate when its gain exceeds unity at 180° can be seen from the Nyquist plot, root locus plot, or Bode plot [8]. Since a Bode plot is a graph of the logarithm of the system transfer function versus frequency, open-loop gain and phase are plotted with a log-frequency axis, and the gain and phase margins can be seen from this plot. In this case, tangential approximations for gain and phase are often used. A Nyquist plot is represented by a graph in polar coordinates, in which the gain and phase of a frequency response are plotted. The plot of these phasor quantities shows the phase as the angle and the magnitude as the distance from the origin. This plot combines the two types of Bode plot—magnitude and phase—on a single graph, with frequency as a parameter along the curve. In this case, if the locus does not surround the point $-1 + j0$, so poles of $H(s)$ are in the left half of the s-plane and that corresponds to the stable region. In the root locus technique, which determines the position of system poles and zeros in the s-plane, the graph or plot illustrates how the closed loop poles (roots of the characteristic equation) vary with loop gain and at what value of the loop gain they enter the right half of the s-plane. The relationship of the system poles and zeros determines the degree of stability. The root locus contour can be determined by using special set of rules [41,42].

There are two sources for stability limit in a charge-pump PLL [12,43]. First, it has a critical stability limitation due to the discrete nature of the PFD and CP output. As a sampled system, the charge-pump PLL will become unstable if the loop gain is made so large that the bandwidth becomes comparable to the sampling frequency. Limited loop gain sets upper boundary of the loop bandwidth obtainable for a given input reference frequency ω_{ref}. Stability limit requires that

$$\omega_n^2 < \frac{\omega_{ref}^2}{\pi\left(\pi + \dfrac{\omega_{ref}}{\omega_1}\right)} \tag{9.60}$$

where ω_n is the natural frequency and $\omega_1 = 1/\tau_1$ is shown in Figure 9.16. The relationship between ω_n and loop bandwidth defined by ω_c is approximately obtained by

$$\omega_c \approx \frac{\omega_n^2}{\omega_1} \tag{9.61}$$

for critically damped and overdamped system. Substituting Eq. (9.61) into Eq. (9.60) results in

$$\omega_c < \frac{\omega_{ref}}{\pi\left(1 + \dfrac{\pi\omega_1}{\omega_{ref}}\right)} \tag{9.62}$$

which indicates that the loop bandwidth has to be significantly lower than the frequency of the input reference signal. Commonly, ω_c is chosen below one-tenth of ω_{ref} to guarantee stability.

The second stability limit comes from the open-loop transfer function. As it is seen from Eq. (9.41), the open-loop transfer function of a third-order charge-pump PLL has two poles at the origin, which makes the loop inherently unstable. A zero should be placed at a lower frequency than the crossover frequency to make the phase margin large enough (>45°). Since the zero reduces the slope of the magnitude response, an additional pole at a higher frequency than the crossover frequency is also required to maintain adequate spurious signal rejection. When a zero is located at ¼ of the crossover frequency and a pole is placed at 4 times of ω_c, the loop is critically damped with a damping ratio of 1 and a phase margin of 63° can be achieved. Normally, a critically damped loop works best for a typical PLL design.

9.6.3 Transient Response

One important aspect of the PLL performance is the manner in which it responds to a change to input frequency or phase, and one of the most important measures is its response to a step change of the input. The step error response of the PLL in the time domain represents the phase error in response to a unit phase step at the input or the frequency error in response to a unit frequency step at the input. In this case, the transient response will be characterized by the *rise time* t_r (the time for transient curve to first reach the value that will eventually be steady-state value), the *peaking time* t_p (the time for transient curve to reach the first peak), the *settling time* t_s (the time for the envelope of the response to reach and remain less than a specified fraction of the final steady-state value, for example, within 10%), and the *peak-overshoot ratio POR* (the ratio of the maximum value of the transient curve to its steady-state value), as shown in Figure 9.25 [41].

In the slow-acquisition PLL, the transient response of the VCO does not affect acquisition time. However, in the high-speed PLL, the transient response of the VCO does affect PLL acquisition performance because the time constants of the LF are set close to those of the VCO [44]. In this case, minimum settling time for a third-order charge-pump PLL can be provided by placing the three closed-loop poles given by the root locus at equal distances from the origin of the axis in the *s*-plane [45]. When the bandwidth is less than 1/20th of the input reference frequency, a good estimation of the settling time t_s can be defined as

$$t_s = \frac{1}{\varsigma\omega_c}\ln\left(\frac{\Delta\omega_{out}}{\Delta\omega_e}\right) \tag{9.63}$$

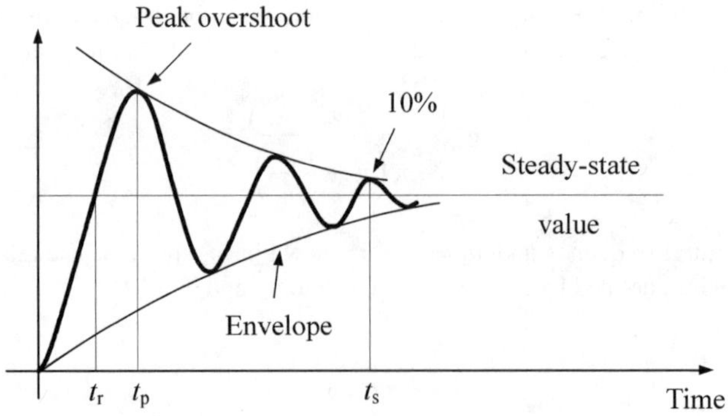

FIGURE 9.25 Transient response to step input change.

where $\Delta\omega_{out}$ is the frequency step and $\Delta\omega_e$ is the required frequency accuracy. As the loop bandwidth frequency ω_c increases, the settling time t_s gets shorter if the damping ratio ζ is fixed.

The effect of the damping ratio on settling time for a third-order charge-pump PLL is plotted in Figure 9.26, which shows that the settling time is fastest when the loop is critically damped, and further underdamping does not improve the settling time [43]. However, underdamping is not desirable, since it increases overshoot in transient response, which can be limited within the dynamic range of the charge-pump PLL. The overshoot in transient response also translates into the gain peaking in the frequency domain, which amplifies the phase noise of the reference signal at the PLL output. An overdamped condition can be recommended to use for better stability with narrow loop bandwidth. Generally, loop bandwidth and damping ratio have to be determined carefully, depending on the system requirements, since they improve some aspects of the performance, but deteriorate others at the same time.

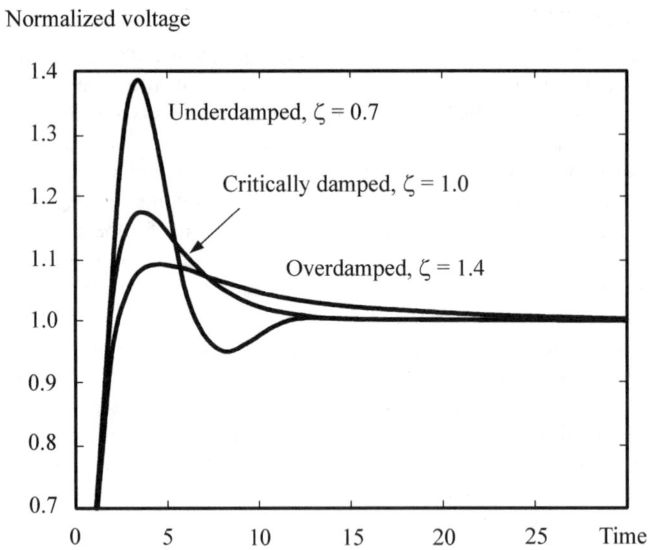

FIGURE 9.26 Transient responses of third-order PLL.

9.6.4 Noise

In general, the system noise bandwidth can be defined as the bandwidth B_n of a rectangular filter that has that nominal power gain (or unity in particular cases) within the passband and zero gain outside the passband and passes the same total power as does the actual filter when they are both excited by a uniformly distributed power density. As a result,

$$B_n = \frac{1}{2\pi}\int_0^\infty |H(j\omega)|^2 d\omega \tag{9.64}$$

where $H(j\omega)$ is the closed-loop transfer function.

For a first-order PLL,

$$B_n = \frac{1}{2\pi}\int_0^\infty \left|\frac{K/j\omega}{1+K/j\omega}\right|^2 d\omega = \frac{K}{4} \tag{9.65}$$

which means that the equivalent noise bandwidth extends beyond the 3-dB bandwidth by almost 60% for the single-pole response of the first-order loop since the ratio between the dc loop gain K and cutoff frequency ω_c is written as $\omega_c = 2\pi f_c = K$.

For the second-order PLL with a lag–lead filter, the noise bandwidth is calculated from

$$B_n = \frac{\omega_n}{2}\left(\varsigma + \frac{1}{4\varsigma}\right) \tag{9.66}$$

where ω_n and ζ are determined by Eqs. (9.57) and (9.58) [6].

For a first-order PLL, no LF exists and the PD is usually implemented using an analog multiplier or an XOR gate. Assuming no DIV and ideal PD, the phase noise at the PLL output will be represented by a combination of the input reference noise and VCO noise. As a result, the fundamental property of the PLL states that its phase noise is dominated by the input source noise at frequency offsets below the loop bandwidth and by the VCO noise at frequency offsets above the loop bandwidth. This is because the PLL is unable to react fast enough to fast random changes in the VCO output, but compensates the slow random variations produced by the VCO by adjusting its control voltage. Therefore, the PLL having a noisy VCO and a clean reference input should be designed to have a large loop bandwidth. Since the loop bandwidth is inversely related to the PLL settling time, such a PLL takes little time for locking and has a large noise reduction of the VCO noise. However, if the loop bandwidth is small, the PLL takes longer time for locking and leaves much of the VCO noise unsuppressed.

Higher order PLL employs the PFD, CP, and LF to ensure zero static phase error. The major internal noise sources of the charge-pump PLL include the DIV noise, charge-pump noise, LF noise, and VCO noise. In addition, additive noise, predominantly thermal, within the PFD gives rise to timing jitter in the edges of the output pulses, which is translated to equivalent phase jitter, thus injecting this noise into the system [46]. The LF and VCO are subject to high-pass filtering by loop action and are thus significant sources of out-of-band phase noise, whereas the DIV and CP are subject to low filtering by loop action and are thus significant sources of in-band phase noise. The excess noise of the DIV and CP can be modeled by an additive $1/f$ noise, whereas the VCO noise and low-pass LF noise can be mainly represented by $1/f^2$ noise characteristic. If the loop bandwidth is large, the PLL has a large reduction in $1/f^2$ noise, but cannot provide a sufficient suppression of $1/f$ noise. However, if the loop bandwidth is small, the charge-pump PLL provides a large $1/f$ noise reduction, but leaves much of the $1/f^2$ noise unsuppressed. Since the transfer functions of the DIV and CP noise sources to the output represent the low-pass characteristics and the transfer functions of the VCO and LF to the output are described by the high-pass characteristics, an optimum phase noise spectrum of the system can be achieved for the 3-dB bandwidth for low- and high-pass transfer functions when power spectrum densities for both $1/f$ and $1/f^2$ noises sources intersect [47].

Consequently, the main contributors to the output phase noise response at offsets lower than the loop bandwidth or in-band noise sources are the reference oscillator noise, DIV noise, and CP noise. In this case, by using a clean reference oscillator (such as a crystal or dielectric resonator oscillator), a large CP gain, and a low frequency division ratio the phase noise and jitter at the PLL output can be reduced at low frequency offsets. However, an increased CP gain may affect the VCO control voltage due to a small net current injection in a locked condition causing a shift in VCO frequency. This in turn produces a phase error at the charge-pump input due to the feedback loop that tends CP to inject a corrective current pulse to bring the VCO frequency back to its previous value. As a result, the VCO frequency toggles between two values under the locked condition. In a frequency domain, this effect will produce spurs at the reference frequency or multiples of the reference frequency. The magnitude of these spurs is directly proportional to the ripple in the control voltage and hence the charge-pump gain. Reference spurs should be at least 50 dB below the PLL output frequency to ensure a clean spectrum. At frequency offsets larger the loop bandwidth, the main contributor is the VCO noise. In this case, a low K_{VCO} is required for a low phase noise PLL. However, reducing K_{VCO} will make the loop dynamics slower, and the PLL will take a long time to correct abrupt changes in frequency and require a large time to lock during start-up. In addition, the effect of the thermal noise of the resistor in the LF should be minimized, for example, by optimizing the physical dimensions of the filter MOS capacitors or in conjunction with op-amp noise [48].

9.7 PHASE MODULATION USING PHASE-LOCKED LOOPS

By recognizing the loop transfer function properties, the VCO can be used as a phase or frequency modulator. The VCO control port varies the VCO frequency, so the VCO can be used as both the carrier generator and the modulator. If the modulating signal is inserted into the circuit in front of the LF, the result is a VCO phase-modulated signal, provided that the signal bandwidth is below the cutoff frequency of the LPF transfer function. If the modulating signal is inserted after the LF, then the transfer function to the VCO output represents a high-pass response. In this case, frequency modulation is generated, and the modulating signal bandwidth must not extend below the loop cutoff frequency. Thus, no frequency offset can be applied, because the rejection at dc is infinite. A VCO with a linear tuning characteristic is required in this latter case, and highly suggested in the former one. However, this type of the VCO tuning characteristic requires a more complicated and carefully designed oscillator circuit.

When a PLL is operated as an open-loop modulator, modulation is disabled just prior to a modulating signal transmission burst in order to allow the PLL circuit to stabilize. Once the PLL is stabilized, modulation is applied for a short time period. Therefore, the open-loop modulators operate only in short bursts between tuning cycles. As a result, the open-loop nature of the modulator results in a PLL free-running during bursts, thus increasing its phase noise. In addition, the PLL conventionally requires precise calibration to ensure proper operation when used in an open-loop configuration. In a closed-loop modulator scheme, the output of a VCO is fed back for minimizing the phase difference between the VCO output and a reference signal input. In this case, frequency modulation can be performed by directly modulating the input of a VCO with a narrowband modulating signal when the bandwidth of the modulating signal is less than the closed-loop bandwidth of the VCO. The closed-loop dynamics of a PLL do not appreciably distort modulation present in the VCO feedback signal as long as the modulation bandwidth is less than the closed-loop bandwidth of the PLL. Signal distortion occurs when a wideband modulation signal is directly applied to a VCO, the output of which is fed back to an input of a closed-loop PLL. The closed-loop dynamics of a PLL, in particular the limited bandwidth of the LF, result in attenuation of the high frequency components of a wideband

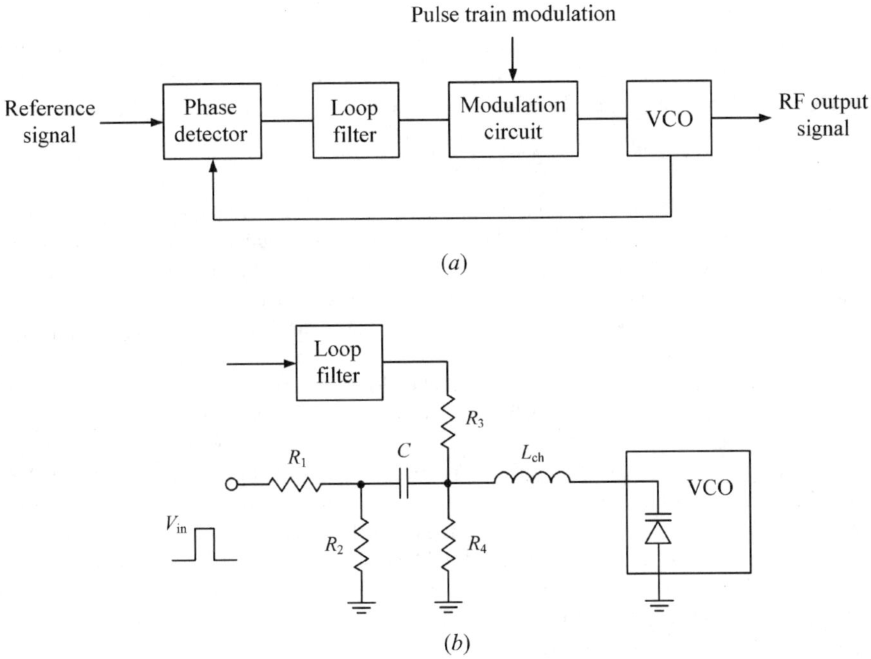

FIGURE 9.27 Schematics of phase modulation PLL circuit.

modulating signal and distort the group delay of the modulating signal. One possibility for achieving wideband modulation using a closed-loop PLL involves injecting some components of a wideband modulating signal into one part of the PLL and other components into a different part of the PLL. For example, the high-frequency components can directly modulate the input of a VCO, while the low-frequency components alter the instantaneous division ratio of the PLL feedback path. However, equalization is needed in this case to cancel overlap between the high and low frequency components in order to prevent the growth of signal distortion.

Figure 9.27(*a*) shows the schematic of a PLL with phase modulation where the modulating signal is applied to the VCO through the modulation circuit inserted between the LF and the VCO and the PLL is used only for frequency stabilization [49]. When the modulating signal, whose frequency is much higher than the loop bandwidth, is introduced to modulate the VCO, the loop does not respond to this modulating signal. The schematic of the modulation circuit is shown in Figure 9.27(*b*), where the modulating signal is capacitively coupled to the VCO input, converting the modulating square wave signal into pulse signal. The modulating signal and error voltage from the LF are added together and applied to the VCO. The phase change due to the pulse modulation can be written as

$$\Delta\theta = \int_0^t \Delta\omega(t)\, dt \tag{9.67}$$

where $\Delta\omega(t)$ is the frequency change due to the modulating signal.

By assuming that the oscillation frequency of the VCO changes instantly with the tuning voltage, the frequency change can be represented in terms of an exponential function as

$$\Delta\omega(t) = K_{\mathrm{VCO}}\Delta V(t) = \pm K_{\mathrm{VCO}} V_{\mathrm{p}} \exp\left(-\frac{t}{RC}\right) \tag{9.68}$$

where $R = R_1' + R_2'$, $V_p = (R_2'/R)(R_1'/R_1)\,V_{in}$, $R_1' = R_1R_2/(R_1 + R_2)$, and $R_2' = R_3R_4/(R_3 + R_4)$. Substituting Eq. (9.68) into Eq. (9.67) and using the conditions that the time constant RC is much smaller than the period of the modulating signal and that the modulating frequency is much higher than the loop bandwidth, the phase shift due to the modulating signal can be simplified to

$$\Delta\theta(t) = \pm K_{VCO} V_p RC \tag{9.69}$$

from which it follows that the modulated phase is determined by the frequency tuning sensitivity of the VCO, modulating signal amplitude, and time constant of the modulation circuit. In this case, the input data stream is restricted to have a zero mean value; otherwise any nonzero mean value would cause the loop to pull back the desired phase shift due to the modulation.

The block diagram of an open-loop modulation shown in Figure 9.28(*a*) is used in DECT (Digital European Cordless Telephone) system. Its output carrier frequency range is from 1.88 to 1.90 GHz with Gaussian frequency-shift keying (GFSK) modulation and high data rate of 1.152 Mb/s. In this case, the loop is initially closed to the RF output to provide $f_{out} = Nf_{ref}$. By putting the CP output into high-impedance mode, the loop is opened and the modulation data is fed to the Gaussian filter. The modulating voltage then appears at the VCO input where it is multiplied by K_{VCO}. When the data burst finishes, the loop is returned to the closed-loop mode of operation. It should be noted that the voltage drift, and hence the system frequency drift, is directly dependent on the leakage current of the CP, when CP is in the high-impedance mode. The block schematic of the direct Gaussian minimum shift keying (GMSK) modulator is shown in Figure 9.28(*b*), where a low-power MSK-modulated carrier signal is generated as a reference signal for the PLL at the desired output microwave transmit frequency [50]. The PLL allows the VCO to frequency track the stable MSK-modulated input signal, while providing Gaussian filtering to convert the input minimum shift keying (MSK) signal spectrum to the GMSK signal spectrum at the VCO output. To realize Gaussian filtering, the loop response must approximate the Gaussian LPF response. Since MSK is equivalent to CPFSK (continuous phase frequency-shift keying) with a modulation index of 0.5, a CPFSK modulator, which is realized by integrating the baseband binary information signal, can be used to phase modulate the carrier. The CPFSK modulator is capable of providing full 360° linear control of the signal phase.

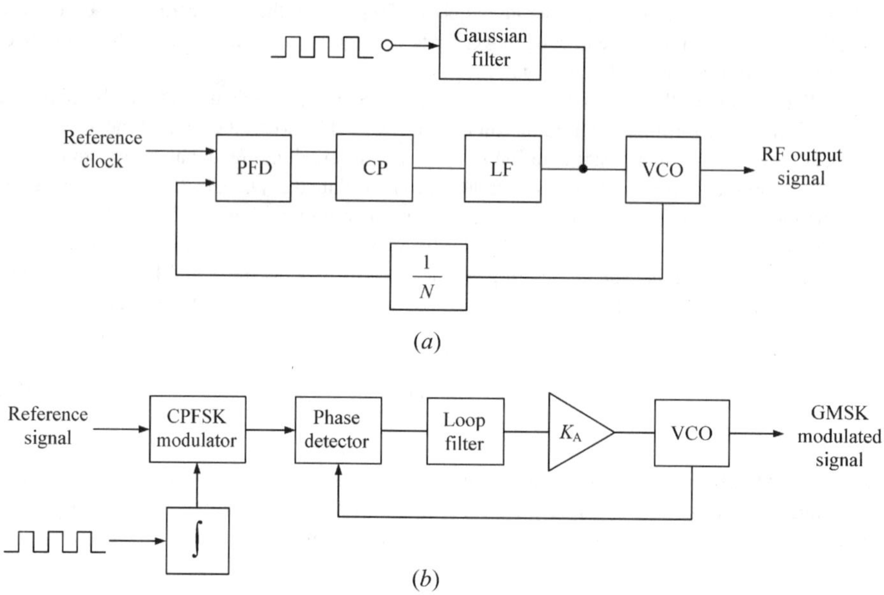

FIGURE 9.28 Block diagrams of phase-modulated PLL.

9.8 FREQUENCY SYNTHESIZERS

Frequency synthesizer techniques can be separated into three general classes: direct analog synthesis, indirect synthesis, and direct digital synthesis [51]. A direct analog synthesizer is based on frequency mixing, multiplication, and switching with appropriate filtering to achieve the desired output performance when the frequency of the reference oscillator or oscillators is translated directly. Indirect synthesizers utilize PLLs with VCOs. In direct digital synthesizers (DDS), the output waveform is directly synthesized digitally from the reference oscillator that is used as a clock for the digital operation. The proper choice of the synthesizer type is based on such parameters as the number of frequencies, frequency spacing, frequency switching time, noise, spurious level, particular technology, and cost.

9.8.1 Direct Analog Synthesizers

This direct analog synthesizer creates its output frequency by mixing two or more signals to produce sum or difference frequency followed by frequency multiplication, division, or phase locking [6,52]. The output signals can be extracted from low-frequency (with crystal or SAW resonators) or high-frequency oscillators (with dielectric, cavity, or sapphire resonators) by frequency multiplication, division, or phase locking. The key advantages of direct frequency synthesizers are fast frequency switching and capability to generate signals with very low phase noise. This low-noise performance is achieved by the selection of topologies and components such that the additive phase noise of all the components is considerably smaller than the multiplied phase noise of the reference oscillators that determine the output phase noise of the synthesizer. The drawbacks of direct synthesizers are that they are hardware intensive and tend to generate an excessive number of spurious signals.

Figure 9.29(*a*) shows the simplest type of the direct analog synthesizer, where two crystal oscillators are applied to a mixer and the resulting output is then filtered to produce the desired frequency [35]. Each oscillator has a number of crystals that are closely spaced in frequency. For example, if oscillator f_1 has 10 crystals in the range of 20.000–20.450 MHz, each frequency differs from the other by 50 kHz. Similarly assume that the other oscillator f_2 has 10 crystals at 5-kHz spacings in the range of 25.000–25.045 MHz. By choosing appropriate combinations, 100 output frequencies are available in the range of 45.000–45.495 MHz, each separated by 5 kHz. The bandpass filter selects only the sum frequency, and it should offer good suppression of the 5-MHz image as well as harmonics of the two oscillators.

Figure 9.29(*b*) shows the block diagram of a direct analog synthesizer that consists of a set of the separate reference oscillators that are used twice to generate the output signal with desired frequency [52,53]. In this case, when one reference oscillator is selected, its signal is sent simultaneously to a mixer directly and through the frequency multiplier with multiplication factor N. The output signal represents any sum or difference frequency $f_{\mathrm{out}} = Nf_{\mathrm{i}} \pm f_{\mathrm{j}}$, where $1 \leq i, j \leq 4$, yielding 16 frequencies at the output for a set of four reference oscillators. Variations on this concept would include the use of more than one multiplier and the use of multipliers in the direct path. An equivalent architecture could use DIVs such that the direct and divided paths to mixer would produce the same result, provided the reference frequencies were appropriately higher. This technique is often used to synthesize frequencies with spacings that are smaller than the reference frequency spacings. Frequency filtering requires the use of narrow-band filters to reduce spurs and harmonic components. However, this can introduce several problems when the output frequency range is restricted to the filter passband, the settling time is inversely proportional to the filter passband, and large group delays are provided by narrow-band filters.

9.8.2 Integer-*N* Synthesizers Using PLL

The most popular technique of frequency synthesis is based on the use of a PLL offering smaller step size, reduced spurious levels, and lower complexity in comparison with direct analog schemes

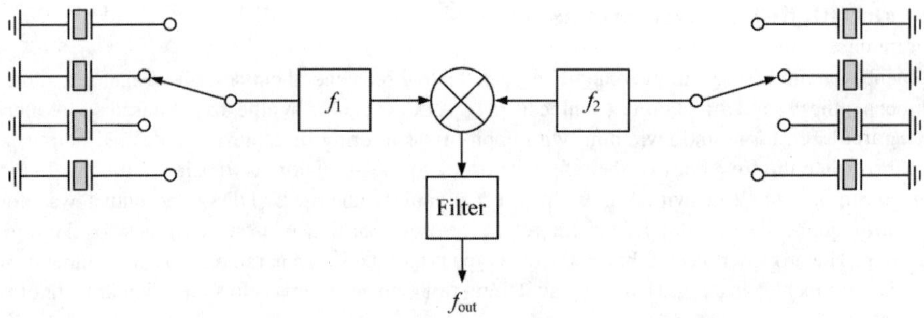

(a)

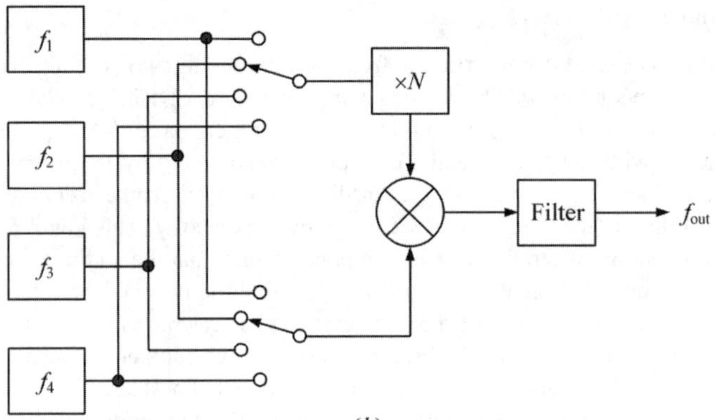

(b)

FIGURE 9.29 Block diagrams of direct frequency synthesizers.

[54,55]. A typical single-loop PLL charge-pump integer-N synthesizer, whose block diagram is shown in Figure 9.30, includes a VCO generating a signal that is fed back to a PFD through a DIV with a variable frequency division ratio N. The reference signal that is set to a desirable step size using a DIV with a division ratio R is applied to the other input of the detector. Once the loop is locked, the output frequency becomes equal to the reference frequency as $f_{out} = (N/R)f_{ref}$. The frequency N divider (or counter) is a programmable component that provides a proper channel selection. In this case, if 200-kHz spacing is required for a 900-MHz output, this desired frequency spacing can be achieved by using a stable crystal-based 10-MHz reference oscillator and dividing it down by 50. Then, the N-value in the feedback path would need to be of the order of 4500, which means that at least a

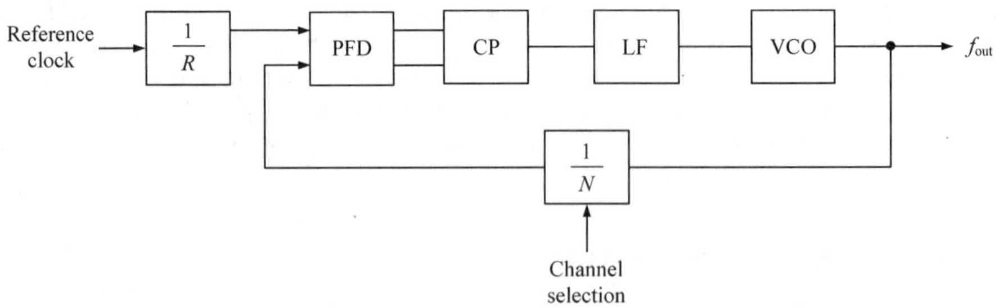

FIGURE 9.30 Block diagram of integer-N synthesizer.

13-bit counter is required. If the required output frequency is too high, it makes sense to precede the programmable counter with a fixed counter element, which is called the *prescaler*, to bring the very high input frequency down to a range at which standard CMOS technology can be used.

Integer-N synthesizer architecture is the preferred solution for minimizing power consumption and die area due to its simplicity. However, it lacks the flexibility of arbitrary choosing f_{ref}, since it is fixed by certain integer division ratios. Also, the drawbacks of this scheme are longer frequency switching time, which is inversely proportional to the loop bandwidth and consequently step size, and considerable higher phase noise in comparison with direct analog techniques. With integer-N frequency synthesis, it is relatively easy to achieve good performance on any single parameter if the other parameters are disregarded. In addition, the following combinations of performance benefits can be achieved at certain costs: small step size and low phase noise (at least at relatively high offset frequencies) at the cost of slow tuning speed; fast tuning speed and low phase noise if large step size is allowed; or fast tuning speed and moderately small step size at the cost of high reference feedthrough spurs. However, it is difficult to have small step size, fast tuning speed, low phase noise, and low reference feedthrough spurs at the same time. A PLL synthesizer with high switching speed, which is composed of digital signal processing (DSP) except for the VCO, is characterized by the acquisition time that is as short as 0.1 ms and the phase noise that is less than -60 dB at the VCO output in the steady-state condition [56].

9.8.3 Fractional-N Synthesizers Using PLL

The fractional-N approach to frequency synthesis enables fast dynamics to be achieved within the PLL by allowing a high reference frequency with fractional feedback ratios [57,58]. The fractional-N synthesizer, whose block diagram is shown in Figure 9.31, includes a dual modulus DIV that can switch its division ratio between N and $N + 1$. By dividing the VCO frequency by N during K cycles of the VCO and by $N + 1$ during $(2^k - K)$ VCO cycles, it is possible to make the average division ration equal to $N + K/2^k$, assuming a k-bit accumulator controlling the prescaler. One of the advantages of this type of synthesizer is that the synthesizer step can be made arbitrary small by adding more bits to the $\Delta\Sigma$ modulator. For example, with a 20-bit $\Delta\Sigma$ fractional-N synthesizer and a 16-MHz reference frequency, the output frequency of the synthesizer is programmable in steps less than 16 Hz [59]. The higher resolution of the frequency division ratio contributes to the use of a higher input frequency and wider loop bandwidth while maintaining channel frequency spacing.

However, if the division modulus is switched periodically, the output is modulated by the beat frequency of the fractional modulus, resulting in the fractional spurs that are very close to the carrier. The fractional spurs can be reduced by breaking the regularity of the division modulus switching period, effectively making the beat frequency randomized. A dithering mechanism using

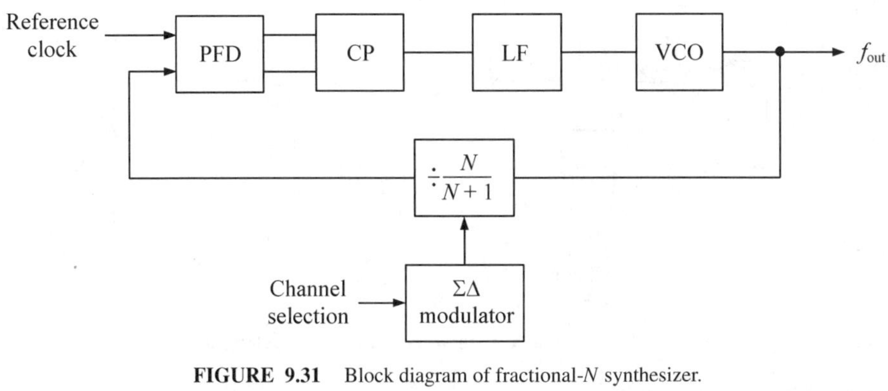

FIGURE 9.31 Block diagram of fractional-N synthesizer.

$\Delta\Sigma$ modulator can not only randomize the beat frequency, but shape the noise spectrum so that it has more power at higher frequency. The high-frequency quantization noise is filtered by the LF of the PLL. A combination of the order of the $\Delta\Sigma$ modulator, its sampling rate, and LF order can reduce the high-frequency quantization noise at levels that make the effect of the noise negligible [43]. A high sampling rate for the $\Delta\Sigma$ modulator results in the quantization noise that is distributed over a wide frequency range. Note that the quantization noise for an nth-order $\Delta\Sigma$ modulator only exhibits an $n-1$ order of noise-shaping characteristic because the quantization phase error is a time-integrated version of the modulator output [60]. The quantization noise in a $\Delta\Sigma$ fractional-N synthesizer can be significantly reduced using an all-digital quantization noise cancellation technique when excellent in-band and out-of-band phase noise performance can be achieved [61]. In the digital synthesis technique, the quantized phase error word is sent to a digital LF, which processes the error word and outputs a VCO control word to a digital-to-analog converter (DAC), which then turns this control word into a control voltage for the VCO. Generally, the PLLs for $\Delta\Sigma$ fractional-N synthesizer can be fully integrated using both bipolar junction transistor (BJT) and CMOS technologies [28,62]. The radio transmitter can be realized using this fractional-N approach that performs phase/frequency modulation in a continuous manner such as GMSK or GFSK by direct modulation of the synthesizer [63].

Offset PLLs offer another method of frequency synthesis [52,53]. Figure 9.32 shows a block diagram of a single-offset analog synthesizer where the VCO frequency is heterodyned to a lower frequency using the signal of an offset generator. The offset signal can be produced using an additional PLL or a chain of frequency multipliers. Frequency agility can be achieved with a multifrequency offset generator in conjunction with a fixed-frequency or a multifrequency reference. An arbitrary frequency resolution can be realized by successively heterodyning the VCO frequency with several multifrequency offset generators with successively finer frequency resolution. One of the problems associated with any frequency-mixing scheme is a possible false lock due to undesired mixing products. This type of failure requires a sufficiently accurate coarse-tuning mechanism. In this case, a digital-to-analog converter may be included to coarse-tune the VCO, the tuning characteristic of which should be linear over operating temperature range. Besides, it is required to provide sufficient isolation between the mixer ports and the synthesizer output.

Figure 9.33 shows the block schematic of a dual-loop synthesizer where the variable offset signal can be generated with another PLL. In this case, assuming a 1 MHz-step synthesizer operating between 9 and 10 GHz, the first loop provides 9–10 GHz frequency coverage with a 100 MHz step size by varying the division ratio N_1 between 90 and 100. At the same time, the output of the first PLL is used as an offset signal for the second loop to keep the mixer output below 100 MHz. Thus, for the desired 1 MHz step size, the maximum division ratio for the second loop does not exceed 100 as well. The phase noise degradation for both loops, which is set by the maximum division ratio, will not exceed 40 dB compared to 80 dB for the single-loop case. Therefore, splitting the design

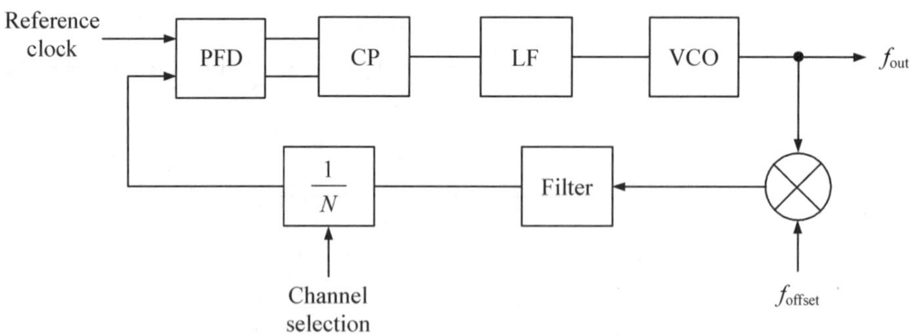

FIGURE 9.32 Single-offset analog synthesizer architecture.

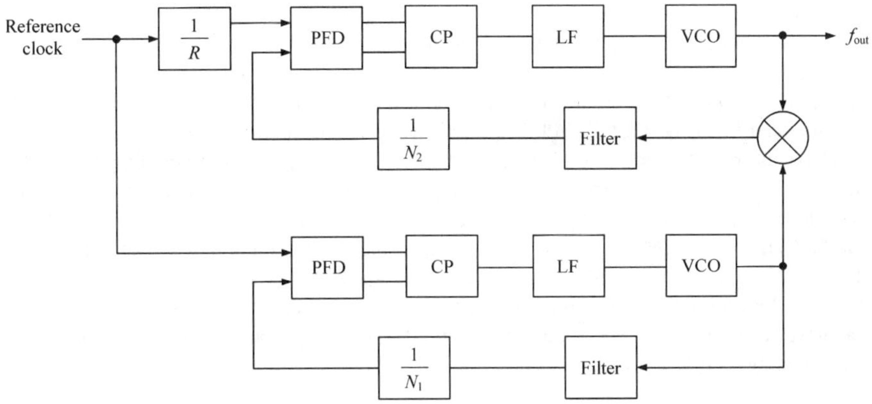

FIGURE 9.33 Dual-loop analog synthesizer architecture.

in two loops can potentially result in overall 40-dB phase noise improvement. Greater phase noise improvement due to smaller step sizes can be achieved using a larger number of loops.

9.8.4 Direct Digital Synthesizers

The DDS appeared in 1971 are the most recent addition to frequency synthesis architectures [64]. Generally, the implementation of the DDS is divided into two basic parts: a discrete phase generator (the accumulator) outputting a phase value and a phase-to-waveform converter outputting the desired signal. The block diagram of the direct digital synthesizer is shown in Figure 9.34 [65,66]. In response to digital commands, an accumulator generates a digital approximation of a linearly increasing phase function, at a rate controlled by a reference oscillator. The output of the accumulator is applied to a ready-only-memory (ROM) lookup table that converts the phase samples into samples of a sinusoidal waveform. The waveform can be defined with up to 2^N phase values, where N is a size (or word length) of the accumulator. The ROM output is then fed into the DAC, which generates an analog approximation of the sinusoidal waveform that, after filtering by LPF, represents the output of the direct digital synthesizer.

The main advantages of the DDS architecture are fast and phase-continuous frequency switching, arbitrarily small frequency spacing, small size, and low cost. For example, the DDS can cover a bandwidth from dc to 75 MHz in steps of 0.035 Hz with a switching speed of 6.7 ns [67]. Its main drawbacks are a limited operating frequency and relatively high noise and spurious signal levels. The clock of the digital circuitry (or sampling frequency f_s) has to be at least twice as high as the output frequency f_{out}, therefore operating a ROM and a linear DAC at 4.8 GHz to generate 2.4 GHz output signals can be challenging using current technologies, and power consumption will be too high. However, from practical point of view, lower output frequency is often preferred, usually

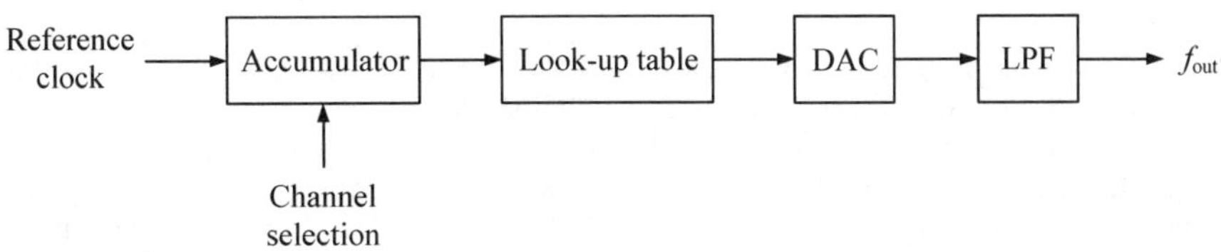

FIGURE 9.34 Block diagram of direct digital synthesizer.

$f_{out(max)} = f_s/4$, for easier analog signal reconstruction. At the same time, the minimum output frequency that can be generated by DDS is defined as $f_{out(min)} = f_s/2^N$. Large quantization noise and harmonic distortion of a high-speed DAC can significantly degrade the spectral purity of the output signal. The DDS power dissipation can be reduced if a nonlinear DAC is used in place of the ROM lookup table and linear DAC [68]. In this case, the function of the nonlinear DAC is to convert the digital phase information from the phase accumulator directly into analog sine output signal. As a result, with 8-bit phase accumulator and sine-weighted DAC, the DDS is capable to operate up to a 32-GHz clock frequency for all frequency control words and synthesize sine-wave outputs from 125 MHz to 16 GHz in 125-MHz steps [69]. By using a 11-bit pipeline phase accumulator, a 10-bit sine-weighted DAC segmented in a 6-bit coarse and eight 3-bit fine DACs, a low-power ultrahigh-speed and high-resolution DDS implemented in a 0.13-μm SiGe BiCMOS technology with $f_{max} = 250$ GHz achieves a maximum clock frequency of 8.6 GHz and a spurious-free dynamic range (SFDR) of approximately 45 dBc with a 4.2958-GHz output [70].

REFERENCES

1. E. V. Appleton, "Automatic Synchronization of Triode Oscillators," *Proc. Camb. Phil. Soc.*, vol. 21, pt. III, p. 231, 1922–1923.
2. H. de Bellescize, "La Reception Synchrone," *L'Onde Electrique*, vol. 11, pp. 230–240, June 1932.
3. F. M. Gardner, *Phaselock Techniques*, New York: John Wiley & Sons, 2005.
4. G.-C. Hsieh and J. C. Hung, "Phase-Locked Loop Techniques – A Survey," *IEEE Trans. Ind. Electron.*, vol. IE-43, pp. 609–615, Dec. 1996.
5. J. Frank, A. Newton, and L. Pugliese, "Solid State, Phase-Locked, Tunable UHF Power Oscillators," *IEEE Trans. Aerospace Navigational Electron.*, pp. 3.3.2/1–3.3.2/8, Oct. 1963.
6. V. Manassewitsch, *Frequency Synthesizers: Theory and Design*, New York: John Wiley & Sons, 1976.
7. D. Banerjee, *PLL Performance, Simulation and Design*, National Semiconductor, 2006.
8. W. F. Egan, *Phase-Lock Basics*, New York: John Wiley & Sons, 2008.
9. S. C. Gupta, "Phase-Locked Loops," *Proc. IEEE*, vol. 63, pp. 291–306, Feb. 1975.
10. B. Razavi, *Monolithic Phase-Locked Loops and Clock Recovery Circuits: Theory and Design*, New York: Wiley-IEEE Press, 2008.
11. A. Carlosena and A. Manuel-Lazaro, "Design of High-Order Phase-Locked Loops," *IEEE Trans. Circuits and Systems–II: Express Briefs*, vol. CAS-II-54, pp. 9–13, Jan. 2007.
12. F. M. Gardner, "Charge-Pump Phase-Locked Loops," *IEEE Trans. Commun.*, vol. COM-28, pp. 1849–1858, Nov. 1980.
13. Z. Wang, "An Analysis of Charge-Pump Phase-Locked Loops," *IEEE Trans. Circuits and Systems–I: Regular Papers*, vol. CAS-I-52, pp. 2128–2138, Oct. 2005.
14. J. Lee, M. Liu, and H. Wang, "A 75-GHz Phase-Locked Loop in 90-nm CMOS Technology," *IEEE J. Solid-State Circuits*, vol. SC-43, pp. 1414–1426, June 2008.
15. S.-R. Han, C.-N. Chuang, and S.-I. Liu, "A Time-Constant Calibrated Phase-Locked Loop with a Fast-Locked Time," *IEEE Trans. Circuits and Systems–II: Express Briefs*, vol. CAS-II-54, pp. 34–37, Jan. 2007.
16. W. C. Lindsey and C. M. Chie, "A Survey of Digital Phase-Locked Loops," *Proc. IEEE*, vol. 69, pp. 410–431, Apr. 1981.
17. R. B. Staszewski and P. T. Balsara, "Phase-Domain All-Digital Phase-Locked Loop," *IEEE Trans. Circuits and Systems–II: Express Briefs*, vol. CAS-II-52, pp. 159–163, Mar. 2005.
18. R. B. Staszewski, J. I. Wallberg, S. Rezeq, C.-M. Hung, O. E. Eliezer, S. K. Vemulapalli, C. Fernando, K. Maggio, R. Staszewski, N. Barton, M.-C. Lee, P. Cruise, M. Entezari, K. Muhammad, and D. Leipold, "All-Digital PLL and Transmitter for Mobile Phones," *IEEE J. Solid-State Circuits*, vol. SC-40, pp. 2469–2482, Dec. 2005.
19. H. L. Krauss, C. W. Bostian, and F. H. Raab, *Solid State Radio Engineering*, New York: John Wiley & Sons, 1980.

20. H. O. Johansson, "A Simple Precharged CMOS Phase Frequency Detector," *IEEE J. Solid-State Circuits*, vol. SC-33, pp. 295–299, Feb. 1998.

21. C. T. Charles and D. J. Allstot, "A Calibrated Phase/Frequency Detector for Reference Spur Reduction in Charge-Pump PLLs," *IEEE Trans. Circuits and Systems–II: Express Briefs*, vol. CAS-II-53, pp. 822–826, Sept. 2006.

22. J. Rogers, C. Plett, and F. Dai, *Integrated Circuit Design for High-Speed Frequency Synthesis*, Norwood, MA: Artech House, 2006.

23. S. Milicevic and L. MacEachern, "A Phase-Frequency Detector and a Charge Pump Design for PLL Applications," *2008 IEEE Int. Symp. Circuits and Systems Dig.*, pp. 1532–1535.

24. W. Rhee, "Design of High-Performance CMOS Charge Pumps in Phase-Locked Loops," *1999 IEEE Int. Symp. Circuits and Systems Dig.*, vol. II, pp. 545–548.

25. W. J. Gruen, "Theory of AFC Synchronization," *Proc. IRE*, vol. 41, pp. 1043–1048, Aug. 1953.

26. D. Byrd, C. Davis, and W. O. Keese, *A Fast Locking Scheme for PLL Frequency Synthesizers*, National Semiconductor, AN1000, 1995.

27. P. K. Hanumolu, M. Brownlee, K. Mayaram, and U.-K. Moon, "Analysis of Charge-Pump Phase-Locked Loops," *IEEE Trans. Circuits and Systems–I: Regular Papers*, vol. CAS-I-51, pp. 1665–1674, Sept. 2004.

28. J. Craninckx and M. S. J. Steyaert, "A Fully Integrated CMOS DCS-1800 Frequency Synthesizer," *IEEE J. Solid-State Circuits*, vol. SC-33, pp. 2054–2065, Dec. 1998.

29. D. G. Tucker, "The Synchronization of Oscillators," *Electronic Engineer*, vol. 15, pp. 412–418, Mar. 1943, pp. 457–461, Apr. 1943, vol. 16, pp. 26–30, June 1943.

30. R. Adler, "A Study of Locking Phenomena in Oscillators," *Proc. IRE*, vol. 34, pp. 351–357, June 1946.

31. B. Razavi, "A Study of Injection Locking and Pulling in Oscillators," *IEEE J. Solid-State Circuits*, vol. SC-39, pp. 1415–1424, Sept. 2004.

32. R. L. Miller, "Fractional-Frequency Generators Utilizing Regenerative Modulation," *Proc. IRE*, vol. 27, pp. 446–457, July 1939.

33. A. Safarian, S. Anand, and P. Heydari, "On the Dynamics of Regenerative Frequency Dividers," *IEEE Trans. Circuits and Systems–II: Express Briefs*, vol. CAS-II-53, pp. 1413–1417, Dec. 2006.

34. C. Kromer, G. von Buren, G. Sialm, T. Morf, F. Ellinger, and H. Jackel, "A 40-GHz Static Frequency Divider with Quadrature Outputs in 80-nm CMOS," *IEEE Microwave and Wireless Components Lett.*, vol. 16, pp. 564–566, Oct. 2006.

35. W. Hayward, *Introduction to Radio Frequency Design*, Newington, CT: ARRL, 1995.

36. L. F. Fei, "Frequency Divider Design Strategies," *RF Design*, pp. 18–26, Mar. 2005.

37. S. Levantino, L. Romano, S. Pellerano, C. Samori, and A. L. Lacaita, "Phase Noise in Digital Frequency Dividers," *IEEE J. Solid-State Circuits*, vol. SC-39, pp. 775–784, May 2004.

38. A. Grebennikov, *RF and Microwave Transistor Oscillator Design*, London: John Wiley & Sons, 2007.

39. E. Hegazi, H. Sjoland, and A. A. Abidi, "A Filtering Technique to Lower *LC* Oscillator Phase Noise," *IEEE J. Solid-State Circuits*, vol. SC-36, pp. 1921–1930, Dec. 2001.

40. R. B. Staszewski, C.-M. Hung, N. Barton, M.-C. Lee, and D. Leipold, "A Digitally Controlled Oscillator in a 90 nm Digital CMOS Process for Mobile Phones," *IEEE J. Solid-State Circuits*, vol. SC-40, pp. 2203–2211, Nov. 2005.

41. C. S. Weaver, "A New Approach to the Linear Design and Analysis of Phase-Locked Loops," *IEEE Trans. Space Electron. Telemetry*, vol. SET-5, pp. 166–178, Dec. 1959.

42. G. Nash, *Phase-Locked Loop Design Fundamentals*, Freescale Semiconductor, AN535, 2006.

43. S. T. Moon, A. Y. Valero-Lopez, and H. Sanchez-Sinencio, "Fully Integrated Frequency Synthesizers: A Tutorial," *Int. J. High Speed Electron. Syst.*, vol. 15, pp. 353–375, June 2005.

44. A. Shibutani, T. Saba, S. Moro, and S. Mori, "Transient Response of Colpitts-VCO and Its Effect on Performance of PLL System," *IEEE Trans. Circuits and Systems–I: Fundamental Theory Appl.*, vol. CAS-I-45, pp. 717–725, July 1999.

45. S. Levantino, M. Milani, C. Samori, and A. L. Lacaita, "Fast-Switching Analog PLL with Finite-Impulse Response," *IEEE Trans. Circuits Syst.–I: Fundam. Theory Appl.*, vol. CAS-I-51, pp. 1697–1701, Sept. 2004.

46. I. Thompson and P. V. Brennan, "Phase Noise Contribution of the Phase/Frequency Detector in a Digital PLL Frequency Synthesizer," *IEE Proc. Circuits Devices Syst.*, vol. 150, pp. 1–5, Feb. 2003.

47. B. Chuang, Z. Zhenyu, and Z. Minxuan, "A Low-Noise PLL Design Achieved by Optimizing the Loop Bandwidth," *J. Semiconduct.*, vol. 30, pp. 085011/1–085011/4, Aug. 2009.
48. L. Lascari, "Accurate Phase Noise Prediction in PLL Synthesizers," *Appl. Microwave Wireless*, vol. 12, pp. 30–38, Apr. 2000, pp. 90–96, May 2000.
49. W. Wang and L. W. Pearson, "Phase Modulation of a Loop Phase-Locked Grid Oscillator Array," *IEEE Microwave Wireless Components Lett.*, vol. 11, pp. 441–443, Nov. 2001.
50. D. M. Klymyshyn, S. Kumar, and A. Mohammadi, "Direct GMSK Modulation with a Phase-Locked Power Oscillator," *IEEE Trans. Vehicular Technol.*, vol. VT-48, pp. 1616–1625, Sept. 1999.
51. V. Reinhardt, K. Gould, K. McNab, and M. Bustamante, "A Short Survey of Frequency Synthesizer Techniques," *Proc. 40th Annual Frequency Control Symp.*, pp. 355–365, 1986.
52. Z. Galani and R. A. Campbell, "An Overview of Frequency Synthesizers for Radars," *IEEE Trans. Microwave Theory Tech.*, vol. MTT-39, pp. 782–790, May 1991.
53. A. Chenakin, "Building a Microwave Frequency Synthesizer," *High Frequency Electronics*, vol. 7, pp. 58–67, May, pp. 18–26, June, pp. 18–26, July, pp.44–51, Aug., pp. 50–58, Sept. 2008.
54. U. L. Rohde, *Microwave and Wireless Synthesizers: Theory and Design*, New York: John Wiley & Sons, 1997.
55. V. Kroupa, *Phase Lock Loops and Frequency Synthesis*, New York: John Wiley & Sons, 2003.
56. A. Kajiwara and M. Nakagawa, "A New PLL Frequency Synthesizer with High Switching Speed," *IEEE Trans. Vehicular Technol.*, vol. VT-41, pp. 407–413, Nov. 1992.
57. T. A. D. Riley, M. A. Copeland, and T. A. Kwasniewski, "Delta-Sigma Modulation in Fractional-*N* Frequency Synthesis," *IEEE J. Solid-State Circuits*, vol. SC-28, pp. 553–559, May 1993.
58. U. L. Rohde, *Synthesizer Design for Microwave Applications*, Synergy Microwave Corporation, 1999.
59. T. Riley and J. Kostamovaara, "A Hybrid $\Sigma\Delta$ Fractional-N Frequency Synthesizer," *IEEE Trans. Circuits and Systems—II: Analog and Digital Processing*, vol. CAS-II-50, pp. 176–180, Apr. 2003.
60. *Basics of Dual Fractional-N Synthesizers/PLLs*, Skyworks Solutions, 101463A, 2001.
61. C.-M. Hsu, M. Z. Straayer, and M. H. Perrott, "A Low-Noise Wide-BW 3.6-GHz Digital $\Delta\Sigma$ Fractional-N Frequency Synthesizer with a Noise-Shaping Time-to-Digital Converter and Quantization Noise Cancellation," *IEEE J. Solid-State Circuits*, vol. SC-43, pp. 2776–2786, Dec. 2008.
62. W.-Z. Chen and J.-T. Wu, "A 2-V, 1.8-GHz BJT Phase-Locked Loop," *IEEE J. Solid-State Circuits*, vol. SC-34, pp. 784–789, June 1999.
63. M. H. Perrott, T. L. Tewksbury III, and C. G. Sodini, "A 27-mW CMOS Fractional-N Synthesizer Using Digital Compensation for 2.5-Mb/s GFSK Modulation," *IEEE J. Solid-State Circuits*, vol. SC-32, pp. 2048–2060, Dec. 1997.
64. J. Tierney, C. Rader, and B. Gold, "A Digital Frequency Synthesizer," *IEEE Trans. Audio Electroacous.*, vol. AE-19, pp. 48–57, Mar. 1971.
65. V. Kroupa, *Direct Digital Frequency Synthesizers*, New York: John Wiley & Sons, 1998.
66. L. Cordesses, "Direct Digital Synthesis: A Tool for Periodic Wave Generation," *IEEE Signal Process. Mag.*, vol. 21, pp. 50–54, July 2004.
67. H. T. Nicholas III and H. Samueli, "A 150-MHz Direct Digital Frequency Synthesizer in 1.25-μm CMOS with −90-dBc Spurious Performance," *IEEE J. Solid-State Circuits*, vol. SC-26, pp. 1959–1969, Dec. 1991.
68. S. Mortezapour and E. F. Lee, "Design of Low-Power ROM-less Direct Digital Frequency Synthesizer Using Nonlinear Digital-to-Analog Converter," *IEEE J. Solid-State Circuits*, vol. SC-34, pp. 1350–1359, Oct. 1999.
69. S. E. Turner and D. E. Kotecki, "Direct Digital Synthesizer with Sine-weighted DAC at 32-GHz Clock Frequency in InP DHBT Technology," *IEEE J. Solid-State Circuits*, vol. SC-41, pp. 2284–2290, Oct. 2006.
70. X. Geng, F. F. Dai, J. D. Irwin, and R. C. Jaeger, "An 11-Bit 8.6 GHz Direct Digital Synthesizer MMIC with 10-Bit Segmented Sine-Weighted DAC," *IEEE J. Solid-State Circuits*, vol. SC-45, pp. 300–313, Feb. 2010.

10 Power Amplifier Design Fundamentals

Generally, the power amplifier design requires accurate active device modeling, impedance matching depending on the technical requirements and operation conditions, stability in operation, and simplicity in practical implementation. The quality of the power amplifier design is evaluated by realized maximum power gain under stable operation condition with minimum amplifier stages, and the requirement of linearity or high efficiency can be considered where it is needed. For a stable operation, it is necessary to evaluate the operating frequency domains where the active device may be potentially unstable. To avoid the parasitic oscillations, the stabilization circuit technique for different frequency domains (from low frequencies up to high frequencies close to the device transition frequency) is discussed. The key parameter of the power amplifier is its linearity, which is very important for many wireless communication applications. The relationships between the output power, 1-dB gain compression point, third-order intercept point, and intermodulation distortions of the third and higher orders are given and illustrated for different active devices. The device bias conditions that are generally different for linearity or efficiency improvement depend on the power amplifier operation class and the type of the active device. The bias circuits for the voltage-controlled metal-oxide-semiconductor field-effect transistor (MOSFET) devices are simple due to its voltage control, because the gate dc current is equal to the gate leakage current. However, for the current-controlled bipolar devices, they differ substantially, depending on the device base current and class of operation. The basic Classes A, AB, B, and C of the power amplifier operation are introduced, analyzed, and illustrated. The principles and design of the push–pull amplifiers using balanced transistors, as well as broadband and distributed power amplifiers are discussed. Harmonic-control techniques for designing microwave power amplifiers are given with description of a systematic procedure of multiharmonic load–pull simulation using the harmonic balance method and active load–pull measurement system. Finally, the concept of thermal resistance is introduced and heatsink design issues are discussed.

10.1 POWER GAIN AND STABILITY

Power amplifier design aims for maximum power gain and efficiency for a given value of output power with a predictable degree of stability. Instability of the power amplifier will lead to undesired parasitic oscillations and, as a result, to the distortion of the output signal. One of the main reasons for amplifier instability is a positive feedback from the device output to its input through the internal feedback capacitance, inductance of a common grounded device electrode, and external circuit elements. Consequently, a stability analysis is crucial to any power amplifier design, especially at high frequencies.

Figure 10.1 shows the basic block schematic of the single-stage power amplifier circuit, which includes an active device, an input matching circuit to match with the source impedance, and an output matching circuit to match with the load impedance. Generally, the two-port active device

RF and Microwave Transmitter Design, First Edition. Andrei Grebennikov.

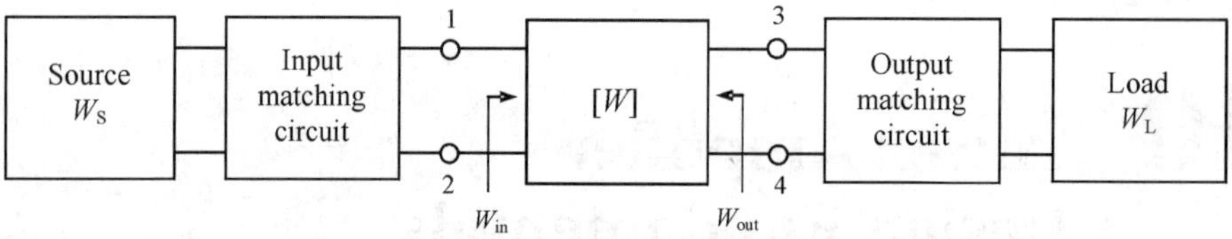

FIGURE 10.1 Generalized single-stage power amplifier circuit.

is characterized by a system of immittance W-parameters, which means any system of impedance Z-parameters or admittance Y-parameters [1,2]. The input and output matching circuits transform the source and load immittances W_S and W_L into specified values between points *1–2* and *3–4*, respectively, by means of which the optimum design operation mode of the power amplifier is realized.

The operating power gain G_P, which represents the ratio of power dissipated in the active load $\mathrm{Re}W_L$ to the power delivered to the input port of the active device, can be expressed in terms of the immittance W-parameters as

$$G_P = \frac{|W_{21}|^2 \,\mathrm{Re}W_L}{|W_{22} + W_L|^2 \,\mathrm{Re}W_{in}} \tag{10.1}$$

where

$$W_{in} = W_{11} - \frac{W_{12}W_{21}}{W_{22} + W_L} \tag{10.2}$$

is the input immittance and W_{ij} $(i, j = 1, 2)$ is the immittance two-port parameters of the active device equivalent circuit.

The transducer power gain G_T, which represents the ratio of power dissipated in the active load $\mathrm{Re}W_L$ to the power available from the source, can be expressed in terms of the immittance W-parameters as

$$G_T = \frac{4\,|W_{21}|^2 \,\mathrm{Re}W_S \,\mathrm{Re}W_L}{|(W_{11} + W_S)(W_{22} + W_L) - W_{12}W_{21}|^2}. \tag{10.3}$$

The operating power gain G_P does not depend on the source parameters and characterizes only the effectiveness of the power delivery from the input port of the active device to the load. This power gain helps to evaluate the power gain of a multistage amplifier when the overall operating power gain $G_{P(total)}$ is equal to the product of each stage G_P. The transducer power gain G_T includes an assumption of conjugate matching of the load and the source.

The bipolar transistor simplified small-signal π-hybrid equivalent circuit shown in Figure 10.2 demonstrates an example of a conjugate-matched bipolar Class A power amplifier. The impedance Z-parameters of the equivalent circuit of the bipolar transistor in a common emitter configuration can be obtained as

$$\begin{aligned} Z_{11} &= r_b + \frac{1}{g_m + j\omega C_\pi} & Z_{12} &= \frac{1}{g_m + j\omega C_\pi} \\ Z_{21} &= -\frac{1}{j\omega C_c}\frac{g_m - j\omega C_c}{g_m + j\omega C_\pi} & Z_{22} &= \left(1 + \frac{C_\pi}{C_c}\right)\frac{1}{g_m + j\omega C_\pi} \end{aligned} \tag{10.4}$$

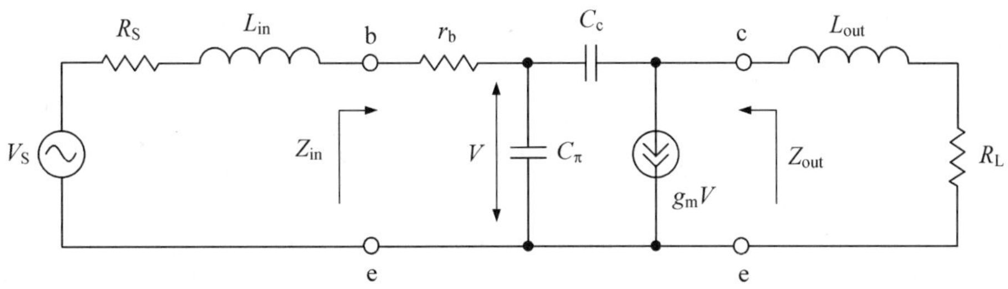

FIGURE 10.2 Simplified equivalent circuit of matched bipolar power amplifier.

where g_m is the device transconductance, r_b is the series base resistance, C_π is the base–emitter capacitance including both diffusion and junction components, and C_c is the feedback collector capacitance.

By setting the device feedback impedance Z_{12} to zero and complex-conjugate matching conditions of $R_S = \mathrm{Re}Z_{in}$ and $L_{in} = -\mathrm{Im}Z_{in}/\omega$ at the input and of $R_L = \mathrm{Re}Z_{out}$ and $L_{out} = -\mathrm{Im}Z_{out}/\omega$ at the output, the small-signal transducer power gain G_T can be calculated from

$$G_T = \left(\frac{f_T}{f}\right)^2 \frac{1}{8\pi f_T r_b C_c} \tag{10.5}$$

where $f_T = g_m/(2\pi C_\pi)$ is the device transition frequency.

Figure 10.3 shows an example of a conjugate-matched field-effect transistor (FET) power amplifier. The admittance Y-parameters of the small-signal equivalent circuit of any FET device in a common source configuration can be obtained by

$$\begin{aligned} Y_{11} &= \frac{j\omega C_{gs}}{1 + j\omega C_{gs}R_{gs}} + j\omega C_{gd} \quad Y_{12} = -j\omega C_{gd} \\ Y_{21} &= \frac{g_m}{1 + j\omega C_{gs}R_{gs}} - j\omega C_{gd} \quad Y_{22} = \frac{1}{R_{ds}} + j\omega\left(C_{ds} + C_{gd}\right) \end{aligned} \tag{10.6}$$

where g_m is the device transconductance, R_{gs} is the gate–source resistance, C_{gs} is the gate–source capacitance, C_{gd} is the feedback gate–drain capacitance, C_{ds} is the drain–source capacitance, and R_{ds} is the differential drain–source resistance.

Since the value of the gate–drain capacitance C_{gd} is normally relatively small, the effect of the feedback admittance Y_{12} can be neglected in a simplified case. Then, it is necessary to set $R_S = R_{gs}$

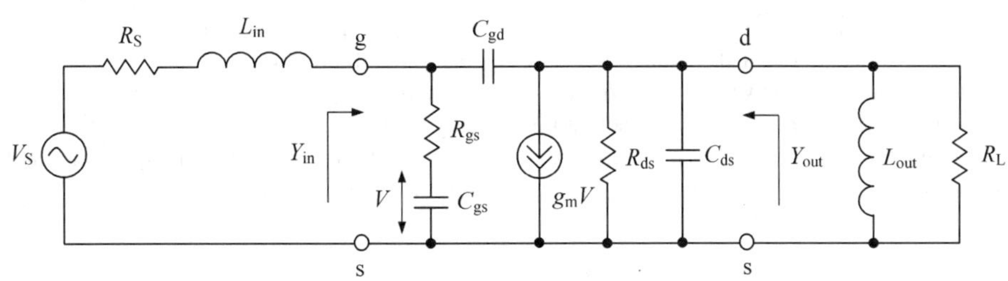

FIGURE 10.3 Simplified equivalent circuit of matched FET power amplifier.

and $L_{in} = 1/\omega^2 C_{gs}$ for input matching, whereas $R_L = R_{ds}$ and $L_{out} = 1/\omega^2 C_{ds}$ for output matching. Hence, the transducer power gain G_T can approximately be calculated from

$$G_T\left(C_{gd} = 0\right) = MAG = \left(\frac{f_T}{f}\right)^2 \frac{R_{ds}}{4R_{gs}} \tag{10.7}$$

where $f_T = g_m/(2\pi C_{gs})$ is the device transition frequency and *MAG* is the maximum available gain representing a theoretical limit on the power gain that can be achieved under complex-conjugate matching conditions.

From Eqs. (10.5) and (10.7) it follows that the small-signal power gain of a conjugate-matched power amplifier for any type of the active device drops off as $1/f^2$ or 6 dB per octave. Therefore, $G_T(f)$ can readily be predicted at a certain frequency f, if its value is known at the transition frequency f_T, by

$$G_T(f) = G_T(f_T)\left(\frac{f_T}{f}\right)^2. \tag{10.8}$$

It should be noted that previous analysis is based on the linear small-signal consideration when generally nonlinear device current source as a function of both the input and output voltages can be characterized by the linear transconductance g_m as a function of the input voltage and the output differential resistance R_{ds} as a function of the output voltage. This is a result of a Taylor series expansion of the output current as a function of the input and output voltages with maintaining only the dc and linear components. Such an approach helps to understand and derive the maximum achievable power-amplifier parameters in a linear approximation. In this case, an active device is operated in a Class A mode when one half of a dc power is dissipated in the device, while the other half is transformed to the fundamental-frequency output power, resulting in a maximum ideal collector efficiency of 50%. The device output resistance R_{out} remains constant and can be calculated as a ratio of the dc supply voltage to the dc current flowing through the active device. In a common case, for a complex-conjugate matching procedure, the device output immittance under large-signal consideration should be calculated using Fourier analysis of the output current and voltage fundamental components. This means that, unlike a linear Class A mode, an active device is operated in a its linear region only for part of the entire period, and its output resistance is defined as a ratio of the fundamental-frequency output voltage to the fundamental-frequency output current. This is not a physical resistance resulting in a corresponding power loss inside the device, but an equivalent resistance required to use for a conjugate matching procedure. In this case, the complex-conjugate matching is valid and required, firstly, to compensate for the reactive part of the device output impedance and, secondly, to provide a proper load resistance resulting in a maximum power gain for a given supply voltage and required output power delivered to the load. Note that this is not the maximum available small-signal power gain that can be achieved in a linear operation mode, but a maximum achievable large-signal power gain that can be achieved for a particular operation mode with a certain conduction angle. Of course, the maximum large-signal power gain is smaller than the small-signal power gain for the same input power, since the output power in a nonlinear operation mode also includes the powers at the harmonic components of the fundamental frequency.

According to the immittance approach to the stability analysis of the active two-port network, it is necessary and sufficient for its unconditional stability if the following system of equations can be satisfied for the given active device with both open-circuit input and output ports:

$$\begin{cases} \mathrm{Re}\left[W_S(\omega) + W_{in}(\omega)\right] > 0 \\ \mathrm{Im}\left[W_S(\omega) + W_{in}(\omega)\right] = 0 \end{cases} \tag{10.9}$$

or

$$\begin{cases} \text{Re}\left[W_{\text{L}}(\omega) + W_{\text{out}}(\omega)\right] > 0 \\ \text{Im}\left[W_{\text{L}}(\omega) + W_{\text{out}}(\omega)\right] = 0. \end{cases} \tag{10.10}$$

In the case of the opposite signs in Eqs. (10.9) and (10.10), the active two-port network can be treated as unstable or potentially unstable.

Analyzing Eq. (10.9) or Eq. (10.10) on the extremum results in a special relationship between the device immittance parameters called the *device stability factor*

$$K = \frac{2\text{Re}W_{11}\text{Re}W_{22} - \text{Re}\left(W_{12}W_{21}\right)}{|W_{12}W_{21}|} \tag{10.11}$$

which shows a stability margin indicating how far from zero value are the real parts in Eqs. (10.9) and (10.10) being positive [3]. An active device is unconditionally stable if $K > 1$ and potentially unstable if $K < 1$.

When the device is potentially unstable, an improvement of the power amplifier stability can be provided with the appropriate choice of the source and load immittances W_{S} and W_{L}. The circuit stability factor K_{T} in this case is defined in the same way as the device stability factor K, but by taking into account $\text{Re}W_{\text{S}}$ and $\text{Re}W_{\text{L}}$ along with the device W-parameters. The circuit stability factor is given by

$$K_{\text{T}} = \frac{2\text{Re}\left(W_{11} + W_{\text{S}}\right)\text{Re}\left(W_{22} + W_{\text{L}}\right) - \text{Re}\left(W_{12}W_{21}\right)}{|W_{12}W_{21}|}. \tag{10.12}$$

If the circuit stability factor $K_{\text{T}} > 1$, the power amplifier is unconditionally stable. However, the power amplifier becomes potentially unstable if $K_{\text{T}} < 1$. The value of $K_{\text{T}} = 1$ corresponds to the border of the circuit unconditional stability. The values of the circuit stability factor K_{T} and device stability factor K become equal if $\text{Re}W_{\text{S}} = \text{Re}W_{\text{L}} = 0$.

When the device stability factor $K > 1$, the operating power gain G_{P} has to be maximized. By analyzing Eq. (10.1) on extremum, it is possible to find optimum values $\text{Re}W_{\text{L}}^{\text{o}}$ and $\text{Im}W_{\text{L}}^{\text{o}}$ when the operating power gain G_{P} is maximal [3,4]. As a result,

$$G_{\text{Pmax}} = \frac{|W_{21}/W_{12}|}{K + \sqrt{K^2 - 1}}. \tag{10.13}$$

The power amplifier with an unconditionally stable active device provides a maximum power gain operation only if the input and output of the active device are conjugate-matched with the source and load impedances, respectively. For the lossless input matching circuit when the power available at the source is equal to the power delivered to the input port of the active device, that is $P_{\text{S}} = P_{\text{in}}$, the maximum operating power gain is equal to the maximum transducer power gain, that is, $G_{\text{Pmax}} = G_{\text{Tmax}}$.

Domains of the device potential instability include the operating frequency ranges where the active device stability factor is equal to $K < 1$. Within the bandwidth of such a frequency domain, parasitic oscillations can occur, defined by internal positive feedback and operating conditions of the active device. The instabilities may also be induced by the RF drive power and remain on its removal. One of the most serious cases of the power amplifier instability can occur when there is a variation of the load impedance. Under these conditions, the transistor may be destroyed almost instantaneously. But even if it is not destroyed, the instability can result in an increased level of the spurious emissions in the output spectrum of the power amplifier tremendously. Generally, the following classification for linear instabilities can be made [5]:

- Low-frequency oscillations produced by thermal feedback effects.
- Oscillations due to internal feedback.
- Negative resistance- or conductance-induced instabilities due to transit-time effects, avalanche multiplication, and so on.
- Oscillations due to external feedback as a result of insufficient decoupling of the dc supply, and so on.

Therefore, it is very important to determine the effect of the device feedback parameters on the origin of the parasitic self-oscillations and to establish possible circuit configurations of the parasitic oscillators. Based on the simplified bipolar equivalent circuit shown in Figure 10.2, the device stability factor can be expressed through the parameters of the transistor equivalent circuit as

$$K = 2r_b g_m \frac{1 + \dfrac{g_\mathrm{m}}{\omega_T C_c}}{\sqrt{1 + \left(\dfrac{g_\mathrm{m}}{\omega C_c}\right)^2}}. \tag{10.14}$$

where $\omega_\mathrm{T} = 2\pi f_\mathrm{T}$ [1,2].

At very low frequencies, the bipolar transistors are potentially stable, and the fact that $K \to 0$ when $f \to 0$ can be explained by simplifying the bipolar equivalent circuit. In practice, at low frequencies, it is necessary to take into account the dynamic base–emitter resistance r_π and Early collector–emitter resistance r_ce, the presence of which substantially increases the value of the device stability factor. This gives the only one unstable frequency domain with $K < 1$ and low boundary frequency f_p1. However, an additional region of possible low frequency oscillations can occur due to thermal feedback where the collector junction temperature becomes frequently dependent, and the common base configuration is especially affected by this [6].

Equating the device stability factor K with unity allows us to determine the high boundary frequency of a frequency domain of the bipolar transistor potential instability as

$$f_\mathrm{p2} = \frac{g_\mathrm{m}/2\pi C_\mathrm{c}}{\sqrt{(2r_\mathrm{b} g_\mathrm{m})^2 \left(1 + \dfrac{g_\mathrm{m}}{\omega_\mathrm{T} C_\mathrm{c}}\right)^2 - 1}}. \tag{10.15}$$

When $r_\mathrm{b} g_\mathrm{m} > 1$ and $g_\mathrm{m} \gg \omega_\mathrm{T} C_\mathrm{c}$, Eq. (10.15) is simplified to

$$f_{p2} \approx \frac{1}{4\pi r_b C_\pi}. \tag{10.16}$$

At higher frequencies, a presence of the parasitic reactive intrinsic transistor parameters and package parasitics can be of great importance in view of power amplifier stability. The parasitic series emitter lead inductance L_e shown in Figure 10.4 has a major effect upon the device stability factor. The presence of L_e leads to the appearance of the second frequency domain of potential instability at higher frequencies. The circuit analysis shows that the second frequency domain of potential instability can be realized only under the particular ratios between the normalized parameters $\omega_\mathrm{T} L_\mathrm{e}/r_\mathrm{b}$ and $\omega_\mathrm{T} r_\mathrm{b} C_\mathrm{c}$ [1,2]. For example, the second domain does not occur for any values of L_e when $\omega_\mathrm{T} r_\mathrm{b} C_\mathrm{c} \geq 0.25$.

An appearance of the second frequency domain of the device potential instability is the result of the corresponding changes in the device feedback phase conditions and takes place only under a

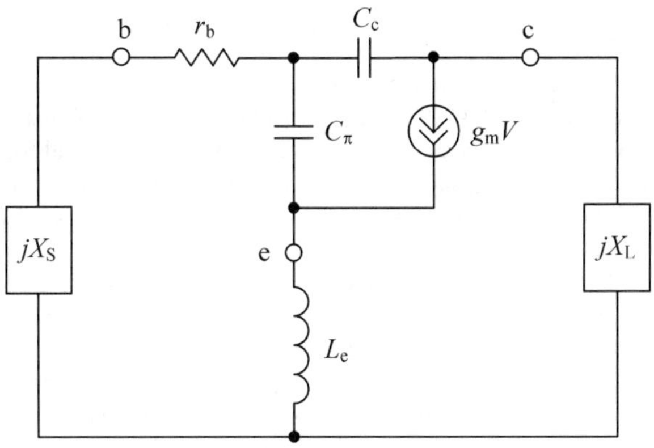

FIGURE 10.4 Simplified bipolar π-hybrid equivalent circuit with emitter lead inductance.

simultaneous effect of the collector capacitance C_c and emitter lead inductance L_e. If the effect of one of these factors is lacking, the active device is characterized by only the first domain of its potential instability.

Figure 10.5 shows the potentially realizable equivalent circuits of the parasitic oscillators. If the value of a series emitter inductance L_e is negligible, the parasitic oscillations can occur only when the values of the source and load reactances are positive, that is, $X_S > 0$ and $X_L > 0$. In this case,

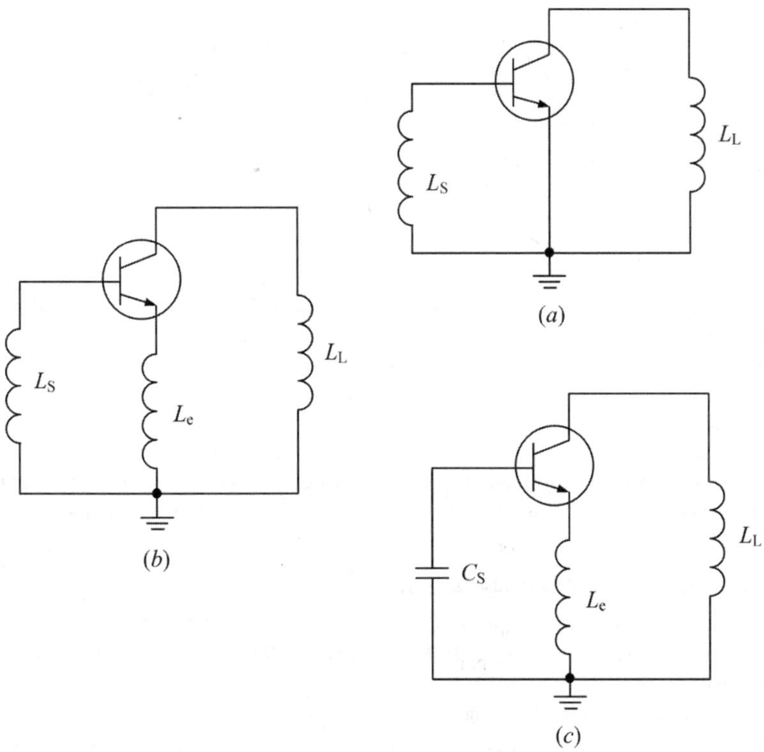

FIGURE 10.5 Equivalent circuits of parasitic oscillators.

the parasitic oscillator shown in Figure 10.5(*a*) represents the inductive three-point circuit, where inductive elements L_S and L_L, in combination with the collector capacitance C_c, form a Hartley oscillator. From a practical point of view, the more the value of the collector dc-feed inductance exceeds the value of the base bias inductance, the more likely are low frequency parasitic oscillators. It was observed that a very low inductance, even a short between emitter and base, can produce very strong and dangerous oscillations that may easily destroy a transistor [5]. Therefore, it is recommended to increase the value of the base choke inductance and to decrease the value of the collector dc-feed inductance.

The presence of L_e leads to narrowing of the first frequency domain of the potential instability, which is limited to the high boundary frequency f_{p2}, and can contribute to appearance of the second frequency domain of the potential instability at higher frequencies. The parasitic oscillator that corresponds to the first frequency domain of the device potential instability can be realized only if the source and load reactances are inductive, that is, $X_S > 0$ and $X_L > 0$, with the equivalent circuit of such a parasitic oscillator shown in Figure 10.5(*b*). The parasitic oscillator corresponding to the second frequency domain of the device potential instability can be realized only if the source reactance is capacitive and load reactance is inductive, that is, $X_S < 0$ and $X_L > 0$, with the equivalent circuit shown in Figure 10.5(*c*). The series emitter inductance L_e is an element of fundamental importance for the parasitic oscillator that corresponds to the second frequency domain of the device potential instability. It changes the circuit phase conditions so that it becomes possible to establish the oscillation phase balance condition at high frequencies. However, if it is possible to eliminate the parasitic oscillations at high frequencies by other means, increasing of L_e will result to narrowing of a low frequency domain of potential instability, thus making the power amplifier potentially more stable, though at the expense of reduced power gain.

Similar analysis of the MOSFET power amplifier shows the two frequency domains of MOSFET potential instability due to internal feedback gate–drain capacitance C_{gd} and series source inductance L_s [1,7]. Because of the very high gate leakage resistance, the value of the low boundary frequency f_{p1} is sufficiently small. For usually available conditions for power MOSFET devices when $g_m R_{ds} =$ 10–30 and $C_{gd}/C_{gs} = 0.1–0.2$, the high boundary frequency f_{p2} can approximately be calculated from

$$f_{p2} \approx \frac{1}{4\pi R_{gs} C_{gs}}. \tag{10.17}$$

It should be noted that power MOSFET devices have a substantially higher value of $g_m R_{ds}$ at small values of the drain current than at its higher values. Consequently, for small drain current, the MOSFET device is characterized by a wider domain of potential instability. This domain is significantly wider than the same first domain of the potential instability of the bipolar transistor. The series source inductance L_s contributes to appearance of the second frequency domain of the device potential instability. The potentially realizable equivalent circuits of the MOSFET parasitic oscillators are the same as for the bipolar device shown in Figure 10.5.

Thus, to prevent parasitic oscillations and to provide a stable operation mode of any power amplifier, it is necessary to take into consideration the following common requirements:

- Use an active device with stability factor $K > 1$.
- If it is impossible to choose an active device with $K > 1$, it is necessary to provide the circuit stability factor $K_T > 1$ by the appropriate choice of the real parts of the source and load immittances.
- Disrupt the equivalent circuits of the possible parasitic oscillators.
- Choose proper reactive parameters of the matching circuit elements adjacent to the input and output ports of the active device, which are necessary to avoid the self-oscillation conditions.

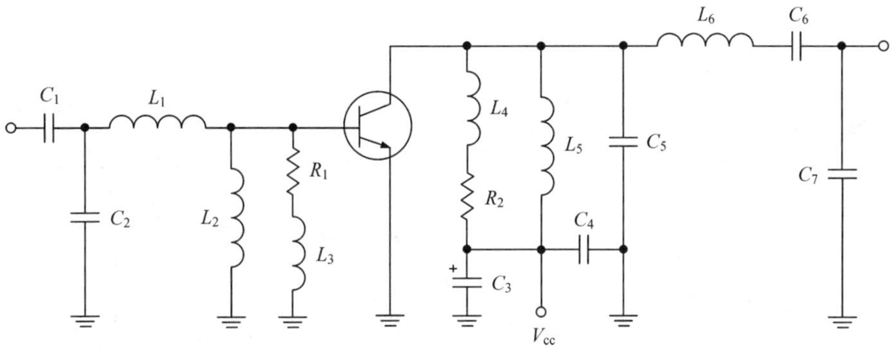

FIGURE 10.6 Stabilized bipolar Class C VHF power amplifier.

Generally, the parasitic oscillations can arise on any frequency within the potential instability domains for particular values of the source and load immittances W_S and W_L. The frequency dependencies of W_S and W_L are very complicated and very often cannot be predicted exactly, especially in multistage power amplifiers. Therefore, it is very difficult to propose a unified approach to provide a stable operation mode of the power amplifiers with different circuit configuration and operation frequencies. In practice, the parasitic oscillations can arise close to the operating frequencies due to the internal positive feedback inside the transistor and at the frequencies sufficiently far from the operating frequencies due to the external positive feedback created by the surface-mounted elements. As a result, the stability analysis of the power amplifier must include the methods to prevent the parasitic oscillations in different frequency ranges.

Figure 10.6 shows an example of a stabilized bipolar very high frequency (VHF) power amplifier configured to operate in a zero-bias Class C mode. Conductive input and output loading due to resistances R_1 and R_2 eliminate a low-frequency instability domain. The series inductors L_3 and L_4 contribute to higher power gain if the resistance values are too small, and can compensate for the capacitive input and output device impedances. To provide a negative-bias Class C mode, the shunt inductor L_2 can be removed. The equivalent circuit of the potential parasitic oscillator at higher frequencies is realized by means of the parasitic reactive parameters of the transistor and external circuitry. The only possible equivalent circuit of such a parasitic oscillator at these frequencies is shown in Figure 10.5(*c*). It can only be realized if the series emitter lead inductance is present. Consequently, the electrical length of the emitter lead should be reduced as much as possible, or, alternatively, the appropriate reactive immittances at the input and output transistor ports are provided. For example, it is possible to avoid the parasitic oscillations at these frequencies if the inductive immittance is provided at the input of the transistor and capacitive reactance is provided at the output of the transistor. This is realized by an input series inductance L_1 and an output shunt capacitance C_5.

The collector efficiency of the power amplifier can be increased by removing the shunt capacitance and series *RL* circuit in the load network. The remaining series *LC* circuit provides high impedances at the second-order and higher order harmonic components of the output current, which are flowing now through the device collector capacitance unlike being grounded by the shunt capacitance. As a result, the bipolar Class C power amplifier, the circuit schematic of which is shown in Figure 10.7, achieved a collector efficiency of 73% and a power gain of 9 dB with an output power of 13.8 W at an operating frequency of 160 MHz [8]. However, special care must be taken to eliminate parasitic spurious oscillations. In this case, the most important element in preventing the potential instability is the base bias resistor R_b. For example, for a relatively large base choke inductor L_b and $R_b = 1\ \text{k}\Omega$, spurious oscillations exist at any tuning. Tuning becomes possible with no parasitic oscillations for output voltage standing wave ratio (*VSWR*) less than 1.3 or supply voltage more than 22 V when R_b is

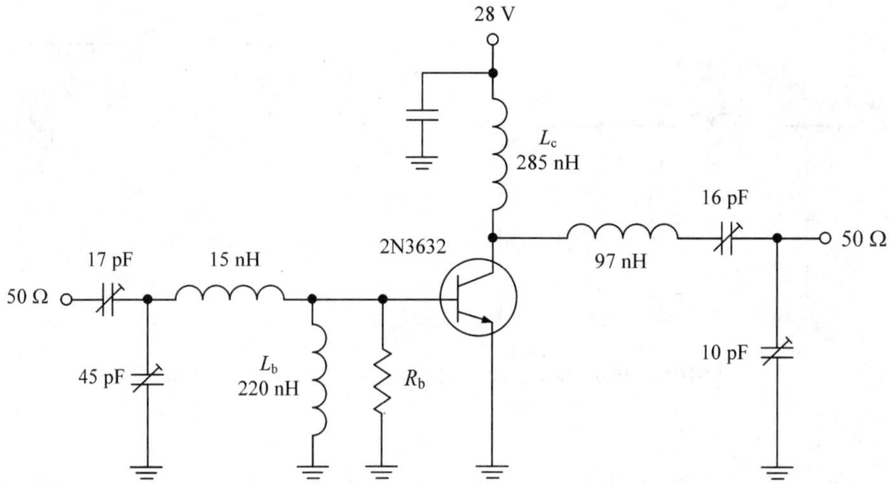

FIGURE 10.7 High efficiency bipolar Class C VHF power amplifier.

reduced to 470 Ω. However, a very small reduction in input drive power causes spurious oscillations. Further reduction of R_b to 47 Ω provides a stable operation for output $VSWR \leq 7$ and supply voltages down to 7 V. Finally, no spurious oscillations occur at any load, supply voltage and drive power level for $R_b = 26$ Ω.

Since the transistor using as an active device in power amplifiers is characterized by a substantially nonlinear behavior, this can result in nonlinear instabilities, which generally provide the parametric generation of both harmonic and subharmonic components. The subharmonics can be explained by parametric varactor-junction action of the collector-to-base voltage-dependent capacitance when the large-signal driving acts like pumping a varactor diode, as in a parametric amplifier [9,10]. Such an amplifier exhibits negative resistance under certain conditions when a circuit starts oscillating at subharmonics or rational fractions of the operating frequency [11]. Generally, the parametric oscillations are the result of the external force impact on the element of the oscillation system by varying its parameter. Understanding of the physical origin of this parametric effect is very important in order to disrupt any potentially realizable parametric oscillator circuits. Especially, it is a serious concern for high-efficiency power amplifiers, in general, and Class E power amplifiers, in particular, with very high voltage peak factor and voltage swing across the device nonlinear output capacitance, since the transistor is operated in pinch-off, active and saturation regions.

Figure 10.8 shows the simplified large-signal equivalent circuit of the (*a*) MOSFET or (*b*) bipolar device with a nonlinear current source $i(t)$, respectively. The most nonlinear capacitances are the bipolar collector capacitance C_c and the MOSFET gate–drain capacitance C_{gd} and drain–source capacitance C_{ds}, which can be modeled as junction capacitances with different sensitivities γ. However, since the drain–source capacitance C_{ds} is normally greater by 8–10 times than the gate–drain capacitance C_{gd}, the parametric effect due to C_{ds} causes a major effect on potential parametric oscillations in a MOSFET power amplifier. The value of the collector capacitance C_c is by the order smaller than the value of the base–emitter capacitance C_π in active mode. In this case, the circuit of a potential parametric oscillator represents a system with one degree of freedom as shown in Figure 10.8(*c*), where V_{cc} is the supply voltage applied to the varying output capacitance C_{out} ($C_{out} \approx C_c$ for a bipolar device and $C_{out} \approx C_{ds}$ for a MOSFET device), while the capacitor C_0 and inductor L_0 represent the series high-Q resonant circuit tuned to the fundamental frequency and having high impedances at the second-order and higher order harmonics [12].

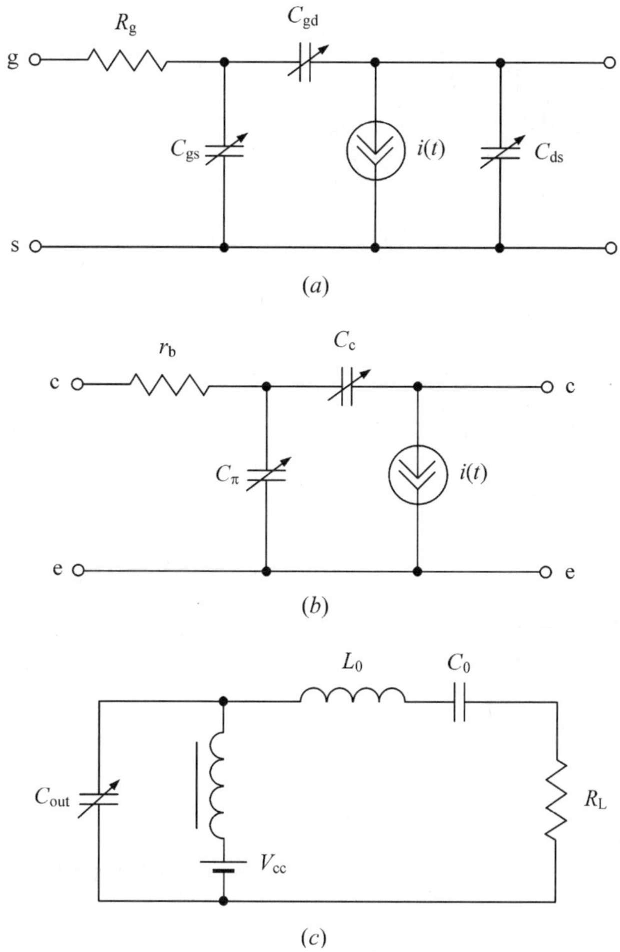

FIGURE 10.8 Simplified nonlinear transistor models and output amplifier circuit.

The most probable parametric oscillations in the nonlinear power amplifier can occur at a subharmonic frequency $\omega_{1/2} = \omega_0/2$, where ω_0 is the operating frequency varying the device capacitance. To eliminate such a parasitic subharmonic parametric oscillation, it is necessary to provide the circuit design solution when the device output can see very high impedance at a subharmonic frequency $\omega_{1/2}$. Alternatively, an additional lossy element in the subharmonic circuit with its proper isolation from the fundamental circuit can be incorporated. In other words, it is necessary to break out any possible resonant conditions at the subharmonic frequency component, which can cause the parametric oscillations.

10.2 BASIC CLASSES OF OPERATION: A, AB, B, AND C

To determine the operation classes of the power amplifier, consider a simple resistive stage shown in Figure 10.9, where L_{ch} is the ideal choke inductor with zero series resistance and infinite reactance at the operating frequency, C_b is the dc-blocking capacitance with infinite value having zero reactance

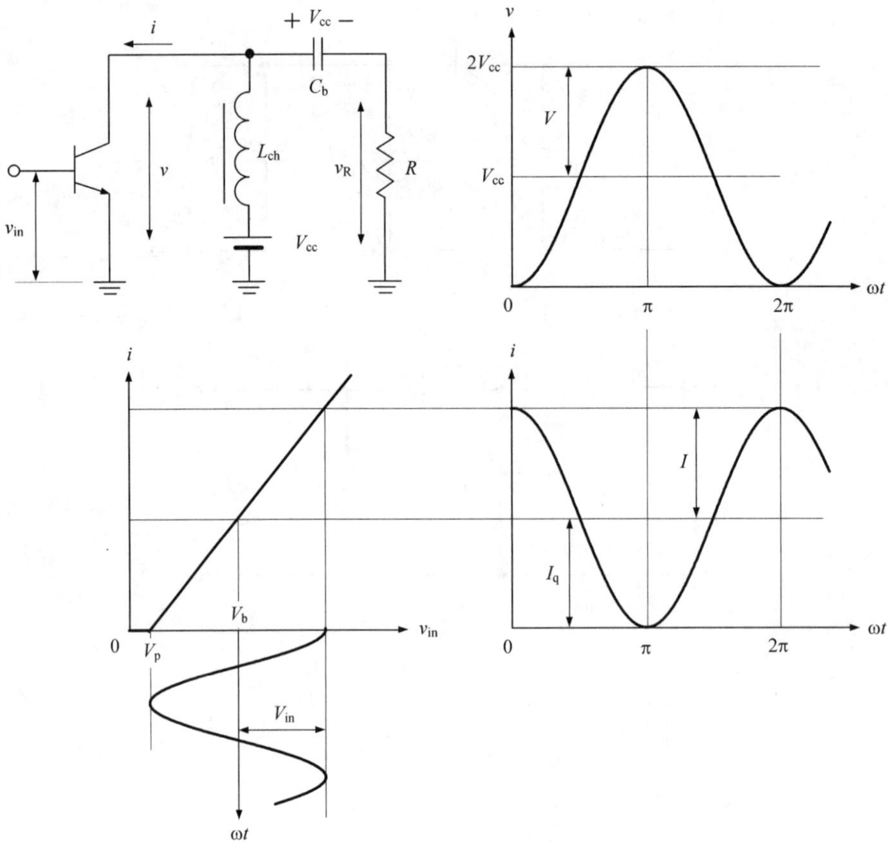

FIGURE 10.9 Voltage and current waveforms in Class A operation.

at the operating frequency, and R is the load resistance. The dc supply voltage V_{cc} is applied to both plates of the dc-blocking capacitor, being constant during the entire signal period. The active device behaves as an ideal voltage-controlled current source having zero saturation resistance.

Let us assume the input signal to be in a cosine form of

$$v_{in} = V_{bias} + V_{in}\cos\omega t \tag{10.18}$$

where V_{bias} is the input dc bias voltage. The operating point must be fixed at the middle point of the linear part of the device transfer characteristic with $V_{in} \le V_{bias} - V_p$, where V_p is the device pinch-off voltage. Normally, to simplify an analysis of the power amplifier operation, the device transfer characteristic is represented by a piecewise-linear approximation. As a result, the output current is cosinusoidal,

$$i = I_q + I\cos\omega t \tag{10.19}$$

with the quiescent current I_q greater or equal to the collector current amplitude I. In this case, the output collector current contains only two components—dc and cosine—and the averaged current magnitude is equal to a quiescent current or dc component I_q.

The output voltage v across the device collector represents a sum of the dc supply voltage V_{cc} and cosine voltage v_R across the load resistance R. Consequently, the greater the output current i, the greater the voltage v_R across the load resistance R and smaller the output voltage v. Thus, for a purely real load impedance when $Z_L = R$, the collector voltage v is shifted by 180° relative to the input voltage v_{in} and can be written as

$$v = V_{cc} + V\cos(\omega t + 180^\circ) = V_{cc} - V\cos\omega t \tag{10.20}$$

where V is the output voltage amplitude.

Substituting Eq. (10.19) into Eq. (10.20) yields

$$v = V_{cc} - \left(i - I_q\right) R. \tag{10.21}$$

Equation (10.21) can be rewritten in the form of

$$i = \left(I_q + \frac{V_{cc}}{R}\right) - \frac{v}{R} \tag{10.22}$$

which determines a linear dependence of the collector current versus collector voltage. Such a combination of the cosine collector voltage and current waveforms is known as a Class A operation mode. In real practice, because of the device nonlinearities, it is necessary to connect a parallel LC circuit with resonant frequency equal to the operating frequency to suppress any possible harmonic components.

Circuit theory prescribes that the collector efficiency η can be written as

$$\eta = \frac{P}{P_0} = \frac{1}{2}\frac{I}{I_q}\frac{V}{V_{cc}} = \frac{1}{2}\frac{I}{I_q}\xi \tag{10.23}$$

where

$$P_0 = I_q V_{cc} \tag{10.24}$$

is the dc output power,

$$P = \frac{IV}{2} \tag{10.25}$$

is the power delivered to the load resistance R at the fundamental frequency f_0, and

$$\xi = \frac{V}{V_{cc}} \tag{10.26}$$

is the collector voltage peak factor.

Then, by assuming the ideal conditions of zero saturation voltage when $\xi = 1$ and maximum output current amplitude when $I/I_q = 1$, from Eq. (10.23) it follows that the maximum collector efficiency in a Class A operation mode is equal to

$$\eta = 50\%. \tag{10.27}$$

However, as it follows from Eq. (10.23), increasing the value of I/I_q can further increase the collector efficiency. This leads to a step-by-step nonlinear transformation of the current cosine waveform to its pulsed waveform when the amplitude of the collector current exceeds zero value

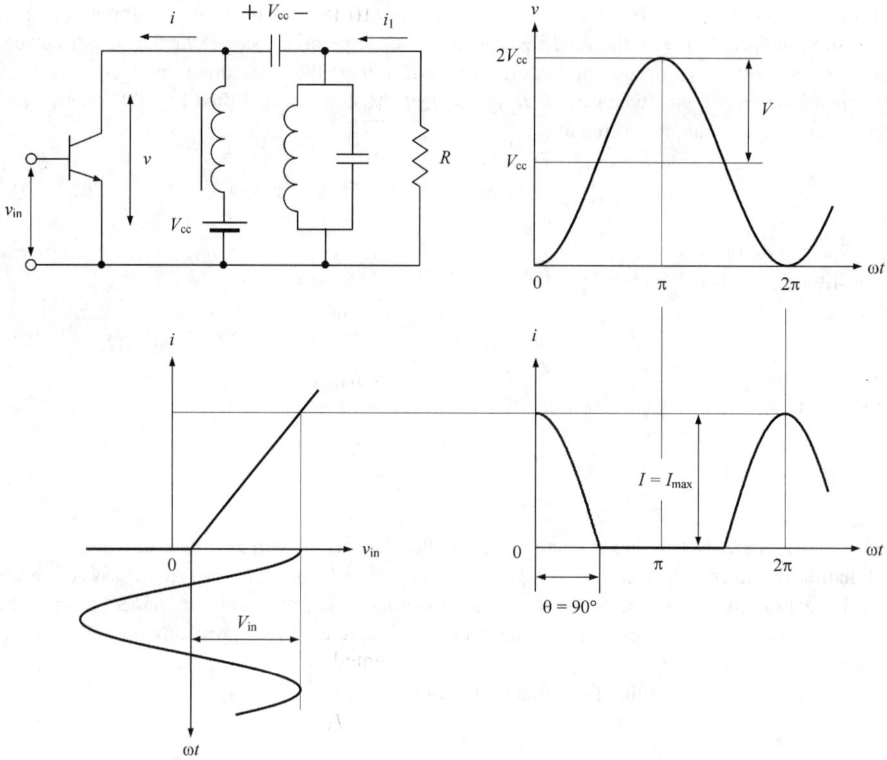

FIGURE 10.10 Voltage and current waveforms in Class B operation.

during only a part of the entire signal period. In this case, an active device is operated in the active region followed by the operation in the pinch-off region when the collector current is zero, as shown in Figure 10.10. As a result, the frequency spectrum at the device output will generally contain the second-order, third-order and higher order harmonics of the fundamental frequency. However, due to high quality factor of the parallel resonant *LC* circuit, only fundamental-frequency signal is flowing into the load, while the short-circuit conditions are fulfilled for higher order harmonic components. Therefore, ideally the collector voltage represents a purely sinusoidal waveform with the voltage amplitude $V \leq V_{cc}$.

Analytically such an operation can be written as

$$i\,(\omega t) = \begin{cases} I_q + I\cos\omega t & -\theta \leq \omega t < \theta \\ 0 & \theta \leq \omega t < 2\pi - \theta \end{cases} \tag{10.28}$$

where the conduction angle 2θ is the angle of current flow indicating the part of the RF current cycle for which device conduction occurs and determines the moment when output current i takes a zero value [13]. At this moment

$$i\,(\theta) = I_q + I\cos\theta = 0 \tag{10.29}$$

and θ can be calculated from

$$\cos\theta = -\frac{I_q}{I}. \tag{10.30}$$

The conduction angle can also be calculated using Eq. (10.18) when voltage $v_{\rm in}$ becomes equal to a pinch-off voltage $V_{\rm p}$ at $\omega t = \theta$ as

$$\cos\theta = -\frac{V_{\rm bias} - V_{\rm p}}{V_{\rm in}}. \tag{10.31}$$

Consequently, in a common case,

$$i\,(\omega t) = I\,(\cos\omega t - \cos\theta)\,. \tag{10.32}$$

When $\omega t = 0$, the output collector current has maximum amplitude of

$$i\,(0) = I\,(1 - \cos\theta) = I_{\rm max}. \tag{10.33}$$

As a result, the basic definitions for nonlinear operation modes of a power amplifier through half the conduction angle can be introduced as follows:

- When $\theta > 90^\circ$, then $\cos\theta < 0$ and $I_{\rm q} > 0$ corresponding to Class AB operation.
- When $\theta = 90^\circ$, then $\cos\theta = 0$ and $I_{\rm q} = 0$ corresponding to Class B operation.
- When $\theta < 90^\circ$, then $\cos\theta > 0$ and $I_{\rm q} < 0$ corresponding to Class C operation.

The periodic pulsed output current $i(\omega t)$ can be represented as a Fourier series expansion

$$i(\omega t) = I_0 + I_1\cos\omega t + I_2\cos 2\omega t + I_3\cos 3\omega t + \cdots \tag{10.34}$$

where the dc, fundamental-frequency, and nth harmonic components are calculated by

$$I_0 = \frac{1}{2\pi}\int_{-\theta}^{\theta} g_{\rm m}V_{\rm in}\,(\cos\omega t - \cos\theta)d\omega t = \gamma_0\,(\theta)\,I \tag{10.35}$$

$$I_1 = \frac{1}{\pi}\int_{-\theta}^{\theta} g_{\rm m}V_{\rm in}\,(\cos\omega t - \cos\theta)\cos\omega t d\omega t = \gamma_1\,(\theta)\,I \tag{10.36}$$

$$I_n = \frac{1}{\pi}\int_{-\theta}^{\theta} g_{\rm m}V_{\rm in}\,(\cos\omega t - \cos\theta)\cos n\omega t d\omega t = \gamma_{\rm n}\,(\theta)\,I \tag{10.37}$$

where $\gamma_n(\theta)$ is called the nth-harmonic coefficient of expansion of the output current cosine waveform or the nth-harmonic current coefficient [1,14]. They can be analytically defined as

$$\gamma_0\,(\theta) = \frac{1}{\pi}\,(\sin\theta - \theta\cos\theta) \tag{10.38}$$

$$\gamma_1\,(\theta) = \frac{1}{\pi}\left(\theta - \frac{\sin 2\theta}{2}\right) \tag{10.39}$$

$$\gamma_n\,(\theta) = \frac{1}{\pi}\left[\frac{\sin\,(n-1)\,\theta}{n\,(n-1)} - \frac{\sin\,(n+1)\,\theta}{n\,(n+1)}\right] \tag{10.40}$$

where $n = 2, 3, \ldots$.

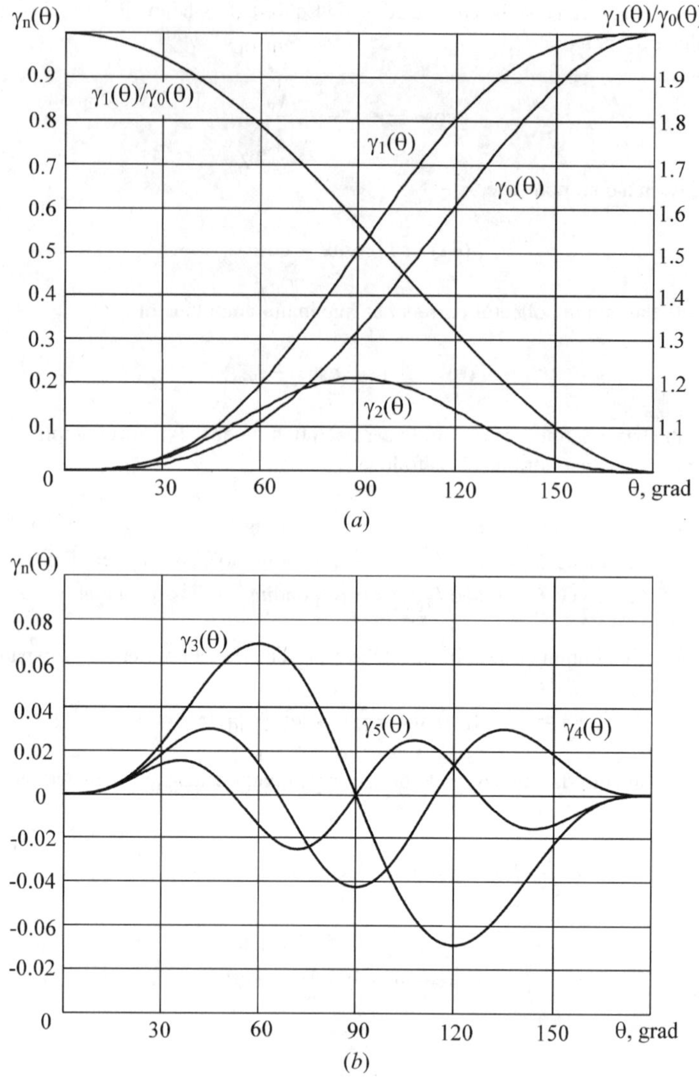

FIGURE 10.11 Dependencies of $\gamma_n(\theta)$ for dc, fundamental and higher order current components.

The dependencies of $\gamma_n(\theta)$ for the dc, fundamental-frequency, second-order and higher order current components are shown in Figure 10.11. The maximum value of $\gamma_n(\theta)$ is achieved when $\theta = 180°/n$. Special case is $\theta = 90°$, when odd current coefficients are equal to zero, that is, $\gamma_3(\theta) = \gamma_5(\theta) = \cdots = 0$. The ratio between the fundamental-frequency and dc components $\gamma_1(\theta)/\gamma_0(\theta)$ varies from 1 to 2 for any values of the conduction angle, with a minimum value of 1 for $\theta = 180°$ and a maximum value of 2 for $\theta = 0°$. It is necessary to pay attention to the fact that, for example, the current coefficient $\gamma_3(\theta)$ becomes negative within the interval of $90° < \theta < 180°$. This implies the proper phase changes of the third current harmonic component when its values are negative and it becomes out of phase relative to the fundamental. Consequently, if the harmonic components, for which $\gamma_n(\theta) > 0$, achieve positive maximum values at the time moments corresponding to the middle points of the current waveform, the harmonic components, for which $\gamma_n(\theta) < 0$, can achieve negative maximum values at these time moments. As a result, combination of different harmonic components with proper loading will result to flattening of the current or voltage waveforms, thus improving efficiency of the power amplifier. The amplitude of corresponding current harmonic component can be obtained by

$$I_n = \gamma_n(\theta)\, I = \gamma_n(\theta)\, g_m V_{in}. \tag{10.41}$$

From Eq. (10.35) it follows that the dc current component is a function of θ in the operation modes with $\theta < 180^\circ$, in contrast to a Class A operation mode where $\theta = 180^\circ$ and the dc current is equal to the quiescent current during the entire period.

The collector efficiency of a resonant power amplifier biased to operate in the nonlinear mode can be obtained from

$$\eta = \frac{P_1}{P_0} = \frac{1}{2}\frac{I_1}{I_0}\xi = \frac{1}{2}\frac{\gamma_1}{\gamma_0}\xi. \tag{10.42}$$

where

$$P_0 = V_{cc} I_0 = V_{cc} I \gamma_0(\theta) \tag{10.43}$$

is the dc output power and

$$P_1 = \frac{VI_1}{2} = \frac{VI\gamma_1(\theta)}{2} \tag{10.44}$$

is the fundamental-frequency output power delivered to the load $P_1 = P_L$.

If $\xi = 1$ and $\theta = \pi/2$ or 90°, then from Eqs. (10.38) and (10.39) it follows that the maximum collector efficiency in a Class B operation mode is equal to

$$\eta = \frac{\pi}{4} \cong 78.5\%. \tag{10.45}$$

Equation (10.42) shows that the collector efficiency of a nonlinear power amplifier with pulsed collector current waveform depends on the conduction angle, where θ corresponds to half the conduction angle. This means that reduction in θ results in lower γ_1, and to increase the fundamental-frequency power P_1, it is necessary to increase the current amplitude I. Since the current amplitude I is determined by the input voltage amplitude V_{in}, the input power P_{in} must be increased. The collector efficiency also increases with reduced value of θ and becomes maximum when $\theta = 0^\circ$ where a ratio γ_1/γ_0 is maximal, as follows from Figure 10.11(*a*). For example, the collector efficiency η increases from 78.5% to 92% when θ reduces from 90° to 60°. However, it requires the input voltage amplitude V_{in} to be increased by 2.5 times, resulting in lower values of the power-added efficiency (*PAE*), which is defined as

$$PAE = \frac{P_1 - P_{in}}{P_0} = \frac{P_1}{P_0}\left(1 - \frac{1}{G_P}\right) \tag{10.46}$$

where

$$G_P = \frac{P_1}{P_{in}}$$

is the operating power gain.

Consequently, to obtain an acceptable tradeoff between high power gain and high *PAE*, the conduction angle should be chosen within the range of $120^\circ \le 2\theta \le 190^\circ$. If it is required to provide high collector efficiency of the active device having high gain capability, it is necessary to choose a Class C operation mode with θ close to 60°. However, when the input power is limited and power gain is not sufficient, a Class AB operation mode with small quiescent current when θ is slightly greater than 90° is recommended. In the latter case, the linearity of the power amplifier can be significantly improved.

Since the parallel *LC* circuit is tuned to the fundamental frequency, the voltage across the load resistor *R* can be considered cosinusoidal. By using Eqs. (10.20), (10.28), and (10.30), the relationship between the collector current i and collector voltage v during a time period of $-\theta \leq \omega t < \theta$ can be expressed by

$$i = \left(I_q + \frac{V_{cc}}{\gamma_1 R} \right) - \frac{v}{\gamma_1 R} \tag{10.47}$$

where the fundamental current coefficient γ_1 as a function of θ is determined by Eq. (10.39) and the load resistance is determined as $R = V/I_1$, where I_1 is the fundamental current amplitude. Equation (10.47) representing the dependence of the collector current on the collector voltage for any values of conduction angle in the form of a straight line function is called the *load line* of the active device. For a Class A operation mode with $\theta = 180°$ when $\gamma_1 = 1$, Eq. (10.47) is identical to Eq. (10.22).

Figure 10.12 shows the idealized active device output *I–V* curves and load lines for different conduction angles according to Eq. (10.47) with the corresponding collector and current waveforms. From Figure 10.12 it follows that the maximum collector current amplitude I_{max} corresponds to the minimum collector voltage V_{sat} when $\omega t = 0$, and is the same for any conduction angle. The slope of the load line defined by its slope angle β is different for the different conduction angles and values of the load resistance, and it can be obtained by

$$\tan \beta = \frac{I}{V(1 - \cos \theta)} = \frac{1}{\gamma_1 R} \tag{10.48}$$

from which it follows that greater the slope angle β of the load line, smaller is the value of the load resistance *R* for the same θ.

In general, the entire load line represents a broken line *PK* including a horizontal part, as shown in Figure 10.12. Figure 10.12(*a*) represents a load line *PNK* corresponding to a Class AB mode with $\theta > 90°$, $I_q > 0$, and $I < I_{max}$. Such a load line moves from point *K*, corresponding to the maximum output current amplitude I_{max} at $\omega t = 0$ and determining the device saturation voltage V_{sat} through the point *N* located at the horizontal axis v where $i = 0$ and $\omega t = \theta$. For a Class AB operation, the conduction angle for the output current pulse between points *N*′ and *N*″ is greater than 180°. Figure 10.12(*b*) represents a load line *PMK* corresponding to a Class C mode with $\theta < 90°$, $I_q < 0$, and $I > I_{max}$. For a Class C operation, the load line intersects a horizontal axis v in a point *M*, and the conduction angle for the output current pulse between points *M*′ and *M*″ is smaller than 180°. Hence, generally the load line represents a broken line with the first section having a slope angle β and another horizontal section with zero current *i*. In a Class B mode, the collector current represents half-cosine pulses with the conduction angle of $2\theta = 180°$ and $I_q = 0$.

Now let us consider a Class B operation with increased amplitude of the cosine collector voltage. In this case, as follows from Figure 10.13, an active device is operated in the saturation, active, and pinch-off regions, and the load line represents a broken line *LKMP* with three linear sections (*LK*, *KM*, and *MP*). The new section *KL* corresponds to the voltage saturation region resulting in the half-cosine output current waveform with a depression in the top part. With further increase of the output voltage amplitude, the output current pulse can be split into two symmetrical pulses containing a significant level of the higher order harmonic components. The same result can be achieved by increasing a value of the load resistance *R* when the load line is characterized by smaller slope angle β.

The collector current waveform becomes asymmetrical for the complex load, the impedance of which represents the load resistance and capacitive or inductive reactances. In this case, the Fourier

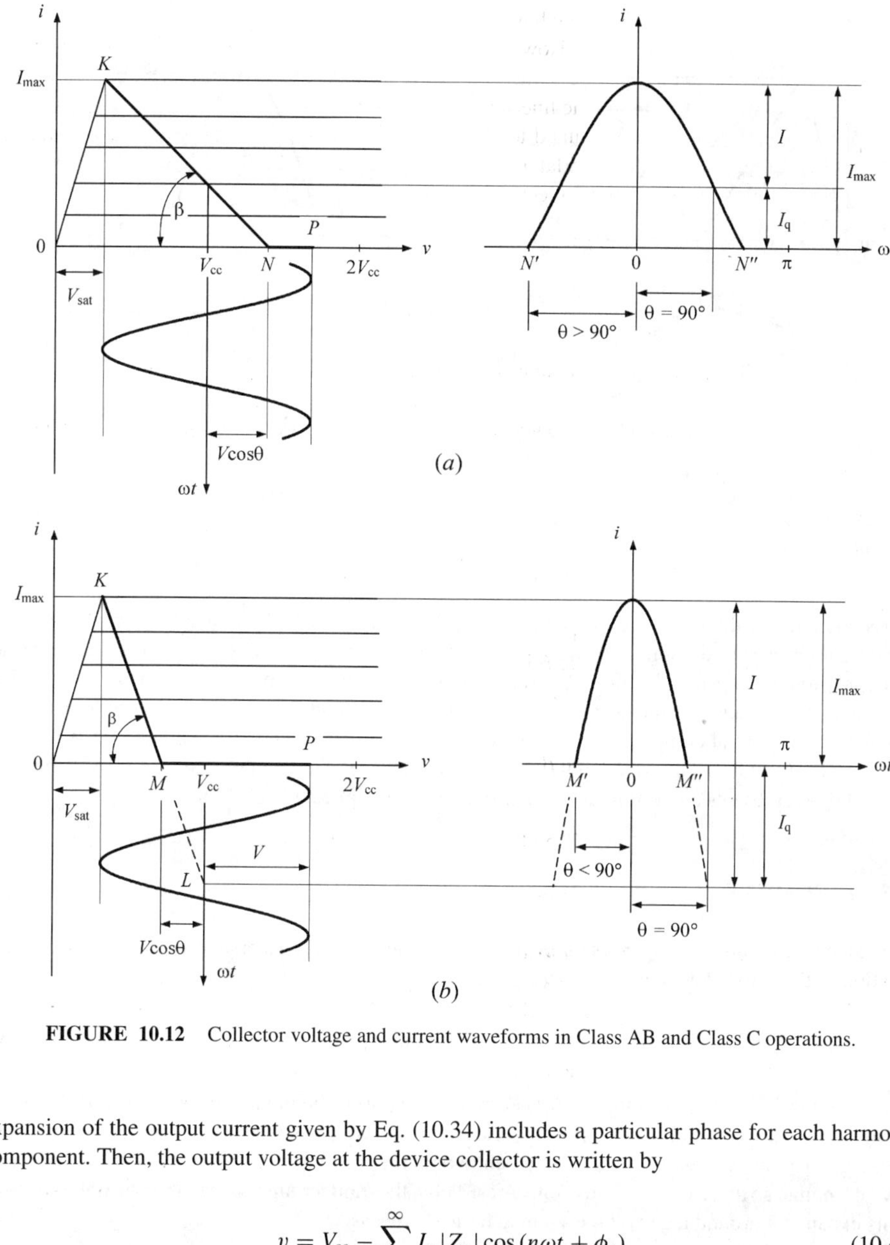

FIGURE 10.12 Collector voltage and current waveforms in Class AB and Class C operations.

expansion of the output current given by Eq. (10.34) includes a particular phase for each harmonic component. Then, the output voltage at the device collector is written by

$$v = V_{cc} - \sum_{n=1}^{\infty} I_n \, |Z_n| \cos(n\omega t + \phi_n) \tag{10.49}$$

where I_n is the amplitude of nth output current harmonic component, $|Z_n|$ is the magnitude of the load network impedance at nth output current harmonic component, and ϕ_n is the phase of nth output current harmonic component. Assuming that Z_n is zero for $n = 2, 3, \ldots$, which is possible for a resonant load network having negligible impedance at any harmonic component except the fundamental, Eq. (10.49) can be rewritten as

$$v = V_{cc} - I_1 \, |Z_1| \cos(\omega t + \phi_1) \,. \tag{10.50}$$

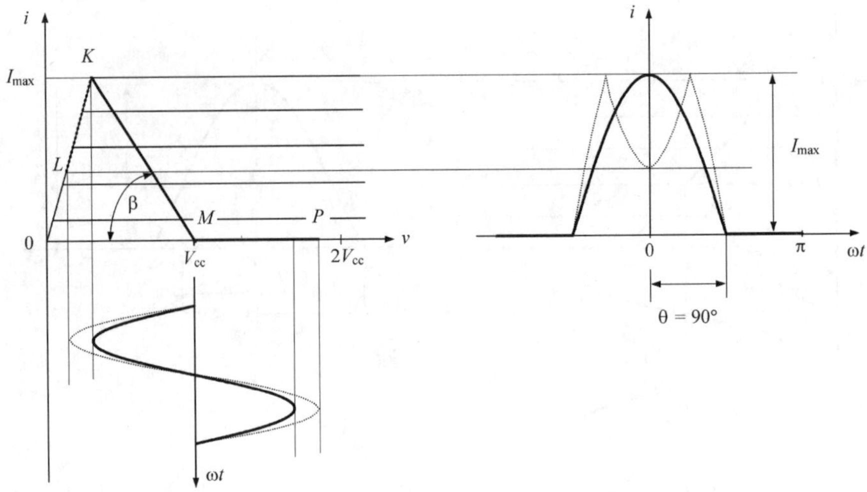

FIGURE 10.13 Collector voltage and current waveforms for the device operating in saturation, active and pinch-off regions.

As a result, for the inductive load impedance, the depression in the collector current waveform reduces and moves to the left side of the waveform, whereas the capacitive load impedance causes the depression to deepen and shift to the right side of the collector current waveform [15]. This effect can simply be explained by the different phase conditions for fundamental and higher order harmonic components in general composing the collector current waveform, and is illustrated by the different load lines for (*a*) inductive and (*b*) capacitive load impedances shown in Figure 10.14. Note that now the load line represents a two-dimensional curve with a complicated behavior.

10.3 LINEARITY

To evaluate the nonlinear properties of the power amplifier, first it is necessary to consider the transfer function of the active device in the form of

$$i\,(\omega t) = f\,[v\,(\omega t)] \tag{10.51}$$

where $i(\omega t)$ is the output collector or drain current and $v(\omega t)$ is the input gate–source or base–emitter voltage. It is convenient to apply a power-series analysis, which is relatively easy to use and which gives a good intuitive sense of the nonlinear behavior of the active device. Assume that the nonlinearity is weak enough so that a power series converges. Then, the transfer function $f(v)$ can be approximated by its expanding around a dc voltage V_0 in a Taylor series as

$$f(v) = f(V_0) + \sum_{n=1}^{\infty} \frac{1}{n!} \left. \frac{\partial^{(n)} f(v)}{\partial v^n} \right|_{v=V_0} (v - V_0)^n. \tag{10.52}$$

The nonlinear properties are usually determined by a two-tone excitation test signal with individual components separated slightly in frequency, which can be represented in a common case of unequal amplitudes as

$$v = V_0 + V_1 \cos\omega_1 t + V_2 \cos\omega_2 t. \tag{10.53}$$

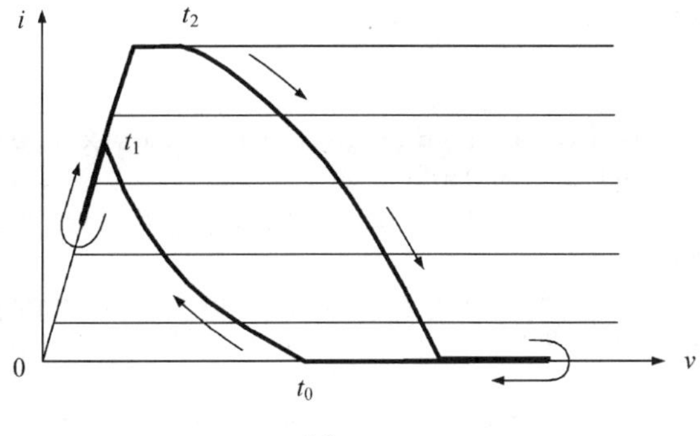

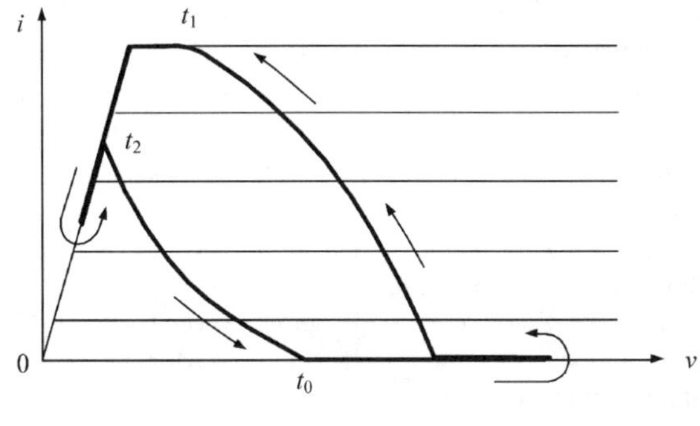

FIGURE 10.14 Load lines for (*a*) inductive and (*b*) capacitive load impedances.

For first three derivatives, the output signal can be represented by a Taylor series expansion with the appropriate equating of the frequency component terms by

$$
\begin{aligned}
i = {} & f(V_0) + \frac{1}{4}\left.\frac{\partial^2 f(v)}{\partial v^2}\right|_{v=V_0}\left(V_1^2 + V_2^2\right) \\
& + \left[\left.\frac{\partial f(v)}{\partial v}\right|_{v=V_0} + \frac{1}{4}\left.\frac{\partial^3 f(v)}{\partial v^3}\right|_{v=V_0}\left(\frac{1}{2}V_1^2 + V_2^2\right)\right] V_1 \cos\omega_1 t \\
& + \left[\left.\frac{\partial f(v)}{\partial v}\right|_{v=V_0} + \frac{1}{4}\left(V_1^2 + \frac{1}{2}V_2^2\right)\right] V_2 \cos\omega_2 t \\
& + \frac{1}{4}\left.\frac{\partial^2 f(v)}{\partial v^2}\right|_{v=V_0}\left(V_1^2 \cos 2\omega_1 t + V_2^2 \cos 2\omega_2 t\right) \\
& + \frac{1}{24}\left.\frac{\partial^3 f(v)}{\partial v^3}\right|_{v=V_0}\left(V_1^3 \cos 3\omega_1 t + V_2^3 \cos 3\omega_2 t\right) \\
& + \frac{1}{2}\left.\frac{\partial^2 f(v)}{\partial v^2}\right|_{v=V_0} V_1 V_2 \cos\left(\omega_1 \pm \omega_2\right) t \\
& + \frac{1}{8}\left.\frac{\partial^3 f(v)}{\partial v^3}\right|_{v=V_0}\left[V_1^2 V_2 \cos\left(2\omega_1 \pm \omega_2\right) t + V_1 V_2^2 \cos\left(\omega_1 \pm 2\omega_2\right) t\right] \cdots
\end{aligned} \tag{10.54}
$$

The following conclusions can be drawn from the above Taylor series expansion of the active device transfer function:

- Variation of the device bias point is directly proportional to the second derivative (in a common case, even derivatives) of the transfer function.
- Device transfer function will be linear only if the third derivative (in a common case, odd derivatives) is equal to zero.
- Even harmonic components are the result of even derivatives of the device transfer function, whereas odd harmonic components are the result of odd derivatives of the device transfer function.
- First-order mixing products (total and differential) are provided by even derivatives of the device transfer function.
- Mixing products of the third order and more are mainly determined by the odd derivatives of the device transfer function.
- Distortions, which are determined by the second derivative (second amplitude degree) or by the third derivative (third amplitude degree) of the device transfer function, are called the *second-order intermodulation distortions* or the *third-order intermodulation distortions*, respectively.

From Eq. (10.54) it follows that the output current amplitude of the fundamental, second, and third harmonic or intermodulation components depends on the first, second, and third degree of the input voltage amplitude, respectively. Consequently, the output power levels of the linear, second-order, and third-order frequency components show a straight-line behavior and vary by 1 dB, 2 dB, and 3 dB, respectively, with an input power level of 1 dB variation. Further analysis of Eq. (10.54) would show that n-order components also vary by n dB with an input power level of 1 dB variation. As a result, these straight lines in terms of dBm intersect at the proper intercept points. Each point is different for each order of intermodulation products. Consequently, if the intercept point is determined, for example, experimentally for a given type of the transistor, it is easy to evaluate the harmonic and n-order intermodulation output power levels at the arbitrary level of input power. Figure 10.15 shows the straight-line dependencies for the fundamental, second harmonic and third-order intermodulation (IM_3) components with the corresponding intercept points IP_2 and IP_3, respectively.

For any straight line, we can write the following equation:

$$P_{\mathrm{IM}_n} = nP_{\mathrm{in}} + P_{n0} \ \mathrm{dBm} \tag{10.55}$$

where P_{n0} is a constant that will be evaluated.

The linear fundamental-frequency output power is equal to

$$P_{\omega 1} = 10\log_{10}\left(G_{\mathrm{p}} P_{\mathrm{in}}\right) = P_{\mathrm{in}} + G_{\mathrm{p}} \ \mathrm{dBm}. \tag{10.56}$$

Equation (10.55) can be rewritten as

$$P_{\mathrm{IM}_n} = nP_{\omega 1} + P_{n0} - nG_{\mathrm{p}} \ \mathrm{dBm}. \tag{10.57}$$

At the intercept point IP_n,

$$P_{\mathrm{IM}_n} = P_{\omega 1} = IP_n \tag{10.58}$$

which yields a ratio

$$(1-n)\,IP_n = P_{\mathrm{no}} - nG_{\mathrm{p}} \ \mathrm{dBm}. \tag{10.59}$$

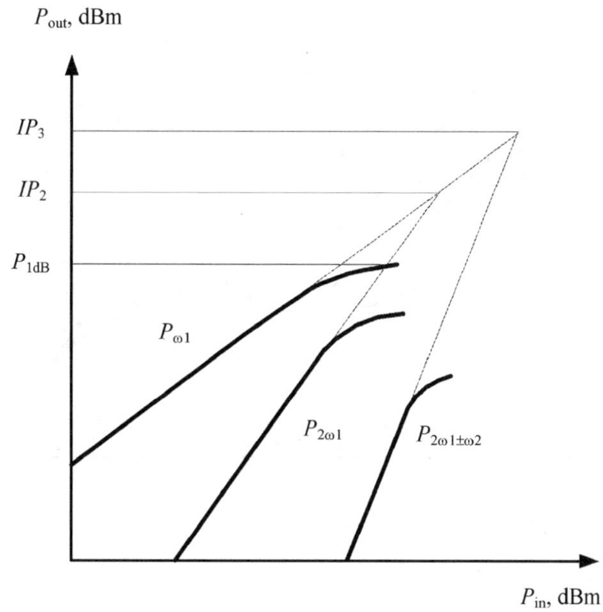

FIGURE 10.15 Straight-line dependencies for harmonic and intermodulation components.

Then,

$$P_{\mathrm{IM}_n} = nP_{\omega_1} - (n-1)\,IP_n \text{ dBm}. \tag{10.60}$$

Equation (10.60) can be used to evaluate the relationship between the fundamental output power P_{ω_1}, the output power corresponding to the n-order component P_{IM_n}, and the n-order intercept point IP_n at the input level below the device voltage saturation. For example, the second harmonic component $P_{2\omega_1}$ and the third-order intermodulation component $P_{2\omega_1-\omega_2}$ can be easily evaluated as

$$P_{2\omega_1} = 2P_{\omega_1} - IP_2 \text{ dBm} \tag{10.61}$$

$$P_{2\omega_1-\omega_2} = 3P_{\omega_1} - 2IP_3 \text{ dBm}. \tag{10.62}$$

The 1-dB compression level of output power, at which a value of the power gain decreases by 1 dB compared to its small-signal value in a linear operation region, can be estimated by comparing the first- and third-order terms in Eq. (10.54) as

$$P_{1\mathrm{dB}} = IP_3 - 9 \text{ dB dBm} \tag{10.63}$$

although, depending on the different devices with different types of their nonlinear transfer function, the difference $IP_3 - P_{1\,\mathrm{dB}}$ may vary between 8 and 15 dB.

Equations (10.62) and (10.63) give a convenient and simple qualitative estimate of the basic nonlinear performance of the real power amplifier. For example, if $IP_3 = 50$ dBm, then $P_{1\mathrm{dB}} =$ 41 dBm with the third-order nonlinear component $P_{2\omega_1-\omega_2} = 23$ dBm. To improve the linearity of the power amplifier, it is necessary to reduce the output power level for a given value of the intercept point. The level of -30 dB (relative to the fundamental-frequency output power P_{ω_1}) for the third-order intermodulation component $P_{2\omega_1-\omega_1}$ can be achieved only when $P_{\omega_1} = 35$ dBm, which means that the output power has to be reduced by 6 dB (or four times).

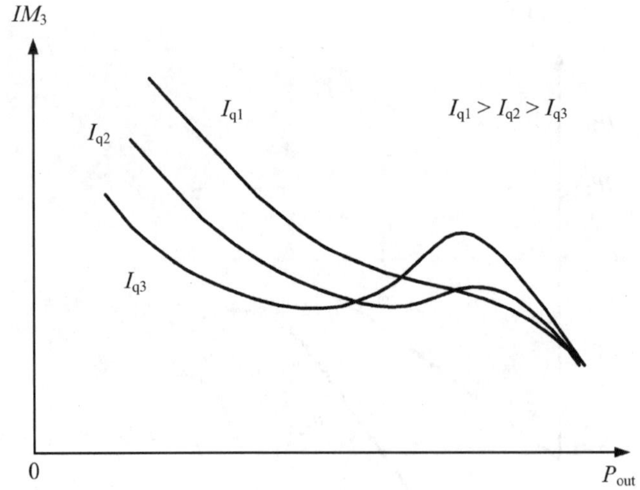

FIGURE 10.16 MESFET power amplifier intermodulation distortions for different quiescent currents.

It is well known that setting the dc drain current of a GaAs metal–semiconductor field-effect transistor (MESFET) device to approximately $0.5I_{dss}$ maximizes not only its gain but also intermodulation intercept points. First, the transfer $I_{ds}(V_{gs})$ curve is clearly more linear near sweet spot of $0.5I_{dss}$ where its slope, that is, the device transconductance, is maximal, and thus mainly influences the first-degree coefficient of its Taylor series expansion. Secondly, the nonlinearity of the $C_{gs}(V_{gs})$ curve significantly decreases with the bias point shifted toward higher values. To provide the high-efficiency operation mode of the MESFET power amplifier, it is necessary to use a value of the gate–source bias voltage quite close to the pinch-off voltage, with the appropriate worsening of its linear properties. However, in this case it is possible to choose the bias point with the drain quiescent current I_q in limits of 0.1–$0.15I_{dss}$, when the carrier-to-third-order intermodulation ratio IM_3 can be minimized at sufficiently high output power, as shown in Figure 10.16. First of all, this is a result of the quadratic dependence of drain current I_{ds} on gate voltage V_{gs} near the pinch-off point. Besides, given a certain value of the gate–source bias voltage, the third-order and the intermodulation components, which are the results of an interaction of the second harmonics $2\omega_1$ and $2\omega_2$ and differential component $\omega_2 - \omega_1$ with the fundamental components ω_1 and ω_2, cancel each other. Since this cancelation depends on the load and source impedances at the frequencies far from the operating frequency bandwidth, it is necessary to provide an additional tuning of the input and output matching circuits to minimize intermodulation distortion.

Figure 10.17 shows an output power spectrum containing only n-order intermodulation components, which are the result of the effect of two-tone input excitation. The amplitude of the higher order intermodulation components decreases significantly when frequency increases. To evaluate the linear behavior of the transistor, it is sufficient to measure the amplitudes of the largest intermodulation components, that is, their third-order and the fifth-order components. The carrier-to-third-order intermodulation ratio or the third-order intermodulation coefficient IM_3, as well as the fifth-order intermodulation coefficient IM_5, are defined as

$$IM_3 = 10\log_{10}\left(\frac{P_{2\omega_1-\omega_2}}{P}\right) = P_{2\omega_1-\omega_2} - P \text{ dBc} \tag{10.64}$$

$$IM_5 = 10\log_{10}\left(\frac{P_{3\omega_1-2\omega_2}}{P}\right) = P_{3\omega_1-2\omega_2} - P \text{ dBc} \tag{10.65}$$

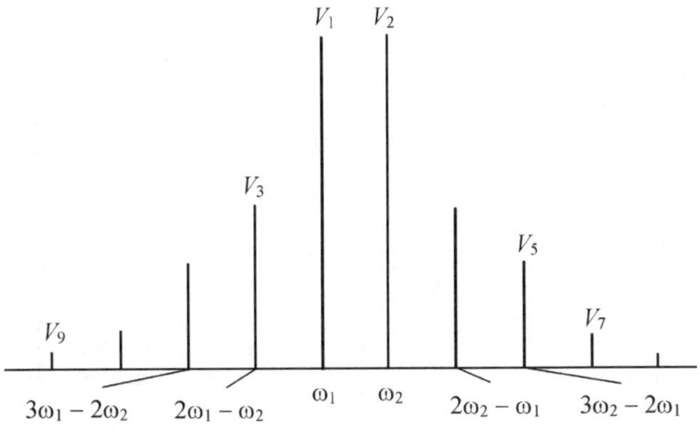

FIGURE 10.17 Typical output power spectrum for two-tone excitation.

where $P = P_{\omega_1} = P_{\omega_2}$ for equal two-tone signal amplitudes. On the other hand, the third-order intermodulation coefficient IM_3 can be directly calculated from Eq. (10.62) by

$$IM_3 = P_{2\omega_1 - \omega_2} - P_{\omega_1} = 2P_{\omega_1} - 2IP_3 \text{ dBc} \qquad (10.66)$$

and, for example, for the power amplifier with $IP_3 = 50$ dBm and $P_{1\text{dB}} = 41$ dBm, the third-order intermodulation coefficient IM_3 is equal to -18 dBc.

The linearity of the MOSFET power amplifiers is strongly sensitive to the bias conditions when a choice of the optimum bias voltage in Class AB operation mode improves the third-order intermodulation distortion by more than 10 dB [16]. This is a result of a quadratic character of the sufficiently long section of the device transfer dependence $I_{\text{ds}}(V_{\text{gs}})$. Minimum power gain flatness over the dynamic power range, linear P_{out} versus P_{in}, corresponds to the best linearity condition.

For bipolar transistors, a similar approach can be used when an improved intermodulation distortion is achieved by optimizing the collector quiescent current with proper base bias condition. In this case, in Class AB operation mode, the minimum level of the third-order intermodulation components is a function of the values of both the output power P_{out} and the collector quiescent current I_{q}. Low values of I_{q} give better linearity at higher power levels, whereas higher values of I_{q} give better linearity at lower power levels. Since the "sweet" point moves when different I_{q} are used, minimum gain flatness over the total dynamic range corresponds to the best linearity that can be achieved by adding a series resistor R with optimum value to the base bias circuit [17]. This series resistor is connected between the base bias voltage supply V_{b} and the bypass circuit C_{bypass}, followed by a quarterwave microstrip line to provide better isolation between RF and dc paths, as shown in Figure 10.18.

The use of a series resistor can minimize the gain flatness and stabilize the level of third-order intermodulation components over the dynamic range. For example, an increase in RF output power causes the appropriate increase in the dc collector current. This leads, in turn, to an increase in the dc base current with the corresponding increase of the voltage drop across the resistor R and decrease in a value of the base bias voltage. To determine the value of R, it needs to set a goal for the level of the intermodulation components at both high and low output power levels. Then, the value of R can be calculated according to

$$R = \frac{V_{\text{b1}} - V_{\text{b2}}}{I_{\text{b2}} - I_{\text{b1}}} \qquad (10.67)$$

where V_{b1} is the dc voltage at low power level, V_{b2} is the dc voltage at high power level, I_{b1} is the dc base current at low power level, and I_{b2} is the dc base current at high power level.

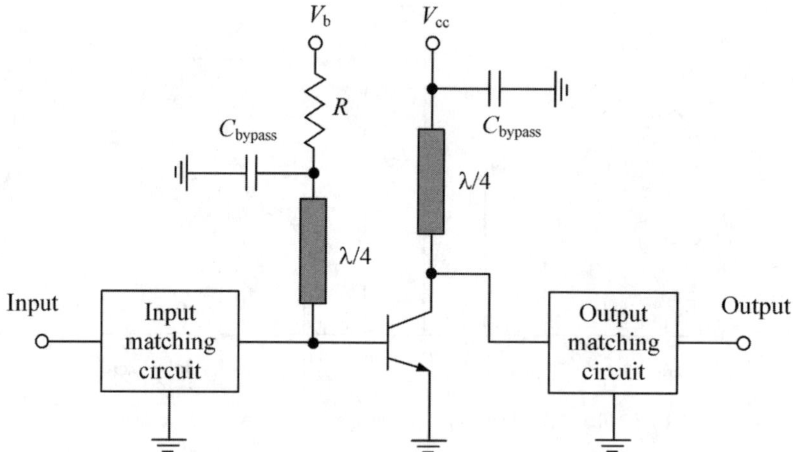

FIGURE 10.18 Bipolar power amplifier with linearizing bias resistor.

In practice, the level of n-order intermodulation component is usually given relative to the peak envelope power P_{PEP} calculated from two-tone excitation signal measurements. In this case, the waveform of the signal dissipated in the load will be significantly different from the sine wave, and the signal voltage can be written for a two-tone signal of equal amplitudes $V_1 = V_2 = V$ as

$$v_L = V_1 \sin\omega_1 t + V_2 \sin\omega_2 t = 2V \cos\Omega t \sin\omega t \tag{10.68}$$

where $\omega = (\omega_1 + \omega_2)/2$ is the center RF signal frequency and $\Omega = (\omega_2 - \omega_1)/2$ is the low intercarrier frequency or pulse envelope.

The waveform of such a two-tone driving signal with equal amplitudes is shown in Figure 10.19, where $T = 2\pi/\Omega$ is the period of envelope. The peak envelope power P_{PEP} that corresponds to the output power with maximum amplitude $2V$ is equal to

$$P_{PEP} = \frac{(2V)^2}{2R_L} \tag{10.69}$$

where R_L is the load resistance. The total output power provided by each sinusoidal tone of a two-tone excitation signal with equal amplitudes is

$$P_{out} = P_{\omega 1} + P_{\omega 2} = \frac{V^2}{R_L}. \tag{10.70}$$

Comparing Eqs. (10.69) and (10.70) yields

$$P_{PEP} = 2P_{out} = 4P \tag{10.71}$$

where $P = P_{\omega 1} = P_{\omega 2}$.

If the output amplitude and/or phase of the two- or multi-tone test signal are affected by the tone difference, the power amplifier exhibits the so-called memory effect. Memory is caused by the storage of energy because of storage elements (capacitors and indictors) that has to be charged or discharged. Smooth memory effects (usually at low frequencies) do not usually affect the linearity of the power amplifier itself. A phase rotation of 10°–20° or an amplitude change of less than 0.5 dB, as a

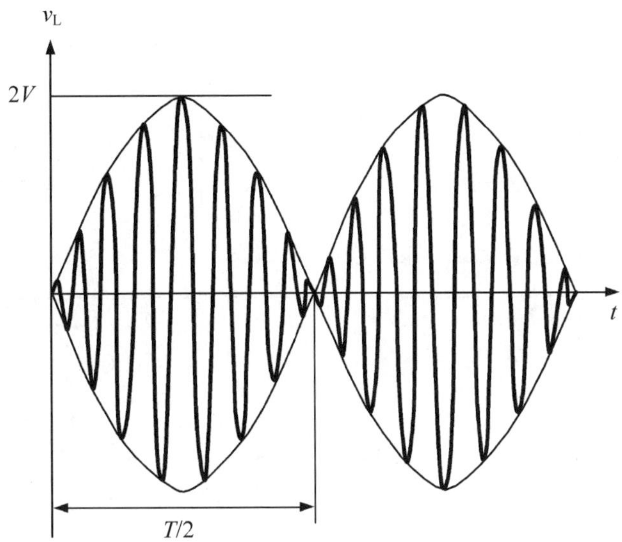

FIGURE 10.19 Two-tone driving signal.

function of modulation frequency, has no dramatic effect on the linearity of the device. There are two memory effects: electrical and thermal. Electrical memory effects are produced by nonconstant node impedances within different frequency ranges corresponding to the dc, fundamental, and its higher order harmonic components. Most of these effects are generated by frequency-dependent envelope impedance, and those close to the dc are the most harmful. Also, the greater the difference between the tones, the more the effect on higher frequency tone can be caused by the device transit time. Thermal memory effects are generated by the junction temperature that is modulated by the applied signal, and are much more prominent in a slow envelope sweep.

10.4 NONLINEAR EFFECT OF COLLECTOR CAPACITANCE

Generally, the dependence of the collector capacitance on the output voltage represents a nonlinear function. To evaluate the influence of the nonlinear collector capacitance on electrical behavior of the power amplifier, let us consider the load network including a series resonant L_0C_0 circuit tuned to the fundamental frequency that provides open-circuit conditions for the second-order and higher order harmonic components of the output current and an L-type matching circuit with series inductor L and shunt capacitor C, as shown in Figure 10.20(a). The matching circuit is required to match the equivalent output resistance R, corresponding to the required output power at the fundamental frequency, with the standard load resistance R_L. Figure 10.20(b) shows the simplified output equivalent circuit of the bipolar power amplifier.

The total output current flowing through the device collector can be written as

$$i = I_0 + \sum_{n=1}^{\infty} I_n \cos(n\omega t + \phi_n) \tag{10.72}$$

where I_n and ϕ_n are the amplitude and phase of the nth harmonic component, respectively.

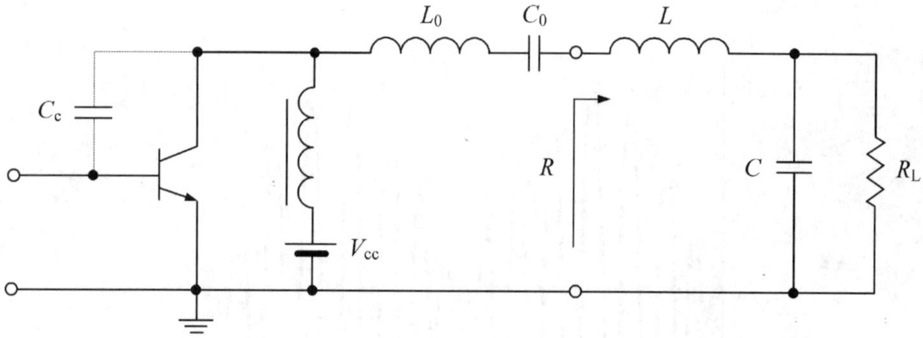

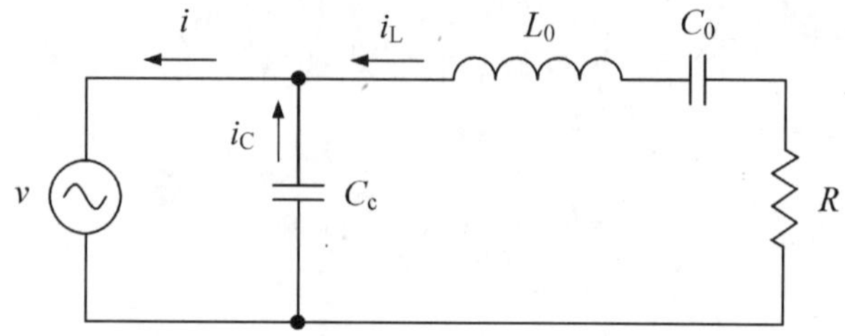

FIGURE 10.20 Resonant power amplifier circuit schematics.

An assumption of a high quality factor of the series resonant circuit allows the only fundamental-frequency current component to flow into the load. The current flowing through the nonlinear collector capacitance C_c, shown in Figure 10.20(*a*) outside of the transistor, consists of the fundamental-frequency and higher order harmonic components,

$$i_C = I_C \cos(\omega t + \phi_1) + \sum_{n=2}^{\infty} I_n \cos(n\omega t + \phi_n) \tag{10.73}$$

where I_C is the fundamental-frequency capacitor current amplitude.

The nonlinear behavior of the collector junction capacitance is described by

$$C_c(v) = C_0 \left(\frac{\varphi + V_{cc}}{\varphi + v} \right)^{\gamma} \tag{10.74}$$

where C_0 is the collector capacitance at $v = V_{cc}$, V_{cc} is the supply voltage, φ is the contact potential, and γ is the junction sensitivity, being equal to 0.5 for abrupt junction.

As a result, the expression for charge flowing through the collector capacitance can be obtained by

$$q = \int_0^v C(v)dv = \int_0^v \frac{C_0 (\varphi + V_{cc})^{\gamma}}{(\varphi + v)^{\gamma}} dv. \tag{10.75}$$

When $v = V_{cc}$, then

$$q_0 = \frac{C_0(\varphi + V_{cc})}{1-\gamma}\left[1 - \left(\frac{\varphi}{\varphi + V_{cc}}\right)^{1-\gamma}\right]. \tag{10.76}$$

Although the dc charge component q_0 is a function of the voltage amplitude, its deviation at maximum voltage amplitude from the small-signal value normally does not exceed 20% for $\gamma = 0.5$. Then, assuming q_0 is determined by Eq. (10.76) as a constant component, the total charge q of the nonlinear capacitance C_c can be represented by the dc component q_0 and ac component Δq written as

$$q = q_0 + \Delta q = q_0\left(1 + \frac{\Delta q}{q_0}\right) = q_0\frac{(\varphi + v)^{1-\gamma} - \varphi^{1-\gamma}}{(\varphi + V_{cc})^{1-\gamma} - \varphi^{1-\gamma}}. \tag{10.77}$$

Since usually $V_{cc} \gg \varphi$, then from Eq. (10.77) it follows that

$$\frac{v}{V_{cc}} = \left(1 + \frac{\Delta q}{q_0}\right)^{\frac{1}{1-\gamma}} \tag{10.78}$$

where $q_0 \cong C_0V_{cc}/(1-\gamma)$.

On the other hand, the linear charge component Δq can be written using Eq. (10.73) as

$$\Delta q = \int i_C(t)\,dt = \frac{I_C}{\omega}\sin(\omega t + \phi_1) + \sum_{n=2}^{\infty}\frac{I_n}{n\omega}\sin(n\omega t + \phi_n). \tag{10.79}$$

As a result, substituting Eq. (10.79) into Eq. (10.78) yields

$$\frac{v}{V_{cc}} = \left[1 + \frac{I_C(1-\gamma)}{\omega C_0V_{cc}}\sin(\omega t + \phi_1) + \sum_{n=2}^{\infty}\frac{I_n(1-\gamma)}{n\omega C_0V_{cc}}\sin(n\omega t + \phi_n)\right]^{\frac{1}{1-\gamma}}. \tag{10.80}$$

After applying a Taylor series expansion to Eq. (10.80), it is sufficient to be limited to its first three terms to reveal the parametric effect. Then, equating the fundamental-frequency collector voltage components gives

$$\frac{v_1}{V_{cc}} = \frac{I_C}{\omega C_0V_{cc}}\sin(\omega t + \phi_1) + \frac{I_CI_2\gamma}{(2\omega C_0V_{cc})^2}\cos(\omega t + \phi_2 - \phi_1) + \frac{I_2I_3\gamma}{12(\omega C_0V_{cc})^2}\cos(\omega t + \phi_3 - \phi_2). \tag{10.81}$$

Consequently, by taking into account that $v_1 = V_1\sin(\omega t + \phi_1)$, the fundamental voltage amplitude V_1 can be obtained from Eq. (10.81) by

$$\frac{V_1}{V_{cc}} = \frac{I_C}{\omega C_0V_{cc}}\left[1 + \frac{I_2\gamma}{4\omega C_0V_{cc}}\cos(90^\circ + \phi_2 - 2\phi_1) + \frac{I_2I_3\gamma}{12\omega C_0V_{cc}I_C}\cos(90^\circ + \phi_3 - \phi_2 - \phi_1)\right]. \tag{10.82}$$

Since a large-signal value of the abrupt-junction collector capacitance usually doesn't exceed 20%, the fundamental-frequency capacitor current amplitude I_C as a first order approximation can be written as $I_C \cong \omega C_0V_1$. As a result, from Eq. (10.82) it follows that, because of the parametric transformation due to the collector capacitance nonlinearity, the fundamental-frequency collector voltage amplitude increases by σ_p times according to

$$\sigma_p = 1 + \frac{I_2\gamma}{4\omega C_0 V_{cc}}\cos(90^\circ + \phi_2 - 2\phi_1) + \frac{I_2 I_3 \gamma}{12(\omega C_0)^2 V_1 V_{cc}}\cos(90^\circ + \phi_3 - \phi_2 - \phi_1) \quad (10.83)$$

where $\sigma_p = \xi_p/\xi$ and ξ_p is the collector voltage peak factor with parametric effect [1,15].

From Eq. (10.83) it follows that, to maximize the collector voltage peak factor and, consequently, the collector efficiency for a given value of the supply voltage V_{cc}, it is necessary to provide the following phase conditions:

$$\phi_2 = 2\phi_1 - 90^\circ \quad (10.84)$$

$$\phi_3 = 3\phi_1 - 180^\circ. \quad (10.85)$$

Then, for $\gamma = 0.5$,

$$\sigma_p = 1 + \frac{I_2}{8\,\omega C_0 V_{cc}} + \frac{I_2 I_3}{24\,(\omega C_0)^2 V_1 V_{cc}}. \quad (10.86)$$

Equation (10.86) demonstrates the theoretical possibility to increase the collector voltage peak factor by 1.1–1.2 times, thus achieving collector efficiency of 85% to 90%. Physically, the improved efficiency can be explained by the transformation of powers corresponding to the second-order and higher order harmonic components into the fundamental-frequency output power because of the nonlinearity of the collector capacitance. However, this becomes effective only in the case of the load network with a series resonant circuit, since it ideally provides infinite impedance at the second-order and higher order harmonic components, unlike the load network with a parallel resonant circuit having ideally zero impedance at these harmonics.

10.5 DC BIASING

The simplest way to provide a proper dc biasing for power MOSFET device in Class A or Class AB operation, is to use the potentiometer-type voltage divider for gate bias and choke inductance in the drain circuit, as shown in Figure 10.21(*a*). However, in this case, any variation of the ambient temperature or bias voltage in a wide range leads to the variations of the quiescent current and, as a result, to appropriate variations of the output power, linearity, efficiency and gain of the power amplifier. The threshold voltage of the MOSFET transistor V_{th} varies with temperature T linearly with the approximate speed of $\Delta V_{th}/\Delta T \cong 2$ mV/°C. But simple adding a diode in series to the variable resistor allows the quiescent current variation to be reduced substantially over temperature. A bias circuit corresponding to this stabilizing condition is shown in Figure 10.21(*b*). For a high value of V_{th}, several diodes can be connected in series. The reason to use such a simple bias circuit for power MOSFET biasing is that its dc gate current is extremely small being equal to the gate–source leakage current only.

In contrast to MOSFET devices, where it can be possible to choose the optimum operating point with practically zero temperature coefficient or to be limited to just an additional diode only, the bipolar transistors require the more complicated approach of dc biasing depending on a class of operation. For example, in a Class A operation mode, to obtain maximum linear power gain in high-power amplifiers with minimum variations in a wide temperature range, the bias circuit shown in Figure 10.22 can be designed [18]. Here, due to the large negative feedback due to the resistor R_3, the operating point of the RF power bipolar transistor is extremely stabilized for wide variations in ambient temperature. For instance, if the dc collector current of the RF power transistor rises, due

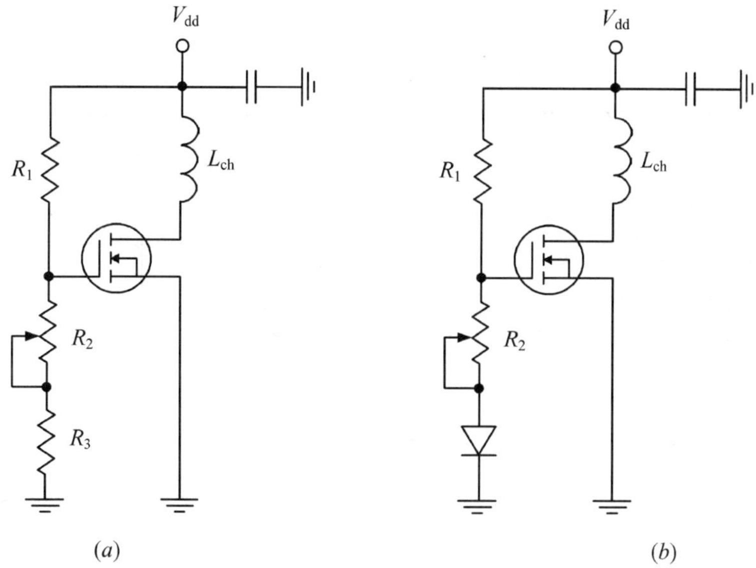

FIGURE 10.21 Typical MOSFET bias circuits.

to an increase in ambient temperature, the collector voltage of this transistor will fall in accordance with the voltage drop across the feedback resistor R_3 causing in turn an appropriate decrease in the transistor collector current. The diode is used to compensate for the temperature coefficient of the base–emitter junction voltage of the *p–n–p* transistor. The variable resistor R_1 in series with this diode serves to adjust the dc collector current of RF transistor in its operating point at the desired value. The resistor R_5 is necessary to reduce the variation in collector current of the *p–n–p* transistor, whereas the resistor R_4 is included to protect the RF power transistor and to reduce the dissipation in the *p–n–p* transistor. The parameters of this bias circuit are given for the collector current of RF power transistor of 0.9 A providing a collector voltage of 25 V for a supply voltage of 28 V.

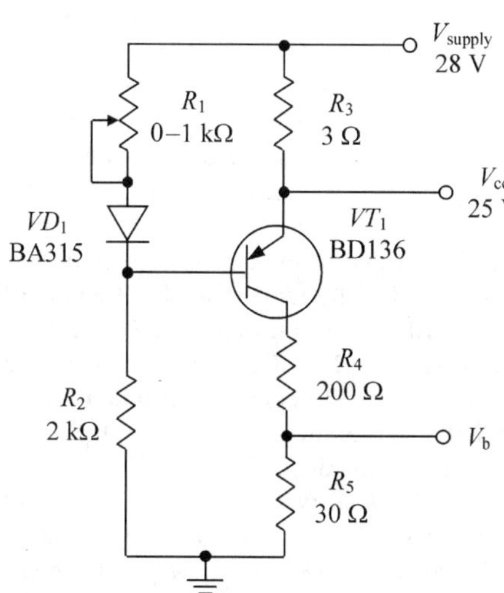

FIGURE 10.22 Typical bipolar bias circuit for Class A operation.

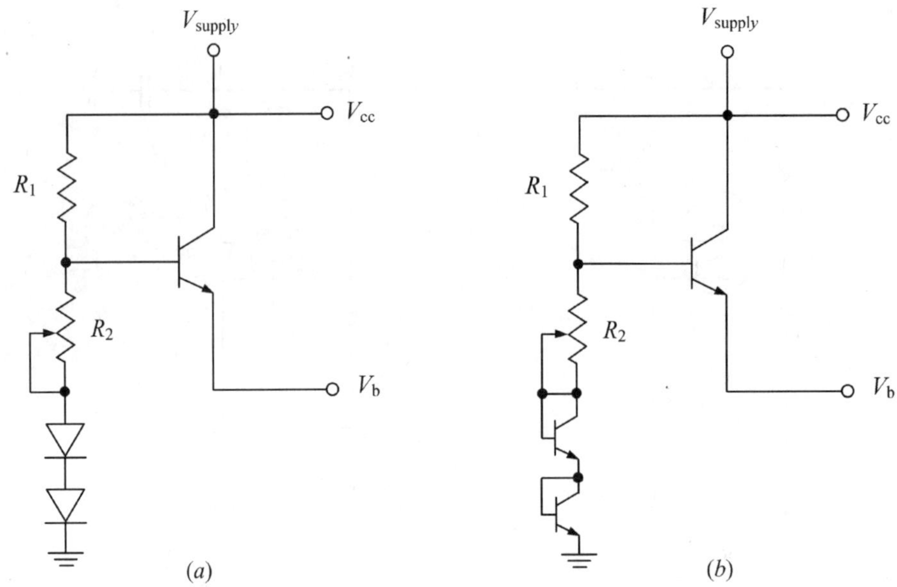

FIGURE 10.23 Simple bipolar bias circuits for Class AB operation.

In a Class AB operation, the bias circuit has to deliver a dc voltage, slightly adjustable approximately in limits of 0.7–0.8 V with a wide range of the current values to stabilize the base current of the RF bipolar transistor. Besides, it is necessary to provide an operation mode of the power amplifier with temperature compensation (collector current stabilization over temperature) and minimum possible reference current (dc current from the reference dc voltage supply). One of the simplest versions of such a bias circuit with silicon diode temperature compensation is shown in Figure 10.23(*a*). By using the emitter follower based on the *n–p–n* transistor, it is possible to increase the base current of the RF power transistor for high power operation. In this bias circuit, each silicon diode can be replaced by the *n–p–n* diode-connected transistor, the collector and the base of which are directly connected to each other, as shown in Figure 10.23(*b*). A better temperature-compensating result can be achieved using similar devices for RF and dc paths, only with reduced area sizes for the bias circuit devices. Such an approach is usually used in monolithic integrated circuit design when the same transistor cells with different area size are used for both RF power device and bias circuit transistors.

Figure 10.24 shows a more complicated bias circuit that is commonly used for biasing of the high-power RF transistors to provide their temperature-stable and reliable operation mode [18]. The temperature stabilization is provided with the parallel connection of the base–emitter diode junction of the transistor VT_1, whereas high value of the bias drive current for the RF power transistor is delivered by the transistor VT_2. If the dc collector current of an RF power transistor is 5 A and a value of its β_F is approximately equal to 10, then the maximum base current of the RF power transistor can be 0.5 A. In this bias circuit, the resistor R_5 is used to reduce the base current variations. At V_b = 0.7 V, for a current of 15 mA, the value of R_5 should be equal to 0.7 V/15 mA = 47 Ω. Suppose that a value of the collector current of VT_1 is 30 mA. As a result, if the base–emitter junction voltage of VT_2 is equal to 0.8 V with a voltage across the resistor R_2 of 28 − 1.5 = 26.5 V, its value is 26.5 V/ 30 mA≅ 820 Ω. The variable resistor R_3 serves to adjust the output voltage in limits of 0.1 V. To limit the maximum collector current of VT_2 by a value of 0.5 A, it is best to use the resistor R_4. Since VT_2 has a value of the saturation voltage of 0.8 V, it follows that the maximum value of R_4 is 26.5 V/ 0.5 A = 53 Ω. It is sufficient to use its value of 47 Ω with a power dissipation of $(0.5\text{ A})^2 \times 47\ \Omega$ = 11.75 W. Such a bias circuit can produce the parasitic oscillations near 1 MHz with highly capacitive

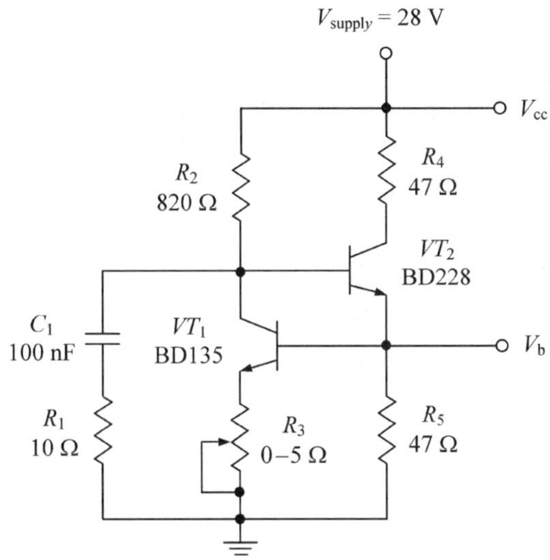

FIGURE 10.24 Typical bipolar bias circuit for Class AB operation.

loads. Therefore, to prevent these oscillations, it is necessary to connect the *RC* circuit between the collector of VT_1 and ground.

In most wireless communication systems, it is preferred that the power amplifier operates with high efficiency, maintaining an acceptable linearity over the desired supply voltage and output power ranges. However, there is a tradeoff between efficiency and linearity, with improvement in one coming at the expense of another. This means that it is necessary to provide an optimum stable fixed or adaptive bias point over wide temperature range and process variations. As a current-controlled device, the bipolar transistor in RF operation requires the dc base driving current, the value of which depends on the output power and device parameters. Because technologically the bipolar device represents a parallel connection of the basic cells, it is important to use the ballast series resistors to avoid current imbalance and possible device collapse at higher current density levels. Another important aspect is to keep the dc base–emitter bias point constant (or properly variable) over any RF input power variations to prevent the linearity worsening at maximum output power for power amplifiers with variable envelope signal (such as EDGE, WCDMA, LTE, or CDMA2000).

The typical temperature-compensation current-mirror bias circuit with one reference transistor Q_1 and one driving transistor Q_2 is shown in Figure 10.25(*a*). This circuit keeps the quiescent current for RF device Q_0 sufficiently constant over temperature variations, and the current flowing through resistor R_2 is sufficiently small. It is very important to provide the proper ratio between ballast resistors R_1 and R_0, equal to the reverse ratio of the device areas Q_0/Q_1. This can minimize the overall performance variation with temperature as well as stabilize the dc bias point. The latter case is very important for the variable envelope signals, as the dc bias voltage $V_{\rm be0}$ establishes the conduction angle and operation class for the RF device. If the dc base–emitter bias voltage reduces with the increase of RF input power, the Class AB mode required for linear operation changes to a nonlinear Class C operation.

Figure 10.25(*b*) shows the dependence of the dc base–emitter bias voltage $V_{\rm be0}$ versus RF input power $P_{\rm in}$ for the second stage of a wideband code-division multiple access (WCDMA) InGaP/GaAs heterojunction bipolar transistor (HBT) power amplifier for different cases, with ballast resistor $R_1 = 0$ (curve 1) and optimum ballast resistor R_1 (curve 2). From Figure 10.25(*b*) it follows that including of the ballast resistor with optimum value results in a more constant base–emitter dc bias voltage over a wider range of input powers, thus improving the linearity performance of the power amplifier at high

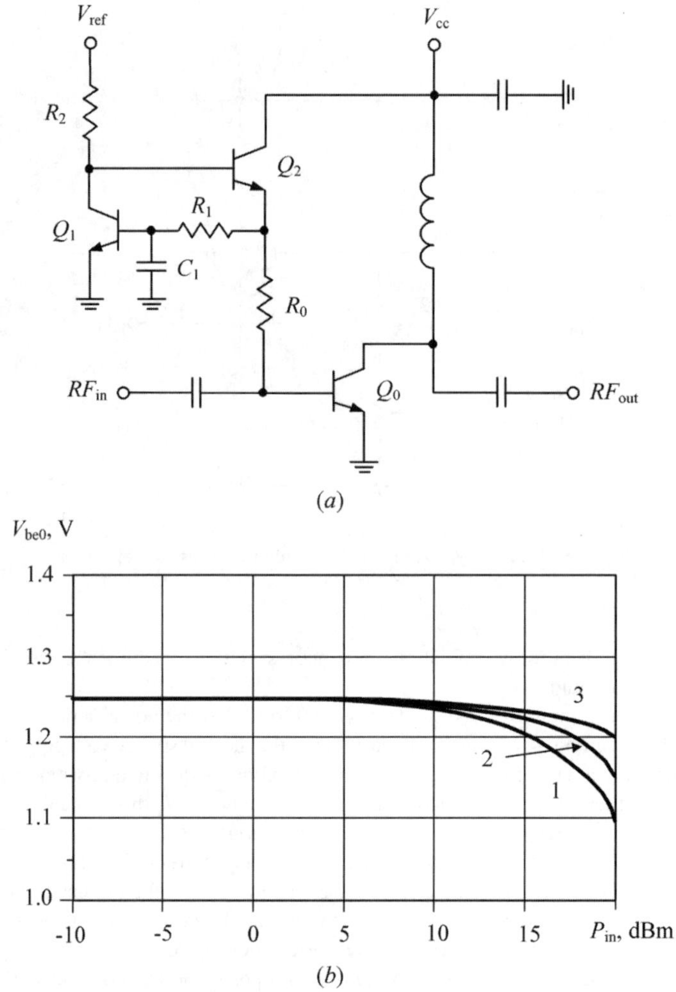

FIGURE 10.25 Current-mirror bipolar bias circuit and its performance.

power levels. In addition, it is best to use a shunt capacitance C_1 connected to the base of the device Q_1 to form a low-pass *RC* filter, which provides better isolation of the bias circuit from RF signal, with a yet more constant base–emitter dc bias voltage (curve 3).

Figure 10.26 shows the emitter follower bias circuit that provides temperature compensation and minimizing reference current requirements [19]. The emitter follower bias circuit requires only several tens of microamperes of reference current, whereas the current mirror bias circuit requires a few milliamperes of reference current. Both the current mirror and emitter follower bias circuits have similar current-voltage behavior but, for the same circuit parameters (R_0, R_1, and R_2) and devices areas for Q_0, Q_1, and Q_2, the emitter follower circuit is less sensitive to the reference voltage variations compared with current mirror bias circuit. Variations of the collector supply voltage V_{cc} in limits of 3.0–5.0 V have no effect on the quiescent current set by the reference voltage V_{ref}. However, for similar temperature performance, the emitter follower bias circuit may supply more dc current from dc power supply.

Figure 10.27 shows the bias circuit configuration stabilizing the quiescent current in some range of variations of the reference voltage V_{ref} [20,21]. This bias circuit incorporates a feedback loop

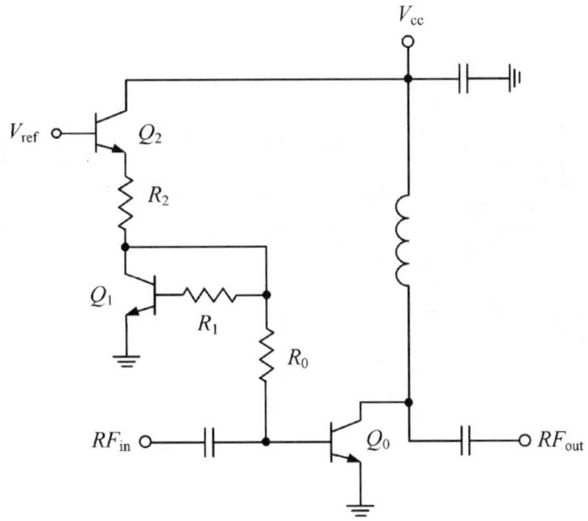

FIGURE 10.26 Bipolar power amplifier stage with emitter follower bias circuit.

in the form of differential amplifier, which stabilizes the dc operating point. The dc base current required for RF device Q_0 is fed by the driver transistor Q_2. For better temperature compensation, the ratio between resistors R_1 and R_0 should be equal to the reverse device area ratio Q_0/Q_1. The base nodes of the transistors Q_3 and Q_4 forming a differential amplifier are at the same potential $V_{b1} = V_{b2}$. The voltage difference $V_{ref} - V_{b2}$ across the resistor R_5 sets the reference current, which is a mirror current for a quiescent current. In this case, if, for some reason, the temperature of the

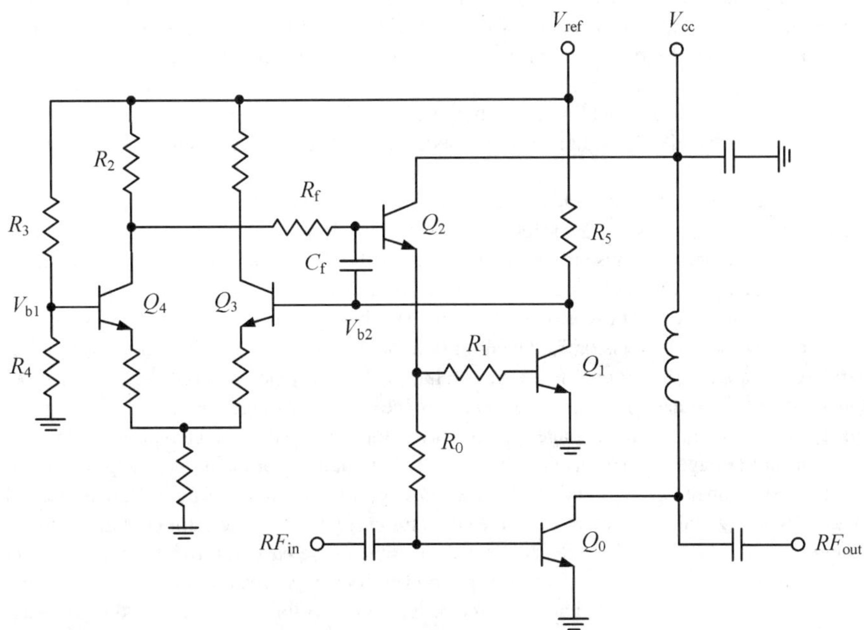

FIGURE 10.27 Bipolar power amplifier stage with feedback loop bias circuit.

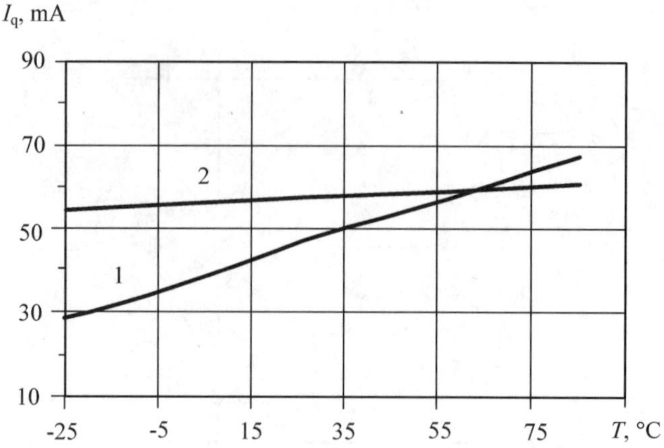

FIGURE 10.28 Quiescent current versus temperature for current mirror and feedback loop bias circuits.

RF power device rises causing the proper increase in a quiescent current, the collector current of a transistor Q_1 increases proportionally causing the decrease of the base voltage V_{b2} due to the increase of voltage drop across the resistor R_5. As a result, as the base voltage of the differential amplifier V_{b2} decreases while the base voltage V_{b1} remains constant, their difference contributes to the decrease of the differential amplifier output voltage resulting in the decrease of base current flowing through the driver transistor Q_2.

Figure 10.28 shows the measured quiescent current I_q of the current mirror and feedback loop bias circuits versus temperature of a wireless handset power amplifier at a supply voltage $V_{cc} = 3.5$ V [21]. The current mirror bias circuit shown in Figure 10.25 demonstrates a significant variation of 36 mA for a quiescent current over whole temperature range from -25 to $+85$°C (curve 1). At the same time, the variation of a quiescent current for the feedback loop bias circuit is negligible (curve 2). However, the feedback loop bias circuit has a tendency to be potentially unstable. To prevent the possible instability, it is necessary to provide a proper choice of feedback elements R_f and C_f acting as a low-pass filter section. In addition, a value of the capacitance C_f may be sufficiently large that increases the die size and cost, which are very important specification parameters for wireless mobile handsets.

To overcome stability problem and to keep similar insensitivity of the bias operating point over the same temperature range, process and reference voltage variations, the bias circuit schematic shown in Figure 10.29(a) can be used. This bias circuit includes a differential amplifier based on two transistors Q_3 and Q_4, located between the diode-connected reference transistor Q_1 and driving transistor Q_2. A value of the resistance R_1 is chosen to minimize the sensitivity to the temperature variations, and its value is scaled with ballast resistor R_0 depending on the device areas. In addition to minimizing the temperature variations, the proper choice of a value of the resistance R_2 contributes to minimizing the sensitivity of the quiescent current to the reference voltage changes. For instance, if, for some reason, a value of the reference voltage V_{ref} increases, then this results in an appropriate increase of the base–emitter voltage of the transistor Q_3. As the base potential of the transistor Q_4 is fixed using diode-connected transistor Q_1, the difference in base–emitter voltages of the differential amplifier transistors Q_3 and Q_4 causes an increase of the collector current of the transistor Q_3 and an increase of the voltage drop across the resistor R_2. As a result, voltage at the base of the transistor Q_2 reduces to keep the same value of a quiescent current of the power device Q_0. In addition, the diode-connected reference transistor Q_1 tracks the changes in the base–emitter voltage variations of the power device over temperature. As shown in Figure 10.29(b), the quiescent current I_q varies only in limits of 10% over a reference voltage range of 2.6–2.9 V. At the same time, the reference current I_{ref} is sufficiently

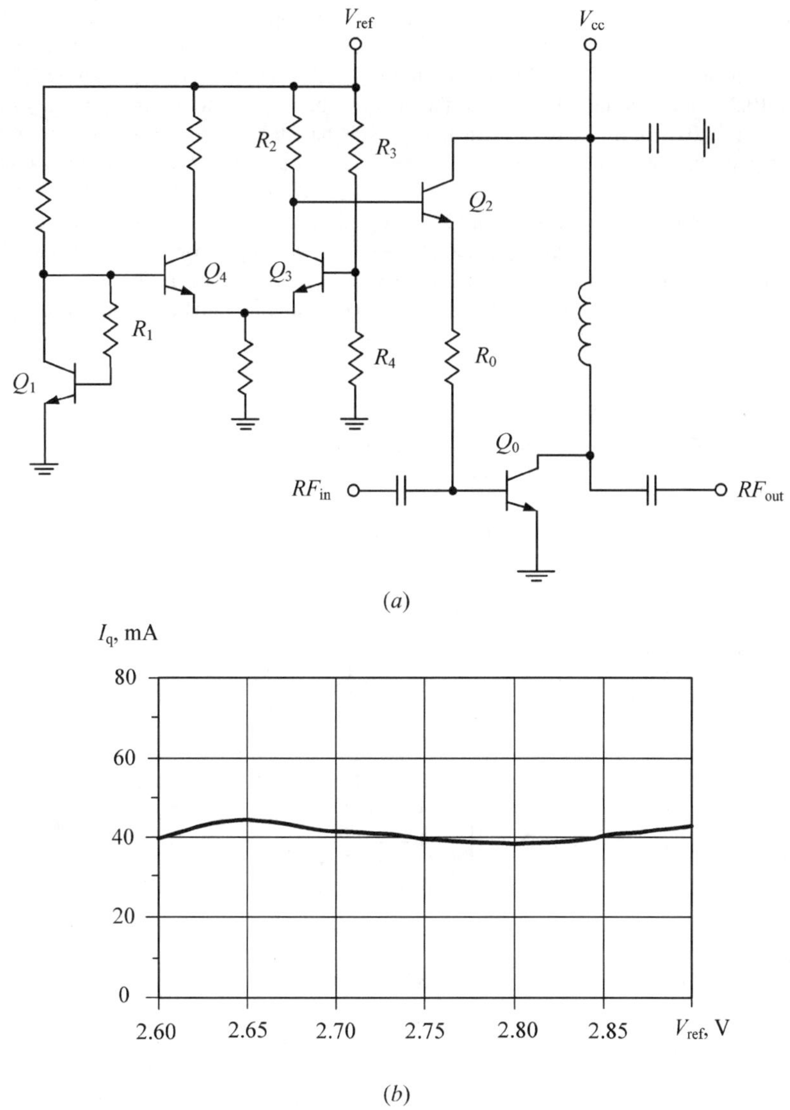

FIGURE 10.29 Bipolar power amplifier stage with differential amplifier bias circuit.

small, in limits of 2–4 mA. The quiescent current is very sensitive to variations of the values of the resistors R_3 and R_4 for both differential amplifier bias circuits shown in Figures 10.27 and 10.29(*a*). To eliminate this sensitivity, it is the best to replace these resistors partly using diode-connected transistors with small areas.

To provide an efficient linear power amplifier operation, it is necessary to minimize the quiescent current at backoff output powers since maximum of the power density function for CDMA2000 or WCDMA standards with nonconstant envelope occurs at the output powers of about 25–30 dB below the saturation output power. As a current controlled device, the bipolar transistor at RF operation requires the dc base driving current, the value of which depends on the output power and device parameters. Unlike the current mirror and emitter follower bias circuits, together with the temperature compensation over wide range of ambient temperatures, the adaptive bias circuit can control dc power consumption as the output power varies by greatly improving a *PAE* when the output power is low and maintaining a high linearity of the power amplifier when its output power is high [22].

Figure 10.30(*a*) shows the schematic of the adaptive bias circuit where the value of the ballast resistance R_1 is chosen to minimize the sensitivity to the base bias current variations and its value is scaled with ballasting resistor R_2 to the ratio of reverse device areas Q_2/Q_1 [23]. The bypass capacitor is needed to isolate dc bias circuit from the RF path. Since the RF transistor Q_1 is biased to a Class AB with a small quiescent current and its collector current is a function of the input power, the collector current of the current mirror device Q_2 increases with input power. This increased current decreases the base voltage of the device Q_3, thus decreasing its collector current. Then, this decreased collector current increases the base voltage of the device Q_4 due to a smaller voltage drop across the

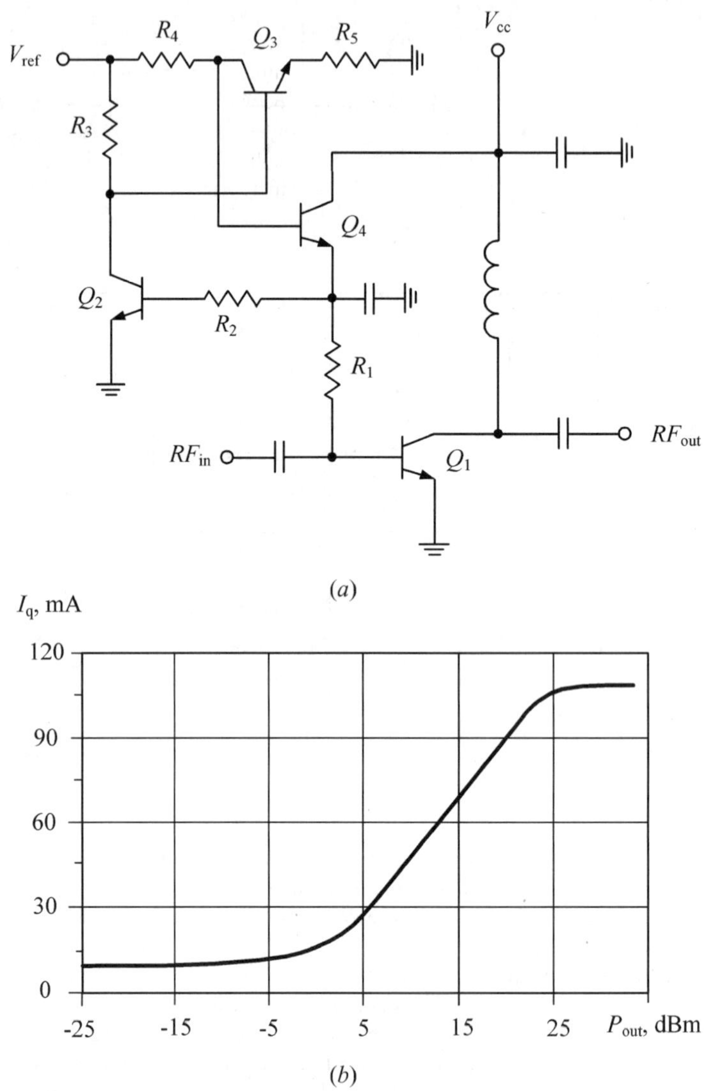

FIGURE 10.30 Adaptive bias circuit and its performance.

resistor R_4 forcing its emitter current and the collector current of the RF device Q_1 to increase. Thus, the quiescent current of the adaptive bias circuit is an increasing function of the input power. This provides high quiescent current at high output powers when high linearity with tradeoff efficiency is achieved and low quiescent current at low output powers when high linearity and increased efficiency are obtained. As an example shown in Figure 10.30(*b*), the quiescent current I_q varies from 110 mA at a maximum $P_{out} = 33$ dBm to 12 mA for low output powers. The minimum value of a quiescent current is defined by the values of resistors R_4 and R_5.

10.6 PUSH–PULL POWER AMPLIFIERS

Generally, if it is necessary to increase the overall output power of the power amplifier, several active devices can be used in parallel or push–pull configurations. In a parallel configuration, the active devices are not isolated from each other that requires a very good circuit symmetry and output impedance becomes too small in the case of high output power. The latter drawback can be eliminated in a push–pull configuration, which provides increased values of the input and output impedances. In this case, for the same output power level, the input impedance Z_{in} and output impedance Z_{out} are approximately four times as high as that of in a parallel connection of the active devices. At the same time, the loaded quality factors of the input and output matching circuits remain unchanged because both the real and reactive parts of these impedances are increased by the factor of four. Very good circuit symmetry can be provided using the balanced active devices with common emitters in a single package. The basic concept of a push–pull operation can be analyzed by using the equivalent circuit shown in Figure 10.31 [24].

It is most convenient to consider an ideal Class B operation, which means that each transistor conducts exactly half a 180° cycle with zero quiescent current. Let us also assume that the number of turns of both primary and secondary windings of the output transformer T_2 is equal when $n_1 = n_2$, and the collector current of each transistor can be represented in the following half-sinusoidal form:

for the first transistor

$$i_{c1} = \begin{cases} +I_c \sin\omega t & 0 \le \omega t < \pi \\ 0 & \pi \le \omega t < 2\pi \end{cases} \tag{10.87}$$

for the second transistor

$$i_{c2} = \begin{cases} 0 & 0 \le \omega t < \pi \\ -I_c \sin\omega t & \pi \le \omega t < 2\pi \end{cases} \tag{10.88}$$

where I_c is the output current amplitude.

Being transformed through the output transformer T_2 with the appropriate out-of-phase conditions, the total current flowing across the load R_L is obtained by

$$i_R(\omega t) = i_{c1}(\omega t) - i_{c2}(\omega t) = I_c \sin\omega t. \tag{10.89}$$

The current flowing into the center tap of the primary windings of the output transformer T_2 is the sum of the collector currents resulting in

$$i_{cc}(\omega t) = i_{c1}(\omega t) + i_{c2}(\omega t) = I_c\,|\sin\omega t|. \tag{10.90}$$

Ideally, even-order harmonics being in phase are canceled out and should not appear at the load. In practice, a level of the second harmonic component of 30–40 dB below the fundamental is allowable. However, it is necessary to connect a bypass capacitor to the center tap of the primary winding to

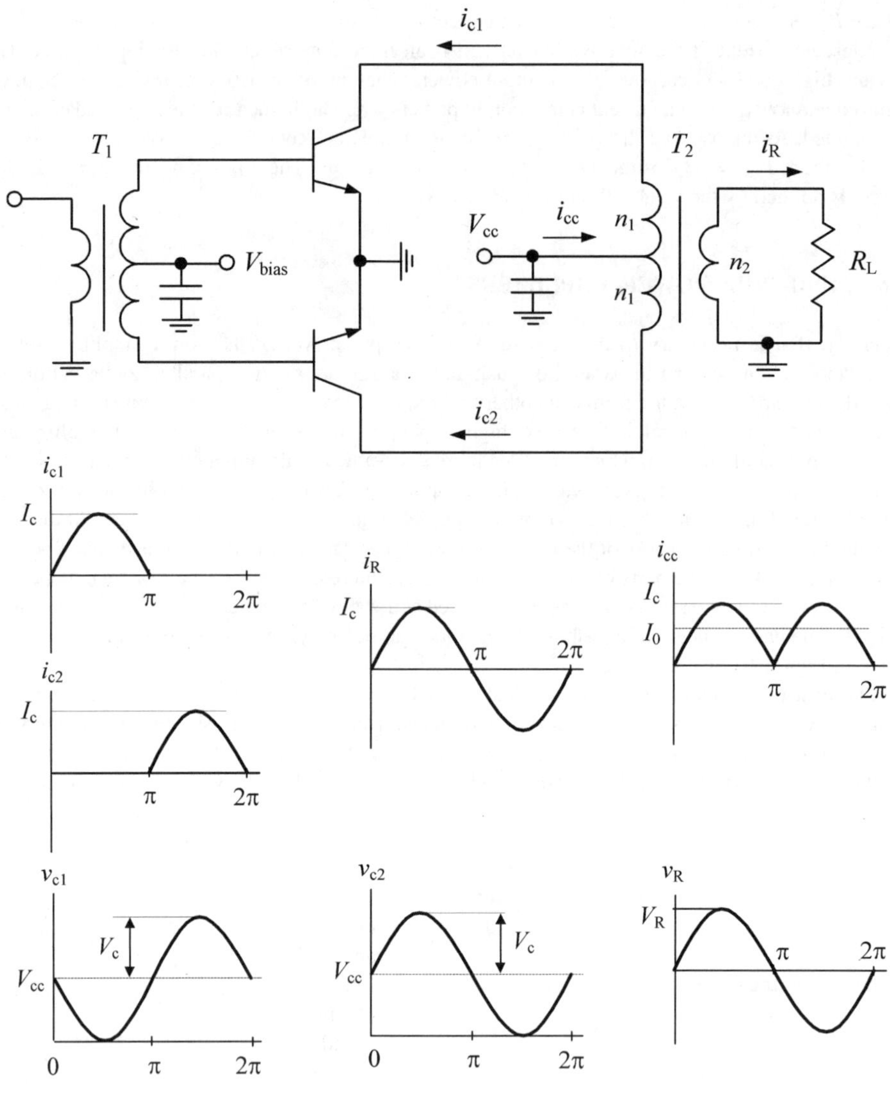

FIGURE 10.31 Basic concept of push–pull operation.

exclude power losses due to even-order harmonics. The current $i_R(\omega t)$ produces the load voltage $v_R(\omega t)$ onto the load R_L as

$$v_R(\omega t) = I_c R_L \sin(\omega t) = V_R \sin(\omega t) \tag{10.91}$$

where V_R is the load voltage amplitude.

The total dc collector current is defined as the average value of $i_{cc}(\omega t)$, which yields

$$I_0 = \frac{1}{2\pi}\int_0^{2\pi} i_{cc}(\omega t)\, d(\omega t) = \frac{2}{\pi} I_c. \tag{10.92}$$

The total dc power P_0 and fundamental-frequency output power P_1, for the ideal case of zero saturation voltage of both transistors when $V_c = V_{cc}$ and taking into account that $V_R = V_c$ for equal turns of windings when $n_1 = n_2$, are calculated from

$$P_0 = \frac{2}{\pi} I_c V_{cc} \tag{10.93}$$

$$P_1 = \frac{I_c V_{cc}}{2}. \tag{10.94}$$

Consequently, the maximum theoretical collector efficiency that can be achieved in a push–pull Class B operation mode is equal to

$$\eta = \frac{P_1}{P_0} = \frac{\pi}{4} \cong 78.5\%. \tag{10.95}$$

In a balanced circuit, identical sides carry 180° out-of-phase signals of equal magnitude. If perfect balance is maintained on both sides of the circuit, the difference between signal magnitudes becomes equal to zero in each midpoint of the circuit, as shown in Figure 10.32. This effect is called the *virtual grounding*, and this midpoint line is referred to as the *virtual ground*. The virtual ground, being actually inside the device package, reduces a common mode inductance and results in better stability and usually higher power gain.

When using a balanced transistor, new possibilities for both internal and external impedance matching procedure emerge. For instance, for a push–pull operation mode of two single-ended transistors, it is necessary to provide reliable grounding for input and output matching circuits for each device, as shown in Figure 10.33(*a*). Using the balanced transistors simplifies significantly the matching circuit topologies with the series inductors and parallel capacitors connected between amplifying paths, as shown in Figure 10.33(*b*), and dc blocking capacitors are not needed.

For a push–pull operation of the power amplifier with a balanced transistor, it is also necessary to provide the unbalanced-to-balanced transformation referenced to the ground both at the input and at the output of the power amplifier. The most suitable approach to solve this problem in the best possible manner at high frequencies and microwaves is to use the transmission-line transformers, as shown in Figure 10.34. If the characteristic impedance Z_0 of the coaxial transmission line is equal to the input impedance at the unbalanced end of the transformer, the total impedance from both devices seen at the balanced end of the transformer will be equal to the input impedance. Hence, such a transmission-line transformer can be used as a 1:1 balun. If $Z_0 = 50\ \Omega$, for the standard input impedance of 50 Ω, the impedance seen at the each balanced part is equal to 25 Ω, which then is

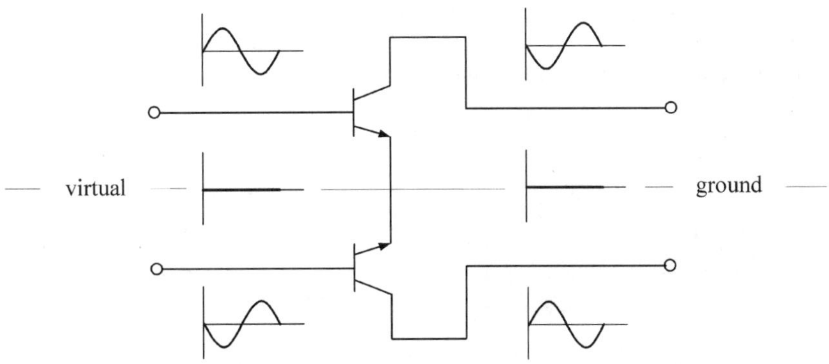

FIGURE 10.32 Basic concept of balanced transistor.

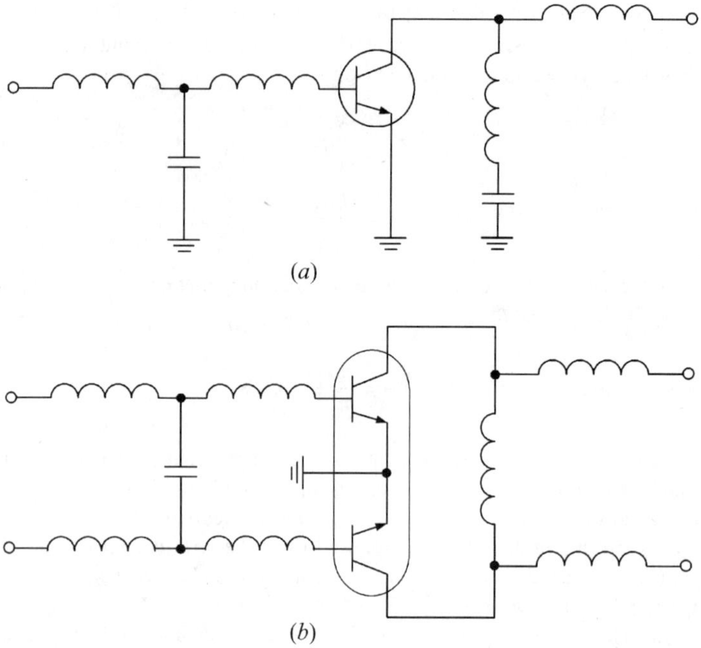

FIGURE 10.33 Matching technique for (*a*) single-ended and (*b*) balanced transistors.

necessary to match with the appropriate input impedance of each part of a balanced transistor. The input and output matching circuits can easily be realized by using the series microstrip lines with parallel capacitors.

The miniaturized compact input unbalanced-to-balanced transformer shown in Figure 10.35 covers the frequency bandwidth up to octave with well-defined rejection-mode impedances [25]. To avoid the parasitic capacitance between the outer conductor and the ground, the coaxial semirigid transformer T_1 is mounted atop microstrip shorted stub l_1 and soldered continuously along its length. The electrical length of this stub is usually chosen from the condition of $\theta \leq \pi/2$ on the high bandwidth frequency depending on the matching requirements. To maintain circuit symmetry on the balanced side of the transformer network, another semirigid coaxial section T_2 with unconnected center conductor is soldered continuously along microstrip shorted stub l_2. The lengths of T_2 and l_2 are equal to the lengths of T_1 and l_1, respectively. Because the input short-circuited microstrip stubs provide inductive impedances, the two series capacitors C_1 and C_2 of the same value are used for matching purposes, thereby forming the first high-pass matching section and providing dc blocking at the same time. The practical circuit realization of the output matching circuit and balanced-to-unbalanced transformer or balun can be the same as for the input matching circuit.

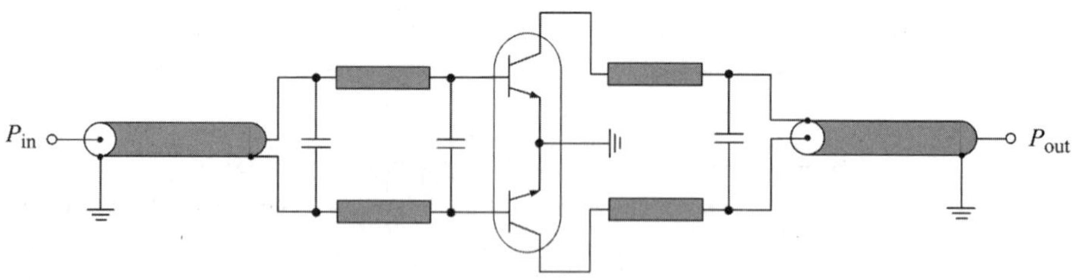

FIGURE 10.34 Push–pull power amplifier with balanced-to-unbalanced transformers.

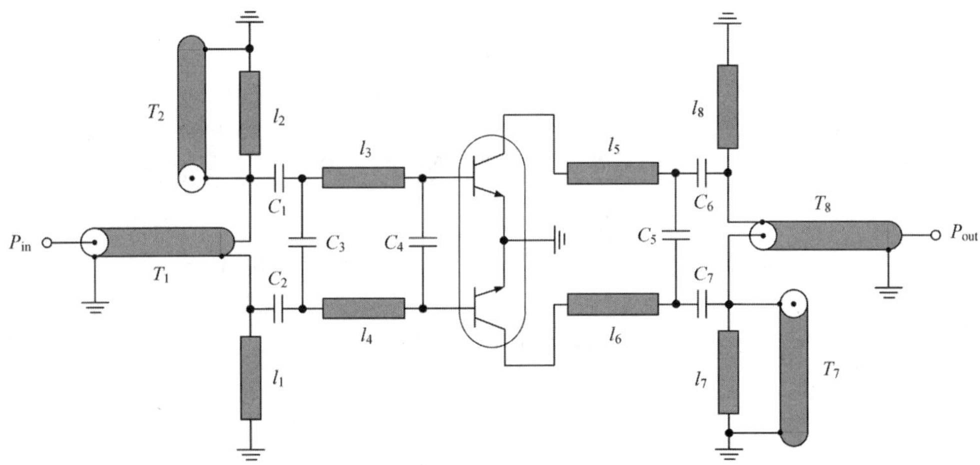

FIGURE 10.35 Push–pull power amplifier with compact balanced-to-unbalanced transformers.

Instead of a passive balun at the input, the active-balun circuit can be used to provide the out-of-phase input signals so that the push–pull power amplifier can be driven from a single unbalanced transmission line, which is very important for a very broadband operation or in a monolithic design. In the former case, the low boundary frequency of operation depends on the values of the dc blocking capacitors only. The simplest active balun or power splitter is a single-stage amplifying stage with two outputs connected to the drain and source separated from the dc voltage supply and the ground by the RF chokes, respectively, as shown in Figure 10.36(*a*). By using a MOSFET device, a broadband operation up to $0.1f_T$ can be easily provided, though the power gain is usually negligible and amplitude and phase accuracy at high frequencies is not sufficiently high. To improve the performance over broader frequency bandwidth, a differential amplifier with identical output paths, which schematic is shown in Figure 10.36(*b*), can be used. The advantage of using a MOSFET device compare to its bipolar counterpart is its high device input impedance so that no need to use the complicated input matching circuit and it is easy to provide symmetry of operation over very wide frequency range.

Figure 10.37(*a*) shows a schematic diagram of an active power splitter or paraphase amplifier based on the differential amplifier with a current source using MESFET devices [26]. The unbalanced input is applied to one of the gates while the other gate is RF shorted to ground by means of a bypass capacitor. For *X*-band operation, a capacitance of 8 pF or greater (reactance at 10 GHz is approximately equal to 2 Ω) is desirable. The output tuning is identical to that used in the first stage of the push–pull amplifier. The series and shunt inductors at the input are used for impedance matching with standard 50-Ω input transmission line. The transistor that functions as a current source is realized by 1.5 times larger device. To provide high impedance at the common source connection to the differential pair, the gate of the current source transistor is RF shorted to ground, and a shunt RF grounded inductor is used at the drain to parallel resonate the device drain–source capacitance. As a result, in a frequency range from 8.4 to 8.65 GHz, the amplitudes were within 1 dB and the phase difference was between 140° and 180°.

Figure 10.37(*b*) shows the circuit schematic of an active power splitter using both the common gate and common source GaAs FETs which can be used as separate devices or in a dual-gate configuration [27,28]. One gate is grounded through a bypass capacitor for a common gate device, while one source is directly grounded for a common source device. The source of a common gate device and the gate of a common source device are connected in phase at the input port through a phase shift circuit representing a short transmission line and a dc blocking capacitor, respectively. A short transmission

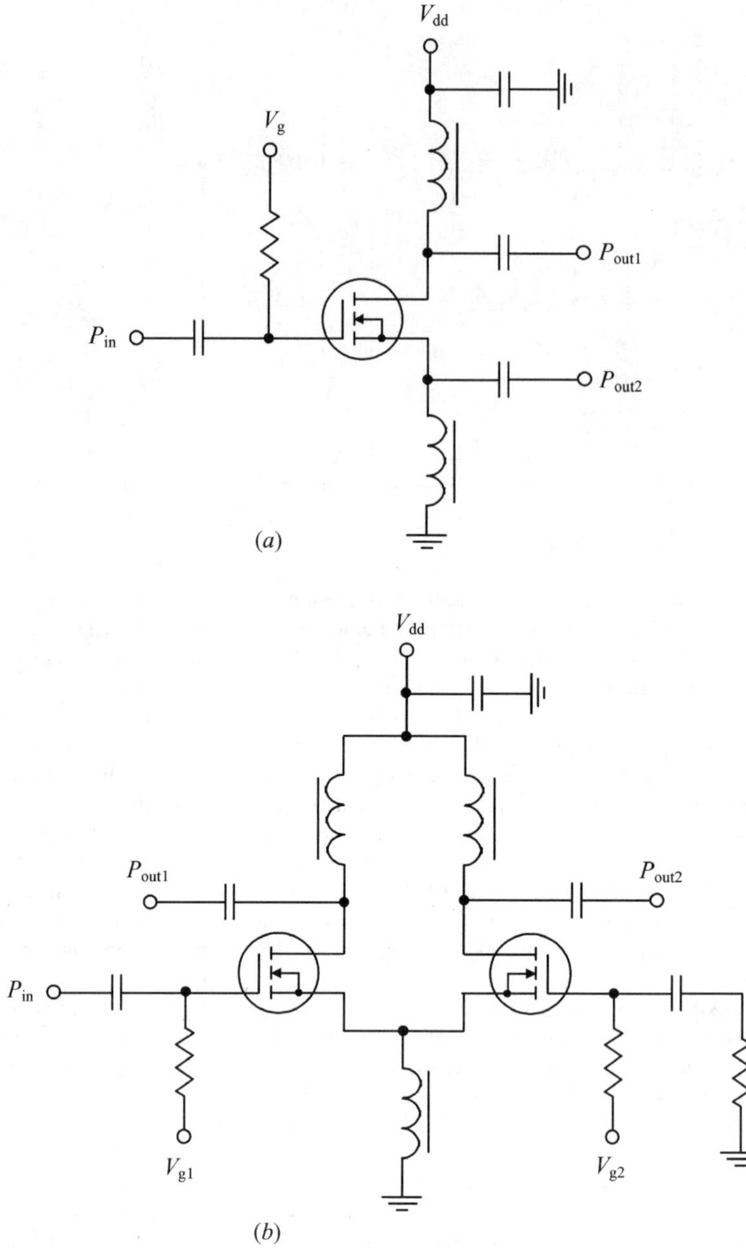

FIGURE 10.36 Broadband active MOSFET power splitters.

line is added to adjust phase difference between the output ports to 180°. The drain bias is supplied through resistors that define the output impedance, though also transmission lines can be used. In addition, the broadband input matching circuit at the input port is necessary to maximize the power gain over wide frequency range. As a result, by using the two 0.15-μm HEMTs with $f_T = 75$ GHz in a monolithic implementation, the measured 1-dB frequency range from 22 to 32 GHz with phase difference within 170° and 180° was achieved [28].

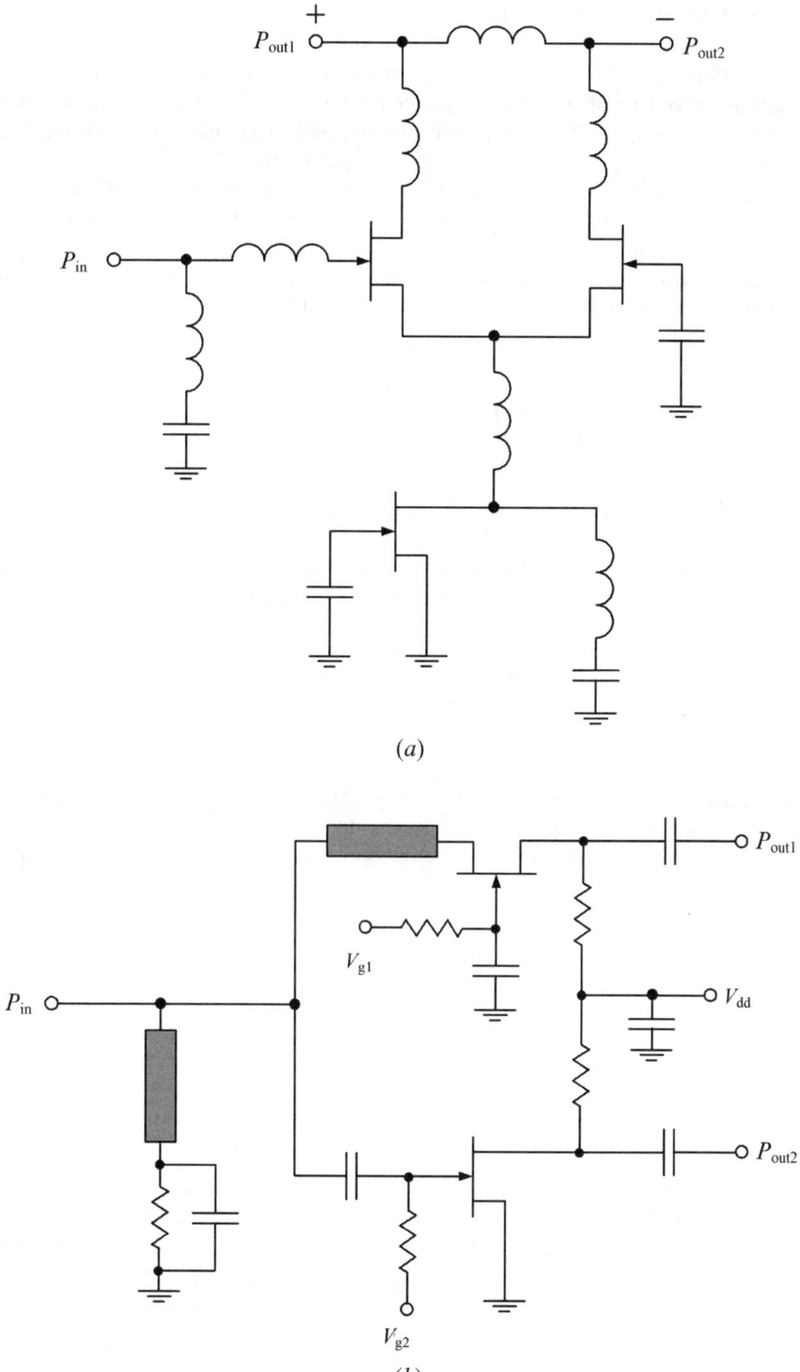

FIGURE 10.37 Active GaAs FET power splitters.

10.7 BROADBAND POWER AMPLIFIERS

Generally, the design for a broadband matching circuit, as a main aspect of the broadband power amplifier design procedure, should solve a problem with contradictory requirements when wider matching bandwidth is required with minimum reflection coefficient, or how to minimize the number of the matching network sections for a given wideband specification. The necessary requirements are determined by the Bode–Fano criterion that gives, for certain canonical types of load impedances, a theoretical limit on the minimum reflection coefficient magnitude that can be obtained with an arbitrary matching network [29,30].

For the lossless matching networks with s parallel RC load impedance shown in Figure 10.38(a) and with series LR load impedance shown in Figure 10.38(b), the Bode-Fano criterion states that

$$\int_0^\infty \ln \frac{1}{|\Gamma(\omega)|} d\omega \leq \frac{\pi}{\tau} \tag{10.96}$$

where $\Gamma(\omega)$ is the reflection coefficient seen looking into the arbitrary lossless matching network and $\tau = RC = L/R$.

For the lossless matching networks with a series RC load impedance shown in Figure 10.38(c) and with parallel LR load impedance shown in Figure 10.38(d), the Bode-Fano integral is written as

$$\int_0^\infty \frac{1}{\omega^2} \ln \frac{1}{|\Gamma(\omega)|} d\omega \leq \pi\tau. \tag{10.97}$$

The mathematical relationships expressed by Eqs. (10.96) and (10.97) reflect the flat responses of an ideal filter over the required frequency bandwidth shown in Figure 10.39 for two different cases. For the same load, both plots illustrate the important tradeoff: the wider the matching network

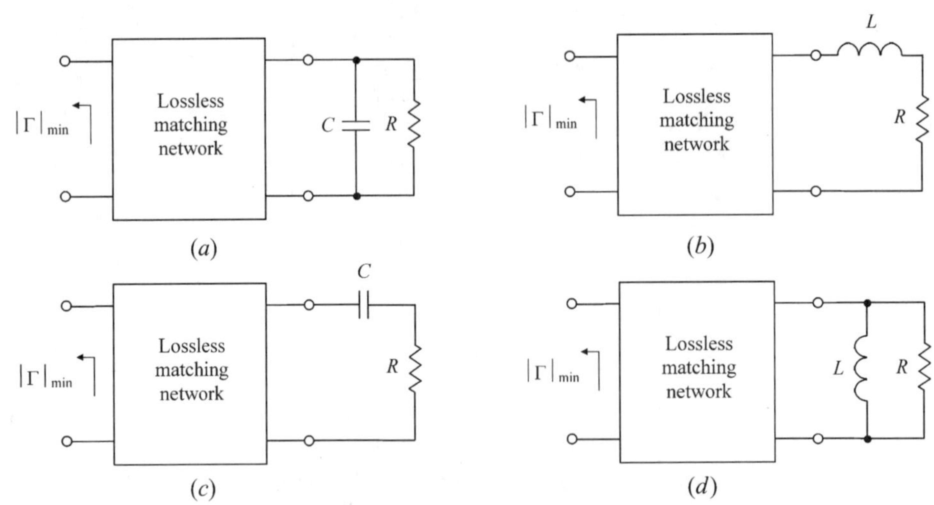

FIGURE 10.38 Loaded lossless matching circuits.

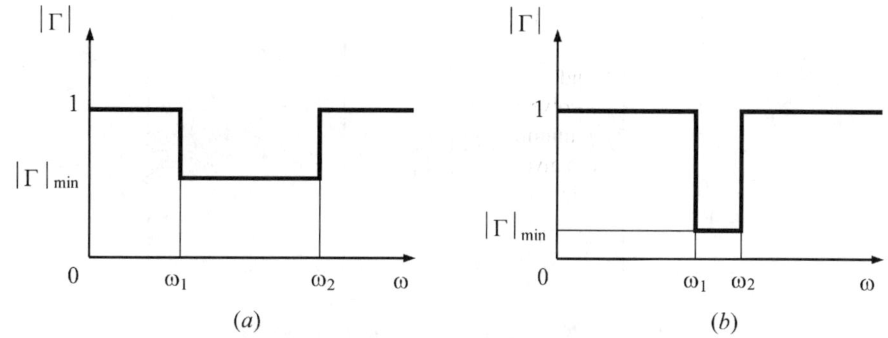

FIGURE 10.39 Ideal filter flat responses.

bandwidth, the worse the reflection coefficient magnitude. From Eq. (10.96) it follows that, when $|\Gamma|$ is constant within the frequency bandwidth and $|\Gamma| = 1$ otherwise,

$$\int_0^\infty \ln \frac{1}{|\Gamma(\omega)|} d\omega = \int_{\omega_1}^{\omega_2} \ln \frac{1}{|\Gamma(\omega)|} d\omega = \Delta\omega \ln \frac{1}{|\Gamma|} \leq \frac{\pi}{\tau}. \tag{10.98}$$

Then,

$$|\Gamma|_{\min} = \exp\left(\frac{-\pi}{\Delta\omega\tau}\right) \tag{10.99}$$

where $\Delta\omega = \omega_2 - \omega_1$.

Similarly, for the lossless network with a series *RC* load impedance and with a parallel *LR* load impedance,

$$|\Gamma|_{\min} = \exp\left(\frac{-\pi\omega_0^2\tau}{\Delta\omega}\right) \tag{10.100}$$

where $\omega_0 = \sqrt{\omega_1\omega_2}$ is the center bandwidth frequency. It should be noted that theoretical bandwidth limits can be realized only with an infinite number of matching network sections. The frequency bandwidth with minimum reflection coefficient magnitude is determined by a loaded quality factor $Q_L = \omega_0\tau$ for the series *RL* or parallel *RC* circuit and $Q_L = 1/(\omega_0\tau)$ for the parallel *RL* or series *RC* circuit, respectively. The Chebyshev matching transformer with a finite number of sections can be considered as a close approximation to the ideal passband network when the ripple of the Chebyshev response is made equal to $|\Gamma|_{\min}$.

Generally, Eqs. (10.99) and (10.100) can be rewritten in a simplified form of

$$|\Gamma|_{\min} = \exp\left(-\pi\frac{Q_0}{Q_L}\right) \tag{10.101}$$

where $Q_0 = \omega_0/\Delta\omega$.

To correctly design the broadband matching circuits for power amplifiers, it is necessary to transform and match the device complex impedances with the source and load impedances, which are usually resistive and equal to 50 Ω. For high-power or low-supply voltage cases, the device impedances may be sufficiently small, therefore it needs to include an ideal transformer *IT* together with a matching circuit, as shown in Figure 10.40. Such an ideal transformer provides only a required

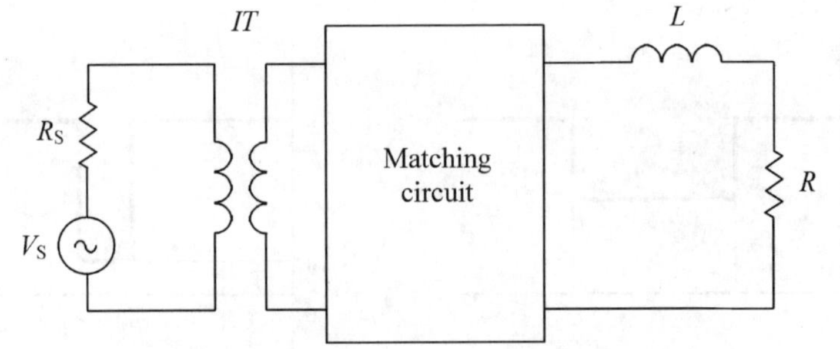

FIGURE 10.40 Matching circuit with ideal transformer.

transformation between the source resistance R_S and input impedance of the matching circuit, and it does not affect the circuit frequency characteristics.

To implement such an ideal transformer to the impedance-transforming circuit design, it is useful to operate with the Norton transform. As a result, an ideal transformer with two capacitors C_1 and C_2 shown in Figure 10.41(a) is replaced by three capacitances $C_{\rm I}$, $C_{\rm II}$ and $C_{\rm III}$ connected in the form of a π-transformer, as shown in Figure 10.41(b). Their values are determined by

$$C_{\rm I} = n_{\rm T}\left(n_{\rm T} - 1\right) C_1 \tag{10.102}$$

$$C_{\rm II} = n_{\rm T} C_1 \tag{10.103}$$

$$C_{\rm III} = C_2 - \left(n_{\rm T} - 1\right) C_1 \tag{10.104}$$

where $n_{\rm T}$ is the transformation coefficient. In this case, all of the parameters of these two-port networks are identical at any frequency. However, such a replacement is possible only if the capacitance $C_{\rm III}$ obtained by Eq. (10.104) is positive and, consequently, physically realizable.

Similarly, an ideal transformer with two inductors L_1 and L_2 shown in Figure 10.42(a) can be replaced by three inductors $L_{\rm I}$, $L_{\rm II}$ and $L_{\rm III}$ connected in the form of a T-transformer in Figure 10.42(b), with values determined by

$$L_{\rm I} = n_{\rm T}\left(n_{\rm T} - 1\right) L_2 \tag{10.105}$$

$$L_{\rm II} = n_{\rm T} L_2 \tag{10.106}$$

$$L_{\rm III} = L_1 - \left(n_{\rm T} - 1\right) L_2. \tag{10.107}$$

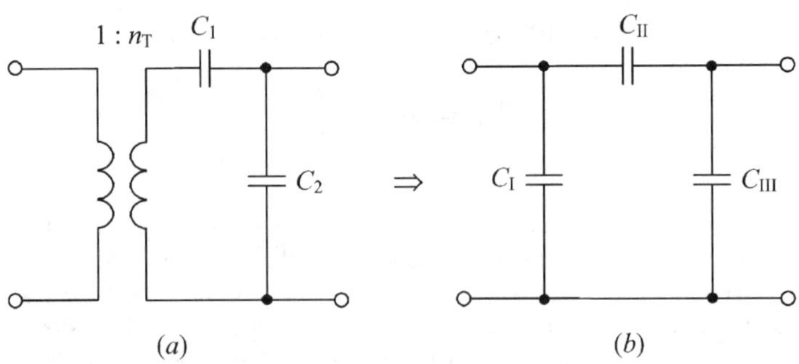

FIGURE 10.41 Capacitive impedance-transforming circuits.

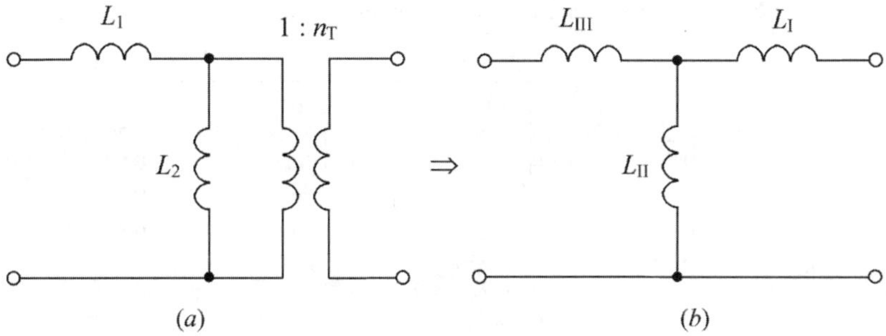

FIGURE 10.42 Inductive impedance-transforming circuits.

Again, this replacement is possible only if the inductance L_{III} defined by Eq. (10.107) is positive and, consequently, physically realizable.

The broadband impedance-transforming circuits generally represent the transforming bandpass filters when the in-band matching requirements with specified ripple must be satisfied. The out-of-band mismatching can be very significant. One of the design methods of such matching circuits is based on the theory of the transforming low-pass filters of a ladder configuration of series inductances alternating with shunt capacitances, whose two-section equivalent representation is shown in Figure 10.43. For a large ratio of R_0/R_5, mismatching at zero frequency is sufficiently high, and such a matching circuit can be treated as a bandpass impedance-transforming filter.

Table 10.1 gives the maximum passband ripples and coefficients g_1 and g_2 needed to calculate the parameters of a two-section low-pass Chebyshev filter for different transformation ratios $r = R_0/R_5$ and frequency bandwidths $w = 2(f_2 - f_1)/(f_2 + f_1)$, where f_2 and f_1 are the high- and low-bandwidth frequencies, respectively [31]. The coefficients g_3 and g_4 are calculated as $g_3 = rg_2$ and $g_4 = g_1/r$, respectively, and the circuit elements can be obtained by

$$C_1 = \frac{g_1}{\omega_0 R_0} \qquad C_3 = \frac{g_3}{\omega_0 R_0} \tag{10.108}$$

$$L_2 = \frac{g_2 R_0}{\omega_0} \qquad L_4 = \frac{g_4 R_0}{\omega_0} \tag{10.109}$$

As an example, consider the design of the broadband input matching circuit in the form of a two-section low-pass transforming filter shown in Figure 10.43(*a*), with a center bandwidth frequency $f_0 = 3$ GHz to match the source impedance $R_S = R_0 = 50\ \Omega$ with the device input impedance $Z_{in} = R_{in} + j\omega_0 L_{in}$, where $R_{in} = R_5 = 2\ \Omega$, $L_{in} = L_4 = 0.223$ nH, and $\omega_0 = 2\pi f_0$. The value of the series

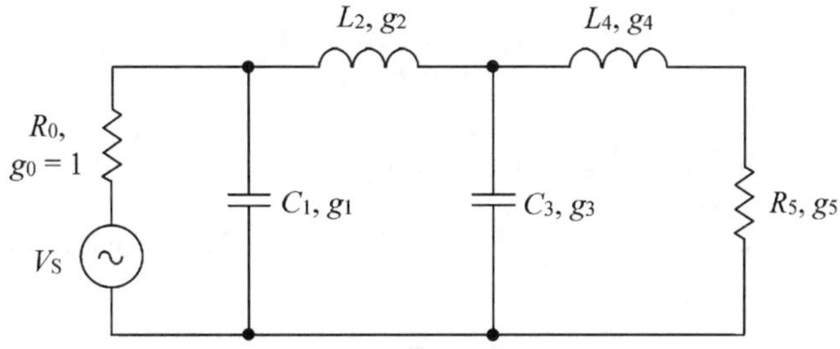

FIGURE 10.43 Two-section impedance-transforming circuit.

TABLE 10.1 Two-Section Low-Pass Chebyshev Filter Parameters.

r	w	Ripple (dB)	g_1	g_2
5	0.1	0.000087	1.26113	0.709217
	0.2	0.001389	1.27034	0.704050
	0.3	0.007023	1.28561	0.695548
	0.4	0.022109	1.30687	0.638859
10	0.1	0.000220	1.60350	0.591627
	0.2	0.003516	1.62135	0.585091
	0.3	0.017754	1.65115	0.574412
	0.4	0.055746	1.69304	0.559894
25	0.1	0.000625	2.11734	0.462747
	0.2	0.009993	2.15623	0.454380
	0.3	0.050312	2.22189	0.440863
	0.4	0.156725	2.31517	0.422868
50	0.1	0.001303	2.57580	0.384325
	0.2	0.020801	2.64380	0.374422
	0.3	0.104210	2.75961	0.358638
	0.4	0.320490	2.92539	0.338129

input device inductance $L_{\rm in} = L_4$ is chosen to satisfy Table 10.1 for $r = 25$, $w = 0.4$, maximum ripple of 0.156725, and $g_1 = 2.31517$. From Eq. (10.109) for $g_2 = 0.422868$, it follows that

$$L_4 = \frac{g_4 R_0}{\omega_0} = \frac{g_1 R_0}{\omega_0 r} = 0.223 \text{ nH}.$$

As a result, the circuit parameters shown in Figure 10.44(a) are calculated from Eqs. (10.108) and (10.109), thus resulting in the corresponding circuit frequency response shown in Figure 10.44(b), with the required passband from 2.6 to 3.4 GHz. The particular value of the inductance $L_{\rm in}$ is chosen for the design convenience. If this value differs from the required value, it means that it is necessary to change the maximum frequency bandwidth, the power ripple or the number of ladder sections.

Another approach is based on the transformation from the low-pass impedance-transforming prototype filters, whose simple L-, T- and π-type equivalent circuits are shown in Figure 10.45, to the bandpass impedance-transforming filters. Table 10.2 gives the parameters of the low-pass impedance-transforming Chebyshev filters-prototypes for different maximum in-band ripples and number of sections n [32]. This transformation can be obtained using the frequency substitution in the form of

$$\omega \rightarrow \frac{\omega_0}{\Delta\omega}\left(\frac{\omega}{\omega_0} - \frac{\omega_0}{\omega}\right) \tag{10.110}$$

where $\omega_0 = \sqrt{\omega_1 \omega_2}$ is the center bandwidth frequency, $\Delta\omega = \omega_2 - \omega_1$ is the passband, ω_1 and ω_2 are the low and high edges of the passband.

As a result, a series inductance $L_{\rm k}$ is transformed into a series LC circuit according to

$$\omega L_{\rm k} = \frac{\omega_0}{\Delta\omega}\left(\frac{\omega}{\omega_0} - \frac{\omega_0}{\omega}\right) L_{\rm k} = \omega L'_{\rm k} - \frac{1}{\omega C'_{\rm k}} \tag{10.111}$$

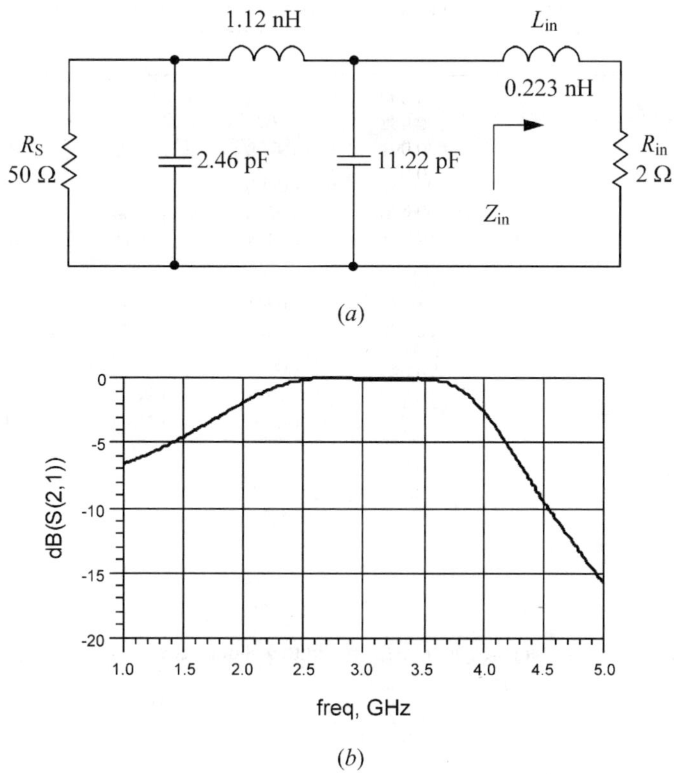

FIGURE 10.44 Two-section broadband low-pass matching circuit and its frequency response.

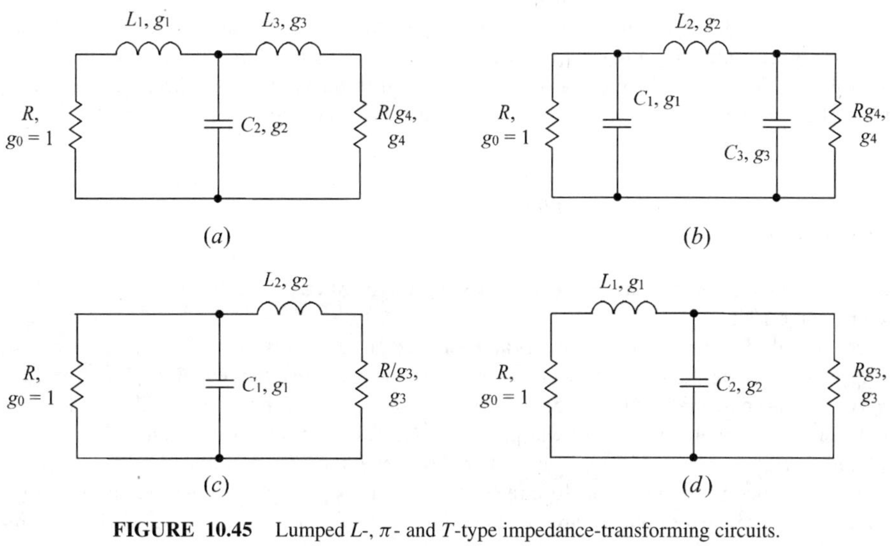

FIGURE 10.45 Lumped L-, π- and T-type impedance-transforming circuits.

TABLE 10.2 Parameters of Low-Pass Chebyshev Filters-Prototypes.

Ripple (dB)	n	g_1	g_2	g_3	g_4
	1	0.0960	1.0000		
0.01	2	0.4488	0.4077	1.1007	
	3	0.6291	0.9702	0.6291	1.0000
	1	0.3052	1.0000		
0.1	2	0.8430	0.6220	1.3554	
	3	1.0315	1.1474	1.0315	1.0000
	1	0.4342	1.0000		
0.2	2	1.0378	0.6745	1.5386	
	3	1.2275	1.1525	1.2275	1.0000
	1	0.6986	1.0000		
0.5	2	1.4029	0.7071	1.9841	
	3	1.5963	1.0967	1.5963	1.0000

where

$$L'_k = \frac{L_k}{\Delta\omega} \qquad C'_k = \frac{\Delta\omega}{\omega_0^2 L_k}. \tag{10.112}$$

Similarly, a shunt capacitance C_k is transformed into a shunt LC-circuit

$$\omega C_k = \frac{\omega_0}{\Delta\omega}\left(\frac{\omega}{\omega_0} - \frac{\omega_0}{\omega}\right) C_k = \omega C'_k - \frac{1}{\omega L'_k} \tag{10.113}$$

where

$$C'_k = \frac{C_k}{\Delta\omega} \qquad L'_k = \frac{\Delta\omega}{\omega_0^2 C_k}. \tag{10.114}$$

The low-pass impedance-transforming prototype filter will be transformed to the bandpass impedance-transforming filter when all its series elements are replaced by the series resonant circuits and all its parallel elements are replaced by the parallel resonant circuits, where each of them are tuned to the center bandwidth frequency ω_0. The bandpass filter elements can be calculated from

$$\Delta\omega C_k = \frac{g_k}{R} \tag{10.115}$$

$$\Delta\omega L_k = g_k R \tag{10.116}$$

where k is an element serial number for low-pass prototype filter, g_k are the appropriate coefficients given by Table 10.2.

Generally, the low-pass prototype filters and obtained on their basis bandpass filters do not perform an impedance transformation. The input and output resistances are either equal for the symmetric T- or π-type filters shown in Figure 10.45(a, b) where $g_4 = 1$ or their ratio is too small for L-type filters as those shown in Figure 10.45(c, d) where $g_3 < 2$. Therefore, in this case, it is necessary to use an ideal transformer concept. This approach is based on using the existing data tables from which the parameters of such impedance-transforming networks can be easily calculated for a given quality factor of the device input or output circuit. However, they can also be very easy verified or optimized by using a computer-aided design (CAD) optimization procedure incorporating in any comprehensive circuit simulator.

Consider the design example of the broadband interstage impedance-transforming filter with the center bandwidth frequency of 1 GHz to match the output driver-stage circuit with the input final-stage

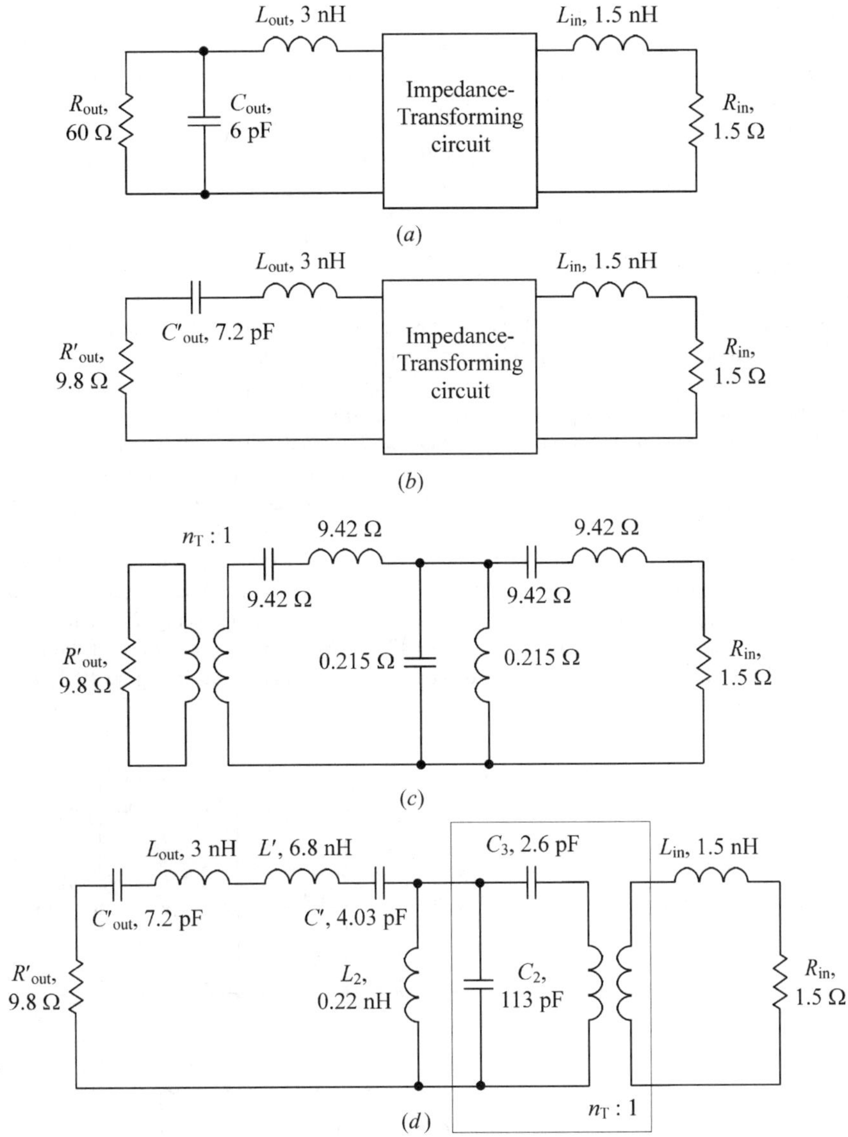

FIGURE 10.46 Impedance-transformer design procedure using low-pass filter-prototype.

circuit of the power amplifier, as shown in Figure 10.46(*a*) [1]. In this case, it is convenient initially to convert the parallel connection of the device output resistance R_{out} and capacitance C_{out} into the appropriate series connection at the center bandwidth frequency ω_0 according to

$$R'_{out} = \frac{R_{out}}{1 + (\omega_0 R_{out} C_{out})^2} \tag{10.117}$$

$$C'_{out} = \frac{1 + (\omega_0 R_{out} C_{out})^2}{(\omega_0 R_{out})^2 \, C_{out}} \tag{10.118}$$

as shown in Figure 10.46(*b*).

For the three-section low-pass impedance-transforming prototype filter shown in Figure 10.45(a) with a maximum in-band ripple of 0.1 dB, we then can obtain $g_1 = g_3 = 1.0315$, $g_2 = 1.1474$, $g_4 = 1$ for $n = 3$ from Table 10.2. According to Eq. (10.116), the relative frequency bandwidth in this case is defined as

$$\frac{\Delta\omega}{\omega_0} = \frac{g_1 R_{\text{in}}}{\omega_0 L_{\text{in}}} = 16.5\%$$

based on a value of which the shunt capacitor C_2 can be calculated using Eq. (10.115) providing a capacitive reactance equal to 0.215 Ω. The inductive reactance corresponding to a series inductance L_{in} is equal to 9.42 Ω.

To convert the low-pass filter to its bandpass prototype, it is necessary to connect the capacitance in series to the input inductance and inductance in parallel to shunt capacitance for and calculate with the same reactances to resonate at the center bandwidth frequency ω_0, as shown in Figure 10.46(c) where an ideal transformer IT with transformation coefficient $n_{\text{T}} = \sqrt{9.8/1.5} = 2.556$ is included. Here, the reactances for each series element are equal to 9.42 Ω, whereas the reactances for each parallel element are equal to 0.215 Ω. Then, in order to apply a Norton transform, move the corresponding elements with transformed parameters (each inductive and capacitive reactance is multiplied by n_{T}^2) to the left-hand side of IT, as shown in Figure 10.46(d) where the required series elements with reactances of $9.42n_{\text{T}}^2$ are realized by existing device output capacitance C'_{out} and inductance L_{out} and additional elements L' and C'. Finally, by using a Norton transform shown in Figure 10.41 with the ideal transformer IT and two capacitances, the resulting impedance matching bandpass filter is obtained, as shown in Figure 10.47(a). The frequency response of the filter with minimum in-band ripple and significant out-of-band suppression is shown in Figure 10.47(b). In the

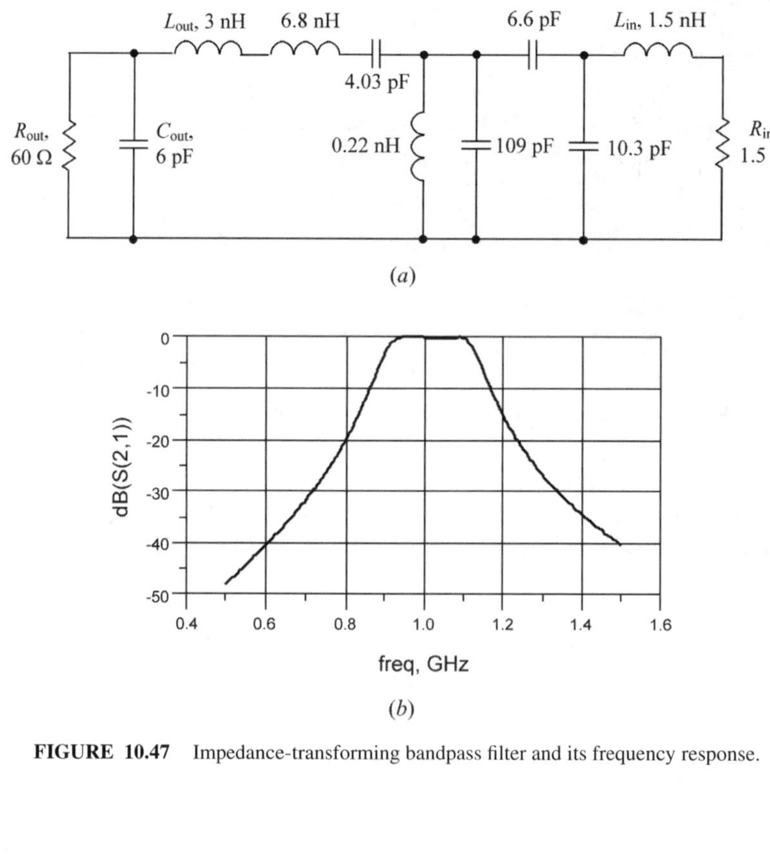

FIGURE 10.47 Impedance-transforming bandpass filter and its frequency response.

case of serious difficulties with practical implementation of very low inductance of 0.22 nH or very high capacitance of 109 pF, it is possible to design a multi-section low-pass impedance-transforming circuit.

An alternative impedance matching technique depends on using the multisection impedance-matching transformers consisting of the stepped transmission-line sections with different characteristic impedances. These transmission-line transformers, in contrast to the continuously tapered transmission-line transformers are significantly shorter and find broader performance. For example, a stepped transmission-line transformer, which consists of a cascaded connection of five uniform sections of equal lengths $l = \lambda_0/4$, where λ_0 is the wavelength corresponding to the central bandwidth frequency f_0, can provide a maximum *VSWR* of 1.021 in an octave frequency bandwidth having a total characteristic impedance change ratio of 8:1 [33]. Such a transformer can be constructed using any type of transmission lines. Such a stepped transmission-line transformer represents an antimetric structure, for which the relationship between the characteristic impedances of its sections may be written in the general form of

$$Z_i Z_{n+1-i} = Z_S Z_L \tag{10.119}$$

where $i = 1, 2, \ldots, n$ and n is the number of sections, Z_S is the source impedance and Z_L is the load impedance [34].

The main drawback of such transformers is their significant total length of $L = n\lambda_0/4$. However, it is possible to reduce the overall transformer length by applying other profiles of its structure. The stepped transformers using n cascaded uniform transmission-line sections of various lengths with alternating impedances are shorter by 1.5–2. In this case, a number of sections n is always an even number and the section impedances may be equal to the source and load impedances to be matched. The total length of such a stepped transmission-line transformer, which four-section structure is shown in Figure 10.48(*a*), can be further reduced by using the structure representing the cascade connection of n transmission line sections of the same length $l < \lambda_0/4$ with

$$\begin{aligned} Z_1 < Z_3 < \cdots < Z_{n-1} \\ Z_2 < Z_4 < \cdots < Z_n \end{aligned} \tag{10.120}$$

where n is an even number and $Z_1 > Z_n$ when $Z_S < Z_L$ [35]. In this case, the total transformer length is shorter by three times compared to the basic structure with the quarterwave sections, and an octave passband from 2 to 4 GHz shown in Figure 10.48(*b*) for the lossless ideal transmission-line sections is provided with input return loss better than 25 dB.

To reduce the high impedance ratio of the stepped transformers with a short total length, their optimized structures can be used with the section length $l_i = l_{n+1-i}$, where $i = 1, 2, \ldots, n/2$, and the same characteristic impedances for odd and even sections according to

$$\begin{aligned} Z_1 = Z_3 = \cdots = Z_{n-1} \\ Z_2 = Z_4 = \cdots = Z_n \end{aligned} \tag{10.121}$$

where $Z_1 Z_2 = Z_S Z_L$, $Z_n < Z_S$, $Z_{n-1} > Z_L$ [36]. In particular situations of high impedance-matching ratio at microwave frequencies when it is necessary to match the standard source load impedance of 50 Ω with the device input and output impedance of 1 Ω and smaller, both the length and width of the microstrip-line sections can be optimized. For example, with the use of a multisection transformer with seven quarter-wavelength microstrip lines of different characteristic impedances, a power gain of 9 ± 1 dB and a *PAE* of 37.5 ± 7.5% over a frequency range of 5–10 GHz was achieved for a 15-W GaAs MESFET power amplifier [37].

The dissipative or lossy gain compensation matching circuits can achieve the important tradeoffs between the power gain, reflection coefficient and frequency bandwidth. Moreover, the resistive nature of such a matching circuit may also improve power amplifier stability and reduce its size and cost

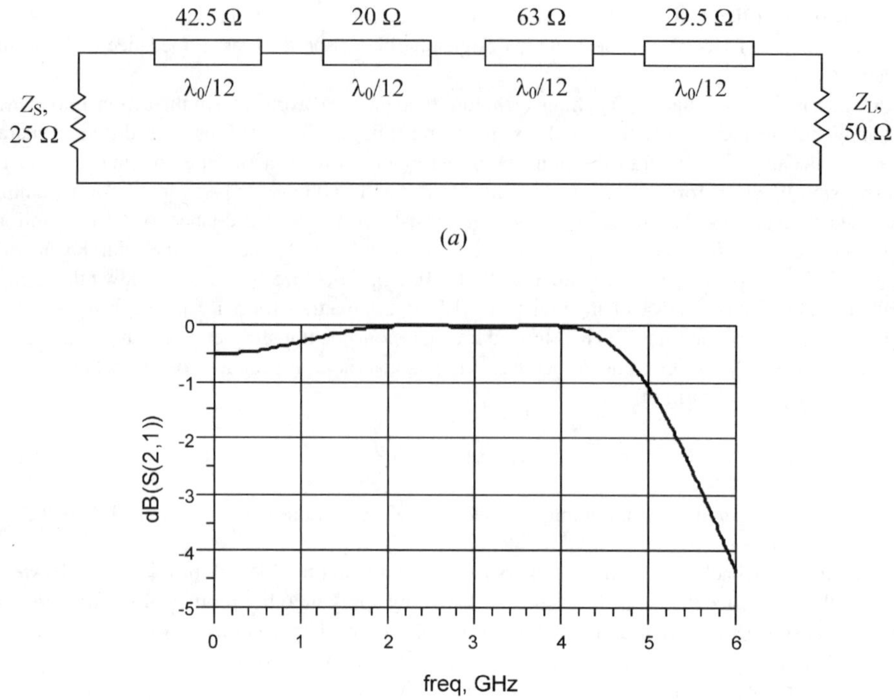

(b)

FIGURE 10.48 Stepped transmission-line transformer with equal-length sections.

because it represents a simple lossy matching circuit structure. In many practical cases, to provide broadband matching with minimum gain flatness and input reflection coefficient, it is sufficient to use the resistive shunt element at the transistor input. An additional matching improvement at high bandwidth frequencies can be achieved by employing inductive reactive elements in series to the resistor [1].

Figure 10.49 shows the circuit schematics of the input, interstage, and output networks intended for microwave broadband power amplifiers. The constant-resistance input network shown in Figure 10.49(*a*) enables the input device impedance to be purely resistive and equal to $Z_{in} = R_{in}$, when

$$R_1 = R_{in} \tag{10.122}$$

$$L_1 = C_{gs} R_{in}^2 \tag{10.123}$$

$$C_1 = \frac{L_g}{R_{in}^2} \tag{10.124}$$

that makes broadband transformation of the input resistance to the source resistance much easier. In the output network shown in Figure 10.49(*b*), a value of the drain inductance is chosen to compensate for the capacitive device output reactance at the center bandwidth frequency. Then, a resonant frequency of the parallel resonant circuit is set to be equal to the same center bandwidth frequency. In this case, for the frequencies where the device drain impedance Z_d is capacitive, the impedance of the parallel resonant circuit is inductive. On the other hand, for the frequencies where the drain impedance Z_d is inductive, the impedance of the parallel resonant circuit is capacitive. As a result, the broadband

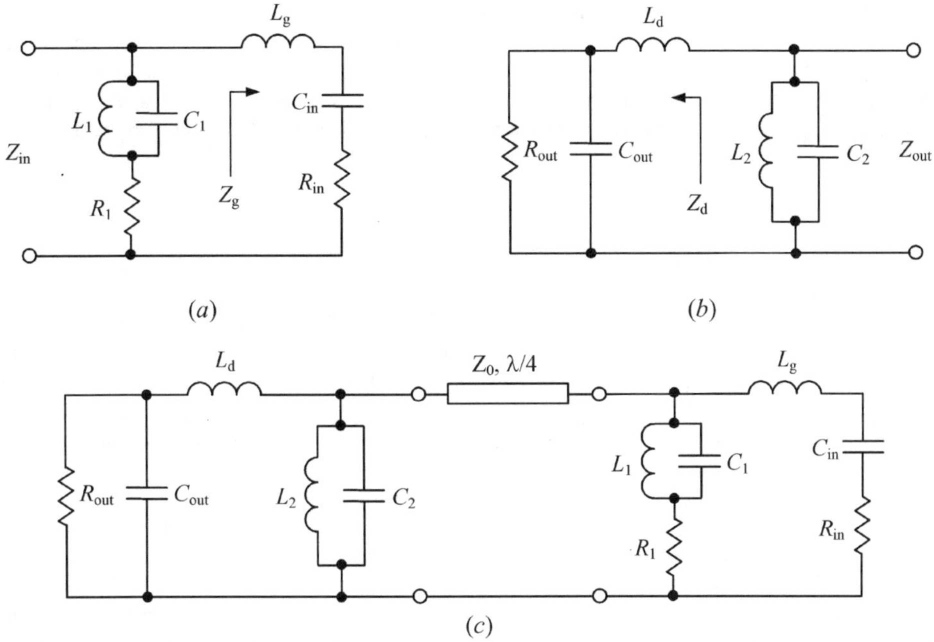

FIGURE 10.49 Schematics of (*a*) input, (*b*) output and (*c*) interstage broadband matching circuits.

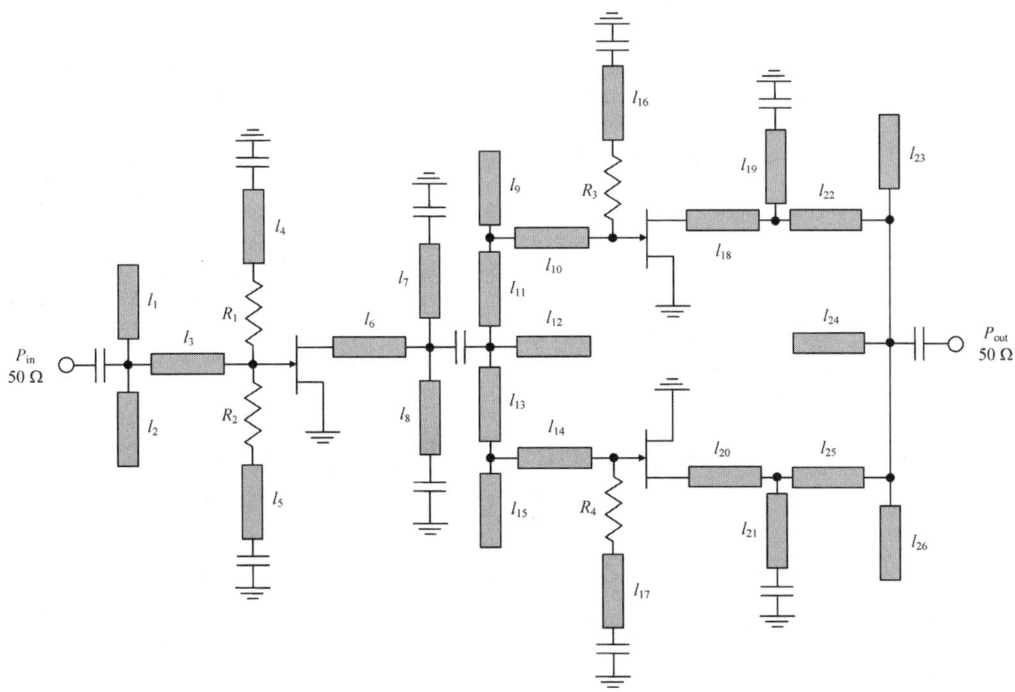

FIGURE 10.50 Microstrip two-stage lossy match MESFET power amplifier.

reactance compensation is realized when the reactive part of the impedance Z_d becomes sufficiently small over a wide frequency bandwidth. In microwave applications, such a parallel resonant circuit is fabricated by using a quarterwave short-circuited stub [38]. The interstage matching network, the schematic diagram of which is shown in Figure 10.49(*c*), is comprised of the input and output networks described above and an additional quarterwave microstrip transformer with the characteristic impedance

$$Z_0 = \sqrt{\frac{R_{in} L_d}{R_{out} C_{out}}}. \tag{10.125}$$

The schematic diagram of a two-stage lossy match MESFET power amplifier is shown in Figure 10.50 [39]. By using a 1.05-mm device in the driver stage and two 1.35-mm devices in the final stage, the saturated output power of 27.7 ± 2.7 dBm with a linear power gain of 8.3 ± 2.8 dB and a drain efficiency of 15.3 ± 8.3% was achieved in a frequency range of 4–25 GHz. The input and interstage constant-resistance networks are each represented by the series connections of a resistance and an inductance, which is fabricated by using a high-impedance microstrip line. When being connected in parallel, two such networks provide the input purely resistive impedance, where l_4 and l_5 are the series microstrip lines, and R_1 and R_2 are the series resistances. The short-circuited microstrip lines l_7 and l_8 in the interstage network and the short-circuited microstrip lines l_{19} and l_{21} in the output network, with quarterwave electrical lengths at the center frequency, provide the parallel resonant circuits connected to the device outputs. The series microstrip lines l_{10} and l_{14} in the interstage network are the quarterwave impedance transformers, which provide matching between the output impedance of the driver-stage device and the input impedance of the final-stage devices connected in parallel. The input and output matching circuits represent the transmission-line T-transformers, where a series microstrip line and a parallel open-circuited microstrip stub replace a series inductance and a parallel capacitance, respectively.

Figure 10.51 shows the circuit schematic of the broadband GaN HEMT microwave power amplifier with the device geometry of 0.7 μm × 1 mm, transition frequency $f_T = 18$ GHz, and maximum frequency $f_{max} = 35$ GHz. The optimized input three-element lossy *LCR* matching circuit provides the power gain up to 11.5 dB and low input reflection less than −10 dB over a frequency range from 3 to 9 GHz [40]. Since the impedance at the input of the lossy matching circuit is only about 10 Ω,

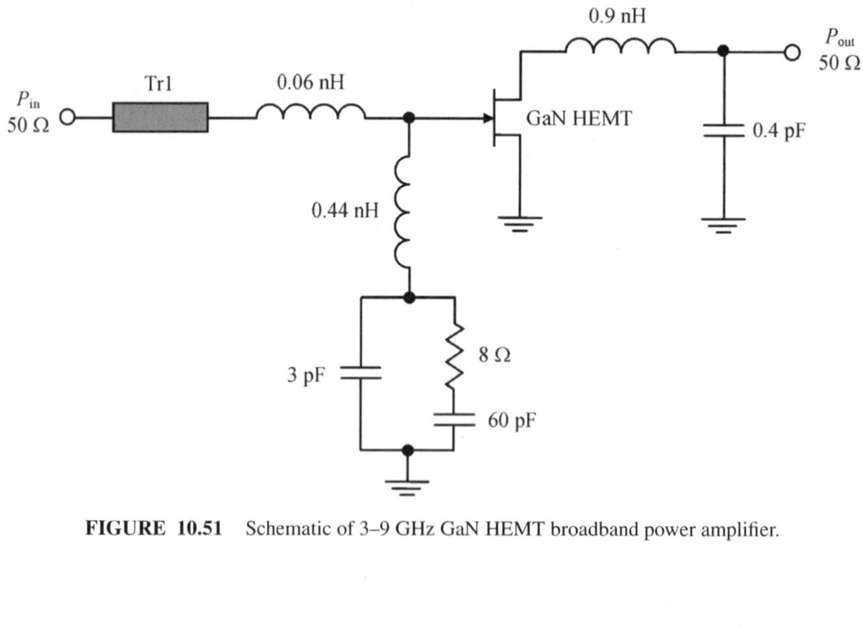

FIGURE 10.51 Schematic of 3–9 GHz GaN HEMT broadband power amplifier.

this necessitates an additional broadband impedance transformation (Tr1) from 50 to 10 Ω, which was realized using a few sections of the quarterwave coplanar transmission lines with decreasing impedance. The output network incorporates an *LC* circuit to compensate for the output device capacitance such that the intrinsic device sees approximately a real load within frequency bandwidth. Because the optimum load for this 1-mm device with supply voltage of 20 V is of about 50 Ω, consequently no output impedance transformation is needed. The output power was of about 1.6 W with the *PAE* ranging from 14 to 24% in a frequency bandwidth from 4 to 8 GHz.

A very broadband power amplifier operation with high power gain can be achieved by using a simple *RL* gain-compensating matching circuit at the input and a compensating series inductor L_{out} at the output of each stage in a multistage power amplifier, which simplified schematic (without bias circuits) is shown in Figure 10.52(*a*). The combination of these compensating elements in interstage matching circuit shown in Figure 10.52(*b*) is necessary to compensate first for the input device reactance due to the gate source capacitance C_{gs} when its input impedance will become purely active equal to R when $R = R_{gs}$ and $L = C_{gs}R^2$ and, secondly, to compensate for the device output drain–source capacitance C_{ds} when the output inductance can be calculated as $L_{out} = C_{ds}R^2$. By using this compensating inductor, the power gain can be increased at high bandwidth frequencies when the effect of the drain–source capacitance C_{ds} becomes significant. The effect of the feedback gate–drain capacitance C_{gd} is neglected because its value is sufficiently small compared with C_{gs} and C_{ds}. For a wider operating bandwidth, it is better to use the transistors of the same gate geometry, though different sizes are also possible. In this case, it is necessary to choose the different parameters of interstage matching circuits and optimize them for maximum bandwidth (the resistance R at the input can be set to 50 Ω if the source impedance is 50 Ω). Then, the overall operating power gain can be estimated from

$$G_P(\text{dB}) = 20n \log_{10}(g_m R) \tag{10.126}$$

where n is the number of stages. As a result, by using the GaAs MESFET devices with transconductances $g_m = 55$ mS in a similar four-stage power amplifier structure, the output power of 10 ± 1 dBm

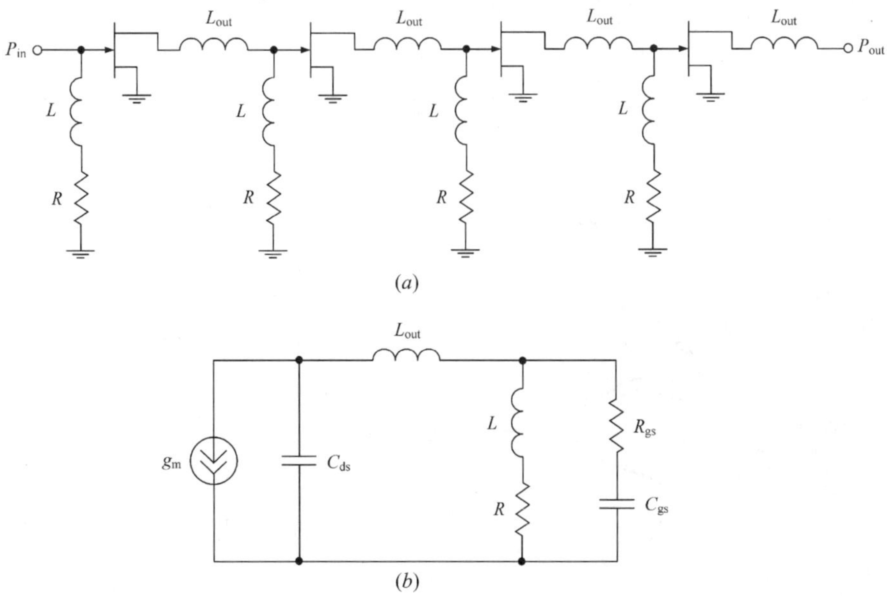

FIGURE 10.52 Schematic of 1–10 GHz four-stage broadband power amplifier.

for the input power of −28 dBm was achieved across the frequency bandwidth from 1 to 10 GHz at the supply voltage of 2 V [41].

Figure 10.53(*a*) shows the circuit schematic of a broadband laterally diffused metal-oxide semiconductor field-effect transistor (LDMOSFET) high power amplifier with the device geometry of 1.25 μm × 40 mm [1]. The optimized input three-element lossy matching circuit allows a very broadband operation to be provided with the minimum power gain flatness. The using of a 1:2 output transformer contributes to an increase in the output power level. The capacitance of 20 pF connected in parallel with the resistance of 27 Ω provides an additional increase of a power gain at higher bandwidth frequencies. The simulation results are plotted in Figure 10.53(*b*), where the output powers of 22–25 W with a power gain of 13.7 ± 0.3 dB in a frequency range of 5–300 MHz can be achieved (curve 1). In this case, input return loss is greater than 8 dB up to 225 MHz (curve 2). The direct connection of the standard 50-Ω load to the device drain through the blocking capacitor provides an output power level within the range of 6–7 W.

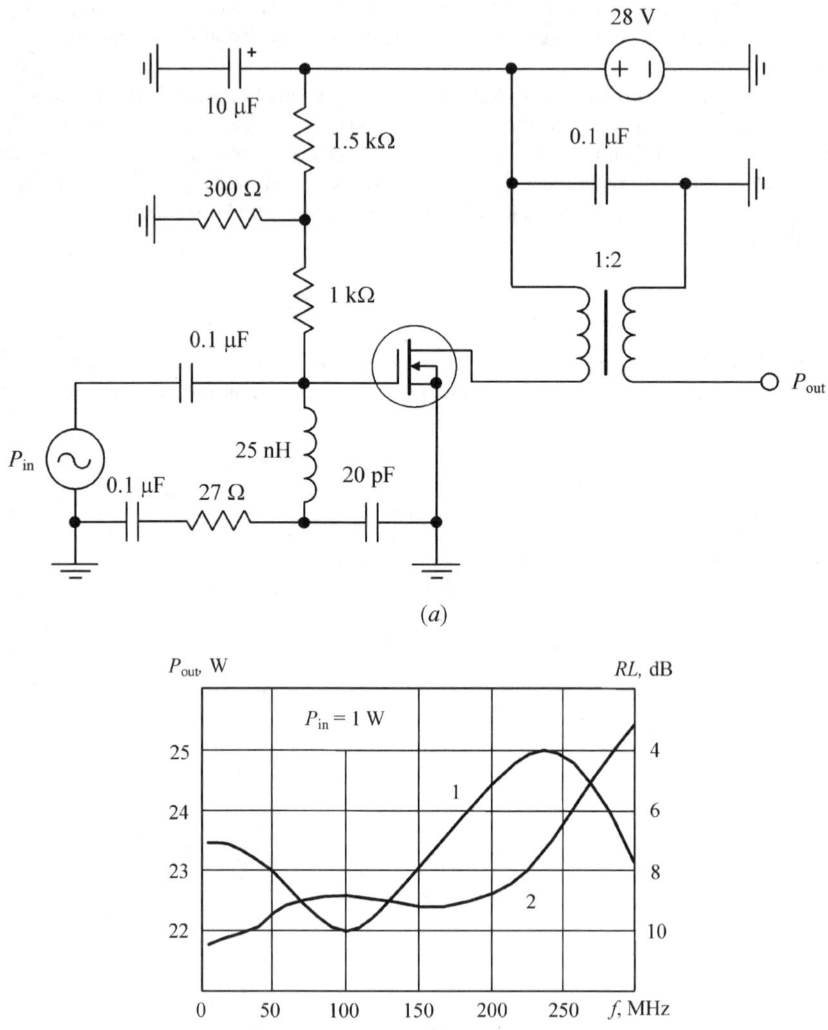

FIGURE 10.53 Circuit schematic of broadband LDMOSFET high power amplifier.

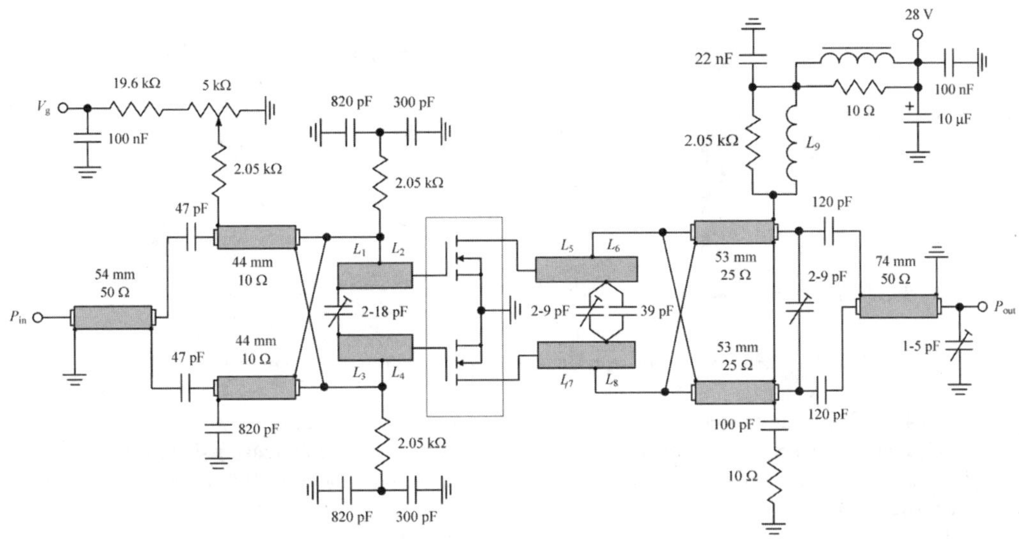

FIGURE 10.54 Circuit diagram of broadband high power VHF MOSFET amplifier.

Figure 10.54 shows the schematic diagram of a two-octave high-power amplifier covering a frequency range of 100–450 MHz [42]. The BLF548 device is a balanced *n*-channel enhancement-mode vertical diffused metal-oxide-semiconductor (VDMOS) transistor designed for use in broadband amplifiers with an output power of 150 W and a power gain of more than 10 dB in a frequency range of up to 500 MHz. In a frequency range of 100–500 MHz, the real part of its input impedance is almost constant and equal to $\text{Re}Z_{\text{in}} = 0.43\ \Omega$, whereas the imaginary part of the input impedance changes its capacitive impedance $\text{Im}Z_{\text{in}} = -4.1\ \Omega$ at 100 MHz to the inductive impedance $\text{Im}Z_{\text{in}} = 0.5\ \Omega$ at 500 MHz. The output impedance is capacitive and equal to $Z_{\text{out}} = (1.1 - j0.8)\ \Omega$ at 500 MHz. Coaxial semirigid baluns are used to transform the unbalanced 50-Ω source and load into two out-of-phase 180° sections of 25 Ω, respectively, followed by the coaxial transformers, with the characteristic impedance of 10 Ω for the input matching and with the characteristic impedance of 25 Ω for the output matching. This yields the lower impedance $R_{\text{in}} = \sqrt{25 \times 10}/4 = 3.95\ \Omega$, which is necessary to transform to the device input impedance of 0.43 Ω, and the higher impedance $R_{\text{out}} = \sqrt{25 \times 25}/4 = 6.25\ \Omega$, which is necessary to transform to device output impedance of 2.8 Ω. Final matching is provided by simple *L*-transformers with series microstrip lines and parallel variable capacitors. The microstrip lines were fabricated on a 30-mil substrate with a dielectric permittivity $\varepsilon_r = 2.2$. In this case, the dimensions of each microstrip line with characteristic impedance of 20 Ω are as follows: L_1 and L_3 are 5 × 8 mm^2, L_2 and L_4 are 2.5 × 8 mm^2, L_5 and L_7 are 11.5 × 8 mm^2, L_6 and L_8 are 4 × 8 mm^2. To compensate for the 6-dB/octave slope, conjugate matching is provided at 450 MHz since, at lower frequencies, a mismatch gives a required decrease of power gain to provide acceptable broadband power gain flatness. As a result, the gain variation of an output power level of 150 W is smaller than 1 dB with an input return loss of more than 12 dB in a frequency range of 100–450 MHz.

10.8 DISTRIBUTED POWER AMPLIFIERS

The potential of traveling-wave or distributed amplification for obtaining power gains over wide frequency bands has been recognized yet in 1937 [43]. However, first theoretical analysis and its practical verification were obtained for very broadband electron-tube amplifiers more than a decade

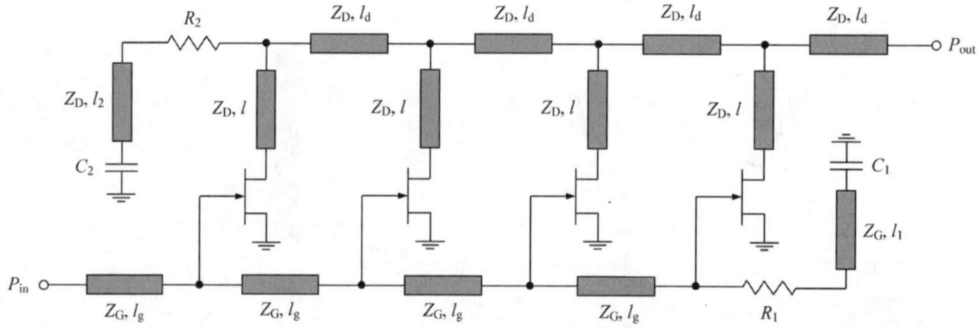

FIGURE 10.55 Schematic representation of four-stage FET traveling-wave amplifier.

later [44,45]. The basic concept was based on the idea to combine the interelectrode capacitances of the electron tubes with series inductors to form two lumped-element artificial transmission lines coupled by the active device transconductances. As a result, it became possible to provide amplification over much wider bandwidths than was achievable with conventional power amplifiers.

In actual circuits, however, the extreme bandwidths predicted by a first-order theory are modified by several factors, such as capacitive and inductive couplings, loading of the lumped-element transmission lines due to grid and coil losses, lead inductance, and parasitic capacitances associated with the coil inductance. Therefore, it is only recently, with the availability of good-quality microwave GaAs FETs, the distributed power amplifiers has again become popular at microwave frequencies. In this case, GaAs FETs are used as the active devices, and the input and output lines are periodically loaded microstrip transmission lines. With such an arrangement, the factors mentioned above as degrading the expected performance are either completely eliminated or their effect is included in the design. In addition, the resultant amplifiers exhibit very low sensitivities to process variations and are relatively easy to design and simulate.

The simplified schematic representation of a four-stage FET traveling-wave amplifier is shown in Figure 10.55, where microstrip lines are periodically loaded with the complex gate and drain impedances of the devices, thus forming lossy transmission-line structures of different characteristic impedance and propagation constant [46]. An RF signal applied at the input end of the gate line travels down the line to the other end, where it is absorbed by the terminating impedance connecting at the end of the gate line and including the resistor R_1. However, a significant portion of the signal is proportionally dissipated by the gate circuits of the individual FETs along the way. The input signal sampled by the gate circuits at different phases (and generally at different amplitudes) is transferred to the drain line through the FET transconductances. If the phase velocity of the signal at the drain line is identical to the phase velocity of the gate line, then the signals on the drain line add. The addition will be in phase only for the forward-traveling signal. Any signal which travels backward, and is not fully canceled by the out-of-phase additions, will be absorbed by the terminating impedance connecting at the end of the drain line and including the resistor R_2.

In conventional power amplifiers, it is impossible to increase the gain-bandwidth product by just paralleling the FETs because the resulting increase in transconductance g_m is compensated for by the corresponding increase in the input and output capacitances. The distributed power amplifier overcomes this problem by adding the individual device transconductances without adding their input and output capacitances, which are now the parts of the artificial gate and drain transmission lines, respectively. If the spacing between FETs is small compared to the wavelength, the characteristic impedances of the gate and drain lines shown in Figures 10.56(*a*) and 10.56(*b*), respectively, can be approximated as

$$Z_G = \sqrt{\frac{L_g}{C_g + \dfrac{C_{gs}}{l_g}}} \tag{10.127}$$

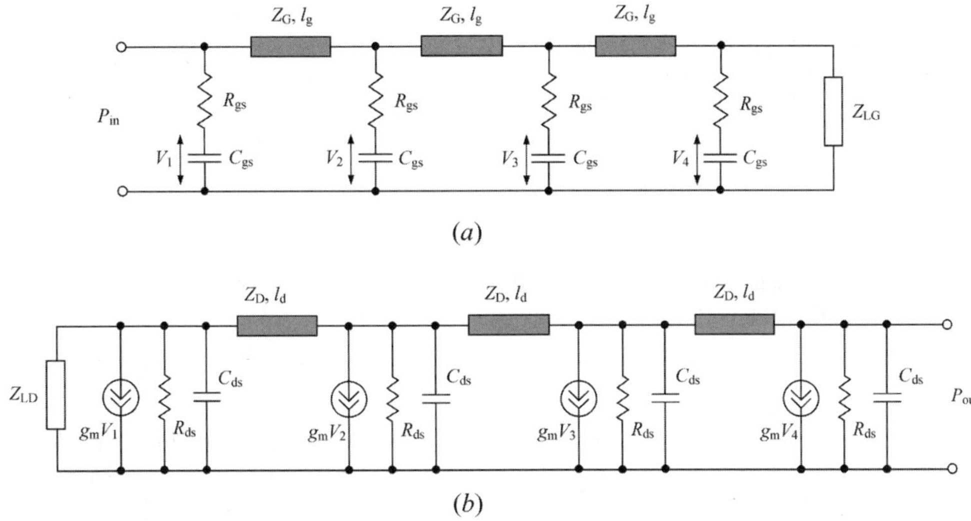

FIGURE 10.56 Simplified equivalent-circuit diagram of FET traveling-wave amplifier.

and

$$Z_D = \sqrt{\frac{L_d}{C_d + \dfrac{C_{ds}}{l_d}}} \tag{10.128}$$

where C_{gs} is the gate–source capacitance, C_{ds} is the drain–source capacitances of the unit FET cell, l_g and l_d are the lengths of the unit gate- and drain-line sections, and L_g, C_g and L_d, C_d are the per-unit-length inductance and capacitance of the gate and drain lines, respectively. Here, the effect of the gate resistance R_{gs} and drain resistances R_{ds} are neglected. Note that the impedance expressions in Eqs. (10.127) and (10.128) are clearly independent of the number of FETs used in the circuit.

As a result, the operating power gain for n-cell circuit by approximating the gate and drain lines as continuous structures can be written as

$$G_P = g_m^2 Z_G Z_D \left| \frac{\gamma_g l_g \left[\exp\left(-n\gamma_g l_g\right) - \exp\left(-n\gamma_d l_d\right)\right]}{\gamma_g^2 l_g^2 - \gamma_d^2 l_d^2} \right|^2. \tag{10.129}$$

where

$$\gamma_g = \frac{\omega^2 C_{gs}^2 R_{gs}}{2l_g} Z_G + j\omega\sqrt{L_g\left(C_g + \frac{C_{gs}}{l_g}\right)} \equiv \alpha_g + j\beta_g$$

$$\gamma_d = \frac{1}{2R_{ds}l_d} Z_D + j\omega\sqrt{L_d\left(C_d + \frac{C_{ds}}{l_d}\right)} \equiv \alpha_d + j\beta_d.$$

Under normal operating conditions, the signals in the gate and drain lines are near synchronism when $\beta_g l_g \cong \beta_d l_d$ and Eq. (10.129) can be simplified for $Z_G \cong Z_D \equiv Z_0$ to

$$G_P = g_m^2 Z_0^2 \frac{\left[\exp\left(-n\alpha_g l_g\right) - \exp\left(-n\alpha_d l_d\right)\right]^2}{\left(\alpha_g l_g - \alpha_d l_d\right)^2} \tag{10.130}$$

from which it follows that, as the number of unit cells n is increased, the power gain does not increase monotonically and approaches zero in limit as n gets large.

For values of $n\alpha_g l_g \leq 1$ and when the drain-line losses are negligible compared to the gate-line losses, Eq. (10.130) can be rewritten as

$$G_P \cong \frac{n^2 g_m^2 Z_0^2}{4} \left(1 - \frac{n\alpha_g l_g}{2} - \frac{n^2 \alpha_g^2 l_g^2}{6}\right)^2 \tag{10.131}$$

which means that, in this operating conditions, the power gain can be made proportional to n^2. In this case, by using the expression for the gate-line attenuation constant α_g,

$$n Z_0 R_{gs} \omega^2 C_{gs}^2 \leq 2 \tag{10.132}$$

that defines the upper limit to the total gate periphery that can be used in practical traveling-wave power amplifier or the maximum number of unit cells for a given FET device. Similar results can be obtained by applying a theoretical analysis based on matrix technique by employing a finite number of active and passive circuit elements [47].

Figure 10.57 shows the frequency dependence of a power gain for different numbers of FET unit cells or stages [46]. As we can see, there is an optimum number n which provides maximum frequency bandwidth with minimum gain variations and reasonable power gain. For example, a power gain of 9 ± 1 dB over a bandwidth of 1–13 GHz obtained for a four-stage traveling-wave amplifier with total GaAs FET gate periphery of 4×300 μm. It should be mentioned that the resistive part of gate loading typically results in 3-dB gain reduction. Effect of drain loading is not so significant, however the power gain can be increased by about 1 dB for increased values of the drain loading resistance.

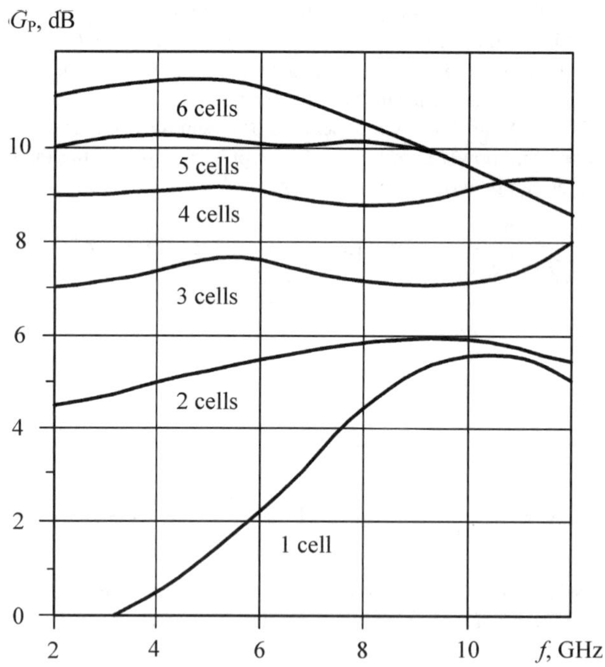

FIGURE 10.57 Power gain versus frequency for different numbers of cells.

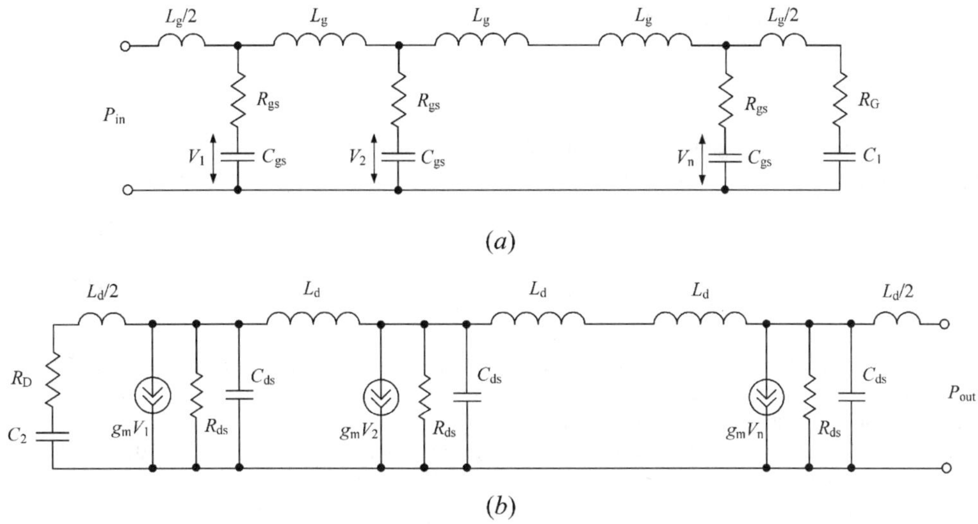

FIGURE 10.58 Simplified equivalent-circuit diagram of FET distributed amplifier with lumped inductors.

The equivalent gate and drain transmission lines based on lumped inductors and capacitors are shown in Figures 10.58(*a*) and 10.58(*b*), respectively [48]. If the gate circuit cutoff frequency is defined as $\omega_g = 1/R_{gs}C_{gs}$ and drain circuit cutoff frequency is defined as $\omega_d = 1/R_{ds}C_{ds}$, the propagation characteristics of the transmission-line sections with constant characteristic impedance are known. By requiring the phase shift between each gate-line and drain-line section to be equal, the cutoff frequency ω_c for both transmission lines must also be equal. As a result, the power gain of the lumped traveling-wave amplifier can be written as

$$G_P = \frac{g_m^2 Z_{01} Z_{02} \sinh^2\left[\frac{n}{2}\left(\alpha_d - \alpha_g\right)\right] \exp\left[-n\left(\alpha_d + \alpha_g\right)\right]}{4\left[1 + \left(\frac{\omega}{\omega_g}\right)^2\right]\left[1 - \left(\frac{\omega}{\omega_c}\right)^2\right] \sinh^2\left[\frac{1}{2}\left(\alpha_d - \alpha_g\right)\right]} \tag{10.133}$$

where

$$\omega_c = \frac{2}{\sqrt{L_g C_{gs}}} = \frac{2}{\sqrt{L_d C_{ds}}}$$

and

$$Z_{01} = \sqrt{\frac{L_g}{C_{gs}}} \quad Z_{02} = \sqrt{\frac{L_d}{C_{ds}}}$$

are the characteristic impedances of the gate and drain lines, respectively.

By maximizing Eq. (10.133) at a given frequency, the optimum number of unit cells $n = N_{opt}$ can be shown to be

$$N_{opt} = \frac{1}{\alpha_d - \alpha_g} \ln\left(\frac{\alpha_d}{\alpha_g}\right). \tag{10.134}$$

The maximum gain-bandwidth product of the distributed power amplifier can be estimated by

$$\sqrt{G_{P0}} f_{1dB} \approx 0.8 f_{max} \tag{10.135}$$

where G_{P0} is the power gain of the amplifier in the low-frequency limit, f_{1dB} is the frequency at which the power gain of the amplifier falls below G_{P0} by 1 dB, and f_{max} is the maximum frequency of oscillation of the FET given by

$$f_{max} = \frac{g_m}{4\pi C_{gs}} \sqrt{\frac{R_{ds}}{R_{gs}}}. \tag{10.136}$$

The attenuation of the gate line α_g increases rapidly with frequency, resulting in a lower power gain at high bandwidth frequencies. In addition, at a given frequency the number of devices is limited. In this case, if the gate-line attenuation can be made very small, the input signal is nearly evenly applied to all FETs in the amplifier and the power gain will remain constant over a wide frequency band. However, since the gate-line attenuation is directly proportional to the gate–source capacitance C_{gs}, then it is possible to reduce its effect by adding a series capacitance $C = qC_{gs}$ connected to each gate, as shown in Figure 10.59 [49]. As a result, since the effective gate capacitance is reduced by a factor of $q/(1 + q)$, the gate-line attenuation α_g decreases by a factor of $q/(1 + q)$ and the gate circuit cutoff frequency ω_g is increased by a factor of $(1 + q)/q$ at a fixed frequency. The gate voltage, however, divides between C and C_{gs}. The FET can now be considered as a modified device having an effective gate–source capacitance of $C'_g = qC_{gs}/(1 + q)$ and an effective transconductance of $g'_m = qg_m/(1 + q)$.

By using a cascode connection of transistors, it is possible to significantly improve the isolation between the input and output artificial transmission lines. Figure 10.60(*a*) shows the circuit schematic of a three-stage cascode distributed amplifier based on bipolar devices [50]. The emitter degeneration resistor increases the device input impedance and helps to reduce the output distortion. Instead of a line section consisting of a series lumped inductor L and a shunt capacitor C with the frequency-independent characteristic impedance $Z_0 = \sqrt{L/C}$ shown in Figure 10.60(*b*), the m-derived T-section shown in Figure 10.60(*c*) can be used by adding a parallel inductor to provide an additional degree of freedom [51]. Both sections still maintain the same input and output impedances, but the m-derived T-section has an LC series resonance in its shunt arm. This resonance provides the ability to modify the passband attenuation. As a result, the m-derived T-section has a flatter passband and a better input reflection coefficient than its corresponding T-section with constant Z_0. Being implemented in SiGe (refers to integration of bipolar junction transistors and CMOS technology) or HBT technology, such a cascode distributed power amplifier can achieve a measured passband from 100 MHz to 50 GHz with a 1-dB compression power gain varying from 6 to 8.5 dB with output power of 4.2 ± 2 dBm [50].

For a typical distributed amplifier with the drain-line termination, there is a serious limitation in efficiency, which is lower by about 25% compared to a conventional single-ended power amplifier [52].

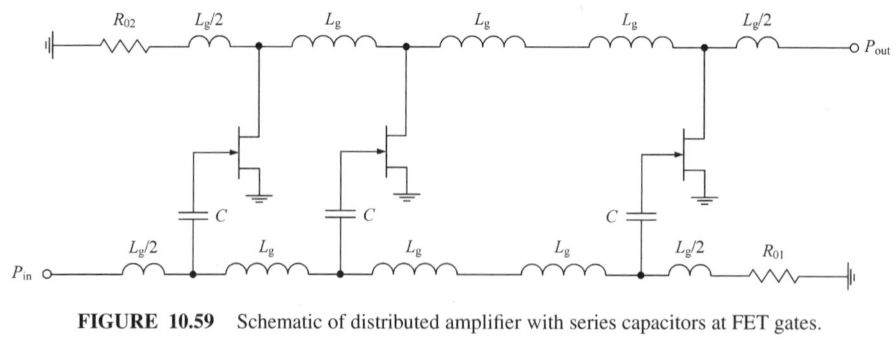

FIGURE 10.59 Schematic of distributed amplifier with series capacitors at FET gates.

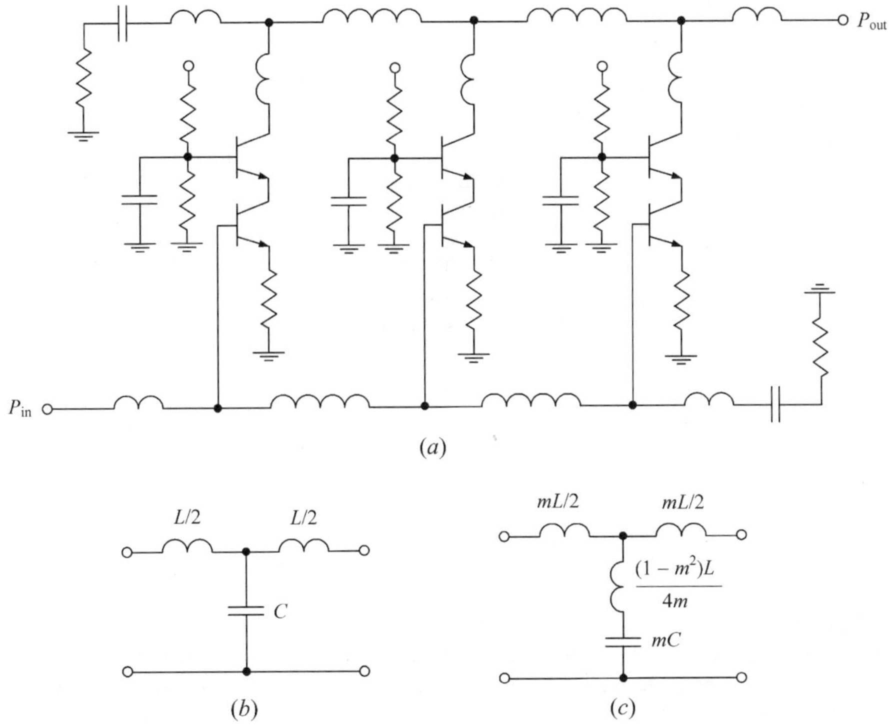

FIGURE 10.60 Schematic of three-stage bipolar cascode distributed amplifier.

However, the effects of this drain-line termination can be mitigated when the resistive termination is neglected and, by tapering the drain-line characteristic impedance, the optimum load can be presented to each active device. In this case, the efficiency can be nearly as twice as high for the same multi-octave frequency bandwidth. The output power and efficiency over wide range of microwave frequencies can also be increased by using a 0.25-μm GaN on SiC device technology [53].

10.9 HARMONIC TUNING USING LOAD–PULL TECHNIQUES

Generally, the transistor is operated as a nonlinear active device under large-signal conditions, resulting in a significant amount of harmonic components. Since the second harmonic is the largest harmonic component, it was found experimentally that just providing the proper source and load second-harmonic-terminating impedances at the input and output ports of the transistor can significantly improve the power amplifier efficiency At microwave frequencies, the special network based on three coupled bars positioned between two ground planes can be used to realize a wide range of impedance matching at the fundamental frequency, while simultaneously presenting reactive impedance at the second-harmonic component, thus decoupling the second-harmonic signal from 50-Ω load [54].

To maximize the amplifier efficiency, the proper harmonic impedances must be provided at the transistor output port. At microwave frequencies, those impedances must be measured or simulated accurately, taking into account the transistor parasitic elements and defining accurately the values of the device equivalent-circuit parameters under large-signal operation. Since it is necessary to realize zero or infinite harmonic impedance conditions seen by the internal device current source required for ideal conventional Class F or inverse Class F modes, the corresponding optimum reactances at the harmonics must be presented at the device external output port which can then be translated to

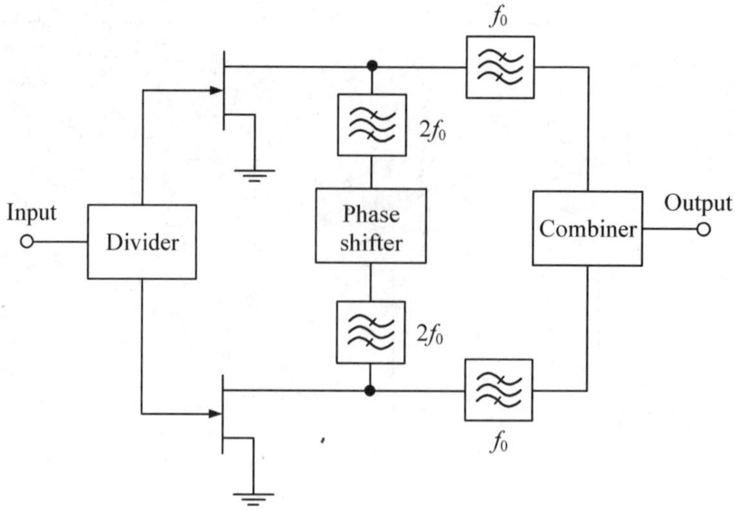

FIGURE 10.61 Circuit configuration of harmonic-reaction amplifier.

those internal required for switched-mode operation. Once the optimum impedances at the device output for the second and third harmonics are established by accurate measurements or simulations, the next step is to properly design the load network using lumped or transmission-line elements to achieve those impedances.

Figure 10.61 shows the circuit configuration of a high-efficiency MESFET power amplifier using a harmonic-reaction technique based on the effect of the second-harmonic injection [55]. Unlike the balanced power amplifier, this configuration includes a second-harmonic path between the output ports of the devices. The fundamental-frequency output paths and the second-harmonic path are designed to provide the independent matching of the device output impedances, complex-conjugate impedance matching with 50-Ω load at the fundamental frequency and optimum reactance matching at the second harmonic for maximum drain efficiency. Since the second-harmonic path has well-matched impedance characteristics, each MESFET device mutually injects without reflection a second-harmonic component into the other MESFET device through the second-harmonic path. However, it should be noted that high-efficiency operation is possible only if both devices are driven with phase-coherent and equal-amplitude input signals. As a result, an output power of 5 W with a *PAE* of 70% and a drain efficiency of 75% was obtained at the operating frequency of 2 GHz and dc-supply voltage of 7 V.

Figure 10.62(*a*) shows the circuit diagram of a single-ended *X*-band MESFET power amplifier designed to provide simultaneously the optimum load impedance at the fundamental frequency and zero impedance at the second harmonic at the device output port [56]. The load network includes the series parasitic drain lead inductance, which was optimized to provide the proper reactance to the device. An open-circuit stub with electrical length of 45° was used for the second-harmonic short circuit. It becomes capacitive at the fundamental frequency and its capacitance is a part of the output matching circuit. The output power of 450 mW and maximum *PAE* of 61% with a drain efficiency of 76% were obtained at the operating frequency of 10 GHz, using a MESFET device with the gate geometry of 0.5 μm × 1200 μm. The same power amplifier without the second-harmonic tuning circuit had demonstrated a *PAE* of only 50%. Figure 10.62(*b*) shows an *X*-band 0.5 W MESFET power amplifier with the same device geometry having a load network with lumped elements tuned to optimally terminate the second and third harmonics [57]. As a result, the measured efficiency of the power amplifier with optimal harmonic tuning was 49.3%, while the maximum efficiency measured for a device tuned to only the fundamental was 44.5%.

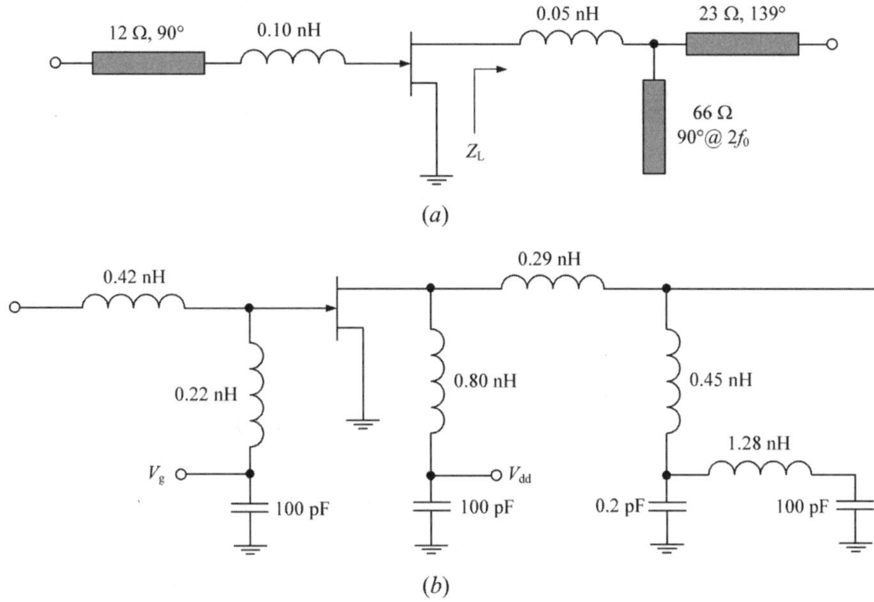

FIGURE 10.62 Circuit diagrams of Class B power amplifier with harmonic tuning.

One popular and effective way to measure the fundamental optimum device output impedances is to use the passive or active load–pull measurement system [58,59]. The objective of the active load–pull measurement technique is to find the best output load in terms of the optimum value of the load reflection coefficient $\Gamma_L(f_0)$ presented to the transistor output at the fundamental frequency, such that the power gain and *PAE* are optimized. Although cumbersome and labor intensive, such an experimental technique can provide accurate information regarding the nonlinear operation of the active device. It should be mentioned that the transistor large-signal *S*-parameters measured at a certain bias condition and output power provide insufficient information for the design of strongly driven power amplifiers with strongly nonlinear behavior of the active device. Therefore, when the transistor is operated in a nonlinear mode, measurement techniques are required that reproduce large-signal operation as input and output impedances or a nonlinear active device model, that are valid for all used bias conditions and frequency ranges.

Figure 10.63 shows the basic components of an active load–pull measurement system based on two six-port reflectometers [60]. At the input of the device under test (DUT), the six-port reflectometer SP1 provides a measurement of the large-signal reflection coefficient $\Gamma_{in}(f_0)$ and the input power $P_{in}(f_0)$ absorbed by the DUT. At the output of the DUT, the measurements of the large-signal reflection coefficient $\Gamma_L(nf_0)$ and the output power $P_L(nf_0)$ for $n = 1, 2$ are achieved by the six-port reflectometer SP2. The evaluation of $\Gamma_L(nf_0)$ and $P_L(nf_0)$ at a particular frequency requires that the power at that frequency be extracted from the output spectrum by means of splitting and filtering circuits inserted between all detection ports and the power sensors. Performing a fundamental load-pulling consists of varying the load $\Gamma_L(f_0)$ over the entire Smith chart with RF and dc measurements for each load, while the load $\Gamma_L(2f_0)$ is maintained constant and near 50 Ω. However, the fundamental load–pull technique does not include the effects of harmonic loading. The second-harmonic load–pull characterization consists in varying the load $\Gamma_L(2f_0)$, while maintaining the load $\Gamma_L(f_0)$ at a desired value. As a result, it was found that the best performance in terms of power gain and *PAE* is obtained for purely reactive second-harmonic loads, since this allows the elimination of the resistive power losses. For a specific case of the MESFET power amplifier designed to operate at 3.5 GHz, an improvement of *PAE* by about 8% was achieved by using a second harmonic active load–pull characterization technique. The

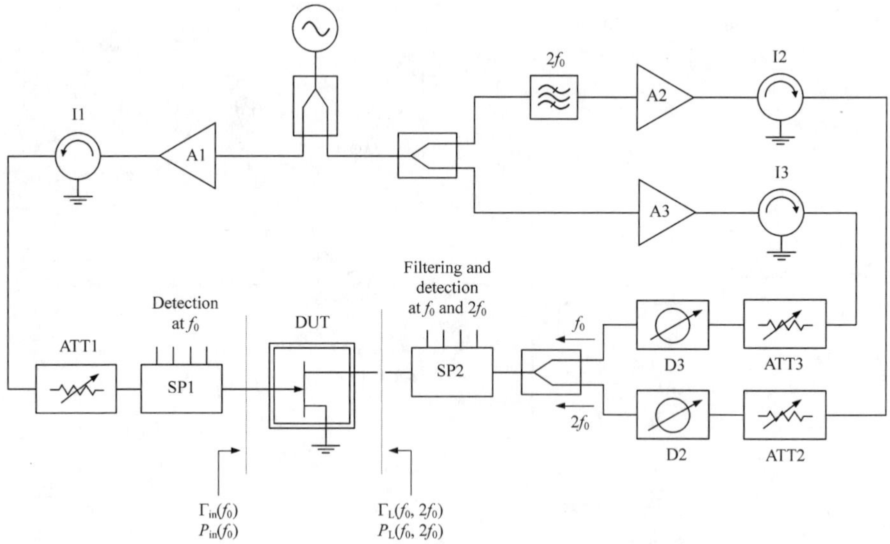

FIGURE 10.63 Harmonic active load–pull measurement system schematic.

active load–pull system based on a six-port reflectometer can also be used to optimize both linearity and output power or *PAE* for modulated signal [61].

In addition to harmonic loading at the device output, the harmonic tuning at the device input can also be effective. For Class AB power amplifiers, both the linearity and overall efficiency can be improved by the suppression of even-order harmonic components at the device input. Figure 10.64 shows an example of the circuit schematic of a microwave MESFET power amplifier with source second-harmonic tuning designed to operate in a Class AB mode [62]. To terminate even harmonics at the device input, a quarterwave transmission line at the fundamental frequency was short-circuited at its far end using a large-value capacitance C_2. The series transmission line with varying electrical length θ is necessary to provide a symmetrical gate-voltage waveform by compensating for the phase of the second-harmonic termination, taking into account the device input shunt capacitance and parasitic series gate inductance. The results of nonlinear circuit simulations show a linear (at 1 dB gain compression point) output power of 37.4 dBm at an operating frequency of 6 GHz with a *PAE* improved by 7% by using source harmonic tuning. It should be noted that changes in the source second-harmonic impedance can vary *PAE* significantly, from 30 to 80% for a 5 GHz 28.4-dBm MESFET power amplifier [63].

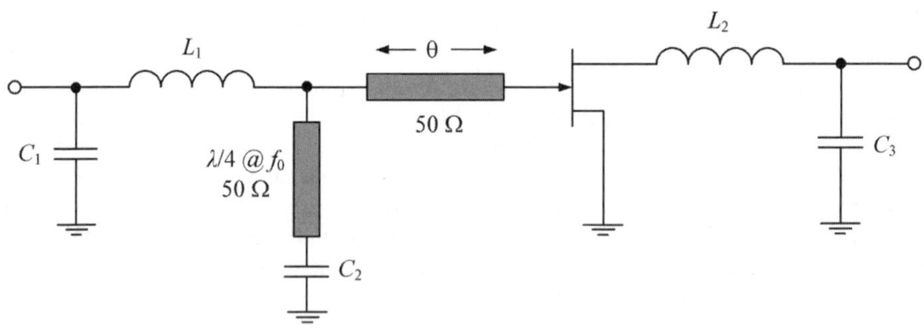

FIGURE 10.64 Circuit schematic of Class AB power amplifier with source harmonic tuning.

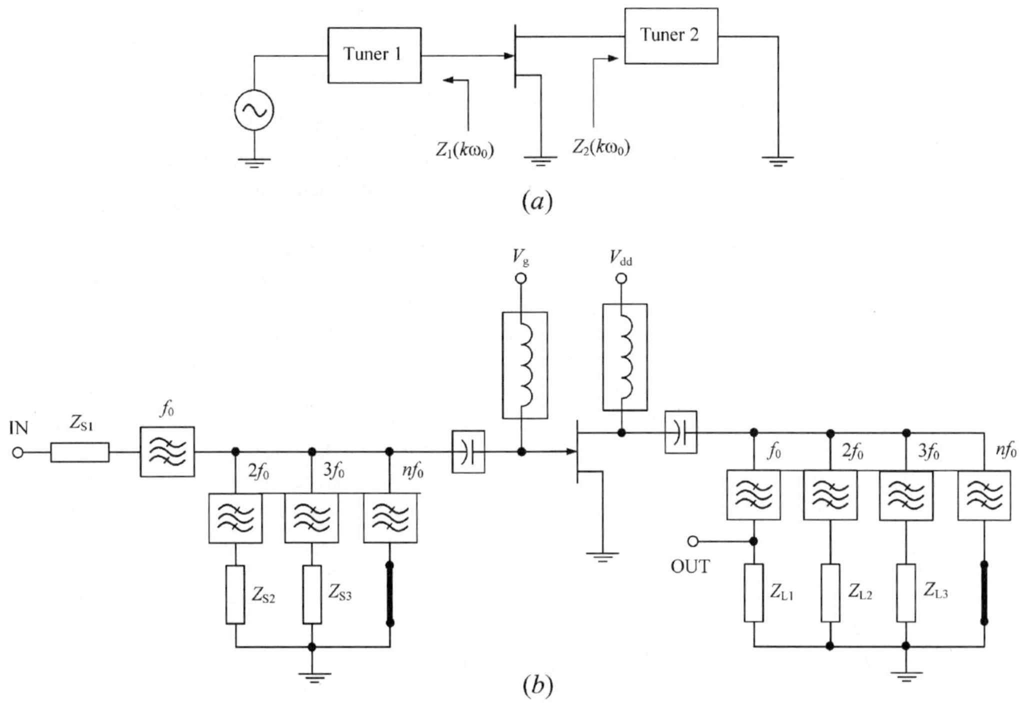

FIGURE 10.65 Circuit topologies for multi-harmonic load–pull design procedure.

A systematic procedure of multi-harmonic load–pull simulation using the harmonic-balance method includes two basic steps: finding the optimal loading at each harmonic component, and checking the power levels of higher harmonics [64]. An impedance-sampling method can be used to find the optimal harmonic loading. Generally, only one impedance component $Z_i(k\omega_0)$ at kth harmonic and ith external port is sampled at the defined impedance plane, while the others are kept constant at each step. For a specific case of a MESFET power amplifier with input and output tuners, as shown in Figure 10.65(*a*), the design procedure steps are as follows:

1. Start load–pull simulation by sampling the impedance at the fundamental frequency $Z_i(\omega_0)$ and find the optimal load $Z_{iopt}(\omega_0)$.
2. Fix $Z_i(\omega_0)$ at $Z_{iopt}(\omega_0)$ and check the output spectrum to observe the effects of higher harmonic loads on the power amplifier response.
3. Perform load–pull simulation by sampling the kth harmonic impedance $Z_i(k\omega_0)$ and find the optimal load $Z_{iopt}(k\omega_0)$.
4. Fix $Z_i(k\omega_0)$ at $Z_{iopt}(k\omega_0)$ and check the output spectrum to observe the effects of higher harmonic loads on the power amplifier response.

It should be noted that, if checking the output power spectrum shows that the power at higher harmonics is sufficiently small, the design procedure is stopped.

For a 900 MHz power amplifier using a GaAs MESFET device with a gate periphery of 0.7 μm × 600 μm biased at 3 V with a quiescent current of approximately 10% of total dc drain current, it was shown that, with sinusoidal driving signal, the best power-added efficiencies are achieved at load phase angles close to ±180° for the second harmonic, corresponding to short-circuit condition, and at load phase angles close to zero at the third harmonic, corresponding to open-circuit condition, approximating a conventional Class F mode [65]. The simulation results demonstrated a strong dependence of efficiency on the second harmonic termination, while the effect of the third harmonic was much weaker. The effect of the fourth- and fifth-harmonic terminations, stronger in the former case and weaker in the latter case, results in efficiency variation of approximately 2.5%. It is necessary to minimize the resistive losses in the harmonic terminations, intended to be purely reactive, that could be changed from reactive to resistive if the harmonic terminations are excessively lossy.

The properly phased input and output voltage harmonic components using a multi-harmonic terminating scheme shown in Figure 10.65(*b*) can be achieved by choosing the optimum impedances for each harmonic component [66,67]. For example, in a Class F mode these output impedances can be purely resistive (low at even harmonics and high at odd harmonics), since the drain (collector) current second harmonic is always in-phase and the drain (collector) current third harmonic is always out-of-phase with respect to the fundamental component for conduction angles ranging from 180° (Class B biasing) to 0° (Class C biasing). Under slightly overdriven device operation when the drain current waveform represents a truncated half-sinusoid, the third harmonic becomes out-of-phase also for Class AB biasing with conduction angles greater than 180°. The same result can be obtained by using a driver stage in a Class F mode. However, in an inverse Class F mode, the drain current second harmonic must be out of phase with respect to the fundamental-frequency component, which can be realized by reactive input termination of the second harmonic providing the required phase shift. Fortunately, the main contribution to the harmonic-generating mechanism at the device input is provided by its nonlinear input gate–source capacitor by generating the second and third harmonic components with proper phasing. As a result, optimum half-sinusoidal drain voltage waveform corresponding to an inverse Class F mode is achieved by optimizing the input and output impedances at the second and third harmonics simultaneously.

Generally, it is very difficult to provide accurate source- and load–pull measurements at very high microwave frequencies where de-embedding of waveguide transitions and parasitic discontinuities leads to significant measurement errors. It can be done accurately enough by using on-wafer measurements for an active device with limited output power capability at the fundamental frequency. At the same time, the *S*-parameters of tuner structures can be accurately predicted using both linear and electromagnetic circuit simulators. The determination of the optimum impedances at the second harmonic can be done by computer source- and load–pull analysis. In this case, there is no problem to isolate the fundamental and second-harmonic paths to determine their optimum impedances. Figure 10.66 shows the schematic of the source- and load–pull simulation setup for a *V*-band InP HEMT power amplifier, where the ideal output duplexer separates the fundamental-frequency from the second-harmonic components [68]. By changing the ideal transformer turns ratio and the reactive component values, the output power, *PAE*, power gain, and stability circle contours can be simulated for various bias voltages and operating frequencies around a center bandwidth frequency of 60 GHz.

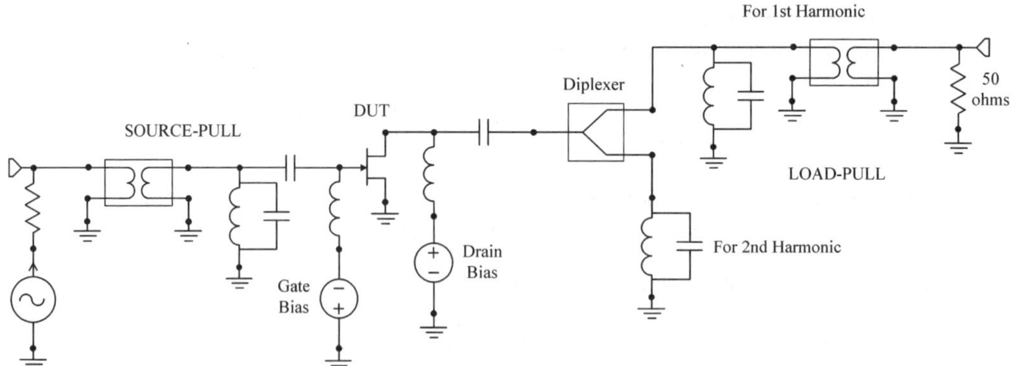

FIGURE 10.66 Schematic of source- and load–pull simulation setup.

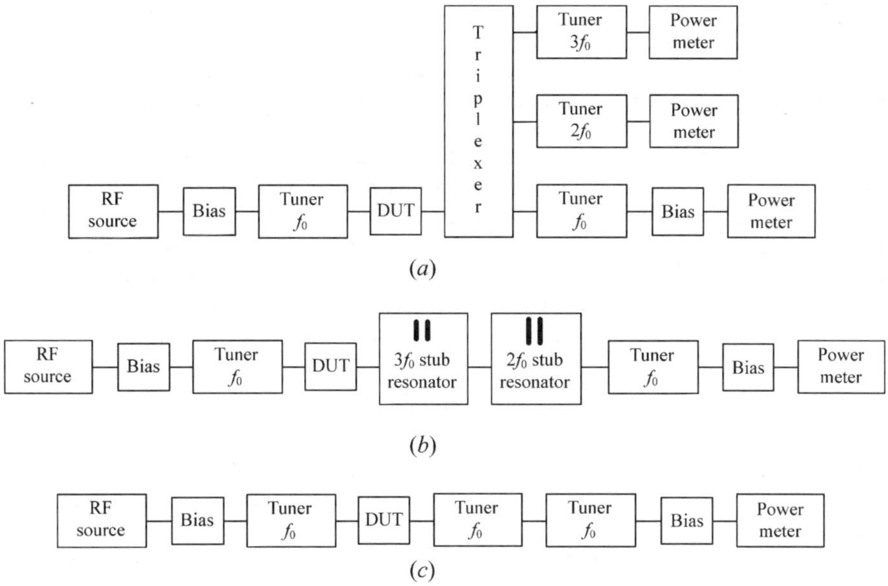

FIGURE 10.67 Circuit topologies for multi-harmonic load–pull design procedure.

The optimum second-harmonic termination resulted in a theoretical 6% improvement in *PAE* at the 3 dB gain-compression point under Class AB bias. However, in practice, a smaller improvement is expected because low-loss reactive harmonic loading at 120 GHz is difficult to physically implement without degrading the fundamental-frequency load match at 60 GHz.

Three basic methods of harmonic tuning have been offered commercially for load pull system with passive automated tuners [69]. The *triplexer tuning method* uses filters to separate the fundamental and harmonic signals so they may be tuned separately. The block diagram corresponding to this method is shown in Figure 10.67(*a*) where a triplexer includes a low-pass filter for the fundamental frequency f_0, a bandpass filter for the second harmonic $2f_0$, and a bandpass or high-pass filter for the third harmonic $3f_0$. The *stub resonator method* is based on using open-circuit stubs with quarterwave lengths at the second and third harmonic components, respectively, connected to the center conductor with a sliding contact according to the block diagram shown in Figure 10.67(*b*). The *cascaded tuner method* uses the two cascaded tuners shown in Figure 10.67(*c*) with 625 states each, producing nearly 400,000 available impedance states at the fundamental frequency with a variety of the impedances at the second harmonic. This allows the possibility to optimize the impedance at $2f_0$ with approximately constant impedance at f_0. Each method has its own advantages and disadvantages. For example, the triplexer method is the only approach with high tuning isolation, however with slightly more power loss at harmonic frequency, especially at the third harmonic, compared to the stub resonator method. At the same time, the stub resonator method with dual stubs is operated over a very narrow bandwidth. The cascaded tuner method has an advantage here, because no hardware needs to be changed; however, its tuning isolation is very poor. Thus, the triplexer method looks the best choice because of the major advantages over the other methods, with typical return-loss isolation over 100 dB and insertion loss of 0.2–0.3 dB.

10.10 THERMAL CHARACTERISTICS

The reliability of such active semiconductor components as transistors and diodes decreases with increasing their junction temperature. Since RF power transistors very often dissipate a significant amount of power, it is important to provide them with heatsinks to dispose of the unwanted heat. Usually, the data sheets specify a maximum junction temperature that should not be exceeded, and it is desirable to maintain as low a junction temperature as possible (compromising with size, weight, and cost of entire system) because many of the semiconductor failure mechanisms are temperature-dependent.

The heat flow P which is transferred through a medium composed of dissimilar materials, as shown in Figure 10.68(*a*), can be modeled by the thermal model given in Figure 10.68(*b*) through the analogy to an electrical circuit shown in Figure 10.68(*c*) where heat flow is represented by current, temperatures are represented by voltages, heat sources are represented by constant current sources, thermal resistances are represented by resistors, and thermal capacitances by capacitors. Thermal capacitors can also be added to model time constants due to heat storage [24]. The property of various elements to transfer the heat is characterized by the thermal resistance

$$R_{\mathrm{th}} = \frac{\Delta T}{P} \tag{10.137}$$

where P is the heat flow through the surface of the element, $\Delta T = T_1 - T_2$ is the temperature difference, T_1 and T_2 are the surface temperatures at the beginning and end of the element, respectively. Thermal resistance is defined as the temperature difference across a structure when a unit of heat energy flows through it in unit time. Similar to electrical circuits, the thermal model can include both

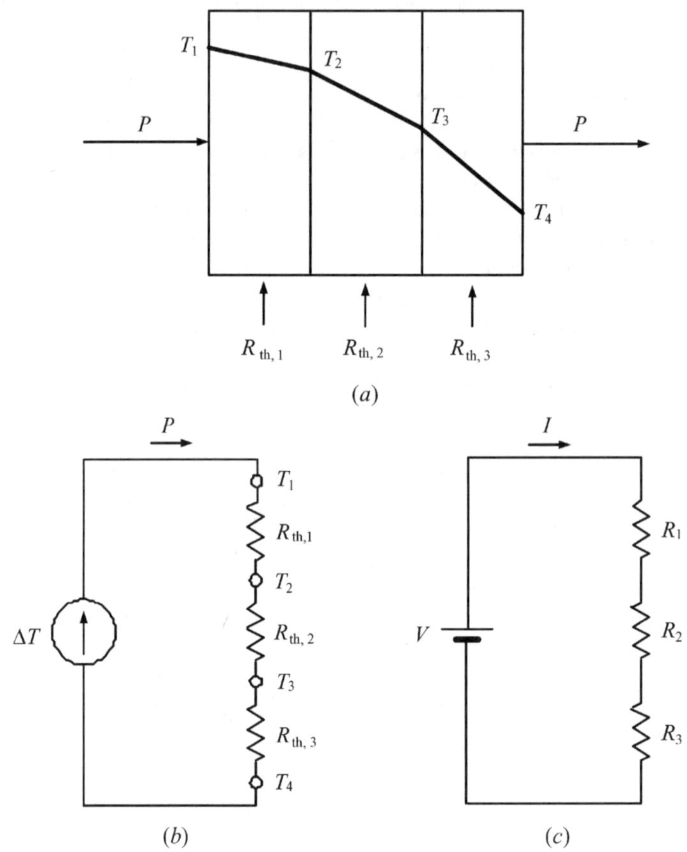

FIGURE 10.68 Thermal resistance model and its electrical equivalent.

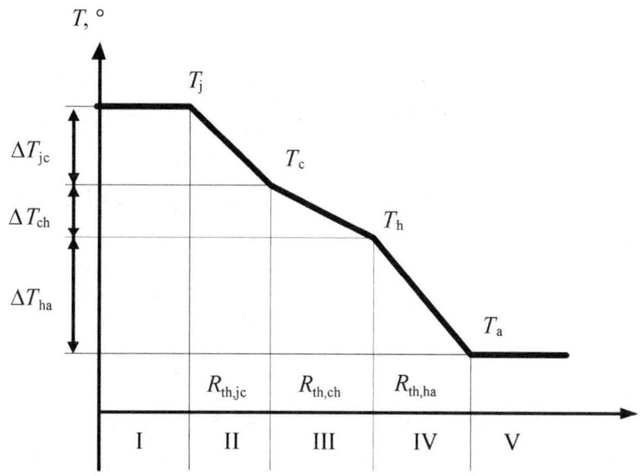

FIGURE 10.69 Model of heat transfer from transistor to ambient.

series and parallel connections of the thermal resistances. In a thermal model shown in Figure 10.68(*b*), the thermal resistances are connected in series, hence its total resistance is equal to

$$R_{\text{th},\Sigma} = \sum_{i=1}^{n} R_{\text{th},i}.$$

The model of heat transfer from junction to ambient is shown in Figure 10.69 where the heat generated in the power transistor chip flows from junction with a temperature T_j to the mounting base, which is in close contact with the case characterized by a temperature T_c, then it flows via contact pressure to heatsink with a temperature T_h, and finally the heatsink dissipates heat to ambient with a temperature T_a by conduction, convection, and radiation [15]. In high-power applications, forced air cooling by fans or forced water cooling may be provided. The total thermal resistance $R_{\text{th, ja}}$ between junction and ambient is given by

$$R_{\text{th, ja}} = R_{\text{th, jc}} + R_{\text{th, ch}} + R_{\text{th, ha}} \tag{10.138}$$

where $R_{\text{th, jc}}$ is the thermal resistance between junction and case, $R_{\text{th, ch}}$ is the thermal resistance between case and heatsink, and $R_{\text{th, ha}}$ is the thermal resistance between heatsink and ambient.

According to the heat transfer model shown in Figure 10.69, the junction temperature T_j can be defined as

$$T_j = T_a + \Delta T_{jc} + \Delta T_{ch} + \Delta T_{ha} \tag{10.139}$$

where T_a is the ambient temperature and $\Delta T_{\text{th, jc}}$, $\Delta T_{\text{th, ch}}$ and $\Delta T_{\text{th, ha}}$ are the temperatures differences between junction and case, case and heatsink, and heatsink and ambient, respectively. From Eqs. (10.137)–(10.139) it follows that

$$T_j = T_a + P_{\text{diss}} \left(R_{\text{th, jc}} + R_{\text{th, ch}} + R_{\text{th, ha}} \right) \tag{10.140}$$

where P_{diss} is the power dissipated by the semiconductor device [70].

When several transistors and diodes with the total dissipated power $P_{\mathrm{diss},\,\Sigma}$ are attached to the same heatsink, Eq. (10.140) can be rewritten as

$$T_{\mathrm{j}} = T_{\mathrm{a}} + P_{\mathrm{diss}}\left(R_{\mathrm{th,\,jc}} + R_{\mathrm{th,\,ch}}\right) + \frac{P_{\mathrm{diss},\,\Sigma}R_{\mathrm{th,\,ha}}}{K_{\mathrm{th}}} \tag{10.141}$$

where $K_{\mathrm{th}} = 0.7$–1.0 is the temperature coefficient taking into account the uneven nature of the heatsink temperature. Similarly, the case temperature can be written as

$$T_{\mathrm{c}} = T_{\mathrm{a}} + P_{\mathrm{diss}}R_{\mathrm{th,\,ch}} + \frac{P_{\mathrm{diss},\,\Sigma}R_{\mathrm{th,\,ha}}}{K_{\mathrm{th}}}. \tag{10.142}$$

When any significant amount of power is being dissipated, something must be done to fill the air voids between mating surfaces in the thermal path. Otherwise, the interface thermal resistance will be unnecessarily high and quite dependent upon the surface finishes. In this case, the thermal compounds based on silicone oil or other synthetic base fluids with conductive particles or dry conductive pads can be used [70]. The part of the heatsink that is in contact with the heat source must be perfectly flat that allows a thinner layer of thermal compound to be used, which will reduce the thermal resistance between heatsink and heat source. Heatsinks must be designed in a way that air can easily and quickly float through the cooler, and reach all cooling fins. Especially heatsinks having a very large amount of fine fins, with small distances between the fins may not allow good air flow. A compromise between high surface (many fins with small gaps between them) and good aerodynamics must be found. This also depends on the fan the heatsink is used with. A powerful fan can force air even through a heatsink with lots of fine fins with only small gaps for air flow, whereas there should be fewer cooling fins with more space between them on a passive heatsink. Large cooling fins are pointless if the heat cannot reach them, so the heatsink must be designed to allow good thermal transfer from the heat source to the fins. Thicker fins have better thermal conductivity, which means that a compromise between high surface (many thin fins) and good thermal transfer (thicker fins) must be found.

To design a heatsink, first it is necessary to define the powers dissipated by each transistor and diode and to know the maximum value (worst case) for the ambient temperature T_{a}. Then, by setting the heatsink overheating temperature $\Delta T_{\mathrm{ha}} = 10$–$25^{\circ}$ relatively to the ambient temperature, the required value for the thermal resistance $R_{\mathrm{th,\,ha}}$ between heatsink and ambient is calculated according to Eqs. (10.139) and (10.141) as $R_{\mathrm{th,\,ha}} = \Delta T_{\mathrm{ha}}K_{\mathrm{th}}/P_{\mathrm{diss},\,\Sigma}$. Finally, the junction temperature T_{j} or case temperature T_{c} is calculated for each component from Eq. (10.141) or Eq. (10.142), respectively, based on the value for $R_{\mathrm{th,\,jc}}$ obtained from data sheets and the value for $R_{\mathrm{th,\,ch}}$ estimated for particular mounting details. Check that the values for T_{j} or T_{c} do not exceed the absolute maximum rating for the components. Then, choose an appropriately sized heatsink for the calculated $R_{\mathrm{th,\,ha}}$ using heatsink nomograms provided by manufacturers.

REFERENCES

1. A. Grebennikov, *RF and Microwave Power Amplifier Design*, New York: McGraw-Hill, 2004.
2. V. M. Bogachev and V. V. Nikiforov, *Transistor Power Amplifiers* (in Russian), Moskva: Energiya, 1978.
3. J. M. Rollett, "Stability and Power Gain Invariants of Linear Two-Ports," *IRE Trans. Circuit Theory*, vol. CT-9, pp. 29–32, Jan. 1962.
4. J. G. Linvill and L. G. Schimpf, "The Design of Tetrode Transistor Amplifiers," *Bell Syst. Technol. J.*, vol. 35, pp. 813–840, Apr. 1956.
5. O. Muller and W. G. Figel, "Stability Problems in Transistor Power Amplifiers," *Proc. IEEE*, vol. 55, pp. 1458–1466, Aug. 1967.
6. O. Muller, "Internal Thermal Feedback in Fourpoles, Especially in Transistors," *Proc. IEEE*, vol. 52, pp. 924–930, Aug. 1964.

7. V. V. Nikiforov, E. P. Stroganova, and I. V. Shevnin, "An Influence of Parasitic Lead Inductances on Stability of Power MOSFET Amplifiers (in Russian)," *Radiotekhnika*, vol. 45, pp. 100–102, May 1990.
8. J. Vidkjaer, "Instabilities in RF-Power Amplifiers Caused by a Self-Oscillation in the Transistor Bias Network," *IEEE J. Solid-State Circuits*, vol. SC-11, pp. 703–712, Oct. 1976.
9. D. R. Lohrmann, "Parametric Oscillations in VHF Transistor Power Amplifiers," *Proc. IEEE*, vol. 54, pp. 409–410, Mar. 1966.
10. D. R. Lohrmann, "Amplifiers Has 85% Efficiency while Providing up to 10 Watts Power over a Wide Frequency Band," *Electron. Des.*, vol. 14, pp. 38–43, Mar. 1966.
11. P. Penfield and R. P. Rafuse, *Varactor Applications*, Cambridge: The MIT Press, 1962.
12. A. Grebennikov and N. O. Sokal, *Switchmode RF Power Amplifiers*, New York: Newnes, 2007.
13. F. E. Terman and J. H. Ferns, "The Calculation of Class C Amplifier and Harmonic Generator Performance of Screen-Grid and Similar Tubes," *Proc. IRE*, vol. 22, pp. 359–373, Mar. 1934.
14. A. I. Berg, *Theory and Design of Vacuum-Tube Generators* (in Russian), Moskva: Gosenergoizdat, 1932.
15. V. I. Kaganov, *Transistor Radio Transmitters* (in Russian), Moskva: Energiya, 1976.
16. A. Wood, "LDMOS Transistor Powers PCS Base-Station Amplifiers," *Microw. RF*, vol. 37, pp. 69–80, Mar. 1998.
17. K. Vennema, "Two-Tone Linearity in a 900 MHz Silicon Bipolar Class AB Amplifier," *Microwave J.*, vol. 39, pp. 88–93, Feb. 1996.
18. *RF Transmitting Transistor and Power Amplifier Fundamentals*, Philips Semiconductors, 1998.
19. T. Sato and C. Grigorean, "Design Advantages of CDMA Power Amplifiers Built with MOSFET Technology," *Microwave J.*, vol. 45, pp. 64–78, Oct. 2002.
20. E. Jarvinen, "Radio Frequency Amplifiers," U.S. Patent 6,052,032, Apr. 18, 2000.
21. E. Jarvinen, S. Kalajo, and M. Matilainen, "Bias Circuits for GaAs HBT Power Amplifiers," *2001 IEEE MTT-S Int. Microwave Symp. Dig.*, vol. 1, pp. 507–510.
22. Y. S. Noh and C. S. Park, "An Intelligent Power Amplifier MMIC Using a New Adaptive Bias Control Circuit for W-CDMA Applications," *IEEE J. Solid-State Circuits*, vol. SC-39, pp. 967–970, June 2004.
23. A. Grebennikov, B. Zogl, H. Hermann, C. Roth, and W. Thomann, "High-Efficiency Balanced Switch-Path Monolithic SiGe HBT Power Amplifiers for Wireless Applications," *Proc. 37th Europ. Microwave Conf.*, 2007, pp. 1189–1192.
24. H. L. Krauss, C. W. Bostian, and F. H. Raab, *Solid State Radio Engineering*, New York: John Wiley & Sons, 1980.
25. B. M. Lee, "Apply Wideband Techniques to Balanced Amplifiers," *Microwaves*, vol. 19, pp. 83–88, Apr. 1980.
26. V. Sokolov and R. E. Williams, "Development of GaAs Monolithic Power Amplifiers in *X*-Band," *IEEE Trans. Electron Dev.*, vol. ED-27, pp. 1164–1171, June 1980.
27. T. Hiraoka, T. Tokumitsu, and M. Akaike, "A Miniaturized Broad-Band MMIC Frequency Doubler," *IEEE Trans. Microwave Theory Tech.*, vol. MTT-38, pp. 1932–1937, Dec. 1990.
28. T. Tokumitsu, A. Oya, K. Sakai, and Y. Hasegawa, "A K-Band Bi-Phase Modulator MMIC for UWB Application," *IEEE Microwave Wireless Compon. Lett.*, vol. 15, pp. 159–161, Mar. 2005.
29. H. W. Bode, *Network Analysis and Feedback Amplifier Design*, New York: Van Nostrand, 1945.
30. R. M. Fano, "Theoretical Limitations on the Broad-Band Matching of Arbitrary Impedances," *J. Franklin Inst.*, vol. 249, pp. 57–83, Jan. 1950, pp. 139–154, Feb. 1950.
31. G. L. Matthaei, "Tables of Chebyshev Impedance-Transforming Networks of Low-Pass Filter Form," *Proc. IEEE*, vol. 52, pp. 939–963, Aug. 1964.
32. G. L. Matthaei, L. Young, and E. M. T. Jones, *Microwave Filters, Impedance-Matching Networks, and Coupling Structures*, New York: Artech House, 1980.
33. S. B. Cohn, "Optimum Design of Stepped Transmission-Line Transformers," *IRE Trans. Microwave Theory Tech.*, vol. MTT-3, pp. 16–21, Apr. 1955.
34. H. J. Riblet, "General Synthesis of Quarter-Wave Impedance Transformers," *IRE Trans. Microwave Theory Tech.*, vol. MTT-5, pp. 36–43, Apr. 1957.
35. G. L. Matthaei, "Short-Step Chebyshev Impedance Transformers," *IEEE Trans. Microwave Theory Tech.*, vol. MTT-14, pp. 372–383, Aug. 1966.

36. V. P. Meschanov, I. A. Rasukova, and V. D. Tupikin, "Stepped Transformers on TEM-Transmission Lines," *IEEE Trans. Microwave Theory Tech.*, vol. MTT-44, pp. 793–798, June 1996.
37. Y. Ito, M. Mochizuki, M. Kohno, H. Masuno, T. Takagi, and Y. Mitsui, "A 5-10 GHz 15-W GaAs MESFET Amplifier with Flat Gain and Power Responses," *IEEE Microwave Guided Wave Lett.*, vol. 5, pp. 454–456, Dec. 1995.
38. K. B. Niclas, "On Design and Performance of Lossy Match GaAs MESFET Amplifiers," *IEEE Trans. Microwave Theory Tech.*, vol. MTT-30, pp. 1900–1906, Nov. 1982.
39. Y. Ito, M. Nii, Y. Kohno, M. Mochizuki, and T. Takagi, "A 4 to 25 GHz 0.5 W Monolithic Lossy Match Amplifier," *1994 IEEE MTT-S Int. Microwave Symp. Dig.*, pp. 257–260.
40. Y.-F. Wu, R. A. York, S. Keller, B. P. Keller, and U. K. Mishra, "3-9 GHz GaN-Based Microwave Power Amplifiers with L-C-R Broad-Band Matching," *IEEE Microwave Guided Wave Lett.*, vol. 9, pp. 314–316, Aug. 1999.
41. B. Y. Banyamin and M. Berwick, "Analysis of the Performance of Four-Cascaded Single-Stage Distributed Amplifiers," *IEEE Trans. Microwave Theory Tech.*, vol. MTT-48, pp. 2657–2663, Dec. 2000.
42. Application Note AN98021, *100-450 MHz 250 W Power Amplifier with the BLF548 MOSFET*, Philips Semiconductors, March 1998.
43. W. S. Percival, "Thermionic Valve Circuits," British Patent 460562, Jan. 1937.
44. E. L. Ginzton, W. R. Hewlett, J. H. Jasberg, and J. D. Noe, "Distributed Amplification," *Proc. IRE*, vol. 36, pp. 956–969, Aug. 1948.
45. W. H. Horton, J. H. Jasberg, and J. D. Noe, "Distributed Amplifiers: Practical Considerations and Experimental Results," *Proc. IRE*, vol. 38, pp. 748–753, July 1950.
46. Y. Ayasli, R. L. Mozzi, J. L. Vorhaus, L. D. Reynolds, and R. A. Pucel, "A Monolithic GaAs 1-13 GHz Traveling-Wave Amplifier," *IEEE Trans. Microwave Theory Tech.*, vol. MTT-30, pp. 976–981, July 1982.
47. K. B. Niclas, W. T. Wilser, T. R. Kritzer, and R. R. Pereira, "On Theory and Performance of Solid-State Microwave Distributed Amplifiers," *IEEE Trans. Microwave Theory Tech.*, vol. MTT-31, pp. 447–456, June 1983.
48. J. B. Beyer, S. N. Prasad, R. C. Becker, J. E. Nordman, and G. K. Hohenwarter, "MESFET Distributed Amplifier Design Guidelines," *IEEE Trans. Microwave Theory Tech.*, vol. MTT-32, pp. 268–275, Mar. 1984.
49. S. N. Prasad, J. B. Beyer, and I.-K. Chang, "Power-Bandwidth Considerations in the Design of MESFET Distributed Amplifiers," *IEEE Trans. Microwave Theory Tech.*, vol. MTT-36, pp. 1117–1123, July 1988.
50. J. Aguirre and C. Plett, "50-GHz SiGe HBT Distributed Amplifiers Employing Constant-k and m-Derived Filter Sections," *IEEE Trans. Microwave Theory Tech.*, vol. MTT-52, pp. 1573–1579, May 2004.
51. Y. Chen, J. B. Beyer, V. Sokolov, and J. P. Culp, "A 11 GHz Hybrid Paraphase Amplifier," *1986 IEEE Int. Solid-State Circuits Conf. Dig.*, pp. 236–237.
52. L. Zhao, A. Pavio, B. Stengel, and B. Thompson, "A 6 Watt LDMOS Broadband High Efficiency Distributed Power Amplifier Fabricated Using LTCC Technology," *2002 IEEE MTT-S Int. Microwave Symp. Dig.*, pp. 897–900.
53. C. Campbell, C. Lee, V. Williams, M.-Y. Kao, H.-Q. Tserng, P. Saunier, and T. Balisteri, "A Wideband Power Amplifier MMIC Utilizing GaN on SiC HEMT Technology," *IEEE J. Solid-State Circuits*, vol. SC-44, pp. 2640–2647, Oct. 2009.
54. S. R. Mazumder, A. Azizi, and F. E. Gardiol, "Improvement of a Class-C Transistor Power Amplifier by Second-Harmonic Tuning," *IEEE Trans. Microwave Theory Tech.*, vol. MTT-27, pp. 430–433, May 1979.
55. T. Nojima and S. Nishiki, "High Efficiency Microwave Harmonic Reaction Amplifier," *1988 IEEE MTT-S Int. Microwave Symp. Dig.*, pp. 1007–1010.
56. M. A. Khatibzadeh and H. Q. Tserng, "Harmonic Tuning of Power FETs at *X*-Band," *1990 IEEE MTT-S Int. Microwave Symp. Dig.*, pp. 989–992.
57. B. Kopp and D. D. Heston, "High-Efficiency 5-Watt Power Amplifier with Harmonic Tuning," *1988 IEEE MTT-S Int. Microwave Symp. Dig.*, pp. 839–842.
58. J. M. Cusak, S. M. Perlow, and B. S. Perlman, "Automatic Load Contour Mapping for Microwave Power Transistors," *IEEE Trans. Microwave Theory Tech.*, vol. MTT-22, pp. 1146–1152, Dec. 1974.
59. Y. Takayama, "A New Load-Pull Characterization Method for Microwave Power Transistors," *1976 IEEE MTT-S Int. Microwave Symp. Dig.*, pp. 218–220.

60. P. Berini, M. Desgagne, F. M. Ghannouchi, and R. G. Bosisio, "An Experimental Study of the Effects of Harmonic Loading on Microwave MESFET Oscillators and Amplifiers," *IEEE Trans. Microwave Theory Tech.*, vol. MTT-42, pp. 943–950, June 1994.

61. S. Bensmida, E. Bergeault, G. I. Abib, and B. Huyart, "Power Amplifier Characterization: An Active Load-Pull System Based on Six-Port Reflectometer using Complex Modulated Carrier," *IEEE Trans. Microwave Theory Tech.*, vol. MTT-54, pp. 2707–2712, June 2006.

62. S. Watanabe, S. Takatsuka, K. Takagi, H. Kuroda, and Y. Oda, "Effect of Source Harmonic Tuning on Linearity of Power GaAs FET under Class AB Operation," *IEICE Trans. Electron.*, vol. E79-C, pp. 611–616, May 1996.

63. L. C. Hall and R. J. Trew, "Maximum Efficiency Tuning of Microwave Amplifiers," *1991 IEEE MTT-S Int. Microwave Symp. Dig.*, pp. 123–126.

64. Q. Cai, J. Herber, and S. Peng, "A Systematic Scheme for Power Amplifier Design Using a Multi-Harmonic Loadpull Simulation Technique," *1998 IEEE MTT-S Int. Microwave Symp. Dig.*, pp. 161–165.

65. J. Staudinger, "Multiharmonic Load Termination Effects on GaAs MESFET Power Amplifiers," *Microwave J.*, vol. 39, pp. 60–77, Apr. 1996.

66. P. Colantonio, F. Giannini, G. Leuzzi, and E. Limiti, "Multiharmonic Manipulation for Highly Efficient Microwave Power Amplifiers," *Int. J. RF Microwave Computer-Aided Eng.*, vol. 11, pp. 366–384, Nov. 2001.

67. P. Colantonio, F. Giannini, E. Limiti, and V. Teppati, "An Approach to Harmonic Load- and Source-Pull Measurements for High-Efficiency PA Design," *IEEE Trans. Microwave Theory Tech.*, vol. MTT-52, pp. 191–198, Jan. 2005.

68. O. S. A. Tang, S. M. J. Liu, P. C. Chao, W. M. T. Kong, K. C. Hwang, K. Nichols, and J. Heaton, "Design and Fabrication of a Wideband 56- to 63-GHz Monolithic Power Amplifier with Very High Power-Added Efficiency," *IEEE J. Solid-State Circuits*, vol. SC-35, pp. 1298–1306, Sept. 2000.

69. G. Simpson, "A Comparison of Harmonic Tuning Methods for Load Pull Systems," *Maury Microw. Corp.*, 2004.

70. Application Note 1040, *Mounting Considerations for Power Semiconductors,* Freescale Semiconductor, 1993.

11 High-Efficiency Power Amplifiers

High efficiency of the power amplifier can be obtained by using Class D, Class F, or Class E operation modes and their subclasses, depending on the technical requirements. In all cases, an efficiency improvement in practical implementation is achieved by providing the nonlinear operation conditions when an active device can simultaneously operate in pinch-off, active, and saturation regions resulting in nonsinusoidal collector current and voltage waveforms, symmetrical for Class D and Class F and asymmetrical for Class E (DE, FE) operation modes. In Class F amplifiers analyzed in frequency domain, the fundamental-frequency and harmonic load impedances are optimized by short-circuit termination and open-circuit peaking to control the voltage and current waveforms at the device output to obtain maximum efficiency. In Class E amplifiers analyzed in time domain, an efficiency improvement is achieved by realizing the on/off active device switching operation (the pinch-off and saturation modes) with special current and voltage waveforms so that high voltage and high current do not concur at the same time.

11.1 CLASS D

The efficiency of a power amplifier can be maximized if the active device is ideally operated as a switch. When the transistor is turned on, the voltage is nearly zero and high current is flowing through the device; that is, the transistor acts as a low resistance (closed switch) during this part of a period. When the transistor is turned off, the current is zero and there is high voltage across the device; that is, the transistor acts as an open switch during the other part of a period. Conceptually, a Class D power amplifier employs a pair of active devices operating in a push–pull mode and a tuned output circuit [1,2]. The switched-mode power amplifiers with output filter tuned to the fundamental frequency transform ideally the total dc power into a fundamental-frequency power delivered to the load without power losses at the harmonics. The output circuit is tuned to the switching frequency and removes ideally its all harmonic components resulting in a purely sinusoidal signal delivered to the load. Consequently, the theoretical efficiency of an idealized Class D power amplifier achieves 100%. Let us consider the basic principles, circuit schematics, and voltage/current waveforms corresponding to the different types of a Class D power amplifier with output filter [3–5].

11.1.1 Voltage-Switching Configurations

Figure 11.1(*a*) shows the simplified circuit schematic of a complementary voltage-switching Class D bipolar power amplifier consisting of the same type of the active devices, fundamentally tuned series L_0C_0 filter, and load resistance R_L. The large-value bypass capacitor C_b is necessary to isolate the dc power supply by bypassing all radio frequency (RF) signals to ground. The input transformer causes both active devices to be driven with currents that are 180° out of phase by reversing one secondary winding on the transformer. However, there is no need in phase reversing if the transistors are true-complementary with different base or channel majority-carrier type that simplifies the circuit

RF and Microwave Transmitter Design, First Edition. Andrei Grebennikov.

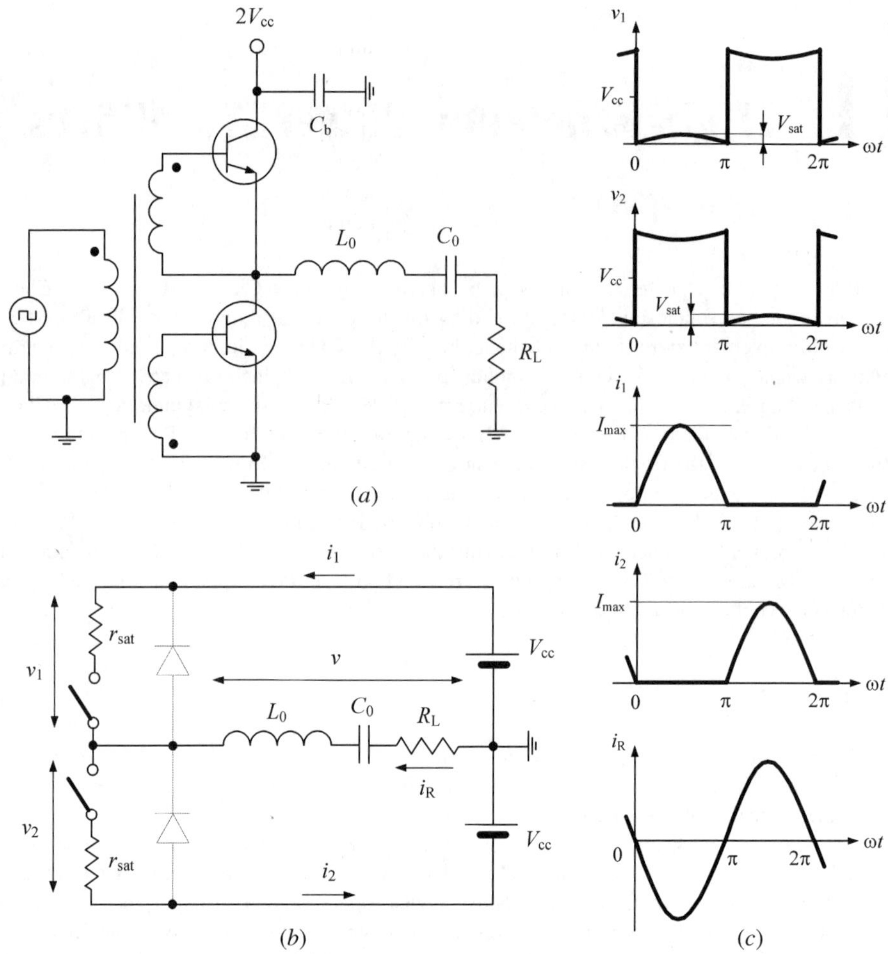

FIGURE 11.1 Complementary voltage-switching configuration with series filter.

schematic. Note that any type of the vacuum tubes, bipolar, and MOSFET transistors can be used in this circuit if suitable drive is applied. Due to the grounding effect of a bypass capacitor C_b, the RF connection of the transistor outputs is parallel, thus resulting in an equivalent load resistance equal to $2R_L$ for each device.

To determine the collector voltage and current waveform and to calculate the output power and collector efficiency, the following assumptions are taken into account:

- Power loss due to flow of leakage current during transistor pinch-off is negligible.
- Power loss due to nonideal tuning is negligible.
- Power loss during switching transitions is negligible.

In this case, each active device is considered a switch with the saturation resistance r_{sat} shown in Figure 11.1(*b*) that is driven in such a way in order to provide an alternating switching between the on- and off-state operation conditions of the transistor under an assumption of a 50% duty cycle. The alternating half-period device switching between its pinch-off mode and saturation mode results in rectangular collector voltage pulses with a maximum amplitude of $2V_{cc}$, as shown in Figure 11.1(*c*). The collector–voltage pulses, which contain the odd harmonic components only, are applied to the series L_0C_0 filter with high-loaded quality factor $Q_L = \omega L_0/R_L \gg 1$ tuned to the fundamental frequency $\omega_0 = 1/\sqrt{L_0C_0}$. This results in the fundamental-frequency sinusoidal current $i_R = -I_R \sin\omega t$ flowing to the load R_L, where I_R is the load current amplitude. The half waves of this current flow in turn through the transistors representing the half-sinusoidal collector–current pulses that contain the fundamental-frequency, second-order and higher order even-harmonic components only. The shape of the saturation voltage with maximum amplitude V_{sat} is fully determined by the collector–current waveform when $i(\omega t) > 0$ according to $v_{sat}(\omega t) = r_{sat} i(\omega t)$, where $i(\omega t) = i_1(\omega t)$ or $i_2(\omega t)$ for a symmetrical circuit with identical transistors. The collector voltage peak factor is equal to $v/V_{cc} = 2$.

It should be noted that for an operation with conduction angles less than 180°, when both active devices are turned on for duty cycles less than 50%, there is a time period when both active devices are turned off simultaneously. Therefore, in order that the load current $i_R(\omega t)$ could flow continuously, it is necessary to include a diode in parallel to each device, as shown in Figure 11.1(*b*). The operation conditions with conduction angles greater than 180° are unacceptable, since both dc power supplies are connected to each other through the small saturation resistances of the identical transistors equal to $2r_{sat}$, thus resulting in the significant efficiency reduction due to an increased total current flowing through both transistors.

Now let us determine the voltage $v(\omega t) = v_1(\omega t) - V_{cc}$ at the input of a series L_0C_0 circuit and collector current $i_1(\omega t)$ for the first transistor operating as a switch with a saturation resistance r_{sat}. When switch is turned on for $0 \le \omega t \le \pi$,

$$v(\omega t) = -V_{cc} + r_{sat} I_R \sin \omega t \tag{11.1}$$

$$i_1(\omega t) = I_R \sin \omega t. \tag{11.2}$$

When switch is turned off for $\pi \le \omega t \le 2\pi$,

$$v(\omega t) = V_{cc} + r_{sat} I_R \sin \omega t \tag{11.3}$$

$$i_1(\omega t) = 0. \tag{11.4}$$

The fundamental-frequency voltage amplitude V of the voltage $v(\omega t)$ can be calculated by applying a Fourier transform to Eqs. (11.1) and (11.2) from

$$V = -\frac{1}{\pi} \int_0^{2\pi} v(\omega t) \sin \omega t \, d\omega t = \frac{4}{\pi} V_{cc} - r_{sat} I_R. \tag{11.5}$$

Similarly, the dc current I_0 can be obtained from Eq. (11.2) by

$$I_0 = \frac{1}{2\pi} \int_0^{\pi} i_1(\omega t) \, d\omega t = \frac{I_R}{\pi}. \tag{11.6}$$

Taking into account that $I_R = V/R_L$ and the fact that the sinusoidal output current flows through either one or another transistor depending on which device is turned on and having a half-sinusoidal waveform, using Eq. (11.5) will result in

$$I_R = I_{max} = \frac{4}{\pi} \frac{V_{cc}}{R_L} \frac{1}{1 + \dfrac{r_{sat}}{R_L}}. \tag{11.7}$$

The dc power P_0 and fundamental-frequency output power P can be obtained using Eqs. (11.5) and (11.6), respectively, by

$$P_0 = 2V_{cc}I_0 = \frac{8}{\pi^2}\frac{V_{cc}^2}{R_L}\frac{1}{1+\dfrac{r_{sat}}{R_L}} \tag{11.8}$$

$$P = \frac{1}{2}I_R^2R_L = \frac{8}{\pi^2}\frac{V_{cc}^2}{R_L}\frac{1}{\left(1+\dfrac{r_{sat}}{R_L}\right)^2}. \tag{11.9}$$

As a result, the collector efficiency η of a complementary voltage-switching push–pull power amplifier with a series filter can be written as

$$\eta = \frac{P}{P_0} = \frac{1}{1+\dfrac{r_{sat}}{R_L}}. \tag{11.10}$$

From Eq. (11.10), it follows that the collector efficiency is equal to 100% for an idealized case of the lossless active devices with zero saturation resistance. It should be mentioned that the dc current drawn from the power supply represents the form of a half-sinusoidal pulse train. Therefore, it is very important to provide the proper RF bypassing circuit representing either a single large-value capacitor or an additional filter with a low cutoff frequency in the power supply line. In addition, the loaded quality factor for a simple series-tuned L_0C_0 circuit must be chosen greater than five to provide a good compromise between the prevention of harmonic current and coil losses. Additional harmonic suppression can be obtained by inserting standard filters between the series-tuned circuit and the load. Note that the collector efficiency degrades with increasing operating frequency where the switching transitions become an appreciable fraction of the signal period. In practice, it was found that high efficiency of a Class D power amplifier can be maintained to frequencies of the order of $0.1f_T$ for low-power transistors and to $0.01f_T$ for high-power transistors rated at greater than 10 W, where f_T is the device transition frequency [6].

The broadband center-tapped and balanced-to-unbalanced transformers can also be used in voltage-switching Class D power amplifiers in much the same manner as they are used in conventional Class B push–pull power amplifiers. In this case, such a configuration is called the *transformer-coupled voltage-switching Class D power amplifier*. Figure 11.2 shows the simplified circuit schematic of a transformer-coupled voltage-switching Class D bipolar power amplifier including an output

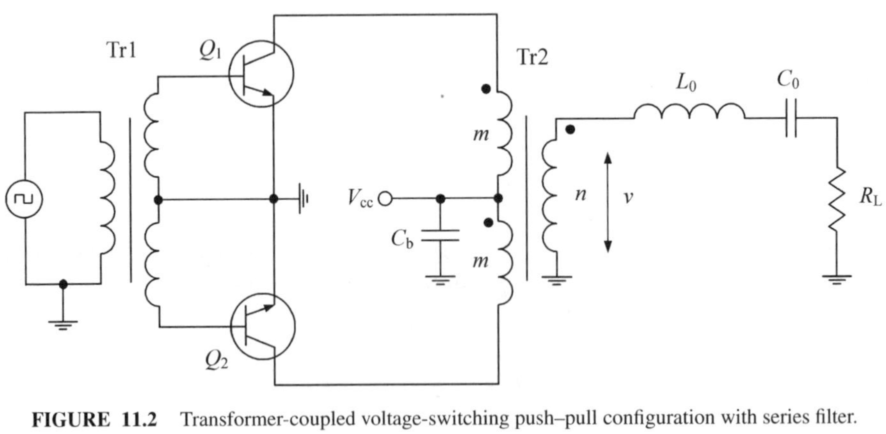

FIGURE 11.2 Transformer-coupled voltage-switching push–pull configuration with series filter.

series-tuned L_0C_0 circuit and a load resistance R_L. As in the complementary voltage-switching configuration, the input transformer Tr1 causes both active devices Q_1 and Q_2 to be driven with currents that are 180° out of phase to switch on and off alternately. During the first half-cycle when transistor Q_1 is turned on, its collector voltage $v_1(\omega t)$ is equal to zero, assuming zero saturation resistance. As a result, dc supply voltage V_{cc} is placed across one-half of the primary winding of the transformer Tr2, being then transformed to the voltage $(-n/m)V_{cc}$ on its secondary winding. When transistor Q_2 is turned on, dc supply voltage V_{cc} is applied to the other half of the primary winding, causing voltage $(n/m)V_{cc}$ to appear on the secondary winding. The resulting secondary voltage $v(\omega t)$ represents a square wave with levels of $\pm(n/m)V_{cc}$, while the collector voltages are square waves with levels of 0 and $+2V_{cc}$. The collector currents of each transistor are the half-sinusoidal waveforms with opposite phase of 180° between each other whose peak amplitudes are $(4/\pi)(V_{cc}/R)$ resulting in a sinusoidal current flowing onto the load R_L.

11.1.2 Current-Switching Configurations

Figure 11.3(*a*) shows the simplified circuit schematic of a symmetrical current-switching Class D bipolar power amplifier consisting of the same type of the active devices, fundamentally tuned parallel L_0C_0 filter, and load resistance R_L. The RF choke L_{ch} connected to the center point of the inductor L_0 is necessary to isolate the dc power supply and make the circuit symmetrical. Each active device is considered to be a switch with the saturation resistance r_{sat} shown in Figure 11.3(*b*) that is driven in such a way in order to provide an alternating switching between the on- and off-state operation conditions of the transistor under an assumption of a 50% duty cycle. Since the only dc current I_0 is flowing through the RF choke L_{ch}, the alternating half-period device switching between their pinch-off and saturation modes results in rectangular collector current pulses with a maximum amplitude of $2I_0$, as shown in Figure 11.3(*c*).

The collector current pulses, which contain the odd harmonic components only, are applied to the parallel L_0C_0 filter with high-loaded quality factor $Q_L = \omega L_0/R_L \gg 1$ tuned to the fundamental frequency $\omega_0 = 1/\sqrt{L_0C_0}$ resulting in the fundamental-frequency sinusoidal voltage $v_R = V_R \sin\omega t$ across the load R_L. During half a period when one transistor is turned on, the half wave of this voltage is applied to the other transistor representing the half-sinusoidal collector voltage pulses that contain the fundamental-frequency, second-order and higher order even-harmonic components only. The flat shape of the saturation voltage with maximum amplitude V_{max} is fully determined by the collector–current waveform when $i(\omega t) > 0$ according to $v_{sat}(\omega t) = r_{sat}\, i_1(\omega t)$ or $r_{sat}\, i_2(\omega t)$ for a symmetrical circuit with identical transistors. Unlike a complementary voltage-switching configuration, the RF connection of the transistor outputs is series, thus resulting in an equivalent load resistance equal to $R_L/2$ for each device.

It should be noted that the symmetrical current-switching Class D power amplifier, the collector voltage and current waveforms of which are shown in Figure 11.3(*c*), is the dual of the complementary voltage-switching Class D power amplifier, the collector voltage and current waveforms of which are shown in Figure 11.1(*c*), because the voltage and current waveforms are interchanged. However, if, in the case of a voltage-switching configuration, the collector peak voltage is defined by the dc supply voltage V_{cc} only, then, in the latter case of a current-switching configuration, the transistors represent current switches with the current amplitude defined by the dc voltage supply V_{cc}, saturation resistance r_{sat}, and load resistance R_L. For an operation mode with conduction angles greater than 180°, there is a time period when both active devices are turned on simultaneously, and parallel-tuned circuit is shunted by a small resistance $2r_{sat}$. Therefore, to eliminate this shunting effect accompanied by power losses in both transistors, it is necessary to include a diode in series to each device collector, as shown in Figure 11.3(*b*). However, the operation conditions with conduction angles less than 180° are unacceptable, since there are time intervals when both transistors are turned off simultaneously causing the significant increase in the collector voltage amplitude due to the growth of the current flowing through RF choke.

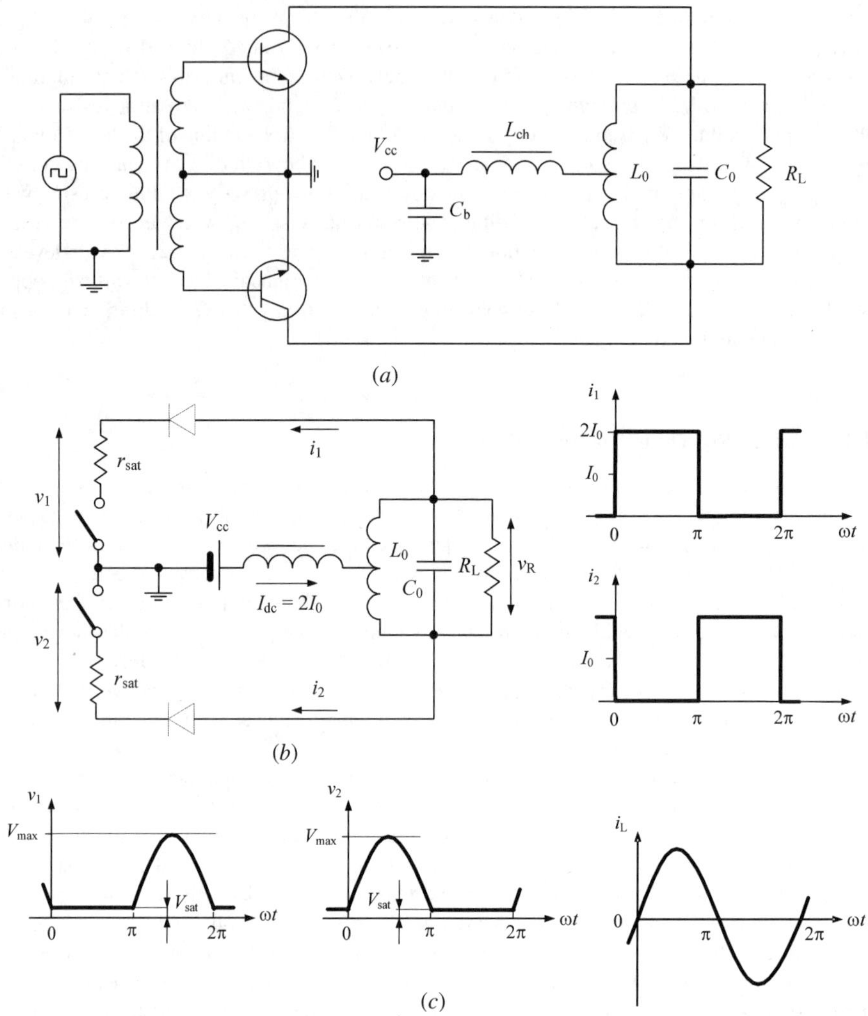

FIGURE 11.3 Symmetrical current-switching configuration with parallel filter.

Now let us determine the current $i(\omega t) = i_1(\omega t) - I_0$ in a parallel L_0C_0 circuit and collector voltage $v_1(\omega t)$ for the first transistor operating as a current switch with a saturation resistance r_{sat}. When first switch is turned on for $0 \leq \omega t \leq \pi$, $i_1(\omega t) = 2I_0$ resulting in

$$i\,(\omega t) = I_0 \tag{11.11}$$

$$v_1\,(\omega t) = V_{sat} = 2r_{sat}I_0 \tag{11.12}$$

where I_0 is the dc supply current of each device.

When first switch is turned off for $\pi \leq \omega t \leq 2\pi$, $i_1(\omega t) = 0$ resulting in

$$i\,(\omega t) = -I_0 \tag{11.13}$$

$$v_1\,(\omega t) = -\,(V_{max} - 2r_{sat}I_0)\sin\omega t + 2r_{sat}I_0 \tag{11.14}$$

where V_{max} is the collector peak voltage.

The fundamental-frequency current amplitude I of the current $i(\omega t)$ can be calculated by applying a Fourier transform to Eqs. (11.11) and (11.13) from

$$I = \frac{1}{\pi} \int_0^{2\pi} i(\omega t) \sin \omega t d\omega t = \frac{4}{\pi} I_0. \tag{11.15}$$

Similarly, the dc supply voltage V_{cc} can be written from Eqs. (11.12) and (11.14) by

$$V_{cc} = \frac{1}{2\pi} \int_0^{\pi} v_1(\omega t) d\omega t = \frac{1}{\pi} (V_{max} - 2r_{sat} I_0). \tag{11.16}$$

Taking into account that $V_1 = V_{max} - 2r_{sat}I_0 = IR_L$, the dc current I_0 and fundamental-frequency collector voltage V_1 can be obtained by

$$I_0 = \left(\frac{\pi}{2}\right)^2 \left(1 + \frac{\pi^2}{2} \frac{r_{sat}}{R_L}\right)^{-1} \frac{V_{cc}}{R_L} \tag{11.17}$$

$$V_1 = \pi \left(1 + \frac{\pi^2}{2} \frac{r_{sat}}{R_L}\right)^{-1} V_{cc}. \tag{11.18}$$

The dc power P_0 and fundamental-frequency output power P and can be obtained using Eqs. (11.17) and (11.18) and taking into account that $V_1 = V_R$ by

$$P_0 = 2V_{cc} I_0 = \frac{\pi^2}{2} \frac{V_{cc}^2}{R_L} \frac{1}{1 + \frac{\pi^2}{2} \frac{r_{sat}}{R_L}} \tag{11.19}$$

$$P = \frac{1}{2} \frac{V_R^2}{R_L} = \frac{\pi^2}{2} \frac{V_{cc}^2}{R_L} \frac{1}{\left(1 + \frac{\pi^2}{2} \frac{r_{sat}}{R_L}\right)^2}. \tag{11.20}$$

As a result, the collector efficiency η of a symmetrical current-switching push–pull power amplifier with parallel filter can be written as

$$\eta = \frac{P}{P_0} = \frac{1}{1 + \frac{\pi^2}{2} \frac{r_{sat}}{R_L}}. \tag{11.21}$$

Figure 11.4 shows the circuit schematic of a current-switching Class D power amplifier, where the output balanced-to-unbalanced transformer is used to connect to a standard load. In this case, such a configuration is called the *transformer-coupled current-switching Class D power amplifier*, which is the dual of the transformer-coupled voltage-switching Class D power amplifier because the collector voltage and current waveforms are interchanged. As in the transformer-coupled voltage-switching configuration, the input transformer Tr1 is necessary to drive both active devices with currents having opposite phases for the on-to-off alternate device switching. The output transformer Tr2 is considered as ideal, having m turns in each half of the primary winding and n turns in the secondary winding. However, the dc current supply is connected to the center tap of the transformer primary winding through the RF choke. The load network of the transformer-coupled current-switching configuration requires a parallel fundamentally tuned resonant L_0C_0 circuit connected in parallel to the load R_L, instead of a series fundamentally tuned resonant L_0C_0 circuit required for the transformer-coupled

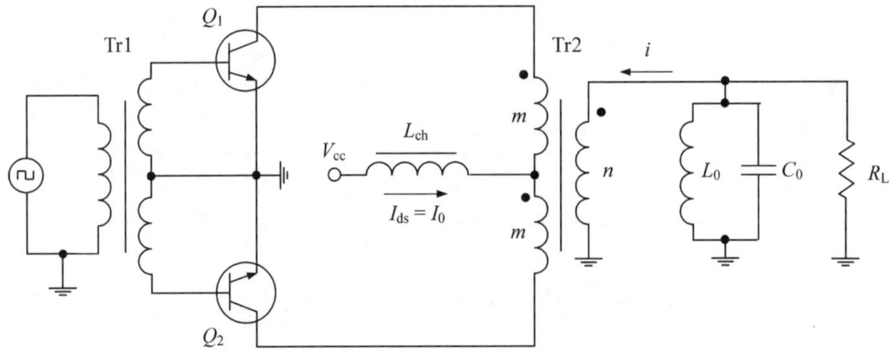

FIGURE 11.4 Transformer-coupled current-switching configuration with parallel filter.

voltage-switching configuration. Whichever device is turned on during the first half-cycle, it takes the entire dc current I_0 resulting in a rectangular collector current waveform with levels of 0 and I_0. Its collector voltage is equal to zero when it turned on, assuming zero saturation resistance. Transformation of the rectangular collector currents from half of the primary winding to the secondary winding produces a rectangular current $i(\omega t)$ with levels of $\pm(m/n)I_0$ resulting in a sinusoidal current flowing onto the load R_L.

11.1.3 Drive and Transition Time

The driving circuitry of a Class D power amplifier is very similar to that of a Class B power amplifier and must provide the driving signal sufficient to ensure that the active devices are alternately saturated or pinched off during the proper time period. Generally, if a current-switching Class D power amplifier requires a rectangular driven signal, a voltage-switching Class D power amplifier can be driven by either a rectangular- or sinusoidal-driven signal. The sinusoidal current driven through the primary winding of the input transformer causes the half-sinusoidal currents to be driven into each base from the transformer secondary winding. The current flowing into the corresponding base results in a rise of the base-to-emitter voltage to V_{bmax}, which is approximately equal to its threshold voltage of 0.7 V for a bipolar device and corresponds to the maximum value required to saturate the MOSFET device. At the same time, the current flowing into the base of the other transistor is shifted by 180° causing the fall of its base-to-emitter voltage to $-V_{bmax}$, ensuring that the transistor is pinched off.

Figure 11.5 shows the different configurations of the input driving circuits for a current-switching Class D bipolar power amplifier. In the case of the driver with a rectangular current waveform, the active devices can be connected either in parallel, as shown in Figure 11.5(*a*), or in series, as shown in Figure 11.5(*b*), where L_{ch} is the choke inductor. For a parallel connection, the driving current is distributed between bipolar transistors being reversely proportional to their input impedances. The input impedance of the transistor at saturation is significantly less that that of in the active and pinch-off regions. Since the transistor cannot change instantly its saturation mode to pinch-off conditions under negative driving signal, it absorbs the most part of the driving current, thus preventing the other transistor to go into saturation under positive driving signal. Therefore, the series connection of the bipolar transistors is more effective because the driving current flows through both devices. However, the voltage across the reverse-biased base–emitter junction will be increased, which value must be controlled to prevent the device failure.

When the driver with a rectangular voltage waveform is used, the driving current flows through the forward-biased base–emitter junction of one transistor and the reverse-biased base–emitter junction of the other transistor when they are connected in series, as shown in Figure 11.5(*c*). If a shunt resistor R is connected in parallel to the transistor input, the effect of switching losses can be reduced because

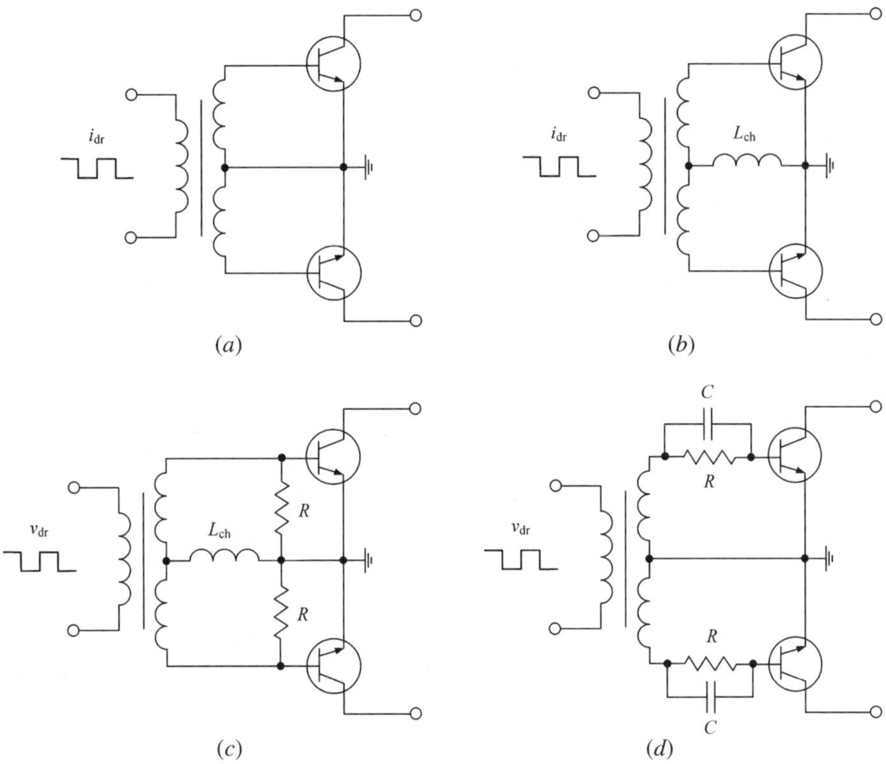

FIGURE 11.5 Input driving circuits.

the device input capacitance is faster discharged through this resistor. A similar correction effect can be achieved by using a parallel *RC* circuit connected in series to each transistor when both transistors are connected in parallel, as shown in Figure 11.5(*d*). However, an increased current is required from the driver to saturate the transistor.

Generally, the input circuit of the transistor, either bipolar or MOSFET, can be represented as a purely capacitance only at low frequencies. In switching applications, however, the finite rise and fall times are the result of the effect of the much higher frequency components than the fundamental. For example, if, at 30-MHz carrier, the switching time of 4 ns can be tolerated at amplitude of 80% that represents roughly a 100-MHz sine wave [7]. Besides, the input capacitance is usually nonlinear and varies over bias conditions. For example, the MOSFET gate–source capacitance varies with gate and drain voltages. At increased gate voltage, it goes down to its lowest value, just before reaching the threshold voltage, and then goes up to be constant at saturation. At the same time, when the MOSFET device begins to draw drain current, the drain voltage is lowered resulting in a reduction of the depletion area and causing an overlap between the gate and bulk material. This in turn increases the value of the gate–drain capacitance, which takes maximum value at zero drain voltage and positive gate voltage corresponding to the maximum device transconductance. In bipolar transistors, the base–emitter nonlinear capacitance represents the large diffusion capacitance in the active and saturation regions and much lower junction capacitance in the pinch-off region.

However, at higher frequencies, the device equivalent input circuit must include also a series resistance, which is the base ohmic resistance for bipolar transistor or effective gate resistance consisting of the distributed channel and gate electrode resistances for MOSFET device. In this case, the power losses occur due to the finite voltage rise time τ_r and fall time τ_f as a function of the

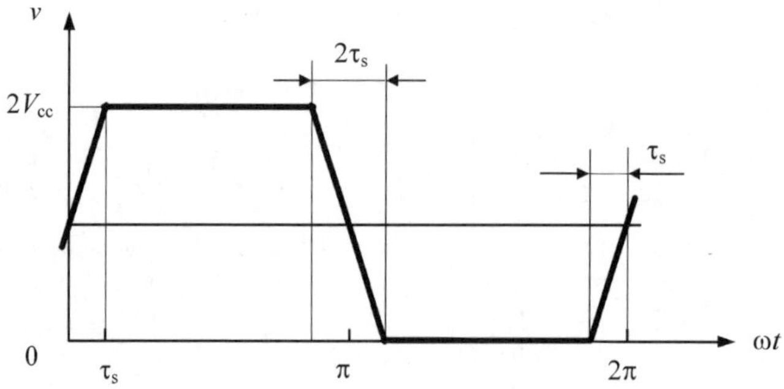

FIGURE 11.6 Transition time in Class D power amplifier.

device input time constant $\tau_{in} = R_{in}C_{in}$. An advantage of the sinusoidal driving signal compared with the rectangular drive is that there is no need to use a broadband input transformer, and all lead and leakage inductances in the gate drive circuit can be absorbed into the input resonant circuit. To minimize power losses for a sinusoidal input drive, the voltage amplitude of the driving signal must be increased since the voltage amplitude across the input capacitance decreases due to the increased voltage drop across the input resistance R_{in}.

Generally, an exact analysis that includes real transition waveforms and effects of elements of the complete device equivalent circuit is very complicated. This analysis of power losses can be substantially simplified by assuming that the resultant collector or drain voltage waveform of a complementary voltage-switching Class D power amplifier is trapezoidal when the transitions produce ramp voltage waveforms, as shown in Figure 11.6 [4]. Usually, under sinusoidal drive with increased voltage amplitude, the shapes of the rise time τ_r and fall time τ_f are close to ramp. The transition time required by a single transistor to complete the entire switching process is shown in Figure 11.6, as being converted to the angular time τ_s. This time can include an effect of the output capacitance and other device parasitics. Both transistors are then assumed to have zero saturation resistances, and switching is completed within $2\tau_s$.

The fundamental-frequency collector voltage amplitude V is obtained by a Fourier integral of the trapezoidal waveform taking into account the dc supply voltage of $2V_{cc}$ as

$$V = \frac{4}{\pi} V_{cc} \frac{\sin\tau_s}{\tau_s} \approx \frac{4}{\pi} V_{cc} \left(1 - \frac{\tau_s^2}{6}\right) \tag{11.22}$$

where the linear approximation is valid only for small values of τ_s.

For the fundamental-frequency output power $P = V^2/(2R_L)$ and dc current $I_0 = V/(\pi R_L)$, where R_L is the load resistance, the collector efficiency can be calculated from

$$\eta = \frac{\pi}{4} \frac{V}{V_{cc}} = \frac{\sin\tau_s}{\tau_s}. \tag{11.23}$$

The same results may be obtained also for the symmetrical current-switching and both voltage-switching and current-switching transformer-coupled Class D power amplifier configurations [8].

In addition, the presence of the parasitic collector capacitances C_{c1} and C_{c2} cause the additional power losses due to finite charge storage process. When the transistor Q_1 is turned on and the transistor Q_2 is turned off, the capacitor C_{c1} is discharged through the Q_1 and the capacitor C_{c2} is charged instantaneously to $2V_{cc}$. However, when the transistor Q_2 is turned on and the transistor Q_1 is turned off, then the capacitor C_{c2} is discharged instantaneously through Q_2 and the capacitor C_{c1} is charged instantaneously to $2V_{cc}$. Since this occurs twice during each period and the power losses due to energy dissipated in the both transistors with charging and discharging processes are equal, the total power losses due to parasitic collector capacitances for a complementary voltage-switching Class D power amplifier with supply voltage of $2V_{cc}$ and zero saturation resistance r_{sat} can be written as

$$P_s = (C_{c1} + C_{c2})(2V_{cc})^2 f_0 = 4C_c V_{cc}^2 f_0 \tag{11.24}$$

where $C_c = C_{c1} + C_{c2}$ and f_0 is the operating frequency. The total switching power losses described by Eq. (11.24) can also characterize the transformer-coupled voltage-switching Class D power amplifier with a supply voltage of V_{cc}, in which the parasitic collector capacitances C_{c1} and C_{c2} are charged and discharged between 0 and $2V_{cc}$.

In the current-switching Class D power amplifier, the parasitic capacitances do not provide charging and discharging losses since the collector currents flowing through the transistor have fixed constant values (either zero or I_{ds}). However, there is another mechanism of power losses associated with parasitic inductances L_{c1} and L_{c2} (due to finite lead length and leakage inductance in transformer) connected in series with the transistor collectors. In this case, the currents flowing through the active devices jump when switching occurs, since they must be changed from zero to maximum instantaneously, twice during each period. Therefore, the total power losses due to the parasitic collector series inductances for a current-switching Class D power amplifier with zero saturation resistance can be written as

$$P_s = L_c I_{dc}^2 f_0 \tag{11.25}$$

where $L_c = L_{c1} + L_{c2}$.

11.2 CLASS F

The possibility to improve efficiency by approximating the anode voltage waveform to square wave to minimize the value of the saturation voltage compared to the supply voltage over half an entire interval $0 \leq \omega t \leq 2\pi$ was discussed yet in the early 1920s. As a practical solution it was proposed to use a load network with a third-harmonic trap in series to the anode, as shown in Figure 11.7(*a*) [9,10]. However, the effect of the inclusion of a third-harmonic resonator was described and analyzed

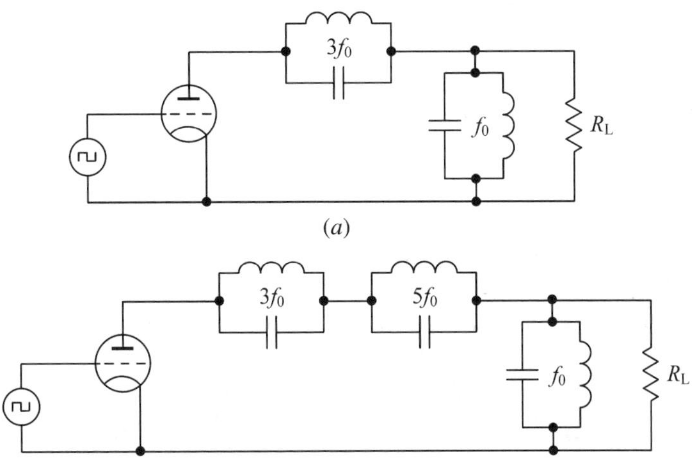

FIGURE 11.7 Polyharmonic power amplifiers.

in detail 1.5 decade later [11,12]. It was shown that the symmetrical anode voltage waveform and level of its depression can be provided with opposite phase conditions between the fundamental and third harmonic and optimum value of the ratio between their voltage amplitudes. In addition, it was noted that high operation efficiency can be achieved even when impedance of the third-harmonic resonator is equal or slightly greater than impedance of the fundamental tank. To maximize efficiency of the vacuum-tube amplifier with better approximating a square voltage anode waveform, it was also suggested to use an additional resonator tuned to the fifth harmonic, as shown in Figure 11.7(*b*) [13].

Figure 11.8 shows that the shapes of the voltage and current waveforms can be significantly changed with increasing fundamental voltage amplitude by adding even one additional harmonic component being properly phased. For example, the combination of the fundamental and third-harmonic components being out of phase at center point results in a flattened voltage waveform with depression in its center. It is clearly seen from Figure 11.8(*a*) that the proper ratio between the amplitudes of the fundamental and third-harmonic components can provide the flattened voltage waveform with minimum depression and maximum difference between its peak amplitude and amplitude of the fundamental component. Similarly, the combination of the fundamental-frequency and second-harmonic components being in-phase at the center point flattens the current waveform corresponding

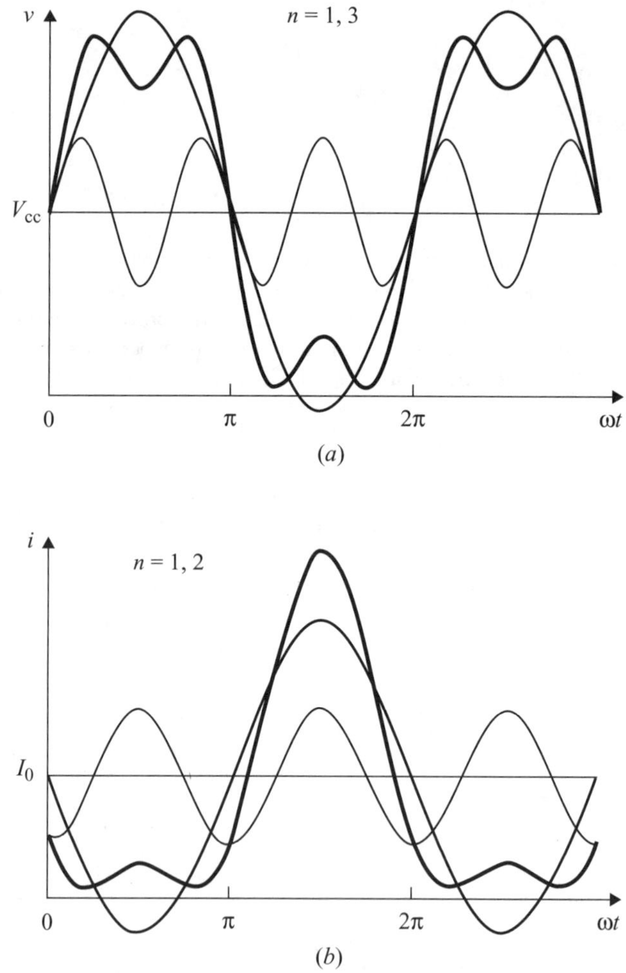

FIGURE 11.8 Fourier voltage and current waveforms with third and second harmonics.

to the maximum values of the voltage waveform and sharpens the current waveform corresponding to the minimum values of the voltage waveform, as shown in Figure 11.8(*b*). The optimum ratio between the amplitudes of the fundamental-frequency and second current harmonic components can maximize a peak value of the current waveform with its minimized value determined by the device saturation resistance in a practical circuit. Thus, power loss due to the active device can be minimized since the results of the integration over period when minimum voltage corresponds to maximum current will give small value compared with the power delivered to the load.

11.2.1 Idealized Class F Mode

Generally, an infinite number of odd-harmonic tank resonators can maintain a square collector voltage waveform, also providing a half-sinusoidal current waveform. Figure 11.9(*a*) shows such a Class F power amplifier with a multiple-resonator output filter to control the harmonic content of its collector (anode or drain) voltage and current waveforms, thereby shaping them to reduce dissipation and to increase efficiency [4,14].

To simplify an analysis of a Class F power amplifier, a simple equivalent circuit of which is shown in Figure 11.9(*b*), the following several assumptions are introduced:

- Transistor has zero saturation voltage, zero saturation resistance, infinite off-resistance, and its switching action is instantaneous and lossless.
- RF choke allows only a dc current and has no resistance.
- Quality factors of all parallel resonant circuits have infinite impedance at the corresponding harmonic and zero impedance at other harmonics.
- There are no losses in the circuit except only into the load R_L.
- Operation mode with a 50% duty cycle.

To determine the idealized collector voltage and current waveforms, let us consider the distribution of voltages and currents in the load network assuming the sinusoidal fundamental current flowing into the load as $i_R(\omega t) = I_R \sin(\omega t)$, where I_R is its amplitude. The voltage $v(\omega t)$ across the switch can be represented as a sum of the dc voltage V_{cc}, fundamental voltage $v_R = i_R R_L$ across the load resistor, and voltage v_{odd} across the odd-harmonic resonators,

$$v(\omega t) = V_{cc} + v_{odd}[(2n+1)\omega t] + v_R(\omega t). \tag{11.26}$$

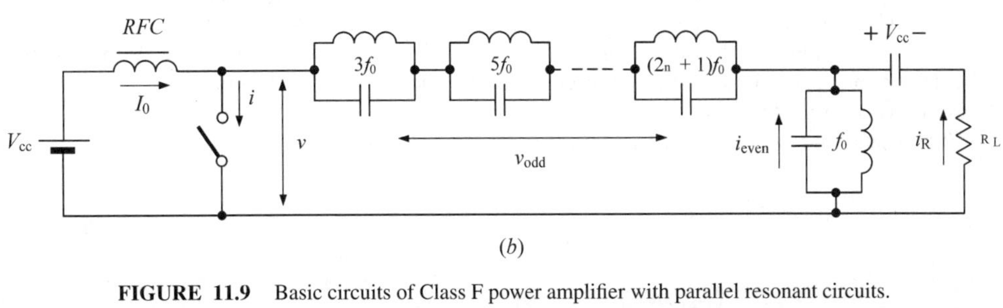

FIGURE 11.9 Basic circuits of Class F power amplifier with parallel resonant circuits.

Since the time moment t was chosen arbitrarily, by introducing a phase shift of π, Eq. (11.26) can be rewritten for periodical sinusoidal functions as

$$v(\omega t + \pi) = V_{cc} - v_{odd}[(2n+1)\omega t] - v_R(\omega t). \tag{11.27}$$

Then, the summation of Eqs. (11.26) and (11.27) yields

$$v(\omega t) = 2V_{cc} - v(\omega t + \pi). \tag{11.28}$$

From Eq. (11.28), it follows that maximum value of the collector voltage cannot exceed a value of $2V_{cc}$ and the time duration with maximum voltage of $v = 2V_{cc}$ coincides with the time duration with minimum voltage of $v = 0$. Since the collector voltage is zero when the switch is closed, the only possible waveform for the collector voltage is a square wave composing of only dc, fundamental-frequency, and odd-harmonic components.

During the interval $0 < \omega t \leq \pi$ when the switch is closed, the current $i(\omega t)$ flowing through the switch can be written as

$$i(\omega t) = I_0 + i_{even}(2n\omega t) + i_R(\omega t) \tag{11.29}$$

whereas during the interval $\pi < \omega t \leq 2\pi$ when the switch is open, the current $i(\omega t + \pi)$ is equal to zero resulting in

$$0 = I_0 + i_{even}(2n\omega t) - i_R(\omega t). \tag{11.30}$$

Then, by substituting Eq. (11.30) into Eq. (11.29), we can rewrite Eq. (11.29) as

$$i(\omega t) = 2i_R(\omega t) = 2I_R \sin(\omega t) \tag{11.31}$$

from which it follows that the amplitude of the current flowing through the switch during interval $0 < \omega t \leq \pi$ is two times greater than the amplitude of the fundamental current. Thus, in a general case of entire interval, Eq. (11.29) can be rewritten as

$$i(\omega t) = I_R(\sin \omega t + |\sin \omega t|) \tag{11.32}$$

which means that the switch current represents half-sinusoidal pulses with the amplitude equal to double load current amplitude.

Consequently, for a purely sinusoidal current flowing into the load as shown in Figure 11.10(*a*), the ideal collector voltage and current waveforms can be represented by the appropriate normalized waveforms shown in Figures 11.10(*b*) and 11.10(*c*), respectively. Here, a sum of the fundamental and odd harmonics approximates a square voltage waveform, and a sum of the fundamental and even harmonics approximates a half-sinusoidal collector current waveform. As a result, the shapes of the collector current and voltage waveforms provide a condition when the current and voltage do not overlap simultaneously. Such a condition, with symmetrical collector voltage and current waveforms, corresponds to an idealized Class F operation mode with 100% collector efficiency.

A Fourier analysis of the current and voltage waveforms allows us to obtain the equations for the dc current, fundamental voltage and current components in the collector voltage and current waveforms, as explained in the following.

The dc current I_0 can be calculated from Eq. (11.32) as

$$I_0 = \frac{1}{2\pi} \int_0^{\pi} 2I_R \sin \omega t \, d\omega t = \frac{2I_R}{\pi} \tag{11.33}$$

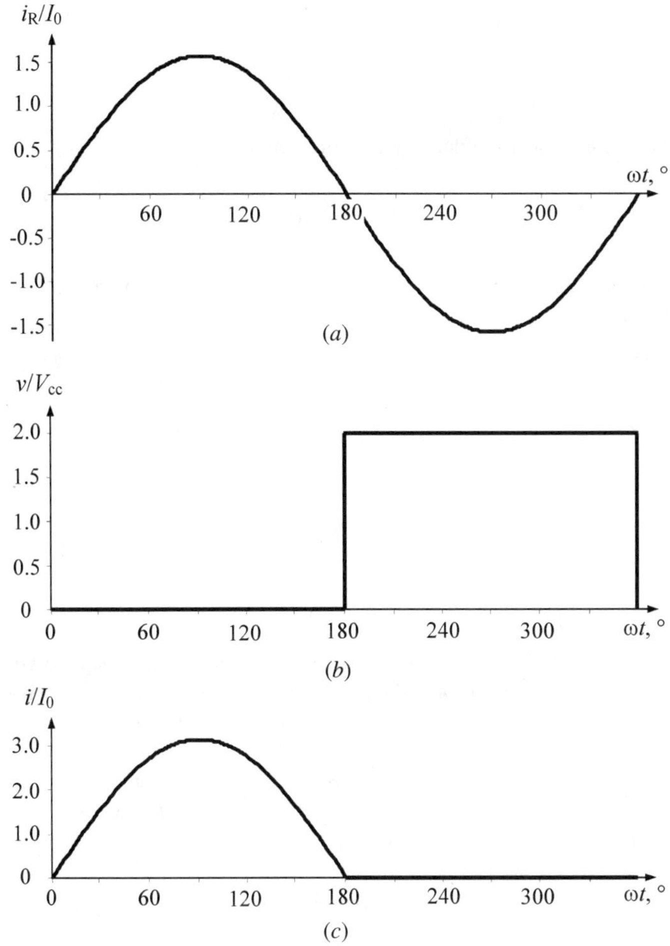

FIGURE 11.10 Ideal waveforms of Class F power amplifier.

the fundamental current component can be calculated from Eq. (11.32) as

$$I_1 = \frac{1}{\pi} \int_0^{\pi} 2I_R \sin^2 \omega t \, d\omega t = I_R \tag{11.34}$$

the fundamental voltage component can be calculated using Eq. (11.28) as

$$V_1 = V_R = \frac{1}{\pi} \int_{\pi}^{2\pi} 2V_{cc} \sin(\omega t + \pi) \, d\omega t = \frac{4V_{cc}}{\pi} \tag{11.35}$$

where $V_R = I_R R_L$ is the fundamental voltage amplitude across the load resistor R_L.

Then, the dc power and output power at the fundamental frequency are calculated from

$$P_0 = V_{cc} I_0 = \frac{2V_{cc} I_R}{\pi} \tag{11.36}$$

and

$$P_1 = \frac{V_1 I_1}{2} = \frac{2V_{cc} I_R}{\pi} \tag{11.37}$$

resulting in a theoretical collector efficiency with maximum value of

$$\eta = \frac{P_1}{P_0} = 100\%. \tag{11.38}$$

In this case, the impedance conditions seen by the device collector for an idealized Class F mode must be equal to

$$Z_1 = R_1 = \frac{8}{\pi^2} \frac{V_{cc}}{I_0} \tag{11.39}$$

$$Z_{2n} = 0 \quad \text{for even harmonics} \tag{11.40}$$

$$Z_{2n+1} = \infty \quad \text{for odd harmonics} \tag{11.41}$$

which are similar to that of derived from the limiting case of the optimum efficiency Class B mode [15].

11.2.2 Class F with Quarterwave Transmission Line

Ideally, a control of an infinite number of the harmonics maintaining a square voltage waveform and a half-sinusoidal current waveform at the drain can be provided by using a serious quarterwave transmission line and a parallel-tuned resonant circuit, as shown in Figure 11.11. Such type of a Class F power amplifier was initially proposed to use at very high frequencies where implementation of the load networks with lumped elements only is difficult and the parasitic device output (lead or package) inductor is sufficiently small [4]. In this case, the quarterwave transmission line transforms the load impedance according to

$$R = \frac{Z_0^2}{R_L} \tag{11.42}$$

where Z_0 is the characteristic impedance of a transmission line [16]. For even harmonics, the short circuit on the load side of the transmission line is repeated, thus producing a short circuit at the drain. However, the short circuit at the load produces an open circuit at the drain for odd harmonics with resistive load at the fundamental.

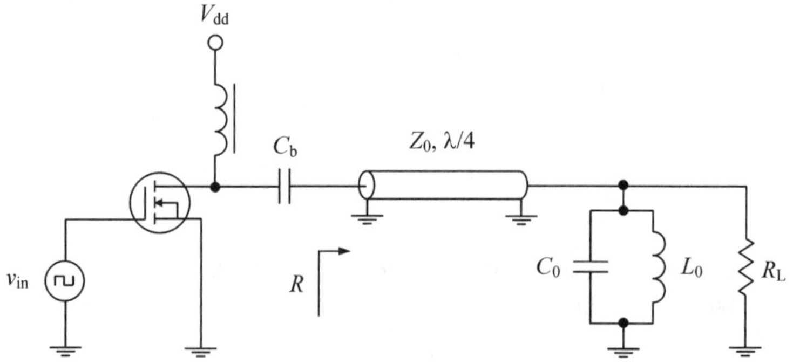

FIGURE 11.11 Class F power amplifier with series quarterwave transmission line.

Generally, at low drive level, the active device acts as a current source (voltage-controlled in the case of the MOSFETs or MESFETs and current-controlled in the case of bipolar transistors). As drive input increases, the active device enters voltage saturation resulting in a harmonic-generation process. Since the transmission line presents the high impedance conditions to all odd harmonics, all odd harmonics provide a proper contribution to the output voltage waveform. As a result, at high drive level, the output voltage waveform becomes a complete square wave and the active device is voltage-saturated for a full half-cycle. In this case, the transistor acts as a switch rather than a saturating current source.

An alternative configuration of the Class F power amplifier with a shunt transmission line located in between the dc power supply and device collector is shown in Figure 11.12(*a*). In this case, there is no need to use an RF choke and a series blocking capacitor because a series fundamentally tuned resonant L_0C_0 circuit is used instead of a parallel fundamentally tuned resonant circuit. However, unlike the case with a series quarterwave transmission line, such a Class F load network configuration with a shunt quarterwave transmission line does not provide an impedance transformation. Therefore, the load resistance R, which is equal to the equivalent active device output resistance at the fundamental frequency, must then be transformed to the standard load resistance R_L. Let us now derive analytically some basic fundamental properties of a quarterwave transmission line. The transmission line in time domain can be represented as an element with finite delay time depending on its electrical length. Consider a simple load network of the Class F power amplifier shown in Figure 11.12(*b*) consisting of a parallel quarterwave transmission line grounded at the end through power supply, a series

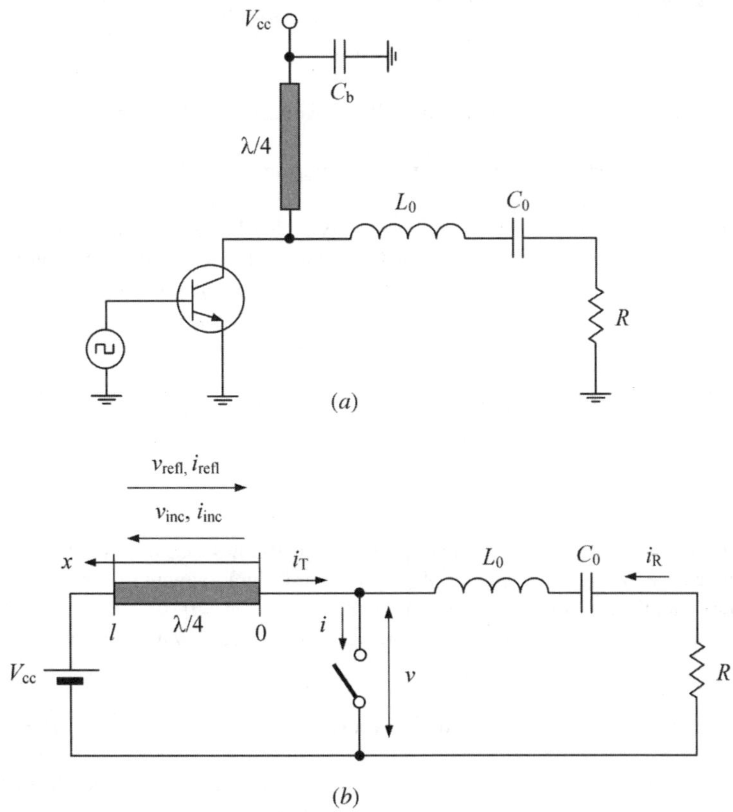

FIGURE 11.12 Class F power amplifier with shunt quarterwave transmission line.

fundamentally tuned L_0C_0 circuit, and a load resistance R. In an idealized case, the intrinsic device output capacitance is assumed to be negligible to affect the power amplifier RF performance. The loaded quality factor Q_L of the series resonant L_0C_0 circuit is high enough to provide the sinusoidal output current i_R flowing into the load R.

To define the collector voltage and current waveforms, consider the electrical behavior of a homogeneous lossless quarterwave transmission line connected to the dc voltage supply with RF grounding [3,17]. In this case, the voltage $v(t, x)$ in any cross section of such a transmission line can be represented as a sum of the incident and reflected voltages $v_{inc}(\omega t - 2\pi x/\lambda)$ and $v_{refl}(\omega t + 2\pi x/\lambda)$, generally with an arbitrary waveform. When $x = 0$, the voltage $v(t, x)$ is equal to the collector voltage

$$v(\omega t) = v(t, 0) = v_{inc}(\omega t) + v_{refl}(\omega t). \tag{11.43}$$

At the same time, at another end of the transmission line when $x = \lambda/4$, the voltage is constant and equal to

$$V_{cc} = v(t, \pi/2) = v_{inc}(\omega t - \pi/2) + v_{refl}(\omega t + \pi/2). \tag{11.44}$$

Since the time moment t was chosen arbitrarily, let us rewrite Eq. (11.44) using a phase shift of $\pi/2$ for each voltage as

$$v_{inc}(\omega t) = V_{cc} - v_{refl}(\omega t + \pi). \tag{11.45}$$

Substituting Eq. (11.45) into Eq. (11.43) yields

$$v(\omega t) = v_{refl}(\omega t) - v_{refl}(\omega t + \pi) + V_{cc}. \tag{11.46}$$

Consequently, for the phase shift of π, the collector voltage can be obtained by

$$v(\omega t + \pi) = v_{refl}(\omega t + \pi) - v_{refl}(\omega t + 2\pi) + V_{cc}. \tag{11.47}$$

For an idealized operation condition with a 50% duty cycle when during half a period the transistor is turned on and during another half a period the transistor is turned off with overall period of 2π, the voltage $v_{refl}(\omega t)$ can be considered the periodical function with a period of 2π,

$$v_{refl}(\omega t) = v_{refl}(\omega t + 2\pi). \tag{11.48}$$

As a result, the summation of Eqs. (11.46) and (11.47) results in the basic expression for collector voltage in the form of

$$v(\omega t) = 2V_{cc} - v(\omega t + \pi). \tag{11.49}$$

From Eq. (11.49), which is similar to Eq. (11.28), it follows that the maximum value of the collector voltage cannot exceed a value of $2V_{cc}$ and the time duration with maximum voltage of $v = 2V_{cc}$ coincides with the time duration with minimum voltage of $v = 0$.

Similarly, an equation for the current i_T flowing into the quarterwave transmission line can be obtained as

$$i_T(\omega t) = i_T(\omega t + \pi) \tag{11.50}$$

which means that the period of a signal flowing into the quarterwave transmission line is equal to π because it contains only even harmonics, since a shorted quarterwave transmission line has an infinite impedance at odd harmonics at its input.

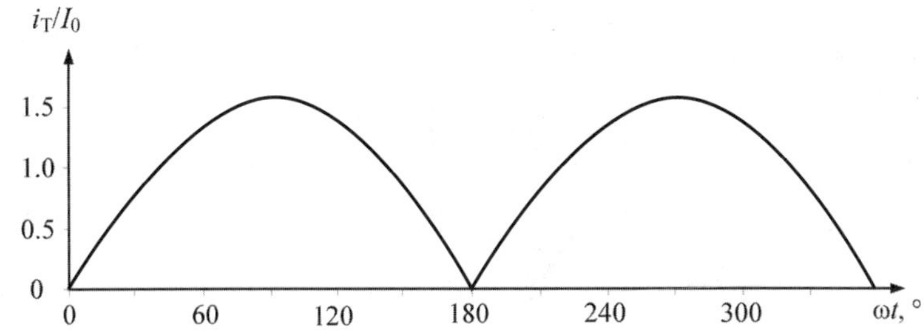

FIGURE 11.13 Ideal current waveform in quarterwave transmission line.

Let the transistor operates as an ideal switch: it is closed during the interval $0 < \omega t \leq \pi$ where $v = 0$ and open during the interval $\pi < \omega t \leq 2\pi$, where $v = 2V_{cc}$ according to Eq. (11.49). During the interval $\pi < \omega t \leq 2\pi$ when switch is open, the load is directly connected to the transmission line and $i_T = -i_R = -I_R \sin\omega t$. Consequently, during the interval $0 < \omega t \leq \pi$ when switch is closed, $i_T = I_R \sin\omega t$ according to Eq. (11.50). Hence, the current flowing into the quarterwave transmission line at any ωt can be represented by

$$i_T(\omega t) = I_R|\sin \omega t| \tag{11.51}$$

where I_R is the amplitude of current flowing into the load.

Since the collector current is defined as $i = i_T + i_R$, then

$$i(\omega t) = I_R(\sin \omega t + |\sin \omega t|) \tag{11.52}$$

which means that the collector current represents half-sinusoidal pulses with the amplitude equal to double load current amplitude.

Consequently, for a purely sinusoidal current flowing into the load due to infinite loaded quality factor of the series fundamentally tuned L_0C_0 circuit shown in Figure 11.10(*a*), the ideal collector voltage and current waveforms can be represented by the corresponding normalized square and half-sinusoidal waveforms shown in Figures 11.10(*b*) and 11.10(*c*), respectively, where I_0 is the dc current. Here, a sum of the fundamental and odd harmonics approximates a square voltage waveform and a sum of the fundamental and even harmonics approximates a half-sinusoidal collector current waveform. The waveform corresponding to the normalized current flowing into the quarterwave transmission line, which is shown in Figure 11.13, represents a sum of even harmonics only. As a result, the shapes of the collector current and voltage waveforms provide a condition when the current and voltage do not overlap simultaneously.

11.2.3 Effect of Saturation Resistance

In a real transistor, the saturation or on-resistance r_{sat} is not equal to zero, and transistor dissipates some amount of power due to the collector current flowing through this resistance when the transistor is turned on. The simplified equivalent circuit of a Class F power amplifier with a quarterwave transmission line where the transistor is represented by a nonideal switch with saturation resistance r_{sat} and parasitic output capacitance C_{out} is shown in Figure 11.14. During the interval $0 < \omega t \leq \pi$ when the switch is closed, the saturation voltage v_{sat} due to the current $i(\omega t)$ flowing through the switch can be written as

$$v_{sat}(\omega t) = V_{sat} \sin \omega t = 2I_R r_{sat} \sin \omega t \tag{11.53}$$

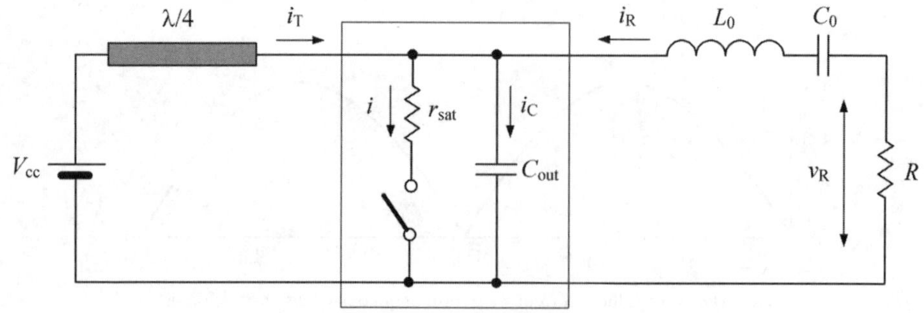

FIGURE 11.14 Effect of parasitic on-resistance and shunt capacitance.

where, by using Eq. (11.35), the saturation voltage amplitude V_{sat} can be obtained by

$$V_{sat} = 2V_R \frac{r_{sat}}{R} = \frac{8V_{cc}}{\pi} \frac{r_{sat}}{R}. \tag{11.54}$$

The collector current and voltage waveforms are shown in Figure 11.15, where the half-sinusoidal current flowing through the saturation resistance r_{sat} causes the deviation of the voltage waveform from the ideal square waveform. In this case, the bottom part of the voltage waveform becomes sinusoidal with the amplitude V_{sat} during the interval $0 < \omega t \leq \pi$. From Eq. (11.49), it follows that the same sinusoidal behavior will correspond to the top part of the voltage waveform during the interval $\pi < \omega t \leq 2\pi$.

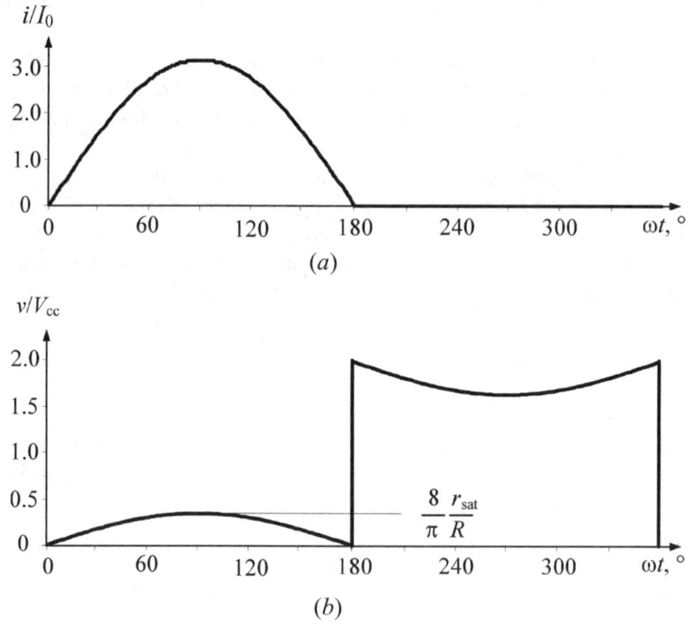

FIGURE 11.15 Idealized collector current and voltage waveforms with nonzero on-resistance.

Now let us evaluate the power losses and collector efficiency due to presence of the saturation resistance r_{sat}. Using Eqs. (11.31), (11.33), and (11.35) results in

$$\frac{P_{sat}}{P_0} = \frac{1}{2\pi}\int_0^{2\pi} \frac{i^2(\omega t)\, r_{sat}}{I_0 V_{cc}} d\omega t = \frac{r_{sat}}{2\pi I_0 V_{cc}}\int_0^{2\pi} (2I_R)^2 \sin^2 \omega t\, d\omega t = \frac{r_{sat} I_R}{V_{cc}}\frac{I_R}{I_0} = \frac{r_{sat}}{R}\frac{I_R}{I_0}\frac{V_R}{V_{cc}} = \frac{2r_{sat}}{R}. \tag{11.55}$$

Hence, the collector efficiency can be calculated from

$$\eta = 1 - \frac{P_{sat}}{P_0} = 1 - \frac{2r_{sat}}{R}. \tag{11.56}$$

In practice, the idealized collector voltage and current waveforms can be realized at low frequencies when effect of the device output capacitance is negligible. At higher frequencies, the effect of the output capacitance contributes to a nonzero switching time resulting in the time periods when the collector voltage and collector current exist at the same time when simultaneously $v > 0$ and $i > 0$. Consequently, such a load network with shunt capacitance cannot provide the switched-mode operation with an instantaneous transition from the device pinch-off to saturation mode. Therefore, during a nonzero time interval, the device operates in active region as a nonlinear current source.

11.2.4 Load Networks with Lumped and Distributed Parameters

Theoretical results show that the proper control of the second and third harmonics only can significantly increase the collector efficiency of the power amplifier by flattening the output voltage waveform. Since practical realization of a multielement high-order *LC* resonant circuit can cause a serious implementation problem, especially at higher frequencies, it is sufficient to be confined to a three- or four-element resonant circuit composing the load network of the power amplifier. In addition, it is necessary to take into account that, in practice, both extrinsic and intrinsic transistor output capacitance has a substantial effect on the efficiency. The output capacitance C_{out} can represent the collector capacitance C_c in the case of the bipolar transistor or drain–source capacitance plus gate–drain capacitance $C_{ds} + C_{gd}$ in the case of the FET device.

For a lumped-circuit amplifier, a special three-element load network can be used to approximate to ideal Class F by providing both high impedance at the fundamental and third harmonics and zero impedance at the second harmonic at the collector (or drain) by compensating for the influence of C_{out}. Examples of such load networks with additional parallel and series resonant circuits located between the dc power supply and device output are shown in Figure 11.16 [18,19]. Here, the output circuit of the active device is represented by a multiharmonic current source, and R_{out} is the equivalent output resistance at the fundamental frequency defined as a ratio of the fundamental voltage at the device output to the fundamental current flowing into the device.

The reactive part of the output admittance or susceptance $B_{net} = \mathrm{Im}(Y_{net})$ of the load network with parallel resonant tank shown in Figure 11.16(*b*) including the device output capacitance C_{out} can be written by

$$B_{net} = \omega C_{out} - \frac{1 - \omega^2 L_2 C_2}{\omega L_1 \left(1 - \omega^2 L_2 C_2\right) + \omega L_2}. \tag{11.57}$$

By applying three-harmonic impedance conditions at the device collector (or drain), open-circuited for fundamental-frequency and third harmonic when $B_{net}(\omega_0) = B_{net}(3\omega_0) = 0$ and short-circuited for

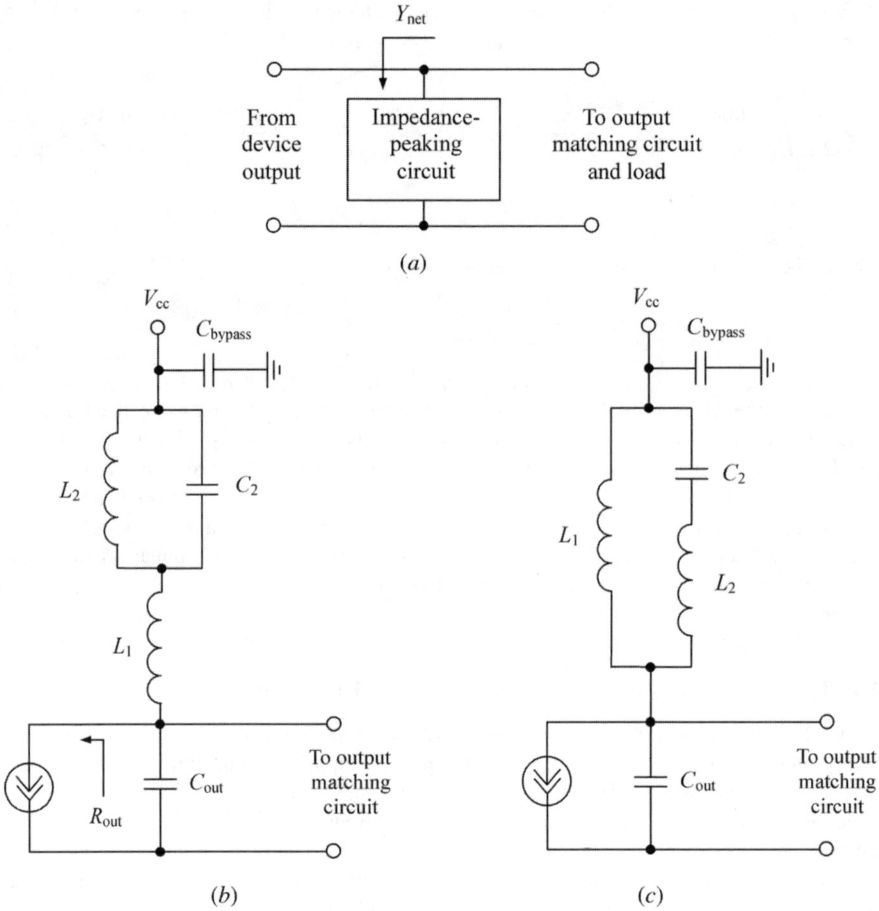

FIGURE 11.16 Load networks with parallel and series resonant circuits.

second harmonic when $B_{net}(2\omega_0) = \infty$, the values of elements of this impedance-peaking circuit can be derived as

$$L_1 = \frac{1}{6\omega_0^2 C_{out}} \quad L_2 = \frac{5}{3}L_1 \quad C_2 = \frac{12}{5}C_{out} \tag{11.58}$$

where the sum of the reactance of the parallel resonant tank, consisting of an inductor L_2 and a capacitor C_2, and inductor L_1 creates resonances at the fundamental and third-harmonic components, and the series capacitance of the tank circuit in series with L_1 creates a low-impedance series resonance at the second-harmonic component [18].

Applying the same conditions for the load network with series resonant circuit shown in Figure 11.16(*c*) results in the ratios between elements given by

$$L_1 = \frac{4}{9\omega_0^2 C_{out}} \quad L_2 = \frac{9}{15}L_1 \quad C_2 = \frac{15}{16}C_{out} \tag{11.59}$$

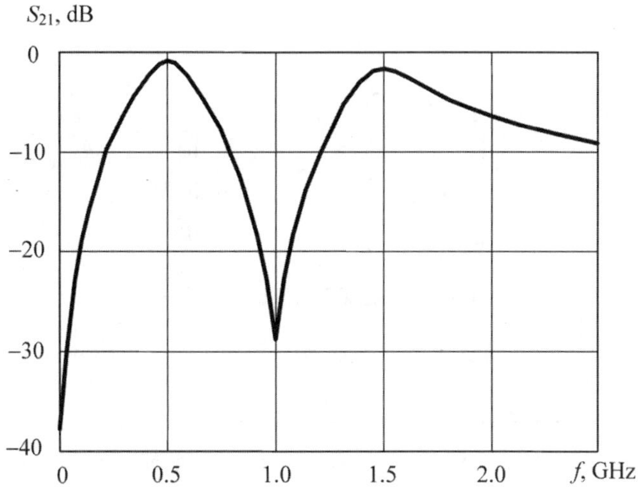

FIGURE 11.17 Frequency response of load network with parallel resonant circuit.

where an inductance L_2 and a capacitance C_2 create a short-circuit condition at the second harmonic, and all elements create the parallel resonant tanks for fundamental-frequency and third-harmonic components [19].

To determine the amplitude characteristics of the load network in frequency domain, it is best to represent the load network as shown in Figure 11.16(a) and simulate the small-signal S-parameters. Then, it is just necessary to plot a magnitude of S_{21} in decibels. As an example, the frequency-response characteristic of the load network with a parallel resonant circuit, the parameters of which are calculated based on the fundamental frequency $f_0 = 500$ MHz, is shown in Figure 11.17. The circuit parameters are $C_{\text{out}} = 2.2$ pF, $R_{\text{out}} = 200\ \Omega$, $C_2 = 5.3$ pF, $L_1 = 7.7$ nH, and $L_2 = 12.8$ nH with inductor quality factor $Q_{\text{ind}} = 20$. In this case, the power-amplifier efficiency will be effectively increased if the first element of the output matching circuit adjacent to the transistor output is series and inductive to provide high impedance at harmonics.

As a first approximation for comparison between different operation modes, the output device resistance R_{out} at the fundamental frequency required to realize a Class F operation mode with third-harmonic peaking can be estimated as the equivalent resistance $R_{\text{out}} = R_1^{(\text{F})} = V_1/I_1$ determined at the fundamental frequency for an ideal Class F operation, where V_1 and I_1 are the fundamental voltage and current at the device output. Assuming zero saturation voltage and using Eq. (11.35) yield

$$R_1^{(\text{F})} = \frac{4}{\pi}\frac{V_{\text{cc}}}{I_1} = \frac{4}{\pi}R_1^{(\text{B})} \tag{11.60}$$

where $R_1^{(\text{B})} = V_{\text{cc}}/I_1$ is the fundamental output resistance in an ideal Class B.

The ideal Class F power amplifier with all even-harmonic short-circuit termination and third-harmonic peaking achieves a maximum drain efficiency of 88.4% [20]. Such an operation mode is easy to realize by using the transmission lines in the load network circuit. The load network impedance-peaking circuit topology of such a transmission-line power amplifier is shown in Figure 11.18 [3,18].

In this case, a quarterwave transmission line TL_1 located between the dc power supply and drain terminal provides short-circuit termination for even voltage harmonics. The electrical length θ_3 of an open-circuit stub TL_3 is chosen to have a quarter wavelength at the third-harmonic component to realize short-circuit condition at the end of the series transmission line TL_2, whose electrical length θ_2 should provide an inductive reactance to resonate with the device output capacitance C_{out} at the

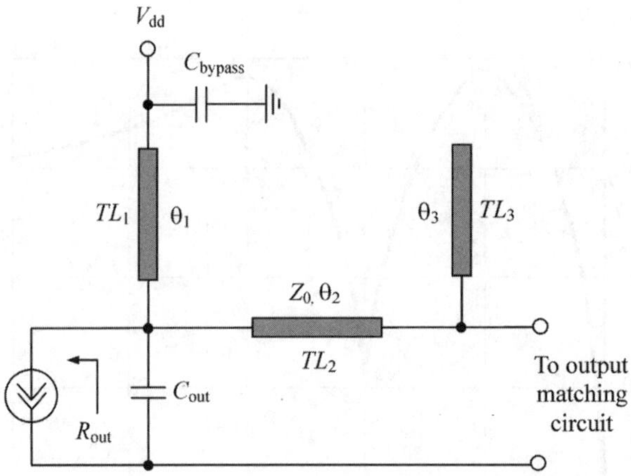

FIGURE 11.18 Transmission-line impedance-peaking circuit.

third voltage harmonic. As a result, the electrical lengths of the transmission lines on fundamental frequency can be obtained from

$$\theta_1 = \frac{\pi}{2} \quad \theta_2 = \frac{1}{3}\tan^{-1}\left(\frac{1}{3Z_0\omega_0 C_{out}}\right) \quad \theta_3 = \frac{\pi}{6} \tag{11.61}$$

where Z_0 is the characteristic impedance of the series transmission line TL_2 and ω_0 is the radial operating frequency. Figure 11.19 shows an example of the frequency-response characteristic of the microstrip impedance-peaking circuit using alumina substrate for the device output resistance $R_{out} = 50\ \Omega$ and output capacitance $C_{out} = 2.2$ pF, characteristic impedance of microstrip lines $Z_0 = 50\ \Omega$ and electrical length $\theta_2 = 15°$. From Figure 11.19, it follows that, for short-circuited conditions for all even harmonics and third-harmonic peaking, an additional output impedance matching at the

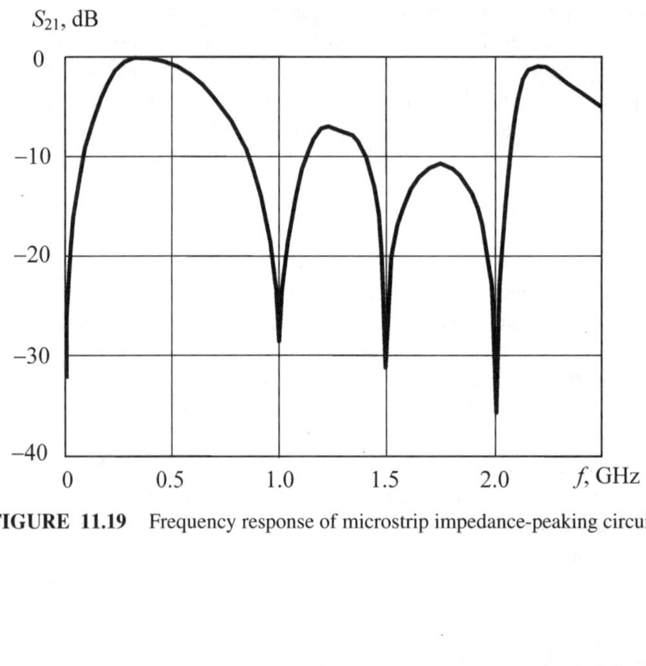

FIGURE 11.19 Frequency response of microstrip impedance-peaking circuit.

operating fundamental frequency $f_0 = 500$ MHz is required to compensate for the reactive part and to match the real part of the realized output impedance with the standard load impedance of 50 Ω.

11.3 INVERSE CLASS F

The effect of the inclusion of the second-harmonic resonator in series to the anode, as shown in Figure 11.20(*a*), was described and analyzed yet in the early 1940s [12,21]. It was shown that the symmetrical anode current waveform and level of its depression can be provided with the opposite phase conditions between the fundamental-frequency and second-harmonic components and an optimum value of the ratio between their voltage amplitudes. It was noted that high operation efficiency could be achieved even when impedance of the parallel circuit to second harmonic was equal or slightly greater than impedance of the tank circuit to fundamental frequency. In practical vacuum-tube power amplifiers intended for operation at very high frequencies, the peak output power and anode efficiency can be increased by 1.15–1.2 times [22]. In addition, it was later suggested to use an additional resonator, tuned to the fourth harmonic and connected in series with the second-harmonic resonator, as shown in Figure 11.20(*b*), to maximize efficiency of the vacuum-tube amplifier with approximate square voltage driving waveform [23].

The simple solution to realize out-of-phase conditions between the voltage fundamental-frequency and second-harmonic components at the device output is to use a second-harmonic parallel resonator connected in series to the device input [3,20]. Such an approach gives a possibility to flatten the anode voltage waveform in active region avoiding the device saturation mode. In this case, the driver stage is loaded by the nonlinear diode-type input grid impedance of the final-stage device providing a flattened grid voltage waveform, which includes the fundamental-frequency and second-harmonic components. The presence of the strong second-harmonic component results in a second-harmonic voltage drop across the resonator. The loaded quality factor of the second-harmonic resonator must be high enough to neglect the voltage drop at the fundamental frequency. As a result, the second-harmonic resonator has no effect on the voltage fundamental-frequency component, however it provides a phase shift of 180° for the second-harmonic component, since increasing of the voltage drop across

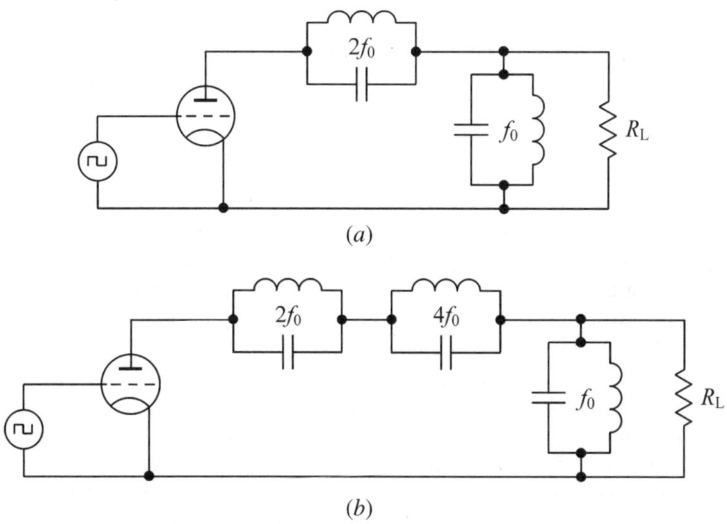

FIGURE 11.20 Polyharmonic power amplifiers.

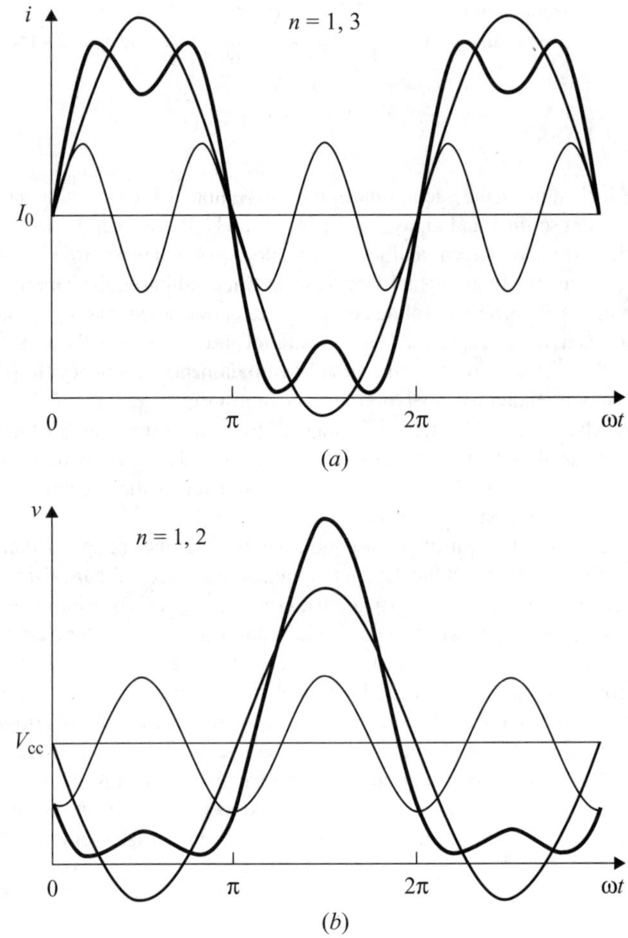

FIGURE 11.21 Fourier current and voltage waveforms with third and second harmonics.

the resonator results in decreasing of the voltage drop across the grid-cathode (base–emitter or gate–source) terminals.

Figure 11.21 shows that the shapes of the voltage and current waveforms can be significantly transformed with increased voltage peak factor by adding even one additional harmonic component being properly phased. For example, the combination of the fundamental-frequency and third-harmonic components being out-of-phase results in a flattened current waveform with depression in its center shown in Figure 11.21(*a*), which can be minimized by using the proper ratio between the amplitudes of the fundamental and third harmonics. Similarly, the combination of the fundamental and second harmonics being in phase sharpens the voltage waveform corresponding to minimum values of the voltage waveform, as shown in Figure 11.21(*b*). The optimum ratio between the amplitudes of the fundamental and second current harmonics can maximize the current waveform in one-half of period and minimize the current waveform during the other half of period determined by the device saturation resistance in a practical circuit. This means that the power loss due to the active device can be minimized since the results of the integration over period when minimum current corresponds to maximum voltage will give small value compared with the power delivered to the load.

11.3.1 Idealized Inverse Class F Mode

Generally, an infinite number of even-harmonic tank resonators can maintain a square current waveform, also providing a half-sinusoidal voltage waveform at the collector. Figure 11.22(*a*) shows such an inverse Class F power amplifier with a multiple-resonator output filter to control the harmonic content of its collector (anode or drain) voltage and current waveforms, thereby shaping them to reduce dissipation and to increase efficiency.

The term "inverse" means that collector voltage and current waveforms are interchanged compared to a conventional case under the same idealized assumptions. Consequently, for a purely sinusoidal current flowing into the load shown in Figure 11.23(*a*), the ideal collector current waveform is composed by a fundamental component and odd harmonics approximating a square waveform, as shown in Figure 11.23(*b*). The collector voltage waveform is composed by the fundamental component and even harmonics approximating a half-sinusoidal waveform, as shown in Figure 11.23(*c*). As a result, the shapes of the collector current and voltage waveforms provide a condition when the current and voltage do not overlap simultaneously similar to a conventional Class F mode. Such a condition, with symmetrical collector voltage and current waveforms, corresponds to an idealized inverse Class F operation mode with 100% collector efficiency.

Similar analysis of the distribution of voltages and currents in the inverse Class F load network as for a conventional Class F mode results in equations for the collector current and voltage waveforms in the inverse form of

$$i(\omega t) = 2I_0 - i(\omega t + \pi) \tag{11.62}$$

where I_0 is the dc current, and

$$v(\omega t) = V_R(\sin \omega t + |\sin \omega t|) \tag{11.63}$$

where V_R is the load fundamental-frequency amplitude. From Eq. (11.62), it follows that maximum value of the collector current cannot exceed a value of $2I_0$, and the time duration with maximum amplitude of $i = 2I_0$ coincides with the time duration with minimum amplitude of $i = 0$. Since the collector current is zero when the switch is open, the only possible waveform for the collector current is a square wave composed of only dc, fundamental-frequency, and odd-harmonic components.

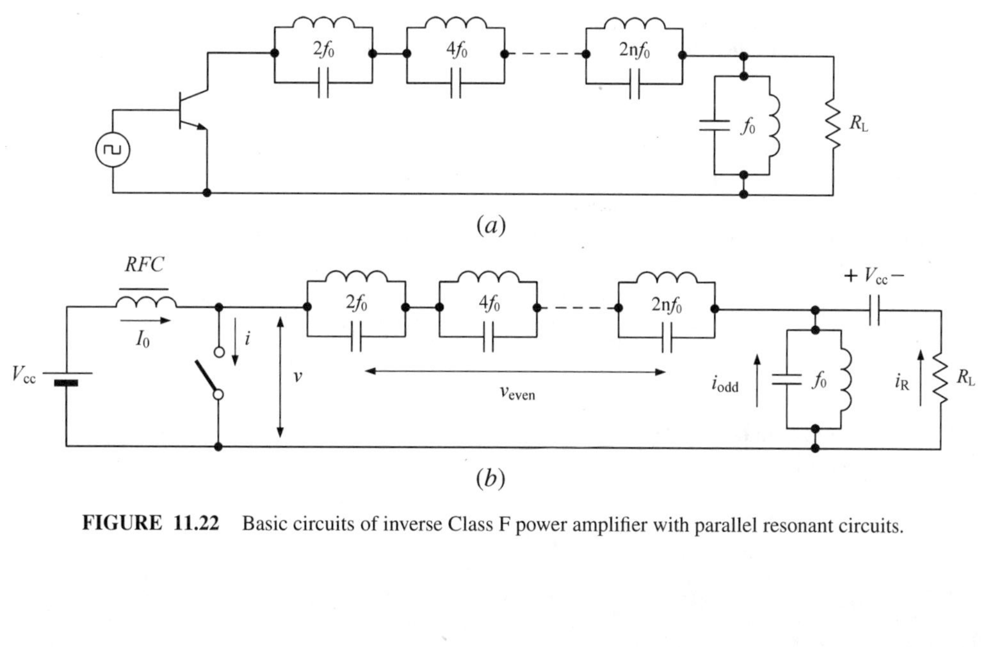

FIGURE 11.22 Basic circuits of inverse Class F power amplifier with parallel resonant circuits.

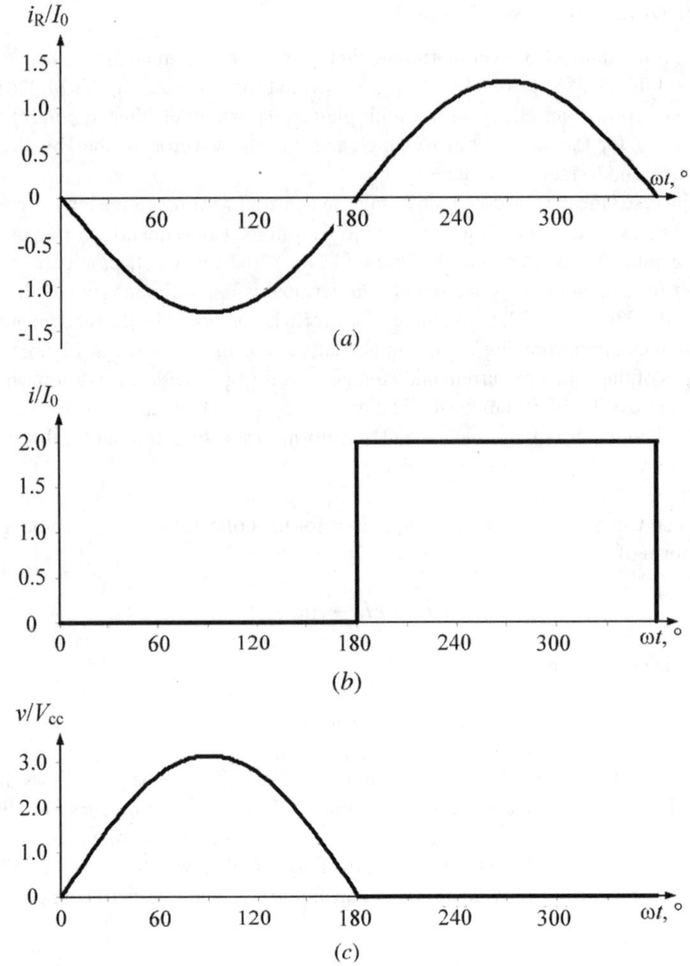

FIGURE 11.23 Ideal waveforms of inverse Class F power amplifier.

By using a Fourier analysis of the current and voltage waveforms, the following equations for the dc voltage, fundamental voltage, and current components in the collector voltage and current waveforms can be obtained:

The fundamental current component can be calculated using Eq. (11.62) as

$$I_1 = I_R = \frac{1}{\pi}\int_{\pi}^{2\pi} 2I_0 \sin(\omega t + \pi)\, d\omega t = \frac{4I_0}{\pi} \tag{11.64}$$

the dc voltage V_{cc} can be calculated from Eq. (11.63) as

$$V_{cc} = \frac{1}{2\pi}\int_{0}^{\pi} 2V_R \sin\omega t\, d\omega t = \frac{2V_R}{\pi} \tag{11.65}$$

the fundamental voltage component can be calculated from Eq. (11.63) as

$$V_1 = \frac{1}{\pi}\int_0^{\pi} 2V_R \sin^2 \omega t \, d\omega t = V_R. \tag{11.66}$$

Then, the ratio between the dc and output power at the fundamental frequency, P_0 and P_1, can be given by

$$P_1 = \frac{V_1 I_1}{2} = \frac{1}{2}\frac{\pi V_{cc}}{2}\frac{4I_0}{\pi} = P_0 \tag{11.67}$$

resulting in a theoretical collector efficiency with maximum value of 100%.

The impedance conditions seen by the device collector for an idealized inverse Class F mode must be equal to

$$Z_1 = R_1 = \frac{\pi^2}{8}\frac{V_{cc}}{I_0} \tag{11.68}$$

$$Z_{2n+1} = 0 \quad \text{for odd harmonics} \tag{11.69}$$

$$Z_{2n} = \infty \quad \text{for even harmonics.} \tag{11.70}$$

11.3.2 Inverse Class F with Quarterwave Transmission Line

An idealized inverse Class F operation mode can also be represented by using a sequence of the series resonant circuits tuned to the fundamental and odd harmonics, as shown in Figure 11.24(*a*). In this case, it is assumed that each resonant circuit has zero impedance at the corresponding fundamental

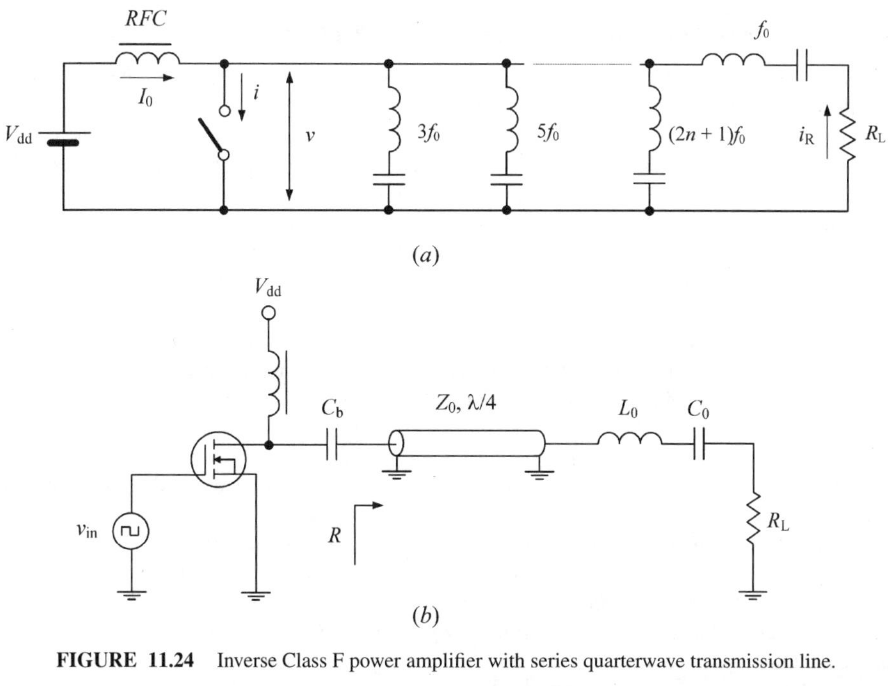

FIGURE 11.24 Inverse Class F power amplifier with series quarterwave transmission line.

frequency f_0 and its odd-harmonic components $(2n + 1)f_0$ and infinite impedance at even harmonics $2nf_0$ realizing the idealized inverse Class F square current and half-sinusoidal voltage waveforms at the device output terminal. As a result, the active device that is driven to operate as a switch sees the load resistance R_L at the fundamental frequency, while the odd harmonics are shorted by the series resonant circuits.

An infinite set of the series resonant circuits tuned to the odd harmonics can be effectively replaced by a quarterwave transmission line with the same operating capability. Such a circuit representation of an inverse Class F power amplifier with a series quarterwave transmission line loaded by the series resonant circuit tuned to the fundamental is shown in Figure 11.24(*b*) [13,24]. The series-tuned output circuit presents to the transmission line a load resistance at the frequency of operation. At the same time, the quarterwave transmission line transforms the load impedance according to

$$R = \frac{Z_0^2}{R_L} \tag{11.71}$$

where Z_0 is the characteristic impedance of a transmission line. For even harmonics, the open circuit on the load side of the transmission line is repeated, thus producing an open circuit at the drain. However, the quarterwave transmission line converts the open circuit at the load to a short circuit at the drain for odd harmonics with resistive load at the fundamental.

Consequently, for a purely sinusoidal current flowing into the load due to infinite loaded quality factor of the series fundamentally tuned circuit, the ideal drain current and voltage waveforms can be represented by the corresponding normalized square and half-sinusoidal waveforms shown in Figures 11.23(*b*) and 11.23(*c*), respectively. Here, a sum of the fundamental and odd harmonics approximates a square current waveform and a sum of the fundamental and even harmonics approximates a half-sinusoidal drain voltage waveform. As a result, the shapes of the drain current and voltage waveforms provide a condition when the current and voltage do not overlap simultaneously. The quarterwave transmission line causes the output voltage across the load resistor R_L to be phase-shifted by 90° relative to the fundamental-frequency components of the drain voltage and current.

11.3.3 Load Networks with Lumped and Distributed Parameters

Theoretical results show that the proper control of the second harmonic can significantly increase the collector efficiency of the power amplifier by flattening of the output current waveform and minimizing the product of integration of the voltage and current waveforms. Practical realization of a multielement high-order *LC* resonant circuit can cause a serious implementation problem, especially at higher frequencies and in monolithic integrated circuits, when only three-harmonic components can be effectively controlled. Therefore, it is sufficient to be confined to the three- or four-element resonant circuit composing the load network of the power amplifier. In this case, the operation with a second-harmonic open circuit and third-harmonic short circuit is a promising concept for low voltage power amplifiers [25].

In addition, it is necessary to take into account that, in practice, both extrinsic and intrinsic transistor parasitic elements like output shunt capacitance or serious inductance have a substantial effect on the efficiency. The output capacitance C_{out} can represent the collector capacitance C_c in the case of the bipolar transistor or drain–source capacitance plus gate–drain capacitance $C_{ds} + C_{gd}$ in the case of the FET device. The output inductance L_{out} is generally composed of the bond-wire and lead inductances for a packaged transistor, effect of which becomes significant at higher frequencies.

The equivalent circuit of the second-harmonic impedance-peaking circuit is shown in Figure 11.25. Here, the series circuit consisting of an inductor L_1 and a capacitor C_1 creates a resonance at the second harmonic. Since the device output inductor L_{out} and capacitor C_{out} are tuned to create an open-circuited condition at the second harmonic, the device collector sees resultant high impedance at the second harmonic. To achieve a second harmonic high impedance, an external inductance may

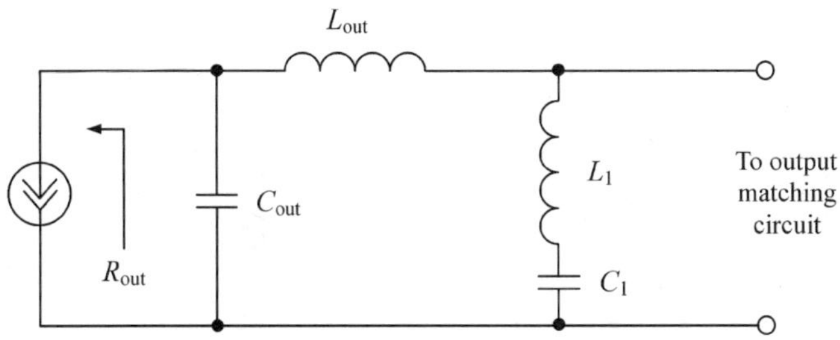

FIGURE 11.25 Second-harmonic impedance-peaking circuit.

be added to interconnect the device output inductance L_{out} directly at the output terminal (collector or drain) if its value is not accurate enough. As a result, the values of the network parameters are

$$L_{out} = \frac{1}{4\omega_0^2 C_{out}} \quad L_1 = \frac{1}{4\omega_0^2 C_1}. \tag{11.72}$$

As a first approximation for numerical calculation, the output device resistance R_{out} at the fundamental frequency required to realize an inverse Class F operation mode with second-harmonic peaking can be estimated as an equivalent resistance $R_{out} = R_1^{(invF)} = V_1/I_1$ determined at the fundamental frequency for an ideal inverse Class F operation. Assuming zero saturation voltage and using Eq. (11.65) yield

$$R_1^{(invF)} = \frac{\pi}{2}\frac{V_{cc}}{I_1} = \frac{\pi^2}{8}R_1^{(F)} = \frac{\pi}{2}R_1^{(B)} \tag{11.73}$$

where $R_1^{(F)}$ is the fundamental output resistance in a conventional Class F mode and $R_1^{(B)} = V_{cc}/I_1$ is the fundamental output resistance in an ideal Class B.

The ideal inverse class F power amplifier with all voltage third-order and higher order odd-harmonic short-circuit termination cannot be provided by using a single parallel transmission line that can be easily realized by a quarterwave transmission line for even harmonics in the conventional class F power amplifier. In this case, with a sufficiently simple circuit schematic convenient for practical realization, applying the current second-harmonic peaking and voltage third-harmonic termination can result in a maximum drain efficiency of more than 80% [3,26]. The output impedance-peaking load network of such a microstrip power amplifier is shown in Figure 11.26, and its circuit structure is similar to the one used to provide a conventional Class F operation mode.

As it follows from Eq. (11.73), the equivalent output resistance for an ideal inverse Class F mode is higher by more than 1.5 times compared to a conventional Class B operation. Therefore, using an inverse Class F operation mode simplifies the subsequent output matching circuit design by minimizing the impedance transformation ratio. This is very important for high output power level when the output resistance is sufficiently small. However, maximum amplitude of the output voltage waveform can exceed the supply voltage by about three times. In practice, it is required to increase the device breakdown voltage or to reduce the supply voltage. The latter is not desirable because of a decrease in power gain and efficiency.

For such an inverse Class F microstrip power amplifier, it is necessary to provide the following electrical lengths for the transmission lines at the fundamental frequency:

$$\theta_1 = \frac{\pi}{3} \quad \theta_2 = \frac{1}{2}\tan^{-1}\left[\left(2Z_0\omega_0 C_{out} + \frac{1}{\sqrt{3}}\right)^{-1}\right] \quad \theta_3 = \frac{\pi}{4} \tag{11.74}$$

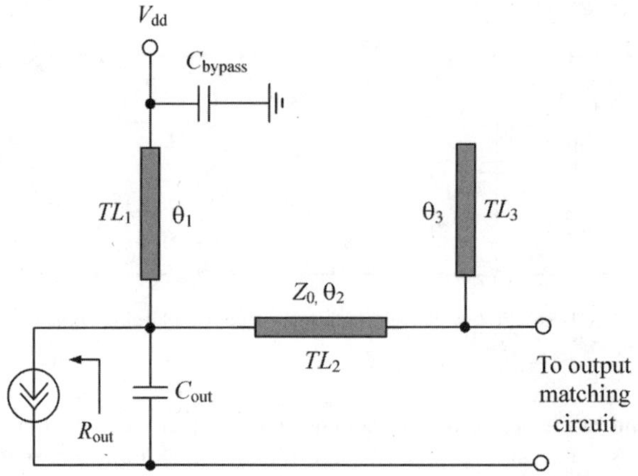

FIGURE 11.26 Transmission-line impedance-peaking circuit.

where Z_0 is the characteristic impedance of the microstrip lines. The transmission line TL_1 with electrical length $\theta_1 = 60°$ provides a short-circuited condition for the third harmonic and introduces a capacitive reactance at the second harmonic. The open-circuit stub TL_3 with electrical length $\theta_3 = 45°$ creates a short-circuited condition at the end of the transmission line TL_2 at the second harmonic. Thus, the transmission line TL_2 should provide an inductive reactance at the second harmonic to compensate for the capacitive reactances of the transmission line TL_1 and device output capacitance C_{out} to create a second-harmonic tank.

As an example, Figure 11.27 shows the frequency-response characteristic of the microstrip impedance-peaking circuit using an alumina substrate for the device output resistance $R_{out} = 50\ \Omega$ and output capacitance $C_{out} = 2.2$ pF, characteristic impedance of microstrip lines $Z_0 = 50\ \Omega$ and electrical length $\theta_2 = 19°$. From Figure 11.27, it follows that, for the second-harmonic peaking and third-harmonic short-circuit termination, an additional output matching at the fundamental

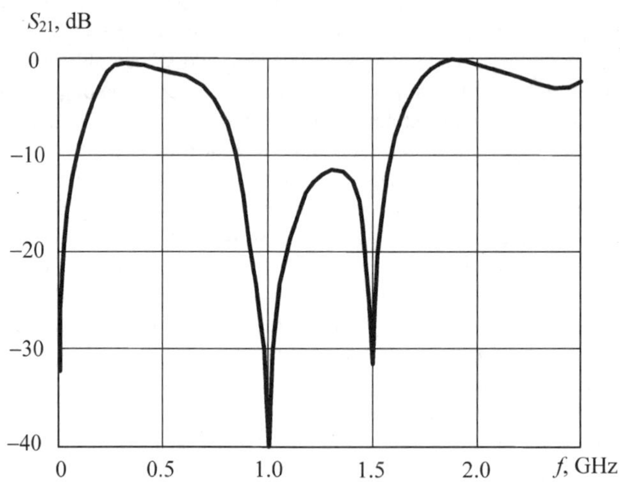

FIGURE 11.27 Frequency response of the microstrip impedance-peaking circuit.

frequency $f_0 = 500$ MHz is required, taking into account the reactance introduced by the impedance-peaking circuit.

11.4 CLASS E WITH SHUNT CAPACITANCE

Using resonant circuits tuned to the odd or even harmonics of the fundamental frequency in the load network by realizing biharmonic or polyharmonic operation modes of the vacuum-tube power amplifiers is very effective to increase their operating efficiency. This implies ideally the in-phase or out-of-phase harmonic conditions when symmetrical flattened voltage or current waveforms can be formed. However, as it turned out, this is not the only way to improve the power amplifier efficiency. Figure 11.28 shows the circuit schematic of the vacuum-tube power amplifier with a parallel-tuned *LC* circuit inserted between the anode and the output matching circuit, with a resonant frequency equal to about 1.5 times the carrier frequency of the signal to be amplified [27]. In other words, if the carrier signal is transmitting at a fundamental frequency f_0, the parallel resonant circuit will have a resonant frequency of about $1.5f_0$ followed by a filter or output matching circuit to suppress the harmonics of the fundamental frequency and to maximize the output power at the fundamental frequency delivered to the standard load. As a result, an efficiency of 89% was achieved for a 3.2-MHz vacuum-tube high-power amplifier.

Although it was assumed that such a parallel resonant circuit introduces considerable impedance to its own second harmonic, which is the third harmonic $3f_0$ of the carrier frequency and could result in a flattened anode voltage waveform, another interesting and nontrivial conclusion could be derived from this circuit topology. In this case, provided the output π-type matching circuit has purely resistive impedance at the fundamental frequency and capacitive reactances at the harmonic components, the anode of the device sees inductive impedance at the fundamental frequency and capacitive reactances at the second-order and higher order harmonic components. This means that the voltage and current waveforms are not symmetrical anymore representing an alternative mechanism of the efficiency improvement. Such an effect of increasing efficiency when the output resonant circuit of the vacuum-tube Class C power amplifier is detuned relatively to the carrier frequency was firstly described seven years earlier [28]. In this case, the anode efficiencies of about 92–93% were achieved for the phase angles of the output load network in limits of 30–40° resulting in the proper inductive reactance at the fundamental frequency and capacitive reactances at the harmonic components seen by the anode of the active device.

Later, it was discovered that very high efficiencies could be obtained with a series resonant *LC* circuit connected to the transistor [29]. The reasons for the high efficiencies is that, due to a proper

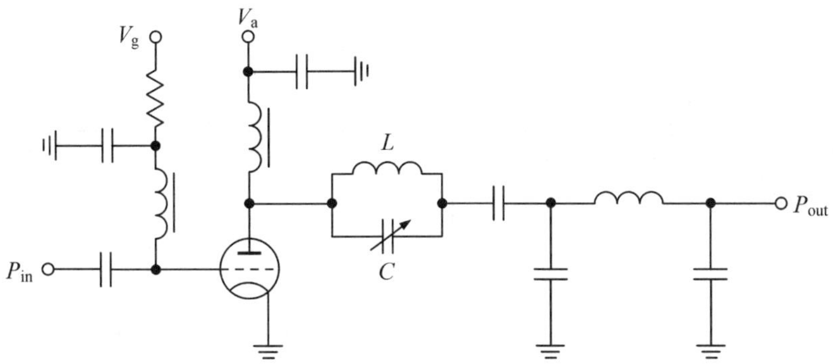

FIGURE 11.28 Class C power amplifier with detuned resonant circuit.

choice of transistor and circuit parameters, the transistor operates in a pure switching mode, and the voltage across the transistor and the current flowing through it can both be made equal to zero during the switching transient interval. To satisfy this condition, the current and voltage must be zero at time just prior to the conduction interval when transistor goes to the saturation mode and the series-tuned circuit must appear inductive at the operating frequency. In this case, a loaded quality factor of the series-tuned circuit of about 10 will give a good sinusoidal shape to the load current. As a result, a 20-W 500-kHz bipolar power amplifier was built having a collector efficiency of 94% with a conduction angle of 180°. The exact theoretical analysis of the single-ended switching-mode power amplifier with a shunt capacitance and a series *LC* circuit was then given by Kozyrev [30].

11.4.1 Optimum Load Network Parameters

However, the single-ended switched-mode power amplifier with a shunt capacitor as a Class E power amplifier was introduced by Sokals in 1975 and has found widespread application due to its design simplicity and high operation efficiency [31,32]. This type of high-efficiency power amplifiers is widely used in different frequency ranges and output power levels ranging from several kilowatts at low RF frequencies up to about one watt at microwaves [33]. The characteristics of a Class E power amplifier can be determined by finding its steady-state collector voltage and current waveforms. The basic circuit of a Class E power amplifier with a shunt capacitance is shown in Figure 11.29(*a*), where the load network consists of a capacitor C shunting the transistor, a series inductor L, a series fundamentally tuned L_0C_0 circuit, and a load resistor R. In a common case, a shunt capacitor C can represent the intrinsic device output capacitor and external circuit capacitor added by the load network. The collector of the transistor is connected to the supply voltage by an RF choke with high reactance at the fundamental frequency. The active device is considered an ideal switch that is driven in such a way as to provide the device switching between its on- and off-state operation conditions.

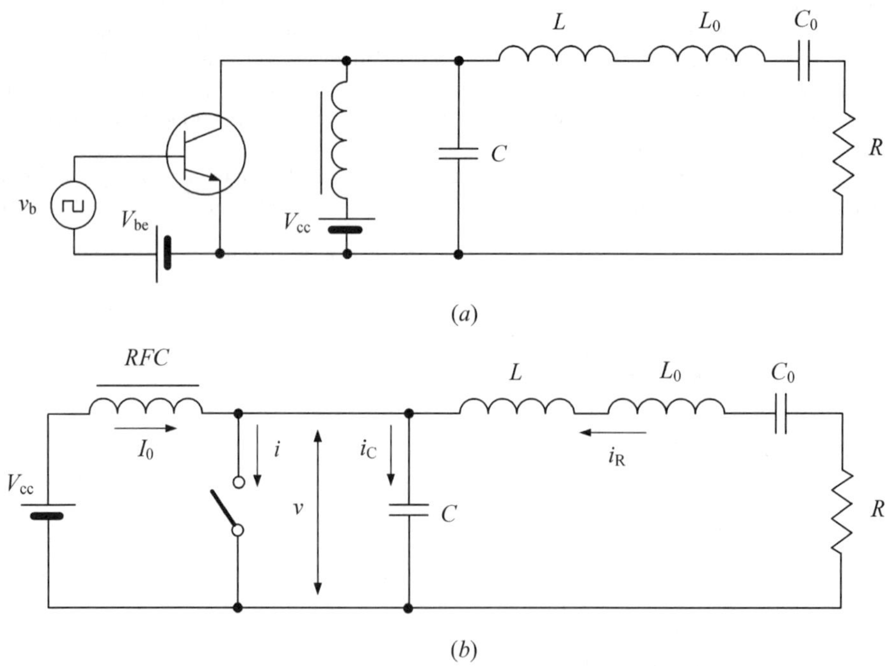

FIGURE 11.29 Basic circuits of Class E power amplifier with shunt capacitance.

As a result, the collector voltage waveform is determined by the switch when it is turned on and by the transient response of the load network when the switch is turned off.

To simplify an analysis of a Class E power amplifier, a simple equivalent circuit of which is shown in Figure 11.29(*b*), the following several assumptions are introduced:

- The transistor has zero saturation voltage, zero saturation resistance, infinite off-resistance, and its switching action is instantaneous and lossless.
- The total shunt capacitance is independent of the collector and is assumed linear.
- The RF choke allows only a constant dc current and has no resistance.
- The loaded quality factor $Q_L = \omega L_0/R = 1/\omega C_0 R$ of the series resonant L_0C_0 circuit tuned to the fundamental frequency is high enough for the output current to be sinusoidal at the switching frequency.
- There are no losses in the circuit except only in the load R.
- For an optimum operation mode, a 50% duty cycle is used.

For a lossless operation mode, it is necessary to provide the following optimum conditions for voltage across the switch just prior to the start of switch on at the moment $\omega t = 2\pi$, when transistor is saturated:

$$v\left(\omega t\right)\big|_{\omega t=2\pi} = 0 \tag{11.75}$$

$$\left.\frac{dv\left(\omega t\right)}{d\omega t}\right|_{\omega t=2\pi} = 0 \tag{11.76}$$

where $v(\omega t)$ is the voltage across the switch.

The detailed theoretical analysis of a Class E power amplifier with shunt capacitance for any duty cycle is given in [34], where the load current is assumed to be sinusoidal,

$$i_R(\omega t) = I_R \sin\left(\omega t + \varphi\right) \tag{11.77}$$

where φ is the initial phase shift.

When the switch is turned on for $0 \leq \omega t < \pi$, the current through the capacitance

$$i_C\left(\omega t\right) = \omega C \frac{dv\left(\omega t\right)}{d\omega t} = 0 \tag{11.78}$$

and, consequently,

$$i(\omega t) = I_0 + I_R \sin\left(\omega t + \varphi\right) \tag{11.79}$$

under the initial on-state condition $i\left(0\right) = 0$. Hence, the dc current can be defined as

$$I_0 = -I_R \sin\varphi \tag{11.80}$$

and the current through the switch can be rewritten by

$$i(\omega t) = I_R \left[\sin\left(\omega t + \varphi\right) - \sin\varphi\right]. \tag{11.81}$$

When the switch is turned off for $\pi \leq \omega t < 2\pi$, the current through the switch $i\left(\omega t\right) = 0$, and the current flowing through the capacitor C can be written as

$$i_C(\omega t) = I_0 + I_R \sin\left(\omega t + \varphi\right) \tag{11.82}$$

producing the voltage across the switch by the charging of this capacitor according to

$$v(\omega t) = \frac{1}{\omega C} \int_{\pi}^{\omega t} i_C(\omega t) d\omega t = -\frac{I_R}{\omega C} \left[\cos(\omega t + \varphi) + \cos\varphi + (\omega t - \pi)\sin\varphi \right]. \tag{11.83}$$

Applying the first optimum condition given by Eq. (11.75) enables the phase angle φ to be determined as

$$\varphi = \tan^{-1}\left(-\frac{2}{\pi}\right) = -32.482^\circ. \tag{11.84}$$

Consideration of trigonometric relationships shows that

$$\sin\varphi = \frac{-2}{\sqrt{\pi^2 + 4}} \qquad \cos\varphi = \frac{\pi}{\sqrt{\pi^2 + 4}}. \tag{11.85}$$

By using Fourier-series expansion and Eqs. (11.80) and (11.85), the expression to determine the supply voltage V_{cc} can be written as

$$V_{cc} = \frac{1}{2\pi} \int_{0}^{2\pi} v(\omega t) d\omega t = \frac{I_0}{\pi \omega C}. \tag{11.86}$$

As a result, the normalized steady-state collector voltage waveform for $\pi \leq \omega t < 2\pi$ and current waveform for period of $0 \leq \omega t < \pi$ are

$$\frac{v(\omega t)}{V_{cc}} = \pi \left(\omega t - \frac{3\pi}{2} - \frac{\pi}{2} \cos\omega t - \sin\omega t \right) \tag{11.87}$$

$$\frac{i(\omega t)}{I_0} = \frac{\pi}{2} \sin\omega t - \cos\omega t + 1. \tag{11.88}$$

Figure 11.30 shows the normalized (*a*) load current, (*b*) collector voltage waveform, and (*c*) collector current waveforms for an idealized optimum Class E with shunt capacitance. From collector voltage and current waveforms it follows that, when the transistor is turned on, there is no voltage across the switch and the current $i(\omega t)$ consisting of the load sinusoidal current and dc current flows through the device. However, when the transistor is turned off, this current flows through the shunt capacitance C. The jump in the collector current waveform at the instant of switching off is necessary to obtain nonzero output power at the fundamental frequency delivered to the load, which can be defined as an integration of the product of the collector voltage and current derivatives over the entire period [35].

As a result, there is no nonzero voltage and current simultaneously, which means a lack of the power losses and gives an idealized collector efficiency of 100%. This implies that the dc power and fundamental-frequency output power delivered to the load are equal,

$$I_0 V_{cc} = \frac{I_R^2}{2} R. \tag{11.89}$$

Consequently, the value of dc supply current I_0 can be determined using Eqs. (11.80) and (11.85) by

$$I_0 = \frac{V_{cc}}{R} \frac{8}{\pi^2 + 4} = 0.577 \frac{V_{cc}}{R}. \tag{11.90}$$

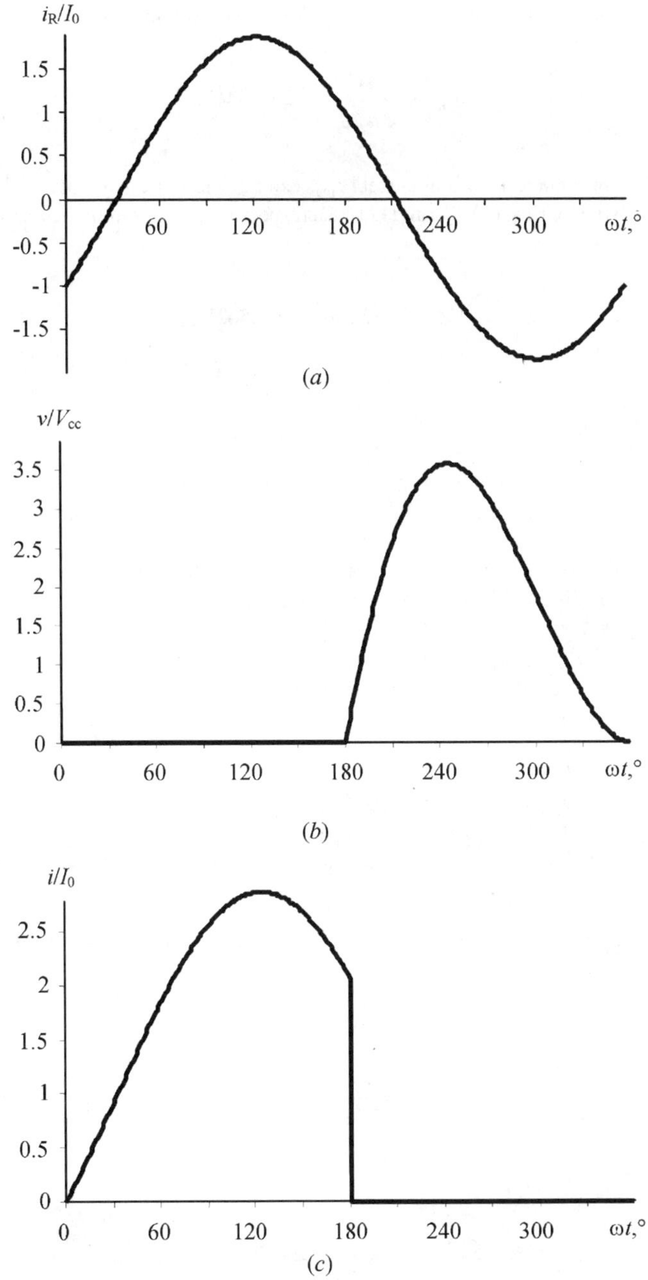

FIGURE 11.30 Normalized (*a*) load current and collector (*b*) voltage and (*c*) current waveforms for idealized optimum Class E with shunt capacitance.

Then, the amplitude of the output voltage $V_R = I_R R$ can be obtained from

$$V_R = \frac{4V_{cc}}{\sqrt{\pi^2 + 4}} = 1.074V_{cc}. \tag{11.91}$$

The peak collector voltage V_{max} and current I_{max} can be determined by differentiating the appropriate waveforms given by Eqs. (11.87) and (11.88), respectively, and setting the results equal to zero, which gives

$$V_{max} = -2\pi\varphi V_{cc} = 3.562V_{cc} \tag{11.92}$$

and

$$I_{max} = \left(\frac{\sqrt{\pi^2 + 4}}{2} + 1\right) I_0 = 2.8621I_0. \tag{11.93}$$

The fundamental-frequency voltage $v_1(\omega t)$ across the switch consists of the two quadrature components, as shown in Figure 11.31, whose amplitudes can be found using Fourier formulas and Eq. (11.87) by

$$V_R = -\frac{1}{\pi}\int_0^{2\pi} v(\omega t)\sin(\omega t + \varphi)\, d\omega t = -\frac{I_R}{\pi\omega C}\left(\frac{\pi}{2}\sin 2\varphi + 2\cos 2\varphi\right) \tag{11.94}$$

$$V_L = -\frac{1}{\pi}\int_0^{2\pi} v(\omega t)\cos(\omega t + \varphi)\, d\omega t = \frac{I_R}{\pi\omega C}\left(\frac{\pi}{2} + \pi\sin^2\varphi + 2\sin 2\varphi\right). \tag{11.95}$$

As a result, the optimum series inductance L and shunt capacitance C can be calculated from

$$\frac{\omega L}{R} = \frac{V_L}{V_R} = 1.1525 \tag{11.96}$$

$$\omega CR = \frac{\omega C}{I_R} V_R = 0.1836. \tag{11.97}$$

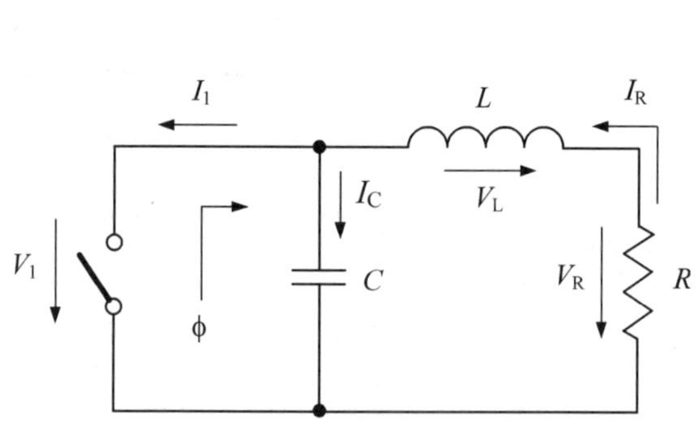

FIGURE 11.31 Equivalent Class E load network at fundamental frequency.

The optimum load resistance R can be obtained using Eqs. (11.89) and (11.91) for the supply voltage V_{cc} and fundamental-frequency output power P_{out} delivered to the load as

$$R = \frac{8}{\pi^2 + 4} \frac{V_{cc}^2}{P_{out}} = 0.5768 \frac{V_{cc}^2}{P_{out}}. \tag{11.98}$$

Finally, the phase angle of the load network seen by the switch at the fundamental and required for an idealized optimum Class E with shunt capacitance can be determined through the load network parameters using Eqs. (11.96) and (11.97) by

$$\phi = \tan^{-1}\left(\frac{\omega L}{R}\right) - \tan^{-1}\left(\frac{\omega C R}{1 - \dfrac{\omega L}{R}\omega C R}\right) = 35.945^\circ. \tag{11.99}$$

When realizing an optimum Class E operation mode, it is very important to know up to which maximum frequency such an idealized efficient operation mode can be extended. In this case, it is possible to establish a relationship between a maximum frequency f_{max}, a parallel shunt capacitance C, and a supply voltage V_{cc} by using Eqs. (11.86) and (11.93) when

$$f_{max} = \frac{1}{\pi^2} \frac{1}{\sqrt{\pi^2 + 4} + 2} \frac{I_{max}}{C_{out} V_{cc}} = \frac{I_{max}}{56.5 C_{out} V_{cc}} \tag{11.100}$$

where $C = C_{out}$ is the device output capacitance limiting the maximum operation frequency of an ideal Class E circuit [36].

The high-Q_L assumption for the series resonant L_0C_0 circuit can lead to considerable errors if its value is substantially small in real circuits [37]. For example, for a 50% duty cycle, the values of the circuit parameters for the loaded quality factor less than unity can differ by several tens of percents. At the same time, for $Q_L \geq 7$, the errors are found to be less than 10% and become less than 5% for $Q_L \geq 10$. To match the optimum Class E load network resistance R with standard load impedance $R_L = 50\ \Omega$, the series resonant L_0C_0 circuit should be followed or fully replaced by the matching circuit, in which the first element must represent the series inductor to provide high impedance at harmonics [3].

11.4.2 Saturation Resistance and Switching Time

In practical power amplifier design, especially when a value of the supply voltage is sufficiently small, it is very important to predict the overall degradation of power amplifier efficiency due to the finite value of the transistor saturation resistance. Figure 11.32 shows the simplified equivalent circuit of a Class E power amplifier with shunt capacitance including the on-resistance or saturation resistance r_{sat} connected in series to the ideal switch. To obtain a quantitative estimate of the power losses due to the contribution of r_{sat}, the saturated output power P_{sat} can be obtained with a simple approximation

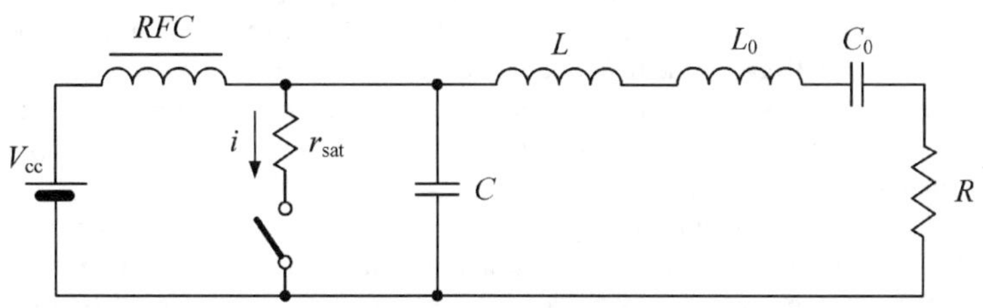

FIGURE 11.32 Equivalent Class E load network with saturation resistance.

when the current $i(\omega t)$ flowing through the saturation resistance r_{sat} is determined in an ideal case by Eq. (11.88).

An analytical expression to calculate the power losses due to the saturation resistance r_{sat}, whose value is assumed constant, can be represented in the normalized form according to

$$\frac{P_{sat}}{P_0} = \frac{r_{sat}}{2\pi I_0 V_{cc}} \int_0^{\pi} i^2(\omega t)\, d\omega t \tag{11.101}$$

where $P_0 = I_0 V_{cc}$ is the dc power.

By taking into account that

$$\int_0^{\pi} \left(\frac{\pi}{2}\sin\omega t - \cos\omega t + 1\right)^2 d\omega t = \frac{\pi}{8}\left(\pi^2 + 28\right) I_0^2 \tag{11.102}$$

Eq. (11.101) can be rewritten using Eq. (11.90) as

$$\frac{P_{sat}}{P_0} = \frac{r_{sat}}{2\pi}\frac{I_0}{V_{cc}}\frac{\pi}{8}\left(\pi^2 + 28\right) = \frac{r_{sat}}{2R}\frac{\pi^2 + 28}{\pi^2 + 4} = 1.365\frac{r_{sat}}{R}. \tag{11.103}$$

The collector efficiency η can be calculated from

$$\eta = \frac{P_{out}}{P_0} = \frac{P_0 - P_{sat}}{P_0} = 1 - \frac{P_{sat}}{P_0}. \tag{11.104}$$

Consequently, the presence of the saturation resistance results in the finite value of the saturation voltage V_{sat}, which can be defined from

$$\frac{V_{sat}}{V_{cc}} = 1 - \frac{1}{1 + 1.365\dfrac{r_{sat}}{R}} \tag{11.105}$$

being normalized to the dc supply voltage V_{cc} [38].

More detailed theoretical analysis of the time-dependent behavior of the collector voltage and current waveforms shows that, for a finite value of the saturation resistance r_{sat}, the optimum conditions for idealized operation mode given by Eqs. (11.75) and (11.76) do not correspond anymore to the minimum dissipated power losses, and there are optimum nonzero values of the collector voltage and its derivative at switching time instant corresponding to minimum overall power losses [39]. For example, even for small losses with the normalized loss parameter $\omega C r_{sat} = 0.1$ for a duty cycle of 50%, the optimum series inductance L is almost two times greater, while the optimum shunt capacitance C is of about 20% greater than those obtained under optimum conditions given by Eqs. (11.75) and (11.76). However, for collector efficiencies of 90% and greater, both the optimum inductance and optimum capacitance differ by less than 20% from their optimum values for $r_{sat} = 0$ [40]. Thus, generally the switching conditions given by Eqs. (11.75) and (11.76) can be considered optimum only for an idealized case of a Class E load network with zero saturation resistance providing the switched-mode transistor operation when it operates in pinch-off and saturation regions only. However, they can be considered as a sufficiently accurate initial guess for further design and optimization of the real high-efficiency power amplifier circuits.

For an ideal active device without any memory effects due to intrinsic phase delays, the switching time is equal to zero when the rectangular input drive results in a rectangular output response. Such an ideal case assumes zero device feedback capacitance and zero device input resistance. However, in a real situation, especially at higher frequencies, it is very difficult to realize the driving signal close to the rectangular form as it leads to the significant circuit complexity. Fortunately, to realize high-efficiency operation conditions, it is sufficient to drive the power amplifier simply with a sinusoidal signal. The finite-time transition from the saturation mode to the pinch-off mode through the device active mode takes place at the point of the intersection of the curve corresponding to the base (or channel) charge process (curve 1) and the curve corresponding to the required ideal collector current waveform provided by the load network (curve 2), as shown in Figure 11.33(*a*). To minimize the switching time interval, it is sufficient to slightly overdrive the active device with signal amplitude by 20–30% higher than it is required for a conventional Class B power amplifier, as shown in Figure 11.33(*b*) [3]. As an alternative, the second-harmonic component (approximation of a half-sinusoidal waveform) or third-harmonic component (approximation of a rectangular waveform with close to trapezoidal waveform) with proper phasing can be added to the input driving signal. In both cases, the overall driving waveform will be steeper compared with simply sinusoidal driving signal, thus resulting in a faster switching operation time.

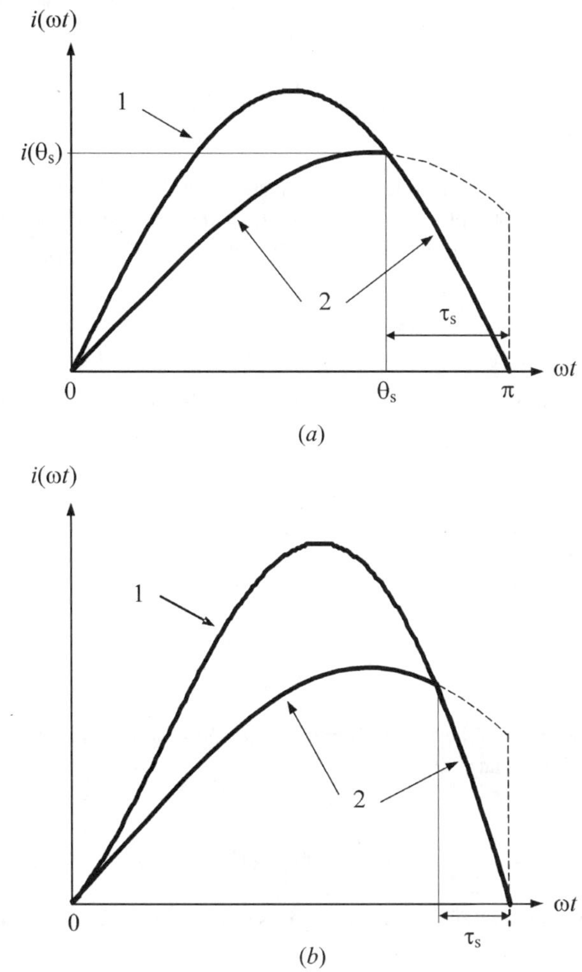

FIGURE 11.33 Collector current waveforms due to finite switching time.

The power dissipated during this on-to-off transition can be calculated assuming zero on-resistance as

$$P_s = \frac{1}{2\pi} \int_{\theta_s}^{\pi} i(\omega t)\, v(\omega t)\, d\omega t \qquad (11.106)$$

where the collector voltage during the transition time $\tau_s = \pi - \theta_s$ is given by

$$v(\theta_s) = \frac{1}{\omega C} \int_{\theta_s}^{\pi} i_C(\omega t)\, d\omega t. \qquad (11.107)$$

The short duration of the switching time and the proper behavior of the resulting collector (or drain) waveform allows us to make an additional assumption of a linearly decreasing collector current during fall time $\tau_s = \pi - \theta_s$ starting at $i(\theta_s)$ at time θ_s and decaying to zero at time π, which can be written by

$$i(\omega t) = i(\theta_s)\left(1 - \frac{\omega t - \theta_s}{\tau_s}\right) \qquad (11.108)$$

where $i(\theta_s)$ corresponds to the peak collector current shown in Figure 11.33(*a*) [38,40]. In this case, the capacitor-charging current $i_C(\omega t) = i(\theta_s) - i(\omega t)$, being zero during saturation mode, varies linearly between zero and $i(\theta_s)$ during on-to-off transition according to

$$i_C(\omega t) = i(\theta_s)\, \frac{\omega t - \theta_s}{\tau_s}. \qquad (11.109)$$

The collector voltage produces a parabolic voltage waveform during the switching interval according to Eq. (11.107) given by

$$v(\omega t) = \frac{i(\theta_s)}{2\omega C \tau_s} (\omega t - \theta_s)^2. \qquad (11.110)$$

As a result, the power dissipated during transition according to Eq. (11.106) is then

$$P_s = \frac{i^2(\theta_s)\, \tau_s^2}{48\pi \omega C}. \qquad (11.111)$$

For an optimum power amplifier by assuming that, in view of a short transition time, $i(\theta_s) = i(\pi)$, from Eq. (11.88) it follows that $i(\pi) = 2I_0$, hence

$$P_s = \frac{I_0^2 \tau_s^2}{12\pi \omega C}. \qquad (11.112)$$

Taking into account Eq. (11.86), the switching loss power P_s normalized to the dc power P_0 can be obtained from

$$\frac{P_s}{P_0} = \frac{I_0 \tau_s^2}{12\pi \omega C V_{cc}} = \frac{\tau_s^2}{12}. \qquad (11.113)$$

The collector efficiency η can be estimated as

$$\eta = 1 - \frac{P_s}{P_0} = 1 - \frac{\tau_s^2}{12}. \tag{11.114}$$

As follows from Eq. (11.113), the power losses due to the finite switching time are sufficiently small and, for example, for $\tau_s = 0.35$ or 20° they are only about 1%, whereas for $\tau_s = 60^\circ$ they are approximately equal to 10%. A more exact analysis assuming a linear variation of the collector current during on-to-off transition results in similar results when efficiency degrades to 97.72% for $\tau_s = 30^\circ$ and to 90.76% for $\tau_s = 60^\circ$ [41]. Considering an exponential collector current decay rather than linear during the fall time shows the similar result for $\tau_s = 30^\circ$ when $\eta = 96.8\%$, but the collector efficiency degrades more significantly at longer fall times when, for example, $\eta = 86.6\%$ for $\tau_s = 60^\circ$ [42].

11.4.3 Load Network with Transmission Lines

The transmission lines are often preferred over lumped inductors at microwave frequencies because of the convenience of their practical implementation, more predictable performance, less insertion loss, and less effect of the parasitic elements. For example, the matching circuit can be composed with any types of the transmission lines including open-circuit or short-circuit stubs to provide the required matching and harmonic-suppression conditions. In this case, to approximate the idealized Class E operation mode of the microwave power amplifier or oscillator, it is necessary to design the transmission-line load network satisfying the required idealized optimum impedances at the fundamental-frequency and harmonic components. The device output capacitance can fully represent the required shunt capacitance whose optimum value is defined by Eq. (11.97). Consequently, the main problem is to satisfy the optimum requirements on the fundamental-frequency impedance $Z_L(\omega_0)$ shown in Figure 11.34(*a*) and harmonic-component impedances $Z_L(n\omega_0)$ shown in Figure 11.34(*b*), which can be written using Eq. (11.96) at fundamental frequency f_0 as

$$Z_L(\omega_0) = R + j\omega L = R\left(1 + j\frac{\omega L}{R}\right) = R\left(1 + j\tan 49.052^\circ\right) \tag{11.115}$$

and at harmonic components nf_0, where $n = 2, 3, \ldots, \infty$, as

$$Z_L(n\omega_0) = \infty. \tag{11.116}$$

Generally, it is practically impossible to realize these conditions for infinite number of the harmonic components by using a finite number of the transmission lines only. However, as it turned out from the Fourier-series analysis, a good approximation to Class E mode may be obtained with the dc, fundamental-frequency, and second-harmonic components of the voltage waveform across the switch [36,43]. Figure 11.34(*c*) shows the collector (drain) voltage waveform containing these two-harmonic components (dotted line) plotted along with an ideal voltage waveform (solid line). In this case, the Class E load network designed for microwave applications will include the series microstrip line l_1 and open-circuit stub l_2, as shown in Figure 11.35(*a*). The electrical lengths of lines l_1 and l_2 are chosen to be of about 45° at the fundamental frequency to provide an open-circuit condition seen from the device output at the second harmonic according to Eq. (11.116). Their characteristic impedances are calculated to satisfy the required inductive impedance condition at the fundamental frequency given by Eq. (11.115). In the case of a packaged active device, its output lead inductance can be accounted for by shortening the length of l_1.

In some case, a value of the device output capacitance exceeds the required optimum value for a Class E mode with shunt capacitance. In this situation, it is possible to approximate Class E mode with high efficiency by setting an adequate optimum load at the fundamental frequency and

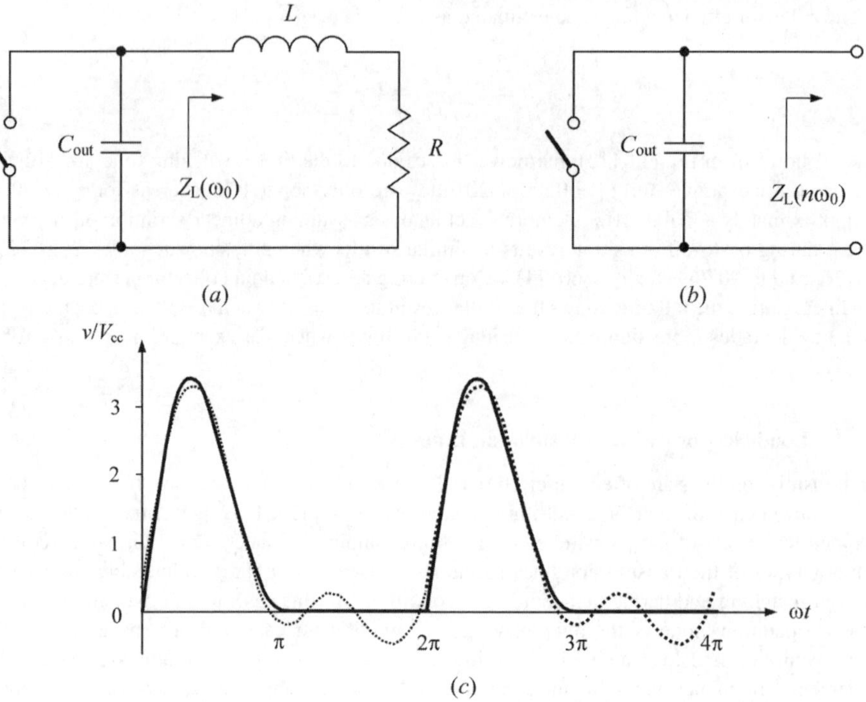

FIGURE 11.34 Optimum load impedance and two-harmonic Class E voltage waveform.

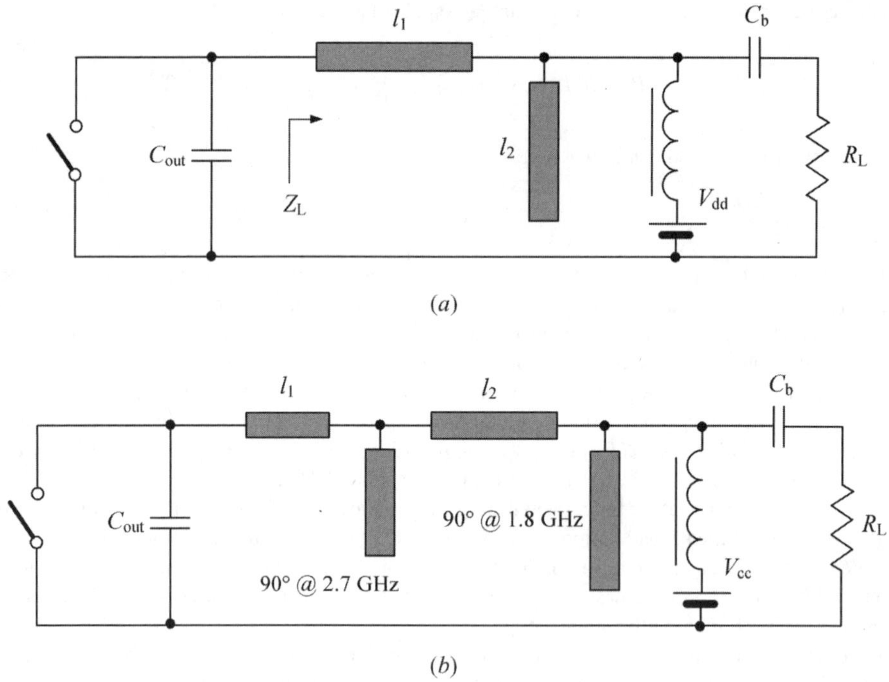

FIGURE 11.35 Equivalent circuits of Class E power amplifiers with transmission lines.

strong reactive load at the second- and third-harmonic components [44]. Such a harmonic control network consists of the open-circuit quarterwave stubs at the second- and third-harmonic components separately, as shown in Figure 11.35(*b*), where the third-harmonic quarterwave stub is located before the second-harmonic quarterwave stub. As a result, very high collector efficiency can be achieved even with values of the device output capacitance higher than it is conventionally required at the expense of lower output power, keeping the load at the second and third harmonics strictly inductive. The maximum collector efficiency over 90% with the output power of 1.5 W for a test power amplifier using the commercial bipolar transistor MRF557 was measured at a carrier frequency of 900 MHz.

An analysis of an optimum Class E mode in frequency domain shows that the combined effect of the saturation resistance r_{sat} and second-harmonic loading is characterized by a shift of the optimum load fundamental-frequency impedance [45]. Besides, parasitic effects due to packaging and microstrip junction discontinuities may upset the transmission-line open-circuit requirement. Therefore, it is practically important to predict the power amplifier performance due to nonideal harmonic terminations. For example, the voltage peak factor increases when the second-harmonic load varies from capacitive to inductive, and the excessive current flow can be realized under capacitive load reactance. It appears that, while a capacitive second-harmonic load achieves higher output power, an inductive second-harmonic load provides better efficiency. In this case, to approximate a switched-mode Class E operation mode, the maximum collector efficiency with lower output power is achieved for nonzero collector voltage and negative collector voltage-derivative conditions [46]. Generally, by providing the open-circuit termination of the second- and third-harmonic components, the collector efficiency can be increased by 10% [47].

11.5 CLASS E WITH FINITE DC-FEED INDUCTANCE

In real practice, it is impossible to realize an RF choke with infinite impedance at the fundamental frequency and its harmonic components. Moreover, using a finite dc-feed inductance has an advantage of minimizing size, cost, and complexity of the overall circuit. The detailed approach to analyze the effect of a finite dc-feed inductance on the idealized Class E mode with shunt capacitance and series filter was firstly described in [48]. It was based on Laplace-transform technique to solve a second-order differential equation describing the behavior of a Class E load network with finite dc-feed inductance. Later this approach was extended to the load network with finite Q_L-factor of the series filter and finite device saturation resistance [49,50]. However, since the results of excessive analytical and numerical calculations are given only for a few particular cases, it is difficult to figure out the basic behavior of the load network elements and define simple equations for their parameters. Generally, based on the composing of the circuit equations in the form of a system of the first-order differential equations for currents and voltages and setting the design specifications, the optimum Class E load network parameters can be numerically calculated taking into account the finite dc-feed inductance, drain current fall time, finite Q_L-factor, nonzero device saturation resistance, and nonlinear operation of any passive element simultaneously [51]. Also, it was analytically shown for a duty cycle of 50% based on the Class E optimum conditions that the series excessive reactance can be either inductive or capacitive depending on the values of the dc-feed inductance and shunt capacitance [52,53]. Based on the certain numbers of cases, a Lagrange polynomial interpolation was used to obtain explicit and directly usable design equations for an idealized Class E with finite dc-feed inductance and series inductive reactance [54].

11.5.1 General Analysis and Optimum Circuit Parameters

The generalized second-order load network of a switched-mode Class E power amplifier with finite dc-feed inductance is shown in Figure 11.36(*a*) [3,55]. The load network consists of a shunt capacitance C, a parallel inductance L, a series reactance X, a series resonant L_0C_0 circuit tuned to the fundamental frequency, and a load resistance R. In a common case, a shunt capacitance C can represent the

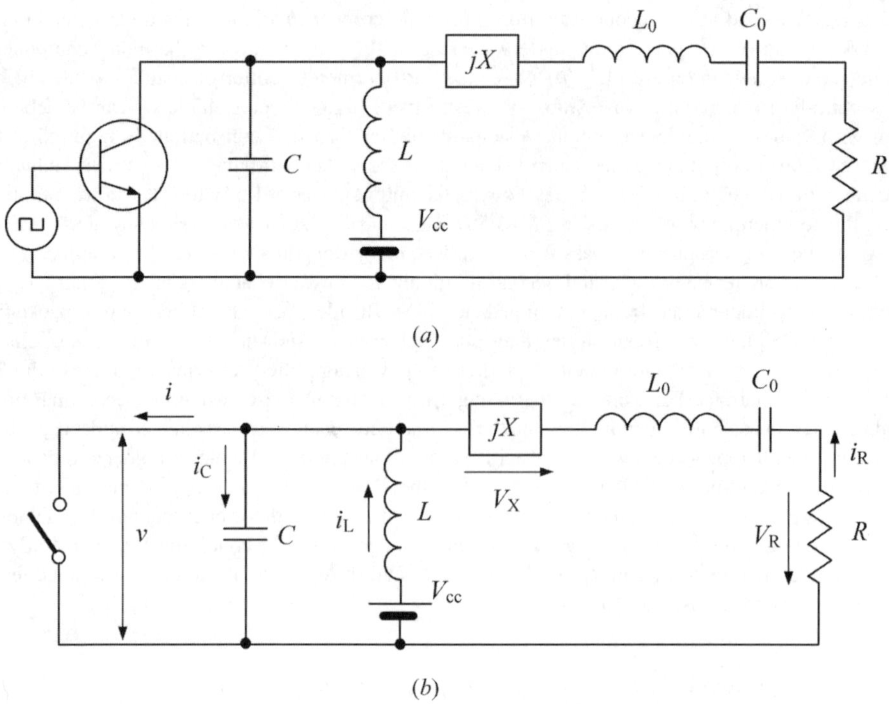

FIGURE 11.36 Equivalent circuits of the Class E power amplifiers with generalized load network.

intrinsic device output capacitance and external circuit capacitance added by the load network, a parallel inductance L represents the finite dc-feed inductance and a series reactance X can be positive (inductance), negative (capacitance), or zero depending on the Class E mode. The active device is considered an ideal switch that is driven in such a way in order to provide the device switching between its on- and off-state operation conditions. To simplify an analysis of the general-circuit Class E power amplifier, a simplified equivalent circuit of which is shown in Figure 11.36(*b*), it needs to introduce the preliminary assumptions similar to those for the Class E power amplifier with shunt capacitance. Assume that the losses in the reactive circuit elements are negligible, duty cycle is 50%, and loaded quality factor of the series L_0C_0 circuit is sufficiently high. For a lossless operation mode, it is necessary to provide the optimum zero voltage and zero voltage-derivative conditions for voltage $v(\omega t)$ across the switch just prior to the start of switch given by Eqs. (11.75) and (11.76).

Let the output current flowing through the load be written as sinusoidal by

$$i_{\rm R}\,(\omega t) = I_{\rm R} \sin\,(\omega t + \varphi) \tag{11.117}$$

where $I_{\rm R}$ is the load current amplitude and φ is the initial phase shift.

When the switch is turned on for $0 \leq \omega t < \pi$, the voltage on the switch $v(\omega t) = V_{\rm cc} - v_{\rm L}(\omega t) = 0$, the current flowing through the capacitance $i_{\rm C}(\omega t) = \omega C(di_{\rm L}/d\omega t) = 0$, and

$$i\,(\omega t) = i_{\rm L}\,(\omega t) + i_{\rm R}\,(\omega t) = \frac{1}{\omega L}\int_0^{\omega t} V_{\rm cc} d\omega t + i_{\rm L}\,(0) + I_{\rm R} \sin\,(\omega t + \varphi) = \frac{V_{\rm cc}}{\omega L}\omega t$$
$$+\, I_{\rm R}\,[\sin\,(\omega t + \varphi) - \sin\varphi] \tag{11.118}$$

where the initial value for the current $i_L(\omega t)$ flowing through the dc-feed inductance L at $\omega t = 0$ can be found using Eq. (11.117) for $i(0) = 0$ as $i_L(0) = -I_R \sin\varphi$.

When the switch is turned off for $\pi \leq \omega t < 2\pi$, the switch current $i(\omega t) = 0$, and the current $i_C(\omega t) = i_L(\omega t) + i_R(\omega t)$ flowing through the capacitance C can be rewritten as

$$\omega C \frac{dv(\omega t)}{d(\omega t)} = \frac{1}{\omega L} \int_{\pi}^{\omega t} [V_{cc} - v(\omega t)]\, d(\omega t) + i_L(\pi) + I_R \sin(\omega t + \varphi) \tag{11.119}$$

under the initial off-state conditions $v(\pi) = 0$ and

$$i_L(\pi) = i(\pi) - i_R(\pi) = \frac{V_{cc}\pi}{\omega L} - \omega L I_R \sin\varphi.$$

Equation (11.119) can be represented in the form of the linear nonhomogeneous second-order differential equation as

$$\omega^2 LC \frac{d^2 v(\omega t)}{d(\omega t)^2} + v(\omega t) - V_{cc} - \omega L I_R \cos(\omega t + \varphi) = 0 \tag{11.120}$$

the general solution of which can be obtained in the normalized form of

$$\frac{v(\omega t)}{V_{cc}} = C_1 \cos(q\omega t) + C_2 \sin(q\omega t) + 1 - \frac{q^2 p}{1 - q^2} \cos(\omega t + \varphi) \tag{11.121}$$

where

$$q = \frac{1}{\omega\sqrt{LC}} \tag{11.122}$$

$$p = \frac{\omega L I_R}{V_{cc}} \tag{11.123}$$

and the coefficients C_1 and C_2 are determined from the initial off-state conditions by

$$C_1 = -(\cos q\pi + q\pi \sin q\pi) - \frac{qp}{1 - q^2}\left[q\cos\varphi \cos q\pi - \left(1 - 2q^2\right)\sin\varphi \sin q\pi\right] \tag{11.124}$$

$$C_2 = (q\pi \cos q\pi - \sin q\pi) - \frac{qp}{1 - q^2}\left[q\cos\varphi \sin q\pi + \left(1 - 2q^2\right)\sin\varphi \cos q\pi\right]. \tag{11.125}$$

The dc supply current I_0 can be found using Fourier formula and Eq. (11.118) by

$$I_0 = \frac{1}{2\pi} \int_0^{2\pi} i(\omega t)\, d(\omega t) = \frac{I_R}{2\pi}\left(\frac{\pi^2}{2p} + 2\cos\varphi - \pi \sin\varphi\right). \tag{11.126}$$

In an idealized Class E operation mode, there is no nonzero voltage and current simultaneously that means a lack of the power losses and gives an idealized collector efficiency of 100%. This implies that the dc power P_0 and fundamental output power P_{out} are equal,

$$I_0 V_{cc} = \frac{V_R^2}{2R} \tag{11.127}$$

where $V_R = I_R R$ is the fundamental voltage amplitude across the load resistance R.

As a result, by using Eqs. (11.126) and (11.127) and taking into account that $R = V_R^2/2P_{out}$, the optimum load resistance R for the specified values of a supply voltage V_{cc} and a fundamental output power P_{out} can be obtained by

$$R = \frac{1}{2}\left(\frac{V_R}{V_{cc}}\right)^2 \frac{V_{cc}^2}{P_{out}} \tag{11.128}$$

where

$$\frac{V_R}{V_{cc}} = \frac{1}{\pi}\left(\frac{\pi^2}{2p} + 2\cos\varphi - \pi\sin\varphi\right). \tag{11.129}$$

The normalized load network inductance L and capacitance C can be appropriately defined using Eqs. (11.122), (11.123), and (11.126) by

$$\frac{\omega L}{R} = p \Big/ \left(\frac{\pi}{2p} + \frac{2}{\pi}\cos\varphi - \sin\varphi\right) \tag{11.130}$$

$$\omega CR = 1 \Big/ \left(q^2 \frac{\omega L}{R}\right). \tag{11.131}$$

The series reactance X, which may generally have an inductive, capacitive, or zero reactance, depending on the load network parameters, can be calculated using the two quadrature fundamental-frequency voltage Fourier components of

$$V_R = -\frac{1}{\pi}\int_0^{2\pi} v(\omega t)\sin(\omega t + \varphi)\, d\omega t \tag{11.132}$$

$$V_X = -\frac{1}{\pi}\int_0^{2\pi} v(\omega t)\cos(\omega t + \varphi)\, d\omega t. \tag{11.133}$$

The fundamental-frequency current flowing through the switch consists of the two quadrature components, the amplitudes of which can be found using Fourier formulas and Eq. (11.118) by

$$I_R = \frac{1}{\pi}\int_0^{2\pi} i(\omega t)\sin(\omega t + \varphi)\, d\omega t = \frac{I_R}{\pi}\left[\frac{\pi\cos\varphi - 2\sin\varphi}{p} + \frac{\pi}{2} - \sin 2\varphi\right]$$

$$I_X = -\frac{1}{\pi}\int_0^{2\pi} i(\omega t)\cos(\omega t + \varphi)\, d\omega t = \frac{I_R}{\pi}\left[\frac{\pi\sin\varphi + 2\cos\varphi}{p} - 2\sin^2\varphi\right].$$

Generally, Eq. (11.121) for a normalized collector voltage contains the three unknown parameters q, p, and φ, which must be analytically or numerically determined. In a common case, the parameter q can be considered a variable, and the other two parameters p and φ are calculated from a system of the two equations resulting from applying of the two optimum zero voltage and zero voltage-derivative conditions given by Eqs. (11.75) and (11.76) to Eq. (11.121). Figure 11.37 shows the dependencies of the optimum parameters p and φ versus q for a Class E with finite dc-feed inductance.

Based on the calculated optimum parameters p and φ as the functions of q, the optimum load network parameters of the Class E load network with a finite dc-feed inductance can be determined

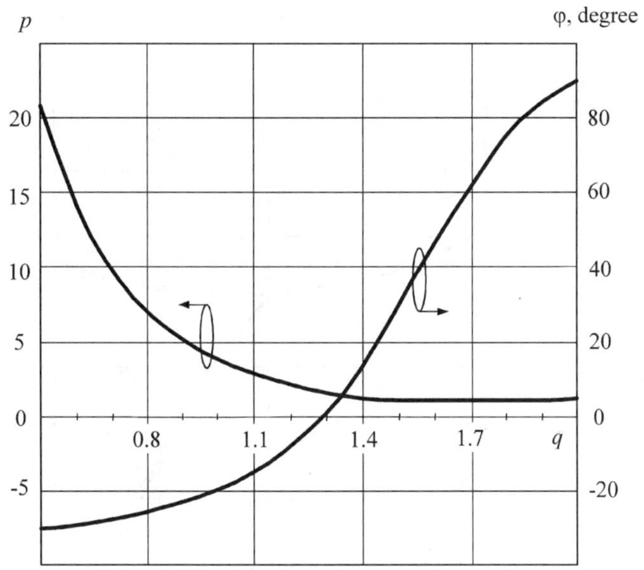

FIGURE 11.37 Optimum Class E parameters p and φ versus q.

using Eqs. (11.128) to (11.131). The series reactance X can be calculated through the ratio of the two quadrature fundamental-frequency voltage Fourier components given in Eqs. (11.132) and (11.133) as

$$\frac{X}{R} = \frac{V_X}{V_R}. \tag{11.134}$$

The dependences of the normalized optimum dc-feed inductance $\omega L/R$ and series reactance X/R are shown in Figure 11.38(a), while the dependences of the normalized optimum shunt capacitance ωCR and load resistance RP_{out}/V_{cc}^2 are plotted in Figure 11.38(b). Here, we can see that the subharmonic case of $q = 0.5$ is very close to a Class E mode with shunt capacitance, since the value of the normalized inductance $\omega L/R$ is sufficiently high and the variations of normalized values of ωCR and RP_{out}/V_{cc}^2 are insignificant. The value of the series reactance X changes its sign from positive to negative, which means that the inductive reactance is followed by the capacitive reactance. As a result, there is special case of a load network with a parallel circuit and a load resistance only when $X = 0$ at $q = 1.412$. In this case, the maximum value of the optimum load resistance R is provided for the same supply voltage and output power, thus simplifying the matching with the standard load of 50 Ω. Also, the values of a dc-feed inductance L become sufficiently small, thus making Class E very attractive for monolithic applications. The maximum operation frequency f_{max} is realized at $q =$ 1.468 where the normalized optimum shunt capacitance ωCR reaches its maximum.

The graphical solutions for the optimum load network parameters can be replaced by the analytical design equations represented in terms of the simple second- and third-order polynomial functions given by Tables 11.1 and 11.2 for different ranges of the parameter q [56]. The maximum difference between the polynomial approximations and exact numerical solutions given in the graphic form is of about 2%.

11.5.2 Parallel-Circuit Class E

The theoretical analysis of a switched-mode parallel-circuit Class E power amplifier using a series filter, which basic circuit is shown in Figure 11.39(a), was done firstly by Kozyrev with the calculation

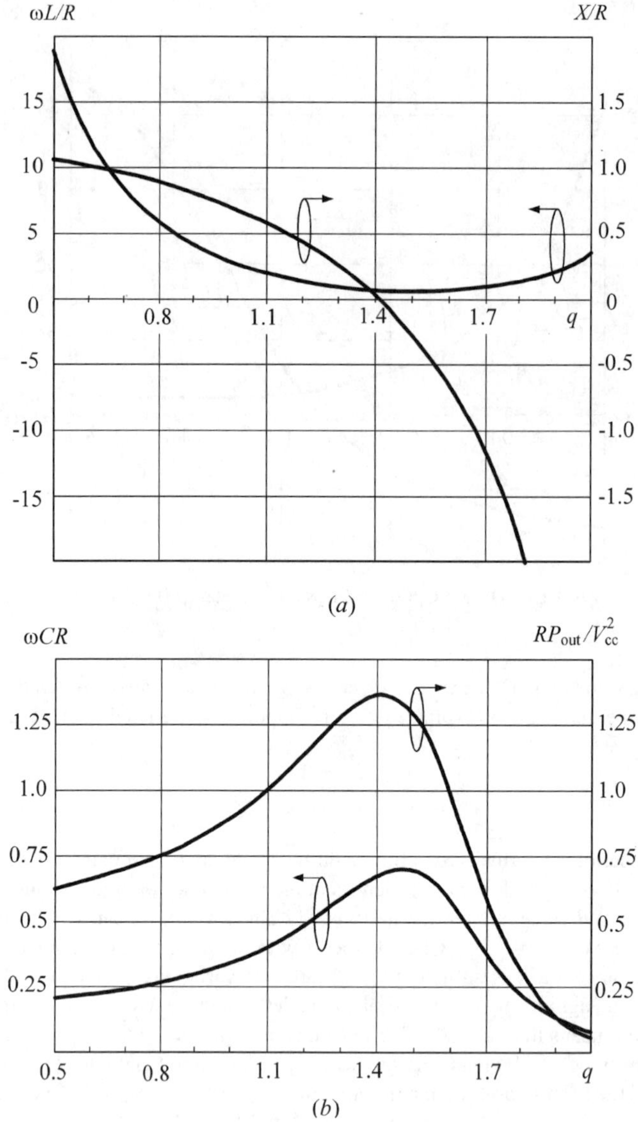

FIGURE 11.38 Normalized optimum Class E load network parameters.

TABLE 11.1 Load Network Parameters for 0.6 < q < 1.0.

Parameter	Design Equation
$\frac{\omega L}{R}$	$44.93q^2 - 94.32q + 52.46$
$\omega C R$	$0.426q^2 - 0.379q + 0.3$
$\frac{X}{R}$	$-0.73\,q^2 + 0.411\,q + 1.03$
$\frac{P_{out}R}{V_{cc}^2}$	$0.74q^2 - 0.6q + 0.76$

TABLE 11.2 Load Network Parameters for 1.0 < q < 1.65.

Parameter	Design Equation
$\frac{\omega L}{R}$	$8.085q^2 - 24.53q + 19.23$
ωCR	$-6.97q^3 + 25.93q^2 - 31.071q + 12.48$
$\frac{X}{R}$	$-2.9q^3 + 8.8q^2 - 10.2q + 5.02$
$\frac{P_{out}R}{V_{cc}^2}$	$-11.9q^3 + 42.753q^2 - 49.63q + 19.7$

of the voltage and current waveforms and some graphical results [30,57]. The load network consists of a finite dc-feed inductor L, a shunt capacitor C, a series L_0C_0 resonant circuit tuned to the fundamental frequency, and a load resistor R. In this case, the switch sees a parallel connection of the load resistor R and parallel LC circuit at the fundamental frequency, as shown in Figure 11.39(*b*), where also the real and imaginary collector fundamental-frequency current components I_X and I_R and real collector fundamental-frequency voltage component V_R are indicated.

In the case of a parallel-circuit Class E load network without series phase-shifting reactance, since the parameter q is unknown a priori, generally it is necessary to solve a system of three equations to define the three unknown parameters q, p, and φ. The two equations are the result of applying of the two optimum zero voltage and zero voltage-derivative conditions given by Eqs. (11.75) and (11.76) to Eq. (11.121). Since the fundamental-frequency collector voltage is fully applied to the load, this means that its reactive part must have zero value resulting in an additional equation

$$V_X = -\frac{1}{\pi}\int_0^{2\pi} v(\omega t)\cos(\omega t + \varphi)d\omega t = 0. \tag{11.135}$$

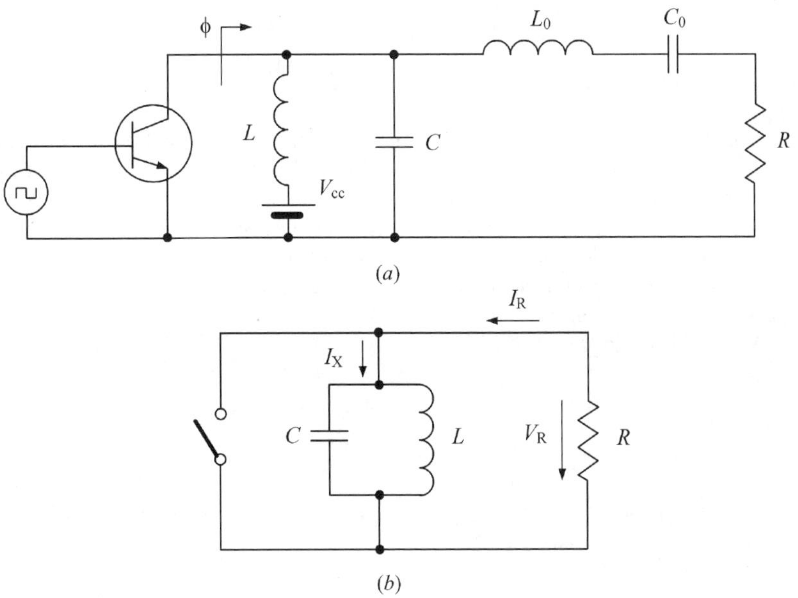

FIGURE 11.39 Equivalent circuits of the parallel-circuit Class E power amplifier.

Solving the system of three equations with three unknown parameters numerically gives the following values [3,58,59]:

$$q = 1.412 \tag{11.136}$$

$$p = 1.210 \tag{11.137}$$

$$\varphi = 15.155^\circ. \tag{11.138}$$

Figure 11.40 shows the normalized (*a*) load current and collector (*b*) voltage, and (*c*) current waveforms for an idealized optimum parallel-circuit Class E operation. From collector voltage and current waveforms it follows that there is no nonzero voltage and current simultaneously. When this happens, there no power loss occurs and an idealized collector efficiency of 100% is achieved.

By using Eqs. (11.128) to (11.131), the optimum load resistance R, parallel inductance L, and parallel capacitance C can be appropriately obtained by

$$R = 1.365\frac{V_{cc}^2}{P_{out}} \tag{11.139}$$

$$L = 0.732\frac{R}{\omega} \tag{11.140}$$

$$C = \frac{0.685}{\omega R}. \tag{11.141}$$

The dc supply current I_0 can be calculated from Eq. (11.126) as

$$I_0 = 0.826 I_R. \tag{11.142}$$

The phase angle ϕ seen from the device collector at the fundamental frequency can be represented either through the two quadrature fundamental-frequency current Fourier components I_X and I_R or as a function of load network elements by

$$\phi = \tan^{-1}\left(\frac{R}{\omega L} - \omega RC\right) = 34.244^\circ. \tag{11.143}$$

If the calculated value of the optimum Class E resistance R is too small or differs significantly from the required load impedance, it is necessary to use an additional matching circuit to deliver maximum output power to the load. It should be noted that, among a family of the Class E load networks, a parallel-circuit Class E load network offers the largest value of R, thus simplifying the final matching design procedure. In this case, the first series element of such matching circuits should be the inductor to provide high impedance conditions for harmonics, as shown in Figure 11.41.

The peak collector current I_{max} and peak collector voltage V_{max} can be determined from Eqs. (11.118), (11.121), and (11.142) as

$$I_{max} = 2.647 I_0 \tag{11.144}$$

$$V_{max} = 3.647 V_{cc}. \tag{11.145}$$

The maximum frequency f_{max} can be calculated using Eqs. (11.139) and (11.141) when $C = C_{out}$, where C_{out} is the device output capacitance, as

$$f_{max} = 0.0798\frac{P_{out}}{C_{out}V_{cc}^2} \tag{11.146}$$

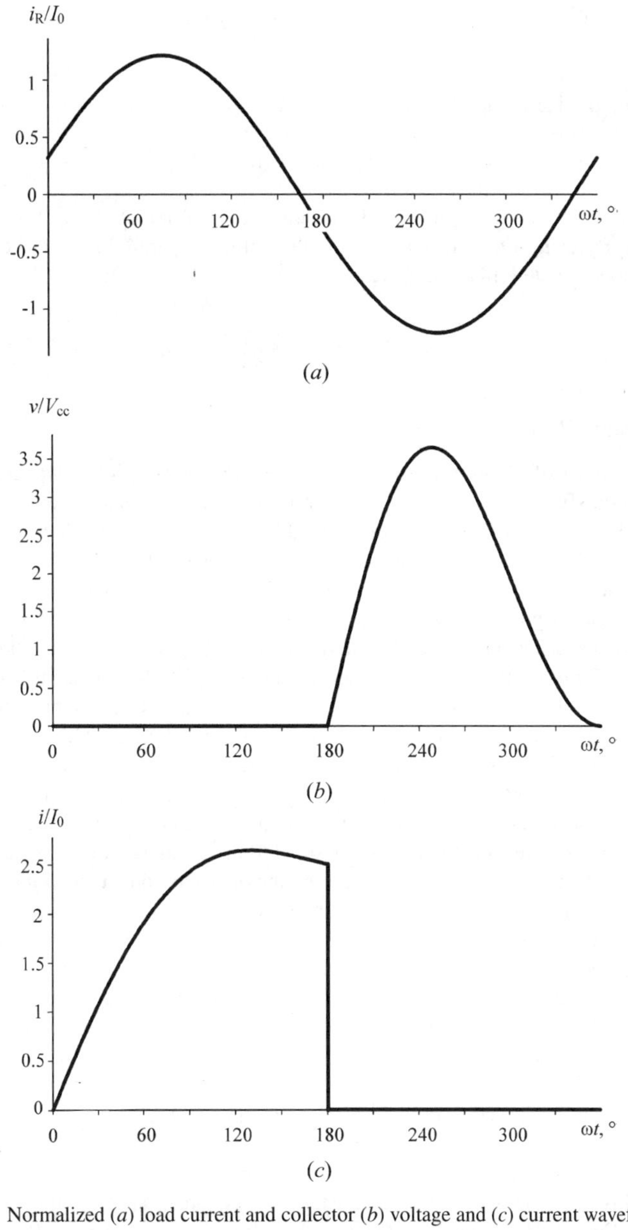

FIGURE 11.40 Normalized (*a*) load current and collector (*b*) voltage and (*c*) current waveforms for idealized optimum parallel-circuit Class E.

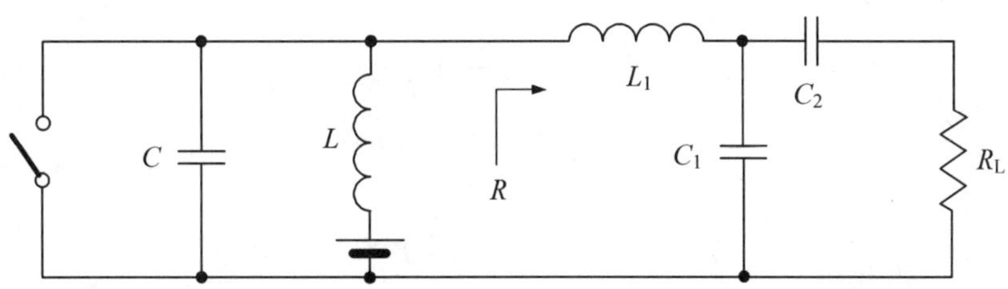

FIGURE 11.41 Parallel-circuit class E power amplifier with lumped matching circuit.

which is 1.4 times higher than maximum operation frequency for an optimum Class E power amplifier with shunt capacitance [60].

At microwave frequencies, the parallel inductance L can be replaced by a short-length short-circuited transmission line TL according to

$$Z_0 \tan\theta = \omega L \tag{11.147}$$

where Z_0 and θ are the characteristic impedance and electrical length of such a transmission line, respectively [61]. By using Eq. (11.140) defining the optimum parallel inductance L for a parallel-circuit Class E mode, Eq. (11.147) can be rewritten as

$$\tan\theta = 0.732 \frac{R}{Z_0}. \tag{11.148}$$

11.5.3 Broadband Class E

The conventional design of a high-efficiency switched-mode tuned Class E power amplifier requires a high Q_L-factor to satisfy the necessary harmonic impedance conditions at the output device terminal. However, if a sufficiently small value of the loaded quality factor Q_L is chosen, a high-efficiency broadband operation of the Class E power amplifier can be realized. For example, a simple network consisting of a series resonant LC circuit tuned to the fundamental frequency and a parallel inductor provides a constant load phase angle of 50° in a frequency range of about 50% [62]. For the first time, such a reactance compensation technique using a single-resonant circuit had been applied to the varactor tuned Gunn oscillator and parametric amplifier [63]. Moreover, it became possible to increase the tuning range of an oscillator by adding more stages of reactance compensation. For instance, for a resonant circuit having a 50-Ω load, an improvement of 4% in the tuning range can theoretically be achieved as a result of applying a double-resonant circuit reactance compensation, whereas, for a resonant circuit operating into 100-Ω load, an increase in the tuning range is 17% [64].

To describe reactance compensation circuit technique, let us consider the simplified equivalent load network with a series resonant L_0C_0 circuit tuned to the fundamental frequency and a shunt LC circuit providing a constant load phase angle seen by the device output, as shown in Figure 11.42(*a*). The reactances of the series and shunt resonant circuits vary with frequency, increasing in the case of a series resonant circuit and reducing in the case of a loaded parallel resonant circuit near the radial resonant frequency ω_0, as shown in Figure 11.42(*b*) by curve 1 and curve 2, respectively. Near the resonant frequency ω_0 of the series circuit with positive slope of its reactance, the slope of a shunt circuit reactance is negative. This reduces the overall reactance slope of the load network (curve 3). With a proper choice of the circuit elements, a constant load angle over a wide frequency bandwidth is established.

This technique can be directly applied to the parallel-circuit Class E power amplifier because its load-network configuration has an exactly the same structure [65,66]. The parallel-circuit configuration exactly matches the broadband operation conditions, unlike the Class E mode with shunt capacitance. In this case, the optimum load resistance R and phase angle ϕ of the load network can be obtained from Eqs. (11.139) and (11.143), respectively.

The load network input admittance $Y_{in} = 1/Z_{in}$ can be written by

$$Y_{in}(\omega) = \left(j\omega C + \frac{1}{j\omega L} + \frac{1}{R + j\omega' L_0} \right) \tag{11.149}$$

where

$$\omega' = \omega \left(1 - \frac{\omega_0^2}{\omega^2} \right) \tag{11.150}$$

and $\omega_0 = 1/\sqrt{L_0C_0}$ is the resonant frequency.

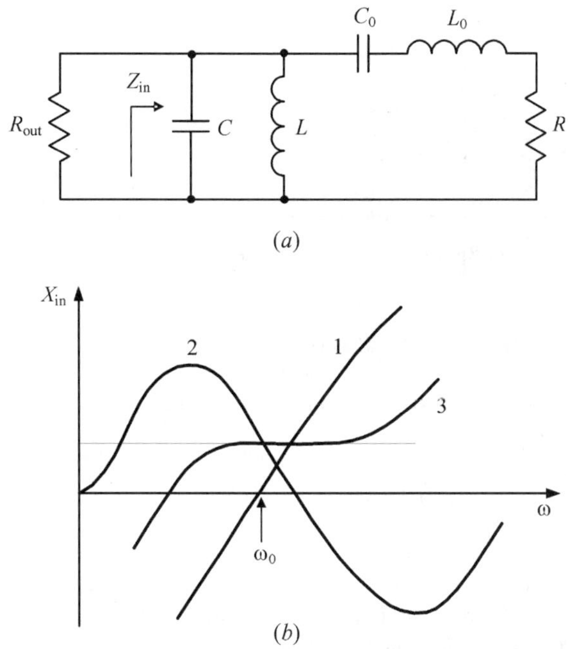

FIGURE 11.42 Single reactance compensation.

The parallel inductance L and shunt capacitance C required for an optimum switched-mode Class E operation are calculated as the functions of the load resistance R and radial frequency ω from Eqs. (11.140) and (11.141), respectively. The parameters of the series-resonant L_0C_0 circuit must be chosen to provide a constant phase angle of the load network over a required wide frequency bandwidth.

The frequency bandwidth will be maximized if, at a resonant frequency ω_0,

$$\left.\frac{dB_{in}(\omega)}{d\omega}\right|_{\omega=\omega_0} = 0 \tag{11.151}$$

where

$$B_{in}(\omega) = \text{Im}Y_{in}(\omega) = \omega C - \frac{1}{\omega L} - \frac{\omega' L_0}{R^2 + (\omega' L_0)^2} \tag{11.152}$$

is the load-network input susceptance. The concept of a susceptance compensation technique, which is equivalent to a reactance compensation technique, is used to simplify the calculation procedure. Thus, an additional equation can be written as

$$C + \frac{1}{\omega_0^2 L} - \frac{2L_0}{R^2} = 0. \tag{11.153}$$

As a result, by substituting Eqs. (11.140) and (11.141) into Eq. (11.153), the series capacitance C_0 and inductance L_0 can be calculated at the resonant frequency ω_0 by

$$L_0 = 1.026\frac{R}{\omega_0} \tag{11.154}$$

$$C_0 = 1/\omega_0^2 L_0. \tag{11.155}$$

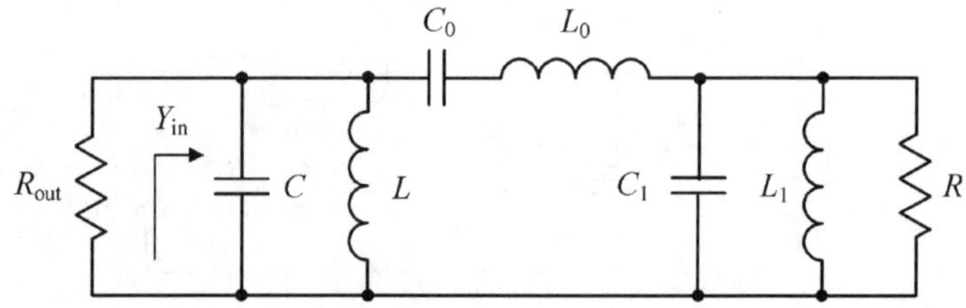

FIGURE 11.43 Double reactance compensation circuit.

Wider frequency bandwidth can be achieved using a double-resonant reactance compensation circuit shown in Figure 11.43, where L_0C_0 and L_1C_1 are the series and parallel resonant circuits, respectively. In this case, a system of the two additional equations can be used and solved according to

$$\left.\frac{dB_{in}(\omega)}{d\omega}\right|_{\omega=\omega_0} = \left.\frac{d^3B_{in}(\omega)}{d\omega^3}\right|_{\omega=\omega_0} = 0 \tag{11.156}$$

as the second derivative cannot provide an appropriate analytical expression.

To determine the load-network parameters for a double-resonant circuit reactance compensation with the input load network susceptance $B_{in} = \text{Im}Y_{in}$ given by

$$B_{in}(\omega) = \omega C - \frac{1}{\omega L} + \omega' \frac{C_1R^2\left[1-(\omega')^2L_0C_1\right] - L_0}{R^2\left[1-(\omega')^2L_0C_1\right]^2 + (\omega' L_0)^2} \tag{11.157}$$

it is necessary to solve simultaneously two following equations at a resonant frequency ω_0,

$$C + \frac{1}{\omega_0^2 L} - 2\frac{C_1R^2 - L_0}{R^2} = 0 \tag{11.158}$$

$$\frac{1}{\omega_0^2 L} + \frac{C_1R^2 - L_0}{R^2} - 8\omega_0^2 L_0\left[C_1^2 + \frac{\left(C_1R^2 - L_0\right)\left(L_0 - 2C_1R^2\right)}{R^4}\right] = 0. \tag{11.159}$$

As a result, the parameters of the series and shunt resonant circuits with the corresponding loaded quality factors $Q_0 = \omega_0 L_0/R$ and $Q_1 = \omega_0 C_1 R$ close to unity and greater – as a starting point for circuit optimization – can be calculated from

$$L_0 = \frac{R}{\omega_0}\frac{2}{\sqrt{5}-1} \qquad C_0 = \frac{1}{\omega_0^2 L_0} \tag{11.160}$$

$$C_1 = \frac{L_0}{R^2}\frac{3-\sqrt{5}}{2} \qquad L_1 = \frac{1}{\omega_0^2 C_1}. \tag{11.161}$$

The circuit simulations for these two types of reactance compensation load networks were performed at a resonant frequency $f_0 = 150$ MHz for a standard load resistance $R = 50\ \Omega$. Figure 11.44 shows the frequency dependencies of the load network phase angle ϕ for the single-reactance (curve 1) and double-reactance (curve 2) compensation circuits. It is apparent that the reactance-compensation technique provides very broadband operation conditions. Using just a single reactance load network yields a significant widening of the operating frequency bandwidth with minimum deviations of the magnitude and phase of the load-network input impedance. A double-reactance compensation load

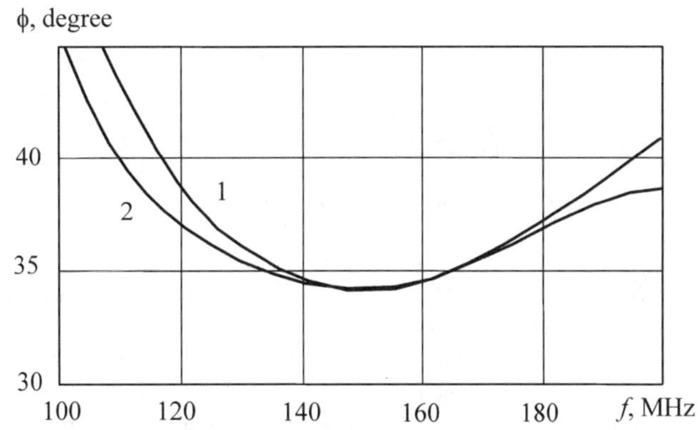

FIGURE 11.44 Reactance compensation load network broadband performance.

network obtains a maximum deviation from the optimum value of about 34° by only 3° in a frequency range of 120–180 MHz.

11.5.4 Power Gain

The load network corresponding to any type of Class E mode, in order to realize the idealized switching conditions, is tuned to provide inductive or capacitive impedance at the fundamental frequency, thus violating the conjugate matching conditions required for conventional Class B operation to provide maximum power delivery to the load. This means that generally the output voltage and current waves consist of both incident and reflected components. Besides, the power gain in a switching mode is normally lower than in a conventional mode because higher driving voltage is required to realize a device saturation mode. In this case, the ratio of the power gain in a Class E power amplifier to that in a Class B power amplifier, for the same output power, is inversely proportional to a squared ratio of their voltage peak factors [67]. For a Class E with one inductor and one capacitor, a maximum operating power gain $G_{P(E)max}$ of a single-stage bipolar power amplifier can be estimated by

$$G_{P(E)max} \approx \frac{\pi}{2}\frac{f_T}{f}\frac{V_{cc}}{V_{th}} \tag{11.162}$$

where f is the operating frequency, V_{th} is the threshold voltage, and it is assumed that the effect of a finite fall time is negligible [68].

However, it is very important to qualitatively compare the power gains of the Class B and Class E power amplifiers as the functions of the device and load network parameters. In this case, a parallel-circuit Class E mode looks very attractive since it provides the highest value of the load resistance compared with others Class E alternatives. The operating power gain G_P expressed through the active device Y-parameters and load can be obtained by

$$G_P = \frac{|Y_{21}|^2}{\mathrm{Re}Y_{in}}\frac{G}{|Y_{22} + Y_L|^2} \tag{11.163}$$

where Y_{21} and Y_{22} are the device transfer and driving-point output admittances, Y_{in} is the input admittance of the loaded device, and $Y_L = G + jB$ is the load admittance [4]. Since the power amplifier is operated in a nonlinear mode, the admittance Y-parameters of the active device are considered as linearized at the fundamental frequency. For example, for a power amplifier operating at the same conduction angle over various bias and drive conditions, these Y-parameters remain constant and operating power gain becomes a function of only the load admittance.

Equation (11.163) can be rewritten as

$$G_{\mathrm{P}} = \frac{|Y_{21}|^2}{\mathrm{Re}Y_{\mathrm{in}}} \frac{R}{\left(1 + \dfrac{G_{22}}{G}\right)^2 + \left(\dfrac{B_{22} + B}{G}\right)^2} \tag{11.164}$$

where $G_{22} = \mathrm{Re}Y_{22}$, $B_{22} = \mathrm{Im}Y_{22}$, and $R = 1/G$ is the load resistance. Without significant loss of accuracy, the output conjugate-matching condition between imaginary parts of the device output admittance and the load admittance can be replaced by a simple condition of $B_{22} + B = 0$, which means a resonance tuning of the load network including the device output susceptance B_{22}. Also, normally the device output conductance G_{22} is significantly smaller than the load conductance G for both MOSFET and bipolar transistors, especially at frequencies well below the device transition frequency f_{T}.

Consequently, the simplified ratio between the operating power gain $G_{\mathrm{P(E)}}$ of a parallel-circuit Class E power amplifier and the operating power gain $G_{\mathrm{P(B)}}$ of a conventional Class B power amplifier with conjugate-matched load can be written by

$$\frac{G_{\mathrm{P(E)}}}{G_{\mathrm{P(B)}}} = \frac{1}{1 + (B_{22} + B)^2 R_{(\mathrm{E})}^2} \frac{R_{(\mathrm{E})}}{R_{(\mathrm{B})}} \tag{11.165}$$

where $R_{(\mathrm{E})}$ is the load resistance of a Class E power amplifier and $R_{(\mathrm{B})}$ is the load resistance of a Class B power amplifier.

For an ideal optimum Class E operation mode with 100% collector efficiency, from Eqs. (11.128) and (11.139) it follows that

$$V_{\mathrm{R(E)}} = 1.652 V_{\mathrm{cc}}. \tag{11.166}$$

For the same output power P_{out} and taking into account that $V_{\mathrm{R(B)}} = V_{\mathrm{cc}}$ in a conventional Class B with zero saturation voltage, we can write

$$\frac{R_{(\mathrm{E})}}{R_{(\mathrm{B})}} = \frac{V_{\mathrm{R(E)}}^2}{V_{\mathrm{R(B)}}^2} = 2.729 \tag{11.167}$$

that shows the significantly higher value for the load resistance in an optimum parallel-circuit Class E operation mode.

As a result, the power gain ratio given by Eq. (11.165) can be rewritten in the form of

$$\frac{G_{\mathrm{P(E)}}}{G_{\mathrm{P(B)}}} = \frac{2.729}{1 + \tan^2\phi} = 1.865 \tag{11.168}$$

where $\phi = 34.244^\circ$ is the phase angle between the fundamental-frequency voltage and current components at the device output required for the optimum parallel-circuit Class E mode.

The result given by Eq. (11.168) means that, ideally, the operating power gain for a switched-mode parallel-circuit Class E mode compared to a conventional Class B mode is almost the same, and even slightly greater despite the mistuning of the load network. This can be explained by the larger value of the load resistance required for the optimum parallel-circuit Class E load network. For example, for a Class E with shunt capacitance, the operating power gain is smaller compared to a Class B power amplifier because its optimum Class E load resistance is about 2.4 times smaller than that of a parallel-circuit Class E mode [69]. The idealized conditions for a switched-mode operation can be achieved with instant on/off active device switching, which requires the rectangular input driving signal compared with sinusoidal driving signal for a conventional Class B mode. However, the power losses due to the finite switching time are sufficiently small and, for example, for switching time of $\tau_s = 0.35$ or 20° are only of about 1% [59]. Consequently, a slight overdrive of the active device is needed when the input power should be increased by 1–2 dB to minimize the switching time and maximize the collector efficiency of a switched-mode parallel-circuit Class E power amplifier. As a result, its resulting operating power gain becomes approximately equal to the operating power gain of a conventional Class B power amplifier.

11.6 CLASS E WITH QUARTERWAVE TRANSMISSION LINE

The ideal Class F load network with a quarterwave transmission line and a series L_0C_0 filter tuned to the fundamental frequency can provide a collector efficiency of 100% when the open-circuit conditions for odd-harmonic components and short-circuit conditions for even-harmonic components are realized. However, in practice, the idealized collector rectangular voltage and half-sinusoidal current waveforms cannot be realized at sufficiently high frequencies when an effect of the device output shunt capacitance C_{out} and series output inductance L_{out}, which can generally include the lead and bondwire device inductances, is significant. In this case, the device is operated during a finite time interval in active region as a current source, and the required optimum conditions can be provided only for the fundamental frequency and several higher order harmonic components. Fortunately, as it will be further demonstrated, the collector efficiency can be increased and effect of the output capacitance can be compensated with a proper value of this output series inductance realizing the switching Class E operation conditions. The obvious advantage of such a load network shown in Figure 11.45 is a combination of the high-operating efficiency corresponding to a Class E mode and an even-harmonic suppression due to a quarterwave transmission line used in a Class F mode.

11.6.1 General Analysis and Optimum Circuit Parameters

Figure 11.46 shows a Class E load network consisting of a shunt capacitor C, a series inductor L, a quarterwave transmission line, a series reactance X, a series resonant L_0C_0 circuit tuned to the fundamental frequency, and a load resistor R. The bottom end of a quarterwave transmission line is connected between a series inductor L and a series reactance X, while its top end is connected to a dc power supply being RF grounded through the bypass capacitor [3]. In a common case, a shunt capacitance C can represent the intrinsic device output capacitance and external circuit capacitance added by the load network. The series reactance X generally can be positive (inductance), negative (capacitance) or zero depending on the values of the shunt capacitance C and series inductance L. The active device is considered an ideal switch that is driven in such a way to provide the device switching between its on- and off-state operation conditions. To simplify an analysis of a Class E power amplifier

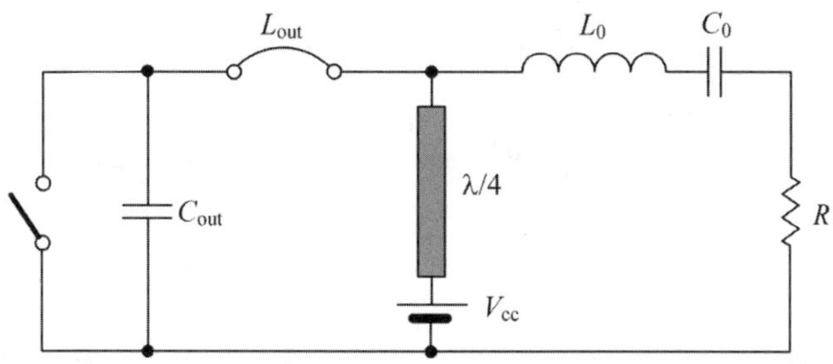

FIGURE 11.45 Class F load network with parasitic shunt capacitance and bondwire inductance.

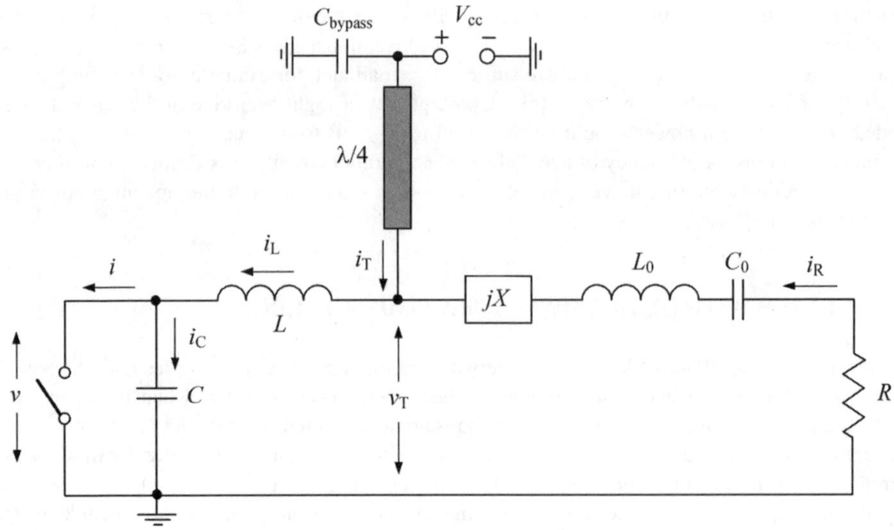

FIGURE 11.46 Equivalent circuit of Class E power amplifier with quarterwave transmission line.

with a quarterwave transmission line, it needs to introduce the preliminary assumptions similar to those for the Class E power amplifier with shunt capacitance. Assume that the losses in the reactive circuit elements are negligible, duty cycle is 50%, and loaded quality factor of the series L_0C_0 circuit is sufficiently high. For a lossless operation mode, it is necessary to provide the optimum zero voltage and zero voltage-derivative conditions for voltage $v(\omega t)$ across the switch just prior to the start of switch given by Eqs. (11.75) and (11.76).

Let the output current flowing into the load be sinusoidal as

$$i_R(\omega t) = I_R \sin(\omega t + \varphi) \tag{11.169}$$

where φ is the initial phase shift due to the shunt capacitance and series inductance.

When the switch is turned on for $0 \leq \omega t < \pi$, the current flowing through the shunt capacitance $i_C(\omega t) = 0$ and, consequently,

$$i(\omega t) = i_L(\omega t) = i_T(\omega t) + i_R(\omega t). \tag{11.170}$$

When the switch is turned off for $\pi \leq \omega t < 2\pi$, there is no current flowing through the switch when $i(\omega t + \pi) = 0$, and the current flowing through the shunt capacitance C is

$$i_C(\omega t + \pi) = i_L(\omega t + \pi) = i_T(\omega t + \pi) + i_R(\omega t + \pi). \tag{11.171}$$

To link up both Eqs. (11.170) and (11.171), each corresponding to one-half a period, it is necessary to use a basic equation for the current flowing into the quarterwave transmission line given by Eq. (11.50) as

$$i_T(\omega t) = i_T(\omega t + \pi) \tag{11.172}$$

which means that the period of a signal flowing into the quarterwave transmission line is equal to π because it contains only even harmonics.

Then,

$$i_{\mathrm{L}}(\omega t)=i_{\mathrm{T}}(\omega t+\pi)+i_{\mathrm{R}}(\omega t)=i_{\mathrm{L}}(\omega t+\pi)+i_{\mathrm{R}}(\omega t)-i_{\mathrm{R}}(\omega t+\pi) \tag{11.173}$$

resulting in

$$i_{\mathrm{L}}(\omega t)-i_{\mathrm{L}}(\omega t+\pi)=2i_{\mathrm{R}}(\omega t). \tag{11.174}$$

The current $i_{\mathrm{L}}(\omega t+\pi)=i_{\mathrm{C}}(\omega t+\pi)$ can be expressed through the voltages $v_{\mathrm{T}}(\omega t+\pi)$ and

$$v_{\mathrm{L}}(\omega t+\pi)=\omega L\frac{di_{\mathrm{L}}(\omega t+\pi)}{d\omega t} \tag{11.175}$$

as

$$i_{\mathrm{L}}(\omega t+\pi)=\omega C\frac{dv(\omega t+\pi)}{d\omega t}=\omega C\frac{d}{d\omega t}\left[v_{\mathrm{T}}(\omega t+\pi)-\omega L\frac{di_{\mathrm{L}}(\omega t+\pi)}{d\omega t}\right]. \tag{11.176}$$

Now we can use the equation for a voltage at the input of the quarterwave transmission line corresponding to each of a period given by Eq. (11.49) as

$$v_{\mathrm{T}}(\omega t)=2V_{\mathrm{cc}}-v_{\mathrm{T}}(\omega t+\pi). \tag{11.177}$$

Hence, when the switch is turned on resulting in $v_{\mathrm{T}}(\omega t)=v_{\mathrm{L}}(\omega t)$,

$$v_{\mathrm{T}}(\omega t+\pi)=2V_{\mathrm{cc}}-\omega L\frac{di_{\mathrm{L}}(\omega t)}{d\omega t}. \tag{11.178}$$

Substituting Eq. (11.178) into Eq. (11.176) and using Eqs. (11.179) and (11.174) yield a second-order nonhomogeneous differential equation corresponding to half a period of $\pi\leq\omega t<2\pi$ when $i_{\mathrm{L}}(\omega t+\pi)=i_{\mathrm{C}}(\omega t+\pi)$ in the form of

$$\frac{d^2 i_{\mathrm{C}}(\omega t+\pi)}{d(\omega t)^2}+\frac{q^2}{2}i_{\mathrm{C}}(\omega t+\pi)-I_{\mathrm{R}}\sin(\omega t+\varphi)=0 \tag{11.179}$$

or

$$\frac{d^2 i_{\mathrm{C}}(\omega t)}{d(\omega t)^2}+\frac{q^2}{2}i_{\mathrm{C}}(\omega t)+I_{\mathrm{R}}\sin(\omega t+\varphi)=0 \tag{11.180}$$

where

$$q=\frac{1}{\omega\sqrt{LC}}. \tag{11.181}$$

The general solution of Eq. (11.180) in the normalized form can be written by

$$\frac{i_{\mathrm{C}}(\omega t)}{I_{\mathrm{R}}}=C_1\cos\left(\frac{q\omega t}{\sqrt{2}}\right)+C_2\sin\left(\frac{q\omega t}{\sqrt{2}}\right)+\frac{2}{2-q^2}\sin(\omega t+\varphi) \tag{11.182}$$

where the coefficients C_1 and C_2 are determined from the initial off-state conditions.

The first initial condition is obtained from Eq. (11.174) as

$$i_{\mathrm{C}}(\omega t)|_{\omega t=\pi}=2i_{\mathrm{R}}(\pi) \tag{11.183}$$

taking into account that $i_{\mathrm{L}}(\pi)=i_{\mathrm{C}}(\pi)$ and $i_{\mathrm{L}}(2\pi)=i_{\mathrm{C}}(2\pi)=0$.

To obtain a second initial condition, let us substitute Eq. (11.175) in Eq. (11.178) and use Eq. (11.174). As a result,

$$\omega L\frac{di_L(\omega t+\pi)}{d\omega t} = 2V_{cc} - \omega L\frac{di_L(\omega t)}{d\omega t} = 2V_{cc} - \omega L\frac{di_L(\omega t+\pi)}{d\omega t} - 2\omega L\frac{di_R(\omega t)}{d\omega t}. \tag{11.184}$$

Then, by taking into account that $i_L(\pi) = i_C(\pi)$, we can write

$$\left.\frac{di_C(\omega t)}{d\omega t}\right|_{\omega t=\pi} = \frac{V_{cc}}{\omega L} - I_R\cos\varphi. \tag{11.185}$$

As a result, applying the initial conditions given by Eqs. (11.183) and (11.185) to Eq. (11.182) yields

$$C_1 = -\frac{\sqrt{2}}{qp}\sin\left(\frac{q\pi}{\sqrt{2}}\right) - \frac{q\sqrt{2}}{2-q^2}\sin\left(\frac{q\pi}{\sqrt{2}}\right)\cos\varphi - 2\frac{1-q^2}{2-q^2}\cos\left(\frac{q\pi}{\sqrt{2}}\right)\sin\varphi \tag{11.186}$$

$$C_2 = \frac{\sqrt{2}}{qp}\cos\left(\frac{q\pi}{\sqrt{2}}\right) + \frac{q\sqrt{2}}{2-q^2}\cos\left(\frac{q\pi}{\sqrt{2}}\right)\cos\varphi - 2\frac{1-q^2}{2-q^2}\sin\left(\frac{q\pi}{\sqrt{2}}\right)\sin\varphi \tag{11.187}$$

where

$$p = \frac{\omega L I_R}{V_{cc}}. \tag{11.188}$$

The dc supply current I_0 can be written using the Fourier formula and Eqs. (11.171) and (11.172) by

$$I_0 = \frac{1}{2\pi}\int_0^{2\pi} i_T(\omega t)\,d\omega t = \frac{1}{\pi}\int_0^{\pi} i_T(\omega t+\pi)\,d\omega t = \frac{1}{\pi}\int_0^{\pi}\left[i_C(\omega t+\pi) - i_R(\omega t+\pi)\right]d\omega t. \tag{11.189}$$

Then, substituting Eqs. (11.169) and (11.182) into Eq. (11.189) results in

$$I_0 = \frac{I_R}{\pi}\left\{C_1\frac{\sqrt{2}}{q}\left[\sin\left(q\pi\sqrt{2}\right) - \sin\left(\frac{q\pi}{\sqrt{2}}\right)\right] - C_2\frac{\sqrt{2}}{q}\left[\cos\left(q\pi\sqrt{2}\right) - \cos\left(\frac{q\pi}{\sqrt{2}}\right)\right] - \frac{2q^2}{2-q^2}\cos\varphi\right\}. \tag{11.190}$$

The voltage $v(\omega t)$ across the switch is produced by charging of the shunt capacitor C by the current given by Eq. (11.182) according to

$$\begin{aligned} v(\omega t) = \frac{1}{\omega C}\int_{\pi}^{\omega t} i_C(\omega t)\,d\omega t = \frac{I_R\sqrt{2}}{q\omega C}\left\{C_1\left[\sin\left(\frac{q\omega t}{\sqrt{2}}\right) - \sin\left(\frac{q\pi}{\sqrt{2}}\right)\right]\right. \\ \left. - C_2\left[\cos\left(\frac{q\omega t}{\sqrt{2}}\right) - \cos\left(\frac{q\pi}{\sqrt{2}}\right)\right] - \frac{q\sqrt{2}}{2-q^2}\left[\cos(\omega t+\varphi) + \cos\varphi\right]\right\}. \end{aligned} \tag{11.191}$$

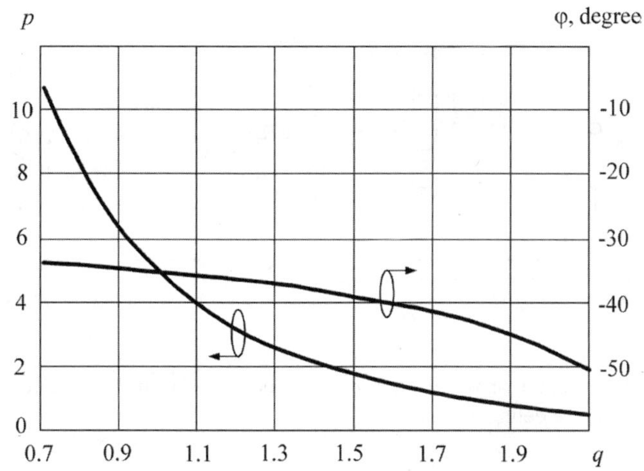

FIGURE 11.47 Optimum quarterwave-line Class E parameters p and φ versus q.

Generally, Eq. (11.191) for the collector voltage contains the three unknown parameter q, p, and φ, which must be determined. In a common case, the parameter q can be considered a variable, and the other two parameters p and φ are determined from a system of the two equations resulting from applying of the two optimum zero voltage and zero voltage-derivative conditions given by Eq. (11.75) and (11.76) to Eq. (11.191). Figure 11.47 shows the dependences of the optimum parameters p and φ versus q for a Class E with a parallel quarterwave transmission line.

The current $i(\omega t) = i_L(\omega t)$ flowing through the inductor L and switch when switch is turned on for $0 \leq \omega t < \pi$ can be written using Eqs. (11.170) to (11.172) and (11.182) in a normalized form as

$$\frac{i(\omega t)}{I_R} = C_1 \cos\left[\frac{q}{\sqrt{2}}(\omega t + \pi)\right] + C_2 \sin\left[\frac{q}{\sqrt{2}}(\omega t + \pi)\right] + 2\frac{1-q^2}{2-q^2}\sin(\omega t + \varphi). \tag{11.192}$$

Then, the normalized voltage $v_T(\omega t)$ during this period can be obtained by

$$\frac{v_T(\omega t)}{V_{cc}} = \frac{\omega L I_R}{V_{cc}}\frac{di_L(\omega t)}{d\omega t} = p\left\{-C_1\frac{q}{\sqrt{2}}\sin\left[\frac{q}{\sqrt{2}}(\omega t + \pi)\right]\right.$$
$$\left. + C_2\frac{q}{\sqrt{2}}\cos\left[\frac{q}{\sqrt{2}}(\omega t + \pi)\right] + 2\frac{1-q^2}{2-q^2}\cos(\omega t + \varphi)\right\}. \tag{11.193}$$

However, when switch is turned off for $\pi \leq \omega t < 2\pi$, it can be calculated from Eq. (11.177).

To calculate the optimum load network parameters for a Class E with a parallel quarterwave transmission line, first it is necessary to define the reactive quadrature fundamental-frequency Fourier component of the voltage $v_T(\omega t)$ according to

$$V_X = -\frac{1}{\pi}\int_0^{\pi} v_T(\omega t)\cos(\omega t + \varphi)\,d\omega t - \frac{1}{\pi}\int_{\pi}^{2\pi}[2V_{cc} - v_T(\omega t - \pi)]\cos(\omega t + \varphi)\,d\omega t. \tag{11.194}$$

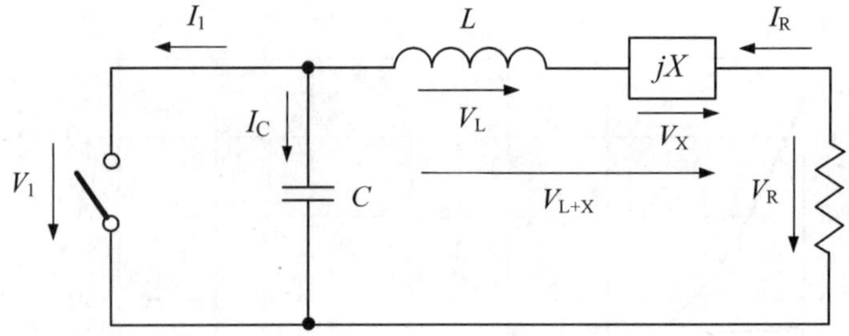

FIGURE 11.48 Equivalent quarterwave-line Class E load network at fundamental frequency.

Then, the two active and reactive quadrature fundamental-frequency Fourier components of the collector voltage $v(\omega t)$ are defined from

$$V_R = -\frac{1}{\pi}\int_0^{2\pi} v(\omega t)\sin(\omega t + \varphi)\,d\omega t \tag{11.195}$$

$$V_{L+X} = -\frac{1}{\pi}\int_0^{2\pi} v(\omega t)\cos(\omega t + \varphi)\,d\omega t. \tag{11.196}$$

Finally, as it follows from Figure 11.48, the optimum normalized series reactance X, series inductance L, and shunt capacitance C can be calculated from

$$\frac{X}{R} = \frac{V_X}{V_R} \tag{11.197}$$

$$\frac{\omega L}{R} = \frac{V_{L+X}}{V_R} - \frac{X}{R} \tag{11.198}$$

$$\omega CR = \frac{1}{q^2 \dfrac{\omega L}{R}}. \tag{11.199}$$

By taking into account that $R = V_R^2/2P_{out}$, the optimum load resistance R for the specified values of a supply voltage V_{cc} and an output power P_{out} delivered to the load can be obtained by

$$R = \frac{1}{2}\left(\frac{V_R}{V_{cc}}\right)^2 \frac{V_{cc}^2}{P_{out}} \tag{11.200}$$

or, using Eq. (11.188), by

$$R = \frac{1}{2}\left(\frac{p}{l}\right)^2 \frac{V_{cc}^2}{P_{out}} \tag{11.201}$$

where $l = \omega L/R$.

The dependences of the normalized optimum series inductance $\omega L/R$ and series reactance X/R are shown in Figure 11.49(*a*), while the dependences of the normalized optimum shunt capacitance ωCR and load resistance RP_{out}/V_{cc}^2 are plotted in Figure 11.49(*b*). Here, we can see that the greater values of the normalized series inductance $\omega L/R$ can be achieved for the lower values of q. To compensate for such an increased inductive value, the reactance X should have a negative capacitive

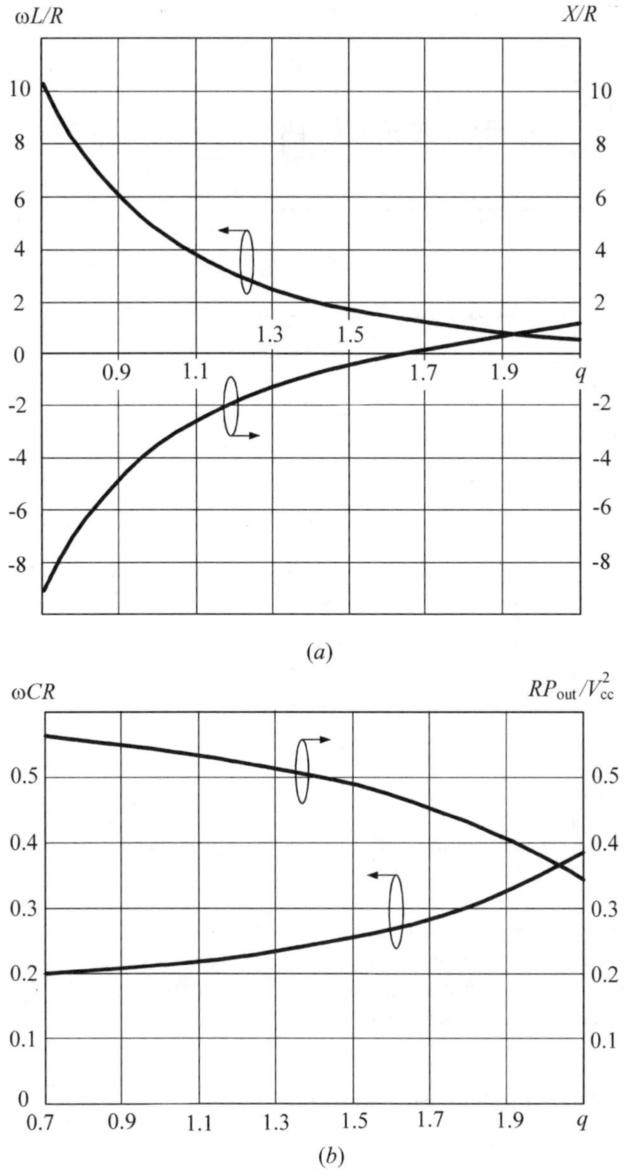

FIGURE 11.49 Normalized optimum quarterwave-line Class E load network parameters.

value. Generally, the value of the series reactance X changes its sign from negative to positive, which means that the capacitive reactance is followed by the inductive reactance, and it is required to add an additional inductance at higher values of q. As a result, there is special case of a load network with $X = 0$ when there is no need for additional phase compensation. The variations of normalized values of ωCR and RP_{out}/V_{cc}^2 versus q are not so significant.

Figure 11.50 shows the circuit schematics of the Class E power amplifiers with a parallel quarterwave transmission lines corresponding to the different values of the optimum parameter q. The inclusion of a series capacitance C_x is necessary to compensate for the excessive inductive reactance at the fundamental frequency when the series reactance X is negative, as shown in Figure 11.50(a).

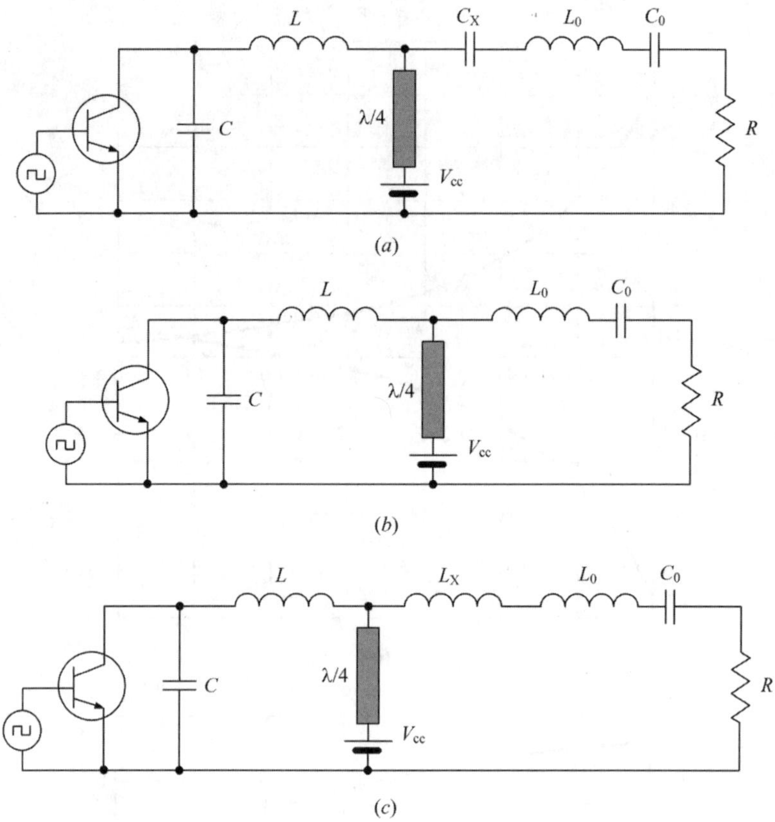

FIGURE 11.50 Schematic of Class E power amplifiers with quarterwave line.

However, when the series reactance X is positive, an additional series inductance L_X shown in Figure 11.50(*c*) is necessary to increase the total series inductance at the fundamental frequency. The special case is a zero series reactance X corresponding to a circuit schematic shown in Figure 11.50(*b*), which we will further consider in more detail.

11.6.2 Load Network with Zero Series Reactance

In this case, the Class E load network consists of a shunt capacitor C, a series inductor L, a parallel quarterwave transmission line connected to a voltage supply, a series L_0C_0 resonant circuit tuned to the fundamental frequency, and a load resistor R. Because the parameter q corresponding to a zero reactance X is unknown a priori, generally it is necessary to solve a system of three equations to define the three unknown parameters q, p, and φ. The two equations are the result of applying of the two optimum zero voltage and zero voltage-derivative conditions given by Eqs. (11.75) and (11.76) to Eq. (11.191). Since the fundamental component of the voltage $v_T(\omega t)$ is fully applied to the load, this means that its reactive part must have zero value resulting in an additional equation

$$V_X = -\frac{1}{\pi}\int_0^{\pi} v_T(\omega t)\cos(\omega t + \varphi)\,d\omega t - \frac{1}{\pi}\int_{\pi}^{2\pi}[2V_{cc} - v_T(\omega t - \pi)]\cos(\omega t + \varphi)\,d\omega t = 0. \tag{11.202}$$

As a result, the exact values, presented in [3,55], are obtained numerically for the unknown parameters as

$$q = 1.649 \tag{11.203}$$

$$p = 1.302 \tag{11.204}$$

$$\varphi = -40.8^\circ. \tag{11.205}$$

Figure 11.51 shows the normalized (*a*) load current, (*b*) collector voltage, and (*c*) current waveforms for idealized optimum Class E mode with a parallel quarterwave transmission line. From the collector voltage and current waveforms it follows that, when the transistor is turned on, there is no voltage across the switch, and the collector current consisting of the load sinusoidal current shown in Figure 11.51(*a*) and transmission-line current shown in Figure 11.52(*a*) flows through the switch. However, when the transistor is turned off, this current with the waveform shown in Figure 11.52(*b*) now flows through the shunt capacitance, the charging process of which produces the collector voltage.

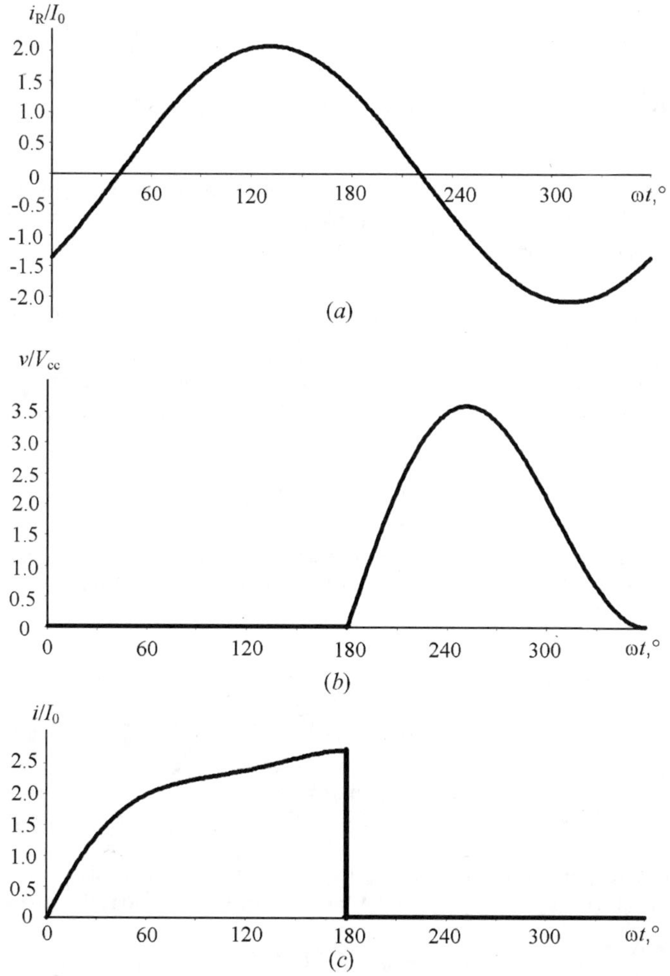

FIGURE 11.51 Voltage and current waveforms of quarterwave-line Class E power amplifier.

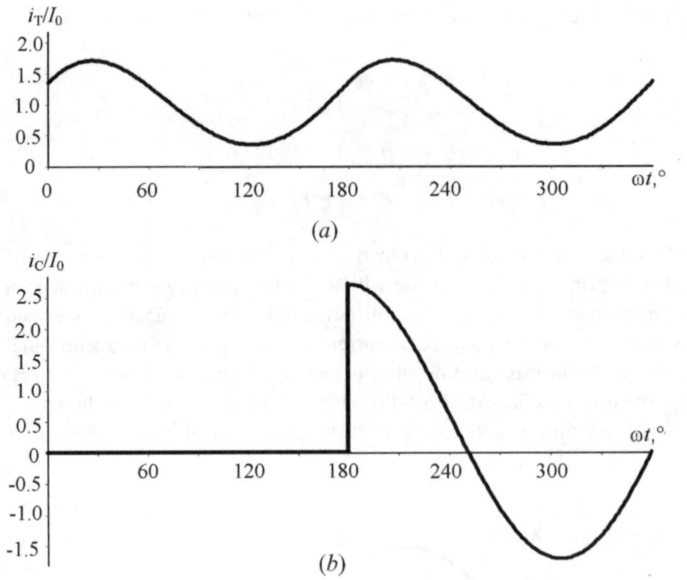

FIGURE 11.52 Current waveforms of quarterwave-line Class E power amplifier.

In an idealized Class E operation mode, there is no nonzero voltage and current simultaneously that means a lack of the power losses and gives an idealized collector efficiency of 100%. This implies that the dc power and fundamental output power are equal,

$$I_0 V_{cc} = \frac{I_R V_R}{2} \tag{11.206}$$

where $V_R = I_R R$ is the fundamental voltage amplitude across the load resistance R.

By using Eqs. (11.188) and (11.206), the normalized inductance can be defined as

$$\frac{\omega L}{R} = \frac{p}{2}\left(\frac{I_0}{I_R}\right)^{-1}. \tag{11.207}$$

As a result, the exact values of the optimum series inductance L, shunt capacitance C, and load resistance R can be calculated by using Eqs. (11.190), (11.199), and (11.201) as

$$L = 1.349\frac{R}{\omega} \tag{11.208}$$

$$C = \frac{0.2725}{\omega R} \tag{11.209}$$

$$R = 0.465\frac{V_{cc}^2}{P_{out}}. \tag{11.210}$$

The peak collector current I_{max} and peak collector voltage V_{max} can be determined directly from Eqs. (11.191) and (11.192) using Eq. (11.190) from numerical calculations that gives

$$I_{max} = 2.714\, I_0 \tag{11.211}$$

$$V_{max} = 3.589\, V_{cc}. \tag{11.212}$$

TABLE 11.3 Optimum Impedances at Fundamental and Harmonics for Different Class E Load Networks.

Class E load network	f_0 (Fundamental)	$2nf_0$ (Even Harmonics)	$(2n+1)f_0$ (Odd Harmonics)
Class E with shunt capacitance	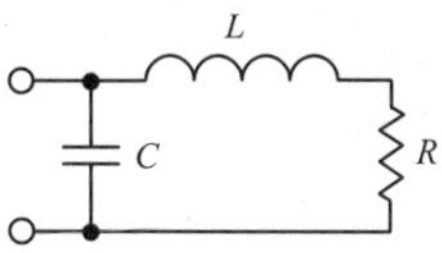	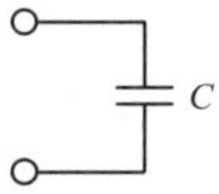	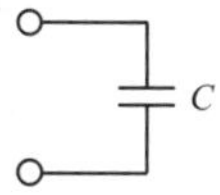
Class E with parallel circuit	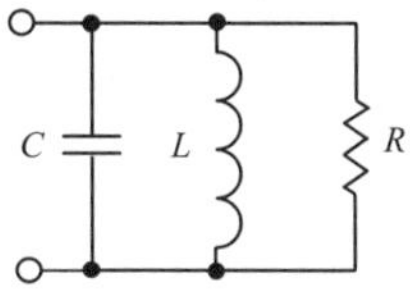	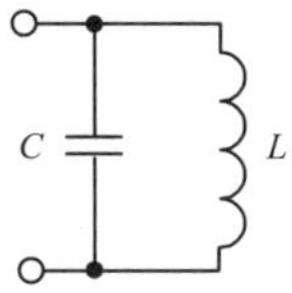	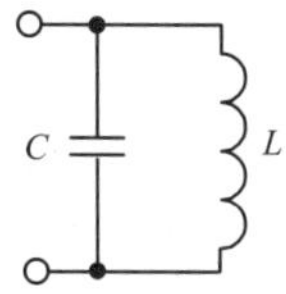
Class E with quarterwave transmission line	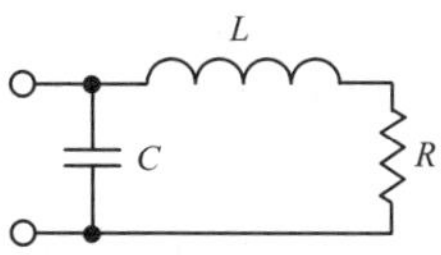	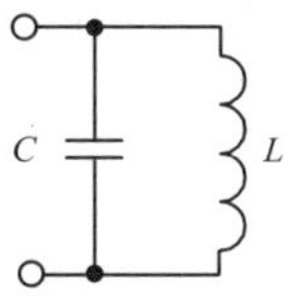	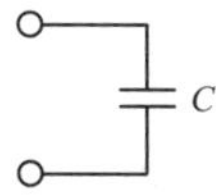

Using Eqs. (11.209) and (11.210) with $C = C_{out}$, where C_{out} is the device output capacitance, gives the value of a maximum operation frequency f_{max} of

$$f_{max} = 0.093 \frac{P_{out}}{C_{out} V_{cc}^2} \tag{11.213}$$

which is 1.63 times higher than the maximum operation frequency for an optimum Class E mode with shunt capacitance [60].

In Table 11.3, the optimum impedances seen by the device collector at the fundamental-frequency and higher order harmonic components are illustrated by the appropriate circuit configurations. It can be seen that Class E mode with a quarterwave transmission line shows different impedance properties at even and odd harmonics. At odd harmonics, the optimum impedances can be established by the shunt capacitance that is required for all harmonic components in Class E with a shunt capacitance At even harmonics, the optimum impedances are realized by using a parallel LC circuit that is required for all harmonic components in Class E with a parallel circuit, which is a subclass of a Class E with finite dc-feed inductance. Thus, the frequency properties of a grounded parallel quarterwave transmission line with its open-circuit conditions at odd harmonics and short-circuit conditions at even harmonics enable Class E with a quarterwave transmission line to combine simultaneously the harmonic impedance conditions typical for both Class E with a shunt capacitance and Class E with a parallel circuit.

11.6.3 Matching Circuits with Lumped and Distributed Parameters

The theoretical results obtained for the Class E power amplifier with a quarterwave transmission line show that it is enough to use a very simple load network to realize the optimum impedance conditions even for four harmonics. As follows from Figure 11.52(a), the current flowing into the quarterwave transmission line is very close to the sinusoidal second-harmonic current, which means that the level of fourth-order and higher order harmonic components is negligible because of the significant shunting effect of the capacitance C. In this case, as the shunt capacitor C and series inductor L both provide

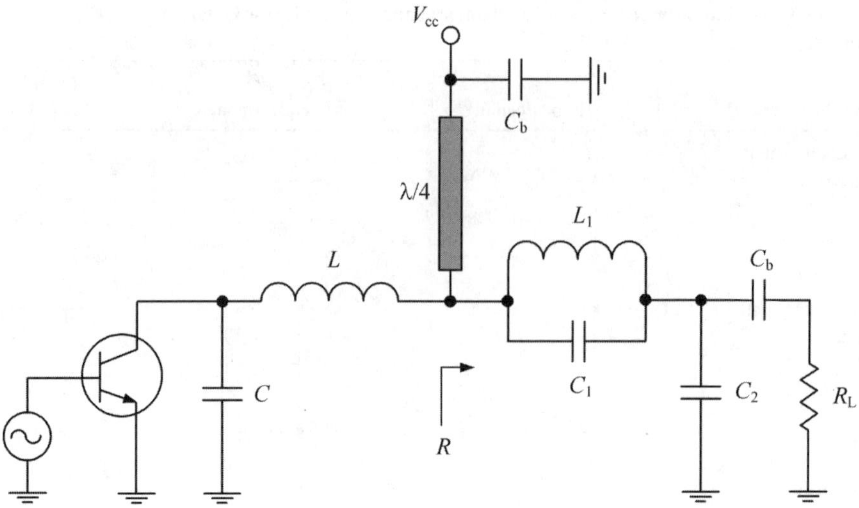

FIGURE 11.53 Schematic of quarterwave-line Class E power amplifier with lumped matching circuit.

optimum inductive impedance at the fundamental frequency and the quarterwave transmission line realizes the shorting of even harmonics, it is only required to provide an open-circuit condition at the third-harmonic component. Consequently, when the ideal series L_0C_0 circuit is replaced by the output matching circuit, the optimum impedance conditions for Class E load network with a quarterwave transmission line can be practically fully realized by simply providing an open-circuit condition at the third-harmonic component.

Figure 11.53 shows the circuit schematic of a lumped Class E power amplifier with a parallel quarterwave transmission line, where the parallel resonant L_1C_1 circuit is tuned to the third-harmonic component and C_b represents the blocking or bypass capacitor. Since the reactance of the parallel third-harmonic tank circuit is inductive at the fundamental, it is enough to use the shunt capacitance C_2 composing the L-type low-pass matching circuit to provide the required impedance matching of the optimum Class E load resistance R with the standard load impedance of $R_L = 50\ \Omega$. In this case, it is assumed that $R < R_L$, which is normally the case for the high-power or low-voltage power amplifiers.

To calculate the parameters of the matching-circuit elements, it is convenient to consider the loaded quality factor $Q_L = \omega C_2 R_L$, which can also be expressed through the resistances R and R_L as

$$Q_L = \sqrt{\left(\frac{R_L}{R}\right) - 1}. \tag{11.214}$$

As a result, the matching circuit parameters can be calculated at the radial fundamental frequency ω_0 from

$$C_2 = \frac{Q_L}{\omega_0 R_L} \tag{11.215}$$

$$L_1 = \frac{8}{9}\frac{Q_L R}{\omega_0} \tag{11.216}$$

$$C_1 = \frac{1}{9\omega_0^2 L_1}. \tag{11.217}$$

At microwaves, the series lumped inductor L can be replaced by the short-length series transmission line. In this case, when the shunt capacitance C represents a fully internal active device output

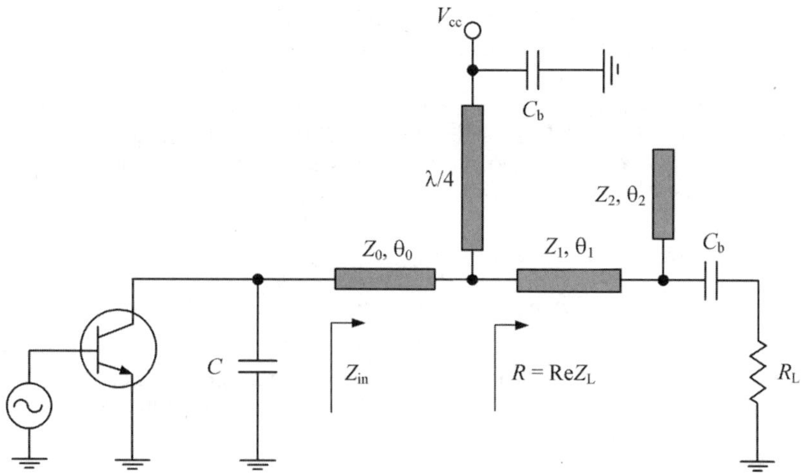

FIGURE 11.54 Schematic of transmission-line Class E power amplifier.

capacitance, the bondwire and lead inductances can be taken into account because they provide an inductive reactance and make the series transmission line shorter. Figure 11.54 shows the circuit schematic of a transmission-line Class E power amplifier with a parallel quarterwave transmission line. The output matching circuit represents the *L*-type low-pass matching circuit consisting of a series transmission line with a short electrical length θ_1, which provides an inductive reactance, and a shunt open-circuit stub with an electrical length θ_2 of less than 90°, which provides a capacitive reactance.

Usually, the characteristic impedance Z_0 of a transmission line (in most practical cases equal to 50 Ω) is much higher than the required optimum Class E load network resistance *R*. Consequently, the input impedance Z_{in} of the loaded series transmission line with the characteristic impedance Z_0 and electrical length θ_0 under the condition of $(R/Z_0)\tan\theta_0 \ll 1$ when the electrical length of a transmission line is less than 45° at the fundamental is determined by

$$Z_{in} = Z_0 \frac{R + jZ_0 \tan\theta_0}{Z_0 + jR\tan\theta_0} = Z_0 \frac{\dfrac{R}{Z_0} + j\tan\theta_0}{1 + j\dfrac{R}{Z_0}\tan\theta_0} \cong R + jZ_0\tan\theta_0. \tag{11.218}$$

As a result, the required optimum value of θ_0 for Class E mode with a quarterwave transmission line using Eqs. (11.208) and (11.218) can be obtained from

$$\theta_0 = \tan^{-1}\left(1.349\frac{R}{Z_0}\right). \tag{11.219}$$

The output matching circuit is necessary to match the required optimum Class E resistance *R* calculated in accordance with Eq. (11.210) to a standard load resistance of 50 Ω. In addition, it is required to provide an open-circuit condition at the third-harmonic component. This can be easily done using the output matching topology in the form of an *L*-type transformer with the series transmission line and open-circuit stub. In this case, the electrical lengths of the series transmission line and open-circuit stub should be chosen to be 30° each. The load impedance Z_L seen by a quarterwave transmission line can be written by

$$Z_L = Z_1 \frac{R_L\left(Z_2 - Z_1\tan^2\theta\right) + jZ_1Z_2\tan\theta}{Z_1Z_2 + j\left(Z_1 + Z_2\right)R_L\tan\theta} \tag{11.220}$$

where Z_1 and θ_1 are the characteristic impedance and electrical length of the series transmission line, and Z_2 and θ_2 are the characteristic impedance and electrical length of the open-circuit stub, and $\theta = \theta_1 = \theta_2 = 30°$.

Hence, the complex-conjugate matching with the load can be provided by proper choice of the characteristic impedances Z_1 and Z_2. Separating Eq. (11.220) into real and imaginary parts and taking into account that $\text{Re}Z_L = R$ and $\text{Im}Z_L = 0$, the following system of two equations with two unknown parameters is obtained:

$$Z_1^2 Z_2^2 - R_L^2 (Z_1 + Z_2) \left(Z_2 - Z_1 \tan^2\theta\right) = 0 \tag{11.221}$$

$$(Z_1 + Z_2)^2 R_L^2 R \tan^2\theta - Z_1^2 Z_2^2 \left[R_L \left(1 + \tan^2\theta\right) - R\right] = 0 \tag{11.222}$$

which enables the characteristic impedances Z_1 and Z_2 to be properly calculated.

This system of two equations can be explicitly solved as a function of the parameter $r = R_L/R$ resulting in

$$\frac{Z_1}{R_L} = \frac{\sqrt{4r - 3}}{r} \tag{11.223}$$

$$\frac{Z_1}{Z_2} = 3\left(\frac{r-1}{r}\right). \tag{11.224}$$

Consequently, for the specified value of the parameter r with the required Class E optimum load resistance R and standard load $R_L = 50\ \Omega$, the characteristic impedance Z_1 is calculated from Eq. (11.223) and then the characteristic impedance Z_2 is calculated from Eq. (11.224). For example, if the required Class E optimum load resistance R is equal to 12.5 Ω resulting in $r = 4$, the characteristic impedance of the series transmission line Z_1 is equal to 45 Ω and the characteristic impedance of the open-circuit stub Z_2 is equal to 20 Ω.

Unlike the transmission-line Class E load network approximations with two-harmonic control [43] and with three-harmonic control [44], the Class E load network with a quarterwave transmission line, which can provide the optimum impedance conditions for at least four-harmonic components, is very simple in circuit implementation and does not require an additional lumped RF choke element. In addition, there is no need to use the special computer simulation tools which are generally required to calculate the parameters of the Class E transmission-line load-network topologies, since all parameters of the Class E load network with a quarterwave transmission line and the output matching circuit parameters are easily calculated explicitly from simple analytical equations. Besides, such a Class E load network with a quarterwave transmission line is very useful in practical design by providing a significant higher order harmonic suppression.

11.7 CLASS FE

In practice, the idealized collector voltage and current waveforms corresponding to Class F mode can be realized at sufficiently low frequencies when effect of the device collector capacitance is negligible. At higher frequencies, the effect of the collector capacitance contributes to a finite switching time resulting in the time periods when the collector voltage and collector current exist at the same time, that is, simultaneously $v > 0$ and $i > 0$. As a result, such a load network with shunt capacitance cannot provide the switched-mode operation with an instantaneous transition from the device pinch-off to saturation mode. Therefore, during a finite time interval the device operates in active region as a nonlinear current source with the reverse-biased collector–base junction and the collector current is provided by this current source.

To minimize power losses during transients, the optimum zero voltage and zero voltage-derivative Class E conditions given by Eqs. (11.75) and (11.76) can be applied to Class F circuit with additionally connected shunt capacitance providing soft-switching operation mode [70]. Experimental results demonstrate that high drain efficiency of more than 80% can be achieved with Class C biasing at operating frequency of 13.56 MHz with maximum peak factor of about two, which is significantly lower than that of in a switched-mode Class E mode.

It should be mentioned that the rectangular voltage and half-sine current collector waveform can be provided by a push–pull voltage-switching Class D power amplifier with the fundamentally tuned series L_0C_0 filter, circuit schematic of which is shown in Figure 11.55(*a*). In this case, the

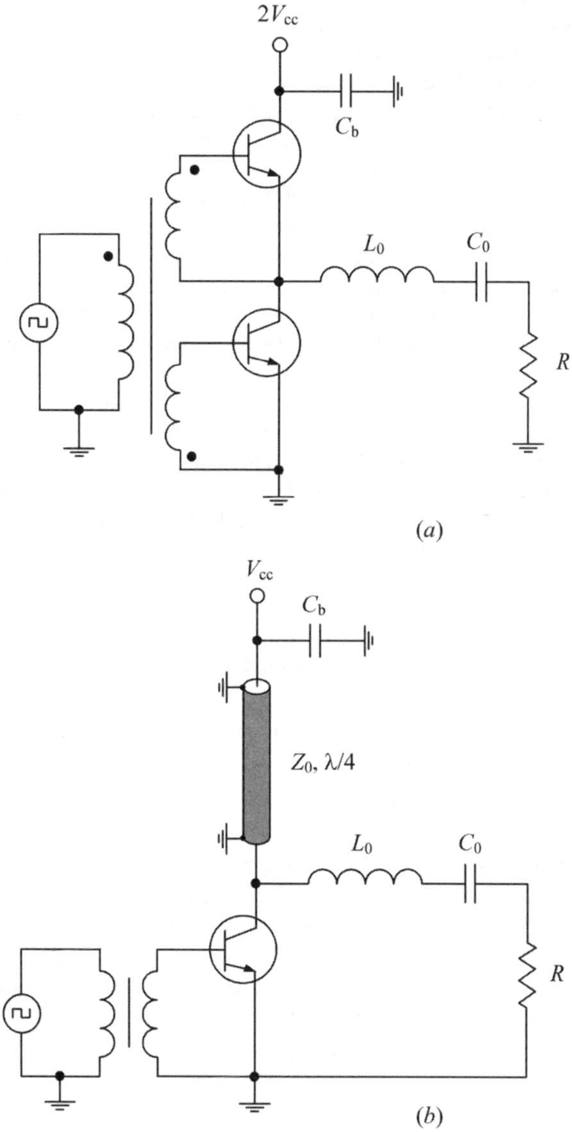

(*a*)

(*b*)

FIGURE 11.55 Voltage-switching Class D and Class F power amplifiers.

bipolar transistors must be driven in such a way as to provide an alternate voltage switching between the on- and off-state modes. However, in a single-ended power amplifier with a parallel quarterwave transmission line shown in Figure 11.55(*b*), the symmetrizing action of the transmission-line behavior can easily provide a function of the voltage inverter resulting in the same collector waveforms with voltage peak factor of two. In this case, the quarterwave line stores the voltage waveform in a traveling wave along its length, which returns delayed by one-half fundamental period and inverted, because of power-reflection condition at the short-circuit termination. However, the presence of parasitic collector capacitances degrades the performances of both power amplifiers due to finite time of the charging and discharging processes. As a result, the frequency limitations of both power amplifiers are provided by the device parasitic collector shunt capacitances, resulting in the increased switching transition times due to the capacitor charging and discharging processes, and the fact that transistor switches charge and discharge the shunt capacitance, dissipating power in the charging and discharging processes.

A possible way to eliminate these power losses and to extend the voltage-switching class D mode to higher frequencies is to introduce a dead time during the period when one device has already turned off but the other has not turned on yet, and the inductive load network is used to charge and discharge the shunt capacitances. This can be done by introducing the Class E switching conditions when the switching loss during off-to-on transition is reduced to zero by the operating requirements of zero voltage at zero voltage slope at the end of period. Since the shunt capacitor must be discharged at that time instant, an additional series inductor L with optimum value should be included into the load network, as shown in Figure 11.56(*a*). As a result, the switching current and voltage waveforms have the characteristics of both Class D and Class E operation modes, and such an operation mode is called Class DE. Such a Class DE power amplifier was firstly described by Zhukov and Kozyrev and has found some particular applications due to its high operation efficiency at higher operating frequencies for driving signals with duty cycles $D < 0.5$ [3,71]. Similarly, a quarterwave transmission line can be used as a voltage inverter resulting in a single-ended Class FE (or Class EF) power amplifier, which combines a class F operating mode with Class E switching conditions.

The optimum parameters of a Class FE power amplifier can be determined based on its steady-state collector voltage and current waveforms [72]. Figure 11.57 shows the two equivalent Class FE switching circuits that occur during a switching cycle, the load network of which consists of a grounded parallel quarterwave transmission line, a shunt capacitance C, a series inductor L, a series fundamentally tuned L_0C_0 circuit, and a load resistor R. An active device is considered an ideal switch that is driven in such a way as to provide the device switching between its on- and off-state operation conditions.

To simplify the analysis of a Class FE power amplifier, the following assumptions are introduced:

- The transistor has zero saturation voltage, zero saturation resistance, infinite off-resistance, and its switching is instantaneous and lossless.
- The shunt capacitance is assumed to be linear.
- The loaded quality factor $Q_L = \omega L_0/R = 1/\omega C_0 R$ of the series resonant L_0C_0 circuit tuned to the fundamental frequency is high enough for the output current to be sinusoidal at the switching frequency.
- There are no losses in the circuit except only into the load R.

The Class E switching conditions for the transistor switch can be written as

$$v\left(\omega t\right)\big|_{\omega t=2\pi} = 0 \tag{11.225}$$

$$\left.\frac{dv\left(\omega t\right)}{d\omega t}\right|_{\omega t=2\pi} = 0 \tag{11.226}$$

where v is the voltage across the switch.

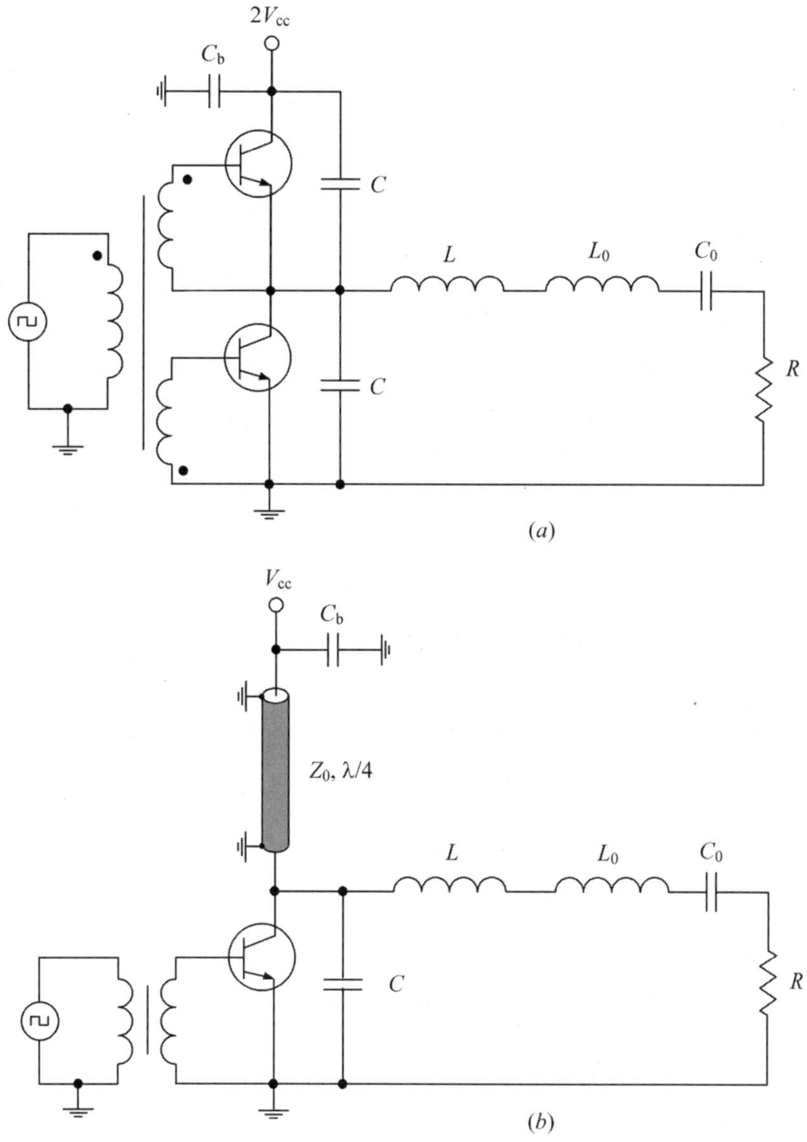

FIGURE 11.56 Voltage-switching Class DE and Class FE power amplifiers.

Let the output current flowing through the load be written as sinusoidal,

$$i_R(\omega t) = I_R \sin(\omega t + \varphi) \tag{11.227}$$

where I_R is the current amplitude and φ is the initial phase shift.

The basic Kirchhoff equations characterizing the electrical behavior of the equivalent circuits of a Class FE power amplifier are

$$i(\omega t) + i_C(\omega t) = i_T(\omega t) + i_R(\omega t) \tag{11.228}$$

$$i_C(\omega t) = \omega C \frac{dv(\omega t)}{d\omega t}. \tag{11.229}$$

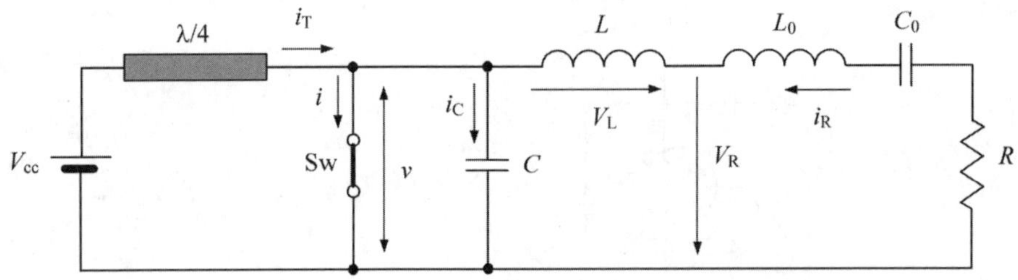

(*a*)

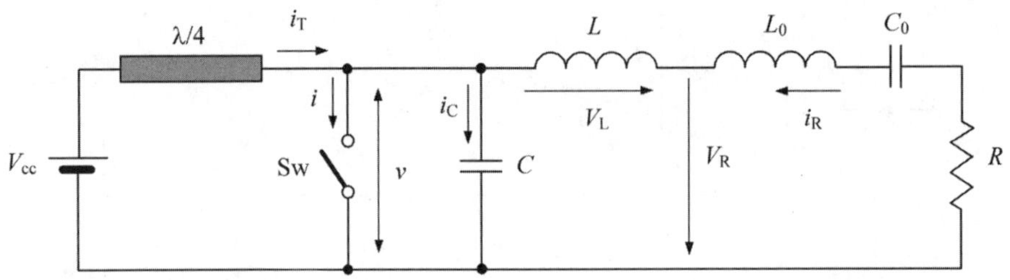

(*b*)

FIGURE 11.57 Equivalent circuits of Class FE power amplifier.

During the interval of $0 \le \omega t \le \pi - \tau_d$ before the switching instant, where τ_d is the dead time, the switch is turned on conducting the load current, as shown in Figure 11.57(*a*), resulting in the current and voltage conditions of

$$i_C(\omega t) = 0 \tag{11.230}$$

$$i(\omega t) = i_R(\omega t) + i_T(\omega t) \tag{11.231}$$

$$v(\omega t) = 0. \tag{11.232}$$

During the interval of $\pi - \tau_d \le \omega t \le \pi$ or dead time, the switch is turned off, as shown in Figure 11.57(*b*), and currents continue to flow through the shunt capacitance charging the capacitor. Using Eq. (11.229) and initial condition of $v(\pi - \tau_d) = 0$ yield

$$i(\omega t) = i_T(\omega t) = 0 \tag{11.233}$$

$$i_C(\omega t) = i_R(\omega t) \tag{11.234}$$

$$v(\omega t) = \frac{1}{\omega C} \int_{\pi-\tau_d}^{\omega t} i_C(\omega t) d\omega t + v(\pi - \tau_d) = -\frac{I_R}{\omega C} \left[\cos(\tau_d + \varphi) + \cos(\omega t + \varphi)\right]. \tag{11.235}$$

During the interval of $\pi \le \omega t \le 2\pi - \tau_d$ before the switching instant, the switch is still turned off, resulting in

$$i_T(\omega t) = -i_R(\omega t) \tag{11.236}$$

$$i(\omega t) = i_C(\omega t) = 0 \tag{11.237}$$

$$v(\omega t) = 2V_{cc}. \tag{11.238}$$

During the interval of $2\pi - \tau_d \le \omega t \le 2\pi$ or dead time, the switch is turned off, and currents flow through the shunt capacitance discharging the capacitor. Using Eq. (11.229) and initial condition of $v(2\pi - \tau_d) = 2V_{cc}$ yield

$$i(\omega t) = i_T(\omega t) = 0 \tag{11.239}$$

$$i_C(\omega t) = i_R(\omega t) \tag{11.240}$$

$$v(\omega t) = \frac{1}{\omega C}\int_{2\pi-\tau_d}^{\omega t} i_C(\omega t)d\omega t + v(2\pi - \tau_d) = \frac{I_R}{\omega C}(\cos\tau_d - \cos\omega t) + 2V_{cc}. \tag{11.241}$$

Equation (11.240) can also be written in a differential form using Eqs. (11.237) and (11.239) as

$$\frac{dv(\omega t)}{d\omega t} = \frac{I_R}{\omega C}\sin(\omega t + \varphi). \tag{11.242}$$

Applying the switching condition given in Eq. (11.236) to Eq. (11.242) results in

$$\sin(2\pi + \varphi) = 0 \tag{11.243}$$

from which it follows that the initial phase φ can be set to zero,

$$\varphi = 0. \tag{11.244}$$

Figure 11.58 shows the ideal current and voltage waveforms of a Class FE power amplifier in an idealized optimum operation mode during the whole interval $0 \le \omega t \le 2\pi$ corresponding to its equivalent circuits shown in Figure 11.57. The quarterwave transmission line has a half-wave symmetry, which means that the line attempts to do the same work in the first and second halves of the cycle. Therefore, the transmission-line half-repeating current must necessarily instantaneously fall to zero at $\pi - \tau_d$ and $2\pi - \tau_d$, being zero during both collector voltage edges. Note that, for the sinusoidal driving signals, the transistor must be biased in Class C mode to provide duty cycle D less than 50% ($D < 0.5$) because the period of time when the device is turned off must exceed the period of time when the device is turned on.

From the boundary condition $v(\pi) = 2V_{cc}$ or $v(2\pi) = 0$ by using the corresponding Eq. (11.235) or (11.241), we can write

$$\frac{\omega C V_{cc}}{I_R} = \frac{1 - \cos\tau_d}{2}. \tag{11.245}$$

The fundamental-frequency component of the voltage $v(\omega t)$ across the switch can be represented by the two quadrature components as shown in Figure 11.57, the amplitudes of which can be found using Fourier formulas by

$$V_R = -\frac{1}{\pi}\int_0^{2\pi} v(\omega t)\sin\omega t\, d\omega t = \frac{2(1 + \cos\tau_d)}{\pi}V_{cc} = I_R R \tag{11.246}$$

$$V_L = -\frac{1}{\pi}\int_0^{2\pi} v(\omega t)\cos\omega t\, d\omega t = \frac{\tau_d - \sin\tau_d\cos\tau_d}{\pi}\frac{I_R}{\omega C}. \tag{11.247}$$

In an idealized Class FE operation mode, there is no nonzero voltage and current simultaneously that gives an idealized collector efficiency of 100% when the dc power P_0 and fundamental output

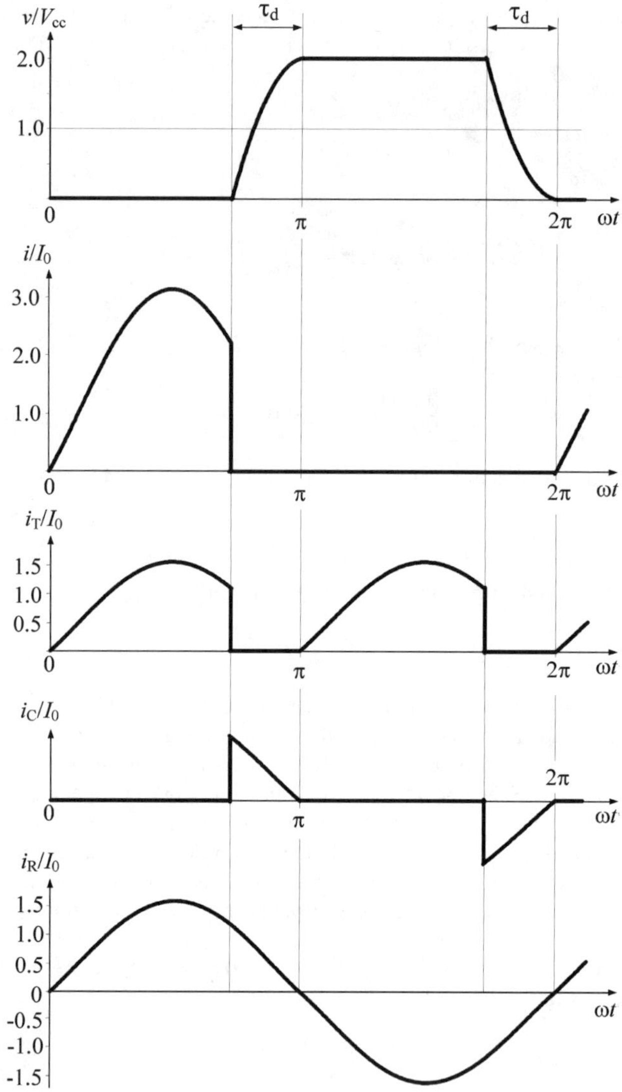

FIGURE 11.58 Waveforms for optimum Class FE operation mode.

power P_{out} are equal. Consequently, the load current amplitude I_R as a function of the dead time τ_d can be obtained using Eqs. (11.127) and (11.246) as

$$I_R = \frac{\pi}{1 + \cos \tau_d} I_0. \tag{11.248}$$

As a result, the optimum normalized shunt capacitance C and series inductance L as the functions of a dead time τ_d can be calculated using Eq. (11.245) from

$$\omega CR = \frac{\sin^2 \tau_d}{\pi} \tag{11.249}$$

$$\frac{\omega L}{R} = \frac{V_L}{V_R} = \frac{\tau_d - \sin\tau_d \cos \tau_d}{\sin^2\tau_d}. \tag{11.250}$$

The optimum load resistance R can be obtained using Eq. (11.246) for the constant dc supply voltage V_{cc} and fundamental-frequency output power P_{out} delivered to the load as

$$R = \frac{1}{2}\frac{V_R^2}{P_{out}} = \frac{2(1+\cos\tau_d)^2}{\pi^2}\frac{V_{cc}^2}{P_{out}}. \tag{11.251}$$

The maximum fundamental-frequency output power P_{out} delivered to the load as the functions of the dead time τ_d and capacitance C can be obtained from Eqs. (11.249) and (11.251) for the supply voltage V_{cc} as

$$P_{out} = \frac{2}{\pi}\left(\frac{1+\cos\tau_d}{\sin\tau_d}\right)^2 \omega C V_{cc}^2. \tag{11.252}$$

The maximum operating frequency f_{max} in an optimum Class FE mode is limited by the device output (collector or drain–source) capacitance C_{out}. From Eq. (11.252) it follows that

$$f_{max} = 0.25\left(\frac{\sin\tau_d}{1+\cos\tau_d}\right)^2 \frac{P_{out}}{C_{out}V_{cc}^2}. \tag{11.253}$$

Figure 11.59 shows the dependence of the normalized maximum operating frequency $\left(f_{max}C_{out}V_{cc}^2\right)/P_{out}$ on the dead time τ_d. From Eq. (11.253) and Figure 11.59, it follows that the maximum operating frequency increases with smaller dead times, and, for a $\tau_d = 45°$,

$$f_{max} = 0.043\frac{P_{out}}{C_{out}V_{cc}^2} \tag{11.254}$$

which is the same as for a Class DE and close to the maximum operating frequency of a Class E with 50% duty cycle [60]. However, for a saturation period of $\pi - \tau_d = (180 - 45)° = 135°$ when the transistor is turned on, the maximum operating frequency f_{max} of a Class E power amplifier is more than three times greater [30,73]. At the same time, for a saturation period of 135°, the collector voltage peak of a Class E power amplifier is three times larger than the applied supply voltage V_{cc} [74].

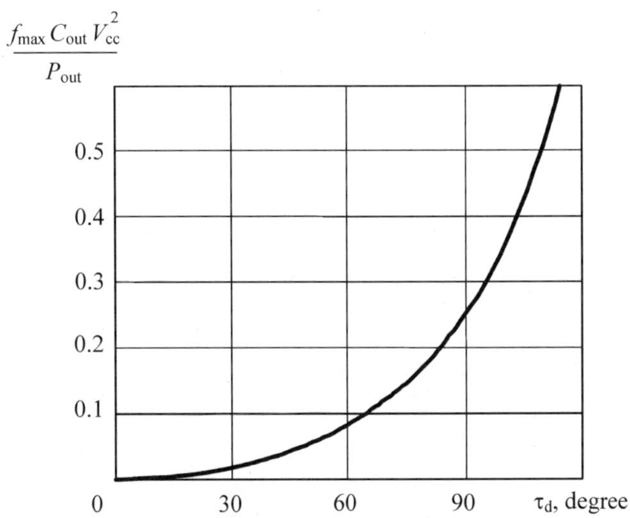

FIGURE 11.59 Normalized maximum operating frequency versus dead time.

TABLE 11.4 Optimum Impedances at Fundamental and Harmonics.

High-efficiency Mode	f_0 (Fundamental)	$2nf_0$ (Even Harmonics)	$(2n+1)f_0$ (Odd Harmonics)
Class F with quarterwave line	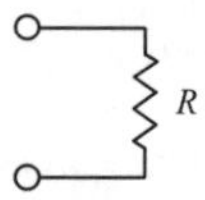	Short	Open
Class E with shunt capacitance	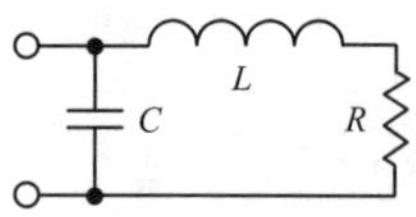	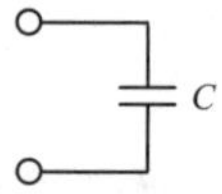	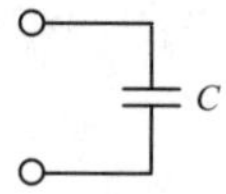
Class FE with quarterwave line	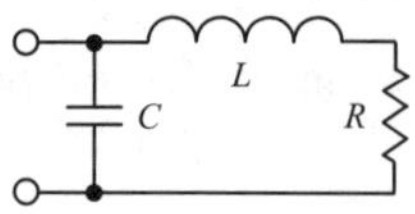	Short	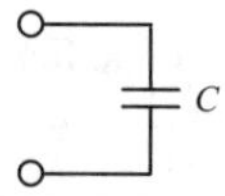

Note that the maximum collector voltage peak factor of a Class FE power amplifier never exceeds a factor of two for any saturation periods.

In Table 11.4, the optimum impedances seen from the device collector at the fundamental and higher order harmonic components are illustrated by the appropriate circuit configurations. It can be seen that Class F mode with a quarterwave transmission line shows purely resistive impedance at the fundamental, short-circuit conditions at even harmonics and open-circuit conditions at odd harmonics. As opposed to Class F mode, Class E mode with shunt capacitance requires inductive impedance at the fundamental and capacitive reactance at any other harmonic component. As a result, the frequency properties of a grounded quarterwave transmission line with its input open-circuit conditions at odd harmonics and short-circuit conditions at even harmonics enables Class FE with a quarterwave transmission line to combine simultaneously the harmonic impedance conditions typical for both Class F with a quarterwave transmission line and Class E with a shunt capacitance.

Figure 11.60 shows a Class FE power amplifier schematic with a series quarterwave transmission line consisting of a shunt capacitor C, a series quarterwave transmission line loaded by a parallel

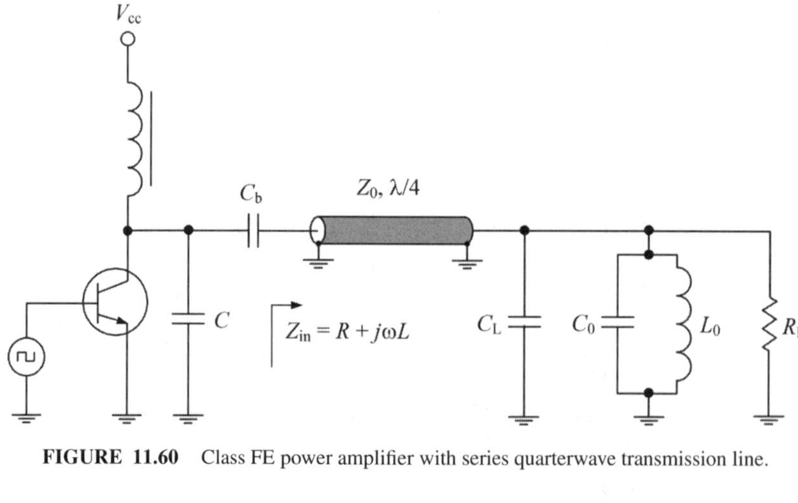

FIGURE 11.60 Class FE power amplifier with series quarterwave transmission line.

resonant L_0C_0 circuit tuned to the fundamental frequency, a shunt capacitor C_L, and a load resistor R_L. In a common case, a shunt capacitance C can represent the intrinsic device output capacitance and external circuit capacitance added by the load network. The RF choke is necessary to isolate dc power supply from RF circuit, while the blocking capacitor C_b is necessary to separate dc supply circuit from the load. The operation principle of such a Class FE power amplifier with a series quarterwave line is similar to that of with a parallel quarterwave line, assuming that the parallel L_0C_0 filter is ideal having infinite impedance at the fundamental frequency and zero impedances at the second-order and higher order harmonic components. In this case, the short circuit on the load side of the quarterwave transmission line produces a short circuit at its input for even harmonics and open circuit at its input for odd harmonics with resistive load at the fundamental frequency.

Unlike the parallel quarterwave transmission line, a series quarterwave transmission line can serve as an impedance transformer at the fundamental frequency. In this case, there is no need in an additional output impedance matching circuit required for Class FE power amplifier with a parallel quarterwave transmission line. The impedance Z_{in} at the input of the loaded quarterwave transmission line with the characteristic impedance Z_0 and electrical length θ at the fundamental frequency can be written as

$$Z_{in} = Z_0 \frac{Z_L + jZ_0 \tan\theta}{Z_0 + jZ_L \tan\theta}\bigg|_{\theta=90^\circ} = \frac{Z_0^2}{Z_L} \tag{11.255}$$

where

$$Z_L = \frac{R_L}{1 + j\omega C_L R_L}. \tag{11.256}$$

Separating Eq. (11.255) into real and imaginary parts enables the standard load resistance R_L and shunt capacitance C_L to be expressed through the optimum load resistance R and series inductance L corresponding to the Class FE mode according to Eqs. (11.250) and (11.251) for a particular value of the transmission-line characteristic impedance Z_0 as

$$R_L = \frac{Z_0^2}{R} \quad C_L = \frac{L}{Z_0^2}. \tag{11.257}$$

At lower frequencies, the length of a quarterwave transmission line becomes too long, so it is possible to approximate the Class FE operation mode with a second-harmonic termination only, by using an additional series resonant circuit L_2C_2, which provides a short circuit for the second harmonic at the device drain, as shown in Figure 11.61(*a*) [75,76]. Such a load network represents a subclass of Class FE and is called Class EF_2 (or Class F_2E) when, for a second-harmonic short-circuit termination, the idealized optimum Class E zero-voltage and zero-voltage-derivative conditions are satisfied at the end of period, as shown in Figure 11.61(*b*) for the duty cycle $D = 0.35$.

From comparison of the theoretical output voltage and current waveforms shown in Figure 11.58 for a general Class FE and in Figure 11.61(*b*) for a second-harmonic Class EF_2, respectively, it follows that if the voltage peak factors for both cases are practically the same, however the peak current value for the latter case is substantially higher. In addition, the higher duty cycle is used, the higher voltage peak factor of the Class EF_2 mode has been achieved, approaching the factor of 4 for the duty cycles close to $D = 0.5$ [75]. As a result, for a duty cycle $D = 0.3$, over 500 W of an output power with a drain efficiency above 92% was achieved at the switching frequency of 30 MHz using a 500-V vertical MOSFET device [76]. However, for the duty cycles greater than 0.5, it is preferable to use a Class E/F_3 mode where the series resonant circuit L_2C_2 represents a third-harmonic short circuit, resulting in a third-harmonic approximation of a Class E/F where the Class E idealized optimum conditions are applied to an inverse Class F. For example, when $D = 0.55$, the peak factor of the flattened drain voltage waveform is only slightly exceeds the factor of 3 [75].

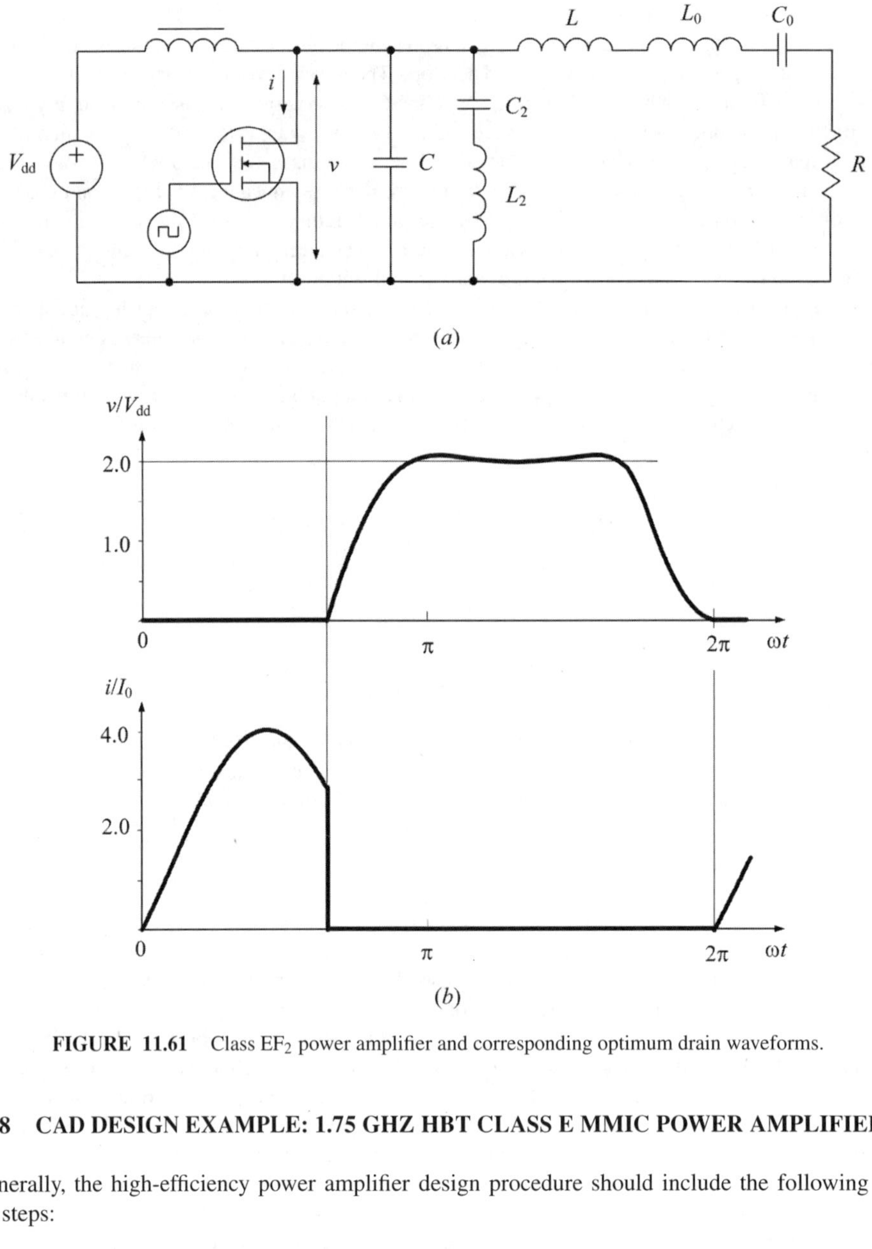

FIGURE 11.61 Class EF_2 power amplifier and corresponding optimum drain waveforms.

11.8 CAD DESIGN EXAMPLE: 1.75 GHZ HBT CLASS E MMIC POWER AMPLIFIER

Generally, the high-efficiency power amplifier design procedure should include the following basic steps:

- The proper selection of the active device, accurate measurements of its small-signal S-parameters under different bias conditions in a wide frequency range, and then transformation of the measured S-parameters to the impedance Z- and admittance Y-parameters to describe the device electrical behavior through the nonlinear model in the form of its equivalent circuit with generally nonlinear parameters.
- The analytical calculation of the optimum parameters of the proper load network in a high-efficiency mode for given output power, supply voltage, and device output capacitance to provide the maximum collector/drain efficiency and required level of harmonic suppression.

- The choice of a proper bias circuits to minimize the reference current and optimize the quiescent current for maximum power-added efficiency and minimum output power variations over wide temperature range.
- The design of the input and interstage matching circuits to provide the minimum input return loss, maximum power gain and power-added-efficiency, and stable operating conditions.
- The final circuit parameter optimization to maximize power-added efficiency.

As an example, suppose that our design objective is to design a high-efficiency monolithic parallel-circuit Class E power amplifier operating at 1.75 GHz with an output fundamental-frequency power $P_{out} = 33$ dBm, a power gain $G_P = 27$ dB and a supply voltage $V_{cc} = 5$ V [3]. To satisfy these requirements, our decision is to choose a two-stage topology and InGaP GaAs HBT devices with the minimum saturation voltage at low supply voltage and the transition frequency $f_T > 25$ GHz. Figure 11.62 shows the nonlinear equivalent circuit of the HBT transistor with the circuit parameters corresponding to the device emitter area of $A = 90$ (3×30) μm^2. Here, the input parameters as R_{be} and C_{be} represent the input diode forward-biased junction where the diffusion capacitor C_{be} has a sufficiently high value, thus shunting the differential resistance R_{be}. Since the voltage swing is very high in a high efficiency mode, the junction collector capacitance C_{bc} can be chosen at some middle bias point, between the base and collector dc voltages. In our case, $C_{bc} = 4.25 \times 10^{-14}\,{}^{*}A$.

Ideally, in a switched-mode operation, the active device should act as an ideal switch, driven to be turned on or off by the input RF signal. By assuming the collector efficiency of the second-stage transistor of 80%, the dc power can be calculated to be equal to $P_0 = 2$ W/0.8 = 2.5 W with the dc supply current $I_0 = P_0/V_{cc} = 2.5$ W/5 V = 500 mA. Since, for an HBT device selected, the recommended dc current density for a reliable operation over a wide temperature range should not exceed 15 mA per 90 μm^2, the overall emitter area was chosen with some margin to be equal to 5400 μm^2. The dc output characteristics $I_{ce}(V_{ce})$ of such an HBT transistor, given in amperes and volts, for different base bias voltage V_{be}, varying from 1.2 to 1.5 V, are shown in Figure 11.63. In this case, the peak collector voltage for a parallel-circuit Class E mode according to Eq. (11.144) is equal to $I_{max} = 2.647I_0 = 2.15$ A. The idealized Class E load line (with instant transition between pinch-off and saturation regions) is shown as a broken solid line in two sections: horizontal at zero current (transistor is pinched off) and slanted upwards to the right (transistor in saturation) with instant transition between pinch-off and saturation regions. It can be seen that the operating point moves along the horizontal V_{ce} axis (pinch-off region) and then along the collector current saturated line (saturation

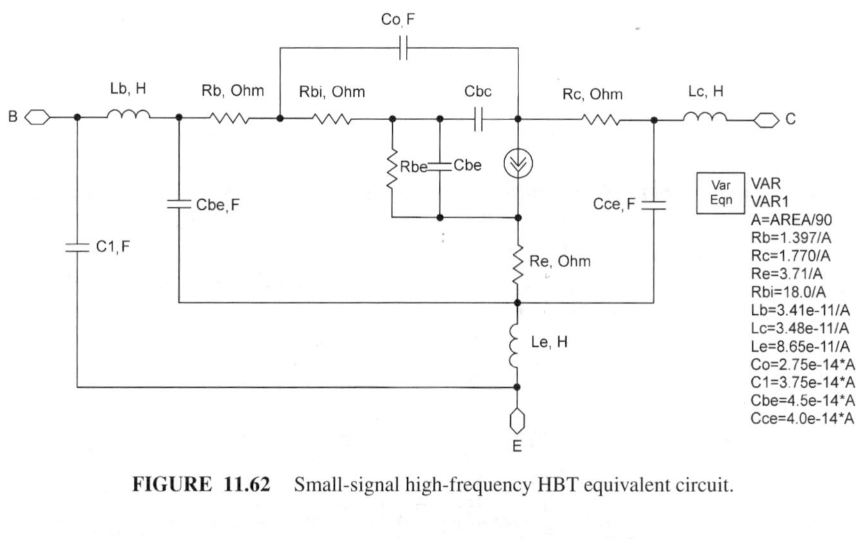

FIGURE 11.62 Small-signal high-frequency HBT equivalent circuit.

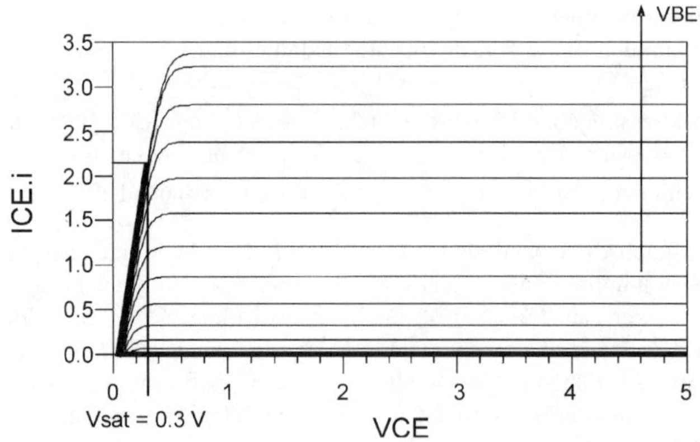

FIGURE 11.63 Device output current-voltage characteristics and load line.

region) until $I_{ce} = 2.15$ A. At this final point, the saturation voltage can be found as equal to $V_{sat} = 0.3$ V. This means that the power loss due to the finite device saturation resistance can be calculated as $P_{sat} = I_0 V_{sat} = 150$ mW with degradation in the collector efficiency of $(1 - P_{sat}/P_0) = 0.06$ or 6%.

First, we shall design the second stage for a maximum collector efficiency with some intermediate value of the source resistance much less than 50 Ω. The idealized optimum parameters of a parallel-circuit Class E load network can be calculated using Eqs. (11.139) to (11.141), by taking into account the saturation voltage V_{sat}, as

$$R = 1.365\frac{(V_{cc} - V_{sat})^2}{P_{out}} = 15\ \Omega$$

$$L = 0.732\frac{R}{\omega} = 1.0\,\text{nH}$$

$$C = \frac{0.685}{\omega R} = 3.9\,\text{pF}.$$

It should be noted that, in view of a saturation voltage, the calculated parameters of the Class E load network generally cannot be considered optimum unlike the ideal case of zero saturation voltage. However, they can be considered as a sufficiently accurate initial guess for final design and optimization when efficiency is sufficiently high. In addition, the collector capacitance can be larger than required, especially at higher frequencies and for the transistor with large emitter area required for high output power. In our case, the total output capacitance can be estimated as $C_{out} = C_{ce} + C_0 + C_{bc} = 6.6$ pF for total emitter area of 5400 μm^2, which is 1.5 times greater than required for the optimum parallel-circuit Class E mode. The excessive output capacitance of (6.6 – 3.9) = 2.7 pF is responsible for additional switching losses which occur during the transitions from the saturation to pinch-off modes of the device operation. To compensate for this capacitance at the fundamental frequency, it is necessary to connect the corresponding inductance in parallel, which value is equal to 3 nH. Hence, the parallel connection of the two inductors with values of 1 nH and 3 nH results in a final value of (1 × 3)/(1 + 3) = 0.75 nH.

Assuming the quality factor of the series filter of $Q_L = 10$, the series capacitance C_0 and inductance L_0 are calculated as

$$C_0 = \frac{1}{\omega R Q_L} = 0.6\ \text{pF}$$

$$L_0 = \frac{1}{\omega^2 C_0} = 13.8\ \text{nH}.$$

Since the input device impedance is sufficiently low, the intermediate source resistance is chosen to be equal to 5 Ω. In this case, it will be enough to use only one input matching section. The input matching circuit can be composed in the form of a high-pass L-type section with a shunt inductor and a series capacitor. The parameters of this matching circuit can finally be optimized in a large-signal mode to minimize input return loss as a criterion by using Smith chart tuning procedure.

Figure 11.64 shows the simulation setup of the second-power amplifier stage designed to operate in a parallel-circuit Class E mode. To simulate the electrical characteristics and waveforms, the corresponding current probes and voltage wire labels are incorporated into the circuit. As a current-controlled device, the bipolar transistor requires the dc base driving current, the value of which depends on the input power and device parameters. Because technologically the bipolar device represents a parallel connection of the basic cells (fingers), the important issue is to use the ballasting series resistors to avoid the current imbalance and possible device collapse at higher current density levels. Generally, different types of the current-mirror bias circuits can be effectively used to bias a transistor. In our case, the emitter follower bias circuit that provides the temperature compensation and minimizes the reference current is used. It is very important to provide the proper ratio between the ballasting resistors R_{34} and R_{33}, which preferably should be equal to the ratio of the corresponding device areas of 5400/270 = 20. The emitter follower bias circuit normally requires only several tens of microamperes of reference current.

Figure 11.65 shows the measurement equations required to plot the small- and large-signal power amplifier characteristics. The small-signal frequency dependence of the stability factor K is shown in Figure 11.66(a) demonstrating the stable conditions over entire frequency range with $K > 1$. However, it is not enough to just simulate the small-signal performance. It is important also to verify the potential instability that can occur in the form of injection-locking effect at large-signal mode near operating frequency. The dependencies of the real parts of the device impedance at the base and the input impedance seen by the source are shown in Figure 11.66(b). It is seen that the real part of the base impedance is slightly negative. In order to compensate for this negative resistance, it is necessary to connect the resistance of 0.35 Ω in series to the base compromising the stable operation and sufficient power gain. The real part of the input impedance seen by the source is close to the required 5 Ω at nominal large-signal operation.

As a result, the second-power amplifier stage exhibits a linear power gain of about 13 dB and saturated output power of 33.1 dBm, which dependencies are shown in Figure 11.67(a), with a maximum collector efficiency of 85.3% and a maximum power-added efficiency of 60.5%, which dependencies are shown in Figure 11.67(b). It is seen that the collector efficiency improvement is achieved at the expense of the significant power gain reduction. However, the maximum power-added efficiency occurs when the power gain is reduced by about 3 dB, which is a result of the contribution of the driving power.

Figure 11.68 shows the collector and base voltage waveform corresponding to a Class E approximation. The collector voltage waveform with a peak factor of about 18/5 = 3.6 was achieved for a maximum collector efficiency with the base voltage waveform having a flattened top part and causing the device to operate in a switching mode with minimum switching loss.

To match the Class E load resistance $R = 15\ \Omega$ with a standard load resistance $R_L = 50\ \Omega$, it is necessary to use a matching circuit with the series inductor as a first element to provide high impedance at the harmonics. Since the ratio of impedances is not so high, we can use the low-pass L-type matching section with design equations, resulting in

$$Q_L = \sqrt{\frac{50}{15} - 1} \cong 1.5$$

$$L = \frac{Q_L R}{\omega} = 2.1\ \text{nH}$$

$$C = \frac{Q_L}{\omega R_L} = 2.9\ \text{pF}.$$

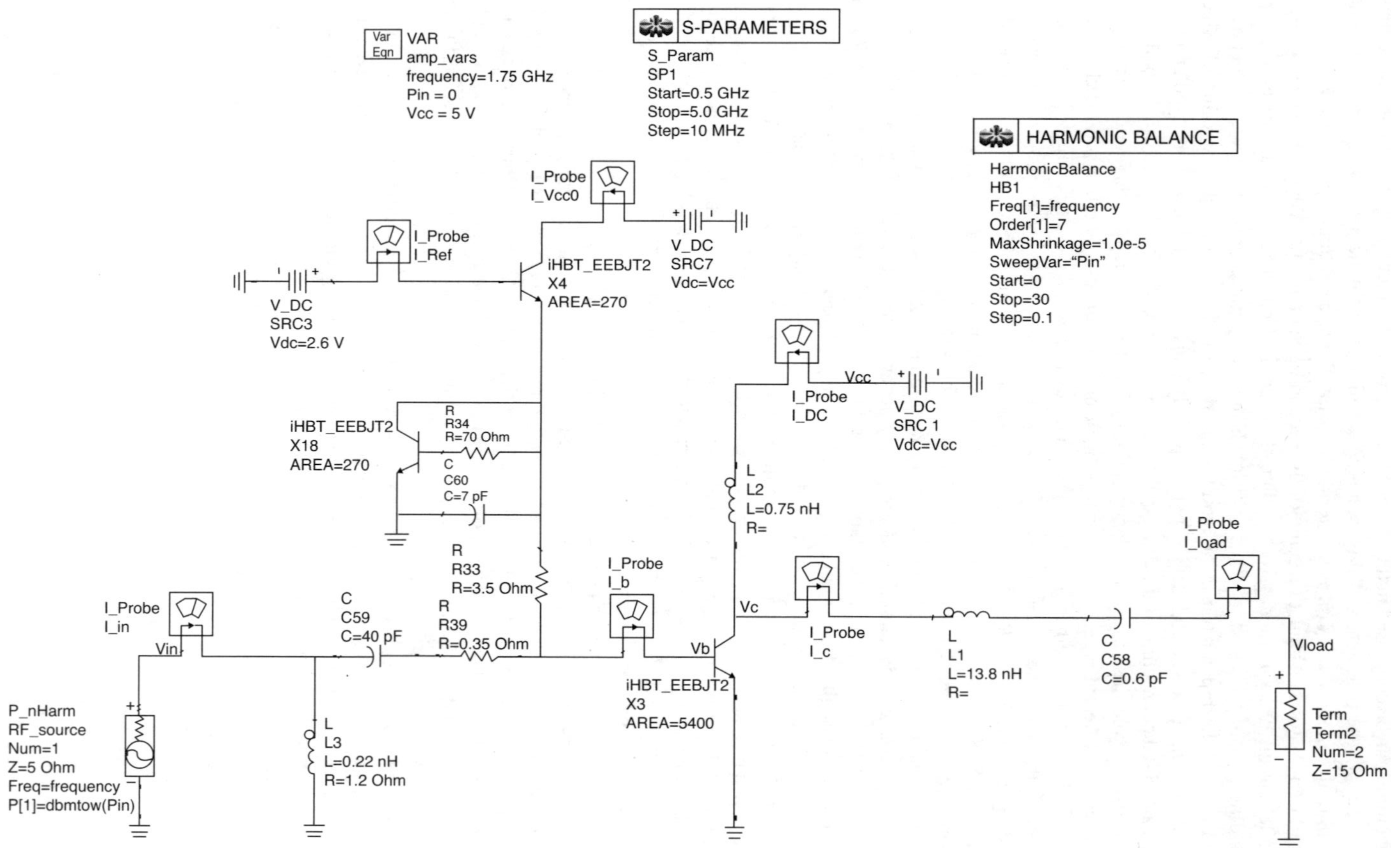

FIGURE 11.64 Simulation setup for Class E second stage.

```
Eqn P_DC=Vcc[0]*(I_Ref.i[::,0]+I_Vcc0.i[::,0]+I_DC.i[::,0])
Eqn Pload_W=mag(Vload[::,1])*mag(I_load.i[::,1])/2
Eqn Pload_dBm=10*log(Pload_W)+30
Eqn Efficiency=100*Pload_W/(Vcc[0]*I_DC.i[::,0])
Eqn Gain=Pload_dBm-Pin
Eqn K=stab_fact(S)
Eqn Pin_W=mag(Vin[::,1])*mag(I_in.i[::,1])/2
Eqn PAE=100*(Pload_W-Pin_W)/P_DC
```

FIGURE 11.65 Measurement equations.

Figure 11.69 shows the simulation setup of the Class E second stage with output matching circuit. Since the loaded quality facto Q_L is sufficiently small, the conditions at the fundamental-frequency and harmonic components will be slightly different compared to the ideal case. In this case, the easiest and fastest way to maximize collector efficiency is to tune manually the parameters of the basic elements such as the parallel inductor. The maximum collector efficiency of 84.8% and power gain of about 13 dB were achieved, as shown in Figure 11.70, by tuning the value of the parallel

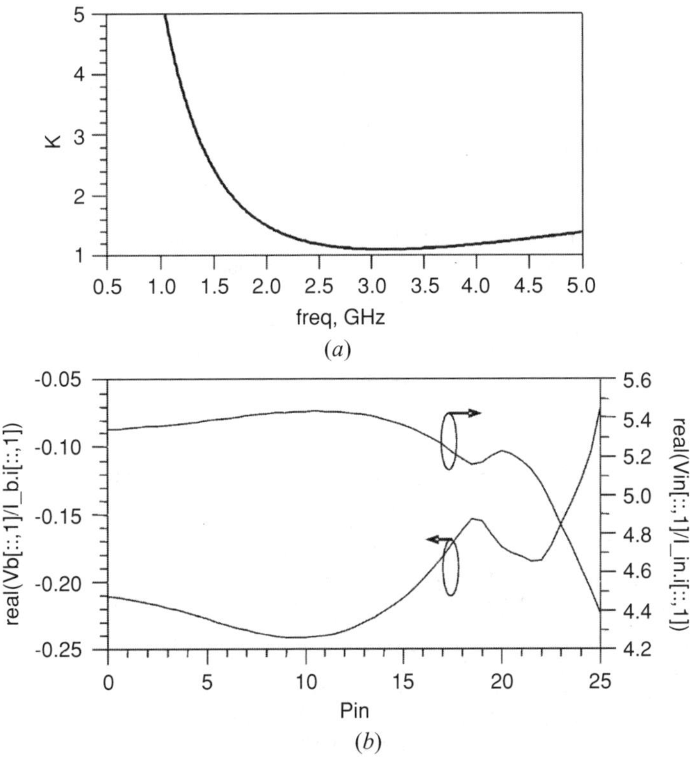

FIGURE 11.66 Input impedance and stability.

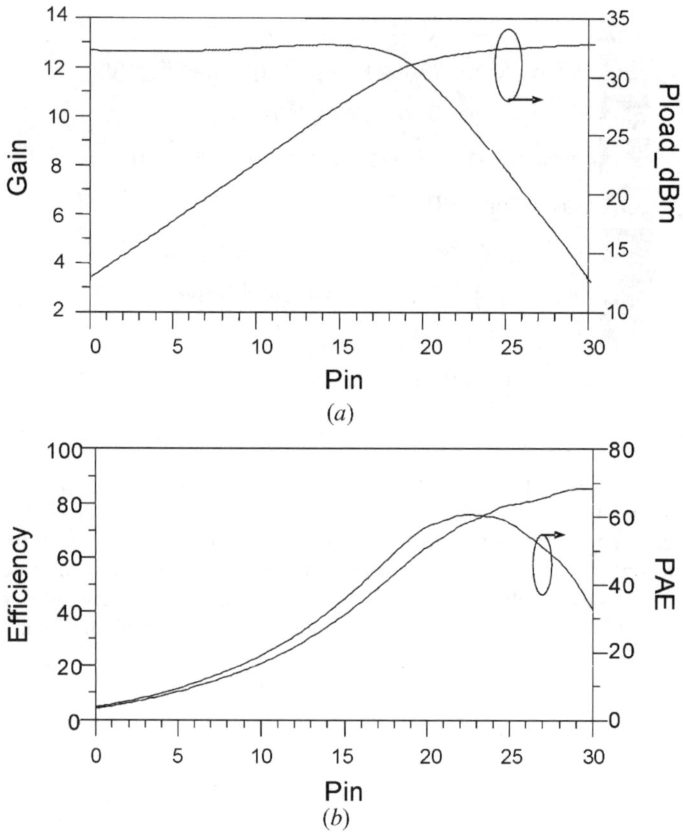

FIGURE 11.67 Power gain, output power and efficiency.

inductor from 0.75–1.0 nH. This inductor is shown in Figure 11.69 inside the circle. However, a single low-pass matching section can provide a second-harmonic suppression of about 20 dB. Consequently, to improve the spectral performance, it is necessary to use two or more low-pass sections in succession, better with equal quality factors to provide a wider frequency bandwidth.

Figure 11.71 shows the simulation setup of the first power amplifier stage designed to operate in a Class AB mode. Since it is sufficient to provide an output power from the first stage of not more

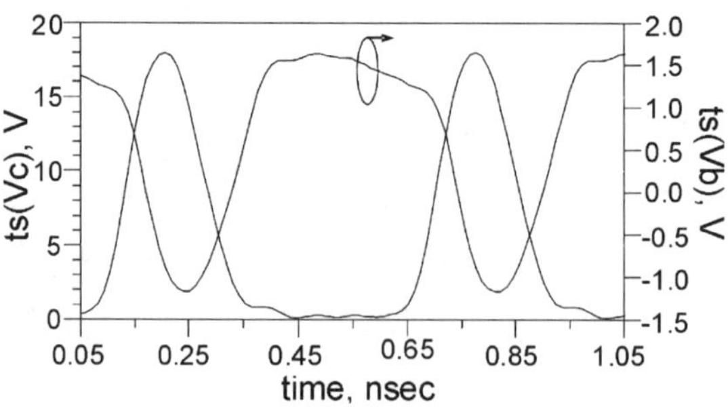

FIGURE 11.68 Collector and base voltage waveforms.

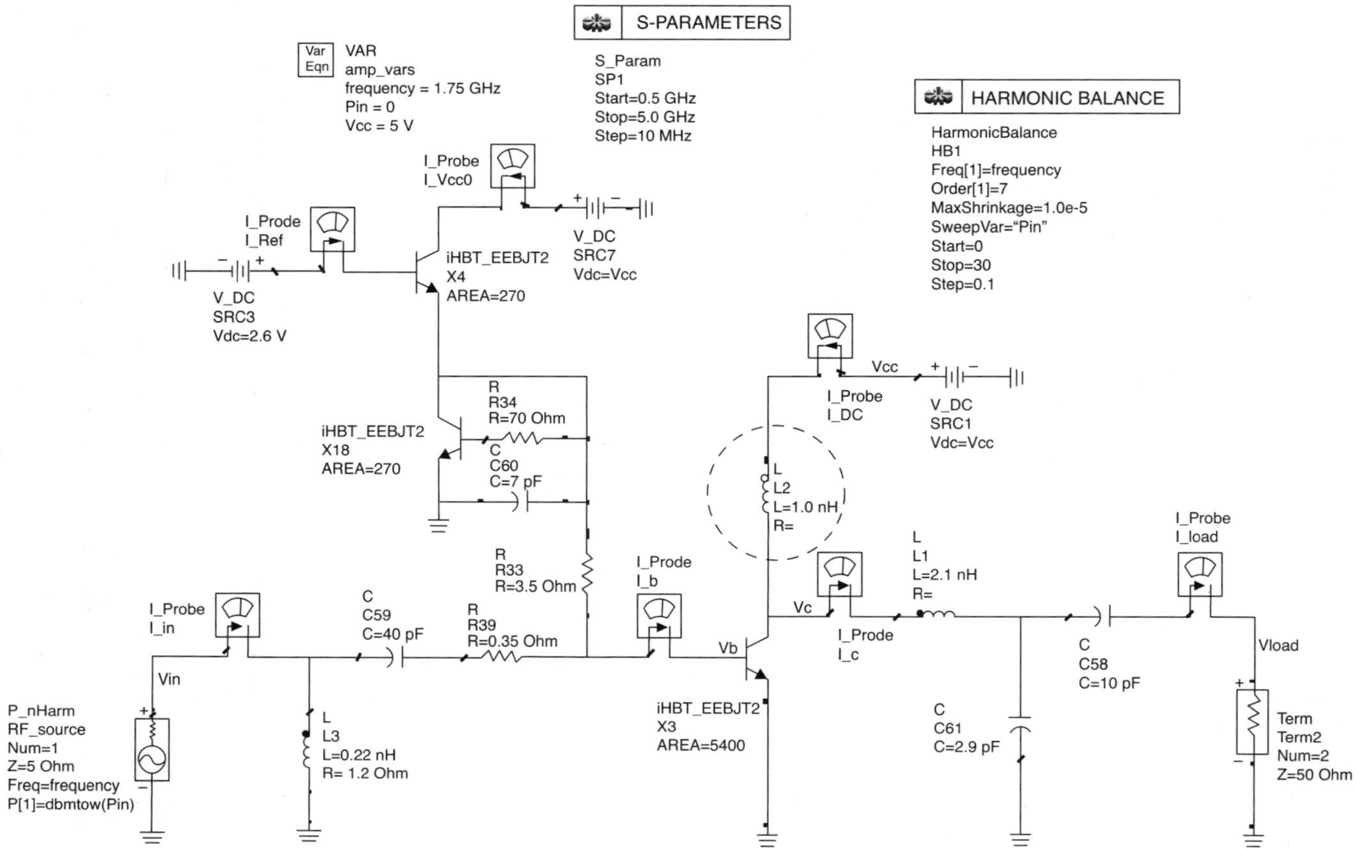

FIGURE 11.69 Simulation setup for Class E second stage with output matching circuit.

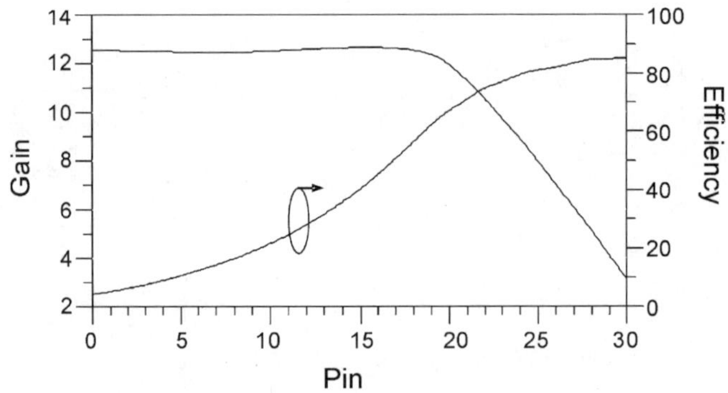

FIGURE 11.70 Power gain and collector efficiency.

than 200 mW, the device emitter area size was chosen to be 720 μm^2 and the output resistance is assumed to be 50 Ω, which is close to the calculated value of $R_{\text{out}} = (V_{\text{cc}} - V_{\text{sat}})^2/2P_{\text{out}} = (4.7 \times 4.7)/0.4 = 55.2\ \Omega$. Then, conjugately match the intermediate load resistance of 5 Ω with the chosen output resistance $R_{\text{out}} = 50\ \Omega$, it is convenient to use a high-pass L-type matching section with the shunt inductor as a first element followed by the series capacitor. In this case, the shunt inductor with a bypassing capacitor at its end in practical implementation also can serve as a dc power supply path. The parameters of the output matching circuit are calculated as

$$Q_{\text{L}} = \sqrt{\frac{50}{5} - 1} = 3$$

$$L = \frac{50}{\omega Q_{\text{L}}} = 1.5\,\text{nH}$$

$$C = \frac{1}{5\omega Q_{\text{L}}} = 6.1\,\text{pF}.$$

To obtain a final value of the shunt inductor, it is necessary to take into account the device collector capacitance. In our case, the total output capacitance can be estimated as $C_{\text{out}} = 6.6/7.5 = 0.88$ pF for emitter area of 720 μm^2. To compensate for this capacitance at the fundamental frequency, it is necessary to connect the inductance of 9.4 pF in parallel. Hence, the parallel connection of the two inductors with values of 9.4 nH and 1.5 nH results in a final value of 1.29 nH. The input matching circuit is composed in the form of a high-pass L-type section, and the parameters of its elements can be simulated in a large-signal mode to minimize input return loss as a criterion by using Smith chart tuning procedure. To improve stability factor and simplify the input matching, the resistance of 8 Ω, connected in series to the transistor base terminal, was included.

The small-signal frequency dependence of the stability factor K and magnitude of S_{11} in decibels are shown in Figure 11.72(a) demonstrating the stable conditions over entire frequency range with $K > 1$ and input return loss better than 20 dB at the operating frequency. Under large-signal operation, the first power amplifier stage exhibits a linear power gain of about 15 dB and a maximum collector efficiency of about 72%, the dependencies of which on the input power are shown in Figure 11.72(b). It is seen that the collector efficiency improvement is achieved at the expense of the significant power gain reduction. However, the sufficiently high efficiency of about 65% can be achieved at a 3-dB compression point when the power gain is reduced by 3 dB from its linear value.

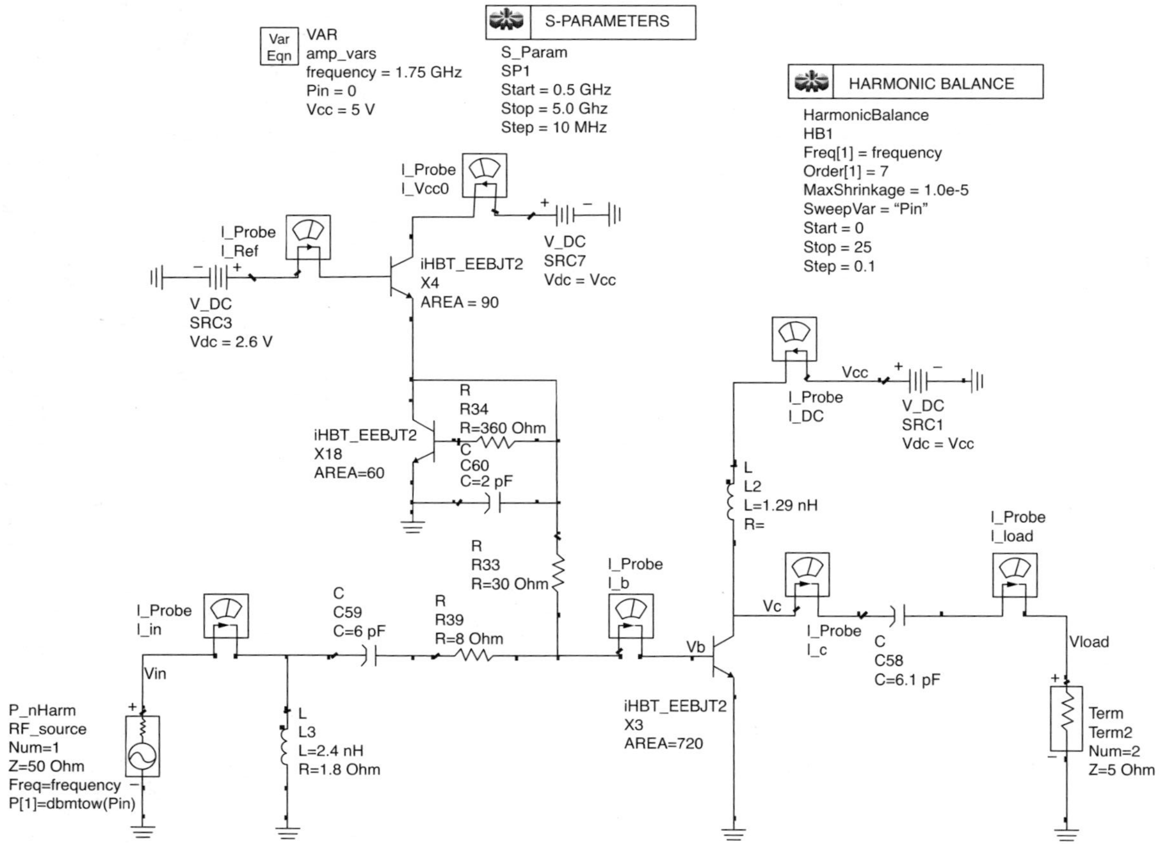

FIGURE 11.71 Simulation setup for Class AB first stage.

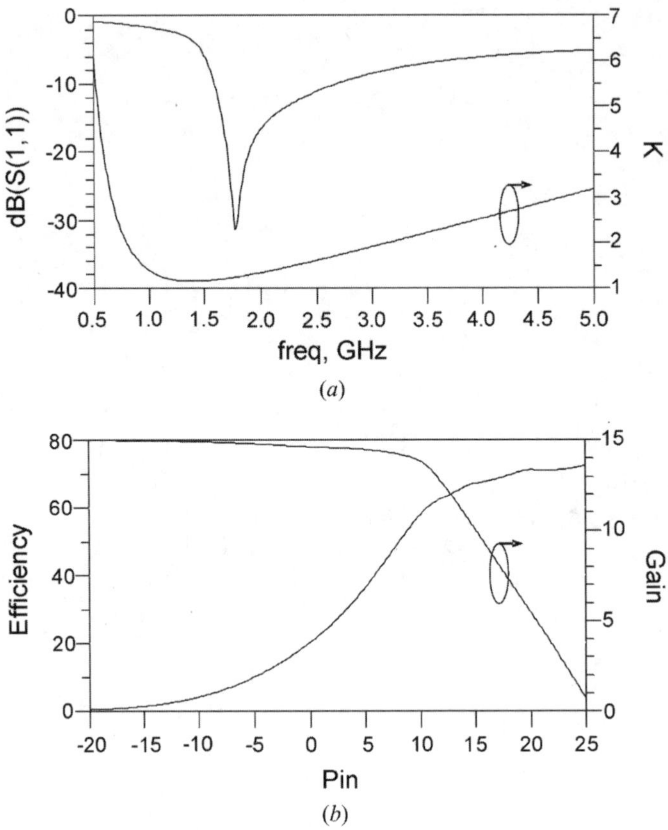

FIGURE 11.72 Electrical performance of first power amplifier stage.

The next step is to combine the first power amplifier stage with the load impedance of 5 Ω and the second-power amplifier stage with the source impedance of 5 Ω in a two-stage power amplifier. Figure 11.73 shows the simulation setup of the two-stage Class E power amplifier with lumped elements. The main attention should paid to the interstage matching circuit consisting of the two high-pass matching sections since the harmonic impedances are not exactly the same, having some certain values at the harmonics. In this case, it is necessary to provide some tuning of the elements around their initial values shown in a circle. It can be done sufficiently fast even with manual tuning.

Figure 11.74 shows the measurement equations required to plot the small- and large-signal power amplifier characteristics including all dc power sources. The small-signal frequency dependences of the stability factor K and magnitude of S_{11} in decibels are shown in Figure 11.75(*a*) demonstrating the stable conditions over entire frequency range with $K > 2$ and input return loss better than 12 dB at the operating frequency. Under large-signal operation, the two-stage parallel-circuit Class E power amplifier exhibits a linear power gain of more than 29 dB and a maximum power-added efficiency of about 68%, the dependencies of which on the input power are shown in Figure 11.75(*b*). The sufficiently high efficiency of about 57.5% can be achieved at a 1-dB compression point when the power gain is reduced by 1 dB from its small-signal linear value.

Finally, the lumped inductors must be replaced by the microstrip lines using the standard FR4 substrate. The lumped capacitors in the output matching circuit can be implemented as the MIM

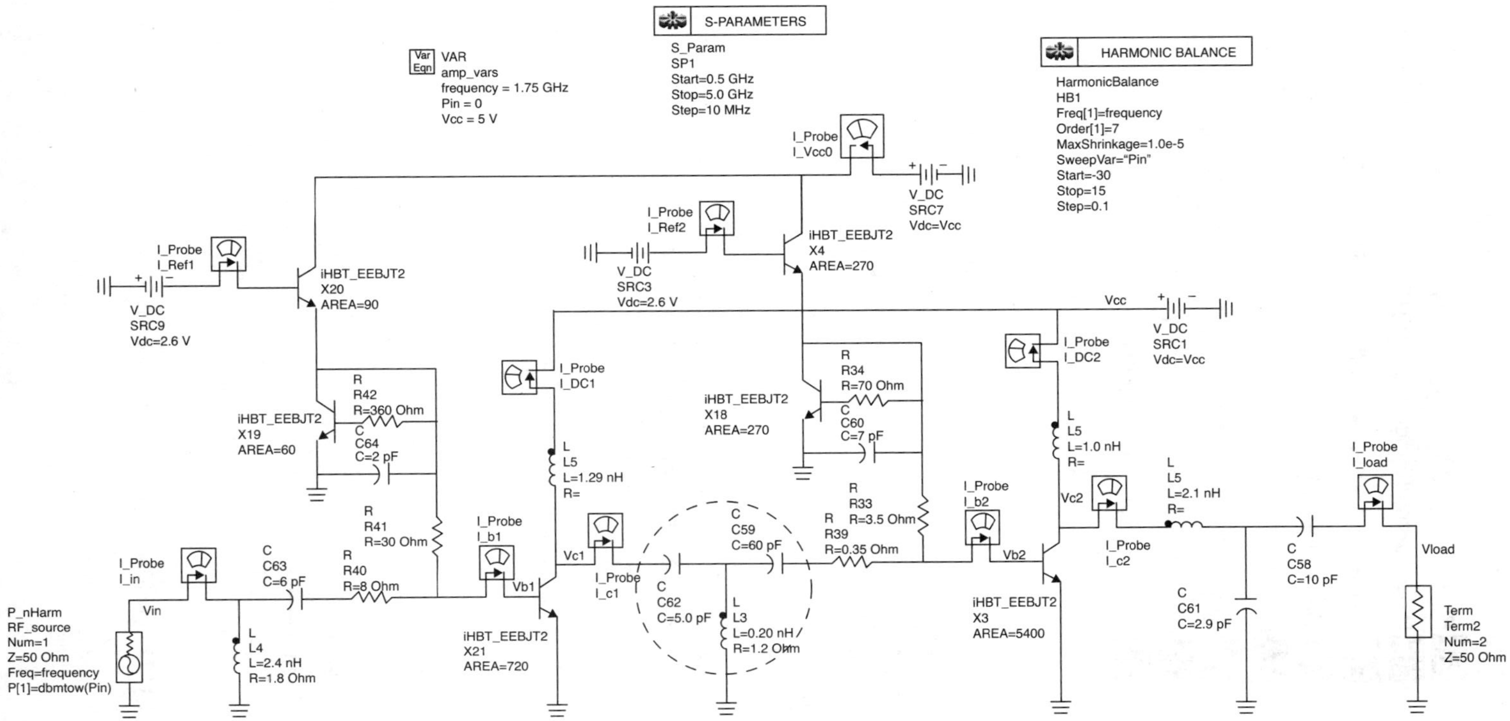

FIGURE 11.73 Simulation setup for two-stage Class E power amplifier with lumped elements.

```
Eqn Pin_W=mag(Vin[::,1])*mag(I_in.i[::,1])/2
Eqn Pload_W=mag(Vload[::,1])*mag(I_load.i[::,1])/2
Eqn Pload_dBm=10*log(Pload_W)+30
Eqn Efficiency=100*Pload_W/(Vcc[0]*I_DC2.i[::,0])
Eqn Gain=Pload_dBm-Pin
Eqn K=stab_fact(S)
Eqn I_DC=I_Ref+I_Vcc0.i[::,0]+I_DC1.i[::,0]+I_DC2.i[::,0]
Eqn PAE=100*(Pload_W-Pin_W)/(Vcc[0]*I_DC)
Eqn I_Ref=I_Ref1.i[::,0]+I_Ref2.i[::,0]
```

FIGURE 11.74 Measurement equations.

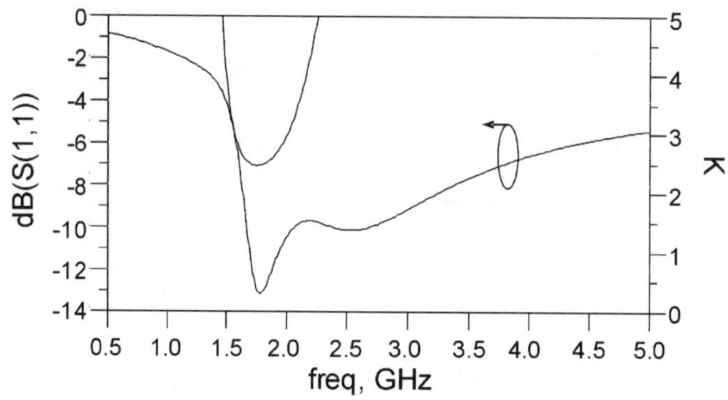

(*a*)

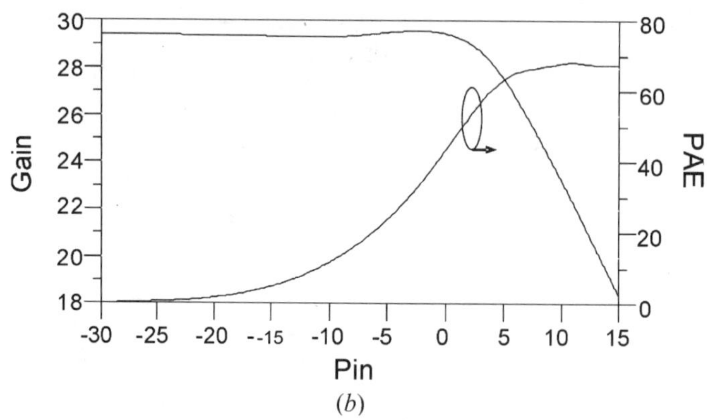

(*b*)

FIGURE 11.75 Electrical performance of two-stage power amplifier with lumped elements.

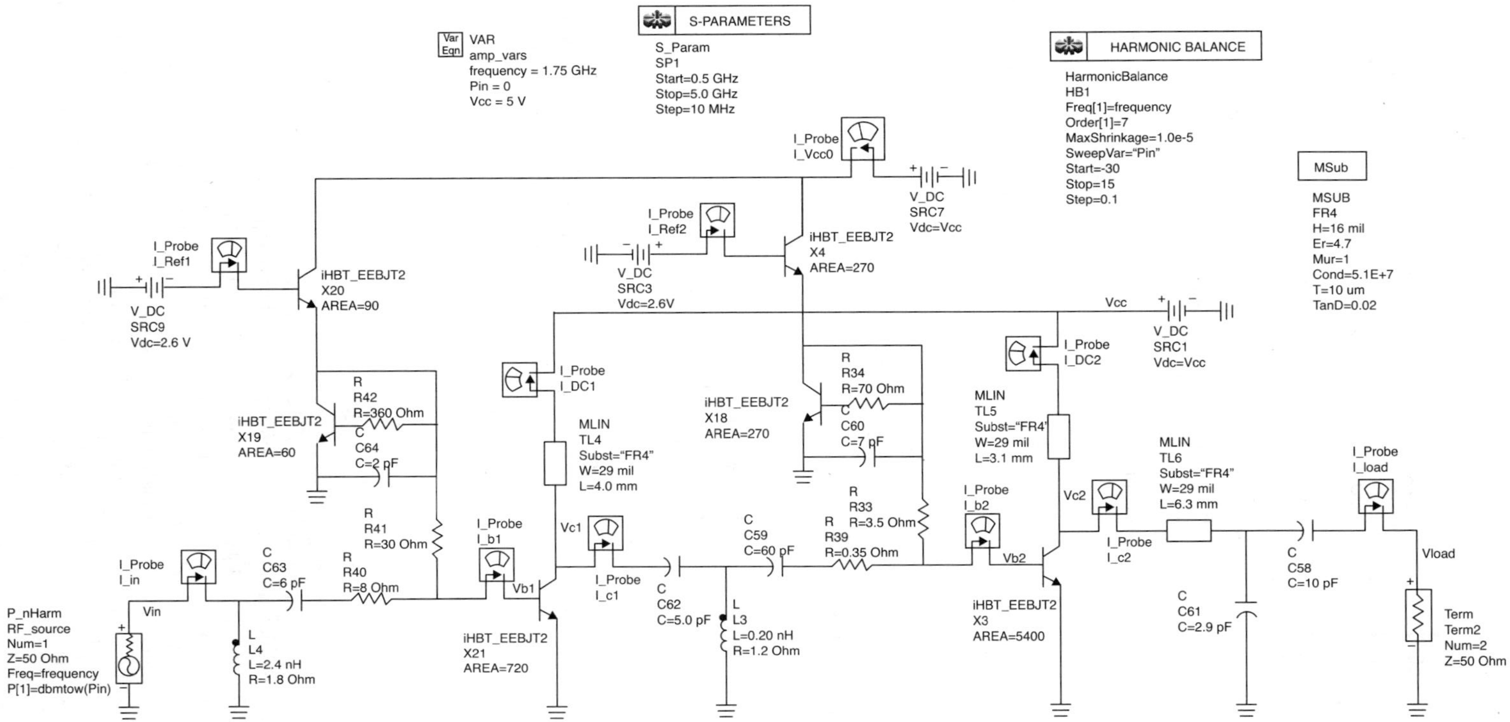

FIGURE 11.76 Simulation setup for two-stage Class E power amplifier with microstrip lines.

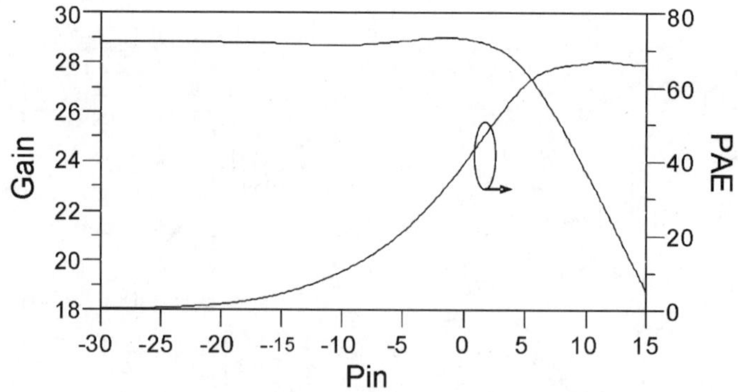

FIGURE 11.77 Electrical performance of two-stage power amplifier with microstrip lines.

capacitors with using a separate die to provide their high quality factors. To calculate the parameters of the microstrip lines, the electrical length θ as a function of the lumped inductance L can be approximately defined by

$$\theta = \tan^{-1}\frac{\omega L}{Z_0}. \tag{11.258}$$

Then, the electrical lengths of the first-stage collector microstrip line, second-stage collector microstrip line, and series microstrip line in the output matching circuit can be respectively calculated as

$$\theta_1 = \tan^{-1}(0.284) = 15.85^\circ$$
$$\theta_1 = \tan^{-1}(0.220) = 12.41^\circ$$
$$\theta_1 = \tan^{-1}(0.462) = 24.79^\circ.$$

Using the LineCalc program, available in Tools displayed by the project menu, results in the following geometrical length of the microstrip lines with 26-mm width (characteristic impedance of approximately 50 Ω) implemented into the 16-mm FR4 substrate with the parameters given in Figure 11.76:

$$l_1 = 4.0\,\text{mm}$$
$$l_2 = 3.1\,\text{mm}$$
$$l_3 = 6.3\,\text{mm}.$$

Figure 11.76 shows the simulation setup of the two-stage parallel-circuit InGaP GaAs HBT Class E MMIC power amplifier with external microstrip lines. In principle, there is no need to tune any circuit element. In this case, the maximum power-added efficiency of 67.1% and power gain of 22.5 dB with an output power of 33.6 dBm can be achieved, as shown in Figure 11.77. At the same time, the output power of 33 dBm and power gain of 27 dB are provided with a power-added efficiency as high as 62.7%. Some small additional tuning of circuit parameters around their nominal values, probably, can slightly improve the performance. However, based on the example described above, you can see how effective a simple analytical approach with quick manual tuning can be, providing high performance and significantly speeding up the entire high-efficiency power amplifier design procedure.

REFERENCES

1. M. R. Osborne, "Design of Tuned Transistor Power Amplifiers," *Electronic Eng.*, vol. 40, pp. 436–443, Aug. 1968.
2. P. J. Baxandall, "Transistor Sine Wave LC Oscillators, Some General Considerations and New Developments," *IEE Proc. Electric Power Appl.*, vol. 106, pp. 748–758, May 1959.
3. A. Grebennikov and N. O. Sokal, *Switchmode RF Power Amplifiers*, New York: Newnes, 2007.
4. H. L. Krauss, C. W. Bostian, and F. H. Raab, *Solid State Radio Engineering*, New York: John Wiley & Sons, 1980.
5. A. D. Artym, *Class D Power Amplifiers and Switching Generators in Radio Communication and Broadcasting* (in Russian), Moskva: Svyaz, 1980.
6. W. J. Chudobiak and D. F. Page, "Frequency Limitations of Class-D Transistor Amplifiers," *IEEE J. Solid-State Circuits*, vol. SC-4, pp. 25–37, Feb. 1969.
7. H. O. Granberg, "Applying Power MOSFETs in Class D/E RF Power Amplifier Design," *RF Design*, vol. 8, pp. 42–47, June 1985.
8. T.-P. Hung, A. G. Metzger, P. J. Zampardi, M. Iwamoto, and P. M. Asbeck, "Design of High-Efficiency Current-Mode Class-D Amplifiers for Wireless Applications," *IEEE Trans. Microwave Theory Tech.*, vol. MTT-53, pp. 144–151, Jan. 2005.
9. D. C. Prince, "Vacuum Tubes as Power Oscillators, Part III," *Proc. IRE*, vol. 11, pp. 527–550, Sept. 1923.
10. J. Zenneck and H. Rukop, *Lehrbuch der Drahtlosen Telegraphie*, Stuttgart: Ferdinand Enke, 1925.
11. I. N. Fomichev, "A New Method to Increase Efficiency of the Radio Broadcasting Station (in Russian)," *Elektrosvyaz*, pp. 58–66, June 1938.
12. R. I. Sarbacher, "Power-Tube Performance in Class C Amplifiers and Frequency Multipliers as Influenced by Harmonic Voltage," *Proc. IRE*, vol. 31, pp. 607–625, Nov. 1943.
13. H. J. Round, "Wireless Telegraph and Telephone Transmission," U.S. Patent 1,564,627, Dec. 1925.
14. F. H. Raab, "An Introduction to Class-F Power Amplifiers," *RF Design*, vol. 19, pp. 79–84, July 1996.
15. D. M. Snider, "A Theoretical Analysis and Experimental Confirmation of the Optimally Loaded and Overdriven RF Power Amplifier," *IEEE Trans. Electron Devices*, vol. ED-14, pp. 851–857, Dec. 1967.
16. F. H. Raab, "FET Power Amplifier Boosts Transmitter Efficiency," *Electronics*, vol. 49, pp. 122–126, June 1976.
17. V. A. Borisov and V. V. Voronovich, "Analysis of Switched-Mode Transistor Amplifier with Parallel Forming Transmission Line (in Russian)," *Radiotekhnika i Elektronika*, vol. 31, pp. 1590–1597, Aug. 1986.
18. A. V. Grebennikov, "Circuit Design Technique for High Efficiency Class F Amplifiers," *IEEE MTT-S Int. Microwave Symp. Dig.*, vol. 2, pp. 771–774, 2000.
19. C. Trask, "Class-F Amplifier Loading Networks: A Unified Design Approach," *IEEE MTT-S Int. Microwave Symp. Dig.*, vol. 1, pp. 351–354, 1999.
20. F. H. Raab, "Class-F Power Amplifiers with Maximally Flat Waveforms," *IEEE Trans. Microwave Theory Tech.*, vol. MTT-45, pp. 2007–2012, Nov. 1997.
21. A. I. Kolesnikov, "A New Method to Improve Efficiency and to Increase Power of the Transmitter (in Russian)," *Master Svyazi*, pp. 27–41, June 1940.
22. E. S. Glazman, L. B. Kalinin, and Y. I. Mikhailov, "Improving VHF Transmitter Efficiency by Using the Biharmonic Mode," *Telecommunications and Radio Engineering*, Part 1, vol. 30, pp. 46–51, July 1975.
23. V. J. Tyler, "A New High-Efficiency High-Power Amplifier," *Marconi Review*, vol. 21, pp. 96–109, 1958.
24. M. K. Kazimierczuk, "A New Concept of Class F Tuned Power Amplifier," *Proc. 27th Midwest Circuits and Systems Symp.*, pp. 425–428, 1984.
25. P. Heymann, R. Doerner, and M. Rudolph, "Harmonic Tuning of Power Transistors by Active Load–Pull Measurement," *Microwave J.*, vol. 43, pp. 22–37, June 2000.
26. C. J. Wei, P. DiCarlo, Y. A. Tkachenko, R. McMorrow, and D. Bartle, "Analysis and Experimental Waveform Study on Inverse Class-F Mode of Microwave Power FETs," *IEEE MTT-S Int. Microwave Symp. Dig.*, vol. 1, pp. 525–528, 2000.
27. J. W. Wood, "High Efficiency Class C Amplifier," U.S. Patent 3,430,157, Feb. 1969.

28. E. P. Khmelnitsky, *Operation of Vacuum-Tube Generator on Detuned Resonant Circuit* (in Russian), Moskva: Svyazizdat, 1962.

29. G. D. Ewing, *High-Efficiency Radio-Frequency Power Amplifiers*, Ph.D. Dissertation, Oregon State University, June 1964.

30. V. B. Kozyrev, "Single-Ended Switching-Mode Tuned Power Amplifier with Filtering Resonant Circuit (in Russian)," *Poluprovodnikovye Pribory v Tekhnike Svyazi*, vol. 6, pp. 152–166, 1971.

31. N. O. Sokal and A. D. Sokal, "Class E – A New Class of High-Efficiency Tuned Single-Ended Switching Power Amplifiers," *IEEE J. Solid-State Circuits*, vol. SC-10, pp. 168–176, June 1975.

32. N. O. Sokal and A. D. Sokal, "High-Efficiency Tuned Switching Power Amplifier," U.S. Patent 3,919,656, Nov. 1975.

33. N. O. Sokal, "Class E High-Efficiency Power Amplifiers, from HF to Microwave," *IEEE MTT-S Int. Microwave Sym. Dig.*, vol. 2, pp. 1109–1112, June 1998.

34. F. H. Raab, "Idealized Operation of the Class E Tuned Power Amplifier," *IEEE Trans. Circuits and Systems*, vol. CAS-24, pp. 725–735, Dec. 1977.

35. B. Molnar, "Basic Limitations on Waveforms Achievable in Single-Ended Switching-Mode Tuned (Class E) Power Amplifiers," *IEEE J. Solid-State Circuits*, vol. SC-19, pp. 144–146, Feb. 1984.

36. T. B. Mader, E. W. Bryerton, M. Marcovic, M. Forman, and Z. Popovic, "Switched-Mode High-Efficiency Microwave Power Amplifiers in a Free-Space Power-Combiner Array," *IEEE Trans. Microwave Theory Tech.*, vol. MTT-46, pp. 1391–1398, Oct. 1998.

37. M. Kazimierczuk and K. Puczko, "Exact Analysis of Class E Tuned Power Amplifier at any Q and Switch Duty Cycle," *IEEE Trans. Circuits and Systems*, vol. CAS-34, pp. 149–158, Feb. 1987.

38. F. H. Raab and N. O. Sokal, "Transistor Power Losses in the Class E Tuned Power Amplifier," *IEEE J. Solid-State Circuits*, vol. SC-13, pp. 912–914, Dec. 1978.

39. A. N. Bruevich, "About Optimum Parameters of Switching-Mode Tuned Power Amplifier with Filtering Resonant Circuit (in Russian)," *Poluprovodnikovaya Elektronika v Tekhnike Svyazi*, vol. 18, pp. 43–48, 1977.

40. I. A. Popov (Ed.), *Transistor Generators of Harmonic Oscillations in Switching Mode* (in Russian), Moskva: Radio i Svyaz, 1985.

41. M. Kazimierczuk, "Effect of the Collector Current Fall Time on the Class E Tuned Power Amplifier," *IEEE J. Solid-State Circuits*, vol. SC-18, pp. 181–193, Apr. 1983.

42. J. A. Blanchard and J. S. Yuan, "Effect of Collector Current Exponential Decay on Power Efficiency for Class E Tuned Power Amplifier," *IEEE Trans. Circuits and Systems—I: Fundamental Theory Appl.*, vol. CAS-I-41, pp. 69–72, Jan. 1994.

43. T. B. Mader and Z. B. Popovic, "The Transmission-Line High-Efficiency Class-E Amplifier," *IEEE Microwave and Guided Wave Lett.*, vol. 5, pp. 290–292, Sept. 1995.

44. F. J. Ortega-Gonzalez, J. L. Jimenez-Martin, A. Asensio-Lopez, and G. Torregrosa-Penalva, "High-Efficiency Load–Pull Harmonic Controlled Class-E Power Amplifier," *IEEE Microwave and Guided Wave Lett.*, vol. 8, pp. 348–350, Oct. 1998.

45. Y.-B. Choi and K.-K. M. Cheng, "Generalised Frequency-Domain Analysis of Microwave Class-E Power Amplifiers," *IEE Proc. Microwave Antennas Propag.*, vol. 148, pp. 403–409, Dec. 2001.

46. P. Watson, R. Neidhard, L. Kehias, R. Welch, T. Quach, R. Worley, M. Pacer, R. Pappaterra, R. Schweller, and T. Jenkins, "Ultra-High Efficiency Operation Based on an Alternative Class-E Mode," *IEEE GaAs IC Symp. Dig.*, pp. 53–56, 2000.

47. T. K. Quach, P. M. Watson, W. Okamura, E. N. Kaneshiro, A. Gutierrez-Aitken, T. R. Block, J. W. Eldredge, T. J. Jenkins, L. T. Kehias, A. K. Oki, D. Sawdai, R. J. Welch, and R. D. Worley, "Ultra-High Efficiency Power Amplifier for Space Radar Applications," *IEEE J. Solid-State Circuits*, vol. SC-37, pp. 1126–1134, Sep. 2002.

48. R. E. Zulinski and J. W. Steadman, "Class E Power Amplifiers and Frequency Multipliers with Finite DC-Feed Inductance," *IEEE Trans. Circuits and Systems*, vol. CAS-34, pp. 1074–1087, Sept. 1987.

49. C. P. Avratoglou, N. C. Voulgaris, and F. I. Ioannidou, "Analysis and Design of a Generalized Class E Tuned Power Amplifier," *IEEE Trans. Circuits and Systems*, vol. CAS-36, pp. 1068–1079, Aug. 1989.

50. G. H. Smith and R. E. Zulinski, "An Exact Analysis of Class E Power Amplifiers with Finite DC-Feed Inductance at Any Output Q," *IEEE Trans. Circuits and Systems*, vol. CAS-37, pp. 530–534, Apr. 1990.

51. H. Sekiya, I. Sasase, and S. Mori, "Computation of Design Values for Class E Amplifiers without Using Waveform Equations," *IEEE Trans. Circuits and Systems–I: Fundamental Theory Appl.*, vol. CAS-I-49, pp. 966–978, July 2002.
52. C.-H. Li and Y.-O. Yam, "Maximum Frequency and Optimum Performance of Class E Power Amplifiers," *IEE Proc. Circuits Devices Syst.*, vol. 141, pp. 174–184, June 1994.
53. C. K. Ho, H. Wong, and S. W. Ma, "Approximation of Non-Zero Transistor ON Resistance in Class-E Amplifiers," *Proc. 5th IEEE Int. Caracas Conf. Devices Circuits Syst.*, vol. 1, pp. 90–93, Nov. 2004.
54. D. Milosevic, J. van der Tang, and A. van Roermund, "Explicit Design Equations for Class-E Power Amplifiers with Small DC-Feed Inductance," *Proc. 2005 Europ. Conf. Circuit Theory and Design*, vol. 3, pp. 101–104, Aug./Sept. 2005.
55. A. Grebennikov, "Load Network Design Techniques for Class E RF and Microwave Amplifiers," *High Frequency Electronics*, vol. 3, pp. 18–32, July 2004.
56. M. Acar, A. J. Annema, and B. Nauta, "Generalized Design Equations for Class-E Power Amplifiers with Finite DC Feed Inductance," *Proc. 36th Europ. Microwave Conf.*, pp. 1308–1311, 2006.
57. A. Grebennikov, "Class E High-Efficiency Power Amplifiers: Historical Aspect and Future Prospect," *Applied Microwave & Wireless*, vol. 14, pp. 64–71, July 2002, pp. 64–72, Aug. 2002.
58. A. V. Grebennikov and H. Jaeger, "Class E with Parallel Circuit – A New Challenge for High-Efficiency RF and Microwave Power Amplifiers," *IEEE MTT-S Int. Microwave Sym. Dig.*, vol. 3, pp. 1627–1630, 2002.
59. A. Grebennikov, "Switched-Mode RF and Microwave Parallel-Circuit Class E Power Amplifiers," *Int. J. RF and Microwave Computer-Aided Eng.*, vol. 14, pp. 21–35, Jan. 2004.
60. M. K. Kazimierczuk and W. A. Tabisz, "Class C-E High-Efficiency Tuned Power Amplifier," *IEEE Trans. Circuits and Systems*, vol. CAS-36, pp. 421–428, Mar. 1989.
61. A. V. Grebennikov and H. Jaeger, "High Efficiency Transmission Line Tuned Power Amplifier," U.S. Patent 6,552,610, April 2003.
62. J. K. A. Everard and A. J. King, "Broadband Power Efficient Class E Amplifiers with a Non-Linear CAD Model of the Active MOS Device," *J. IERE*, vol. 57, pp. 52–58, Mar. 1987.
63. C. S. Aitchison and R. V. Gelsthorpe, "A Circuit Technique for Broadbanding the Electronic Tuning Range of Gunn Oscillators," *IEEE J. Solid-State Circuits*, vol. SC-12, pp. 21–28, Feb. 1977.
64. R. V. Gelsthorpe and C. S. Aitchison, "Analytical Evaluation of the Components Necessary for Double Reactance Compensation of an Oscillator," *Electronics Lett.*, vol. 12, pp. 485–486, Sept. 1976.
65. H. Jaeger, A. V. Grebennikov, E. P. Heaney, and R. Weigel, "Broadband High-Efficiency Monolithic InGaP/GaAs HBT Power Amplifiers for 3G Handset Applications," *IEEE MTT-S Int. Microwave Symp. Dig.*, vol. 2, pp. 1035–1038, 2002.
66. H. Jaeger, A. Grebennikov, E. Heaney, and R. Weigel, "Broadband High-Efficiency Monolithic InGaP/GaAs HBT Power Amplifiers for Wireless Applications," *Int. J. RF and Microwave Computer-Aided Eng.*, vol. 13, pp. 496–510, Nov./Dec. 2003.
67. I. A. Popov, "Switching Mode of Single-Ended Transistor Power Amplifier (in Russian)," *Poluprovodnikovye Pribory v Tekhnike Svyazi*, vol. 5, pp. 15–35, 1970.
68. M. Kazimierczuk, "Charge-Control Analysis of Class E Tuned Power Amplifier with Only One Inductor and One Capacitor in Load Network," *IEEE Trans. Electron Devices*, vol. ED-31, pp. 366–373, Mar. 1984.
69. J. Modzelewski, "Power Gain of Class E and Class B VHF Tuned Power Amplifiers," *Proc. 15th Int. Conf. Microwaves Radar Wireless Commun.*, vol. 1, pp. 41–44, 2004.
70. J. W. Phinney, D. J. Perreault, and J. H. Lang, "Radio-Frequency Inverters with Transmission-Line Input Networks," *IEEE Trans. Power Electronics*, vol. PE-22, pp. 1154–1161, July 2007.
71. S. A. Zhukov and V. B. Kozyrev, "Push–Pull Switching-Mode Tuned Power Amplifier without Commutation Losses (in Russian)," *Poluprovodnikovye Pribory v Tekhnike Elektrosvyazi*, vol. 15, pp. 95–106, 1975.
72. A. Grebennikov, "High-Efficiency Class FE Tuned Power Amplifiers," *IEEE Trans. Circuits and Systems–I: Regular Papers*, vol. CAS-I-55, pp. 3284–3292, Nov. 2008.
73. M. Albulet, "An Exact Analysis of Class DE Amplifier at Any Output *Q*," *IEEE Trans. Circuits and Systems–I: Fundamental Theory Appl.*, vol. CAS-I-46, pp. 1228–1239, Oct. 1999.

74. D. J. Kessler and M. K. Kazimierczuk, "Power Losses and Efficiency of Class-E Power Amplifier at Any Duty Ratio," *IEEE Trans. Circuits and Systems–I: Regular Papers*, vol. CAS-I-51, pp. 1675–1689, Sept. 2004.
75. Z. Kaczmarczyk, "High-Efficiency Class E, EF_2, and E/F_3 Inverters," *IEEE Trans. Industrial Electronics*, vol. IE-53, pp. 1584–1593, Oct. 2006.
76. J. M. Rivas, Y. Han, O. Leitermann, A. D. Sagneri, and D. J. Perreault, "A High-Frequency Resonant Inverter Topology with Low-Voltage Stress," *IEEE Trans. Power Electronics*, vol. PE-23, pp. 1759–1771, July 2008.

12 Linearization and Efficiency Enhancement Techniques

In modern wireless communication systems, it is very important to realize simultaneously high-efficiency and linear operation of the power amplifiers. This chapter describes a variety of techniques and approaches that can improve the power amplifier performance. To increase efficiency over power backoff range, the Doherty, outphasing, and envelope-tracking power amplifier architectures, as well as switched multipath power amplifier configurations are discussed and analyzed. There are several linearization techniques that provide linearization of both entire transmitter system and individual power amplifier. Feedforward, cross cancellation, or reflect forward linearization techniques are available technologies for satellite and cellular base station applications with achieving very high linearity levels. The practical realization of these techniques is quite complicated and very sensitive to both the feedback loop imbalance and the parameters of its individual components. Analog predistortion linearization technique is the simplest form of power amplifier linearization and can be used for handset application, although significant linearity improvement is difficult to realize. Different types of the feedback linearization approaches together with digital linearization techniques are potentially attractive to be used in handset or base station transmitters. The choice of a proper high-efficiency approach or linearity correction scheme depends on performance tradeoffs as well as manufacturing capabilities. Finally, the potential semidigital and digital amplification approaches are discussed with their architectural advantages and problems in practical implementation.

12.1 FEEDFORWARD AMPLIFIER ARCHITECTURE

In the middle of 1920s, H. S. Black firstly proposed the method of suppressing even- and odd-order distortion components produced in nonlinear transmitting system [1]. However, interest in this invention was limited at that time due to success of the competitive feedback approach invented later by him as well, due to the simplicity and effectiveness of the latter. Only almost three decades later W. D. Lewis extended this approach to microwave frequencies by using the wave-guide sections for delay lines, branch-line hybrid junctions and directional couplers [2]. Since then, the interest of feedforward correction in RF and microwave applications had become significant to satisfy simultaneously strong requirements in high output powers, extremely high degrees of linearity, good time stability and broad bandwidths. H. Seidel described in detail the application of a feedforward compensated circuit in which the amplified signal is compared with a time-shifted reference signal [3]. In this case, the error component, which includes both noise and distortion components introduced by the main amplifier, is then amplified by means of a high-quality linear subsidiary amplifier and added to the time-shifted amplified signal in such a phase as to minimize the error in the output signal. To minimize errors due to impedance mismatch in the amplifier circuit, hybrid-coupler power dividers can be used. At the same time, to minimize noise in the output signal due to the subsidiary amplifier, the portion of input signal coupled to the subsidiary amplifier must be larger than that of coupled to

RF and Microwave Transmitter Design, First Edition. Andrei Grebennikov.

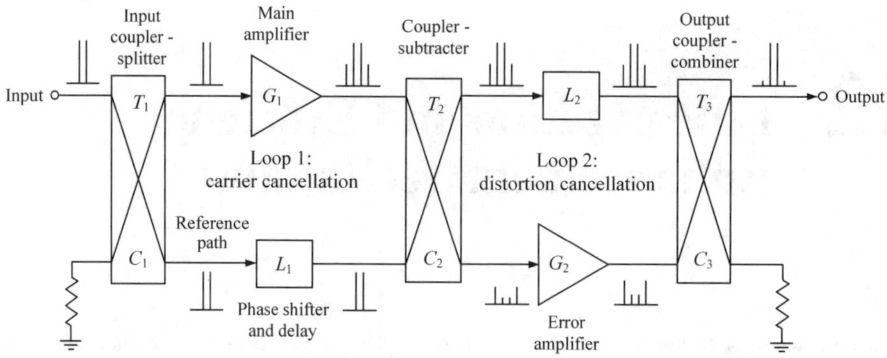

FIGURE 12.1 Basic structure and operation principle of feedforward amplifier.

the main amplifier. Most efficient utilization of the power in the amplified signal and the error signal can be realized by using a reactive three-port network to match the main signal path and the error signal path to the output load. As a part of a test to determine its applicability to coaxial repeaters, a feedforward error-controlled system was applied to a coaxial flat-gain amplifier operating in the frequency range of 0.5–20 MHz. As a result, a modulation product reduction of greater than 35 dB over a 40:1 bandwidth was achieved [4,5]. The use of feedforward architecture could result in up to 20 dB distortion improvement in a feedback amplifier operating over the whole frequency decade 30–300 MHz [6]. In a practical 2.2 GHz feedforward amplifier system with a power gain of 30 dB and an output power of 1.25 W, the suppression of the intermodulation distortion products of at least 50 dB from the carrier level was achieved [7].

Figure 12.1 shows the basic structure and principle of operation of the feedforward amplifier. The linearization feedforward system consists of two cancellation loops and generally includes the main power amplifier, three couplers, two phase shifters, and an auxiliary error amplifier. The operation of the feedforward linearization circuit is based on the subtraction of two equal signals with subsequent cancellation of the error signal in the amplifier output spectrum. Its operation principle can be seen clearly from the two-tone test spectra at various points in the block diagram. The input signal is split by input coupler-splitter into two identical parts, although in a common case the ratio used in the splitting process does not need to be equal, with one part going to the main power amplifier while the other part goes to a delay element. The signal in the top path is amplified by the main amplifier whose inherent nonlinear behavior contributes to the intermodulation and harmonic distortion components that are added to the original signal. This signal is sampled and scaled by the coupler-subtracter before being combined with the delayed distortion-free portion of the input signal. The resulting error signal ideally contains only the distortion components provided by the main amplifier. The error signal is then amplified linearly by low-power high-linearity error amplifier to the level required to cancel the distortion in the main part, and is then fed to the output directional coupler-combiner, on the other input of which a time-delayed and out-of-phase main-path signal is forwarded. The resulting signal at the feedforward linearization system output is an error-free signal in an ideal case or essentially an amplified version of the original input signal in practice.

The operation quality of the feedforward amplifier system obviously depends significantly on cancellation accuracy at the coupler-subtracter and output coupler-combiner. The level of distortion reduction is determined by the cancellation occurring at output coupler-combiner, and cancellation of the fundamental signals at coupler-subtracter is required to prevent subtraction of the fundamentals at output coupler-combiner and consequent gain loss. At the same time, cancellation of the fundamentals at coupler-subtracter is also important in order to prevent large amplitude error signals from entering error amplifier and possibly causing significant distortion in that amplifier. Generally, in the first carrier-cancellation loop the precision in cancellation is required only to such a degree as to avoid any substantial degradation of linearity in the error amplifier [5]. On the other hand, since the second distortion-cancellation loop controls the entire linearity improvement of the feedforward system, the degree of its balance should always be at the highest level [7].

To analyze the effect of imperfect magnitude and phase equalization in the amplifier and delay line paths at any particular frequency, consider the signal in each upper and lower paths of the first cancellation loop to be cosinusoidal in the form of

$$v_1 = V \cos \omega t \tag{12.1}$$

$$v_2 = (V \pm \Delta V) \cos(\omega t \pm \theta) \tag{12.2}$$

where ΔV is the amplitude imbalance and θ is the phase imbalance.

After subtraction of these signals in a coupler-subtracter, we have in a normalized form

$$\frac{\Delta v}{V} = \cos \omega t - \left(1 \pm \frac{\Delta V}{V}\right) \cos(\omega t \pm \theta) = (1 - \alpha\cos\theta) \cos \omega t \pm \alpha \sin\theta \sin \omega t \tag{12.3}$$

where $\Delta v = v_1 - v_2$ and

$$\alpha = 1 \pm \frac{\Delta V}{V}. \tag{12.4}$$

As a result, for a total imbalance magnitude,

$$\frac{|\Delta v|}{V} = \sqrt{(1 - \alpha\cos\theta)^2 + (\alpha \sin\theta)^2}. \tag{12.5}$$

Since cancellation achieved by the second loop can be analyzed ideally in a similar way, the cancellation result achieved by the first and second loops independently can be rewritten in the corresponding forms of

$$CANC_1(\text{dB}) = 10 \log_{10}\left(1 + \alpha_1^2 - 2\alpha_1 \cos\theta_1\right) \tag{12.6}$$

$$CANC_2(\text{dB}) = 10 \log_{10}\left(1 + \alpha_2^2 - 2\alpha_2 \cos\theta_2\right) \tag{12.7}$$

where α_1 and θ_1 are the amplitude and phase imbalance in the first loop, and α_2 and θ_2 are the amplitude and phase imbalance in the second loop, respectively [8].

Figure 12.2 shows the distortion cancellation as a function of amplitude and phase imbalance. From these curves it follows that, to obtain a high degree of cancellation accuracy, it is necessary to maintain extremely small amounts of amplitude imbalance. For example, a 40 dB of cancellation would require a phase imbalance of less than 1° and an amplitude imbalance of less than 0.1 dB. However, a demand for a high degree of linearity improvement will cause the system to become sensitive to circuit parameter variations due to temperature change. To achieve temperature stability in a practical system, the degree of linearity improvement should be kept at a reasonably low level. For example, a 30 dB of cancellation would require only an amplitude imbalance of 0.25 dB and a phase imbalance of 1.8°. To improve the temperature stability characteristic, it is better to realize both main and error amplifiers using the same technology, similar components and assembly techniques. However, if a higher degree of balance is to be maintained at all times, an automatic adaptive control system must be employed. Besides, an additional transmission-line delay mismatch can be taken into account when using transmission lines in high-frequency feedforward linearization systems. For example, if the difference in wavelength between the transmission lines in upper and lower paths at the center bandwidth frequency f_0 is equal to $0.1f_0$, then, in order to obtain a 30 dB of cancellation

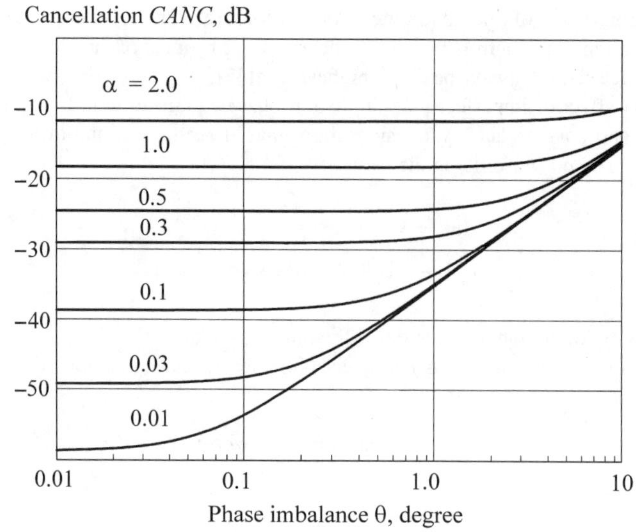

FIGURE 12.2 Cancellation as function of amplitude and phase imbalance.

with 30 MHz bandwidth at 800 MHz for $\alpha = 0.1$ dB, the phase imbalance should be maintained within approximately 1.0° [9].

It is of great importance for telecommunication system to minimize its nonlinear distortion level, and the main factor of its linearity is a level of the third-order intermodulation products at the system output. In this case, consider the cancellation provided by both fist and second loop through the parameters of the feedforward system [10]. At the output of the coupler-subtracter with suppressed carrier P_{supp}, the cancellation of the first loop is defined as

$$CANC_1 = \frac{P_{\text{supp}}}{C_2 P_{\text{main}}} \tag{12.8}$$

where C_2 is the coupling coefficient of the second coupler-subtracter and P_{main} is the carrier power level of the main amplifier. On the other hand, the cancellation achieved in the second loop is

$$CANC_2 = \frac{P_{\text{IM3supp}}}{P_{\text{IM3main}}} \frac{1}{T_2 L_2 T_3} \tag{12.9}$$

where P_{IM3main} is the power level of the third-order intermodulation component, P_{IM3supp} is the power level of the third-order intermodulation component of the main amplifier suppressed at the linearizer output due to the corrective action of the second loop, L_2 is the delay-line loss in the second loop, T_2 and T_3 are the transmission losses in the coupler-subtracter and output coupler-combiner, respectively.

The effective cancellation of the overall feedforward linearization system is the ratio of the power level of all intermodulation components at the feedforward system output over the power level of the intermodulation products for open-loop configuration. As a result, for in-phase addition of the intermodulation components of the main and error amplifiers, the effective cancellation can be expressed by

$$CANC_{\text{eff}}(\text{dB}) = 20\log_{10}\left[\sqrt{CANC_2} + \sqrt{CANC_1^3 \left(\frac{IP_{3\text{main}}}{IP_{3\text{error}}}\right)^2 \frac{T_2^2 L_2^2}{\alpha_2^3} \left(\frac{T_3}{C_3}\right)^2}\right] \tag{12.10}$$

where the amplitude imbalance α_2 is defined as the ratio of the power gains of the two paths

$$\alpha_2 = \frac{T_2 L_2 T_3}{C_2 G_2 C_3} \tag{12.11}$$

where G_2 is the power gain of the error amplifier, C_3 is the coupling coefficient of the output coupler-combiner, and $IP_{3\text{main}}$ and $IP_{3\text{error}}$ are the third-order intercept points of the main and error amplifiers, respectively. The first term in Eq. (12.10) depends on the balance level achieved in the second loop, whereas the second term defines the possible imbalance created by the first loop and some other feedforward circuit parameters. In particularly, an error amplifier with sufficiently low power capabilities having too small value of $IP_{3\text{error}}$ or too big coupling coefficient C_3 of the output coupler-combiner and loss $(T_2L_2T_3)$ through the main path increases the effect of the amplitude and phase imbalance.

The relationship between the overall feedforward system efficiency η and the efficiencies of the two amplifiers, η_{main} for main amplifier and η_{error} for error amplifier, when the losses $(T_2L_2T_3)$ through the main path are considered negligible, can be written as

$$\eta = \frac{\eta_{\text{main}}\eta_{\text{error}}C_3\left(1 - C_3\right)}{\eta_{\text{error}}C_3 + \eta_{\text{main}} f_{\text{main}}\left(1 - C_3\right)} \tag{12.12}$$

where $\log_{10} f_{\text{main}} = -(C/I)_{\text{main}}/10$, $(C/I)_{\text{main}}$ is the ratio of carrier to third-order intermodulation product of the main amplifier [11,12]. Provided the optimum value of C_3, which maximizes the overall efficiency η when the other system parameters are fixed, the maximum η_{max} can be obtained by

$$\eta_{\text{max}} = \eta_{\text{main}} \Big/ \left(1 + \sqrt{\frac{\eta_{\text{main}}}{\eta_{\text{error}}} f_{\text{main}}}\right)^2 \tag{12.13}$$

which shows the efficiency degradation due to the linearization system [13]. For example, for a typical 10 dB coupling ratio of the output coupler-combiner, only 10% of the power from the error amplifier reaches the load, which means that the error amplifier must produce ten times the power of the distortion in the main amplifier. In this case, it should operate in an inefficient linear mode in order not to disturb the error signal. As a result, the dc power consumed by the error amplifier can represent a significant part of that of the main amplifier. In addition, it needs to take into account the fact that, despite its excellent distortion cancellation property, generally the feedforward amplifier system requires well-equalized circuitry and is characterized by substantially increased complexity and cost.

The efficiency of the conventional feedforward linearization system using a balanced configuration of the main amplifier shown in Figure 12.3(*a*) can be improved by providing some restructuring of the system. As a result, the modified feedforward system consists of three major loops shown in Figure 12.3(*b*): carrier cancelation loop, balanced power amplifier loop, and error-injection loop [14]. In this case, the carrier cancelation loop extracts the error signal from the amplified signal at the output of the top power amplifier, whereas the error-injection loop provides an injection of the amplitude-adjusted and properly phased distortion into the output of the bottom power amplifier. Finally, the amplified signals in balanced paths are combined in the output hybrid combiner with corresponding distortion cancellation. Unlike the conventional feedforward system, in its balanced version each power amplifier sees only one coupler, either coupler-subtracter or output coupler-combiner, which means that there is no additional insertion loss due to output coupler-combiner as in the conventional feedforward system. As a result, for the four-carrier wideband code division multiple access (WCDMA) signal with peak-to-average ratio of 10 dB, an efficiency improvement by 2% at an average output power of 40 dBm with an improvement in adjacent channel leakage power ratio (*ACLR*) (5 MHz offset) of about 18.6 dB by cancellation at the center bandwidth frequency of 2.14 GHz.

However, it is a serious problem for the conventional feedforward linearization system to maintain the great accuracy in amplitude and phase balance over time, temperature, supply voltage, or input source and load variations. In practice, some forms of the gain and phase adjustments are essential to

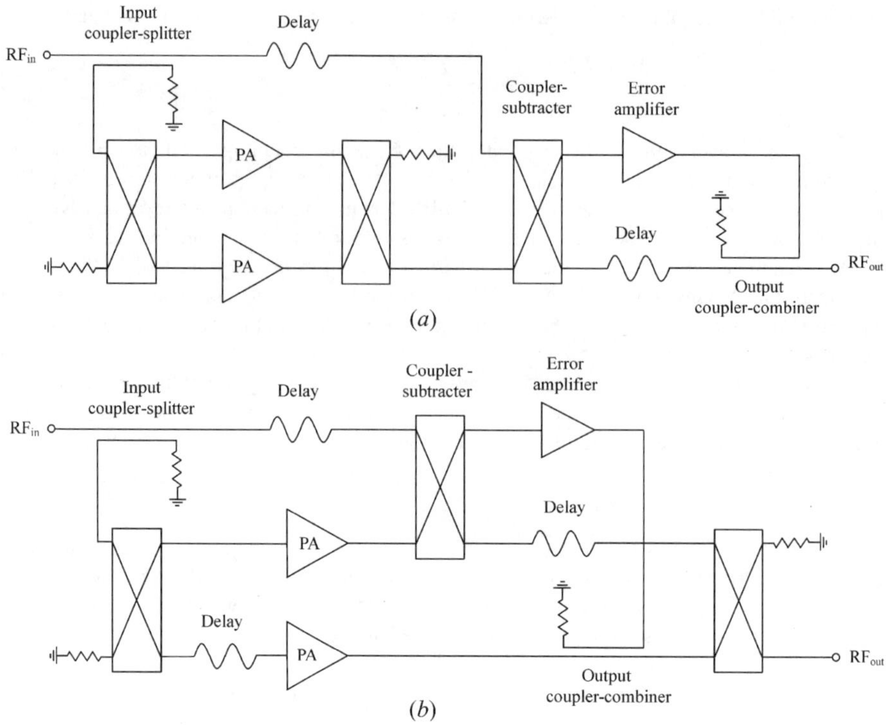

FIGURE 12.3 Balanced feedforward amplifier topologies.

achieve acceptably low level of intermodulation distortion. Figure 12.4(*a*) shows a block schematic of the analog adaptive feedforward linearization system that includes a feedback network for adaptively adjusting the performance of the overall feedforward system to compensate for uncontrolled variations of its component parameters [15]. The feedback network provides a control of the carrier and distortion cancellation loops by comparing the signals sampled at their inputs and outputs and adaptively adjusts the corresponding vector modulators to minimize the amplitude and phase imbalance when it is necessary. Different adaptation algorithms using optimization techniques can be implemented to improve the cancellation results for an analog adaptive feedforward linearization system [16].

Using a digital signal processing (DSP) creates a good opportunity to provide a correction in the amplitude and phase imbalance in the feedforward linearization system at baseband level, thus making this procedure more predictable and fast and overcoming the problems with mixer dc offset and masking of strong signals by weaker ones than can compromise analog adaptive implementations [17]. To compensate for the component frequency response and the nonadaptive nature of the delay lines, a hybrid of the conventional feedforward linearizer and a digital signal processor can be used, as shown in Figure 12.4(*b*), where both the amplifier input signal and the reference signal are generated by DSP [18]. The reference signal is then used to cancel the linearly amplified component of the distorted amplifier output signals, leaving an error signal containing only the main amplifier distortion. By generating the reference signal in the DSP, rather than using an analog splitter, some of the analog hardware can be moved into a simpler digital implementation, with independent control of the main amplifier and reference signals by using equalizers. In this case, the amplitude and reference equalizers correct the phase shift, time delay and nonideal response of the analog components to achieve the proper distortion cancellation. By improving the cancellation of the first loop, a more accurate error signal is generated that consists only of the distortion from the main amplifier. Generally, the use of

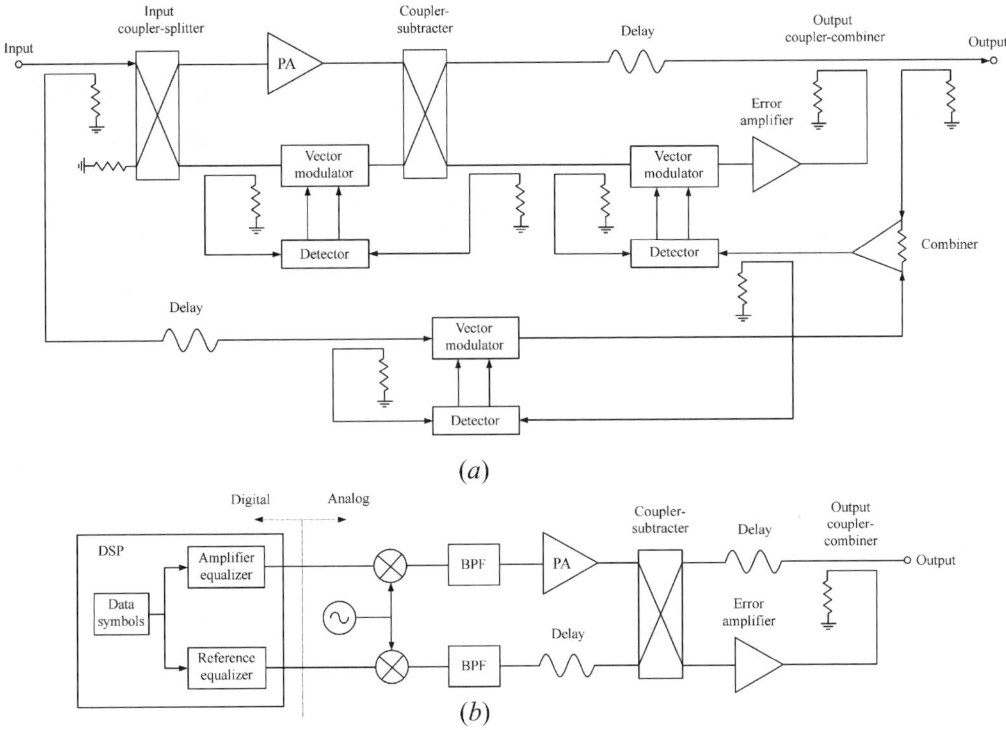

FIGURE 12.4 Adaptive analog and digital feedforward amplifier linearizers.

amplifier and reference equalizers in the first loop has an advantage in that the tuning, previously done manually, has now been moved back into the DSP where it can be done adaptively.

12.2 CROSS CANCELLATION TECHNIQUE

As an alternative approach proposed in the middle of 1930s that provides higher efficiency: distortion in nonlinear power amplifiers can be eliminated by using an auxiliary amplifier in which a fraction of the main-amplifier input signal combined with a fraction of the distorted main-amplifier output signal produces a correcting component that, combined with the total output, restores this to the same shape as the input [19,20]. The approach is now known as a cross cancellation technique that combines the high efficiency of a parallel or balanced power amplifier with linearization capabilities of the predistortion linearizers.

The basic cross cancellation scheme shown in Figure 12.5(a) includes the two identical power amplifiers connected in parallel configuration under equal input drive conditions. Balancing of signal levels in both amplifying paths is only possible by employing the input divider with unequal division ratio where the more power goes to the lower amplifying path, in which the sampled output signal by the output directional coupler from the upper amplifying path is delivered through the input directional coupler with proper coupling coefficient. To equalize the signal phases in both amplifying paths, the phase delay elements are included in the input and output circuits of the corresponding signal paths. As a result, the lower power amplifier is operated as a predistorted power amplifier with the predistortion signal created by the upper power amplifier. In this case, distortion neutralization is obtained by the injection of the distortion components to the lower amplifying path in such a manner

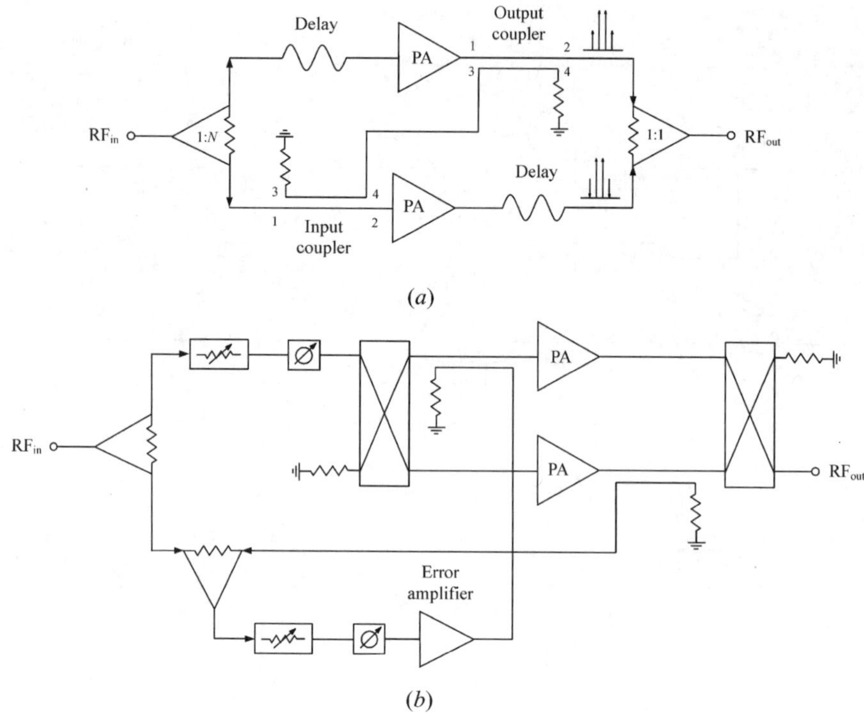

FIGURE 12.5 Cross cancellation linearizer diagrams.

that they will be 180° out of phase with those being created by the upper power amplifier at the input of the output combiner.

To derive some analytical relationships between parameters of the cross cancellation linearization system shown in Figure 12.5(*a*), consider an idealized approach where the system parameters are normalized to the input power and power ratio of the input unequal divider is equal to 1:*N*. Then, the output powers of lower and upper amplifying paths at the corresponding inputs of the output combiner can be written as

$$\begin{aligned} G_P(1 - C_{31}) + IM &= \left[N(1 - C_{31}) - (G_P + IM)C_{31}^2\right]G_P + IM \\ &= \left[N(1 - C_{31}) - G_P C_{31}^2\right]G_P + \left(1 - G_P C_{31}^2\right)IM \end{aligned} \tag{12.14}$$

where G_P is the power gain of each power amplifier (PA), *IM* is the intermodulation distortion introduced by each PA, and C_{31} is the coupling factor of each directional coupler calculated as the ratio of power at the output port 3 relative to the input port 1 equal to the coupling factor C_{24} of the input directional coupler when its port 4 becomes an input port. From comparison between the left- and right-hand sides of Eq. (12.14) it follows that the out-of-phase conditions for intermodulation components at the corresponding inputs of the output in-phase combiner can be obtained when

$$G_P C_{31}^2 = 2 \tag{12.15}$$

resulting in

$$N = \frac{3 - C_{31}}{1 - C_{31}}. \tag{12.16}$$

As an example, if the power gain of each PA is $G_P = 200$ or 23 dB, then from Eq. (12.15) it follows that it is necessary to choose the input and output directional couplers with coupling factor $C_{31} = 0.1$ or -10 dB and the input power divider with $N = 3.2$ (about 5 dB) resulting then from Eq. (12.16). In this case, the power gain of the overall system reduces to 19.5 dB. However, it was found that the linearity improvement is not as high as in a feedforward linearizer where the distortions are subtracted at its output. This is because the amplifying paths are not really identical. To make the cross cancellation system more symmetrical, it is necessary to equalize the insertion losses in the output circuits of both amplifying paths by introducing a required attenuation in a lower path, which in turn results in reduced system efficiency. Generally, in practical applications, with varying input drive levels and temperature, it is necessary to use phase shifters and variable attenuators that are controlled by a power-minimization loop controller that serves to minimize the distortion components in a composite output signal [21].

Figure 12.5(*b*) show the cross cancellation technique based on a balanced power amplifier configuration where the distortion generated in one balanced path, which is identical to the other path, are used to cancel the distortions generated by the whole balanced power amplifier [22]. This approach provides a control of the error signal separately, as in the feedforward technique. However, the main difference between these two techniques is that the error signal is added to the input of the amplifying path, not to the output, thus improving the system efficiency. In this case, samples of the signal and distortion from lower amplifying path are combined with a portion of the reference signal delivered from the input splitter such that the linear components of these two signals are cancel each other leaving the distortion components only from the sampled path of the balanced power amplifier. The gain and phase of the distortion are then adjusted using variable attenuator, phase shifter and linear error amplifier so that, when it is coupled into the input of the other path of the balanced power amplifier, the distortions generated by both paths of the balanced power amplifier are cancelled.

12.3 REFLECT FORWARD LINEARIZATION AMPLIFIER

The reflect forward adaptive linearizer (RFAL) technique uses the input reflected signal from one power amplifier to develop an intermodulation-correcting signal forwarded to the input of the other power amplifier, which are connected in parallel or balanced configuration [23,24]. Figure 12.6 shows

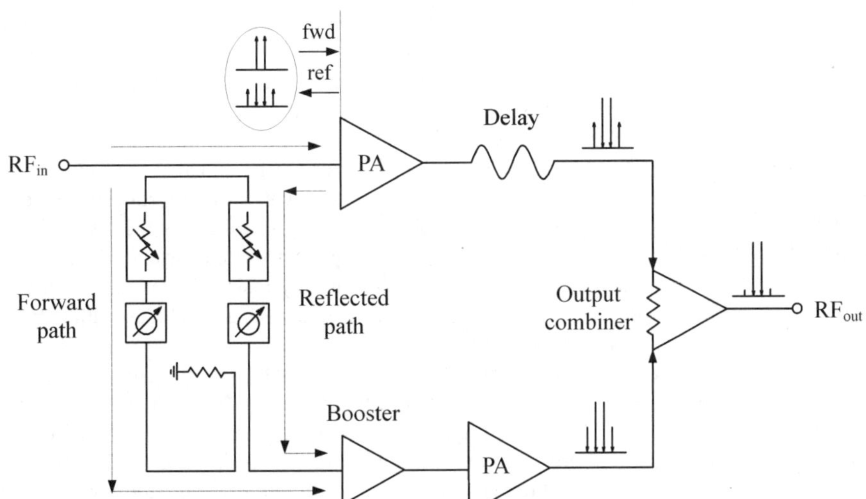

FIGURE 12.6 Block schematic of reflect forward linearization amplifier.

a block diagram of the RFAL amplifier that includes the two identical power amplifiers, low-distortion booster amplifier, forward and reflected paths with two directional couplers at the input, and delay line and in-phase combiner at the output. With RFAL technique, the power combining efficiency of the two power amplifiers approaches that of the conventional parallel power amplifier configuration, while the intermodulation products can be improved from 20 to 30 dBc at the center bandwidth frequency.

The basic principle of RFAL operation can be described as follows: when the two-tone forward fundamental signal is reflected from the transistor input of upper power amplifier, the resulting reflected two-tone fundamental signal is out-of-phase relative to the input forward signal and is in-phase relative to the output signal. In this case, the intermodulation components appeared at the upper power amplifier input as a result of its active device nonlinearity are in-phase relative to the output intermodulation components. The input reflected composite signal containing the fundamental and intermodulation components is used as a correcting signal for the lower power amplifier. This correcting signal, when properly amplified and phased in the reflected path, cancels the output intermodulation distortions produced by the lower power amplifier when it is combined with the input signal flowing through the forward path. The booster amplifier in a lower path is necessary to equalize the drive signal levels for both upper and lower power amplifiers. In this case, the signals from the upper and lower amplifying paths have the in-phase fundamental and out-of-phase intermodulation components at the corresponding inputs of the output in-phase combiner, thus resulting in a distortion cancellation in the combined signal flowing into the load. The laterally diffused metal-oxide-semiconductor (LDMOS) RFAL amplifier with output power of 43 dBm achieves an improvement of the third-order intermodulation products by over 15 dB and a total efficiency of more than 20% over the frequency range from 865 to 895 MHz [25]. For very high cancelation requirements over wide temperature and drive conditions an adaptive feedback network can be used to monitor the relative amplitude and phase at the outputs of the power amplifiers and drive the voltage variable attenuators and phase shifters to maintain the optimum conditions.

12.4 PREDISTORTION LINEARIZATION

To achieve simultaneously high-efficiency and low-distortion operation conditions of the power amplifier when the linearity requirements are not extremely high, it is possible to use a predistortion linearizer that provides the positive amplitude and negative phase deviations for input RF signal to compensate for the active device nonlinearity whose nonlinear behavior (when a power amplifier is operated close to saturation) usually represents the opposite behavior of its amplitude and phase characteristics. Historically, the initial idea to compensate for the third-order intermodulation products arising in vacuum-tube amplifier was to use the linearization scheme where the nonlinear amplifier having a compressing characteristic is followed by a nonlinear element having an expanding characteristic and producing the third-order distortion of opposite sign to that of the amplifier [26].

The block diagram of a linearized power amplifier system with a predistortion linearizer with indication of the appropriate amplitude and phase dependences at each stage of the system is shown in Figure 12.7, where also a variable attenuator for adjusting the amplitude level of the input signal is included. At microwaves, a linearized power amplifier usually includes two isolators for stable operation conditions. The conventional predistortion linearizer circuits generally use either diodes or transistors as sources of intermodulation [27,28].

As an interesting fact, yet at the beginning of 1920s it was claimed that, by using similar vacuum tubes in both stages of a two-stage power amplifier with similar signals at their inputs, the even harmonics generated by the first amplification are neutralized by the even harmonics generated by the second amplification because they are similar in amplitude and opposite in phase at the output of the second vacuum tube [29]. Indeed, as it turned out with regard to modern transistor power amplifiers using GaAs pseudomorphic high electron mobility transistor (pHEMT) devices, it is enough to proper choose a bias point of a driver-stage device in a two-stage amplifier to provide a negative phase deviation to compensate for the positive phase deviation of the final stage [30]. In this case, the

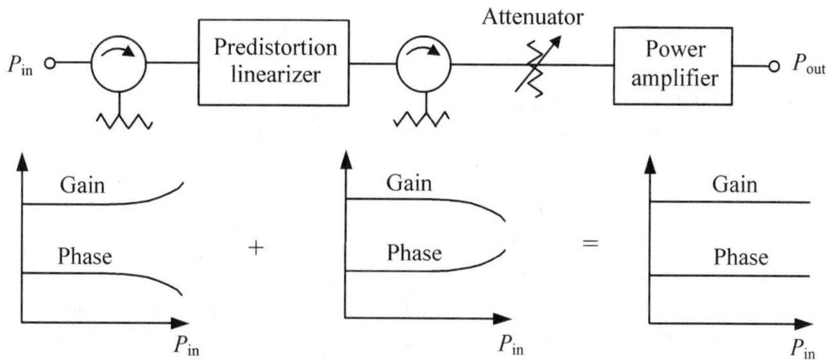

FIGURE 12.7 Block diagram of power amplifier with predistortion linearizer.

quiescent current of the driver-stage device, whose size is three times smaller than that of a final-stage device, is sufficiently small. As a result, for a quiescent current equal to 1.25% of the device dc saturated current, an improvement of more than 5 dB in *ACLR* of a whole high-efficiency two-stage cellular-phone WCDMA power amplifier operating at 1.95 GHz can be achieved at backoff output powers close to the saturation power.

Figure 12.8(*a*) shows the schematic of a simple diode linearizer composed of a series Schottky diode and a parallel capacitor with two RF chokes for dc feed and two blocking capacitors, which provides positive amplitude and negative phase deviations when input power increases [31]. The equivalent circuit of the series diode is shown in Figure 12.8(*b*), where R is the diode equivalent resistance and C_j is the junction capacitance. With the increase of an incident input signal power, the forward diode current increases that leads to the decrease of the diode resistance R. In this case, the positive amplitude and negative phase deviations can be achieved under low forward-bias conditions when the diode current ranges from 0.1 to 1.0 mA, and, in the latter case, the phase deviation can reach a value of $-30°$. Applying such a linearizer to a 1.9 GHz monolithic microwave integrated circuit (MMIC) power amplifier with saturated power of 22.5 dBm, an improvement of adjacent channel power ratio (*ACPR*) of 5 dB can be achieved for the quadrature phase shift keying (QPSK) modulated signal when output powers are less than 15 dBm.

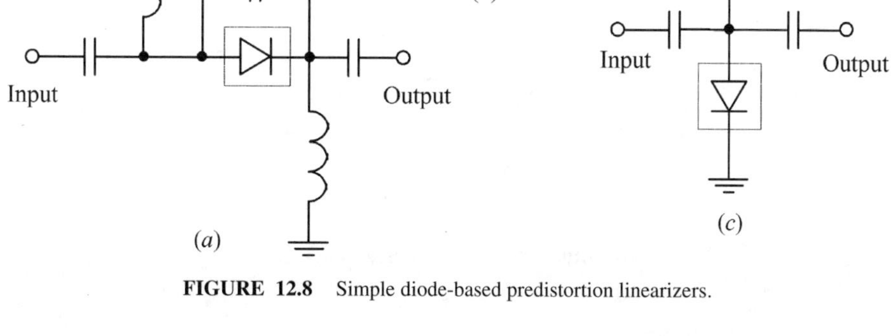

FIGURE 12.8 Simple diode-based predistortion linearizers.

A similar improvement of *ACPR* can be achieved by using a linearizer based on a parallel Schottky diode with the bias feed resistor R_b, which is shown in Figure 12.8(*c*) [32]. With the increase of input power, the bias point of a diode changes due to the voltage drop across the resistor R_b caused in turn by the increased diode forward current. As a result, due to the decreased diode resistance R, the linearizer achieves positive gain and negative phase deviations. By applying such a linearizer to a 2.7 GHz power amplifier, a maximum improvement of 5 dB was achieved for low quiescent current conditions at output power of 34 dBm.

Positive amplitude deviation with negative phase deviation can also be achieved using a series-feedback GaAs metal semiconductor field-effect transistor (MESFET) amplifier with a large source inductance L_s, a block diagram of which (including matching circuits) is shown in Figure 12.9(*a*) [33]. The required amplitude and phase deviations are due to nonlinearities of the device transconductance g_m, gate–source capacitance C_{gs}, and differential drain–source resistance R_{ds}. For the device with a gate width of 1.2 mm, a nonlinearity of g_m contributes to the positive amplitude deviation when $L_s =$ 20 nH. At the same time, nonlinearities of both g_m and R_{ds} contribute to the negative phase deviation when $L_s \geq 3$ nH. A nonlinearity of C_{gs} has a negligibly small effect on both the amplitude and phase deviations. As a result, for a linearizer with $L_s = 16$ nH at an operating frequency of 1.9 GHz, the positive amplitude and negative phase deviations were obtained across the input power dynamic range from 5 to 18 dBm, with amplitude deviation of 2.5 dB and phase deviation of 30° at 18 dBm input power. The GaAs MESFET device was biased in Class AB mode with a drain–source supply voltage of 2 V providing a quiescent current of 78 mA. By applying this linearizing technique to a 1.9 GHz MMIC power amplifier with 1-dB compressed power of 17 dBm, an improvement of *ACPR* up to 7 dB was achieved for a $\pi/4$-shifted QPSK signal.

As an alternative, it is also possible to achieve positive amplitude and negative phase deviations using a source-grounded MESFET device with zero drain–source supply voltage [34]. The schematic diagram of such a linearizer is shown in Figure 12.9(*b*). In this case, for the device with a gate width of 240 μm at the saturation power of 20 mW under the gate bias condition of $V_g = -0.4$ V, the

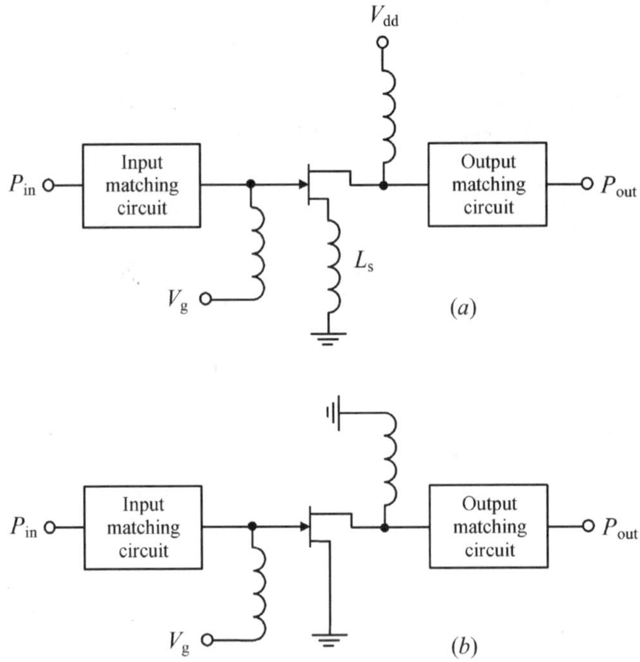

FIGURE 12.9 Transistor-based linearizers.

3-dB increased power gain and of about 30-degree negative phase were achieved due to the varying drain–source resistance. Because of its simplicity, such a linearizer can operate from 2 to 12 GHz with good thermal stability. When it was implemented into a 50 W solid-state power amplifier system at operating frequency of 7 GHz, the system noise power ratio was improved over 15-dB dynamic range, in particular by 2 dB at the 3-dB output power backoff point.

More advanced configurations of the predistortion linearizer is based on the splitting of the input signal into nonlinear and linear paths using a directional coupler or a hybrid divider with subsequent subtraction of resulting signals in the output coupler-subtracter. The block diagram of such a predistortion linearizer that employs two power amplifiers in a balanced configuration using two 90-degree hybrids is shown in Figure 12.10(*a*) [28]. In this case, the upper power amplifier is operated in a linear Class A mode while the lower power amplifier is biased in a nonlinear Class AB or B mode to generate the proper intermodulation products by controlling the input power and device bias conditions. The phase shifter in a lower amplifying path is necessary to optimize the level of the fundamental components in the resulting output spectrum. Since both devices present approximately the same input impedances, a low input return loss is provided because the most of the reflected power flows into the isolated port. Figure 12.10(*b*) shows the practical microwave microstrip implementation of a two-path predistortion linearizer with an input 90-degree branch-line hybrid coupler, a nonlinear power amplifier in a lower path and an output directional coupler [35]. The microstrip transmission line in a lower amplifying path having a required electrical length compensates for the additional phase shift provided by the active device, whereas the required amplitude conditions are realized with the coupling coefficient of an output coupler-subtracter to be chosen. As a result, for a *Ku*-band multicarrier 4.5-W power amplifier, the phase deviation of a 12-dBm signal at the linearizer output up to -10° was achieved with a 22-dBm signal at the linearizer input.

Figure 12.11 shows the modified three-path predistortion linearizer structure, where a balanced configuration with a nonlinear power amplifier is adjusted for suppression of the fundamental components with the resulting error signal. Then, the amplitude-adjusted and properly phased error signal is amplified by an error amplifier and added to the linear component in the upper path, which is a

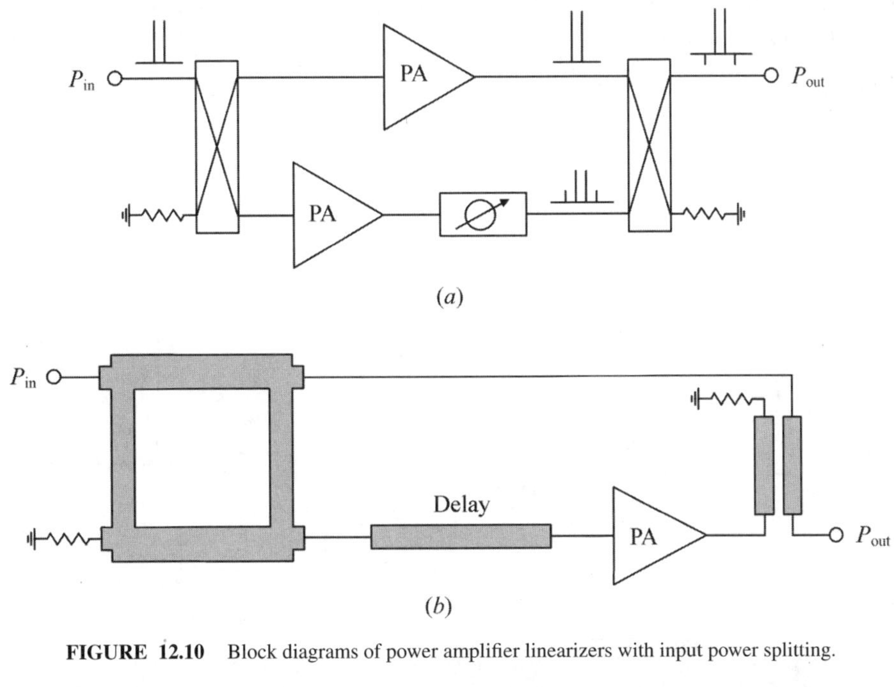

FIGURE 12.10 Block diagrams of power amplifier linearizers with input power splitting.

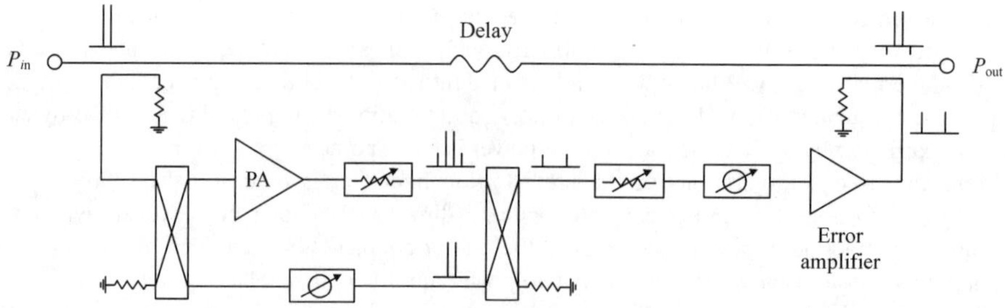

FIGURE 12.11 Modified three-path predistortion linearizer.

delayed portion of the input signal. However, it is very difficult to match the nonlinear characteristics of the predistorter and the main power amplifier because generally they differ both in size and number of stages that can only result to less than 10 dB improvement of adjacent channel leakage power ratio (*ACLR*) at 5 MHz offset from the center bandwidth frequency [36]. Therefore, it is very important for further linearity improvement to use similar devices in the predistorter and in the main power amplifier with a preferred balanced structure.

The block schematic of a power amplifier module, which includes a three-path predistortion linearizer shown in Figure 12.11 and a main power amplifier based on four power amplifiers configured into a balanced structure, is shown in Figure 12.12. In this case, the same Freescale MW6S004NT LDMOS devices were used both in the linearizer and main balanced power amplifier operated in a Class AB mode. As a result, a significant improvement of *ACLR* was achieved, from −42 to −57 dB for a 32-dBm channel power and from −37 to −49 dB for a 36-dBm channel power, with the signal peak-to-average ratio (*PAR*) of 6.5 dB.

The concept of a feedforward loop with its high cancellation performance can be used for a predistortion linearizer implementation. Since the feedforward loop is placed in front of the main amplifier, the linearity and power requirements to the error amplifier are reduced significantly compared to the conventional feedforward system. In this case, the delay-line and coupler losses are not so significant factors affecting the amplifier performance. Figure 12.13 shows the simplified schematic diagram of a power amplifier module with the feedforward distortion linearization using five identical power amplifiers based on MRF5S21090 LDMOS devices [37]. For a forward-link four-carrier WCDMA

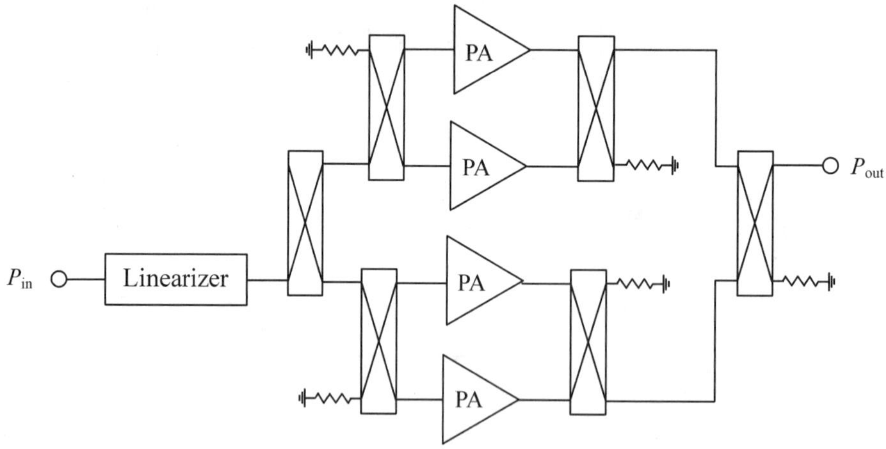

FIGURE 12.12 Power amplifier module with linearizer.

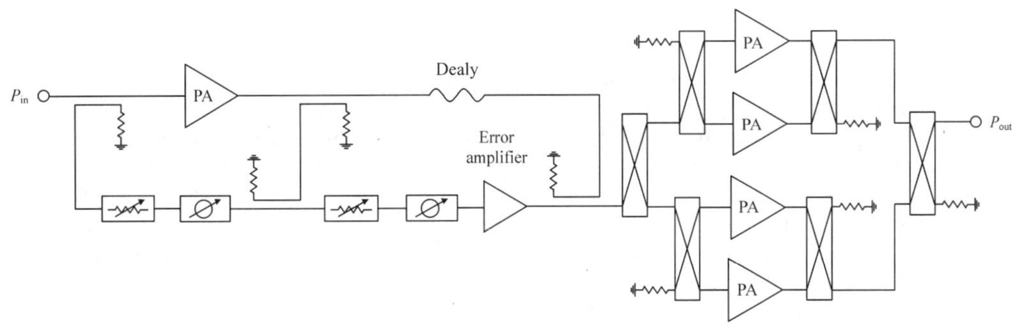

FIGURE 12.13 Power amplifier module with feedforward predistortion linearizer.

signal at 2.35 GHz, the *ACLR* was enhanced by about 7 dB at 5 MHz offset and the total efficiency of 12.7% was achieved at an average output power of 47.8 dBm, backed-off by 10.8 dB from the total peak power of 720 W.

Figure 12.14 shows the block schematic of a digital predistortion linearizer where the predistortion algorithm is based on an initially measured PA amplitude-to-amplitude modulation (*AM–AM*) and amplitude-to-phase modulation (*AM–PM*) response extracted from the *S*-parameter measurements by a vector network analyzer (VNA) [38]. The amplitude and phase characteristics are interpolated using splines, which are continuous piecewise cubic functions with continuous first and second derivatives. The interpolated amplitude and phase characteristics are then used to compute the appropriate predistortion coefficients representing a lookup table (LUT), which are multiplied with the original IS-95 signal to generate the desired predistorted baseband signal. The results show the limitations of this technique when the LDMOSFET power amplifier operation conditions are close to saturation when better than 6-dB improvement in *ACPR* can only be achieved. Generally, an adaptive correction mechanism is required to maintain the linearized power amplifier performance over varying load, supply voltage, or temperature conditions. This means that the LUT needs to be updated continuously to keep difference between the source signal and the transmitted signal sufficiently small. This can be realized by downconverting the portion of the transmitted signal and comparing it with source signal. In this case, it is important to provide the optimization of the wordlengths required in different parts of the predistortion linearizer to reduce power consumption and increase bandwidth for the required adjacent channel interference level [39]. The feedback complexity can be reduced with special adaptation algorithm when a single mixer and analog-to-digital converter (ADC) may be used in the feedback path instead of the full quadrature demodulation [40]. In addition, a noniterative adaptation method can be used to eliminate the convergence constrains usual for iterative methods [41].

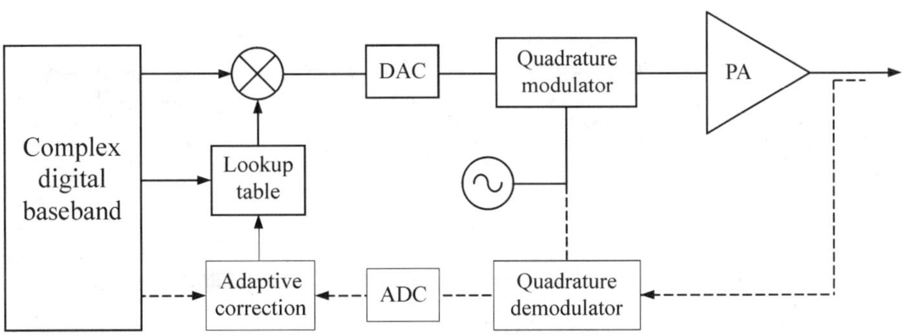

FIGURE 12.14 Digital predistortion system.

12.5 FEEDBACK LINEARIZATION

The principle of a feedback linearization of the power amplifier at the carrier frequency was invented by Harold S. Black in 1927. A year later he filed the patent application on vacuum-tube feedback amplifier [42]. Black recognized that by using a large amount of feedback in an amplifier comprising several vacuum-tube stages in cascade to yield a very high open-loop gain gives a glorious opportunity to make the resulting negative feedback amplifier increased in bandwidth and insensitive to nonlinearity and uncertainty in the characteristics of the vacuum tube [43]. The gain of the negative feedback amplifier decreases by amount of the feedback or loop gain, so do the nonlinear components. In this case, the negative feedback amplifier becomes insensitive to the gain or phase variations as long as its stability conditions are satisfied. In the latter case, if each of three tuned circuits of a three-stage vacuum-tube RF feedback amplifier can be assumed to have the phase-gain characteristics of the interstage circuits very nearly to infinite Q-tuned circuits, the maximum amount of feedback will allow a phase margin of 30° from a total phase shift of 180° [44]. Unfortunately, the significance of this invention was not fully understood at that time, as well as the operation principle of a negative feedback amplifier. For instance, Black's director of research insisted that a negative feedback amplifier would never work, similarly the Patent Office initially did not believe it would work and took over nine years to decide to issue the patent [45]. However, even nowadays his pioneering role for further achievements in feedback theory and practice is not well known.

The basic structure of a negative feedback amplifier at microwave frequencies is shown in Figure 12.15(*a*) where the power amplifier, bandpass filter (BPF) and feedback loop elements are chosen to provide a loop transmission greater than one with a phase shift of 180° within the operating bandwidth [46]. In this case, the BPF must be a single tuned resonator, so that the phase shift introduced will be less than 90°. By using a 50-dB gain power amplifier, an improvement of the

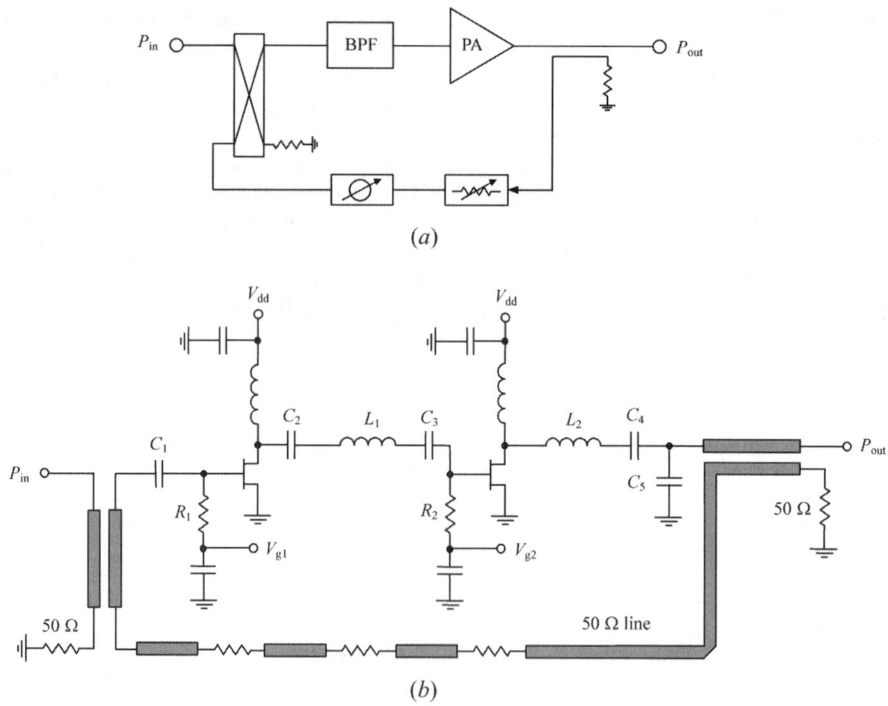

FIGURE 12.15 Negative feedback power amplifier schematic.

third-order intermodulation components by 10 dB (from −30 down to −40 dBc) at midband frequency of 2 GHz with the power gain of the closed loop system of about 34 dB and output power of 33 dBm was achieved [47]. It should be mentioned that the feedback power amplifier designed to operate at 835 MHz in communication system can provide a distortion improvement of 10 dB at saturated region, if the feedback gain of at least 20 dB is achieved [48]. Although, a feedback gain of 10 dB results in approximately 10-dB improvement in adjacent channel leakage power at 3 dB power backoff from saturation. In contrast, for the simple resistive feedback single-stage power amplifier, the level of the third-order IMD components at 2 GHz can be reduced by about 5 dB for the medium-signal levels only [49].

Figure 12.15(*b*) shows the circuit schematic of the two-stage power amplifier designed for a 3.4–4.2 GHz frequency bandwidth [50]. The output power of 27 dBm was achieved by using a power MESFET device with a gate length of 0.5 μm in the second stage. Loop gain was adjusted by changing the coupling in the output directional coupler, whereas a phase shift of 180° in the feedback path at midband was obtained using a microstrip line with the proper electrical length. The predicted open-loop gain was 19 dB at 3.7 GHz, decreasing to 13 dB at the upper bandwidth frequency. The shunt resistor at the input of the first MESFET device is necessary to improve the input return loss. As a result, for the close-loop configuration, an improvement of the third-order intermodulation components of 7–9 dB over a 750 MHz bandwidth at –3 dB power backoff with a power gain of about 10 dB and an input return loss of more than 10 dB.

To improve the gain capability of the negative feedback amplifier, it is possible to combine its simplicity with a feedforward technique to provide a separate path for the error signal only. Figure 12.16 shows the schematic of the power amplifier module with a feedback-feedforward linearization when the intermodulation distortion products at the main PA input, where they are symbolically plotted as out-of-phased, should be phased in a way that to cancel the intermodulation distortion products at its output. This modified negative feedback circuit can provide a reduction in the intermodulation distortion products equal to the traditional negative feedback topology without the usual reduction in overall amplifier gain [51]. According to the intermodulation distortion analysis based on Volterra series, the reduction of the power amplifier linear gain is dependent on the feedback at the fundamental only, while the third-order intermodulation components are reduced due to effect of the feedback factor both at the fundamental and at the intermodulation frequencies when a reduction is equal to the loop gain in the latter case [52]. This means that, if the fundamentals were removed from the feedback loop, there would still be a reduction of the intermodulation products due to the feedback at the intermodulation frequencies with unaffected amplifier gain. In this case, the feedback signal is coupled from the output of the power amplifier, properly attenuated and phase shifted to form the error signal with cancelled fundamental by combining with a lower portion of undistorted input signal. The resulting error signal is then amplified, scaled in the amplitude and phase, and

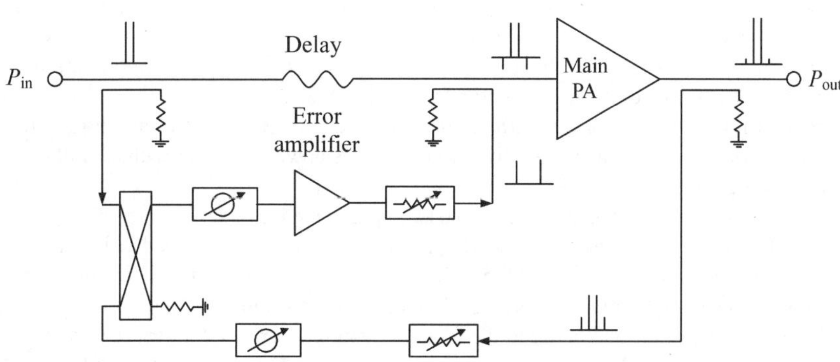

FIGURE 12.16 Power amplifier module with feedback–feedforward linearization.

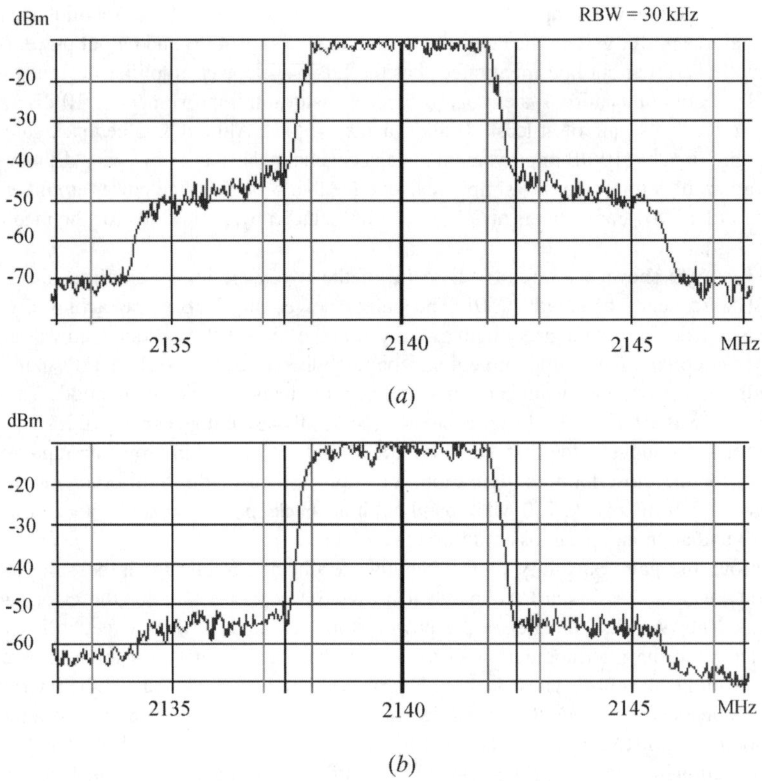

FIGURE 12.17 Measured spectrum of WCDMA 3GPP FWD output signal (*a*) before and (*b*) after feedback–feedforward linearization.

finally combined with a delayed upper portion of the undistorted signal to form the composite signal at the amplifier input with out-of-phased distortion products required to cancel the intermodulation products at the amplifier output. The stability analysis shows that, for narrow-band amplifiers with minimum loop delay, the reduction of third-order intermodulation products can exceed 20 dB [51].

In this case, when four Freescale MW6S004NT LDMOS devices operated in Class AB mode are used in the main PA that represents a balanced configuration of the four amplifying stages similar to shown in Figure 12.12, a 7-dB improvement of *ACLR* from −38 to −47 dB for a 35-dBm channel power with the signal peak-to-average ratio of 6.5 dB was achieved, as shown in Figure 12.17. The power amplifier gain was 17 dB and coupling coefficient of the microstrip output coupler was 10 dB. This indicates that the effect of the negative feedback on the intermodulation products is independent of the cancellation of the fundamentals in the loop, and circuit can successfully reduce the intermodulation distortion without affecting the gain of the power amplifier. For further intermodulation reduction, it is necessary to use a multistage power amplifier with overall high power gain to increase the loop gain.

The feedback linearization technique can also be implemented reversely to a feedforward scheme, as shown in Figure 12.18, where the linearizing circuit consists of the feedback, canceling and feeding blocks [53]. At the canceling block, the amplitude-corrected and phase-shifted feedback signal is combined with the sampled input signal to form the error signal, which is then amplified and added with proper amplitude and phase to the input undistorted signal within the feeding block. As a result, for a 38-dB main amplifier gain, the cancellation of the third-order intermodulation products from −22 to −42 dBc were achieved at the output power P_{1dB} = 27 dBm and operating frequency

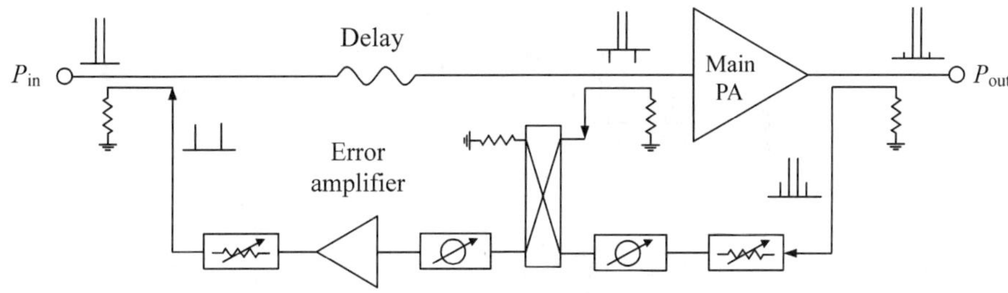

FIGURE 12.18 Power amplifier with feedback predistortion linearization.

of 1.85 GHz. However, the critical problems of this analog feedback predistortion scheme are the operation at bandwidth limitation caused by the loop delay and an oscillation tendency caused by the feedback nature. By employing a digital LUT technique, these limitations can be overcome, while maintaining the advantages of the feedback circuit [54]. As a result, the distortion is corrected in a digital domain and further enhanced by the feedback linearization. In this case, the structure of a digital feedback linearizer is the same as the analog feedback counterpart, except that the feedback signal in the cancellation loop constructs a LUT in the digital domain, and the gain factors of the signal canceling and feedback paths are adjusted by the DSP instead of using the variable attenuators and phase shifters. The predistortion signal is extracted directly from the LUT, which has been updated using the error signal extracted at the signal cancellation loop beforehand. Thus, the time delay through the loop is eliminated, and the bandwidth limitation does not exist anymore. At the same time, the oscillation tendency of the feedback circuit can be suppressed easily by digital control of the feedback loop parameters. Compared to the conventional digital predistortion technique, less number of iterations is required for convergence with simpler algorithm.

The envelope-feedback linearization is a simple and popular technique to improve the distortion associated with the amplitude nonlinearity of the power amplifier [44,55]. The basis of this technique, the block diagram of which is shown in Figure 12.19(*a*), is to compare the envelopes of the input and output signals, and to control the instantaneous gain of the power amplifier so as to minimize the difference between them. In this case, the RF input and output signals are sampled by the corresponding input and output couplers and then each fed to the proper input of the differential amplifier. The difference signal, representing the error between the input and output envelopes, is used to drive a variable gain amplifier to modify the envelope of the input signal that drives the power amplifier. The linearization effect depends on the nonlinearity of the detectors, especially at low signal levels, the bandwidth, time delay, and phase-gain characteristic of the feedback loop and the sensitivity of the variable gain amplifier. Therefore, where the *AM–AM* distortion is dominant, the two-tone intermodulation products are typically reduced by up to 10 dB. To provide a significant improvement in the nonlinear distortion cancellation, both RF feedback and envelope feedback methods can be combined [44].

Figure 12.19(*b*) shows the envelope-feedback power amplifier module with a digital adaptive predistortion developed for code division multiple access (CDMA) handset application in a frequency range of 887–925 MHz [56]. The block diagram of the power amplifier module includes additionally a surface acoustic wave (SAW) filter, a phase-controlling block and a CMOS integrated circuit incorporating two LUTs to linearize *AM–AM* and *AM–PM* characteristics. The variable gain amplifier is based on a dual-gate MOSFET to linearize *AM–AM* characteristic, which can easily vary the power gain by controlling its second-gate bias voltage in a range of more than 10 dB. In the phase-controlling block, a varactor diode was used with the phase range of more than 10°. By producing appropriate *AM–PM* predistortion data including phase characteristics of the dual-gate MOSFET and the following PA block, the total *AM–PM* can be linearized. Because the variations of phase with supply voltage and temperature according to the measurement results were insignificant, it was enough to use the LUT with modified initial data to minimize phase variations versus input power

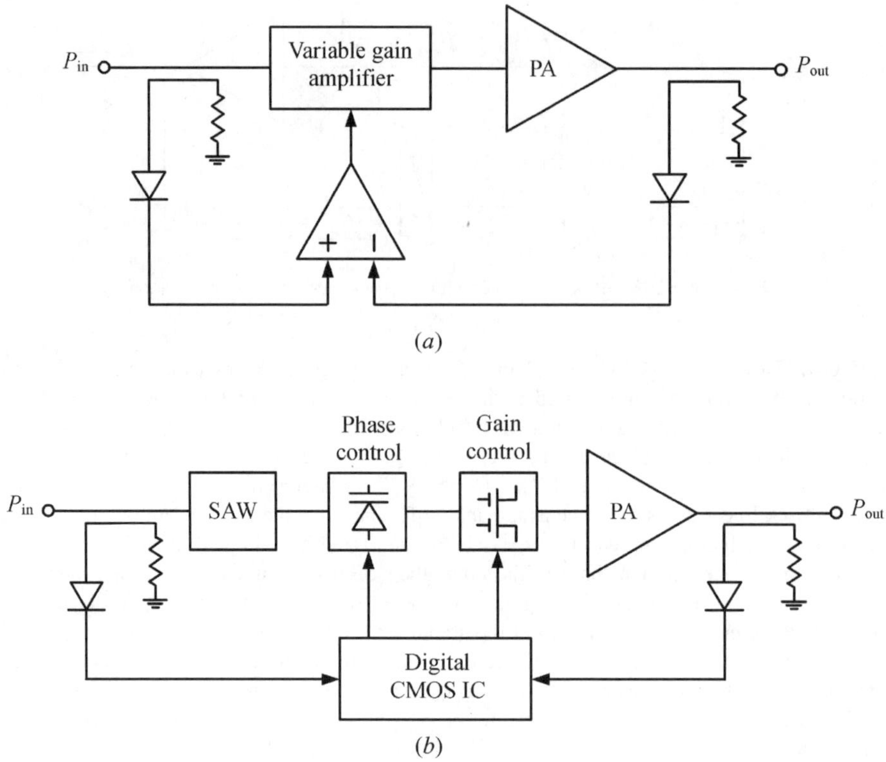

FIGURE 12.19 Block diagrams of power amplifier with adaptive predistortion.

that were predictable in advance. To linearize *AM–AM* characteristic, the adaptive predistortion was used by modifying the data in LUT during linearization process. It was found that the allowable time delay must be less than 40 ns for CDMA signal. As a result, by using a digital adaptive predistortion mechanism for only *AM–AM* characteristic, the *PAE* was increased up to 48% for the output power of 27.5 dBm and *ACPR* of −49 dBc. The complementary metal-oxide-semiconductor (CMOS) integrated circuit, whose size is 2.5×2.5 mm^2, consumed of about 15 mA.

The *AM–AM* linearization can also be achieved by dynamically varying the gate bias voltage of the final-stage transistor. In such an adaptive double-envelope feedback two-stage MESFET power amplifier the gain variations are detected directly while the phase variations are detected through a 90° branch-line coupler [57]. This is possible since when the gate goes more negative, the MESFET is closer to pinch-off and its gain is reduced. On the other hand, as the gate goes more positive, the MESFET will approach Class A, where its gain is maximal. The dynamic bias on the gate of the MESFET resulted in an increase in P_{1dB} by 1 dB, which improved the *PAE* by 5%.

To overcome the limitation of the envelope-feedback technique to correct for *AM–PM* distortion, the polar-loop technique can be used where a phased-lock loop is added to the envelope feedback system resulting in a polar-loop feedback [58]. However, the key disadvantage of a polar feedback occurs in the generally different bandwidths required for the amplitude and phase feedback paths, as well as locking capability of the phased-lock loop is limited at low signal level in the presence of interference coupled to the transmitter output from antenna that leads to a poorer overall performance. The Cartesian-feedback technique can solve at some extent the problems associated with the wide bandwidth of the signal phase by applying modulation feedback in the in-phase (I) and quadrature (Q) Cartesian baseband components. Figure 12.20(a) shows the basic block diagram of the Cartesian

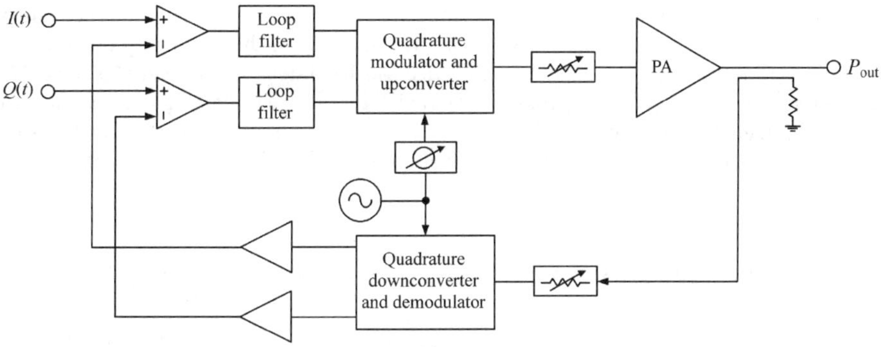

(a)

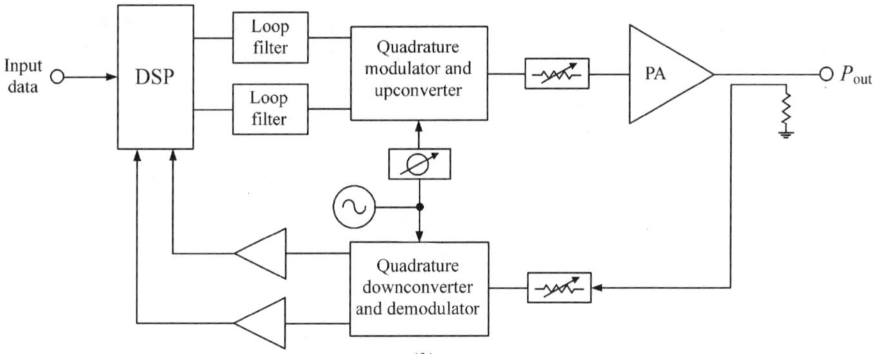

(b)

FIGURE 12.20 Digital predistortion system.

loop transmitter with two identical feedback loops for *I* and *Q* channels [12,59]. In this case, the sampled output signal is quadrature downconverted and demodulated to Cartesian co-ordinate signals, amplified and then subtracted from the corresponding input signals to form a distorted error signal that is complementary to the distortion signal at the PA output after being filtered and upconverted. The amount of the distortion reduction is equal to the loop gain and is high inside the passband of the loop filters. The phase shifter is necessary to control the phase delays that cause rotation of the signal constellation. The bandwidth of the Cartesian components is narrower than that of the RF signal, however errors in the feedback loop are not corrected and feedback components must be as linear as the desired loop linearity. The Cartesian feedback can be used to linearize multicarrier power amplifiers, improving output spectrum by 10–30 dB [60].

Further linearity improvement can be achieved by using a combined Cartesian loop and adaptive baseband predistortion linearization techniques using DSP, as shown in Figure 12.20(*b*) [61]. Adaptive predistortion provides a continuous adjusting of the loop time delays and updating of the predistorter LUT with information of the PA nonlinearity, which changes with supply voltage, temperature, load voltage standing wave ratio (*VSWR*), output power, and other environmental effects. For example, by using the predistorter implemented in DSP with 5-bit effective resolution digital-to-analog converters at its output and optimized using a direct search algorithm, a further 11-dB reduction of the third-order intermodulation components was achieved when a predistorted component has been added to the error signal [62]. However, as the bandwidth of transmitting signal gets wider, the feedback loop delay mismatch becomes increasingly detrimental to the convergence of LUT adaptation algorithm. In this case, the analog Cartesian feedback, which is separated from LUT, can be used only for LUT training, thus resulting in the energy-efficient and low-complexity adaptive linearization [63]. Then, after the completion of LUT training, the analog Cartesian is turned off so that open-loop predistortion is performed using a compact Cartesian LUT to linearize wideband signals. The renewal of the LUT can happen without interrupting ongoing communication because the LUT training needs a millisecond and most communication protocols implement enough buffering for an error control method such as automatic repeat request. How often the LUT would need a renewal depends on PA characteristics and the environmental conditions.

12.6 DOHERTY POWER AMPLIFIER ARCHITECTURES

The Doherty power amplifier technique was introduced in the middle of 1930s as a more efficient alternative to the conventional amplitude-modulation techniques having low average efficiency [64]. The classical two-stage Doherty power amplifier architecture shown in Figure 12.21 incorporates two power amplifiers, normally called the *main (carrier)* and *auxiliary (peaking)* amplifiers, separated by a quarterwave transmission line in the main amplifier path. The quarterwave transmission line on the input of the auxiliary amplifier is required to compensate for the 90° phase shift caused by the quarterwave transmission line at the output of the main amplifier. The main amplifier is biased in Class B mode, while the auxiliary amplifier is biased in Class C mode.

The condition of power conservation for a lossless output transmission line results in

$$I_3 = I_1\sqrt{\frac{R_1}{R_3}}. \tag{12.17}$$

The current division ratio is defined by

$$\beta = \frac{I_3}{I_2 + I_3}. \tag{12.18}$$

As a result, the overall output power P_{out} is the sum of the main amplifier output power $P_1 = \beta P_{\text{out}}$ and auxiliary amplifier output power $P_2 = (1 - \beta)P_{\text{out}}$. The impedance seen at the output of the 50-Ω transmission line is

$$R_3 = \frac{I_2 + I_3}{I_3}\frac{Z_1^2}{R_L} = \frac{Z_1^2}{\beta R_L} \tag{12.19}$$

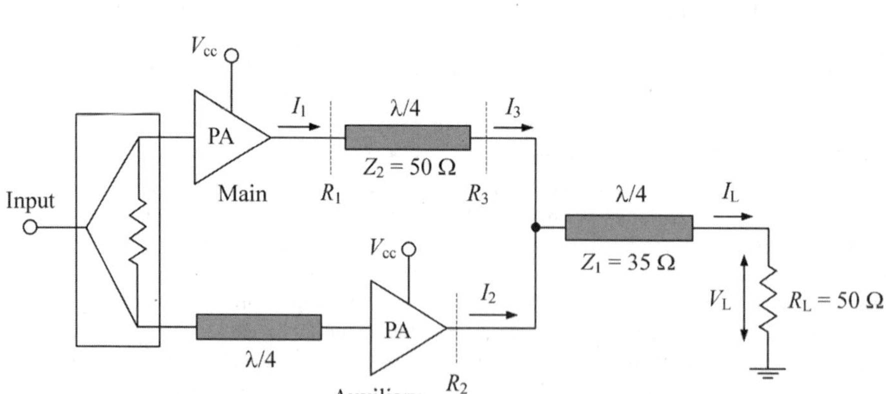

FIGURE 12.21 Classical two-stage Doherty power amplifier architecture.

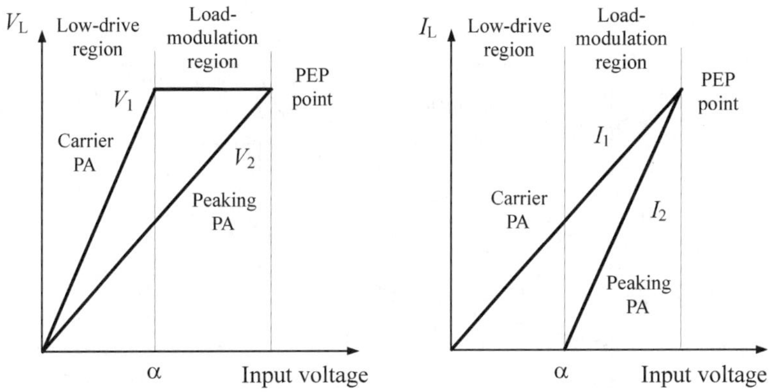

FIGURE 12.22 Ideal voltage and current behavior of Doherty amplifier.

whereas the impedance seen by the auxiliary power amplifier is

$$R_2 = \frac{I_2 + I_3}{I_2}\frac{Z_1^2}{R_L} = \frac{Z_1^2}{(1-\beta)\,R_L}. \tag{12.20}$$

The basic operation principle of the Doherty power amplifier architecture can be analyzed for low, medium and peak output power regions separately [65]. Figure 12.22 shows the current and voltage behavior for ideal transistors and lossless matching circuits where V_L is the load voltage and I_L is the load current. At peak output power P_{PEP} when both main and auxiliary amplifiers are saturated, the resultant collector efficiency is equal to the maximum achievable efficiency of $\eta = \pi/4$ for an ideal Class B operation. For the classical Doherty power amplifier architecture shown in Figure 12.21 with the current and power division ratios of $\beta = \alpha = 0.5$ when both main and auxiliary amplifiers produce equal output powers, their load impedances are equal to $R_1 = R_3 = R_2 = Z_2 = 2Z_1^2/R_L$. If the characteristic impedance of the output transmission line is chosen equal to $Z_1 = 35\ \Omega$, then $R_1 = R_3 = R_2 = Z_2 = R_L = 50\ \Omega$.

At lower power levels in a low-drive region, the auxiliary (peaking) amplifier is turned off because the instantaneous amplitude of the input signal is insufficient to overcome the negative Class C bias and appears as an open circuit, whereas the main (carrier) amplifier operates in active region. In this case, the load impedance seen by the main amplifier is

$$R_1 = \left(\frac{Z_2}{Z_1}\right)^2 R_L \tag{12.21}$$

resulting in $R_1 = 2R_L = 100\ \Omega$ when $Z_1 = 35\ \Omega$ and $Z_2 = R_L = 50\ \Omega$. Because the output power of the main amplifier in saturation is four times less than the peak output power P_{PEP}, the collector efficiency of the main amplifier in an ideal Class B mode will be twice than that of a conventional Class B power amplifier, achieving maximum of 78.5% at backoff power level of −6 dB, as shown in Figure 12.23.

At medium power levels in a load-modulation region, the main amplifier is saturated whereas the auxiliary amplifier is turned on and operates in the active region. Since the output voltage of the main amplifier $V_1 = I_1R_1$ is constant under saturation conditions, from Eq. (12.17) it follows that current I_3 is constant in the medium power region as well. The collector efficiency of the main amplifier remains at its maximum value, whereas the collector efficiency of the auxiliary amplifier increases up to its maximum value for Class B operation at peak output power P_{PEP}. As a result, the Doherty

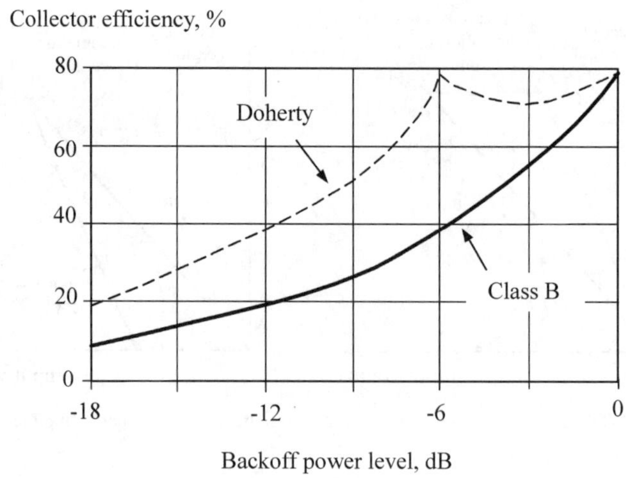

FIGURE 12.23 Collector efficiencies of ideal two-stage Doherty architecture and Class B power amplifier.

power amplifier architecture achieves maximum efficiency at both the transition −6 dB backoff point and the peak output power, and remains relatively high in between, as seen from Figure 12.23.

There is a possibility to extend the collector efficiency over a wider range of output powers if we choose the main and auxiliary amplifiers with different output power capabilities, smaller for the carrier (main) amplifier and larger for the peaking (auxiliary) amplifier. For instance, for a power ratio $\alpha = 0.25$, the transition point with maximum collector efficiency corresponds to the backoff power level of −12 dB from peak output power [65]. At peak output power when the carrier and peaking amplifiers are saturated, from consideration of their output powers it follows that $R_1 = Z_2 = R_3 = 3R_2$. As a result, $I_2 = 3I_3$ and $\beta = 0.25$. The output impedances R_2 and R_3 as functions of the load resistance R_L and characteristic impedance Z_1 can be obtained from Eqs. (12.19) and (12.20). For example, if we choose the characteristic impedance of the output transmission line and load resistance equal to $Z_1 = 15\ \Omega$ and $R_L = 50\ \Omega$, respectively, then the characteristic impedance of the quarterwave transformer and output impedance of the main amplifier are $Z_2 = R_1 = 18\ \Omega$, whereas the output impedance of the peaking amplifier is equal to $R_2 = R_1/3 = 6\ \Omega$.

Since from Eqs. (12.19) and (12.20) it follows that

$$R_1 = \frac{Z_2^2}{\beta R_3} \tag{12.22}$$

hence, at lower power levels when the peaking amplifier is turned off, the output impedance R_1 is four times higher than that at peak output power where $R_1 = Z_2 = R_3$. The scaling ratio of 4:1 was used for InGaP/GaAs HBT devices with total emitter areas of 3360 and 840 μm^2 for the peaking and carrier amplifiers, respectively, to implement the extended Doherty technique into the monolithic power amplifier developed for CDMA handset applications As a result, the power-added efficiencies of 45% and 23% was measured at the highest output power of 25 dBm and at 10-dB backoff level, respectively [66]. Note that the conventional Class AB power amplifiers designed for the same application normally have the power-added efficiency of about four times lower at this backoff power. For the packaged devices when it is difficult to choose the proper power ratio between the devices, it is convenient to use the identical power amplifiers that can compose ideally the N-way Doherty architecture where one carrier power amplifier is in parallel with $N - 1$ numbers of the peaking amplifiers. This is the simplest hybrid approach to acquire an $N - 1$ times larger-sized peaking amplifier compared with the carrier amplifier [67].

The loss due to effect of the noninfinite impedance of the peaking power amplifier when it is turned off depends on the impedance seen from its output end, especially at microwaves. Therefore, to increase this output impedance by rotating the phase, the output matching circuit, which is required to transform the device output impedance at maximum power to 50 Ω when it is turned on, should be followed by a particular section of a transmission line with the characteristic impedance of 50 Ω [68]. From the other hand, in order to avoid an increase in size that is very important for monolithic design, the proper structure of the output matching circuit can be selected depending on the output power [69].

An asymmetrical Doherty architecture exhibits a significant drop in efficiency in the region between the power at the transition backoff point and peak output power. However, it is possible to use more than two power amplifiers in order to maintain the efficiency without significant dropping at the backoff output power levels. The basic multistage Doherty power amplifier architecture shown in Figure 12.24(*a*) uses more than one peaking amplifiers, with quarterwave transmission lines to combine their output powers [70]. The characteristic impedances of each output quarterwave transmission line depend on the levels of backoff and can be calculated from

$$Z_{0i} = R_L \prod_{j=1}^{i} \gamma_j \tag{12.23}$$

$$\prod_{j=k}^{(i+k)/2} \gamma_{(2j-k)} = 10^{(B_i/20)} \tag{12.24}$$

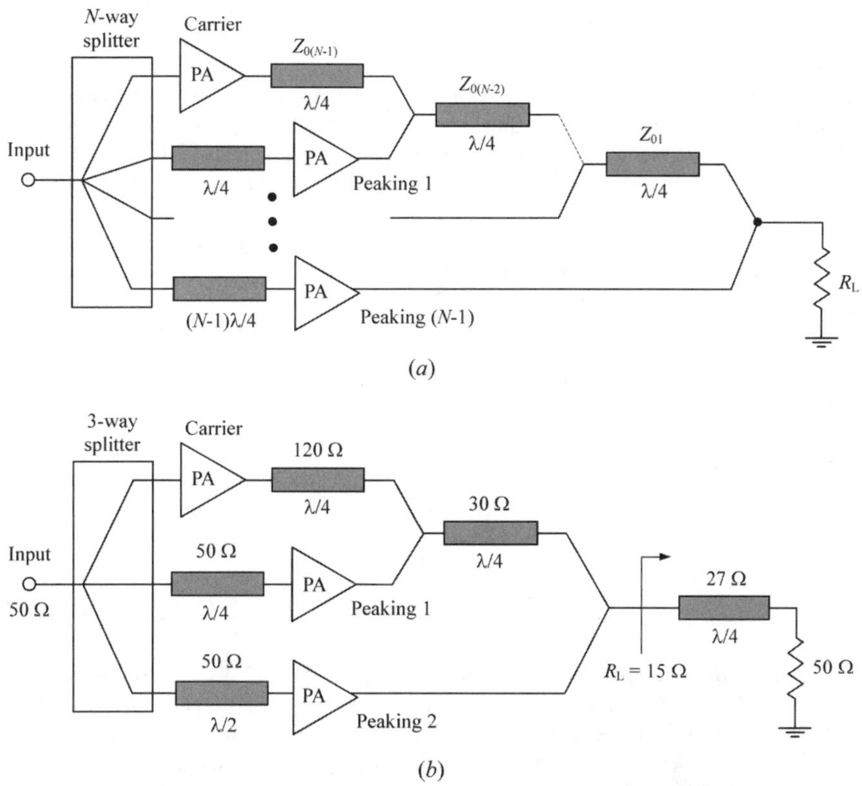

FIGURE 12.24 Multistage Doherty power amplifier architectures.

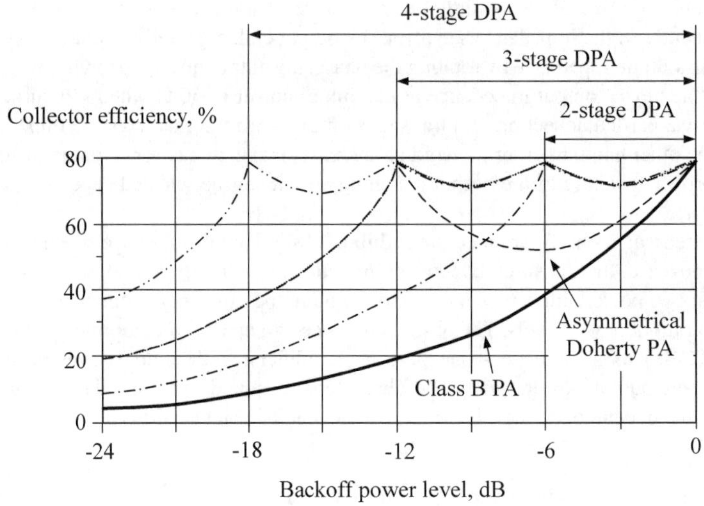

FIGURE 12.25 Efficiencies of the different Doherty power amplifier architectures.

where $i = 1, 2, \ldots, N-1$, $k = 1$ (for odd i) or 2 (for even i), N is the total number of amplifier stages, and B_i is the backoff level (positive value in decibels) from maximum output power of the system at which the efficiency will peak. The maximum level of backoff B_{N-1} is set by the carrier amplifier and the number of efficiency peaking points is directly proportional to the number of amplifier stages used in the design.

Figure 12.25 shows the instantaneous collector efficiencies of the multistage Doherty power amplifier (DPA) architectures for two, three, and four stages having maximum efficiency at the transition points of −6 dB, −12 dB, and −18 dB backoff output power levels, respectively. From Figure 12.25, it follows that the multistage architecture provides higher efficiencies at backoff levels in between the efficiency peaking points compared with asymmetrical Doherty architecture and significantly higher efficiency at all backoff output power levels compared with the conventional Class B power amplifiers. For the most practical case of a three-stage Doherty power amplifier, the block schematic of which is shown in Figure 12.24(*b*), the characteristic impedances of each output quarterwave transmission line can be found from Eqs. (12.23) and (12.24) to be

$$Z_{01} = \gamma_1 R_L \tag{12.25}$$

$$Z_{02} = \gamma_1 \gamma_2 R_L \tag{12.26}$$

where

$$\gamma_1 = 10^{(B_1/20)} \tag{12.27}$$

$$\gamma_2 = 10^{(B_2/20)} \tag{12.28}$$

where $B_1 = 6$ and $B_2 = 12$ dB for peak efficiencies at −6 dB and −12 dB backoff, respectively.

For a 1.95-GHz WCDMA application, such a three-stage Doherty power amplifier structure using GaAs FET devices with the device periphery ratio of 1:2:4 and microstrip power combining elements provides the power-added efficiency of 48.5% and power gain of 12 dB at $P_{1dB} = 33$ dBm. The peak power-added efficiencies of 42% and 27% were measured at −6 dBc and −12 dBc backoff levels [70]. Efficiencies of the three-stage Doherty amplifier at the maximum power and backoff transition points, in particular at -12 dB backoff power, can be significantly increased up to more than 60% by using the highly effective GaN HEMT devices and apply digital predistortion technique for linearization [71].

It is difficult to apply directly the Doherty technique to the design of the power amplifier integrated circuits with high level of integration, since the physical size of the quarterwave transmission lines is too large. For example, for a FR-4 substrate with effective dielectric permittivity of $\varepsilon_r = 3.48$, the geometrical lengths of the quarterwave transmission lines are 48 mm, 19 mm, and 8.7 mm at the operating frequencies of 900 MHz, 2.4 GHz, and 5.2 GHz, respectively [72]. Therefore, the best solution in this situation is to replace the quarterwave transmission lines by the lumped elements using the single-frequency equivalence between the circuits with lumped elements and a quarterwave transmission line.

Consider thc transmission matrix $[ABCD]_L$ for a quarterwave transmission line shown in Figure 12.26(a) and transmission matrix $[ABCD]_T$ for a π-type low-pass lumped circuit, consisting of the series inductance and two shunt capacitances shown in Figure 12.26(b) given respectively by

$$[ABCD]_T = \left.\begin{bmatrix} \cos\theta & jZ_0\sin\theta \\ j\dfrac{\sin\theta}{Z_0} & \cos\theta \end{bmatrix}\right|_{\theta=\pi/2} = \begin{bmatrix} 0 & jZ_0 \\ \dfrac{j}{Z_0} & 0 \end{bmatrix} \tag{12.29}$$

$$[ABCD]_L = \begin{bmatrix} 1 & 0 \\ j\omega C & 1 \end{bmatrix}\begin{bmatrix} 1 & j\omega L \\ 0 & 1 \end{bmatrix}\begin{bmatrix} 1 & 0 \\ j\omega C & 1 \end{bmatrix} = \begin{bmatrix} 1-\omega^2 LC & j\omega L \\ j\omega L\left(2-\dfrac{1}{\omega^2 LC}\right) & 1-\omega^2 LC \end{bmatrix}. \tag{12.30}$$

Equating the corresponding elements of both matrices in Eqs. (12.29) and (12.30) yields the following relationships between the circuit elements:

$$Z_0\omega C = \frac{Z_0}{\omega L} = 1. \tag{12.31}$$

The same relationships can be established between the characteristic impedance of the three-quarterwave transmission line and π-type high-pass lumped circuit consisting of the series capacitance and two shunt inductances shown in Figure 12.26(c). Similarly, the input branch-line hybrid coupler or

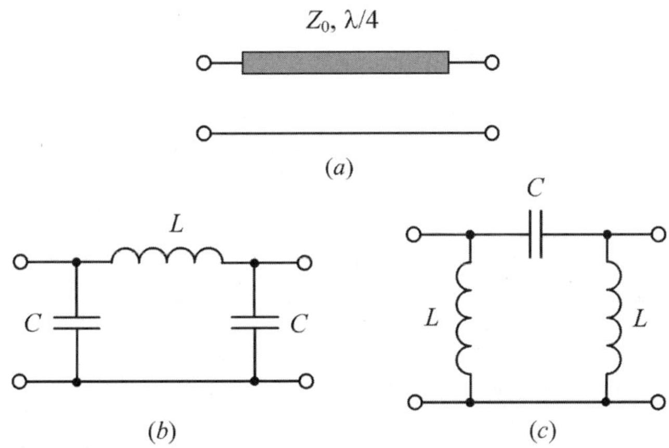

FIGURE 12.26 Quarterwave transmission line and its single-frequency lumped equivalent circuits.

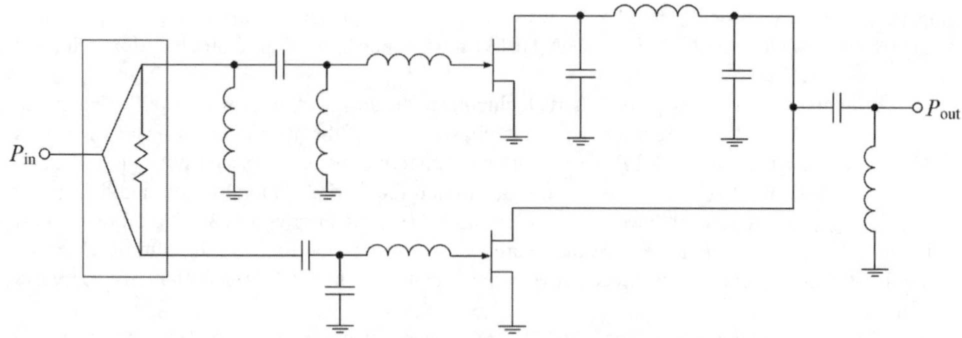

FIGURE 12.27 A 900-MHz Doherty amplifier implemented with lumped elements.

in-phase Wilkinson divider can be replaced by using a π-type low-pass lumped circuit in each their branch. Figure 12.27 shows a 900-MHz GaAs MESFET lumped two-stage Doherty power amplifier architecture where the output quarterwave transmission line connected to the carrier amplifier is replaced by a π-type low-pass lumped circuit and the input quarterwave transmission line (providing phase shifting) connected to the peaking amplifier is replaced by a π-type high-pass lumped circuit, connected to the carrier amplifier input. In addition, the output quarterwave transformer connected to the load is replaced by an L-type high-pass matching circuit, whereas the two L-type low-pass matching circuits are used to provide the input matching of the carrier and peaking amplifiers. At the carrier amplifier input path, the right-hand shunt inductance as a part of the equivalent three-quarterwave phase shifter and shunt capacitance as a part of the input L-type high-pass matching circuit can be combined into a single shunt inductance. For such a lumped Doherty power amplifier with power-added efficiency of about 52% at maximum output power, the power usage efficiency according to a probability density function (PDF) for the CDMA signal is 14.1%, which is more than three times higher compared to the conventional Class AB power amplifier with an efficiency of 4.4% [72].

However, the nonideal power gain and phase in a high power nonlinear region can cause a linearity problem for transmitting signals with nonconstant envelope in communication transmitters. To solve this problem, an improved Doherty power amplifier architecture can be used employing an envelope tracking technique to control gate bias voltage of the peaking amplifier in accordance with the input signal envelope. Such an approach can provide also higher efficiency realizing a lower bias voltage for higher output power. This compromises both high efficiency and good linearity requirements over a wide range of output powers. Figure 12.28 shows a block diagram of a 2.14-GHz LDMOSFET microstrip Doherty power amplifier with adaptive gate bias control [73]. In this case, the device output impedance transformed by the lower-path offset line becomes high enough to block the output power leakage of the peaking amplifier at low power mode. For the same average output powers of 32.7 dBm, such a two-stage Doherty power amplifier demonstrates an improvement in *PAE* of 15.2% at *ACLR* of -30 dBc compared to its Class AB counterpart. This is because in the Doherty power amplifier the quiescent current is maintained constant only for the carrier amplifier, whereas the bias point of the peaking amplifier is varied according to the input signal envelope. Further efficiency improvement can be achieved by applying a Class F or a Class E mode to the both carrier and peaking power amplifiers [74].

The linearity of the power amplifier can be improved by using the digital signal processing to provide more accurate gate bias control combined with digital predistortion correcting the gain and phase characteristics in a high power region simultaneously. Figure 12.29 shows an 840-MHz MESFET Doherty power amplifier architecture using a hybrid technology with the DSP implemented

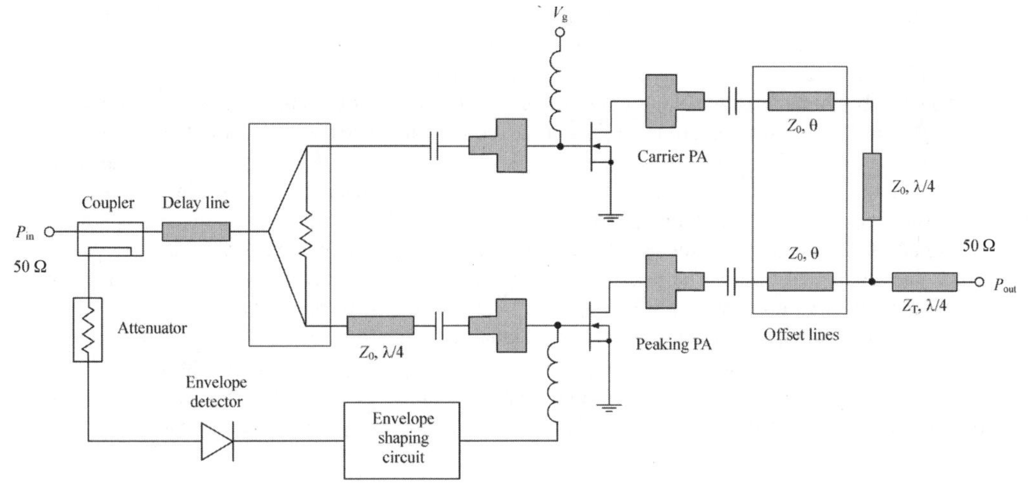

FIGURE 12.28 Block diagram of LDMOSFET microstrip Doherty power amplifier with adaptive gate bias control.

externally on a board controlled by a personal computer [75]. In this case, the DSP generates both the baseband I and Q signals, which are then upconverted to an RF signal using the quadrature modulator, and the voltage signal V_{g2}, which is applied to the peaking amplifier as the gate bias. This results in an efficiency improvement by the dynamic gate biasing of the peaking amplifier according to the instantaneous envelope of the input signal. At the same time, the phase performance is corrected by the phase predistortion at baseband level based on the dynamic gate bias-voltage values from the gain correction, thus resulting in a linearity improvement. An overall improvement of *PAE* of 3–5% and *ACPR* of about 10 dB at an average output power of 23 dBm can be achieved by utilizing such a DSP technique [72,75]. Significant linearity improvement can also be achieved by using a digital feedback predistortion linearization technique [76].

12.7 OUTPHASING POWER AMPLIFIERS

The outphasing modulation technique was invented in the middle of 1930s in order to improve both the efficiency and linearity of AM-broadcast transmitters [77]. Substantially later in 1970s, its

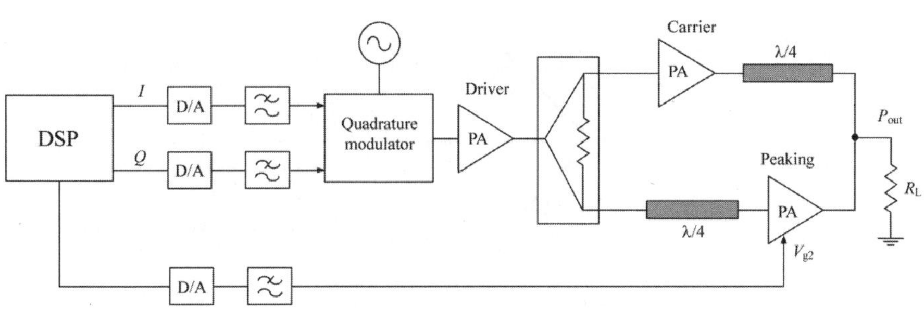

FIGURE 12.29 Doherty power amplifier architecture with DSP control.

application was extended up to microwave frequencies under the name LINC (*linear amplification using nonlinear components*) [78]. An outphasing transmitter operates as a linear power amplifier system for amplitude-modulated signal having a linear transfer function over a wide range of the input signal levels by combining the outputs of two nonlinear power amplifiers that are driven with signals of constant amplitude but different time-varying phases corresponding to the envelope of the input signal.

A simple outphasing power amplifier system is shown in Figure 12.30(*a*) [79]. The signal component separator (SCS) generates from the input amplitude-modulated signal two sinewave signals of constant envelopes with different phases $+\phi(t)$ and $-\phi(t)$. These two signals are then separately amplified by the identical nonlinear power amplifiers each and combined to produce the output amplitude-modulated signal. The peak output power is obtained with $\phi = 90^\circ$ when currents from power amplifiers with equal amplitudes $I_L = I_1 = I_2$ are added in phase, similar to a push–pull operation. Zero output power corresponds to the signal with $\phi = 0^\circ$ when equal currents from power amplifiers cancel each other resulting in $I_L = 0$. Intermediate values of phase in between $0^\circ < \phi < 90^\circ$ produce intermediate values of the output voltage amplitude. As shown in Figure 12.30(*b*), the time-varying phase ϕ can be written using the vector sum of the output voltages V_1 and V_2 by

$$\phi = \sin^{-1}\left(\frac{V_L}{V_{LPEP}}\right) \tag{12.32}$$

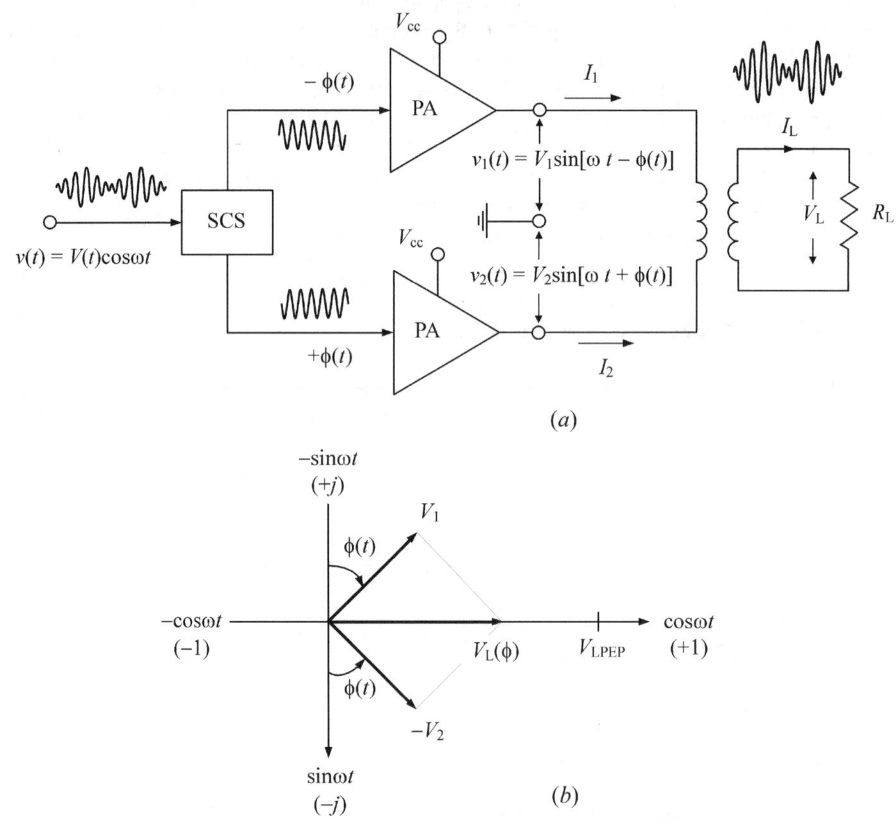

FIGURE 12.30 Simple outphasing power amplifier system.

where $V_L = I_L R_L$ is the output voltage amplitude across the load resistance R_L and V_{LPEP} is the maximum output voltage amplitude at peak envelope power.

The instantaneous collector efficiency of a simple outphasing system with Class B power amplifiers can be calculated from

$$\eta = \frac{\pi}{4} \frac{V_L}{V_{LPEP}} \tag{12.33}$$

having maximum value of 78.5% in saturation when $V_L = V_{LPEP}$ with $\phi = 90^\circ$ and zero value when $V_L = 0$ with $\phi = 0^\circ$. Thus, the efficiency of a simple power amplifier outphasing system is the same as that of an ideal Class B power amplifier, reducing linearly with the output voltage amplitude. In this case, to perform accurately the required signal component separation, it is necessary to use the digital signal processing (DSP) technique [80].

The outphasing power amplifier systems used at higher frequencies and microwaves employ hybrid combiners to isolate the two power amplifiers from each other, allowing them to see resistive loads at all signal levels as shown in Figure 12.31(a). Typical structures of the hybrid combiners represent a quadrature branch-line coupler or an in-phase Wilkinson combiner that are fully matched and lossy combining structures with high isolation between the combined amplifying paths for properly phased and equal signal powers. However, since both power amplifiers deliver full power all of the time, the efficiency of such a hybrid-coupled microwave LINC transmitter varies with the output power resulting in its significant degradation at lower power levels. This is because most of the output power is dissipated in the ballast resistor R_0 of the combining network when the two power amplifiers are operated substantially out-of-phase.

Figure 12.31(b) shows the power recycling schematic where RF-to-dc converter is implemented with high-speed Schottky diodes and optimized matching networks [81]. To achieve better efficiency, the diodes should be switched fully having minimal series on-resistances. As a result, at the operating

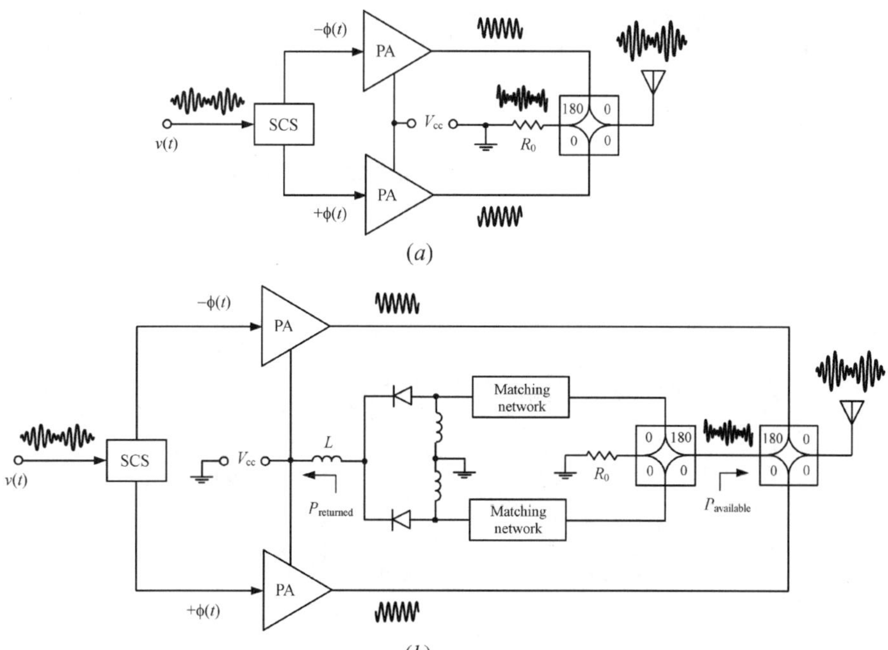

FIGURE 12.31 Outphasing power amplifier system with hybrid combiner.

frequency of 1.95 GHz, the measured power reuse efficiency, which is defined as a ratio of the returned power P_{returned} to the power available from the hybrid $P_{\mathrm{available}}$, was found to be approximately 63% at power levels varied with supply voltage. The amount of the overall system efficiency improvement depends on the modulation scheme and could be compromised for modulation schemes that exhibit very deep variations in envelope power on a regular basis. By eliminating the ballast resistor in a Wilkinson combiner resulting in a simple lossless T-type combiner, efficiency can also be increased due to varying load impedance seen by each device [82]. However, there is a tradeoff between efficiency and linearity when a matched combiner provides greater linearity performance compared to the lossless combiner, while the lossless combiner shows equal or better efficiency performance compared to the matched combiner.

The efficiency at lower output voltages can be significantly improved by using a lossless Chireix outphasing power amplifier system shown in Figure 12.32(a), which includes additional quarterwave transmission lines and shunt reactances. Phasor analysis of the load network for the time-varying phase ϕ with impedance-transforming quarterwave transmission line results in

$$Y_3 = \frac{2R_{\mathrm{L}}}{Z_0^2}\frac{V_{\mathrm{L}}}{V_{\mathrm{LPEP}}}(\sin\phi + j\cos\phi) \tag{12.34}$$

$$Y_4 = \frac{2R_{\mathrm{L}}}{Z_0^2}\frac{V_{\mathrm{L}}}{V_{\mathrm{LPEP}}}(\sin\phi - j\cos\phi) \tag{12.35}$$

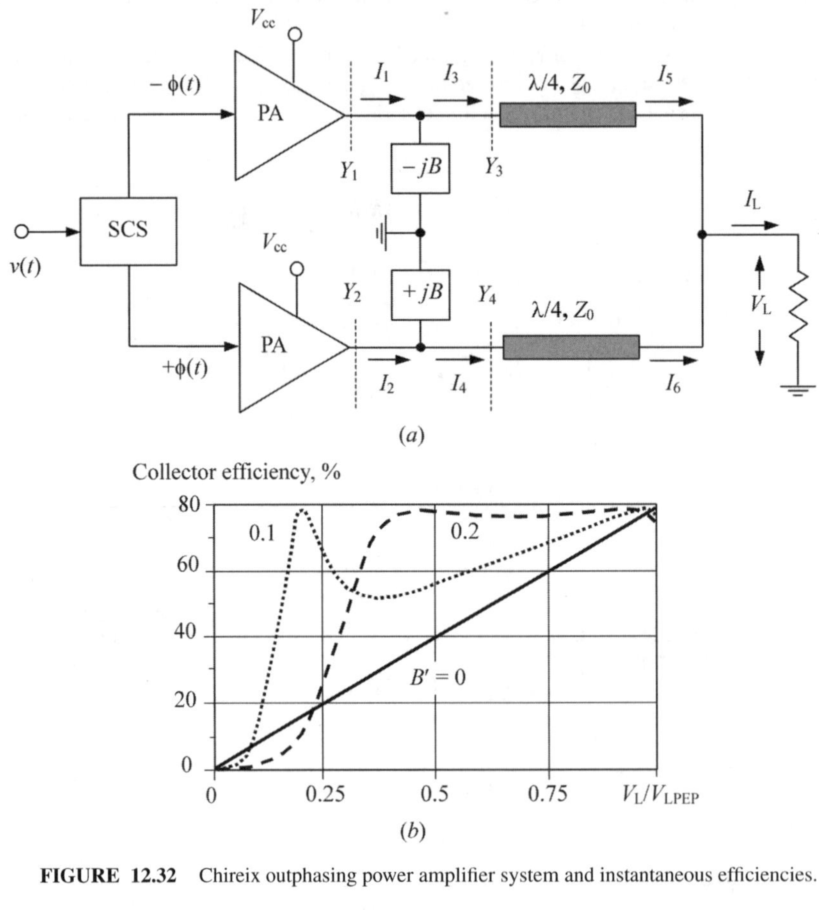

FIGURE 12.32 Chireix outphasing power amplifier system and instantaneous efficiencies.

where Z_0 is the characteristic impedance of the transmission lines [79]. From Eqs. (12.34) and (12.35) it follows that the admittances Y_3 and Y_4 are purely resistive only for $\phi = 90^\circ$ corresponding to the case of in-phase output currents. However, for most values of phase ϕ, the power amplifiers have highly reactive loads that become completely reactive when $\phi = 0^\circ$ with out-of-phase output currents. The effect of the reactive loads can be partially compensated by adding the corresponding shunt susceptances $-B$ and $+B$, respectively. In this case, the reactive parts of the admittances $Y_1 = Y_3 - jB$ and $Y_2 = Y_4 + jB$ can be zeroed at one specific output voltage amplitude by setting

$$B = \frac{2R_\mathrm{L}}{Z_0^2}\frac{V_\mathrm{L}}{V_\mathrm{LPEP}}\sqrt{1 - \left(\frac{V_\mathrm{L}}{V_\mathrm{LPEP}}\right)^2} \tag{12.36}$$

which can be obtained by substituting Eq. (12.32) into Eqs. (12.34) and (12.35). As a result, for the case of a purely resistive load, the instantaneous collector efficiency of a Chireix outphasing system with ideal Class B power amplifiers can reach the maximum value of

$$\eta = \frac{\pi}{4}. \tag{12.37}$$

The instantaneous efficiencies of the Chireix outphasing system for different values of the normalized shunt susceptance $B' = BZ_0^2/2R_\mathrm{L}$ are shown in Figure 12.32(*b*), from which it follows that the selection of a proper value of B increases efficiency at a specified medium level of the output voltage amplitude, however it is degraded at low and high amplitudes. Using a value of $B' = 0.2$ can provide high efficiency over the upper 6 dB of the output voltage range. The case of $B' = 0$ corresponds to the collector efficiency variations of an ideal Class B power amplifier. An improvement in the average efficiency calculated over a wide range of output voltages for various amplitude-modulated signals of up to a factor of two over that of an ideal Class B power amplifier can be achieved by properly selecting the shunt susceptances in outphasing power amplifier system. On the whole, to design such an outphasing system, it is necessary to consider simultaneously such factors as a complexity of the SCS circuit and sensitivity of the power amplifiers to the wide range of load impedances.

Further efficiency improvement can potentially be achieved by using the push–pull voltage-switching Class D, transmission-line Class F, or switching-mode Class E power amplifiers [83,84]. Using a push–pull configuration for each saturated Class B power amplifier based on pHEMT devices in a Chireix outphasing system had contributed to the increased efficiency of 42.2% at -7 dB backoff, which is more that two times improvement over the Wilkinson combiner system [85]. The wideband efficient operation of a LINC transmitter over frequency range from 30 to 450 MHz can be achieved by using the push–pull resistive-load Class D power amplifiers with harmonic filter bank [86]. At microwave frequencies, it can be realized by replacing the single quarterwave transmission lines in Class F power amplifiers with either lumped low-pass *LC* ladder networks or cascades of the quarterwave transmission lines, having a smaller transformation ratio for each line [87].

In a practical LINC transmitter there are three main mechanisms that degrade the overall performance: The power gain and phase imbalance between two RF amplifying paths and the different nonlinear characteristics of both power amplifiers [88]. For example, if the gain imbalance between paths is about 1%, the output rejection may reduce to 45 dB and even further to 28 dB only depending on the relative level of the input modulating signal amplitude. On the other hand, a phase error between amplifying paths as low as 2° diminishes the undesired response rejection to only 33 dB for the best case. At the same time, effect of the imbalance of the power amplifier nonlinearities is less meaningful. In terms of the Chireix power amplifier parameters, it is very important to minimize the combiner impedance mismatching, especially at microwave frequencies where the Chireix outphasing combiner should consist of two shunt stubs of equal and opposite reactances and two series quarterwave transmission lines [89].

Therefore, the practical implementation of the entire LINC or Chireix outphasing power amplifier system is very difficult because of its inherent strong sensitivity to the amplitude and phase errors

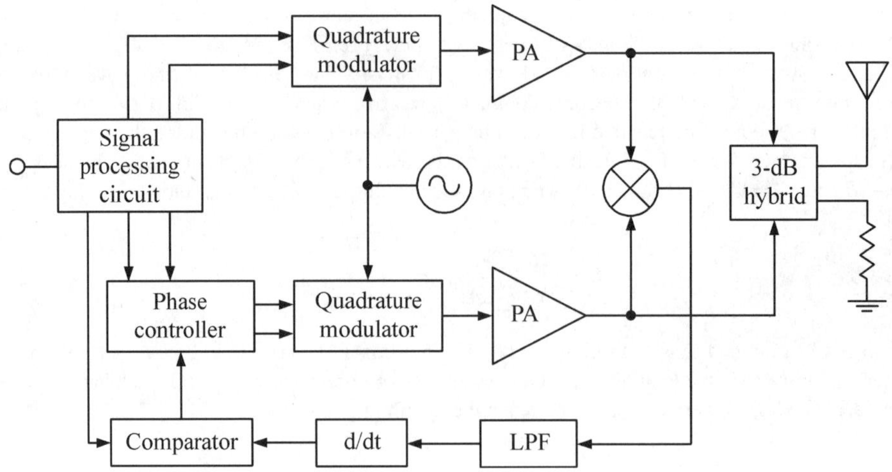

FIGURE 12.33 LINC transmitter with phase error compensating loop.

caused, first of all, by the difference in electrical lengths between the two power amplifier branches. Figure 12.33 shows the block diagram of the LINC architecture incorporating a feedback loop to compensate for phase errors [90]. In this case, the phase difference between the two branches is detected by a multiplier that allows a phase control of one branch by adding or subtracting a certain phase increment. As a result, the output power of 7.5 W with power amplifier efficiency of 21% (including hybrid and isolator losses) was achieved at 900 MHz. An additional linearity improvement can be achieved by providing an adaptive amplitude adjustment as well, by using a feedback loop from the output to measure output signal in adjacent channels and find optimal gain and phase correction by optimization algorithm [91]. To provide high amplitude and phase accuracy of the LINC system, a DSP-based architecture can be developed where the compensation of the amplitude and phase imbalances can be accomplished using calibration schemes [92]. As an example, for a LINC transmitter amplifying a $\pi/4$-shifted differential QPSK signal, an *ACPR* of −65 dBc can be obtained without predistortion when the amplifier branch phase imbalance range is less than $\pm 0.6^\circ$ and the gain imbalance range is less than ± 0.07 dB. With adaptive digital predistortion, the same *ACPR* level can be achieved over a phase imbalance range of $\pm 7^\circ$ and a gain imbalance range of ± 1 dB [93]. The average efficiency of 30% with *ACLR* of −45 dBc can be achieved for a 3GPP WCDMA signal in a frequency range of 2.11–2.17 GHz, measured at a channel output power of 20 W [94].

The CALLUM (combined analogue locked loop universal modulator) transmitter contains such a feedback control loop that incorporates the component separation, upconversion, amplification and summation all within the loop, as shown in Figure 12.34 [95]. The control signals are generated from the error signal between the Cartesian input components and the demodulated output of the summation port. If the demodulation process is exact with perfect gain and phase matching, and if the loop gain is infinite, then the action of the feedback control loop will error to be zero. The system can be viewed as a pair of analog cross-coupled phase-locked, excited by the modulation waveform represented with Cartesian components. Consequently, the stable region of lock covers only one quadrant, thus limiting the basic CALLUM transmitter to such modulation formats as full amplitude modulation and amplitude shift keying. In order to achieve the stable operation in all four quadrants of the input signal phase, it is necessary to include the switching matrix in the VCO control path to control the sign of the feedback components as a function of the input vector [96]. However, an ultimate limitation of the feedback compensation employed in the CALLUM transmitter is the restriction on loop gain and modulation bandwidth that limits the potential CALLUM transmitter implementation to such narrow-band systems as TETRA [97].

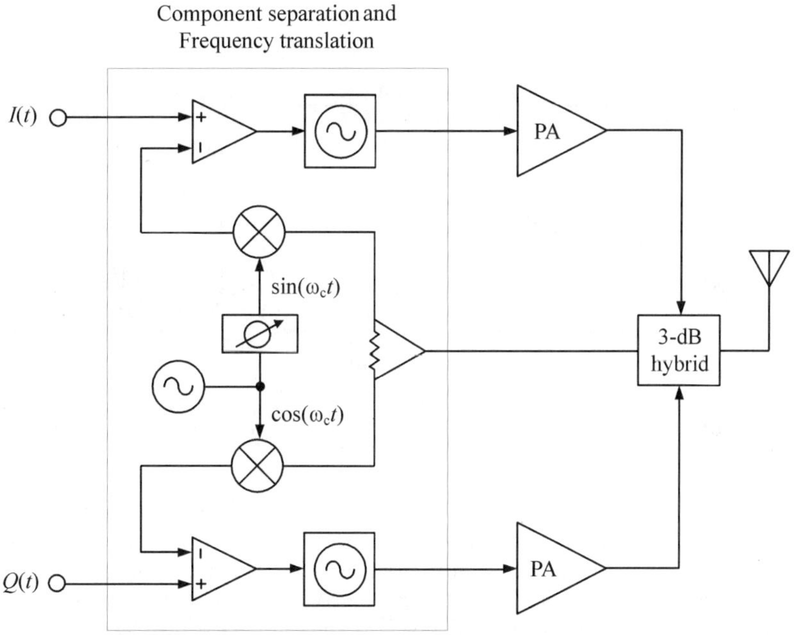

FIGURE 12.34 Basic structure of CALLUM transmitter.

12.8 ENVELOPE TRACKING

For a convenience of further consideration, it is important to represent the collector (drain) efficiency of a power amplifier in an analytical form (excluding losses in the output load network), which can be expressed as

$$\eta = \frac{P_1}{P_0} = \frac{1}{2}\frac{I_1}{I_0}\frac{V}{V_{cc}} \tag{12.38}$$

where $P_1 = I_1\ V$ is the output power at fundamental frequency, $P_0 = I_0 V_{cc}$ is the dc supply power, I_1 is the fundamental current amplitude, I_0 is the dc current, V is the fundamental collector voltage amplitude, and V_{cc} is the collector supply voltage. For the operation conditions with the same conduction angle (for example, in Class B, the conduction angle is equal to 180° regardless of the collector current waveform amplitude), the current ratio between fundamental and dc components I_1/I_0 keeps a constant value. Generally, depending on the conduction angle, the current ratio I_1/I_0 varies from 1 in Class A with a conduction angle of 360° to 2 for ideal limiting case of Class C with a conduction angle of 0°. For Class B operation, the current ratio I_1/I_0 is equal to 1.57. Consequently, for a nearly Class B operation when the value of the conduction angle deviates from 180° by not too much, the current ratio I_1/I_0 varies within a small range of 10–20%. Thus, as follows from Eq. (14.38), the main factor of collector efficiency improvement at backoff output power levels is the voltage ratio V/V_{cc}, which should be kept constant for different output power levels. This can be achieved by reducing the supply voltage V_{cc} using envelope tracking technique or increasing the load resistance.

Since the collector efficiency is proportional to the ratio of fundamental amplitude to dc supply voltage, it becomes extremely small already at the output powers less than peak power by 10 dB, provided the load and dc supply voltage are kept constant. However, normally for wireless CDMA2000 or WCDMA mobile transmitters, the output power can vary in a wide dynamic range of about 80 dB

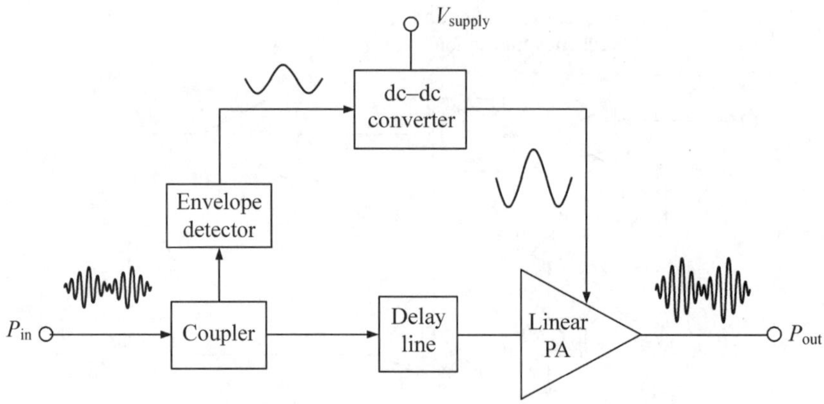

FIGURE 12.35 Envelope-tracking power amplifier architecture with analog control.

with maximum statistically averaged transmitting power required to deliver signal to the base station of about 15–30 dB less than the peak output power. Therefore, the average envelope-tracking technique can be very useful to increase power amplifier efficiency in a wide range of output powers of mobile transmitters by varying the dc supply voltage according to the RF signal envelope. When the supply voltage tracks the instantaneous output envelope, it is known as a *wideband envelope* or *envelope-follower tracking*. However, when the supply voltage tracks the long-term average of the output envelope, it is known as an *average envelope* or *power tracking*. The envelope-follower tracking is necessary to increase efficiency at fixed output power, while the power tracking is required to increase efficiency at backoff powers.

Figure 12.35 shows the envelope-tracking power amplifier architecture with analog control where the envelope detector is included at the input to detect the input signal envelope [98,99]. Here, a dc–dc converter is used to provide the dynamically controlled supply voltage to a linear power amplifier. While both buck (step-down) or boost (step-up) dc–dc converters can be used, the latter allows operation of the power amplifier from a supply voltage higher than the dc-supply voltage. Such a dc–dc boost converter implemented using AlGaAs/GaAs HBT process with 10 MHz switching (clock) frequency can provide the efficiency of 74% at maximum dc power [98]. The ripple in the dc–dc converter output results in the spurious in the output signal spectrum of approximately 60 dB lower than the fundamental spectral component. The delay line is necessary to compensate for the phase misalignment between the envelope and RF signal paths.

The linearity of a power amplifier with envelope tracking is usually worse than that of the power amplifier with fixed supply voltage because of the gain and phase variations with supply voltage. In this case, it is possible to use a fixed analog predistortion or digital control to minimize the increased nonlinearity to a considerable extent. Figure 12.36 shows the envelope-tracking power amplifier architecture with a digital control [100]. In addition to providing a proper control voltage for the dc–dc converter according to the signal envelope, the DSP system also computes a predistorted input signal for both the in-phase (I) and quadrature (Q) channels using amplifier amplitude and phase characteristics. The practical results show that, in this case, the *ACPR* of the IS-95 CDMA input signal can be improved by 8 dB at an output power of 28 dBm. However, the *ACPR* is quite sensitive to the timing relationships between the varying supply voltage and the input signal. For a GaAs MESFET power amplifier with boost converter operated at 950 MHz with maximum output power of 1 W, the power usage efficiency calculated in accordance with probability distribution function (PDF) of its output power is 1.64 times higher than the one for just battery operation [99]. However, to meet CDMA IS-95 specifications for *ACPR*, a fast feedback loop regulation scheme and dynamic gate biasing are needed to reduce the intermodulation distortions caused by the gain variation (due to the significant variation of the device transconductance with the supply voltage) and parasitic

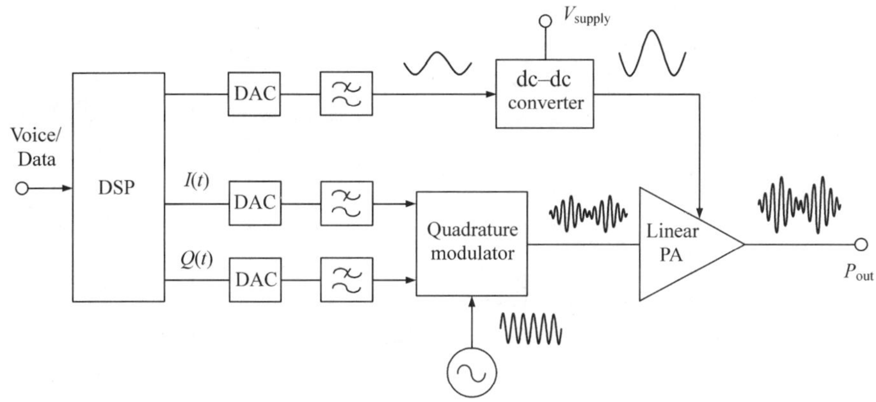

FIGURE 12.36 Envelope-tracking power amplifier architecture with digital control.

phase modulation (due to the inherently nonlinear behavior of the intrinsic device capacitances). Efficiency improvement can be achieved if to use a power amplifier with harmonic tuning instead of linear power amplifier operating in Class AB using high-performance GaN HEMT device [101]. In modern transmitters for wireless applications, the DSP technique are used to provide both the separate envelope signal and predistortion linearization using the feedback loop from the output.

It is much easier to provide a power tracking since power control dynamically changes the supply voltage and current as a function of power at a much slower rate compared to envelope. The Class-S modulator that can provide a long-term dynamically controlled supply voltage is similar in form to a buck dc–dc converter where the width or duty cycle of the pulses is proportional to an input control voltage. The control voltage corresponds to the long-term root mean square (rms) value of the modulated signal. The high switching frequency provides several advantages: reduced value and size of the low-pass filter (LPF), better suppression of the switching frequency and fast dynamic response. Maximizing the quality factors of the elements of the LPF and minimizing the on-resistances of the switching nMOS and pMOS transistors using larger size of their gate channel widths can result in a Class-S modulator efficiency of approximately 90% at the switching frequency of 5 MHz and slightly less at that of 10 MHz [102]. Figure 12.37 shows the envelope-tracking power amplifier architecture with a Class-S modulator.

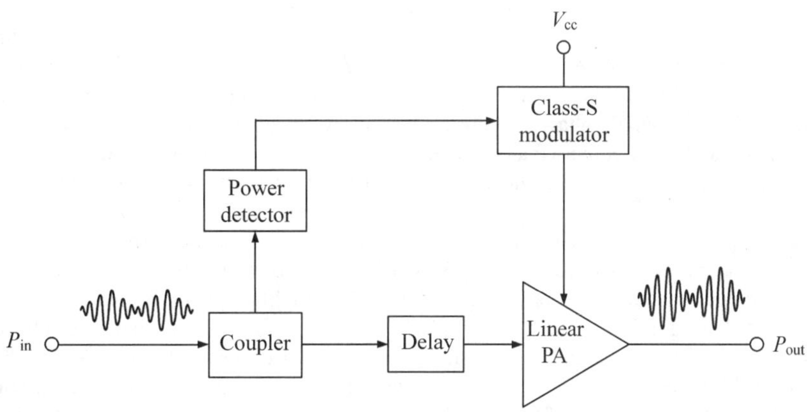

FIGURE 12.37 Envelope-tracking power amplifier architecture with Class-S modulator.

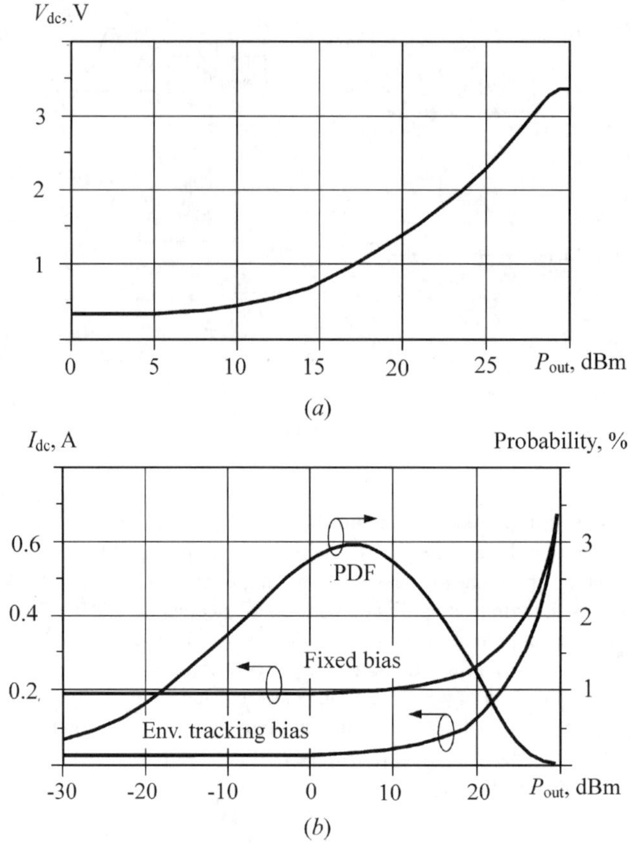

FIGURE 12.38 Envelope-tracking power-amplifier dc supply (*a*) voltage and (*b*) current.

The experimental results for a 2 W envelope-tracking MESFET power amplifier using a high-efficiency Class-S modulator intended for CDMA cellular handset applications in a frequency range of 824–849 MHz are shown in Figure 12.38 [102]. The phase deviations of the output signal across the dynamic range of the supply voltage were less than 3° up to the 2-dB gain compression point resulting in adjacent and alternate channel power ratios of −46 dBc and −56 dBc, respectively. Each *ACPR* is measured as the ratio of power in a 30-kHz bandwidth offset from the carrier by 885 kHz for adjacent channel and 1.98 MHz for alternate channel. The supply voltage is dynamically stepped down via the Class-S modulator to approximately 0.3 V in a deep backoff with maximum voltage of about 3.3 V due to small voltage drop from the battery dc supply voltage of 3.5 V, as shown in Figure 12.38(*a*). In this case, more than five times improvement in power usage efficiency, as shown in Figure 12.38(*b*), was obtained compared to a power amplifier with fixed bias voltage when the significant amount of the quiescent current is still kept over a wide range of output powers.

In handset wireless applications, the dc–dc converters should provide high efficiency, small size, and low cost operation. In this case, a deep submicron SiGe BiCMOS process technology can be used to fabricate the monolithic supply-modulated power amplifier where the power transistor can be implemented using the SiGe HBT process [103]. The size of the passive elements can be reduced to practical values for integration by increasing the switching frequencies to the order of 100 MHz. Switches can be implemented using nMOS and pMOS devices with optimum channel widths to minimize their power losses. The filter capacitance is realized as a MOS capacitance having higher specific capacitance compared with a metal–insulator–metal (MIM) capacitance. To increase the quality factor of the filter inductance, it is implemented using a thick last metal layer far above the substrate. Such a monolithic 900 MHz power amplifier with a high speed synchronous buck dc–dc converter can provide substantially higher efficiency compared with the similar power amplifier using constant voltage supply at lower power levels. By using a buck-boost converter with high efficiency over a wide loading range, the system efficiency can be further improved [104].

However, the system efficiency is slightly lower at high power than that of the conventional power amplifier with constant supply voltage due to certain effect of the dc–dc converter. For an 802.11 g OFDM application at 2.4 GHz, the overall efficiency of the complete wideband envelope-tracking system with SiGe HBT power amplifier is achieved 28% with an EVM (error vector magnitude) of 5% at an output power of 20 dBm [105]. Since the wideband envelope-tracking system has a significant nonlinearity associated with the collector modulation, as well as the intrinsic nonlinearity of the power amplifier, off-chip baseband digital predistortion with a LUT scheme was implemented to improve the system linearity, in particular EVM.

12.9 SWITCHED MULTIPATH POWER AMPLIFIERS

The power amplifiers in wireless communications systems operating in WCDMA or CDMA2000 standards are required to cover linearly a dynamic range of the transmitting output powers up to 80 dB. As a result, being designed for the highest power level with maximum achievable efficiency, the power amplifier tends to operate less efficiently at lower power levels, which leads to shortening the life of a battery. Figure 12.39(*a*) shows the transmitter architecture including a variable gain amplifier, a power amplifier to provide a high output power level, a bypass line for bypassing the smaller output power level, and a two-pole switch between two signal paths [106]. Normally, the power amplifiers for wireless handset transmitters are designed to achieve the power gain of about 25–30 dB. Therefore, it is very important to provide an efficient operation condition for the maximum probable transmitting power required to deliver signal to the base station, which is of about 12–15 dB less than the maximum peak power. The output power of the widely used variable gain amplifiers is usually less than 10 dBm, otherwise it is difficult to realize their linear operation. Besides, the variable gain amplifiers usually have a sufficiently high value for their noise figure. This contributes to the degradation in nonlinear distortion and excessive noise level in receiver bandwidth, which can only be improved by additional filtering, which increases the cost and size of the transmitter. A possible solution to improve the performance of the wireless handset transmitter is to use two

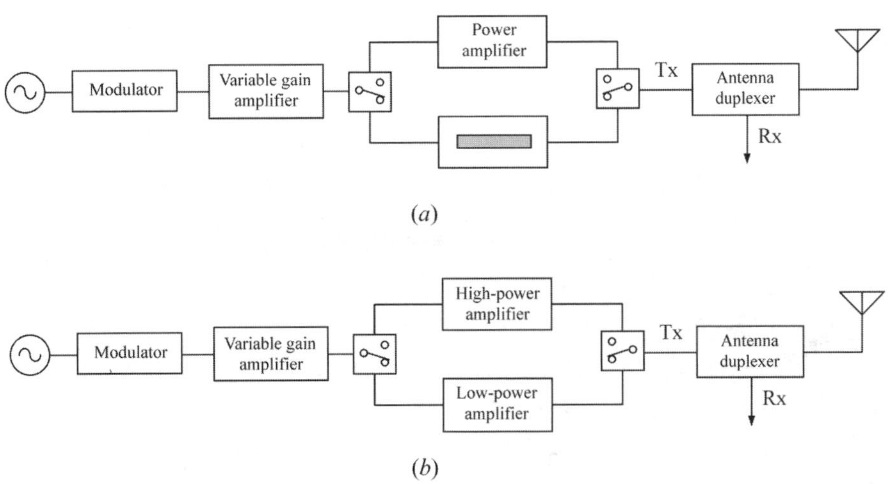

FIGURE 12.39 Transmitter architectures with dual-path power amplifier.

power amplifier paths with different output power levels. This can result in a significant reduction of power consumption, because the low-power amplifier provides higher efficiency at output power level corresponding to maximum PDF. The block schematic of the dual-path transmitter architecture is shown in Figure 12.39(*b*). When it is required to transmit the signal with an output power between the maximum level P_{max} and the statistically averaged power P_{avg}, the low-power amplifier is turned off. When it is enough to transmit the signal with output power equal or less than P_{avg}, the high power amplifier is turned off. So, at any moment the only one power amplifier is switched on. To further improve the efficiency of the transmitter, it may need to provide more than two power amplifier paths with different output power levels connected in parallel to the multipole switch.

In another approach, a mechanism for switching the output path between two or several power devices is used [107]. In this switchable-path power amplifier circuit shown in Figure 12.40, a combining network with a Schottky diode is used to switch the output path between the first active device, designed for maximum output power level, and the second active device, designed for the specified output power level with increased efficiency. As a result, the combining network is operable such that either the first power device Q_1 or the second power device Q_2 drives the power amplifier output.

An efficiency improvement can also be achieved by bypassing the power amplifier stages. For example, the power amplifier topology shown Figure 12.41(*a*) provides the possibility of bypassing the second stage [108]. At lower output power levels, the signal amplification can be achieved using only the first stage PA_1 with switch S_1 being turned on, whereas the maximum output power level is achieved using the two-stage power amplifier configuration with switches S_2 and S_3 being turned on. To eliminate any additional impedance matching, both stages should be designed to operate in a 50-Ω environment at the input and output. An improvement in average efficiency of greater than four times at backoff output power levels, compared with the conventional two-stage architecture, is

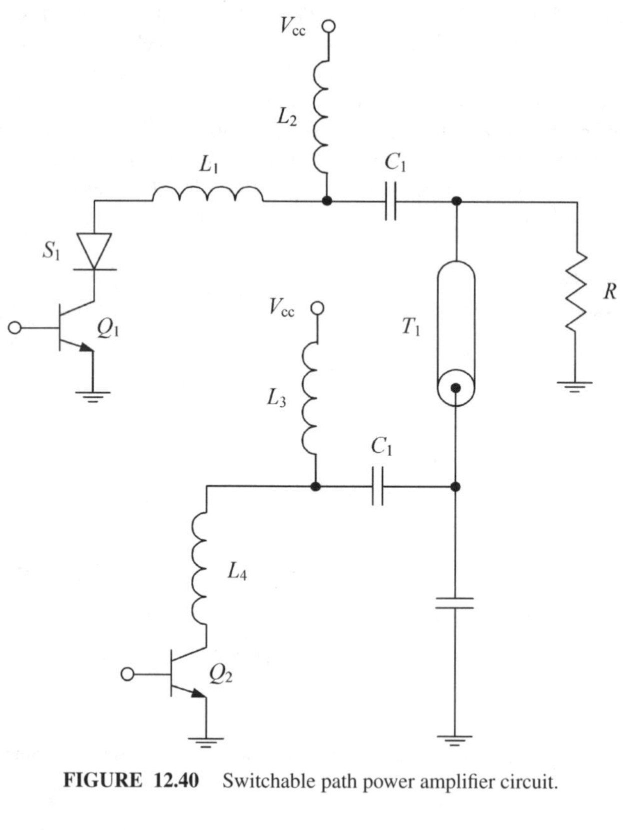

FIGURE 12.40 Switchable path power amplifier circuit.

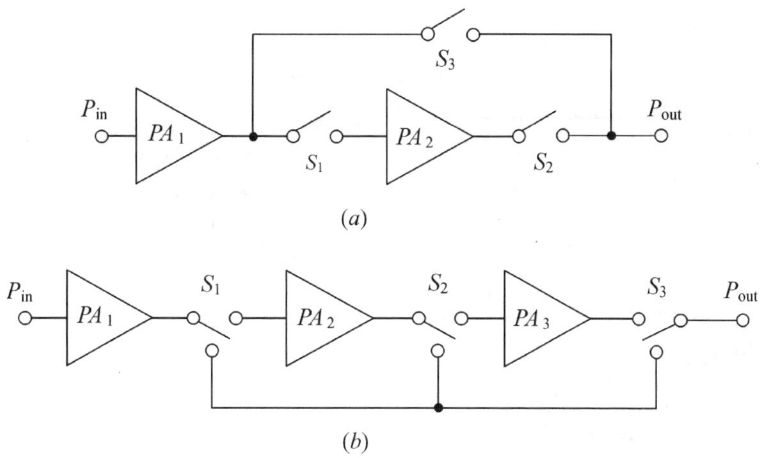

FIGURE 12.41 Switched-stage power amplifier configurations.

realized for a CDMA power amplifier operating over frequency bandwidth of 825–849 MHz with maximum output power of 29 dBm [109].

The power amplifier schematic shown in Figure 12.41(*b*) includes three amplifier stages with each stage having its own configuration so that a selected number of desired output power levels may be obtained directly from a selected power amplifier stage [110]. The amplifier stages also can be configured in tandem to deliver maximum output power when the output of the first power amplifier PA_1 provides an input to the second power amplifier PA_2, while the output of the second power amplifier PA_2 provides an input to the third power amplifier PA_3. As a result, a three-stage power amplifier can provide three output power levels with maximum achievable efficiency.

As an alternative to switching of the power amplifier paths, it is possible to improve efficiency of the power amplifier operated at different output power levels with a fixed supply voltage by providing the impedance transformation between the load and active device using switched-circuit arrangements of the load network [111]. The load network may include the series transmission line and parallel capacitors that can properly be connected by the on-to-off switching of a corresponding *p–i–n* diode to provide the impedance matching at the maximum and specified reduced power levels. For practical implementation, it is necessary to choose *p–i–n* diodes with minimal series resistance and minimize an influence of the diode biasing circuitry on RF performance. The operational principle of the diode-switched or variable load network configuration is illustrated in Figure 12.42. To maximize efficiency of the power amplifier, the load resistance for different output power levels should be different, so as to provide a collector voltage amplitude close to the value of the supply voltage V_{cc}, in accordance with Eq. (12.38). This means that the load-line angle at lower output power levels becomes smaller, so that the smaller collector current amplitude corresponds to approximately the same collector voltage amplitude as for the higher output power level. Moreover, for lower power levels, the saturation voltage becomes smaller, as seen from the collector voltage amplitude corresponding to the collector current amplitude I'' in Figure 12.42, and peak collector voltage is even higher. This smaller load-line angle corresponds to the higher value of the load resistance seen by the device collector.

To provide higher efficiency over a wide range of output power levels, the load network configurations with variable circuit elements can be used [112]. Figure 12.43(*a*) shows a possible load network topology with variable inductances. The variable inductance can be practically implemented by means of the series connection of a quarterwave transmission line and a varactor diode, as shown in Figure 12.43(*b*). The characteristic impedance of a quarterwave transmission line must be sufficiently low, so that the voltage swing across the varactor capacitance is substantially reduced. Using commercially available varactor diodes, it is possible to achieve an approximately constant value of maximum power-added efficiency over the 5 dB dynamic range and more than double improvement

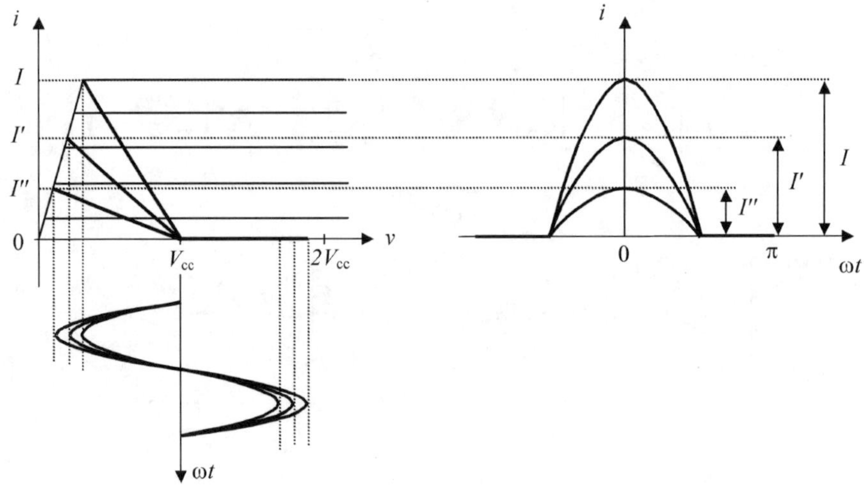

FIGURE 12.42 Collector voltage and current waveforms for different load lines.

in efficiency at −15 dB backoff output power level. The reverse-bias voltage applied to the varactor diodes should be high enough to minimize the insertion losses and parasitic phase distortion.

Using the diode switches in the load network results in increased size, cost, circuit complexity, and additional power losses. To improve the performance of the wireless handset transmitter, it might be possible to use its architecture with two power amplifier paths with different output power levels and a single three-port nonswitchable output load network [113]. Such an approach provides high-efficiency operation at low and medium output power levels with a significant reduction of the overall transmitter power consumption. The basic dual-path two-stage power amplifier schematic with a three-port output matching circuit is shown in Figure 12.44(*a*), which also includes a common first stage, a common three-port interstage matching circuit and a dual-path second stage. When it is necessary to transmit a signal with output power between the maximum output power P_{max} and some averaged backoff output power P_{avg}, the low-power amplifier stage with active device Q_3 is turned off. When it is enough to transmit the signal with output power equal or less than P_{avg}, the high-power amplifier stage with active device Q_2 is turned off. Both transistors are biased in Class AB with a small quiescent current to provide a linear operation. The area of the smaller device Q_3 corresponds to the output power P_{avr} required to deliver to the antenna. The three-port output matching circuit should be configured so that it provides a lower load impedance seen by the collector of the transistor Q_2 to deliver maximum output power P_{max} with maximum achievable collector efficiency and a higher load impedance seen by the collector of the transistor Q_3 to maximize collector efficiency at

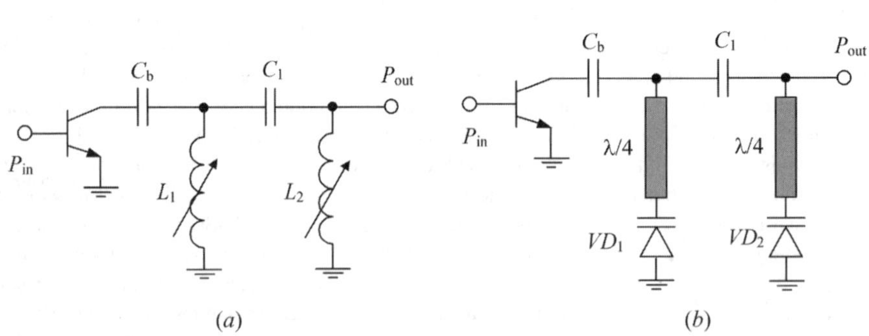

FIGURE 12.43 Load network configurations with variable circuit elements.

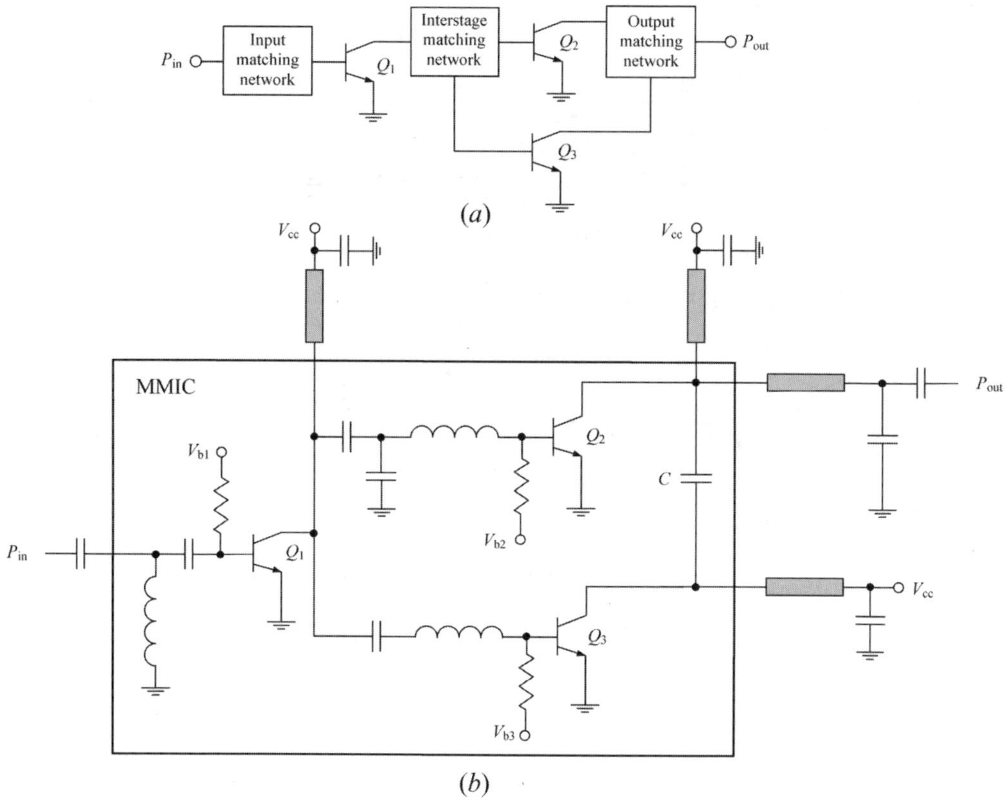

FIGURE 12.44 Three-port load network configurations of dual-path power amplifier.

the most probable output power P_{avr}. Since such a dual-path configuration with a single three-port load network does not require additional components rather than transistors, it is very practical for single-chip integration without serious increasing in manufacturing cost and size.

Figure 12.44(*b*) shows the circuit schematic of a two-stage InGaP/GaAs HBT MMIC power amplifier intended for WCDMA handset applications [113]. The emitter areas of the transistors for driver stage Q_1, power stage Q_2, and dual stage Q_3 were chosen of 480 μm^2, 1920 μm^2, and 480 μm^2, respectively. The output impedance-transforming circuit is constructed with a series capacitance *C*. The current-mirror circuits located within the integrated circuit were used for transistor biasing. As a result, by using a Class AB mode with quiescent currents of 10 mA for the devices Q_1 and Q_3 and 40 mA for the device Q_2 at supply voltage of $V_{cc} = 3.3$ V, the *PAE* of 16.4% was obtained in low-power mode with $P_{1dB} = 16.7$ dBm. In high-power mode with $P_{1dB} = 27.6$ dBm, the *PAE* was 34.2%. Generally, an overall efficiency improvement of at least 1.81 times over a wide power range was provided, compared to a conventional two-stage Class AB power amplifier.

Figure 12.45 shows an alternative approach to realize the dual-path power amplifier using a chain configuration of the output impedance-transforming circuit [114]. The dual-chain two-stage InGaP/GaAs HBT MMIC power amplifier with common input and output matching circuits is implemented by parallel connection of two amplifying chains having different output powers. Either the low-power amplifying chain with $P_{1dB} = 16$ dBm or the high-power amplifying chain with $P_{1dB} = 28$ dBm is activated through the bias selection. The matching circuit between the collectors of the output transistors, with a series inductance and a shunt capacitance composing a simple low-pass *L*-type matching circuit, allows the power-added efficiency at backoff output powers to be increased

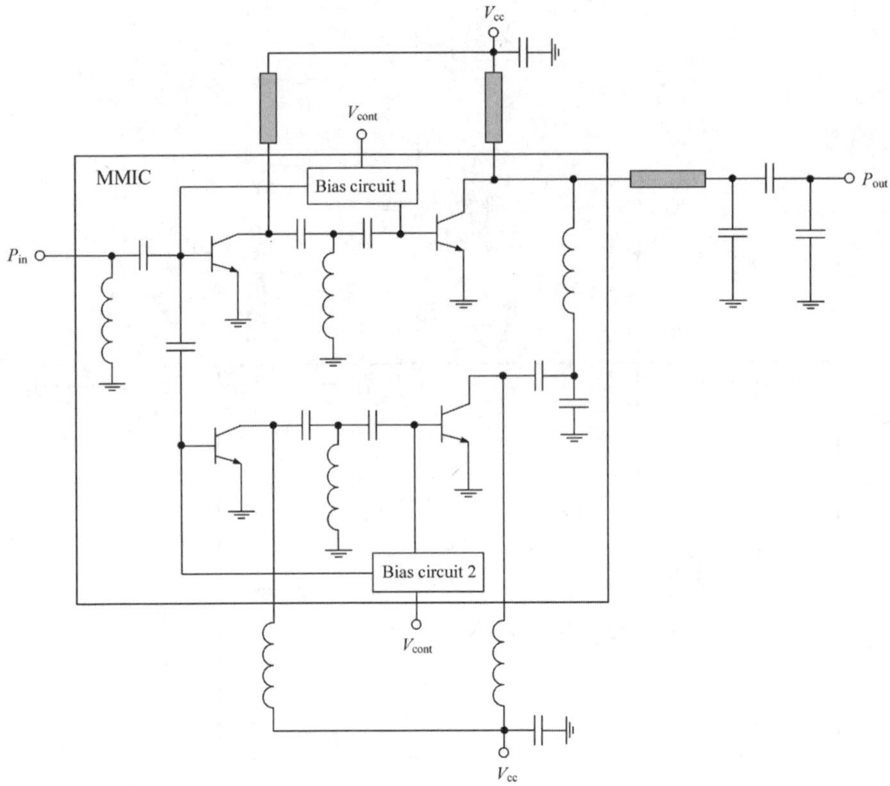

FIGURE 12.45 Schematic of dual-chain MMIC power amplifier.

by more accurate impedance matching of the higher output impedance of the low-power device to the 50-Ω load. Using such a chain power amplifier configuration, it is possible to obtain the *PAE* of 21% at $P_{1dB} = 16$ dBm in a low-power mode and *PAE* of 40% at $P_{1dB} = 28$ dBm in a high-power mode.

Figure 12.46(*a*) shows the block diagram of the power amplifier with high efficiency at different output power levels, which incorporates an additional separate low-power amplifying path connected between the isolated ports of a quadrature input hybrid power divider and a quadrature output hybrid power combiner [115]. In a high power mode, the bias control unit provides proper biasing of the transistors in both balanced paths and switches off the biasing in a low-power path 4-to-4′. In this case, the output powers from both high-power balanced paths 2-to-2′ and 3-to-3′ are combined in the output hybrid combiner and the low-power path is isolated from them, thus providing a high efficiency at high output power. When all ports of a hybrid branch-line combiner are matched, power entering input port 1 is divided between the output ports 2 and 3 with a phase shift of 90° between these outputs and no power is delivered to the isolated port 4. In a low power mode, the bias control unit switches off the biasing of transistors in both balanced paths and provides proper biasing for a low power amplifying path. A quadrature hybrid has an important advantage compared to the in-phase dividers that, at equal values of reflection coefficients from loads connected to the output ports 2 and 3, the reflection wave is lacking at the input port 1 and, consequently, input *VSWR* of a quadrature hybrid does not depend on the equal-load mismatch level. In this case, all reflected power in a quadrature divider is dissipated in a 50-Ω ballast resistor connected to the isolated port 4. Similarly, for a quadrature combiner, the reflection waves from two inputs will combine and flow to the load. As a result, the isolated port of the input hybrid becomes the input for a low-power amplifying path, while the isolated port of the output hybrid operates as its output 50-Ω port. Hence, the high efficiency at lower output power levels

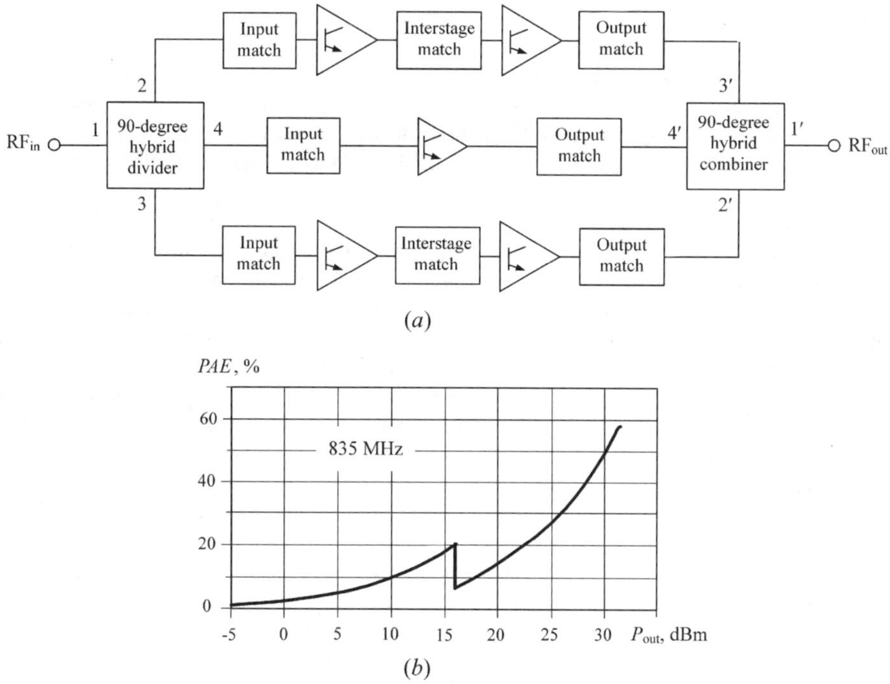

FIGURE 12.46 Block diagram and test result of balanced switched-path MMIC power amplifier.

can be achieved, and no need to include any additional switches to isolate high-power balanced paths and low-power path from each other.

Practical results for high-efficiency linear multiband and multimode two-stage balanced switched-path SiGe HBT MMIC power amplifiers intended to operate across DCS1800/PCS1900 and WCDMA850/1900/2100 frequency bands had demonstrated the power-added efficiency of $\geq$50% at maximum output power and $\geq$20% at backoff output power of 16 dBm. For example, the maximum output power of 31.5 dBm was achieved with a maximum power-added efficiency of 58% at operating frequency of 835 MHz, while, for an average WCDMA power of 28.5 dBm, a *PAE* of 41.5% was measured. In a low-power mode at $P_{out} = 16$ dBm when $ACLR = -34$ dBc, the efficiency is increased by greater than a factor of two reaching 20% by turning off the bias current for the high-power balanced paths and switching on the low-power path, as shown in Figure 12.46(*b*) [115].

To provide an efficient linear power amplifier operation, it is necessary to minimize the quiescent current at backoff output powers because maximum of the power density function for WCDMA standard occurs at an output power of about 25 dB below its maximum level. As a current-controlled device, the bipolar transistor at RF operation requires the dc base driving current, whose value depends on the output power and device parameters. From technology point of view, since a bipolar device represents a parallel connection of the basic cells, important issue is to use the ballasting series resistors to avoid the current imbalance and possible device collapse at higher current density levels. The basic current-mirror and emitter-follower bias circuits can provide the temperature compensation over a wide range of ambient temperatures, with very small reference current for the latter case [116]. However, an adaptive bias circuit can additionally control dc power consumption with varying output power, thus greatly improving power-added efficiency when the output power is low and maintaining high linearity of the power amplifier when its output power is high [117].

In wireless base-station applications, it is possible to use a combiner design technology for dual-state operation [118,119]. In such a dual-state power amplifier, the power amplifier consisting of a four-amplifier array and a 50-Ω matched transmission-line combiner is operated under two modes depending upon the number of activated power amplifiers. In a low-state operation, the only one power amplifier is activated, while the other three power amplifiers are switched off. It is assumed that, when power amplifier is switched off, its output impedance is nearly infinite (open circuit) and, due to the effect of the transmission lines with a quarterwave length each, the three deactivated amplifiers are properly isolated from the activated amplifier. In a high-state operation, all four power amplifiers are evenly delivering maximum power to the 50-Ω load. As a result, maximum efficiency is provided at the maximum and 6-dB backoff output powers. However, in practice at high frequencies, the output impedance of a deactivated power amplifier is not an open circuit due to effect of the active device parasitic elements and output matching circuit. Also, additional losses occur due to the asymmetrical structure of the output combiner, and two quarterwave transmission lines are necessary at the input so that the power can be added in phase at the output. As a result, the overall efficiency significantly degrades at both maximum and 6-dB output powers.

The switched-line architecture of a dual-state power amplifier shown in Figure 12.47 can improve the power amplifier efficiency at the maximum and 6-dB backoff power levels by using of two switches connected at the ground ends of two additional parallel quarterwave transmission lines [120]. In a high-power mode when all four power amplifiers are activated, as shown in Figure 12.47(*a*), the switches are turned on providing open-circuit conditions at the opposite input ends of these transmission lines, and all four power amplifiers are evenly delivering maximum power to the 50-Ω load. In this case, the power combining structure is symmetrical, and each power amplifying path contains only two series output quarterwave transmission lines. In a low-state mode when the power amplifier 1 is activated only, the switches are turned off providing short-circuit conditions at the input ends of these quarterwave transmission lines, as shown in Figure 12.47(*b*), resulting in a full isolation of the deactivated power amplifiers from the activated one. Consequently, the two peak high efficiencies can be achieved at the maximum and 6-dB backoff power levels. Since the switches are located between the ground and the ends of the parallel quarterwave transmission lines at the opposite sides of the power combiner, they consume a very small amount of dc power.

12.10 KAHN EER TECHNIQUE AND DIGITAL POWER AMPLIFICATION

12.10.1 Envelope Elimination and Restoration

Modern wireless communication systems require that the multiple-carrier signal with a nonconstant envelope is fed through the power amplifier linearly and efficiently. However, there is a tradeoff between power amplifier efficiency and linearity, with improvement in one coming at the expense of the other. In a traditional analog envelope elimination and restoration (EER) approach, where special devices are required to separate amplitude (envelope) and RF phase-modulated signals, one type of power amplifier is responsible for envelope signal amplification, while another type of power amplifier is fed by a constant-envelope RF signal, as shown in Figure 12.48(*a*). The limiter removes the envelope component to form a purely phase-modulated RF signal. This constant-envelope signal is then amplified efficiently by a nonlinear power amplifier (PA) using either conventional Classes AB, B, or C with certain conduction angle or switching-mode Classes D, E, or F, depending on the operating frequency, frequency bandwidth and output power. At the same time, the envelope component is detected, isolated from the phase-modulated component and efficiently processed using a Class S modulator based on pulse-width modulation (PWM) technique to modulate the phase-modulated signal at the final stage of the power amplifier. Amplitude modulation of the final stage of the power amplifier restores the envelope to the phase-modulated carrier signal creating an amplitude replica of the input signal.

This technique was firstly developed by Kahn in 1950s to improve the efficiency of short-wave broadcast transmitters [121]. In contrast to linear power amplifiers, a Kahn EER transmitter can operate with high efficiency over a wide dynamic range of backoff output power levels, thus producing an average efficiency that is three to five times higher [122,123]. To minimize misalignment between

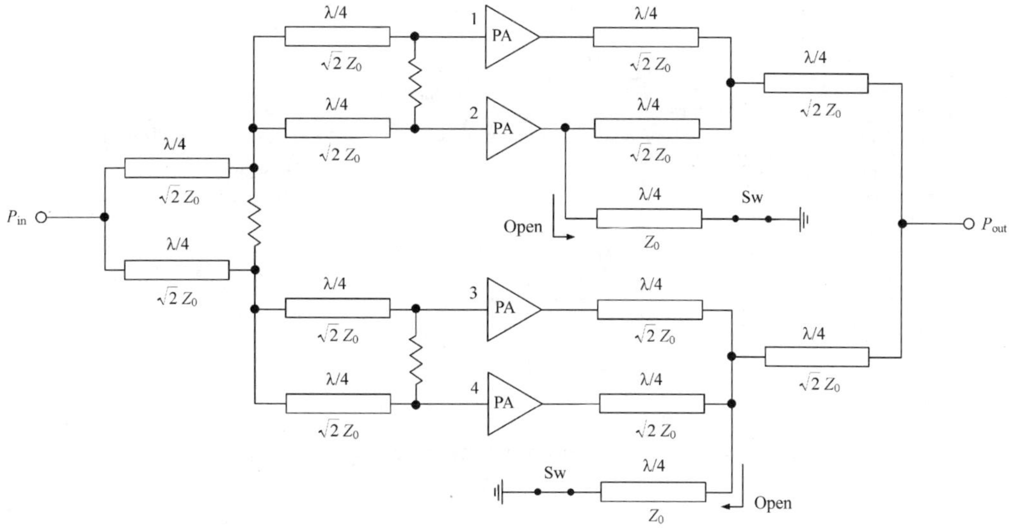

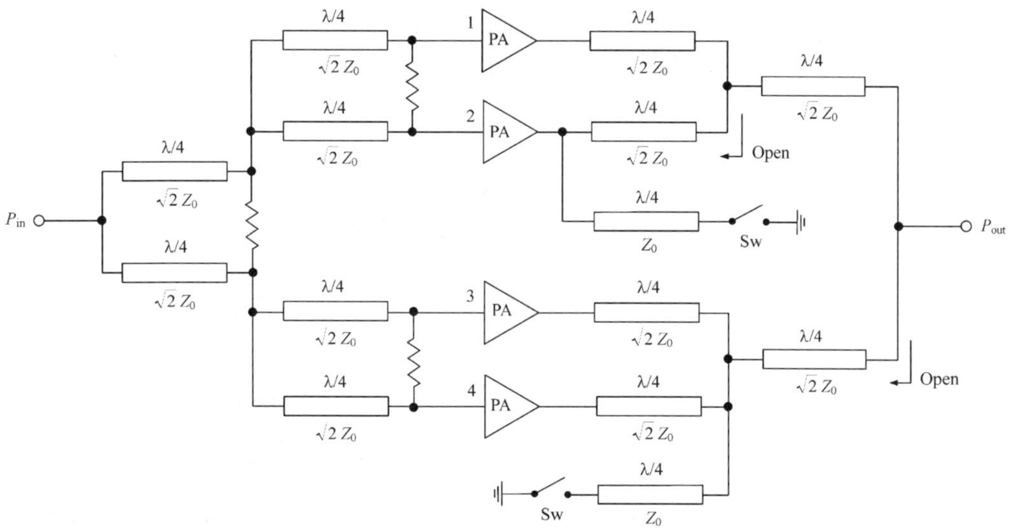

FIGURE 12.47 Schematic of switched-line dual-state power amplifier architecture.

phase and amplitude, the delay line is required. Adding the output envelope feedback circuit with envelope detector allows the intermodulation distortion to be reduced [124]. In modern transmitters for wireless applications, the both envelope and phase-modulated signals can be easily generated separately using DSP technique, resulting in a system architecture shown in Figure 12.48(*b*). In this case, the phase-modulated signal with constant envelope is upconverted to the desired output RF frequency using a direct or double conversion scheme. For a direct conversion scheme, the baseband signal containing the phase information directly modulates the RF carrier. The experimental results for mobile RF polar transmitters with CDMA IS-95 signals show that the overall system efficiency more than 30% can be achieved at moderate average output power levels satisfying the corresponding linearity requirements [125,126].

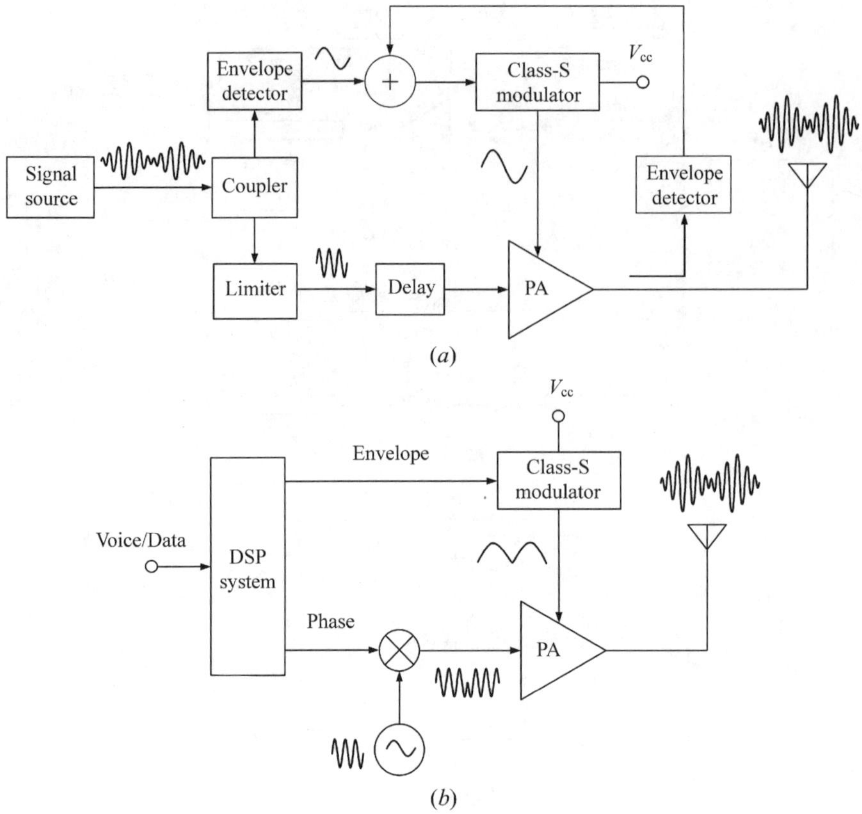

FIGURE 12.48 Block diagrams of Kahn EER transmitters.

Nevertheless, there are several serious limitations of a traditional Kahn EER architecture for practical implementation. The output voltage characteristic of the RF power transistor should be linear enough to restore accurately the input envelope. However, in practice this is not the case, especially for high power devices. Besides, the size of a Class-S modulator is sufficiently large for MMIC implementation, and it becomes too inefficient at high frequencies. In addition, a high degree of amplitude and phase tracking accuracy is required to achieve a high depth of modulation. For instance, a combination of 0.2 dB of amplitude and 3° of phase tracking error will result in a maximum modulation dynamic range of 20 dB. To obtain the level of the third-order intermodulation components better than −30 dBc, the differential time delay (Δt) between the envelope and phase-modulated paths defined as $\Delta t \leq 0.1/B_{RF}$, where B_{RF} is the bandwidth of the RF signal, must be sufficiently small [127]. Finally, there should be a tradeoff between the switching frequency in the Class-S modulator and the order of the low-pass filter in order to minimize the intermodulation distortion.

12.10.2 Pulse-Width Carrier Modulation

Figure 12.49(a) shows the transmitter architecture with pulse-width carrier modulation, which is an alternative to a conventional Kahn EER technique [128]. In this case, similarly to the Kahn EER technique, the phase-modulated input signal with a nonconstant envelope is split by a coupler into separate envelope and phase-modulated components by means of the envelope detector and limiter, respectively. The envelope is converted to duty factor variations of pulse train, typically at a sampling rate at least an order of magnitude higher in frequency than the maximum frequency of the envelope.

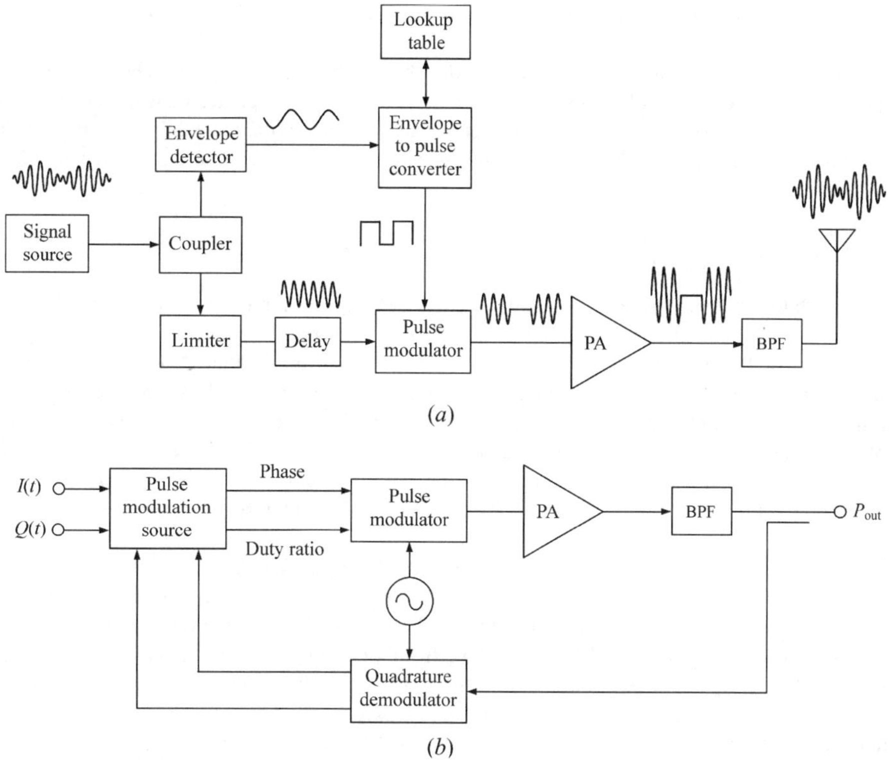

FIGURE 12.49 Transmitter architecture with pulse-width carrier modulation.

However, the pulse train is then applied as pulse-width modulation onto a low-power carrier phase-modulated signal using a pulse modulator. The combined signal is transmitted to a high-efficiency broadband nonlinear power amplifier. Unwanted sidebands caused by the sampling process are then rejected by a bandpass filter (BPF) at the amplifier output to pass only the carrier and its modulated signal bandwidth and hence recover a linear replica of the original source signal. Pulse modulation source can operate using Cartesian coordinate I and Q baseband data determining the phase values and duty ratios for modulating the pulsed envelope signal, as shown in Figure 12.49(*b*) [129,130]. Using a LUT in the envelope-to-pulse modulation circuitry is important to maximize the envelope dynamic range by proper adjustment of the pulse width and the sampling period. The feedback loop with a quadrature demodulator reproducing the baseband in-phase and quadrature feedback signals may be necessary to correct the nonlinearity effects introduces by components within the circuit and improve linearity [129].

Generally, the both envelope and phase-modulated signals can also be generated separately using DSP technique, and digitization of the envelope signal can be provided by a delta-sigma modulator with a sampling rate of ten times of the RF bandwidth [131,132]. The main disadvantage of such transmitter architectures is a sufficiently high level of insertion loss of the narrowband BPF, the quality factor of which should be very high to minimize the switching frequency sidebands. Nevertheless, in some particular cases, it is possible to realize a compromise solution between BPF insertion loss and switching frequency, thus resulting in a higher average efficiency for signals with nonconstant envelope than can be achieved by the systems at backoff output power levels using linear power amplifiers. For example, by using a second-order delta-sigma modulator with 200 MHz switching frequency and a Class E power amplifier, an output power of 18 dBm with a power-added efficiency of

32% for a three-tone signal was achieved at operating frequency of 2.47 GHz with a BPF insertion loss of 2 dB [132]. In this case, to minimize the quantization noise level within the bandwidth of interest close to the carrier, such as the receive band in wireless base station or mobile transceiver, the switching frequency must be chosen as high as possible [133]. However, the higher the switching frequency, the broader must be the power amplifier to operate in a purely switching mode, thus transmitting the overall spectrum including harmonics of the switching frequency without introducing significant nonlinearities.

12.10.3 Class S Amplifier

An approach to efficiently transmit the audio signals by means of a pulse-width modulation, which also was variously called *pulse-duration modulation* or *two-state amplification*, was invented yet in the middle of 1920s. In that proposed case, the efficiency of the multistage transmission system is increased by producing a series of electrical impulses varying in length in accordance with the amplitude of the transmitting audio-frequency signal. To obtain these impulses, the first vacuum-tube amplifying stage was considered a modulator with varying grid potential according to the audio-frequency signal amplitude, thus providing the output current impulses with different amplitude, with the second and third amplifying stages operating as the amplifier and the limiter, respectively [134]. The bandwidth of the output low-pass filter is chosen to transmit freely the audio frequencies but attenuate the high-frequency carrier signal. A few years later, the triangular high-frequency carrier signal was proposed to apply to the grid of a current-switching push–pull amplifier resulting in the rectangular-wave output currents at anodes to form a pulse-width modulation transmission, which was then called the *Class S amplification* [135]. In this case, the original audio-frequency signal was restored by using a two-section low-pass filter at the circuit output. With further introduction of the integrated circuit, the pulse-width modulation is widely used based on a comparator when a triangular waveform of a high frequency is fed into one input of a comparator, whereas the audio signal is applied to the other comparator input.

Figure 12.50(*a*) shows the simplified circuit schematic of a complementary voltage-switching Class S amplifier where the slowly varying average or dc component of the voltage waveform $v_2(\omega t)$ is coupled to the load R_L through low-pass filter L_0C_0, producing the output voltage $v_R(\omega t)$ and output current $i_R(\omega t)$ that is a replica of the original low-frequency signal [123]. The low-pass filter has an input series inductor and therefore high impedance to the switching frequency and its harmonics. The output current $i_R(\omega t)$ is drawn through the low-pass filter and, at any given time, through one of the transistors and diodes with waveforms shown in Figure 12.50(*b*). In this case, diodes in parallel with the transistors are necessary to provide a reverse-direction current path for the output current [136]. However, precaution must be taken to make sure that the two transistors are never turned on together causing a very high short-circuit current to flow. The Class S amplifier, similar to other switching amplifiers, provides less than 100% efficiency because of the effects of finite saturation resistance, shunt capacitance and transition time. To provide the maximum allowable distortion in the forms of spurious products in the output spectrum, some trade-offs must be made between a lower switching frequency and more number of sections in a low-pass filter.

12.10.4 Digital RF Amplification

The low-frequency Class S amplification architecture can also be implemented to digitally transmit the signals at radio and microwave frequencies. In this case, the transmitter block schematic shown in Figure 12.51(*a*), which corresponds to a high-frequency Class S amplification architecture, employs the bandpass delta-sigma (BPDS) modulator to encode the desired signal into a single-bit digital data stream, a switching-mode power amplifier to amplify the two-level transmitting signal efficiently and a bandpass filter (BPF) that is required to significantly suppress the unwanted spectral components [137]. Here, the signals at different points are illustrated in time and frequency domains demonstrating the effect of a BPDS modulator, a power amplifier and a BPF on the transmitting signal from RF

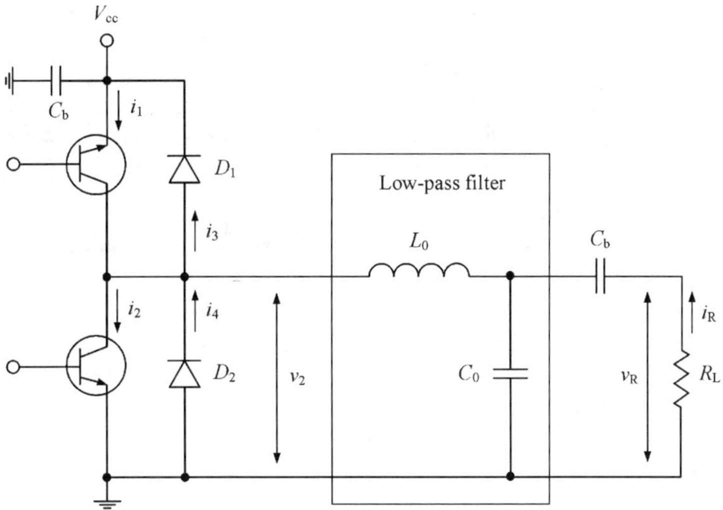

(a)

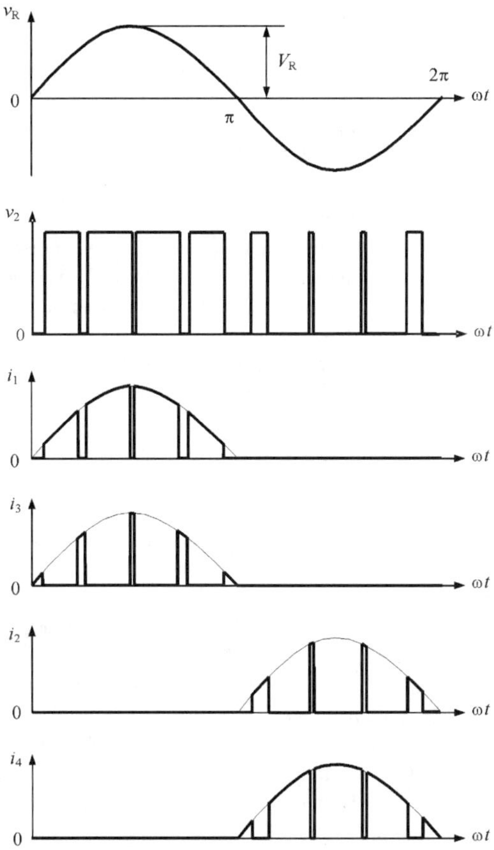

(b)

FIGURE 12.50 Class S amplifier with voltage and current waveforms.

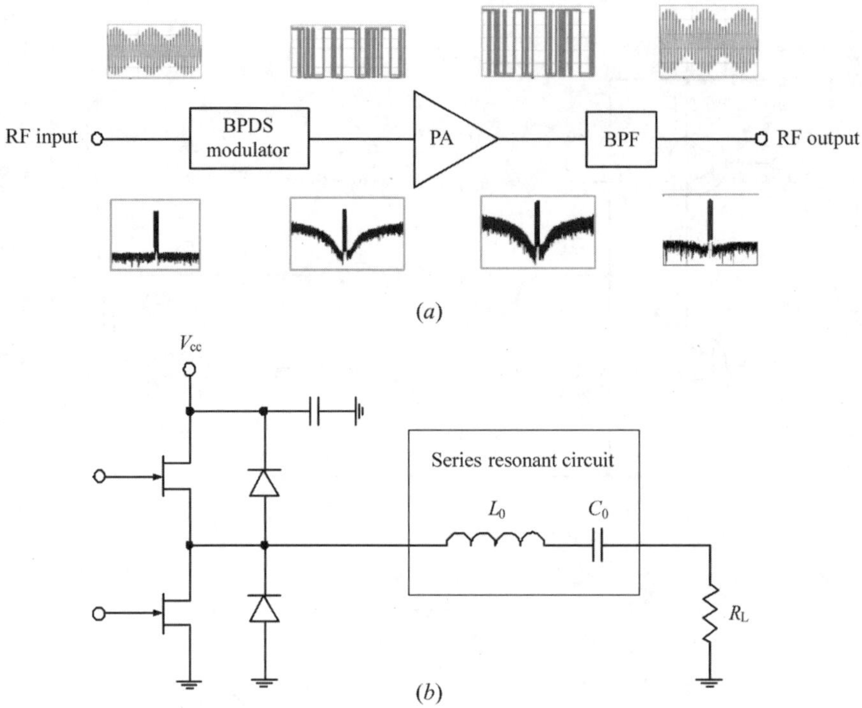

FIGURE 12.51 High-frequency Class S amplification with bandpass filter.

input to RF output. The incoming amplitude and phase-modulated signal is passed through a BPDS modulator becoming a binary signal, in which the quantization noise associated with the digitization is spectrally shaped so that its level was minimized within the band of interest.

The switching-mode power amplifier can be implemented as a voltage-switching Class D power amplifier with a series resonant circuit L_0C_0, the schematic of which is shown in Figure 12.51(*b*). Here, the diodes in parallel to the transistors are necessary to provide the current flow during intervals when the transistors are turned off, because the amplifier is driven by signals with a duty ratio in general not equal to a nominal duty ratio of 50%. The output spectrum also contains images of the input signal at frequencies $f_0 \pm nf_s$, where f_0 is the carrier frequency, f_s is the switching frequency and n is an integer, that are suppressed by the output bandpass filter, together with the excessive level of the quantization noise. In this case, the power associated with the undesired out-of-band spectral components is reflected, rather than absorbed in the filter, in a way that preserves high efficiency of the Class S amplification.

Figure 12.52 shows the circuit diagram of a DSP-based high-frequency transmitter including a Class D power amplifier and a bandpass filter [138]. The I and Q signals generated at baseband are provided by samples at a baseband frequency f_{bb}, which is an exact submultiple of f_s and which typically is chosen to be higher than the desired signal bandwidth by a factor of at least 10 with moderate resolution. The upconverter and BPDS modulator operate with the highest clock rate or switching frequency f_s when it is selected to correspond to a multiple of f_0 (typically $f_s = 4f_0$) in order to simplify the computations. To generate signals with different carrier frequencies, a phased locked loop with frequency control to generate f_s is assumed. The BPDS modulator comprises one or more digital resonators, a single-bit quantizer, a dither source, and a feedback loop, in which the output signal is fed back to the input. Then, the BPDS output signal is used to drive a switching-mode high-efficiency Class D power amplifier.

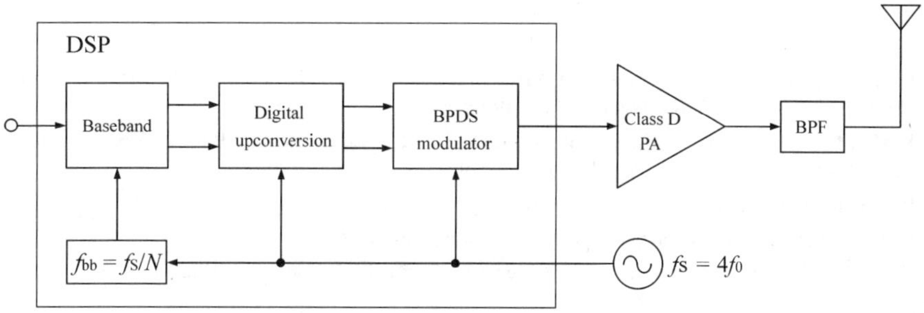

FIGURE 12.52 DSP-based Class S transmitter architecture.

The performance of the BPDS modulator depends mainly on two parameters: the order of the modulator and the oversampling ratio (OSR) of the input signal. The OSR is defined as a ratio of sampling (switching) frequency to two times the input signal bandwidth. The higher the order is, the better the quantization noise is filtered at the output bandwidth, but the modulator structure becomes more complicated. For example, for a 2.14-GHz WCDMA carrier, the switching frequency must be equal to $f_s = 4f_0 = 8.56$ GHz, which means that, for a 60 MHz bandwidth, the OSR = 8.56 GHz/(2 × 60 MHz) = 71.33, resulting in the *ACLR* of at least −50 dBc [139]. However, the sampling rate of 8.56 GHz is definitely too high for modern technology, especially in implementation of a high-efficiency high-power switching-mode amplifier, the active devices of which must be characterized by a transition frequency f_T much higher than f_s, of at least three times to include its third harmonic component in the transmitting spectrum for increased efficiency and linearity. For a CDMA IS-95 signal with a bandwidth of 1.23 MHz and a *PAR* of 5.5 dB at a carrier frequency of 800 MHz, a drain efficiency of 31% was achieved for a voltage-switching CMOS Class D power amplifier with an output power of 15 dBm and an *ACPR* of −43 dBc using a clock rate for a two-level BPSD of 3.2 GHz [140]. However, to calculate a power-added efficiency of the high-frequency Class S amplification path, it is necessary to operate with a coding efficiency that represents a ratio of the in-band power to the total pulse train power [141]. In this case, a high level of out-of-band spectral power is wasteful and reduces the utility of the pulse train to generate a load in-band power may significantly affect the overall system efficiency. Although out-of-band spectral components do not dissipate any power in the load network of an ideal Class-D power amplifier, however they are dissipated in its input circuit and require the excessive dc power for a driver amplifier, thus significantly reduces the operating power gain and, as a result, the power-added efficiency, compared to the linear amplification architecture. Therefore, in practice it is necessary to choose the active devices such as a pHEMT or a GaN HEMT with the minimum saturation resistances, high gain capability and maximum transition frequency.

REFERENCES

1. H. S. Black, "Translating System," U.S. Patent 1,686,792, Oct. 1928.
2. W. D. Lewis, "Self-Correcting Amplifier," U.S. Patent 2,592,716, Apr. 1952.
3. H. Seidel, "Feed-Forward Amplifier," U.S. Patent 3,471,798, Oct. 1969.
4. H. Seidel, "A Feedforward Experiment Applied to an L-4 Carrier System Amplifier," *IEEE Trans. Commun. Technol.*, vol. COM-19, pp. 320–325, June 1971.
5. H. Seidel, "A Microwave Feedforward Experiment," *Bell System Tech. J.*, vol. 50, pp. 2879–2916, Nov. 1971.
6. R. G. Meyer, R. Eschenbach, and W. M. Edgerley, "A Wide-Band Feedforward Amplifier," *IEEE J. Solid-State Circuits*, vol. SC-9, pp. 422–428, Dec. 1974.
7. C.-C. Hsieh and S.-P. Chan, "A Feedforward *S*-Band MIC Amplifier System," *IEEE J. Solid-State Circuits*, vol. SC-11, pp. 271–278, Apr. 1976.

8. R. J. Wilkinson and P. B. Kenington, "Specification of Error Amplifiers for Use in Feedforward Transmitters," *IEE Proc.-G*, vol. 139, pp. 447–480, Aug. 1992.
9. S.-G. Kang and I.-K. Lee, "Cancellation Performance of a Linearisation Loop of a Feedforward Amplifier," *Electronics Lett.*, vol. 33, pp. 444–446, Mar. 1997.
10. K. Konstantinou and D. K. Paul, "Analysis and Design of Broadband High Efficiency Feedforward Amplifiers," *IEEE MTT-S Int. Microwave Symp. Dig.*, vol. 2, pp. 867–870, 1996.
11. P. B. Kenington, "Efficiency of Feedforward Amplifiers," *IEE Proc.-G*, vol. 139, pp. 591–593, Oct. 1992.
12. P. B. Kenington, *High Linearity RF Amplifier Design*, Boston: Artech House, 2000.
13. E. E. Eid, F. M. Ghannouchi, and F. Beauregard, "Optimal Feedforward Linearization System Design," *Microwave J.*, vol. 38, pp. 78–86, Nov. 1995.
14. H. Choi, Y. Jeong, J. S. Kenney, and C. D. Kim, "Cross Cancellation Technique Employing an Error Amplifier," *IEEE Microwave and Wireless Components Lett.*, vol. 18, pp. 488–490, July 2008.
15. R. M. Bauman, "Adaptive Feed-Forward System," U.S. Patent 4,389,618, June 1983.
16. B. Shi, W. Shan, and L. Sundstrom, "An Analog Adaptive Feedforward Amplifier Linearizer," *Proc. 34th European Microwave Conf.*, pp. 1065–1068, 2004.
17. S. J. Grant, J. K. Cavers, and P. A. Goud, "A DSP Controlled Adaptive Feedforward Amplifier Linearizer," *5th IEEE Int. Conf. Record. Universal Personal Commun.*, pp. 788–792, 1996.
18. S. G. Randall, J. G. McRory, and R. H. Johnston, "Broadband DSP Based Feedforward Amplifier Linearizer," *Electronics Lett.*, vol. 38, pp. 1470–1471, Nov. 2002.
19. W. W. Macalpine, "Distortionless Transmission System," U.S. Patent 2,043,587, June 1936.
20. P. O. Pedersen, "A Distortion-Free Amplifier," *Proc. IRE*, vol. 28, pp. 59–66, Feb. 1940.
21. L. T. Mucenieks, C. S. Robertson, B. L. Irvine, and N. Salvador, "RF Power Amplifier Linearization Using Parallel RF Power Amplifiers Having Intermod-Complementing Predistortion Paths," U.S. Patent 6,111,462, Aug. 2000.
22. M. Billsberry, "Balanced Distortion Reduction Circuit," U.S. Patent 6,734,726, May 2004.
23. R. Gutierrez, "Reflect Forward Adaptive Linearizer Amplifier," U.S. Patent 6,573,793, June 2003.
24. R. Gutierrez, "The RFAL Technique for Cancellation of Distortion in Power Amplifiers," *High Frequency Electronics*, vol. 4, pp. 18–28, June 2004.
25. R. Gutierrez, "High-Efficiency Linearized LDMOS Amplifiers Utilize the RFAL Architecture," *High Frequency Electronics*, vol. 6, pp. 22–28, Feb. 2006.
26. E. Peterson, "Control of Distortion," U.S. Patent 2,233,061, Feb. 1941.
27. T. Nojima and T. Kohno, "Cuber Predistortion Linearizer for Relay Equipment in 800 MHz Band Land Mobile Telephone System," *IEEE Trans. Vehicular Technol.*, vol. VT-34, pp. 169–177, Nov. 1985.
28. R. C. Tupynamba and E. Camargo, "MESFET Nonlinearities Applied to Predistortion Linearizer Design," *IEEE MTT-S Int. Microwave Symp. Dig.*, pp. 955–958, 1992.
29. H. S. Read, "Electric Circuits," U.S. Patent 1,464,111, Aug. 1923.
30. Y. Ikeda, K. Mori, S. Shinjo, F. Kitabayashi, A. Ohta, T. Takagi, and O. Ishida, "An L-Band High Efficiency and Low Distortion Multi-Stage Amplifier Using Self Phase Distortion Compensation Technique," *IEICE Trans. Electron.*, vol. E85-C, pp. 1967–1972, Dec. 2002.
31. K. Yamauchi, K. Mori, M. Nakayama, Y. Itoh, Y. Mitsui, and O. Ishida, "A Novel Series Diode Linearizer for Mobile Radio Power Amplifiers," *IEEE MTT-S Int. Microwave Symp. Dig.*, vol. 2, pp. 831–834, 1996.
32. K. Yamauchi, K. Mori, M. Nakayama, Y. Mitsui, and T. Takagi, "A Microwave Miniaturized Linearizer Using a Parallel Diode with a Bias Feed Resistance," *IEEE Trans. Microwave Theory Tech.*, vol. MTT-45, pp. 2431–2435, Dec. 1997.
33. M. Nakayama, K. Mori, K. Yamauchi, Y. Itoh, and Y. Mitsui, "An Amplitude and Phase Linearizing Technique for Linear Power Amplifiers," *Microwave J.*, vol. 39, pp. 96–104, Mar. 1996.
34. S. Ogura, K. Seino, T. Ono, A. Kamikokura, and H. Hirose, "Development of a Compact Broadband FET Linearizer for Satellite Use," *IEEE MTT-S Int. Microwave Symp. Dig.*, vol. 3, pp. 1195–1198, 1997.
35. J. Boix, "Ku-Band Solid State Power Amplifiers for the New European Satellites," *Proc. 23rd Europ. Microwave Conf.*, pp. 11–14, 1993.

36. J. Yi, Y. Yang, M. Park, W. Kang, and B. Kim, "Analog Predistortion Linearizer for High-Power RF Amplifiers," *IEEE Trans. Microwave Theory Tech.*, vol. 48, pp. 2709–2713, Dec. 2000.

37. I. Kim, J. Cha, S. Hong, Y. Y. Woo, J. Kim, and B. Kim, "Predistortion Power Amplifier for Base-Station Using a Feedforward Loop Linearizer," *Proc. 36th Europ. Microwave Conf.*, pp. 141–144, 2006.

38. F. Zavosh, D. Runton, and C. Thron, "Digital Predistortion Linearizes CDMA LDMOS Amps," *Microwaves & RF*, vol. 39, pp. 55–61, Mar. 2000.

39. L. Sundstrom, M. Faulkner, and M. Johansson, "Quantization Analysis and Design of a Digital Predistortion Linearizer for RF Power Amplifiers," *IEEE Trans. Vehicular Technol.*, vol. VT-45, pp. 707–719, Nov. 1996.

40. A. R. Mansell and A. Bateman, "Adaptive Predistortion with Reduced Feedback Complexity," *Electronics Lett.*, vol. 32, pp. 1153–1154, June 1996.

41. N. Nascas and Y. Papananos, "A New Non-Iterative, Adaptive Baseband Predistortion Method for High Power RF Amplifiers," *IEEE Circuits and Systems Symp. Dig.*, vol. I, pp. 413–416, 2003.

42. H. S. Black, "Wave Translation System," U.S. Patent 2,003,282, June 1935 (filed Aug. 1923).

43. H. S. Black, "Stabilized Feed-Back Amplifiers," *Proc. IEEE*, vol. 87, pp. 379–385, Feb. 1999; reprinted from *Electronic Engineering*, vol. 53, pp. 114–120, Jan. 1934.

44. W. B. Bruene, "Distortion Reducing Means for Single-Sideband Transmitters," *Proc. IRE*, vol. 44, pp. 1760–1765, Dec. 1956.

45. H. S. Black, "Inventing the Negative Feedback Amplifier," *IEEE Spectrum*, vol. 14, pp. 55–60, Dec. 1977.

46. H. A. Rosen and A. T. Owens, "Power Amplifier Linearity Studies for SSB Transmissions," *IEEE Trans. Commun. Syst.*, vol. CS-12, pp. 150–159, June 1964.

47. C.-C. Hsieh and E. Strid, "A S-Band High Power Feedback Amplifier," *IEEE MTT-S Int. Microwave Symp. Dig.*, pp. 182–184, June 1977.

48. K. Oosaki and Y. Akaiwa, "Nonlinearity Compensation of Linear Power Amplifier for Mobile Communication," *4th IEEE Int. Conf. Record. Universal Personal Commun.*, pp. 302–305, Nov. 1995.

49. R. J. Gilmore, R. Kiehne, and F. J. Rosenbaum, "Circuit Design to Reduce 3rd Order Intermodulation Distortion in FET Amplifiers," *IEEE MTT-S Int. Microwave Symp. Dig.*, pp. 413–416, June 1985.

50. A. K. Ezzeddine, H.-L. Hung, and H. C. Huang, "An MMIC *C*-Band FET Feedback Power Amplifier," *IEEE Trans. Microwave Theory Tech.*, vol. MTT-38, pp. 350–357, Apr. 1990.

51. J. G. McRory and R. H. Johnston, "An RF Amplifier for Low Intermodulation Distortion," *IEEE MTT-S Int. Microwave Symp. Dig.*, vol. 3, pp. 1741–1744, 1994.

52. S. Narayanan, "Application of Volterra Series to Intermodulation Distortion Analysis of Transistor Feedback Amplifiers," *IEEE Trans. Circuit Theory*, vol. CT-17, pp. 518–527, Nov. 1970.

53. Y. Kim, Y. Yang, S. Kang, and B. Kim, "Linearization of 1.85 GHz Amplifier Using Feedback Predistortion Loop," *IEEE MTT-S Int. Microwave Symp. Dig.*, vol. 3, pp. 1675–1678, 1998.

54. Y. Y. Woo, J. Kim, J. Yi, S. Hong, I. Kim, J. Moon, and B. Kim, "Adaptive Digital Feedback Predistortion Technique for Linearizing Power Amplifiers," *IEEE Trans. Microwave Theory Tech.*, vol. MTT-55, pp. 932–940, May 2007.

55. T. Arthanayake and H. B. Wood, "Linear Amplification Using Envelope Feedback," *Electronics Lett.*, vol. 7, pp. 145–146, Apr. 1971.

56. S. Kusunoki, K. Yamamoto, T. Hatsugai, H. Nagaoka, K. Tagami, N. Tominaga, K. Osawa, K. Tanabe, S. Sakurai, and T. Iida, "Power-Amplifier Module with Digital Adaptive Predistortion for Cellular Phones," *IEEE Trans. Microwave Theory Tech.*, vol. MTT-50, pp. 2979–2986, Dec. 2002.

57. J.-S. Cardinal and F. Ghannouchi, "A New Adaptive Double Envelope Feedback (ADEF) Linearizer for Solid State Power Amplifiers," *IEEE Trans. Microwave Theory Tech.*, vol. MTT-43, pp. 1508–1515, July 1995.

58. V. Petrovic and W. Gosling, "Polar-Loop Transmitter," *Electronics Lett.*, vol. 15, pp. 286–288, May 1979.

59. V. Petrovic, "Reduction of Spurious Emission from Radio Transmitters by Means of Modulation Feedback," *Proc. IEE Conf. Radio Spectrum Conservation Technol.*, pp. 44–49, 1983.

60. M. Johansson and L. Sundstrom, "Linearisation of RF Multicarrier Amplifiers Using Cartesian Feedback," *Electronics Lett.*, vol. 30, pp. 1110–1112, July 1994.

61. Y. Nagata, "Linear Amplification Technique for Digital Mobile Communications," *Proc. 39th IEEE Vehicular Technol. Conf.*, pp. 159–164, 1989.

62. A. R. Mansell and A. Bateman, "Transmitter Linearisation Using Composite Modulation Feedback," *Electronics Lett.*, vol. 32, pp. 2120–2121, Nov. 1996.

63. S. W. Chung, J. W. Holloway, and J. L. Dawson, "Energy-Efficient Digital Predistortion with Lookup Table Training Using Analog Cartesian Feedback," *IEEE Trans. Microwave Theory Tech.*, vol. MTT-56, pp. 2248–2258, Oct. 2008.

64. W. H. Doherty, "A New High Efficiency Power Amplifier for Modulated Waves," *Proc. IRE*, vol. 24, pp. 1163–1182, Sept. 1936.

65. F. H. Raab, "Efficiency of Doherty RF Power-Amplifier Systems," *IEEE Trans. Broadcast.*, vol. BC-33, pp. 77–83, Sept. 1987.

66. M. Iwamoto, A. Williams, P -F. Chen, A. G. Metzger, L. E. Larsson, and P. M. Asbeck, "An Extended Doherty Amplifier with High Efficiency Over a Wide Power Range," *IEEE Trans. Microwave Theory Tech.*, vol. MTT-49, pp. 2472–2479, Dec. 2001.

67. Y. Yang, J. Cha, B. Shin, and B. Kim, "A Fully Matched N-Way Doherty Amplifier with Optimized Linearity," *IEEE Trans. Microwave Theory Tech.*, vol. MTT-51, pp. 986–993, Mar. 2003.

68. Y. Yang, J. Yi, Y. Y. Woo, and B. Kim, "Optimum Design for Linearity and Efficiency of a Microwave Doherty Amplifier Using a New Load matching Technique," *Microwave J.*, vol. 44, pp. 20–36, Dec. 2001.

69. Y. Takayama, T. Harada, T. Fujita, and K. Maenaka, "Design Method of Microwave Doherty Power Amplifiers and Its Application to Si Power MOSFET Amplifiers," *Electronics and Communications in Japan*, part 2, vol. 88, pp. 9–17, Apr. 2005.

70. N. Srirattana, A. Raghavan, D. Heo, P. E. Allen, and J. Laskar, "Analysis and Design of a High-Efficiency Multistage Doherty Power Amplifier for Wireless Communications," *IEEE Trans. Microwave Theory Tech.*, vol. MTT-53, pp. 852–860, Mar. 2005.

71. M. J. Pelk, W. C. E. Neo, J. R. Gajadharsing, R. S. Pengelly, and L. C. N. de Vreede, "A High-Efficiency 100-W GaN Three-Way Doherty Amplifier for Base-Station Applications," *IEEE Trans. Microwave Theory Tech.*, vol. MTT-56, pp. 1582–1591, July 2008.

72. Y. Zhao, M. Iwamoto, D. Kimball, L. E. Larsson, and P. M. Asbeck, "A 900 MHz Doherty Amplifier Implemented with Lumped Elements," *IEEE Topical Workshop on Power Amplifiers for Wireless Commun.*, San Diego, 2003.

73. Y. Yang, J. Cha, B. Shin, and B. Kim, "A Microwave Doherty Amplifier Employing Envelope Tracking Technique for High Efficiency and Linearity," *IEEE Microwave and Wireless Components Lett.*, vol. 13, pp. 370–372, Sept. 2003.

74. Y.-S. Lee, M.-W. Lee, and Y.-H. Jeong, "Highly Efficient Doherty Amplifier Based on Class-E topology for WCDMA Applications," *IEEE Microwave and Wireless Components Lett.*, vol. 18, pp. 608–610, Sept. 2008.

75. Y. Zhao, M. Iwamoto, L. E. Larsson, and P. M. Asbeck, "Doherty Amplifier with DSP Control to Improve performance in CDMA Operation," *IEEE MTT-S Int. Microwave Symp. Dig.*, vol. 2, pp. 687–690, 2003.

76. J. Moon, Ja. Kim, I. Kim, Ju. Kim, and B. Kim, "Highly efficiency Three-Way Saturated Doherty Amplifier with Digital Feedback Predistortion," *IEEE Microwave and Wireless Components Lett.*, vol. 18, pp. 539–541, Aug. 2008.

77. H. Chireix, "High Power Outphasing Modulation," *Proc. IRE*, Vol. 23, pp. 1370–1392, Nov. 1935.

78. D. C. Cox, "Linear Amplification with Nonlinear Components," *IEEE Trans. Commun.*, vol. COM-22, pp. 1942–1945, Dec. 1974.

79. F. H. Raab, "Average Efficiency of Outphasing Power-Amplifier Systems," *IEEE Trans. Commun.*, vol. COM-33, pp. 1094–1099, Oct. 1985.

80. S. A. Hetzel, A. Bateman, and J. P. McGeehan, "A LINC Transmitter," *Proc. 41st IEEE Vehicular Technol. Conf.*, pp. 133–137, 1991.

81. R. Langridge, T. Thornton, P. M. Asbeck, and L. E. Larson, "A Power Re-Use Technique for Improved Efficiency of Outphasing Microwave Power Amplifiers," *IEEE Trans. Microwave Theory Tech.*, vol. MTT-47, pp. 1467–1470, Aug. 1999.

82. C. P. Conradi, R. H. Johnston, and J. G. McRory, "Evaluation of a Lossless Combiner in a LINC Transmitter," *Proc. 1999 IEEE Canadian Conf. Electrical and Computer Eng.*, vol. 1, pp. 105–110, May 1999.
83. B. Shi and L. Sundstrom, "Investigation of a Highly Efficient LINC Amplifier Topology," *Proc. 54th IEEE Vehicular Technol. Conf.*, pp. 1215–1219, 2001.
84. K. Tom, V. Bassoo, M. Faulkner, and T. Lejon, "Load-Pull Analysis of Outphasing Class-E Power Amplifier," *Proc. 2nd Int. Conf. Wireless Broadband and Ultra Wideband Commun.*, pp. 52–55, 2007.
85. I. Hakala, D. K. Choi, L. Gharavi, N. Kajakine, J. Koskela, and R. Kaunisto, "A 2.14-GHz Chireix Outphasing Transmitter," *IEEE Trans. Microwave Theory Tech.*, vol. MTT-53, pp. 2129–2138, June 2005.
86. G. Hegazi, T. Chu, S. Heibel, J. Jordan, and H. Sasmazer, "Linear Wideband VHF/UHF Quad LINC Transmitter System," *Proc. 2007 IEEE Military Commun. Conf.*, pp. 1–6, Oct. 2007.
87. S. Hamedi-Hagh and C. A. T. Salama, "Wideband CMOS Integrated RF Combiner for LINC Transmitters," *IEEE MTT-S Int. Microwave Symp. Dig.*, vol. 1, pp. 41–43, 2003.
88. F. Casadevall and J. J. Olmos, "On the Behavior of the LINC Transmitter," *Proc. 40th IEEE Vehicular Technol. Conf.*, pp. 29–34, May 1990.
89. A. Birafane and A. B. Kouki, "On the Linearity and Efficiency of Outphasing Microwave Amplifiers," *IEEE Trans. Microwave Theory Tech.*, vol. MTT-52, pp. 1702–1708, July 2004.
90. S. Tomisato, K. Chiba, and K. Murota, "Phase Error Free LINC Modulator," *Electronics Lett.*, vol. 25, pp. 576–577, Apr. 1989.
91. L. Sundstrom, "Automatic Adjustment of Gain and Phase Imbalances in LINC Transmitters," *Electronics Lett.*, vol. 31, pp. 155–156, Feb. 1995.
92. X. Zhang, L. E. Larson, P. M. Asbeck, and P.Nanawa, "Gain/Phase Imbalance-Minimization Techniques for LINC Transmitters," *IEEE Trans. Microwave Theory Tech.*, vol. MTT-49, pp. 2507–2515, Dec. 2001.
93. C. P. Conradi and J. G. McRory, "Predistorted LINC Transmitter," *Electronics Lett.*, vol. 38, pp. 301–302, Mar. 2002.
94. A. Huttunen and R. Kaunisto, "A 20-W Chireix Outphasing Transmitter for WCDMA Base Stations," *IEEE Trans. Microwave Theory Tech.*, vol. MTT-55, pp. 2709–2718, Dec. 2007.
95. A. Bateman, "The Combined Analogue Locked Loop Universal Modulator (CALLUM)," *Proc. 42nd IEEE Vehicular Technol. Conf.*, pp. 759–763, 1992.
96. A. Bateman, M. Li, and K. Chan, "Implementation of the LINC Transmitter Using the Combined Analogue Locked Loop Universal Modulator (CALLUM)," *Proc. 7th IEE Europ. Conf. Mobile and Personal Commun.*, pp. 31–37, 1993.
97. D. J. Jennings and J. P. McGeehan, "A High-Efficiency RF Transmitter Using VCO-Derived Synthesis: CALLUM," *IEEE Trans. Microwave Theory Tech.*, vol. MTT-47, pp. 715–721, June 1999.
98. G. Hannigton, P.-F. Chen, V. Radisic, T. Itoh, and P. M. Asbeck, "Microwave Power Amplifier Efficiency Improvement with a 10 MHz HBT DC-DC Converter," *IEEE MTT-S Int. Microwave Symp. Dig.*, pp. 313–316, 1998.
99. G. Hannigton, P.-F. Chen, P. M. Asbeck, and L. E. Larson, "High-Efficiency Power Amplifier Using Dynamic Power-Supply Voltage for CDMA Applications," *IEEE Trans. Microwave Theory Tech.*, vol. MTT-47, pp. 1471–1476, Aug. 1999.
100. M. Ranjan, K. H. Koo, G. Hannigton, C. Fallesen, and P. M. Asbeck, "Microwave Power Amplifiers with Digitally-Controlled Power Supply Voltage for High Efficiency and High Linearity," *IEEE MTT-S Int. Microwave Symp. Dig.*, vol. 1, pp. 493–496, 2000.
101. D. F. Kimball, J. Jeong, C. Hsia, P. Draxler, S. Lanfranco, W. Nagy, K. Linthicum, L. E. Larson, and M. Asbeck, "High-Efficiency Envelope-Tracking W-CDMA Base-Station Amplifier Using GaN HFETs," *IEEE Trans. Microwave Theory Tech.*, vol. MTT-54, pp. 3848–3856, Nov. 2006.
102. J. Staudinger, B. Gilsdorf, D. Newman, G. Norris, G. Sadowniczak, R. Sherman, and T. Quach, "High Efficiency CDMA RF Power Amplifier Using Dynamic Envelope Tracking Technique," *IEEE MTT-S Int. Microwave Symp. Dig.*, vol. 2, pp. 873–876, 2000.
103. S. Abedinpour, I. Deligoz, J. Desai, M. Figiel, and S. Kiaei, "Monolithic Supply Modulated RF Power Amplifier and DC–DC Power Converter IC," *IEEE MTT-S Int. Microwave Symp. Dig.*, vol. 1, pp. 89–92, 2003.

104. B. Sahu and G. A. Rincon-Mora, "A High-Efficiency Linear RF Power Amplifier with a Power-Tracking Dynamically Adaptive Buck-Boost Supply," *IEEE Trans. Microwave Theory Tech.*, vol. MTT-52, pp. 112–120, Jan. 2004.

105. F. Wang, D. F. Kimball, D. Y. Lie, P. M. Asbeck, and L. E. Larsson, "A Monolithic High-Efficiency 2.4-GHz 20-dBm SiGe BiCMOS Envelope-Tracking OFDM Power Amplifier," *IEEE J. Solid-State Circuits*, vol. SC-42, pp. 1271–1281, June 2007.

106. Y. Aihara, "Transmitter Power Varying Device Having a Bypass Line for a Power Amplifier," U.S. Patent 5,909,643, June 1999.

107. J. R. King and G. A. Olsen, "Switchable Path Power with Schottky Diode Combining Network," U.S. Patent 6,181,208, Jan. 2001.

108. S. Brozovich, "High Efficiency Multiple Power Level Amplifier Circuit," U.S. Patent 5,661,434, Aug. 1997.

109. J. Staudinger, "Applying Switched Gain Stage Concepts to Improve Efficiency and Linearity of Mobile CDMA Power Amplification," *Microwave J.*, vol. 43, pp. 152–162, Sept. 2000.

110. H. X. Wu, "High-Efficient Configurable Power Amplifier for Use in a Portable Unit," U.S. Patent 5,758,269, May 1998.

111. A. V. Grebennikov, H. Jaeger, and E. P. Heaney, "High-Efficiency Cellular Phone Transmitters with Reduced Power Consumption," *Proc. 2002 European Conf. Wireless Technol.*, pp. 115–118.

112. G. Leuzzi and C. Micheli, "Variable-Load Constant-Efficiency Power Amplifier for Mobile Communications Applications," *Proc. 11th European GAAS Symp.*, pp. 481–484, 2003.

113. H.-M. Park, S.-H. Cheon, J.-W. Park, and S. Hong, "Demonstration of On-Chip Appended Power Amplifier for Improved Efficiency at Low Power Region," *IEEE MTT-S Int. Microwave Symp. Dig.*, vol. 2, pp. 691–694, 2003.

114. J. H. Kim, J. H. Kim, Y. S. Noh, and C. S. Park, "An InGaP-GaAs HBT MMIC Smart Power Amplifier for W-CDMA Mobile Handsets," *IEEE J. Solid-State Circuits*, vol. SC-38, pp. 905–910, June 2003.

115. A. Grebennikov, B. Zogl, H. Hermann, C. Roth, and W. Thomann, "High-Efficiency Balanced Switched-Path Monolithic SiGe HBT Power Amplifiers for Wireless Applications," *Proc. 37th Europ. Microwave Conf.*, pp. 1189–1192, 2007.

116. T. Sato and C. Grigorean, "Design Advantages of CDMA Power Amplifiers Built with MOSFET Technology," *Microwave J.*, vol. 45, pp. 64–78, Oct. 2002.

117. Y. S. Noh and C. S. Park, "An Intelligent Power Amplifier MMIC Using a New Adaptive Bias Control Circuit for W-CDMA Applications," *IEEE J. Solid-State Circuits*, vol. SC-39, pp. 967–970, June 2004.

118. C. Y. Hang, Y. Wang, and T. Itoh, "A New Amplifier Power Combining Scheme with Optimum Efficiency under Variable Outputs," *IEEE MTT-S Int. Microwave Symp. Dig.*, vol. 2, pp. 913–916, 2002.

119. Y. Chung, J. Jeong, Y. Wang, D. Ahn, and T. Itoh, "Power Level-Dependent Dual-Operating Mode LDMOS Power Amplifier for CDMA Wireless Base-Station Applications," *IEEE Trans. Microwave Theory Tech.*, vol. MTT-53, pp. 739–746, Feb. 2005.

120. A. Grebennikov, "High-Efficiency Dual-State Power Amplifier," Europ. Patent Appl. EP 2,169,825, Mar. 2010.

121. L. R. Kahn, "Single Sideband Transmission by Envelope Elimination and Restoration," *Proc. IRE*, vol. 40, pp. 803–806, July 1952.

122. F. H. Raab, "High Efficiency Amplification Techniques," *IEEE Circuits and Systems J.*, vol. 7, pp. 3–11, Dec. 1975.

123. H. L. Krauss, C. W. Bostian, and F. H.Raab, *Solid State Radio Engineering*, New York: John Wiley & Sons, Ltd. 1980.

124. F. H. Raab and D. J. Rupp, "Class-S High-Efficiency Amplitude Modulator," *RF Design*, vol. 17, pp. 3–11, May 1994.

125. D. K. Su and W. J. McFarland, "An IC for Linearizing RF Power Amplifiers Using Envelope Elimination and Restoration," *IEEE J. Solid-State Circuits*, vol. SC-33, pp. 2252–2258, Dec. 1998.

126. J. Choi, J. Yim, J. Yang, J. Kim, J. Cha, D. Kang, D. Kim, and B. Kim, "A $\Delta\Sigma$-Digitized Polar RF Transmitter," *IEEE Trans. Microwave Theory Tech.*, vol. MTT-55, pp. 2679–2690, Dec. 2007.

127. F. H. Raab, "Intermodulation Distortion in Kahn-Technique Transmitters," *IEEE Trans. Microwave Theory Tech.*, vol. MTT-44, pp. 2273–2278, Dec. 1996.

128. R. Posner and Z. Regev, "Transmission of Multiple Carrier Signals in a Nonlinear System," U.S. Patent 5,249,201, Sept. 1993.

129. P. Midya, R. M. Jackson, and D. Anderson, "Method and Apparatus for Generating a Modulated Signal," U.S. Patent 5,838,210, Nov. 1998.

130. P. Wagh and P. Midya, "High-Efficiency Switched-Mode RF Power Amplifier," *Proc. 42nd Midwest Symp. Circuits and Systems*, vol. 2, pp. 1044–1047, 1999.

131. Y. Wang, "An Improved Kahn Transmitter Architecture Based on Delta-Sigma Modulation," *IEEE MTT-S Int. Microwave Symp. Dig.*, vol. 3, pp. 1327–1300, 2003.

132. A. Dupuy and Y. E. Wang, "High Efficiency Power Transmitter Based on Envelope Delta-Sigma Modulation (EDSM)," *Proc. 60th IEEE Vehicular Technol. Conf.*, pp. 2092–2095, 2004.

133. C. Berland, I. Hibon, J. F. Bercher, M. Villegas, D. Belot, D. Pache, and V. Le Goascoz, "A Transmitter Architecture for Nonconstant Envelope Modulation," *IEEE Trans. Circuits and Systems–II: Express Briefs*, vol. CAS-II-53, pp. 13–17, Jan. 2006.

134. R. A. Heising, "Transmission System," U.S. Patent 1,655,543, Jan. 1928.

135. B. D. Bedford, "Electric Amplifying Circuits," U.S. Patent 1,874,159, Aug. 1932.

136. H. R. Camenzind, "Modulated Pulse Audio Power Amplifiers for Integrated Circuits," *IEEE Trans. Audio and Electroacoustics*, vol. AU-14, pp. 136–140, Sept. 1966.

137. A. Jayaraman, P. F. Chen, G. Hanington, L. Larson, and P. Asbeck, "Linear High-Efficiency Microwave Power Amplifiers Using Bandpass Delta-Sigma Modulators," *IEEE Microwave and Guided Wave Lett.*, vol. 8, pp. 121–123, Mar. 1998.

138. J. Keyzer, J. Hinrichs, A. Metzger, M. Iwamoto, I. Galton, and P. Asbeck, "Digital Generation of RF Signals for Wireless Communications with Band-Pass Delta-Sigma Modulation," *IEEE MTT-S Int. Microwave Symp. Dig.*, vol. 3, pp. 2127–2130, 2001.

139. J. Ketola, J. Sommarek, J. Vankka, and K. Halonen, "Transmitter Utilizing Bandpass Delta-Sigma Modulator and Switching Mode RF Power Amplifier," *IEEE Int. Sym. Circuits and Systems Dig.*, vol. 1, pp. 633–636, 2004.

140. T.-P. Hung, J. Rode, L. E. Larson, and P. M. Asbeck, "Design of H-Bridge Class-D Power Amplifiers for Digital Pulse Modulation Transmitters," *IEEE Trans. Microwave Theory Technol.*, vol. MTT-55, pp. 2845–2855, Dec. 2007.

141. T. Johnson and S. P. Stapleton, "RF Class-D Amplification with Bandpass Sigma-Delta Modulator Drive Signals," *IEEE Trans. Circuits and Systems–I: Regular Papers*, vol. CAS-I-53, pp. 2507–2520, Dec. 2006.

13 Control Circuits

This chapter discusses the circuit schematics and main properties of the semiconductor control circuits that are usually characterized by small size, low power consumption, high-speed performance, and operating life. Generally, they can be built based on the *p–i–n* diodes, silicon MOSFET, or GaAs MESFET transistors and can be divided into two basic parts: amplitude and phase control circuits. The control circuits are necessary to protect high power devices from excessive peak voltage or dc current conditions. They also used as switching elements for directing signal between different transmitting paths, as variable gain amplifiers to stabilize transmitter output power, as attenuators and phase shifters to changing the amplitude and phase of the transmitting signal paths in array systems, or as limiters to protect power-sensitive components.

13.1 POWER DETECTOR AND *VSWR* PROTECTION

In wireless communication systems based on GSM/EDGE, CDMA2000, WCDMA, or OFDM standards with nonconstant envelope signals, it is required that the power amplifier could operate with high power, high efficiency, high linearity, and high spectral purity simultaneously under significant load variation where the values of the load voltage standing wave ratio (*VSWR*) may be very high. For conventional power amplifiers, the performance degrades significantly with high *VSWR* when the entire active device sees the high impedance conditions corresponding to the device operation in a highly nonlinear saturation mode. At the same time, when the entire device sees the low impedance conditions corresponding to the device operation in an active linear mode that is far from saturation, efficiency degrades significantly. In addition, reactive loads can lead to high second-breakdown stresses and increased dissipation in the transistors, both of which lead to reduced reliability [1]. Protection against low-impedance loads can be achieved by placing a current-limiting regulator in the dc feed to the final power amplifier stage. However, protection against load variations in general requires more complicated circuitry.

For a load impedance $Z_L = R_L + jX_L$, the *VSWR* will vary according to

$$VSWR = \frac{1 + |\Gamma|}{1 - |\Gamma|} \tag{13.1}$$

where the magnitude of the reflection coefficient Γ is written as

$$|\Gamma| = \sqrt{\frac{(R_L - Z_0)^2 + X_L^2}{(R_L + Z_0)^2 + X_L^2}} \tag{13.2}$$

and Z_0 is nominal load resistance. In this case, the condition $VSWR = 1$ corresponds to a perfectly matched condition, while $VSWR = \infty$ corresponds to a complete mismatch. The same value of

RF and Microwave Transmitter Design, First Edition. Andrei Grebennikov.

VSWR can arise from numerous combinations of resistance and reactance, with the locus of all points corresponding to a given value of *VSWR* appeared on the Smith chart as a circle whose center is at 50 Ω (or the center of the chart) and which passes through impedances 50 Ω divided by *VSWR* and 50 Ω multiplied by *VSWR*.

Figure 13.1(*a*) shows the *VSWR* protection approach based on an active feedback circuit that is placed between the collector and base terminals of the power transistor, and the large collector voltage swing is directly and drastically fed back to the base terminal [2]. Under an oversupply and mismatch load condition, the maximum collector voltage of the power transistor grossly increases that results in an appropriate increase of the overall collector–emitter voltage across the active feedback circuit. When this voltage exceeds a turn-on voltage of the active feedback circuit, large current flows through this protection circuit and the collector voltage of the power transistor is prevented from reaching the breakdown limit. In this case, such an active feedback circuit operating as a voltage limiter can

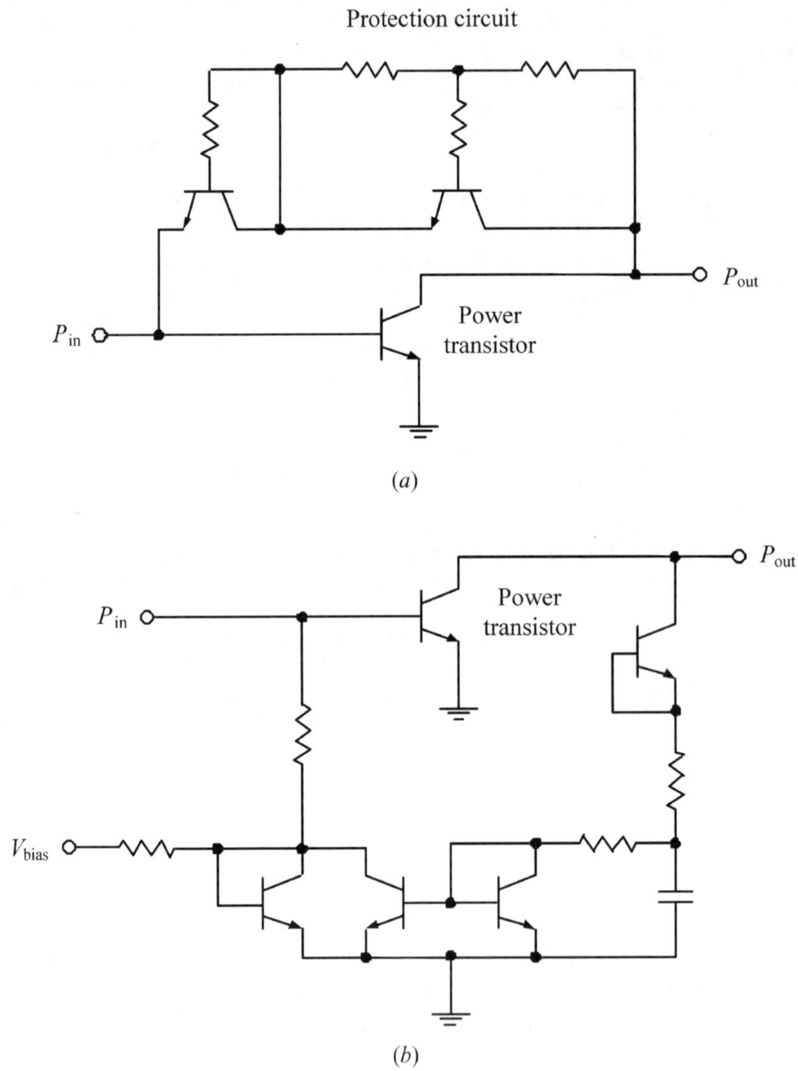

FIGURE 13.1 Schematics of *VSWR* protection feedback circuits.

protect the output stage from an excessive collector voltage swing for a mismatched load with *VSWR* up to 10:1 without seriously affecting the circuit performance for a matched 50-Ω load. In the other feedback approach, circuit schematic of which is shown in Figure 13.1(*b*), the *VSWR* protection is based on a current-mirror circuit in feedback path when less current flows to the base terminal of the power transistor for higher collector voltage approaching the breakdown voltage, thus reducing the base bias voltage of the power transistor [3].

Control of the output power that is usually needed in wireless communication system to optimize the propagation link budget can also be performed by a closed-loop approach. In this case, as shown in Figure 13.2(*a*), the radio frequency (RF) power is sensed at the amplifier output using a directional coupler and is detected by a diode [4,5]. The resulting signal is compared to a reference voltage through an error amplifier whose output controls the power amplifier bias voltage. The loop forces the sensed output voltage and the reference voltage to be equal. Therefore, power control can be accomplished by changing the reference voltage. However, in this case the dynamic range is limited to that of the detector diode, usually about 20 dB without compensation, and the loop gain can vary considerably over the dynamic range causing stability problems because of the nonlinear transfer characteristic of the power amplifier. An alternative approach, particularly applicable to amplifiers utilizing square law devices such as MOSFETs, uses supply voltage to control the output power of the amplifier.

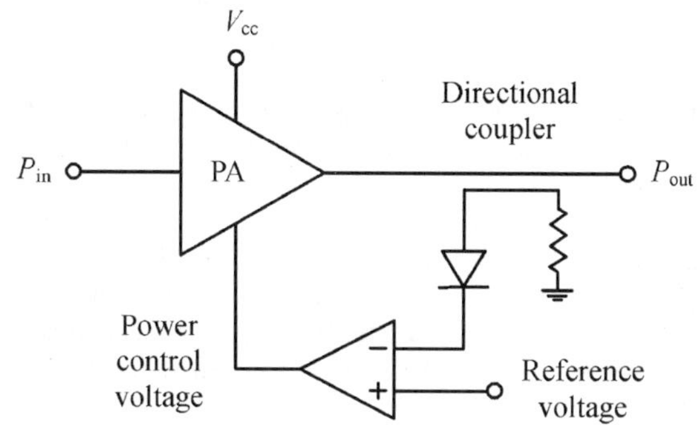

(*a*)

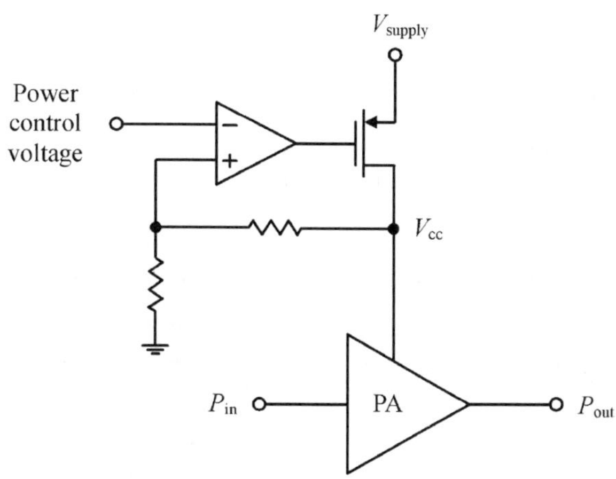

(*b*)

FIGURE 13.2 Block diagrams of power control circuits.

By using a simple linear regulator that is implemented through a power pMOS transistor and an operational amplifier, as shown in Figure 13.2(*b*), the collector (or drain) voltage can be varied linearly over wide range of supply voltages. Reducing the collector voltage effectively limits the output voltage swing and, hence, limits the output power with very fast response time [6]. In this case, output power is predictable after a simple calibration and burst shaping with associate switching transients can be easily controlled. However, a very low saturation resistance is needed for the pMOS transistor to preserve the overall efficiency under maximum output power levels. The transistor protection circuit can be generally based on an adaptive output power control that includes sensing of both dc current and output voltage swing under severe load mismatched conditions [7].

Two topologies for capacitively-coupled power detectors are shown in Figure 13.3 [8]. In Figure 13.3(*a*), the value of the coupled capacitor is usually adjusted so that the sampled power is 15–20 dB below the output power, while the value of the shunt resistor is selected to obtain the largest possible detected voltage and bandwidth. The capacitive-type sampling has a larger bandwidth potential than resistive or inductive. When the shunt resistor is replaced by a diode, as shown in Figure 13.3(*b*), the input power can be significantly reduced to obtain the same detected voltage.

For a nonconstant envelope RF signal, an envelope detector can be used that takes a high-frequency signal as input, and provides an output that ideally should be an envelope of the original signal. The shunt capacitor in the circuit stores up charge on the rising edge, and releases it slowly through the resistor when the signal falls. The diode in series rectifies the incoming signal, allowing current flow only when the positive input terminal is at a higher potential than the negative input terminal. Filtering is then used to smooth the final result. The simple low-pass filtering is often not sufficient since some ripple is likely to remain on the detected envelope. More complicated filter circuit gives a smoother result, but decreases the responsiveness of the design, so practical solutions are usually a compromise. The simplest form of envelope detector is the diode that is located between the input and output of a circuit, connected to a resistor and capacitor in parallel from the output of the circuit to the ground. If the resistor and capacitor are correctly chosen, the output of this circuit should approximate a voltage-shifted version of the original (baseband) signal.

The systems described that sense voltage or current associated with the power output only account for the amplitude of the output voltage or current. However, for accurate power detection, it needs to account for both amplitude and phase. The system that senses incident power only may fail to accurately sense the transmitting or true output power because it does not account for any reflected power if there is an impedance mismatch at the load. In addition, the control of reflected power is required to exactly measure the load *VSWR* because the reflection coefficient Γ is a ratio between the reflected (V_r) and incident (V_i) voltages. Figure 13.4 shows the power amplifier architecture with *VSWR* detection and protection mechanism based on the transmission-line directional couplers and variable-gain amplifier (VGA). To protect from high *VSWR* condition, the power amplifier can also be powered down into a safe operating mode using bias or dc supply current control of the power transistor [9,10].

When the output signal traveling along the transmission line encounters a load with impedance Z_L equal to the transmission-line characteristic impedance Z_0, all of the available signal power is

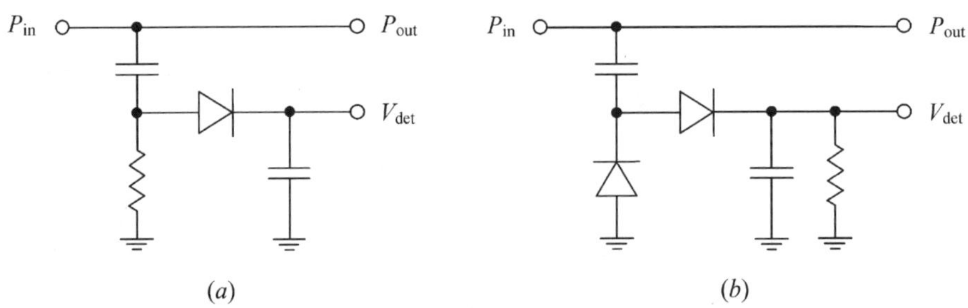

FIGURE 13.3 Schematics of capacitively-coupled voltage detectors.

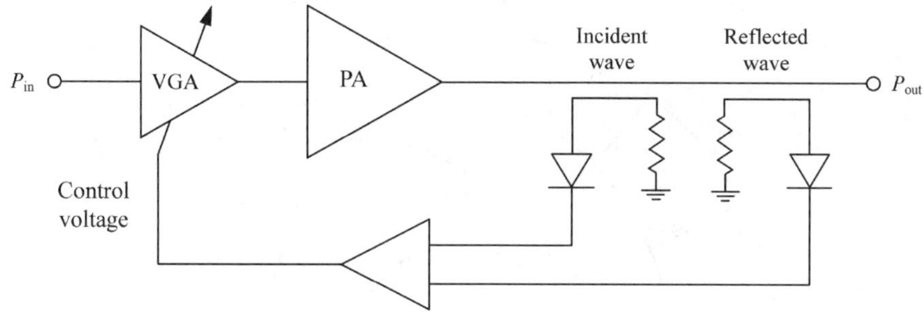

FIGURE 13.4 Block diagram of *VSWR*-protected power amplifier circuit with power control.

delivered to the load. However, any mismatch, when $Z_L < Z_0$ or $Z_L > Z_0$, causes reflected currents and voltages along the line, creating a standing wave. In this case, the *VSWR* represents a ratio of maximum (V_{max}) and minimum (V_{min}) voltage magnitudes measured along the transmission line. To accurately calculate the reflection coefficient, the directivity D of the directional coupler should be taken into account [10]. Once the incident and reflected signals have been sampled and isolated, the magnitudes of these signals need to be detected, which require dual detectors. Accuracy of the detection method will determine the accuracy of *VSWR* that usually requires detectors with high dynamic range.

Implementation of the 45° delay lines into each parallel path of the power amplifier and using the in-phase Wilkinson combiners in pairs and a hybrid quadrature combiner in the output combining circuit, as shown in Figure 13.5, can improve the overall power amplifier characteristics that become more insensitive to variations of the load *VSWR* when the *ACLR* (adjacent channel leakage power ratio) reduces by more than 12 dB and efficiency increases by more than 10% for *VSWR* = 3 [11]. The basic idea is to spread the different impedances seen by the device outputs when phase delay of the reflected signals varies between 0 and 180° with a step of 45°, thus creating different impedances along the corresponding load *VSWR* circle on the Smith chart.

In this case, each from four devices sees different load impedance, as seen from the Smith chart (normalized to 50 Ω) shown in Figure 13.6, where the load variation circle is given for *VSWR* = 2, as an example. As a result, the only one device from entire four sees the high impedance of 100 Ω, while the impedances seen by the other three devices are spread along the Smith chart circle, with real parts being in between 100 and 25 Ω. This means that the output power is reduced only for one device and linearity is disturbed significantly only by one device, unlike a conventional parallel-operation power amplifier where it is distorted by all four devices. In a practical implementation, for a 3.5-V 29-dBm GaAs MESFET power amplifier designed to operate in a 900-MHz digital cellular phone system with $\pi/4$-DQPSK modulation, the *ACLR* below −45 dBc with an efficiency of over 45% can be obtained for a load *VSWR* ≤ 3.

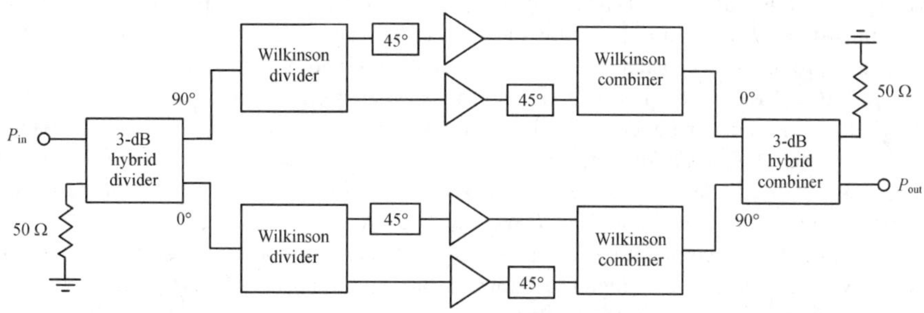

FIGURE 13.5 Balanced power amplifier topology with 45° delay lines.

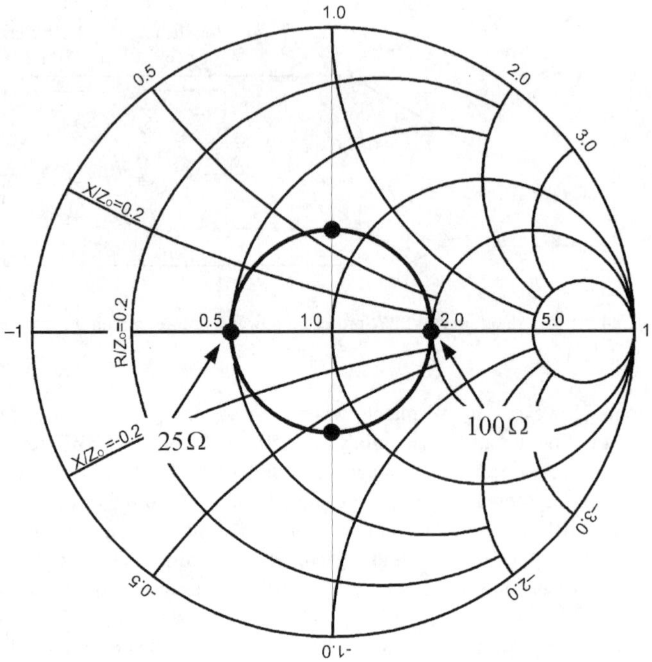

FIGURE 13.6 Smith-chart impedances for $VSWR = 2$.

13.2 SWITCHES

Semiconductor diodes have been extensively used for many years as RF and microwave switching elements for directing signal or power flow between different ports [12,13]. In these applications, the RF impedance may be switched from one value (usually small) in a forward-biased or conducting state to another (usually large) in a reverse-biased or nonconducting state. The impedance changes are provided by changing the dc bias from a high forward current to a small back current that is applied through leads, which are isolated from the transmitting signal by RF chokes and bypass capacitors. A large forward bias will decrease the diode resistance and results in high effective impedance that will absorb little power. On the other hand, a large reverse bias will result in low impedance that both absorbs power and causes high reflection.

Generally, diodes are inherently nonlinear and some care must be taken in establishing the limits on current, voltage, and dissipation to prevent significant harmonic generation and diode damage at high power levels. The diode operating performance can be described by a simple equivalent circuit that consists of the series inductance L, which is usually assumed to be constant, the junction capacitance C_j, which is variable and represents a derivative of the stored charge with respect to voltage, the series spreading resistance R_s, which is a function of the conductivity of the semiconductor and the contact area, being approximately constant for reverse-bias voltages, and the junction nonlinear differential resistance R_j that varies with amplitude and polarity of the applied dc bias. In a reverse-bias mode when switch is closed, it can be represented by a series combination of the inductance L, capacitance C_j, and reverse series R_s. In a forward-bias mode when switch is open, the junction resistance R_j is small, thereby shunting the junction capacitance C_j, and the switch can be equivalently represented by a series combination of the inductance L and resistance R_s.

The *p–i–n* diodes are commonly used as switching elements to control RF or microwave signals because they are characterized by high breakdown voltages, are capable of high-speed operation, and can be easily integrated with planar technology. A *p–i–n* diode can be used in either a series or a shunt configuration to form a single-pole single-through (SPST) RF switch. In the series configuration shown in Figure 13.7(*a*), the switch is turned on when the diode is forward biased, while in the shunt configuration shown in Figure 13.7(*b*) the switch is turned on when the diode is reverse biased. In both cases, input power is reflected when the switch is turned off. The dc block capacitors should have very high values at the RF operating frequency, while the RF choke inductors should have very high RF impedance. In microwave circuit designs, the high impedance quarter-wavelength lines can replace the RF chokes in order to provide RF blocking. The shunt SPST switch offers higher isolation over a wide frequency range, approximately 20 dB for a singled diode switch, and can result in a design capable of handling more power since it is easier to connect the diode to heatsink. However, the series SPST switch is commonly used when minimum insertion loss is required over a wide frequency range. In series connected switches, maximum isolation depends primarily on the junction capacitance of the *p–i–n* diode, while the insertion loss and power dissipation are functions of the diode series resistance.

To improve isolation between input and output ports, the switch design can employ combinations of series and shunt diodes (compound switches). The two most common compound SPST switch configurations represent a series-shunt circuit shown in Figure 13.7(*c*) and a *T*-type circuit shown in Figure 13.7(*d*). In the insertion loss state for a compound switch, the series diode is forward biased and the shunt diode is at zero or reverse bias. These circuits offer improved overall performance but the added circuit complexity degrades the *VSWR* and insertion loss. Since all diodes are not simultaneously biased in one state or the other, there is an increase in complexity of the bias circuitry. A summary of design equations used for calculating insertion loss and isolation for simple and compound SPST switches is given in Table 13.1, where $X_C = 1/\omega C_j$ and Z_0 is the input and output characteristic impedance [14].

When it is necessary to switch multiple ports in the system, multithrough switches can be used, the simplest example of which is a single-pole double-through (SPDT) switch, in which the signal from the input port can be delivered to either of two output ports. If the SPDT switch is symmetrical,

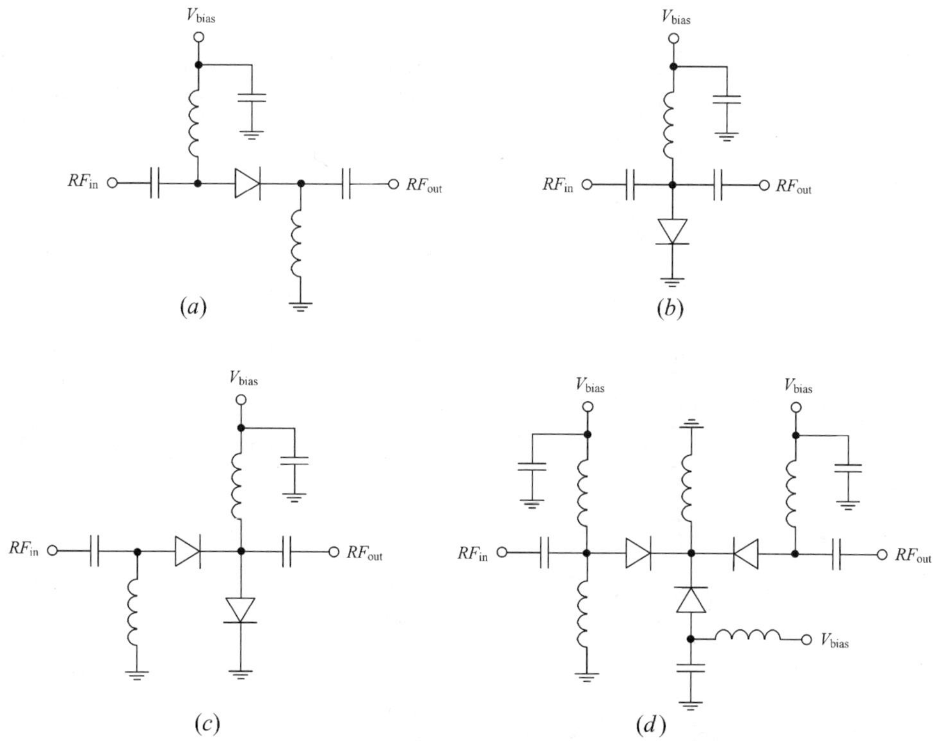

FIGURE 13.7 Basic schematics of SPST *p–i–n* diode switches.

TABLE 13.1 Summary of Design Equations for SPST Switches.

Type	Isolation, dB	Insertion Loss, dB
Series	$10\log_{10}\left[1+\left(\frac{X_c}{2Z_0}\right)^2\right]$	$20\log_{10}\left(1+\frac{R_c}{2Z_0}\right)$
Shunt	$20\log_{10}\left(1+\frac{Z_0}{2R_s}\right)$	$10\log_{10}\left[1+\left(\frac{Z_0}{2X_c}\right)^2\right]$
Series-shunt	$10\log_{10}\left[\left(1+\frac{Z_0}{2R_s}\right)^2+\left(\frac{X_c}{2Z_0}\right)^2\left(1+\frac{Z_0}{R_s}\right)^2\right]$	$10\log_{10}\left[\left(1+\frac{R_s}{2Z_0}\right)^2+\left(\frac{Z_0+R_s}{2X_c}\right)^2\right]$
T-type	$10\log_{10}\left[1+\left(\frac{X_c}{Z_0}\right)^2\right]+10\log_{10}\left[\left(1+\frac{Z_0}{2R_s}\right)^2+\left(\frac{X_c}{2R_s}\right)^2\right]$	$20\log_{10}\left(1+\frac{R_s}{2Z_0}\right)+10\log_{10}\left[1+\left(\frac{Z_0+R_s}{2X_c}\right)^2\right]$

each switch branch performs like the SPST equivalent, but the isolation of a double-through switch is increased by 6 dB. This effect occurs because the branch with turned-off diode is shunted by the branch with turned-on diode and its 50-Ω termination, causing the RF voltage across the turned-off diode to be 50% less than would be the case for the equivalent SPST switch.

Figure 13.8 shows the three basic circuit configurations for the SPDT switches [15,16]. The circuit schematic of a shunt SPDT *p–i–n* diode switch is shown in Figure 13.8(*a*), where the

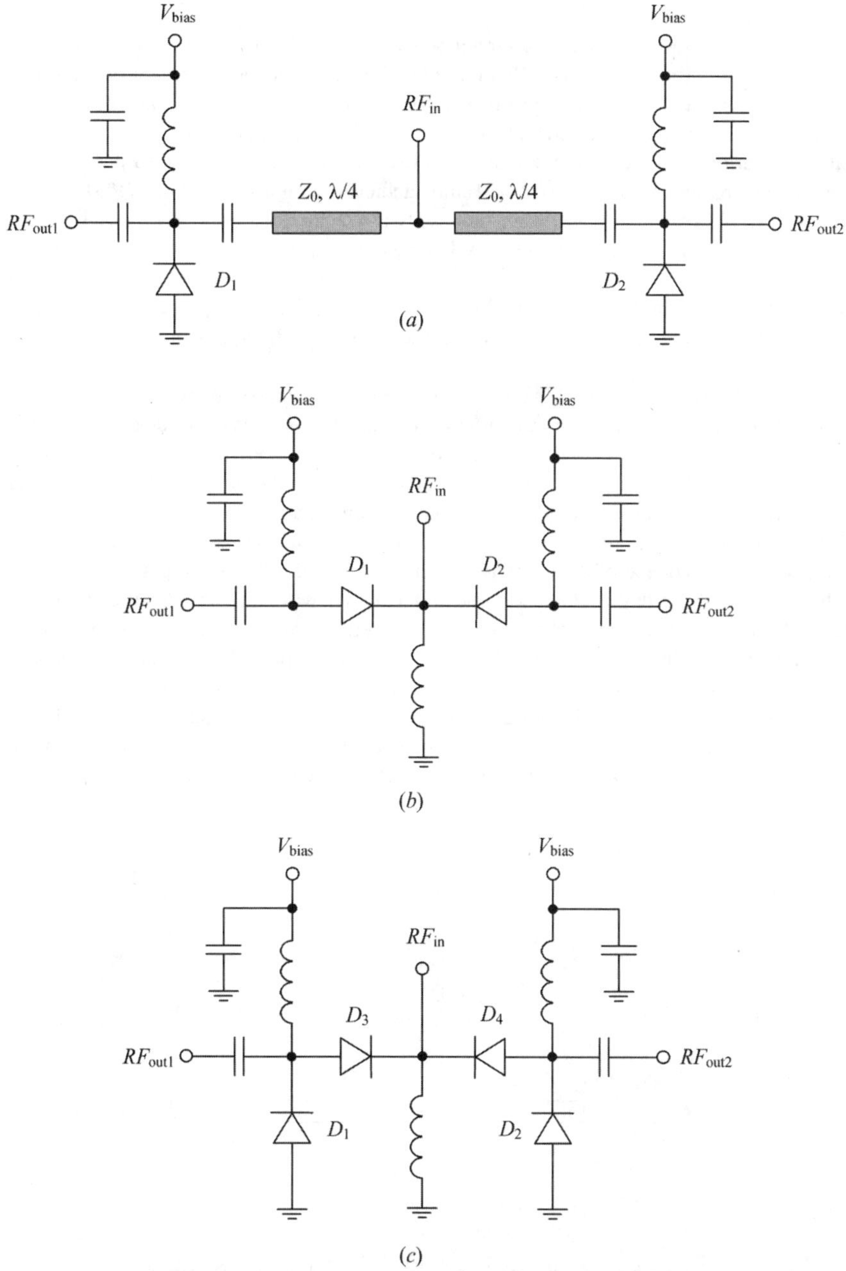

FIGURE 13.8 Basic schematics of SPDT *p–i–n* diode switches.

quarter-wavelength transmission lines are used between common junction (input port) and each shunt diode. When diode D_1 is forward biased and diode D_2 is zero or reverse biased, the RF power will flow from the input port RF_{in} to the output port RF_{out2}, and the output port RF_{out1} will be isolated. In this case, the quarterwave line will transform the short circuit at diode D_1 into an open circuit at the common junction, thus eliminating any reactive loading of the closed port and improving isolation between the output ports. However, as the frequency is changed from the center bandwidth frequency f_0, the transmission lines will change in electrical length, creating a mismatch at the common junction. For example, the *VSWR* of this structure will be 1.43:1 when the frequency ratio f/f_0 or f_0/f is equal to 1.2 (40% bandwidth).

The series SPDT diode switch, the circuit schematic of which is shown in Figure 13.8(*b*), is capable of very large (multioctave) bandwidth, limited only by the parameters of the bias inductors, blocking capacitors, and length of any transmission line between the diode and common junction. This structure can be easily fabricated with beam-lead diodes on alumina substrate. However, parasitic capacitances can result in poor isolation at microwave frequencies, with a 6 dB per octave roll-off as a function of frequency. Better isolation can be achieved with a series-shunt SPDT diode switch shown in Figure 13.8(*c*). In this case, when diodes D_2 and D_3 are forward biased into a low-resistance state, while diodes D_1 and D_4 are reverse biased into a high-resistance state, RF power flows from the input port RF_{in} to the output port RF_{out1}. Diode D_4 acts as an open circuit to isolate the short circuit at diode D_2 from the common junction. However, it is complicated to properly implement such a switch in practice because it is difficult to mount the D_3 and D_4 diode junctions electrically close to the common point.

The bandwidth of the shunt SPDT diode switch can be improved by the simple impedance matching technique. In this case, a third quarterwave transmission line is placed between the common junction and input port RF_{in}, as shown in Figure 13.9 [16]. The specific value of the transmission-line characteristic impedances will determine the *VSWR* and bandwidth of the switch. For example, setting the characteristic impedance Z_0 of all three transmission lines to 35 Ω will result in a 1.43:1 *VSWR* frequency bandwidth of 100% (3:1). Additional bandwidth improvement can be achieved if the characteristic impedance of the quarterwave line connected to the input port RF_{in} is selected to be slightly different from that for the other lines. The quarter-wavelength transmission lines can also be used as RF chokes in the bias network of each *p–i–n* diode arm [17]. Higher isolation over wide range at microwave frequencies can be achieved by using additional shunt diodes and transmission-line sections [18]. The ladder lumped resonance circuits in parallel to the *p–i–n* diodes can be used to simplify the SPDT diode switch configurations, minimize their size, and reduce cost [19].

Different types of SPDT diode switches can be used to separately connect the transmitting and receiving paths to antenna in radio communication system. When *p–i–n* diodes are used as elements in

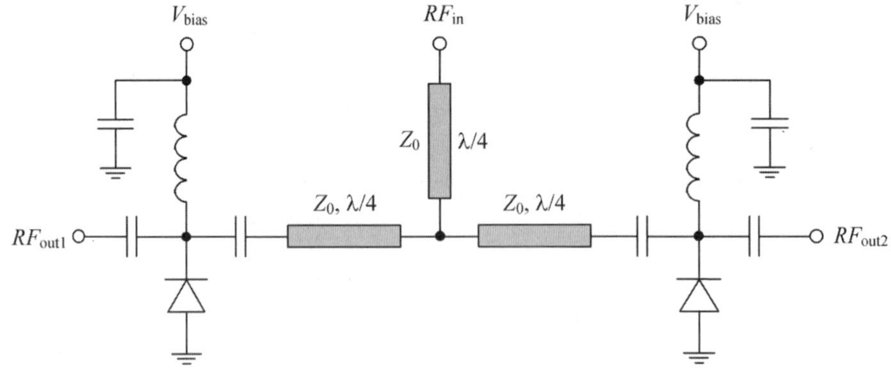

FIGURE 13.9 Basic schematic of broadband shunt SPDT switch.

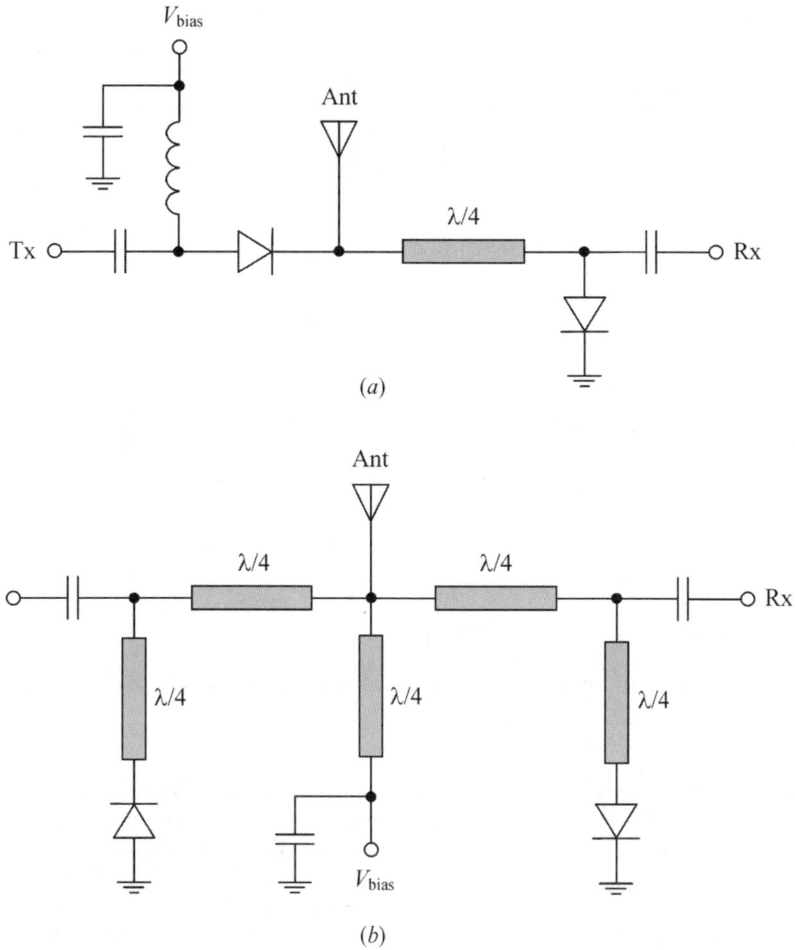

FIGURE 13.10 Basic schematics of T/R SPDT switches.

these switches, they offer higher reliability, better mechanical ruggedness, and faster switching speed than their electro-mechanical counterparts. Figure 13.10(*a*) shows the typical transmit-receive (T/R) series-shunt SPDT diode switch configuration with a quarterwave transmission line. In this case, when both diodes are forward biased, the transmitter is connected to the antenna, and the receiver is protected by the low series resistance R_s of the shunt diode, terminating the quarter-wavelength line and creating high impedance at its opposite end connected to antenna. For $R_s = 1\ \Omega$ or less, the isolation greater than 30 dB and insertion loss less than 0.3 dB can be expected over a 10% bandwidth. However, when both diodes are reverse biased, the transmitter port is isolated by high reactance of the series diode, and low junction capacitance of the shunt diode contributes to a low direct insertion loss between the antenna and the receiver. The biasing scheme is very simple, requiring only one RF choke and two dc blocking capacitors. The circuit schematic of a shunt SPDT diode T/R switch convenient to implement at microwave frequencies is shown in Figure 13.10(*b*), where one diode is forward biased when the other diode is reverse biased, with alternate switching of the transmitting and receiving paths.

In some cases, especially for integrated circuit (IC) design when low dc power consumption is required, the GaAs FET switches can be used to control power flow between the input and output

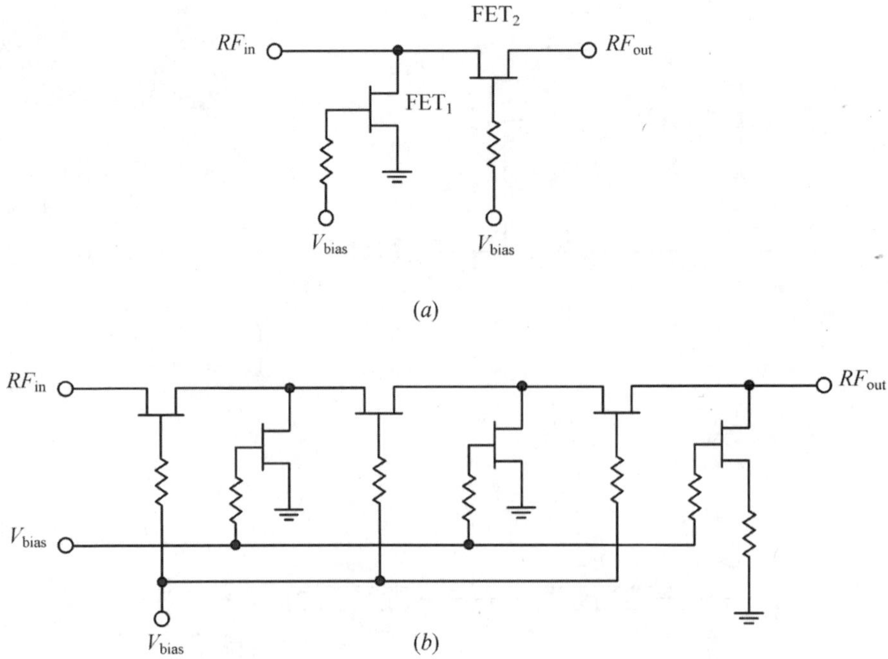

FIGURE 13.11 Basic configurations of GaAs MESFET switches.

ports. The simplest SPST configuration of such a GaAs MESFET switch is shown in Figure 13.11(*a*) where a sufficiently low insertion loss during on-state and high isolation during off-state can be achieved [20]. In on-state, when the FET_1 is open-circuited and the FET_2 is short-circuited, an RF signal is transferred from the input port RF_{in} to the output port RF_{out}. In off-state, when the FET_1 is short-circuited and the FET_2 is open-circuited, an RF signal can be isolated from the output port RF_{out}. Higher isolation can be achieved by cascading the GaAs MESFET SPST switches that can result in the isolation of 60 dB and insertion loss of 1.6 dB at a frequency of 1.9 GHz for a three-stage switch IC circuit shown in Figure 13.11(*b*), with the total chip size of $0.6 \times 0.9\ \text{mm}^2$ [21]. Very high switching speed of the GaAs FET switches can be provided using HEMT technology [22]. Potential benefit of using CMOS switches in wireless communication systems rather than GaAs FET switches are the fact that CMOS switches do not require a negative control voltage. Besides, if they can be implemented in a standard digital CMOS process, then these switches can be integrated with the other RF blocks. In this case, a series-shunt CMOS SPDT T/R switch can provide over 40-dB isolation with less than 0.8-dB insertion loss for frequencies up to 1 GHz and higher than 21-dB isolation with less than 2-dB insertion loss up to 20 GHz and transmitting power level up to 30 dBm [23,24].

13.3 PHASE SHIFTERS

RF and microwave phase shifters play an important role in many wireless applications. One of them is the phased-array radars where changing the phase of each array element allows the main antenna beam to be steered. Generally, the basic elements of the phase shifters are the *p–i–n* diodes, GaAs FET devices, transmission-line segments, and microwave hybrids. The final choice of a phase-shifter circuitry depends on the required frequency bandwidth, insertion loss, switching speed, phase resolution, or power handling capability.

13.3.1 Diode Phase Shifters

Several types of the phase shifters, such as reflection-type, switched line, low-pass/high-pass, or loaded-line can be constructed with *p–i–n* diode switching or varactor tuning elements. Compared to the ferrite phase shifters, diode phase shifters are characterized by the substantially smaller size, integration capability with planar circuitry, and high speed. Figure 13.12(*a*) shows a typical circuit of the *reflection-type phase shifter* consisting of a circulator with diode controllable reflective phase termination [25,26]. In this case, a controllable time delay of arbitrary length is achieved if the diode is assumed to switch between short and open impedances. Generally, such a reflection-type phase shifter can be composed of a shunt diode with a short-circuited transmission line, a series diode with an open-circuited transmission line, or a lumped circuit including diode parasitics terminating the transmission line. A typical nonperfect diode switch will be inductive for conduction and capacitive for reverse bias. However, proper selection of diode parasitics can compensate for the errors causing by these reactances. To provide a wideband phase shifting, the diode is connected to the circulator or hybrid without the transmission line with fixed electrical length. For a perfect switch with 0 Ω when diode is closed and ∞ Ω when diode is open, these impedances result in a maximum phase shift $\Delta\phi = 2\pi(\Delta l/\lambda) = \pi$ or 180° with $\Delta l = \lambda/2$. The frequency bandwidth for reflection-type phase shifter is limited by the bandwidth of a circulator or 3-dB hybrid.

Figure 13.12(*b*) shows the reflection-type phase shifter where separating the incident and reflected signals is realized by a quadrature hybrid, such as the 3-dB branch-line or directional coupler. In operation, the input signal is divided into two quadrature components with equal amplitudes among

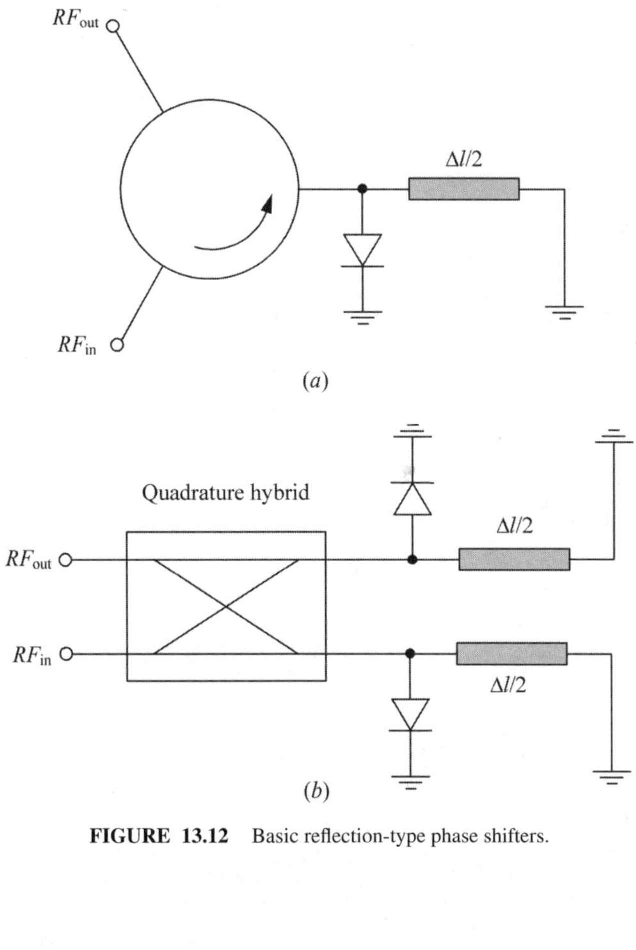

FIGURE 13.12 Basic reflection-type phase shifters.

the right-hand ports of the hybrid. The diodes are both biased in the same state, either forward or reverse, so the signals reflected from the identical terminations will add in phase at the indicated output port. Turning the diodes on or off changes the total path length for both reflected waves by Δl, thus produces a phase shift of $\Delta\phi$ at the output. There are infinite numbers of choices of line lengths that give the desired $\Delta\phi$, but the bandwidth is optimized if the reflection coefficients for the two diode states are phase conjugates.

A continuous variation of phase shift can be realized by using varactor diodes whose capacitance value depends on the reverse bias voltage [27]. Ideally, the magnitude of the diode reflection coefficient is fixed at unity, and the phase varies from zero (zero capacitance) to -180° (maximum capacitance). As a result, the phase of the output signal is varied continuously by changing the bias voltage on the varactor diodes. However, in practice, the phase-shift range is limited by the capacitance ratio of the varactor diodes. For example, a full 360° phase shift within the frequency range from 16 to 18 GHz was realized by cascading three phase-shifter chips, each of which was based on a microstrip 90° branch-line hybrid coupler and two abrupt varactor diodes, as shown in Figure 13.13(*a*) [28]. The phase shifter with hyperabrupt varactors can provide a larger phase shift with better linearity as a function of bias voltage, and it is feasible to realize a 360° phase shift using two such chips with approximately 3-dB insertion loss. In this case, a very linear 200° phase-shift response can be achieved within the limited bias voltage range [29]. The reflection-type phase shifter using conventional hybrid ring requires a 90° line section added to one of its reflected ports, as shown in Figure 13.13(*c*) [30]. However, an additional quarterwave line can be eliminated by using an asymmetric hybrid ring that generally allows it to be terminated by arbitrary impedances [31]. Both branch-line and ring hybrids are characterized by the frequency bandwidth limited to about 10%. The substantially broader frequency bandwidth (up to octave) can be achieved with a coupled-line hybrid, the circuit diagram of

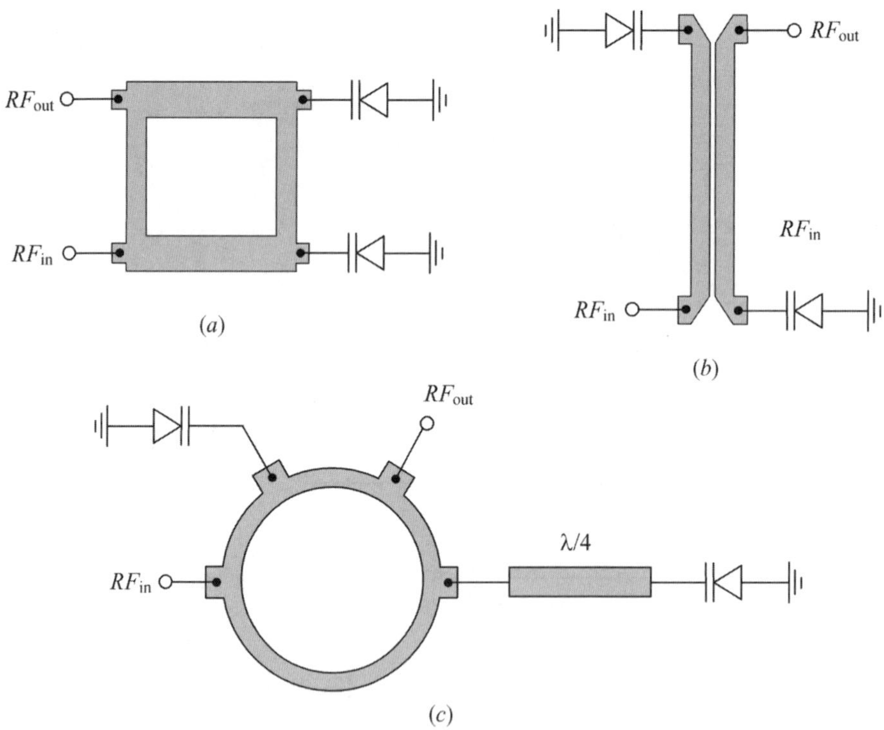

FIGURE 13.13 Hybrid reflection-type phase shifters.

which is shown in Figure 13.13(*b*). For the two cascaded 3-dB multilayer folded directional couplers with abrupt varactors, the phase shift up to 98° was measured over two octaves, from 4.4 to 16.1 GHz [32].

Hyperabrupt varactor diodes have a nonlinear doping profile that facilitates a larger capacitance as the reverse bias voltage is increased and a larger tuning ratio occurs. Therefore, these diodes have better capacitance-voltage linearity and tunability. However, the maximum linear power from a single varactor is limited due to the second- and third-order nonlinear distortions. In this case, to eliminate the second-order and other even-order distortions, two varactor diodes can be connected in an antiseries configuration. As a result, the branch-line reflection-type phase shifter with parallel resonant circuits, each of which is comprised of an antiseries varactor pair and an inductor and connected to the corresponding reflected port of the branch-line hybrid, as shown in Figure 13.14(*a*), significantly improves the maximum linear output power and provides a continuous phase shift of 180° at an operating frequency of 1.88 GHz [33]. For a given varactor or varactor pair with a limited capacitance tuning ratio, the phase-shift range can be increased by using an impedance-transforming branch-line coupler that also provides an impedance transformation between the varactor loads and the input and output ports, thus making the entire circuit more compact [34]. A full 360° phase shift with a silicon varactor of a 1.4-to-8-pF capacitance ratio can be obtained at an operating frequency of 2 GHz when each reflection load for an impedance-transforming branch-line coupler consists of two varactors interconnected with a quarter-wavelength transmission line, as shown in Figure 13.14(*b*) [35].

The *loaded-line phase shifter* is very useful for obtaining small phase shifts up to approximately 45° due to its simplicity and low insertion loss [25,36]. The circuit consists of two equal two-state switchable admittances Y_i (i = 1, 2) connected in shunt with a transmission-line section, and the transition between two admittance states produces a certain change $\Delta\phi$ in the phase of the transmission

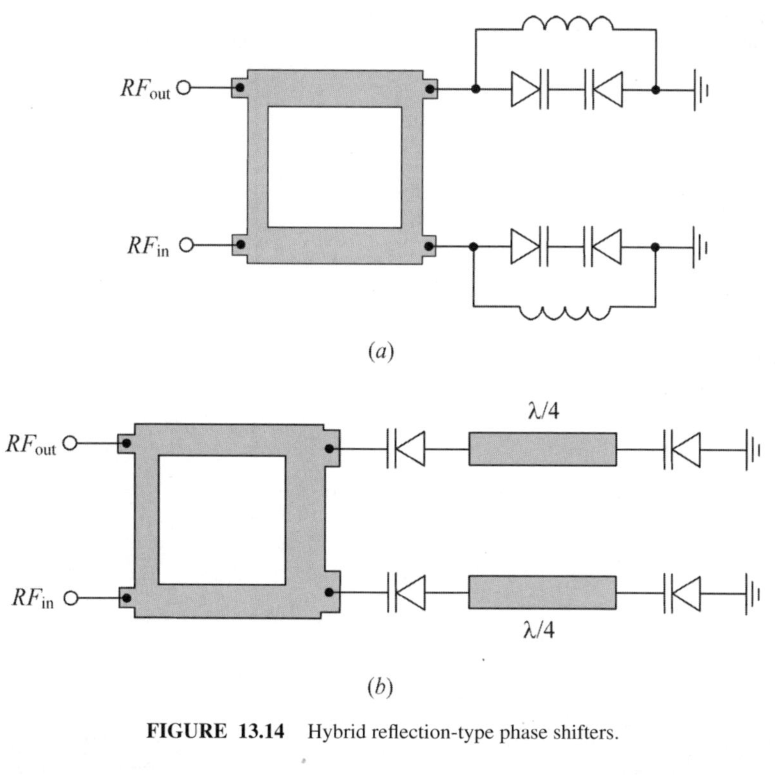

FIGURE 13.14 Hybrid reflection-type phase shifters.

scattering coefficient S_{21}. The scattering *S*-matrix can directly be obtained by transformation from the *ABCD*-matrix that can be derived in the idealized lossless case as

$$[ABCD] = \begin{bmatrix} 1 & 0 \\ jB_i & 1 \end{bmatrix} \begin{bmatrix} \cos\theta & jZ_T \sin\theta \\ j\dfrac{\sin\theta}{Z_T} & \cos\theta \end{bmatrix} \begin{bmatrix} 1 & 0 \\ jB_i & 1 \end{bmatrix}$$

$$= \begin{bmatrix} \cos\theta - B_i Z_T \sin\theta & jZ_T \sin\theta \\ \dfrac{j}{Z_T}\left[2B_i Z_T \cos\theta + \left(1 - B_i^2 Z_T^2\right)\sin\theta\right] & \cos\theta - B_i Z_T \sin\theta \end{bmatrix} \tag{13.3}$$

where $B_i = \mathrm{Im}Y_i$ is the shunt susceptance, Z_T is the characteristic impedance, and θ is the electrical length of the transmission-line section [26,37]. The transformation to the scattering matrix for a reciprocal, symmetrical two-port network is given by

$$S_{11} = S_{22} = \frac{BY_0 - CZ_0}{2A + BY_0 + CZ_0} \tag{13.4}$$

$$S_{21} = S_{12} = \frac{2}{2A + BY_0 + CZ_0} \tag{13.5}$$

where $Z_0 = 1/Y_0$ is the characteristic impedance of the circuit into which the phase shifter is connected.

The required phase shift $\Delta\phi$ corresponds to the change in the argument ϕ_i of $S_{21}(B_i)$ when the shunt elements make the transition from reactive admittances jB_1 to jB_2 by switching the *p–i–n* diodes, as shown in Figure 13.15(*a*) with lumped inductors and capacitors. In this case, the elimination of reflected waves from the input of the phase shifter by input matching reduces the return loss and removes possibility of phase errors that can be caused by interaction of the reflected waves with adjacent cascaded stages. The condition of the input match is specified by requiring $S_{11} = 0$ in Eq. (13.4). As a result,

$$BY_0 = CZ_0 \tag{13.6}$$

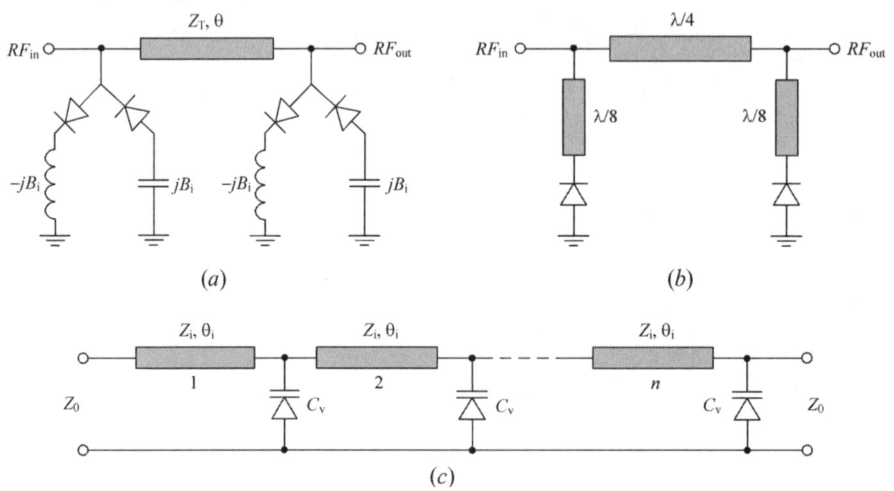

FIGURE 13.15 Basic circuits of loaded-line phase shifters.

and the transmission scattering element S_{21} becomes

$$S_{21} = \frac{1}{A + BY_0} = \frac{1}{(\cos\theta - B_i Z_T \sin\theta) + jZ_T Y_0 \sin\theta}. \tag{13.7}$$

Since the magnitude of S_{21} is equal to unity in the input-matched lossless case, from Eq. (13.7) it follows that

$$\cos\phi = \cos\theta - B_i Z_T \sin\theta \tag{13.8}$$

$$\sin\phi = -Z_T Y_0 \sin\theta \tag{13.9}$$

where ϕ is the phase angle of S_{21}. It can be seen from Eq. (13.9) that $\sin\phi$ remains constant during load switching, while $\cos\phi$ in Eq. (13.8) assumes two values with two loading admittance states. Consequently, the phase shift $\Delta\phi$ due to the switching between the susceptance values B_1 and B_2 can be written from Eq. (13.8) as

$$\Delta\phi = \cos^{-1}(\cos\theta - B_1 Z_T \sin\theta) - \cos^{-1}(\cos\theta - B_2 Z_T \sin\theta). \tag{13.10}$$

The spacing between diodes of $\lambda/4$ when the electrical length of the transmission line is equal to 90° provides the widest frequency bandwidth and is mostly used in practice. However, a phase shift greater than 45° is difficult to achieve without causing excessively high *VSWR* [26]. In this case, the phase is switched symmetrically about 90° by increments of $\pm\Delta\phi/2$ by the line loading. Substituting $\phi = (90^\circ \pm \Delta\phi/2)$ into Eqs. (13.8) and (13.9) results in

$$Z_T = Z_0 \cos(\Delta\phi/2) \tag{13.11}$$

$$B_i Z_0 = \pm\tan(\Delta\phi/2) \tag{13.12}$$

representing the design equations for the lossless loaded-line phase shifter with $\Delta\phi$. Hence, the phase shift $\Delta\phi$ can be properly set by switching between two susceptances B_1 and B_2 as

$$\Delta\phi = 2\tan^{-1}\frac{B_1 - B_2}{2Y_0}. \tag{13.13}$$

From Eq. (13.8), it follows that $d\phi/dB_i$ is positive for $\theta \leq 90^\circ$. Therefore, the addition of capacitive susceptance loading increases the phase angle ϕ. For complex-conjugate loading when $B_1 = -B_2$,

$$\Delta\phi = 2\tan^{-1}\frac{B_1}{Y_0} = 2\tan^{-1}\left(-\frac{B_2}{Y_0}\right) \tag{13.14}$$

where $B_1 = \omega C$ and $B_2 = -1/\omega L$ corresponding to Figure 13.15(*a*) or $B_1 = Y_s$ and $B_2 = -Y_s$, where $Z_s = 1/Y_s$ is the stub characteristic impedance, corresponding to Figure 13.15(*b*) [37,38].

The phase shift can be substantially extended using a transmission line that is periodically loaded with varactor diodes, as shown in Figure 13.15(*c*) [39]. In this case, the maximum possible phase shift is defined by $\Delta\phi = n\theta_i\left(\sqrt{1 + x/\theta_i} - \sqrt{1 + xy/\theta_i}\right)$, where n is the number of sections, θ_i is the electrical length of the transmission-line section measured in radians, $x = Z_i\omega C_{vmax}$, $y = C_{vmin}/C_{vmax}$, C_{vmin} and C_{vmax} are the minimum and maximum varactor capacitances, and the transmission-line characteristic impedance Z_i is calculated from $Z_i = Z_0\sqrt{1 + x/\theta_i}$, where Z_0 is the loaded characteristic impedance usually equal to 50 Ω. As a result, the variable phase shift up to 360° can be achieved at 20 GHz with the maximum insertion loss of 4.2 dB and return loss better than −12 dB for such a distributed phase shifter based on a coplanar-waveguide line.

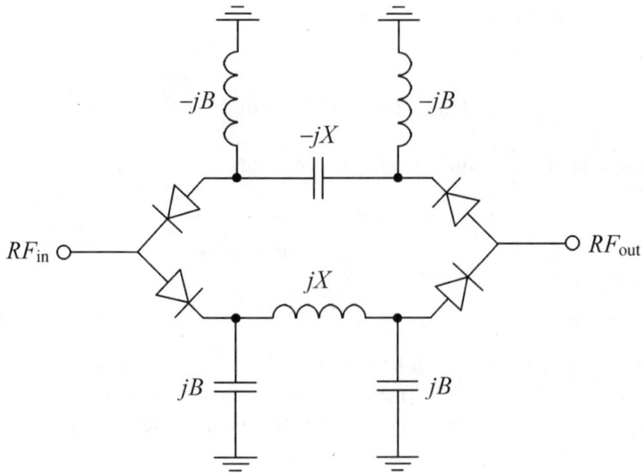

FIGURE 13.16 Basic circuit of low-pass/high-pass phase shifter.

In a *low-pass/high-pass phase shifter*, the low-pass filter consisting of the series inductors and shunt capacitors provides phase delay to the signals passing through it, while the high-pass filter consisting of the series capacitors and shunt inductors provides phase advance. By arranging diode switches (usually SPDT-type) to permit switching between low- and high-pass modes, it is possible to make such a phase shifter smaller in size and similar in bandwidth than the other types of lumped-element phase shifters [26]. This type of phase shifter is mostly applicable at lower frequencies, particularly in the VHF and UHF frequency bands. The *ABCD*-matrix of the low-pass π-section of a low-pass/high-pass phase shifter, the basic schematic of which is shown in Figure 13.16 can be written as

$$[ABCD] = \begin{bmatrix} 1 & 0 \\ jB & 1 \end{bmatrix} \begin{bmatrix} 1 & jX \\ 0 & 1 \end{bmatrix} \begin{bmatrix} 1 & 0 \\ jB & 1 \end{bmatrix} = \begin{bmatrix} 1 - XB & jX \\ j\left(2B - XB^2\right) & 1 - XB \end{bmatrix} \tag{13.15}$$

where X is the series reactance and B is the shunt susceptance due to the inductance and capacitance, respectively.

The transmission scattering element S_{21} defined by transformation from the *ABCD*-matrix given by Eq. (13.5) is equal to

$$S_{21} = \frac{2}{2\,(1 - XB) + j\,[XY_0 + BZ_0\,(2 - XB)]} \tag{13.16}$$

whose phase ϕ is written as

$$\phi = \tan^{-1}\left[-\frac{XY_0 + BZ_0\,(2 - XB)}{2\,(1 - XB)}\right]. \tag{13.17}$$

For a condition of the input match when $S_{11} = 0$ in Eq. (13.4) and lossless operation when the magnitude of S_{21} is equal to unity, from Eq. (13.16) it follows that

$$X = \frac{2BZ_0^2}{1 + (BZ_0)^2} \tag{13.18}$$

and substitution of Eq. (13.18) into Eq. (13.17) results in

$$\phi = \tan^{-1}\left[-\frac{2BZ_0}{1-(BZ_0)^2}\right]. \tag{13.19}$$

When both B and X change signs that corresponds to the high-pass π-section, the phase remains the same but changes sign. Therefore, the phase shift $\Delta\phi$ caused by switching between low- and high-pass sections is twice the value of the phase ϕ in Eq. (13.19),

$$\Delta\phi = 2\tan^{-1}\left[-\frac{2BZ_0}{1-(BZ_0)^2}\right]. \tag{13.20}$$

Consequently, the susceptance B and reactance X of a low-pass/high-pass phase shifter can be expressed in terms of the phase shift $\Delta\phi$ from Eqs. (13.20) and (13.18), respectively, as

$$B = Y_0 \tan\left(\frac{\Delta\phi}{4}\right) \tag{13.21}$$

$$X = Z_0 \sin\left(\frac{\Delta\phi}{2}\right). \tag{13.22}$$

For a T-section filter instead of a π-section, it is necessary to exchange B and X given in Eqs. (13.21) and (13.22). Each T-section or π-section phase shifter configuration can provide a phase shift of 90°±2° over almost an octave, while the smaller phase shifts are available over more than octave. However, the amplitude difference varies with frequency, since the band-edge slopes of the two filters are rolling in opposite directions.

The *switched-line phase shifter*, the basic schematic of which is shown in Figure 13.17(*a*), is based on a very simple principle when two transmission lines of different lengths are switched by two SPDT *p–i–n* diode switches to route the signal flow [26,40]. When the *p–i–n* diodes D_1 and D_2 are turned on while the *p–i–n* diodes D_3 and D_4 are turned off, the reference lower-path transmission line is connected into the circuit. However, when the *p–i–n* diodes D_1 and D_2 are turned off while the *p–i–n* diodes D_3 and D_4 are turned on, the upper-path transmission line is connected into the circuit. By switching a signal between two lines of different lengths, it is possible to provide a specific phase shift. The difference in physical length Δl between these two transmission-line paths results in the phase shift $\Delta\phi$ defined by

$$\Delta\phi = \beta(l_2 - l_1) = 2\beta\frac{\Delta l}{2} = 2\pi\frac{\Delta l}{\lambda} \tag{13.23}$$

where β is the propagation constant of the transmission line, $\lambda = 2\pi/\beta$ is the transmission-line wavelength, $\Delta l = l_2 - l_1$, l_2 is the length of the upper-path line and l_1 is the length of the lower-path line. Because the phase shift in the switched-line configuration is well defined by difference in the transmission-line lengths only, such a phase shifter does not need any additional adjustment and is very stable over time and temperature.

Switched-line digital phase shifters providing a certain phase increment can be made by a cascade connection of individual binary switched-line phase shifters [41]. For a switched-line phase shifter comprised of the nth consecutive stages, the smallest phase increment is defined by

$$\Delta\phi = \frac{\Delta\phi_{\max}}{2^n} \tag{13.24}$$

where $\Delta\phi_{\max}$ is the maximum phase shift. Generally, multibit switched-line phase shifters can vary the phase shift up to 360°. Figure 13.17(*b*) shows the basic circuit schematics of the 4-bit microstrip

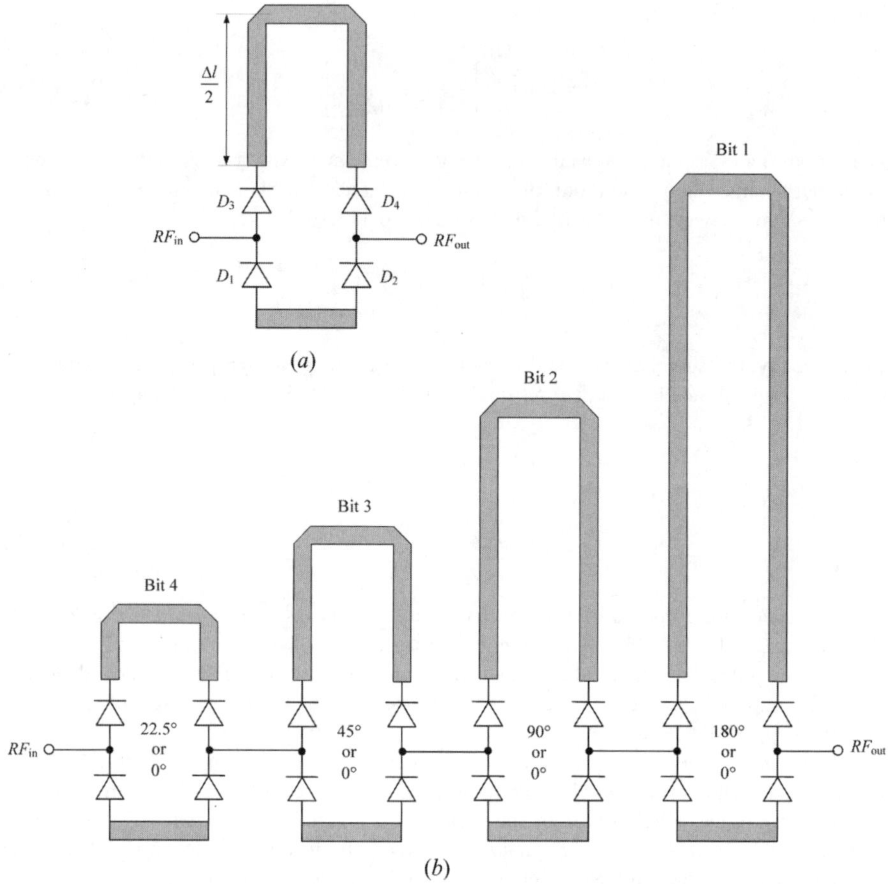

FIGURE 13.17 Schematics of switched-line phase shifters.

switched-line *p–i–n* diode phase shifter consisting of four binary phase shifters and covering 180° in nominal steps of 22.5°. The main problem with multibit phase shifters is a large variation in the return loss across the phase states that can limit their phase resolution and accuracy.

13.3.2 Schiffman 90° Phase Shifter

The basic differential Schiffman 90° phase shifter consists of two separate sections, one of which is a reference uncoupled transmission line while the other is a single *C*-section representing a pair of parallel coupled transmission lines directly connected to each other at one end, as shown in Figure 13.18(*a*) [42].

The phase shift ϕ_1 of the coupled-line section is defined as

$$\phi_1 = \cos^{-1}\left(\frac{\dfrac{Z_{0e}}{Z_{0o}} - \tan^2\theta}{\dfrac{Z_{0e}}{Z_{0o}} + \tan^2\theta}\right) \tag{13.25}$$

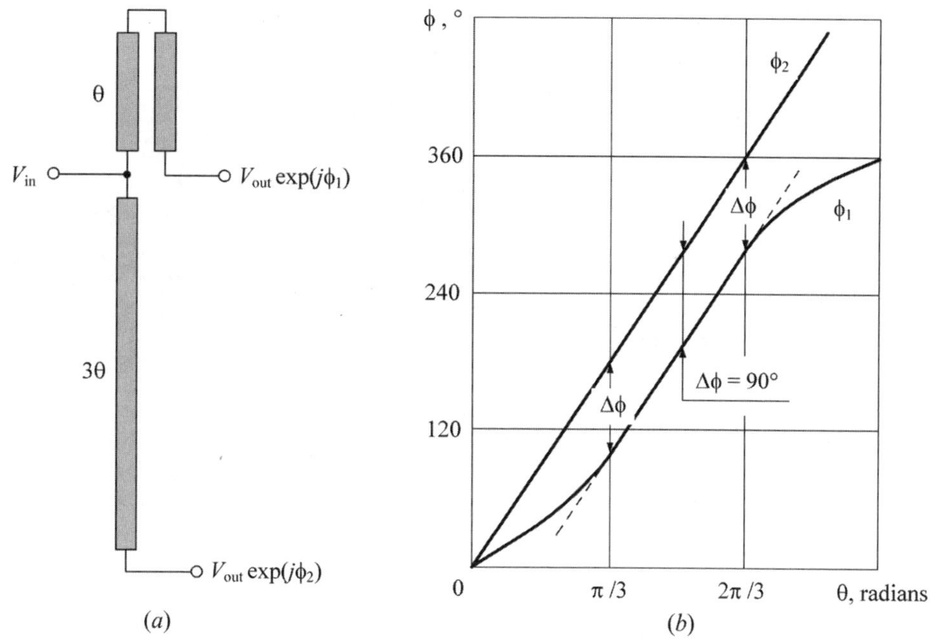

FIGURE 13.18 Schiffman differential 90° phase shifter.

where θ is the electrical length, Z_{0e} and Z_{0o} are the even- and odd-mode impedances of the coupled lines, respectively. In this case, provided that the phase shift ϕ_1 can be properly specified by appropriate choice of the ratio $\rho = Z_{0e}/Z_{0o}$ and electrical length θ, it follows that a network with a desirable difference phase response can be obtained by connecting in parallel a coupled-line section and a suitable length of uniform transmission line. As a result, this section can provide a nearly constant phase shift $\Delta\phi = \phi_2 - \phi_1 = 90^\circ$ compared to a reference line with electrical length 3θ. When $\theta = \pi/2$ or 90° and $\rho = 3$, the differential phase shift $\Delta\phi = 90\pm4.8^\circ$ over a 2.34:1 bandwidth can be obtained, as shown in Figure 13.18(*b*). Coupling coefficient between two coupled lines can be reduced by using a parallel or double parallel Schiffman phase shifter configurations [43].

To provide a multioctave operation, the differential 90° phase shifter can be constructed by a cascade of pairs of coupled transmission lines having different coupling coefficients and connected together at their far ends, as shown in Figure 13.19(*a*) [44]. However, the broader bandwidth and smaller coupling coefficients can be achieved with a structure containing a set of two coupled lines with optimized coupling coefficients, which is shown in Figure 13.19(*b*) for the case of four coupled pairs [45]. In this case, the matching condition is assumed for each coupled-line section,

$$Z_{0i} = \sqrt{Z_{0oi} Z_{0ei}} \tag{13.26}$$

where Z_{0i} is the characteristic impedance of the *i*th coupled section, Z_{0oi} and Z_{0ei} are the odd- and even-mode impedances, respectively. The coupling coefficient of each coupled-line section can be specified as

$$C_i = \frac{Z_{0ei} - Z_{0oi}}{Z_{0ei} + Z_{0oi}}. \tag{13.27}$$

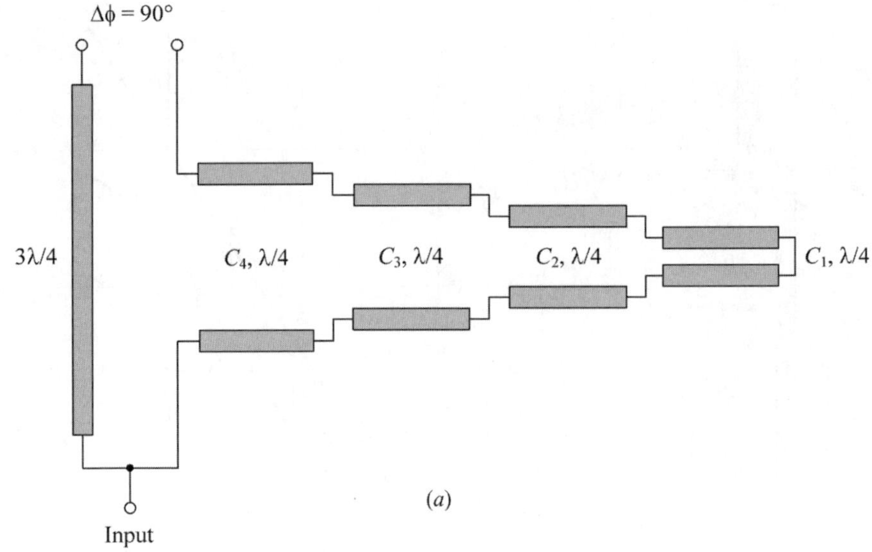

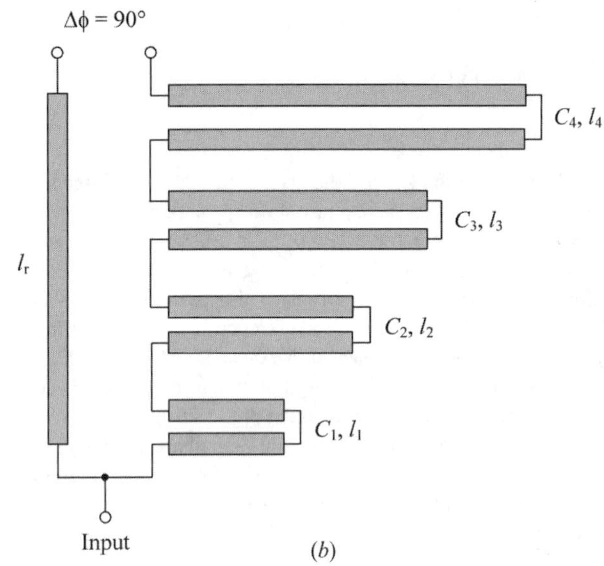

FIGURE 13.19 Structures of multioctave differential 90° phase shifters.

resulting in the differential phase shift $\Delta\phi = 90\pm1.2^\circ$ over a 6:1 bandwidth for $l_r = 4.0006\lambda$, $C_1 = 0.815$ and $l_1 = 0.167\lambda$, $C_2 = 0.63$ and $l_2 = 0.349\lambda$, $C_3 = 0.266$ and $l_3 = 0.627\lambda$, $C_4 = 0.121$ and $l_4 = 0.78\lambda$, where λ is the line wavelength corresponding to the center operating frequency.

Schiffman phase shifters can be very useful to achieve the sufficiently flat 90° relative-phase differences between two output ports. Ideally, it can be done for equal even- and odd-mode characteristics, corresponding to stripline realization in a homogeneous substrate. Therefore, since even- and odd-mode phase velocities in the coupled-line section are not equal in the microstrip Schiffman phase shifter, it is necessary to apply special techniques to equalize phase velocities corresponding to these two modes.

13.3.3 MESFET Phase Shifters

Significant progress in monolithic microwave integrated circuits (MMICs) up to millimeter waves has made the GaAs MESFET devices very attractive for realizing active phase-shifting components intended to widely use in active phased-array antennas for broadband satellite and wireless communication systems. Although the MESFET itself is a three-terminal device, the switching mechanism takes place between the drain and the source with the control voltage applied to the gate. In this configuration, the FET behaves as a passive element, and only dc power consumption is due to gate leakage current. With this switching configuration, the GaAs MESFETs can replace *p–i–n* diodes to be effectively used in the monolithic broadband load-line and reflection-type phase shifters [46,47]. As a result, a 180° reflection-type phase-shifter MMIC can demonstrate a phase shifter of 180±7° over frequency range of 0.5–20 GHz. In a dual-gate GaAs MESFET, the phase shift is obtained by changing the dc voltage applied to the control gate of the FET, and a linear phase shift of 100° can be obtained at an operating frequency of 12 GHz [48]. There are several approaches to realize active phase shifters using dual-gate FETs. One of them is to use them as switchable amplifiers that allow different phase length paths to be chosen to achieve the desired phase shift [49]. The phase shift over an octave band from 4 to 8 GHz can be achieved by using a dual-gate FET as a variable gain amplifier and phase shift is obtained by complex addition of two orthogonal variable vectors [50]. Using HBT or HEMT technology, application of the vector sum method for active phase shifters can be extended up to 20 GHz and higher [51].

Figure 13.20(*a*) shows the 90° phase shifter basic configuration where the two dual-gate FET amplifiers are excited in quadrature phase through a 90° hybrid splitter and the outputs of both FET amplifiers are then combined through an in-phase combiner to produce a phase-controlled output signal. The gain of each dual-gate FET amplifier can be controlled from +10 to −30 dBm

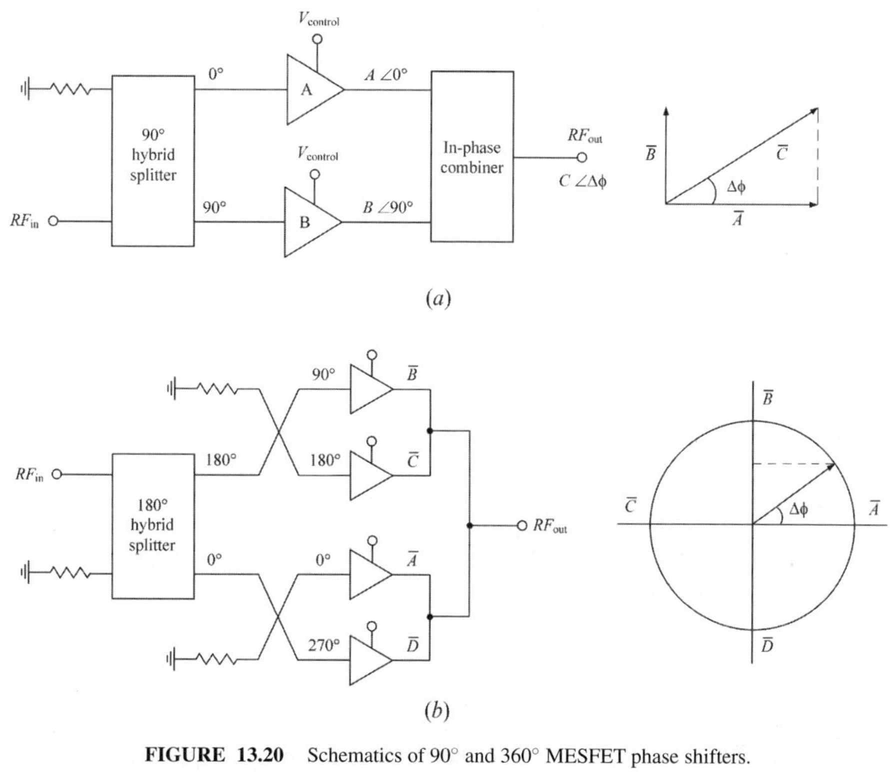

FIGURE 13.20 Schematics of 90° and 360° MESFET phase shifters.

by controlling the second gate bias voltage. In this case, the output phase shift is defined as $\Delta\phi = \tan^{-1}(B/A)$ and the output amplitude as $C = \sqrt{A^2 + B^2}$, which can be kept constant by independently adjusting the gain of each of the dual-gate FET amplifiers. The block schematic of a phase shifter continuously variable from 0 to 360° is shown in Figure 13.20(*b*). The 360° phase shift is achieved by the sum of four quadrature vectors $A\angle 0°$, $B\angle 90°$, $C\angle 180°$, and $D\angle 270°$ with properly controlled amplitudes *A*, *B*, *C*, and *D*. These four quadrature vectors can be realized by a 180° power divider, two 90° hybrids, four dual-gate FET amplifiers, and an in-phase four-way power combiner. In this case, the four quadrants of 360° phase shift are obtained using a combination of two vectors at a time. Thus, by controlling the bias voltages of two amplifiers at one time, while the other two are switched off, the total of 360° continuous phase shift is achieved.

The multibit phase shifter MMICs can be realized based on a low-pass/high-pass phase shifting concept. Generally, four bits can be designed with 22.5°, 45°, 90°, and 180° values. In this case, an optimum performance of each bit over wide frequency range can be achieved by selecting a suitable topology. However, for a narrow-band operation, a simple phase bit topology shown in Figure 13.21(*a*) can be used to provide 22.5°, 45°, or 90° phase shift [52]. Here, for reference state

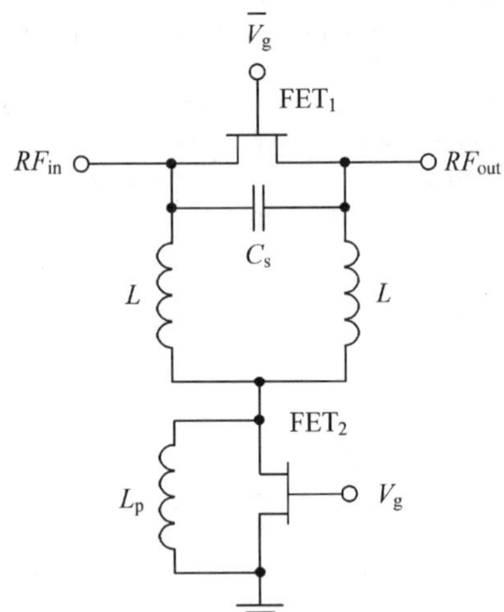

(*a*)

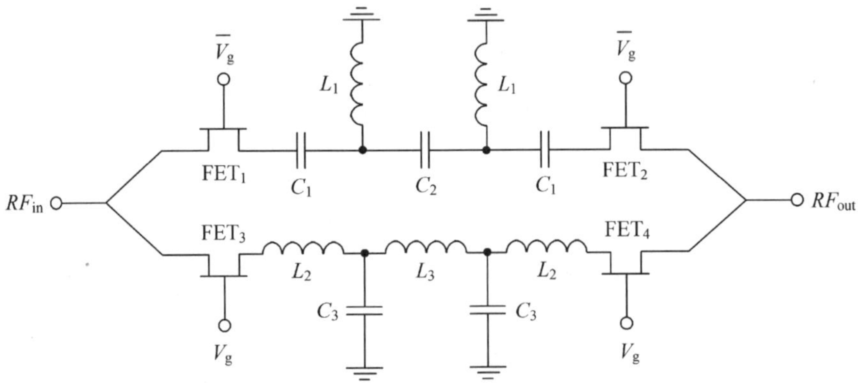

(*b*)

FIGURE 13.21 Basic circuit of MESFET low-pass/high-pass phase shifters.

operation, FET_1 is switched on and FET_2 is switched off. The inductor L_p is sized to parallel resonate the drain–source capacitance of FET_2 that contributes to a series combination of inductances L in parallel with the on-state resistance of FET_1. To activate the phase shift state, FET_1 is switched off and FET_2 is switched on, resulting in the approximate equivalent high-pass π-network, where the series capacitance C is represented by the parallel combination of C_s and drain–source capacitance of FET_1.

The transmission *ABCD*-matrix of the high-pass *LC*-type π-section can be written as

$$[ABCD] = \begin{bmatrix} 1 - \dfrac{1}{\omega^2 LC} & \dfrac{1}{j\omega C} \\ \dfrac{2}{j\omega L} - \dfrac{1}{j\omega^3 L^2 C} & 1 - \dfrac{1}{\omega^2 LC} \end{bmatrix} \tag{13.28}$$

that results in design equations for L and C as a function of the phase shift $\Delta\phi$ in a matched condition with $S_{11} = S_{22} = 0$ given by

$$L = \frac{Z_0}{\omega \tan(\Delta\phi/2)} \tag{13.29}$$

$$C = \frac{1}{\omega Z_0 \sin \Delta\phi} \tag{13.30}$$

where Z_0 is the loaded characteristic impedance.

The 180° phase shifter can be designed by switching of the low- and high-pass filters, as shown in Figure 13.21(*b*) [53]. In this case, to switch the signal between these filters, the FET-based SPDT switches are used. As a result, the phase error less than 3° can be achieved across the operating frequency bands of 1.4–2.4 GHz and 2.3–3.8 GHz. Generally, the low-pass/high pass phase shifter configuration is very useful when wider bandwidth and compact size are required. Here, the phase shift is realized by using the phase lag characteristics of the low- and high-pass sections that are connected between two SPDT switches. However, this topology has relatively higher insertion loss.

13.4 ATTENUATORS

Analog and digital attenuators that are capable of a large attenuation range with low insertion loss and *VSWR* can be designed using *p–i–n* diodes where a *p–i–n* diode is operated as an electronically variable resistor controlled by dc current [54]. In a switching *p–i–n* diode operation, to provide low phase shift and high dynamic step attenuation range up to 70 dB, several two-path resistive attenuator sections can be cascaded by means of a couple of SPDT *p–i–n* diode switches [55]. Monolithic integration of a *p–i–n* attenuator can reduce the size, complexity, and overall system cost. Generally, to select the *p–i–n* diode for an attenuator application, it is necessary to be aware of the range of its resistance that will define the dynamic range of the attenuation, and nonlinear behavior, since its operating bias point often occurs at a low value of forward bias current.

Figure 13.22 shows the basic circuits of the bridged-*T* and π-type *p–i–n* diode attenuators [14]. In the bridged-*T* attenuator shown in Figure 13.22(*a*), the required attenuation is provided by absorbing RF power in the *p–i–n* diodes D_1 and D_2 and can be defined in decibels as

$$A\,(\text{dB}) = 20\log_{10}\left(1 + \frac{Z_0}{R_{s1}}\right) \tag{13.31}$$

where $Z_0 = \sqrt{R_{s1} R_{s2}}$ is the loaded characteristic impedance, R_{s1} and R_{s2} are the series resistances of the forward-biased *p–i–n* diodes D_1 and D_2, respectively.

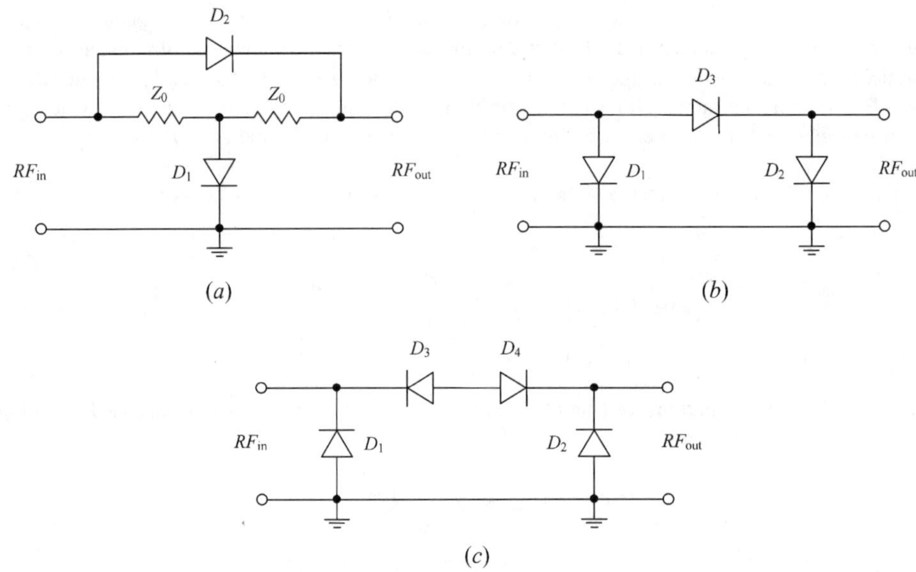

FIGURE 13.22 Basic circuits of T- and π-type *p–i–n* diode attenuators.

The expression for attenuation in the π-type attenuator shown in Figure 13.22(*b*) is given by

$$A\,(\text{dB}) = 20\log_{10}\left(\frac{R_{s1} + Z_0}{R_{s1} - Z_0}\right) \tag{13.32}$$

where $R_{s1} = R_{s2}$ and $R_{s3} = 2R_{s1}Z_0^2/\left(R_{s1}^2 - Z_0^2\right)$ is the series resistance of the forward-biased *p–i–n* diode D_3. From Eq. (13.32) it follows that the minimum values of equal diode resistances R_{s1} and R_{s2} are equal to Z_0.

The *p–i–n* diodes in both attenuator configurations are biased at two different resistance points simultaneously to achieve proper attenuation performance. These lumped-element structures are very compact and can be used at relatively low frequencies up to 1 GHz. Their input and output port impedances remain constant, while the overall attenuation can be varied over a range of less than 1 dB to greater than 20 dB. The benefit of the π-attenuator that includes four *p–i–n* diodes, as shown in Figure 13.22(*c*), is its symmetry that allows for a simpler bias network and a reduction of distortion due to cancellation of harmonic components in the back-to-back configuration of the series diodes D_3 and D_4 [56].

Figure 13.23(*a*) shows the reflection-type *p–i–n* diode attenuator based on a quadrature branch-line hybrid or a Lange coupler where two diodes are connected to the reflected ports. In this circuit, the input power is divided equally between two reflected ports and passes to the two *p–i–n* diodes that should be spaced equally and used as mismatched elements. Because the incident power is divided into two parts, such a reflection-type attenuator is capable of handling twice the power of the simple diode configuration. The reflected power from the diodes, which depends on the bias voltage, flows to the output port and ideally should not affect the input port, resulting in a good input matching. For zero or infinite impedance at the reflected ports, the RF signal from reflected ports is fully delivered to the output port. In this case, the attenuation is equal to 0 dB. Infinite or maximum attenuation occurs when the reflected ports are loaded by the 50 Ω, providing a zero reflection.

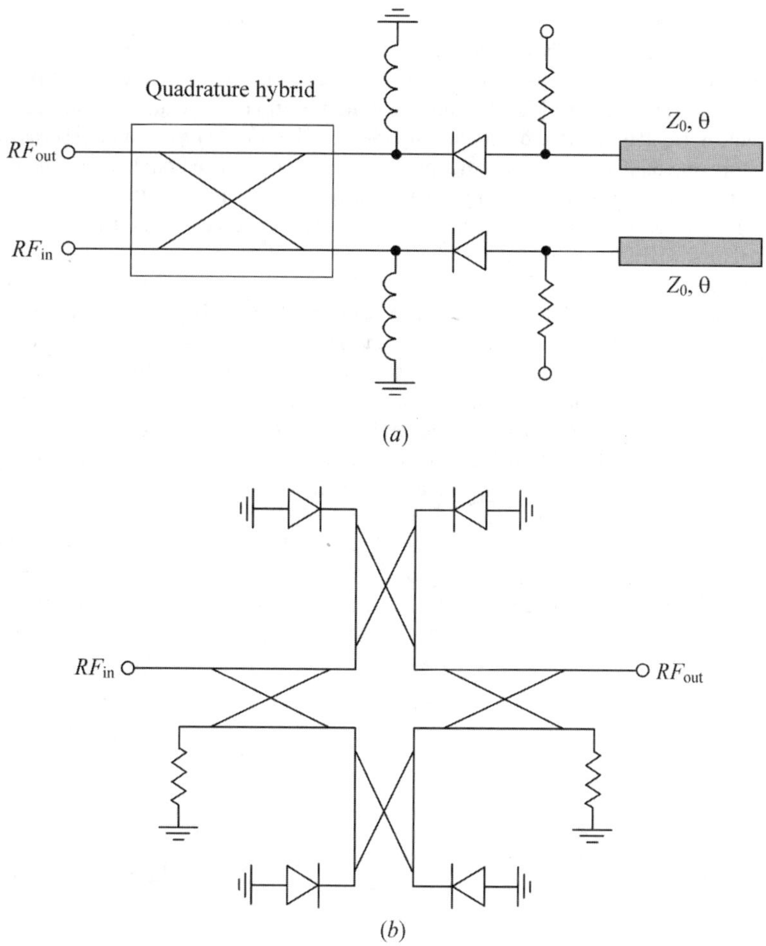

FIGURE 13.23 Reflection-type low-phase shift diode attenuators.

At any *p–i–n* diode bias condition, the phase characteristic of the reflection-type attenuator shown in Figure 13.23(*a*) can be written as

$$\Delta\phi = \tan^{-1}\left(\frac{\omega L_s - \dfrac{\omega R_p^2 C_j}{1+\omega^2 R_p^2 C_j^2} - \dfrac{Z_0}{\tan\theta}}{R_s - Z_0 + \dfrac{R_p}{1+\omega^2 R_p^2 C_j^2}}\right) - \tan^{-1}\left(\frac{\omega L_s - \dfrac{\omega R_p^2 C_j}{1+\omega^2 R_p^2 C_j^2} - \dfrac{Z_0}{\tan\theta}}{R_s + Z_0 + \dfrac{R_p}{1+\omega^2 R_p^2 C_j^2}}\right) \qquad (13.33)$$

where L_s is the parasitic series inductance, R_s is the series *i*-layer resistance, C_j is the diode junction capacitance shunted by a parallel resistance R_p, Z_0 and θ are the characteristic impedance and electrical length of the open-circuit stub, respectively [57]. In this case, a low phase shift attenuator can be realized by adjusting the value of θ to minimize the effects of the parasitic series inductance L_s and varying resistance R_j. As a result, at a center bandwidth frequency of 1855 MHz, this reflection-type attenuator can achieve less than 3° phase shift with up to 30 dB attenuation and return loss below 17 dB at both input and output ports.

The frequency bandwidth of a reflection-type variable attenuator is defined by the bandwidth properties of the quadrature hybrid. The microstrip 3-dB branch-line coupler in conjunction with two varactor diodes instead of *p–i–n* diodes can vary the attenuation from 2.2 to 17 dB over a 40% bandwidth from 2.8 to 4.2 GHz [58]. To minimize the variation of attenuation with frequency and improve return loss, two or more quadrature hybrids can be used. For example, by cascading two balanced bi-phase modulators with opposite phase where each balanced modulator consists of a pair of reflection-type attenuators connected by means of input and output quadrature directional couplers, as shown in Figure 13.23(*b*), the phase error less than 5° over a 32 dB attenuation range in a frequency range from 6 to 18 GHz was achieved [59].

The performance of the variable attenuators can be improved in terms of frequency bandwidth, linearity, power handling, and integration level by using MOSFET, MESFET, or HEMT technology. The basic mechanism of the circuit is the change in the low-field drain–source resistance of a zero-biased field-effect transistor controlled by the gate–source voltage [60]. In this case, the variable attenuator operating as a controlled resistor represents a passive component and does not dissipate any dc power. The value of resistance can generally vary from the open-gate resistance to infinite when the gate voltage is changed from the built-in voltage (positive) of the gate barrier to the pinch-off voltage (negative). The attenuator nonlinear distortion can be reduced by connecting two FETs in antiparallel (or antiseries) configuration or using a single FET with negative feedback [61].

The classic bridged-*T* attenuator shown in Figure 13.24(*a*) becomes a continuously variable attenuators when the series resistance R_1 and the shunt resistance R_2 are allowed to vary according to $Z_0 = \sqrt{R_1 R_2}$, where Z_0 is the desired characteristic impedance. As a result, the attenuation in a

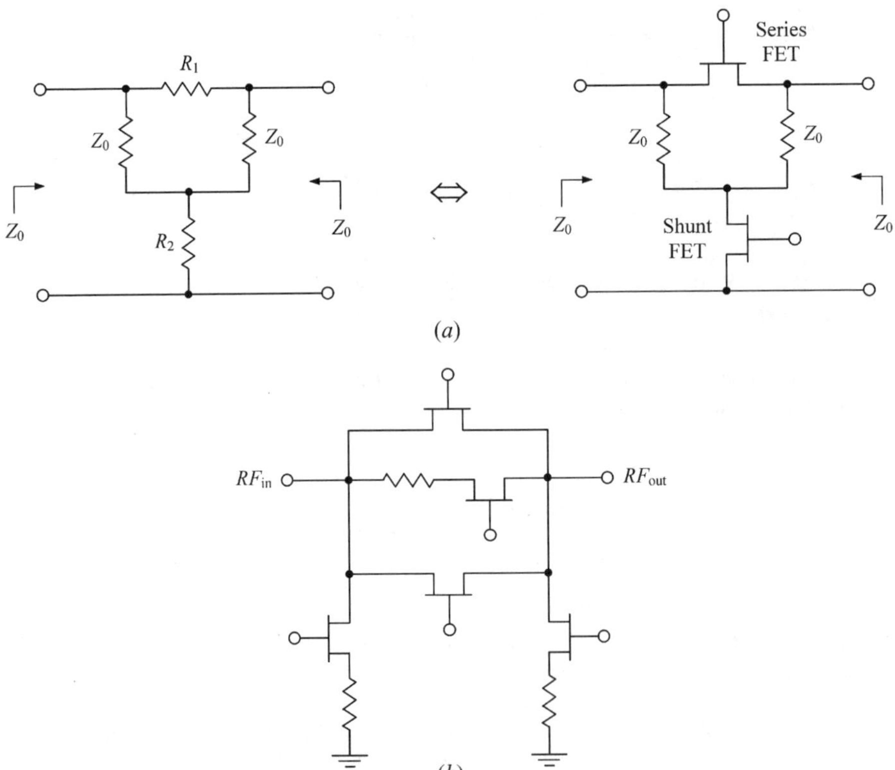

FIGURE 13.24 Resistive and MESFET variable attenuators.

matched condition is defined as

$$A\,(\text{dB}) = 20\log_{10}\left(1+\sqrt{\frac{R_1}{R_2}}\right) = 20\log_{10}\left(1+\frac{R_1}{R_2}\right). \tag{13.34}$$

By replacing the series and shunt resistors R_1 and R_2 with MESFET devices that generally can operate in a forward- or zero-biased conditions, as shown in Figure 13.24(*a*), a monolithic implementation of the bridged-T attenuator can be realized [62]. Several factors have to be considered when using FETs for the shunt and series elements. In this case, the FETs width must be chosen wide enough for low insertion loss in the minimum attenuation state, but small enough to limit parallel drain–source capacitance C_{ds} so that the isolation at higher frequencies is sufficient in the maximum attenuation state. The maximum attenuation is most dependent on the C_{ds} of the series FET. By adding additional FET devices, a digital step attenuator can be designed that provides precise and well-controlled attenuation ranges over wide frequency range. Figure 13.24(*b*) shows the modified π-type MESFET attenuator circuit, based on which the broadband low-loss 5- and 6-bit digital attenuators can be built [63].

At low frequencies when the effect of the MESFET capacitances can be neglected, the attenuation in a matched condition for three-FET T- and π-type attenuators shown in Figures 13.25(*a*) and 13.25(*b*), respectively, can be written as

$$A\,(\text{dB}) = 20\log_{10}\left(\frac{R_2}{R_1+R_2+Z_0}\right) \tag{13.35}$$

where R_1 is the series-FET resistance, R_2 is the shunt-FET resistance, and

$$Z_0 = \sqrt{R_1^2+2R_1R_2}. \tag{13.36}$$

However, generally these T- and π-type variable attenuators can be represented through the equivalent circuits consisting of the series and shunt resistances and capacitances, as shown in

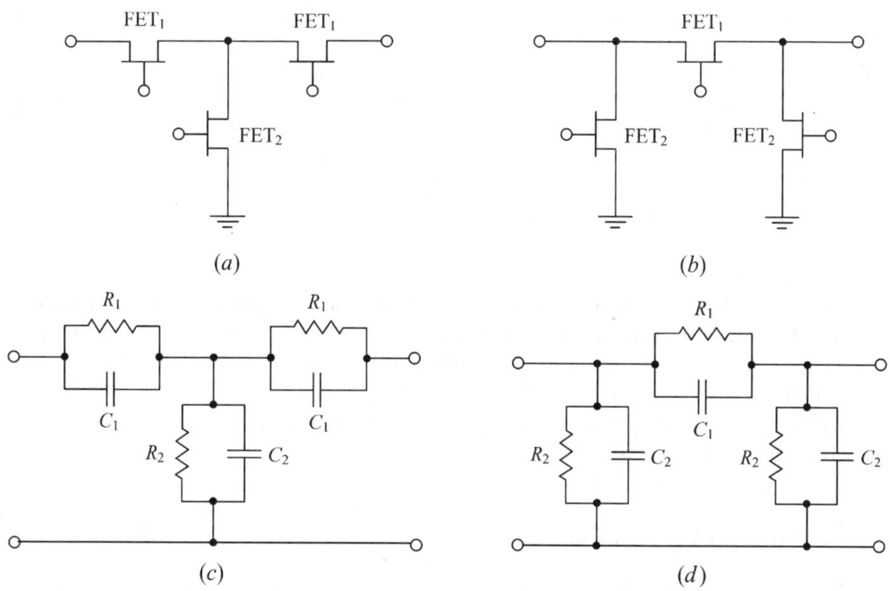

FIGURE 13.25 MESFET attenuators and their equivalent circuits.

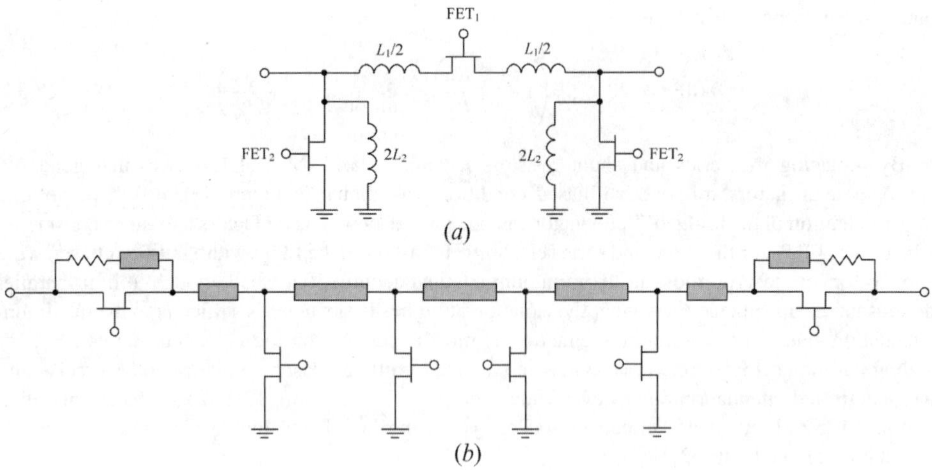

FIGURE 13.26 Schematics of MESFET bandpass attenuators.

Figures 13.25(*c*) and 13.25(*d*), respectively [64]. For the same R_1, the π-type circuit is characterized by less insertion loss than the T-type circuit. The value of R_1 can be reduced by increasing the number of gate fingers, although the parasitic capacitance C_1 will also increase limiting the dynamic range of attenuation at high frequencies. In terms of the dynamic range, T-type attenuators become advantageous over π-type attenuators. Besides, the T-network is more linear than the π-network, especially at higher attenuation settings [65]. Since power-handling capability is limited by the performance of the shunt FET, dual-gate devices can be used for the shunt FET to increase its knee and breakdown voltages. To achieve better linearity, the two nonidentical cascaded T-type stages that are activated consecutively can be used [66].

In order to absorb the device parasitic capacitances C_1 and C_2 representing the device drain–source capacitances C_{ds} (or generally equivalent output capacitances C_{out}) and eliminate their effect on attenuator performance, a bandpass filter topology can be superimposed onto the π-type MMIC variable attenuator, as shown in Figure 13.26(*a*) [67]. In this case, the bandpass filter parameters are determined as $L_1 = Z_0^2 C_1$ and $L_2 = 1/(2\omega_1\omega_2 C_2)$, where $Z_0 = \sqrt{L_1/2C_2} = \sqrt{L_2/C_1}$ is the filter image impedance corresponding to the loaded characteristic impedance equal to 50 Ω, ω_1 and ω_2 are the lower and upper cutoff frequencies in radians with the center bandwidth frequency $\omega_0 = \sqrt{\omega_1\omega_2}$, respectively. Figure 13.26(*b*) shows the T-type MMIC variable attenuator where a shunt FET is distributed into four cells interconnected by high-impedance transmission-line sections [68]. At a minimum attenuation setting when the shunt FETs are equivalently represented as capacitances, the distributed structure absorbs the FET capacitances into an artificial 50-Ω transmission line. At a maximum attenuation setting with the shunt FETs effectively operated as low resistances, the distributed LR structure yields an increased attenuation with frequency, thus compensating for degraded isolation due to parasitic capacitances of the series FETs. Moreover, a 50-Ω resistor connected in parallel with each series FET minimizes the parasitic capacitances at the maximum attenuation setting by allowing the series FETs to be biased well below their pinch-off voltage. As a result, a 30-dB dynamic range of attenuation in a frequency range from dc up to 50 GHz with a minimum insertion loss of 1.8 dB at 26.5 GHz and 2.6 dB at 40 GHz was achieved.

13.5 VARIABLE GAIN AMPLIFIERS

For mobile phone or satellite communication transmitters with power control schemes, a variable gain amplifier with wide dB-linear gain control and high transmission linearity is required. To achieve high dynamic gain control with low input/output *VSWR*, the low-cost variable gain amplifier (VGA) can consist of three stages in a cascade connection where the first stage represents the input buffer to keep constant input *VSWR*, the second stage provides the variable gain control by varying the base bias current, and the third stage is operated as a fixed gain output buffer to amplify the RF signal [69]. A shunt feedback scheme can be employed to reduce nonlinear distortion and instability. Using SiGe HBT technology, a constant *VSWR* of less than 1.6, a maximum gain of 23 dB, and a dynamic gain control range of 30 dB was achieved in 824–849 MHz band. However, linearity of the gain control characteristic (of about 15 dB) is limited when the bias voltage of a single bipolar device is varied. Substantially higher linear dynamic range (of about 40 dB) can be achieved by using the VGA consisting of a linear π-type MESFET attenuator with sequential bias control of the series and shunt devices and a two-stage amplifier [70]. To improve an overall system transmission linearity, the input predistorter based on a common gate FET device with pair of shunt inductors can be inserted, which is designed to have gain expansion and negative phase shift characteristics with an increased input power when biased near pinch-off mode [71].

The variable gain amplifier can be realized in the form of the cascode amplifier, a typical circuit schematic of which is shown in Figure 13.27(*a*). Such a cascode VGA based on GaAs HBT technology can operate as an attenuator with a linear power dynamic range of about 20 dB, as shown in Figure 13.27(*b*). Its gain control is achieved by adjusting the base bias of the common base device. In this case, the dc current consumption is sufficiently small (less than 8 mA at maximum output power for $V_{contr} = 2.2$ V) because a common voltage supply is required for both transistors. The

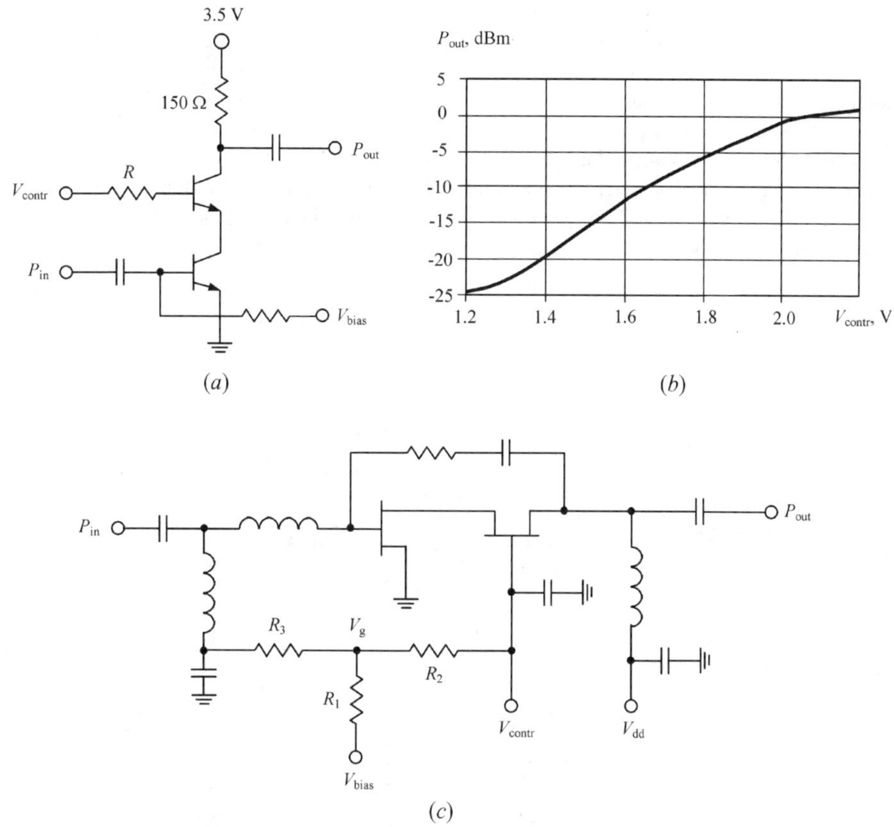

FIGURE 13.27 Cascode VGA schematics and power tuning curve.

maximum output power with a good linearity can be realized when the values of the resistor R are chosen within the range of 0.5 to1.0 kΩ. However, the phase difference over entire control voltage range can be sufficiently high, exceeding 25° [72]. The main benefit of using a cascode configuration is an inherent reduction of the parasitic feedback capacitance, thus enhancing the high frequency response and providing a compact two-stage structure.

In active phased-array radar antennas, it is important to independently control the amplitude and phase of the signals. Therefore, the transmission phase deviation over the gain control dynamic range should be minimized. Figure 13.27(*c*) shows the circuit schematic of a cascode-connected FET VGA where the two gate bias voltages are controlled differently to improve the transmission phase deviation [73]. Since the main contribution to the phase variation in a cascode-connected FET amplifier is caused by nonlinearity of the gate–drain capacitance C_{gd} of a common source FET, then the gate bias voltage of a common source FET has to be reduced simultaneously with decreasing the gate bias voltage of a common gate FET. As a result, by choosing the optimum resistance values $R_1 = 150\ \Omega$, $R_2 = 600\ \Omega$, and $R_3 = 3\ \text{k}\Omega$, the phase deviation is suppressed from between 0° and −30° to between +3° and −5° with a 25-dB gain control at an operating frequency of 1.9 GHz for fixed $V_{bias} = -1.43$ V and varying V_{contr} from 0.75 to −2.75 V.

The dual-gate FET can be considered as two separate FETs connected in cascode configuration where the current characteristics of the bottom FET are determined by the top FET. Consequently, the gain of the dual-gate FET can easily be varied only by the second gate bias voltage. Therefore, not only can the circuit construction be simplified, but also power dissipation can be decreased by using dual-gate FETs in microwave variable-gain amplifiers. The gain control characteristic of the dual-gate FET is due to the fact that its transconductance can be varied by adjusting the gate bias voltage. Although the transconductance can be varied by the first gate bias voltage as well as the second gate bias voltage, it does not monotonically change with the first gate bias and reaches maximum, while the transconductance smoothly decreases with the second gate bias voltage within the specified range without degradation of bandpass performance [74]. Therefore, the second gate bias dependence of the transconductance can be preferably used for a variable-gain amplifier. Figure 13.28 shows the broadband microwave two-stage dual-gate FET VGA, where the first stage is necessary to provide an input broadband impedance matching and the gain control is provided by the second stage [75]. As a result, a 15-dB gain control (between +3 and −13 dBm) with an input *VSWR* less than 3.5 was achieved over the ultra-wide frequency range from 100 MHz to 17 GHz by varying the second gate bias voltage from 0 to −1.7 V.

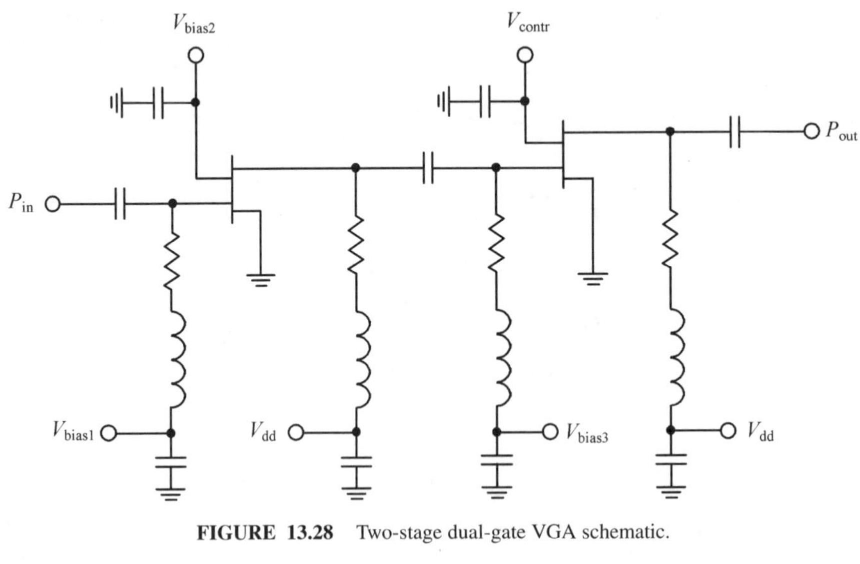

FIGURE 13.28 Two-stage dual-gate VGA schematic.

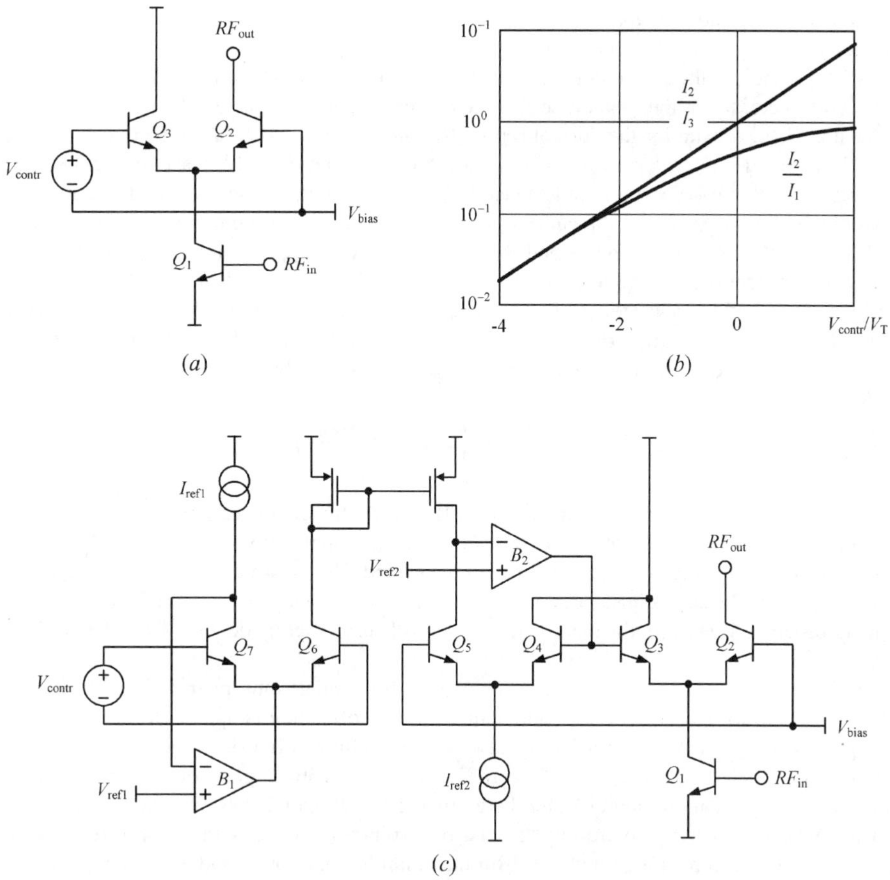

FIGURE 13.29 Simplified schematics of differential VGA and gain tuning curves.

Figure 13.29(a) shows the conventional differential amplifier where the signal current from a common emitter tail transistor Q_1 is split between the two top transistors Q_2 and Q_3 with the collector current from Q_2 being used as the output and the collector current from Q_3 being shunted to a supply, and a voltage V_{contr} between the bases of the transistors determines the attenuation of the transmitting signal [76]. The gain-control characteristic can be found by solving for the current ratio I_2/I_1, where the current I_{n} corresponds to the transistor Q_{n}, $n = 1, 2, \ldots$. A linear change in the difference between the base–emitter voltages of bipolar transistors changes the ratio of their current exponentially as

$$\frac{I_2}{I_3} = \exp\left(\frac{V_{\text{contr}}}{V_{\text{T}}}\right) \tag{13.37}$$

where V_{T} is the thermal voltage. Solving for the gain-control characteristic by determining the current ratio I_2/I_1 results in

$$\frac{I_2}{I_1} = \frac{1}{1 + \exp\left(-\dfrac{V_{\text{contr}}}{V_{\text{T}}}\right)}. \tag{13.38}$$

Figure 13.29(*b*) shows the curves that are plotted from Eqs. (13.37) and (13.38), respectively. Here, it can be seen that when the control voltage V_{contr} is applied as a voltage across the bases of transistors Q_2 and Q_3, the gain-control characteristic is fairly linear-in-dB at high attenuation levels, but the error increases at higher gains so that the maximum gain is expected at a 6-dB loss. The true exponential characteristic for the current ratio I_2/I_1 can be achieved based on the circuit schematic shown in Figure 13.29(*c*), where all corresponding bipolar and pMOS transistors are assumed to be identical. In this case, a transconductance buffer amplifier B_1 is used to supply current to the transistors Q_6 and Q_7 such that the current I_7 is equal to the reference current I_{ref1}. The current I_6, which is a function of the current I_7 and the control voltage V_{contr}, is mirrored by the two pMOS devices. The current I_5 is forced to be equal to the current I_7 by a buffer B_2 that controls the voltage between the bases of the transistors Q_4 and Q_5, and the same voltage is applied between the transistors Q_2 and Q_3. This sets the current ratio I_5/I_{ref2} equal to the current ratio I_2/I_1, and the reference current I_{ref2} is chosen to be equal to the reference current I_{ref1}. Consequently,

$$\frac{I_2}{I_1} = \frac{I_5}{I_{ref2}} = \frac{I_6}{I_7} = \exp\left(\frac{V_{contr}}{V_T}\right) \tag{13.39}$$

which demonstrates the most desirable linear-in-dB gain-control characteristic. As a result, the gain-control range can be measured of about 50 dB for 900-MHz applications using a BiCMOS process. It should be noted that the equivalence between the dc and RF performances can be achieved if the device transconductance is constant over the entire bias range and there is no effect of the device output impedance. Therefore, the measured gain-control curve slightly deviates from the ideal linear characteristic.

It is well known that a fully balanced (or double differential) multiplier is based on the three differential transistor pairs where the modulating (or control) signal is applied to the inputs of the two top transistor pairs and the carrier signal is applied to the tail transistor pair [77]. In this case, using a silicon bipolar technology with $f_T = 10$ GHz results in a 50-dB gain control range with 24 dB of maximum gain and input/output *VSWR* better than 2:1 at all gain levels over the frequency range from 100 MHz to 4 GHz by controlling the base bias voltages of the top transistor differential pairs [78]. Similar gain control range with 24-dBm maximum output power and efficiency greater than 27% at an operating frequency of 1.95 GHz for mobile WCDMA applications can be achieved using a GaAs HBT technology [79].

13.6 LIMITERS

Generally, limiter represents a two-port network that limits the output power level to a constant value at high input power levels to protect power-sensitive components. A typical single-stage limiter circuit shown in Figure 13.30(*a*) uses the limiter diode, which is a special type of a *p–i–n* diode [80,81]. For incoming signals that are below the threshold level in amplitude, the diode acts as an ordinary unbiased *p–i–n* diode. In this case, the electrical field of the incoming signal is not large enough to force carriers into the diode *i*-layer so its resistance remains high and it appears to be a capacitor of relatively small value. When the incident signal exceeds the threshold power level, the diode *i*-layer is flooded with carriers during the positive half-cycle of the incoming RF signal. The dc current begins to flow in the loop formed by the limiter diode and RF choke, and the limiter diode biases itself to a low resistance value in a matter of nanoseconds, thus shorting out the signal transmission at high power levels. In this case, the limiter circuit operates as a reflective switch, reflecting the large signal back to its source and protecting the subsequent circuitry. When the large amplitude of the incoming signal reduces to the level below the threshold voltage, the carriers in the diode *i*-layer rapidly recombine, and the diode junction resistance becomes very high, allowing small signals to pass. The inductance of the RF choke is chosen to have a sufficiently high reactance and out-of-band series resonance so that it also produces negligible in-band reflection loss.

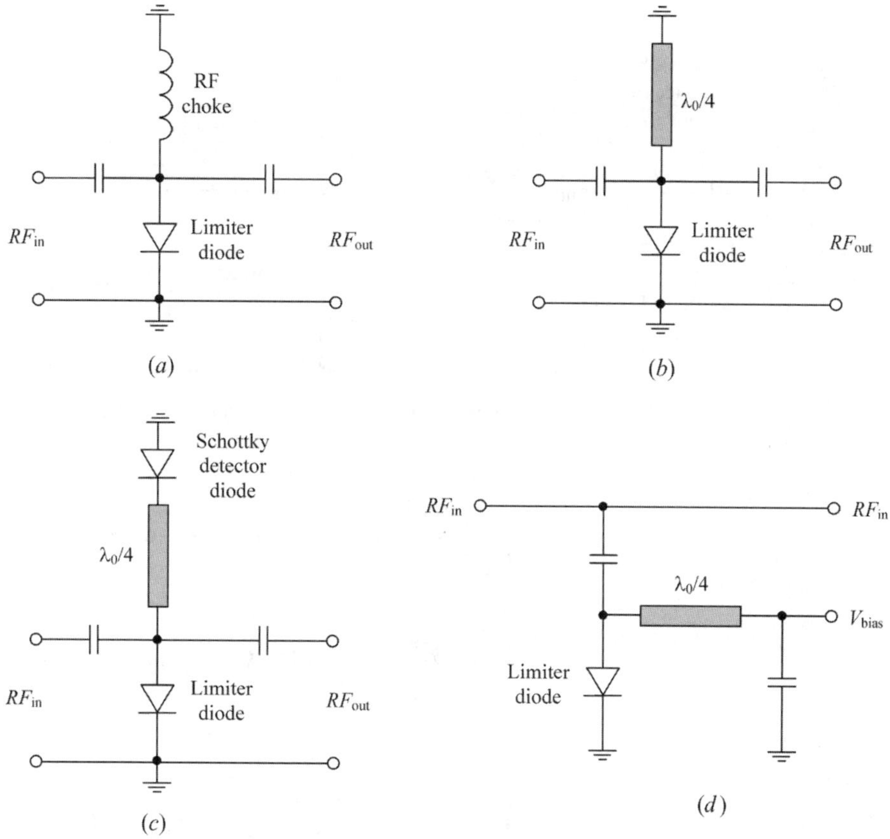

FIGURE 13.30 Basic circuits of *p–i–n* diode limiters.

There are three basic types of a power limiter circuit, usually based on the diode shunt configurations [41]. In a passive (or self-bias) limiter, lumped and transmission-line configurations of which are shown in Figures 13.30(*a*) and 13.30(*b*), respectively, the limited output power depends on the variable diode resistance under zero-bias condition. Since the limiter diodes are turned on by the incoming RF signal itself, a passive limiter is a self-activating switch that is activated by a high-level incident power. When the self-rectified bias is increasing with increasing input power, the diode impedance is decreasing, and speed of this process depends on the diode lifetime (20–200 ns). This relatively slow process leads to a high dissipation power on the diode, causing spike leakage power nearly equal to the incident power level. The maximum incoming RF power that a limiter diode can handle is limited by either diode breakdown voltage or its power-dissipation capability. The quasi-active limiter shown in Figure 13.30(*c*) includes a high-power Schottky-barrier diode detector that supplies dc current to the limiter diode. The Schottky diode rectification process is much faster than the limiter diode self-bias. In this case, the limiter diode is turned on primarily by the external bias current from the Schottky detector diode that is sensitive to the incident power. The active limiter shown in Figure 13.30(*d*) operates using a separate dc current source, and it is used in the sensitive time control systems.

The single-stage limiter can typically produce 20–30 dB of isolation, depending on the input signal frequency and the characteristics of the diode. However, much more isolation can be required to protect sensitive components. Figure 13.31 shows the basic circuit of a two-stage *p–i–n* diode

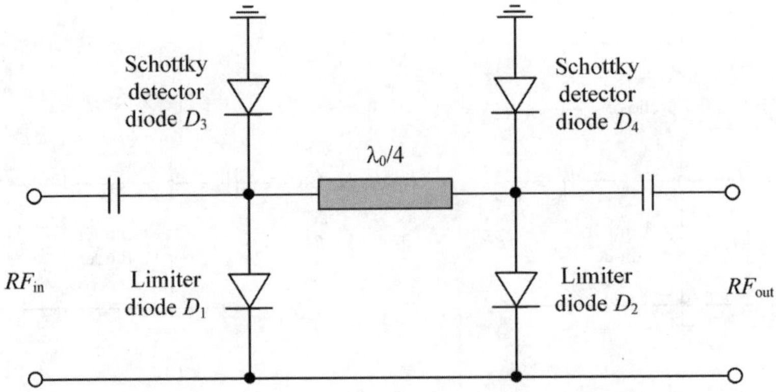

FIGURE 13.31 Basic circuit of two-stage *p–i–n* diode limiter.

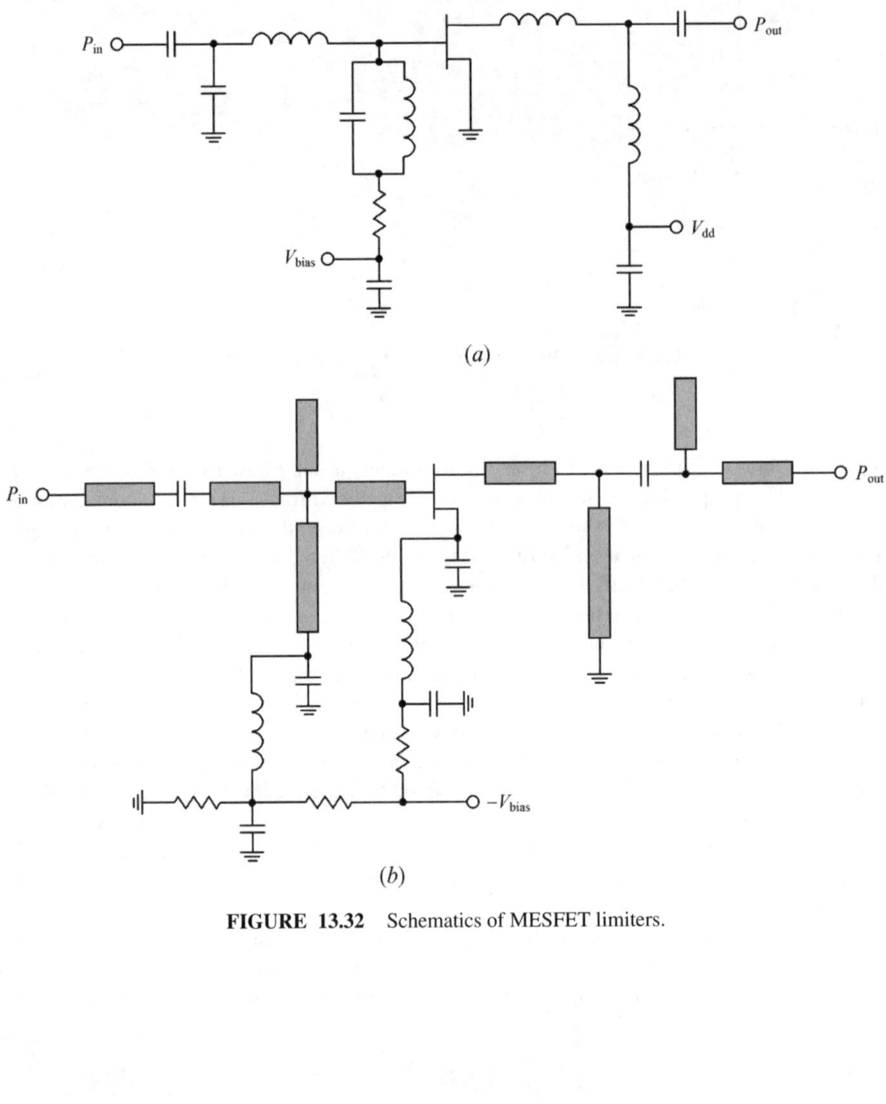

FIGURE 13.32 Schematics of MESFET limiters.

microwave power limiter with limiter diodes D_1 and D_2 and two Schottky detector diodes D_3 and D_4, where the Schottky diodes D_3 and D_4 are coupled with the main 50-Ω transmitting path through high-impedance quarterwave transmission lines [41]. These Schottky diodes detect the incident power and rapidly apply self-rectified current to the limiter diodes D_1 and D_2, causing a rapid increase in attenuation. The first high-power limiter diode D_1 is used as a prelimiter of the high-incident power, while the second low-power limiter diode D_2 provides protection during the initial stages of pulse rise time as having a faster response. Ideally, the electrical spacing between two limiter diodes should be of a one-quarter wavelength or an odd multiple of a one-quarter wavelength. The parasitic reactances of the limiter diodes can be compensated for by choosing the optimum spacing and characteristic impedance of the transmission line between them. If the limiter is required to handle very large input signals, a third stage can be added at the limiter input, spaced another one-quarter wavelength from the second diode, which now becomes the intermediate limiter stage. The double-stacked diode configurations can be used to handle about 6 dB more of the input power and to reduce the parasitic diode shunt capacitance by a factor of 2 [82,83]. Variable attenuation levels are achieved through biasing of the limiter diodes since the diode impedance is current-dependent and results in a continuously variable attenuation [84].

Instead of the *p–i–n* diode limiter, a GaAs MESFET limiter can be used where the device is operated beyond its compression point so that the RF level at its output cannot exceed the saturation level, irrespective of the incoming signal amplitude. In this case, to reduce the phase shift that depends on the amplitude of the RF signal and mainly caused by the device input gate–source capacitance variations occurring during a fraction of the RF cycle when the gate becomes forward biased, an additional series resistor with optimum value is included into the gate bias circuit to create the voltage drop that results in the increased reverse gate–source voltage [85]. As a result, for a given input power, the fraction of the RF cycle corresponding to a forward-biased gate is smaller and the variations of the gate–source capacitance also diminish. Figure 13.32(*a*) shows the circuit schematic of a 8-GHz 12-dBm MMIC MESFET limiter that incorporates a special gate biasing scheme with optimized series resistor and makes use of appropriate load conditions resulting in a phase variations less than 8° over a 22-dB input power range [86]. The hybrid topology of a microwave MESFET limiter with a single bias circuit shown in Figure 13.32(*b*) is based on the microstrip-line design and provides low phase distortions [87]. The dual-gate MESFET can be used as an excellent wideband limiter because it offers the design advantages of lower power output limiting, a sharper saturation knee, and the ability to set the output power level by changing the dc bias point [88].

REFERENCES

1. H. L. Krauss, C. W. Bostian, and F. H. Raab, *Solid State Radio Engineering*, New York: John Wiley & Sons, 1980.
2. K. Yamamoto, S. Suzuki, K. Mori, T. Asada, T. Okuda, A. Inoue, T. Miura, K. Chomei, R. Hattori, M. Yamanouchi, and T. Shimura, "A 3.2-V Operation Single-Chip Dual-Band AlGaAs/GaAs HBT MMIC Power Amplifier with Active Feedback Circuit Technique," *IEEE J. Solid-State Circuits*, vol. SC-35, pp. 1109–1120, Aug. 2000.
3. D. Dening and K. Muhonen, "Power Amplifier Protection Circuit," U. S. Patent 6,720,831, Apr. 2004.
4. Ang. Scuderi, L. La Paglia, Ant. Scuderi, F. Carrara, and G. Palmisano, "A VSWR-Protected Silicon Bipolar RF Power Amplifier with Soft-Slope Power Control," *IEEE J. Solid-State Circuits*, vol. SC-40, pp. 611–621, Mar. 2005.
5. F. Carrara, C. D. Presti, Ant. Scuderi, C. Santagati, and G. Palmisano, "A Methodology for Fast VSWR Protection Implemented in a Monolithic 3-W 55% PAE RF CMOS Power Amplifier," *IEEE J. Solid-State Circuits*, vol. SC-43, pp. 2057–2066, Sept. 2008.
6. Application Note 1599, *Power Control with the MRFIC0913 GaAs IPA and MC33169 Support IC*, Freescale Semiconductor, 2000.

7. A. van Bezooijen, F. van Straten, R. Mahmoudi, and A. H. M. van Roermund, "Avalanche Breakdown Protection by Adaptive Output Power Control," *IEEE Radio and Wireless Symp. Dig.*, pp. 519–522, 2006.
8. I. J. Bahl, "Broadband Power Detectors," *IEEE Microwave Mag.*, vol. 8, pp. 82–86, June 2007.
9. D. W. Bowman, "Circuit to Protect RF Output Amplifier Against Mismatch Damage," U. S. Patent 3,852,669, Dec. 1974.
10. A. Rachakonda and L. Hawkins, "Log Amps and Directional Couplers Enable VSWR Detection," *RF Design*, pp. 28–34, Jan. 2007.
11. H. Ikeda, H. Kosugi, and T. Uwano, "A Low Distortion and High Efficiency Parallel-Operation Power Amplifier Combined in Different Phases in Wide Range of Load Impedances," *IEEE MTT-S Int. Microwave Symp. Dig.*, pp. 535–538, 1996.
12. M. R. Millet, "Microwave Switching by Crystal Diodes," *IEEE Trans. Microwave Theory Tech.*, vol. MTT-6, pp. 284–290, July 1958.
13. M. E. Hines, "Fundamental Limitations in RF Switching and Phase Shifting Using Semiconductor Diodes," *Proc. IEEE*, vol. 52, pp. 697–708, June 1964.
14. Application Note, *Design with PIN Diodes*, Skyworks Solutions, Apr. 2009.
15. Application Note 922, *Applications of PIN Diodes*, Agilent Technologies, 1999.
16. Application Note 957–1, *Broadbanding the Shunt PIN Diode SPDT Switch*, Agilent Technologies, 1996.
17. K. W. Kobayashi, A. K. Oki, D. K. Umemoto, S. K. Z. Claxton, and D. C. Streit, "Monolithic GaAs HBT *p–i–n* Diode Variable Gain Amplifiers, Attenuators, and Switches," *IEEE Trans. Microwave Theory Tech.*, vol. MTT-41, p. 2295–2302, Dec. 1993.
18. J.-L. Lee, D. Zych, E. Reese, and D. M. Drury, "Monolithic 2–18 GHz Low Loss, On-Chip Biased PIN Diode Switches," *IEEE Trans. Microwave Theory Tech.*, vol. MTT-43, pp. 250–256, Feb. 1995.
19. S. Tanaka, S. Horiuchi, T. Kimura, and Y. Atsumi, "Design and Fabrication of Multiband *p–i–n* Diode Switches with Ladder Circuits," *IEEE Trans. Microwave Theory Tech.*, vol. MTT-54, pp. 1561–1568, Apr. 2006.
20. K. Miyatsuji and D. Ueda, "A GaAs High Power RF Single Pole Dual Throw Switch IC for Digital Mobile Communication System," *IEEE J. Solid-State Circuits*, vol. SC-30, pp. 979–983, Sept. 1995.
21. Y. Ota, M. Sakakura, K. Fujimoto, S. Yamamoto, and H. Fujimoto, "High Isolation and Low Insertion Loss Switch IC Using GaAs MESFET's," *IEEE Trans. Microwave Theory Tech.*, vol. MTT-43, pp. 2175–2177, Sept. 1995.
22. J. Frounchi and S. Harrold, "Very Fast GaAs HEMT Analogue Switch," *Electronics Lett.*, vol. 33, pp. 1499–1501, Aug. 1997.
23. K. F.-J. Huang O, "A 0.5-μm CMOS T/R Switch for 900-MHz Wireless Applications," *IEEE J. Solid-State Circuits*, vol. SC-36, pp. 486–492, Mar. 2001.
24. Q. Li and Y. P. Zhang, "CMOS T/R Switch Design: Towards Ultra-Wideband and Higher Frequency," *IEEE J. Solid-State Circuits*, vol. SC-42, pp. 563–570, Mar. 2007.
25. J. F. White, "High-Power, *p–i–n* Diode Controlled, Microwave Transmission Phase Shifters," *IEEE Trans. Microwave Theory Tech.*, vol. MTT-13, pp. 233–242, Mar. 1965.
26. R. V. Carver, "Broad-Band Diode Phase Shifters," *IEEE Trans. Microwave Theory Tech.*, vol. MTT-20, pp. 314–323, May 1972.
27. R. H. Hardin, E. J. Downey, and J. Munushian, "Electronically-Variable Phase Shifters Utilizing Variable Capacitance Diodes," *Proc. IRE*, vol. 48, pp. 944–945, May 1960.
28. C.-L. Chen, W. E. Courtney, L. J. Mahoney, M. J. Manfra, A. Chu, and H. A. Atwater, "A Low-Loss *Ku*-Band Monolithic Analog Phase Shifter," *IEEE Trans. Microwave Theory Tech.*, vol. MTT-35, pp. 315–320, Mar. 1987.
29. E. C. Niehenke, V. V. DiMarco, and A. Friedberg, "Linear Analog Hyperabrupt Varactor Diode Phase Shifter," *IEEE MTT-S Int. Microwave Symp. Dig.*, pp. 657–660, 1985.
30. J. F. White, "Diode Phase Shifters for Array Antennas," *IEEE Trans. Microwave Theory Tech.*, vol. MTT-22, pp. 658–674, June 1974.
31. H.-R. Ahn and I. Wolff, "Asymmetric Ring-Hybrid Phase Shifters and Attenuators," *IEEE Trans. Microwave Theory Tech.*, vol. MTT-50, pp. 1146–1155, Apr. 2002.

32. S. Lucyszyn and I. D. Robertson, "Two-Octave Bandwidth Monolithic Analog Phase Shifter," *IEEE Microwave and Guided Wave Lett.*, vol. 2, pp. 343–345, Aug. 1992.

33. A. Keerti, J. Xiang, and A.-V. Pham, "High Power Linearized RF Phase Shifter Using Anti-Series Diodes," *IEEE Microwave and Wireless Components Lett.*, vol. 16, pp. 200–202, Apr. 2006.

34. C.-S. Lin, S.-F. Chang, C.-C. Chang, and Y.-H. Shu, "Design of a Reflection-Type Phase Shifter with Wide Relative Phase Shift and Constant Insertion Loss," *IEEE Trans. Microwave Theory Tech.*, vol. MTT-55, pp. 1862–1868, Sept. 2007.

35. C.-S. Lin, S.-F. Chang, and W.-C. Hsiao, "A Full-360° Reflection-Type Phase Shifter with Constant Insertion Loss," *IEEE Microwave and Wireless Components Lett.*, vol. 18, pp. 106–108, Feb. 2008.

36. W. A. Davis, "Design Equations and Bandwidth of Loaded-Line Phase Shifters," *IEEE Trans. Microwave Theory Tech.*, vol. MTT-22, pp. 561–563, May 1974.

37. H. A. Atwater, "Circuit Design of the Loaded-Line Phase Shifter," *IEEE Trans. Microwave Theory Tech.*, vol. MTT-33, pp. 626–634, July 1985.

38. I. J. Bahl and K. C. Gupta, "Design of Loaded-Line *p–i–n* Diode Phase Shifter Circuit," *IEEE Trans. Microwave Theory Tech.*, vol. MTT-28, pp. 219–224, Mar. 1980.

39. A. S. Nagra and R. A. York, "Distributed Analog Phase Shifters with Low Insertion Loss," *IEEE Trans. Microwave Theory Tech.*, vol. MTT-47, pp. 1705–1711, Sept. 1999.

40. G. D. Lynes, G. E. Johnson, B. E. Huckleberry, and N. H. Forrest, "Design of a Broad-Band 4-Bit Loaded Switched-Line Phase Shifter," *IEEE Trans. Microwave Theory Tech.*, vol. MTT-22, pp. 693–697, June 1974.

41. L. G. Maloratsky, *Passive RF & Microwave Integrated Circuits*, New York: Newnes, 2004.

42. B. M. Schiffman, "A New Class of Broad-Band Microwave 90-Degree Phase Shifters," *IEEE Trans. Microwave Theory Tech.*, vol. MTT-6, pp. 232–237, Apr. 1958.

43. J. L. R. Quirarte and J. P. Starski, "Novel Schiffman Phase Shifters," *IEEE Trans. Microwave Theory Tech.*, vol. MTT-41, pp. 9–14, Jan. 1993.

44. B. M. Schiffman, "Multisection Microwave Phase-Shift Networks," *IEEE Trans. Microwave Theory Tech.*, vol. MTT-14, p. 209, Apr. 1966.

45. V. P. Meschanov, I. V. Metelnikova, V. D. Tupikin, and G. G. Chumaevskaya, "A New Structure of Microwave Ultrawide-Band Differential Phase Shifter," *IEEE Trans. Microwave Theory Tech.*, vol. MTT-42, pp. 762–765, May 1994.

46. C. Andricos, I. J. Bahl, and E. L. Griffin, "*C*-Band 6-Bit GaAs Monolithic Phase Shifter," *IEEE Trans. Microwave Theory Tech.*, vol. MTT-33, pp. 1591–1596, Dec. 1985.

47. K. Miyaguchi, M. Hieda, K. Nakahara, H. Kurusu, M. Nii, M. Kasahara, T. Takagi, and S. Urasaki, "An Ultra-Broad-Band Reflection-Type Phase-Shifter MMIC with Series and Parallel *LC* Circuits," *IEEE Trans. Microwave Theory Tech.*, vol. MTT-49, pp. 2446–2452, Dec. 2001.

48. C. Tsironis and P. Harrop, "Dual-Gate MESFET Phase Shifter with Gain at 12 GHz," *Electronics Lett.*, vol. 16, pp. 553–554, July 1980.

49. J. L. Vorhaus, R. A. Pucel, and Y. Tajima, "Monolithic Dual-Gate GaAs FET Digital Phase Shifter," *IEEE Trans. Microwave Theory Tech.*, vol. MTT-30, pp. 982–992, July 1982.

50. M. Kumar, R. J. Menna, and H.-C. Huang, "Broad-Band Active Phase Shifter Using Dual-Gate MESFET," *IEEE Trans. Microwave Theory Tech.*, vol. MTT-29, pp. 1098–1102, Oct. 1981.

51. P.-Y. Chen, T.-W. Huang, H. Wang, Y.-C. Wang, C.-H. Chen, and P.-C. Chao, "K-Band HBT and HEMT Monolithic Active Phase-Shifters Using Vector Sum Method," *IEEE Trans. Microwave Theory Tech.*, vol. MTT-52, pp. 1414–1424, May 2004.

52. C. F. Campbell and S. A. Brown, "A Compact 5-Bit Phase-Shifter MMIC for *K*-Band Satellite Communication Systems," *IEEE Trans. Microwave Theory Tech.*, vol. MTT-48, pp. 2652–2656, Dec. 2000.

53. I. J. Bahl and D. Conway, "*L*- and *S*-Band Compact Octave Bandwidth 4-Bit MMIC Phase Shifters," *IEEE Trans. Microwave Theory Tech.*, vol. MTT-56, pp. 293–299, Feb. 2008.

54. J. K. Hunton and A. G. Ryals, "Microwave Variable Attenuators and Modulators Using *PIN* Diodes," *IEEE Trans. Microwave Theory Tech.*, vol. MTT-10, pp. 262–273, July 1962.

55. F. G. Atanasso, "A Low Phase Shift Step Attenuator Using *p–i–n* Diodes Switches," *IEEE Trans. Microwave Theory Tech.*, vol. MTT-28, pp. 774–776, July 1980.

56. Application Note 1003, *A Wideband General Purpose PIN Diode Attenuator*, Skyworks Solutions, 2005.

57. W.-T. Kang, I.-S. Chang, and M.-S Kang, "Reflection-Type Low-Phase-Shift Attenuators," *IEEE Trans. Microwave Theory Tech.*, vol. MTT-46, pp. 1019–1021, July 1998.

58. C. R. Trent and T. M. Weller, "S-Band Reflection Type Variable Attenuator," *IEEE Microwave and Wireless Components Lett.*, vol. 12, pp. 243–245, July 2002.

59. D. Adler and P. Mariato, "Broadband Phase Invariant Attenuator," *IEEE MTT-S Int. Microwave Symp. Dig.*, pp. 673–676, 1988.

60. W. Gosling, "Voltage Controlled Attenuation Using Field Effect Transistors," *IEEE Trans. Audio*, vol. AU-13, pp. 112–120, Sept. 1965.

61. H. P. von Ow, "Reducing Distortion in Controlled Attenuators Using FET," *Proc. IEEE*, vol. 56, pp. 1718–1719, Oct. 1968.

62. G. S. Barta, K. E. Jones, G. C. Herrick, and E. W. Strid, "Surface-Mounted GaAs Active Splitters and Attenuator MMIC's Used in a 1–10-GHz Leveling Loop," *IEEE Trans. Microwave Theory Tech.*, vol. MTT-34, pp. 1569–1575, Dec. 1986.

63. D. Krafcsik, F. Ali, and S. Bishop, "Broadband, Low-Loss 5- and 6-Bit Digital Attenuators," *IEEE MTT-S Int. Microwave Symp. Dig.*, pp. 673–676, 1988.

64. Y. Tajima, T. Tsukii, R. Mozzi, E. Tong, L. Hanes, and B. Wrona, "GaAs Monolithic Wideband (2–18 GHz) Variable Attenuators," *IEEE MTT-S Int. Microwave Symp. Dig.*, pp. 479–481, 1982.

65. H. Dogan and R. G. Meyer, "Intermodulation Distortion in CMOS Attenuators and Switches," *IEEE J. Solid-State Circuits*, vol. SC-42, pp. 529–539, Mar. 2007.

66. H. Dogan, R. G. Meyer, and A. M. Niknejad, "Analysis and Design of RF CMOS Attenuators," *IEEE J. Solid-State Circuits*, vol. SC-43, pp. 2269–2283, Oct. 2008.

67. S. M. Daoud and P. N. Shastry, "A Novel Wideband MMIC Voltage Controlled Attenuator with a Bandpass Filter Topology," *IEEE Trans. Microwave Theory Tech.*, vol. MTT-54, pp. 2576–2583, June 2006.

68. H. Kondoh, "DC-50 GHz MMIC Variable Attenuator with a 30dB Dynamic Range," *IEEE MTT-S Int. Microwave Symp. Dig.*, pp. 499–502, 1988.

69. C.-W. Kim and Y.-G. Kim, "A 2.7-V SiGe HBT Variable Gain Amplifier for CDMA Applications," *IEEE Microwave and Wireless Components Lett.*, vol. 13, pp. 502–504, Dec. 2003.

70. M. Inamori, K. Motoyoshi, T. Kitazawa, K. Tara, and M. Hagio, "A New GaAs Variable-Gain Amplifier MMIC with a Wide-Dynamic-Range and Low-Voltage-Operation Linear Attenuation Circuit," *IEEE Trans. Microwave Theory Tech.*, vol. MTT-48, pp. 182–186, Feb. 2000.

71. G. Hau, T. B. Nishimura, and N. Iwata, "High Efficiency, Wide Dynamic Range Variable Gain and Power Amplifier MMICs for Wide-Band CDMA Handsets," *IEEE Microwave and Wireless Components Lett.*, vol. 11, pp. 13–15, Jan. 2001.

72. Q. Chaudhry, R. Alidio, G. Sakamoto, and T. Cisco, "A SiGe MMIC Variable Gain Cascode Amplifier," *IEEE Microwave and Wireless Components Lett.*, vol. 12, pp. 424–425, Nov. 2002.

73. H. Hayashi and M. Muragushi, "An MMIC Variable-Gain Amplifier Using a Cascode-Connected FET with Constant Phase Deviation," *IEICE Trans. Electron.*, vol. E81-C1, pp. 70–77, Jan. 1998.

74. M. Maeda and Y. Minai, "Microwave Variable-Gain Amplifier with Dual-Gate GaAs FET," *IEEE Trans. Microwave Theory Tech.*, vol. MTT-22, pp. 1226–1230, Dec. 1974.

75. N. Mamodaly, P. Quentin, P. Dueme, and J. Obregon, "100 MHz to 17 GHz Dual-Gate Variable-Gain Amplifier," *IEEE Trans. Microwave Theory Tech.*, vol. MTT-30, pp. 918–919, June 1982.

76. D. Coffing, E. Main, M. Randol, and G. Szklarz, "A Variable Gain Amplifier with 50-dB Control Range for 900-MHz Applications," *IEEE J. Solid-State Circuits*, vol. SC-37, pp. 1169–1175, Sept. 2002.

77. A. Bilotti, "Applications of a Monolithic Analog Multipliers," *IEEE J. Solid-State Circuits*, vol. SC-3, pp. 373–380, Dec. 1968.

78. I. Kipnis, J. F. Kukielka, J. Wholey, and C. P. Snapp, "Silicon Bipolar Fixed and Variable Gain Amplifier MMICs for Microwave and Lightwave Applications up to 6 GHz," *IEEE MTT-S Int. Microwave Symp. Dig.*, pp. 109–112, 1989.

79. V. T. S. Vintola, M. J. Matilainen, S. J. K. Kalajo, and E. A. Jarvinen, "Variable-Gain Power Amplifier for Mobile WCDMA Applications," *IEEE Trans. Microwave Theory Tech.*, vol. MTT-49, pp. 2464–2471, Dec. 2001.

80. Application Note, *PIN Limiter Diodes in Receiver Protectors*, Skyworks Solutions, Aug. 2008.

81. Application Note 1959, *Low Cost Surface Mount Power Limiters*, Agilent Technologies, 1999.

82. D. J. Seymour, D. D. Heston, and R. E. Lehmann, "X-Band and Ka-Band Monolithic GaAs PIN Diode Variable Attenuation Limiters," *IEEE MTT-S Int. Microwave Symp. Dig.*, pp. 255–258, 1988.

83. D. G. Smith, D. D. Heston, and D. L. Allen, "Designing High-Power Limiter Circuits with GaAs PIN Diodes," *IEEE MTT-S Int. Microwave Symp. Dig.*, pp. 329–332, 1999.

84. D. G. Smith, D. D. Heston, J. Heston, B. Heimer, and K. Decker, "Designing Reliable High-Power Limiter Circuits with GaAs PIN Diodes," *IEEE MTT-S Int. Microwave Symp. Dig.*, vol. 3, pp. 1245–1247, 2002.

85. T. Parra, M. Gayral, O. Llopis, M. Pouysegur, J. Sautereau, and J. Graffeuil, "Design of a Low Phase Distortion GaAs FET Power Limiter," *IEEE Trans. Microwave Theory Tech.*, vol. MTT-39, pp. 1059–1062, June 1991.

86. T. Parra, J. Dienor, M. Gayral, M. Pouysegur, J. Sautereau, and J. Graffeuil, "*X*-Band Low Phase Distortion MMIC Power Limiter," *IEEE Trans. Microwave Theory Tech.*, vol. MTT-41, pp. 876–879, May 1993.

87. S. Fukuda, M. Kitamura, Y. Ara, and I. Haga, "A New Microwave Amplitude Limiter Using GaAs Field Effect Transistor," *IEEE MTT-S Int. Microwave Symp. Dig.*, pp. 240–242, 1977.

88. R. J. Hamilton, "Wide Band, Dual-Gate GaAs F.E. T. Output Limiters," *IEEE MTT-S Int. Microwave Symp. Dig.*, pp. 256–258, 1979.

14 Transmitter Architectures

This chapter describes the different types of radio transmitter architectures, history of radio communication, conventional types of radio transmission, and modern communication systems. Amplitude-modulated transmitters representing the oldest technique for radio communication are based on high- or low-level modulation methods, with the particular case of an amplitude keying. Single-sideband transmitters as the next-generation transmitters could provide higher efficiency because of the transmission of a single sideband only. Frequency-modulated transmitters then became a revolutionary step to improve the quality of a broadcast transmission. TV transmitters include different modulation techniques for transmitting audio and video information, both analog and digital. Wireless communication transmitters as a part of the cellular technologies provide a worldwide wireless radio access. Radar transmitters are required for many commercial and military applications such as phased-array radars, automotive radars, or electronic warfare systems. Satellite transmission systems contribute to worldwide transmission of any communication signals through satellite transponders and offer communication for areas with any population density and location. Ultra-wideband transmission is very attractive for its low-cost and low-power communication applications, occupying a very wide frequency range.

14.1 AMPLITUDE-MODULATED TRANSMITTERS

Amplitude-modulated transmitters whose extensive development began in about 1927 based on vacuum tubes for generating radio-frequency signals are used in high-frequency communication, broadcasting, aircraft, and navigation systems. Early continuous-wave vacuum-tube transmitters used a power oscillator as the generator of the RF power. However, the requirements for better frequency stability soon caused a significant change in the transmitter configuration because of success in the development in crystal stability when the requirements for piezoelectric crystal control of frequency began establishing in 1929 [1]. Currently, amplitude-modulated radio transmissions are used worldwide at medium frequencies (530–1700 kHz) and high-frequency bands (3–26 MHz) where the carrier signals are amplitude modulated with audio bandwidths of 5 kHz [2]. In short-wave broadcasting, transmitters use both solid-state (5-kW modules) or vacuum-tube RF power amplifiers (PAs), with power levels ranging from 10 kW to 2 MW. Full-carrier double-sideband amplitude modulation can be accomplished by several techniques, including high-level modulation by variation of the final RF PA supply voltage by transformer coupling and low-level modulation by controlling the gate bias voltage. An advantage of the high-level modulation is that only the final stage needs to be modulated when all preliminary stages can be driven at a constant RF level. In low-level modulation, a small audio stage is used to modulate a low-power RF carrier stage, the output of which is then amplified using a linear RF PA. However, the serious disadvantage of this system is that the amplifier chain is less efficient because it has to be linear to preserve modulation. In order to use Class-C PA as a final stage, the feedback linearization scheme is required.

RF and Microwave Transmitter Design, First Edition. Andrei Grebennikov.

14.1.1 Collector Modulation

The traditional method of accomplishing high-quality amplitude modulation (AM) uses a transformer-coupled push–pull Class B audio-frequency (AF) amplifier to vary the supply collector (drain) voltage of a Class RF PA designed in a conventional (with shunt resonant circuit) or mixed-mode (with series resonant circuit) configuration, as shown in Figure 14.1(*a*) [3]. When efficiency is not a critical parameter, a Class A audio amplifier can be used. In this case, all components in a modulating path, namely the transformer, AF amplifier, and AF drivers if required, are collectively called the *modulator*. The linearity of high-level modulation technique depends on the linearity of the modulator and the modulation characteristic of the RF PA. In the latter case, the nonlinear capacitive coupling between

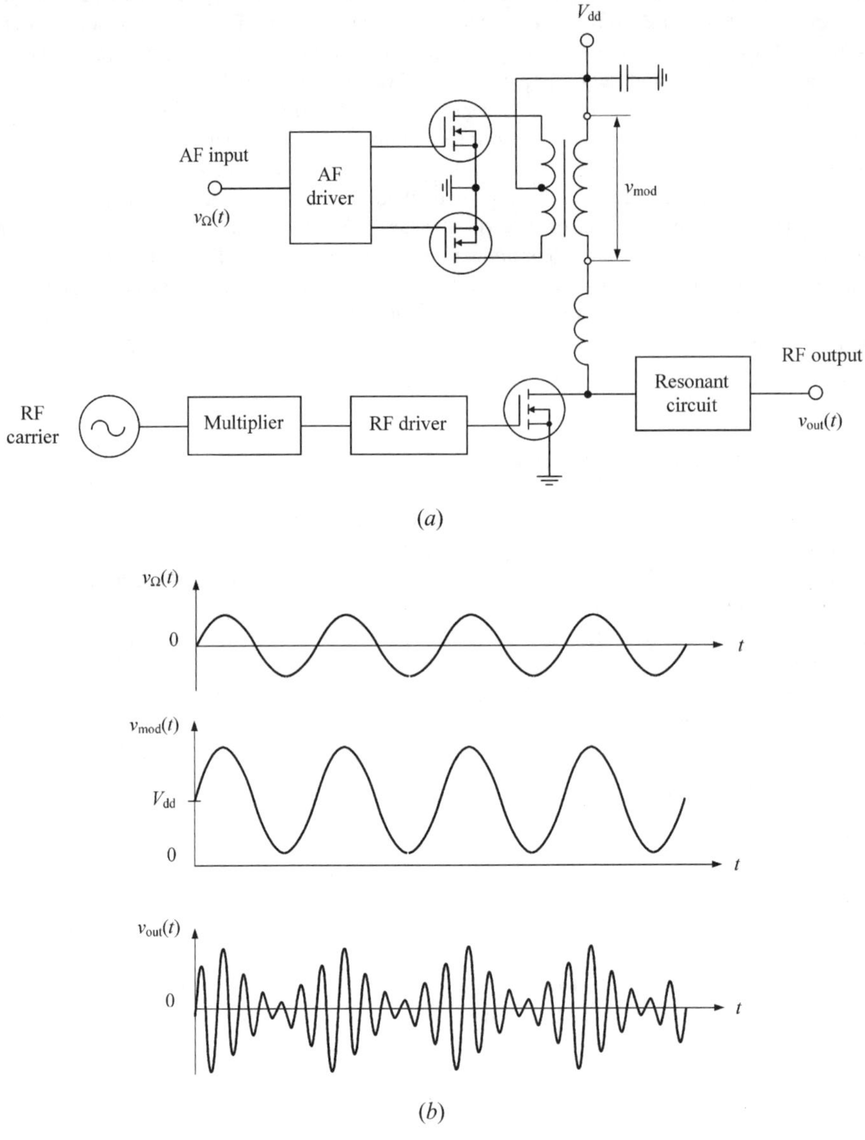

FIGURE 14.1 Basic transmitter structure and operation waveforms for collector modulation.

gate and drain of the field-effect transistor (or between base and collector of the bipolar device) produces a minimum output or feedthrough signal that is present even when $V_{dd} = 0$, resulting in both amplitude and phase modulation (PM) nonlinearity at low signal level. Consequently, too much drive results in nonlinearity at low outputs, while too little drive results in nonlinearity at high outputs when device saturation does not occur at high values of V_{dd}.

An analysis of waveforms shown in Figure 14.1(*b*) shows that the drain (or collector) supply voltage of the final RF power amplifying stage swings above and below the unmodulated carrier level, which is equal to V_{dd}. In this case, the final RF power amplifying stage must deliver the peak envelope power when its supply voltage is equal to $2V_{dd}$. Since the drain dc and fundamental current amplitudes change almost linearly with the supply voltage, the final RF amplifying stage represents approximately constant resistive load and efficiency close to the saturated efficiency in Class B. As a result, the low-level nonlinear distortions and high overall efficiency can generally be achieved. The overall efficiency of the transformer-coupled collector modulation by a single sinusoid can be calculated (assuming that only the final RF PA is modulated and the power required by the AF and RF drivers is ignored) as

$$\eta = \eta_{RF}\eta_{AF}\frac{1+\dfrac{m^2}{2}}{\eta_{AF}+\dfrac{m^2}{2}} \tag{14.1}$$

where m is the modulation index, η_{RF} is the efficiency of the final RF PA, and η_{AF} is the efficiency of the final audio amplifier. If both the AF and RF final amplifiers had 78.5% efficiency of an ideal Class B mode, the overall efficiency η of the two final amplifiers would be 72% for $m = 1$. If the final AF amplifier operates in Class A with 50% efficiency, the overall efficiency is 58.9%.

To minimize the amplitude nonlinear distortions at low voltages at the device collector when the base–collector junction can be forward biased resulting in low junction impedance and feedthrough signal, an additional collector modulation of the driver stage (with modulation index $m < 1$) by using the modulating signal coupled with the AF amplifier output transformer can be implemented, as shown in Figure 14.2. This provides the lower signal power at the input of the final RF amplifying stage at low voltages at the device collector, thus minimizing the feedthrough signal through it and

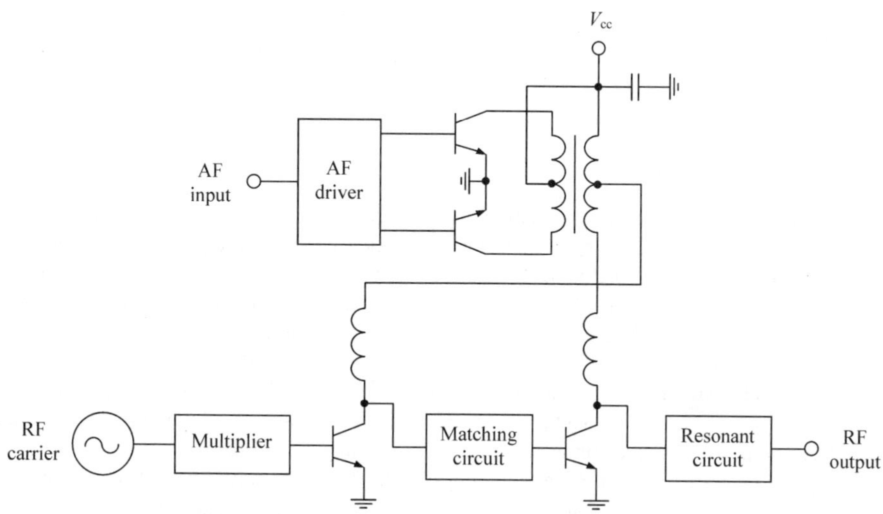

FIGURE 14.2 Basic transmitter structure for dual collector modulation.

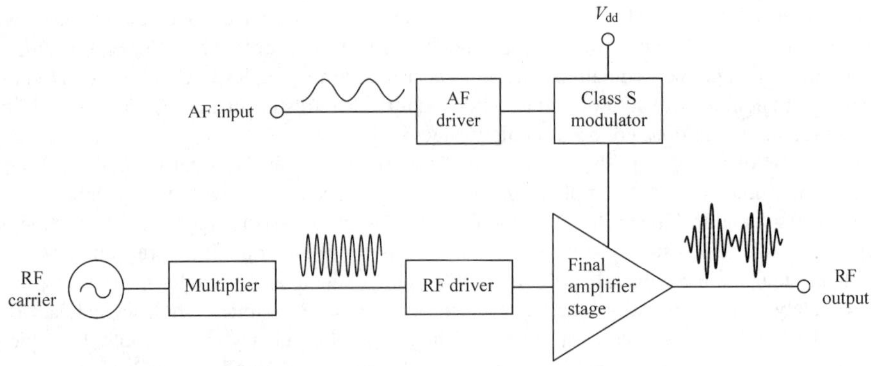

FIGURE 14.3 Basic transmitter structure with Class-S modulator.

linearizing its modulation characteristic. Modulation can also be accomplished by supplying audio-frequency power simultaneously to the plate of the modulating amplifier and to the grid of the driver stage in the vacuum-tube transmitter [4]. An automatic gain control can be applied to the modulator circuit to maintain the average percentage of modulation at a relatively high value irrespective of the average level of the incoming modulating signal. It is particularly helpful in the communication systems where the average voice power will vary over a wide range from speaker to speaker.

Figure 14.3 shows the AM transmitter structure with corresponding waveforms where the collector modulation is accomplished by using a Class S modulator, which is based on a series pulse-width modulation principle. In this case, since the modulator output voltage cannot exceed V_{dd}, the peak envelope output power must be produced with V_{dd} (rather than $2V_{dd}$) supplied to the final RF amplifier stage, with the unmodulated carrier level equal to $V_{dd}/2$. Therefore, the Class S modulator is biased to produce an output from the modulator of $V_{dd}/2$ in the absence of an input AF signal. Then, applying the AF signal causes the modulator to produce a voltage that varies between 0 and V_{dd}. The overall efficiency of the modulator and final RF power stage is obtained as

$$\eta = \eta_{RF}\eta_{AF} \tag{14.2}$$

where the efficiency η_{AF} of a Class S modulator is assumed to be constant. When using high-efficiency Class S modulator with final Class D PA, the overall efficiency of the short-wave high-power broadcasting transmitters of around 70% can be achieved [5,6].

14.1.2 Base Modulation

AM can also be accomplished by variation of the gate (or base) bias voltage in a Class B PA in accordance with AF modulating signal that produces a corresponding variation in the voltage amplitude of the RF carrier signal at the device drain (or collector). Although some distortion of the audio signal is inherent in this method of AM, its circuit implementation is simple and suitable for low-cost AM transmitters. The gate modulation of the field-effect transistor (or base modulation of the bipolar device) is similar to the grid modulation of the vacuum-tube PAs [7]. Figure 14.4(a) shows the basic transmitter structure with gate–bias modulation of the final RF amplifying stage based on a MOSFET device. In this case, the voltage applied to the gate represents the sum of the RF driving, dc bias, and AF modulating signal voltages, and the conduction angle for the RF carrier signal varies in accordance with the time-varying AF modulating signal. As an alternative, the AF modulating signal can be provided directly from the output of the AF driver, with separate gate bias circuit for final RF amplifying device.

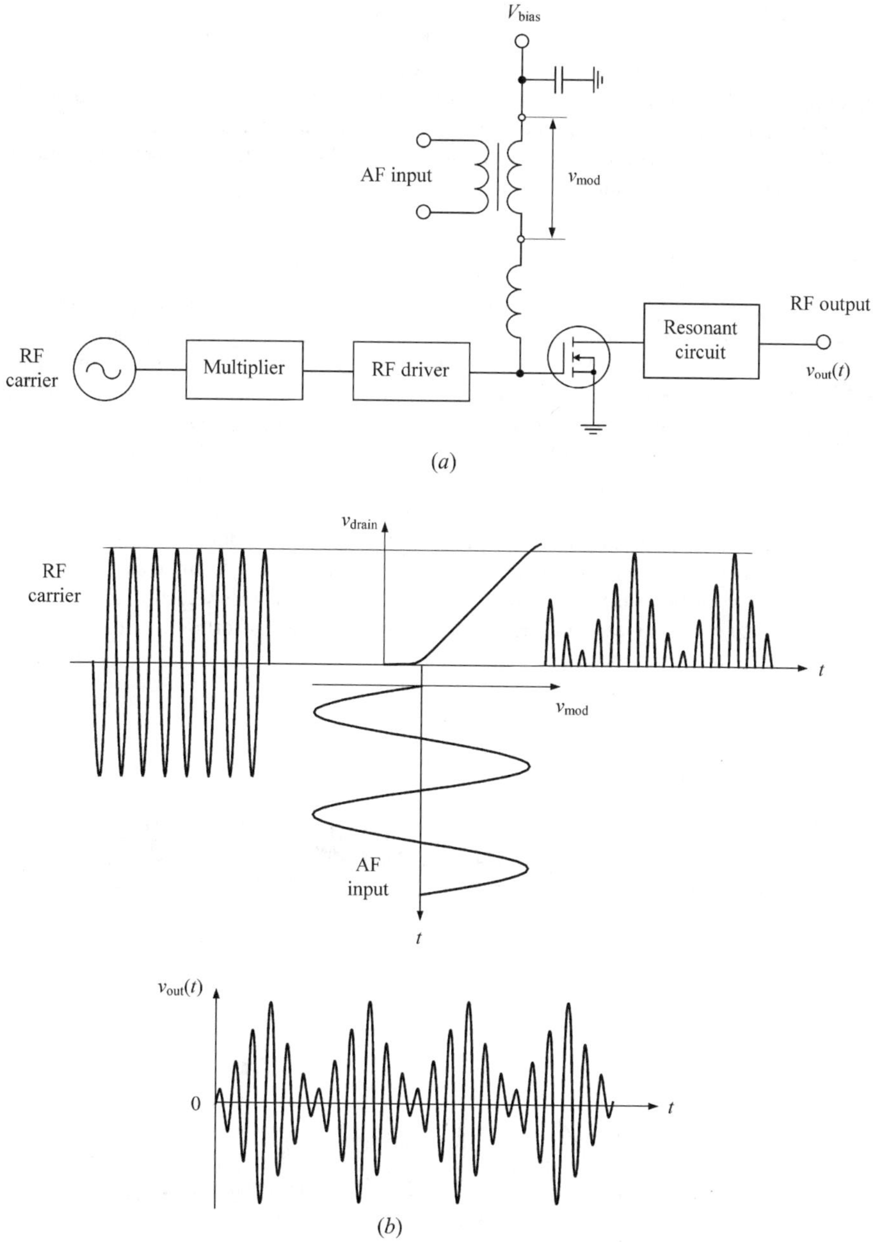

FIGURE 14.4 Basic transmitter structure and operation waveforms for base modulation.

The amplitude-modulated RF output signal is shown in Figure 14.4(*b*), where the dc bias and the modulating signal are chosen to produce a Class B operation at positive modulation peaks and a zero conduction angle at negative modulation peaks, based on the sufficiently linear device transfer characteristic. The simple overall efficiency approximation assumes a linear variation of the instantaneous time-varying efficiency from zero value with $v_{drain} = 0$ to maximum (peak-envelope) value η_{max} with $v_{drain} = V_{dd}$, where V_{dd} is the drain supply voltage. As a result, the efficiency for an unmodulated carrier signal is $\eta_{max}/2$ and the average efficiency for 100% modulation by a single sinusoid is $3\eta_{max}/4$. However, due to significant nonlinearity of the device transfer characteristic close to cutoff and saturation regions, the modulation index is usually chosen to be less than 70–80% to minimize amplitude (AM–AM) distortions. At the same time, nonlinearity of the input device nonlinear capacitance, which can be modeled as a junction capacitance, can cause the unwanted phase (AM–PM) distortions of the transmitting amplitude-modulated signal if its phase characteristic is considered an important parameter.

14.1.3 Low-Level Modulation

The block diagram of a typical amplitude-modulated transmitter, in which an amplitude modulator is followed by a linear PA, is shown in Figure 14.5 with corresponding waveforms. This is a low-level modulation system where the modulation takes place at a power level less than transmitter output, and the linear amplification of the amplitude-modulated signals is often used in low-power broadcasting and communication transmitters. In this case, to minimize nonlinear distortions of the transmitting signal, the input level of the amplitude-modulated signal and the gate (or base) bias condition with constant conduction angle must be chosen so that only the linear part of the transfer characteristic of the linear PA can be used, thus providing its active (linear current source) mode and avoiding saturated mode. Since existing bipolar or field-effect power transistors operated in Class B do not change abruptly from a cutoff mode to an active mode that would result to so-called *crossover distortion*, the active device must be biased in a Class AB with small quiescent current in the collector or drain. The amount of a quiescent current required for minimum distortion is usually determined

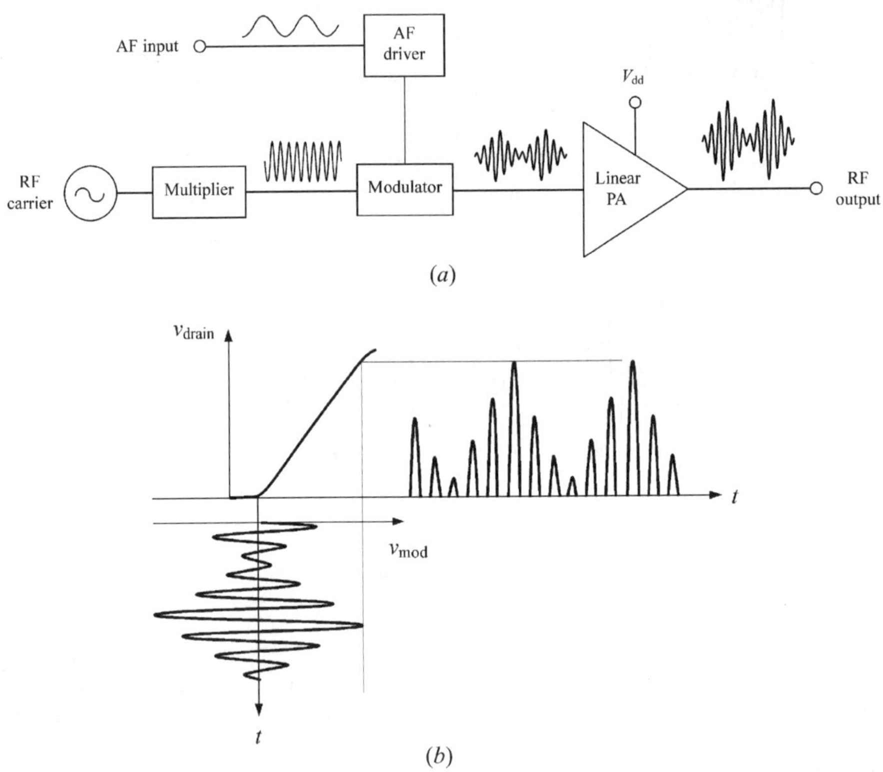

FIGURE 14.5 Basic transmitter structure and operation waveforms for low-level modulation.

by simulation or measurement (typically, in between 1% and 10% of the collector or drain dc current under nominal operation), since theoretical prediction is quite complicated.

For a sinusoidal full-carrier AM with $m = 1$ in a Class B PA, the dc input current is proportional to the RF output current (or voltage) amplitude. Consequently, the average dc power for an AM signal is the same as the dc power for the carrier without modulation, with an unmodulated carrier level equal to $V_{dd}/2$. The average RF output power is equal to $P_{avr} = P_0\,(1 + m^2/2) = (3/2)P_0$, where P_0 is the unmodulated carrier power that is four times less than the peak envelope power P_{PEP}. As a result, the average drain efficiency can be written as

$$\eta = \frac{(3/2)\,(P_{PEP}/4)}{(1/2)\,(4/\pi)\,P_{PEP}} = \frac{3\pi}{16} \tag{14.3}$$

which is approximately equal to 58.9% [3].

14.1.4 Amplitude Keying

The continuous-wave radio transmitters are often operated in an amplitude-shift-keying (ASK) mode required in remote control circuits, security systems, wireless data transmission, or in accordance to dots and dashes to produce radio telegraph signals in a simple case of on-off keying. This amplitude keying is usually performed at a low or moderate power level, and the required transmitter power is then obtained by the high-power amplifying stage with a fixed bias. The block diagram of a simple amplitude-keyed transmitter is shown in Figure 14.6(a), where the gate of the active device is turned on and off in accordance with the rectangular modulating signal delivered directly through the gate bias circuit. In a 60-GHz high data rate transmitter, the ASK modulator block consists of a 60-GHz fundamental-mode oscillator, a buffer, and a switch (using a 0.1-μm GaAs HEMT technology) followed by the driver, bandpass filter (BPF), PA, and antenna fabricated with a low-temperature co-fired ceramic (LTCC) [8].

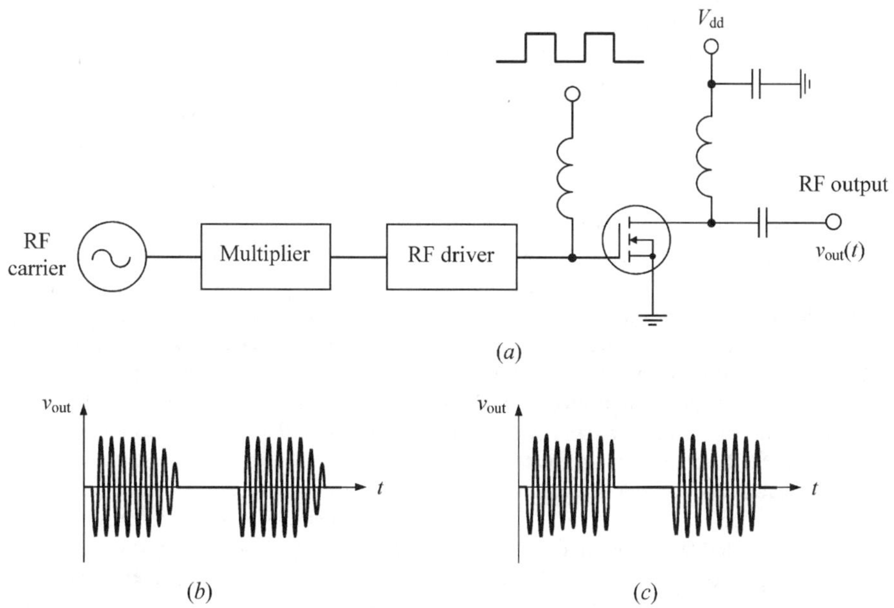

FIGURE 14.6 Basic transmitter structure and operation waveforms for amplitude keying.

The optional inclusion of a frequency multiplier improves stability by allowing the frequencies of the oscillator and high-power amplifiers (HPAs) to be different. In amplitude keying, the RF pulses should have rounded edges, as shown in Figure 14.6(*b*). Otherwise, if the voltage rises and falls very rapidly, the abrupt changes in amplitude will introduce high-order sidebands, at frequencies differing significantly from the carrier frequency. However, when the rise and fall times are too long, this will lead to the nonlinear distortions of the transmitted signal that may significantly complicate its identification after detection. In this case, it is necessary to choose the active devices with low input time constant and high transition frequency. In addition, it needs to carefully design the bias circuit in order to maximize isolation between modulating signal path and the carrier signal path and minimize low-frequency transient response shown in Figure 14.6(*c*).

14.2 SINGLE-SIDEBAND TRANSMITTERS

In mid-1915, during the radio–telephone experiments conducted with a powerful vacuum-tube transmitter in Arlington, Virginia, it was recognized by Harold Arnold that the antenna could be tuned to one side of the carrier frequency that contained all the signal elements necessary to reproduce the original speech [9]. In addition to suppressing one sideband, John Carson proposed suppression of the carrier as well [10]. As a result, a transatlantic reliable one-way speech transmission was demonstrated in January 1923 using a powerful experimental single-sideband (SSB) transmitter, operating at a mid-band frequency of 57 kHz [11]. The first short-wave SSB equipment provided only a single speech channel on one side of the carrier, but it was quickly determined that a common PA could be made sufficiently linear to permit adding a second channel on the opposite of the carrier [12]. In this case, two frequency tones of equal amplitudes were then used for testing the transmitting signal quality, each frequency tone being half the peak sideband amplitude. With introduction of automotive tuning capability of the output resonant circuit and high-power wideband ferrite balun at the end of 1950s, the broadband operation of a short-wave vacuum-tube SSB transmitter was extended to cover a frequency range from 2 to 25 MHz. One of the first solid-state transmitters that used eight wideband RF modules in a hybrid arrangement to produce an output power of 1 kW was introduced in the early 1970s [13]. Modern SSB transmitters are widely used in long-range military, marine, aircraft, and amateur communication systems where frequency of operation can vary over several octaves.

An SSB can be obtained by passing the output of a carrier-suppression system through filter circuits that are sufficiently selective to transmit one sideband while suppressing the other. However, the requirements that filter must meet are sufficiently strong, especially if the carrier frequency is too high. For example, the standard frequency band of the telephone channel is from 300 to 3400 Hz, which means that very effective rejection can be done for the lowest offset frequency of 2×300 Hz. Even with well-designed filters such sharpness of discrimination is possible only if the carrier frequency is low enough so that 600 Hz is not too small a percentage of the carrier frequency. As a result, the filter method is suitable only for obtaining SSBs corresponding to low or moderate carrier frequencies. If the sideband must be located in a high-frequency part of spectrum, then it must be translated in frequency by heterodyne or superheterodyne action, resulting generally to multiple-conversion transmitter architectures mostly based on the discrete-band and broadband methods.

The discrete-band frequency translation technique has evolved from vacuum-tube transmitters and is well suited for operation in several small assigned bands. Figure 14.7(*a*) shows the discrete-band SSB transmitter architecture with two balanced modulators. For such an SSB communication transmitter, the SSB signal at intermediate frequency can be achieved using a balanced modulator and a BPF (which can represent a quartz or electromechanical filter) for fixed intermediate carrier frequency f_1 in range of 100 to 500 kHz. In this case, the modulating signal with frequencies from f_1 + 300 Hz to f_1 + 3400 Hz applies to the second balanced modulator, and filter must suppress one of the sidebands separated by $2(f_1 + 300\text{ Hz})$, or more by about 1 MHz for $f_1 = 500$ kHz than without frequency upconversion. After filtering, which can be done with simple multisection *LC* filters, the resulting SSB signal then translated to the assigned small frequency band within the high frequency

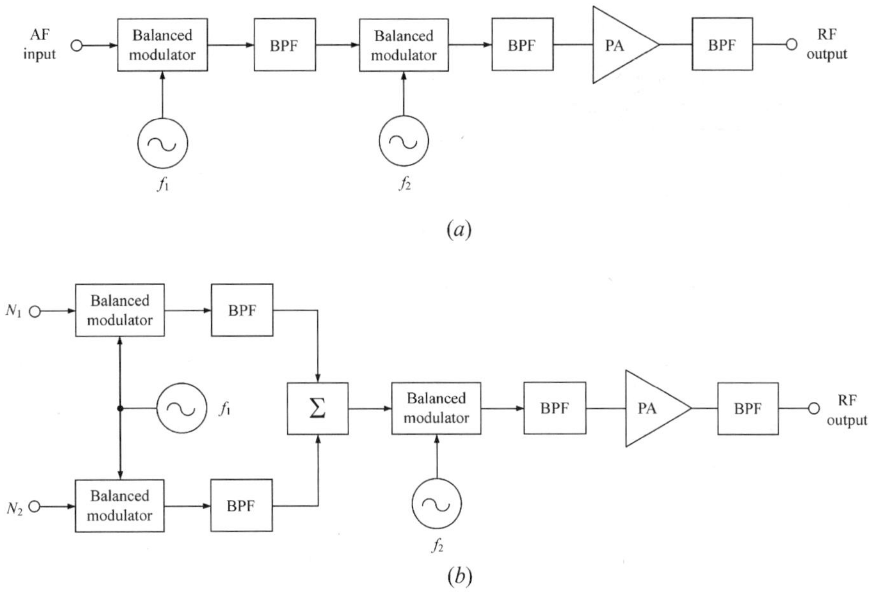

FIGURE 14.7 Basic structures of multiple-conversion SSB transmitters.

range from 1.6 to 30 MHz intended for HF communication service. The BPF generally represents a bank of switched BPFs, and band selection is provided by activating the corresponding filter. Some applications, particularly military communications, require transmitters capable of operating on any frequency in the HF band. In this case, the second carrier source (f_2) in Figure 14.7(a) will represent a frequency synthesizer followed by a low-pass filter (LPF) and a broadband power [3]. The SSB transmitter can generally provide a capability of multichannel transmission. Its two-channel structure is shown in Figure 14.7(b), where N_1 and N_2 are the audio modulating signals of the first and second channels, respectively. At the output of the first balanced modulators, the high sideband is selected by one filter and the low sideband is selected by another filter. Then, both these sidebands are combined and translated to higher frequencies by the second balanced modulator followed by the filter whose passband must include both telephone channels.

The direct-conversion SSB transmitters, the basic structure of which is shown in Figure 14.8(a), are intended for use on a single band of frequencies in low-cost applications. SSB modulation using a phasing method to provide the direct conversion is based on two balanced modulators arranged with RF inputs that differ in phase by 90°, and with AF modulating inputs that are likewise identical except that each frequency component of the modulating voltage applied to the second balanced modulator differs in phase by 90° from the phase of the corresponding component of the modulating voltage applied to the first balanced modulator. As an example, the simplified circuit of the transformer-coupled MOSFET balanced modulator is shown in Figure 14.8(b). When the outputs of these two balanced modulators are added, one of the sidebands is canceled out, resulting (together with the fact that the carrier is suppressed by the balanced modulators) in an output that contains only a single sideband. The first practical implementation of this principle described in 1948 included a balanced-modulator system using screen-grid-modulated Class C amplifiers [14]. The quadrature signals at the carrier frequency f_0 can be obtained by digital frequency division of the output of a voltage-controlled oscillator (VCO) operating at $4f_0$. As shown in Figure 14.8(a), the signal at $4f_0$ is divided by a flip-flop whose outputs are at frequency $2f_0$ and 180° out of phase. The subsequent division of each of these signals by a pair of flip-flops then produces the corresponding outputs at frequency f_0 that differ in phase by 90°.

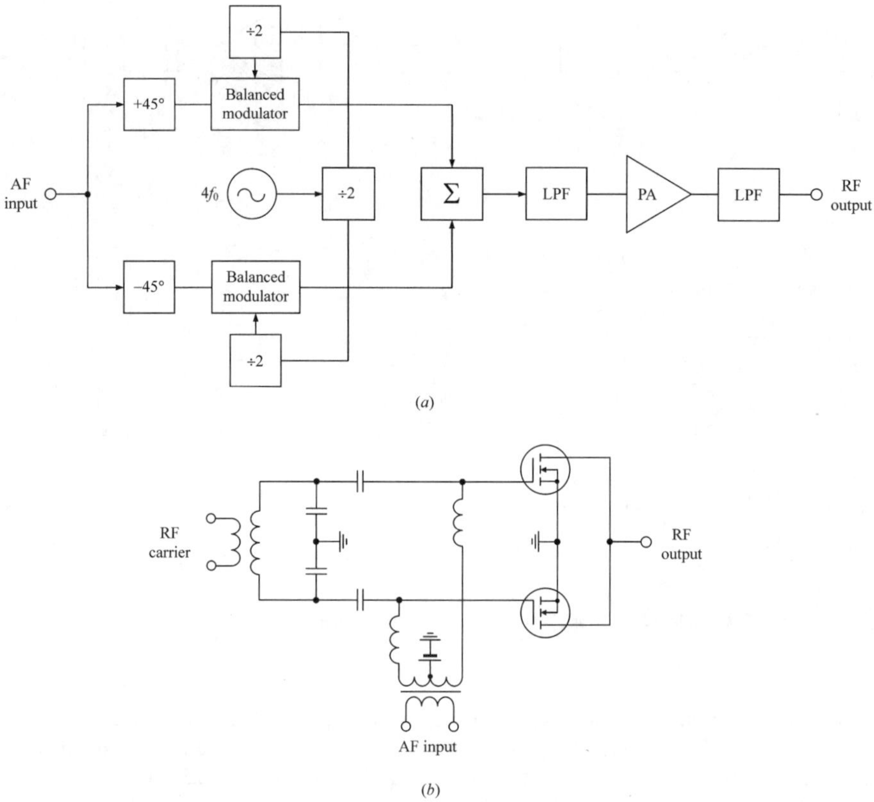

FIGURE 14.8 Basic structure of direct-conversion SSB transmitter.

For a two-tone signal with equal amplitudes commonly used to test an SSB transmitter, the average RF output power is equal to $P_{avr} = P_{PEP}/2$, because each of the tone with half of the peak envelope amplitude would dissipate $P_{PEP}/4$. The dc input current (and dc power) will vary with the output current, forming a full-wave rectified shape whose peak value is the current required by an amplifier delivering P_{PEP}. The average dc input current is therefore $2/\pi$ times the peak dc input current. As a result, the average drain efficiency of the PA can be written as

$$\eta = \frac{(1/2)P_{PEP}}{(2/\pi)(4/\pi)P_{PEP}} = \frac{\pi^2}{16} \tag{14.4}$$

which is approximately equal to 61.7% [15].

14.3 FREQUENCY-MODULATED TRANSMITTERS

Despite the fact that the subject of frequency modulation (FM) is very old and was proposed before the introduction of the vacuum-tube oscillators, Edwin Armstrong was the first who developed and installed a working frequency-modulated transmitter and began regular broadcasts in 1939 [16]. A few years later, to provide the center frequency stability of a high-order for wideband frequency-modulation systems, direct crystal control of the center frequency was proposed as an ideal solution [17]. Since then frequency-modulated transmitters have found extensive use at frequencies above

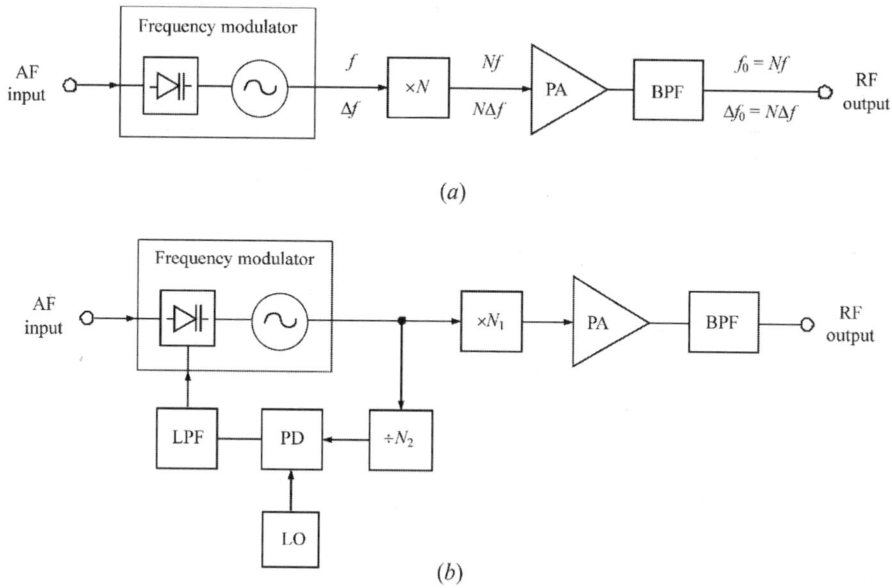

FIGURE 14.9 Basic structures of FM transmitters.

40 MHz in such applications as frequency-modulated broadcasting, television sound transmission, mobile radio, or radio-relay systems. For example, frequency-modulated broadcasting is now used worldwide in the 88–108 MHz band and offers greatly reduced noise and interference with lower distortion and larger audio bandwidth. The first frequency-modulated transmitter utilizing GaAs power MESFETs for a 7-GHz frequency-modulated radio-relay equipment was developed in mid-1970s [18].

Generally, FM can be accomplished at low-power level either *directly* by variation of the frequency of an oscillator by the audio input signal, or *indirectly* by PM of the RF signal by the time-integrated audio input signal. Figure 14.9(*a*) shows the block schematic of a frequency-modulated transmitter with direct FM where the audio signal is applied to the varactor included to the resonant circuit of a crystal oscillator. To achieve high linearity, most frequency modulators operate at a lower than required operating frequency producing a small modulation index or frequency deviation. To achieve the required operating frequency and to increase the frequency deviation, one or several multiplier stages with multiplication factor N are used, then followed by a PA and a BPF. High stability of the operating frequency can be achieved by using a crystal varactor-controlled oscillator or a phase-locked loop with local crystal oscillator (LO), as shown in Figure 14.9(*b*). In this case, the frequency-modulated signal is coupled to pass through the frequency divider to reduce the modulation index and to increase the amplitude of the average frequency compared to sideband amplitudes. Then, it is delivered to the phase detector (PD) to compare with the frequency of the highly stable reference LO signal. When the average frequency of a frequency-modulated signal is not equal to the frequency of the LO signal, then the PD output signal represents a low-frequency error voltage, the frequency and phase of which is defined by the difference of the divided average frequency-modulated and reference LO signals. This voltage is applied to the varactor through the LPF, which has a bandwidth to pass the dc voltage and very low frequencies but to remove the modulating-frequency feedback. As a result, the loop responds to the injected error voltage by adjusting the VCO to produce the specified phase shift between its output and the reference signal.

For higher operating frequencies, frequency-modulated transmitter configuration with direct modulation can include mixer to provide frequency upconversion, as shown in Figure 14.10(*a*). In this case, unlike frequency multiplication, frequency upconversion changes neither the modulation index

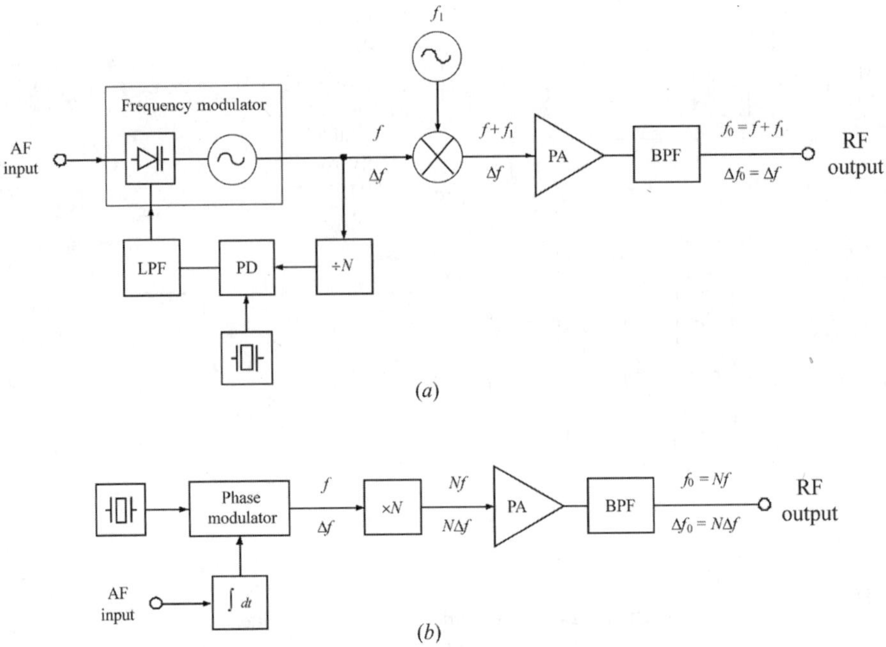

FIGURE 14.10 Block schematics of FM transmitter with direct and indirect modulation.

nor the frequency deviation. High frequency stability can easily be achieved by using indirect method to accomplish FM in frequency-modulated transmitter where the crystal oscillator is not modulated. Indirect FM is commonly produced in portable and mobile transmitter by the phase-shift or reactance modulator. Since this modulator directly varies the phase of the signal, rather than its frequency, it is necessary to time-integrate the input audio signal, as shown in Figure 14.10(*b*). The phase-modulated type of frequency-modulated transmitter has an advantage that the carrier frequency is derived directly from a crystal oscillator, and so is inherently very stable in frequency. The disadvantage of PM is that the maximum frequency deviation obtainable from the modulator, particular at low modulating frequencies, is small. In this case, the phase modulator must be followed by the frequency multiplier because modulation index $\Delta\phi$ produced by the phase modulator usually does not exceed 30°–70° (or 0.5–1.2 rad), which corresponds to low frequency deviation $\Delta f = f_m \Delta\phi$ for fixed modulating audio-signal frequency f_m. Larger linear phase deviations of up to ±180° can be achieved using a phase-locked loop configured to act as a phase modulator where the time-integrated audio-frequency signal is injected into the loop as an error voltage [3].

The output power for commercial frequency-modulated broadcast transmitters typically varies from 1 to 50 kW, with solid-state power amplifiers (SSPAs) up to 20 kW and vacuum-tube amplifying stage for higher power. The modulating signal represents a composite of the primary audio and subcarriers. In stereo broadcasting with two independent audio channels (left channel L and right channel R), these signals are not themselves transmitted. Instead they are combined to form a sum $(L + R)$ and a difference $(L - R)$ channel. The primary sum channel has a 15-kHz bandwidth and modulates the RF carrier with a 75-kHz deviation. Stereo transmission is accomplished by adding a 19-kHz pilot signal and 38-kHz subcarrier with double-sideband suppressed-carrier (DSB/SC) modulation by the difference channel. The 38-kHz signal is derived from the same source as the 19-kHz pilot tone, which provides a reference signal for the stereo decoder. Additional subcarriers are used for background music and the radio data system (RDS). The resultant bandwidth and channel spacing are 200 kHz. Since the high-frequency sounds in a typical audio signal produce less deviation of the carrier, this directly affects the noise performance of the received signal due to poorer noise immunity at these frequencies. To avoid this effect, high frequencies are boosted before being transmitted by pre-emphasis.

Since audio signal transmission in an analog form results in unnecessary expense and degradation in signal quality, the modern frequency-modulated broadcast systems provide a capability of digital audio transmission and distribution. The advantages of digital audio distribution include transparent audio quality, noise immunity, and the ability to diversely route the signal through a multi-node network. Figure 14.11(*a*) shows a typical 1-kW digital VHF/FM transmitter block schematic with an integrated digital exciter (including control unit, signal processor, and digital synthesizer), an upconverter, and an HPA module where each of four PAs provides a 300 W output power. Transmitters with higher output power up to 20 kW can be developed by parallel combining of the basic 1-kW PA modules. Figure 14.11(*b*) shows the circuit schematic of a 300-W MOSFET PA operating over the frequency range of 88 to 108 MHz and providing a power gain of more than 19 dB with a drain efficiency of more than 70% at a supply voltage of 50 V using the balanced RF MOSFET SD2932 device [19]. The input matching network to transform the unbalanced standard 50 Ω to the balanced (gate-to-gate) device input impedance of 5.56 Ω consists of a 1:1 balun based on a 50-Ω coaxial cable

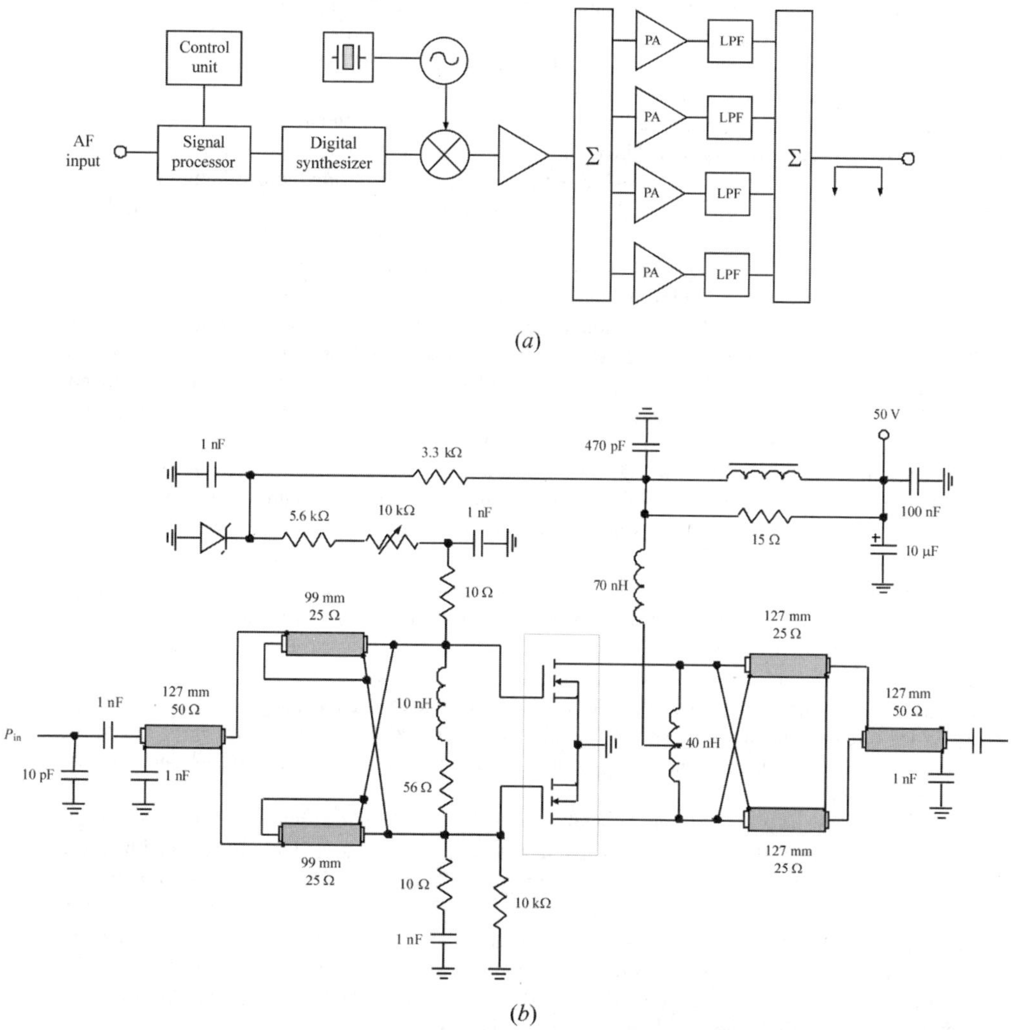

FIGURE 14.11 Schematics of high-power VHF/FM transmitter.

and a 9:1 balanced impedance transformer based on 25-Ω flexible coaxial cables with ferrite core. The output matching network to transform the balanced (drain-to-drain) equivalent output device resistance of about 12 Ω (calculated in Class B for each part as $R_{out} = (0.85V_{dd})^2/2P_{out} = (0.85 \times 50\ V)^2/(2 \times 150\ W) = 6.02\ \Omega$) to the standard 50-Ω load impedance consists of a 4:1 balanced impedance transformer based on 25-Ω flexible coaxial cables and a 1:1 balun based on a 50-Ω coaxial cable. A 10-nH inductor is connected between two gates to compensate for the device input capacitive reactance, whereas a 40-nH inductor connected between two drains is used to compensate for the device output capacitance over whole frequency range.

14.4 TELEVISION TRANSMITTERS

The history of electronic television (TV) transmission is originated from pioneering works of Philo Farnsworth who invented an electronic scanning system (image dissector) and Vladimir Zworykin who introduced an iconoscope camera tube and a kinescope picture tube that together formed the first electronic television system and replaced the cumbersome mechanical system of whirling perforated discs that had dominated the early development of television [20,21]. Zworykin studied at the St. Petersburg Institute of Technology, Russia, being a student of Prof. Boris Rosing, who had filed his first patent on a television system in 1907 [22]. In May 1911, Rosing exhibited a television system, using a mechanical scanner in the transmitter and the electronic Braun tube in the receiver, which was among the first demonstrations of TV of any kind. Fully electronic television broadcasting systems were introduced in Europe, USA, and Japan at the end of 1930s, and then after World War II became commercially available worldwide. A microwave point-to-point radio system for TV channel transmission intended to supplement local wire distribution of television was tested in 1946 [23]. TV broadcasting uses both VHF (in various regions between 48 to 100 MHz and 174 to 230 MHz) and UHF (470–890 MHz) frequency bands. VHF and lower power UHF television transmitters are mostly based on SSPAs, while high-power UHF transmitters employ vacuum tubes.

There are three basic analog television standards with their own history used in different countries such as North American NTSC (National Television System Committee), SECAM (Sequentiel Couleur Avec Memoire), originating in France, and PAL (Phase Alternation Line), developed in Germany. An NTSC television channel occupies a total bandwidth of 6 MHz in the frequency spectrum, as shown in Figure 14.12(*a*). The sound that accompanies the picture is transmitted via a frequency-modulated carrier with a 25-kHz deviation located 4.5 MHz above the picture carrier. The picture carrier (brightness signal) is amplitude-modulated by the composite video signal representing a vestigial-sideband (VSB) transmission that is DSB/SC with the lower sideband removed by filtering before transmission in order to save bandwidth. The seven elements of the composite video signal include horizontal line sync pulse, color reference burst, reference black level, picture luminance information, color saturation information, color hue information, and vertical sync pulse. Since the image is divided into 525 lines (in a picture with a width-to-height ratio of 4:3) that are scanned in an interlaced fashion at 30 Hz, the nominal bandwidth becomes equal to 4.2 MHz [3]. The color subcarrier is 3.579545 MHz above the video carrier, and the I and Q signals of the color information that correspond roughly to the difference between brightness and the red-orange and blue-green components of the image are transmitted by DSB/SC modulation of the color subcarrier.

According to the SECAM standard with a total channel bandwidth of 8 MHz, the image is divided into 625 lines that are scanned in an interlaced fashion at 25 Hz with the nominal bandwidth of 6.0 MHz, and is similar to PAL in these respects. To transmit the two-color information, two different subcarrier frequencies are used (4.25 and 4.406 MHz), as shown in Figure 14.12(*b*), that requires a one-line delay in the receiver to recombine the color information. These two subcarriers are frequency-modulated, instead of AM of a color subcarrier occurring at approximately 3.58 MHz for NTSC and at approximately 4.43 MHz for PAL. The sound carrier is located 6.5 MHz above the picture carrier and frequency modulated with a 25-kHz deviation.

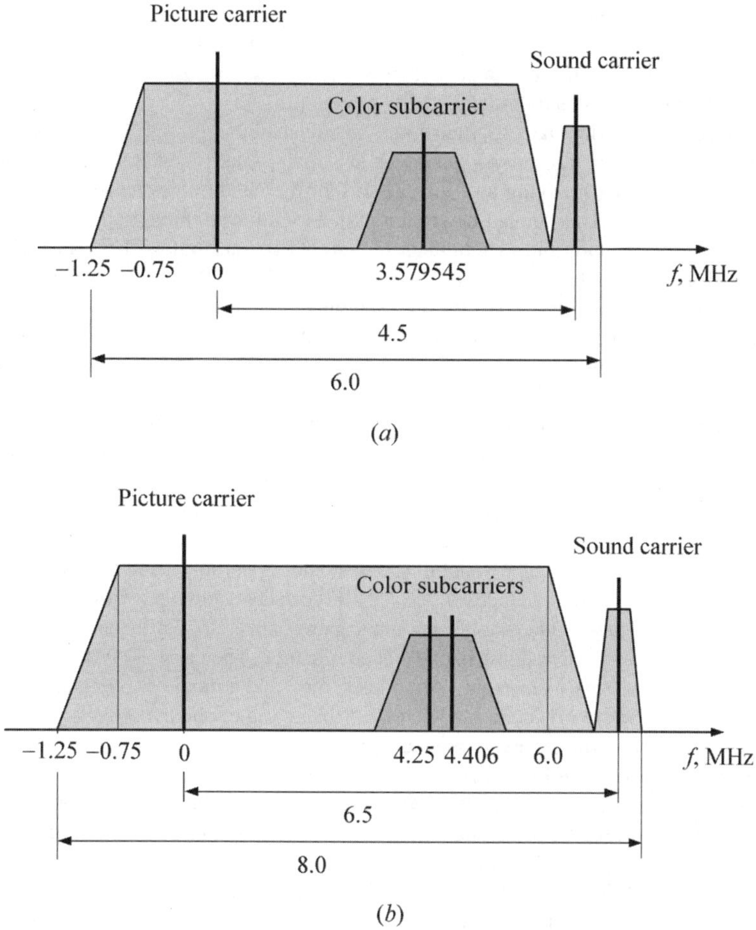

FIGURE 14.12 Spectra of composite TV signals in different analog standards.

Digital television (DTV) supports many different picture formats defined by the combination of size, aspect ratio (width-to-height ratio), and interlacing, and transmitted on radio frequencies that are similar to standard analog television, with the primary difference being the use of multiplex transmitters to allow reception of multiple channels on a single frequency range. In the USA, high-definition television is based on MPEG-2 (Moving Pictures Experts Group) video coding and Dolby AC-3 audio coding as specified by the Advanced Television Systems Committee (ATSC) [24,25]. The image is composed of 1080 lines with 1920 pixels per line, resulting in a data rate of 19.39 Mb/s, which expands to 30.28 Mb/s with the addition of error-correction coding and other overheads. The 8VSB-transmitted signal is based on eight-level DSB/SC modulation followed by filtering to remove most of the lower sideband, allowing it to fit into existing 6-MHz television channels. In the United Kingdom, digital video broadcasting (DVB) is based on MPEG encoding and OFDM using 16 or 64 QAM. The data at rates up to 31.7 Mb/s are distributed among 1704 carriers with 1.1-kHz spacing or 6816 carriers with 4.5-kHz spacing. DTV has several advantages over analog TV, the most significant being that digital channels take up less bandwidth, and the bandwidth needs are continuously variable, at a corresponding reduction in image quality depending on the level of compression as well as the resolution of the transmitted image. This means that digital broadcasters can provide more digital channels in the same space, provide high-definition television (HDTV) service, or provide other non-television services such as multimedia or interactivity. By bringing DTV content to mobile terminals, mobile TV promises to extend, enhance, and enrich conventional TV user experience by allowing consumers not only to watch TV on the move, but also to have access to personalized, time-shifted, and on-demand TV content [26].

The first practical all-solid-state TV transmitter having an output power of 30 kW was developed in Japan in 1985 [27]. This transmitter comprises of a main/stand-by exciter system, two separate solid-state transmitters (each having half the capacity of the rated output power) that are operated in parallel so that each transmitter can be separately released for required maintenance, and output coaxial circuitry for combining and switching the outputs of two transmitters. The transmitter modular design provides high redundancy, while jet air cooling with high efficiency contributes to size and weight reduction. The basic 6-kW UHF PA representing a functional block consists of eight 800-W PA modules, four 28-V power supplies, one 8-way divider at the input, and one 8-way combiner at the output, each employing a suspended substrate stripline that provides an insertion loss of less than 0.15 dB. A total of twelve internally matched 110-W bipolar 2SC2660 A devices operating in a push–pull mode up to 860 MHz are used in parallel in each 800-W PA module, which has no heatsink by adopting jet air cooling. A circulator is employed in each module to achieve its stable operation over load variations. The high temperature protection, reflection detection, and abnormal display latch functions are provided to enhance self-protection function and improve the maintainability.

Figure 14.13 shows the block schematic of a solid-state VHF TV transmitter providing a 30-kW peak visual output power with high reliability, broadband operation, and minimum maintenance requirements [28]. The exciter includes video and IF correction circuitry with sound intercarrier frequency phase locked to the video signal, automatic gain control, SAW filter, and possibility to lock the carrier frequency to an external precision frequency source. The basic 5-kW frame fed from two separate power supplies consists of eight PA modules, fans, and a monitor and control module. The four PAs in each module are first combined using a 3-dB hybrid to obtain 850 W output per module and then the output of two frames are combined by a 16-way full isolation coupling system to obtain 10 kW power. The basic PA uses a balanced bipolar transistor operating in a push–pull Class AB mode. The bipolar transistor is rated at 240 W peak of sync output at the 1-dB gain compression point. With 192 transistors in use, the total capability is approximately 46 kW that provides comfortable linear operation below the compression point.

By using the silicon laterally diffused MOSFET technology, solid-state transmitters based on the LDMOSFET devices become reliable and economically viable alternatives to the traditional

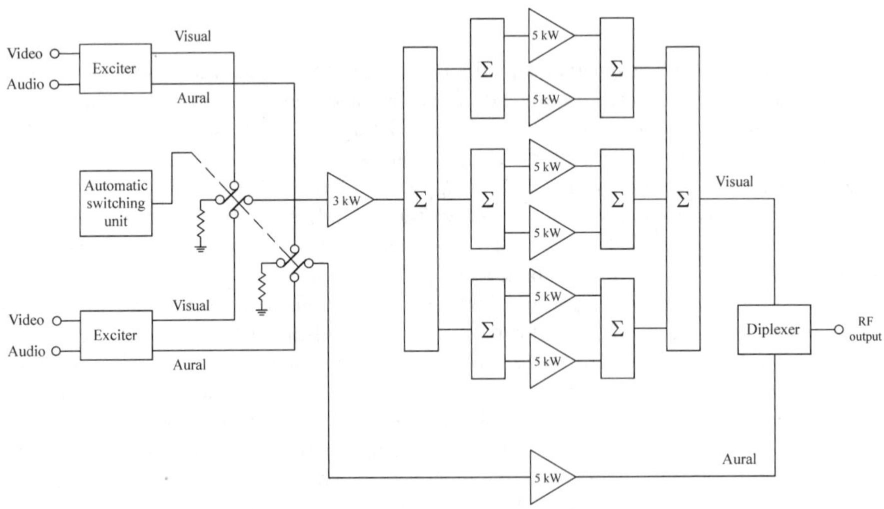

FIGURE 14.13 Block schematic of 30 kW VHF TV transmitter.

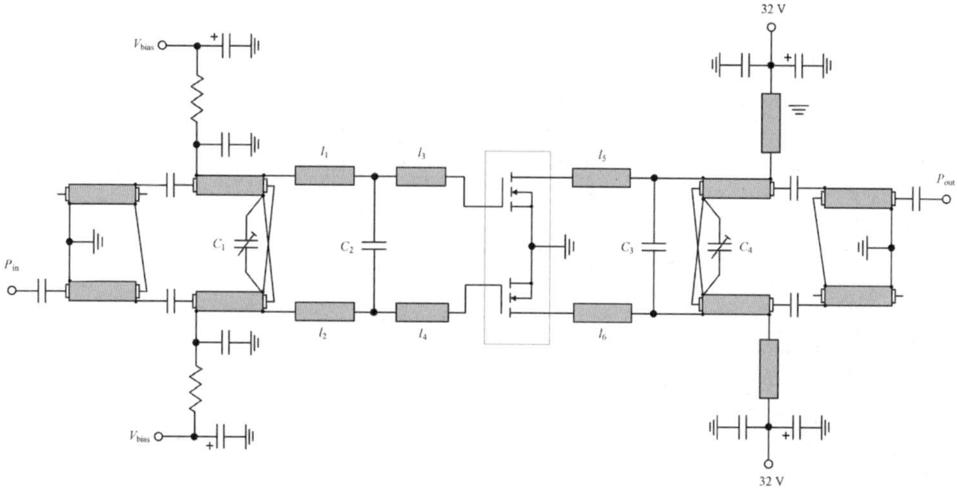

FIGURE 14.14 Circuit schematic of broadband UHF TV power amplifier.

high-power bipolar-transistor or vacuum-tube transmitters in modern UHF DTV service. The LDMOS transistor can provide better thermal and gain-bandwidth performance because the source is connected directly to the ground that can allow the traditional ceramic package to be replaced with a low-cost plastic package. A transmitter that used silicon bipolar transistors would contain more cabinets because the linear output power of an individual bipolar device is less than a similar sized LDMOS transistor [29]. Figure 14.14 shows the circuit schematic of a push–pull 300-W LDMOSFET PA operating over the entire frequency range of 470 to 860 MHz and providing a power gain of more than 18 dB and a third-order intermodulation distortion of better than −35 dBc with a 200-W peak-envelope power at a supply voltage of 32 V using the balanced *n*-channel enhancement-mode lateral MOSFET MRF6P3300 H [30]. The input matching network to transform the unbalanced standard 50 Ω to the balanced (gate-to-gate) device input inductive impedance with real part around 10 Ω (varying from 8.77 Ω at 470 MHz to 13.11 Ω at 710 MHz and 3.73 Ω at 860 MHz) consists of a 1:1 balun based on two 50-Ω coaxial cables and a 4:1 balanced impedance transformer based on two 25-Ω coaxial cables followed by a two-section *LC* impedance transformer with first variable capacitor and series microstrip lines. The output matching network to transform the balanced (drain-to-drain) equivalent output device resistance around 7 Ω to the standard 50-Ω load impedance consists of an *L*-section *LC* impedance transformer, a 4:1 balanced impedance transformer based on 25-Ω coaxial cables, and a 1:1 balun based on two 50-Ω coaxial cables.

The transmitting 8VSB digitally modulated signal in DTV transmitter is characterized by the peak-to-average power ratio of typically 6 to 7 dB, with random peaks and constant average power. In order to maximize adjacent channel interference and maximize in-channel signal-to-noise power ratio, the total spectral re-growth in the transmitting path should be suppressed greater than 35 dB at a 3.4 MHz and 40 dB at 6 MHz offsets from the center frequency. In this case, each amplifier stage must introduce the minimum amount of nonlinearity that can be accomplished by operating early stages in Class A and final stage in Class AB to achieve efficiency better than 35% with IF precorrection circuitry in the exciter [29]. Adaptive precorrection can be used to compare a sample of the transmitter output demodulated signal to the IF input. After referring to a look-up table in memory, the adaptive precorrector adjusts the coefficients of the precorrection algorithm used to linearize the transmitter.

Common amplification of the visual and aural signals generally provides lower dc power to RF power efficiency compared to separate amplification with completely independent paths of amplification for the visual and aural RF signals. This can be attributed to the higher peak-envelope power when more power backoff is needed for linear operation to keep the intermodulation products under

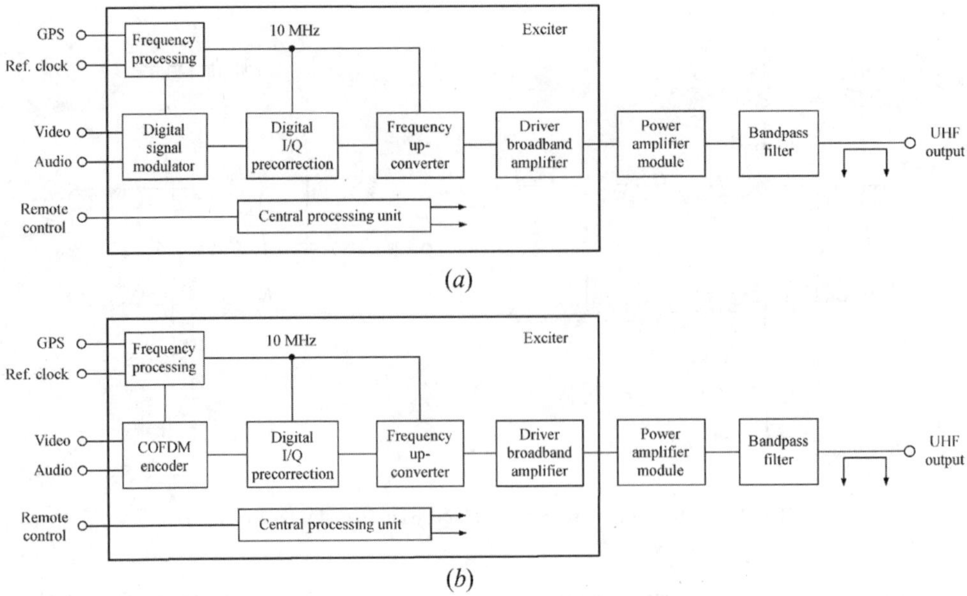

FIGURE 14.15 Block schematics of UHF TV transmitters with common amplification.

control. However, modern digital precorrection techniques allow very linear transmitter performance and freedom from visual-to-aural cross modulation. As a result, the difference between two systems in terms of efficiency becomes very small [31]. In this case, common amplification provides an extremely attractive alternative to external diplexing when it comes to upgrade an analog TV transmitter for digital operation. Figure 14.15(*a*) shows the block diagram for the analog transmitter configuration, including the digital signal modulator that processes the video and audio analog baseband and digital *I*/*Q* precorrector. To modify the transmitter system for DTV service, the only significant hardware change required is the replacement of the digital signal modulator with the coded OFDM (COFDM) encoder, as shown in Figure 14.15(*b*). All other items remain the same as for the analog TV transmitter. A BPF that is placed on the output of a DTV transmitter will slightly degrade the TV output signal by introducing group delay in the transmit chain. The group delay will become substantial if the slope of the bandpass skirts is very steep. In order to minimize group delay, the BPF should be designed with the minimum amount of required attenuation to provide a trade-off between PA linearity and BPF skirt requirements.

14.5 WIRELESS COMMUNICATION TRANSMITTERS

Decades before the cellular technologies began to provide a worldwide wireless radio access, the two-way radio systems met the need for mobile communication. In 1927, a number of bands in the frequency range of 1.5 to 6.0 MHz were allocated to mobile aircraft two-way communication systems [32]. The first two-way municipal police duplex radio communication systems operating between 30 to 42 MHz were introduced in 1930s [33]. The system transmitter that was capable of 100% modulation either by voice or by a 1-kHz calling tone included a vacuum-tube crystal oscillator, several frequency multipliers, an intermediate PA as a driver, and a final push–pull PA module. Microwave radio communication system using a pulse-position modulation that provides eight duplex voice channels through time division multiplexing and operating at 4.5 GHz was developed in mid-1940s [34]. The applications for modern mobile radio include land, marine, and airborne use for both civilian and military organizations. Analog systems use AM or FM and include a control channel for digital information. Each service is assigned to a specific set of frequencies in the VHF and UHF bands. For example, terrestrial trunked radio (TETRA) as a digital land-mobile radio based on time division multiple access is commonly developed to use in different frequency bands, mainly between 350 and 470 MHz, marine communications using narrowband FM is provided in the 156–174 MHz band, civilian aeronautical communications uses AM in the 108–137 MHz band, and military airborne communications uses the frequency range from 225 to 400 MHz [2]. The multimedia millimeter-wave wireless communication service with high-speed modulation capability is provided at 60 GHz and can be based on an ultra-miniature monolithic FM/FSK transmitter/receiver module utilizing a complete MMIC chip set that includes a DR-stabilized fundamental-frequency oscillator and based on a 0.15-μm double heterojunction GaAs FET technology [35].

The first commercial mobile telephone service operated on six channels in the 152–162 MHz band with 60-kHz channel spacing was developed in the USA in 1946 [36]. In 1960, the world's first partly automatic car phone system was launched in Sweden. The first commercially automated cellular 1G (first generation) network was introduced in Japan in 1979. The initial launch network covered the full metropolitan area of Tokyo's over 20 million inhabitants with a cellular network of 23 base stations. Within 5 years, this network had been expanded to cover the whole population of Japan and became the first nation-wide 2G (second generation) network. The second launch of the 1G networks was the simultaneous deployment in Scandinavian countries in 1981 of the Nordic Mobile Telephone (NMT) system, which was the first mobile phone network featuring international roaming. The first network technology on digital 2G cellular technology was introduced in 1991 in Finland on the basis of the GSM standard. In 2001, the first commercial launch of 3G (third generation) system was developed in Japan based on the WCDMA standard. The next evolution to offer broadband wireless access with higher data transfer speeds and capacity represents the 4G (fourth generation) systems based on 3GPP LTE (Long Term Evolution) cellular standard, which can offer peak bit rates of 326.4 Mb/s.

Several architectures can be proposed for wireless communication mobile transmitters depending on the performance requirements [37,38]. In a Cartesian-type transmitter, the quadrature I and Q baseband components are mixed with a quadrature LO signals. In this case, the baseband components are usually generated in the digital domain followed by the digital-to-analog converters (DAC) and LPF to provide delivery of the analog baseband signals to low-frequency inputs of the mixers. The upconversion can be done in one or two steps. Figure 14.16(*a*) shows the superheterodyne transmitter with two-step upconversion where the modulation is done at IF frequency and then subsequently upconverted to the transmitting frequency by an RF mixer. This architecture requires a significant amount of circuitry, which usually includes both the low-pass IF and bandpass RF filters to properly suppress noise and spurious components. In addition, to provide a significant dynamic range of output powers up to 80 dB depending on the communication standard, it is necessary to apply the variable power control scheme directly at RF prior to the PA. The direct-conversion architecture shown in Figure 14.16(*b*) minimizes the number of circuit blocks providing the best compromise of performance versus power consumption and circuit complexity [39]. However, it may require an RF SAW filter to suppress noise floor in the receiver band before feeding the signal into the PA. Also, direct-conversion transmitters suffer from any mismatch in the phase and amplitude of the baseband signals, and any dc offset in the baseband signal can result in LO leakage to the output, thus degrading the transmit signal quality. In both architectures, the PA requires a good linearity, compromising the overall transmitter efficiency and complicating output power control.

Despite various shielding techniques that attempt to isolate the LO in a direct-conversion transmitter, the noisy PA output can still corrupt the oscillator spectrum through injection-pulling (or injection-locking) effect, tending the LO frequency ω_0 to shift towards the frequency of an external stimulus. If the frequency of the injected noise is close to the oscillator natural frequency, then the LO output is disturbed increasingly as the noise magnitude rises, eventually locking to the noise frequency. In practice, noise level as low as 40 dB below the oscillation level may create tremendous disturbance. This unwanted phenomenon can be alleviated if the PA output spectrum is sufficiently higher or lower than the LO frequency. For example, the LO frequency can be offset by subtracting or adding the

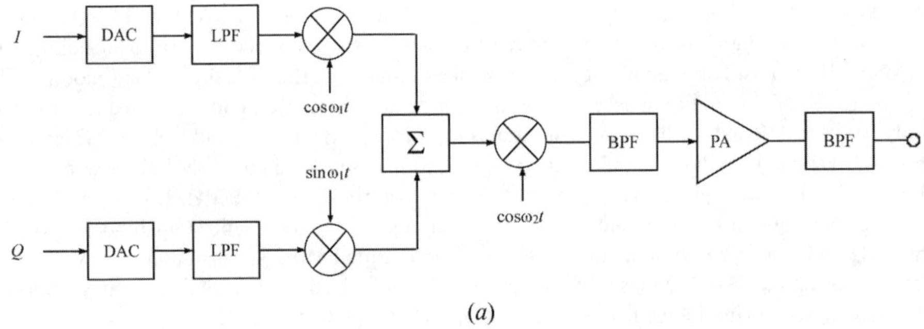

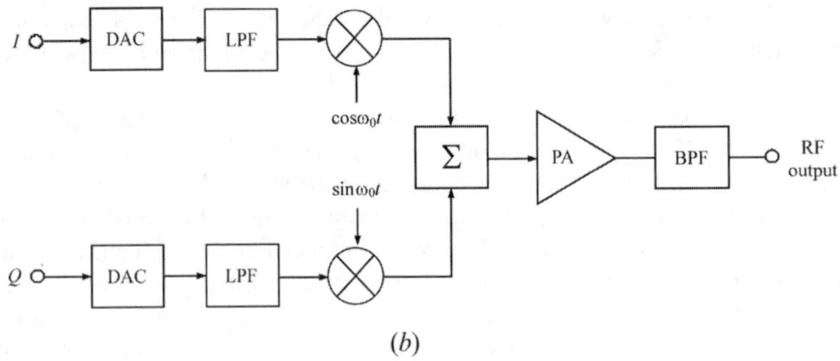

FIGURE 14.16 Double-conversion and direct-conversion transmitter architectures.

output frequency of another LO with subsequent bandpass filtering, as shown in Figure 14.17 [40]. In this case, the BPF selectivity should provide a significantly low level of intermodulation frequency components $m\omega_1 \pm n\omega_2$ resulting from the nonlinearities in the offset mixer.

The easiest way to transmit a constant-envelope signal is to use the direct modulation of a VCO with baseband data, as shown in Figure 14.18(*a*), which has appeared in designs for DECT communication standard [41]. Here, initially the VCO is operated with the phase-locked loop (PLL) to accurately set the carrier frequency and then disconnected so that modulation can be fed into the control input of the VCO unperturbed by its dynamics. Since the VCO performs both frequency translation and modulation, this approach allows a significant reduction in components (with no mixers and only one DAC to produce modulation signal) and low power consumption. However, because the loop is inactive during modulation, the nominal frequency setting of the VCO tends to drift, resulting in a perturbation of the output frequency, which can also be seriously affected by turn-on transients. In addition, the isolation requirements between VCO and PA exclude the possibility of a one-chip solution that limits high integration capability of this approach. The offset-PLL architecture is shown

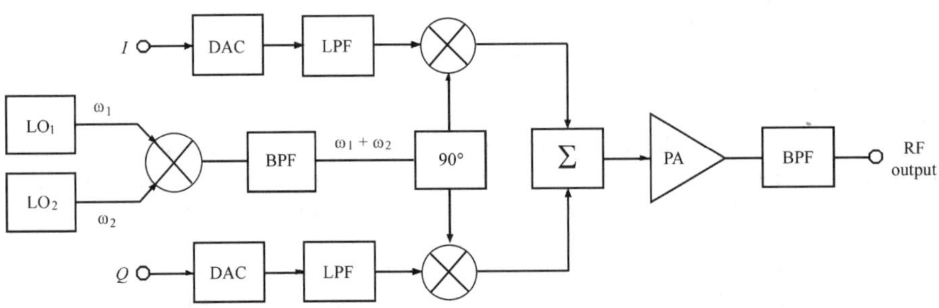

FIGURE 14.17 Direct-conversion transmitter architecture with offset VCO.

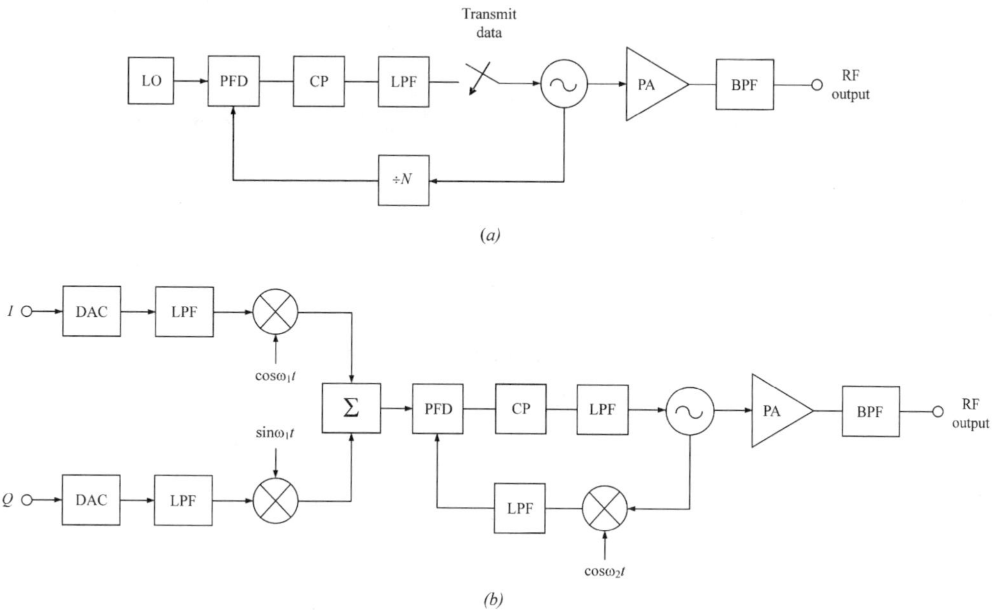

FIGURE 14.18 PLL-based direct-modulated and reference-modulated transmitter architectures.

in Figure 14.18(*b*) where the baseband signal is translated to IF frequency using quadrature modulators and a PLL serves to upconvert the IF signal to RF and to reduce the filtering requirements on the output signal by the inherent loop action with properly chosen bandwidth [42]. The PLL incorporates an offset mixer driven by an additional LO so as to lower the frequency presented to the PFD. This architecture is well suited to low-cost and high-performance systems using constant-envelope modulation. However, injection pulling still is an issue, requiring higher isolation between loop VCO and PA.

In modern communication mobile phone systems, the transmit and receive bands are different, with translation performed at the base station. In a full-duplex system where reception and transmission occur simultaneously through a single antenna, the transmit and receive paths are separated by a narrowband front-end filter block, which is called the *duplexer*. Figure 14.19 shows the typical block schematic of the RF transceiver with a heterodyne receiver and a direct-conversion transmitter [37]. Here, the receive and transmit LOs are embedded in a frequency synthesizer that selects the proper carrier frequency for each of two communication channels (for receive and transmit) according to the digital input when a call is initiated. In the receive path, the downconverted signal is finally applied to an ADC, which is required for such baseband operations as equalization, matched filtering, and dispreading that are performed with high precision in digital domain. In the transmit path, the digitized voice stream, which is subsequently oversampled and subdivided into multi-bit words, is applied to two DACs. The frequency synthesizer in an RF transceiver generates precisely spaced carrier frequencies according to a digital input. For example, in NADC standard, the receive channels are 30 kHz apart and range from 869 to 894 MHz, indicating the very high precision required for defining each channel frequency. If the frequency error is 10 parts per million, then each channel is offset by 9 kHz, which is an appreciable value with respect to the bandwidth of 30 kHz. In this case, the division factor in the synthesizer with a frequency divider varies from 29,000 to 29,800 and is changed by means of the channel-selected bits. The LNA gain must be chosen according to the noise and linearity of the mixer. If this gain is too low, the mixer noise dominates the overall noise figure, and if it is too high, the input signal to the mixer creates high intermodulation products.

For RF signals with a non-constant envelope, the PA must compromise simultaneously both linearity and efficiency performance to satisfy standard requirements and maximize talk time. Recent

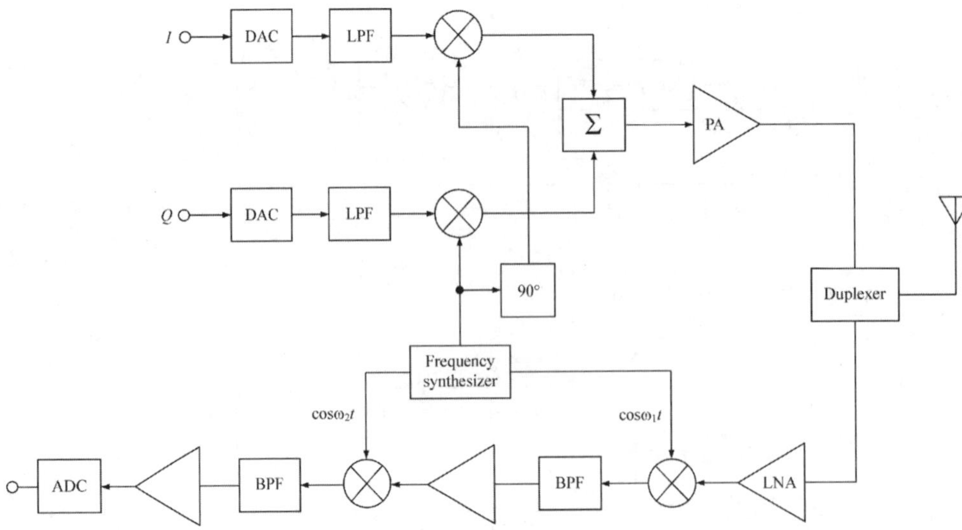

FIGURE 14.19 RF transceiver architecture.

trends in the integration of entire communication system require the development of homodyne radios with direct-conversion transmit and receive paths [39]. The PLL-based or translational loop architecture, where sampled output of the VCO is fed back to the *I*/*Q* modulator included in the loop and which is well-suited for constant-envelope signals, can eliminate the need for RF SAW filter at the PA input. In this case, a single-chip quad-band GSM direct-conversion transceiver can be developed using a BiCMOS technology where the only required external elements (excluding the impedance matching and loop filter components) are the receive-band SAW filters, a switchplexer, and a PA [43].

Polar transmitter architecture was chosen to minimize the dc power consumption and number of external RF filters [44]. In this type of transmitter the modulated signal $A(t)\cos[\omega_0 t + \phi(t)]$ is decomposed into its amplitude and phase polar components. As a result, the PM can be first applied using a PLL-based phase modulator to generate a constant-envelope signal $A_0 \cos[\omega_0 t + \phi(t)]$. Then, the AM can be added by modulating the envelope of the RF signal using a variable gain amplifier (VGA) whose output amplitude is a function of $A(t)$. Figure 14.20(*a*) shows the simplified block diagram of a GSM/EDGE transmitter where an *I*/*Q* modulator generates the modulated IF output from the baseband inputs that contain both AM and PM information [45]. The PM path limits the IF signal to remove AM envelope, and this limited signal is then used as the reference input to a translational loop PLL. The PLL locks the VCO to the transmit carrier frequency, transfers the PM of the reference input onto the VCO, operating as a tunable high-*Q* filter for input noise contributors in the transmitter. The AM signal is extracted from the IF signal using an AM detector followed by the LPF and then directly applied to the VGA where both AM and PM signals are finally recombined.

In GSM mode with constant envelope, the AM path is disabled and the transmitter reverts to a conventional nonlinear offset-PLL transmitter. In order to reconstruct the modulated signal correctly, the AM and PM signals need to be aligned. For example, a delay mismatch of about 80 ns can increase the output spectrum mask at 400 kHz offset to about −54 dBc, which is a minimum level required by the standard. To minimize this misalignment in practice, either a feedback loop or a precision calibration scheme can be used. When AM signal is combined with PM signal, AM-to-PM conversion due to PA nonlinearity can corrupt the signal quality, degrading both spectral mask and error vector magnitude (EVM). Therefore, as an alternative, either polar modulation can be applied to the low-power linearized VGA prior to the high-power PA, or feedback control of both phase and amplitude of the transmitted signal from the output of PA is necessary to correct its nonlinearities under large-signal operation [46].

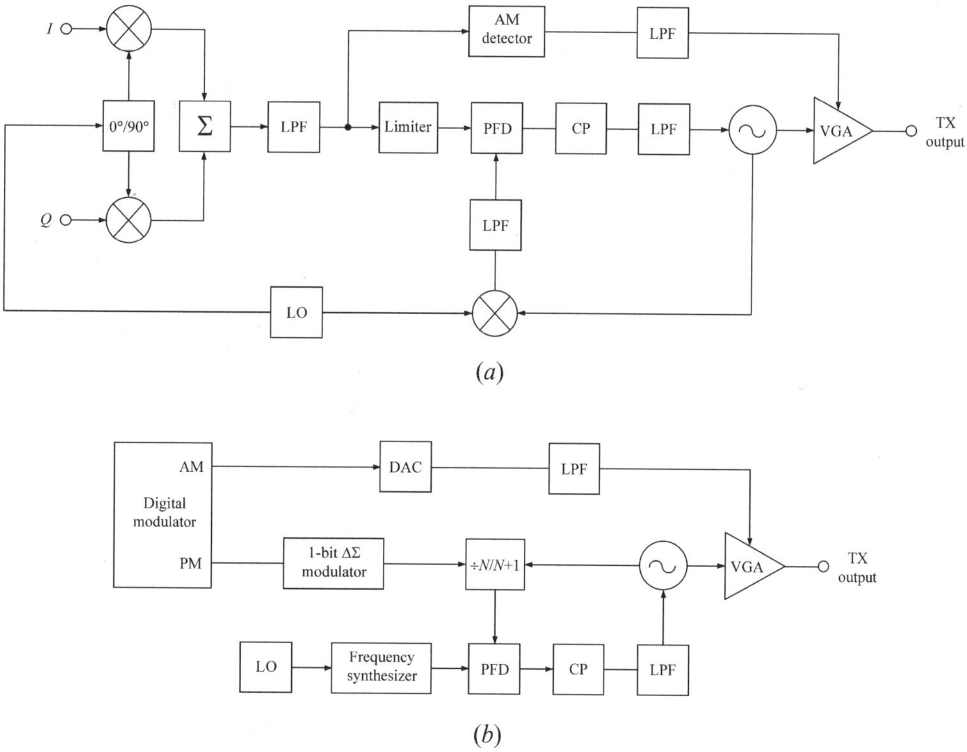

FIGURE 14.20 Polar transmitter architectures.

Figure 14.20(*b*) shows the simplified digital polar transmitter architecture that can be incorporated in a fully integrated single-chip GSM/EDGE radio together with the receiver using a low-cost submicron CMOS technology [44]. The GMSK transmitter is based on the fractional-*N* approach to frequency synthesis, which enables fast dynamics to be achieved within the PLL by allowing a high reference frequency, and low spurious noise can be obtained by using a high-order $\Delta\Sigma$ modulator to perform high resolution [41]. The binary PM data stream is first convolved with a digital finite impulse response (FIR) filter that has a Gaussian shape and fed into the input of a $\Delta\Sigma$ modulator, the output of which controls the instantaneous division value of the PLL. The nominal division value sets the carrier frequency, and its variation causes the output frequency to be modulated according to the input data. Sufficient reduction of the $\Delta\Sigma$-modulator quantization noise can be accomplished through the proper choice of the sample rate, which is usually assumed to be equal to the reference frequency, and the PLL transfer function. To support EDGE functionality, the digital modulator creates a separate AM path that directly modulates the amplitude of RF VGA. By adding a frequency divider by two to create an additional transmitting path, the low-frequency band including 824–849 MHz and 880–915 MHz and high-frequency band ranging from 1850 to 1910 MHz for PCS standard and from 1710 to 1785 MHz for DCS standard can be provided.

Next generation wireless networks will support a wide range of data rates over multiple frequency bands and require adaptive and programmable system resources. Low-cost and low-power designs at the circuit, architectural, and overall system levels implementing in high data rate programmable OFDM multiple-input/multiple-output (MIMO) radio transceivers will cover different frequency bands, support multiple standard, and enable longer battery life for mobile devices [47]. In this case, the architecture and frequency planning must be optimized to avoid coupling in a multistandard MIMO transceiver that may arise from existing or harmonic frequencies of the different standards

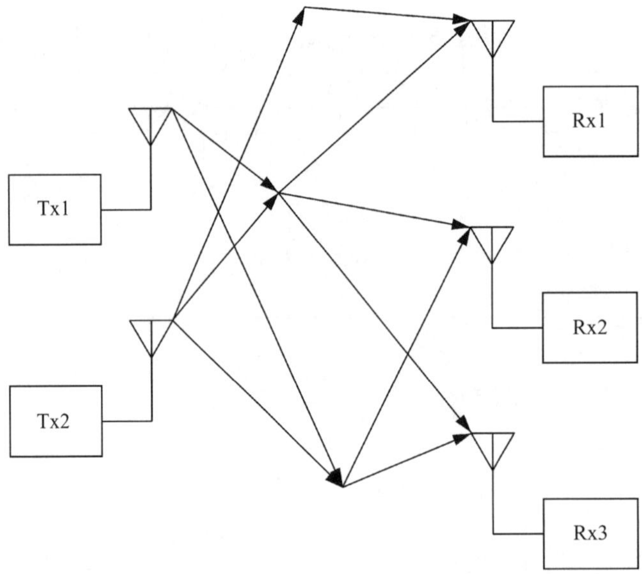

FIGURE 14.21 A 2 × 3 matrix MIMO radio.

and to optimize the cost of the radio by sharing the maximum number of blocks to avoid the need for separate full radio blocks for each standard. For example, the frequency synthesizer and clock generator would generate clocks for all frequency bandwidths (such as 2.4 GHz and 5.5 GHz for WLAN). Once LO signals are generated from the LO generator, they must be routed to different radios at different frequencies, and, in most case, they are routed differentially for noise immunity and in pairs for image rejection. In wireless communications, multiple transmitter and receiver paths are subject to reflections from the surrounding environment. Once a signal is transmitted, the receiver may see several different versions of this signal due to the reflections from walls and elsewhere. To overcome these problems through MIMO support, multiple transmitters and receivers can be used, as shown in Figure 14.21 for two transmitters and three receivers. In this case, multiple streams of data are transmitted simultaneously on the multiple slices of each transmitter. By optimally combining the multiple-received channels at the digital baseband, a higher throughout, a longer communication range, and immunity to fading and interference are all achieved.

14.6 RADAR TRANSMITTERS

Radar is used in many commercial and military applications and represents an object detection system that uses electromagnetic waves to identify the range, altitude, direction, or speed of both moving and fixed objects such as aircraft, ships, motor vehicles, weather formations, and terrain [48]. The first person to use radio waves to detect "the presence of distant metallic objects" was Christian Huelsmeyer, who in 1904 demonstrated the feasibility of detecting the presence of a ship in dense fog, but not its distance. In August 1917, Nikola Tesla first established principles regarding frequency and power level for the first primitive radar units stating that by electromagnetic waves "we may determine the relative position or course of a moving object, such as a vessel at sea, the distance traversed by the same, or its speed." Before the World War II, developments by the British, German, French, Russian, and American scientists led to the modern version of radar. However, Robert Watson-Watt and Arnold Wilkins from Great Britain were the first to fully exploit it as a defense against aircraft attack. The term "radar" is derived from the description of its first primary role as a RAdio Detection And Ranging system. Early radar equipment was adapted from the radio communication field using HF, VHF, and UHF vacuum tubes and antenna techniques [49]. Solid-state transmitters were first introduced into radar applications with availability of the high-power silicon bipolar transistors in the 1970s.

14.6.1 Phased-Array Radars

Early radar transmitters used multiple radiating elements to achieve desired antenna radiation patterns. A rapid electronic phasing of the individual array antenna elements introduced to steer the radar beam with the flexibility and speed of electronics rather than with much slower and less flexible mechanical steering led to the first practical implementation of a 900-MHz 16-element linear-array structure in 1959 [50]. One of the fundamental difficulties in designing a phased array is that significant portions of the microwave power transmitted by one element of the array can be received by the surrounding array antenna elements. This effect, which is known as *array mutual coupling*, can result in a substantial or total loss of transmitted or received radar signal, depending on the coherent combination of all of the mutual-coupling signals in the array. The amplitudes and phases of the array mutual-coupling signals depend primarily on the shape of the radiating antenna elements, the spacing between the array elements, and the number of radiating elements. Substantial amounts of mutual coupling can result in blind spots in the radar-scan sector unless care is taken in the design of the array. These blind spots are angles where the element pattern has a null and the reflection coefficient of the array has a peak close to unity. Generally, the multi-element planar array consists of hundreds or thousands of individual antenna elements located on a flat metallic ground-plane with about one-half wavelength spacing between adjacent elements [51]. Phased-array antennas require accurate calibration of their multiplicity of transmit/receive (T/R) channels, so that the radar main beam can be pointed in the correct direction and the sidelobe levels of the radar antenna can be controlled.

Conventional radars fall into two broad categories independent of what functions they perform [52]. The first category has fixed antenna patterns produced by passive array antennas with centralized transmitters and receivers, as shown in Figure 14.22(*a*). In this case, for fixed main beam, scanning can only be achieved by physically moving the antenna. For example, tracking radar will have a pencil beam that is used to track targets by the use of a mechanical tracking mount. Because of the limitations imposed on such radars by their design, they are considered as the single-function radars. The second category of radars is the passive phased array, incorporating electronic beam scanning or beam shaping by the use of phase shifters, switching elements, as shown in Figure 14.22(*b*), or frequency scanning methods. These features result in more complex radar systems having capability to carry out more than one radar function. Generally, however, the functions of the radar are preprogramed and not adaptable as the radar environment or the threat changes. These systems traditionally use centralized transmitters and receivers and their limited amount of multifunctionality is bought at the price of considerable beamforming losses.

Assume that a flat antenna array is composed of regularly-spaced transmitter elements radiating the signals of the same frequency with equal amplitudes but different initial phase. When all elements transmit in the same phase (as happens with mechanical-scan antennas), the plane of the electromagnetic field or wave front is parallel to that of antenna, and the vector of the main lobe of the transmitted signal (or the main radar beam) is always perpendicular to the wave front of the electromagnetic field. However, when the phase difference between adjacent elements is ϕ, as shown in Figure 14.22(*b*), and the *n*th antenna element is supplied with a signal that is phase shifted by $\phi_n = n\phi$ relative to the first element with zero initial phase, then the length difference l_n can be written as

$$l_n = nd\sin\theta \qquad (14.5)$$

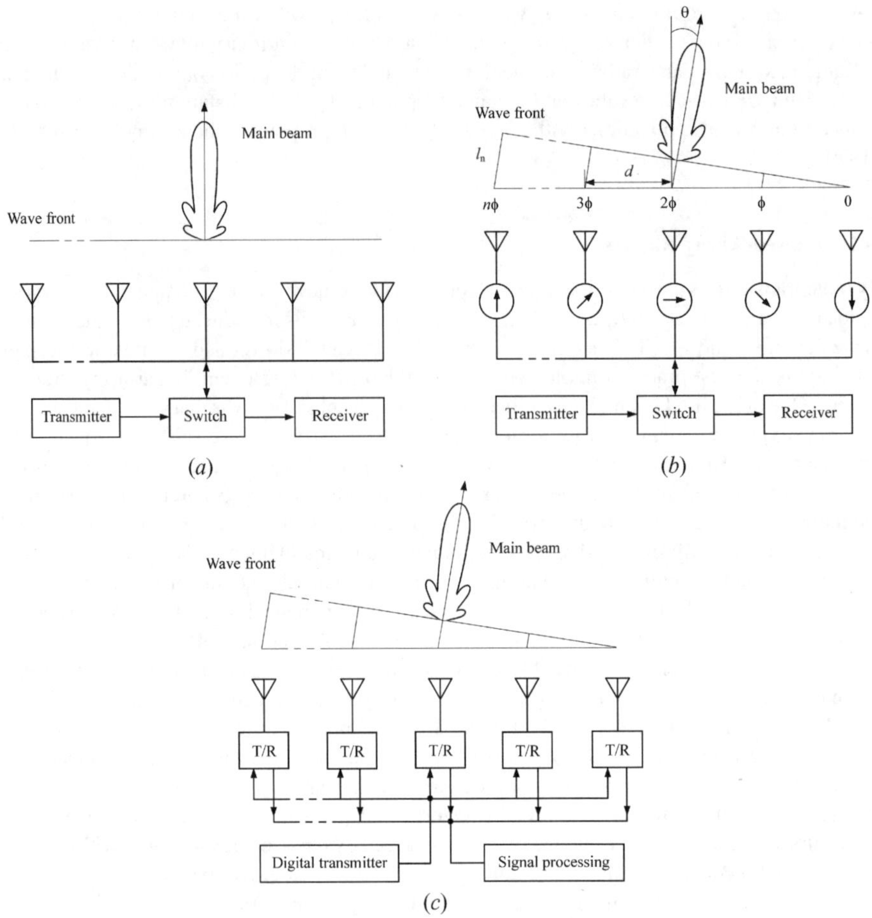

FIGURE 14.22 Beamformer architectures for phased-array radars.

where d is the spacing between adjacent elements and θ is the maximum incident angle determining the main beam direction. The maximum incident angle θ can be defined from a condition when the length difference is compensated by the phase difference, according to

$$\frac{2\pi}{\lambda} l_n - \phi_n = \frac{2\pi}{\lambda} nd \sin\theta - n\phi = 0 \tag{14.6}$$

where λ is the wavelength. As a result, from Eqs. (14.5) and (14.6) it follows that

$$\theta = \sin^{-1} \frac{\lambda\phi}{2d\pi} \tag{14.7}$$

and the main beam direction can be controlled by the incremental phase shift ϕ.

In order to improve the multifunction capability, the adaptive active phased-array radar can be used, where the T/R modules for each antenna element are mounted at the antenna face, as shown in Figure 14.22(*c*), employing adaptive beamforming, radar management, and control techniques resulting in a multifunctional capability [52,53]. These techniques require a much higher level of electronics integration and is more expensive; however, the presence of multiple T/R modules allows the forming of multiple independent main beams, each assigned to its own operation task. Ultra-wideband (UWB) design of the active phased-array radars supporting their multiband and multifunction operations from 3 to 12 GHz or from 8 to 20 GHz is based on the cross-polarization suppressed array architecture, piezoelectric true-time-delay phase shifters, and broadband high-power monolithic amplifiers [54].

A functional diagram of a typical T/R module is shown in Figure 14.23(*a*), where the same phase shifter and attenuator are used for receive and transmit paths depending on a switch position [55,56].

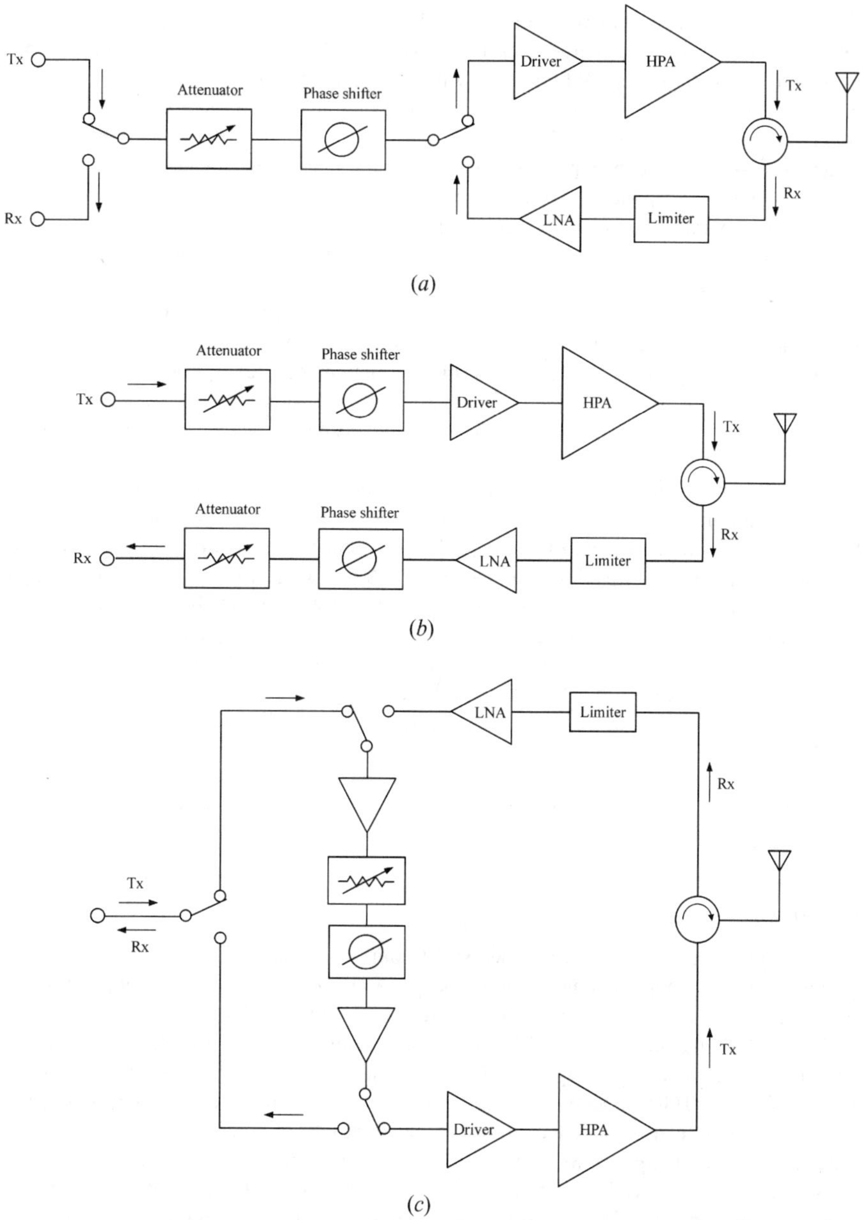

FIGURE 14.23 Block schematics of T/R radar module.

In transmit mode, an RF signal is phase-shifted and amplitude adjusted at each module to produce the desired radiated beam. This signal is then amplified by approximately 50 dB by driver and HPA and finally routed through circulator operating as a duplexer to the radiating antenna element. To improve efficiency of an *L*-band T/R module, an HPA can be designed based on a high-efficiency switched-mode Class E or Class F mode [57]. The circulator provides isolation for the receiver, as well as *VSWR* protection for the PA. Alternative T/R module architectures result from the possibility to improve either cost or performance of the module. Figure 14.23(*b*) shows the configuration where the transmitter and receiver are totally isolated except through the duplexer. Since there are no T/R switches, the dead time required to switch between transmit and receive paths is minimized, with a resulting improvement in radar efficiency. However, separate phase shifter and attenuator for transmit and receive paths are required, thus making such a T/R module configuration more expensive. Figure 14.23(*c*) shows another configuration with reduced number of components where the phase shifter and attenuator are shared between the transmit path and the receive path. However, in this configuration, since the transmit and receive signals flow in the same direction through the phase shifter and attenuator, it becomes possible to use common amplifying stages, thus further reducing the overall number of components. Some disadvantage of this architecture is an increased number of switches, timing complexity, and the potential for oscillations caused by signal leakage.

The packaging technology for a high-power *X*-band T/R module can be based on either low temperature co-fired ceramic (LTCC) or multi-chip-module (MCM) structure [58,59]. For example, a 6-bit phase shifter and a 5-bit attenuator using *p—i–n* diodes and implemented in a SiGe technology can represent separate MMICs [60]. The single-chip phase shifter with six digital bits of 180°, 90°, 45°, 22.5°, 11.25°, and 5.625°, cascaded in a linear arrangement, provides 64 phase states between 0 and 360° in increments of 5.625°. A 5-bit two-chip attenuator consists of 1-, 2-, 4-, 8-, and 16-dB attenuation bits and is divided into two cascadable designs, when a 3-bit attenuator and a 2-bit attenuator to be used are interdispersed between amplifier gain stages.

Figure 14.24 shows the circuit solution for an *X*-band radar GaAs MMIC chipset based on a multifunction-self-aligned-gate (MSAG) MESFET process that provides an output power of about 10 W in a frequency range from 8 to 11 GHz [61]. In this process, which does not use air bridges and provides multilevel plating capability, the active devices use 0.4-μm gates deposited by employing low-cost optical lithography. The digital 5-bit attenuator (32 steps with 0.75 dB), 6-bit phase shifter (64 steps with 5.6°), and buffer amplifier are incorporated in a single chip and fully matched to 50 Ω on both the input and the output. High power-added efficiency of the PA over 40% can be achieved using a 0.35-μm pHEMT or 0.25-μm GaN HEMT technology [62,63]. Typically, the HPA consists of two amplifying stages where the first stage is composed of 4 unit cells and the final stage is composed of 8 or 16 unit cells that are matched and connected in parallel in pairs using quarter-wavelength microstrip lines. The broadband *Ka*-band GaN HEMT MMIC PA can provide an output power of more than 2 W with a power gain of 13 ± 1 dB across the bandwidth from 26 to 36 GHz [64]. At 35 GHz, the measured saturated output power was 36 dBm with a power-added efficiency of 23%.

14.6.2 Automotive Radars

Automotive radars have found a great interest for comfort and safety applications in car industry. Long-range radars for applications including adaptive cruise control (ACC), collision avoidance, and pedestrian detection use the frequency band from 76 to 77 GHz, while the frequency band from 77 to 81 GHz has been allocated for short-range applications such as blind spot monitoring, lane departure warning, and parking aid [65,66]. Recent advances in a SiGe HBT technology (with device f_{max} of over 300 GHz, good noise figure, and high breakdown voltage) enable the realization of 77-GHz radar systems using silicon technology. At the same time, popular CMOS technology is being advanced rapidly and its applications are now expanding into the millimeter-wave region allowing the transceiver chip for 77-GHz automotive radar to be designed in 90-nm CMOS [67]. As a result, the reduced semiconductor cost, together with packaging and system optimization, will ultimately allow deployment of these systems in economy car models.

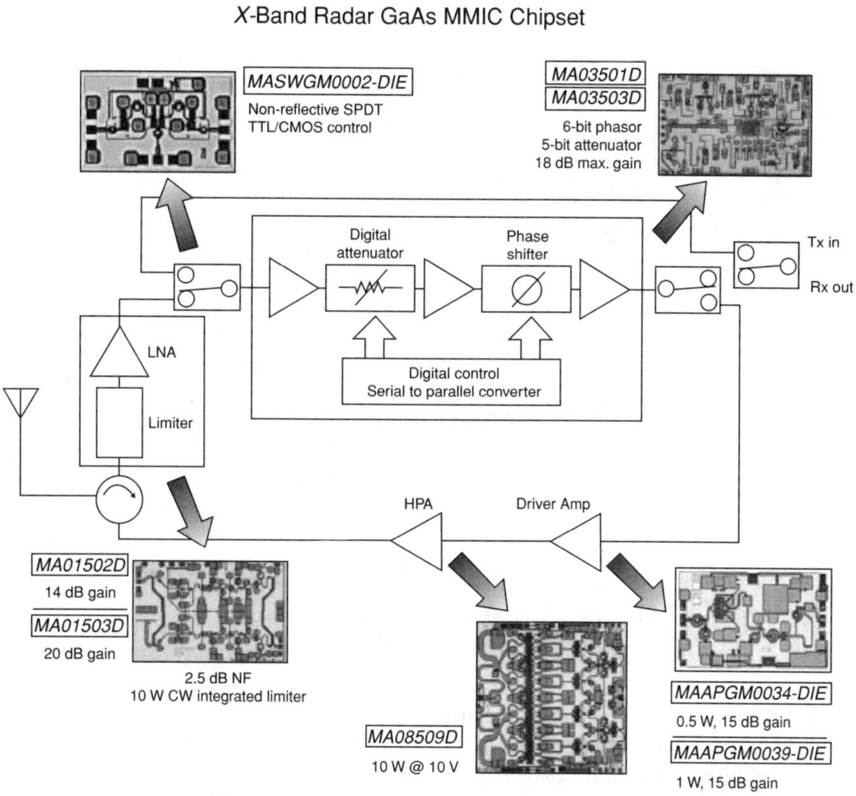

FIGURE 14.24 *X*-band radar GaAs MMIC chipset (courtesy of Inder Bahl, Cobham) [61].

Generally, the basic radar module for a single-beam configuration that can be adapted to multiple-beam operation represents either the monostatic configuration with duplexing between T/R signals at the single antenna, as shown in Figure 14.25(*a*) [68,69], or the bistatic configuration with separate antennas for transmitter and receiver, as shown in Figure 14.25(*b*) [67,70]. If the target has a velocity component along the line-of-site of the radar, the returned signal will be shifted in frequency relative to the transmitted frequency due to the Doppler effect. If the transmitted frequency is f_0 and the target velocity is v, then the shift in frequency, or Doppler frequency, will be $f_D = 2vf_0/c$, where c is the velocity of light. In this case, the received frequency is then $f_0 \pm f_D$, where the plus sign corresponds to an approaching target and the minus sign corresponds to a receiving target.

In terms of integration, a 77-GHz four-element phased-array transceiver system, in which a beamforming capability is achieved by varying the relative phase shifts in each element, can be implemented in a single chip with on-chip antennas [71]. To effectively linearize the VCO tuning characteristic and significantly reduce the VCO phase noise, the PLL can be used to obtain linear FM ramp required to achieve fine target resolution [70,72]. A reconfigurable single-chip T/R structure can be used as the basic building block for a complete four-channel T/R chip, which allows the implementation of the MIMO radar [73]. Its main difference compared to the conventional radar is the capability of transmitting different signals on multiple Tx antennas while keeping these signals separable at reception. In this case, the target can be illuminated from different directions and the number of resolvable targets can be increased in the same direction by improving angular resolution.

Typical monostatic automotive radar front-end module including an oscillator, a frequency divider, a PA, and a differential (or quadrature) hybrid combiner, whose block schematic is shown in

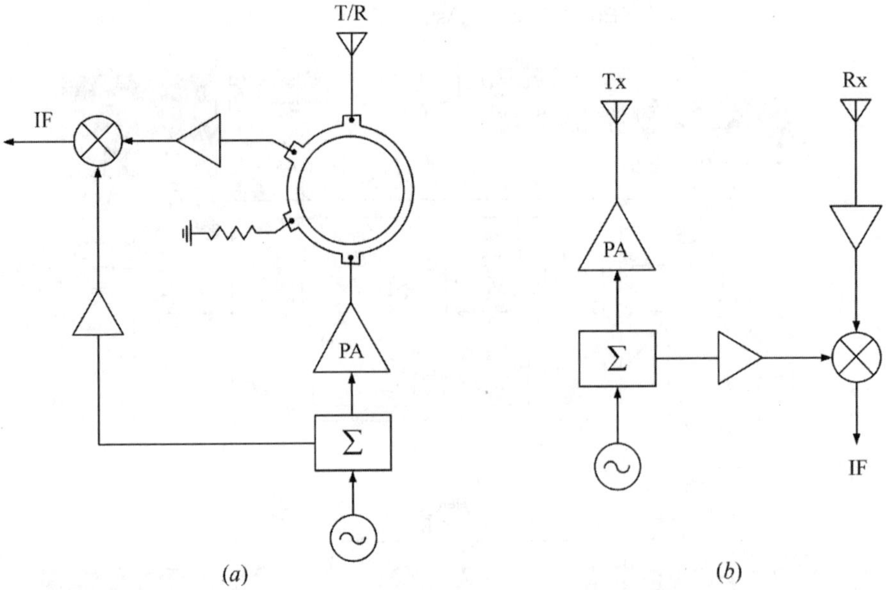

FIGURE 14.25 Block schematic of monostatic and bistatic radar front-ends.

Figure 14.25(*a*), uses a single antenna to reduce the overall size of the radar system. However, the large leakage signal from transmitter due to the low isolation of the antenna circulator (or hybrid coupler) or the reflection from the antenna can significantly reduce the sensitivity of the receiver, and even saturate the receiver components, so that the dynamic range of the receiver is reduced. For example, the isolation of the branch-line hybrid coupler at these frequencies is about 20 dB. To cancel Tx leakage due to poor isolation between antenna and isolated ports of the hybrid, the balanced radar topology can be used that includes an out-of-phase divider and a PA to supply the two Lange couplers with two equal-amplitude and anti-phase signals, as shown in Figure 14.26 [74]. The coupled port of the first Lange coupler is connected to the antenna that is matched to 50 Ω; however, the coupled port of the second Lange coupler is terminated with 50-Ω load. The isolated ports of both Lange couplers are combined with an in-phase Wilkinson combiner. As a result, Tx leakage signals are canceled out at the output of the Wilkinson combiner because they are balanced and 180° out of phase at its input ports.

However, such an automotive radar system is not very efficient since only half an output power flows to the antenna, whereas the other half flows to the ballast resistor of the output hybrid. Besides, half of the received signal power is dissipated too. In this case, a Tx leakage canceller can be introduced using four branch-line hybrid couplers, a 90° microstrip delay line, and a Wilkinson divider [75,76]. Highly efficient differential radar system, whose block schematic is shown in Figure 14.27, incorporates three differential couplers (rat-race hybrids) with three antennas for transmit and receive signals and two mixers [77]. The two transmitting 180° out-of-phase signals initially flow to the left-hand side and right-hand side rat-races where they split into two equal-amplitude signals, one halves flowing to the corresponding left-line and right-line antennas, respectively, and the other halves flowing to the central rat-race from the corresponding output ports of side hybrids, thus resulting in a two times greater power transmission to the central antenna, which is responsible for the front line. No power is flowing to the ballast resistor of the central rat-race hybrid. Generally, there is no power loss in such a radar system except due to certain insertion losses of the hybrids. The receiving signals from the left-line and right-line antennas flow to left-hand side and right-hand side hybrids and mixers, respectively, but not to the central rat-race. The corresponding mixer provides signal

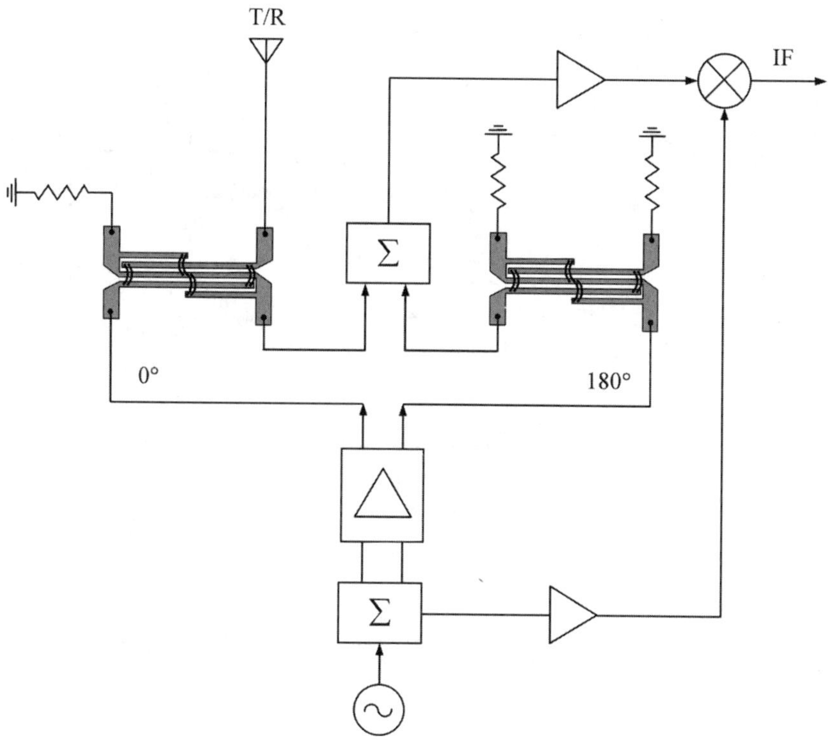

FIGURE 14.26 Balanced automotive radar topology.

detection from the corresponding side. However, the receiving signal flowing to the central rat-race is divided into two parts flowing to the left-side and right-side rat-races, respectively. This means that for a front-line signal, both mixers provide the signal detection simultaneously.

An offset-loop architecture for a frequency modulated continuous wave (FMCW) radar transmitter optimized for low-phase noise and high-sweep bandwidth is shown in Figure 14.28, where a downconverter is based on a dielectric resonator oscillator (DRO) in conjunction with low division factor and a direct digital synthesizer (DDS) acting as a reference clock is used to generate the frequency ramp [78]. In this case, the overall division value is composed of the divider ratio $N = 4$ between the VCO and downconverter and $N = 16$ after the downconverter, thus providing wideband frequency sweeps using only quite a narrow frequency range generated from DDS, where spurious levels at the DDS output are sufficiently low. With a fully integrated dual-band PLL-based frequency synthesizer, the operation in the 22–29 GHz and 77–81 GHz short-range automotive radar bands can be provided [79]. Such a synthesizer consists of two VCOs, one for each radar band, whose outputs are multiplexed into the input of an injection-locked circuit that acts as a divide-by-three circuit for higher band and as a tuned buffer for lower band. The 79-GHz VCO represents the modified differential Colpitts oscillator, with an *LC* degeneration technique, and the 24-GHz VCO is a cross-coupled *LC* oscillator with the tank formed by a spiral inductor and MOS varactors.

One of the important issues in a millimeter-wave PA design is to achieve its high operating efficiency at maximum output power, since generally it is difficult to achieve high efficiency at these frequencies taking into account significant effect of grounding parasitics and component losses. For example, by using a 0.12-μm SiGe BiCMOS process with $f_{\max} = 200$ GHz, the single-ended PA achieves an output power of 17.5 dBm with a power-added efficiency (*PAE*) of 12.8% at 77 GHz [80]. In this case, four stages in a common emitter configuration are biased in Class AB mode where the last

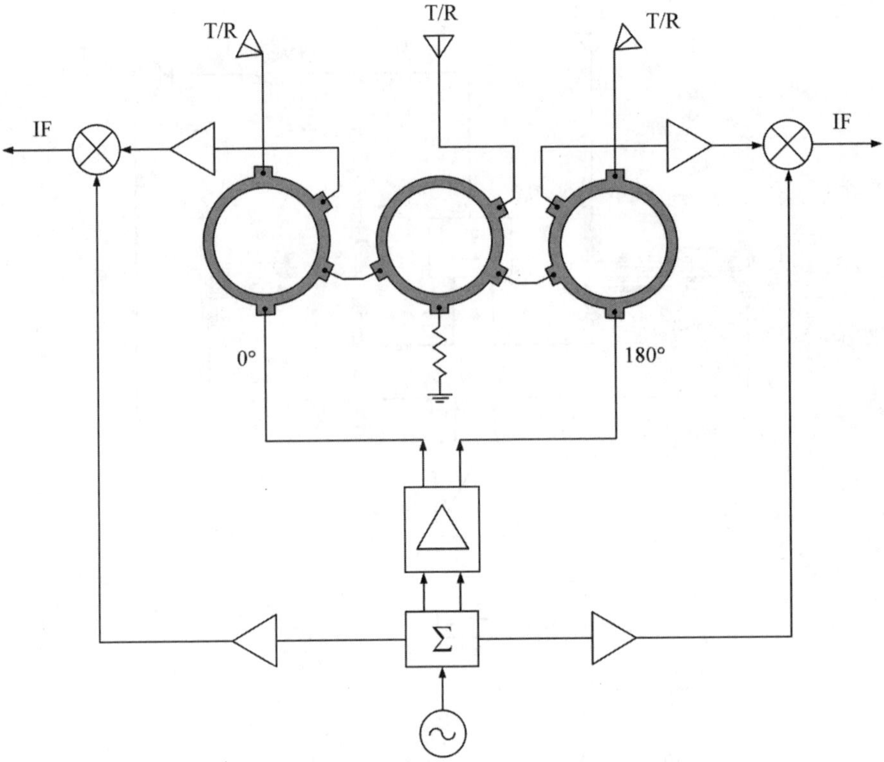

FIGURE 14.27 High-efficiency differential radar system.

three stages represent one, two, and four identical cells, respectively. The series transmission lines and parallel short- or open-circuit stubs were used for impedance matching between stages. Conductor-backed coplanar waveguide (CBCPW) structure was used as the transmission-line structure resulting in a high degree of isolation between adjacent lines and small-size full integration. Balanced three-stage PA based on a similar technology could achieve a saturated power of 19.6 dBm with a *PAE* of 15.4% at 90 GHz [81]. A saturated power of 14.5 dBm with a *PAE* of 15.7% at 77 GHz was achieved by the PA implemented in a 0.13-μm SiGe HBT process with $f_{\max} = 230$ GHz [82]. Usually, the latter power stages operated in Class AB with small quiescent current are configured with common emitters to minimize power consumption and maximize power-added efficiency, while driver stages

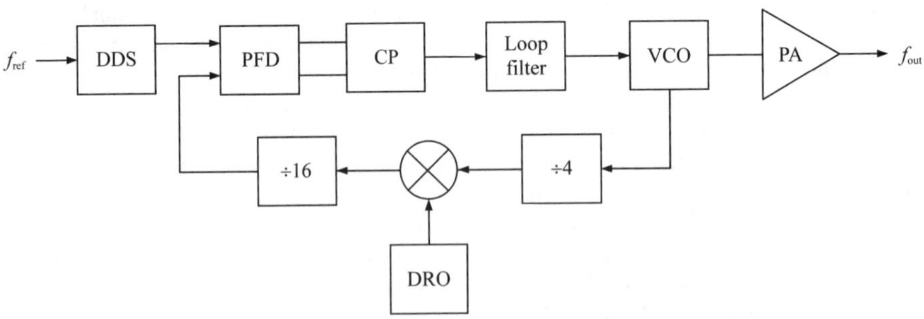

FIGURE 14.28 77-GHz PLL-based automotive radar architecture.

can be designed in a cascode configuration. High breakdown voltage of over 50 V to design a reliable 76-GHz PA can be realized with a 0.15-μm GaN FET process having $f_{\max} = 160$ GHz [83].

14.6.3 Electronic Warfare

Electronic warfare (EW) techniques were initiated during the World War II as radio countermeasures when signal analysis was used to identify the frequencies of enemy systems such as communication, radar, and airborne navigation equipments, which were subsequently disabled with passive countermeasures such as chaff or active noise jamming techniques [84]. Passive countermeasures represent methods to reduce the probability that a target will be detected by the radar, or to present false targets to the radar, and do not involve the emission of any targets and noise. Active countermeasures are based on the radiation of a signal to confuse or deceive the radar or communication system, and generally take the form of jammers. A noise jammer radiates a relatively large power in the operating band of the radar to mask the true target return, or to overload the receiver front-end. There are three basic subsets of the overall discipline of EW such as electronic support measures (ESM), electronic countermeasures (ECM), and electronic counter-countermeasures (ECCM). ESM is based on the use of intercept or warning receivers when actions are taken to search for, intercept, locate, and analyze radiated electromagnetic energy. ECM includes passive and active techniques such as jamming and deception, when actions are taken to prevent or reduce the enemy's use of the electromagnetic spectrum, overload or saturate the data processing of threat systems, and introduce false deceptive data into hostile electronic systems. Since a jamming transmitter may be vulnerable to lethal countermeasures, prudent ECM operational procedures require detection and identification of a target signal before a jamming signal is transmitted. ECCM includes a coding technique that spreads the signal energy over a wide bandwidth to minimize the effect of enemy jamming and interference.

The most common techniques for spreading the signal spectrum are to employ frequency-agile transmissions, either in discrete steps, frequency hopping (FH), or to use a continuously swept FM or phase-shift keying (PSK) of the carrier oscillator. FM is employed in pulsed radar where a linear swept chirp waveform is transmitted and a weighted matched filter is incorporated into the receiver to detect the return echo. FH and PSK, both under control of a pseudo-noise (PN) code, represent two basic modulation techniques for generating the spread-spectrum waveform. Direct-sequence spread spectrum is widely used in satellite communication and navigation systems, where a continuous PN spreading code is modulo-two added with data prior to keying the carrier. FH offers a much flatter transmitted spectrum than PSK or MSK and, for a given RF bandwidth, the FH dwell time is much longer than the equivalent PSK interval. This reduces the PN-code rate with a potential easing of synchronization acquisition in the FH systems. A key advantage of FH is that multiple-access signals in other frequency slots are theoretically non-interfering, regardless of the relative signal strength.

Many EW systems require a high performance front-end covering all or part of the 2–18 GHz band with stringent requirements on the size, weight, power consumption, and reliability. An effective method of meeting all these requirements is to make extensive use of MMICs to allow packaging of major sections of the front-end using a small number of MMIC modules [85]. In this case, the excellent balance and reduced parasitics can minimize the level of spurious products and LO leakage in mixers. In the development of high-frequency VCOs, wider tuning bandwidths can be achieved using MMIC technology due to the elimination of critical bondwire parasitics. In modern ECM systems, VCOs serve as the frequency-agile local oscillators in receiver subsystems and fast-modulation noise sources in active jamming subsystems. Among wideband tunable signal sources such as YIG-tuned oscillators, wideband VCOs using hyperabrupt varactors are preferable because of their small size, low weight, high settling time speed, and capability of fully monolithic integration. Figure 14.29(*a*) illustrates the use of a hyperabrupt VCO as a local oscillator in the receiver system, while a block diagram of an active jamming system using a hyperabrupt VCO as the microwave power source is shown in Figure 14.29(*b*) [86]. In both system applications, the hyperabrupt VCO must exhibit the linear voltage-frequency characteristic, minimum tuning slope variation, low post-tuning drift, fast settling time, and low voltage bias range to allow direct integration with high-speed digital-to-analog

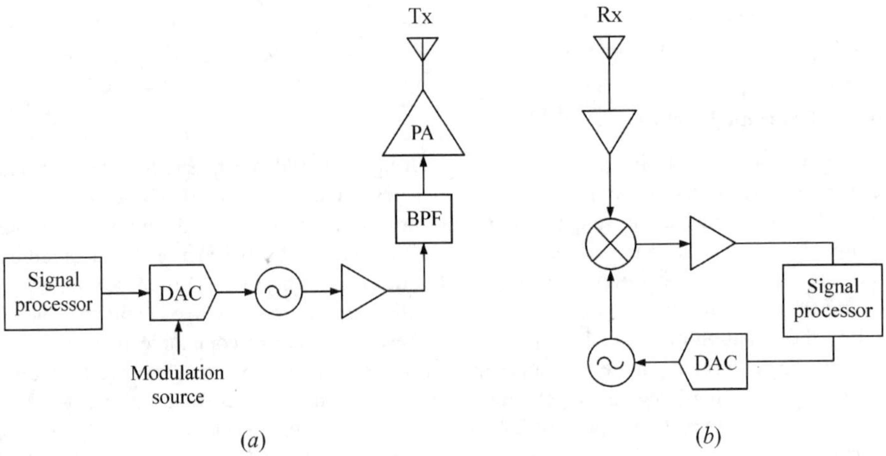

FIGURE 14.29 ECM receiver and transmitter systems.

conversion circuitry [87,88]. For example, the linear tuning characteristic and high-speed tuning capability of the hyperabrupt VCO result in a high intercept probability for the receiver as well as high frequency resolution. Since the hyperabrupt varactor sensitivity γ ranges between 1.0 and 2.0, hence a linear frequency tuning over specified frequency band can be achieved either as a result of simultaneous changes in varactor and VCO bias voltages or using an analog or digital linearizer in front of the VCO [86,89].

To minimize the excessive losses in the resonant circuit under wideband varactor tuning, it is necessary to choose the varactors with maximum quality factors and provide the required broadband matching. For example, the matching of the device output impedance with 50-Ω load can be provided by microstrip transformer of a quarter-wavelength at highest bandwidth frequency. Figure 14.30(*a*) shows the circuit schematic of a wideband FET VCO where hyperabrupt varactors are used in both source and gate circuits. Using the commercially available hyperabrupt GaAs varactor diodes (with a quality factor of 3000 at a bias voltage of 4 V) provides an UWB frequency tuning from 6.5 to 16.1 GHz with tuning linearity in limits of ±1% up to 14 GHz for a varactor bias range of 0 to 25 V [90]. In this case, values for the shunt inductors L_1, L_2, and L_3 together with the series gate inductor L_4 are optimized to meet the required oscillation conditions over the full capacitance range of the varactors.

Using the lumped inductances in parallel to the varactors provides an additional improvement of the circuit sensitivity to the varactor tuning. Improvement occurs when the VCO oscillation frequency is close to the natural resonant frequency of the parallel circuit consisting of the varactor capacitance and parallel inductance. However, to avoid the unstable operation mode accompanied by jumping effects, it is necessary to provide the capacitive reactance of the source circuit and the inductive reactance of the gate circuit in an entire frequency tuning bandwidth. The three-stage balanced buffer amplifier connected directly to the VCO output minimizes the load-pulling effect to less than 1.5 MHz into a 12-dB return loss and increases the output power to more than 18 dBm with a flatness within ±1 dBm.

Figure 14.30(*b*) shows the triple-tuned VCO using an InGaP/GaAs HBT device and hyperabrupt varactors (included into the base, emitter, and collector circuits) with a capacitance ratio of approximately 13.6, which covers the oscillation bandwidth from 5.6 to 16.8 GHz with an output power of 3.4 ± 2.0 dBm at the tuning voltage from −0.35 to +16 V [91]. The phase noise level was lower than −112.0 dBc/Hz at a 1-MHz offset from the carrier.

The main advantage of a push–push oscillator is a high quality factor of the resonant circuit because the fundamental resonator is decoupled from the second-harmonic load that improves both load pulling and noise performances. In a common collector configuration, a series tank circuit

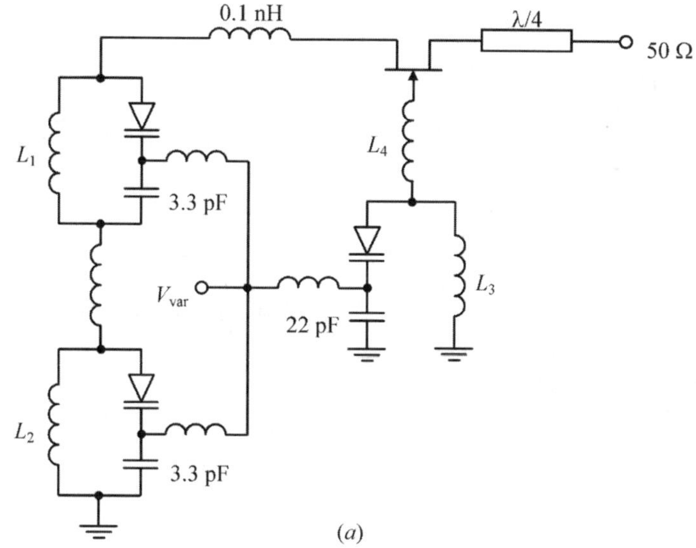

(*a*)

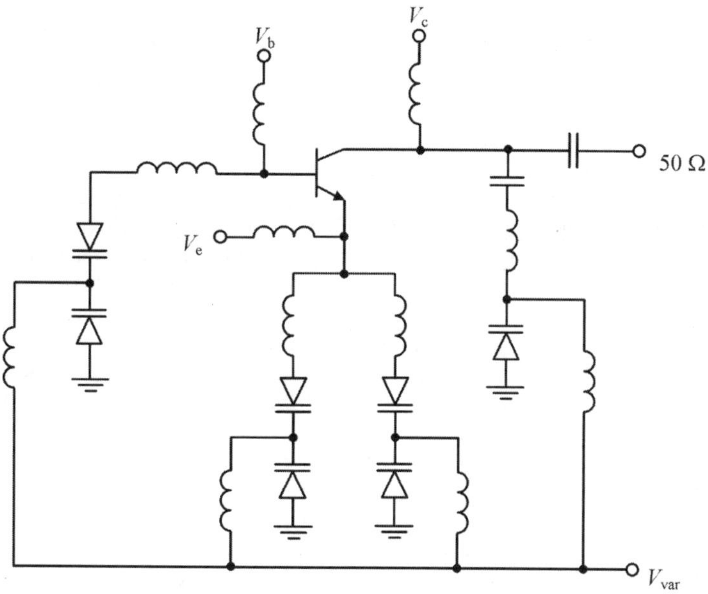

(*b*)

FIGURE 14.30 Microwave monolithic very wideband VCO circuits.

connected between the device bases can be used. The overall resonant circuit includes also transistor collector–base capacitances, varactor capacitances, and bond wire inductances. In this case, the load can be connected conductively to the center point of the resonant-circuit inductor, as shown in Figure 14.31, where odd-harmonic cancelation can be provided. For symmetrical parts in the odd operation mode, this center point becomes a virtual ground. Such a push–push VCO enables to provide very wideband varactor tuning over a frequency range from 6.94 to 19.22 GHz with the fundamental

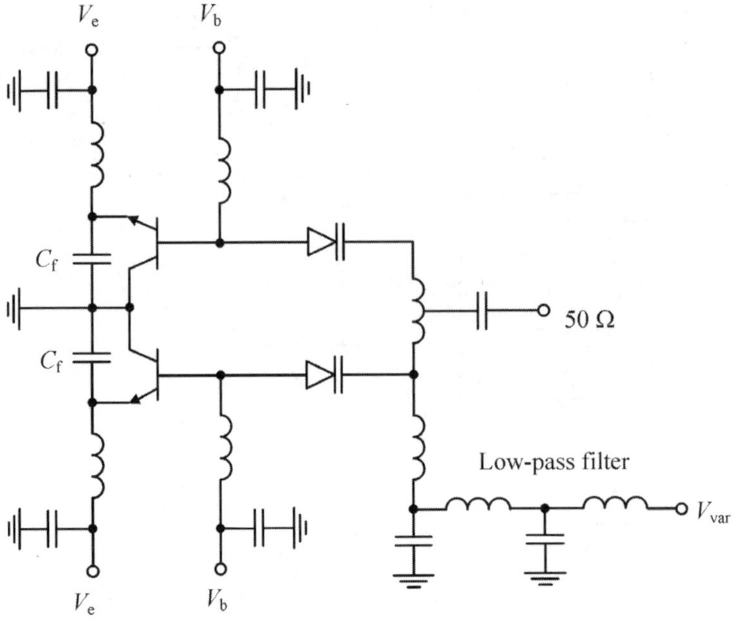

FIGURE 14.31 Microwave very wideband push–push VCO.

harmonic suppression of about 10 dB when additional varactors are used instead of collector–emitter feedback capacitors C_f [92,93]. Such a wideband frequency tuning can be realized using double pairs of GaAs hyperabrupt varactors with a sensitivity of $\gamma = 1.1$ in a bias range of 0 to 20 V.

Microwave SSPAs implemented in MMIC technology are very attractive for EW applications with their broadband capability, small size, low-voltage operation, and high reliability. The first all-monolithic PA MMIC based on a 0.5-μm MESFET technology was designed to operate over the full 6–18 GHz frequency band and to provide more than 1-W output power using a balanced configuration with Lange couplers [94]. The two-stage 6–18 GHz AlGaAs/GaAs HBT PA can achieve an output power varying from 31.0 to 33.9 dBm with a *PAE* varying from 18.4 to 36.7% across the entire frequency band [95]. By using a 0.25-μm power pHEMT process with $f_T = 32$ GHz, the dual-channel three-stage 6–18 GHz MMIC PA provided an output power of 4.3 ± 1.3 W, a small-signal gain of 24 ± 3.5 dB, and a *PAE* of 19.5 ± 7.5% with external output combiner [96]. In this case, the input and output Lange couplers, dividing and combining two amplifier channels, are strongly overcoupled to achieve operation over the 6 to 18 GHz frequency band. An output power varying from 4 to 7.5 W with an average *PAE* of 22% and a power gain of 12 dB from 6 to 18 GHz was achieved by the two-stage MMIC PA, which is implemented in a 0.15-μm pHEMT process and consists of four 800-μm cells driving sixteen 800-μm cells for a total gate periphery of 16 mm [97].

14.7 SATELLITE TRANSMITTERS

Generally, a number of ideas, problems, and potential solutions such as alternative satellite repeater schemes, orbits, mutual visibility and distances, number of satellites, path-loss calculations, and modulation schemes for transoceanic communication by means of satellites were discussed more than 50 years ago [98]. For example, passive and active repeaters or transponders can be used, with smaller ground antennas and transmitter powers in the latter case. The active transponder receives signals on the uplink, transfers them to the downlink frequency, and amplifies them for retransmission to earth. In July 1962, the orbiting Telstar active satellite having a 6-GHz uplink and a 4-GHz downlink with transmitter power of 2.25 W was used for TV transmission, as well as for demonstrations of telephony between American and European cities [99]. Due to demand in wider frequency range and more information capacity, satellite communications have moved from *C*-band (6/4 GHz) to *Ku*-band (14/11 GHz) and then to *Ka*-band (30/20 GHz) frequencies [100]. The potential market for mobile communications provided by satellite systems has resulted in the development of a range of operational systems and conceptual designs, from conventional geosynchronous orbit (GEO) systems to the low earth orbit (LEO) systems [101]. To further improve the performance of high data-rate communication links, millimeter-wave frequencies were considered to use for wireless communications through satellites. However, such effects as atmospheric attenuation and rain absorption increase the channel noise temperature and limit the channel capacity. In this case, the phased-array technique represents a practical solution to compensate for these propagation impairments, since a highly directive antenna array improves the signal-to-noise ratio significantly, thus correspondingly increasing channel capacity. For example, a 16-element phased-array transmitter in a standard 0.18-μm SiGe BiCMOS technology based on 4-bit active phase shifters results in a 12.5 dB of average power gain per channel at 42.5 GHz [102].

Efficient use of communication channels can be achieved through multiple access techniques such as FDMA, which allows more than one earth station to share the bandwidth of a satellite transponder when each earth station is assigned to a specific carrier or set of carriers for signal transmission over a certain frequency bandwidth; TDMA, which is used to provide more than one earth station with access to the same carrier by dividing the transmitting signal into different time slots; and CDMA as a spread-spectrum access technique, in which every earth station may transmit at the same frequency and at the same time when each earth station uses a unique identifying code modulated with the communication signal. In TDMA system, the satellite transponder can operate at saturation, while output backoff of the satellite transponder in an SCPC (single channel per carrier) FDMA system is necessary to reduce intermodulation distortion [103]. The advantage of using FDMA in a wideband frequency transponder is that it allows the simplest ground station configuration requiring no timing by the user, whereas TDMA requires that the earth station complex be synchronized to a high degree of precision and be capable of high peak power transmission. The TDMA modulation equipment can provide a transmission rate of 60 Mb/s and overall capacity of 19 full-duplex 1.544-Mb/s signal channels per transponder, or the equivalent of 456 full-duplex voice-grade channels [104]. CDMA is usually divided into two subclasses: frequency hopping and direct sequence. In the former, a user imposes a preselected frequency hopping sequence on his normal transmission so that mutual interference between two such users occurs only when they are hopped to the same frequency at the same time. In the latter, the transmission is characterized by a phase-modulated carrier with a total bandwidth occupancy much wider than the information bandwidth. The data are added to the multiple-access carrier as either frequency or phase modulation. A PN code is used to spread the spectrum and is typically a maximal length shift register code [105]. The degree of mutual interference between two such users is determined by the cross correlation between the PN codes of two users.

The first completely solid-state transmitter intended for missile and satellite applications to operate as a PPM (pulse position modulation) telemetry transmitter was based on transistor PA stages with a varactor tripler and provided a peak power of 30 W at 234 MHz [106]. Further advances in communications satellite technology have led to the development of different systems that can transmit various kinds of communication signals. In a domestic satellite communications system accommodating numerous small earth stations, it is important that the small earth station equipment remains inexpensive in order to make the system economical. In this case, each station uses a high power FET amplifier, whose output peak power is about 2 W per voice channel, and a low-noise FET amplifier, whose noise temperature is less than 90 K [103]. When more than two voice channels are transmitted, signals are multiplexed using orthogonal carriers in order to keep the transmitting peak power low. These carriers are modulated by different PN codes and amplified simultaneously by a high PA.

Figure 14.32(*a*) shows the typical block schematic of a 6/4 GHz satellite transponder performing the function of a low-noise amplification of the 6-GHz receiving signal, its upconversion to 4-GHz

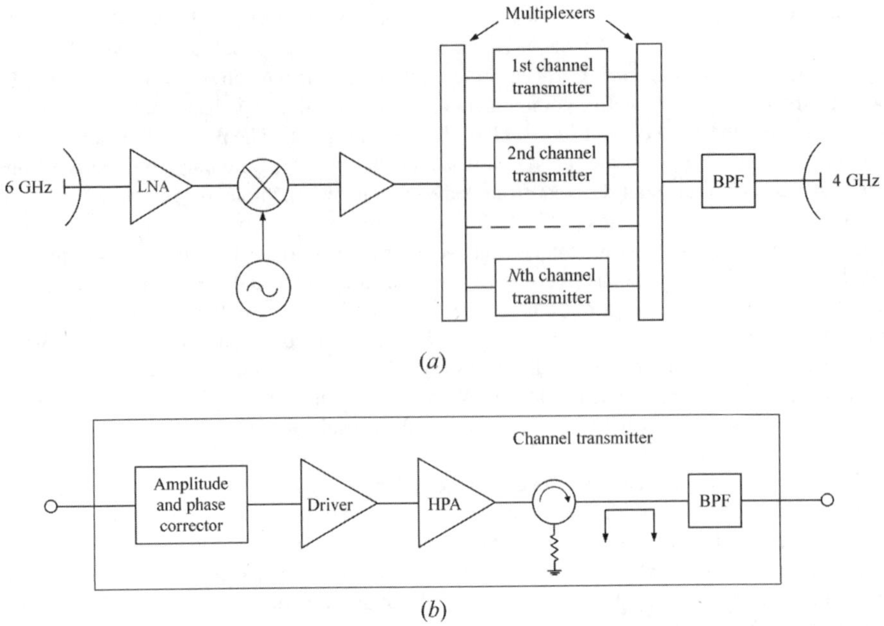

FIGURE 14.32 Block schematics of satellite transponder.

band, and further channel power amplification (with channel bandwidth up to 35–40 MHz) and output filtering. Channelization of communication frequencies after receiver processing and recombination of channels after high-power amplification are provided by input and output multiplexers, respectively. Both input and output multiplexers employ a multiplicity of high-selectivity channel BPFs in combination with an input distribution network and output manifold. Significant achievements in dielectric resonator technology, resulting in dielectric materials of superior stability and reduced loss, have produced filter designs of reduced size, volume, and weight. Due to frequency bandwidth and filter selectivity limitations, the *Ka*-band 30/20 GHz satellite transponders can be designed using 30–12 GHz downconverter and 12–20 GHz upconverter when the intermediate channel amplification is provided in a 12-GHz band [107]. The block schematic of a channel PA is shown in Figure 14.32(*b*), where the amplitude and phase corrector is necessary to linearize the signal amplitude characteristic and to make the signal phase independent of its amplitude. To provide transmitter stable operation with an HPA and sufficient isolation from the other transmitting channel, the circulator operating as an isolator is included at the output of the PA. The output directional coupler is used to measure the levels of incident and reflected powers.

The development of SSPA technology in the late 1970s and early 1980s created a viable alternative to the traveling-wave tube amplifiers (TWTA) to design a channel PA for most *C*-band applications. SSPA designs have utilized power GaAs FETs that became available and space-qualified during that period. As a result, a *C*-band GaAs FET PA could deliver a high output power of 70 W with a high efficiency of more than 50% under 10-V operation at 3.8 GHz, representing a packaged internally matched amplifier with four 35-mm-gate-width heterostructure FETs [108]. A *Ku*-Band GaAs FET SSPA, consisting of four 8-W internally matched power FETs and using a spatial power combiner, demonstrated a 30-W output power across the band of 14.0 to 14.5 GHz with 11% efficiency [109]. By using a 0.35-μm GaAs pHEMT process, a monolithic two-stage SSPA is capable to deliver up to 10 W of output power with a *PAE* of 47% at 11.7 GHz [110]. The MMIC layout, whose dimensions are 4.7×4.0 mm^2, is composed of eight power cells in the final amplifying stage and microstrip combining network.

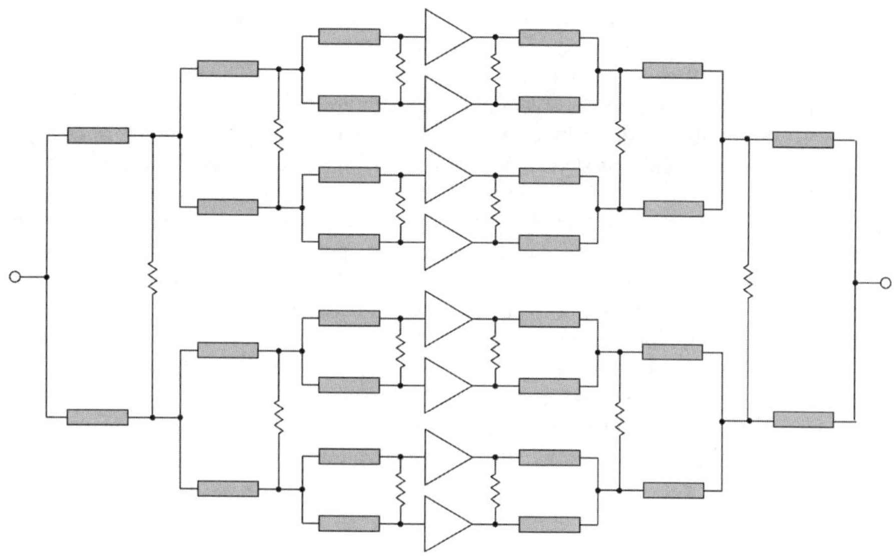

FIGURE 14.33 Block diagram of microwave solid-state high-power amplifier.

Recently, the GaN HEMTs have demonstrated superior microwave power performance over GaAs with output powers of 4 W/mm, which is approximately four times higher compared to the GaAs pHEMTs at *Ku*-band [111]. To achieve a very high output power level, the individual PAs are usually combined in a packaged hybrid module using Wilkinson dividers/combiners [112,113]. Figure 14.33 shows the block diagram of a *C*-band PA with over 340-W output power at 4.8 GHz using 0.8-μm GaN HEMTS and an *X*-band PA with over 100-W output power at 9.8 GHz using 0.25-μm GaN HEMTs obtained under the pulsed condition of 10-μs pulse width and 10% duty cycles [113]. In both cases based on two internally matched GaN HEMT chips and microstrip Wilkinson dividers/combiners, *PAE* exceed 50%.

14.8 ULTRA-WIDEBAND COMMUNICATION TRANSMITTERS

UWB technology was first proposed for communication systems a long ago [114]. Since then it was mainly used for radar-based applications because of the wideband nature of the signal that results in very accurate timing information. By the early 1970s, the system concept and basic components such as pulse train generators and modulators, detection receivers and wideband antennas were available [115]. However, due to recent developments in high-speed switching and narrowband pulse generation technology, UWB has become more attractive for low-cost communications applications, representing now any wireless transmission scheme that occupies a transmission frequency bandwidth of more than 20% of a center frequency, or more than 500 MHz [116]. Such large bandwidths are achieved by using very narrow time-duration baseband pulses of appropriate shape and duration, including the family of Gaussian shaped pulses and their derivatives. Larger transmission bandwidths are preferred to achieve higher data rates without the need to increase transmitting power, resulting in the ability for increasingly fine resolution for multipath arrivals, which leads to reduced fading per resolved path since the impulsive nature of the transmitted waveforms prevents significant overlap and, hence, reduces possibility of destructive combining. A key advantage of UWB designs is that highly linear PAs are generally not required because the UWB pulse generator need only to produce a peak-to-peak voltage swing on the order of 100 mV to meet spectral mask requirements that can be achieved by a suitable UWB waveform choice. The UWB systems are being currently designed for 110 Mb/s at a 10-m range with four collocated clusters that project to spatial capacity of about 1.3 Mb/s/m^2. Thus, UWB represents a tradeoff between lower spectral efficiency for increased power efficiency to achieve a given rate/range operating point with limited transmit power.

In view of UWB pulsed nature with a very sharp rise and fall time for signal transmission, the UWB emissions can potentially interfere with other licensed frequency bands. Therefore, the allowed transmit power spectral density (PSD) is limited to −41.25 dBm/MHz that corresponds to a theoretical maximum total power of −13.9 dBm for a 550-MHz bandwidth signal [117]. The transmitting signal in UWB system can be generally modulated by turning the pulse on and off (OOK), by providing the binary phase-shift keying (BPSK), or by dithering the pulse position modulation (PPM), or the pulse amplitude modulation (PAM). The pulse may have duration on the order of 200 ps and its shape can be designed to concentrate energy over the broad frequency range from 2 to 6 GHz, for example. In this case, a BPF is necessary to use before the antenna to constrain the emissions within this desired frequency band when the filter would have a bandwidth on the order of 4 GHz. UWB systems can also be designed using spread-spectrum codes, which offer better coexistence with other UWB systems. The generation of spread-spectrum UWB signals can be achieved using time hopping or direct sequence methods. In multiband UWB systems when UWB band from 3.1 to 10.6 GHz is divided into 14 subbands with a bandwidth of 528 MHz each, signal transmissions are staggered in time across the constituent subbands, and one of several (typically 3–10) subbands are used sequentially for transmission. The requirement for maximum total power consumption of a UWB transceiver set by specification at 110 Mb/s and 200 Mb/s is 100 mW and 250 mW, respectively.

Generally, low-rate and high-rate pulsed UWB transmitters are based on one of two basic techniques to synthesize pulses in the 3.1–10.6 GHz band [118]. The upconversion pulse (or direct conversion) synthesis technique involves generating a pulse at baseband and upconverting it to center frequency in the UWB band by mixing with a local oscillator, whose output can be either enabled or disabled by a simple switch, thus effectively mixing the RF signal with a rectangular baseband pulse. The carrier-less pulse synthesis involves generating pulses that directly fall in the UWB band without requiring frequency translation where the pulse width is usually defined by delay elements that may be tunable or fixed. In this case, a baseband impulse may excite a filter that shapes the pulse or the pulse may be directly synthesized at RF with no additional filtering required. An advantage of carrier-less techniques over traditional mixed-based architectures is that the carrier frequency generation is inherently duty cycled when RF power is only generated when it is required. However, its main disadvantage is that an integrated downconverting receiver typically cannot share the RF generation circuits, and therefore, must have its own local oscillator. Pulse-based transmitters can additionally be categorized in terms of how pulses are delivered to an antenna, either by analog power amplification or digital buffering. However, in the latter case, linear AM and pulse shaping are more difficult to achieve. Since the direct-conversion UWB transceiver is usually characterized by the large LO leakage to the RF front-end and to antenna because the local oscillator is operated at the same frequency as the RF carrier, the dual-conversion zero-IF transceiver architecture with two-step upconversion can be a suitable alternative [119]. In this case, to reduce the image-rejection problem, which exists in the first conversion, either both local oscillators for channel selection can be variable or an *I*/*Q* local oscillator is properly fixed at desired frequency according to careful frequency planning.

Figure 14.34(*a*) shows the generic UWB pulse-based transmitter architecture with a pulse generator followed by a modulator, both driven by a digital baseband. The modulator is then directly followed by a PA to achieve the required power level, a BPF to reduce spurious emission, and a wideband antenna to radiate pulsed signal. A UWB pulse can be generated by multiplying a sinusoidal signal with an envelope close to zero-order Gaussian shape with a standard deviation of 341 ps and centered in 4.1 GHz to spread energy over a 2-GHz bandwidth [120]. The UWB transmitter based on a 0.18-μm CMOS technology can achieve highest pulse repetition rate of 750 Mb/s with bi-phase modulated pulses of 500-ps duration and 8-GHz center frequency, operating over the band of 6 to 10 GHz and complying with the standard spectrum mask [121]. For a short-range non-coherent communication, the transmitter can use an all-digital architecture and calibration technique to synthesize pulses with programmable width and center frequency using a 90-nm CMOS process [122].

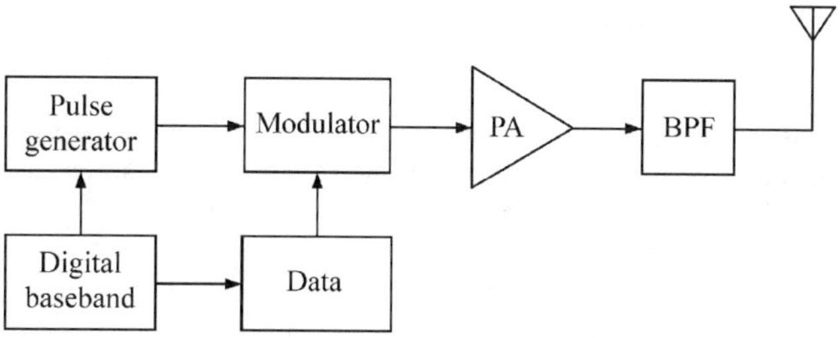

(*a*)

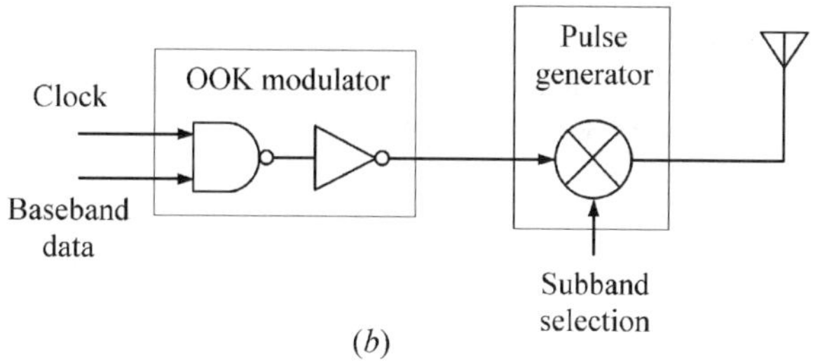

(*b*)

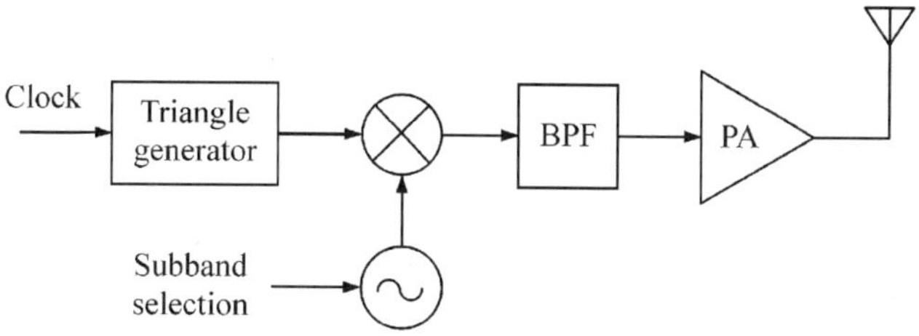

(*c*)

FIGURE 14.34 UWB transmitter architectures.

Figure 14.34(*b*) shows a low-power, low-complexity carrier-based multiband UWB transmitter where OOK modulation is used for low data rate (~100 kb/s) applications [123]. The pulse generator block is capable of generating triangular-enveloped pulses over three 528-MHz subbands in a 3–5 GHz band using a subband selection function. The modulator consists of the NAND and inverter logic components where one input of a NAND block is the baseband data while the other input is the square pulse clock. The clock is provided by the baseband processor with predetermined duration to control the output pulse width to ensure 528-MHz bandwidth for each subband. The input data and clock are combined and the resultant edge-combined square pulse signal drives the pulse generator. As a result, with pulse duration of 3.5 ns, an output pulse spectrum with more than 20 dB of sidelobe rejection can be achieved satisfying spectral mask without additional filtering. This transmitter being implemented in a 0.18-μm CMOS process was designed to operate in a burst mode and dissipates only 18 pJ of energy consumption per pulse. For higher amplitudes of transmitted triangular pulses (for example, with a peak voltage of about 2.8 V under a supply voltage of 3.3 V), in order to satisfy the spectral mask limit, especially in the frequency range from 0.96 to 1.61 GHz assigned for GPS systems, the separate BPF that can represent a single-section coupled-line filter (half-wavelength transmission-line resonator) with a tapered stub can be used [124]. With sufficiently low inductance values, a third-order Bessel bandpass *LC* filter structure can be designed on-chip [125].

The new-generation UWB transmitters, whose typical block schematic is shown in Figure 14.34(*c*), are intended to operate in the 3.1–10.6 GHz band with 100 Mb/s data rate at a link distance of 10 m using BPSK modulated pulses at a maximum pulse repetition frequency of 100 MHz [117]. The transmitter implemented in a 0.18-μm SiGe BiCMOS process uses a direct-conversion architecture to upconvert a shaped baseband pulse train to one of 14 subbands in the UWB band by modulating the bias current of the differential pair with a carrier frequency. The baseband UWB signal is generated using either a programmable arbitrary waveform generator (AWG) or a dedicated discrete pulse generator. Using an AWG enables a large amount of flexibility in the shape of the transmitted pulses, modulation scheme, and duration of transmission. The carrier frequency can be generated either by on-chip VCO or from external LO. In the latter case with a 0.13-μm CMOS technology, the frequency of the external differential input signal from 6 to 22 GHz is divided by the on-chip 2:1 frequency divider to obtain quadrature LO signals from 3 to 11 GHz to cover 14 bands of the UWB system at the same time [126]. With the LO input power of 0 dBm, the quadrature upconversion mixer and balanced PA provide the conversion gain of 1.1 ± 1 dB and power gain of 12.4 ± 2 dB over the frequency range from 3 to 11 GHz, respectively.

A pulse generation principle for a multiband OOK UWB transmitter with on/off switching of a CMOS oscillator, whose circuit schematic is shown in Figure 14.35, can be based on the switched

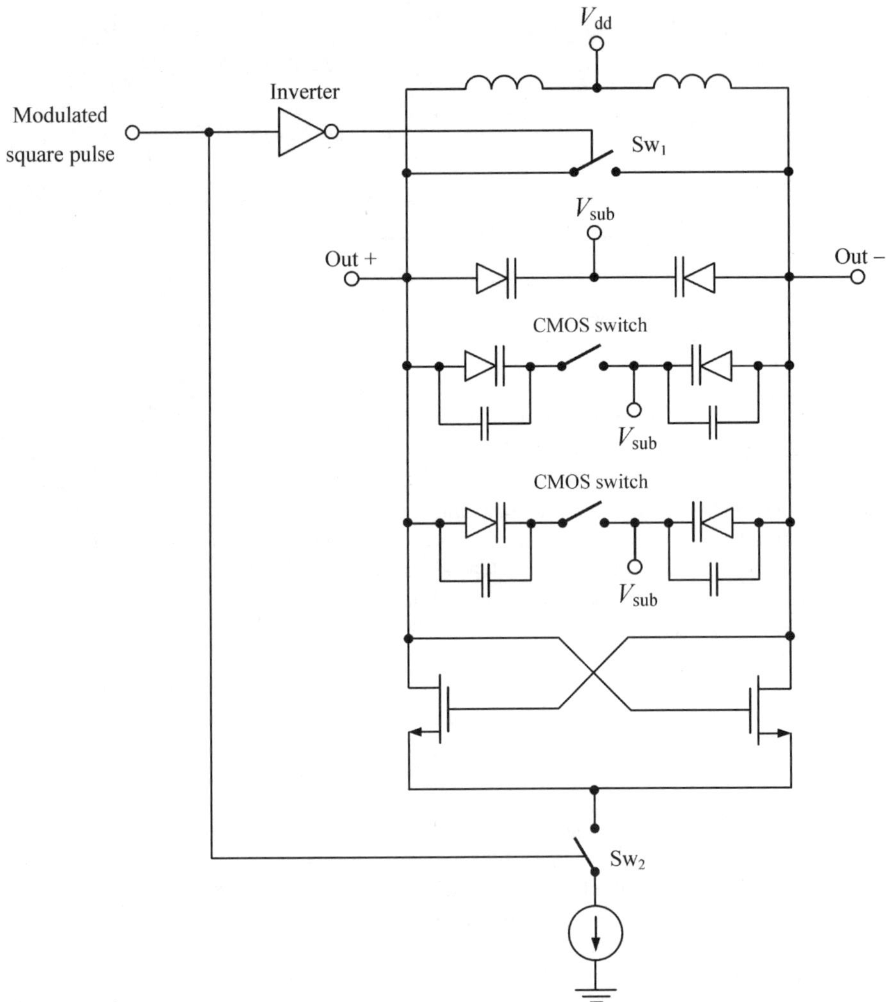

FIGURE 14.35 Schematic of three-band OOK UWB pulse generator.

capacitor bank with subband selection and two-switch approach (with a switch across *LC* tank and a switch in series to current source) [123]. In this case, when the switch Sw_1 is turned off and the switch Sw_2 is turned on, pulses are generated at the oscillator differential outputs and vice versa. An inverter is used to make 180° out-of-phase square pulse train to drive switches. Since the triangular-enveloped pulse is required to provide more than 20 dB of sidelobe rejection, this can be achieved from the square pulse by turning off the oscillations during their rise transient before the pulse enters the steady-state

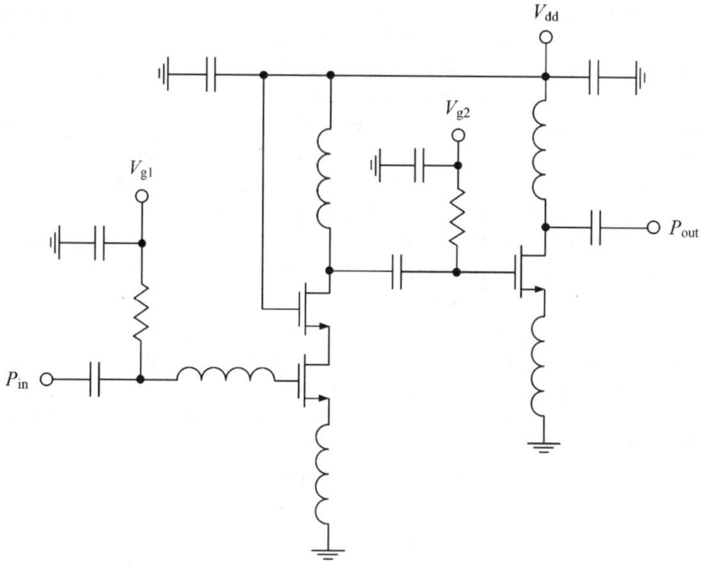

(*a*)

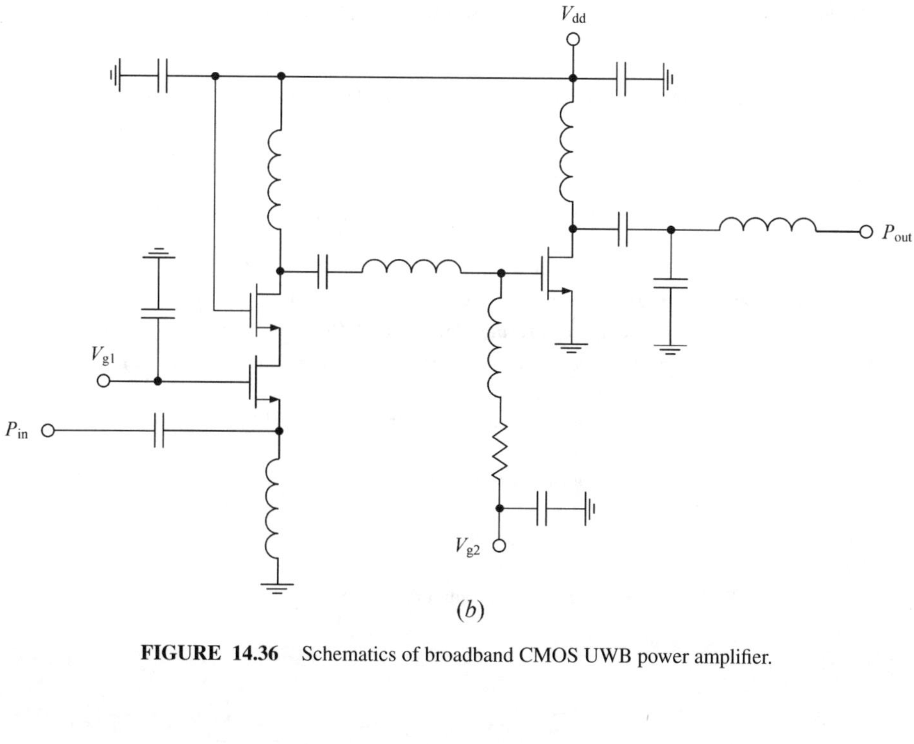

(*b*)

FIGURE 14.36 Schematics of broadband CMOS UWB power amplifier.

mode. By changing the capacitor bank values, the pulse center frequency is shifted, and the three subbands with 528 MHz are centered at 3.2, 3.8, and 4.4 GHz, respectively. In a 0.18-μm SiGe BiCMOS UWB transmitter designed for wireless sensor networks, the VCO switched tank circuit with three capacitor banks controlled by CMOS switches can provide three 500-MHz subbands at center frequencies of 4.6, 6.4, and 8.0 GHz, respectively [127].

The PA in UWB systems should meet several stringent requirements simultaneously such as broadband matching, high power gain, sufficient output power, and reasonable efficiency for low-power operation. By using the lossy input and output *RLC* matching networks and two-stage cascode configuration, it is possible to cover a full UWB band from 3.1 to 10.6 GHz with a power gain above 6 dB and an output power of around 0 dBm using a 0.18-μm CMOS technology [128]. The output power can be increased to more than 5 dBm over a full UWB band using twice as many transistors in a distributed configuration with a wideband bandpass artificial transmission-line topology [129]. Unfortunately, in both cases the efficiency is too low and the power consumption is too high to satisfy such broadband requirements. However, the *PAE* can potentially be increased to more than 30% with proper external broadband output matching for a 0.18-μm CMOS PA, whose circuit schematic is shown in Figure 14.36(*a*) and which is operated in a narrower frequency band from 3 to 5 GHz with an output power over 10 dBm [130]. Here, the source degeneration inductor in the first cascode stage is added for linearity and stability improvement.

Figure 14.36(*b*) shows the circuit schematic of a 0.18-μm CMOS PA with interstage *RLC* broadband impedance matching (including the gate–drain capacitance of the common-gate device and the gate–source capacitance of the common-source device) to provide a flat power gain over a frequency band from 6 to 10 GHz [131]. In this case, a *PAE* of about 15% over most of the bandwidth and a power consumption of only 18 mW can be achieved with an output power of about 5 dBm at a supply voltage of 1.5 V. To simplify the input matching and provide significant interstage isolation, both first-stage nMOS devices are configured with a common gate.

REFERENCES

1. J. O. Weldon, "Transmitters," *Proc. IRE*, vol. 50, pp. 901–906, May 1962.
2. F. H. Raab, R. Caverly, R. Campbell, M. Eron, J. B. Hecht, A. Mediano, D. P. Myer, and J. L. B. Walker, "HF, VHF, and UHF Systems and Technology," *IEEE Trans. Microw. Theory Technol.*, vol. MTT-50, pp. 888–899, Mar. 2002.
3. H. L. Krauss, C. W. Bostian, and F. H. Raab, *Solid State Radio Engineering*, New York: John Wiley & Sons, 1980.
4. N. F. Schlaack and F. A. Polkinghorn, "An Unattended Ultra-Short-Wave Radiotelephone System," *Proc. IRE*, vol. 23, pp. 1275–1285, Nov. 1935.
5. A. Mina, J. O. Weldon, and G. Hulsey, "An Improved High Efficiency 500 Kilowatt Short Wave Broadcast Transmitter with Novel Features," *IEEE Trans. Broadcast.*, vol. BC-34, pp. 134–136, June 1988.
6. M. Yamazoe, K. Haeiwa, S. Hirose, and K. Wakai, "Considerations for Developing a Highly Efficient, High-Power Shortwave AM Transmitter," *Proc. 1996 IEEE Power Conversion Conf.*, pp. 1035–1042.
7. F. E. Terman, *Electronic and Radio Engineering*, New York: McGraw-Hill, 1955.
8. D. Y. Jung, W. I. Chang, and C. S. Park, "A System-on-Package Integration of 60 GHz ASK Transmitter," *Proc. 2006 IEEE Radio and Wireless Conf.*, pp. 151–154.
9. A. A. Oswald, "Early History of Single-Sideband Transmission," *Proc. IRE*, vol. 44, pp. 1672–1679, Dec. 1956.
10. J. R. Carson, "Method and Means for Signaling with High Frequency Waves," U.S. Patent 1,449,382, Mar. 1923 (filed Dec. 1915).
11. H. D. Arnold and L. Espenschied, "Transatlantic Radio Telephony," *Trans. AIEE*, vol. 42, pp. 718–729, Aug. 1923.
12. A. A. Oswald, "A Short-Wave Single-Side-Band Radiotelephone System," *Proc. IRE*, vol. 26, pp. 1431–1454, Dec. 1938.

13. J. Wood, "Developments in the Design and Performance of HF Transmitters," *Electronics & Power*, vol. 33, pp. 392–396, June 1987.

14. O. G. Willard, "A High-Level Single-Sideband Transmitter," *Proc. IRE*, vol. 36, pp. 1419–1425, Nov. 1948.

15. L. R. Kahn, "Comparison of Linear Single-Sideband Transmitters with Envelope Elimination and Restoration Single-Sideband Transmitters," *Proc. IRE*, vol. 44, pp. 1706–1712, Dec. 1956.

16. E. H. Armstrong, "A Method of Reducing Disturbances in Radio Signaling by a System of Frequency Modulation," *Proc. IRE*, vol. 24, pp. 689–740, May 1936.

17. R. J. Pieracci, "A Stabilized Frequency-Modulation System," *Proc. IRE*, vol. 30, pp. 76–80, Feb. 1942.

18. H. Komizo, Y. Arai, and T. Saito, "A 7 GHz FM Transmitter Utilizing GaAs Power MESFETs," *1977 IEEE MTT-S Int. Microwave Symp. Dig.*, pp. 243–245.

19. Application Note 1229, *SD2932 RF MOSFET for 300 W FM Amplifier,* STMicroelectronics, 2007.

20. V. K. Zworykin, "Television System," U.S. Patent 2,141,059, Dec. 1938 (filed Dec. 1923).

21. P. T. Farnsworth, "Television System," U.S. Patent 1,773,980, Aug. 1930 (filed Jan. 1927).

22. J. E. Brittain, "Zworykin on an Electronic Television System," *Proc. IEEE*, vol. 72, p. 723, June 1984.

23. J. F. Wentz and K. D. Smith, "A New Microwave Television System," *Trans. AIEE*, vol. 66, pp. 465–470, Jan. 1947.

24. B. Bhatt, D. Birks, and D. Hermreck, "Digital Television: Making It Work," *IEEE Spectrum*, vol. 34, pp. 19–28, Oct. 1997.

25. G. A. Jones, J. M. Defilippis, H. Hoffmann, and E. A. Williams, "Digital Television Station and Network Implementation," *Proc. IEEE*, vol. 94, pp. 22–36, Jan. 2006.

26. J. Zhou, Z. Ou, M. Rautiainen, T. Koskela, and M. Ylianttila, "Digital Television for Mobile Devices," *IEEE Multimedia*, vol. 16, pp. 60–70, Jan.–Mar. 2009.

27. T. Ikegami, "A Newly Designed UHF Solid-State Television Transmitter," *IEE Colloq. Advances in Solid State Technol. for Broadcast Transmitters*, pp. 2/1–2/17, 1988.

28. P. Corbel, "New Developments in Transistorized TV Transmitters," *IEE Colloq. Advances in Solid State Technol. for Broadcast Transmitters*, pp. 7/1–7/4, 1988.

29. C. Davis, J. Hawkins, and C. Einolf Jr, "Solid State DTV Transmitters," *IEEE Trans. Broadcast.*, vol. BC-43, pp. 252–260, Sept. 1997.

30. Technical Data, *RF Power Field Effect Transistor MRF6P3300 H*, Freescale Semiconductor, 2005.

31. M. J. Horspool, *Analog-to-Digital Upgradeable Transmitters for the Worldwide Market*, Harris Corporation, 2003.

32. R. L. Jones and F. M. Ryan, "Air Transport Communication," *Trans. AIEE*, vol. 49, pp. 187–197, Jan. 1930.

33. S. Becker and L. M. Leeds, "A Modern Two-Way Radio System," *Proc. IRE*, vol. 24, pp. 1183–1206, Sept. 1936.

34. H. S. Black, J. W. Beyer, T. J. Grieser, and F. A. Polkinghorn, "A Multichannel Microwave Radio Relay System," *Trans. AIEE*, vol. 65, pp. 798–805, Dec. 1946.

35. K. Ohata, T. Inoue, M. Funabashi, A. Inoue, Y. Takimoto, T. Kuwabara, S. Shinozaki, K. Maruhashi, K. Hosaya, and H. Nagai, "Sixty-GHz-Band Ultra-Miniature Monolithic T/R Modules for Multimedia Wireless Communication Systems," *IEEE Trans. Microwave Theory Tech.*, vol. MTT-44, pp. 2354–2360, Dec. 1996.

36. H. I. Romnes and R. R. O'Connor, "General Mobile Telephone System," *Trans. AIEE*, vol. 66, pp. 1658–1666, Jan. 1947.

37. B. Razavi, "Challenges in Portable RF Transceiver Design," *IEEE Circuits & Devices Mag.*, vol. 12, pp. 12–25, Sept. 1996.

38. B. Razavi, "RF Transmitter Architectures and Circuits," *Proc. 1999 IEEE Custom Integrated Circuits Conf.*, pp. 197–204.

39. A. Loke and F. Ali, "Direct Conversion Radio for Digital Mobile Phones – Design Issues, Status, and Trends," *IEEE Trans. Microwave Theory Tech.*, vol. MTT-50, pp. 2422–2435, Nov. 2002.

40. T. D. Stetzler, I. G. Post, J. H. Havens, and M. Koyama, "A 2.7–4.5 V Single Chip GSM Transceiver RF Integrated Circuit," *IEEE J. Solid-State Circuits*, vol. SC-30, pp. 1421–1429, Dec. 1995.

41. M. H. Perrott, T. L. Tewksbury III, and C. G. Sodini, "A 27-mW CMOS Fractional-*N* Synthesizer Using Digital Compensation for 2.56-Mb/s GFSK Modulation," *IEEE J. Solid-State Circuits*, vol. SC-32, pp. 2048–2060, Dec. 1997.

42. T. Yamawaki, M. Kokubo, K. Irie, H. Matsui, K. Hori, T. Endou, H. Hagisawa, T. Furiya, Y. Shimizu, M. Katagishi, and J. R. Hildersley, "A 2.7-V GSM RF Transceiver IC," *IEEE J. Solid-State Circuits*, vol. SC-32, pp. 2089–2096, Dec. 1997.

43. R. Magoon, A. Molnar, J. Zachan, G. Hatcher, and W. Rhee, "A Single-Chip Quad-Band (850/900/1800/1900 MHz) Direct Conversion GSM/GPRS RF Transceiver with Integrated VCOs and Fractional-*N* Synthesizer," *IEEE J. Solid-State Circuits*, vol. SC-37, pp. 1710–1720, Dec. 2002.

44. A. Zolfaghari, H. Darabi, H. Jensen, J. Leete, B. Mohammadi, J. Chiu, Q. Li, Z. Zhou, P. Lettieri, Y. Chang, A. Hadji-Abdolhamid, P. Chang, M. Nariman, I. Bhatti, A. Medi, L. Serrano, J. Welz, K. Shoarinejad, S. Hasan, J. Castaneda, J. Kim, H. Tran, P. Kilcoyne, R. Chen, B. lee, B. Zhao, B. Ibrahim, M. Rofougaran, and A. Rofourgaran, "Transmitter Development for Cellular Integrated Circuits," *IEEE Commun. Mag.*, vol. 46, pp. 146–152, Sept. 2008.

45. M. R. Elliott, T. Montalvo, B. P. Jeffries, F. Murden, J. Strange, A. Hill, S. Nandipaku, and J. Harrebek, "A Polar Modulator Transmitter for GSM/EDGE," *IEEE J. Solid-State Circuits*, vol. SC-39, pp. 2190–2199, Dec. 2004.

46. T. Sowlati, D. Rozenblit, R. Pullela, M. Damgaard, E. McCarthy, D. Koh, D. Ripley, F. Balteanu, and G. Gheorghe, "Quad-Band GSM/GPRS/EDGE Polar Loop Transmitter," *IEEE J. Solid-State Circuits*, vol. SC-39, pp. 2179–2189, Dec. 2004.

47. A. Rofougaran, M. Rofourgaran, and A. Behzad, "Radios for Next-Generation Wireless Networks," *IEEE Microwave Mag.*, vol. 6, pp. 38–43, May 2005.

48. M. Skolnik, *Radar Handbook*, New York: McGraw-Hill, 2008.

49. D. K. Barton, "A Half Century of Radar," *IEEE Trans. Microwave Theory Tech.*, vol. MTT-32, pp. 1161–1170, Sept. 1984.

50. A. J. Fenn, D. H. Temme, W. R. Delaney, and W. E. Courtney, "The Development of Phased-Array Radar Technology," *Lincoln Lab. J.*, vol. 12, pp. 321–339, No. 2, 2000.

51. K. F. Molz, "Phased Array Radar Systems," *The Radio and Electronic Eng.*, vol. 28, pp. 331–339, Nov. 1964.

52. J. Holloway, "Design Considerations for Adaptive Active Phased-Array 'Multifunction' Radars," *Electronics & Commun. Eng. J.*, vol. 13, pp. 277–288, Dec. 2001.

53. A. K. Agrawal and E. L. Holzman, "Beamformer Architectures for Active Phased-Array Radar Antennas," *IEEE Trans. Antennas and Propagation*, vol. AP-47, pp. 432–442, Mar. 1999.

54. C. T. Rodenbeck, S. G. Kim, W. H. Tu, M. R. Coutant, S. Hing, M. Li, and K. Chang, "Ultra-Wideband Low-Cost Phased-Array Radars," *IEEE Trans. Microwave Theory Tech.*, vol. MTT-53, pp. 3697–3703, Dec. 2005.

55. D. N. McQuiddy Jr, R. L. Gassner, P. Hull, J. S. Mason, and J. M. Bedinger, "Transmit/Receive Module Technology for X-Band Active Array Radar," *Proc. IRE*, vol. 79, pp. 308–341, Mar. 1991.

56. B. A. Kopp, M. Borkowski, and G. Jerinic, "Transmit/Receive Modules," *IEEE Trans. Microwave Theory Tech.*, vol. MTT-50, pp. 827–834, Mar. 2002.

57. W. N. Edelstein, C. Andricos, and A. Moussessian, "High-Efficiency L-Band Transmit/Receive Module for Synthetic Aperture Radar," *Proc. 2003 IEEE Radar Conf.*, pp. 238–243.

58. B. A. Kopp, "S- and X-Band Radar Transmit/Receive Module Overview," *Proc. 2007 IEEE Radar Conf.*, pp. 948–953.

59. S. K. Yeo, C. H. Kim, J. H. Ahn, J. H. Chun, and Y. S. Kwon, "A Compact High Power T/R Module for X-Band Phased Array Radar Applications Using an Anodized Aluminum Substrate," *Proc. 39th Europ. Microwave Conf.*, pp. 1365–1368, 2009.

60. R. Tayrani, M. A. Teshiba, G. M. Sakamoto, Q. Chaudhry, R. Alidio, Y. Kang, I. S. Ahmad, T. C. Cisco, and M. Hauhe, "Broad-Band SiGe MMICs for Phased-Array Radar Applications," *IEEE J. Solid-State Circuits*, vol. SC-38, pp. 1462–1470, Sept. 2003.

61. I. J. Bahl, *Fundamentals of RF and Microwave Transistor Amplifiers*, New York: John Wiley & Sons, 2009.

62. C. K. Chu, H. K. Huang, H. Z. Liu, C. H. Lin, C. H. Chang, C. L. Wu, C. S. Chang, and Y. H. Wang, "An X-Band High-Power and High-PAE PHEMT MMIC Power Amplifier for Pulse and CW Operation," *IEEE Microwave and Wireless Components Lett.*, vol. 18, pp. 707–709, Oct. 2008.

63. J. Kuehn, F. van Raay, R. Quay, R. Kiefer, T. Maier, R. Stibal, M. Mikulla, M. Seelman-Eggebert, W. Bronner, M. Schlechtweg, O. Ambacher, and M. Thumm, "Design of Highly-Efficient GaN X-Band-Power-Amplifier MMICs," *2009 IEEE MTT-S Int. Microwave Symp. Dig.*, pp. 661–664.

64. A. M. Darwish, K. Boutros, B. Luo, B. Huebschman, E. Viveiros, and H. A. Hung, "4-Watt Ka-Band AlGaN/GaN Power Amplifier MMIC," *2006 IEEE MTT-S Int. Microwave Symp. Dig.*, pp. 730–733.

65. H. Knapp, B. Dehlink, H.-P. Forstner, E. Kolmhofer, K. Aufinger, J. Boeck, and T. F. Meister, "SiGe Circuits for Automotive Radar," *Proc. 2007 IEEE Topical Meeting Silicon Monolithic Integrated Circuits in RF Syst.*, pp. 231–236.

66. W. M. Huang, J. P. John, S. Braithwaite, J. Kirchgessner, I. S. Lim, D. Morgan, Y. B. Park, S. Shams, I. To, P. Welch, R. Reuter, H. Li, A. Ghazinour, P. Wennekers, and Y. Yin, "SiGe 77GHz Automotive Radar Technology," *2007 IEEE Int. Circuits and Syst. Symp. Dig.*, pp. 1967–1970.

67. Y. Kawano, T. Suzuki, M. Sato, T. Hirose, and K. Joshin, "A 77GHz Transceiver in 90 nm CMOS," *Proc. 2009 IEEE Int. Solid-State Circuits Conf.*, pp. 310–311.

68. I. Gresham, N. Jain, T. Budka, A. Alexanian, N. Kinayman, B. Ziegner, S. Brown, and P. Staecker, "A Compact Manufacturable 76–77-GHz Radar Module for Commercial ACC Applications," *IEEE Trans. Microwave Theory Tech.*, vol. MTT-49, pp. 44–58, Jan. 2001.

69. C. Wagner, A. Stelzer, and H. Jaeger, "A Phased-Array Radar Transmitter Based on 77-GHz Cascadable Transceivers," *2009 IEEE MTT-S Int. Microwave Symp. Dig.*, pp. 73–76.

70. T. V. Kerssenbrock, T. Musch, B. Shiek, and P. Heide, "Novel 77 GHz Low-Cost Automotive Radar Module with Fractional-PLL Frequency Linearizer," *Proc. 29th Europ. Microwave Conf.*, pp. 5–8, 1999.

71. A. Natarajan, A. Komijani, X. Guan, A. Babakhani, and A. Hajimiri, "A 77-GHz Phased-Array Transceiver with On-Chip Antennas in Silicon: Transmitter and Local LO-Path Phase Shifting," *IEEE J. Solid-State Circuits*, vol. SC-41, pp. 2807–2819, Dec. 2006.

72. D. Salle, C. Landez, P. Savary, G. Montoriol, R. Gach, H. Li, and A. Ghazinour, "A Fully Integrated 77GHz FMCW Radar Transmitter Using a Fractional-N frequency Synthesizer," *Proc. 6th Europ. Radar Conf.*, pp. 149–152, 2009.

73. R. Feger, C. Wagner, S. Schuster, S. Scheiblhofer, H. Jaeger, and A. Stelzer, "A 77-GHz FMCW MIMO Radar Based on an SiGe Single-Chip Transceiver," *IEEE Trans. Microwave Theory Tech.*, vol. MTT-57, pp. 1020–1035, May 2009.

74. J. G. Kim, S. Ko, S. Jeon, J. W. Park, and S. Hong, "Balanced Topology to Cancel Tx Leakage in CW Radar," *IEEE Microwave and Wireless Components Lett.*, vol. 14, pp. 443–445, Sept. 2004.

75. C.Y. Kim, J. G. Kim, J. H. Oum, J. R. Yang, D.K. Kim, J. H. Choi, S.W. Kwon, S.H. Jeon, J.W. Park, and S. Hong, "Tx Leakage Cancellers for 24 GHz and 77 GHz Vehicular Radar Applications," *2006 IEEE MTT-S Int. Microwave Symp. Dig.*, pp. 1402–1405.

76. C.-Y. Kim, J.-G. Kim, and S. Hong, "A Quadrature Radar Topology with Tx Leakage Canceller for 24-GHz Radar Applications," *IEEE Trans. Microwave Theory Tech.*, vol. MTT-55, pp. 1438–1444, July 2007.

77. A. Grebennikov and G. Haider, "High-Efficiency Differential Radar System," U.S. Patent 7,511,658, Mar. 2009.

78. C. Wagner, A. Stelzer, and H. Jaeger, "PLL Architecture for 77-GHz FMCW Radar Systems with Highly-Linear Ultra-Wideband Frequency Sweeps," *2006 IEEE MTT-S Int. Microwave Symp. Dig.*, pp. 399–402.

79. V. Jain, F. Tzeng, L. Zhou, and P. Heydari, "A Single-Chip Dual-Band 22–29-GHz/77–81-GHz BiCMOS Transceiver for Automotive Radars," *IEEE J. Solid-State Circuits*, vol. SC-44, pp. 3469–3485, Dec. 2009.

80. A. Komijani, and A. Hajimiri, "A Wideband 77-GHz, 17.5-dBm Fully Integrated Power Amplifier in Silicon," *IEEE J. Solid-State Circuits*, vol. SC-41, pp. 1749–1756, Aug. 2006.

81. M. Chang and G. M. Rebeiz, "A Wideband High-Efficiency 79–97 GHz SiGe Linear Power Amplifier with >90 mW Output," *Proc. 2008 IEEE Bipolar/BiCMOS Circuits and Tech. Meeting*, pp. 69–72.

82. S. Nicolson, K. H. K. Yau, S. Pruvost, V. Danelon, P. Chevalier, P. Garcia, A. Chantre, B. Sautreuil, and S. P. Voinigescu, "A Low-Voltage SiGe BiCMOS 77-GHz Automotive Radar Chipset," *IEEE Trans. Microwave Theory Tech.*, vol. MTT-56, pp. 1092–1104, May 2008.

83. S. Yoshida, M. Tanomura, Y. Murase, K. Yamanoguchi, K. Ota, K. Matsunaga, and H. Shimawaki, "A 76 GHz GaN-on-Silicon Power Amplifier for Automotive Radar Systems," *2009 IEEE MTT-S Int. Microwave Symp. Dig.*, pp. 665–668.

84. P. M. Grant and J. H. Collins, "Introduction to Electronic Warfare," *IEE Proc. Commun. Radar & Signal Process.*, vol. 129, pp. 113–132, June 1982.

85. G. Garbe, R. Becker, M. Patel, W. Guthrie, S. Moghe, R. Williams, and G. Giacomino, "A 6.8–10.7 GHz EW Module Using 72 MMICs," *1993 IEEE MTT-S Int. Microwave Symp. Dig.*, pp. 1329–1332.

86. R. Beach, "High Speed Linear Oscillators," *Microwave J.*, vol. 21, pp. 59–65, Dec. 1978.

87. D. A. Boyd, "Design Considerations for Post-Tuning Drift (PTD) Reduction in Modern VCO Subsystems," *Microwave J.*, vol. 29, pp. 121–130, Dec. 1986.

88. D. A. Boyd, "VCO Techniques for EW Systems," *Microwave J.*, vol. 31, pp. 129–138, Feb. 1988.

89. R. N. Buswell, "Behind the Design of VCO Linearizers," *Microwaves*, vol. 15, pp. 36–40, Sept. 1976.

90. K. Adams and C. Oxley, "Broadband FET VCO Design," *Electronic Eng.*, vol. 61, pp. S25–S30, Feb. 1989.

91. M. Tsuru, K. Kawakami, K. Tajima, K. Miyamoto, M. Nakane, K. Itoh, M. Miyazaki, and Y. Isota, "A Triple-Tuned Ultra-Wideband VCO," *IEEE Trans. Microwave Theory Tech.*, vol. MTT-56, pp. 346–354, Feb. 2008.

92. R. G. Winch, "Very Broadband Bipolar VCO," *Electronics Lett.*, vol. 17, pp. 871–873, Nov. 1981.

93. R. G. Winch, "Wide-Band Varactor-Tuned Oscillators," *IEEE J. Solid-State Circuits*, vol. SC-17, pp. 1214–1219, Dec. 1982.

94. A. Platzker, K. T. Hetzler, and J. B. Cole, "Development of Highly Dense Four-Stage Flat-Gain 1-W 6–18-GHz MMIC Power Amplifier Chip," *IEEE J. Solid-State Circuits*, vol. SC-27, pp. 1405–1412, Oct. 1992.

95. M. Salib, A. Gupta, F. Ali, and D. Dawson, "A 1.8-W, 6–18 GHz HBT MMIC Power Amplifier with 10-dB Gain and 37% Peak Power-Added Efficiency," *IEEE Microwave and Guided Wave Lett.*, vol. 3, pp. 325–326, Sept. 1993.

96. A. R. Barnes, M. T. Moore, and M. B. Allenson, "A 6–18 GHz Broadband High Power MMIC for EW Applications," *1997 IEEE MTT-S Int. Microwave Symp. Dig.*, pp. 1429–1432.

97. J. J. Komiak, W. Kong, and K. Nichols, "High Efficiency Wideband 6 to 18 GHz PHEMT Power Amplifier MMIC," *2002 IEEE MTT-S Int. Microwave Symp. Dig.*, pp. 905–907.

98. J. R. Pierce and R. Kompfner, "Transoceanic Communication by Means of Satellite," *Proc. IRE*, vol. 47, pp. 372–380, Mar. 1959; reprinted in *Proc. IEEE*, vol. 85, pp. 1011–1019, June 1997.

99. H. Sobol, "Microwave Communications – An Historical Perspective," *IEEE Trans. Microwave Theory Tech.*, vol. MTT-32, pp. 1170–1181, Sept. 1984.

100. J. E. Keigler and C. E. Profera, "Domestic and Regional Satellite Systems," *Proc. IEEE*, vol. 59, pp. 238–252, Feb. 1971.

101. W. W. Wu, "Satellite Communications," *Proc. IEEE*, vol. 85, pp. 998–1010, June 1997.

102. K.-J. Koh, J. W. May, and G. M. Rebeiz, "A Millimeter-Wave (40–45 GHz) 16-Element Phased-Array Transmitter in 0.18-μm SiGe BiCMOS Technology," *IEEE J. Solid-State Circuits*, vol. SC-44, pp. 1498–1509, May 2009.

103. T. Masamura and T. Inoue, "Satellite Communication System Using TDMA and SSMA," *IEEE Trans. Aerospace Electron. Syst.*, vol. AES-19, pp. 906–914, Nov. 1983.

104. K. A. Miller and A. Schoen, "A Digital Satellite Communications System for Corporate Telecommunications," *IEEE J. Selected Areas Commun.*, vol. SAC-2, pp. 403–413, May 1984.

105. R. Y. Huang and P. Hooten, "Communication Satellite Processing Repeaters," *Proc. IEEE*, vol. 59, pp. 238–252, Feb. 1971.

106. R. J. Bauer and R. Baker, "A Solid State 30 Watt PPM Telemetry Transmitter," *IEEE Trans. Aerospace Navigation. Electron.*, pp. 2.3.1–1 – 2.3.1–15, 1963.

107. D. Cummings, K. J. Sweeney, and N. F. Kington, "The Development of a British Prototype Ks-Band Satellite Transponder," *J. IERE*, vol. 58, pp. 149–157, May 1988.

108. A. Wakejima, T. Asano, T. Hirano, M. Funabashi, and K. Matsunaga, "C-Band GaAs FET Power Amplifiers with 70-W Output Power and 50% PAE for Satellite Communication," *IEEE J. Solid-State Circuits*, vol. SC-40, pp. 2054–2060, Oct. 2005.

109. R. Mallavarpu and M. P. Puri, "High Power 14 GHz SSPA for SATCOM Applications," *1993 IEEE MTT-S Int. Microwave Symp. Dig.*, pp. 1217–1220.

110. F. Scappaviva, R. Cignani, C. Florian, G. Vannini, and M. Feudale, "10 Watt High Efficiency GaAs MMIC Power Amplifier for Space Applications," *Proc. 38th Europ. Microwave Conf.*, pp. 1429–1432, 2008.

111. M. C. Smith and L. P. Dunleavy, "Comparison of Solid State, MPM, and TWT Based Transmitters for Spaceborne Applications," *Proc. 1998 IEEE Southeastcon*, pp. 256–259.

112. S. D. Seymour, "Development of Spacecraft Solid-State High-Power L-Band Amplifiers," *IEE Proc. Commun. Radar & Signal Processing*, vol. 133, pp. 326–338, July 1986.

113. H. Shigematsu, Y. Inoue, A. Akasegawa, M. Yamada, S. Masuda, Y. Kamada, A. Yamada, M. Kanamura, T. Ohki, K. Makiyama, N. Okamoto, K. Imanishi, T. Kikkawa, K. Joshin, and N. Hara, "C-Band 340-W and X-band 100-W GaN Power Amplifiers with over 50% PAE," *2009 IEEE MTT-S Int. Microwave Symp. Dig.*, pp. 1265–1268.

114. C. H. Hoeppner, "Pulse Communication System," U.S. Patent 2,999,128, Sept. 1961 (filed Nov. 1945).

115. G. F. Ross, "Transmission and Reception System for Generating and Receiving Base-Band Duration Pulse Signals without Distortion for Short Base-Band Pulse Communication System," U.S. Patent 3,728,632, Apr. 1973 (filed May 1971).

116. S. Roy, J. R. Foerster, V. S. Somayazulu, and D. G. Leeper, "Ultrawideband Radio Design: The Promise of High-Speed, Short-Range Wireless Connectivity," *Proc. IEEE*, vol. 92, pp. 295–311, Feb. 2004.

117. D. D. Wentzloff, R. Blazquez, F. S. Lee, B. P. Ginsburg, J. Powell, and A. P. Chandrakasan, "System Design Considerations for Ultra-Wideband Communication," *IEEE Commun. Mag.*, vol. 43, pp. 114–121, Aug. 2005.

118. A. P. Chandrakasan, F. S. Lee, D. D. Wentzloff, V. Sze, B. P. Ginsburg, P. P. Mercier, D. C. Daly, and R. Blazquez, "Low-Power Impulse UWB Architectures and Circuits," *Proc. IEEE*, vol. 97, pp. 332–352, Feb. 2009.

119. H. Zheng, S. Lou, D. Lu, C. Shen, T. Chan, and H. S. Luong, "A 3.1 GHz – 8.0 GHz Single-Chip Transceiver for MB-OFDM UWB in 0.18-μm CMOS Process," *IEEE J. Solid-State Circuits*, vol. SC-44, pp. 414–426, Feb. 2009.

120. D. Marchaland, M. Villegas, G. Baudoin, C. Tinella, and D. Belot, "System Concepts Dedicated to UWB Transmitter," *Proc. 2005 Europ. Wireless Tech. Conf.*, pp. 141–144.

121. V. V. Kulkarni, M. Muqsith, K. Niitsu, H. Ishikuro, and T. Kuroda, "A 750 Mb/s, 12 pJ/b, 6-to-10 GHz CMOS IR-UWB Transmitter with Embedded On-Chip Antenna," *IEEE J. Solid-State Circuits*, vol. SC-44, pp. 394–403, Feb. 2009.

122. D. D. Wentzloff, F. S. Lee, D. C. Daly, M. Bhardwaj, P. P. Mercier, and A. P. Chandrakasan, "Energy Efficient Pulsed-UWB CMOS Circuits and Systems," *Proc. 2007 IEEE Int. Ultra-Wideband Conf.*, pp. 282–287.

123. T.-A. Phan, J. Lee, V. Krizhanovskii, S.-K. Han, and S.-G. Lee, "A 18-pJ/Pulse OOK CMOS Transmitter for Multiband UWB Impulse Radio," *IEEE Microwave and Wireless Components Lett.*, vol. 17, pp. 688–690, Sept. 2007.

124. J. Lee, Y. J. Park, M. Kim, C. Yoon, J. Kim, and K. H. Kim, "System-On-Package Ultra-Wideband Transmitter Using CMOS Impulse Generator," *IEEE Trans. Microwave Theory Tech.*, vol. MTT-54, pp. 1667–1674, Apr. 2006.

125. S. Bourdel, Y. Bachelet, J. Gaubert, R. Vauche, O. Fourquin, N. Dehaese, and H. Barthelemy, "A 9-pJ/Pulse 1.42-Vpp OOK CMOS UWB Pulse Generator for the 3.1 – 10.6-GHz FCC Band," *IEEE Trans. Microwave Theory Tech.*, vol. MTT-58, pp. 65–73, Jan. 2010.

126. W.C. Wang, C.P. Liao, Y.K. Lo, Z.D. Huang, F. R. Shahroury, and C.Y. Wu, "The Design of Integrated 3-GHz to 11-GHz CMOS Transmitter for Full-Band Ultra-Wideband (UWB) Applications," *2008 IEEE Int. Circuits and Syst. Symp. Dig.*, pp. 2709–2712.

127. J. Zhao, C. Maxey, A. Narayanan, and S. Raman, "A SiGe BiCMOS Ultra Wide Band RFIC Transmitter Design for Wireless Sensor Networks," *Proc. 2004 IEEE Radio and Wireless Conf.*, pp. 215–218.

128. H. C. Hsu, Z. W. Wang, and G. K. Ma, "A Low Power CMOS Full-Band UWB Power Amplifier Using Wideband RLC Matching Method," *Proc. 2005 IEEE Electron Devices and Solid-State Conf.*, pp. 233–236.

129. C. Lu, A.V. Pham, and M. Shaw, "A CMOS Power Amplifier for Full-Band UWB Transmitters," *2006 IEEE RFIC Symp. Dig.*, pp. 397–400.

130. S. K. Wong, S. Maisurah, M. N. Osman, F. Kung, and J. H. See, "High Efficiency CMOS Power Amplifier for 3 to 5 GHz Ultra-Wideband (UWB) Application," *IEEE Trans. Consumer Electron.*, vol. CE-55, pp. 1546–1550, Aug. 2009.

131. H. W. Chung, C. Y. Hsu, C. Y. Yang, K. F. Wei, and H. R. Chuang, "A 6–10-GHz CMOS Power Amplifier with an Interstage Wideband Impedance Transformer for UWB Transmitters," *Proc. 38th Europ. Microwave Conf.*, pp. 305–308, 2008.

INDEX

RF and Microwave Transmitter Design, First Edition. Andrei Grebennikov.

WILEY SERIES IN MICROWAVE AND OPTICAL ENGINEERING

KAI CHANG, Editor
Texas A&M University

FIBER-OPTIC COMMUNICATION SYSTEMS, Fourth Edition • *Gavind P. Agrawal*

ASYMMETRIC PASSIVE COMPONENTS IN MICROWAVE INTEGRATED CIRCUITS • *Hee-Ran Ahn*

COHERENT OPTICAL COMMUNICATIONS SYSTEMS • *Silvello Betti, Giancarlo De Marchis, and Eugenio Iannone*

PHASED ARRAY ANTENNAS: FLOQUET ANALYSIS, SYNTHESIS, BFNs, AND ACTIVE ARRAY SYSTEMS • *Arun K. Bhattacharyya*

HIGH-FREQUENCY ELECTROMAGNETIC TECHNIQUES: RECENT ADVANCES AND APPLICATIONS • *Asoke K. Bhattacharyya*

RADIO PROPAGATION AND ADAPTIVE ANTENNAS FOR WIRELESS COMMUNICATION LINKS: TERRESTRIAL, ATMOSPHERIC, AND IONOSPHERIC • *Nathan Blaunstein and Christos G. Christodoulou*

COMPUTATIONAL METHODS FOR ELECTROMAGNETICS AND MICROWAVES • *Richard C. Booton, Jr.*

ELECTROMAGNETIC SHIELDING • *Salvatore Celozzi, Rodolfo Araneo, and Giampiero Lovat*

MICROWAVE RING CIRCUITS AND ANTENNAS • *Kai Chang*

MICROWAVE SOLID-STATE CIRCUITS AND APPLICATIONS • *Kai Chang*

RF AND MICROWAVE WIRELESS SYSTEMS • *Kai Chang*

RF AND MICROWAVE CIRCUIT AND COMPONENT DESIGN FOR WIRELESS SYSTEMS • *Kai Chang, Inder Bahl, and Vijay Nair*

MICROWAVE RING CIRCUITS AND RELATED STRUCTURES, Second Edition • *Kai Chang and Lung-Hwa Hsieh*

MULTIRESOLUTION TIME DOMAIN SCHEME FOR ELECTROMAGNETIC ENGINEERING • *Yinchao Chen, Qunsheng Cao, and Raj Mittra*

DIODE LASERS AND PHOTONIC INTEGRATED CIRCUITS • *Larry Coldren and Scott Corzine*

EM DETECTION OF CONCEALED TARGETS • *David J. Daniels*

RADIO FREQUENCY CIRCUIT DESIGN • *W. Alan Davis and Krishna Agarwal*

RADIO FREQUENCY CIRCUIT DESIGN, Second Edition • *W. Alan Davis*

MULTICONDUCTOR TRANSMISSION-LINE STRUCTURES: MODAL ANALYSIS TECHNIQUES • *J. A. Brandão Faria*

PHASED ARRAY-BASED SYSTEMS AND APPLICATIONS • *Nick Fourikis*

SOLAR CELLS AND THEIR APPLICATIONS, Second Edition • *Lewis M. Fraas and Larry D. Partain*

FUNDAMENTALS OF MICROWAVE TRANSMISSION LINES • *Jon C. Freeman*

OPTICAL SEMICONDUCTOR DEVICES • *Mitsuo Fukuda*

MICROSTRIP CIRCUITS • *Fred Gardiol*

HIGH-SPEED VLSI INTERCONNECTIONS, Second Edition • *Ashok K. Goel*

FUNDAMENTALS OF WAVELETS: THEORY, ALGORITHMS, AND APPLICATIONS, Second Edition • *Jaideva C. Goswami and Andrew K. Chan*

HIGH-FREQUENCY ANALOG INTEGRATED CIRCUIT DESIGN • *Ravender Goyal (ed.)*

RF AND MICROWAVE TRANSMITTER DESIGN • *Andrei Grebennikov*

ANALYSIS AND DESIGN OF INTEGRATED CIRCUIT ANTENNA MODULES • *K. C. Gupta and Peter S. Hall*

PHASED ARRAY ANTENNAS, Second Edition • *R. C. Hansen*

STRIPLINE CIRCULATORS • *Joseph Helszajn*

THE STRIPLINE CIRCULATOR: THEORY AND PRACTICE • *Joseph Helszajn*

LOCALIZED WAVES • *Hugo E. Hernández-Figueroa, Michel Zamboni-Rached, and Erasmo Recami (eds.)*

MICROSTRIP FILTERS FOR RF/MICROWAVE APPLICATIONS, Second Edition • *Jia-Sheng Hong*

MICROWAVE APPROACH TO HIGHLY IRREGULAR FIBER OPTICS • *Huang Hung-Chia*

NONLINEAR OPTICAL COMMUNICATION NETWORKS • *Eugenio Iannone, Francesco Matera, Antonio Mecozzi, and Marina Settembre*

FINITE ELEMENT SOFTWARE FOR MICROWAVE ENGINEERING • *Tatsuo Itoh, Giuseppe Pelosi, and Peter P. Silvester (eds.)*

INFRARED TECHNOLOGY: APPLICATIONS TO ELECTROOPTICS, PHOTONIC DEVICES, AND SENSORS • *A. R. Jha*

SUPERCONDUCTOR TECHNOLOGY: APPLICATIONS TO MICROWAVE, ELECTRO-OPTICS, ELECTRICAL MACHINES, AND PROPULSION SYSTEMS • *A. R. Jha*

TIME AND FREQUENCY DOMAIN SOLUTIONS OF EM PROBLEMS USING INTEGTRAL EQUATIONS AND A HYBRID METHODOLOGY • *B. H. Jung, T. K. Sarkar, S. W. Ting, Y. Zhang, Z. Mei, Z. Ji, M. Yuan, A. De, M. Salazar-Palma, and S. M. Rao*

OPTICAL COMPUTING: AN INTRODUCTION • *M. A. Karim and A. S. S. Awwal*

INTRODUCTION TO ELECTROMAGNETIC AND MICROWAVE ENGINEERING • *Paul R. Karmel, Gabriel D. Colef, and Raymond L. Camisa*

MILLIMETER WAVE OPTICAL DIELECTRIC INTEGRATED GUIDES AND CIRCUITS • *Shiban K. Koul*

ADVANCED INTEGRATED COMMUNICATION MICROSYSTEMS • *Joy Laskar, Sudipto Chakraborty, Manos Tentzeris, Franklin Bien, and Anh-Vu Pham*

MICROWAVE DEVICES, CIRCUITS AND THEIR INTERACTION • *Charles A. Lee and G. Conrad Dalman*

ADVANCES IN MICROSTRIP AND PRINTED ANTENNAS • *Kai-Fong Lee and Wei Chen (eds.)*

SPHEROIDAL WAVE FUNCTIONS IN ELECTROMAGNETIC THEORY • *Le-Wei Li, Xiao-Kang Kang, and Mook-Seng Leong*

ARITHMETIC AND LOGIC IN COMPUTER SYSTEMS • *Mi Lu*

OPTICAL FILTER DESIGN AND ANALYSIS: A SIGNAL PROCESSING APPROACH • *Christi K. Madsen and Jian H. Zhao*

THEORY AND PRACTICE OF INFRARED TECHNOLOGY FOR NONDESTRUCTIVE TESTING • *Xavier P. V. Maldague*

METAMATERIALS WITH NEGATIVE PARAMETERS: THEORY, DESIGN, AND MICROWAVE APPLICATIONS • *Ricardo Marqués, Ferran Martín, and Mario Sorolla*

OPTOELECTRONIC PACKAGING • *A. R. Mickelson, N. R. Basavanhally, and Y. C. Lee (eds.)*

OPTICAL CHARACTER RECOGNITION • *Shunji Mori, Hirobumi Nishida, and Hiromitsu Yamada*

ANTENNAS FOR RADAR AND COMMUNICATIONS: A POLARIMETRIC APPROACH • *Harold Mott*

INTEGRATED ACTIVE ANTENNAS AND SPATIAL POWER COMBINING • *Julio A. Navarro and Kai Chang*

ANALYSIS METHODS FOR RF, MICROWAVE, AND MILLIMETER-WAVE PLANAR TRANSMISSION LINE STRUCTURES • *Cam Nguyen*

LASER DIODES AND THEIR APPLICATIONS TO COMMUNICATIONS AND INFORMATION PROCESSING • *Takahiro Numai*

FREQUENCY CONTROL OF SEMICONDUCTOR LASERS • *Motoichi Ohtsu (ed.)*

WAVELETS IN ELECTROMAGNETICS AND DEVICE MODELING • *George W. Pan*

OPTICAL SWITCHING • *Georgios Papadimitriou, Chrisoula Papazoglou, and Andreas S. Pomportsis*

MICROWAVE IMAGING • *Matteo Pastorino*

ANALYSIS OF MULTICONDUCTOR TRANSMISSION LINES • *Clayton R. Paul*

INTRODUCTION TO ELECTROMAGNETIC COMPATIBILITY, Second Edition • *Clayton R. Paul*

ADAPTIVE OPTICS FOR VISION SCIENCE: PRINCIPLES, PRACTICES, DESIGN AND APPLICATIONS • *Jason Porter, Hope Queener, Julianna Lin, Karen Thorn, and Abdul Awwal (eds.)*

ELECTROMAGNETIC OPTIMIZATION BY GENETIC ALGORITHMS • *Yahya Rahmat-Samii and Eric Michielssen (eds.)*

INTRODUCTION TO HIGH-SPEED ELECTRONICS AND OPTOELECTRONICS • *Leonard M. Riaziat*

NEW FRONTIERS IN MEDICAL DEVICE TECHNOLOGY • *Arye Rosen and Harel Rosen (eds.)*

ELECTROMAGNETIC PROPAGATION IN MULTI-MODE RANDOM MEDIA • *Harrison E. Rowe*

ELECTROMAGNETIC PROPAGATION IN ONE-DIMENSIONAL RANDOM MEDIA • *Harrison E. Rowe*

HISTORY OF WIRELESS • *Tapan K. Sarkar, Robert J. Mailloux, Arthur A. Oliner, Magdalena Salazar-Palma, and Dipak L. Sengupta*

PHYSICS OF MULTIANTENNA SYSTEMS AND BROADBAND PROCESSING • *Tapan K. Sarkar, Magdalena Salazar-Palma, and Eric L. Mokole*

SMART ANTENNAS • *Tapan K. Sarkar, Michael C. Wicks, Magdalena Salazar-Palma, and Robert J. Bonneau*

NONLINEAR OPTICS • *E. G. Sauter*

APPLIED ELECTROMAGNETICS AND ELECTROMAGNETIC COMPATIBILITY • *Dipak L. Sengupta and Valdis V. Liepa*

COPLANAR WAVEGUIDE CIRCUITS, COMPONENTS, AND SYSTEMS • *Rainee N. Simons*

ELECTROMAGNETIC FIELDS IN UNCONVENTIONAL MATERIALS AND STRUCTURES • *Onkar N. Singh and Akhlesh Lakhtakia (eds.)*

ANALYSIS AND DESIGN OF AUTONOMOUS MICROWAVE CIRCUITS • *Almudena Suárez*

ELECTRON BEAMS AND MICROWAVE VACUUM ELECTRONICS • *Shulim E. Tsimring*

FUNDAMENTALS OF GLOBAL POSITIONING SYSTEM RECEIVERS: A SOFTWARE APPROACH, Second Edition • *James Bao-yen Tsui*

RF/MICROWAVE INTERACTION WITH BIOLOGICAL TISSUES • *André Vander Vorst, Arye Rosen, and Youji Kotsuka*

InP-BASED MATERIALS AND DEVICES: PHYSICS AND TECHNOLOGY • *Osamu Wada and Hideki Hasegawa (eds.)*

COMPACT AND BROADBAND MICROSTRIP ANTENNAS • *Kin-Lu Wong*

DESIGN OF NONPLANAR MICROSTRIP ANTENNAS AND TRANSMISSION LINES • *Kin-Lu Wong*

PLANAR ANTENNAS FOR WIRELESS COMMUNICATIONS • *Kin-Lu Wong*

FREQUENCY SELECTIVE SURFACE AND GRID ARRAY • *T. K. Wu (ed.)*

ACTIVE AND QUASI-OPTICAL ARRAYS FOR SOLID-STATE POWER COMBINING • *Robert A. York and Zoya B. Popović (eds.)*

OPTICAL SIGNAL PROCESSING, COMPUTING AND NEURAL NETWORKS • *Francis T. S. Yu and Suganda Jutamulia*

ELECTROMAGNETIC SIMULATION TECHNIQUES BASED ON THE FDTD METHOD • *Wenhua Yu, Xiaoling Yang, Yongjun Liu, and Raj Mittra*

SiGe, GaAs, AND InP HETEROJUNCTION BIPOLAR TRANSISTORS • *Jiann Yuan*

PARALLEL SOLUTION OF INTEGRAL EQUATION-BASED EM PROBLEMS • *Yu Zhang and Tapan K. Sarkar*

ELECTRODYNAMICS OF SOLIDS AND MICROWAVE SUPERCONDUCTIVITY • *Shu-Ang Zhou*